JOHN SANTA CRUZ

Essentials of Anatomy and Physiology

Third Edition

Essentials of Anatomy and Physiology

Third Edition

Valerie C. Scanlon, PhD
College of Mount Saint Vincent
Riverdale, New York

Tina Sanders
Medical Illustrator
Castle Creek, New York
Formerly
Head Graphic Artist
Tompkins Courtland Community College
Dryden, New York

F. A. DAVIS COMPANY • Philadelphia

F. A. Davis Company
1915 Arch Street
Philadelphia, PA 19103

Printed in the United States of America

Last digit indicates print number: 10 9 8 7 6 5 4 3 2 1

Nursing Editor: Alan Sorkowitz
Production Editor: Stephen D. Johnson
Cover Designer: Alicia R. Baronsky

As new scientific information becomes available through basic and clinical research, recommended treatments and drug therapies undergo changes. The author and publisher have done everything possible to make this book accurate, up to date, and in accord with accepted standards at the time of publication. The author, editors, and publisher are not responsible for errors or omissions or for consequences from application of the book, and make no warranty, expressed or implied, in regard to the contents of the book. Any practice described in this book should be applied by the reader in accordance with professional standards of care used in regard to the unique circumstances that may apply in each situation. The reader is advised always to check product information (package inserts) for changes and new information regarding dose and contraindications before administering any drug. Caution is especially urged when using new or infrequently ordered drugs.

Library of Congress Cataloging-in-Publication Data

Scanlon, Valerie C., 1946–
Essentials of anatomy and physiology / Valerie C. Scanlon, Tina Sanders.—3nd ed.
p. cm.
ISBN 0-8036-0407-6 (alk. paper)
1. Human physiology. 2. Human anatomy. I. Sanders, Tina, 1943– . II. Title.
QP34.5.S288 1999
612—dc21 98-25201
CIP

Preface

Once again we have the pleasure of speaking for everyone associated with the production of *Essentials of Anatomy and Physiology*, and extending our thanks to all of you who have adopted our textbook. Your response to our second edition was most gratifying, and you have made this third edition possible.

The text has been updated wherever appropriate to include recent research. This will be most apparent in the boxed inserts on clinical applications. Additions to or revisions of basic material include Langerhans cells in Chapter 5, velocity of blood flow in Chapter 13, and nervous regulation of respiration in Chapter 15.

New to this edition are brief end-of-chapter sections describing the effects of aging on the organ systems. Many of the clinical application boxes have been illustrated, some with photographs (including skin disorders, lumbar puncture, and anemia), and others with explanatory diagrams (including diabetes mellitus, emphysema, and renal failure). New illustrations have focused on physiology, and include muscle contraction, physiology of equilibrium, regulation of blood pressure, regulation of respiration, and others. We hope you and your students will find them helpful.

As always, your comments and suggestions will be most welcome, and they may be sent to us in care of the publisher: F. A. Davis Company, 1915 Arch Street, Philadelphia, PA 19103.

Valerie C. Scanlon
Dobbs Ferry, New York

Tina Sanders
Castle Creek, New York

To my students, past and present
VCS

To Brooks, for his encouragement
TS

To the Instructor

Teachers of introductory anatomy and physiology courses face a special challenge: We must distill and express the complexities of human structure and function in a simple way, without losing the essence and meaning of the material. That is the goal of this textbook: to make this material readily accessible to students with diverse backgrounds and varying levels of educational preparation.

No prior knowledge of biology or chemistry is assumed, and even the most fundamental terms are defined thoroughly. Essential aspects of anatomy are presented clearly and reinforced with excellent illustrations. Essential aspects of physiology are discussed simply, yet with accuracy and precision. Again, the illustrations complement the text material and foster comprehension on the part of the student. These illustrations were prepared especially for students for whom this is a first course in anatomy and physiology. As you will see, many are full-page images in which detail is readily apparent. All important body parts have been carefully labeled, but the student is not overwhelmed with unnecessary labels. Wherever appropriate, the legends refer students to the text for further description or explanation.

The text has three unifying themes: the relationship between physiology and anatomy, the interrelations among the organ systems, and the relationship of each organ system to homeostasis. Although each type of cell, tissue, organ, or organ system is discussed simply and thoroughly in itself, applicable connections are made to other aspects of the body or to the functioning of the body as a whole. Our goal is to provide your students with the essentials of anatomy and physiology, and in doing so, to help give them an appreciation for the incredible machine that is the human body.

The sequence of chapters is a very traditional one. Cross-references are used to remind students of what they have learned from previous chapters. Nevertheless, the textbook is very flexible, and, following the introductory four chapters, the organ systems may be covered in almost any order, depending on the needs of your course.

Each chapter is organized internally from the simple to the more complex, with the anatomy followed by the physiology. The *Instructor's Guide* presents modifications of the topic sequences that may be used, again depending on the needs of your course. Certain more advanced topics may be omitted from each chapter without losing the meaning or flow of the rest of the material, and these are indicated, for each chapter, in the *Instructor's Guide.*

Clinical applications are set apart from the text in boxed inserts. These are often aspects of pathophysiology that are related to the normal anatomy or physiology in the text discussion. Each box presents one particular topic and is referenced at the appropriate point in the text. This material is intended to be an integral part of the chapter but is set apart for ease of reference and to enable you to include or omit as many of these topics as you wish. The use of these boxes also enables students to read the text material without interruption and then to focus on specific aspects of pathophysiology. A comprehensive list of the boxes appears inside the book's front and back covers, and another list at the beginning of each chapter cites the boxes within that chapter.

Tables are utilized as summaries of structure and function, to concisely present a sequence of events, or to present additional material that you may choose to include. Each table is referenced in the text and is intended to facilitate your teaching and to help your students learn.

New terms appear in bold type within the text, and all such terms are fully defined in an extensive glossary, with phonetic pronunciations. Bold type may also be used for emphasis whenever one of these terms is used again in a later chapter.

Each chapter begins with a chapter outline and student objectives to prepare the student for the chapter itself. New terminology and related clinical terms are also listed, with phonetic pronunciations. Each of these terms is fully defined in the glossary, with cross-references back to the chapter in which the term is introduced.

At the end of each chapter are a study outline and review questions. The study outline includes all of the essentials of the chapter in a concise outline form. The review questions may be used by the students as a review or self-test. Following each question is a page reference in parentheses. This reference cites the page(s) in the chapter on which the content needed to answer the question correctly can be found. The answers themselves are included in the *Instructor's Guide.*

An important supplementary learning tool for your students is available in the form of a *Student Workbook* that accompanies this text. For each chapter in the textbook, the workbook offers fill-in and matching-column study questions, figure-labeling and figure-coloring exercises, and crossword puzzles based on the chapter's vocabulary list. Also included are comprehensive, multiple-choice chapter tests to provide a thorough review for students. All answers are provided at the end of the workbook.

The instructor's materials for this text include a complete *Instructor's Guide,* a computerized test bank, and a transparency package. The *Instructor's Guide* contains expanded chapter outlines, notes on each chapter's organization and content (useful for modifying the book to your specific teaching needs), topics for class discussions, and answers to the chapter review questions from the textbook. The computerized test bank contains test questions for every chapter of the book, with a total of more than 1500 questions. It uses F. A. Davis's simple but powerful test-generation software, which allows you to select the questions you wish, modify them if you choose, and even add your own questions. The transparency package offers many clear, sharp, full-color transparencies taken from the textbook's illustrations and tables. New to this edition is an instructor's CD-ROM for Windows and Macintosh that provides more illustrations from the text, annotated with suggested points for lecture or class discussion. We hope this PowerPoint presentation software program, which is free to adopters of the text, will give instructors greater flexibility in preparing and presenting classroom materials.

Suggestions and comments from colleagues are always valuable, and yours would be greatly appreciated. When we took on the task of writing and illustrating this textbook, we wanted to make it the most useful book possible for you and your students. Any suggestions that you can give us to help us achieve that goal are most welcome, and they may be sent to us in care of F. A. Davis Company, 1915 Arch Street, Philadelphia, PA 19103.

Valerie C. Scanlon
Dobbs Ferry, New York

Tina Sanders
Castle Creek, New York

To the Student

This is your textbook for your first course in human anatomy and physiology, a subject that is both fascinating and rewarding. That you are taking such a course says something about you: You may simply be curious as to how the human body functions. Or, you may have a personal goal of making a contribution in one of the health-care professions. Whatever your reason, this textbook will help you to be successful in your anatomy and physiology course.

The material is presented simply and concisely, yet with accuracy and precision. The writing style is informal yet clear and specific; it is intended to promote your comprehension and understanding.

ORGANIZATION OF THE TEXTBOOK

To use this textbook effectively, you should know the purpose of its various parts. Each chapter is organized in the following way:

Chapter Outline—This presents the main topics in the chapter, which correspond to the major headings in the text.

Student Objectives—These summarize what you should know after reading and studying the chapter. These are not questions to be answered, but are rather, with the chapter outline, a preview of the chapter contents.

New Terminology and Related Clinical Terminology—These are some of the new terms you will come across in the chapter. Read through these terms before you read the chapter, but do not attempt to memorize them just yet. When you have finished the chapter, return to the list and see how many terms you can define. All of these terms are fully defined in the glossary.

Study Outline—This is found at the end of the chapter. It is a concise summary of the essentials in the chapter. You may find this outline very useful as a quick review before an exam.

Review Questions—These are also at the end of the chapter. Your instructor may assign some or all of them as homework. If not, the questions may be used as a self-test to evaluate your comprehension of the chapter's content. The page number(s) in parentheses following each question refers you to the page(s) in the chapter on which the content needed to answer the question correctly can be found.

OTHER FEATURES WITHIN EACH CHAPTER

Illustrations—These are an essential part of this textbook. Use them. Look at them and study them carefully, and they will be of great help to you as you learn. They are intended to help you develop your own mental picture of the body and its parts and processes. Each illustration is referenced in the text, so you will know just when to consult it.

Boxes—Discussions of clinical applications are in separate boxes in the text so that you may find and refer to them easily. Your instructor may include all or some of these as required reading. If you are planning a career in the health professions, these boxes are an introduction to pathophysiology, and you will find them interesting and helpful.

Bold Type—This is used whenever a new term is introduced, or when an old term is especially important. The terms in bold type are fully defined in the glossary, which includes phonetic pronunciations.

Tables—This format is used to present material in a very concise form. Some tables are summaries of text material and are very useful for a quick review. Other tables present additional material that complements the text material.

To make the best use of your study time, a Student Workbook is available that will help you to focus your attention on the essentials in each chapter. Also included are comprehensive chapter tests to help you determine which topics you have learned thoroughly and which you may have to review. If your instructor has not made the workbook a required text, you may wish to ask that it be ordered and made available in your bookstore. You will find it very helpful.

SOME FINAL WORDS OF ENCOURAGEMENT

Your success in this course depends to a great extent on you. Try to set aside study time for yourself every day; a little time each day is usually much more productive than trying to cram at the last minute.

Ask questions of yourself as you are studying. What kinds of questions? The simplest ones. If you are studying a part of the body such as an organ, ask yourself: What is its name? Where is it? What is it made of? What does it do? That is: name, location, structure, and function. These are the essentials. If you are studying a process, ask yourself: What is happening here? What is its purpose? That is: What is going on? and what good is it? Again, these are the essentials.

We hope this textbook will contribute to your success. If you have any suggestions or comments, we would very much like to hear them. After all, this book was written for you, to help you achieve your goals in this course and in your education. Please send your suggestions and comments to us in care of F. A. Davis Company, 1915 Arch Street, Philadelphia, PA 19103.

Valerie C. Scanlon
Dobbs Ferry, New York

Tina Sanders
Castle Creek, New York

Acknowledgments

We wish to thank the editors and production staff of the F. A. Davis Company, especially:

- Alan Sorkowitz, Nursing Editor, who has been the publishing heart and soul (no, that is not an oxymoron) of this textbook since its inception.
- Stephen D. Johnson, Production Editor, for his meticulous review of the text.
- Herbert J. Powell, Jr., Director of Production, for his high standards of excellence.
- Bill Donnelly, for designing the layout of the book.
- Alicia R. Baronsky, for designing the cover of the book.
- Valerie Scanlon extends special thanks to her sister, Joan Scanlon, for making the computer wonders of the 20th century comprehensible to a 19th century mind.
- In addition, Tina Sanders wishes especially to thank Dolores Lake Taylor for her consultation on the book's art program.

VCS

TS

Contents

CHAPTER 1

Organization and General Plan of the Body

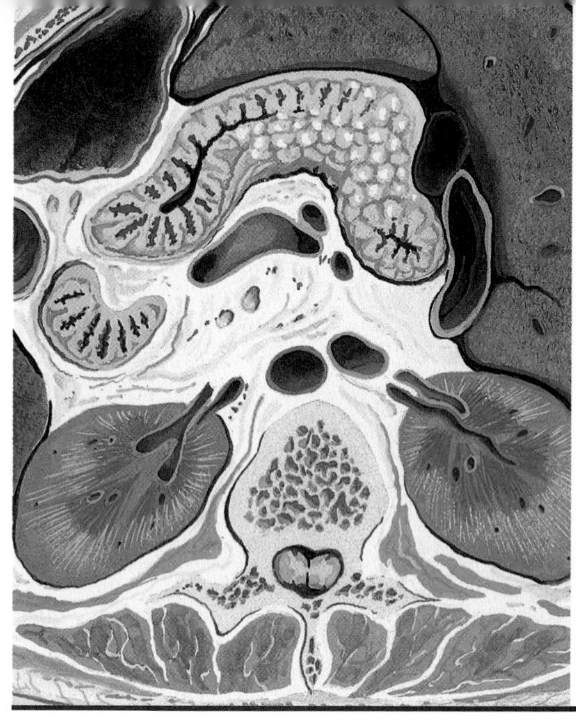

CHAPTER 1

Chapter Outline

Student Objectives

- Define anatomy, physiology, and pathophysiology. Use an example to explain how they are related.
- Name the levels of organization of the body from simplest to most complex, and explain each.
- Define homeostasis, and use an example to explain.
- Explain how a negative feedback mechanism works.
- Describe the anatomical position.
- State the anatomical terms for the parts of the body.
- Use proper terminology to describe the location of body parts with respect to one another.
- Name the body cavities, their membranes, and some organs within each cavity.
- Describe the possible sections through the body or an organ.
- Explain how and why the abdomen is divided into smaller areas. Be able to name organs in these areas.

Organization and General Plan of the Body

New Terminology

Anatomy (uh-**NAT**-uh-mee)
Body cavity (**BAH**-dee **KAV**-i-tee)
Cell (**SELL**)
Homeostasis (HOH-me-oh-**STAY**-sis)
Inorganic chemicals (**IN**-or-GAN-ik **KEM**-i-kuls)
Meninges (me-**NIN**-jeez)
Negative feedback (**NEG**-ah-tiv **FEED**-bak)
Organ (**OR**-gan)
Organ system (**OR**-gan **SIS**-tem)
Organic chemicals (or-**GAN**-ik **KEM**-i-kuls)
Pathophysiology (PATH-oh-FIZZ-ee-**AH**-luh-jee)
Pericardial membranes (PER-ee-**KAR**-dee-uhl **MEM**-brains)
Peritoneum—Mesentery (PER-i-toh-**NEE**-um—**MEZ**-en-TER-ee)
Physiology (FIZZ-ee-**AH**-luh-jee)
Plane (**PLAYN**)
Pleural membranes (**PLOOR**-uhl **MEM**-brains)
Section (**SEK**-shun)
Tissue (**TISH**-yoo)

Related Clinical Terminology

Computed tomography (CT) scan (kom-**PEW**-ted toh-**MAH**-grah-fee SKAN)
Diagnosis (DYE-ag-**NO**-sis)
Disease (di-**ZEEZ**)
Magnetic resonance imaging (MRI) (mag-**NET**-ik **REZ**-ah-nanse **IM**-ah-jing)
Positron emission tomography (PET) (**PAHZ**-i-tron e-**MISH**-un toh-**MAH**-grah-fee)

Terms that appear in ***bold type*** *in the chapter text are defined in the glossary, which begins on p. 528.*

The human body is a precisely structured container of chemical reactions. Have you ever thought of yourself in this way? Probably not, and yet, in the strictly physical sense, that is what each of us is. The body consists of trillions of atoms in specific arrangements and thousands of chemical reactions proceeding in a very orderly manner. That literally describes us, and yet it is clearly not the whole story. The keys to understanding human consciousness and self-awareness are still beyond our grasp. We do not yet know what enables us to study ourselves—no other animals do, as far as we know—but we have accumulated a great deal of knowledge about ourselves. Some of this knowledge makes up the course you are about to take, a course in basic human anatomy and physiology.

Anatomy is the study of body structure, which includes size, shape, composition, and perhaps even coloration. **Physiology** is the study of how the body functions. The physiology of red blood cells, for example, includes what these cells do, how they do it, and how this is related to the functioning of the rest of the body. Physiology is directly related to anatomy. For example, red blood cells contain the mineral iron in molecules of the protein called hemoglobin; this is an aspect of their anatomy. The presence of iron enables red blood cells to carry oxygen, which is their function. All cells in the body must receive oxygen in order to function properly, so the physiology of red blood cells is essential to the physiology of the body as a whole.

Pathophysiology is the study of disorders of functioning, and a knowledge of normal physiology makes such disorders easier to understand. For example, you are probably familiar with the anemia called iron-deficiency anemia. With insufficient iron in the diet, there is not enough iron in the hemoglobin of red blood cells, and less oxygen can be transported throughout the body, resulting in the symptoms of the iron-deficiency disorder. This example shows the relationship between anatomy, physiology, and pathophysiology.

The purpose of this text is to enable you to gain an understanding of anatomy and physiology with the emphasis on normal structure and function. Many examples of pathophysiology have been included, however, to illustrate the relationship of **disease** to normal physiology and to describe some of the procedures used in the **diagnosis** of disease. Many of the examples are clinical applications that will help you begin to apply what you have learned and demonstrate that your knowledge of anatomy and physiology will become the basis for your further study in the health professions.

LEVELS OF ORGANIZATION

The human body is organized in structural and functional levels of increasing complexity. Each higher level incorporates the structures and functions of the previous level, as you will see. We will begin with the simplest level, which is the chemical level, and proceed to cells, tissues, organs, and organ systems. All of the levels of organization are depicted in Fig. 1-1.

CHEMICALS

The chemicals that make up the body may be divided into two major categories: inorganic and organic. **Inorganic chemicals** are usually simple molecules made of one or two elements other than carbon (with a few exceptions). Examples of inorganic chemicals are water (H_2O); oxygen (O_2); one of the exceptions, carbon dioxide (CO_2); and minerals such as iron (Fe), calcium (Ca), and sodium (Na). **Organic chemicals** are often very complex and always contain the elements carbon and hydrogen. In this category of organic chemicals are carbohydrates, fats, proteins, and nucleic acids. The chemical organization of the body is the subject of Chapter 2.

CELLS

The smallest living units of structure and function are **cells.** There are many different types of cells; each is made of chemicals and carries out specific chemical reactions. Cell structure and function are discussed in Chapter 3.

TISSUES

A **tissue** is a group of cells with similar structure and function. There are four groups of tissues:

Epithelial tissues—cover or line body surfaces; some are capable of producing secretions with specific functions. The outer layer of the skin and sweat glands are examples of epithelial tissues.

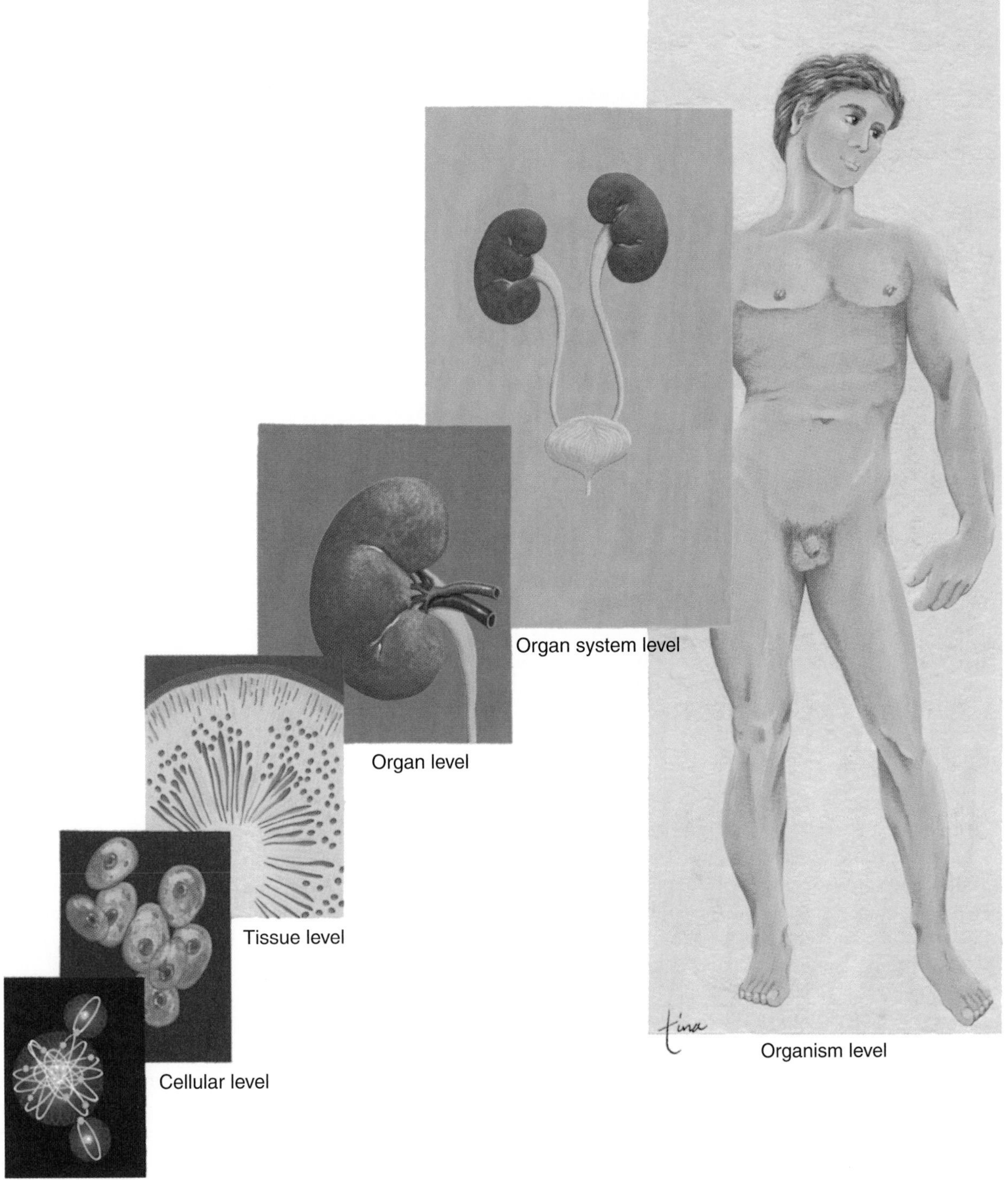

Figure 1–1 Levels of structural organization of the human body, depicted from the simplest (chemical) to the most complex (organism). The organ shown here is the kidney, and the organ system is the urinary system.

Box 1–1 REPLACING TISSUES AND ORGANS

Blood transfusions are probably the most familiar and frequent form of "replacement parts" for people. Blood is a tissue, and when properly typed and cross-matched (blood types will be discussed in Chapter 11) may safely be given to someone with the same or a compatible blood type.

Organs, however, are much more complex structures. When a patient receives an organ transplant, there is always the possibility of rejection (destruction) of the organ by the recipient's immune system (Chapter 14). With the discovery and use of more effective immune-suppressing medications, however, the success rate for many types of organ transplants has increased. Organs that may be transplanted include corneas, kidneys, the heart, the liver, and the lungs.

The skin is also an organ, but skin transplanted from another person will not survive very long. Several kinds of artificial skin are now available to temporarily cover large areas of damaged skin. Patients with severe burns, for example, will eventually need skin grafts from their own unburned skin to form permanent new skin over the burn sites. It is now possible to "grow" a patient's skin in laboratory culture, so that a small patch of skin may eventually be used to cover a large surface. Other cells grown in culture include cartilage, bone, pancreas, and liver. Much research is being done on liver implants (not transplants), clusters of functional liver cells grown in a lab. Such implants would reduce or eliminate the need for human donors.

Many artificial replacement parts have also been developed. These are made of plastic or metal and are not rejected as foreign by the recipient's immune system. Damaged heart valves, for example, may be replaced by artificial ones, and sections of arteries may be replaced by tubular grafts made of synthetic materials. Artificial joints are available for every joint in the body, as is artificial bone for reconstructive surgery. Cochlear implants (tiny, electronic "ears") have provided some sense of hearing for people with certain types of deafness. Work is also progressing on the use of a featherweight computer chip as an artificial retina, and on devices that help damaged hearts pump blood more efficiently.

Connective tissues—connect and support parts of the body; some transport or store materials. Blood, bone, and adipose tissue are examples of this group.

Muscle tissues—specialized for contraction, which brings about movement. Our skeletal muscles and the heart are examples of muscle tissue.

Nerve tissue—specialized to generate and transmit electro-chemical impulses that regulate body functions. The brain and optic nerves are examples of nerve tissue.

The types of tissues in these four groups, as well as their specific functions, are the subject of Chapter 4.

ORGANS

An **organ** is a group of tissues precisely arranged so as to accomplish specific functions. Examples of organs are the kidneys, liver, lungs, and stomach. The stomach is lined with epithelial tissue that secretes gastric juice for digestion. Muscle tissue in the wall of the stomach contracts to mix food with gastric juice and propel it to the small intestine. Nerve tissue carries impulses that increase or decrease the contractions of the stomach (see Box 1–1: Replacing Tissues and Organs).

ORGAN SYSTEMS

An **organ system** is a group of organs that all contribute to a particular function. Examples are the urinary system, digestive system, and respiratory system. In Fig. 1–1 you see the urinary system, which consists of the kidney, ureters, urinary bladder, and urethra. These organs all contribute to the formation and elimination of urine.

As a starting point, Table 1–1 lists the organ systems of the human body with their general functions, and some representative organs (Fig. 1–2); these organ

Table 1–1 THE ORGAN SYSTEMS

System	Functions	Organs*
Integumentary	• Is a barrier to pathogens and chemicals • Prevents excessive water loss	skin, hair, subcutaneous tissue
Skeletal	• Supports the body • Protects internal organs • Provides a framework to be moved by muscles	bones, ligaments
Muscular	• Moves the skeleton • Produces heat	muscles, tendons
Nervous	• Interprets sensory information • Regulates body functions such as movement by means of electro-chemical impulses	brain, nerves, eyes, ears
Endocrine	• Regulates body functions by means of hormones	thyroid gland, pituitary gland
Circulatory	• Transports oxygen and nutrients to tissues and removes waste products	heart, blood, arteries
Lymphatic	• Returns tissue fluid to the blood • Destroys pathogens that enter the body	spleen, lymph nodes
Respiratory	• Exchanges oxygen and carbon dioxide between the air and blood	lungs, trachea, larynx
Digestive	• Changes food to simple chemicals that can be absorbed and used by the body	stomach, colon, liver
Urinary	• Removes waste products from the blood • Regulates volume and pH of blood	kidneys, urinary bladder, urethra
Reproductive	• Produces eggs or sperm • *In women*, provides a site for the developing embryo-fetus	*Female*: ovaries, uterus *Male*: testes, prostate gland

*These are simply representative organs, not an all-inclusive list.

systems make up an individual person. The balance of this text discusses each system in more detail.

HOMEOSTASIS

A person who is in good health is in a state of **homeostasis**. Homeostasis reflects the ability of the body to maintain relative stability and to function normally despite constant changes. Changes may be external or internal, and the body must respond appropriately.

Eating breakfast, for example, brings about an internal change. Suddenly there is food in the stomach, and something must be done with it. What happens? The food is digested or broken down into simple chemicals that the body can use. The protein in a hard-boiled egg is digested into amino acids, its basic chemical building blocks; these can then be used by the body to produce its own specialized proteins.

An example of an external change is a rise in environmental temperature. On a hot day, the body temperature would also tend to rise. However, body temperature must be kept within its normal range of about 97° to 99°F (36° to 38°C), in order to support normal functioning. What happens? One of the body's responses to the external temperature rise is to increase sweating so that excess body heat can be lost by the evaporation of sweat on the surface of the skin. This response, however, may bring about an undesirable internal change, dehydration. What happens? As body water decreases, we feel the sensation of thirst and drink fluids to replace the water lost in sweating. Notice that when certain body responses occur, they reverse the event that triggered them. In the example above, a rising body temperature stimulates increased sweating, which lowers body temperature, which in turn decreases sweating. This is an example of a **negative feedback mechanism**, in which the body's response reverses the stimulus and keeps some aspect of the body within its normal range.

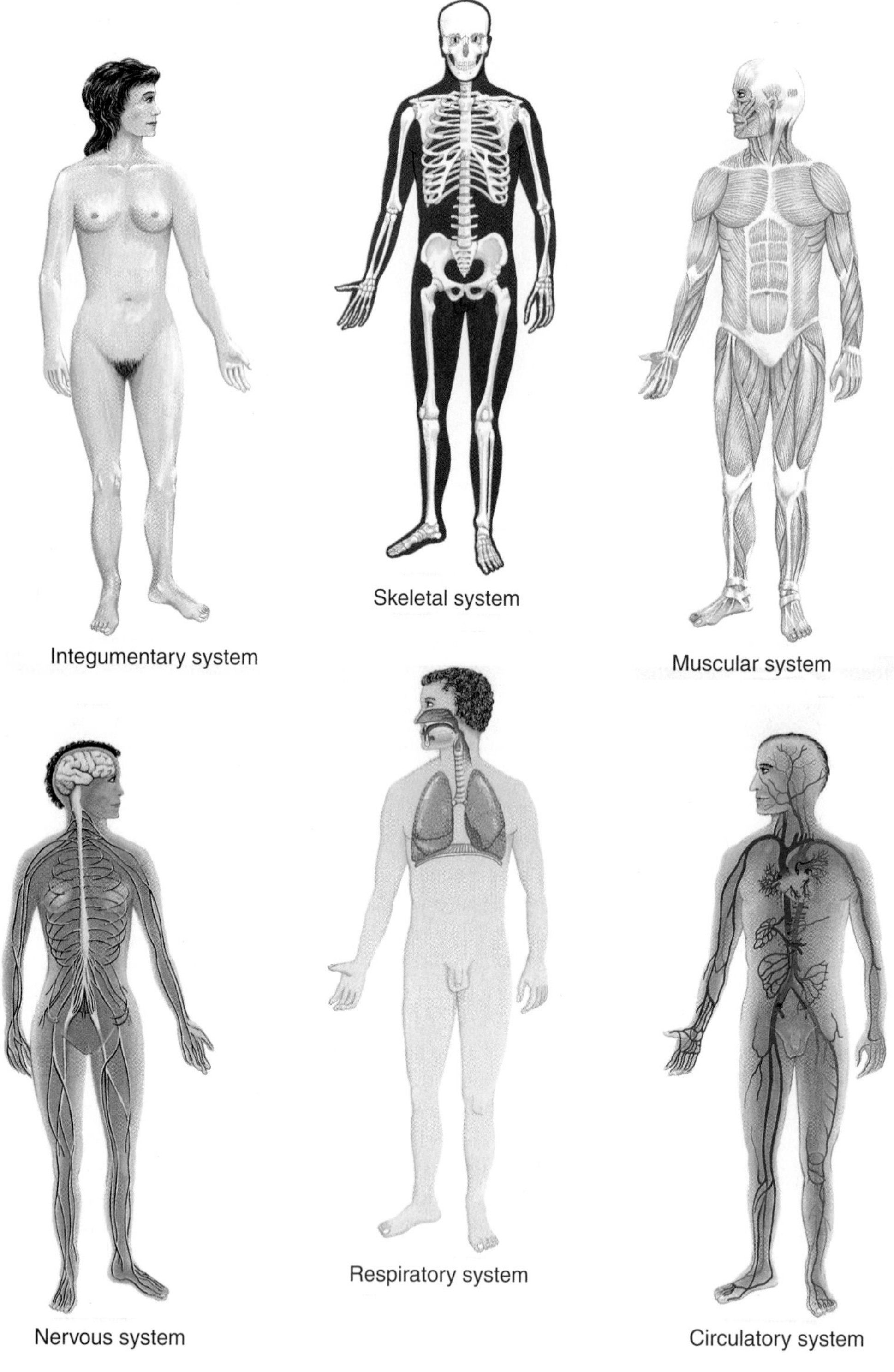

Figure 1–2 Organ systems. Compare the depiction of each system to its description in Table 1–1. Try to name at least one organ shown in each system.

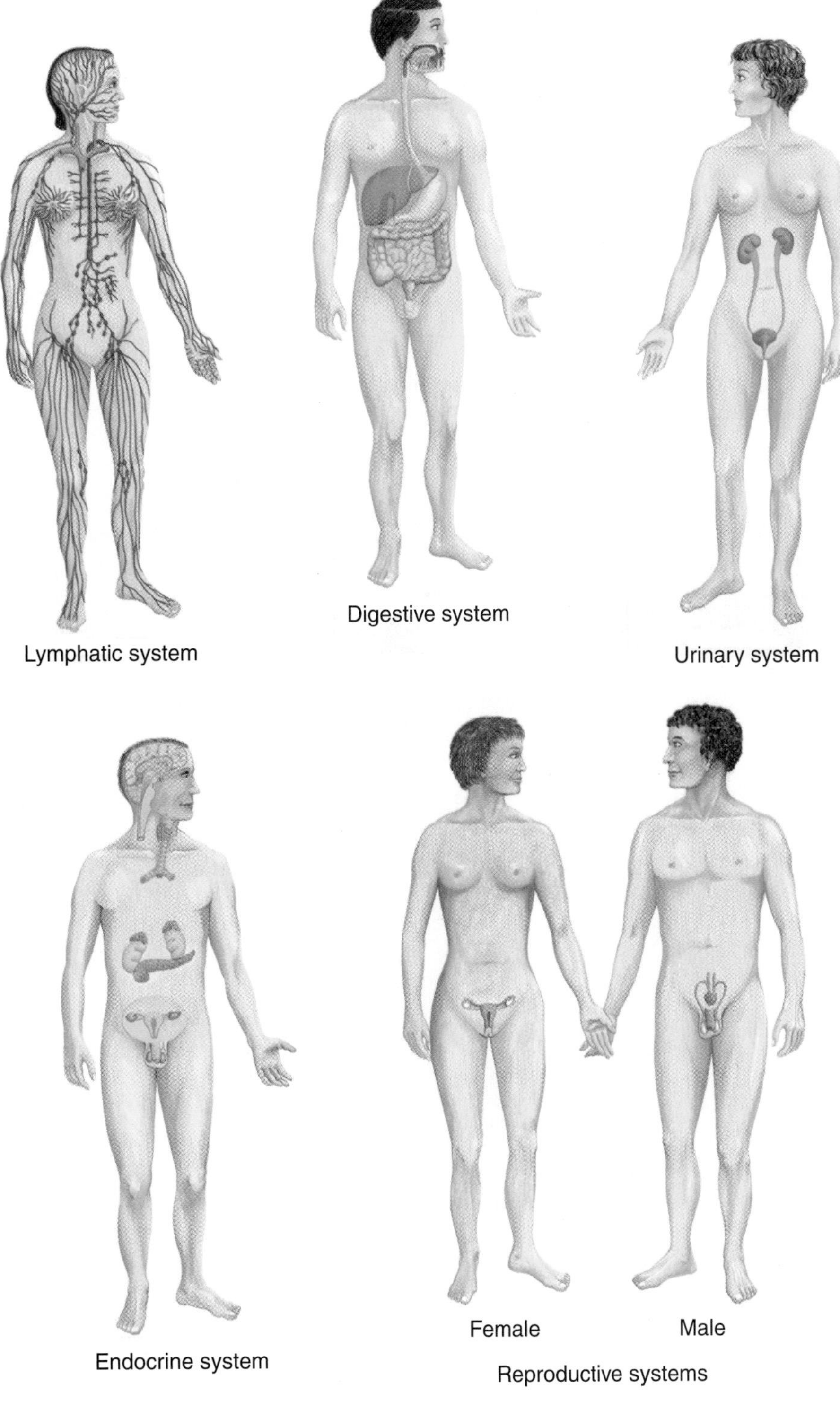

Figure 1–2 (Continued)

You can probably think of many other situations in which your body responds to changes and keeps you alive and healthy—this steady-state or equilibrium is homeostasis. As you continue your study of the human body, keep in mind that the proper functioning of each organ and organ system contributes to homeostasis.

TERMINOLOGY AND GENERAL PLAN OF THE BODY

As part of your course in anatomy and physiology, you will learn many new words or terms. At times you may feel that you are learning a second language, and indeed you are. Each term has a precise meaning, which is understood by everyone else who has learned the language. Mastering the terminology of your profession is essential to enable you to communicate effectively with your co-workers and your future patients. Although the number of new terms may seem a bit overwhelming at first, you will find that their use soon becomes second nature to you.

The terminology presented in this chapter will be used throughout the text in the discussion of the organ systems. This will help to reinforce the meanings of these terms and will transform these new words into knowledge.

Table 1–2 DESCRIPTIVE TERMS FOR BODY PARTS AND AREAS

Term	Definition (Refers to)
Axillary	armpit
Brachial	upper arm
Buccal (oral)	mouth
Cardiac	heart
Cervical	neck
Cranial	head
Cutaneous	skin
Deltoid	shoulder
Femoral	thigh
Frontal	forehead
Gastric	stomach
Gluteal	buttocks
Hepatic	liver
Iliac	hip
Inguinal	groin
Lumbar	small of back
Mammary	breast
Nasal	nose
Occipital	back of head
Orbital	eye
Parietal	crown of head
Patellar	kneecap
Pectoral	chest
Perineal	pelvic floor
Plantar	sole of foot
Popliteal	back of knee
Pulmonary	lungs
Renal	kidney
Sacral	base of spine
Temporal	side of head
Umbilical	naval
Volar (palmar)	palm

BODY PARTS AND AREAS

Each of the terms listed in Table 1-2 and shown in Fig. 1-3 refers to a specific part or area of the body. For example, "femoral" always refers to the thigh. The femoral artery is a blood vessel that passes through the thigh, and the quadriceps femoris is a large muscle group of the thigh.

Another example is "pulmonary," which always refers to the lungs, as in pulmonary artery, pulmonary edema, and pulmonary embolism. Although you may not know the exact meaning of each of these terms now, you do know that each has something to do with the lungs.

TERMS OF LOCATION AND POSITION

When describing relative locations, the body is always assumed to be in **anatomical position**: standing upright facing forward, arms at the sides with palms forward, and the feet slightly apart. The terms of location are listed in Table 1-3, with a definition and example for each. As you read each term, find the body parts used as examples in Figs. 1-3 and 1-4. Notice also that these are pairs of terms and that each pair is a set of opposites. This will help you recall the terms and their meanings.

BODY CAVITIES AND THEIR MEMBRANES

The body has two major cavities: the **dorsal cavity** (posterior) and the **ventral cavity** (anterior). Each of

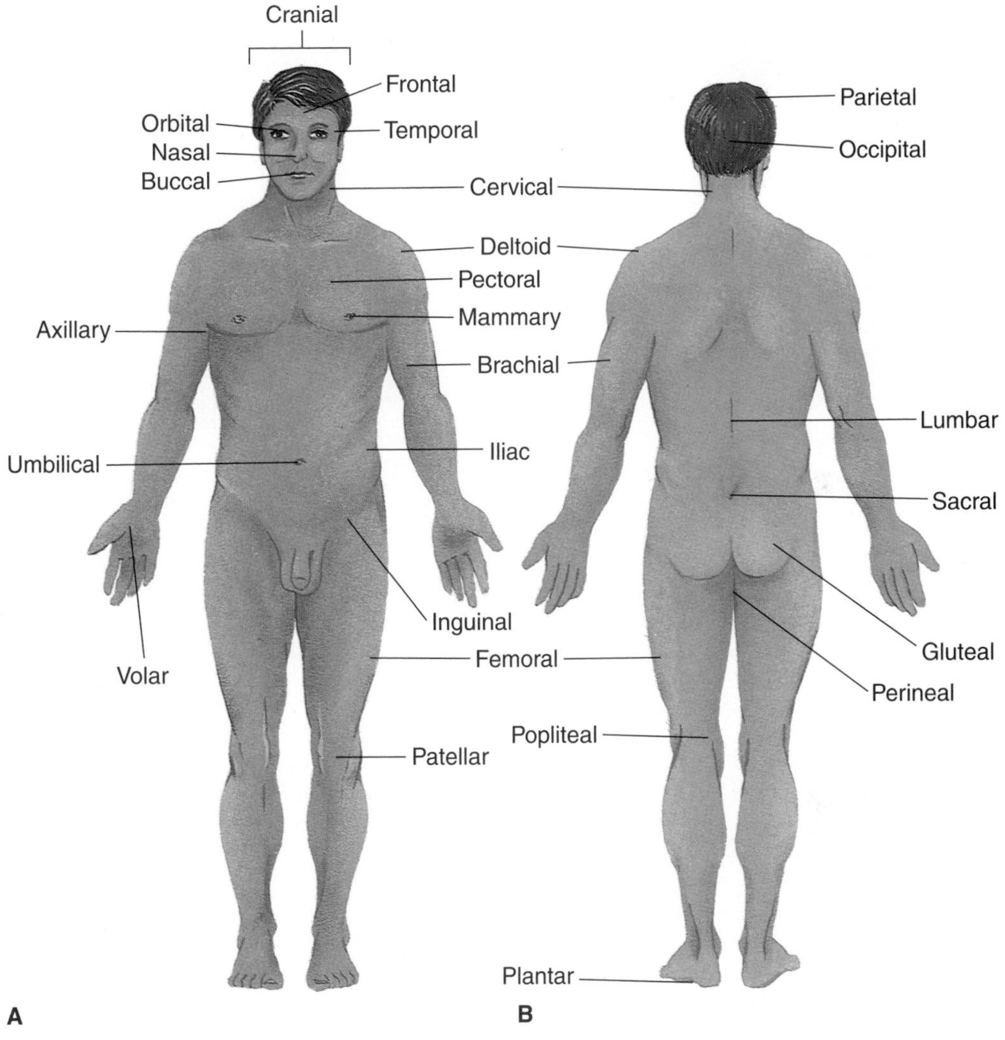

Figure 1–3 Body parts and areas. The body is shown in anatomical position. **(A)** Anterior view. **(B)** Posterior view. (Compare with Table 1–2.)

these cavities has further subdivisions, which are shown in Fig. 1–4.

Dorsal Cavity

The dorsal cavity consists of the cranial cavity and the vertebral or spinal cavity. The **cranial cavity** is formed by the skull and contains the brain. The **spinal cavity** is formed by the backbone (spine) and contains the spinal cord. The membranes that line these cavities and cover the organs of the central nervous system are called the **meninges**.

Ventral Cavity

The ventral cavity consists of two compartments, the thoracic cavity and the abdominal cavity, which are separated by the diaphragm. The pelvic cavity may be

Table 1–3 TERMS OF LOCATION AND POSITION

Term	Definition	Example
Superior	above, or higher	The heart is superior to the liver.
Inferior	below, or lower	The liver is inferior to the lungs.
Anterior	toward the front	The chest is on the anterior side of the body.
Posterior	toward the back	The lumbar area is posterior to the umbilical area.
Ventral	toward the front	The mammary area is on the ventral side of the body.
Dorsal	toward the back	The buttocks are on the dorsal side of the body.
Medial	toward the midline	The heart is medial to the lungs.
Lateral	away from the midline	The shoulders are lateral to the neck.
Internal	within, or interior to	The brain is internal to the skull.
External	outside, or exterior to	The ribs are external to the lungs.
Superficial	toward the surface	The skin is the most superficial organ.
Deep	within, or interior to	The deep veins of the legs are surrounded by muscles.
Central	the main part	The brain is part of the central nervous system.
Peripheral	extending from the main part	Nerves in the arm are part of the peripheral nervous system.
Proximal	closer to the origin	The knee is proximal to the foot.
Distal	farther from the origin	The palm is distal to the elbow.
Parietal	pertaining to the wall of a cavity	The parietal pleura lines the chest cavity.
Visceral	pertaining to the organs within a cavity	The visceral pleura covers the lungs.

considered a subdivision of the abdominal cavity or as a separate cavity.

Organs in the **thoracic cavity** include the heart and lungs. The membranes of the thoracic cavity are serous membranes called the **pleural membranes**. The parietal pleura lines the chest wall, and the visceral pleura covers the lungs. The heart has its own set of serous membranes called the pericardial membranes. The parietal pericardium lines the fibrous pericardial sac, and the visceral pericardium covers the heart muscle.

Organs in the **abdominal cavity** include the liver, stomach, and intestines. The membranes of the abdominal cavity are also serous membranes called the peritoneum and mesentery. The **peritoneum** is the membrane that lines the abdominal wall, and the **mesentery** is the membrane folded around and covering the outer surfaces of the abdominal organs.

The **pelvic cavity** is inferior to the abdominal cavity. Although the peritoneum does not line the pelvic cavity, it covers the free surfaces of several pelvic organs. Within the pelvic cavity are the urinary bladder and reproductive organs such as the uterus in women and the prostate gland in men.

PLANES AND SECTIONS

When internal anatomy is described, the body or an organ is often cut or **sectioned** in a specific way so as to make particular structures easily visible. A **plane** is an imaginary flat surface that separates two portions of the body or an organ. These planes and sections are shown in Fig. 1–5 (see Box 1–2: Visualizing the Interior of the Body).

Frontal (coronal) section—a plane from side to side separates the body into front and back portions.

Sagittal section—a plane from front to back separates the body into right and left portions. A midsagittal section creates equal right and left halves.

Transverse section—a horizontal plane separates the body into upper and lower portions.

Cross-section—a plane perpendicular to the long

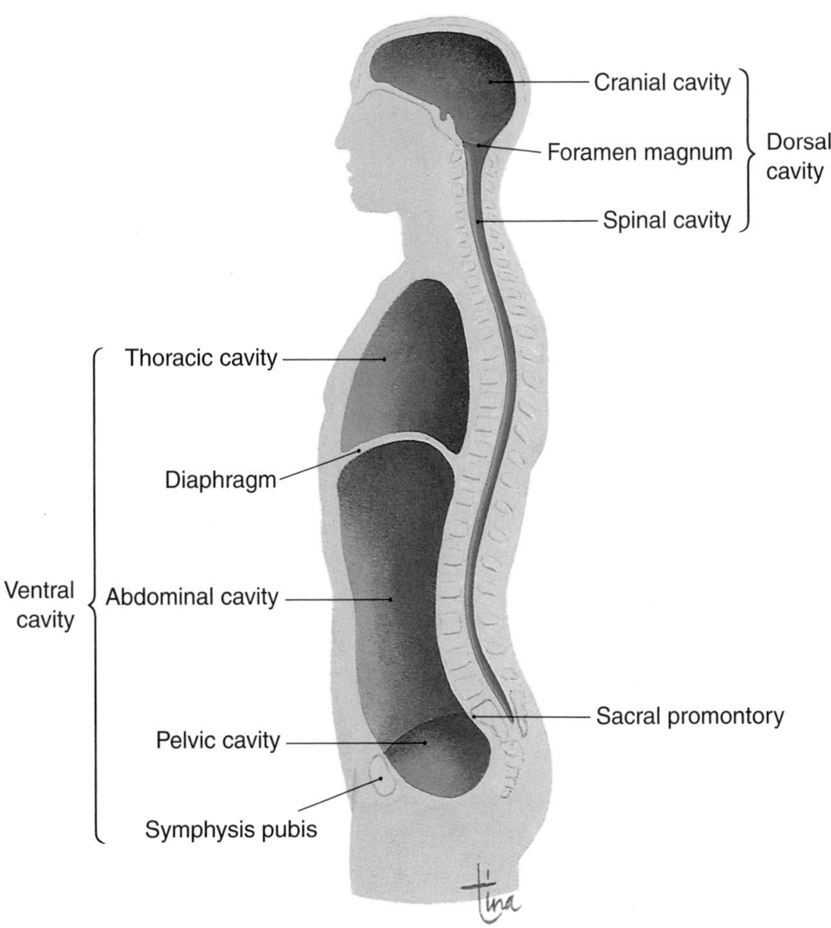

Figure 1–4 Body cavities. (Lateral view from the left side.)

axis of an organ. A cross-section of the small intestine (which is a tube) would look like a circle with the cavity of the intestine in the center.

Longitudinal section—a plane along the long axis of an organ. A longitudinal section of the intestine is shown in Fig. 1-5, and a frontal section of the femur (thigh bone) would also be a longitudinal section (see Fig. 6-1).

AREAS OF THE ABDOMEN

The abdomen is a large area of the lower trunk of the body. If a patient reported "abdominal pain," the physician or nurse would want to know more precisely where the pain was. In order to do this, the abdomen may be divided into smaller regions or areas, which are shown in Fig. 1-6.

Quadrants—a transverse plane and a midsagittal plane that cross at the umbilicus will divide the abdomen into four quadrants. Clinically, this is probably the division used more frequently. The pain of gallstones might then be described as in the right upper quadrant.

Nine Areas—two transverse planes and two sagittal planes divide the abdomen into nine areas.

Upper areas—above the level of the rib cartilages are the left hypochondriac, epigastric, and right hypochondriac.

Middle areas—the left lumbar, umbilical, and right lumbar.

Lower areas—below the level of the top of the pelvic bone are the left iliac, hypogastric, and right iliac. This division is often used in anatomical studies to describe the location of organs. The liver, for example, is located in the epigastric and right hypochondriac areas.

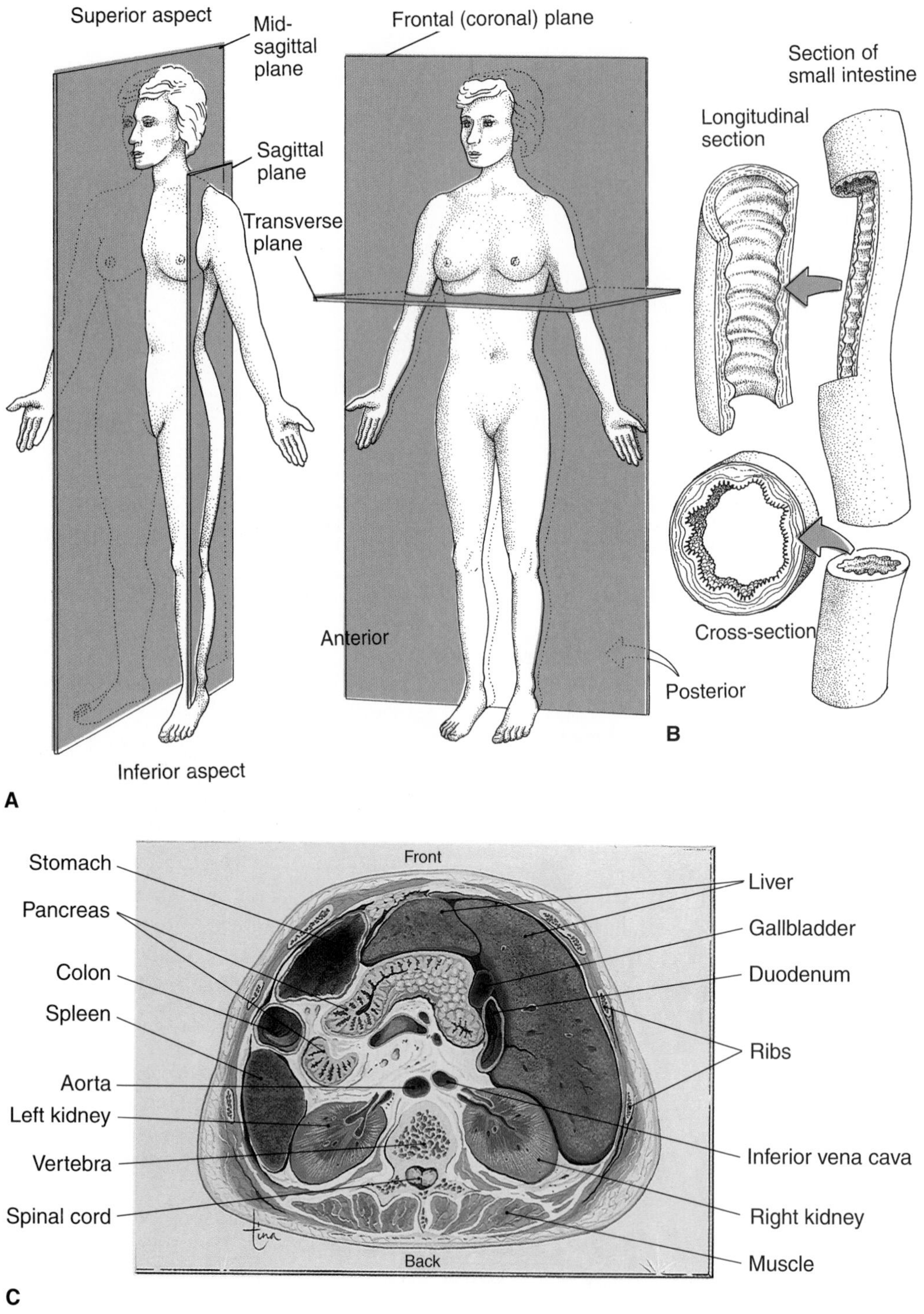

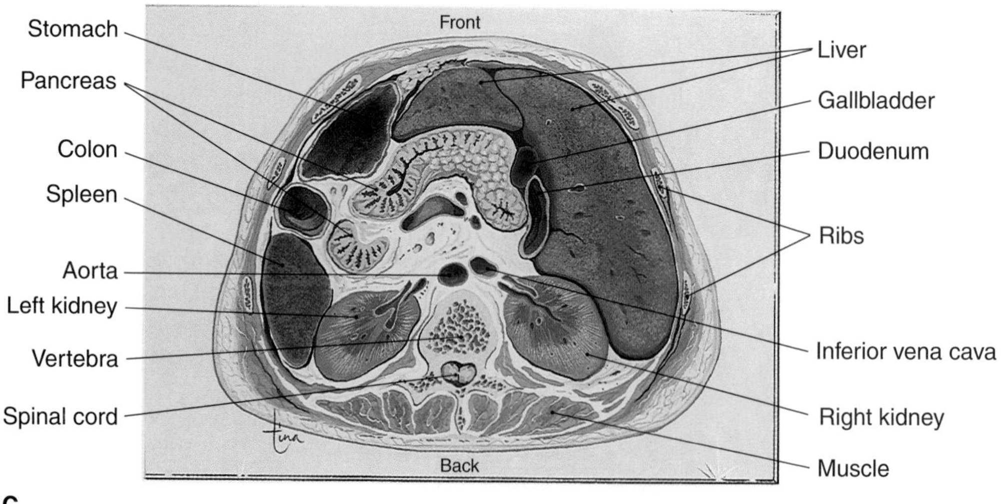

Figure 1–5 **(A)** Planes and sections of the body. **(B)** Cross-section and longitudinal section of the small intestine. **(C)** Transverse section through the upper abdomen.

Box 1–2 VISUALIZING THE INTERIOR OF THE BODY

In the past, the need for exploratory surgery brought with it hospitalization, risk of infection, and discomfort and pain for the patient. Today, however, new technologies and the extensive use of computers permit us to see the interior of the body without surgery.

Computed tomography (CT) scanning uses a narrowly focused x-ray beam that circles rapidly around the body. A detector then measures how much radiation passes through different tissues, and a computer constructs an image of a thin slice through the body. Several images may be made at different levels—each takes only a few seconds—to provide a more complete picture of an organ or part of the body. The images are much more detailed than are those produced by conventional x-rays.

Magnetic resonance imaging (MRI) is another diagnostic tool that is especially useful for visualizing soft tissues, including the brain and spinal cord. Recent refinements have produced images of individual nerve bundles, which had not been possible using any other technique. The patient is placed inside a strong magnetic field, and the tissues are pulsed with radio waves. Because each tissue has different proportions of various atoms, which resonate or respond differently, each tissue emits a characteristic signal. A computer then translates these signals into an image; the entire procedure takes 30 to 45 minutes.

Positron emission tomography (PET) scanning creates images that depict the rates of physiological processes such as blood flow, oxygen usage, or glucose metabolism. The comparative rates are depicted by colors: Red represents the highest rate, followed by yellow, then green, and finally blue representing the lowest rate.

One drawback of the new technologies is their cost; they are expensive. However, the benefits to patients are great: highly detailed images of the body are obtained without the risks of surgery and with virtually no discomfort in the procedures themselves.

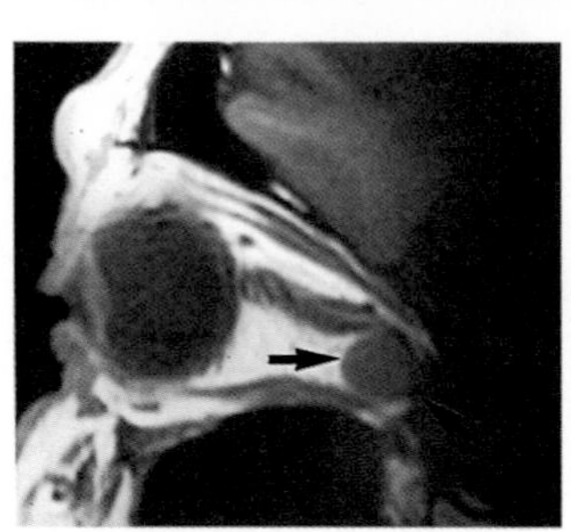
A

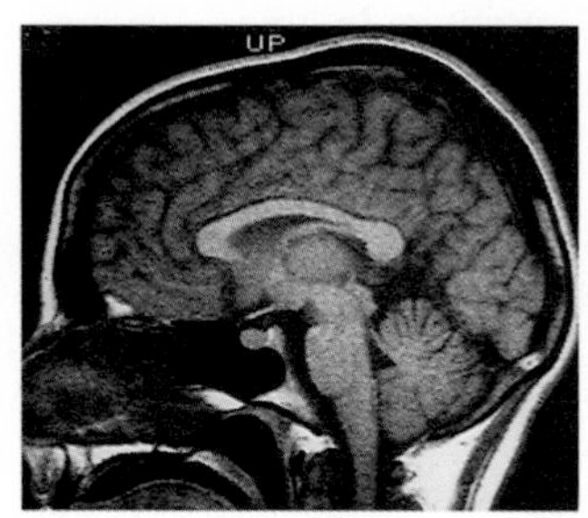

B

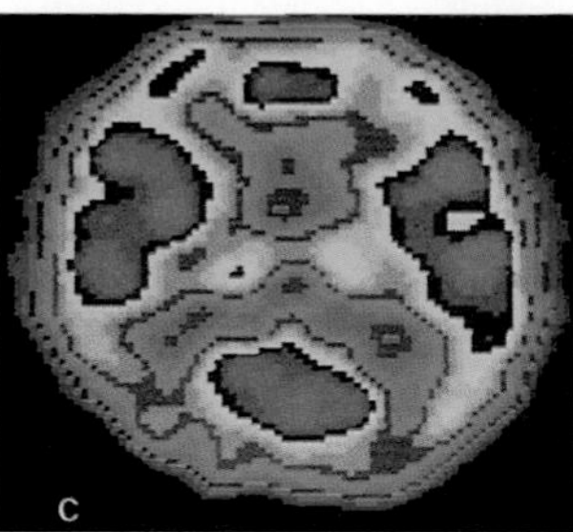

C

Box Figure 1–A Imaging techniques. **(A)** CT scan of eye in lateral view showing a tumor *(arrow)* below the optic nerve. **(B)** MRI of midsagittal section of head (compare with Fig. 8–6 and 15–1). **(C)** PET scan of brain in transverse section (frontal lobes at top) showing glucose metabolism. (From Mazziotta, JC, and Gilman, S: Clinical Brain Imaging: Principles and Applications. FA Davis, Philadelphia, 1992, pp 27 and 298, with permission.)

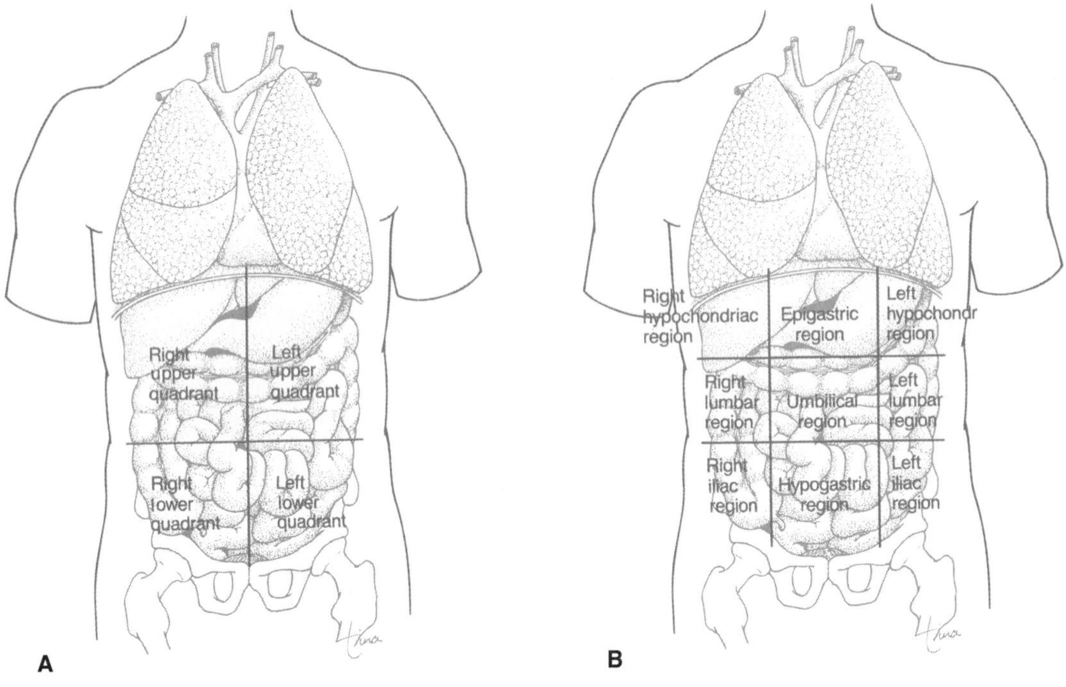

Figure 1–6 Areas of the abdomen. **(A)** Four quadrants. **(B)** Nine regions.

SUMMARY

As you will see, the terminology presented in this chapter is used throughout the text to describe the anatomy of organs and the names of their parts. We will now return to a consideration of the structural organization of the body and to more detailed descriptions of its levels of organization. The first of these, the chemical level, is the subject of the next chapter.

STUDY OUTLINE

Introduction

1. Anatomy—the study of structure.
2. Physiology—the study of function.
3. Pathophysiology—the study of disorders of functioning.

Levels of Organization

1. Chemical—inorganic and organic chemicals make up all matter, both living and non-living.
2. Cells—the smallest living units of the body.
3. Tissues—groups of cells with similar structure and function.
4. Organs—groups of tissues that contribute to specific functions.
5. Organ Systems—groups of organs that work together to perform specific functions (see Table 1-1 and Fig. 1-2).
6. Person—all the organ systems functioning properly.

Homeostasis

1. A state of good health maintained by the normal functioning of the organ systems.
2. The body constantly responds to internal and external changes, yet remains stable.
3. Negative feedback mechanism—a control system in which a stimulus initiates a response that reverses or reduces the stimulus, thereby stopping the response until the stimulus occurs again.

Terminology and General Plan of the Body

1. Body Parts and Areas—see Table 1-2 and Fig. 1-3.
2. Terms of Location and Position—used to describe relationships of position (see Table 1-3 and Figs. 1-3 and 1-4).
3. Body cavities and their membranes (see Fig. 1-4).
   - Dorsal cavity—lined with membranes called meninges; consists of the cranial and vertebral cavities.
     - ◦ Cranial cavity contains the brain.
     - ◦ Vertebral cavity contains the spinal cord.
   - Ventral cavity—the diaphragm separates the thoracic and abdominal cavities; the pelvic cavity is inferior to the abdominal cavity.
     - ◦ Thoracic cavity—contains the lungs and heart.
       - Pleural membranes line the chest wall and cover the lungs.
       - Pericardial membranes surround the heart.
     - ◦ Abdominal cavity—contains many organs including the stomach, liver, and intestines.
       - The peritoneum lines the abdominal cavity; the mesentery covers the abdominal organs.
     - ◦ Pelvic cavity—contains the urinary bladder and reproductive organs.
4. Planes and Sections—cutting the body or an organ in a specific way (see Fig. 1-5).
   - Frontal or Coronal—separates front and back parts.
   - Sagittal—separates right and left parts.
   - Transverse—separates upper and lower parts.
   - Cross—a section perpendicular to the long axis.
   - Longitudinal—a section along the long axis.
5. Areas of the Abdomen—permits easier description of locations:
   - Quadrants—see Fig. 1-6.
   - Nine areas—see Fig. 1-6.

REVIEW QUESTIONS

1. Explain how the physiology of a bone is related to its anatomy. Explain how the physiology of the hand is related to its anatomy. (p. 4)
2. Describe anatomical position. Why is this knowledge important? (p. 10)
3. Name the organ system with each of the following functions: (p. 7)
   a. Moves the skeleton
   b. Regulates body functions by means of hormones
   c. Covers the body and prevents entry of pathogens
   d. Destroys pathogens that enter the body
   e. Exchanges oxygen and carbon dioxide between the air and blood
4. Name the two major body cavities and their subdivisions. Name the cavity lined by the peritoneum, meninges, and parietal pleura. (pp. 10, 12)
5. Name the four quadrants of the abdomen. Name at least one organ in each quadrant. (pp. 13, 16)
6. Name the section through the body that would result in each of the following: equal right and left halves, anterior and posterior parts, superior and inferior parts. (pp. 12-16)
7. Review Table 1-2, and try to find each external area on your own body. (pp. 10-11)
8. Define cell. When similar cells work together, what name are they given? (p. 4)
9. Define organ. When a group of organs works together, what name is it given? (p. 6)
10. Define homeostasis. (pp. 7, 10)
    a. Give an example of an external change and explain how the body responds to maintain homeostasis.
    b. Give an example of an internal change and explain how the body responds to maintain homeostasis.
    c. Briefly explain how a negative feedback mechanism works.

CHAPTER 2

Chapter Outline

Elements
Atoms
Chemical Bonds
Ionic Bonds
Covalent Bonds
Hydrogen Bonds
Chemical Reactions
Inorganic Compounds of Importance
Water
Water Compartments
Oxygen
Carbon Dioxide
Cell Respiration
Trace Elements
Acids, Bases, and pH
    Buffer systems
Organic Compounds of Importance
Carbohydrates
Lipids
Proteins
    Enzymes
Nucleic Acids
    DNA and RNA
    ATP

Box 2–1 BLOOD GASES
Box 2–2 NITRIC OXIDE
Box 2–3 SATURATED AND UNSATURATED FATS

Student Objectives

- Define the terms element, atom, proton, neutron, and electron.
- Describe the formation and purpose of: ionic bonds, covalent bonds, and hydrogen bonds.
- Describe what happens in synthesis and decomposition reactions.
- Explain the importance of water to the functioning of the human body.
- Name and describe the water compartments.
- Explain the roles of oxygen and carbon dioxide in cell respiration.
- State what trace elements are, and name some, with their functions.
- Explain the pH scale. State the normal pH ranges of body fluids.
- Explain how a buffer system limits great changes in pH.
- Describe the functions of monosaccharides, disaccharides, oligosaccharides, and polysaccharides.
- Describe the functions of true fats, phospholipids, and steroids.
- Describe the functions of proteins, and explain how enzymes function as catalysts.
- Describe the functions of DNA, RNA, and ATP.

Some Basic Chemistry

New Terminology

Acid (**ASS**-sid)
Amino acid (ah-**MEE**-noh **ASS**-sid)
Atom (**A**-tum)
Base (**BAYSE**)
Buffer system (**BUFF**-er **SIS**-tem)
Carbohydrates (KAR-boh-**HIGH**-drayts)
Catalyst (**KAT**-ah-list)
Cell respiration (SELL RES-pi-**RAY**-shun)
Covalent bond (ko-**VAY**-lent)
Dissociation-ionization (dih-SEW-see-**AY**-shun; EYE-uh-nih-**ZAY**-shun)
Element (**EL**-uh-ment)
Enzyme (**EN**-zime)
Extracellular fluid (EX-trah-**SELL**-yoo-ler)
Intracellular fluid (IN-trah-**SELL**-yoo-ler)
Ion (**EYE**-on)
Ionic bond (eye-**ON**-ik)
Lipids (**LIP**-ids)
Matter (**MAT**-ter)
Molecule (**MAHL**-e-kuhl)
Nucleic acids (new-**KLEE**-ik **ASS**-sids)
pH and pH scale (Pee-H SKALE)
Protein (**PRO**-teen)
Salt (**SAWLT**)
Solvent-solution (**SAHL**-vent; suh-**LOO**-shun)
Steroid (**STEER**-oyd)
Trace elements (TRAYSE **EL**-uh-ments)

Related Clinical Terminology

Acidosis (ASS-i-**DOH**-sis)
Atherosclerosis (ATH-er-oh-skle-**ROH**-sis)
Hypoxia (high-**POCK**-see-ah)
Saturated (**SAT**-uhr-ay-ted) fats
Unsaturated (un-**SAT**-uhr-ay-ted) fats

*Terms that appear in **bold type** in the chapter text are defined in the glossary, which begins on p. 528.*

When you hear or see the word "chemistry" you may think of test tubes and Bunsen burners in a laboratory experiment. However, literally everything in our physical world is made of chemicals. The paper used for this book, which was once the wood of a tree, is made of chemicals. The air we breathe is a mixture of chemicals in the form of gases. Water, lemonade, and diet soda are chemicals in liquid form. Our foods are chemicals, and our bodies are complex arrangements of thousands of chemicals. Recall from Chapter 1 that the simplest level of organization of the body is the chemical level.

This chapter covers some very basic aspects of chemistry as they are related to living organisms, and most especially as they are related to our understanding of the human body.

ELEMENTS

All matter, both living and not living, is made of elements, the simplest chemicals. An element is a substance made of only one type of atom (therefore, an atom is the smallest part of an element). There are 92 naturally occurring elements in the world around us. Examples are hydrogen (H), iron (Fe), oxygen (O), calcium (Ca), nitrogen (N), and carbon (C). In nature, an element does not usually exist by itself but rather combines with the atoms of other elements to form compounds. Examples of some compounds important to our study of the human body are: water (H_2O), in which two atoms of hydrogen combine with one atom of oxygen; carbon dioxide (CO_2), in which an atom of carbon combines with two atoms of oxygen; and glucose ($C_6H_{12}O_6$), in which six carbon atoms and six oxygen atoms combine with 12 hydrogen atoms.

The elements carbon, hydrogen, oxygen, nitrogen, phosphorus, and sulfur are found in all living things. If calcium is included, these seven elements make up approximately 99% of the human body (weight).

More than 20 different elements are found, in varying amounts, in the human body. Some of these are listed in Table 2–1. As you can see, each element has a standard chemical symbol. This is simply the first (and sometimes the second) letter of the element's English or Latin name. You should know the symbols of the elements in this table, since they are used in textbooks, articles, hospital lab reports, and so on. Notice that if a two-letter symbol is used for an element, the second letter is always lower case, not a capital. For example, the symbol for calcium is "Ca," not "CA." "CA" is an abbreviation often used for "cancer."

Table 2–1 ELEMENTS IN THE HUMAN BODY

Element	Symbol	Atomic Number*	Percent of the Body by Weight
Hydrogen	H	1	9.5
Carbon	C	6	18.5
Nitrogen	N	7	3.3
Oxygen	O	8	65.0
Fluorine	F	9	Trace
Sodium	Na	11	0.2
Magnesium	Mg	12	0.1
Phosphorus	P	15	1.0
Sulfur	S	16	0.3
Chlorine	Cl	17	0.2
Potassium	K	19	0.4
Calcium	Ca	20	1.5
Manganese	Mn	25	Trace
Iron	Fe	26	Trace
Cobalt	Co	27	Trace
Copper	Cu	29	Trace
Zinc	Zn	30	Trace
Iodine	I	53	Trace

*Atomic number is the number of protons in the nucleus of the atom. It also represents the number of electrons that orbit the nucleus.

ATOMS

Atoms are the smallest parts of an element that have the characteristics of that element. An atom consists of three major subunits or particles: protons, neutrons, and electrons (Fig. 2-1). A **proton** has a positive electrical charge and is found in the nucleus (or center) of the atom. A **neutron** is electrically neutral (has no charge) and is also found in the nucleus. An **electron** has a negative electrical charge and is found outside the nucleus orbiting in what may be called an electron cloud or shell around the nucleus.

The number of protons in an atom gives it its **atomic number**. Protons and neutrons have mass and weight; they give an atom its **atomic weight**. In an atom, the number of protons (+) equals the number of electrons (−); therefore, an atom is electrically neutral. The electrons, however, are important in that they may enable an atom to connect or **bond** to other atoms to form **molecules**. A molecule is a combination of atoms (usually of more than one element) which are so tightly bound together that the molecule behaves as a single unit.

Each atom is capable of bonding in only very specific ways. This capability depends on the number and the arrangement of the electrons of the atom. Electrons orbit the nucleus of an atom in shells or **energy levels**. The first, or innermost, energy level can contain a maximum of two electrons and is then considered stable. The second energy level is stable when it contains its maximum of eight electrons. The remaining energy levels, more distant from the nucleus, are also most stable when they contain eight electrons, or a multiple of eight.

A few atoms (elements) are naturally stable, or "uninterested" in reacting, because their outermost energy level already contains the maximum number of electrons. The gases helium and neon are examples of these stable atoms, which do not usually react with other atoms. Most atoms are not stable, however, and tend to gain, lose, or share electrons in order to fill their outermost shell. By doing so, an atom is capable of forming one or more chemical bonds with other atoms. In this way, the atom becomes stable, because its outermost shell of electrons has been filled. It is these reactive atoms that are of interest in our study of anatomy and physiology.

CHEMICAL BONDS

A chemical bond is not a structure, but rather a force or attraction between positive and negative electrical charges that keeps two or more atoms closely associated with each other to form a molecule. By way of comparison, think of gravity. We know that gravity is

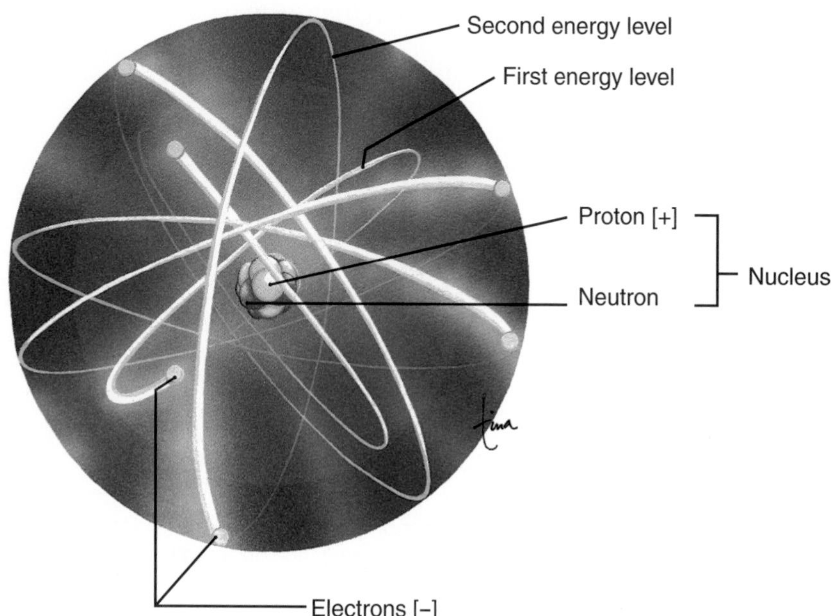

Figure 2–1 An atom of carbon. The nucleus contains six protons and six neutrons (not all are visible here). Six electrons orbit the nucleus, two in the first energy level and four in the second energy level.

not a "thing," but rather the force that keeps our feet on the floor. Molecules formed by chemical bonding then have physical characteristics different from those of the atoms of the original elements. For example, the elements hydrogen and oxygen are gases, but atoms of each may chemically bond to form molecules of water, which is a liquid.

The type of chemical bonding depends upon the tendencies of the electrons of atoms involved, as you will see. Three kinds of bonds are very important to the chemistry of the body: ionic bonds, covalent bonds, and hydrogen bonds.

IONIC BONDS

An **ionic bond** involves the loss of one or more electrons by one atom and the gain of the electron(s) by another atom or atoms. Refer to Fig. 2–2 as you read the following.

An atom of sodium (Na) has one electron in its outermost shell, and in order to become stable, it tends to lose that electron. When it does so, the sodium atom has one more proton than it has electrons. Therefore, it now has an electrical charge (or **valence**) of +1 and is called a sodium **ion** (Na^+). An atom of chlorine has seven electrons in its outermost shell, and in order to become stable tends to gain one electron. When it does so, the chlorine atom has one more electron than it has protons, and now has a charge (valence) of −1. It is called a chloride ion (Cl^-).

When an atom of sodium loses an electron to an atom of chlorine, their ions have unlike charges (positive and negative) and are thus attracted to one another. The result is the formation of a molecule of sodium chloride: NaCl, or common table salt. The bond that holds these ions together is called an ionic bond.

Another example is the bonding of chlorine to calcium. An atom of calcium has two electrons in its outermost shell and tends to lose those electrons in order to become stable. If two atoms of chlorine each gain one of those electrons, they become chloride ions. The positive and negative ions are then attracted to one another, forming a molecule of calcium chloride, $CaCl_2$, which is also a salt. A **salt** is a molecule made of ions other than hydrogen (H^+) ions or hydroxyl (OH^-) ions.

Ions with positive charges are called **cations**. These include Na^+, Ca^{+2}, K^+, Fe^{+2}, and Mg^{+2}. Ions with negative charges are called **anions**, which include Cl^-, SO_4^{-2} (sulfate), and HCO_3^- (bicarbonate). The types of compounds formed by ionic bonding are salts, acids, and bases. (Acids and bases are discussed later in this chapter.)

In the solid state, ionic bonds are relatively strong. Our bones, for example, contain the salt calcium carbonate ($CaCO_3$), which helps give bone its strength. However, in an **aqueous** (water) **solution**, many ionic bonds are weakened. The bonds may become so weak that the bound ions of a molecule separate, creating a solution of free positive and negative ions. For example, if sodium chloride is put in water, it dissolves, then **ionizes**. The water now contains Na^+

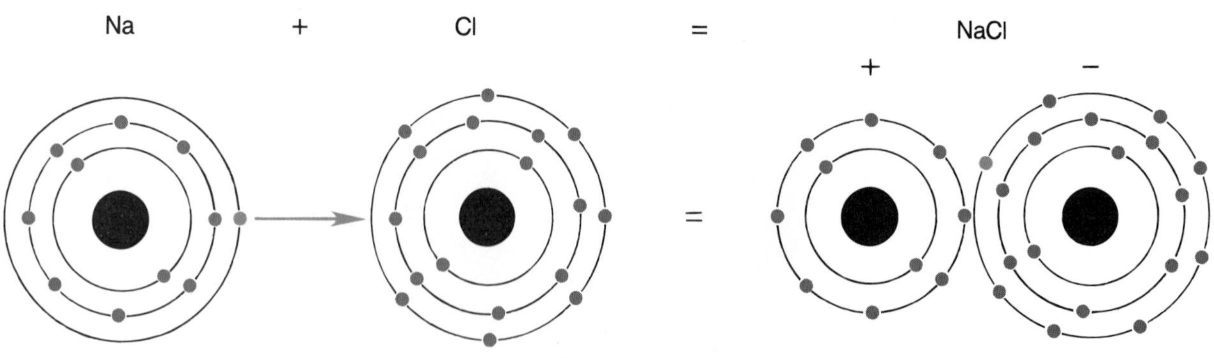

Figure 2–2 Formation of an ionic bond. An atom of sodium loses an electron to an atom of chlorine. The two ions formed have unlike charges, are attracted to one another, and form a molecule of sodium chloride.

ions and Cl^- ions. Ionization, also called **dissociation**, is important to living organisms because once dissociated, the ions are free to take part in other chemical reactions within the body. Cells in the stomach lining produce hydrochloric acid (HCl) and must have Cl^- ions to do so. The chloride in NaCl would not be free to take part in another reaction since it is tightly bound to the sodium atom. However, the Cl^- ions available from ionized NaCl in the cellular water can be used for the **synthesis**, or chemical manufacture, of HCl in the stomach.

COVALENT BONDS

Covalent bonds involve the sharing of electrons between atoms. As shown in Fig. 2–3, an atom of oxygen needs two electrons to become stable. It may share two of its electrons with another atom of oxygen, also sharing two electrons. Together they form a molecule of oxygen gas (O_2), which is the form in which oxygen exists in the atmosphere.

An atom of oxygen may also share two of its electrons with two atoms of hydrogen, each sharing its single electron (see Fig. 2–3). Together they form a molecule of water (H_2O). When writing structural formulas for chemical molecules, a pair of shared electrons is indicated by a single line, as shown in the formula for water; this is a single covalent bond. A double covalent bond is indicated by two lines, as in the formula for oxygen; this represents two pairs of shared electrons.

The element carbon always forms covalent bonds; an atom of carbon has four electrons to share with other atoms. If these four electrons are shared with four atoms of hydrogen, each sharing its one electron, a molecule of methane gas (CH_4) is formed. Carbon may form covalent bonds with other carbons, hydrogen, oxygen, nitrogen, or other elements. Organic

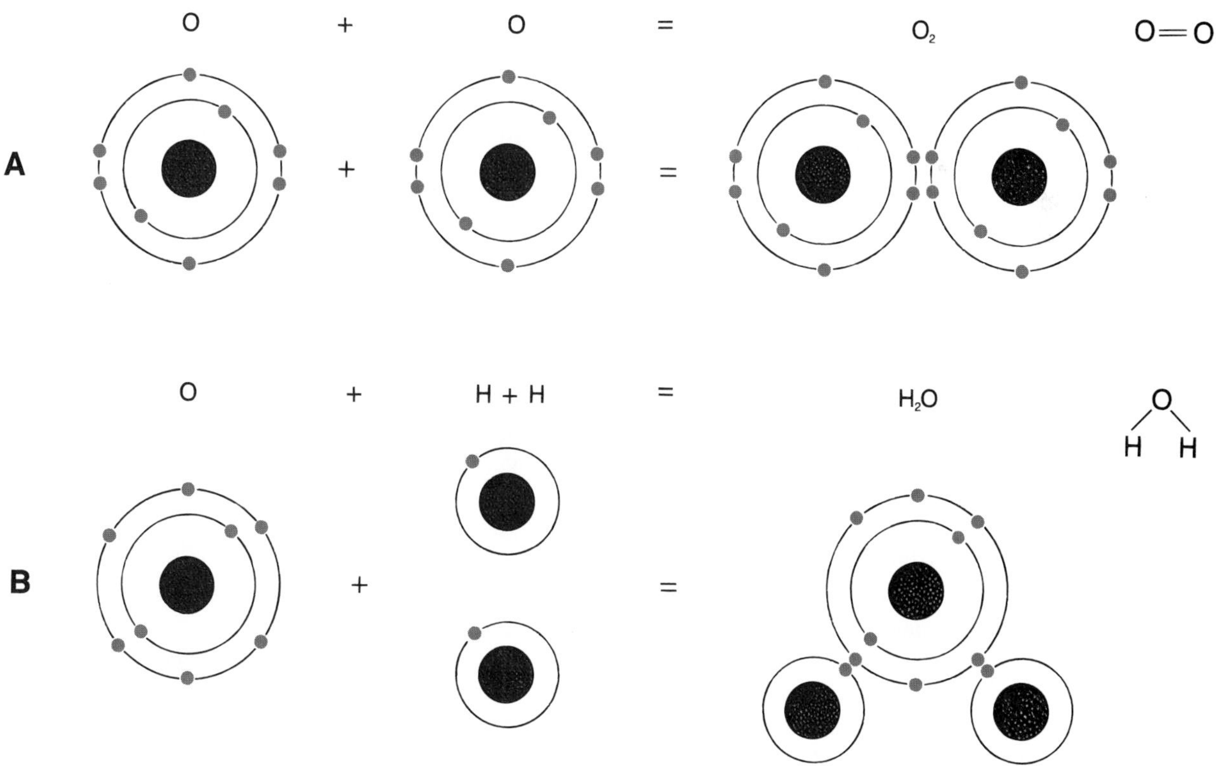

Figure 2–3 Formation of covalent bonds. **(A)** Two atoms of oxygen share two electrons each, forming a molecule of oxygen gas. **(B)** An atom of oxygen shares one electron with each of two hydrogen atoms, each sharing its electron. A molecule of water is formed.

compounds such as proteins and carbohydrates are complex and precise arrangements of these atoms covalently bonded to one another. Covalent bonds are relatively strong and are not weakened in an aqueous solution. This is important because the proteins produced by the body, for example, must remain intact in order to function properly in the water of our cells and blood. The functions of organic compounds will be considered later in this chapter.

HYDROGEN BONDS

A hydrogen bond does not involve the sharing or exchange of electrons, but rather is due to a property of hydrogen atoms. When a hydrogen atom shares its one electron in a covalent bond with another atom, its proton has a slight positive charge and may then be attracted to a nearby oxygen or nitrogen atom, which has a slight negative charge.

Although they are weak bonds, hydrogen bonds are important in several ways. Large organic molecules such as proteins and DNA have very specific functions that depend on their three-dimensional shapes. The shapes of these molecules, so crucial to their proper functioning, are often maintained by hydrogen bonds.

Hydrogen bonds also make water cohesive; that is, each water molecule is attracted to nearby water molecules. Such cohesiveness can be seen if water is dropped onto clean glass; the surface tension created by the hydrogen bonds makes the water form three-dimensional beads. These bonds are also responsible for the important characteristics of water, which are discussed in a later section.

CHEMICAL REACTIONS

A chemical reaction is a change brought about by the formation or breaking of chemical bonds. Two general types of reactions are synthesis reactions and decomposition reactions.

In a synthesis reaction, bonds are formed to join two or more atoms or molecules to make a new compound. The production of the protein hemoglobin in potential red blood cells is an example of a synthesis reaction. Proteins are synthesized by the bonding of many amino acids, their smaller subunits. Synthesis reactions require energy for the formation of bonds.

In a decomposition reaction, bonds are broken, and a large molecule is changed to two or more smaller ones. One example is the digestion of large molecules of starch to many smaller glucose molecules. Some decomposition reactions release energy; this is described in a later section on cell respiration.

In this and future chapters, keep in mind that the term "reaction" refers to the making or breaking of chemical bonds and thus to changes in the physical and chemical characteristics of the molecules involved.

INORGANIC COMPOUNDS OF IMPORTANCE

Inorganic compounds are usually simple molecules that often consist of only one or two different elements. Despite their simplicity, however, some inorganic compounds are essential to normal structure and functioning of the body.

WATER

Water makes up 60% to 75% of the human body, and is essential to life for several reasons.

1. Water is a **solvent**, that is, many substances (called solutes) can dissolve in water. Nutrients such as glucose are dissolved in blood plasma (which is largely water) to be transported to cells throughout the body. The sense of taste depends upon the solvent ability of saliva; dissolved food stimulates the receptors in taste buds. The excretion of waste products is possible because they are dissolved in the water of urine.
2. Water is a lubricant, which prevents friction where surfaces meet and move. In the digestive tract, mucus is a slippery fluid that permits the smooth passage of food through the intestines. Synovial fluid within joint cavities prevents friction as bones move.
3. Water changes temperature slowly. Water will absorb a great deal of heat before its temperature rises significantly, or it must lose a great deal of heat before its temperature drops significantly. This is one of the factors that helps the body maintain a constant temperature. It is also important for the process of sweating. Excess body heat evaporates sweat on the skin surfaces, rather than overheating the body's cells.

WATER COMPARTMENTS

All water within the body is continually moving, but water is given different names when it is in specific body locations, which are called compartments (Fig. 2-4).

Intracellular fluid (ICF)—the water within cells; about 65% of the total body water

Extracellular fluid (ECF)—all the rest of the water in the body; about 35% of the total. More specific compartments of extracellular fluid include:

- **Plasma**—water found in blood vessels
- **Lymph**—water found in lymphatic vessels
- **Tissue fluid or interstitial fluid**—water found in the small spaces between cells
- **Specialized fluids**—synovial fluid, cerebrospinal fluid, aqueous humor in the eye, and others

The movement of water between compartments in the body and the functions of the specialized fluids will be discussed in later chapters.

OXYGEN

Oxygen in the form of a gas (O_2) is approximately 21% of the atmosphere, which we inhale. We all know that without oxygen we wouldn't survive very long, but exactly what does it do? Oxygen is important to us because it is essential for a process called cell respiration, in which cells break down simple nutrients such as glucose in order to release energy. The reason we breathe is to obtain oxygen for cell respiration and to exhale the carbon dioxide produced in cell respiration (this will be discussed in the next section). Biologically useful energy that is released by the reactions of cell respiration is trapped in a molecule called ATP (adenosine triphosphate). ATP can then be used for cellular processes that require energy.

CARBON DIOXIDE

Carbon dioxide (CO_2) is produced by cells as a waste product of cell respiration. You may ask why a waste

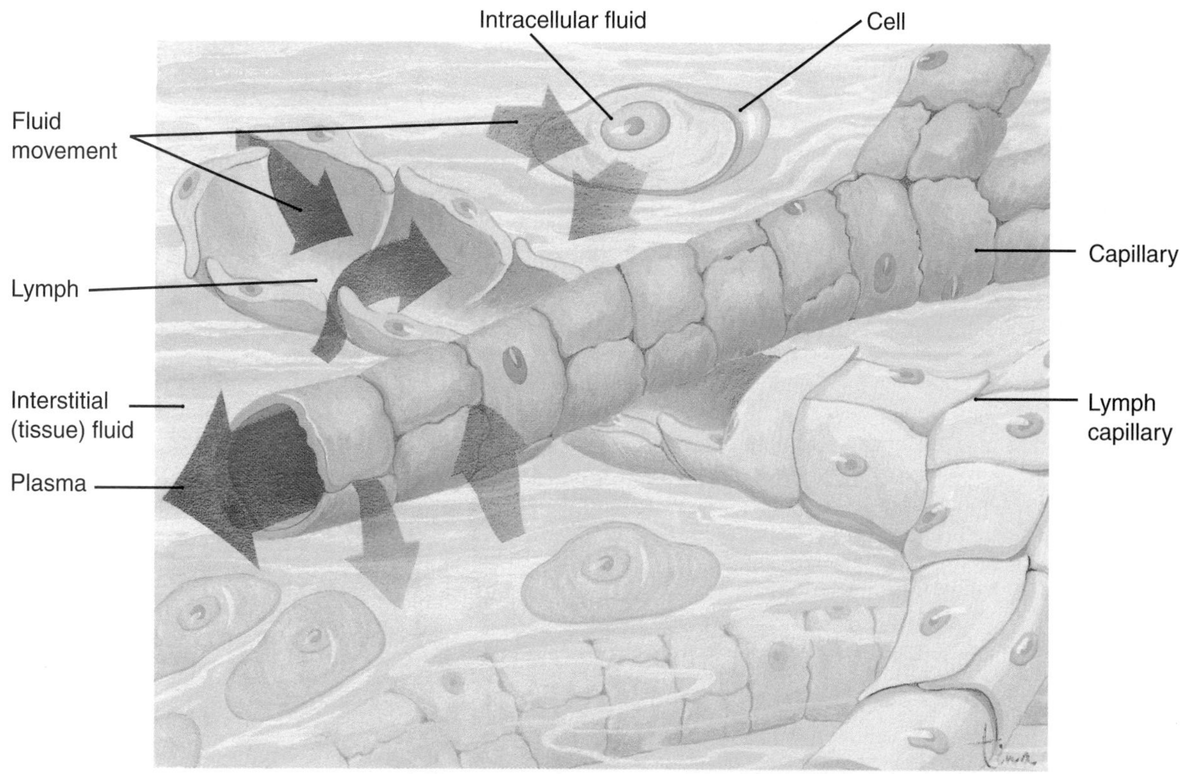

Figure 2–4 Water compartments, showing the names water is given in its different locations and the ways in which water moves between compartments.

product is considered important. Keep in mind that "important" does not always mean "beneficial," but it does mean "significant." If the amount of carbon dioxide in the body fluids increases, it causes these fluids to become too acidic. Therefore, carbon dioxide must be exhaled as rapidly as it is formed to keep the amount in the body within normal limits. Normally this is just what happens, but severe pulmonary diseases such as pneumonia or emphysema decrease gas exchange in the lungs and permit carbon dioxide to accumulate in the blood. When this happens, a person is said to be in a state of **acidosis**, which may seriously disrupt body functioning (see the sections on pH and enzymes later in this chapter; see also Box 2-1: Blood Gases).

CELL RESPIRATION

Cell respiration is the name for energy production within cells and involves both respiratory gases, oxygen and carbon dioxide. There are many chemical reactions involved, but in its simplest form, cell respiration may be summarized by the following equation:

$$\text{Glucose} + 6O_2 \rightarrow 6CO_2 + 6H_2O + \text{ATP} + \text{heat}\ (C_6H_{12}O_6)$$

This reaction shows us that glucose and oxygen combine to yield carbon dioxide, water, ATP, and heat. Food, represented here by glucose, in the presence of oxygen is broken down into the simpler molecules carbon dioxide and water. The potential energy in the glucose molecule is released in two forms: ATP and heat. Each of the four products of this process has a purpose or significance in the body. The carbon dioxide is a waste product that moves from the cells into the blood to be carried to the lungs and eventually exhaled. The water formed is useful and becomes part of the intracellular fluid. The heat produced contributes to normal body temperature. ATP is used for cell processes such as mitosis, protein synthesis, and muscle contraction, all of which require energy and will be discussed a bit further on in the text.

We will also return to cell respiration in later chapters. For now, the brief description above will suffice

Box 2–1 BLOOD GASES

A patient is admitted to the emergency room with a possible heart attack, and the doctor in charge orders "blood gases." Another patient hospitalized with pneumonia has "blood gases" monitored at frequent intervals. What are blood gases, and what does measurement of them tell us? The blood gases are oxygen and carbon dioxide, and their levels in arterial blood provide information about the functioning of the respiratory and circulatory systems. Arterial blood normally has a high concentration of oxygen and a low concentration of carbon dioxide. These levels are maintained by gas exchange in the lungs and by the proper circulation of blood.

A pulmonary disease such as pneumonia interferes with efficient gas exchange in the lungs. As a result, blood oxygen concentration may decrease, and blood carbon dioxide concentration may increase. Either of these changes in blood gases may become life-threatening for the patient, so monitoring of blood gases is important. If blood oxygen falls below the normal range, oxygen will be administered; if blood carbon dioxide rises above the normal range, blood pH will be corrected to prevent serious acidosis.

Damage to the heart may also bring about a change in blood gases, especially oxygen. Oxygen is picked up by red blood cells as they circulate through lung capillaries; as red blood cells circulate through the body, they release oxygen to tissues. What keeps the blood circulating or moving? The pumping of the heart.

A mild heart attack, when heart failure is unlikely, is often characterized by a blood oxygen level that is low but still within normal limits. A more severe heart attack that seriously impairs the pumping of the heart will decrease the blood oxygen level to less than normal. This condition is called **hypoxia**, which means that too little oxygen is reaching tissues. When this is determined by measurement of blood gases, appropriate oxygen therapy can be started to correct the hypoxia.

Table 2–2 TRACE ELEMENTS

Element	Function
Calcium	• Provides strength in bones and teeth • Necessary for blood clotting • Necessary for muscle contraction
Phosphorus	• Provides strength in bones and teeth • Part of DNA and RNA • Part of cell membranes
Iron	• Part of hemoglobin in red blood cells; transports oxygen • Part of myoglobin in muscles; stores oxygen • Necessary for cell respiration
Copper	• Necessary for cell respiration
Sodium and potassium	• Necessary for muscle contraction • Necessary for nerve impulse transmission
Sulfur	• Part of some proteins such as insulin and keratin
Cobalt	• Part of vitamin B_{12}
Iodine	• Part of thyroid hormones—thyroxine

to show that eating and breathing are interrelated; both are essential for energy production.

TRACE ELEMENTS

Trace elements are those that are needed by the body in very small amounts. Although they may not be as abundant in the body as are carbon, hydrogen, or oxygen, they are nonetheless essential. Table 2-2 lists some of these trace elements and their functions (see also Box 2-2: Nitric Oxide).

ACIDS, BASES, AND pH

An **acid** may be defined as a substance that increases the concentration of hydrogen ions (H^+) in a water solution. A **base** is a substance that decreases the concentration of H^+ ions, which in the case of water, has the same effect as increasing the concentration of hydroxyl ions (OH^-).

The acidity or alkalinity (basicity) of a solution is measured on a scale of values called **pH** (parts hydrogen). The values on the **pH scale** range from 0 to 14, with 0 indicating the most acidic level and 14 the most alkaline. A solution with a pH of 7 is neutral because it contains the same number of H^+ ions and OH^- ions. Pure water has a pH of 7. A solution with a higher concentration of H^+ ions than OH^- ions is an acidic solution with a pH below 7. An alkaline solution,

Box 2–2 NITRIC OXIDE

Nitric oxide is a gas with the molecular formula NO. You have probably heard of it as a component of air pollution and cigarette smoke, but it is produced by certain human tissues, and recent research has discovered several functions for this deceptively simple molecule.

Nitric oxide promotes vasodilation of arterioles, permitting greater blood flow and oxygen delivery to tissues. It is involved in nerve impulse transmission in the brain, and may contribute to memory storage. Some immune system cells produce nitric oxide as a cytotoxic (cell-poisoning) agent to help destroy foreign cells.

Investigations are underway to determine if nitric oxide can be used therapeutically. It has already been found useful in the treatment of pulmonary hypertension, to relax abnormally constricted arteries in the lungs to permit normal gas exchange. Another possible therapy is for sickle-cell anemia. Nitric oxide bonds to hemoglobin and may prevent the change of shape characteristic of sickle-cell hemoglobin (see Box 3–2).

Much more research is needed, including a determination of possible harmful side effects of greater than normal amounts of nitric oxide, but the results of clinical trials thus far are very promising.

therefore, has a higher concentration of OH^- ions than H^+ ions and has a pH above 7.

The pH scale, with the relative concentrations of H^+ ions, and OH^- ions, is shown in Fig. 2-5. A change of one pH unit is a 10-fold change in H^+ ion concentration. This means that a solution with a pH of 4 has 10 times as many H^+ ions as a solution with a pH of 5, and 100 times as many H^+ ions as a solution with a pH of 6. Figure 2-5 also shows the pH of some body fluids and other familiar solutions. Notice that gastric juice has a pH of 1 and coffee has a pH of 5. This means that gastric juice has 10,000 times as many H^+ ions as does coffee. Although coffee is acidic, it is a weak acid and does not have the corrosive effect of gastric juice, a strong acid.

The cells and internal fluids of the human body have a pH close to neutral. The pH of intracellular fluid is around 6.8, and the normal pH range of blood is 7.35 to 7.45. Fluids such as gastric juice and urine are technically external fluids, since they are in body tracts that open to the environment. The pH of these fluids may be more strongly acidic or alkaline without harm to the body.

The pH of blood, however, must be maintained within its very narrow, slightly alkaline range. A decrease of only one pH unit, which is 10 times as many H^+ ions, would disrupt the chemical reactions of the blood and cause the death of the individual. Normal metabolism tends to make body fluids more acidic, and this tendency to acidosis must be continually cor-

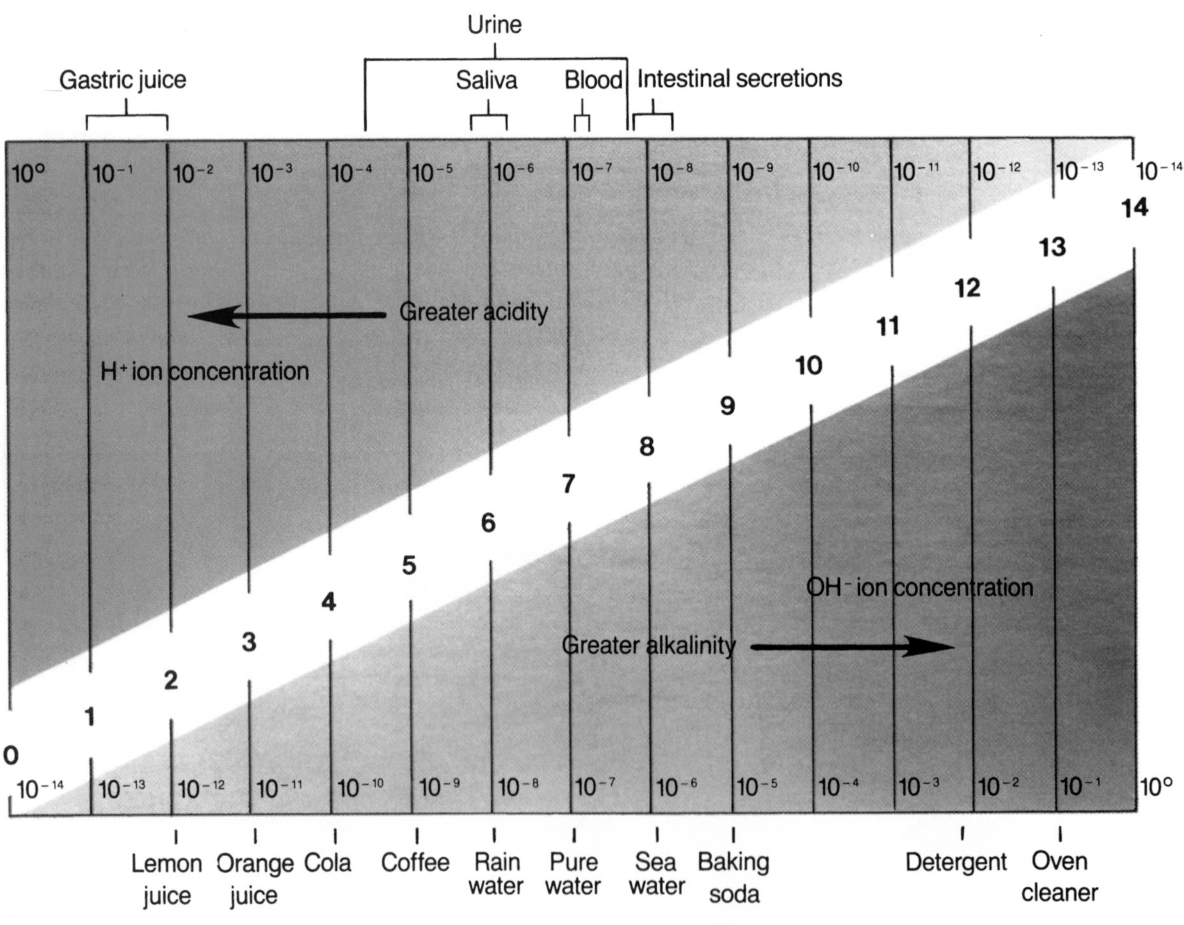

Figure 2-5 The pH scale. The pH values of several body fluids are indicated above the scale. The pH values of some familiar solutions are indicated below the scale.

rected. Normal pH of internal fluids is maintained by the kidneys, respiratory system, and buffer systems. Although acid-base balance will be a major topic of Chapter 19, we will briefly mention buffer systems here.

Buffer Systems

A **buffer system** is a chemical or pair of chemicals that minimize changes in pH by reacting with strong acids or strong bases to transform them into substances that will not drastically change pH. Expressed in another way, a buffer may bond to H^+ ions when a body fluid is becoming too acidic, or release H^+ ions when a fluid is becoming too alkaline.

As a specific example, we will use the bicarbonate buffer system, which consists of carbonic acid (H_2CO_3), a weak acid, and sodium bicarbonate ($NaHCO_3$), a weak base. This pair of chemicals is present in all body fluids but is especially important to buffer blood and tissue fluid.

Carbonic acid ionizes as follows (but remember, because it is a weak acid it does not contribute many H^+ ions to a solution):

$$H_2CO_3 \rightarrow H^+ + HCO_3^-$$

Sodium bicarbonate ionizes as follows:

$$NaHCO_3 \rightarrow Na^+ + HCO_3^-$$

If a strong acid, such as HCl, is added to extracellular fluid, this reaction will occur:

$$HCl + NaHCO_3 \rightarrow NaCl + H_2CO_3$$

What has happened here? Hydrochloric acid, a strong acid that would greatly lower pH, has reacted with sodium bicarbonate. The products of this reaction are NaCl, a salt which has no effect on pH, and H_2CO_3, a weak acid that lowers pH only slightly. This prevents a drastic change in the pH of the extracellular fluid.

If a strong base, such as sodium hydroxide, is added to the extracellular fluid, this reaction will occur:

$$NaOH + H_2CO_3 \rightarrow H_2O + NaHCO_3$$

Sodium hydroxide, a strong base that would greatly raise pH, has reacted with carbonic acid. The products of this reaction are water, which has no effect on pH, and sodium bicarbonate, a weak base that raises pH only slightly. Again, this prevents a drastic change in the pH of the extracellular fluid.

In the body, such reactions take place in less than a second whenever acids or bases are formed that would greatly change pH. Because of the body's tendency to become more acidic, the need to correct acidosis is more frequent. With respect to the bicarbonate buffer system, this means that more $NaHCO_3$ than H_2CO_3 is needed. For this reason, the usual ratio of these buffers is 20:1 ($NaHCO_3:H_2CO_3$).

ORGANIC COMPOUNDS OF IMPORTANCE

Organic compounds all contain covalently bonded carbon and hydrogen atoms and perhaps other elements as well. In the human body there are four major groups of organic compounds: carbohydrates, lipids, proteins, and nucleic acids.

CARBOHYDRATES

A primary function of **carbohydrates** is to serve as sources of energy. All carbohydrates contain carbon, hydrogen, and oxygen and are classified as monosaccharides, disaccharides, oligosaccharides, and polysaccharides. Saccharide means sugar, and the prefix indicates how many are present.

Monosaccharides or single sugar compounds are the simplest sugars. Glucose is a **hexose**, or 6-carbon, sugar with the formula $C_6H_{12}O_6$ (Fig. 2-6). Fructose and galactose also have the same formula, but the physical arrangement of the carbon, hydrogen, and oxygen atoms in each differs from that of glucose. This gives each hexose sugar a different three-dimensional shape. The liver is able to change fructose and galactose to glucose, which is then used by cells in the process of cell respiration to produce ATP.

Another type of monosaccharide is the **pentose**, or 5-carbon, sugar. These are not involved in energy production but rather are structural components of the nucleic acids. Deoxyribose ($C_5H_{10}O_4$) is part of DNA, which is the genetic material of chromosomes. Ribose ($C_5H_{10}O_5$) is part of RNA, which is essential for protein synthesis. We will return to the nucleic acids later in this chapter.

Disaccharides are double sugars, made of two monosaccharides linked by covalent bonds. Examples are sucrose, lactose, and maltose, which are present

A Glucose

B Disaccharide

C Cellulose

D Starch

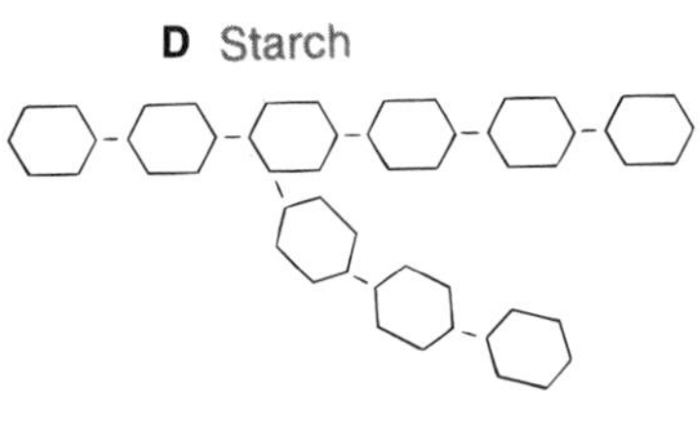

E Glycogen

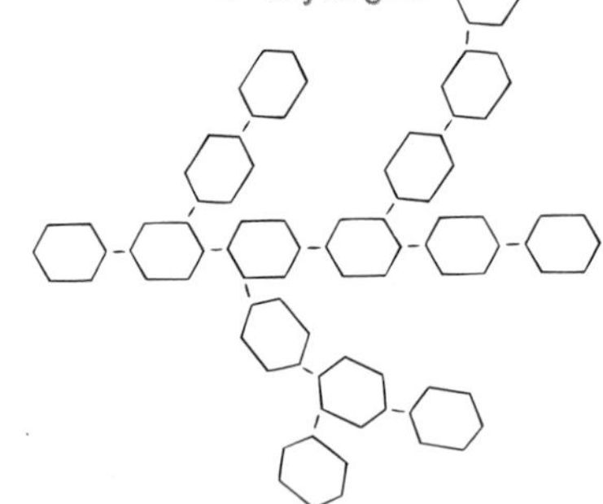

Figure 2–6 Carbohydrates. **(A)** Glucose, depicting its structural formula. **(B)** A disaccharide such as maltose. **(C)** Cellulose, a polysaccharide. **(D)** Starch, a polysaccharide. **(E)** Glycogen, a polysaccharide.

in food. They are digested into monosaccharides and then used for energy production.

The prefix "oligo" means "few"; **oligosaccharides** consist of from 3 to 20 monosaccharides. In human cells, oligosaccharides are found on the outer surface of cell membranes. Here they serve as **antigens**, which are chemical markers (or "sign posts") that identify cells. The A, B, and AB blood types, for example, are the result of oligosaccharide antigens on the outer surface of red blood cell membranes. All of our cells have "self" antigens, which identify the cells that belong in an individual. The presence of "self" antigens on our own cells enables the immune system to recognize antigens that are "non-self." Such foreign antigens include bacteria and viruses, and immunity will be a major topic of Chapter 14.

Polysaccharides are made of thousands of glucose molecules, bonded in different ways, resulting in different shapes (see Fig. 2–6). Starches are branched chains of glucose and are produced by plant cells to store energy. We have digestive enzymes that split the bonds of starch molecules, releasing glucose. The glucose is then absorbed and used by cells to produce ATP.

Glycogen, a highly branched chain of glucose molecules, is our own storage form for glucose. After a meal high in carbohydrates, the blood glucose level rises. Excess glucose is then changed to glycogen and stored in the liver and skeletal muscles. When the blood glucose level decreases between meals, the glycogen is converted back to glucose, which is released into the blood. The blood glucose level is kept within normal limits, and cells can take in this glucose to produce energy.

Cellulose is a nearly straight chain of glucose molecules produced by plant cells as part of their cell walls. We have no enzyme to digest the cellulose we consume as part of vegetables and grains, and it passes through the digestive tract unchanged. Another name for dietary cellulose is "fiber," and although we cannot use its glucose for energy, it does have a function. Fiber provides bulk within the cavity of the large in-

Table 2–3 CARBOHYDRATES

Name	Structure	Function
Monosaccharides—"Single" Sugars		
Glucose	Hexose sugar	• Most important energy source for cells
Fructose and galactose	Hexose sugars	• Converted to glucose by the liver, then used for energy production
Deoxyribose	Pentose sugar	• Part of DNA, the genetic code in the chromosomes of cells
Ribose	Pentose sugar	• Part of RNA, needed for protein synthesis within cells
Disaccharides—"Double" Sugars		
Sucrose, lactose, and maltose	Two hexose sugars	• Present in food; digested to monosaccharides, which are then used for energy production
Oligosaccharides—"Few" Sugars (3–20)		
		• Form "self" antigens on cell membranes; important to permit the immune system to distinguish "self" from foreign antigens (pathogens)
Polysaccharides—"Many" Sugars (Thousands)		
Starches	Branched chains of glucose molecules	• Found in plant foods; digested to monosaccharides and used for energy production
Glycogen	Highly branched chains of glucose molecules	• Storage form for excess glucose in the liver and skeletal muscles
Cellulose	Straight chains of glucose molecules	• Part of plant cell walls; provides fiber to promote peristalsis, especially by the colon

testine. This promotes efficient **peristalsis**, the waves of contraction that propel undigested material through the colon. A diet low in fiber does not give the colon much exercise, and the muscle tissue of the colon will contract weakly, just as our skeletal muscles will become flabby without exercise. A diet high in fiber provides exercise for the colon muscle and may help prevent chronic constipation.

The structure and functions of the carbohydrates are summarized in Table 2-3.

LIPIDS

Lipids contain the elements carbon, hydrogen, and oxygen; some also contain phosphorus. In this group of organic compounds are different types of substances with very different functions. We will consider three types: true fats, phospholipids, and steroids (Fig. 2-7).

True fats are made of one molecule of glycerol and one, two, or three fatty acid molecules. If three fatty acid molecules are bonded to a single glycerol, a **triglyceride** is formed (you can usually find this term on the nutrition labels of some highly processed foods; it means that fat has been added). You have undoubtedly heard of saturated and unsaturated fats; the differences between them are discussed in Box 2-3. True fats are a storage form for excess food, that is, they are stored energy. Any type of food consumed in excess of the body's caloric needs will be converted to fat and stored in adipose tissue. Most adipose tissue is subcutaneous, between the skin and muscles. Some organs, however, such as the eyes and kidneys, are enclosed in a layer of fat that acts as a cushion to absorb shock.

Phospholipids are diglycerides with a phosphate group (PO_4) in the third bonding site of glycerol. Although similar in structure to the true fats, phospholipids are not stored energy but rather structural components of cells. Lecithin is a phospholipid that is part of our **cell membranes**. Another example is **myelin**, which forms the myelin sheath around nerve cells and provides electrical insulation for nerve impulse transmission.

The structure of **steroids** is very different from that of the other lipids. **Cholesterol** is an important ste-

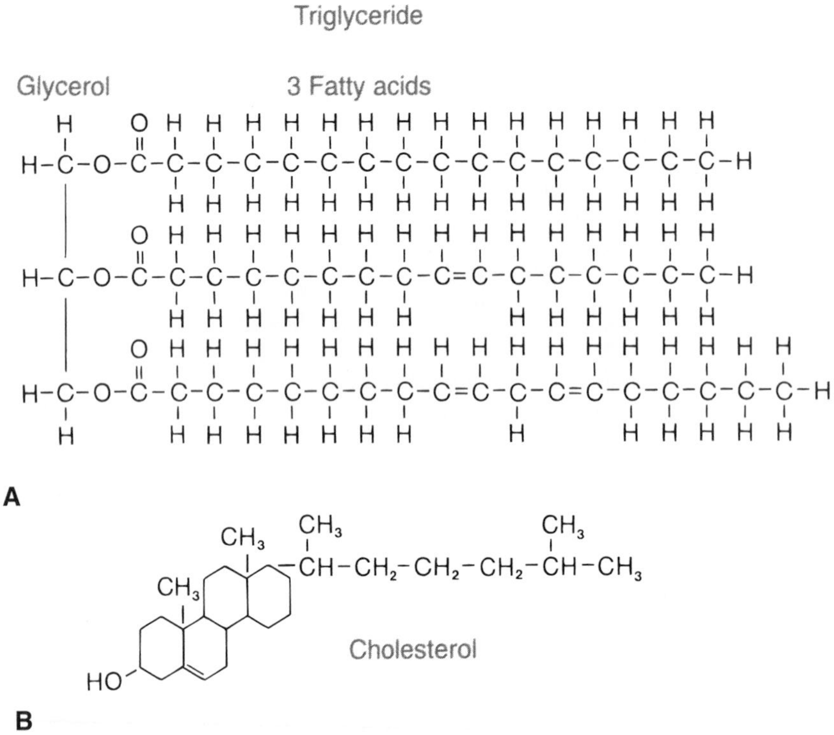

Triglyceride

Glycerol

3 Fatty acids

A

Cholesterol

B

Figure 2–7 Lipids. **(A)** A triglyceride made of one glycerol and three fatty acids. **(B)** The steroid cholesterol. The hexagons represent rings of carbons and hydrogens.

roid; it is made of four rings of carbon and hydrogen (not fatty acids and glycerol) and is shown in Fig. 2–7. The liver synthesizes cholesterol, in addition to the cholesterol we eat in food as part of our diet. Cholesterol is another component of cell membranes and is the precursor (raw material) for the synthesis of other steroids. In the ovaries or testes, cholesterol is used to synthesize the steroid hormones estrogen or testosterone, respectively. A form of cholesterol in the skin is changed to vitamin D on exposure to sunlight. Liver cells use cholesterol for the synthesis of bile salts, which emulsify fats in digestion. Despite its link to coronary artery disease and heart attacks, cholesterol is an essential substance for human beings.

The structure and functions of lipids are summarized in Table 2–4.

PROTEINS

Proteins are made of smaller subunits or building blocks called **amino acids**, which contain the elements carbon, hydrogen, oxygen, nitrogen, and perhaps sulfur. There are about 20 amino acids that make up human proteins. The structure of amino acids is shown in Fig. 2–8. Each amino acid has a central carbon atom covalently bonded to an atom of hydrogen, an amino group (NH_2), and a carboxyl group (COOH). At the fourth bond of the central carbon is the variable portion of the amino acid, represented by R. The R group may be a single hydrogen atom, or a CH_3 group, or a more complex configuration of carbon and hydrogen. This gives each of the 20 amino acids a slightly different physical shape. A bond between two amino acids is called a **peptide bond**, and a short chain of amino acids linked by peptide bonds is a **polypeptide**.

A protein may consist of from 50 to thousands of amino acids. The sequence of the amino acids is specific and unique for each protein. This unique sequence determines the protein's characteristic three-dimensional shape, which in turn determines its function. Our body proteins have many functions; some of these are listed in Table 2–5 and will be mentioned again in later chapters. However, one very important function of proteins will be discussed further here: the role of proteins as enzymes.

Box 2–3 SATURATED AND UNSATURATED FATS

You have probably heard or read that eating foods high in **unsaturated fats** may help prevent heart disease and that a diet high in **saturated fats** may contribute to heart disease. "Heart disease" actually refers to **atherosclerosis** of the coronary arteries, those that supply the heart muscle with oxygen. Atherosclerosis is the deposition of cholesterol and other substances in the walls of these arteries, leading to obstruction or clot formation and a heart attack. These abnormal deposits of cholesterol are more likely to occur when blood cholesterol levels are high, and diet does have an effect on blood cholesterol.

The true fats may be divided into saturated and unsaturated fats. Refer to Fig. 2–7 and notice that one of the fatty acids has single covalent bonds between all its carbon atoms. Each of these carbons is then bonded to the maximum number of hydrogens; this is a saturated fatty acid. The other fatty acids shown have one or more (poly) double covalent bonds between their carbons and less than the maximum number of hydrogens; these are unsaturated fatty acids. At room temperature, saturated fats are often in solid form, while unsaturated fats are often (not always) in liquid form.

Saturated fats tend to be found in animal foods such as beef, pork, eggs, and cheese, but palm oil and coconut oil are also saturated. Unsaturated fats are found in other plant oils such as corn oil, sunflower oil, and safflower oil, but certain fish oils are also unsaturated.

The breakdown products of saturated fats are used by the liver to synthesize cholesterol, which raises the blood cholesterol level. This is not true for the unsaturated fats; they do not contribute to cholesterol formation. This is why a diet high in unsaturated fats may help prevent atherosclerosis. Unsaturated fats, especially polyunsaturated fats, may actually help lower blood cholesterol by decreasing the formation of cholesterol by the liver.

There are other contributing factors to coronary artery disease, such as heredity, smoking, being overweight, and lack of exercise. Diet alone cannot prevent atherosclerosis. However, a diet low in total fat and high in polyunsaturated fats is a good start.

Table 2–4 LIPIDS

Name	Structure	Function
True fats	A triglyceride consists of three fatty acid molecules bonded to a glycerol molecule (some are monoglycerides or diglycerides)	• Storage form for excess food molecules in subcutaneous tissue • Cushion organs such as the eyes and kidneys
Phospholipids	Diglycerides with a phosphate group bonded to the glycerol molecule	• Part of cell membranes (lecithin) • Form the myelin sheath to provide electrical insulation for neurons
Steroids (cholesterol)	Four carbon–hydrogen rings	• Part of cell membranes • Converted to vitamin D in the skin on exposure to UV rays of the sun • Converted by the liver to bile salts, which emulsify fats during digestion • Precursor for the steroid hormones such as estrogen in women (ovaries) or testosterone in men (testes)

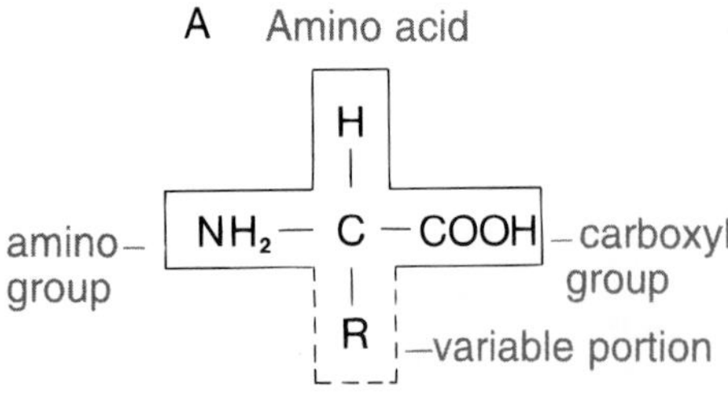

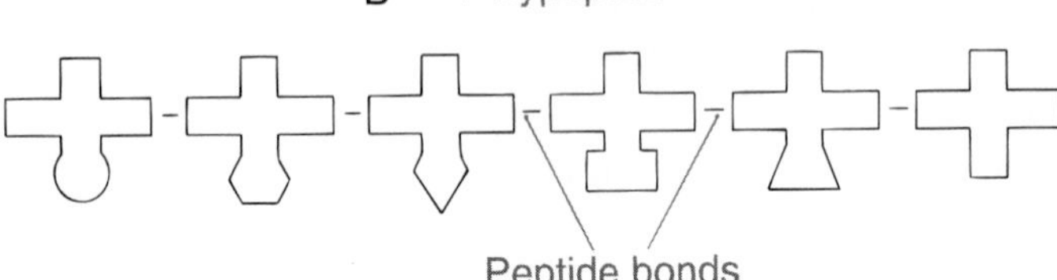

Figure 2–8 Amino acids. **(A)** The structural formula of an amino acid. The "R" represents the variable portion of the molecule. **(B)** A polypeptide. Several amino acids, represented by different shapes, are linked by peptide bonds.

Enzymes

Enzymes are **catalysts**, which means that they speed up chemical reactions without the need for an external source of energy such as heat. The many reactions that take place within the body are catalyzed by specific enzymes; all of these reactions must take place at body temperature.

Table 2–5 FUNCTIONS OF PROTEINS

Type of Protein	Function
Structural proteins	• Form pores and receptor sites in cell membranes • Keratin—part of skin and hair • Collagen—part of tendons and ligaments
Hormones	• Insulin—enables cells to take in glucose; lowers blood glucose level • Growth hormone—increases protein synthesis and cell division
Hemoglobin	• Enables red blood cells to carry oxygen
Antibodies	• Produced by lymphocytes (white blood cells); label pathogens for destruction
Myosin and actin	• Muscle structure and contraction
Enzymes	• Catalyze reactions

The way in which enzymes function as catalysts is called the **Active Site Theory**, which is based on the shape of the enzyme and the shapes of the reacting molecules, called **substrates**. A simple reaction is depicted in Fig. 2–9. Notice that the enzyme has a specific shape, as do the substrate molecules. The active site of the enzyme is the part that matches the shapes of the substrates. The substrates must "fit" into the active site of the enzyme, and temporary bonds may form between the enzyme and the substrate. This is called the enzyme-substrate complex. In this case, two substrate molecules are thus brought close together so that chemical bonds are formed between them, creating a new compound. The product of the reaction, the new compound, is then released, leaving the enzyme itself unchanged and able to catalyze another reaction of the same type.

Each enzyme is specific in that it will catalyze only one type of reaction. An enzyme that digests the protein in food, for example, has the proper shape for that reaction but cannot digest starches. For starch digestion, another enzyme with a differently shaped active site is needed. Thousands of chemical reactions take place within the body, and therefore we have thousands of enzymes, each with its own shape and active site.

The ability of enzymes to function may be limited or destroyed by changes in the intracellular or extracellular fluids in which they are found. Changes in pH and temperature are especially crucial. Recall that the pH of intracellular fluid is approximately 6.8, and that

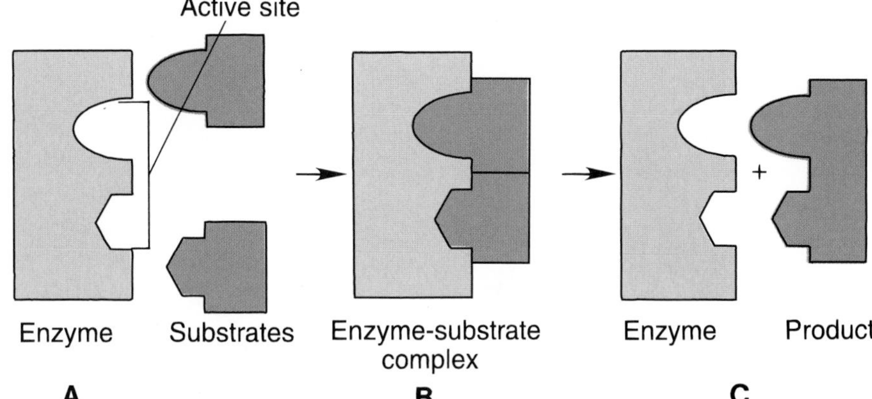

Figure 2–9 Active Site Theory, as shown in a synthesis reaction. **(A)** The enzyme and substrates of this reaction. **(B)** The enzyme–substrate complex. **(C)** The product of the reaction and the intact enzyme.

a decrease in pH means that more H^+ ions are present. If pH decreases significantly, the excess H^+ ions will react with the active sites of cellular enzymes, change their shapes, and prevent them from catalyzing reactions. This is why a state of acidosis may cause the death of cells—the cells' enzymes are unable to function properly.

With respect to temperature, most human enzymes have their optimum functioning in the normal range of body temperature: 97° to 99°F (36° to 38°C). A temperature of 106°F, a high fever, may break the chemical bonds that maintain the shapes of enzymes. If an enzyme loses its shape, it is said to be **denatured**, and a denatured enzyme is unable to function as a catalyst.

NUCLEIC ACIDS

DNA and RNA

The **nucleic acids, DNA** (deoxyribonucleic acid) and **RNA** (ribonucleic acid), are large molecules made of smaller subunits called nucleotides. A **nucleotide** consists of a pentose sugar, a phosphate group, and one of several nitrogenous bases. In DNA nucleotides, the sugar is deoxyribose, and the bases are adenine, guanine, cytosine, or thymine. In RNA nucleotides, the sugar is ribose, and the bases are adenine, guanine, cytosine, or uracil. Small segments of DNA and RNA molecules are shown in Fig. 2–10.

Notice that DNA looks somewhat like a twisted ladder; this is two strands of nucleotides called a double helix (coil). Alternating phosphate and sugar molecules form the uprights of the ladder, and pairs of nitrogenous bases form the rungs. The size of the bases and the number of hydrogen bonds each can create the complementary base pairing of the nucleic acids. In DNA, adenine is always paired with thymine, and guanine is always paired with cytosine.

DNA makes up the chromosomes of cells, and is, therefore, the **genetic code** for hereditary characteristics. The sequence of bases in the DNA strands is actually a code for the many kinds of proteins living things produce; the code is the same in plants, other animals, and microbes. The sequence of bases for one protein is called a gene. Human genes are the codes for the proteins produced by human cells (though some of these genes are also found in all other forms of life). The functioning of DNA will be covered in more detail in the next chapter.

RNA is a single strand of nucleotides (see Fig. 2–10), with uracil nucleotides in place of thymine nucleotides. RNA is synthesized from DNA in the nucleus of a cell but carries out its function in the cytoplasm. This function is protein synthesis, which will also be discussed in the following chapter.

ATP

ATP (adenosine triphosphate) is a specialized nucleotide that consists of the base adenine, the sugar ribose, and three phosphate groups. Mention has already been made of ATP as a product of cell respiration that contains biologically useful energy. ATP is one of several "energy transfer" molecules within cells, transferring the potential energy in food molecules to cell processes. When a molecule of glucose is broken down into carbon dioxide and water with the release of energy, some of this energy is used by

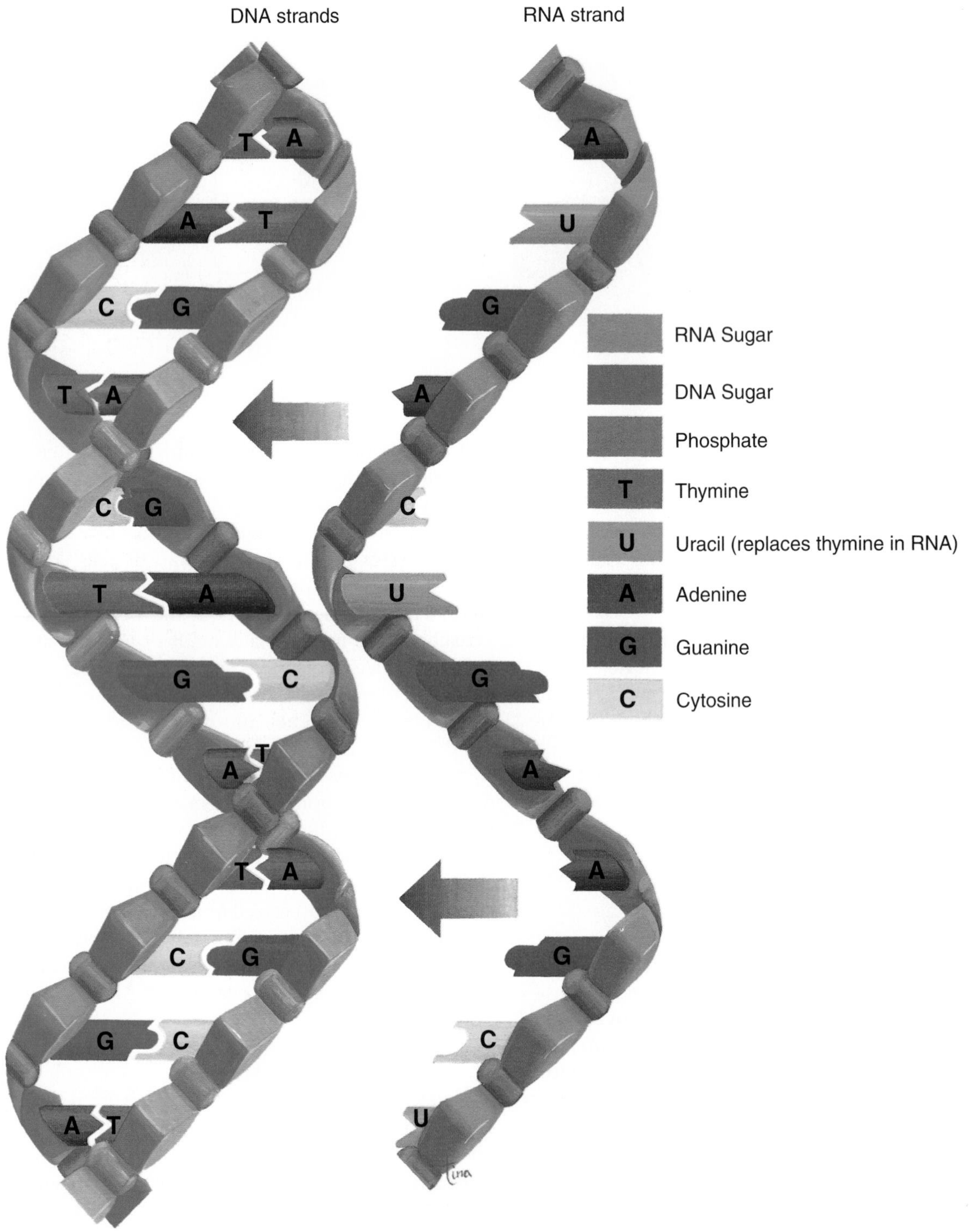

Figure 2–10 DNA and RNA. A small portion of each molecule is shown, with each part of a nucleotide represented by a different color. Note the complementary base pairing of DNA (A–T and G–C). When RNA is synthesized, it is a complementary copy of half the DNA molecule (with U in place of T).

Table 2–6 NUCLEIC ACIDS

Name	Structure	Function
DNA (deoxyribonucleic acid)	A double helix of nucleotides; adenine paired with thymine, and guanine paired with cytosine	• Found in the chromosomes in the nucleus of a cell • Is the genetic code for hereditary characteristics
RNA (ribonucleic acid)	A single strand of nucleotides; adenine, guanine, cytosine, and uracil	• Copies the genetic code of DNA to direct protein synthesis in the cytoplasm of cells
ATP (adenosine triphosphate)	A single adenine nucleotide with three phosphate groups	• An energy-transferring molecule • Formed when cell respiration releases energy from food molecules • Used for energy-requiring cellular processes

the cell to synthesize ATP. Present in cells are molecules of ADP (adenosine diphosphate) and phosphate. The energy released from glucose is used to loosely bond a third phosphate to ADP, forming ATP. When the bond of this third phosphate is again broken and energy is released, ATP then becomes the energy source for cell processes such as mitosis.

All cells have enzymes that can remove the third phosphate group from ATP to release its energy, forming ADP and phosphate. As cell respiration continues, ATP is resynthesized from ADP and phosphate. ATP formation to trap energy from food and breakdown to release energy for cell processes is a continuing cycle in cells.

The structure and functions of the nucleic acids are summarized in Table 2-6.

SUMMARY

All the chemicals we have just described are considered to be non-living, even though they are essential parts of all living organisms. The cells of our bodies are precise arrangements of these non-living chemicals and yet are considered living matter. The cellular level, therefore, is the next level of organization we will examine.

STUDY OUTLINE

Elements

1. Elements are the simplest chemicals, which make up all matter.
2. Carbon, hydrogen, oxygen, nitrogen, phosphorus, sulfur, and calcium make up 99% of the human body.
3. Elements combine in many ways to form molecules.

Atoms (see Fig. 2–1)

1. Atoms are the smallest part of an element that still retain the characteristics of the element.
2. Atoms consist of positively and negatively charged particles and neutral (or uncharged) particles.
   - Protons have a positive charge and are found in the nucleus of the atom.
   - Neutrons have no charge and are found in the nucleus of the atom.
   - Electrons have a negative charge and orbit the nucleus.
3. The number and arrangement of electrons give an atom its bonding capabilities.

Chemical Bonds

1. An ionic bond involves the loss of electrons by one atom and the gain of these electrons by another atom: ions are formed that attract one another (see Fig. 2-2).
   - Cations are ions with positive charges: Na^+, Ca^{+2}.
   - Anions are ions with negative charges: Cl^-, HCO_3^-.

- Salts, acids, and bases are formed by ionic bonding.
- In water, many ionic bonds break; dissociation releases ions for other reactions.

2. A covalent bond involves the sharing of electrons between two atoms (see Fig. 2-3).
   - Oxygen gas (O_2) and water (H_2O) are covalently bonded molecules.
   - Carbon always forms covalent bonds; these are the basis for the organic compounds.
   - Covalent bonds are not weakened in an aqueous solution.
3. A hydrogen bond is the attraction of a covalently bonded hydrogen to a nearby oxygen or nitrogen atom.
   - The three-dimensional shape of proteins and nucleic acids is maintained by hydrogen bonds.
   - Water is cohesive because of hydrogen bonds.

Chemical Reactions

1. A change brought about by the formation or breaking of chemical bonds.
2. Synthesis—bonds are formed to join two or more molecules.
3. Decomposition—bonds are broken within a molecule.

Inorganic Compounds of Importance

1. Water—makes up 60% to 75% of the body.
   - Solvent—for transport of nutrients in the blood and excretion of wastes in urine.
   - Lubricant—mucus in the digestive tract.
   - Prevents sudden changes in body temperature; absorbs body heat in evaporation of sweat.
   - Water compartments—the locations of water within the body (see Fig. 2-4).
     - Intracellular—within cells; 65% of total body water.
     - Extracellular—35% of total body water
       - Plasma—in blood vessels.
       - Lymph—in lymphatic vessels.
       - Tissue fluid—in tissue spaces between cells.
2. Oxygen—21% of the atmosphere.
   - Essential for cell respiration: the breakdown of food molecules to release energy.
3. Carbon Dioxide
   - Produced as a waste product of cell respiration.
   - Must be exhaled; excess CO_2 causes acidosis.
4. Cell Respiration—the energy-producing processes of cells.
   - Glucose + $O_2 \rightarrow CO_2 + H_2O$ + ATP + heat
   - This is why we breathe: to take in oxygen to break down food; to exhale the CO_2 produced.
5. Trace Elements—needed in small amounts (see Table 2-2).
6. Acids, Bases, and pH
   - The pH scale ranges from 0 to 14; 7 is neutral; below 7 is acidic; above 7 is alkaline.
   - An acid increases the H^+ ion concentration of a solution; a base decreases the H^+ ion concentration (or increases the OH^- ion concentration) (see Fig. 2-5).
   - The pH of cells is about 6.8. The pH range of blood is 7.35 to 7.45.
   - Buffer systems maintain normal pH by reacting with strong acids or strong bases to change them to substances that do not greatly change pH.
   - The bicarbonate buffer system consists of H_2CO_3 and $NaHCO_3$.

Organic Compounds of Importance

1. Carbohydrates (see Table 2-3 and Fig. 2-6).
   - Monosaccharides are simple sugars. Glucose, a hexose sugar ($C_6H_{12}O_6$), is the primary energy source for cell respiration.
     - Pentose sugars are part of the nucleic acids DNA and RNA.
   - Disaccharides are made of two hexose sugars. Sucrose, lactose, and maltose are digested to monosaccharides and used for cell respiration.
   - Oligosaccharides consist of from 3 to 20 monosaccharides; they are antigens on the cell membrane that identify cells as "self."
   - Polysaccharides are made of thousands of glucose molecules.
     - Starches are plant products broken down in digestion to glucose.
     - Glycogen is the form in which our bodies store glucose in the liver and muscles.
     - Cellulose, the fiber portion of plant cells, cannot be digested but promotes efficient peristalsis in the colon.
2. Lipids (see Table 2-4 and Fig. 2-7).
   - True fats are made of fatty acids and glycerol; a storage form for energy in adipose tissue. The eyes and kidneys are cushioned by fat.

- Phospholipids are part of cell membranes. An example is myelin, which provides electrical insulation for nerve cells.
- Steroids consist of four rings of carbon and hydrogen. Cholesterol, produced by the liver and consumed in food, is the basic steroid from which the body manufactures others: steroid hormones, vitamin D, and bile salts.

3. Proteins
   - Amino acids are the subunits of proteins; 20 amino acids make up human proteins. Peptide bonds join amino acids to one another (see Fig. 2-8).
   - A protein consists of from 50 to thousands of amino acids in a specific sequence.
   - Protein functions—see Table 2-5.
   - Enzymes are catalysts, which speed up reactions without additional energy. The Active Site Theory is based on the shapes of the enzyme and the substrate molecules: these must "fit" (see Fig. 2-9). The enzyme remains unchanged after the product of the reaction is released. Each enzyme is specific for one type of reaction. The functioning of enzymes may be disrupted by changes in pH or body temperature, which change the shape of the active sites of enzymes.
4. Nucleic Acids (see Table 2-6 and Fig. 2-10).
   - Nucleotides are the subunits of nucleic acids. A nucleotide consists of a pentose sugar, a phosphate group, and a nitrogenous base.
   - DNA is a double strand of nucleotides, coiled into a double helix, with complementary base pairing: A-T and G-C. DNA makes up the chromosomes of cells and is the genetic code for the synthesis of proteins.
   - RNA is a single strand of nucleotides, synthesized from DNA, with U in place of T. RNA functions in protein synthesis.
   - ATP is a nucleotide that is specialized to trap and release energy. Energy released from food in cell respiration is used to synthesize ATP from ADP + P. When cells need energy, ATP is broken down to ADP + P, and the energy is released for cell processes.

REVIEW QUESTIONS

1. State the chemical symbol for each of the following elements: sodium, potassium, iron, calcium, oxygen, carbon, hydrogen, copper, and chlorine. (p. 20)

2. Explain, in terms of their electrons, how an atom of sodium and an atom of chlorine form a molecule of sodium chloride. (pp. 22-23)

3. Explain, in terms of their electrons, how an atom of carbon and two atoms of oxygen form a molecule of carbon dioxide. (pp. 23-24)

4. Name the subunits (smaller molecules) of which each of the following is made: DNA, glycogen, a true fat, and a protein. (pp. 29-32, 34-35)

5. State precisely where in the body each of these fluids is found: plasma, intracellular water, lymph, and tissue fluid. (p. 25)

6. Explain the importance of the fact that water changes temperature slowly. (p. 24)

7. Describe two ways the solvent ability of water is important to the body. (p. 24)

8. Name the organic molecule with each of the following functions: (pp. 31, 34-35, 37)
   a. The genetic code in chromosomes
   b. "Self" antigens in our cell membranes
   c. The storage form for glucose in the liver
   d. The storage form for excess food in adipose tissue
   e. The precursor molecule for the steroid hormones

f. The undigested part of food that promotes peristalsis
g. The sugars that are part of the nucleic acids

9. State the summary equation of cell respiration. (p. 26)
10. State the role or function of each of the following in cell respiration: CO_2, glucose, O_2, heat, and ATP. (pp. 26–27)
11. State a specific function of each of the following in the human body: Ca, Fe, Na, I, and Co. (p. 27)
12. Explain, in terms of relative concentrations of H^+ ions and OH^- ions, each of the following: acid, base, and neutral substance. (pp. 27–29)
13. State the normal pH range of blood. (p. 28)
14. Complete the following equation, and state how each of the products affects pH: $HCl + NaHCO_3 \rightarrow$ ________ + ________. (p. 29)
15. Explain the Active Site Theory of enzyme functioning. (pp. 34–35)
16. Explain the difference between a synthesis reaction and a decomposition reaction (p. 24).

CHAPTER 3

Cells

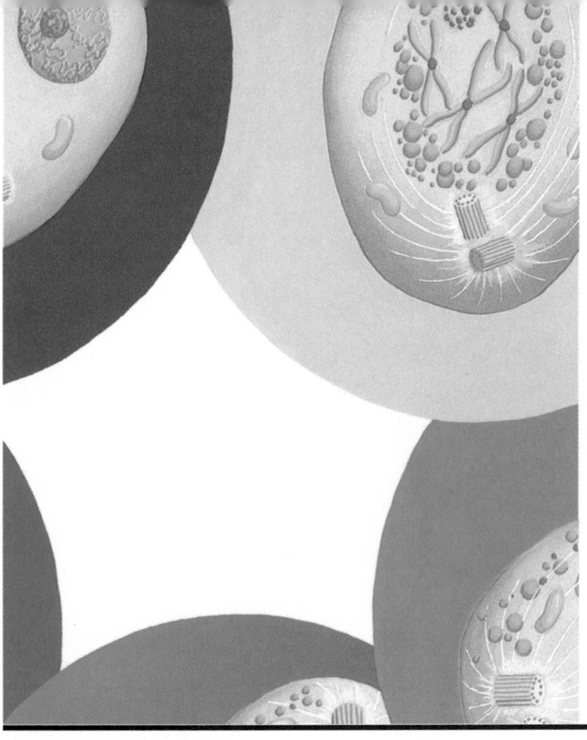

CHAPTER 3

Chapter Outline

Cell Structure
Cell Membrane
Nucleus
Cytoplasm and Cell Organelles
Cellular Transport Mechanisms
Diffusion
Osmosis
Facilitated Diffusion
Active Transport
Filtration
Phagocytosis and Pinocytosis
The Genetic Code and Protein Synthesis
DNA and the Genetic Code
RNA and Protein Synthesis
Cell Division
Mitosis
Meiosis
Aging and Cells

Box 3–1 TERMINOLOGY OF SOLUTIONS
Box 3–2 GENETIC DISEASE—SICKLE-CELL ANEMIA
Box 3–3 ABNORMAL CELLULAR FUNCTIONING—CANCER

Student Objectives

- Name the organic molecules that make up cell membranes and state their functions.
- State the function of the nucleus and chromosomes.
- Describe the functions of the cell organelles.
- Define each of these cellular transport mechanisms and give an example of the role of each in the body: diffusion, osmosis, facilitated diffusion, active transport, filtration, phagocytosis, pinocytosis.
- Describe the triplet code of DNA.
- Explain how the triplet code of DNA is translated in the synthesis of proteins.
- Describe what happens in mitosis and in meiosis.
- Use examples to explain the importance of mitosis.
- Explain the importance of meiosis.

Cells

New Terminology

Active transport (**AK**-tiv **TRANS**-port)
Aerobic (air-**ROH**-bik)
Cell membrane (SELL **MEM**-brayn)
Chromosomes (**KROH**-muh-sohms)
Cytoplasm (**SIGH**-toh-plazm)
Diffusion (di-**FEW**-zhun)
Diploid number (**DIH**-ployd)
Filtration (fill-**TRAY**-shun)
Gametes (**GAM**-eets)
Gene (**JEEN**)
Haploid number (**HA**-ployd)
Meiosis (my-**OH**-sis)
Mitochondria (MY-toh-**KAHN**-dree-ah)
Mitosis (my-**TOH**-sis)
Nucleus (**NEW**-klee-us)
Organelles (OR-gan-**ELLS**)
Osmosis (ahs-**MOH**-sis)
Pinocytosis (PIN-oh-sigh-**TOH**-sis)
Phagocytosis (FAG-oh-sigh-**TOH**-sis)
Selectively permeable (se-**LEK**-tiv-lee **PER**-me-uh-buhl)
Theory (**THEER**-ree)

Related Clinical Terminology

Benign (bee-**NINE**)
Carcinogen (kar-**SIN**-oh-jen)
Chemotherapy (KEE-moh-**THER**-uh-pee)
Genetic disease (je-**NET**-ik di-**ZEEZ**)
Hypertonic (HIGH-per-**TAHN**-ik)
Hypotonic (HIGH-po-**TAHN**-ik)
Isotonic (EYE-so-**TAHN**-ik)
Malignant (muh-**LIG**-nunt)
Metastasis (muh-**TASS**-tuh-sis)
Mutation (mew-**TAY**-shun)

Terms that appear in ***bold type*** *in the chapter text are defined in the glossary, which begins on p. 528.*

All living organisms are made of cells and cell products. This simple statement, called the Cell Theory, was first proposed over 150 years ago. You may think of a **theory** as a guess or hypothesis, and sometimes this is so. A theory, however, is actually the best explanation of all the available evidence. All of the evidence science has gathered so far supports the validity of the Cell Theory.

Cells are the smallest living subunits of a multicellular organism such as a human being. A cell is a complex arrangement of the chemicals discussed in the previous chapter, is living, and carries out specific activities. Microorganisms, such as amoebas and bacteria, are single cells that function independently. Human cells, however, must work together and function interdependently. Homeostasis depends upon the contributions of all of the different kinds of cells.

Human cells vary in size, shape, and function. Most human cells are so small they can only be seen with the aid of a microscope and are measured in units called **microns** (1 micron = 1/25,000 of an inch—see Appendix 1: Units of Measure). One exception is the human ovum or egg cell, which is about 1 millimeter in diameter, just visible to the unaided eye. Some nerve cells, although microscopic in diameter, may be quite long. Those in our arms and legs, for example, are at least 2 feet (60 cm) long.

With respect to shape, human cells vary greatly. Some are round or spherical, others rectangular, still others irregular. White blood cells even change shape as they move.

Cell functions also vary, and since our cells do not act independently, we will cover specialized cell functions in Chapter 4. This chapter is concerned with the basic structure of cells and the cellular activities common to all our cells.

CELL STRUCTURE

Despite their many differences, human cells have several similar structural features: a cell membrane, cytoplasm and cell organelles, and a nucleus. Red blood cells are an exception because they have no nuclei when mature. The cell membrane forms the outer boundary of the cell and surrounds the cytoplasm, organelles, and nucleus.

CELL MEMBRANE

Also called the **plasma membrane**, the **cell membrane** is made of phospholipids, cholesterol, and proteins. The arrangement of these organic molecules is shown in Fig. 3-1. The phospholipids permit lipid-soluble materials to easily enter or leave the cell by diffusion through the cell membrane. The presence of cholesterol decreases the fluidity of the membrane, thus making it more stable. The proteins have several functions: Some form **pores** or openings to permit passage of materials; others are **enzymes** that also help substances enter the cell. Still other proteins, with oligosaccharides on their outer surface, are **antigens**, markers that identify the cells of an individual as "self." Yet another group of proteins serves as **receptor sites** for hormones. Many hormones bring about their specific effects by first bonding to a particular receptor on the cell membrane. This bonding then triggers chemical reactions within the cell membrane or the interior of the cell.

Although the cell membrane is the outer boundary of the cell, it should already be apparent to you that it is not a static or wall-like boundary, but rather an active, dynamic one. The cell membrane is **selectively permeable**, that is, certain substances are permitted to pass through and others are not. These mechanisms of cellular transport will be covered later in this chapter.

NUCLEUS

With the exception of mature red blood cells, all human cells have a nucleus. The **nucleus** is within the cytoplasm and is bounded by a double-layered **nuclear membrane** with many pores. It contains one or more nucleoli and the chromosomes of the cell (Fig. 3-2).

A **nucleolus** is a small sphere made of DNA, RNA, and protein. The nucleoli form a type of RNA called ribosomal RNA, which becomes part of ribosomes (a cell organelle) and is involved in protein synthesis.

The nucleus is the control center of the cell because it contains the chromosomes. The 46 **chromosomes** of a human cell are usually not visible; they are long threads called **chromatin**. When a cell divides, however, the chromatin coils extensively into visible chromosomes. Chromosomes are made of DNA and protein. Remember from our earlier discussion that the

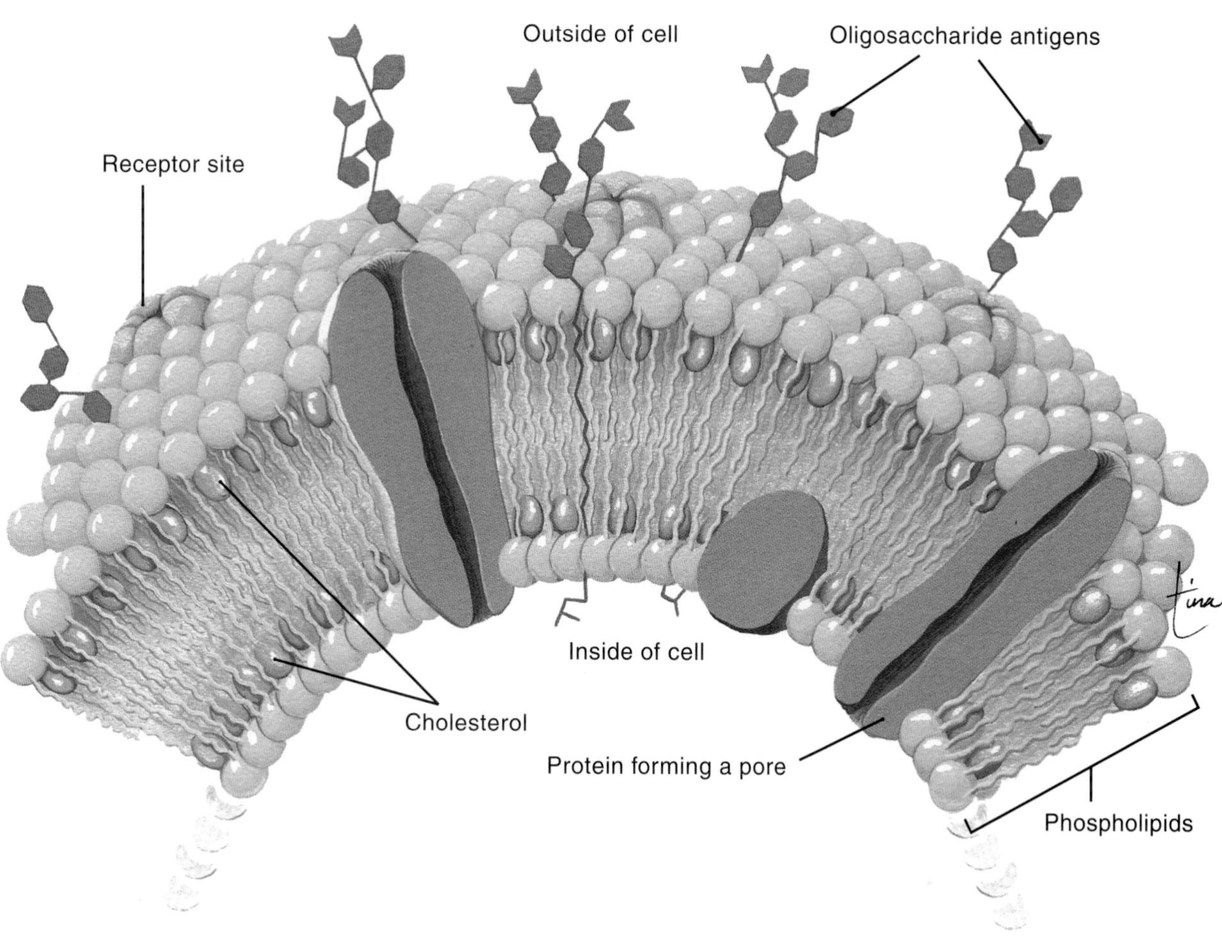

Figure 3–1 The cell (plasma) membrane depicting the types of molecules present.

DNA is the genetic code for the characteristics and activities of the cell. Although the DNA in the nucleus of each cell contains all of the genetic information for all human traits, only a small number of genes (a **gene** is the genetic code for one protein) are actually active in a particular cell. These active genes are the codes for the proteins necessary for the specific cell type. How the genetic code in chromosomes is translated into proteins will be covered in a later section.

CYTOPLASM AND CELL ORGANELLES

Cytoplasm is a watery solution of minerals, gases, and organic molecules that is found between the cell membrane and the nucleus. Chemical reactions take place within the cytoplasm, and many of the cell organelles are found here. Cell **organelles** are intracellular structures, often bounded by their own membranes, that have specific roles in cellular functioning. They are also shown in Fig. 3–2.

The **endoplasmic reticulum** (ER) is an extensive network of membranous tubules that extend from the nuclear membrane to the cell membrane. Rough ER has numerous ribosomes on its surface, whereas smooth ER has no ribosomes at all. As a network of interconnected tunnels, the ER serves as a passageway for the transport of the materials necessary for cell function within the cell. These include proteins synthesized by the ribosomes on the rough ER, and lipids synthesized by the smooth ER.

Ribosomes are very small structures made of protein and ribosomal RNA. Some are found on the sur-

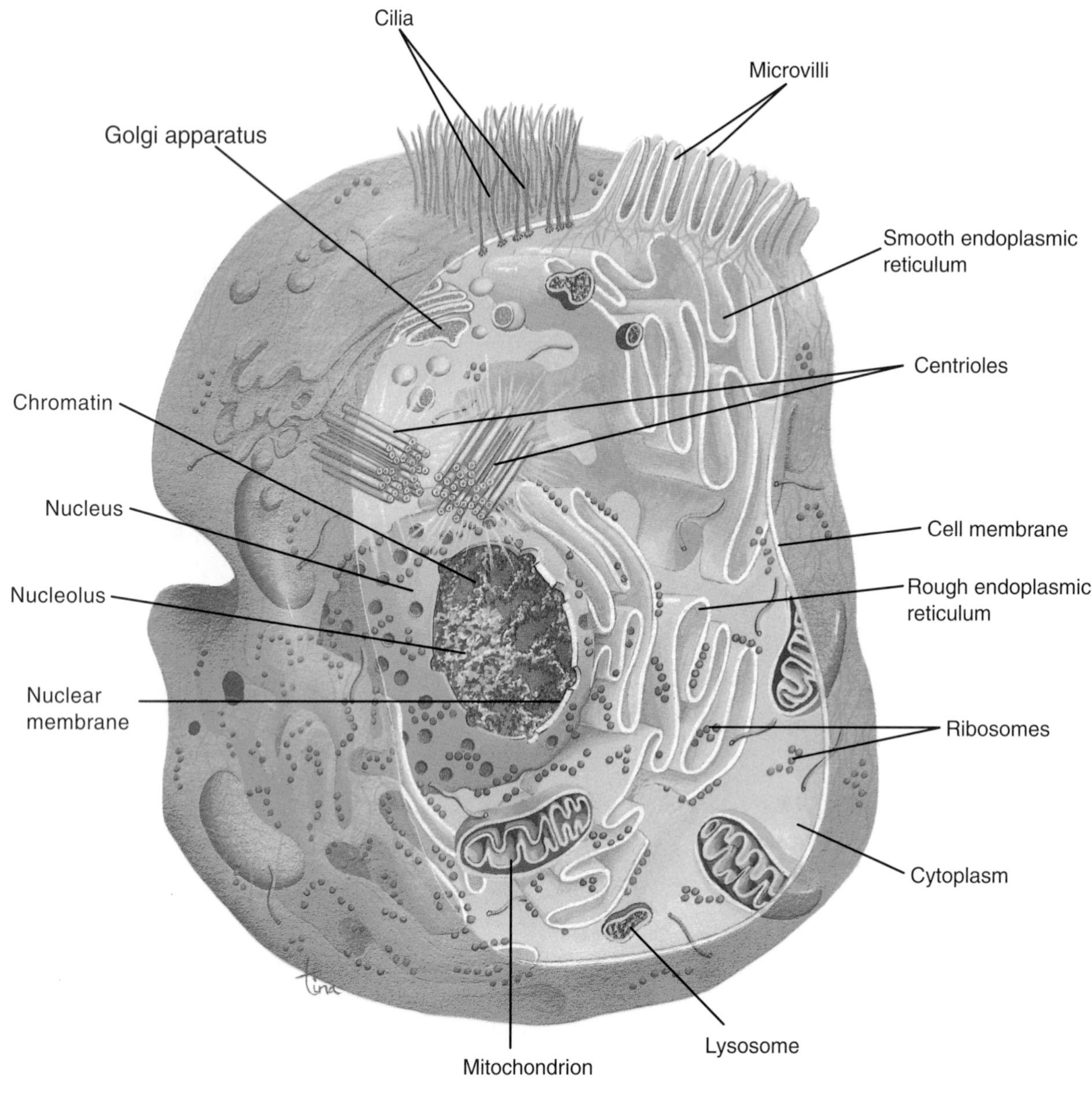

Figure 3–2 Generalized human cell depicting the structural components. See text and Table 3–1 for descriptions.

face of rough ER, while others float freely within the cytoplasm. Ribosomes are the site of protein synthesis.

The **Golgi apparatus** is a series of flat, membranous sacs, somewhat like a stack of saucers. Carbohydrates are synthesized within the Golgi apparatus, and are packaged, along with other materials, for secretion from the cell. To secrete a substance, small sacs of the Golgi membrane break off and fuse with the cell membrane, releasing the substance to the exterior of the cell.

Mitochondria are oval or spherical organelles within the cytoplasm, bounded by a double membrane. The inner membrane has folds called cristae. Within the mitochondria, the **aerobic** (oxygen-requiring) reactions of cell respiration take place. Therefore, mitochondria are the site of ATP (and hence energy) production. Cells that require large amounts of ATP, such as muscle cells, have many mitochondria to meet their need for energy.

Lysosomes are single membrane structures within the cytoplasm that contain digestive enzymes. When certain white blood cells engulf bacteria, the bacteria are digested and destroyed by these lysosomal enzymes. Worn-out cell parts and dead cells are also digested by these enzymes, which is necessary before tissue repair can begin, but which contributes to the process of inflammation in damaged tissues.

Centrioles are a pair of rod-shaped structures perpendicular to one another, located just outside the nucleus. Their function is to organize the spindle fibers during cell division.

Cilia and **flagella** are mobile thread-like projections through the cell membrane. **Cilia** serve the function of sweeping materials across the cell surface. They are usually shorter than flagella, and an individual cell has many of them. Cells lining the fallopian tubes, for example, have cilia to sweep the egg cell toward the uterus. The only human cell with a **flagellum** is the sperm cell. The flagellum provides **motility**, or movement, for the sperm cell.

The functions of the cell organelles are summarized in Table 3-1.

Table 3–1 FUNCTIONS OF CELL ORGANELLES

Organelle	Function(s)
Endoplasmic reticulum (ER)	• Passageway for transport of materials within the cell • Synthesis of lipids
Ribosomes	• Site of protein synthesis
Golgi apparatus	• Synthesis of carbohydrates • Packaging of materials for secretion from the cell
Mitochondria	• Site of aerobic cell respiration—ATP production
Lysosomes	• Contain enzymes to digest ingested material or damaged tissue
Centrioles	• Organize the spindle fibers during cell division
Cilia	• Sweep materials across the cell surface
Flagellum	• Enables a cell to move

CELLULAR TRANSPORT MECHANISMS

Living cells constantly interact with the blood or tissue fluid around them, taking in some substances and secreting or excreting others. There are several mechanisms of transport that enable cells to move materials into or out of the cell: diffusion, osmosis, facilitated diffusion, active transport, filtration, phagocytosis, and pinocytosis. Some of these take place without the expenditure of energy by the cells. But others *do* require energy, often in the form of ATP. Each of these mechanisms is described in the following sections and an example is included to show how each is important to the body.

DIFFUSION

Diffusion is the movement of molecules from an area of greater concentration to an area of lesser concentration (that is, with or along a **concentration gradient**). Diffusion occurs because molecules have free energy, that is, they are always in motion. The molecules in a solid move very slowly; those in a liquid move faster, and those in a gas move faster still, as when ice absorbs heat energy, melts, and then evaporates. In Fig. 3-3, a sugar cube in a glass of water is shown. As the sugar dissolves, the sugar molecules collide with one another. These collisions spread out the sugar molecules until they are evenly dispersed among the water molecules. The molecules are still moving, but as some go to the top others go to the bottom, and so on. Thus, an equilibrium (or steady-state balance) is reached.

Diffusion is a very slow process, but may be an effective transport mechanism across microscopic distances. Within the body, the gases oxygen and carbon dioxide move by diffusion. In the lungs, for example, there is a high concentration of oxygen in the alveoli (air sacs) and a low concentration of oxygen in the

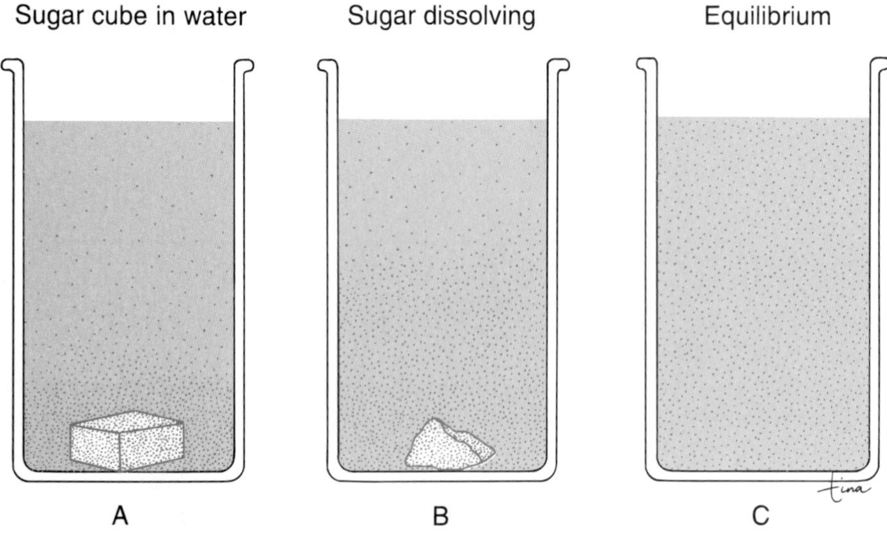

Figure 3–3 Diffusion of sugar in water. **(A)** Sugar cube. **(B)** Partial dissolving and diffusion of sugar molecules. **(C)** Sugar molecules distributed evenly throughout the water (such an equilibrium would take a very long time).

blood in the surrounding pulmonary capillaries. The opposite is true for carbon dioxide: A low concentration in the air in the alveoli and a high concentration in the blood in the pulmonary capillaries. These gases diffuse in opposite directions, each moving from where there is more to where there is less. Oxygen diffuses from the air to the blood to be circulated throughout the body. Carbon dioxide diffuses from the blood to the air to be exhaled.

OSMOSIS

Osmosis may be simply defined as the diffusion of water through a selectively permeable membrane or barrier. That is, water will move from an area with more water present to an area with less water. Another way to say this is that water will naturally tend to move to an area where there is more dissolved material, such as salt or sugar. If a 2% salt solution and a 6% salt solution are separated by a membrane allowing water but not salt to pass through it, water will diffuse from the 2% salt solution to the 6% salt solution. The result is that the 2% solution will become more concentrated, and the 6% solution will become more dilute.

In the body, the cells lining the small intestine absorb water from digested food by osmosis. These cells have first absorbed salts, have become more "salty," and water follows salt into the cells. The process of osmosis also takes place in the kidneys, which reabsorb large amounts of water (many gallons each day) to prevent its loss in urine. In Box 3–1: Terminology of Solutions is some terminology we use when discussing solutions and the effects of various solutions on cells.

FACILITATED DIFFUSION

The word facilitate means to help or assist. In **facilitated diffusion**, molecules move through a membrane from an area of greater concentration to an area of lesser concentration, but they need some help to do this.

In the body, our cells must take in glucose to use for ATP production. Glucose, however, will not diffuse through most cell membranes by itself, even if there is more outside the cell than inside. Diffusion of glucose into most cells requires **carrier enzymes**, proteins that are part of the cell membrane. Glucose bonds to the carrier enzymes, and by doing so be-

Box 3–1 TERMINOLOGY OF SOLUTIONS

Human cells or other body fluids contain many dissolved substances (called **solutes**) such as salts, sugars, acids, and bases. The concentration of solutes in a fluid creates the **osmotic pressure** of the solution, which in turn determines the movement of water through membranes.

As an example here, we will use sodium chloride (NaCl). Human cells have an NaCl concentration of 0.9%. With human cells as a reference point, the relative NaCl concentrations of other solutions may be described with the following terms:

Isotonic—a solution with the same salt concentration as in cells.
The blood plasma is isotonic to red blood cells.

Hypotonic—a solution with a lower salt concentration than in cells.
Distilled water (0% salt) is hypotonic to human cells.

Hypertonic—a solution with a higher salt concentration than in cells.
Sea water (3% salt) is hypertonic to human cells.

Refer now to the diagrams below of red blood cells (RBCs) in each of these different types of solutions, and note the effect of each on osmosis:

- When RBCs are in plasma, water moves into and out of them at equal rates, and the cells remain normal in size and water content.
- If RBCs are placed in distilled water, more water will enter the cells than leave, and the cells will swell and eventually burst.
- If RBCs are placed in seawater, more water will leave the cells than enter, and the cells will shrivel and die.

This knowledge of osmotic pressure is used when replacement fluids are needed for a patient who has become dehydrated. Isotonic solutions are usually used; normal saline and Ringer's solution are examples. These will provide rehydration without causing osmotic damage to cells or extensive shifts of fluid between the blood and tissues.

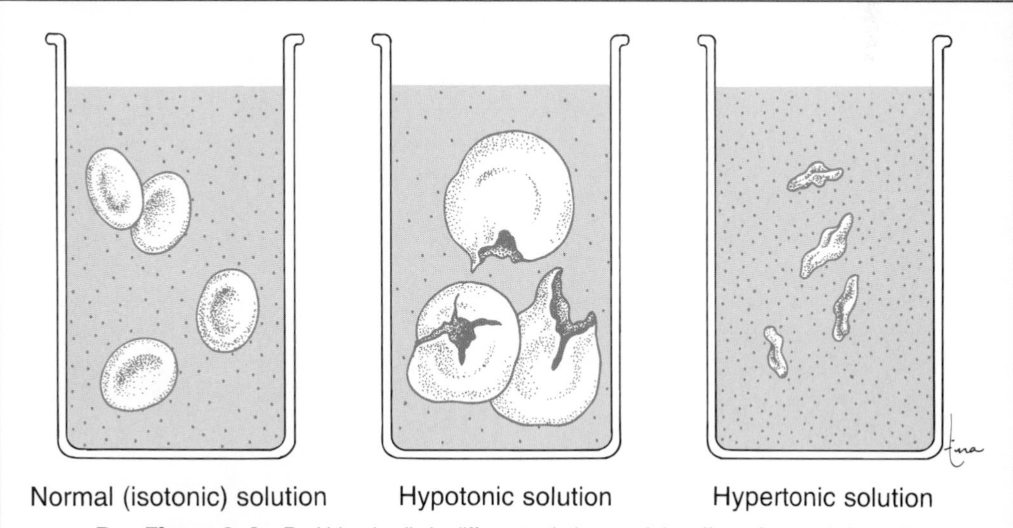

Box Figure 3–A Red blood cells in different solutions and the effect of osmosis in each.

comes soluble in the phospholipids of the cell membrane. The glucose-carrier molecule diffuses through the membrane, and glucose is released to the interior of the cell.

ACTIVE TRANSPORT

Active transport requires the energy of ATP to move molecules from an area of lesser concentration to an area of greater concentration. Notice that this is the opposite of diffusion, in which the free energy of molecules causes them to move to where there are fewer of them. Active transport is therefore said to be movement against a concentration gradient.

In the body, nerve cells and muscle cells have "sodium pumps" to move sodium ions (Na^+) out of the cells. Sodium ions are more abundant outside the cells, and they constantly diffuse into the cell, their area of lesser concentration. Without the sodium pumps to return them outside, the incoming sodium ions would bring about an unwanted nerve impulse or muscle contraction. Nerve and muscle cells constantly produce ATP to keep their sodium pumps working and prevent spontaneous impulses.

Another example of active transport is the absorption of glucose and amino acids by the cells lining the small intestine. The cells use ATP to absorb these nutrients from digested food, even when their intracellular concentration becomes greater than their extracellular concentration.

FILTRATION

The process of **filtration** also requires energy, but the energy needed does not come directly from ATP. It is the energy of mechanical pressure. Filtration means that water and dissolved materials are forced through a membrane from an area of higher pressure to an area of lower pressure.

In the body, **blood pressure** is created by the pumping of the heart. Filtration occurs when blood flows through capillaries, whose walls are only one cell thick and very permeable. The blood pressure in capillaries is higher than the pressure of the surrounding tissue fluid. In capillaries throughout the body, blood pressure forces plasma and dissolved materials through the capillary membranes into the surrounding tissue spaces. This creates more tissue fluid and is how cells receive glucose, amino acids, and other nutrients. Blood pressure in the capillaries of the kidneys also brings about filtration, which is the first step in the formation of urine.

Table 3–2 CELLULAR TRANSPORT MECHANISMS

Mechanism	Definition	Example in the Body
Diffusion	Movement of molecules from an area of greater concentration to an area of lesser concentration.	Exchange of gases in the lungs or body tissues.
Osmosis	The diffusion of water.	Absorption of water by the small intestine or kidneys.
Facilitated diffusion	Carrier enzymes move molecules across cell membranes.	Intake of glucose by most cells.
Active transport	Movement of molecules from an area of lesser concentration to an area of greater concentration (requires ATP).	Absorption of amino acids and glucose from food by the cells of the small intestine.
Filtration	Movement of water and dissolved substances from an area of higher pressure to an area of lower pressure (blood pressure).	Formation of tissue fluid; the first step in the formation of urine.
Phagocytosis	A moving cell engulfs something.	White blood cells engulf bacteria.
Pinocytosis	A stationary cell engulfs something.	Cells of the kidney tubules reabsorb small proteins.

PHAGOCYTOSIS AND PINOCYTOSIS

These two processes are similar in that both involve a cell engulfing something. An example of **phagocytosis** is a white blood cell engulfing bacteria. The white blood cell flows around the bacterium, taking it in and eventually digesting it.

Other cells that are stationary may take in small molecules that become adsorbed or attached to their membranes. The cells of the kidney tubules reabsorb small proteins by **pinocytosis**, so that the protein is not lost in urine.

Table 3-2 summarizes the cellular transport mechanisms.

THE GENETIC CODE AND PROTEIN SYNTHESIS

The structure of DNA, RNA, and protein was described in Chapter 2 but will be reviewed briefly here.

DNA AND THE GENETIC CODE

DNA is a double strand of nucleotides in the form of a **double helix**, very much like a spiral ladder. The rungs of the ladder are made of the four nitrogenous bases, always found in complementary pairs: adenine with thymine (A-T) and guanine with cytosine (G-C). Although DNA contains just these four bases, the bases may be arranged in many different sequences (reading up or down the ladder). It is the sequence of bases that is the **genetic code**. The DNA of our 46 chromosomes is estimated to contain about 6 billion base pairs, which make up as many as 50,000 to 100,000 genes.

Recall that a **gene** is the genetic code for one protein, and a protein is a specific sequence of amino acids. Therefore, a gene, or segment of DNA, is the code for the sequence of amino acids in a particular protein.

The code for a single amino acid consists of three bases in the DNA molecule; this **triplet** of bases may be called a **codon** (Fig. 3-4). There is a triplet of bases in the DNA for each amino acid in the protein. If a protein consists of 100 amino acids, the gene for that protein would consist of 100 triplets, or 300 bases. Some of the triplets will be the same, since the same amino acid may be present in several places within the protein. Also part of the gene are other triplets that start and stop the process of making the protein, rather like punctuation marks.

RNA AND PROTEIN SYNTHESIS

The transcription of the genetic code in DNA into proteins requires the other nucleic acid, **RNA**. DNA is found in the chromosomes in the nucleus of the cell, but protein synthesis takes place on the ribosomes in the cytoplasm. **Messenger RNA (mRNA)** is the intermediary molecule between these two sites.

When a protein is to be made, the segment of DNA that is its gene uncoils, and the hydrogen bonds between the base pairs break (see Fig. 3-4). Within the nucleus are RNA nucleotides (A,C,G,U) and enzymes to construct a single strand of nucleotides that is a complementary copy of half the DNA gene (with uracil in place of thymine). This copy of the gene is mRNA, which then separates from the DNA. The gene coils back into the double helix, and the mRNA leaves the nucleus, enters the cytoplasm, and becomes attached to ribosomes.

As the copy of the gene, mRNA is a series of triplets of bases; each triplet is the code for one amino acid. Another type of RNA, called **transfer RNA (tRNA)**, is also found in the cytoplasm. Each tRNA molecule has an **anticodon**, a triplet complementary to a triplet on the mRNA. The tRNA molecules pick up specific amino acids (which have come from protein in our food) and bring them to their proper triplets on the mRNA. The ribosomes contain enzymes to catalyze the formation of **peptide bonds** between the amino acids. When an amino acid has been brought to each triplet on the mRNA, and all peptide bonds have been formed, the protein is finished.

The protein then leaves the ribosomes and may be transported by the ER to where it is needed in the cell, or it may be packaged by the Golgi apparatus for secretion from the cell. A summary of the process of protein synthesis is found in Table 3-3.

Thus, the expression of the genetic code may be described by the following sequence:

DNA → RNA → Proteins:

Structural Proteins ↙ ↓ Enzymes

Enzymes ↓ Catalyze Reactions

Structural Proteins ↘ Hereditary Characteristics ↙ Catalyze Reactions

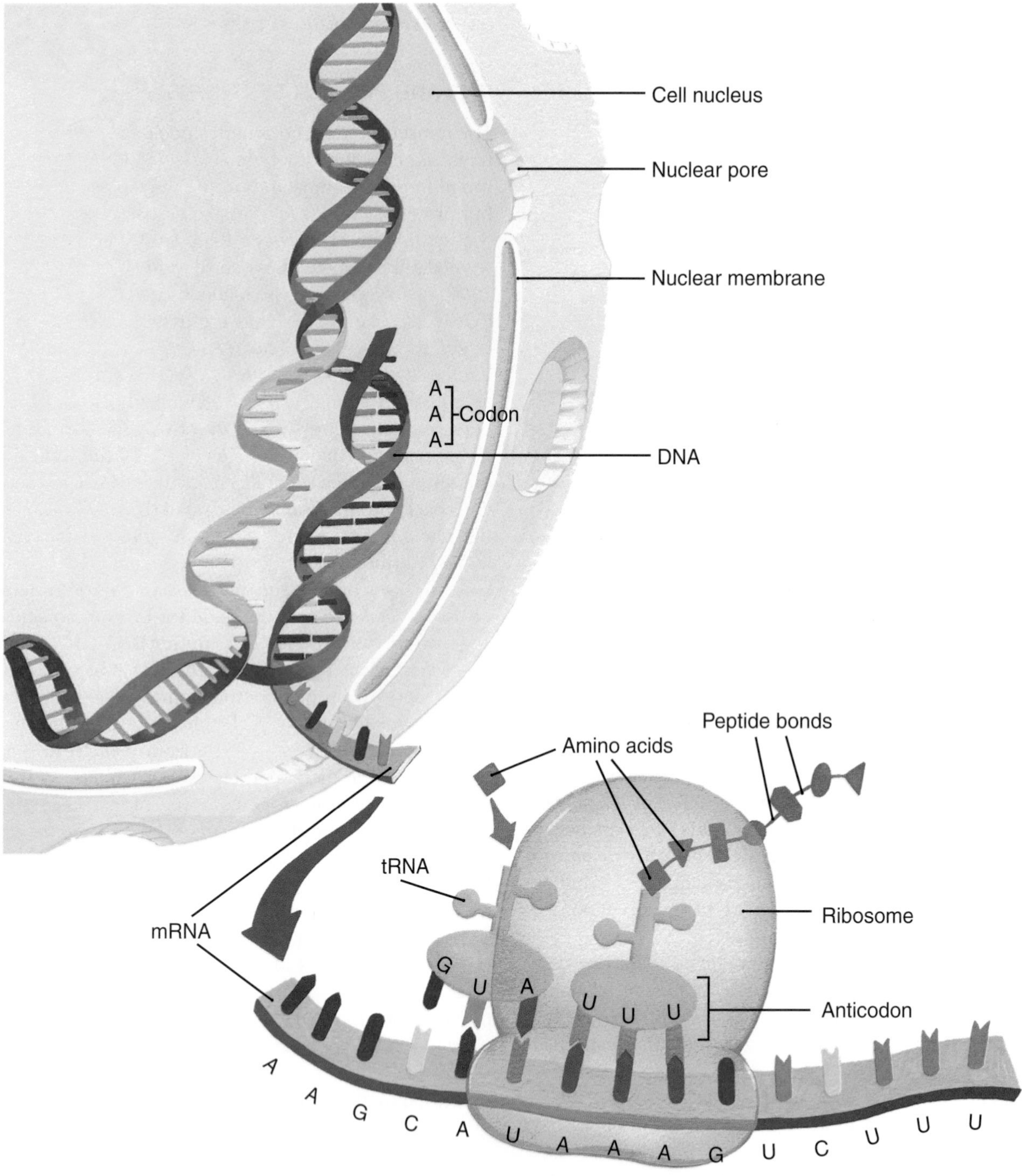

Figure 3–4 Protein synthesis. The mRNA is formed as a copy of a portion of the DNA in the nucleus of a cell. In the cytoplasm, the mRNA becomes attached to ribosomes. See text for further description.

Table 3–3 PROTEIN SYNTHESIS

Molecule or Organelle	Functions
DNA	• A double strand (helix) of nucleotides that is the genetic code in the chromosomes of cells • A gene is the sequence of bases (segment of DNA) that is the code for one protein
mRNA (messenger RNA)	• A single strand of nucleotides formed as a complementary copy of a gene in the DNA • Now contains the triplet code: three bases is the code for one amino acid • Leaves the DNA in the nucleus, enters the cytoplasm of the cell, and becomes attached to ribosomes
Ribosomes	• The cell organelles that are the site of protein synthesis • Attach the mRNA molecule • Contain enzymes to form peptide bonds between amino acids
tRNA (transfer RNA)	• Picks up amino acids (from food) in the cytoplasm and transports them to their proper sites (triplets) along the mRNA molecule

Each of us is the sum total of our genetic characteristics. Blood type, hair color, muscle proteins, nerve cells, and thousands of other aspects of our structure and functioning have their basis in the genetic code of DNA.

If there is a "mistake" in the DNA, that is, incorrect bases or triplets of bases, this mistake will be copied by the mRNA. The result is the formation of a malfunctioning or nonfunctioning protein. This is called a **genetic** or **hereditary disease**, and a specific example is described in Box 3-2: Genetic Disease—Sickle-Cell Anemia.

CELL DIVISION

Cell division is the process by which a cell reproduces itself. There are two types of cell division, mitosis and meiosis. Although both types involve cell reproduction, their purposes are very different.

MITOSIS

Each of us began life as one cell, a fertilized egg. Each of us now consists of billions of cells produced by the process of mitosis. In **mitosis**, one cell with the **dip-**

Box 3–2 GENETIC DISEASE—SICKLE-CELL ANEMIA

A **genetic disease** is a hereditary disorder, one that may be passed from generation to generation. Although there are hundreds of genetic diseases, they all have the same basis: a mistake in DNA. Because DNA makes up the chromosomes that are found in egg and sperm, this mistake may be passed from parents to children.

Sickle-cell anemia is the most common genetic disorder among people of African descent and affects the hemoglobin in red blood cells. Normal hemoglobin, called hemoglobin A (HbA), is a protein made of two identical alpha chains (141 amino acids each) and two identical beta chains (146 amino acids each). In sickle-cell hemoglobin (HbS), the 6th amino acid in each beta chain is incorrect; valine is present instead of the glutamic acid found in HbA. This difference seems minor—only two incorrect amino acids out of a total of over 500—but the consequences for the person are very serious.

HbS has a great tendency to crystallize when oxygen levels are low, as is true in capillaries. When HbS crystallizes, the red blood cells are deformed into crescents (sickles) and other irregular shapes. These irregular, rigid red blood cells clog and rupture capillaries, causing internal bleeding and severe pain. These red blood cells are also fragile, have a short life span, and break up easily, leading to anemia and hypoxia (lack of oxygen).

(Continued)

Box 3–2 GENETIC DISEASE—SICKLE-CELL ANEMIA (Continued)

In the past, children with sickle-cell anemia often died before the age of 20 years. Treatment of this disease has improved greatly, but it is still incurable.

What has happened to cause the formation of HbS rather than HbA? Because hemoglobin is a protein, the genetic code for it is in the DNA (of chromosome 11). One amino acid in the beta chains is incorrect, therefore, one triplet in the DNA gene for the beta chain must be, and is, incorrect. This mistake is copied by mRNA in the cells of the red bone marrow, and HbS is synthesized in red blood cells.

Sickle-cell anemia is a recessive genetic disease, which means that a person with one gene for HbS and one gene for HbA will have "sickle-cell trait." Such a person, sometimes called a carrier, usually will not have the severe effects of sickle-cell anemia, but may pass the gene for HbS to children. It is estimated that 9% of African-Americans have sickle-cell trait and about 1% have sickle-cell anemia.

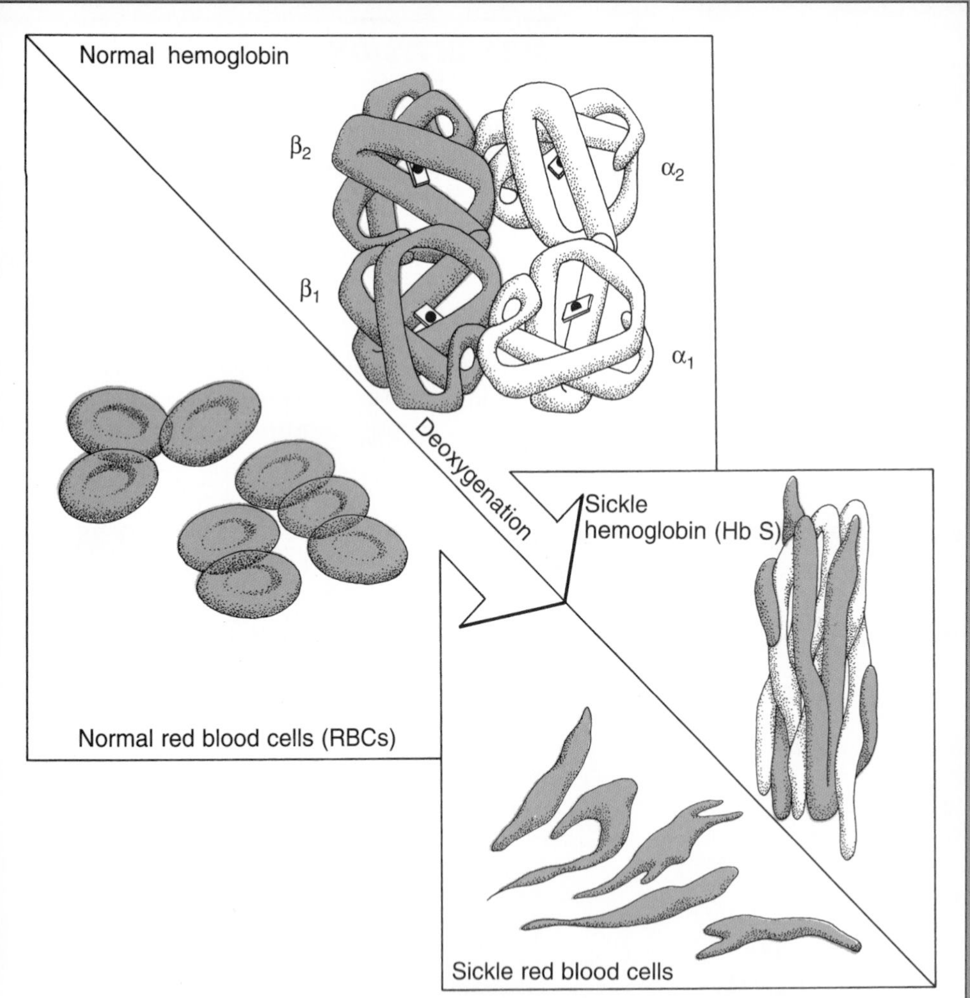

Box Figure 3–B Structure of hemoglobin A and sickle-cell hemoglobin and their effect on red blood cells.

loid number of chromosomes (the usual number, 46 for people) divides into two identical cells, each with the diploid number of chromosomes. This production of identical cells is necessary for the growth of the organism and for repair of tissues (see also Box 3-3: Abnormal Cellular Functioning—Cancer).

Before mitosis can take place, a cell must have two complete sets of chromosomes, because each new cell must have the diploid number. The process of **DNA replication** enables each chromosome to make a copy of itself. The time during which this takes place is called **interphase**, the time between mitotic divisions. Although interphase is sometimes referred to as the resting stage, resting means "not dividing," not "inactive." The cell is quite actively producing a second set of chromosomes and storing energy in ATP.

The stages of mitosis are **prophase, metaphase, anaphase**, and **telophase**. What happens in each of these stages is described in Table 3-4. As you read the events of each stage, refer to Fig. 3-5, which depicts mitosis in a cell with a diploid number of four.

As mentioned previously, mitosis is essential for re-

Box 3-3 ABNORMAL CELLULAR FUNCTIONING—CANCER

There are more than 100 different types of **cancer**, all of which are characterized by abnormal cellular functioning. Normally, our cells undergo mitosis only when necessary and stop when appropriate. A cut in the skin, for example, is repaired by mitosis, usually without formation of excess tissue. The new cells fill in the damaged area, and mitosis slows when the cells make contact with surrounding cells. This is called contact inhibition, which limits the new tissue to just what is needed. **Malignant** (cancer) cells, however, are characterized by uncontrolled cell division. Our cells are genetically programmed to have particular life spans and to divide or die. One gene is known to act as a brake on cell division; another gene enables cells to live indefinitely, beyond their normal life spans, and to keep dividing. Any imbalance in the activity of these genes may lead to abnormal cell division. Such cells are not inhibited by contact with other cells, keep dividing, and tend to spread.

A malignant tumor begins in a primary site such as the colon, then may spread or metastasize. Often the malignant cells are carried by the lymph or blood to other organs such as the liver, where secondary tumors develop. **Metastasis** is characteristic only of malignant cells; **benign** tumors do not metastasize but remain localized in their primary site.

What causes normal cells to become malignant? At present, we have only partial answers. A malignant cell is created by a **mutation**, a genetic change that brings about abnormal cell functions or responses. Environmental substances that cause mutations are called **carcinogens**. One example is the tar found in cigarette smoke, which is definitely linked to lung cancer. Ultraviolet light may also cause mutations, especially in skin that is overexposed to sunlight. For a few specific kinds of cancer, the trigger is believed to be infection with certain viruses that cause cellular mutations. Carriers of hepatitis B virus, for example, are more likely to develop primary liver cancer than are people who have never been exposed to this virus. Recent research has discovered two genes, one on chromosome 2 and the other on chromosome 3, that contribute to a certain form of colon cancer. Both of these genes are the codes for proteins that correct the "mistakes" that may occur when the new DNA is synthesized. When these proteins do not function properly, the mistakes (mutations) in the DNA lead to the synthesis of yet other faulty proteins that impair the functioning of the cell and predispose it to becoming malignant.

Once cells have become malignant, their functioning cannot return to normal. Therefore, the treatments for cancer are directed at removing or destroying the abnormal cells. Surgery to remove tumors, radiation to destroy cells, and **chemotherapy** to stop cell division or interfere with other aspects of cell metabolism are all aspects of cancer treatment.

Table 3–4 STAGES OF MITOSIS

Stage	Events
Prophase	1. The chromosomes coil up and become visible as short rods. Each chromosome is really 2 chromatids (original DNA plus its copy) still attached at a region called the centromere. 2. The nuclear membrane disappears. 3. The centrioles move toward opposite poles of the cell and organize the spindle fibers, which extend across the equator of the cell.
Metaphase	1. The pairs of chromatids line up along the equator of the cell. The centromere of each pair is attached to a spindle fiber. 2. The centromeres now divide.
Anaphase	1. Each chromatid is now considered a separate chromosome; there are two complete and separate sets. 2. The spindle fibers contract and pull the chromosomes, one set toward each pole of the cell.
Telophase	1. The sets of chromosomes reach the poles of the cell and become indistinct as their DNA uncoils to form chromatin. 2. A nuclear membrane re-forms around each set of chromosomes.
Cytokinesis	1. The cytoplasm divides; new cell membrane is formed.

pair of tissues, to replace damaged or dead cells. Some examples may help illustrate this. In several areas of the body, mitosis takes place constantly. These sites include the epidermis of the skin, the stomach lining, and the red bone marrow. For each of these sites, there is a specific reason why this constant mitosis is necessary.

What happens to the surface of the skin? The dead, outer cells are worn off by contact with the environment. Mitosis in the lower living layer replaces these cells, and the epidermis maintains its normal thickness.

The stomach lining, although internal, is also constantly worn away. Gastric juice, especially the hydrochloric acid, is very damaging to cells. Rapid mitosis replaces damaged cells and keeps the stomach lining intact.

One of the functions of red bone marrow is the production of red blood cells. Because red blood cells have a life span of only about 120 days, new ones are needed to replace the older ones that die. Very rapid mitosis in the red bone marrow produces approximately 2 million new red blood cells every second.

It is also important to be aware of the areas of the body where mitosis cannot take place. In an adult, most muscle cells and neurons (nerve cells) cannot reproduce themselves. If they die, their functions are also lost. Someone whose spinal cord has been severed will have paralysis and loss of sensation below the level of the injury. The spinal cord neurons cannot undergo mitosis to replace the ones that were lost, and such an injury is permanent.

The heart is made of cardiac muscle cells, which are also incapable of mitosis. A heart attack (myocardial infarction) means that a portion of cardiac muscle dies because of lack of oxygen. These cells cannot be replaced, and the heart will be a less effective pump. If a large enough area of the heart muscle dies, the heart attack may be fatal.

MEIOSIS

Meiosis is a more complex process of cell division that results in the formation of **gametes**, which are egg and sperm cells. In meiosis, one cell with the diploid number of chromosomes divides twice to form four cells, each with the **haploid number** (half the usual number) of chromosomes. In women, meiosis takes place in the ovaries and is called **oogenesis**. In men, meiosis takes place in the testes and is called **spermatogenesis**. The differences between oogenesis and spermatogenesis will be discussed in Chapter 20, the Reproductive Systems.

The egg and sperm cells produced by meiosis have the haploid number of chromosomes, which is 23 for humans. Meiosis is sometimes called reduction division because the division process reduces the chromosome number in egg or sperm. Then, during **fertilization** in which the egg unites with the sperm, the 23 chromosomes of the sperm plus the 23 chromosomes of the egg will restore the diploid number of 46 in the fertilized egg. Thus the proper chromosome number is maintained in the cells of the new individual.

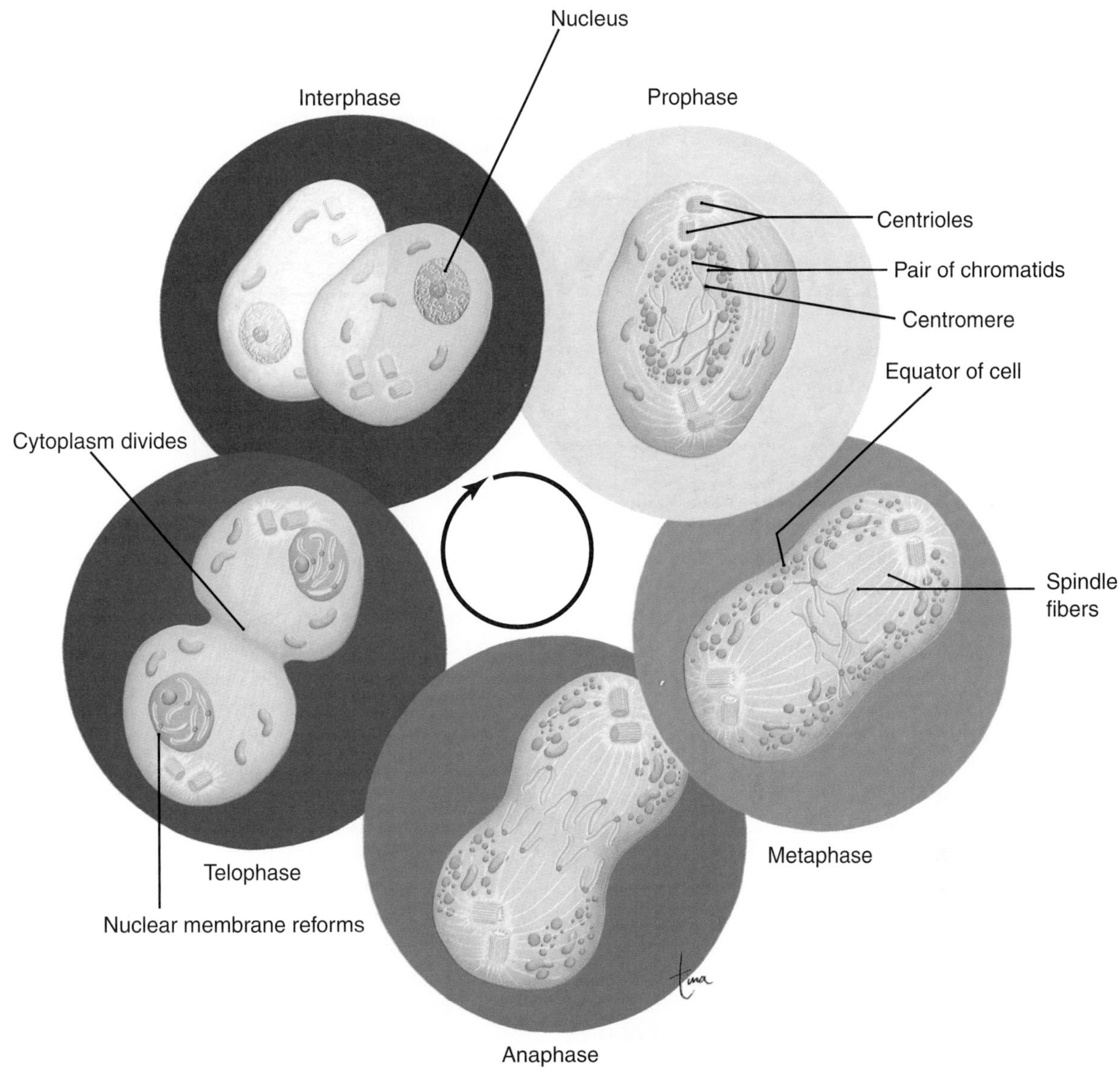

Figure 3–5 Stages of mitosis in a cell with the diploid number of four. See Table 3–4 for descriptions.

AGING AND CELLS

Multicellular organisms, including people, age and eventually die; our cells do not have infinite life spans. It has been proposed that some cells capable of mitosis are limited to a certain number of divisions, that is, every division is sort of a tick-tock off a biological clock. At present, it is not known exactly what this cellular biological clock is.

Cellular aging also involves the inevitable deterioration of chromosomes and cell organelles. Just as the parts of a car break down in time, so too will cells. Unlike cars or machines, cells can often repair themselves, but they do have limits. As cells age, structural proteins break down and are not replaced, or necessary enzymes are not synthesized.

Much about the chemistry of the aging process remains a mystery, though we can describe what happens to organs and to the body as a whole. In each of the following chapters on body systems there is a brief discussion of how aging affects the system. Keep in mind that a system is the sum of its cells, in tissues and organs, and that all aging is ultimately at the cellular level.

SUMMARY

As mentioned at the beginning of this chapter, human cells work closely together and function interdependently. Each type of human cell makes a contribution to the body as a whole. Usually, however, cells do not function as individuals, but rather in groups. Groups of cells with similar structure and function form a tissue, which is the next level of organization.

STUDY OUTLINE

Human cells vary in size, shape, and function. Our cells function interdependently to maintain homeostasis.

Cell Structure—the major parts of a cell are the cell membrane, nucleus (except mature RBCs), cytoplasm, and cell organelles

1. Cell Membrane—the selectively permeable boundary of the cell (see Fig. 3-1).
   - Phospholipids permit diffusion of lipid-soluble materials.
   - Cholesterol provides stability.
   - Proteins form pores, carrier enzymes, "self" antigens, and receptor sites for hormones.
2. Nucleus—the control center of the cell; has a double-layer membrane.
   - Nucleolus—forms ribosomal RNA.
   - Chromosomes—made of DNA and protein; DNA is the genetic code for the structure and functioning of the cell. A gene is a segment of DNA that is the code for one protein. Human cells have 46 chromosomes.
3. Cytoplasm—a watery solution of minerals, gases, and organic molecules; contains the cell organelles; site for many chemical reactions.
4. Cell Organelles—intracellular structures with specific functions (see Table 3-1 and Fig. 3-2).

Cellular Transport Mechanisms—the processes by which cells take in or secrete or excrete materials through the selectively permeable cell membrane

1. Diffusion—movement of molecules from an area of greater concentration to an area of lesser concentration; occurs because molecules have free energy: They are constantly in motion. Oxygen and carbon dioxide are exchanged by diffusion in the lungs.
2. Osmosis—the diffusion of water. Water diffuses to an area of less water, that is, to an area of more dissolved material. The small intestine absorbs water from digested food by osmosis. Isotonic, hypertonic, and hypotonic (see Box 3-1).
3. Facilitated Diffusion—carrier enzymes that are part of the cell membrane permit cells to take in materials that would not diffuse in by themselves. Most cells take in glucose by facilitated diffusion.
4. Active Transport—a cell uses ATP to move substances from an area of lesser concentration to an area of greater concentration. Nerve cells and muscle cells have sodium pumps to return Na^+ ions to the exterior of the cells; this prevents spontaneous impulses. Cells of the small intestine absorb glucose and amino acids from digested food by active transport.
5. Filtration—pressure forces water and dissolved materials through a membrane from an area of higher pressure to an area of lower pressure. Tissue fluid is formed by filtration: Blood pressure forces plasma and dissolved nutrients out of capillaries and into tissues. Blood pressure in the kidney capillaries creates filtration that is the first step in the formation of urine.
6. Phagocytosis—a moving cell engulfs something; white blood cells phagocytize bacteria to destroy them.
7. Pinocytosis—a stationary cell engulfs small mole-

cules; kidney tubule cells reabsorb small proteins by pinocytosis.

The Genetic Code and Protein Synthesis (see Fig. 3–4 and Table 3–3)

1. DNA and the Genetic Code
   - DNA is a double helix with complementary base pairing: A–T and G–C.
   - The sequence of bases in the DNA is the genetic code for proteins.
   - The triplet code: three bases (a codon) is the code for one amino acid.
   - A gene consists of all the triplets that code for a single protein.
2. RNA and Protein Synthesis
   - mRNA is a complementary copy of the sequence of bases in a gene (DNA).
   - mRNA moves from the nucleus to the ribosomes in the cytoplasm.
   - tRNA molecules (in the cytoplasm) have anticodons for the triplets on the mRNA.
   - tRNA molecules bring amino acids to their proper triplets on the mRNA.
   - Ribosomes contain enzymes to form peptide bonds between the amino acids.
3. Expression of the Genetic Code
   - DNA → RNA → Proteins (structural and enzymes) → Hereditary Characteristics.
   - A genetic disease is a "mistake" in the DNA, which is copied by mRNA and results in a malfunctioning protein.

Cell Division

1. Mitosis—one cell with the diploid number of chromosomes divides once to form two cells, each with the diploid number of chromosomes (46 for humans).
   - DNA replication forms two sets of chromosomes during interphase.
   - Stages of mitosis (see Fig. 3–5 and Table 3–4): prophase, metaphase, anaphase, and telophase. Cytokinesis is the division of the cytoplasm following telophase.
   - Mitosis is essential for growth and for repair and replacement of damaged cells.
   - Most adult nerve and muscle cells cannot divide; their loss may involve permanent loss of function.
2. Meiosis—one cell with the diploid number of chromosomes divides twice to form four cells, each with the haploid number of chromosomes (23 for humans).
   - Oogenesis in the ovaries forms egg cells.
   - Spermatogenesis in the testes forms sperm cells.
   - Fertilization of an egg by a sperm restores the diploid number in the fertilized egg.

REVIEW QUESTIONS

1. State the functions of the organic molecules of cell membranes: cholesterol, proteins, and phospholipids. (p. 44)
2. Describe the function of each of these cell organelles: mitochondria, lysosomes, Golgi apparatus, ribosomes, and endoplasmic reticulum. (pp. 45–47)
3. Explain why the nucleus is the control center of the cell. (pp. 44–45)
4. What part of the cell membrane is necessary for facilitated diffusion? Describe one way this process is important within the body. (p. 48)
5. What provides the energy for filtration? Describe one way this process is important within the body. (p. 50)
6. What provides the energy for diffusion? Describe one way this process is important within the body. (pp. 47–48)
7. What provides the energy for active transport? Describe one way this process is important within the body. (p. 50)

8. Define osmosis, and describe one way this process is important within the body. (p. 48)
9. Explain the difference between hypertonic and hypotonic, using human cells as a reference point. (p. 49)
10. In what way are phagocytosis and pinocytosis similar? Describe one way each process is important within the body. (p. 51)
11. How many chromosomes does a human cell have? What are these chromosomes made of? (p. 51)
12. Name the stage of mitosis in which each of the following takes place: (p. 56)
    a. The two sets of chromosomes are pulled toward opposite poles of the cell.
    b. The chromosomes become visible as short rods.
    c. A nuclear membrane re-forms around each complete set of chromosomes.
    d. The pairs of chromatids line up along the equator of the cell.
    e. The centrioles organize the spindle fibers.
    f. Cytokinesis takes place after this stage.
13. Describe two specific ways mitosis is important within the body. Explain why meiosis is important. (pp. 53, 55–56)
14. Compare mitosis and meiosis in terms of: (pp. 53, 56)
    a. Number of divisions
    b. Number of cells formed
    c. Chromosome number of the cell formed
15. Explain the triplet code of DNA. Name the molecule that copies the triplet code of DNA. Name the organelle that is the site of protein synthesis. What other function does this organelle have in protein formation? (pp. 51–53)

CHAPTER 4

Tissues and Membranes

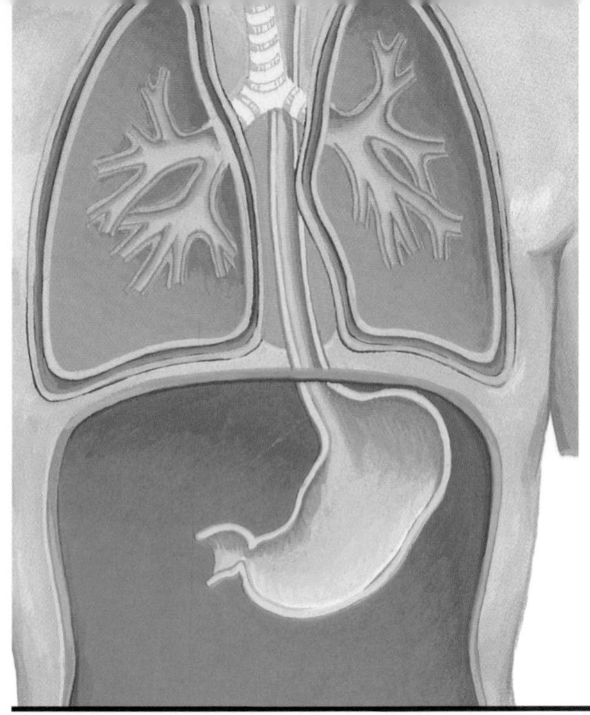

CHAPTER 4

Chapter Outline

Epithelial Tissue
Simple Squamous Epithelium
Stratified Squamous Epithelium
Transitional Epithelium
Simple Cuboidal Epithelium
Simple Columnar Epithelium
Ciliated Epithelium
Glands
- Unicellular glands
- Multicellular glands

Connective Tissue
Blood
Areolar Connective Tissue
Adipose Tissue
Fibrous Connective Tissue
Elastic Connective Tissue
Bone
Cartilage

Muscle Tissue
Skeletal Muscle
Smooth Muscle
Cardiac Muscle

Nerve Tissue

Membranes
Epithelial Membranes
- Serous membranes
- Mucous membranes

Connective Tissue Membranes

Aging and Tissues

Box 4–1 CYSTIC FIBROSIS
Box 4–2 VITAMIN C AND COLLAGEN
Box 4–3 COSMETIC COLLAGEN

Student Objectives

- Describe the general characteristics of each of the four major categories of tissues.
- Describe the functions of the types of epithelial tissues with respect to the organs in which they are found.
- Describe the functions of the connective tissues, and relate them to the functioning of the body or a specific organ system.
- Explain the differences, in terms of location and function, among skeletal muscle, smooth muscle, and cardiac muscle.
- Name the three parts of a neuron and state the function of each. Name the organs made of nerve tissue.
- Describe the locations of the pleural membranes, the pericardial membranes, and the peritoneum–mesentery. State the function of serous fluid in each of these locations.
- State the locations of mucous membranes and the functions of mucus.
- Name some membranes made of connective tissue.
- Explain the difference between exocrine and endocrine glands, and give an example of each.

Tissues and Membranes

New Terminology

Absorption (ab-**ZORB**-shun)
Bone (**BOWNE**)
Cartilage (**KAR**-ti-lidj)
Chondrocyte (**KON**-droh-sight)
Collagen (**KAH**-lah-jen)
Connective tissue (kah-**NEK**-tiv **TISH**-yoo)
Elastin (eh-**LAS**-tin)
Endocrine gland (**EN**-doh-krin GLAND)
Epithelial tissue (EP-i-**THEE**-lee-uhl **TISH**-yoo)
Exocrine gland (**EK**-so-krin GLAND)
Hemopoietic (HEE-moh-poy-**ET**-ik)
Matrix (**MAY**-tricks)
Microvilli (MY-kro-**VILL**-eye)
Mucous membrane (**MEW**-kuss **MEM**-brain)
Muscle tissue (**MUSS**-uhl **TISH**-yoo)
Myocardium (MY-oh-**KAR**-dee-um)
Nerve tissue (NERV **TISH**-yoo)
Neuron (**NYOOR**-on)
Neurotransmitter (NYOOR-oh-**TRANS**-mih-ter)
Osteocyte (**AHS**-tee-oh-sight)
Plasma (**PLAZ**-mah)
Secretion (see-**KREE**-shun)
Serous membrane (**SEER**-us **MEM**-brain)
Synapse (**SIN**-aps)

Terms that appear in ***bold type*** *in the chapter text are defined in the glossary, which begins on p. 528.*

A **tissue** is a group of cells with similar structure and function. The tissue then contributes to the functioning of the organs in which it is found. You may recall that in Chapter 1 the four major groups of tissues were named and very briefly described. These four groups are epithelial, connective, muscle, and nerve tissue.

This chapter presents more detailed descriptions of the tissues in these four categories. For each tissue, its functions are related to the organs of which it is a part. Also in this chapter is a discussion of **membranes**, which are sheets of tissues. As you might expect, each type of membrane has its specific locations and functions.

EPITHELIAL TISSUE

Epithelial tissues are found on surfaces as either coverings (outer surfaces) or linings (inner surfaces). Because they have no capillaries of their own, epithelial tissues receive oxygen and nutrients from the connective tissue beneath them. Many epithelial tissues are capable of secretion and may be called glandular epithelium, or more simply, **glands**.

Classification of the epithelial tissues is based on the type of cell of which the tissue is made, its characteristic shape, and the number of layers of cells. There are three distinctive shapes: **squamous** cells are flat, **cuboidal** cells are cube-shaped, and **columnar** cells are tall and narrow. "**Simple**" is the term for a single layer of cells, and "**stratified**" means that many layers of cells are present (Fig. 4-1 and Table 4-1).

SIMPLE SQUAMOUS EPITHELIUM

Simple squamous epithelium is a single layer of flat cells (Fig. 4-2). These cells are very thin and very smooth—these are important physical characteristics. The alveoli (air sacs) of the lungs are simple squamous epithelium. The thinness of the cells permits the diffusion of gases between the air and blood.

Another location of this tissue is capillaries, the smallest blood vessels. Capillary walls are only one cell thick, which permits the exchange of gases, nutrients, and waste products between the blood and tissue

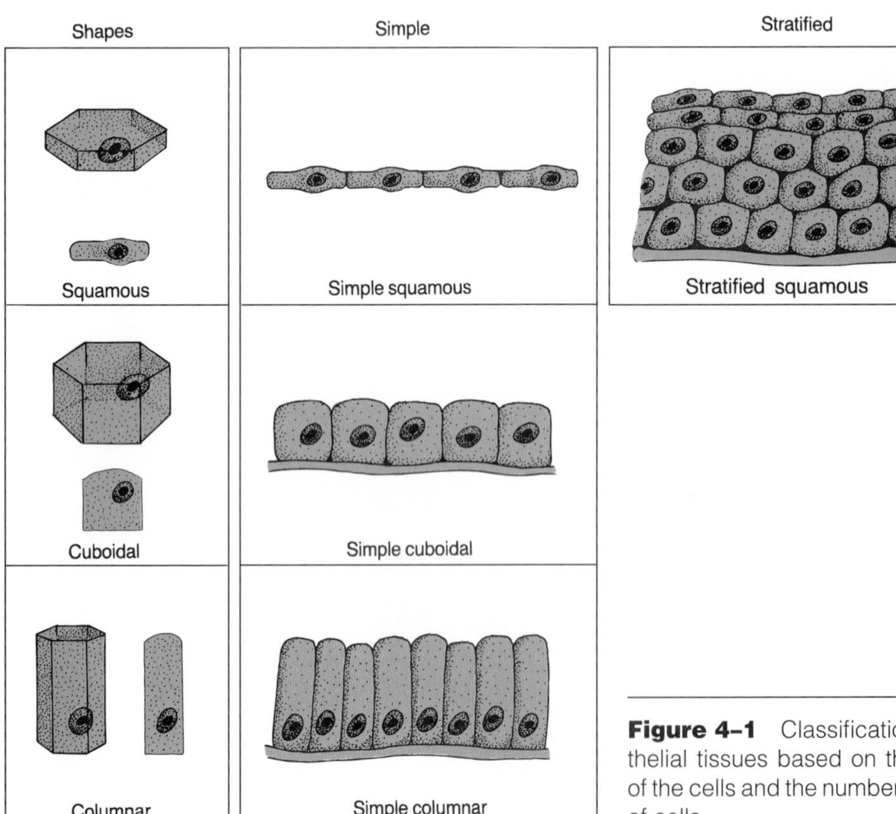

Figure 4-1 Classification of epithelial tissues based on the shape of the cells and the number of layers of cells.

Table 4–1 TYPES OF EPITHELIAL TISSUE

Type	Structure	Location and Function
Simple Squamous	One layer of flat cells	• Alveoli of the lungs—thin to permit diffusion of gases • Capillaries—thin to permit exchanges of materials; smooth to prevent abnormal blood clotting
Stratified Squamous	Many layers of cells; surface cells flat; lower cells rounded; lower layer undergoes mitosis	• Epidermis—surface cells are dead; a barrier to pathogens • Lining of esophagus, vagina—surface cells are living; a barrier to pathogens
Transitional	Many layers of cells; surface cells change from rounded to flat	• Lining of urinary bladder—permits expansion without tearing the lining
Cuboidal	One layer of cube-shaped cells	• Thyroid gland—secretes thyroxine • Salivary glands—secrete saliva
Columnar	One layer of column-shaped cells	• Lining of stomach—secretes gastric juice • Lining of small intestine—secretes enzymes and absorbs end products of digestion (microvilli present)
Ciliated	One layer of columnar cells with cilia on their free surfaces	• Lining of trachea—sweeps mucus and dust to the pharynx • Lining of fallopian tube—sweeps ovum toward uterus

fluid. The interior surface of capillaries is also very smooth (and these cells continue as the lining of the arteries, veins, and heart); this is important because it prevents abnormal blood clotting within blood vessels.

STRATIFIED SQUAMOUS EPITHELIUM

Stratified squamous epithelium consists of many layers of mostly flat cells, although lower cells are rounded. Mitosis takes place in the lowest layer to continually produce new cells to replace those worn off the surface (see Fig. 4–2). This type of epithelium makes up the epidermis of the skin; here the surface cells are dead. Stratified squamous epithelium also lines the oral cavity, esophagus, and, in women, the vagina. In these locations the surface cells are living and make up the mucous membranes of these organs. In all its body locations, this tissue is a barrier to microorganisms because the cells of which it is made are very close together. The more specialized functions of the epidermis will be covered in the next chapter.

TRANSITIONAL EPITHELIUM

Transitional epithelium is a type of stratified epithelium in which the surface cells change shape from round to squamous. The urinary bladder is lined with transitional epithelium. When the bladder is empty, the surface cells are rounded (see Fig. 4–2). As the bladder fills, these cells become flattened. Transitional epithelium enables the bladder to fill and stretch without tearing the lining.

SIMPLE CUBOIDAL EPITHELIUM

Simple cuboidal epithelium is a single layer of cube-shaped cells (Fig. 4–3). This type of tissue makes up the functional units of the thyroid gland and salivary glands; these are examples of **glandular epithelium**. In these glands the cuboidal cells are arranged in small spheres and **secrete** into the cavity formed by the sphere. In the thyroid gland, the cuboidal epithelium secretes the thyroid hormones; thyroxine is an example. In the salivary glands the cuboidal cells secrete saliva.

SIMPLE COLUMNAR EPITHELIUM

Columnar cells are taller than they are wide and are specialized for secretion and absorption. The stomach lining is made of **columnar epithelium** that secretes gastric juice for digestion. The lining of the small intestine (see Fig. 4–3) secretes digestive enzymes, but

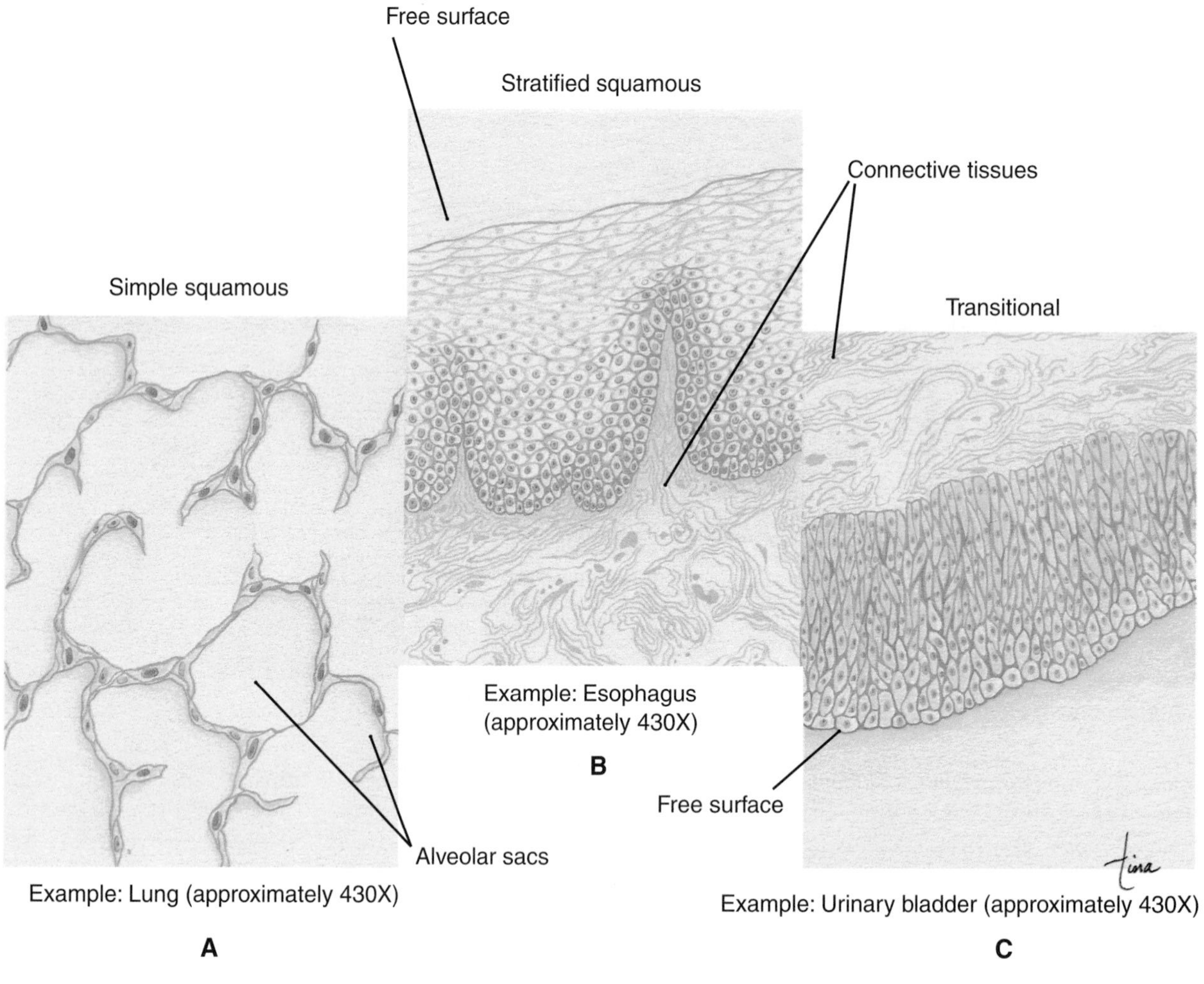

Figure 4–2 Epithelial tissues. **(A)** Simple squamous. **(B)** Stratified squamous. **(C)** Transitional.

these cells also **absorb** the end products of digestion. In order to absorb efficiently, the columnar cells of the small intestine have **microvilli**, which are folds of the cell membrane on their free surfaces (see Fig. 3–2). These microscopic folds greatly increase the surface area for absorption.

Yet another type of columnar cell is the **goblet cell**, which is a unicellular gland. Goblet cells secrete **mucus** and are found in the lining of the intestines and parts of the respiratory tract such as the trachea.

CILIATED EPITHELIUM

Ciliated epithelium consists of columnar cells that have **cilia** on their free surfaces (see Fig. 4–3). Recall from Chapter 3 that the function of cilia is to sweep materials across the cell surface. Ciliated epithelium lines the nasal cavities, larynx, trachea, and large bronchial tubes. The cilia sweep mucus, with trapped dust and bacteria, toward the pharynx to be swallowed. Bacteria are then destroyed by the hydrochloric acid in the stomach.

Another location of ciliated epithelium in women is the lining of the fallopian tubes. The cilia here sweep the ovum, which has no means of self-locomotion, toward the uterus.

GLANDS

Glands are cells or organs that **secrete** something, that is, produce a substance that has a function either at that site or at a more distant site.

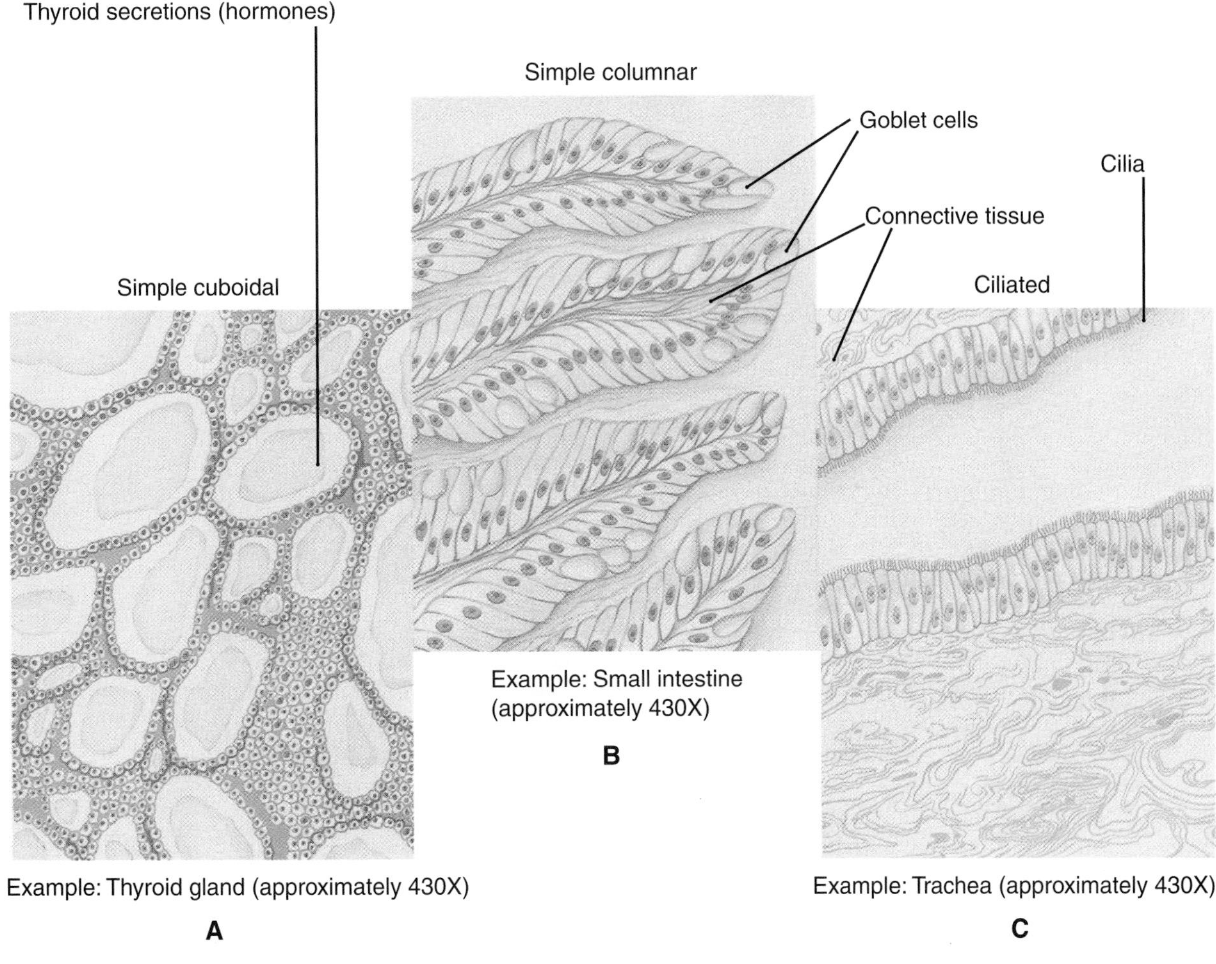

Figure 4–3 Epithelial tissues. **(A)** Simple cuboidal. **(B)** Simple columnar. **(C)** Ciliated.

Unicellular Glands

Unicellular means "one cell." Goblet cells are an example of unicellular glands. As mentioned earlier, goblet cells are found in the lining of the respiratory and digestive tracts. Their secretion is mucus (see also Box 4–1: Cystic Fibrosis).

Multicellular Glands

Most glands are made of many similar cells. **Multicellular** glands may be divided into two major groups: exocrine glands and endocrine glands.

Exocrine glands have **ducts** (tubes) to take the secretion away from the gland to the site of its function. Salivary glands, for example, secrete saliva that is carried by ducts to the oral cavity. Sweat glands secrete sweat that is transported by ducts to the skin surface, where it can be evaporated by excess body heat.

Endocrine glands are ductless glands. The secretions of endocrine glands are a group of chemicals called **hormones**, which enter capillaries and are circulated throughout the body. Hormones then bring about specific effects in their target organs. These will be covered in more detail in Chapter 10. Examples of endocrine glands are the thyroid gland, adrenal glands, and pituitary gland.

The pancreas is an organ that functions as both an exocrine and an endocrine gland. The exocrine portions secrete digestive enzymes that are carried by ducts to the duodenum, their site of action. The endocrine portions of the pancreas, called islets of Lang-

Box 4–1 CYSTIC FIBROSIS

Cystic fibrosis (CF) is a genetic disorder of certain exocrine glands including the salivary glands, sweat glands, the pancreas, and the mucus glands of the respiratory tract.

In the pancreas, thick mucus clogs the ducts and prevents pancreatic enzymes from reaching the small intestine, thus impairing digestion, especially of fats. But the most serious effects of CF are in the lungs. The genetic mistake in CF involves a gene called *cftr*, which codes for protein channels called chloride gateways in epithelial cells. In the lungs, the defective channels prevent salt from entering the cells, creating a very salty tissue fluid around the cells. This salty fluid inactivates defensin, a natural antibiotic produced by lung tissue. In the absence of defensin, a bacterium called *Pseudomonas aeruginosa* stimulates the lung cells to produce copious thick mucus, an ideal growth environment for bacteria. Defensive white blood cells cannot get through the thick mucus, and their activity mistakenly destroys lung tissue. A person with CF has clogged bronchial tubes, frequent episodes of pneumonia, and ultimately, lungs that cannot carry out gas exchange. CF is a chronic, progressive disease that is eventually fatal unless a lung transplant is performed.

CF is one of several disorders believed to be correctable by gene therapy. Current research trials involve the use of non-pathogenic viruses to transfer a normal gene into the cells that line the respiratory airways. Because it involves human subjects, this kind of work proceeds slowly, and it may be several years before conclusive results are obtained as to whether gene therapy offers a cure for CF.

One possible new treatment is an enzyme (to be inhaled in the form of a fine mist) that will break up the thick mucus to a thinner form that may be more easily coughed up. Although this would not be a cure, it may help prolong the lives of people with CF until a cure becomes possible.

erhans, secrete the hormones insulin and glucagon directly into the blood.

CONNECTIVE TISSUE

There are several kinds of **connective tissue**, some of which may at first seem more different than alike. The types of connective tissue include areolar, adipose, fibrous, and elastic tissue as well as blood, bone, and cartilage (Table 4–2). A characteristic that all connective tissues have in common is the presence of a matrix in addition to cells. The **matrix** is a structural network or solution of non-living intercellular material. Each connective tissue has its own specific kind of matrix. The matrix of blood, for example, is blood plasma, which is mostly water. The matrix of bone is made primarily of calcium salts, which are hard and strong. As each type of connective tissue is described in the following sections, mention will be made of the types of cells present as well as the kind of matrix.

BLOOD

Although **blood** is the subject of Chapter 11, a brief description will be given here. The matrix of blood is **plasma**, which is 52% to 62% of the total blood volume in the body. The water of plasma contains dissolved salts, nutrients, and waste products. As you might expect, one of the primary functions of plasma is transport of these materials within the body.

The cells of blood are red blood cells, white blood cells, and platelets, which are actually fragments of cells. These are shown in Fig. 4–4. The blood-forming or **hemopoietic tissues** are the red bone marrow and lymphatic tissue, which includes the spleen and the lymph nodes. Red bone marrow produces red blood cells, the five types of white blood cells, and the platelets. Two kinds of white blood cells are also produced in lymphatic tissue.

The blood cells make up 38% to 48% of the total blood, and each type of cell has its specific function. **Red blood cells** (RBCs) carry oxygen bonded to their hemoglobin. **White blood cells** (WBCs) destroy pathogens and provide us with immunity to some dis-

Table 4–2 TYPES OF CONNECTIVE TISSUE

Type	Structure	Location and Function
Blood	Plasma (matrix) and red blood cells, white blood cells, and platelets	Within blood vessels: • *Plasma*—transports materials • *RBCs*—carry oxygen • *WBCs*—destroy pathogens • *Platelets*—prevent blood loss
Areolar (Loose)	Fibroblasts and a matrix of tissue fluid, collagen, and elastin fibers	Subcutaneous • Connects skin to muscles; WBCs destroy pathogens Mucous membranes (digestive, respiratory, urinary, reproductive tracts) • WBCs destroy pathogens
Adipose	Adipocytes that store fat (little matrix)	Subcutaneous • Stores excess energy Around eyes and kidneys • Cushions
Fibrous	Mostly collagen fibers (matrix) with few fibroblasts	Tendons and ligaments (regular) • Strong to withstand forces of movement of joints Dermis (irregular) • The strong inner layer of the skin
Elastic	Mostly elastin fibers (matrix) with few fibroblasts	Walls of large arteries • Helps maintain blood pressure Around alveoli in lungs • Promotes normal exhalation
Bone	Osteocytes in a matrix of calcium salts and collagen	Bones: • Support the body • Protect internal organs from mechanical injury • Store excess calcium • Contain and protect red bone marrow
Cartilage	Chondrocytes in a flexible protein matrix	Wall of trachea • Keeps airway open On joint surfaces of bones • Smooth to prevent friction Tip of nose and outer ear • Support Between vertebrae • Absorb shock

eases. **Platelets** prevent blood loss; the process of blood clotting involves platelets.

AREOLAR CONNECTIVE TISSUE

The cells of **areolar (or loose) connective tissue** are called **fibroblasts**, which produce protein fibers. **Collagen** fibers are very strong; **elastin** fibers are elastic, that is, able to return to their original length, or recoil, after being stretched. These protein fibers and tissue fluid make up the matrix, or non-living portion, of areolar connective tissue (see Fig. 4–4). Also within the matrix are many white blood cells, which are capable of self-locomotion. Their importance here is related to the locations of areolar connective tissue.

Areolar tissue is found beneath the dermis of the skin and beneath the epithelial tissue of all the body systems that have openings to the environment. Recall that one function of white blood cells is to destroy pathogens. How do pathogens enter the body? Many do so through breaks in the skin. Bacteria and viruses also enter with the air we breathe and the food we eat, and some may get through the epithelial linings of the respiratory and digestive tracts. Areolar con-

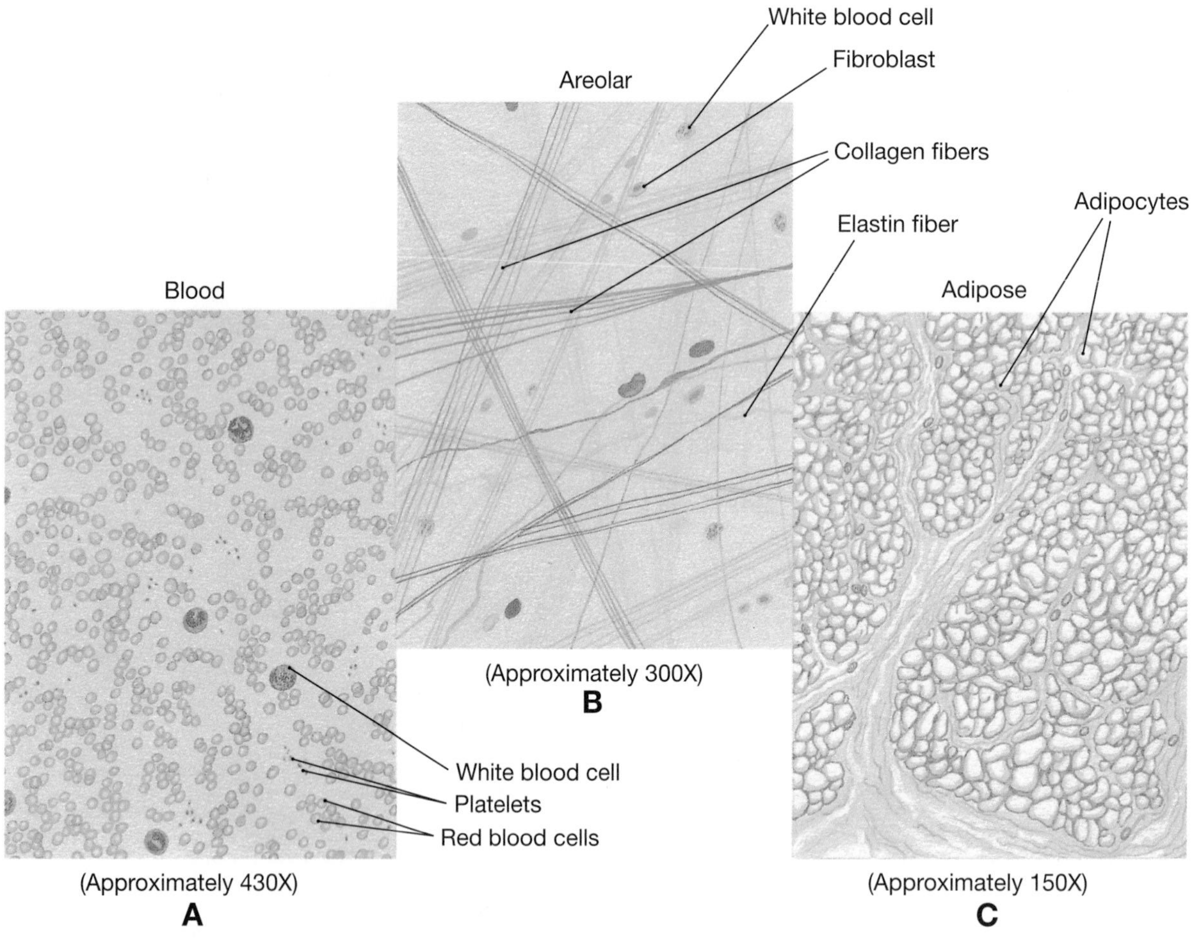

Figure 4–4 Connective tissues. **(A)** Blood. **(B)** Areolar. **(C)** Adipose.

nective tissue with its many white blood cells is strategically placed to intercept pathogens before they get to the blood and circulate throughout the body.

ADIPOSE TISSUE

The cells of **adipose tissue** are called **adipocytes** and are specialized to store fat in microscopic droplets. True fats are the chemical form of long-term energy storage. Excess nutrients have calories that are not wasted but are converted to fat to be stored for use when food intake decreases. The amount of matrix in adipose tissue is small and consists of tissue fluid and a few collagen fibers (see Fig. 4–4).

Most fat is stored subcutaneously in the areolar connective tissue between the dermis and the muscles. This layer varies in thickness among individuals; the more excess calories consumed, the thicker the layer. As was mentioned in Chapter 2, adipose tissue also cushions organs such as the eyes and kidneys.

FIBROUS CONNECTIVE TISSUE

Fibrous connective tissue consists mainly of parallel (regular) collagen fibers with few fibroblasts scattered among them (Fig. 4–5). This parallel arrangement of collagen provides great strength, yet is flexible. The locations of this tissue are related to the need for flex-

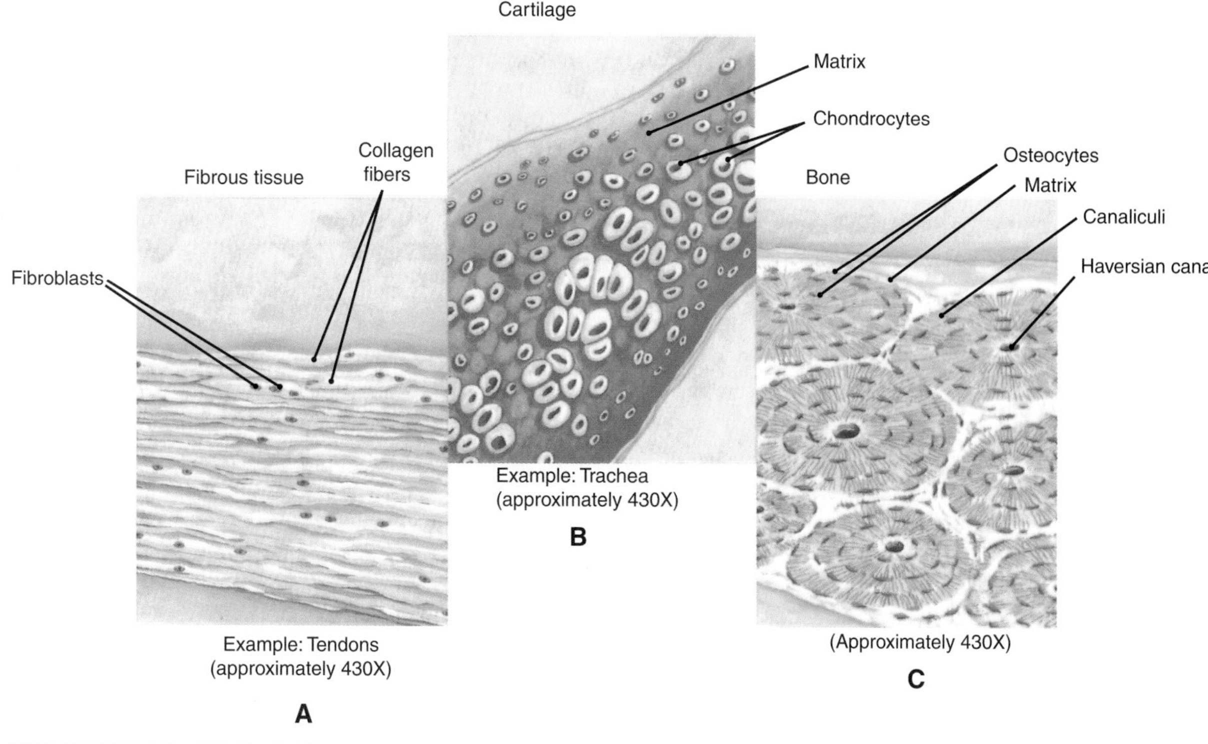

Figure 4–5 Connective tissues. **(A)** Fibrous. **(B)** Cartilage. **(C)** Bone.

ible strength. The outer walls of arteries are reinforced with fibrous connective tissue, because the blood in these vessels is under high pressure. The strong outer wall prevents rupture of the artery (see also Box 4-2: Vitamin C and Collagen). Tendons and ligaments are made of fibrous connective tissue. Tendons connect muscle to bone; ligaments connect bone to bone. When the skeleton is moved, these structures must be able to withstand the great mechanical forces exerted upon them.

Fibrous connective tissue has a relatively poor blood supply, which makes repair a slow process. If you have ever had a severely sprained ankle (which means the ligaments have been overly stretched), you know that complete healing may take several months.

An irregular type of fibrous connective tissue forms the dermis of the skin and the fascia (membranes) around muscles. Although the collagen fibers here are not parallel to one another, the tissue is still strong. The dermis is different from other fibrous connective tissue in that it has a good blood supply (see also Box 4-3: Cosmetic Collagen).

ELASTIC CONNECTIVE TISSUE

As its name tells us, **elastic connective tissue** is primarily elastin fibers. One of its locations is the walls of large arteries. These vessels are stretched when the heart contracts and pumps blood, then recoil when the heart relaxes; this is important to maintain normal blood pressure.

Elastic connective tissue is also found surrounding the alveoli of the lungs. The elastic fibers are stretched during inhalation, then recoil during exhalation to squeeze air out of the lungs. If you pay attention to your breathing for a few moments, you will notice that normal exhalation does not require "work" or energy. This is because of the normal elasticity of the lungs.

Box 4–2 VITAMIN C AND COLLAGEN

In the last two decades, many claims have been made for the effectiveness of vitamin C in preventing or treating the common cold and even cancer. There is no conclusive evidence for these claims, but one function of vitamin C is known with certainty, and that is its role in the synthesis of collagen.

Imagine the protein collagen as a ladder with three uprights and rungs that connect adjacent uprights. Vitamin C is essential for forming the "rungs," without which the uprights will not stay together as a strong unit. Collagen formed in the absence of vitamin C is weak, and the effects of weak collagen are dramatically seen in the disease called scurvy.

In 1753 James Lind, a Scottish surgeon, recommended to the British Navy that lime juice be taken on long voyages to prevent scurvy among the sailors. Scurvy is characterized by bleeding gums and loss of teeth, poor healing of wounds, fractures, and bleeding in the skin, joints, and elsewhere in the body. The lime juice did prevent this potentially fatal disease, as did consumption of fresh fruits and vegetables, although at the time no one knew why. Vitamin C was finally isolated in the laboratory in 1928.

BONE

The prefix that designates bone is "osteo," so bone cells are called **osteocytes**. The matrix of **bone** is made of calcium salts and collagen and is strong, hard, and not flexible. In the shafts of long bones such as the femur, the osteocytes, matrix, and blood vessels are in very precise arrangements called **haversian systems** (see Fig. 4-5). Bone has a good blood supply, which enables it to serve as a storage site for calcium and to repair itself relatively rapidly after a simple fracture. Some bones, such as the sternum (breastbone) and pelvic bone, contain red bone marrow, one of the hemopoietic tissues that produces blood cells.

Other functions of bone tissue are related to the strength of bone matrix. The skeleton supports the body, and some bones protect internal organs from

Box 4–3 COSMETIC COLLAGEN

Collagen is the protein that makes tendons, ligaments, and other connective tissues strong. In 1981, the Food and Drug Administration (FDA) approved the use of cattle collagen by injection for cosmetic purposes, to minimize wrinkles and scars. Indeed, collagen injected below the skin will flatten out deep facial wrinkles and make them less prominent, and millions of people in the United States have had this seemingly simple cosmetic surgery.

There are, however, drawbacks. Injected collagen lasts only a few months; the injections must be repeated several times a year, and they are expensive. Some people experience allergic reactions to the cattle collagen, which is perceived by the immune system as foreign tissue. More seriously, an autoimmune response may be triggered in some individuals, and the immune system may begin to destroy the person's own connective tissue.

In an effort to avoid these problems, some cosmetic surgeons now use the person's own collagen and fat, which may be extracted from the thigh, hip, or abdomen. This technique is still relatively new, and long-term consequences and outcomes have yet to be evaluated. We might remember that for many years the use of silicone injections had been considered safe. Silicone injections are now banned by the FDA, since we now know that they carry significant risk of serious tissue damage.

mechanical injury. A more complete discussion of bone is found in Chapter 6.

CARTILAGE

The protein matrix of **cartilage** differs from that of bone in that it is firm, yet smooth and flexible. Cartilage is found on the joint surfaces of bones, where its smooth surface helps prevent friction. The tip of the nose and external ear are supported by flexible cartilage. The wall of the trachea, the airway to the lungs, contains rings of cartilage to maintain an open air passageway. Discs of cartilage are found between the vertebrae of the spine. Here the cartilage absorbs shock and permits movement.

Within the cartilage matrix are the **chondrocytes**, or cartilage cells (see Fig. 4–5). There are no capillaries within the cartilage matrix, so these cells are nourished by diffusion through the matrix, a slow process. This becomes clinically important when cartilage is damaged, for repair will take place very slowly or not at all. Athletes sometimes damage cartilage within the knee joint. Such damaged cartilage is usually surgically removed in order to preserve as much joint mobility as possible.

MUSCLE TISSUE

Muscle tissue is specialized for contraction. When muscle cells contract, they shorten and bring about some type of movement. There are three types of muscle tissue: skeletal, smooth, and cardiac (Table 4–3). The movements each can produce have very different purposes.

SKELETAL MUSCLE

Skeletal muscle may also be called **striated** muscle or **voluntary** muscle. Each name describes a particular aspect of this tissue, as you will see. The skeletal muscle cells are cylindrical, have several nuclei each, and appear striated, or striped (Fig. 4–6). The striations are the result of the precise arrangement of the contracting proteins within the cells.

Skeletal muscle tissue makes up the muscles that are attached to bones. These muscles are supplied with motor nerves, and thus move the skeleton. They also produce a significant amount of body heat. Each muscle cell has its own motor nerve ending. The nerve impulses that can then travel to the muscles are essential to cause contraction. Although we do not have to consciously plan all our movements, the nerve impulses for them originate in the cerebrum, the "thinking" part of the brain.

Let us return to the three names for this tissue: "skeletal" describes its location, "striated" describes its appearance, and "voluntary" describes how it functions. The skeletal muscles and their functioning are the subject of Chapter 7.

SMOOTH MUSCLE

Smooth muscle may also be called **visceral** muscle or **involuntary** muscle. The cells of smooth muscle

Table 4–3 TYPES OF MUSCLE TISSUE

Type	Structure	Location and Function	Effect of Nerve Impulses
Skeletal	Large cylindrical cells with striations and several nuclei each	Attached to bones • Moves the skeleton and produces heat	Essential to cause contraction (voluntary)
Smooth	Small tapered cells with no striations and one nucleus each	Walls of arteries • Maintains blood pressure Walls of stomach and intestines • Peristalsis Iris of eye • Regulates size of pupil	Bring about contraction or regulate the rate of contraction (involuntary)
Cardiac	Branched cells with faint striations and one nucleus each	Walls of the chambers of the heart • Pumps blood	Regulate only the rate of contraction

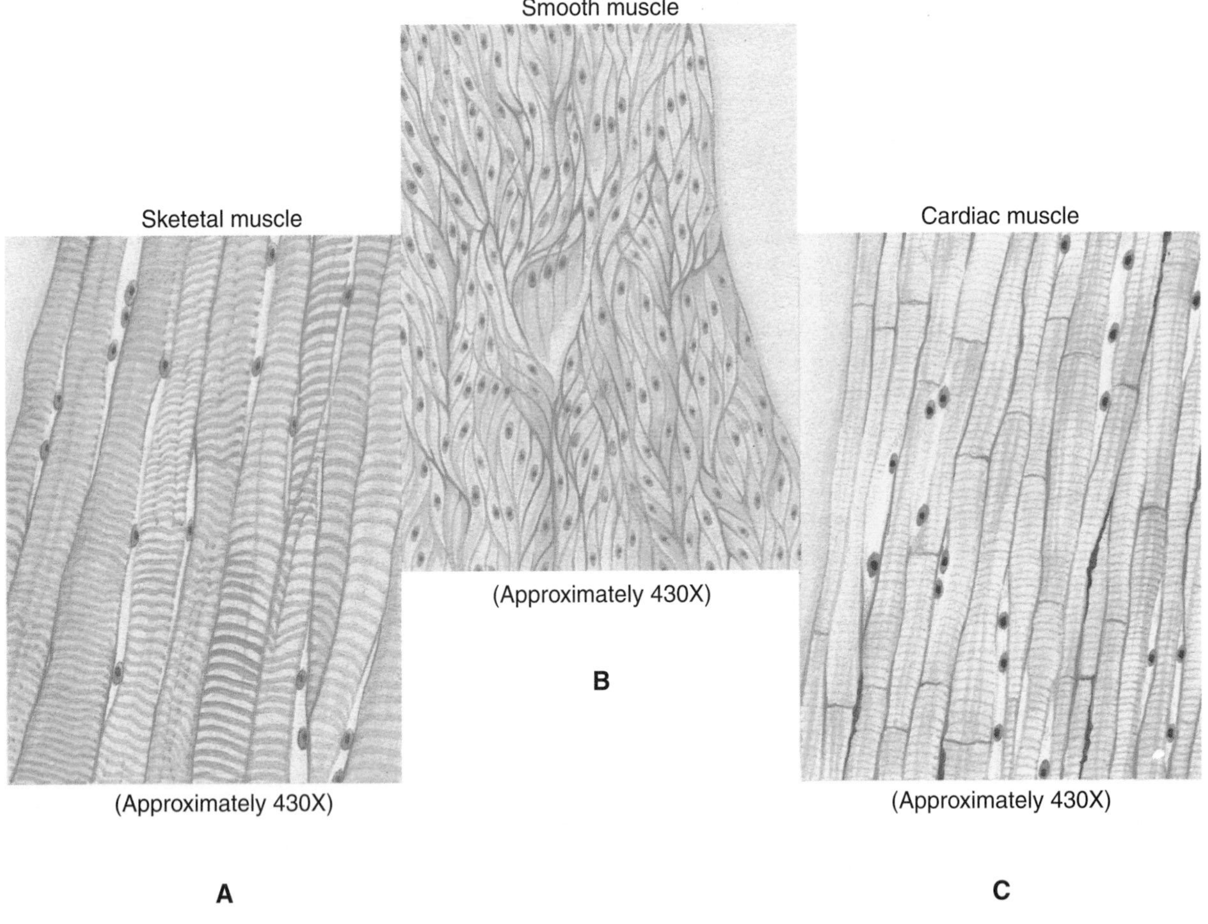

Figure 4–6 Muscle tissues. **(A)** Skeletal. **(B)** Smooth. **(C)** Cardiac.

have tapered ends, a single nucleus, and no striations (see Fig. 4–6). Although nerve impulses do bring about contractions, this is not something most of us can control. The term "visceral" refers to internal organs, many of which contain smooth muscle. The functions of smooth muscle are actually functions of the organs in which the muscle is found.

In the stomach and intestines, smooth muscle contracts in waves called peristalsis to propel food through the digestive tract.

In the walls of arteries and veins, smooth muscle constricts or dilates the vessels to maintain normal blood pressure. The iris of the eye has two sets of smooth muscle fibers to constrict or dilate the pupil, which regulates the amount of light that strikes the retina.

Other functions of smooth muscle are mentioned in later chapters. This is an important tissue that you will come across again and again in our study of the human body.

CARDIAC MUSCLE

The cells of **cardiac muscle** are shown in Fig. 4–6. They are branched, have one nucleus each, and have faint striations. Cardiac muscle, called the **myocardium**, forms the walls of the chambers of the heart. Its function, therefore, is the function of the heart, to pump blood. The contractions of the myocardium create blood pressure and keep blood circulating throughout the body, so that the blood can carry out its many functions.

Cardiac muscle cells have the ability to contract by themselves. Thus the heart maintains its own beat. The role of nerve impulses is to increase or decrease the heart rate, depending upon whatever is needed by the body in a particular situation. We will return to the heart in Chapter 12.

NERVE TISSUE

Nerve tissue consists of nerve cells called **neurons** and some specialized cells found only in the nervous system. The nervous system has two divisions: the central nervous system (CNS) and the peripheral nervous system (PNS). The brain and spinal cord are the organs of the CNS. They are made of neurons and specialized cells called neuroglia. The CNS and the neuroglia are discussed in detail in Chapter 8. The PNS consists of all the nerves that emerge from the CNS and supply the rest of the body. These nerves are made of neurons and specialized cells called Schwann cells. The Schwann cells form the myelin sheath to electrically insulate neurons (Table 4-4).

Neurons are capable of generating and transmitting electrochemical impulses. There are many different kinds of neurons, but they all have the same basic structure (Fig. 4-7). The **cell body** contains the nucleus and is essential for the continuing life of the neuron. An **axon** is a process (cellular extension) that carries impulses away from the cell body; a neuron has only one axon. **Dendrites** are processes that carry impulses toward the cell body; a neuron may have several dendrites. A nerve impulse along the cell membrane of a neuron is electrical, but where neurons meet there is a small space called a **synapse**, which an electrical impulse cannot cross. At a synapse, between the axon of one neuron and the dendrite or cell body of the next neuron, impulse transmission depends upon chemicals called **neurotransmitters**. Each of these aspects of nerve tissue is covered in more detail in Chapter 8.

Nerve tissue makes up the brain, spinal cord, and peripheral nerves. As you can imagine, each of these organs has very specific functions. For now, we will just summarize the functions of nerve tissue. These functions include sensation, movement, the rapid regulation of body functions such as heart rate and breathing, and the organization of information for learning and memory.

MEMBRANES

Membranes are sheets of tissue that cover or line surfaces or separate organs or parts (lobes) of organs from one another. Many membranes produce secretions that have specific functions. The two major categories of membranes are epithelial membranes and connective tissue membranes.

Table 4–4 NERVE TISSUE

Part	Structure	Function
Neuron (nerve cell)		
Cell body	• Contains the nucleus	• Regulates the functioning of the neuron
Axon	• Cellular process (extension)	• Carries impulses away from the cell body
Dendrites	• Cellular process (extension)	• Carry impulses toward the cell body
Synapse	• Space between axon of one neuron and the dendrite or cell body of the next neuron	• Transmits impulses from one neuron to others
Neurotransmitters	• Chemicals released by axons	• Transmit impulses across synapses
Neuroglia	• Specialized cells in the central nervous system	• Form myelin sheaths and other functions
Schwann cells	• Specialized cells in the peripheral nervous system	• Form the myelin sheaths around neurons

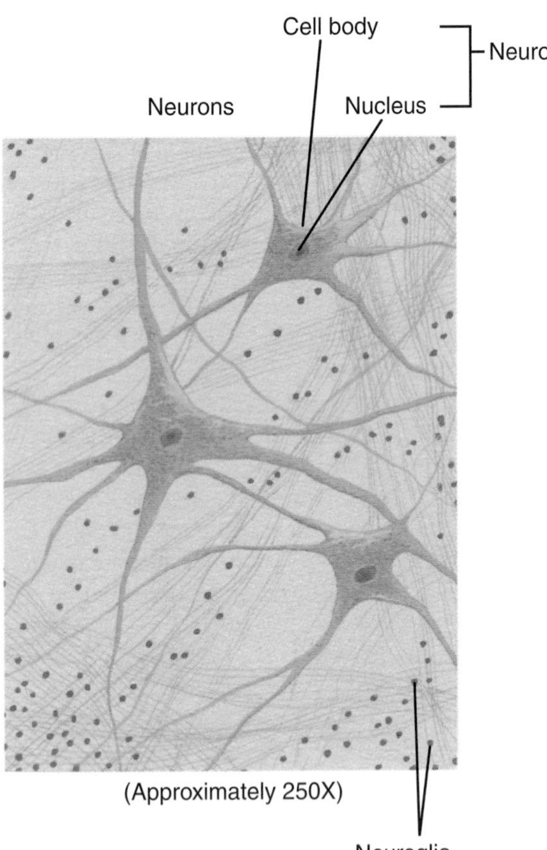

Figure 4–7 Nerve tissue of the central nervous system (CNS).

EPITHELIAL MEMBRANES

There are two types of epithelial membranes, serous and mucous. Each type is found in specific locations within the body and secretes a fluid. These fluids are called serous fluid and mucus.

Serous Membranes

Serous membranes are sheets of simple squamous epithelium that line some closed body cavities and cover the organs in these cavities (Fig. 4–8). The **pleural membranes** are the serous membranes of the thoracic cavity. The parietal pleura lines the chest wall and the visceral pleura covers the lungs. (Notice that "line" means "on the inside" and "cover" means "on the outside." These terms cannot be used interchangeably, because each indicates a different location.) The pleural membranes secrete **serous fluid**, which prevents friction between them as the lungs expand and recoil during breathing.

The heart, in the thoracic cavity between the lungs, has its own set of serous membranes. The parietal **pericardium** lines the fibrous pericardium (a connective tissue membrane), and the visceral **pericardium**, or epicardium, is on the surface of the heart muscle. Serous fluid is produced to prevent friction as the heart beats.

In the abdominal cavity, the **peritoneum** is the serous membrane that lines the cavity. The **mesentery**, or visceral peritoneum, is folded over and covers the abdominal organs. Here, the serous fluid prevents friction as the stomach and intestines contract and slide against other organs (see also Fig. 16–4).

Mucous Membranes

Mucous membranes line the body tracts (systems) that have openings to the environment. These are the respiratory, digestive, urinary, and reproductive tracts. The epithelium of a mucous membrane **(mucosa)**

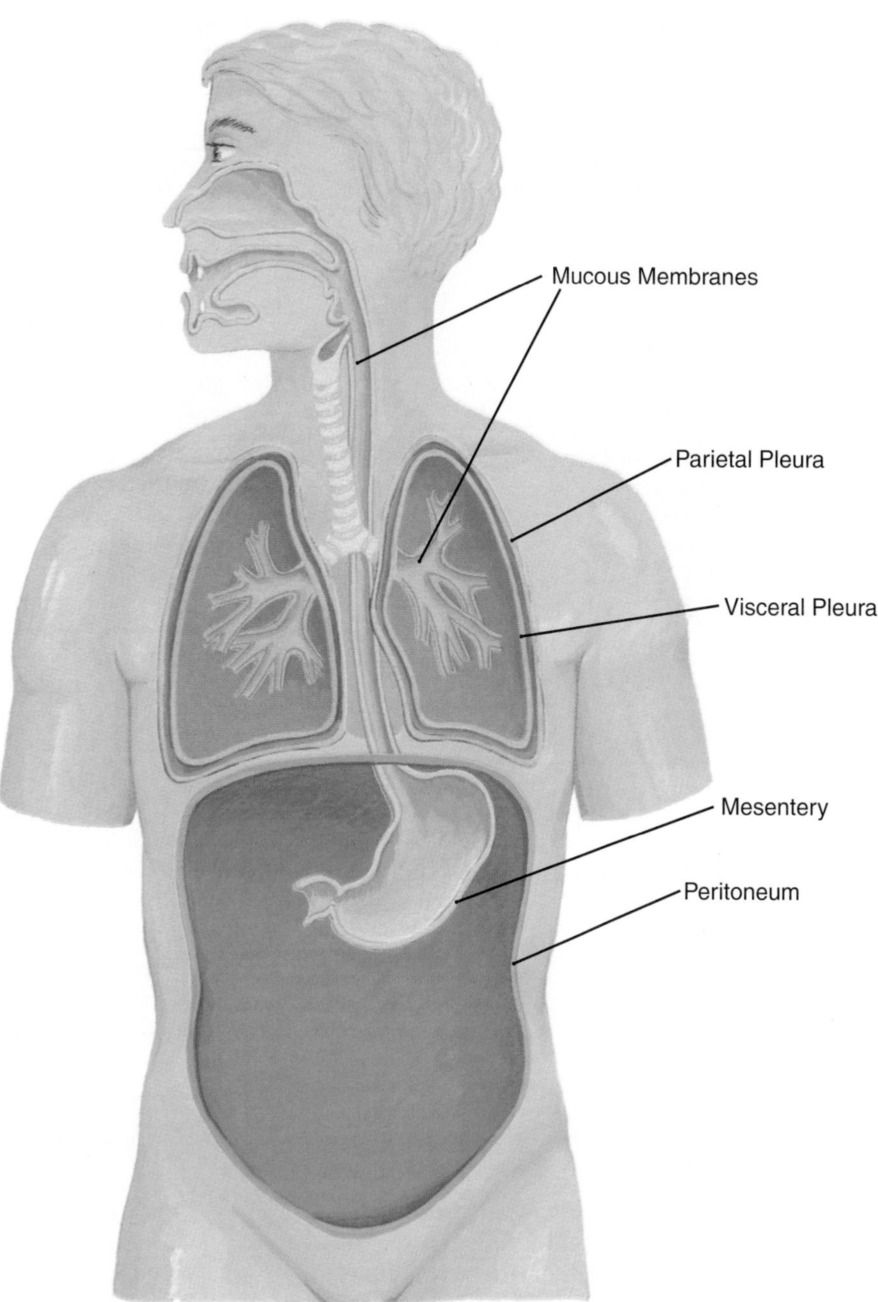

Figure 4–8 Epithelial membranes. Mucous membranes line body tracts that open to the environment. Serous membranes are found within closed body cavities such as the thoracic and abdominal cavities.

varies with the different organs involved. The mucosa of the esophagus and of the vagina is stratified squamous epithelium; the mucosa of the trachea is ciliated epithelium; the mucosa of the stomach is columnar epithelium.

The **mucus** secreted by these membranes keeps the lining epithelial cells wet. Remember that these are living cells, and if they dry out, they will die. In the digestive tract, mucus also lubricates the surface to permit the smooth passage of food. In the respiratory

Table 4–5 CONNECTIVE TISSUE MEMBRANES

Membrane	Location and Function
Superficial Fascia	• Between the skin and muscles; adipose tissue stores fat
Periosteum	• Covers each bone; contains blood vessels that enter the bone • Anchors tendons and ligaments
Perichondrium	• Covers cartilage; contains capillaries, the only blood supply for cartilage
Synovial	• Lines joint cavities; secretes synovial fluid to prevent friction when joints move
Deep Fascia	• Covers each skeletal muscle; anchors tendons
Meninges	• Cover the brain and spinal cord; contain cerebrospinal fluid
Fibrous Pericardium	• Forms a sac around the heart

tract the mucus traps dust and bacteria, which are then swept to the pharynx by ciliated epithelium.

CONNECTIVE TISSUE MEMBRANES

Many membranes are made of connective tissue. Because these will be covered with the organ systems of which they are a part, their locations and functions are summarized in Table 4–5.

AGING AND TISSUES

As mentioned in the previous chapter, aging takes place at the cellular level, but of course is apparent in the groups of cells we call tissues. In muscle tissue, for example, the proteins that bring about contraction deteriorate and are not repaired or replaced. The same is true of collagen and elastin, the proteins of connective tissue. Other aspects of the aging of tissues will be more meaningful to you in the context of the functions of organs and systems, so we will save those for the following chapters.

SUMMARY

The tissues and membranes described in this chapter are more complex than the individual cells of which they are made. However, we have only reached an intermediate level with respect to the structural and functional complexity of the body as a whole. The following chapters are concerned with the organ systems, the most complex level. In the descriptions of the organs of these systems, you will find mention of the tissues and their contributions to each organ and organ system.

STUDY OUTLINE

A tissue is a group of cells with similar structure and function. The four main groups of tissues are: epithelial, connective, muscle, and nerve.

Epithelial Tissue—found on surfaces; have no capillaries; some are capable of secretion; classified as to shape of cells and number of layers of cells (see Table 4–1 and Figs. 4–1, 4–2, and 4–3)

1. Simple Squamous—one layer of flat cells; thin and smooth. Sites: alveoli (to permit diffusion of gases); capillaries (to permit exchanges between blood and tissues).
2. Stratified Squamous—many layers of mostly flat cells; mitosis takes place in lowest layer. Sites: epidermis, where surface cells are dead (a barrier to pathogens); lining of mouth; esophagus; and vagina (a barrier to pathogens).
3. Transitional—stratified, yet surface cells are rounded and flatten when stretched. Site: urinary bladder (to permit expansion without tearing the lining).
4. Simple Cuboidal—one layer of cube-shaped cells. Sites: thyroid gland (to secrete thyroid hormones); salivary glands (to secrete saliva).
5. Simple Columnar—one layer of column-shaped cells. Sites: stomach lining (to secrete gastric juice); small intestinal lining (to secrete digestive enzymes

and absorb nutrients—microvilli increase surface area for absorption).

6. Ciliated—columnar cells with cilia on free surfaces. Sites: trachea (to sweep mucus and bacteria to the pharynx); fallopian tubes (to sweep ovum to uterus).
7. Glands—Epithelial tissues that produce secretions.
   - Unicellular—one-celled glands. Goblet cells secrete mucus in the respiratory and digestive tracts.
   - Multicellular—many-celled glands.
     - Exocrine glands have ducts; salivary glands secrete saliva into ducts that carry it to the oral cavity.
     - Endocrine glands secrete hormones directly into capillaries (no ducts); thyroid gland secretes thyroxine.

Connective Tissue—all have a non-living intercellular matrix and specialized cells (see Table 4–2 and Figs. 4–4 and 4–5)

1. Blood—the matrix is plasma, mostly water; transports materials in the blood. Red blood cells carry oxygen; white blood cells destroy pathogens and provide immunity; platelets prevent blood loss, as in clotting.
2. Areolar (loose)—cells are fibroblasts, which produce protein fibers: collagen is strong, elastin is elastic; the matrix is collagen, elastin, and tissue fluid. White blood cells are also present. Sites: below the dermis and below the epithelium of tracts that open to the environment (to destroy pathogens that enter the body).
3. Adipose—cells are adipocytes that store fat; little matrix. Sites: between the skin and muscles (to store energy); around the eyes and kidneys (to cushion).
4. Fibrous—mostly matrix, strong collagen fibers; cells are fibroblasts. Regular fibrous sites: tendons (to connect muscle to bone); ligaments (to connect bone to bone). Irregular fibrous sites: dermis of the skin and the fascia around muscles.
5. Elastic—mostly matrix, elastin fibers. Sites: walls of large arteries (to maintain blood pressure); around alveoli (to promote normal exhalation).
6. Bone—cells are osteocytes; matrix is calcium salts and collagen, strong and not flexible. Sites: bones of the skeleton (to support the body and protect internal organs from mechanical injury).
7. Cartilage—cells are chondrocytes; protein matrix is firm yet flexible; no capillaries in matrix. Sites: joint surfaces of bones (to prevent friction); tip of nose and external ear (to support); wall of trachea (to keep air passage open); discs between vertebrae (to absorb shock).

Muscle Tissue—specialized to contract and bring about movement (see Table 4–3 and Fig. 4–6)

1. Skeletal—also called striated or voluntary muscle. Cells are cylindrical, have several nuclei, and have striations. Each cell has a motor nerve ending; nerve impulses are essential to cause contraction. Site: skeletal muscles attached to bones (to move the skeleton and produce heat).
2. Smooth—also called visceral or involuntary muscle. Cells have tapered ends, one nucleus each, and no striations. Contraction is not under voluntary control. Sites: stomach and intestines (peristalsis); walls of arteries and veins (to maintain blood pressure); iris (to constrict or dilate pupil).
3. Cardiac—cells are branched, have one nucleus each, and faint striations. Site: walls of chambers of the heart (to pump blood; nerve impulses regulate the rate of contraction).

Nerve Tissue—neurons are specialized to generate and transmit impulses (see Table 4–4 and Fig. 4–7)

1. Cell body contains the nucleus; axon carries impulses away from the cell body; dendrites carry impulses toward the cell body.
2. A synapse is the space between two neurons; a neurotransmitter carries the impulse across a synapse.
3. Specialized cells in nerve tissue are neuroglia in the CNS and Schwann cells in the PNS.
4. Sites: brain; spinal cord; and peripheral nerves (to provide sensation, movement, regulation of body functions, learning, and memory).

Membranes—sheets of tissue on surfaces, or separating organs or lobes

1. Epithelial Membranes (see Fig. 4-8)
   - Serous membranes—in closed body cavities; the serous fluid prevents friction between the two layers of the serous membrane.
     - Thoracic cavity—partial pleura lines chest wall; visceral pleura covers the lungs.
     - Pericardial sac—parietal pericardium lines fibrous pericardium; visceral pericardium (epicardium) covers the heart muscle.

  - Abdominal cavity—peritoneum lines the abdominal cavity; mesentery covers the abdominal organs.
- Mucous membranes—line body tracts that open to the environment: respiratory, digestive, urinary, and reproductive. Mucus keeps the living epithelium wet; provides lubrication in the digestive tract; traps dust and bacteria in the respiratory tract.

2. Connective Tissue Membranes—see Table 4-5.

REVIEW QUESTIONS

1. Explain the importance of each tissue in its location: (pp. 64-65, 66, 73)
   a. Simple squamous epithelium in the alveoli of the lungs
   b. Ciliated epithelium in the trachea
   c. Cartilage in the trachea
2. Explain the importance of each tissue in its location: (pp. 70-71)
   a. Bone tissue in bones
   b. Cartilage on the joint surfaces of bones
   c. Fibrous connective tissue in ligaments
3. State the functions of red blood cells, white blood cells, and platelets. (p. 72)
4. Name two organs made primarily of nerve tissue, and state the general functions of nerve tissue. (p. 75)
5. State the location and function of cardiac muscle. (pp. 74-75)
6. Explain the importance of each of these tissues in the small intestine: smooth muscle and columnar epithelium. (pp. 65-66, 73-74)
7. State the precise location of each of the following membranes: (p. 76)
   a. Peritoneum
   b. Visceral pericardium
   c. Parietal pleura
8. State the function of: (pp. 68, 76, 77)
   a. Serous fluid
   b. Mucus
   c. Blood plasma
9. State two functions of skeletal muscles. (p. 73)
10. Name three body tracts lined with mucous membranes. (pp. 76-77)
11. Explain how endocrine glands differ from exocrine glands. (p. 67)
12. State the function of adipose tissue: (pp. 69, 71)
    a. Around the eyes
    b. Between the skin and muscles
13. State the location of: (p. 78)
    a. Meninges
    b. Synovial membranes
14. State the important physical characteristics of collagen and elastin, and name the cells that produce these protein fibers (p. 69).

CHAPTER 5

The Integumentary System

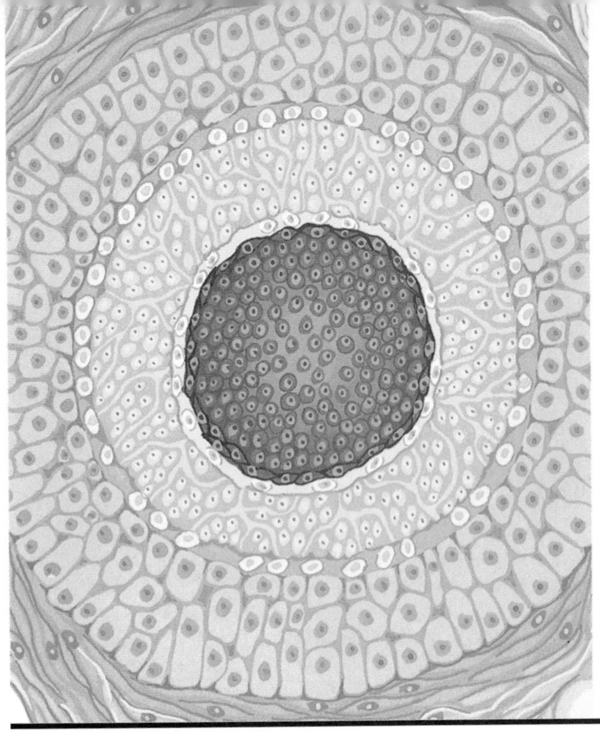

CHAPTER 5

Chapter Outline

The Skin
Epidermis
- Stratum germinativum
- Stratum corneum
- Langerhans cells
- Melanocytes

Dermis
- Hair follicles
- Nail follicles
- Receptors
- Glands
- Blood vessels

Other Functions of the Skin
Subcutaneous Tissue
Aging and the Integumentary System

Box 5–1 BURNS
Box 5–2 PREVENTING SKIN CANCER: COMMON SENSE AND SUNSCREENS
Box 5–3 COMMON SKIN DISORDERS

Student Objectives

- Name the two major layers of the skin and the tissue of which each is made.
- State the locations and describe the functions of the stratum corneum and stratum germinativum.
- Describe the function of Langerhans cells.
- Describe the function of melanocytes and melanin.
- Describe the functions of hair and nails.
- Name the cutaneous senses and explain their importance.
- Describe the functions of the secretions of sebaceous glands, ceruminous glands, and eccrine sweat glands.
- Describe how the arterioles in the dermis respond to heat, cold, and stress.
- Name the tissues that make up the subcutaneous tissue, and describe their functions.

The Integumentary System

New Terminology

Arterioles (ar-**TEER**-ee-ohls)
Ceruminous gland/Cerumen (suh-**ROO**-mi-nus GLAND/suh-**ROO**-men)
Dermis (**DER**-miss)
Eccrine sweat gland (**EK**-rin SWET GLAND)
Epidermis (EP-i-**DER**-miss)
Hair follicle (HAIR **FAH**-li-kull)
Keratin (**KER**-uh-tin)
Melanin (**MEL**-uh-nin)
Melanocyte (muh-**LAN**-oh-sight)
Nail follicle (NAIL **FAH**-li-kull)
Papillary layer (**PAP**-i-LAR-ee **LAY**-er)
Receptors (ree-**SEP**-turs)
Sebaceous gland/sebum (suh-**BAY**-shus GLAND/ **SEE**-bum)
Stratum corneum (**STRA**-tum **KOR**-nee-um)
Stratum germinativum (**STRA**-tum JER-min-ah-**TEE**-vum)
Subcutaneous tissue (SUB-kew-**TAY**-nee-us **TISH**-yoo)
Vasoconstriction (VAY-zoh-kon-**STRIK**-shun)
Vasodilation (VAY-zoh-dye-**LAY**-shun)

Related Clinical Terminology

Acne (**AK**-nee)
Alopecia (AL-oh-**PEE**-she-ah)
Biopsy (**BYE**-op-see)
Carcinoma (KAR-sin-**OH**-mah)
Circulatory shock (**SIR**-kew-lah-TOR-ee SHAHK)
Contusion (kon-**TOO**-zhun)
Decubitis ulcer (dee-**KEW**-bi-tuss **UL**-ser)
Dehydration (DEE-high-**DRAY**-shun)
Dermatology (DER-muh-**TAH**-luh-gee)
Eczema (**EK**-zuh-mah)
Erythema (ER-i-**THEE**-mah)
Histamine (**HISS**-tah-meen)
Hives (**HIGH**-VZ)
Inflammation (IN-fluh-**MAY**-shun)
Melanoma (MEL-ah-**NO**-mah)
Nevus (**NEE**-vus)
Pruritus (proo-**RYE**-tus)
Septicemia (SEP-tih-**SEE**-mee-ah)

Terms that appear in ***bold type*** *in the chapter text are defined in the glossary, which begins on p. 528.*

The **integumentary system** consists of the skin, its accessory structures such as hair and sweat glands, and the subcutaneous tissue below the skin. The **skin** is made of several different tissue types and is considered an organ. Because the skin covers the surface of the body, one of its functions is readily apparent: It separates the body from the external environment and prevents the entry of many harmful substances. The **subcutaneous tissue** directly underneath the skin connects it to the muscles and has other functions as well.

THE SKIN

The two major layers of the skin are the outer **epidermis** and the inner **dermis**. Each of these layers is made of different tissues and has very different functions.

EPIDERMIS

The **epidermis** is made of stratified squamous epithelial tissue and is thickest on the palms and soles. The cells are called **keratinocytes**, and there are no capillaries present between them. Although the epidermis may be further subdivided into four or five sublayers, two of these are of greatest importance: the innermost layer, the stratum germinativum, and the outermost layer, the stratum corneum (Fig. 5-1).

Stratum Germinativum

The **stratum germinativum** is the inner epidermal layer in which **mitosis** takes place. New cells are continually being produced, pushing the older cells toward the skin surface. These cells produce the protein **keratin**, and as they get farther away from the capillaries in the dermis, they die. As dead cells are worn off the skin's surface, they are replaced by cells from within.

Stratum Corneum

The **stratum corneum**, the outermost epidermal layer, consists of many layers of dead cells; all that is left is their **keratin**. The protein keratin is relatively waterproof and prevents evaporation of body water. Also of importance, keratin prevents the entry of water. Without a waterproof stratum corneum, it would be impossible to swim in a pool or even take a shower without damaging our cells.

The stratum corneum is also a barrier to pathogens and chemicals. Most bacteria and other microorganisms cannot penetrate unbroken skin. Most chemicals, unless they are corrosive, will not get through unbroken skin to the living tissue within. One painful exception is the sap of poison ivy. This resin does penetrate the skin and initiates an allergic reaction in susceptible people. The inflammatory response that characterizes allergies causes blisters and severe itching. The importance of the stratum corneum becomes especially apparent when it is lost (see Box 5-1: Burns).

Certain minor changes in the epidermis are undoubtedly familiar to you. When first wearing new shoes, for example, the skin of the foot may be subjected to friction. This will separate layers of the epidermis, or separate the epidermis from the dermis, and tissue fluid may collect, causing a **blister**. If the skin is subjected to pressure, the rate of mitosis in the stratum germinativum will increase and create a thicker epidermis; we call this a **callus**. Although calluses are more common on the palms and soles, they may occur on any part of the skin.

Langerhans Cells

Within the epidermis are **Langerhans cells** (also called dendritic cells because of their branched appearance) that originated in the bone marrow. These cells are quite mobile, and are able to phagocytize foreign material, such as bacteria that enter the body through breaks in the skin. With such ingested pathogens, the Langerhans cells migrate to lymph nodes and present the pathogen to lymphocytes, a type of white blood cell. This triggers an immune response such as the production of antibodies. Recent research indicates that the skin is much more active in immunity than was previously believed, though most of the exact aspects of this have yet to be determined. Immunity is covered in Chapter 14.

Melanocytes

Another type of cell found in the lower epidermis is the melanocyte. **Melanocytes** produce another pro-

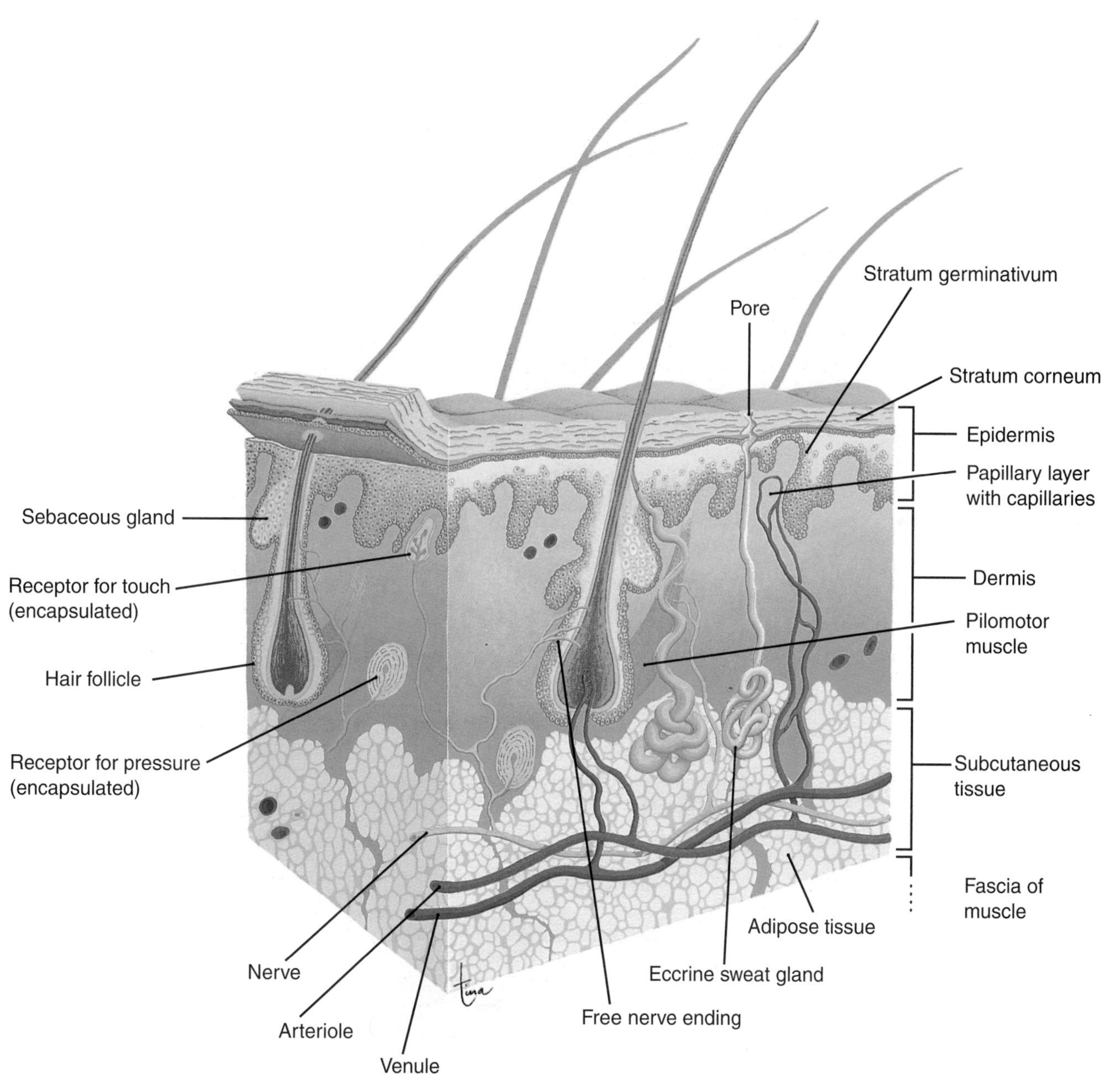

Figure 5–1 Skin. Structure of the skin and subcutaneous tissue.

tein, a pigment called **melanin**. People of the same size have approximately the same number of melanocytes. In people with dark skin, the melanocytes continuously produce large amounts of melanin. The melanocytes of light-skinned people produce less melanin. The activity of melanocytes is genetically regulated; skin color is one of our hereditary characteristics.

In all people, melanin production is increased by exposure of the skin to ultraviolet rays, which are part

Box 5–1 BURNS

Burns of the skin may be caused by flames, hot water or steam, sunlight, electricity, or corrosive chemicals. The severity of burns ranges from minor to fatal, and the classification of burns is based on the extent of damage.

FIRST-DEGREE BURN—only the superficial epidermis is burned, and is painful but not blistered. Light-colored skin will appear red due to localized **vasodilation** in the damaged area. Vasodilation is part of the inflammatory response that brings more blood to the injured site.

SECOND-DEGREE BURN—deeper layers of the epidermis are affected. Another aspect of **inflammation** is that damaged cells release **histamine**, which makes capillaries more

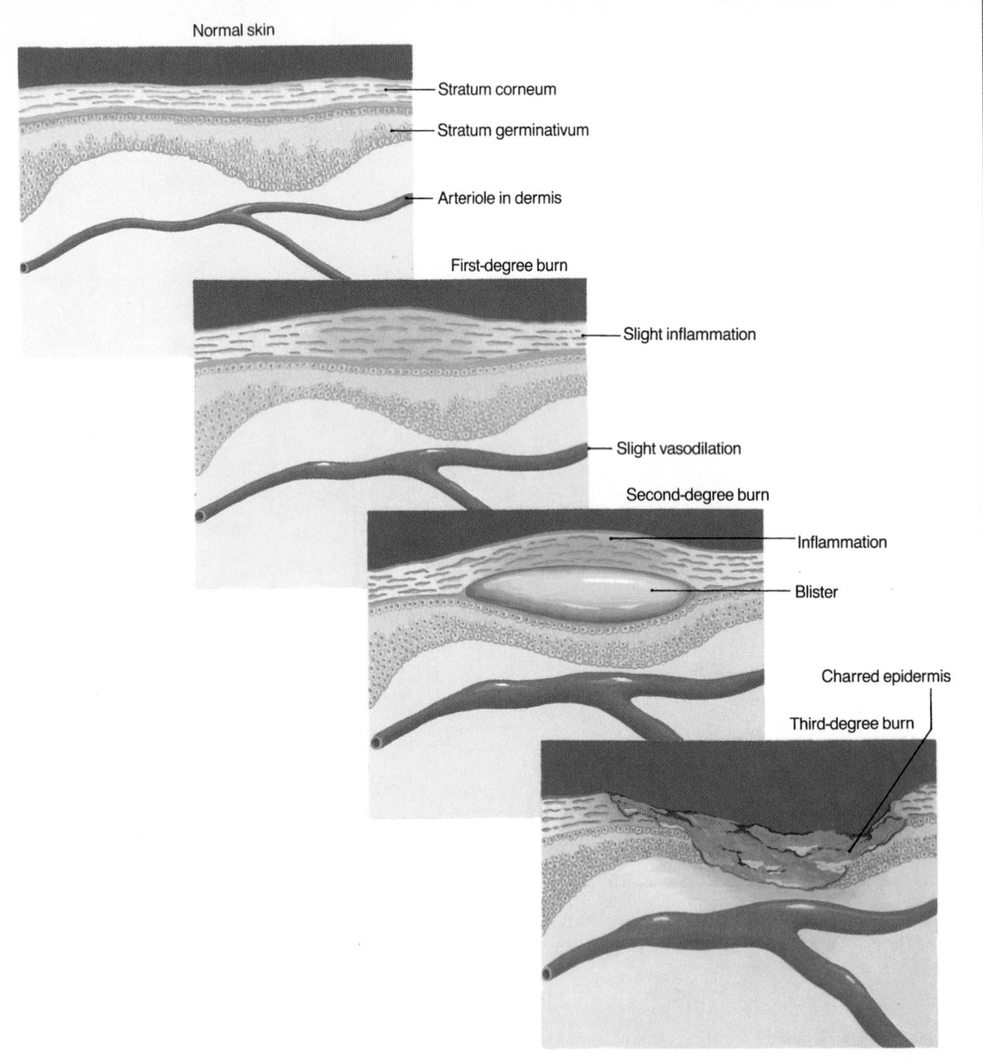

Box Figure 5–A Normal skin section and representative sections showing first-degree, second-degree, and third-degree burns.

Box 5–1 BURNS (Continued)

permeable. More plasma leaves these capillaries and becomes tissue fluid, which collects at the burn site, creating blisters. The burned skin is often very painful.

THIRD-DEGREE BURN—the entire epidermis is charred or burned away, and the burn may extend into the dermis or subcutaneous tissue. Often such a burn is not painful at first, if the receptors in the dermis have been destroyed.

Extensive third-degree burns are potentially life-threatening because of the loss of the stratum corneum. Without this natural barrier, living tissue is exposed to the environment and is susceptible to infection and dehydration.

Bacterial infection is a serious problem for burn patients; the pathogens may get into the blood **(septicemia)** and quickly spread throughout the body. Dehydration may also be fatal if medical intervention does not interrupt and correct the following sequence: Tissue fluid evaporates from the burned surface, and more plasma is pulled out of capillaries into the tissue spaces. As more plasma is lost, blood volume and blood pressure decrease. This is called **circulatory shock**; eventually the heart simply does not have enough blood to pump, and heart failure is the cause of death. To prevent these serious consequences, third-degree burns are covered with donor skin or artificial skin until skin grafts of the patient's own skin can be put in place.

of sunlight. As more melanin is produced, it is taken in by the epidermal cells as they are pushed toward the surface. This gives the skin a darker color, which prevents further exposure of the living stratum germinativum to ultraviolet rays. People with dark skin already have good protection against the damaging effects of ultraviolet rays; people with light skin do not (see Box 5-2: Preventing Skin Cancer: Common Sense and Sunscreens). The functions of the epidermis are summarized in Table 5-1.

DERMIS

The **dermis** is made of an irregular type of fibrous connective tissue. Fibroblasts produce both collagen and elastin fibers. Recall that **collagen** fibers are strong, and **elastin** fibers are able to recoil after being stretched. Strength and elasticity are two characteristics of the dermis. With increasing age, however, the deterioration of the elastin fibers causes the skin to lose its elasticity. We can all look forward to at least a few wrinkles as we get older.

The uneven junction of the dermis with the epidermis is called the **papillary layer** (see Fig. 5-1). Capillaries are abundant here to nourish not only the dermis but also the stratum germinativum. This epidermal layer has no capillaries of its own and depends on the blood supply in the dermis for oxygen and nutrients.

Within the dermis are the accessory skin structures: hair and nail follicles, sensory receptors, and several types of glands. Some of these project through the epidermis to the skin surface, but their active portions are in the dermis.

Hair Follicles

Hair follicles are made of epidermal tissue, and the growth process of hair is very similar to growth of the epidermis. At the base of a follicle is the **hair root**, where mitosis takes place (Fig. 5-2). The new cells produce keratin, get their color from melanin, then die and become incorporated into the **hair shaft**. The hair that we comb and brush every day consists of dead, keratinized cells.

Compared to some other mammals, humans do not have very much hair. The actual functions of human hair are quite few. Eyelashes and eyebrows help to keep dust and perspiration out of the eyes, and the hairs just inside the nostrils help to keep dust out of the nasal cavities. Hair of the scalp does provide insulation from cold for the head. The hair on our bodies, however, no longer serves this function, but we have the evolutionary remnants of it. Attached to each hair follicle is a small, smooth muscle called the **pilomotor** or arrector pili muscle. When stimulated by cold or emotions such as fear, these muscles pull the

Box 5–2 PREVENTING SKIN CANCER : COMMON SENSE AND SUNSCREENS

Anyone can get skin cancer, and the most important factor is exposure to sunlight. Light-skinned people are, of course, more susceptible to the effects of ultraviolet (UV) rays, which may trigger **mutations** in living epidermal cells.

Squamous cell **carcinoma** and basal cell carcinoma (see A below) are the most common forms of skin cancer. The lesions are visible as changes in the normal appearance of the skin, and a **biopsy** (microscopic examination of a tissue specimen) is used to confirm the diagnosis. These lesions usually do not metastasize rapidly, and can be completely removed using simple procedures.

Malignant melanoma (see B below) is a more serious form of skin cancer, which begins in melanocytes. Any change in a pigmented spot or mole **(nevus)** should prompt a person to see a doctor. Melanoma is serious not because of its growth in the skin, but because it may **metastasize** very rapidly to the lungs, liver, or other vital organ. Researchers are currently testing a vaccine for people who have had melanoma. The purpose of the vaccine is to stimulate the immune system strongly enough to prevent a second case, for such recurrences are often fatal.

Although the most common forms of skin cancer are readily curable, prevention is a much better strategy. We cannot, and we would not want to, stay out of the sun altogether, but we may be able to do so when sunlight is most damaging. During the summer months, UV rays are especially intense between 10 A.M. and 2 P.M.. If we are or must be outdoors during this time, use of a sunscreen is recommended by dermatologists.

Sunscreens contain chemicals such as PABA (para-amino benzoic acid) that block UV rays and prevent them from damaging the epidermis. An SPF (sun protection factor) of 15 or higher is considered good protection. Use of a sunscreen on exposed skin not only helps prevent skin cancer but prevents sunburn and its painful effects. It is especially important to prevent children from getting severely sunburned, because such burns have been linked to the development of skin cancer years later.

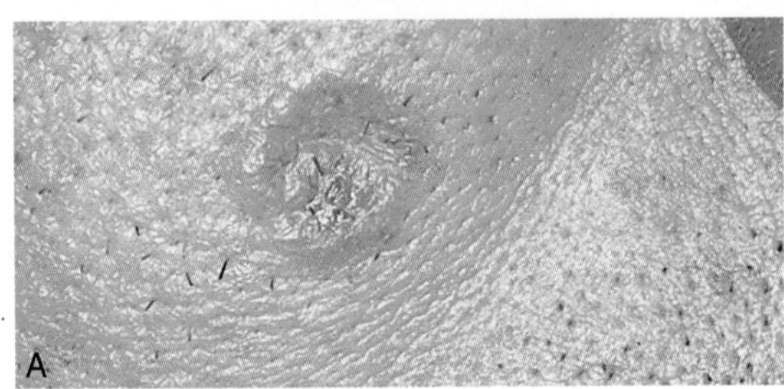

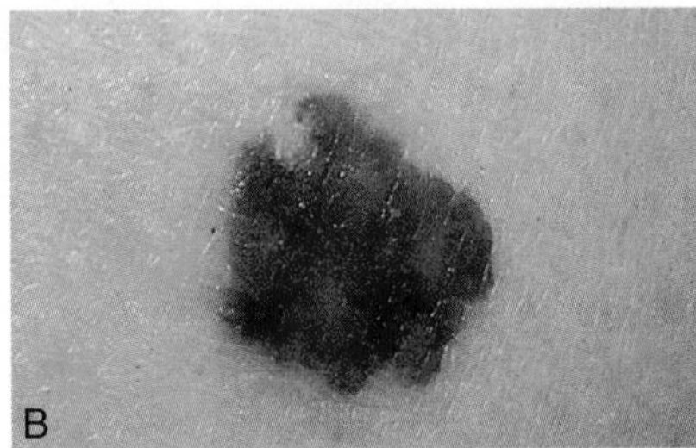

Box Figure 5–B (A) Classic basal cell carcinoma on face. **(B)** Melanoma with typical irregular border. (From Reeves, JRT, and Maibach, HI: Clinical Dermatology Illustrated: A Regional Approach, ed 3. FA Davis, Philadelphia, 1998, pp 329 and 338, with permission.)

hair follicles upright. For an animal with fur, this would provide greater insulation. Because people do not have thick fur, all this does for us is give us "goosebumps."

Nail Follicles

Found on the ends of fingers and toes, **nail follicles** produce nails just as hair follicles produce hair. Mitosis takes place in the **nail root** (Fig. 5-3), and the new

Table 5–1 EPIDERMIS

Part	Function
Stratum corneum (keratin)	• Prevents loss or entry of water • If unbroken, prevents entry of pathogens and most chemicals
Stratum germinativum	• Continuous mitosis produces new cells to replace worn-off surface cells
Langerhans cells	• Phagocytize foreign material and stimulate an immune response by lymphocytes
Melanocytes	• Produce melanin on exposure to ultraviolet (UV) rays
Melanin	• Protects living skin layers from further exposure to UV rays

cells produce keratin (a stronger form of this protein than is found in hair) and then die. Although the nail itself consists of keratinized dead cells, the flat nail bed is living tissue. This is why cutting a nail too short can be quite painful. Nails function to protect the ends of the fingers and toes from mechanical injury and to give the fingers greater ability to pick up small objects.

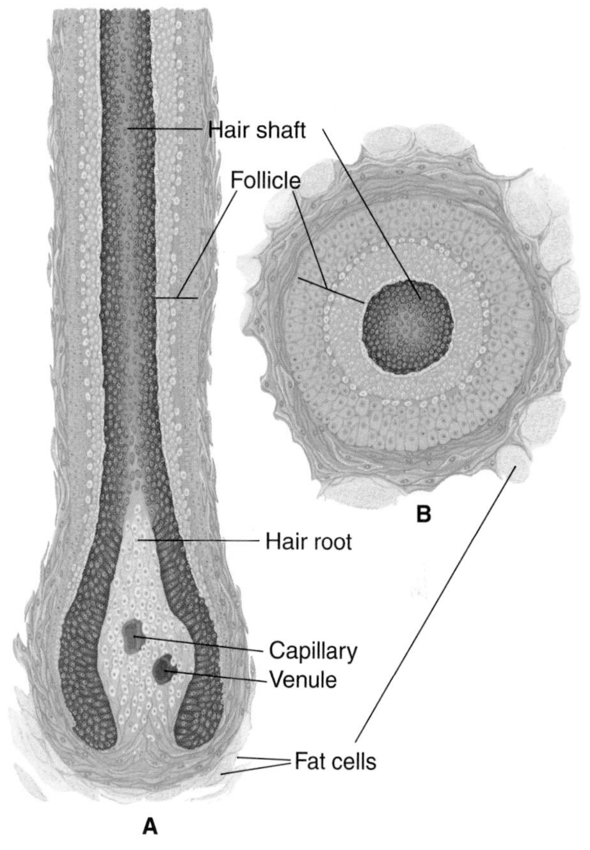

Figure 5–2 Structure of a hair follicle. **(A)** Longitudinal section. **(B)** Cross-section.

Receptors

The sensory **receptors** in the dermis are for the cutaneous senses: touch, pressure, heat, cold, and pain. For each sensation there is a specific type of receptor, which is a structure that will detect a particular change. For pain, the receptors are **free nerve endings**. For the other cutaneous senses, the receptors are called **encapsulated nerve endings**, which means there is a cellular structure around the sensory nerve ending (see Fig. 5–1). The purpose of these receptors and sensations is to provide the central nervous system with information about the external environment and its effect on the skin. This information may stimulate responses; a simple example would be responding to a feeling of cold by putting on a sweater.

The sensitivity of an area of skin is determined by how many receptors are present. The skin of the fingertips, for example, is very sensitive to touch because

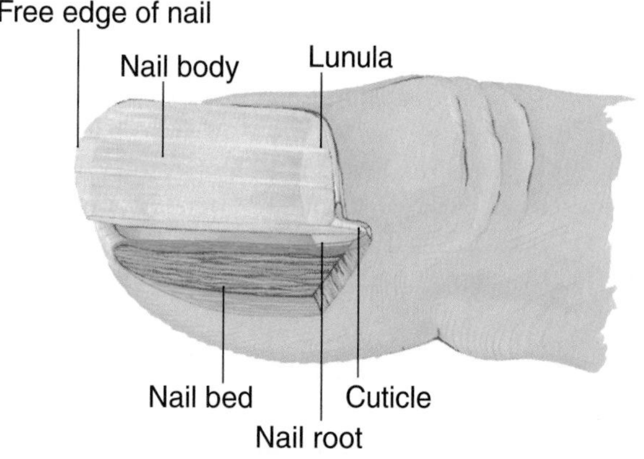

Figure 5–3 Structure of a fingernail shown in longitudinal section.

there are many receptors per square inch. The skin of the upper arm, with few touch receptors per square inch, is less sensitive.

When receptors detect changes, they generate nerve impulses that are carried to the brain, which interprets the impulses as a particular sensation. Sensation, therefore, is actually a function of the brain (we will return to this in Chapters 8 and 9).

Glands

Glands are made of epithelial tissue. The exocrine glands of the skin have their secretory portions in the dermis. Some of these are shown in Fig. 5-1.

Sebaceous Glands The ducts of **sebaceous glands** open into hair follicles or directly to the skin surface. Their secretion is **sebum**, a lipid substance that we commonly refer to as oil. The function of sebum is to prevent drying of skin and hair. The importance of this may not be readily apparent, but skin that is dry tends to crack more easily. Even very small breaks in the skin are potential entryways for bacteria. Decreased sebum production is another consequence of getting older, and elderly people often have dry and more fragile skin.

Adolescents may have the problem of overactive sebaceous glands. Too much sebum may trap bacteria within hair follicles and create small infections. Because sebaceous glands are more numerous around the nose and mouth, these are common sites of pimples in young people (see also Box 5-3: Common Skin Disorders).

Ceruminous Glands These are modified sebaceous glands located in the dermis of the ear canals. Their secretion is called **cerumen** or ear wax. Cerumen keeps the outer surface of the eardrum pliable and prevents drying. However, if excess cerumen accumulates in the ear canal, it may become impacted against the eardrum. This might diminish the acuity of hearing by preventing the eardrum from vibrating properly.

Sweat Glands There are two types of sweat glands, apocrine and eccrine. **Apocrine glands** are most numerous in the axillae (underarm) and genital areas and are most active in stress and emotional situations. Although their secretion does have an odor, it is barely perceptible to other people. However, animals, such as dogs, can tell people apart by their individual scents. If the apocrine secretions are allowed to accumulate on the skin, bacteria metabolize the chemicals in the sweat and produce waste products that have distinct odors that many people find unpleasant.

Eccrine glands are found all over the body but are especially numerous on the forehead, upper lip, palms, and soles. The secretory portion of these glands is simply a coiled tube in the dermis. The duct of this tube extends to the skin's surface, where it opens into a **pore**.

The sweat produced by eccrine glands is important in the maintenance of normal body temperature. In a warm environment, or during exercise, more sweat is secreted onto the skin surface, where it is then evaporated by excess body heat. Although this is a very effective mechanism of heat loss, it has a potentially serious disadvantage. Loss of too much body water in sweat may lead to **dehydration**, as in heat exhaustion. Increased sweating during exercise or on a hot day should always be accompanied by increased fluid intake.

Blood Vessels

Besides the capillaries in the dermis, the other blood vessels of great importance are the arterioles. **Arterioles** are small arteries, and the smooth muscle in their walls permits them to constrict (close) or dilate (open). This is important in the maintenance of body temperature, because blood carries heat, which is a form of energy.

In a warm environment the arterioles dilate **(vasodilation)**, which increases blood flow through the dermis, and brings excess heat close to the body surface to be radiated to the environment. In a cold environment, however, body heat must be conserved if possible, so the arterioles constrict. The **vasoconstriction** decreases the flow of blood through the dermis and keeps heat within the core of the body. This adjusting mechanism is essential for maintaining homeostasis. Regulation of the diameter of the arterioles in response to external temperaure changes is controlled by the nervous system. These changes can often be seen in light-skinned people. Flushing, especially in the face, may be observed in hot weather. In cold, the skin of the extremities may become even paler as blood flow through the dermis decreases. In people with dark skin, such changes are not as readily apparent because they are masked by melanin in the epidermis.

Box 5–3 COMMON SKIN DISORDERS

Impetigo—a bacterial infection often caused by *streptococci* or *staphylococci.* The characteristic pustules (pus-containing lesions) crust as they heal; the infection is contagious to other children.

Eczema—a symptom of what is more properly called atopic dermatitis. This is an allergic reaction more common in children than adults; the rash is itchy (pruritus), and may blister or ooze. Eczema is often related to foods such as fish, eggs, or milk products, or to inhaled allergens such as dust, pollens, or animal dander. Prevention depends upon determining what the child is allergic to and eliminating or at least limiting exposure.

Warts—caused by a virus that makes epidermal cells divide abnormally, producing a growth on the skin that is often raised and has a rough or pitted surface. Warts are probably most common on the hands, but they may be anywhere on the skin. Plantar warts on the sole of the foot may become quite painful because of the constant pressure of standing and walking.

Fever blisters (cold sores)—caused by the Herpes simplex virus, to which most people are exposed as children. An active lesion, usually at the edge of the lip, is painful and oozes. If not destroyed by the immune system, the virus "hides out" and becomes dormant in nerves of the face. Another lesion, weeks or months later, may be triggered by stress or another illness.

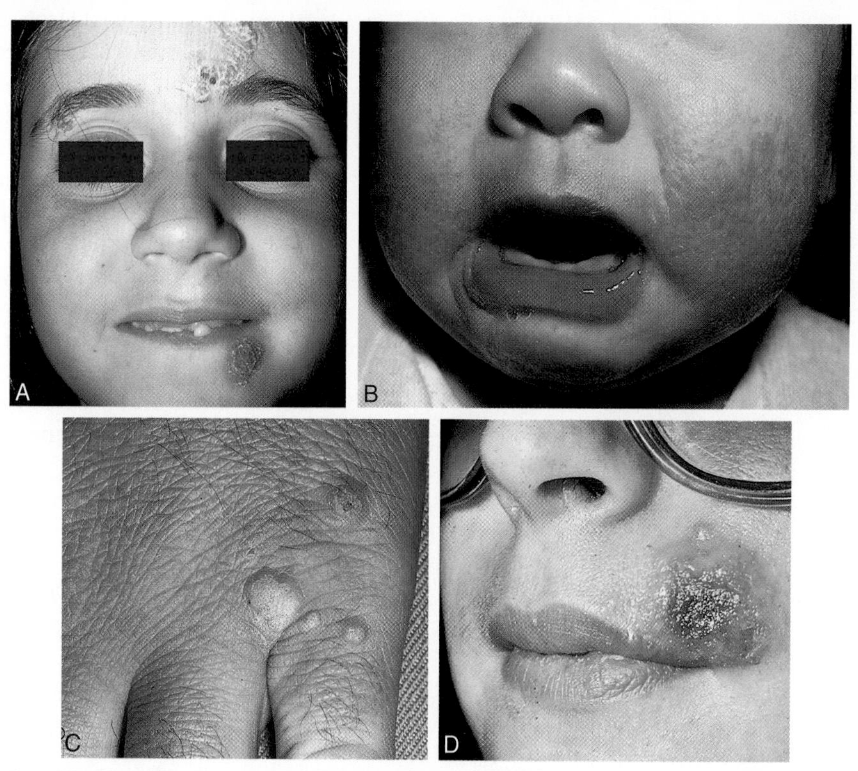

Box Figure 5–C (A) Impetigo. **(B)** Eczema of atopic dermatitis. **(C)** Warts on back of hand. **(D)** Fever blister, localized but severe. (From Reeves, JRT, and Maibach, HI: Clinical Dermatology Illustrated: A Regional Approach, ed 3. FA Davis, Philadelphia, 1998, pp 65, 165, 250, and 304, with permission.)

Vasoconstriction in the dermis may also occur during stressful situations. For our ancestors, stress usually demanded a physical response: Either stand and fight or run away to safety. This is called the "fight or flight response." Our nervous systems are still programmed to respond as if physical activity were necessary to cope with the stress situation. Vasoconstriction in the dermis will shunt, or redirect, blood to more vital organs such as the muscles, heart, and brain. In times of stress, the skin is a relatively unimportant organ and can function temporarily with a minimal blood flow.

OTHER FUNCTIONS OF THE SKIN

Excretion—small amounts of **urea** (a waste product of protein metabolism) and sodium chloride are excreted in sweat. This is a very minor function of the skin; the kidneys are primarily responsible for removing waste products from the blood.

Formation of **vitamin D**—there is a form of cholesterol in the skin that, on exposure to ultraviolet light, is changed to vitamin D. This is why vitamin D is sometimes referred to as the "sunshine vitamin." People who do not get much sunlight depend more on nutritional sources of vitamin D, such as fortified milk. Vitamin D is important for the absorption of calcium and phosphorus from food in the small intestine. The functions of dermal structures are summarized in Table 5-2.

Table 5–2 DERMIS

Part	Function
Papillary Layer	• Contains capillaries that nourish the stratum germinativum
Hair (Follicles)	• Eyelashes and nasal hair keep dust out of eyes and nasal cavities • Scalp hair provides insulation from cold for the head
Nails (Follicles)	• Protect ends of fingers and toes from mechanical injury
Receptors	• Detect changes that are felt as the cutaneous senses: touch, pressure, heat, cold, and pain
Sebaceous Glands	• Produce sebum, which prevents drying of skin and hair
Ceruminous Glands	• Produce cerumen, which prevents drying of the eardrum
Eccrine Sweat Glands	• Produce watery sweat that is evaporated by excess body heat to cool the body
Arterioles	• Dilate in response to warmth to increase heat loss • Constrict in response to cold to conserve body heat • Constrict in stressful situations to shunt blood to more vital organs
Cholesterol	• Converted to vitamin D on exposure to UV rays

SUBCUTANEOUS TISSUE

The **subcutaneous tissue** may also be called the **superficial fascia**, one of the connective tissue membranes. Made of areolar connective tissue and adipose tissue, the superficial fascia connects the dermis to the underlying muscles. Its other functions are those of its tissues, as you may recall from Chapter 4.

Areolar connective tissue contains collagen and elastin fibers and many white blood cells that have left capillaries to wander around here. These migrating white blood cells destroy pathogens that enter the body through breaks in the skin.

The cells (adipocytes) of adipose tissue are specialized to store fat, and our subcutaneous layer of fat stores excess nutrients as a potential energy source. This layer also cushions bony prominences, such as when sitting, and provides some insulation from cold. For people, this last function is relatively minor, because we do not have a thick layer of fat, as do animals

Table 5–3 SUBCUTANEOUS TISSUE

Part	Function
Areolar Connective Tissue	• Connects skin to muscles • Contains many WBCs to destroy pathogens that enter breaks in the skin
Adipose Tissue	• Contains stored energy in the form of true fats • Cushions bony prominences • Provides some insulation from cold

such as whales and seals. The functions of subcutaneous tissue are summarized in Table 5-3.

AGING AND THE INTEGUMENTARY SYSTEM

The effects of age on the integumentary system are often quite visible. Both layers of skin become thinner and more fragile as mitosis in the epidermis slows and fibroblasts in the dermis die and are not replaced. The skin becomes wrinkled as collagen and elastin fibers in the dermis deteriorate. Sebaceous glands and sweat glands become less active; the skin becomes dry; and temperature regulation in hot weather becomes more difficult. Hair follicles become inactive and hair on the scalp and body thins. Melanocytes die and the hair that remains becomes white. There is often less fat in the subcutaneous tissue, which may make an elderly person more sensitive to cold. It is important for elderly people (and those who care for them) to realize that extremes of temperature may be harmful and to take special precautions in very hot or very cold weather.

SUMMARY

The integumentary system is the outermost organ system of the body. You have probably noticed that many of its functions are related to this location. The skin protects the body against pathogens and chemicals, minimizes loss or entry of water, and blocks the harmful effects of sunlight. Sensory receptors in the skin provide information about the external environment, and the skin helps regulate body temperature in response to environmental changes.

STUDY OUTLINE

The integumentary system consists of the skin and its accessory structures and the subcutaneous tissue. The two major layers of the skin are the outer epidermis and the inner dermis.

Epidermis—made of stratified squamous epithelium; cells called keratinocytes (see Fig. 5–1 and Table 5–1)

1. Stratum Germinativum—the innermost layer where mitosis takes place; new cells produce keratin and die as they are pushed toward the surface.
2. Stratum Corneum—the outermost layers of dead cells; keratin prevents loss and entry of water and resists entry of pathogens and chemicals.
3. Langerhans Cells—phagocytize foreign material and stimulate an immune response by lymphocytes.
4. Melanocytes—in the lower epidermis, produce melanin. UV rays stimulate melanin production; melanin prevents further exposure of the stratum germinativum to UV rays by darkening the skin.

Dermis—made of irregular fibrous connective tissue; collagen provides strength, and elastin provides elasticity; capillaries in the papillary layer nourish the stratum germinativum (see Table 5–2 and Fig. 5–1)

1. Hair Follicles—mitosis takes place in the hair root; new cells produce keratin, die, and become the hair shaft. Hair of the scalp provides insulation from cold for the head; eyelashes keep dust out of eyes; nostril hairs keep dust out of nasal cavities (see Figs. 5-1 and 5-2).
2. Nail Follicles—at the ends of fingers and toes; mitosis takes place in the nail root; the nail itself is dead, keratinized cells. Nails protect the ends of the fingers and toes and enable the fingers to pick up small objects (see Fig. 5-3).
3. Receptors—detect changes in the skin: touch, pressure, heat, cold, and pain; provide information about the external environment that initiates appropriate responses; sensitivity of the skin depends on the number of receptors present.
4. Sebaceous Glands—secrete sebum into hair follicles or to the skin surface; sebum prevents drying of skin and hair.
5. Ceruminous Glands—secrete cerumen in the ear canals; cerumen prevents drying of the eardrum.
6. Apocrine Sweat Glands—modified scent glands in axillae and genital area; activated by stress and emotions.

7. Eccrine Sweat Glands—most numerous on face, palms, soles. Activated by high external temperature or exercise; sweat on skin surface is evaporated by excess body heat; potential disadvantage is dehydration.
8. Arterioles—smooth muscle permits constriction or dilation. Vasoconstriction in cold temperatures decreases dermal blood flow to conserve heat in the body core. Vasodilation in warm temperatures increases dermal blood flow to bring heat to the surface to be lost. Vasoconstriction during stress shunts blood away from the skin to more vital organs, such as muscles, to permit a physical response, if necessary.

Other Functions of the Skin

1. Excretion of small amounts of urea and NaCl (minor function).
2. Formation of vitamin D from cholesterol on exposure to UV rays of sunlight.

Subcutaneous Tissue—also called the superficial fascia; connects skin to muscles (see Fig. 5–1 and Table 5–3)

1. Areolar Tissue—contains WBCs that destroy pathogens that get through breaks in the skin.
2. Adipose Tissue—stores fat as potential energy; cushions bony prominences; provides some insulation from cold.

REVIEW QUESTIONS

1. Name the parts of the integumentary system. (p. 84)
2. Name the two major layers of skin, the location of each, and the tissue of which each is made. (pp. 84, 87)
3. In the epidermis: (p. 84)
   a. Where does mitosis take place?
   b. What protein do the new cells produce?
   c. What happens to these cells?
   d. What is the function of Langerhans cells?
4. Describe the functions of the stratum corneum. (p. 84)
5. Name the cells that produce melanin. What is the stimulus? Describe the function of melanin. (pp. 84–85)
6. Where, on the body, does human hair have important functions? Describe these functions. (p. 87)
7. Describe the functions of nails. (pp. 88–89)
8. Name the cutaneous senses. Describe the importance of these senses. (pp. 89–90)
9. Explain the functions of sebum and cerumen. (p. 90)
10. Explain how sweating helps maintain normal body temperature. (p. 90)
11. Explain how the arterioles in the dermis respond to cold or warm external temperatures and to stress situations. (p. 90)
12. What vitamin is produced in the skin? What is the stimulus for the production of this vitamin? (p. 92)
13. Name the tissues of which the superficial fascia is made. Describe the functions of these tissues. (p. 92)

CHAPTER 6

The Skeletal System

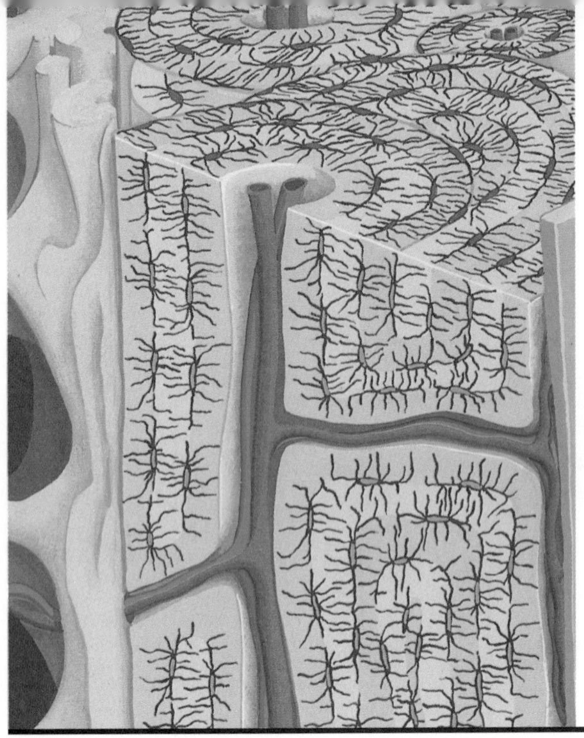

CHAPTER 6

Chapter Outline

Functions of the Skeleton
Types of Bone Tissue
Classification of Bones
Embryonic Growth of Bone
Factors That Affect Bone Growth and Maintenance
The Skeleton
Skull
Vertebral Column
Rib Cage
The Shoulder and Arm
The Hip and Leg
Joints—Articulations
The Classification of Joints
Synovial Joints
Aging and the Skeletal System

Box 6–1 FRACTURES AND THEIR REPAIR
Box 6–2 OSTEOPOROSIS
Box 6–3 HERNIATED DISC
Box 6–4 ABNORMALITIES OF THE CURVES OF THE SPINE
Box 6–5 ARTHRITIS

Student Objectives

- Describe the functions of the skeleton.
- Explain how bones are classified, and give an example of each type.
- Describe how the embryonic skeleton model is replaced by bone.
- Name the nutrients necessary for bone growth, and explain their functions.
- Name the hormones involved in bone growth and maintenance, and explain their functions.
- Explain what is meant by "exercise" for bones, and explain its importance.
- Name all the bones of the human skeleton (be able to point to each on diagrams, skeleton models, or yourself).
- Describe the functions of the skull, vertebral column, rib cage, scapula, and pelvic bone.
- Explain how joints are classified. For each type, give an example, and describe the movement possible.
- Describe the parts of a synovial joint, and explain their functions.

The Skeletal System

New Terminology

Appendicular (AP-en-**DIK**-yoo-lar)
Articulation (ar-TIK-yoo-**LAY**-shun)
Axial (**AK**-see-uhl)
Bursa (**BURR**-sah)
Diaphysis (dye-**AFF**-i-sis)
Epiphysis (e-**PIFF**-i-sis)
Epiphyseal disc (e-**PIFF**-i-SEE-al DISK)
Fontanel (FON-tah-**NELL**)
Haversian system (ha-**VER**-zhun **SIS**-tem)
Ligament (**LIG**-uh-ment)
Ossification (AHS-i-fi-**KAY**-shun)
Osteoblast (**AHS**-tee-oh-BLAST)
Osteoclast (**AHS**-tee-oh-KLAST)
Paranasal sinus (PAR-uh-**NAY**-zuhl **SIGH**-nus)
Periosteum (PER-ee-**AHS**-tee-um)
Suture (**SOO**-cher)
Symphysis (**SIM**-fi-sis)
Synovial fluid (sin-**OH**-vee-al **FLOO**-id)

Related Clinical Terminology

Autoimmune disease (AW-toh-im-**YOON** di-**ZEEZ**)
Bursitis (burr-**SIGH**-tiss)
Cleft palate (KLEFT **PAL**-uht)
Fracture (**FRAK**-chur)
Herniated disc (**HER**-nee-ay-ted DISK)
Kyphosis (kye-**FOH**-sis)
Lordosis (lor-**DOH**-sis)
Osteoarthritis (AHS-tee-oh-ar-**THRY**-tiss)
Osteomyelitis (AHS-tee-oh-my-uh-**LYE**-tiss)
Osteoporosis (AHS-tee-oh-por-**OH**-sis)
Rheumatoid arthritis (**ROO**-muh-toyd ar-**THRY**-tiss)
Rickets (**RIK**-ets)
Scoliosis (SKOH-lee-**OH**-sis)

*Terms that appear in **bold type** in the chapter text are defined in the glossary, which begins on p. 528.*

Imagine for a moment that people did not have skeletons. What comes to mind? Probably that each of us would be a little heap on the floor, much like a jellyfish out of water. Such an image is accurate and reflects the most obvious function of the skeleton: to support the body. Although it is a framework for the body, the skeleton is not at all like the wooden beams that support a house. Bones are living organs that actively contribute to the maintenance of the internal environment of the body.

The **skeletal system** consists of bones and other structures that make up the joints of the skeleton. The types of tissue present are bone tissue, cartilage, and fibrous connective tissue, which forms the ligaments that connect bone to bone.

FUNCTIONS OF THE SKELETON

1. Provides a framework that supports the body; the muscles that are attached to bones move the skeleton.
2. Protects some internal organs from mechanical injury; the rib cage protects the heart and lungs, for example.
3. Contains and protects the red bone marrow, one of the hemopoietic (blood-forming) tissues.
4. Provides a storage site for excess calcium. Calcium may be removed from bone to maintain a normal blood calcium level, which is essential for blood clotting and proper functioning of muscles and nerves.

TYPES OF BONE TISSUE

Bone was described as a tissue in Chapter 4. Recall that bone cells are called **osteocytes**, and the **matrix** of bone is made of calcium salts and collagen. The **calcium salts** are calcium carbonate ($CaCO_3$) and calcium phosphate ($Ca_3(PO_4)_2$), which give bone the strength for its supportive and protective functions. The function of osteocytes is to regulate the amount of calcium that is deposited in, or removed from, the bone matrix.

In bone as an organ, two types of bone tissue are present (Fig. 6–1). **Compact bone** is made of **haversian systems:** cylinders of bone matrix with osteocytes in concentric rings around central **haversian canals**. In the haversian canals are blood vessels; the osteocytes are in contact with these blood vessels and with one another through microscopic channels **(canaliculi)** in the matrix.

The second type of bone tissue is **spongy bone**, which does look rather like a sponge. Osteocytes, matrix, and blood vessels are present but are not arranged in haversian systems. The cavities in spongy bone often contain **red bone marrow**, which produces red blood cells, platelets, and the five types of white blood cells.

CLASSIFICATION OF BONES

1. **Long bones**—the bones of the arms, legs, hands, and feet (but not the wrists and ankles). The shaft of a long bone is the **diaphysis**, and the ends are called **epiphyses** (see Fig. 6–1). The diaphysis is made of compact bone and is hollow, forming a canal within the shaft. This **marrow canal** (or medullary cavity) contains **yellow bone marrow**, which is mostly adipose tissue. The epiphyses are made of spongy bone covered with a thin layer of compact bone. Although red bone marrow is present in the epiphyses of children's bones, it is largely replaced by yellow bone marrow in adult bones.
2. **Short bones**—the bones of the wrists and ankles.
3. **Flat bones**—the ribs, shoulder blades, hipbones, and cranial bones.
4. **Irregular bones**—the vertebrae and facial bones.

Short, flat, and irregular bones are all made of spongy bone covered with a thin layer of compact bone. Red bone marrow is found within the spongy bone.

The joint surfaces of bones are covered with **articular cartilage**, which provides a smooth surface. Covering the rest of the bone is the **periosteum**, a fibrous connective tissue membrane whose collagen fibers merge with those of the tendons and ligaments that are attached to the bone. The periosteum anchors

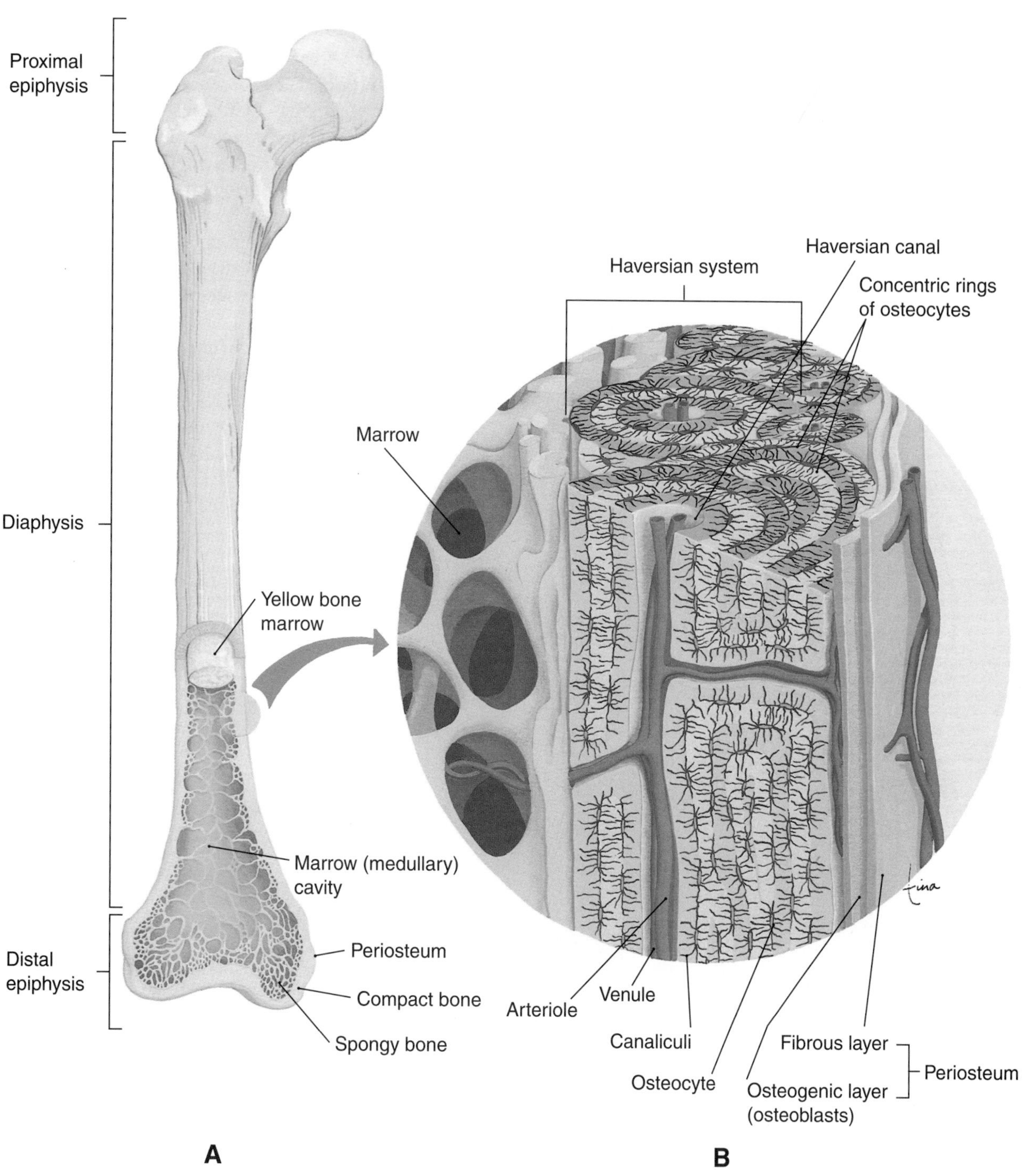

Figure 6–1 Bone tissue. **(A)** Femur with distal end cut in longitudinal section. **(B)** Compact bone showing haversian systems.

these structures and also contains the blood vessels that enter the bone itself.

EMBRYONIC GROWTH OF BONE

During embryonic development, the skeleton is first made of cartilage and fibrous connective tissue, which are gradually replaced by bone. Bone matrix is produced by cells called **osteoblasts** (a blast cell is a "producing" cell, and "osteo" means bone). In the embryonic model of the skeleton, osteoblasts differentiate from the fibroblasts that are present. The production of bone matrix, called **ossification**, begins in a **center of ossification** in each bone.

The cranial and facial bones are first made of fibrous connective tissue. In the third month of fetal development, fibroblasts (spindle-shaped connective tissue cells) become more specialized and differentiate into osteoblasts, which produce bone matrix. From each center of ossification, bone growth radiates outward as calcium salts are deposited in the collagen of the model of the bone. This process is not complete at birth; a baby has areas of fibrous connective tissue remaining between the bones of the skull. These are called **fontanels** (Fig. 6–2), which permit compression of the baby's head during birth without breaking the still thin cranial bones. You may have heard fontanels referred to as "soft spots," and indeed they are. A baby's skull is still quite fragile and must be protected from trauma. By the age of 2 years, all the fontanels have become ossified, and the skull becomes a more effective protective covering for the brain.

The rest of the embryonic skeleton is first made of cartilage, and ossification begins in the third month of gestation in the long bones. Osteoblasts produce bone matrix in the center of the diaphyses of the long bones and in the center of short, flat, and irregular bones. Bone matrix gradually replaces the original cartilage (Fig. 6–3).

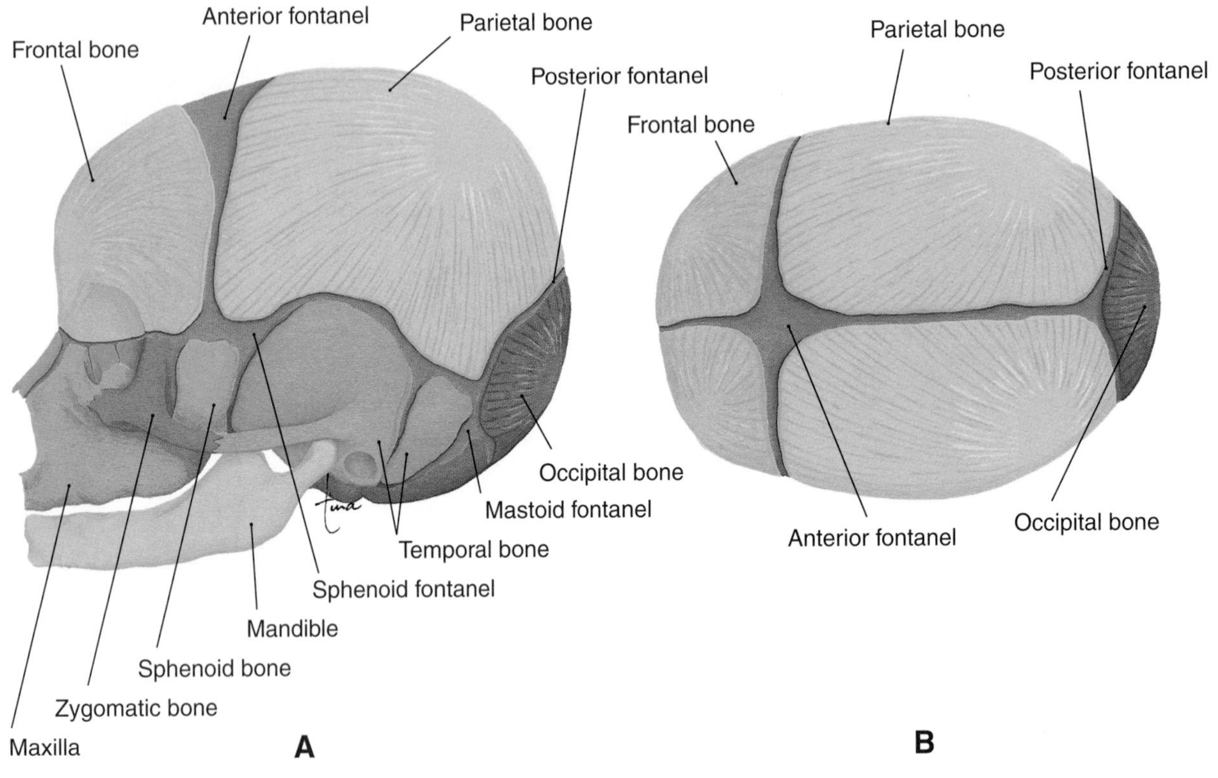

Figure 6–2 Infant skull with fontanels. **(A)** Lateral view of left side. **(B)** Superior view.

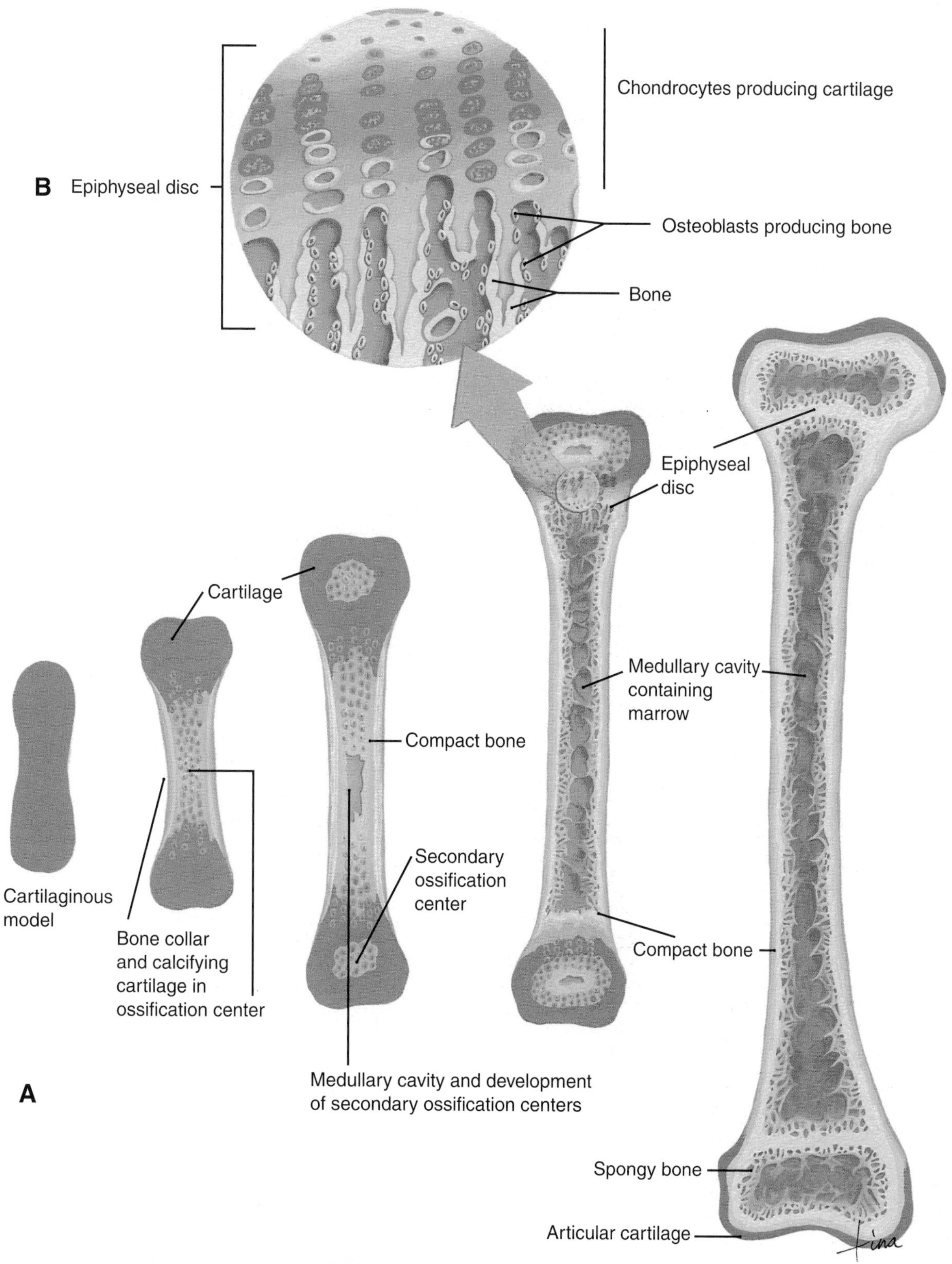

Figure 6–3 The ossification process in a long bone. **(A)** Progression of ossification from the cartilage model of the embryo to the bone of a young adult. **(B)** Microscopic view of an epiphyseal disc showing cartilage production and bone replacement.

Box 6–1 FRACTURES AND THEIR REPAIR

A fracture means that a bone has been broken. There are different types of fractures classified as to extent of damage.

Simple (closed)—the broken parts are still in normal anatomical position; surrounding tissue damage is minimal (skin is not pierced).

Compound (open)—the broken end of a bone has been moved, and it pierces the skin; there may be extensive damage to surrounding blood vessels, nerves, and muscles.

Greenstick—the bone splits longitudinally. The bones of children contain more collagen than do adult bones and tend to splinter rather than break completely.

Comminuted—two or more intersecting breaks create several bone fragments.

Impacted—the broken ends of a bone are forced into one another; many bone fragments may be created.

Spontaneous (pathologic)—a bone breaks without apparent trauma; may accompany bone disorders such as osteoporosis.

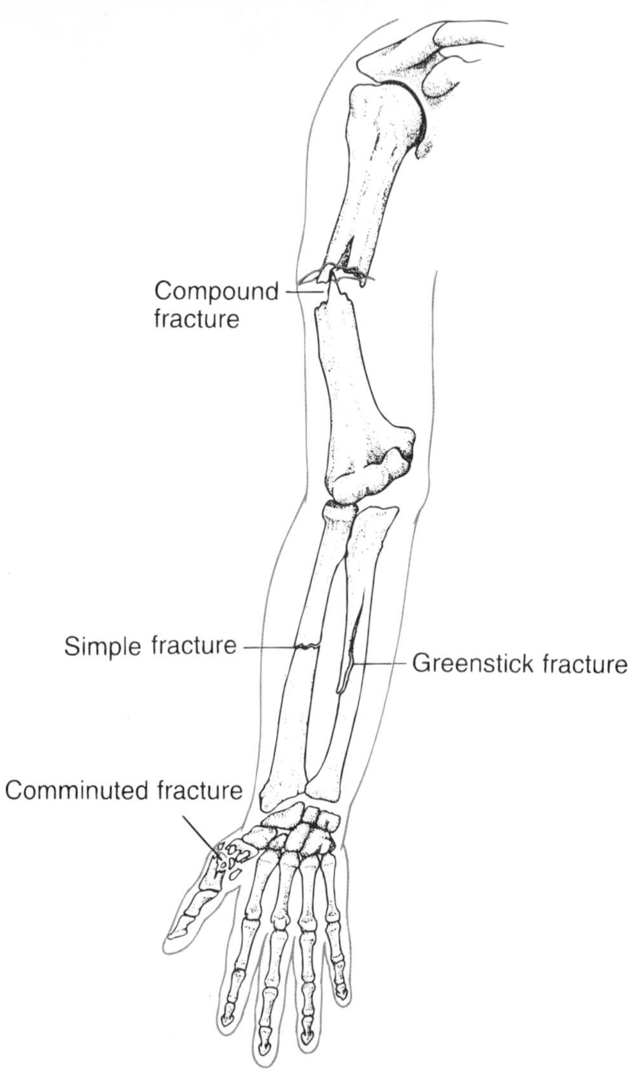

Box Figure 6–A Types of fractures. Several types of fractures are depicted in the right arm.

The Repair Process

Even a simple fracture involves significant bone damage that must be repaired if the bone is to resume its normal function. Fragments of dead or damaged bone must first be removed. This is accomplished by osteoclasts, which dissolve and reabsorb the calcium salts of bone matrix. Imagine a building that has just collapsed; the rubble must be removed before reconstruction can take place. This is what the osteoclasts do. Then, new bone must be produced. The inner layer of the periosteum contains osteoblasts that are activated when bone is damaged. The osteoblasts produce bone matrix to knit the broken ends of the bone together.

Because most bone has a good blood supply, the repair process is usually relatively rapid, and a simple fracture often heals within 6 weeks. Some parts of bones, however, have a poor blood supply, and repair of fractures takes longer. These areas are the neck of the femur (the site of a "fractured hip") and the lower third of the tibia.

Other factors that influence repair include the age of the person, general state of health, and nutrition. The elderly and those in poor health often have slow healing of fractures. A diet with sufficient calcium, phosphorus, vitamin D, and protein is also important. If any of these nutrients is lacking, bone repair will be a slower process.

The long bones also develop centers of ossification in their epiphyses. At birth, ossification is not yet complete and continues throughout childhood. In long bones, growth occurs in the **epiphyseal discs** at the junction of the diaphysis with each epiphysis. An epiphyseal disc is still cartilage, and the bone grows in length as more cartilage is produced on the epiphysis side (see Fig. 6–3). On the diaphysis side, osteoblasts produce bone matrix to replace the cartilage. Between the ages of 16 and 25 years, all of the cartilage of the ephiphyseal discs is replaced by bone. This is called closure of the epiphyseal discs, and the bone lengthening process stops.

Also in long bones are specialized cells called **osteoclasts**. These cells reabsorb bone matrix in the center of the diaphysis to form the **marrow canal**. Blood vessels grow into the marrow canals of embryonic long bones, and red bone marrow is established. After birth, the red bone marrow is replaced by yellow bone marrow. Red bone marrow remains in the spongy bone of short, flat, and irregular bones. For other functions of osteoclasts and osteoblasts, see Box 6–1: Fractures and Their Repair.

FACTORS THAT AFFECT BONE GROWTH AND MAINTENANCE

1. Heredity—each person has a genetic potential for height, with genes inherited from both parents. There are many genes involved, and their interactions are not well understood. Some of these genes are probably those for the enzymes involved in cartilage and bone production, for this is how bones grow.
2. Nutrition—nutrients are the raw materials of which bones are made. Calcium, phosphorus, and protein become part of the bone matrix itself. Vitamin D is needed for the efficient absorption of calcium and phosphorus by the small intestine. Vitamins A and C do not become part of bone but are necessary for the process of bone matrix formation (ossification).

   Without these and other nutrients, bones cannot grow properly. Children who are malnourished grow very slowly and may not reach their genetic potential for height.
3. Hormones—endocrine glands produce hormones that stimulate specific effects in certain cells. Several hormones have important roles in bone growth and maintenance. These include growth hormone, thyroxine, parathyroid hormone, and insulin, which help regulate cell division, protein synthesis, calcium metabolism, and energy production. The hormones and their specific functions are listed in Table 6–1.
4. Exercise or "Stress"—for bones, exercise means bearing weight, which is just what bones are specialized to do. Without this stress (which is normal), bones will lose calcium faster than it is replaced. Exercise need not be strenuous, it can be as simple as the walking involved in everyday activities. Bones that do not get this exercise,

Table 6–1 HORMONES INVOLVED IN BONE GROWTH AND MAINTENANCE

Hormone (Gland)	Functions
Growth Hormone (anterior pituitary gland)	• Increases the rate of mitosis of chondrocytes and osteoblasts • Increases the rate of protein synthesis (collagen, cartilage matrix, and enzymes for cartilage and bone formation)
Thyroxine (thyroid gland)	• Increases the rate of protein synthesis • Increases energy production from all food types
Insulin (pancreas)	• Increases energy production from glucose
Parathyroid Hormone (parathyroid glands)	• Increases the reabsorption of calcium from bones to the blood (raises blood calcium level) • Increases the absorption of calcium by the small intestine and kidneys (to the blood)
Calcitonin (thyroid gland)	• Decreases the reabsorption of calcium from bones (lowers blood calcium level)
Estrogen (ovaries) or Testosterone (testes)	• Promotes closure of the epiphyses of long bones (growth stops) • Helps retain calcium in bones to maintain a strong bone matrix

Box 6–2 OSTEOPOROSIS

Bone is an active tissue; calcium is constantly being removed to maintain normal blood calcium levels. Usually, however, calcium is replaced in bones at a rate equal to its removal, and the bone matrix remains strong.

Osteoporosis is characterized by excessive loss of calcium from bones without sufficient replacement. Recent research has suggested that a certain gene for bone buildup in youth is an important factor; less buildup would mean earlier bone thinning. Contributing environmental factors include smoking, insufficient dietary intake of calcium, inactivity, and lack of the sex hormones. Osteoporosis is most common among elderly women, because estrogen secretion decreases sharply at menopause (in older men, testosterone is still secreted in significant amounts). Factors such as bed rest or inability to get even minimal exercise will make calcium loss even more rapid.

As bones lose calcium and become thin and brittle, fractures are much more likely to occur. Among elderly women, a fractured hip (the neck of the femur) is an all-too-common consequence of this degenerative bone disorder. Such a serious injury is not inevitable, however, and neither is the thinning of the vertebrae that bows the spines of some elderly people. After menopause, women may wish to have a bone density test to determine the strength of their bone matrix. Several medications are available that diminish the rate of bone loss. A diet high in calcium and vitamin D is essential for both men and women, as is moderate exercise. Also an option for women is estrogen replacement therapy, which slows bone loss. Young women and teenagers should make sure they get adequate dietary calcium to form strong bone matrix, because this will delay the serious effects of osteoporosis later in life.

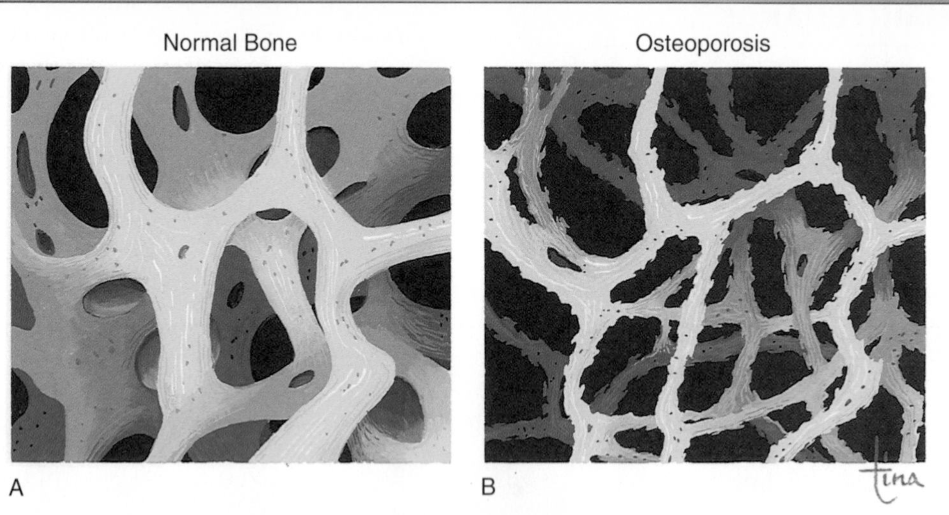

Box Figure 6–B (A) Normal spongy bone, as in the body of a vertebra. **(B)** Spongy bone thinned by osteoporosis.

such as those of bed-ridden patients, will become thinner and more fragile. This condition is discussed further in Box 6-2: Osteoporosis.

THE SKELETON

The human skeleton has two divisions: the **axial skeleton**, which forms the axis, and the **appendicular skeleton**, which supports the appendages or limbs. The axial skeleton consists of the skull, vertebral column, and rib cage. The bones of the arms and legs and the shoulder and pelvic girdles make up the appendicular skeleton. There are 206 bones in total, and the complete skeleton is shown in Fig. 6-4.

SKULL

The **skull** consists of 8 cranial bones and 14 facial bones. Also in the head are three small bones in each middle ear cavity and the hyoid bone that supports the base of the tongue. The **cranial bones** form the braincase that encloses and protects the brain, eyes, and ears. The names of some of these bones will be familiar to you; they are the same as the terminology used (see Chapter 1) to describe areas of the head. These are the **frontal bone, parietal bones** (two), **temporal bones** (two), and **occipital bone**. The **sphenoid bone** and **ethmoid bone** are part of the floor of the braincase and the orbits (sockets) for the eyes. All the joints between cranial bones are immovable joints called **sutures**. It may seem strange to refer to a joint without movement, but the term joint is used for any junction of two bones. The classification of joints will be covered later in this chapter. All the bones of the skull, as well as the large sutures, are shown in Figs. 6-5 through 6-8. Their anatomically important parts are described in Table 6-2.

Of the 14 **facial bones**, only the **mandible** (lower jaw) is movable; it forms a **condyloid joint** with each temporal bone. The other joints between facial bones are all sutures. The **maxillae** are the upper jaw bones, which also form the anterior portion of the hard palate (roof of the mouth). Sockets for the roots of the teeth are found in the maxillae and the mandible. The other facial bones are described in Table 6-2.

Paranasal sinuses are air cavities located in the maxillae, and frontal, sphenoid, and ethmoid bones (Fig. 6-9). As the name "paranasal" suggests, they open into the nasal cavities and are lined with **ciliated epithelium** continuous with the mucosa of the nasal cavities. We are aware of our sinuses only when they become "stuffed up," which means that the mucus they produce cannot drain into the nasal cavities. This may happen during upper respiratory infections such as colds, or with allergies such as hay fever. These sinuses, however, do have functions: They make the skull lighter in weight, because air is lighter than bone, and they provide resonance for the voice.

The **mastoid sinuses** are air cavities in the mastoid process of each temporal bone; they open into the middle ear. Before the availability of antibiotics, middle ear infections often caused mastoiditis, infection of these sinuses.

Within each middle ear cavity are three **auditory bones:** the malleus, incus, and stapes. As part of the hearing process, these bones transmit vibrations from the eardrum to the receptors in the inner ear.

VERTEBRAL COLUMN

The **vertebral column** (spinal column or backbone) is made of individual bones called **vertebrae**. The names of vertebrae indicate their location along the length of the spinal column. There are 7 cervical vertebrae, 12 thoracic, 5 lumbar, 5 sacral fused into 1 sacrum, and 4 to 5 small coccygeal vertebrae fused into 1 coccyx (Fig. 6-10).

The seven **cervical vertebrae** are those within the neck. The first vertebra is called the **atlas**, which supports the skull and forms a **pivot joint** with the odontoid process of the **axis**, the second cervical vertebra. This pivot joint allows us to turn our heads from side to side. The remaining five cervical vertebrae do not have individual names.

The **thoracic vertebrae** articulate (form joints) with the ribs on the posterior side of the trunk. The **lumbar vertebrae**, the largest and strongest bones of the spine, are found in the small of the back. The **sacrum** permits the articulation of the two hipbones: the **sacroiliac joints**. The **coccyx** is the remnant of tail vertebrae, and some muscles of the perineum (pelvic floor) are anchored to it.

All of the vertebrae articulate with one another in sequence to form a flexible backbone that supports the trunk and head. They also form the **vertebral canal**, a continuous tunnel within the bones that contains the spinal cord and protects it from mechanical

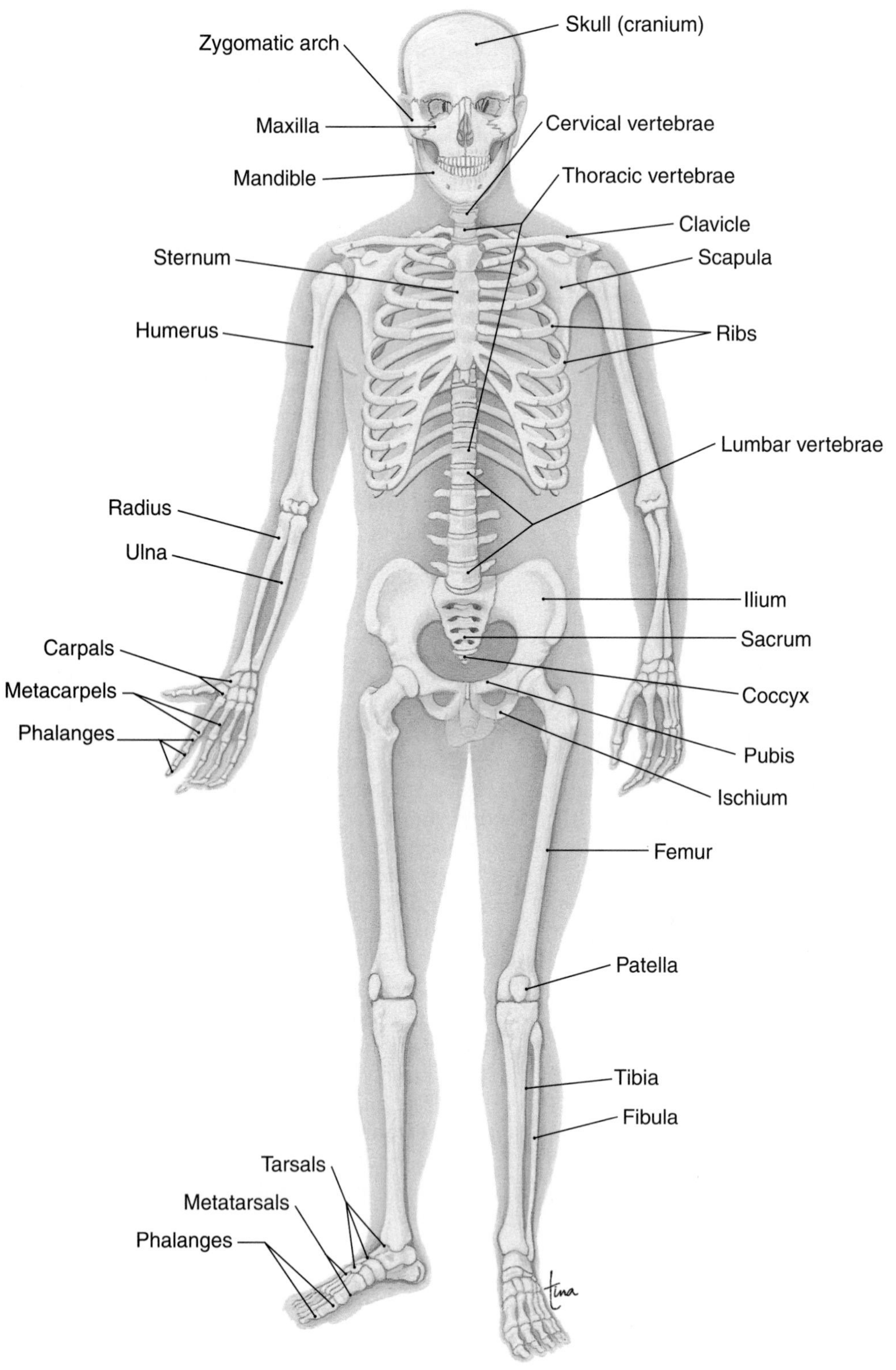

Figure 6–4 Skeleton. Anterior view.

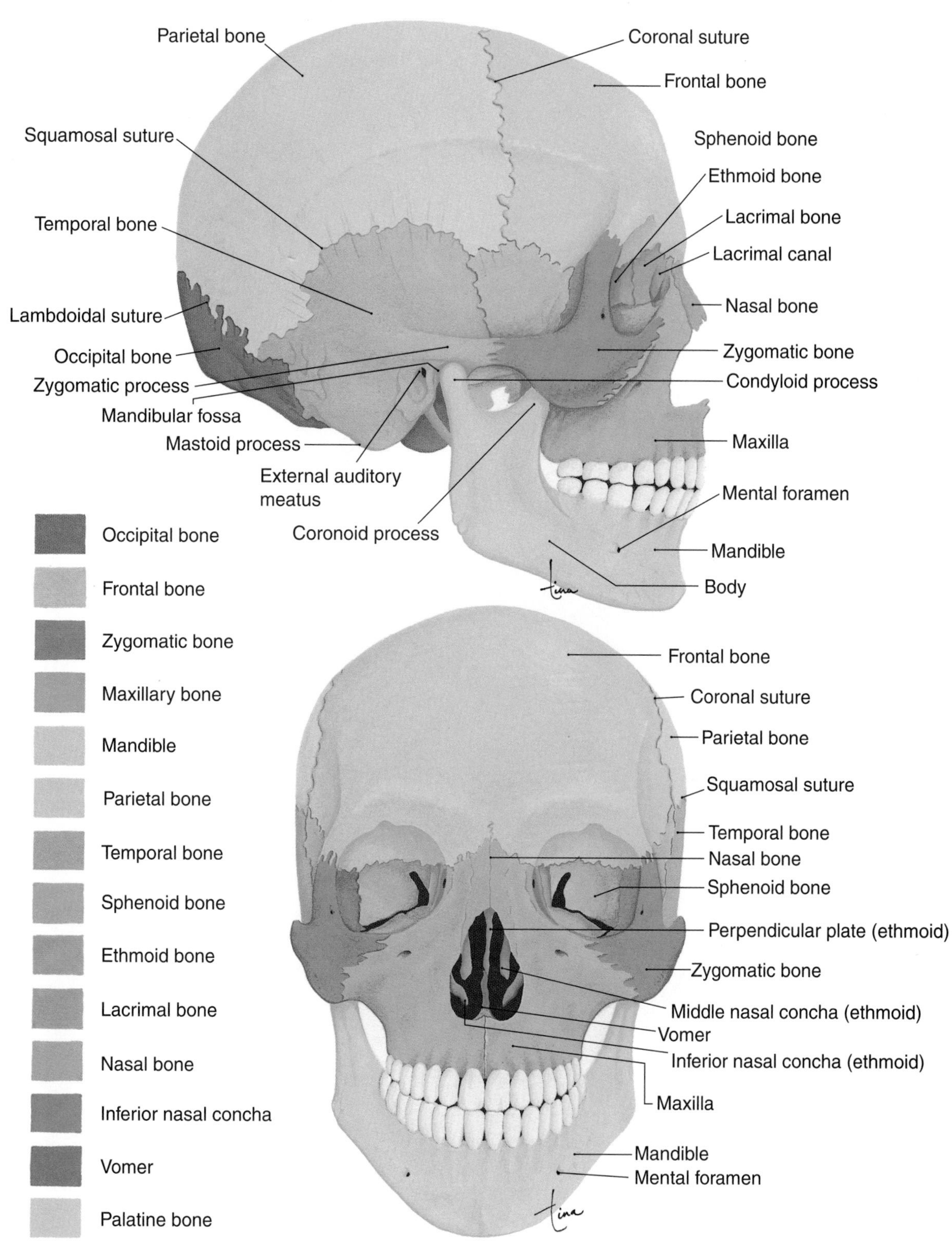

Figure 6–5 Skull. Lateral view of right side.

Figure 6–6 Skull. Anterior view.

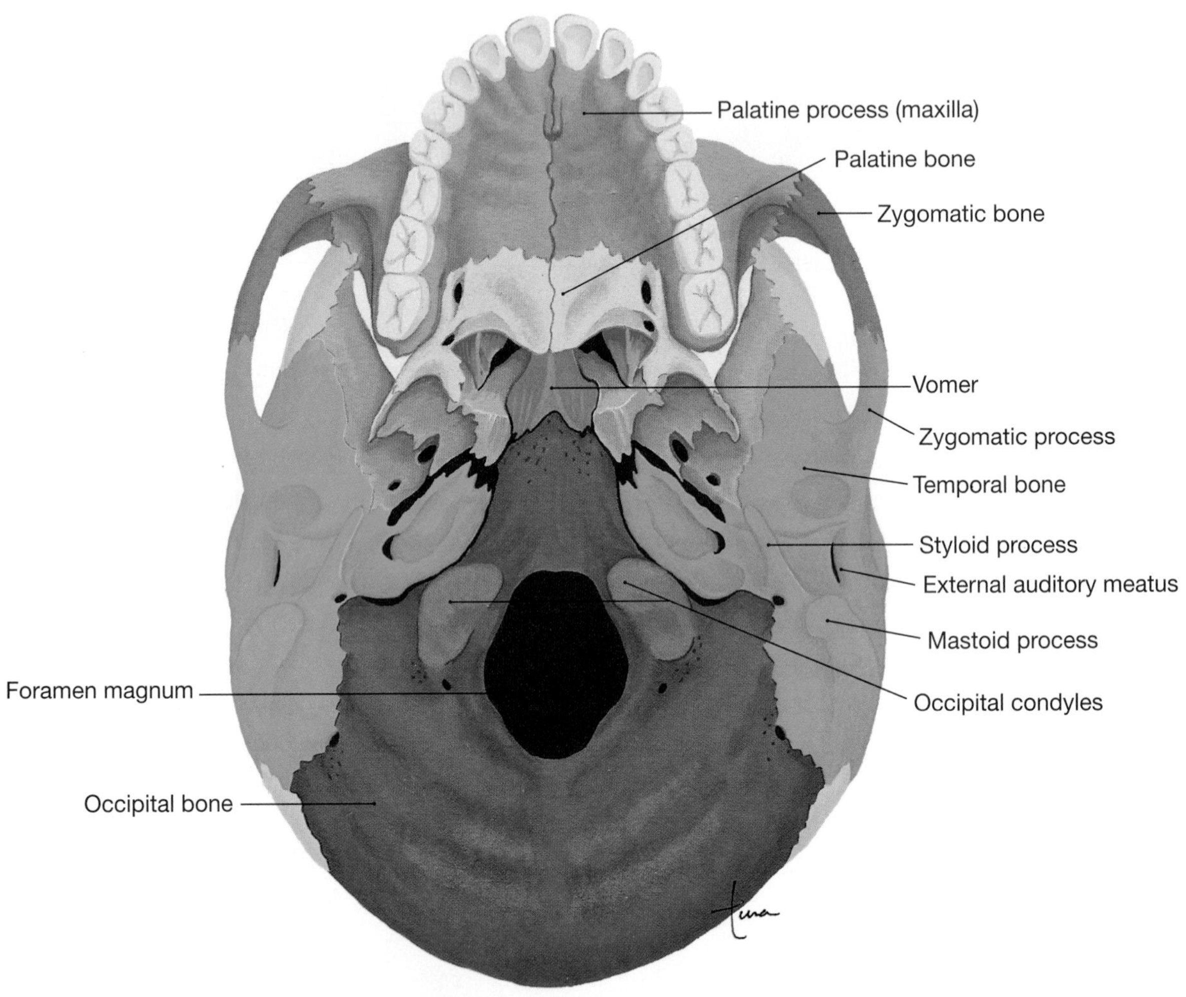

Figure 6–7 Skull. Inferior view with mandible removed.

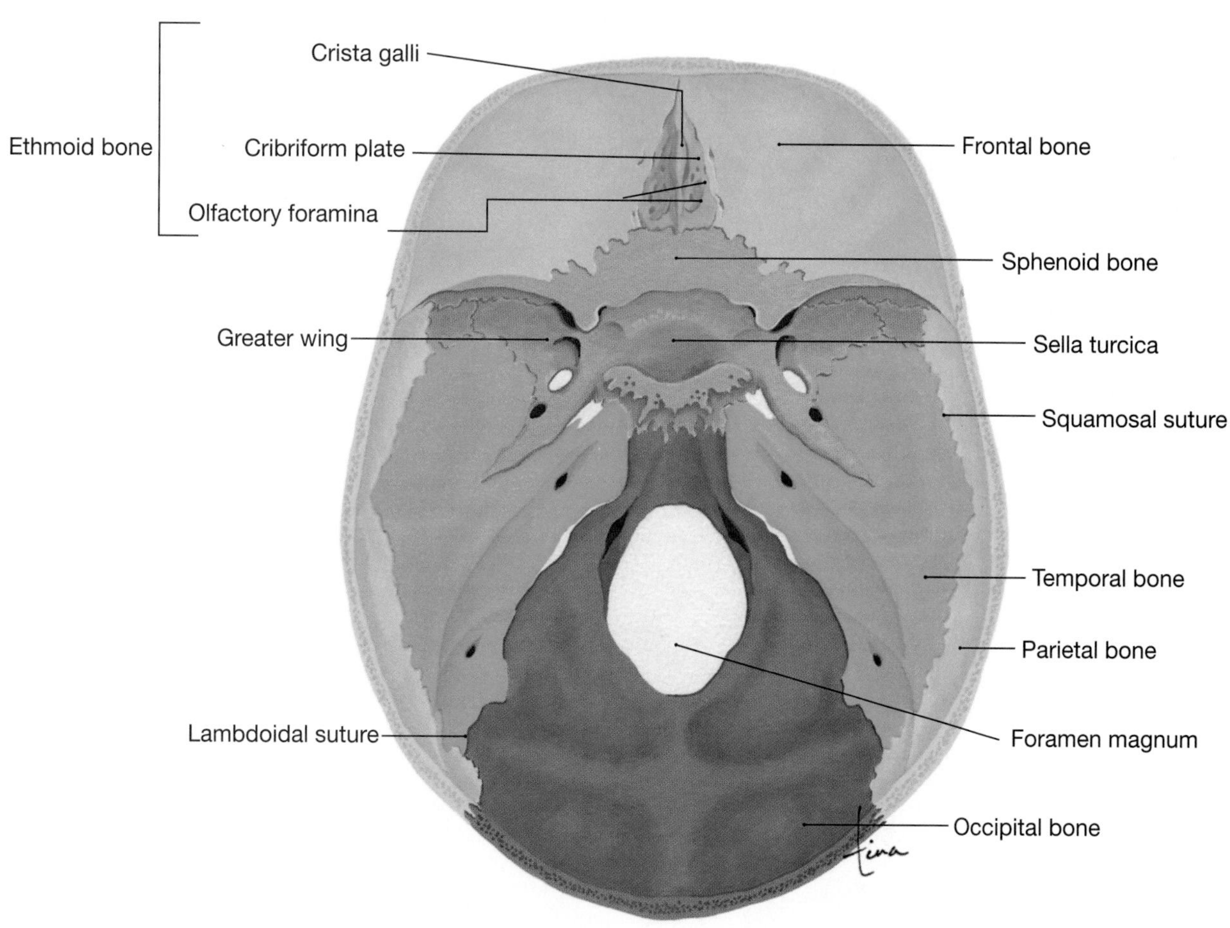

Figure 6–8 Skull. Superior view with the top of cranium removed.

Table 6–2 BONES OF THE SKULL—IMPORTANT PARTS

Terminology of Bone Markings

Foramen—a hole or opening	Meatus—a tunnel-like cavity	Condyle—a rounded projection
Fossa—a depression	Process—a projection	Plate—a flat projection

Bone	Part	Description
Frontal	• Frontal sinus • Coronal suture	• Air cavity that opens into nasal cavity • Joint between frontal and parietal bones
Parietal (2)	• Sagittal suture	• Joint between the 2 parietal bones
Temporal (2)	• Squamosal suture • External auditory meatus • Mastoid process • Mastoid sinus • Mandibular fossa • Zygomatic process	• Joint between temporal and parietal bone • The tunnel-like ear canal • Oval projection behind the ear canal • Air cavity that opens into middle ear • Oval depression anterior to the ear canal; articulates with mandible • Anterior projection that articulates with the zygomatic bone
Occipital	• Foramen magnum • Condyles • Lambdoidal suture	• Large opening for the spinal cord • Oval projections on either side of the foramen magnum; articulate with the atlas • Joint between occipital and parietal bones
Sphenoid	• Greater wing • Sella turcica • Sphenoid sinus	• Flat, lateral portion between the frontal and temporal bones • Central depression that encloses the pituitary gland • Air cavity that opens into nasal cavity
Ethmoid	• Ethmoid sinus • Crista galli • Cribriform plate and Olfactory foramina • Perpendicular plate • Conchae (4 are part of ethmoid; 2 inferior are separate bones)	• Air cavity that opens into nasal cavity • Superior projection for attachment of meninges • On either side of base of crista galli; olfactory nerves pass through foramina • Upper part of nasal septum • Shelf-like projections into nasal cavities which increase surface area of nasal mucosa
Mandible	• Body • Condyles • Sockets	• U-shaped portion with lower teeth • Oval projections that articulate with the temporal bones • Conical depressions that hold roots of lower teeth
Maxilla (2)	• Maxillary sinus • Palatine process • Sockets	• Air cavity that opens into nasal cavity • Projection that forms anterior part of hard palate • Conical depressions that hold roots of upper teeth
Nasal (2)	—	• Forms the bridge of the nose
Lacrimal (2)	Lacrimal canal	• Opening for nasolacrimal duct to take tears to nasal cavity
Zygomatic (2)	—	• Form point of cheek; articulate with frontal, temporal, and maxillae
Palatine (2)	—	• Forms the posterior part of hard palate
Vomer	—	• Lower part of nasal septum

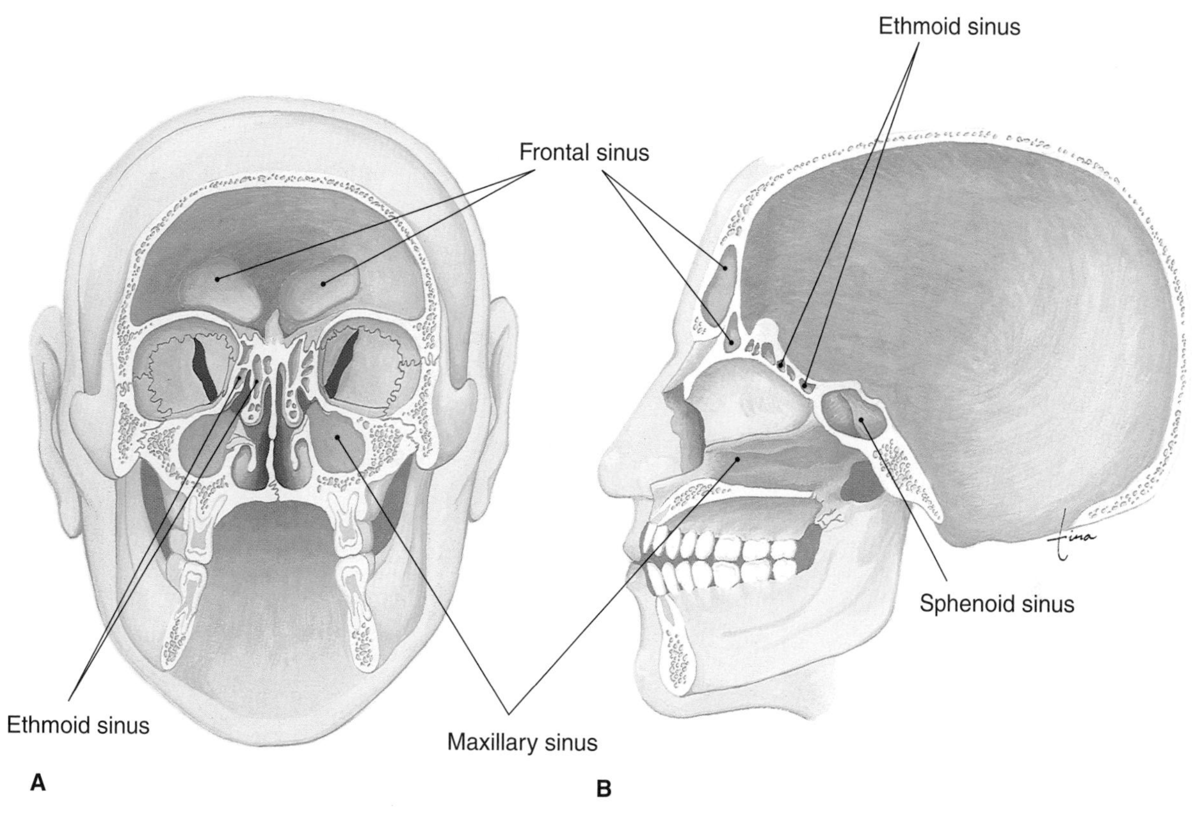

Figure 6–9 Paranasal sinuses. **(A)** Anterior view of skull. **(B)** Left lateral view of skull.

injury. The spinous and transverse processes are projections for the attachment of the muscles that bend the vertebral column.

The supporting part of a vertebra is its body; the bodies of adjacent vertebrae are separated by **discs** of fibrous cartilage. These discs cushion and absorb shock and permit some movement between vertebrae **(symphysis joints)**. Since there are so many joints, the backbone as a whole is quite flexible (see also Box 6-3: Herniated Disc).

The normal spine in anatomical position has four natural curves, which are named after the vertebrae that form them. Refer to Fig. 6-10, and notice that the cervical curve is forward, the thoracic curve backward, the lumbar curve forward, and the sacral curve backward. These curves center the skull over the rest of the body, which enables a person to more easily walk upright (see Box 6-4: Abnormalities of the Curves of the Spine).

RIB CAGE

The **rib cage** consists of the 12 pairs of ribs and the sternum, or breast bone. The three parts of the **sternum** are the upper **manubrium**, the central **body**, and the lower **xiphoid process** (Fig. 6-11).

All the **ribs** articulate posteriorly with the thoracic vertebrae. The first seven pairs of ribs are called **true ribs**; they articulate directly with the manubrium and body of the sternum by means of costal cartilages. The next three pairs are called **false ribs**; their cartilages join the 7th rib cartilage. The last two pairs are called **floating ribs** because they do not articulate with the sternum at all (see Fig. 6-10).

An obvious function of the rib cage is that it encloses and protects the heart and lungs. Keep in mind, though, that the rib cage also protects organs in the upper abdominal cavity, such as the liver and spleen. The other important function of the rib cage depends upon its flexibility: The ribs are pulled upward and

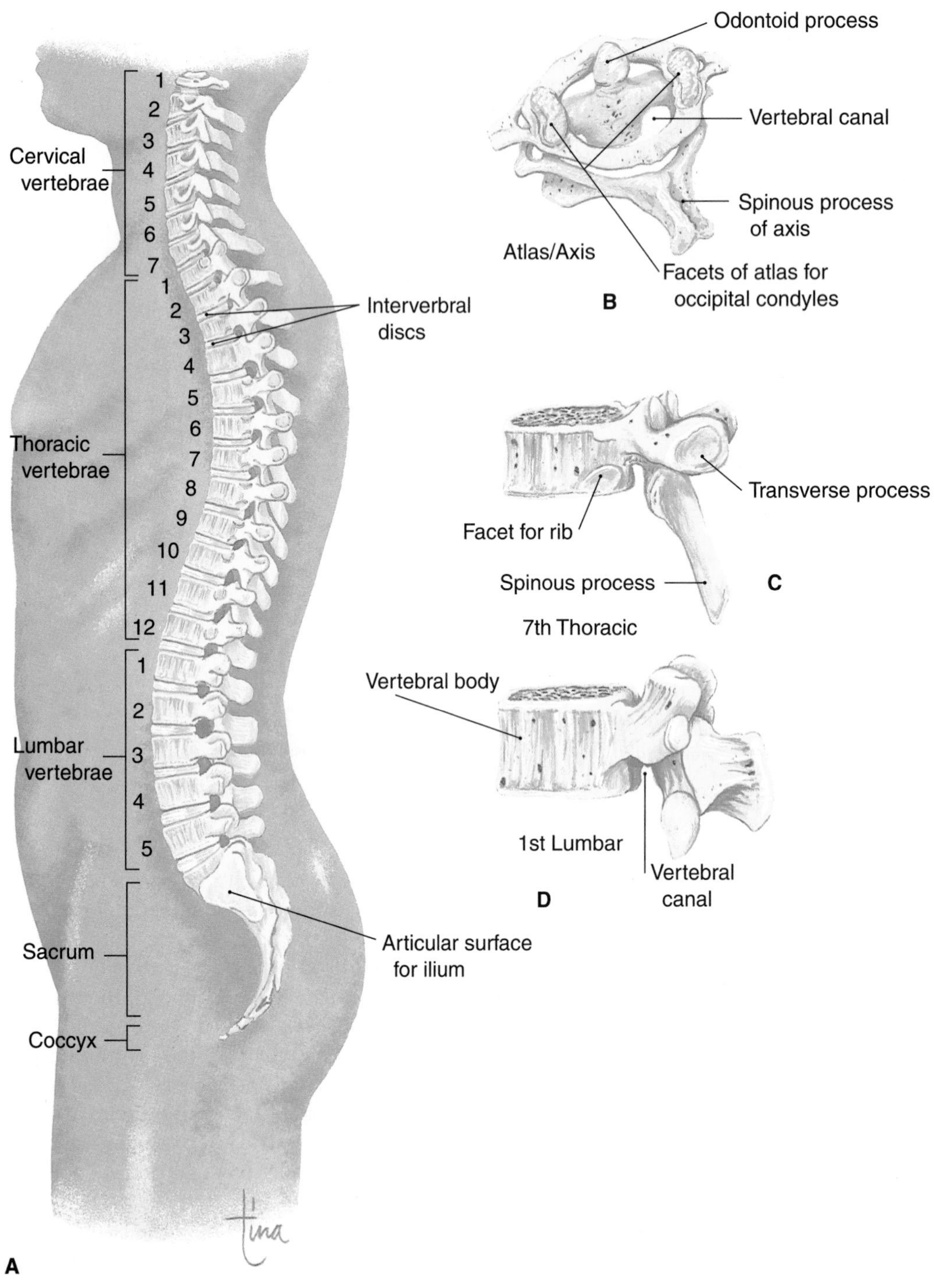

Figure 6–10 Vertebral column. **(A)** Lateral view of left side. **(B)** Atlas and axis, superior view. **(C)** 7th thoracic vertebra, left lateral view. **(D)** 1st lumbar vertebra, left lateral view.

Box 6–3 HERNIATED DISC

The vertebrae are separated by discs of fibrous cartilage that act as cushions to absorb shock. An intervertebral disc has a tough outer covering and a soft center called the nucleus pulposus. Extreme pressure on a disc may rupture the outer layer and force the nucleus pulposus out. This may occur when a person lifts a heavy object improperly, that is, using the back rather than the legs and jerking upward, which puts sudden, intense pressure on the spine. Most often this affects discs in the lumbar region.

Although often called a "slipped disc," the affected disc is usually not moved out of position. The terms **"herniated"** or **"ruptured" disc** more accurately describe what happens. The nucleus pulposus is forced out, usually posteriorly, where it puts pressure on a spinal nerve. For this reason a herniated disc may be very painful or impair function in the muscles supplied by the nerve.

Healing of a herniated disc may occur naturally if the damage is not severe and the person rests and avoids activities that would further compress the disc. Surgery may be required, however, to remove the portion of the nucleus pulposus that is out of place and disrupting nerve functioning.

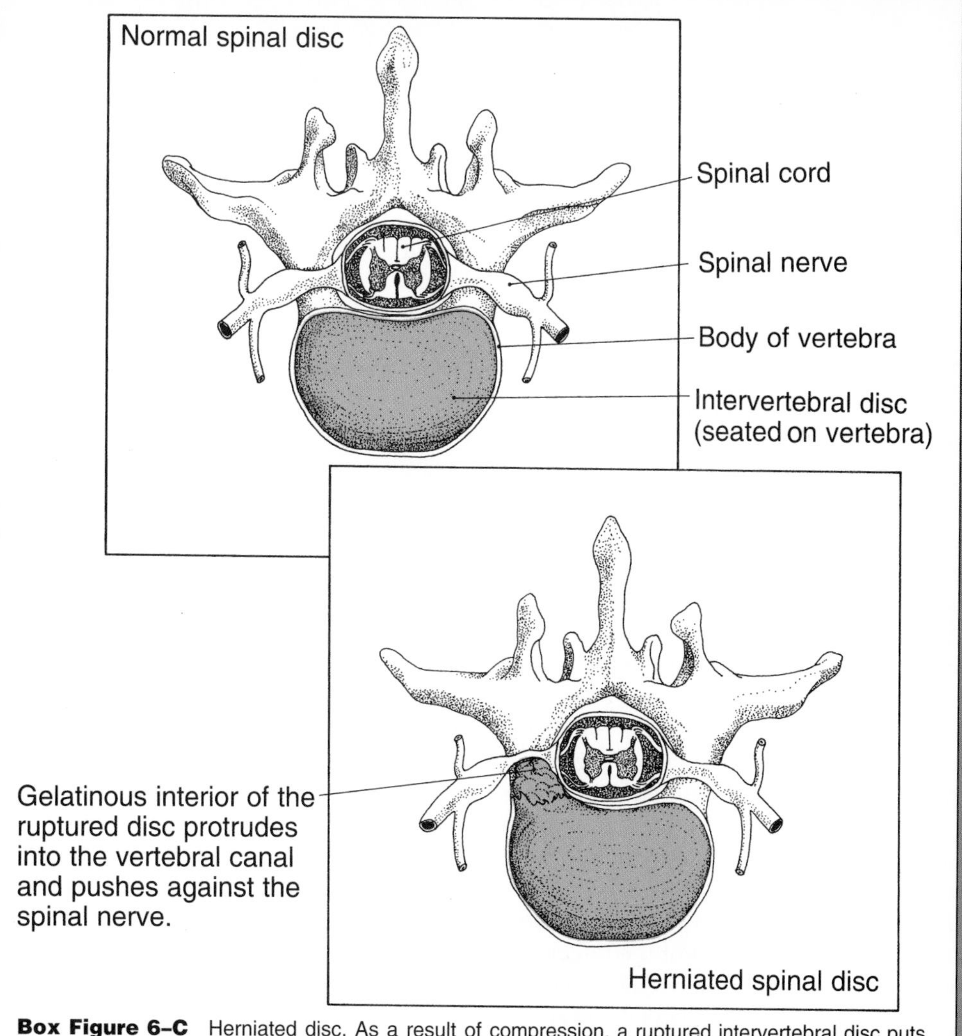

Box Figure 6–C Herniated disc. As a result of compression, a ruptured intervertebral disc puts pressure on a spinal nerve.

Box 6–4 ABNORMALITIES OF THE CURVES OF THE SPINE

Scoliosis—an abnormal lateral curvature, which may be congenital, the result of having one leg longer than the other, or the result of chronic poor posture during childhood while the vertebrae are still growing. Usually the thoracic vertebrae are affected, which displaces the rib cage to one side. In severe cases, the abdominal organs may be compressed, and the expansion of the rib cage during inhalation may be impaired.

Kyphosis*—an exaggerated thoracic curve; sometimes referred to as hunchback.

Lordosis*—an exaggerated lumbar curve; sometimes referred to as swayback.

These abnormal curves are usually the result of degenerative bone diseases such as osteoporosis or tuberculosis of the spine. If osteoporosis, for example, causes the bodies of the thoracic vertebrae to collapse, the normal thoracic curve will be increased. Most often the vertebral body "settles" slowly (rather than collapses suddenly) and there is little, if any, damage to the spinal nerves. The damage to the vertebrae, however, cannot be corrected, so these conditions should be thought of in terms of prevention rather than cure.

*Although descriptive of normal anatomy, the terms kyphosis and lordosis, respectively, are commonly used to describe the abnormal condition associated with each.

outward by the external intercostal muscles. This enlarges the chest cavity, which expands the lungs and contributes to inhalation.

THE SHOULDER AND ARM

The shoulder girdles attach the arms to the axial skeleton. Each consists of a scapula (shoulder blade) and clavicle (collarbone). The **scapula** is a large, flat bone that anchors some of the muscles that move the upper arm. A shallow depression called the glenoid fossa forms a **ball and socket joint** with the humerus, the bone of the upper arm (Fig. 6-12).

Each **clavicle** articulates laterally with a scapula and medially with the manubrium of the sternum. In this position the clavicles act as braces for the scapulae and prevent the shoulders from coming too far forward. Although the shoulder joint is capable of a wide range of movement, the shoulder itself must be relatively stable if these movements are to be effective.

The **humerus** is the long bone of the upper arm. Proximally, the humerus forms a **ball and socket joint** with the scapula. Distally, the humerus forms a **hinge joint** with the ulna of the forearm. This hinge joint, the elbow, permits movement in one plane, that is, back and forth with no lateral movement.

The forearm bones are the **ulna** on the little finger side and the **radius** on the thumb side. The radius and ulna articulate proximally to form a **pivot joint** which permits turning the hand palm up to palm down. You can demonstrate this yourself by holding your arm palm up in front of you, and noting that the radius and ulna are parallel to each other. Then turn your hand palm down, and notice that your upper arm does not move. The radius crosses over the ulna, which permits the hand to perform a great variety of movements without moving the entire arm.

The **carpals** are eight small bones in the wrist; **gliding joints** between them permit a sliding movement. The carpals also articulate with the distal ends of the ulna and radius, and with the proximal ends of the **metacarpals**, the five bones of the hand.

The **phalanges** are the bones of the fingers. There are two phalanges in each thumb and three in each of the fingers. Between phalanges are **hinge joints**, which permit movement in one plane. The thumb, however, is more movable than the fingers because of its carpometacarpal joint. This is a **saddle joint**, which enables the thumb to cross over the palm, and permits gripping. Important parts of these bones are described in Table 6-3.

THE HIP AND LEG

The pelvic girdle consists of the two **hip bones** (coxae or innominate bones), which articulate with

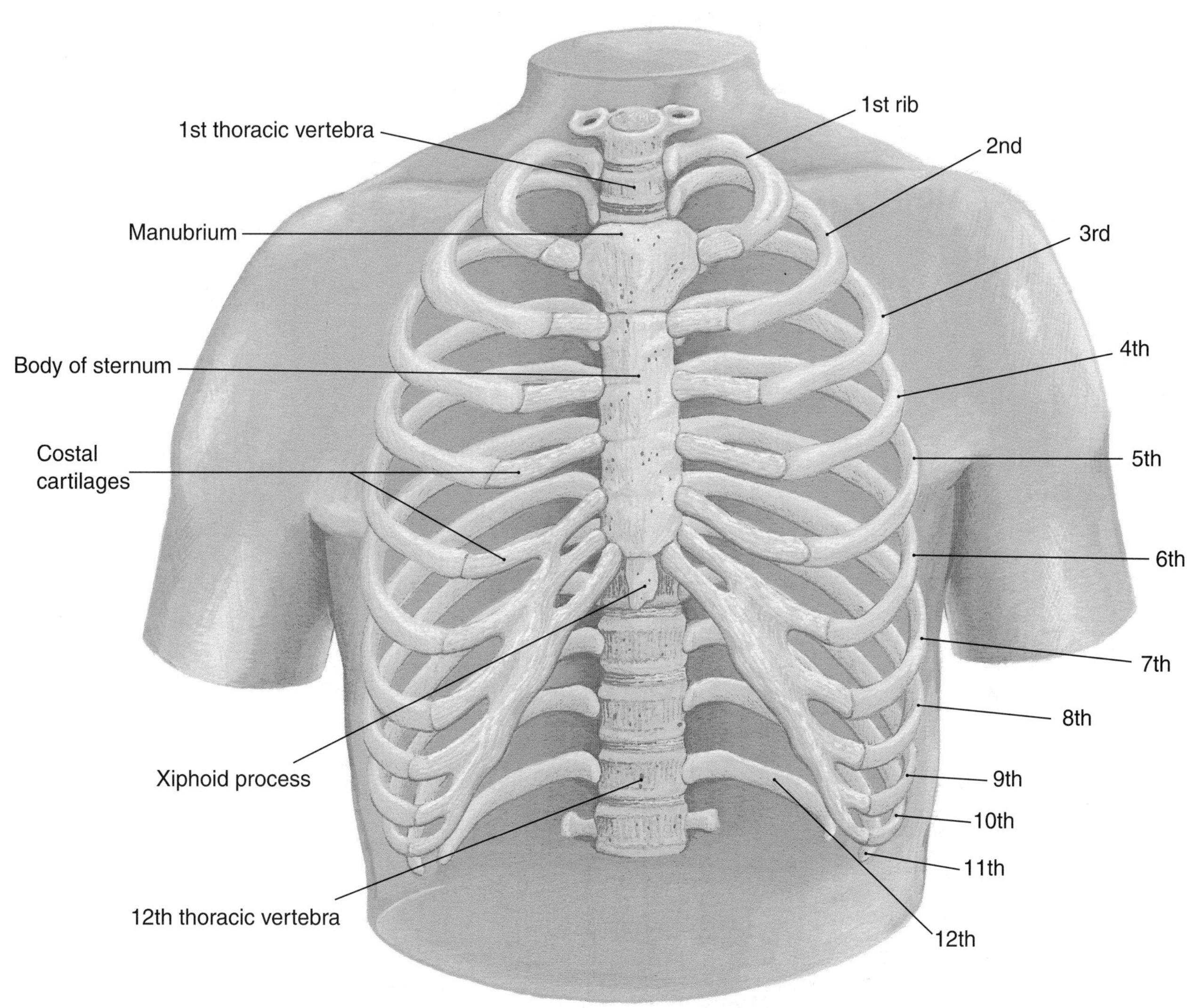

Figure 6–11 Rib cage. Anterior view.

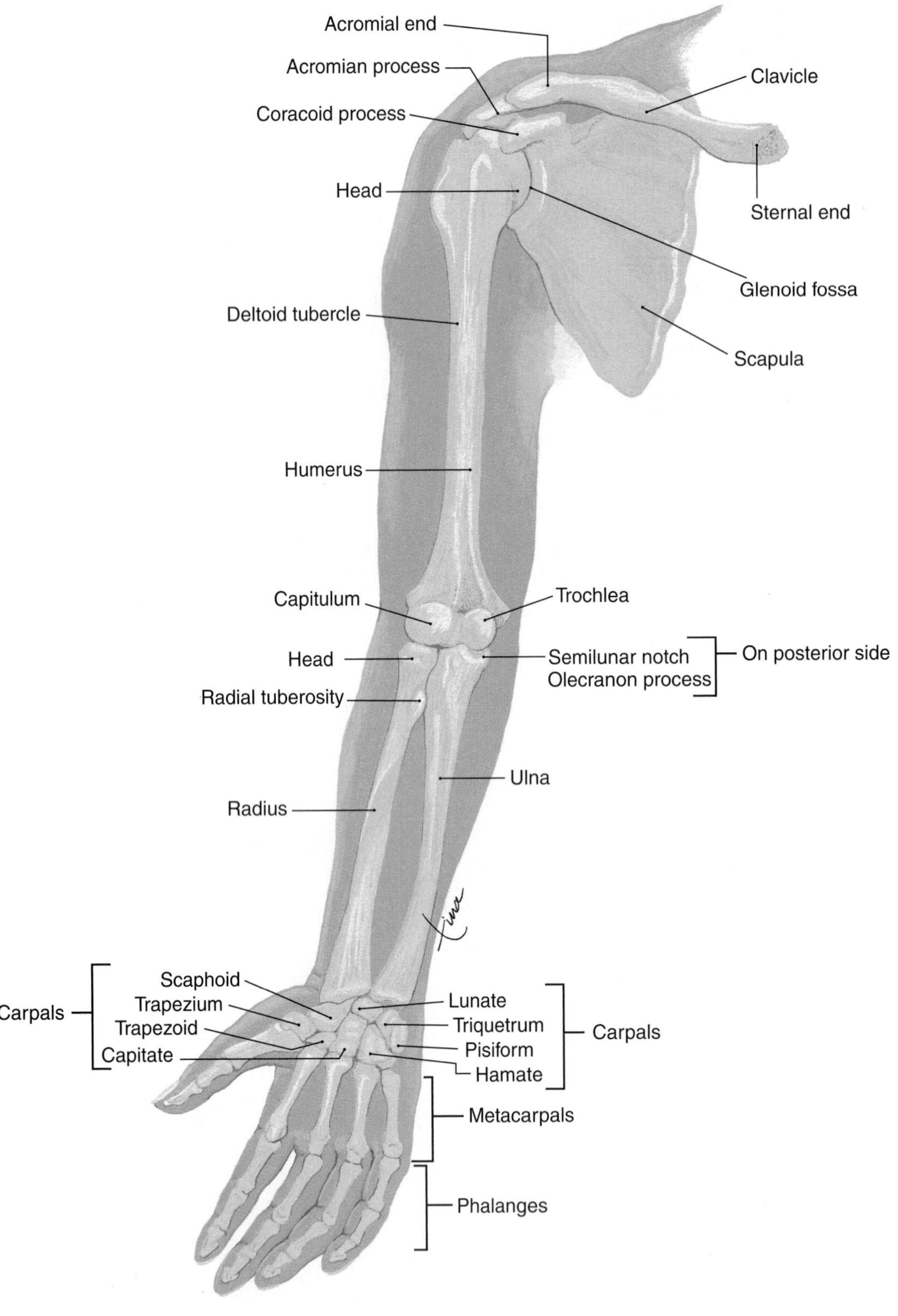

Figure 6–12 Bones of arm and shoulder girdle. Anterior view of right arm.

Table 6–3 BONES OF THE SHOULDER AND ARM—IMPORTANT PARTS

Bone	Part	Description
Scapula	• Glenoid fossa • Spine • Acromion process	• Depression that articulates with humerus • Long, posterior process for muscle attachment • Articulates with clavicle
Clavicle	• Acromial end • Sternal end	• Articulates with scapula • Articulates with manubrium of sternum
Humerus	• Head • Olecranon fossa • Capitulum • Trochlea	• Round process that articulates with scapula • Posterior, oval depression for the olecranon process of the ulna • Round process superior to radius • Concave surface that articulates with ulna
Radius	• Head	• Articulates with the ulna
Ulna	• Olecranon process • Semilunar notch	• Fits into olecranon fossa of humerus • "Half-moon" depression that articulates with the trochlea of ulna
Carpals (8)	• Scaphoid • Lunate • Triquetrum • Pisiform • Trapezium • Trapezoid • Capitate • Hamate	• Proximal row • Distal row

the axial skeleton at the sacrum. Each hip bone has three major parts (Fig. 6–13): the ilium, ischium, and pubis. The **ilium** is the flared, upper portion that forms the sacroiliac joint. The **ischium** is the lower, posterior part that we sit on. The **pubis** is the lower, most anterior part. The two **pubic bones** articulate with one another at the **pubic symphysis**, with a disc of fibrous cartilage between them.

The **acetabulum** is the socket in the hip bone that forms a **ball and socket joint** with the femur. Compared to the glenoid fossa of the scapula, the acetabulum is a much deeper socket. This has great functional importance because the hip is a weight-bearing joint, whereas the shoulder is not. Because the acetabulum is deep, the hip joint is not easily dislocated, even by activities such as running and jumping (landing) which put great stress on the joint.

The **femur** is the long bone of the thigh. As mentioned, the femur forms a very movable ball and socket joint with the hip bone. At its distal end, the femur forms a **hinge joint**, the knee, with the tibia of the lower leg. The **patella**, or knee cap, is anterior to the knee joint, enclosed in the tendon of the quadriceps femoris, a large muscle group of the thigh.

The **tibia** is the weight-bearing bone of the lower leg. Notice in Fig. 6–14 that the **fibula** is not part of the knee joint and does not bear weight. The fibula is important, however, in that leg muscles are attached and anchored to it, and it helps stabilize the ankle. The tibia and fibula do not form a pivot joint as do the radius and ulna in the arm. This makes the lower leg and foot more stable, and thus able to support the body.

The **tarsals** are the seven bones in the ankle. The largest is the **calcaneus**, or heel bone; the **talus** transmits weight between the calcaneus and the tibia. **Metatarsals** are the five long bones of each foot, and **phalanges** are the bones of the toes. There are two phalanges in the big toe and three in each of the other toes. The phalanges of the toes form hinge joints with each other. Because there is no saddle joint in the foot, the big toe is not as movable as is the thumb. Important parts of these bones are described in Table 6–4.

JOINTS—ARTICULATIONS

A joint is where two bones meet, or **articulate**.

THE CLASSIFICATION OF JOINTS

The classification of joints is based on the amount of movement possible. A **synarthrosis** is an immovable joint, such as a suture between two cranial bones. An **amphiarthrosis** is a slightly movable joint, such as

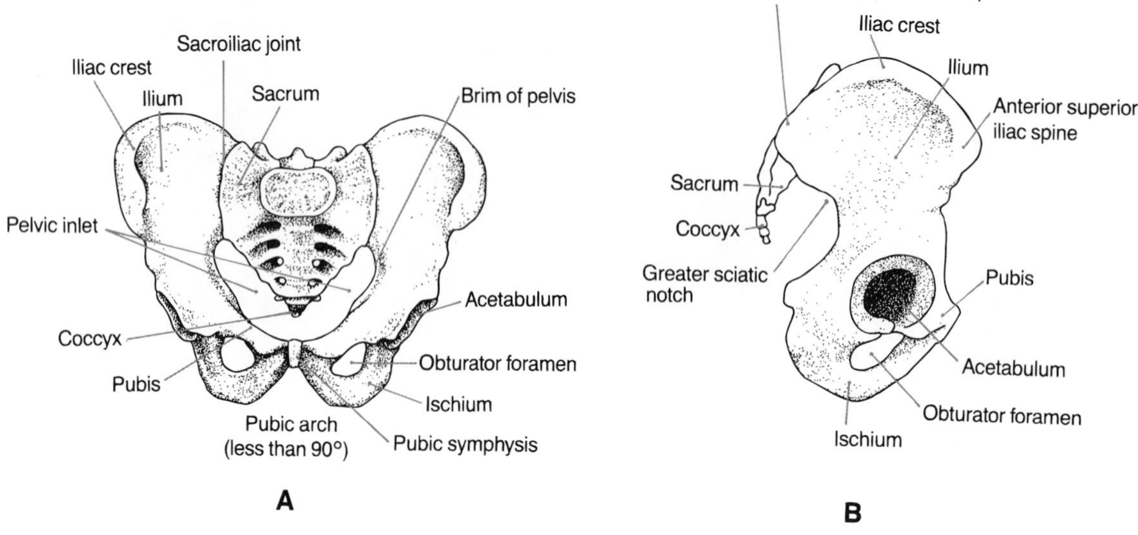

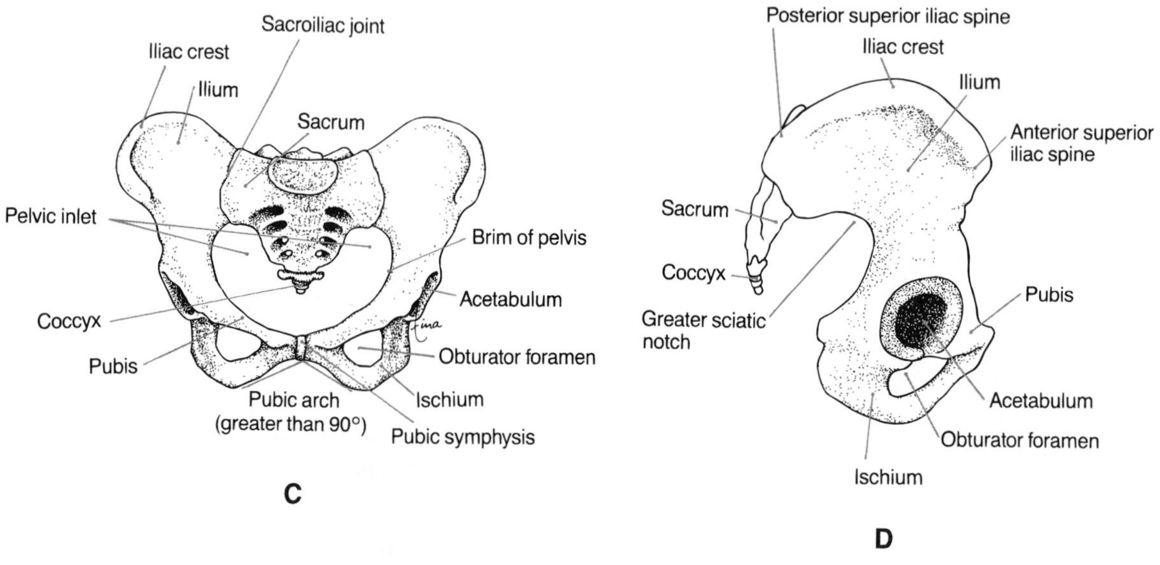

Figure 6–13 Hip bones and sacrum. **(A)** Male pelvis, anterior view. **(B)** Male pelvis, lateral view of right side. **(C)** Female pelvis, anterior view. **(D)** Female pelvis, lateral view of right side.

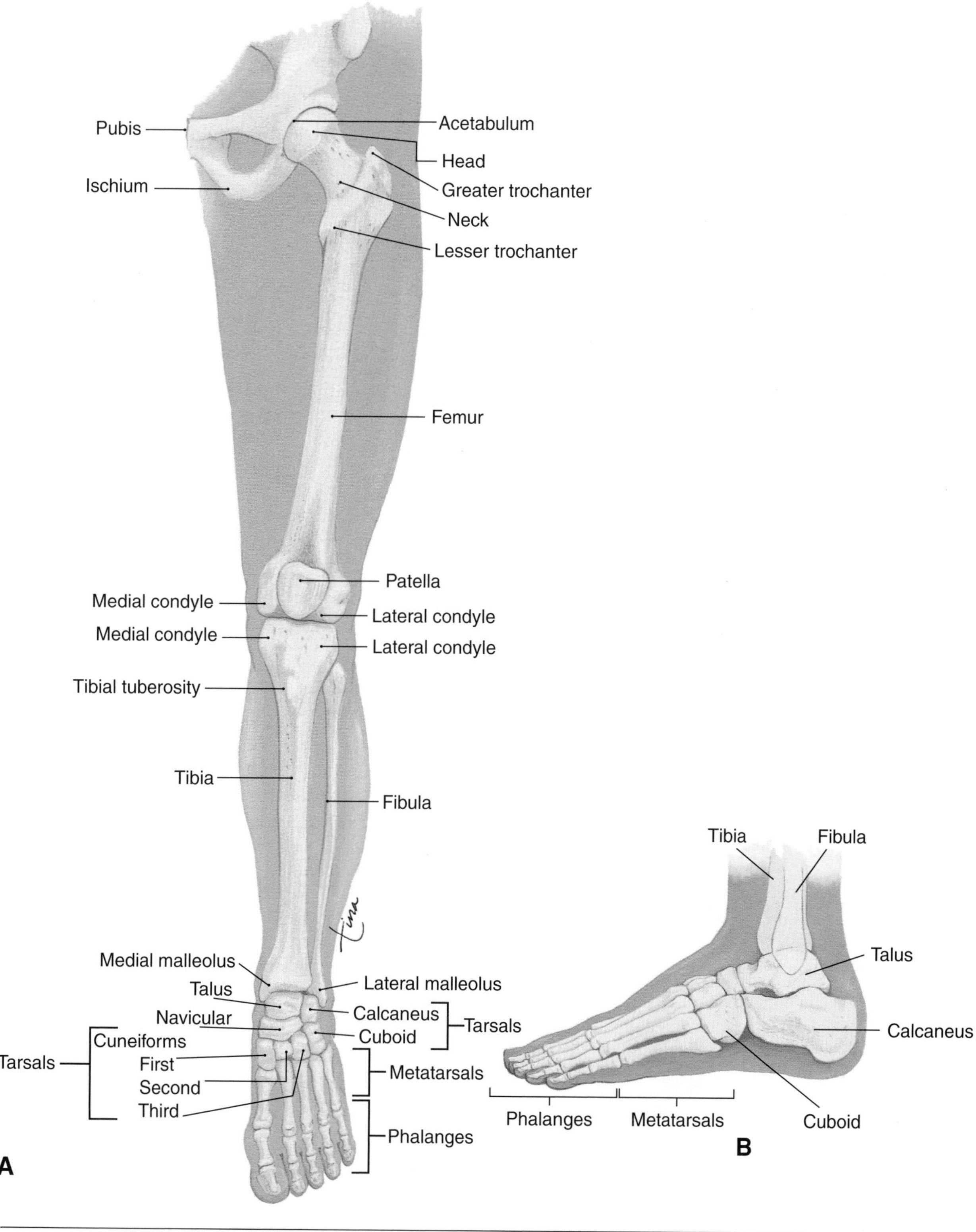

Figure 6–14 **(A)** Bones of the leg and portion of hip bone, anterior view of left leg. **(B)** Lateral view of left foot.

Table 6–4 BONES OF THE HIP AND LEG—IMPORTANT PARTS

Bone	Part	Description
Pelvic (2 hip bones)	• Ilium	Flared, upper portion
	• Iliac crest	Upper edge of ilium
	• Posterior superior iliac spine	Posterior continuation of iliac crest
	• Ischium	Lower, posterior portion
	• Pubis	Anterior, medial portion
	• Pubic symphysis	Joint between the 2 pubic bones
	• Acetabulum	Deep depression that articulates with femur
Femur	• Head	Round process that articulates with hip bone
	• Neck	Constricted portion distal to head
	• Greater trochanter	Large lateral process for muscle attachment
	• Lesser trochanter	Medial process for muscle attachment
	• Condyles	Rounded processes that articulate with tibia
Tibia	• Condyles	Articulate with the femur
	• Medial malleolus	Distal process; medial "ankle bone"
Fibula	• Head	Articulates with tibia
	• Lateral malleolus	Distal process; lateral "ankle bone"
Tarsals (7)	• Calcaneus	Heel bone
	• Talus	Articulates with calcaneus and tibia
	• Cuboid, Navicular	—
	• Cuneiform: 1st, 2nd, 3rd	—

the symphysis joint between adjacent vertebrae. A **diarthrosis** is a freely movable joint. This is the largest category of joints and includes the ball and socket joint, the pivot, hinge, and others. Examples of each type of joint are described in Table 6–5, and many of these are illustrated in Fig. 6–15.

SYNOVIAL JOINTS

All diarthroses, or freely movable joints, are **synovial joints** because they share similarities of structure. A typical synovial joint is shown in Fig. 6–16. On the joint surface of each bone is the **articular cartilage**, which provides a smooth surface. The **joint capsule**,

Table 6–5 TYPES OF JOINTS

Category	Type and Description	Examples
Synarthrosis (immovable)	Suture—fibrous connective tissue between bone surfaces	• Between cranial bones; between facial bones
Amphiarthrosis (slightly movable)	Symphysis—disc of fibrous cartilage between bones	• Between vertebrae; between pubic bones
Diarthrosis (freely movable)	Ball and socket—movement in all planes	• Scapula and humerus; pelvic bone and femur
	Hinge—movement in one plane	• Humerus and ulna; femur and tibia; between phalanges
	Condyloid—movement in one plane with some lateral movement	• Temporal bone and mandible
	Pivot—rotation	• Atlas and axis; radius and ulna
	Gliding—side to side movement	• Between carpals
	Saddle—movement in several planes	• Carpometacarpal of thumb

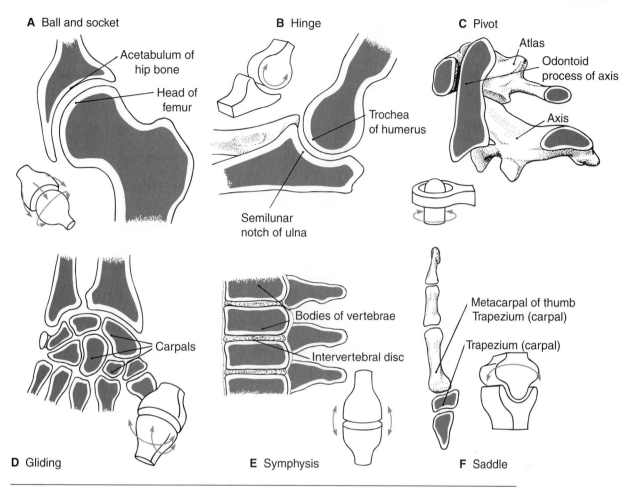

A Ball and socket
Acetabulum of hip bone
Head of femur

B Hinge
Trochea of humerus
Semilunar notch of ulna

C Pivot
Atlas
Odontoid process of axis
Axis

Carpals
D Gliding

Bodies of vertebrae
Intervertebral disc
E Symphysis

Metacarpal of thumb
Trapezium (carpal)
Trapezium (carpal)
F Saddle

Figure 6–15 Types of joints. For each type, a specific joint is depicted, and a simple diagram shows the position of the joint surfaces. **(A)** Ball and socket. **(B)** Hinge. **(C)** Pivot. **(D)** Gliding. **(E)** Symphysis. **(F)** Saddle.

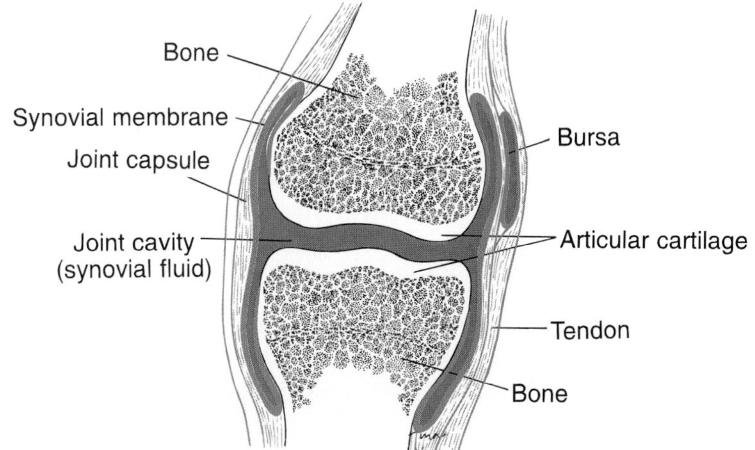

Figure 6–16 Structure of a synovial joint.

Box 6–5 ARTHRITIS

The term **arthritis** means inflammation of a joint. Of the many types of arthritis, we will consider two: osteoarthritis and rheumatoid arthritis.

Osteoarthritis is a natural consequence of getting older. In joints that have borne weight for many years, the articular cartilage is gradually worn away. The once smooth joint surface becomes rough, and the affected joint is stiff and painful. As you might guess, the large, weight-bearing joints are most often subjected to this form of arthritis. If we live long enough, most of us can expect some osteoarthritis in knees, hips, or ankles.

Rheumatoid arthritis can be a truly crippling disease that may begin in early middle age or, less commonly, during adolescence. It is believed to be an **autoimmune disease,** which means that the immune system mistakenly directs its destructive capability against part of the body. Exactly what triggers this abnormal response by the immune system is not known with certainty, but certain bacterial and viral infections have been suggested as possibilities.

Rheumatoid arthritis often begins in joints of the extremities, such as those of the fingers. The autoimmune activity seems to affect the synovial membrane, and joints become painful and stiff. Sometimes the disease progresses to total destruction of the synovial membrane and calcification of the joint. Such a joint is then fused and has no mobility at all.

Treatment of rheumatoid arthritis is directed at reducing inflammation as much as possible, for it is the inflammatory process that causes the damage. Therapies being investigated involve selectively blocking specific aspects of the immune response, and limiting this blockage only to joints. At present there is no cure for autoimmune diseases.

made of fibrous connective tissue, encloses the joint in a strong sheath, like a sleeve. Lining the joint capsule is the **synovial membrane**, which secretes synovial fluid into the joint cavity. **Synovial fluid** is thick and slippery and prevents friction as the bones move.

Many synovial joints also have **bursae** (or bursas), which are small sacs of synovial fluid between the joint and the tendons that cross over the joint. Bursae permit the tendons to slide easily as the bones are moved. If a joint is used excessively, the bursae may become inflamed and painful; this condition is called **bursitis**. Some other disorders of joints are described in Box 6–5: Arthritis.

AGING AND THE SKELETAL SYSTEM

With age, bone tissue tends to lose more calcium than is replaced. Bone matrix becomes thinner, the bones themselves more brittle, and spontaneous fractures are more likely to occur.

Erosion of the articular cartilages of joints is also a common consequence of aging. Joints affected include weight-bearing joints such as the knees, and active, small joints such as those of the fingers.

Although the normal wear and tear of joints cannot be prevented, elderly people can preserve their bone matrix with exercise and diets high in calcium and vitamin D.

SUMMARY

Your knowledge of the bones and joints will be useful in the next chapter as you learn the actions of the muscles that move the skeleton. It is important to remember, however, that bones have other functions as well. As a storage site for excess calcium, bones contribute to the maintenance of a normal blood calcium level. The red bone marrow found in flat and irregular bones produces the blood cells: red blood cells, white blood cells, and platelets. Some bones protect vital organs such as the brain, heart, and lungs. As you can see, bones themselves may also be considered vital organs.

STUDY OUTLINE

The skeleton is made of bone and cartilage and has these functions:

1. Is a framework for support, moved by muscles.
2. Protects internal organs from mechanical injury.
3. Contains and protects red bone marrow.
4. Stores excess calcium; important to regulate blood calcium level.

Bone Tissue (see Fig. 6–1)

1. Osteocytes (cells) are found in the matrix of calcium phosphate, calcium carbonate, and collagen.
2. Compact Bone—haversian systems are present.
3. Spongy Bone—no haversian systems; red bone marrow present.
4. Articular Cartilage—smooth, on joint surfaces.
5. Periosteum—fibrous connective tissue membrane; anchors tendons and ligaments; has blood vessels that enter the bone.

Classification of Bones

1. Long—arms, legs; shaft is the diaphysis (compact bone) with a marrow cavity containing yellow bone marrow (fat); ends are epiphyses (spongy bone) (see Fig. 6-1).
2. Short—wrists, ankles (spongy bone covered with compact bone).
3. Flat—ribs, pelvic bone, cranial bones (spongy bone covered with compact bone).
4. Irregular—vertebrae, facial bones (spongy bone covered with compact bone).

Embryonic Growth of Bone

1. The embryonic skeleton is first made of other tissues that are gradually replaced by bone. Ossification begins in the third month of gestation; osteoblasts differentiate from fibroblasts and produce bone matrix.
2. Cranial and facial bones are first made of fibrous connective tissue; osteoblasts produce bone matrix in a center of ossification in each bone; bone growth radiates outward; fontanels remain at birth, permit compression of infant skull during birth; fontanels are calcified by age 2 (see Fig. 6-2).
3. All other bones are first made of cartilage; in a long bone the first center of ossification is in the diaphysis, other centers develop in the epiphyses. After birth a long bone grows at the epiphyseal discs: cartilage is produced on the epiphysis side, and bone replaces cartilage on the diaphysis side. Osteoclasts form the marrow cavity by reabsorbing bone matrix in the center of the diaphysis (see Fig. 6-3).

Factors That Affect Bone Growth and Maintenance

1. Heredity—many pairs of genes contribute to genetic potential for height.
2. Nutrition—calcium, phosphorus, and protein become part of the bone matrix; vitamin D is needed for absorption of calcium in the small intestine; vitamins C and A are needed for bone matrix production (calcification).
3. Hormones—produced by endocrine glands; concerned with cell division, protein synthesis, calcium metabolism, and energy production (see Table 6-1).
4. Exercise or Stress—weight-bearing bones must bear weight or they will lose calcium and become brittle.

The Skeleton—206 bones in total (see Fig. 6–4)

1. Axial—skull, vertebrae, rib cage.
   - Skull—see Figs. 6-5 through 6-8 and Table 6-2.
     - Eight cranial bones form the braincase, which also protects the eyes and ears; 14 facial bones make up the face; the immovable joints between these bones are called sutures.
     - Paranasal sinuses are air cavities in the maxillae, frontal, sphenoid, and ethmoid bones; lighten the skull and provide resonance for voice (see Fig. 6-9).
     - Three auditory bones in each middle ear cavity transmit vibrations for the hearing process.
   - Vertebral Column—see Fig. 6-10.
     - Individual bones are called vertebrae: 7 cervical, 12 thoracic, 5 lumbar, 5 sacral (fused into one sacrum), 4 to 5 coccygeal (fused into one coccyx). Supports trunk and head, encloses and protects the spinal cord in the vertebral canal. Discs of fibrous cartilage absorb shock between the bodies of adjacent vertebrae, also permit slight movement. Four natural curves center head over body for walking upright (see Table 6-5 for joints).

- Rib Cage—see Fig. 6-11.
  - Sternum and 12 pairs of ribs; protects thoracic and upper abdominal organs from mechanical injury and is expanded to contribute to inhalation. Sternum consists of manubrium, body, and xiphoid process. All ribs articulate with thoracic vertebrae; true ribs (first seven pairs) articulate directly with sternum by means of costal cartilages; false ribs (next three pairs) articulate with 7th costal cartilage; floating ribs (last two pairs) do not articulate with the sternum.

2. Appendicular—bones of the arms and legs and the shoulder and pelvic girdles.
   - Shoulder and Arm—see Fig. 6-12 and Table 6-3.
     - Scapula—shoulder muscles are attached; glenoid fossa articulates with humerus.
     - Clavicle—braces the scapula.
     - Humerus—upper arm; articulates with the scapula and the ulna (elbow).
     - Radius and ulna—forearm—articulate with one another and with carpals.
     - Carpals—eight—wrist; Metacarpals—five—hand; Phalanges—14—fingers (for joints, see Table 6-5).
   - Hip and Leg—see Figs. 6-13 and 6-14 and Table 6-4.
     - Pelvic bone—two hip bones; ilium, ischium, pubis; acetabulum articulates with femur.
     - Femur—thigh; articulates with pelvic bone and tibia (knee).
     - Patella—kneecap; in tendon of quadriceps femoris muscle.
     - Tibia and fibula—lower leg; tibia bears weight; fibula does not bear weight, but does anchor muscles and stabilizes ankle.
     - Tarsals—seven—ankle; calcaneus is heel bone.
     - Metatarsals—five—foot; phalanges—14—toes (see Table 6-5 for joints).

Joints—articulations

1. Classification based on amount of movement:
   - Synarthrosis—immovable.
   - Amphiarthrosis—slightly movable.
   - Diarthrosis—freely movable (see Table 6-5 for examples; see also Fig. 6-15).
2. Synovial Joints—all diarthroses have similar structure (see Fig. 6-16):
   - Articular Cartilage—smooth on joint surfaces.
   - Joint Capsule—strong fibrous connective tissue sheath that encloses the joint.
   - Synovial Membrane—lines the joint capsule; secretes synovial fluid that prevents friction.
   - Bursae—sacs of synovial fluid that permit tendons to slide easily across joints.

REVIEW QUESTIONS

1. Explain the differences between compact bone and spongy bone, and state where each type is found. (p. 98)
2. State the locations of red bone marrow, and name the blood cells it produces. (p. 98)
3. Name the tissue of which the embryonic skull is first made. Explain how ossification of cranial bones occurs. (p. 100)
4. State what fontanels are, and explain their function. (p. 100)
5. Name the tissue of which the embryonic femur is first made. Explain how ossification of this bone occurs. Describe what happens in epiphyseal discs to produce growth of long bones. (p. 103)
6. Explain what is meant by "genetic potential" for height, and name the nutrients a child must have in order to attain genetic potential. (p. 103)
7. Explain the functions of calcitonin and parathyroid hormone with respect to bone matrix and to blood calcium level. (p. 103)

8. Explain how estrogen or testosterone affects bone growth, and when. (p. 103)
9. State one way each of the following hormones helps promote bone growth: insulin, thyroxine, growth hormone. (p. 103)
10. Name the bones that make up the braincase. (p. 105)
11. Name the bones that contain paranasal sinuses and explain the functions of these sinuses. (p. 105)
12. Name the bones that make up the rib cage, and describe two functions of the rib cage. (p. 111)
13. Describe the functions of the vertebral column. State the number of each type of vertebra. (pp. 105, 111)
14. Explain how the shoulder and hip joints are similar and how they differ. (pp. 114, 117)
15. Give a specific example (name two bones) for each of the following types of joints: (p. 120)
    a. Hinge
    b. Symphysis
    c. Pivot
    d. Saddle
    e. Suture
    f. Ball and socket
16. Name the part of a synovial joint with each of the following functions: (pp. 120, 122)
    a. Fluid within the joint cavity that prevents friction
    b. Encloses the joint in a strong sheath
    c. Provides a smooth surface on bone surfaces
    d. Lines the joint capsule and secretes synovial fluid
17. Refer to the diagram (Fig. 6–4) of the full skeleton, and point to each bone on yourself. (p. 106)

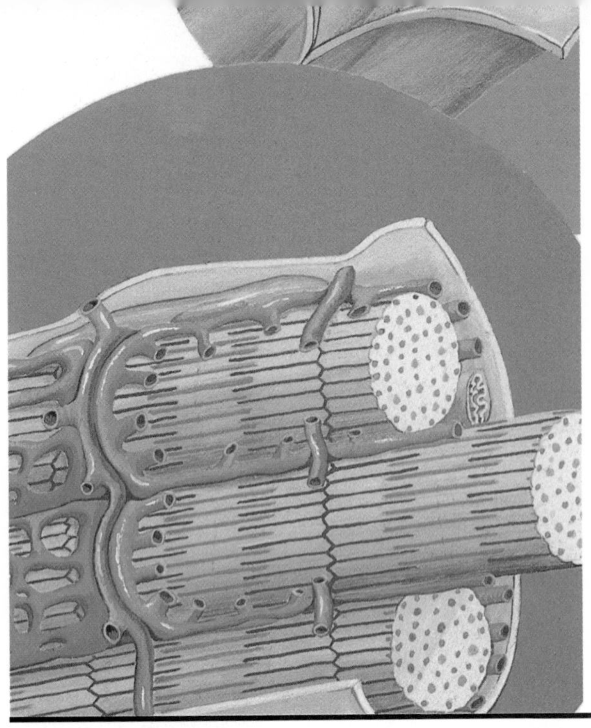

CHAPTER 7

Chapter Outline

Student Objectives

- Name the organ systems directly involved in movement, and state how they are involved.
- Describe muscle structure in terms of muscle cells, tendons, and bones.
- Describe the difference between antagonistic and synergistic muscles, and explain why such arrangements are necessary.
- Explain the role of the brain with respect to skeletal muscle.
- Define muscle tone and explain its importance.
- Explain the difference between isotonic and isometric exercise.
- Define muscle sense and explain its importance.
- Name the energy sources for muscle contraction, and state the simple equation for cell respiration.
- Explain the importance of hemoglobin and myoglobin, oxygen debt, and lactic acid.
- Describe the neuromuscular junction and state the function of each part.
- Describe the structure of a sarcomere.
- Explain in terms of ions and charges: polarization, depolarization, repolarization.
- Describe the Sliding Filament Theory of muscle contraction.
- Describe some of the body's responses to exercise and explain how each maintains homeostasis.
- Learn the major muscles of the body and their functions.

The Muscular System

New Terminology

Actin (**AK**-tin)
Antagonistic muscles (an-**TAG**-on-ISS-tik **MUSS**-uhls)
Creatine phosphate (**KREE**-ah-tin **FOSS**-fate)
Depolarization (DE-poh-lahr-i-**ZAY**-shun)
Fascia (**FASH**-ee-ah)
Insertion (in-**SIR**-shun)
Isometric (EYE-so-**MEH**-trik)
Isotonic (EYE-so-**TAHN**-ik)
Lactic acid (**LAK**-tik **ASS**-id)
Muscle fatigue (**MUSS**-uhl fah-**TEEG**)
Muscle sense (**MUSS**-uhl SENSE)
Muscle tone (**MUSS**-uhl TONE)
Myoglobin (**MYE**-oh-GLOW-bin)
Myosin (**MYE**-oh-sin)
Neuromuscular junction (NYOOR-oh-**MUSS**-kyoo-ler **JUNK**-shun)
Origin (**AHR**-i-jin)
Oxygen debt (**OKS**-ah-jen DET)
Polarization (POH-lahr-i-**ZAY**-shun)
Prime mover (PRIME **MOO**-ver)
Sarcolemma (SAR-koh-**LEM**-ah)
Sarcomeres (**SAR**-koh-meers)
Synergistic muscles (**SIN**-er-JIS-tik **MUSS**-uhls)
Tendon (**TEN**-dun)

Related Clinical Terminology

Anabolic steroids (an-a-**BOLL**-ik **STEER**-oyds)
Atrophy (**AT**-ruh-fee)
Botulism (**BOTT**-yoo-lizm)
Hypertrophy (high-**PER**-truh-fee)
Intramuscular injection (IN-trah-**MUSS**-kyoo-ler in-**JEK**-shun)
Muscular dystrophy (**MUSS**-kyoo-ler **DIS**-truh-fee)
Myalgia (my-**AL**-jee-ah)
Myasthenia gravis (MY-ass-**THEE**-nee-yuh **GRAH**-viss)
Myopathy (my-**AH**-puh-thee)
Paralysis (pah-**RAL**-i-sis)
Range-of-motion exercises (RANJE-of-**MOH**-shun **EKS**-err-sigh-zez)
Sex-linked trait (SEX LINKED **TRAYT**)
Tetanus (**TET**-uh-nus)

*Terms that appear in **bold type** in the chapter text are defined in the glossary, which begins on page 528.*

Do you like to dance? Most of us do, or, we may simply enjoy watching good dancers. The grace and coordination involved in dancing result from the interaction of many of the organ systems, but the one you think of first is probably the muscular system.

There are more than 600 muscles in the human body. Most of these muscles are attached to the bones of the skeleton by tendons, although a few muscles are attached to the undersurface of the skin. The primary function of the **muscular system** is to move the skeleton. The other body systems directly involved in movement are the nervous, respiratory, and circulatory systems. The nervous system transmits the electrochemical impulses that cause muscle cells to contract. The respiratory system exchanges oxygen and carbon dioxide between the air and blood. The circulatory system brings oxygen to the muscles and takes carbon dioxide away.

These interactions of body systems are covered in this chapter, which focuses on the **skeletal muscles**. You may recall from Chapter 4 that there are two other types of muscle tissue: smooth muscle and cardiac muscle. These types of muscle tissue will be discussed in other chapters in relation to the organs of which they are part. Before you continue, you may find it helpful to go back to Chapter 4 and review the structure and characteristics of skeletal muscle tissue. In this chapter we will begin with the gross (large) anatomy and physiology of muscles, then discuss the microscopic structure of muscle cells and the biochemistry of muscle contraction.

MUSCLE STRUCTURE

All muscle cells are specialized for contraction. When these cells contract, they shorten and pull a bone in order to produce movement. Each skeletal muscle is made of thousands of individual muscle cells, which also may be called **muscle fibers** (see Fig. 7-3). Depending on the work a muscle is required to do, variable numbers of muscle fibers contract. When picking up a pencil, for example, only a small portion of the muscle fibers in a muscle will contract. If the muscle has more work to do, such as picking up a book, more muscle fibers will contract to accomplish the task.

Muscles are anchored firmly to bones by **tendons**. Tendons are made of fibrous connective tissue, which, you may remember, is very strong and merges with the **fascia** that covers the muscle and with the **periosteum**, the fibrous connective tissue membrane that covers bones. A muscle usually has at least two tendons, each attached to a different bone. The more immobile or stationary attachment of the muscle is its **origin**; the more movable attachment is called the **insertion**. The muscle itself crosses the joint of the two bones to which it is attached, and when the muscle contracts it pulls on its insertion and moves the bone in a specific direction.

MUSCLE ARRANGEMENTS

Muscles are arranged so as to bring about a variety of movements. The two general types of arrangements are the opposing **antagonists** and the cooperative **synergists**.

Antagonistic Muscles

Antagonists are opponents, so we use the term **antagonistic muscles** for muscles that have opposing or opposite functions. An example will be helpful here—refer to Fig. 7-1 as you read the following. The biceps brachii is the muscle on the front of the upper arm. The origin of the biceps is on the scapula (there are actually two tendons, hence the name "biceps"), and the insertion is on the radius. When the biceps contracts it **flexes** the forearm, that is, bends the elbow (see Table 7-2). Recall that when a muscle contracts it gets shorter and pulls. Muscles cannot push, for when they relax they exert no force. Therefore, the biceps can bend the elbow but cannot straighten it; another muscle is needed. The triceps brachii is located on the back of the upper arm. Its origins (the prefix "tri" tells you that there are three of them) are on the scapula and humerus, and its insertion is on the ulna. When the triceps contracts and pulls, it **extends** the forearm, that is, straightens the elbow.

Joints that are capable of a variety of movements have several sets of antagonists. Notice how many ways you can move your upper arm at the shoulder, for instance. Abducting (laterally raising) the arm is the function of the deltoid. Adducting the arm is brought about by the pectoralis major and latissimus dorsi. Flexion of the arm (across the chest) is also a function of the pectoralis major, and extension of the arm (behind the back) is also a function of the lattissimus dorsi. All of these muscles are described and depicted in the tables and figures later in the chapter. Without antagonistic muscles, this variety of movements would be impossible.

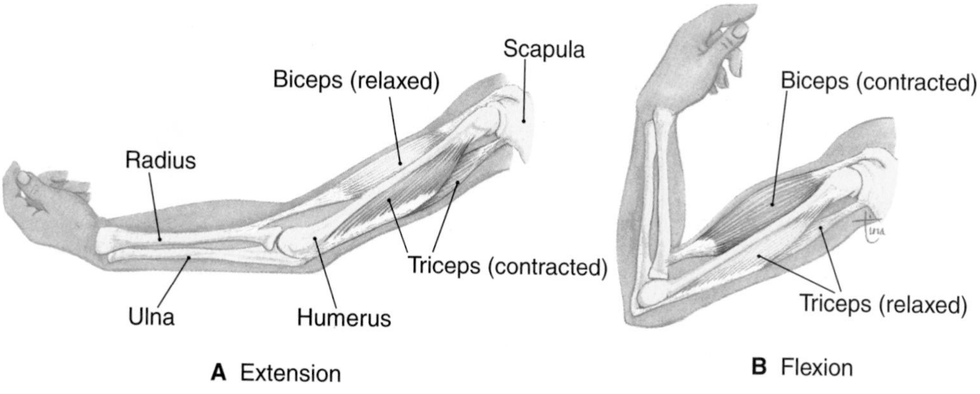

Figure 7–1 Antagonistic muscles. **(A)** Extension of the forearm. **(B)** Flexion of the forearm.

You may be familiar with **range-of-motion** or ROM exercises that are often recommended for bedridden patients. Such exercises are designed to stretch and contract the antagonistic muscles of a joint to preserve as much muscle function and joint mobility as possible.

Synergistic Muscles

Synergistic muscles are those with the same function, or those that work together to perform a particular function. Recall that the biceps brachii flexes the forearm. The brachioradialis, with its origin on the humerus and insertion on the radius, also flexes the forearm. There is even a third flexor of the forearm, the brachialis. You may wonder why we need three muscles to perform the same function, and the explanation lies in the great mobility of the hand. If the hand is palm up, the biceps does most of the work of flexing and may be called the **prime mover**. When the hand is thumb up, the brachioradialis is in position to be the prime mover, and when the hand is palm down, the brachialis becomes the prime mover. If you have ever tried to do chin-ups, you know that it is much easier with your palms toward you than with palms away from you. This is because the biceps is a larger, and usually much stronger, muscle than is the brachialis.

Muscles may also be called synergists if they help to stabilize or steady a joint to make a more precise movement possible. If you drink a glass of water, the biceps brachii may be the prime mover to flex the forearm. At the same time, the muscles of the shoulder keep that joint stable, so that the water gets to your mouth, not over your shoulder or down your chin. The shoulder muscles are considered synergists for this movement because their contribution makes the movement effective.

THE ROLE OF THE BRAIN

Even our simplest movements require the interaction of many muscles, and the contraction of skeletal muscles depends on the brain. The nerve impulses for movement come from the **frontal lobes** of the **cerebrum**. The cerebrum is the largest part of the brain; the frontal lobes are beneath the frontal bone. The **motor areas** of the frontal lobes generate electrochemical impulses that travel along motor nerves to muscle fibers, causing the muscle fibers to contract.

For a movement to be effective, some muscles must contract while others relax. This is what we call coordination, and it is regulated by the **cerebellum**, which is located below the occipital lobes of the cerebrum.

MUSCLE TONE

Except during certain stages of sleep, most of our muscles are in a state of slight contraction; this is what is known as **muscle tone**. When sitting upright, for example, the tone of your neck muscles keeps your head up, and the tone of your back muscles keeps your back straight. This is an important function of muscle tone for human beings, because it helps us to maintain an upright posture. In order for a muscle to remain

Box 7–1 ANABOLIC STEROIDS

Anabolic steroids are synthetic drugs very similar in structure and action to the male hormone **testosterone**. Normal secretion of testosterone, beginning in males at puberty, increases muscle size and is the reason men usually have larger muscles than do women.

Some athletes, both male and female, both amateur and professional, take anabolic steroids to build muscle mass and to increase muscle strength. There is no doubt that the use of anabolic steroids will increase muscle size, but there are hazards, some of them very serious. Side effects of such self-medication include liver damage, kidney damage, disruption of reproductive cycles, and mental changes such as irritability and aggressiveness.

Women athletes may develop increased growth of facial and body hair and may become sterile as a result of the effects of a male hormone on their own hormonal cycles.

slightly contracted, only a few of the muscle fibers in that muscle must contract. Alternate fibers contract so that the muscle as a whole does not become fatigued. This is similar to a pianist continuously rippling her fingers over the keys of the piano—some notes are always sounding at any given moment, but the notes that are sounding are always changing.

Muscle fibers need the energy of ATP in order to contract. When they produce ATP in the process of cell respiration, muscle fibers also produce heat. The heat generated by normal muscle tone is approximately 25% of the total body heat at rest. During exercise, of course, heat production increases significantly.

EXERCISE

Good muscle tone improves coordination. When muscles are slightly contracted, they can react more rapidly if and when greater exertion is necessary. Muscles with poor tone are usually soft and flabby, but exercise will improve muscle tone.

There are two general types of exercise: isotonic and isometric. In **isotonic exercise**, muscles contract and bring about movement. Jogging, swimming, and weight lifting are examples. Isotonic exercise improves muscle tone, muscle strength, and if done repetitively against great resistance (as in weight lifting), muscle size. This type of exercise also improves cardiovascular and respiratory efficiency, because movement exerts demands on the heart and respiratory muscles. If done for 30 minutes or longer, such exercise may be called "aerobic," because it strengthens the heart and respiratory muscles as well as the skeletal muscles.

Isometric exercise involves contraction without movement. If you put your palms together and push one hand against the other, you can feel your arm muscles contracting. If both hands push equally, there will be no movement; this is isometric contraction. Such exercises will increase muscle tone and muscle strength but are not considered aerobic. Without movement, heart rate and breathing do not increase nearly as much as they would during an equally strenuous isotonic exercise.

Many of our actions involve both isotonic and isometric contractions. Pulling open a door requires isotonic contractions of arm muscles, but if the door is then held open for someone else, those contractions become isometric. Picking up your books is isotonic; holding them in your arm is isometric. Both kinds of contractions are needed for even the simplest activities. (With respect to increasing muscle strength, see Box 7-1: Anabolic Steroids.)

MUSCLE SENSE

When you walk up a flight of stairs, do you have to look at your feet to be sure each will get to the next step? Most of us don't (an occasional stumble doesn't count), and for this freedom we can thank our muscle sense. **Muscle sense** is the brain's ability to know where our muscles are and what they are doing, without our having to consciously look at them.

Within muscles are receptors called **stretch receptors** (proprioceptors or muscle spindles). The general function of all sensory receptors is to detect changes. The function of stretch receptors is to detect changes in the length of a muscle as it is stretched. The sensory impulses generated by these receptors are interpreted by the brain as a mental "picture" of where the muscle is.

We can be aware of muscle sense if we choose to be, but usually we can safely take it for granted. In fact, that is what we are meant to do. Imagine what life would be like if we had to watch every move to be sure that a hand or foot performed its intended action. Even simple activities such as walking or eating would require our constant attention.

There are times when we may become aware of our muscle sense. Learning a skill such as typing or playing the guitar involves very precise movements of the fingers, and beginners will often watch their fingers to be sure they are moving properly. With practice, however, muscle sense again becomes unconscious, and the experienced typist or guitarist need not watch every movement.

All sensation is a function of brain activity, and muscle sense is no exception. The impulses for muscle sense are integrated in the **parietal lobes** of the cerebrum (conscious muscle sense) and in the cerebellum (unconscious muscle sense) to be used to promote coordination.

ENERGY SOURCES FOR MUSCLE CONTRACTION

Before discussing the contraction process itself, let us look first at how muscle fibers obtain the energy they need to contract. The direct source of energy for muscle contraction is **ATP**. ATP, however, is not stored in large amounts in muscle fibers and is depleted in a few seconds.

The secondary energy sources are creatine phosphate and glycogen. **Creatine phosphate** is, like ATP, an energy-transferring molecule. When it is broken down (by an enzyme) to creatine, phosphate, and energy, the energy is used to synthesize more ATP. Most of the creatine formed is used to resynthesize creatine phosphate, but some is converted to **creatinine**, a waste product that is excreted by the kidneys.

The most abundant energy source in muscle fibers is **glycogen**. When glycogen is needed to provide energy for sustained contractions (more than a few seconds), it is first broken down into the **glucose** molecules of which it is made. Glucose is then further broken down in the process of cell respiration to produce ATP, and muscle fibers may continue to contract.

Recall from Chapter 2 our simple reaction for cell respiration:

$$\text{Glucose} + O_2 \rightarrow CO_2 + H_2O + \text{ATP} + \text{heat}$$

Look first at the products of this reaction. ATP will be used by the muscle fibers for contraction. The heat produced will contribute to body temperature, and if exercise is strenuous, will increase body temperature. The water becomes part of intracellular water, and the carbon dioxide is a waste product that will be exhaled.

Now look at what is needed to release energy from glucose: oxygen. Muscles have two sources of oxygen. The blood delivers a continuous supply of oxygen, which is carried by the **hemoglobin** in red blood cells. Within muscle fibers themselves there is another protein called **myoglobin**, which stores some oxygen within the muscle cells. Both hemoglobin and myoglobin contain the mineral iron, which enables them to bond to oxygen. (Iron also makes both molecules red, and it is myoglobin that gives muscle tissue a red or dark color.)

During strenuous exercise, the oxygen stored in myoglobin is quickly used up, and normal circulation may not deliver oxygen fast enough to permit the completion of cell respiration. Even though the respiratory rate increases, the muscle fibers may literally run out of oxygen. This state is called **oxygen debt**, and in this case, glucose cannot be completely broken down into carbon dioxide and water. If oxygen is not present (or not present in sufficient amounts), glucose is converted to an intermediate molecule called **lactic acid**, which causes **muscle fatigue**.

In a state of fatigue, muscle fibers cannot contract efficiently, and contraction may become painful. To be in oxygen debt means that we owe the body some oxygen. Lactic acid from muscles enters the blood and circulates to the liver, where it is converted back into glucose. This conversion requires ATP, and oxygen is needed to produce the necessary ATP in the liver. This is why, after strenuous exercise, the respiratory rate and heart rate remain high for a time and only gradually return to normal.

MUSCLE FIBER—MICROSCOPIC STRUCTURE

We will now look more closely at a muscle fiber, keeping in mind that there are thousands of these cylindrical cells in one muscle. Each muscle fiber has its own motor nerve ending; the **neuromuscular junction** is where the motor neuron terminates on the muscle fiber (Fig. 7–2). The **axon terminal** is the enlarged tip of the motor neuron; it contains sacs of the neurotransmitter **acetylcholine** (ACh). The membrane of the muscle fiber is the **sarcolemma**, which contains receptor sites for acetylcholine, and an in-

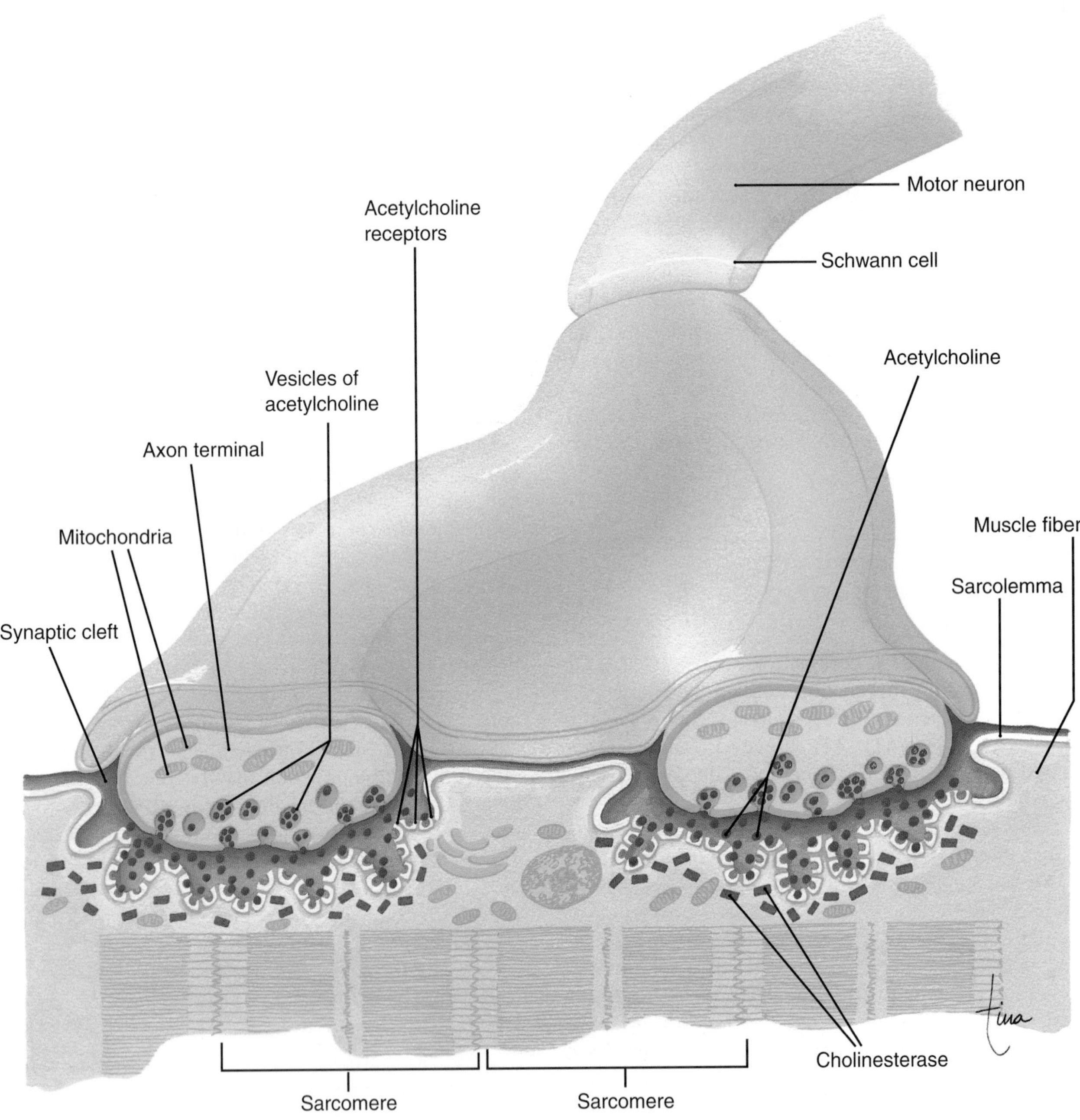

Figure 7–2 Structure of the neuromuscular junction, showing an axon terminal adjacent to the sarcolemma of a muscle fiber.

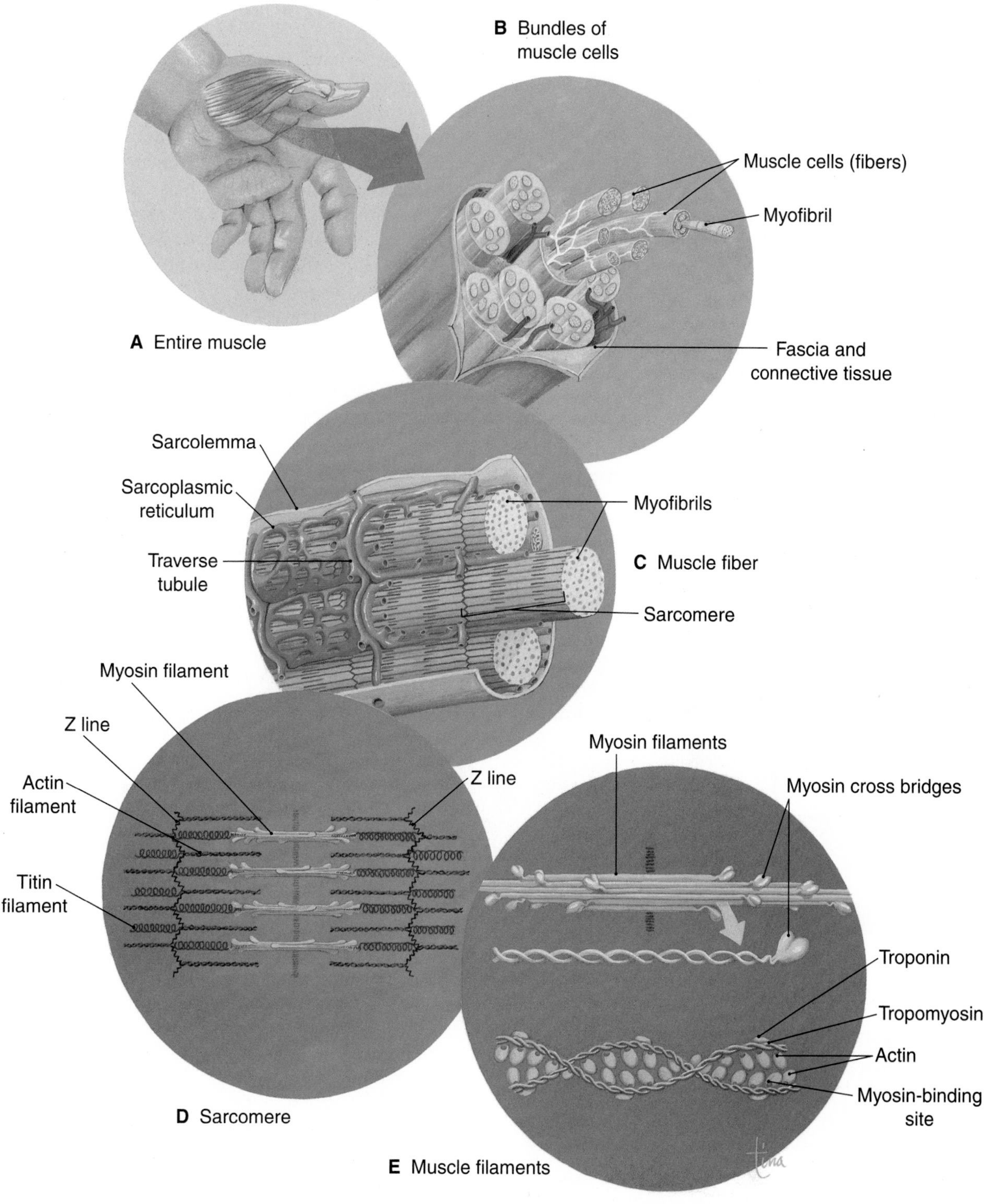

Figure 7–3 Structure of skeletal muscle. **(A)** Entire muscle. **(B)** Bundles of muscle cells within a muscle. **(C)** Single muscle fiber, microscopic structure. **(D)** A sarcomere. **(E)** Structure of muscle filaments.

activator called **cholinesterase**. The **synapse** (or synaptic cleft) is the small space between the axon terminal and the sarcolemma.

Within the muscle fiber are thousands of individual contracting units called **sarcomeres**, which are arranged end to end in cylinders called **myofibrils**. The structure of a sarcomere is shown in Fig. 7–3: The Z lines are the end boundaries of a sarcomere. Filaments of the protein **myosin** are in the center of the sarcomere, and filaments of the protein **actin** are at the ends, attached to the Z lines. Myosin and actin are the contractile proteins of a muscle fiber. Their interactions produce muscle contraction. Also present are two inhibitory proteins, **troponin** and **tropomyosin**, which prevent the sliding of myosin and actin when the muscle fiber is relaxed.

Surrounding the sarcomeres is the **sarcoplasmic reticulum**, the endoplasmic reticulum of muscle cells. The sarcoplasmic reticulum is a reservoir for calcium ions (Ca^{+2}), which are essential for the contraction process.

All of these parts of a muscle fiber are involved in the contraction process. Contraction begins when a nerve impulse arrives at the axon terminal and stimulates the release of acetylcholine. Acetylcholine generates electrical changes (the movement of ions) at the sarcolemma of the muscle fiber. These electrical changes initiate a sequence of events within the muscle fiber that is called the **Sliding Filament Theory** of muscle contraction. We will begin our discussion with the sarcolemma.

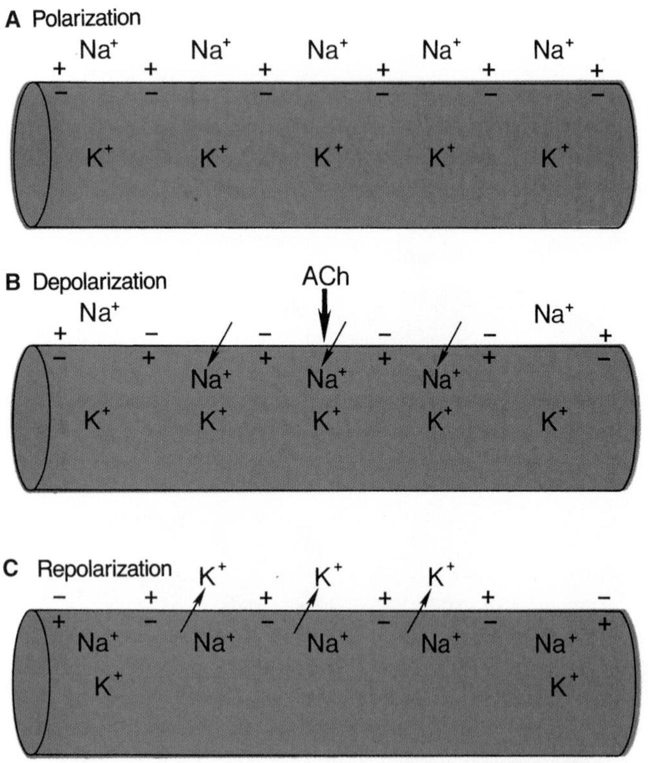

Figure 7–4 Electrical charges and ion concentrations at the sarcolemma. **(A)** Polarization, when the muscle fiber is relaxed. **(B)** Depolarization in response to acetylcholine. **(C)** Repolarization.

Table 7–1 SARCOLEMMA—ELECTRICAL CHANGES

State or Event	Description
Resting Potential	
Polarization	• Sarcolemma has a (+) charge outside and a (−) charge inside. • Na^+ ions are more abundant outside the cell; as they diffuse inward, the sodium pump returns them outside. • K^+ ions are more abundant inside the cell; as they diffuse out the potassium pump returns them inside.
Action Potential	
Depolarization	• ACh makes the sarcolemma very permeable to Na^+ ions, which rush into the cell. • Reversal of charges on the sarcolemma: now (−) outside and (+) inside. • The reversal of charges spreads along the entire sarcolemma. • Cholinesterase at the sarcolemma inactivates ACh.
Repolarization	• Sarcolemma becomes very permeable to K^+ ions, which rush out of the cell. • Restoration of charges on the sarcolemma: (+) outside and (−) inside. • The sodium and potassium pumps return Na^+ ions outside and K^+ ions inside. • The muscle fiber is now able to respond to ACh released by another nerve impulse arriving at the axon terminal.

SARCOLEMMA—POLARIZATION

When a muscle fiber is relaxed, the sarcolemma is polarized (has a resting potential), which is a difference in electrical charges between the outside and the inside. During **polarization**, the outside of the sarcolemma has a positive charge relative to the inside, which is said to have a negative charge. Sodium ions (Na^+) are more abundant outside the cell, and potassium ions (K^+) and negative ions are more abundant inside (Fig. 7-4).

The Na^+ ions outside tend to diffuse into the cell, and the **sodium pump** transfers them back out. The K^+ ions inside tend to diffuse outside, and the **potassium pump** returns them inside. Both of these pumps are active transport mechanisms which, you may recall, require ATP. Muscle fibers use ATP to maintain a high concentration of Na^+ ions outside the cell and a high concentration of K^+ inside. The pumps, therefore, maintain polarization and relaxation until a nerve impulse stimulates a change.

SARCOLEMMA—DEPOLARIZATION

When a nerve impulse arrives at the axon terminal, it causes the release of acetylcholine, which diffuses across the synapse and bonds to **ACh receptors** on the sarcolemma. By doing so, acetylcholine makes the sarcolemma very permeable to Na^+ ions, which rush into the cell. This makes the inside of the sarcolemma positive relative to the outside, which is now considered negative. This reversal of charges is called **depolarization**. The electrical impulse thus generated (called an action potential) then spreads along the entire sarcolemma of a muscle fiber. Depolarization initiates changes within the cell that bring about contraction. The electrical changes that take place at the sarcolemma are summarized in Table 7-1 and shown in Fig. 7-4.

MECHANISM OF CONTRACTION—SLIDING FILAMENT THEORY

All of the parts of a muscle fiber and the electrical changes described earlier are involved in the contraction process, which is a precise sequence of events.

In summary, a nerve impulse causes depolarization of a muscle fiber, and this electrical change enables the myosin filaments to pull the actin filaments toward the center of the sarcomere, making the sarcomere shorter. All of the sarcomeres shorten and the muscle fiber contracts. A more detailed description of this process is the following:

1. A nerve impulse arrives at the axon terminal; acetylcholine is released and diffuses across the synapse.
2. Acetylcholine makes the sarcolemma more permeable to Na^+ ions, which rush into the cell.
3. The sarcolemma depolarizes, becoming negative outside and positive inside.
4. Depolarization stimulates the release of Ca^{+2} ions from the sarcoplasmic reticulum. Ca^{+2} ions bond to the troponin-tropomyosin complex, which shifts it away from the actin filaments.
5. Myosin splits ATP to release its energy; bridges on the myosin attach to the actin filaments and pull them toward the center of the sarcomere, thus making the sarcomere shorter (Fig. 7-5).
6. All the sarcomeres in a muscle fiber shorten—the entire muscle fiber contracts.
7. The sarcolemma repolarizes: K^+ ions leave the cell, restoring a positive charge outside and a negative charge inside. The pumps then return Na^+ ions outside and K^+ ions inside.
8. Cholinesterase in the sarcolemma inactivates acetylcholine.
9. Subsequent nerve impulses will prolong contraction (more acetycholine is released).
10. When there are no further impulses, the muscle fiber will relax and return to its original length.

The above sequence (1-8) describes a single muscle fiber contraction (called a "twitch") in response to a single nerve impulse. Since all of this takes place in less than a second, useful movements would not be possible if muscle fibers relaxed immediately after contracting. Normally, however, nerve impulses arrive in a continuous stream and produce a sustained contraction called **tetanus**, which is a normal state not to be confused with the disease tetanus (see Box 7-2: Tetanus and Botulism). When in tetanus, muscle fibers remain contracted and are capable of effective movements. In a muscle such as the biceps brachii that flexes the forearm, an effective movement means that many of its thousands of muscle fibers are in tetanus.

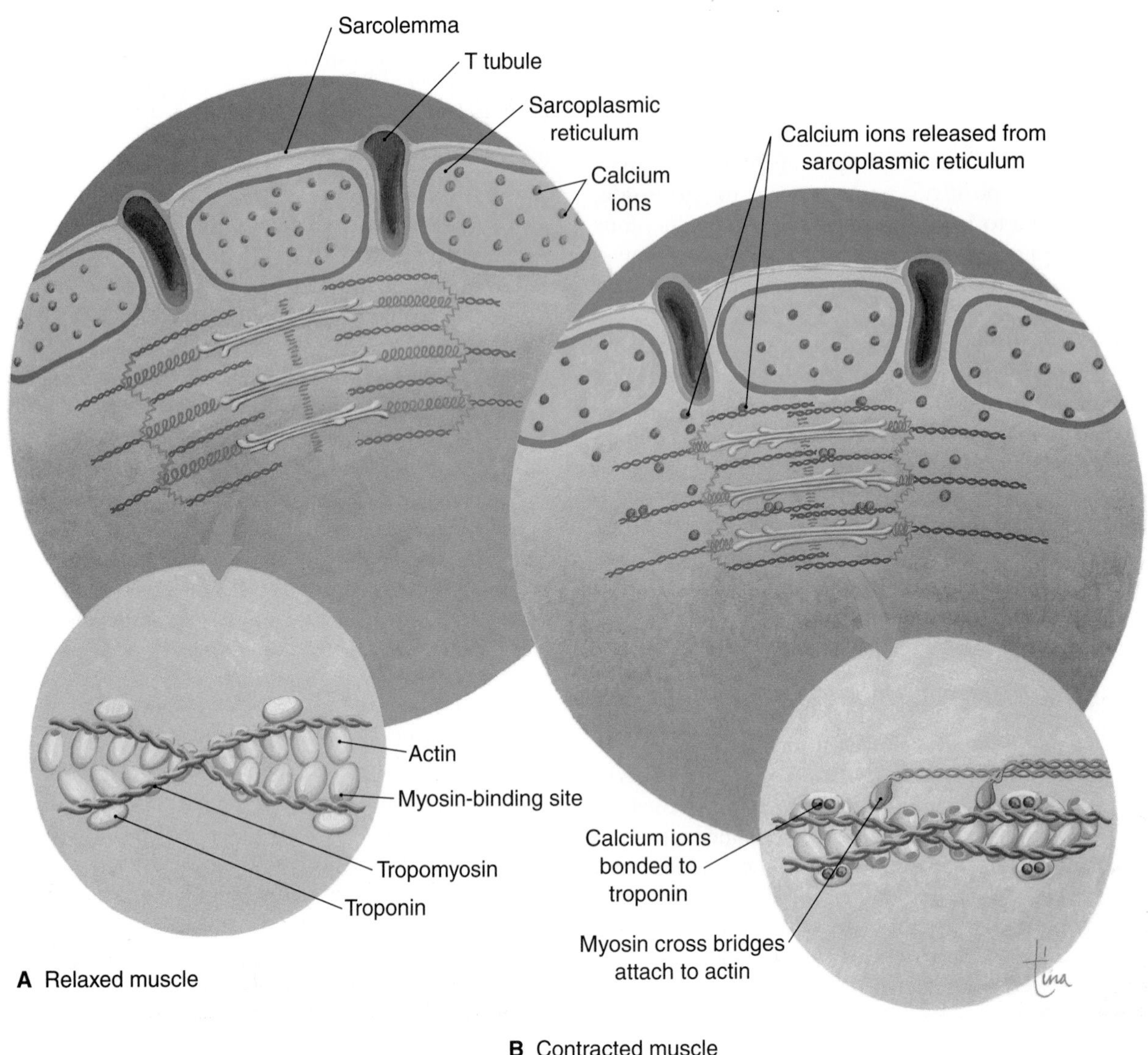

Figure 7–5 Sliding Filament Theory. **(A)** Sarcomere in relaxed muscle fiber. **(B)** Sarcomere in contracted muscle fiber. See text for description.

As you might expect with such a complex process, there are many ways muscle contraction may be impaired. Perhaps the most obvious is the loss of nerve impulses to muscle fibers, as occurs when nerves or the spinal cord are severed, or when a **stroke (cerebrovascular accident)** occurs in the frontal lobes of the cerebrum. Without nerve impulses, skeletal muscles become **paralyzed**, unable to contract. Paralyzed muscles eventually **atrophy**, that is, become smaller from lack of use. Other disorders that affect muscle functioning are discussed in Box 7–3: Muscular Dystrophy, and Box 7–4: Myasthenia Gravis.

RESPONSES TO EXERCISE—MAINTAINING HOMEOSTASIS

Although entire textbooks are devoted to exercise physiology, we will discuss it only briefly here as an

Box 7–2 TETANUS AND BOTULISM

Some bacteria cause disease by producing toxins. A **neurotoxin** is a chemical that in some way disrupts the normal functioning of the nervous system. Because skeletal muscle contraction depends on nerve impulses, the serious consequences for the individual may be seen in the muscular system.

Tetanus is characterized by the inability of muscles to relax. The toxin produced by the tetanus bacteria *(Clostridium tetani)* affects the nervous system in such a way that muscle fibers receive too many impulses, and muscles go into spasms. Lockjaw, the common name for tetanus, indicates one of the first symptoms, which is difficulty opening the mouth because of spasms of the masseter muscles. Treatment requires the antitoxin (an antibody to the toxin) to neutralize the toxin. In untreated tetanus the cause of death is spasm of the respiratory muscles.

Botulism is usually a type of food poisoning, but it is not characterized by typical food poisoning symptoms such as diarrhea or vomiting. The neurotoxin produced by the botulism bacteria *(Clostridium botulinum)* prevents the release of acetylcholine at neuromuscular junctions. Without acetylcholine, muscle fibers cannot contract, and muscles become paralyzed. Early symptoms of botulism include blurred or double vision and difficulty speaking or swallowing. Weakness and paralysis spread to other muscle groups, eventually affecting all voluntary muscles. Without rapid treatment with the antitoxin (the specific antibody to this toxin), botulism is fatal because of paralysis of the respiratory muscles.

Box 7–3 MUSCULAR DYSTROPHY

Muscular dystrophy is really a group of genetic diseases in which muscle tissue is replaced by fibrous connective tissue or by fat. Neither of these tissues is capable of contraction, and the result is progressive loss of muscle function. The most common form is Duchenne's muscular dystrophy, in which the loss of muscle function affects not only skeletal muscle but also cardiac muscle. Death usually occurs before the age of 20 due to heart failure, and at present there is no cure.

Duchenne's muscular dystrophy is a **sex-linked** (or x-linked) **trait**, which means that the gene for it is on the X chromosome and is recessive. The female sex chromosomes are XX. If one X chromosome has a gene for muscular dystrophy, and the other X chromosome has a dominant gene for normal muscle function, the woman will not have muscular dystrophy but will be a carrier who may pass the muscular dystrophy gene to her children. The male sex chromosomes are XY, and the Y has no gene at all for muscle function, that is, no gene to prevent the expression of the gene on the X chromosome. If the X chromosome has a gene for muscular dystrophy, the male will have the disease. This is why Duchenne's muscular dystrophy is more common in males; the presence of only one gene means the disease will be present.

The muscular dystrophy gene on the X chromosome has been located, and the protein the gene codes for has been named dystrophin. Dystrophin seems to be necessary for the stability of the sarcolemma and the proper movement of ions. Determining the precise function of dystrophin, and exactly what goes wrong when this protein malfunctions, may provide a basis or starting point for therapy. Treatments being investigated include the injection of normal muscle cells into affected muscles, and the insertion (using viruses) of normal genes for dystrophin into affected muscle cells.

Box 7–4 MYASTHENIA GRAVIS

Myasthenia gravis is an **autoimmune** disorder characterized by extreme muscle fatigue even after minimal exertion. Women are affected more often than are men, and symptoms usually begin in middle age. Weakness may first be noticed in the facial or swallowing muscles and may progress to other muscles. Without treatment, the respiratory muscles will eventually be affected, and respiratory failure is the cause of death.

In myasthenia gravis, the autoantibodies (self-antibodies) destroy the **acetylcholine receptors** on the sarcolemma. These receptors are the sites to which acetylcholine bonds and stimulates the entry of Na^+ ions. Without these receptors, the acetylcholine released by the axon terminal cannot cause depolarization of a muscle fiber.

Treatment of myasthenia gravis may involve anticholinesterase medications. Recall that cholinesterase is present in the sarcolemma to inactivate acetylcholine and prevent continuous, unwanted impulses. If this action of cholinesterase is inhibited, acetylcholine remains on the sarcolemma for a longer time and may bond to any remaining receptors to stimulate depolarization and contraction.

example of the body's ability to maintain homeostasis. Engaging in moderate or strenuous exercise is a physiological stress situation, a change that the body must cope with and still maintain a normal internal environment, that is, homeostasis.

Some of the body's responses to exercise are diagrammed in Fig. 7-6; notice how they are related to cell respiration.

As you can see, the respiratory and cardiovascular systems make essential contributions to exercise. The integumentary system also has a role, since it eliminates excess body heat. Although not shown below, the nervous system is also directly involved, as we have seen. The brain generates the impulses for muscle contraction and regulates heart rate, breathing rate, and the diameter of blood vessels. The next time you run up a flight of stairs, hurry to catch a bus, or just go swimming, you might reflect a moment on all of the things that are actually happening to your body . . . after you catch your breath.

AGING AND THE MUSCULAR SYSTEM

With age, muscle cells die and are replaced by fibrous connective tissue or by fat. Regular exercise, however,

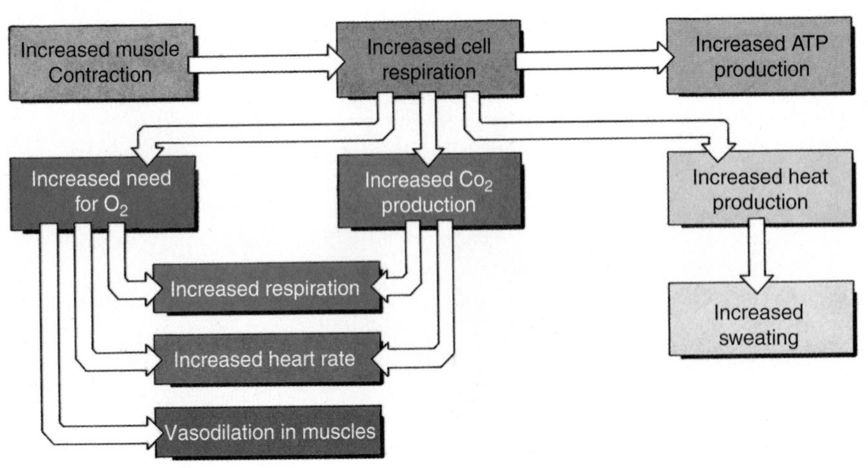

Figure 7–6 Responses of the body during exercise.

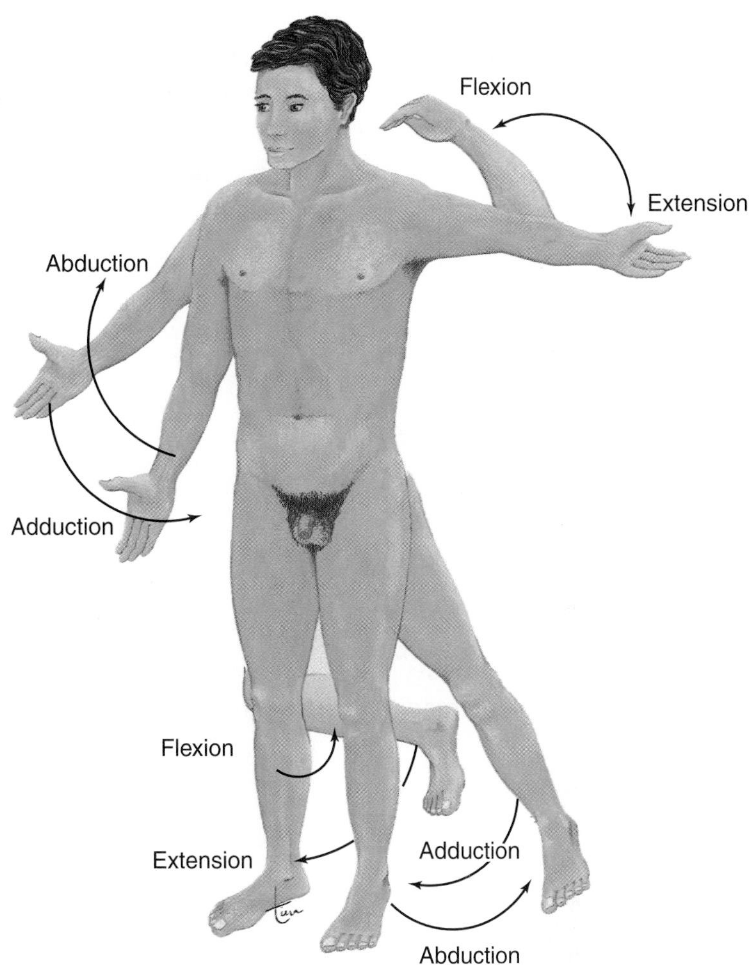

Figure 7–7 Actions of muscles.

delays atrophy of muscles. Although muscles become slower to contract and their maximal strength decreases, exercise can maintain muscle functioning at a level that meets whatever a person needs for daily activities. Such exercise also benefits the cardiovascular, respiratory, and skeletal systems.

MAJOR MUSCLES OF THE BODY

The actions that muscles perform are shown in Fig. 7-7 and are listed in Table 7-2. Most are in pairs as antagonistic functions.

The major muscles are shown in Fig. 7-8. They are listed, according to body area, in Tables 7-3 through 7-6, with associated Figs. 7-9 through 7-12, respec-

Table 7–2 ACTIONS OF MUSCLES

Action	Definition
Flexion	• To decrease the angle of a joint
Extension	• To increase the angle of a joint
Adduction	• To move closer to the midline
Abduction	• To move away from the midline
Pronation	• To turn the palm down
Supination	• To turn the palm up
Dorsiflexion	• To elevate the foot
Plantar flexion	• To lower the foot (point the toes)
Rotation	• To move a bone around its longitudinal axis

Most are grouped in pairs of antagonistic functions.

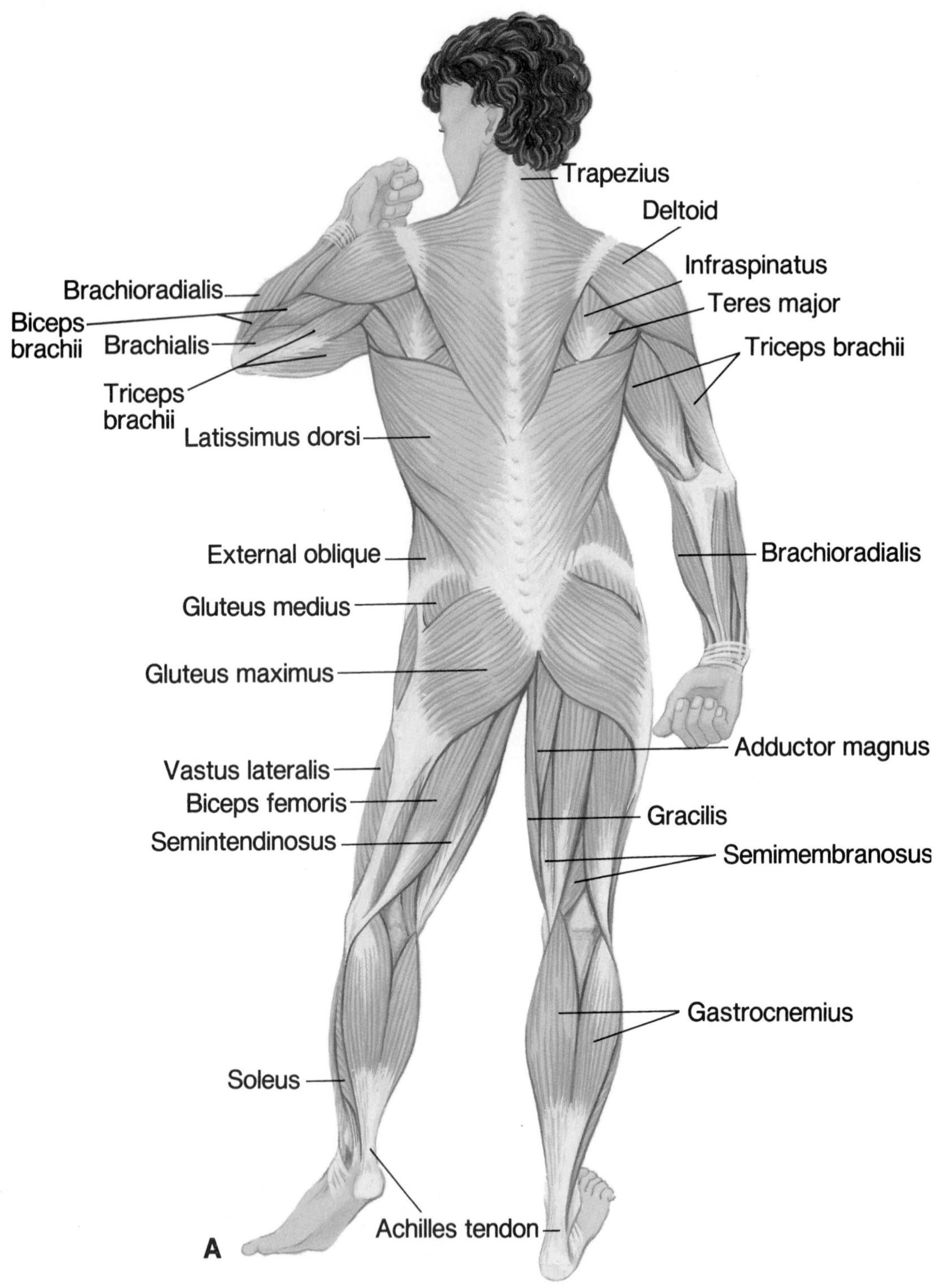

Figure 7–8 Major muscles of the body. **(A)** Posterior view.

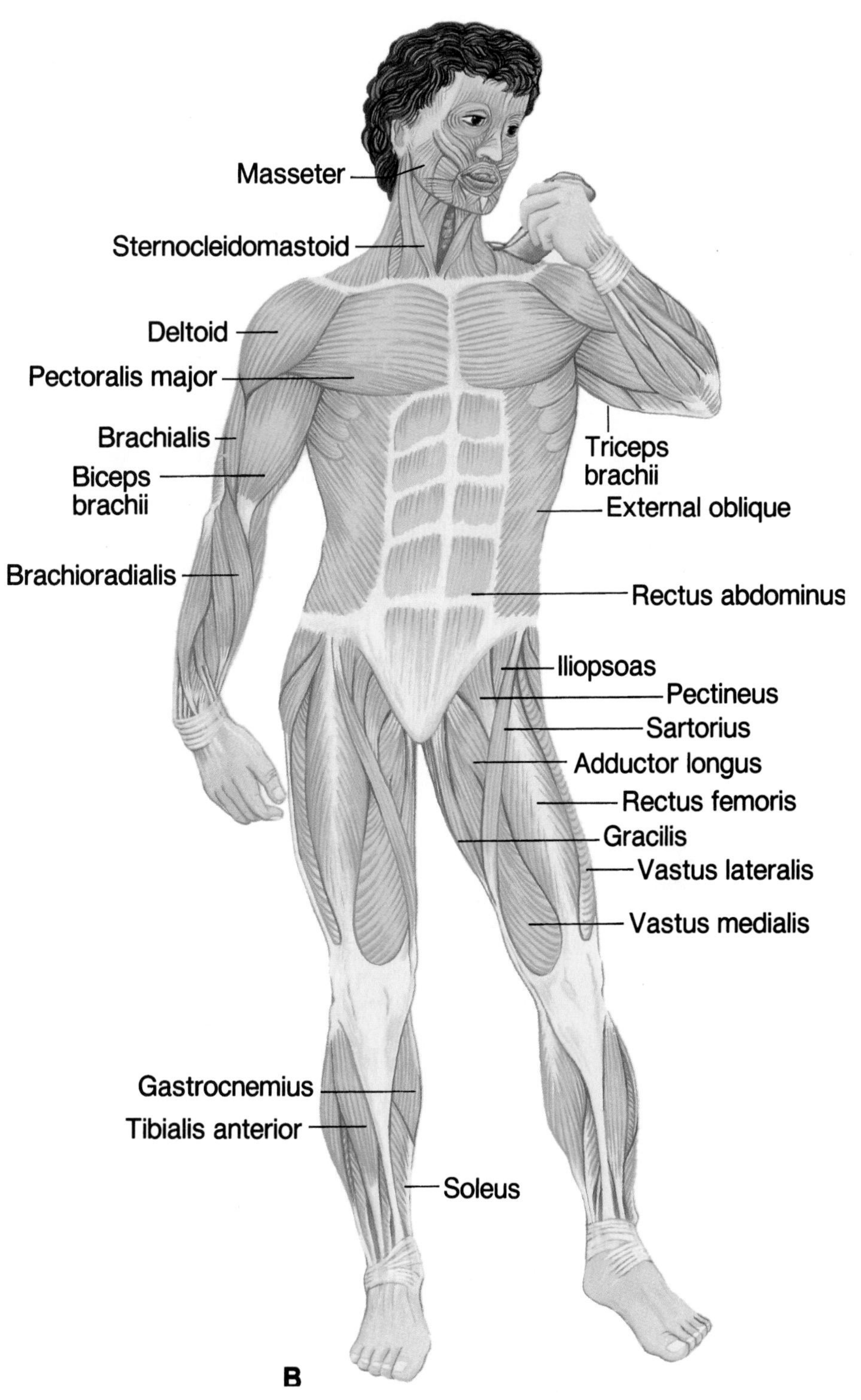

Figure 7–8 Continued. **(B)** Anterior view.

Table 7–3 MUSCLES OF THE HEAD AND NECK

Muscle	Function	Origin	Insertion
Orbicularis oculi	Closes eye	• medial side of orbit	• encircles eye
Orbicularis oris	Puckers lips	• encircles mouth	• skin at corners of mouth
Masseter	Closes jaw	• maxilla and zygomatic	• mandible
Buccinator	Pulls corners of mouth laterally	• maxillae and mandible	• orbicularis oris
Sternocleidomastoid	Turns head to opposite side (both—flex head and neck)	• sternum and clavicle	• temporal bone (mastoid process)
Semispinalis capitis (a deep muscle)	Turns head to same side (both—extend head and neck)	• 7th cervical and first 6 thoracic vertebrae	• occipital bone

Table 7–4 MUSCLES OF THE TRUNK

Muscle	Function	Origin	Insertion
Trapezius	Raises, lowers, and adducts shoulders	• occipital bone and all thoracic vertebrae	• spine of scapula and clavicle
External intercostals	Pull ribs up and out (inhalation)	• superior rib	• inferior rib
Internal intercostals	Pull ribs down and in (forced exhalation)	• inferior rib	• superior rib
Diaphragm	Flattens (down) to enlarge chest cavity for inhalation	• last 6 costal cartilages and lumbar vertebrae	• central tendon
Rectus abdominus	Flexes vertebral column, compresses abdomen	• pubic bones	• 5th–7th costal cartilages and xiphoid process
External oblique	Rotates and flexes vertebral column, compresses abdomen	• lower 8 ribs	• iliac crest and linea alba
Sacrospinalis group (a deep group of muscles)	Extends vertebral column	• ilium, lumbar, and some thoracic vertebrae	• ribs, cervical, and thoracic vertebrae

Table 7–5 MUSCLES OF THE SHOULDER AND ARM

Muscle	Function	Origin	Insertion
Deltoid	Abducts the humerus	• scapula and clavicle	• humerus
Pectoralis major	Flexes and adducts the humerus	• clavicle, sternum, 2nd–6th costal cartilages	• humerus
Latissimus dorsi	Extends and adducts the humerus	• last 6 thoracic vertebrae, all lumbar vertebrae, sacrum, iliac crest	• humerus
Teres major	Extends and adducts the humerus	• scapula	• humerus
Triceps brachii	Extends the forearm	• humerus and scapula	• ulna
Biceps brachii	Flexes the forearm	• scapula	• radius
Brachioradialis	Flexes the forearm	• humerus	• radius

Table 7–6 MUSCLES OF THE HIP AND LEG

Muscle	Function	Origin	Insertion
Iliopsoas	Flexes femur	• ilium, lumbar vertebrae	• femur
Gluteus maximus	Extends femur	• iliac crest, sacrum, coccyx	• femur
Gluteus medius	Abducts femur	• ilium	• femur
Quadriceps femoris group: Rectus femoris Vastus lateralis Vastus medialis Vastus intermedius	Flexes femur and extends lower leg	• ilium and femur	• tibia
Hamstring group: Biceps femoris Semimembranosus Semitendinosus	Extends femur and flexes lower leg	• ischium	• tibia and fibula
Adductor group	Adducts femur	• ischium and pubis	• femur
Sartorius	Flexes femur and lower leg	• ilium	• tibia
Gastrocnemius	Plantar flexes foot	• femur	• calcaneus (Achilles tendon)
Soleus	Plantar flexes foot	• tibia and fibula	• calcaneus (Achilles tendon)
Tibialis anterior	Dorsiflexes foot	• tibia	• metatarsals

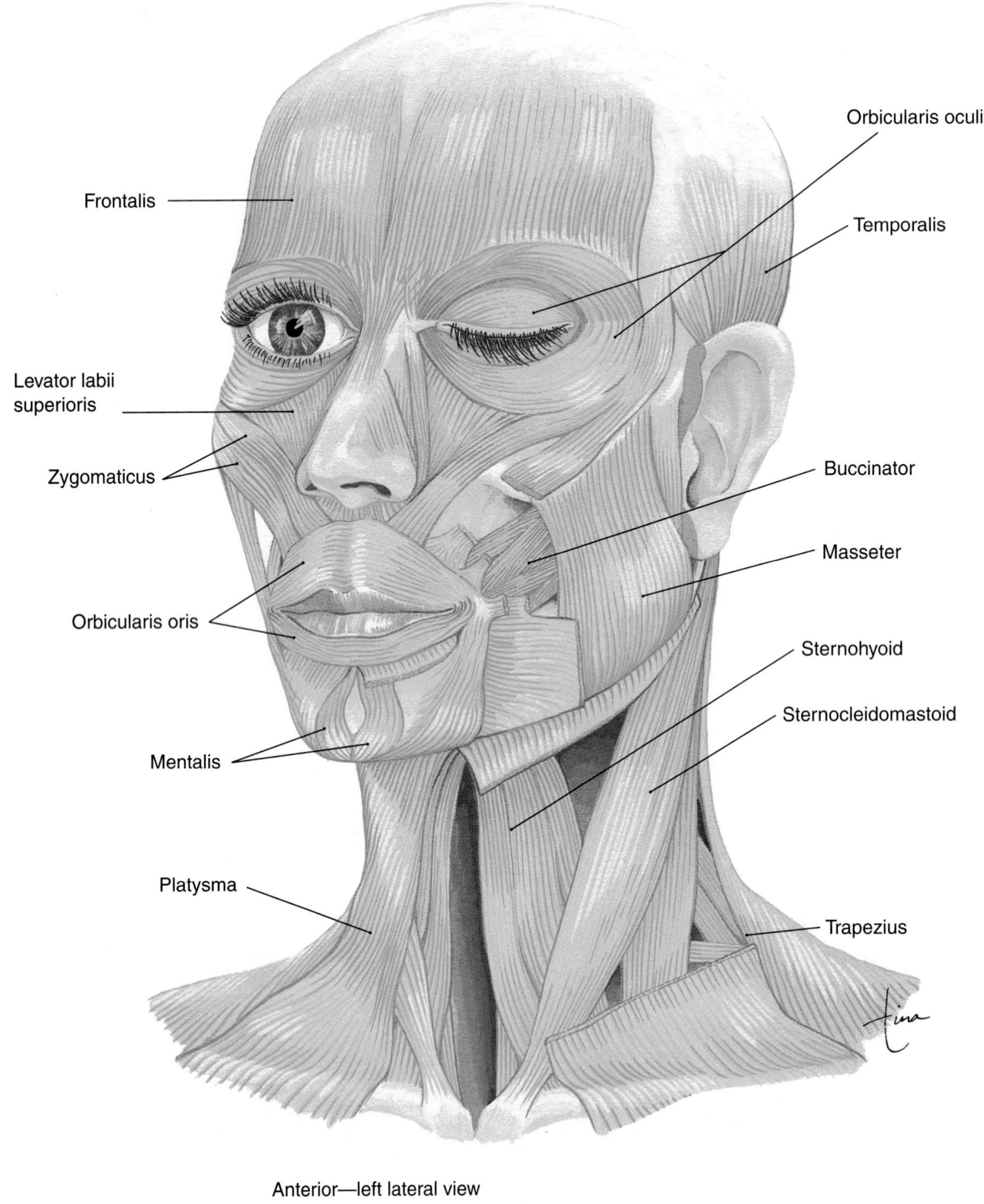

Figure 7–9 Muscles of the head and neck in anterior, left-lateral view.

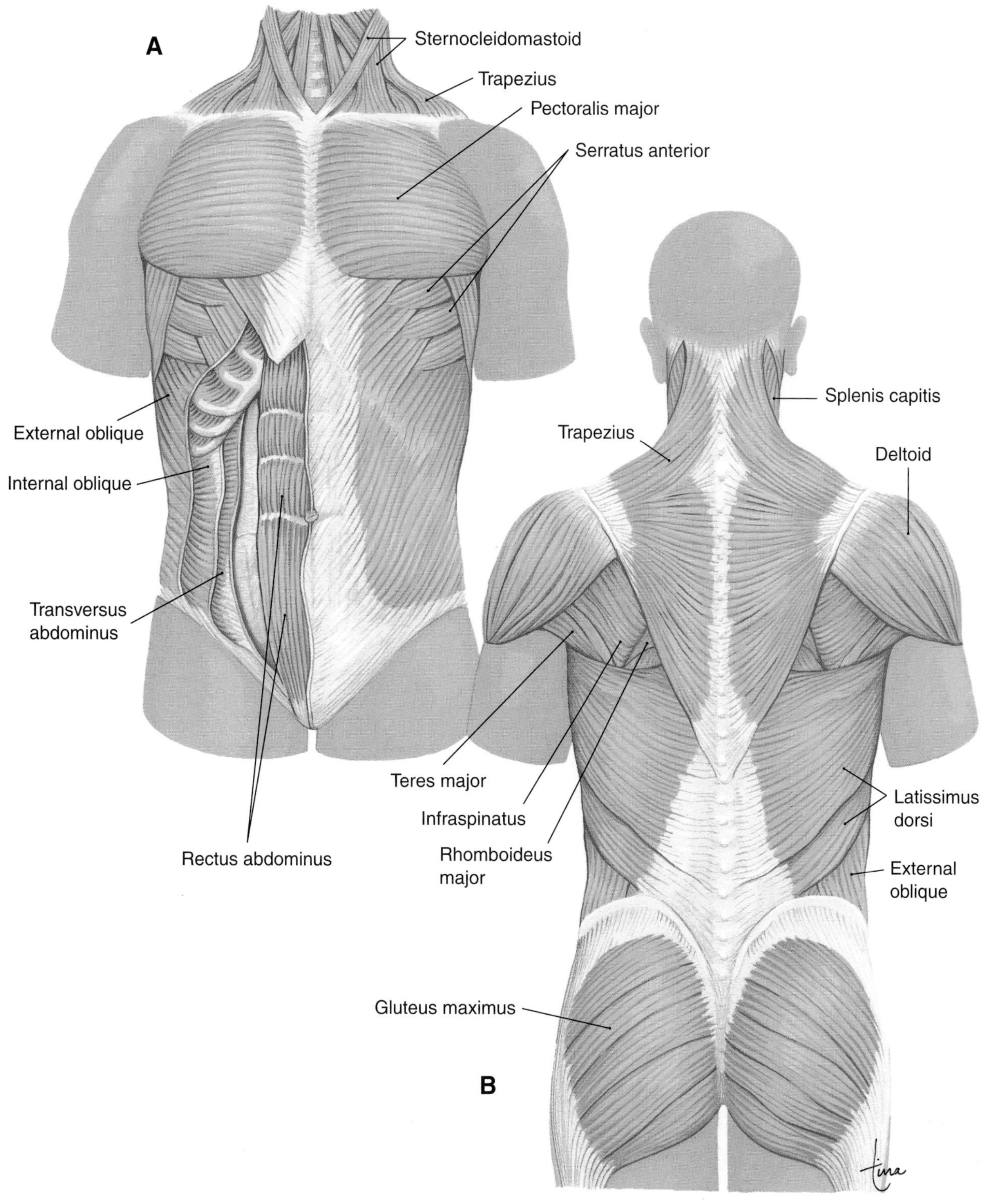

Figure 7–10 Muscles of the trunk. **(A)** Anterior view. **(B)** Posterior view.

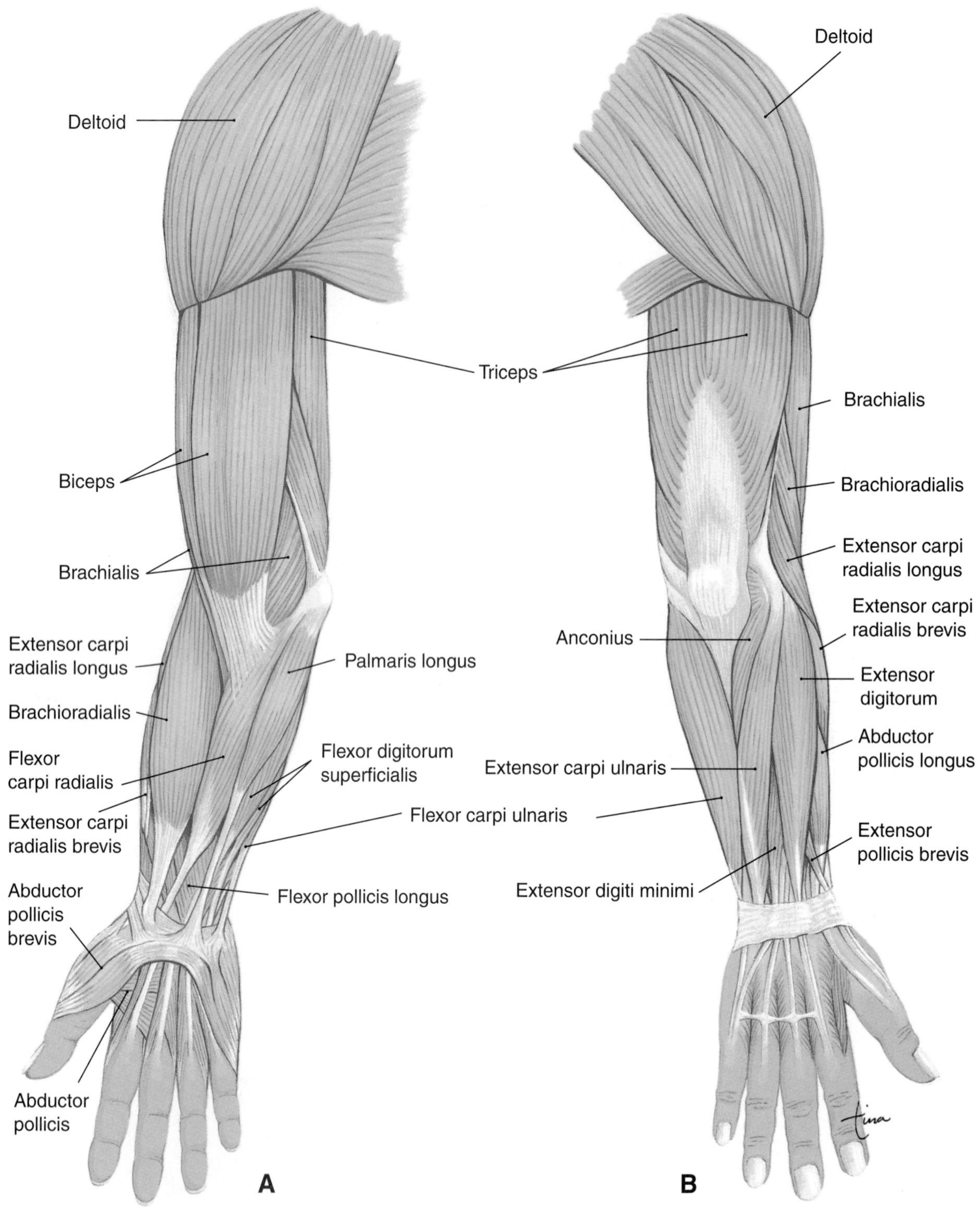

Figure 7–11 Muscles of the arm. **(A)** Anterior view. **(B)** Posterior view.

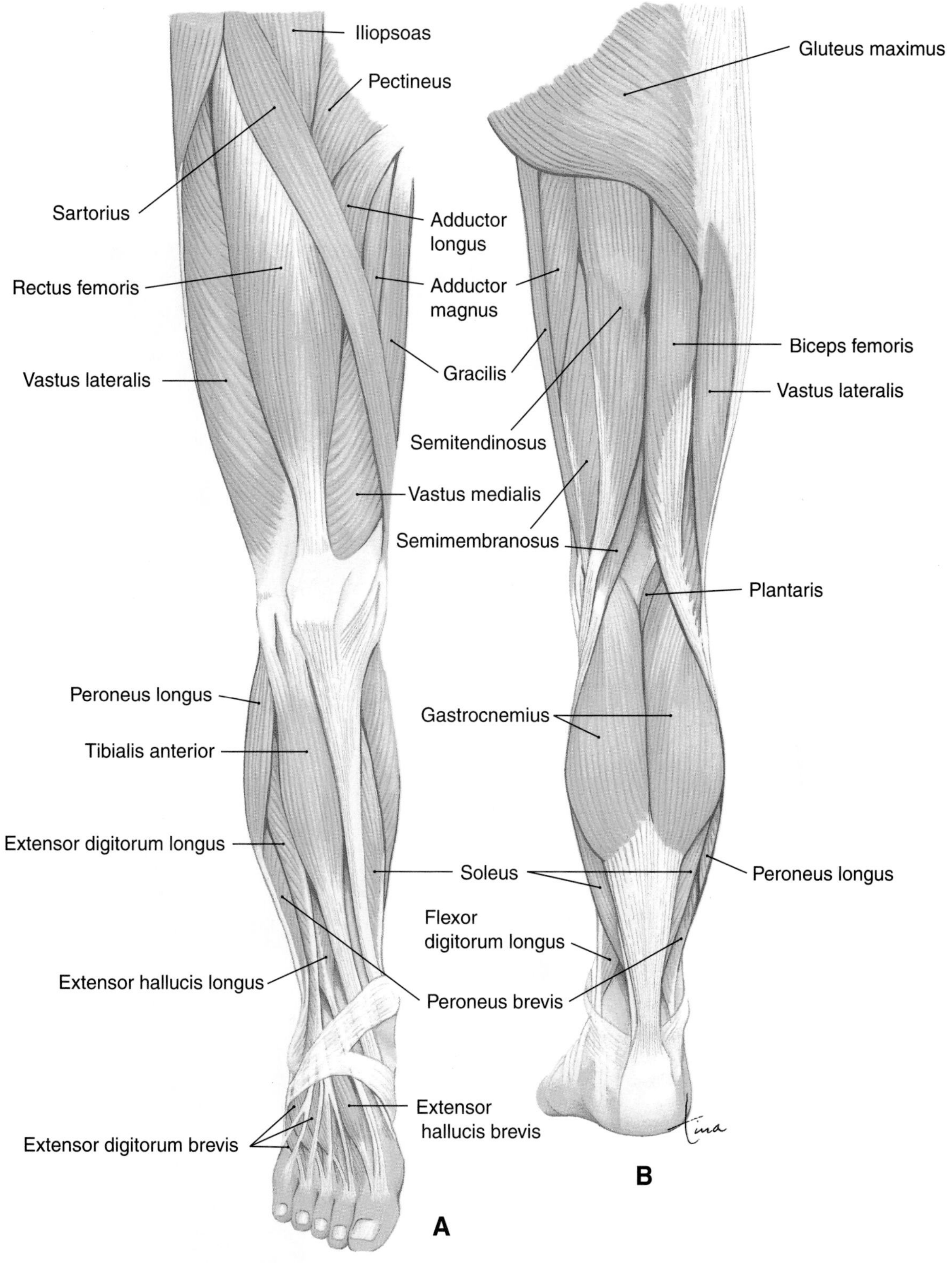

Figure 7–12 Muscles of the leg. **(A)** Anterior view. **(B)** Posterior view.

tively. Learning the muscles and their functions does involve memorization, but the bones you have already learned will help you. For each muscle, note its origin and insertion. If you know the bones to which a muscle is attached, you can determine the joint the muscle affects when it contracts.

The name of the muscle may also be helpful, and again, many of the terms are ones you have already learned. Some examples: "abdominus" refers to an abdominal muscle, "femoris" to a thigh muscle, "brachii" to a muscle of the upper arm, "oculi" to an eye muscle, and so on.

Muscles that are sites for intramuscular injections are shown in Box 7-5.

Box 7–5 COMMON INJECTION SITES

Intramuscular injections are used when rapid absorption is needed, because muscle has a good blood supply. Common sites are the buttock (*gluteus medius*), lateral thigh (*vastus lateralis*), and the shoulder (*deltoid*). These sites are shown below; also shown are the large nerves to be avoided when giving such injections.

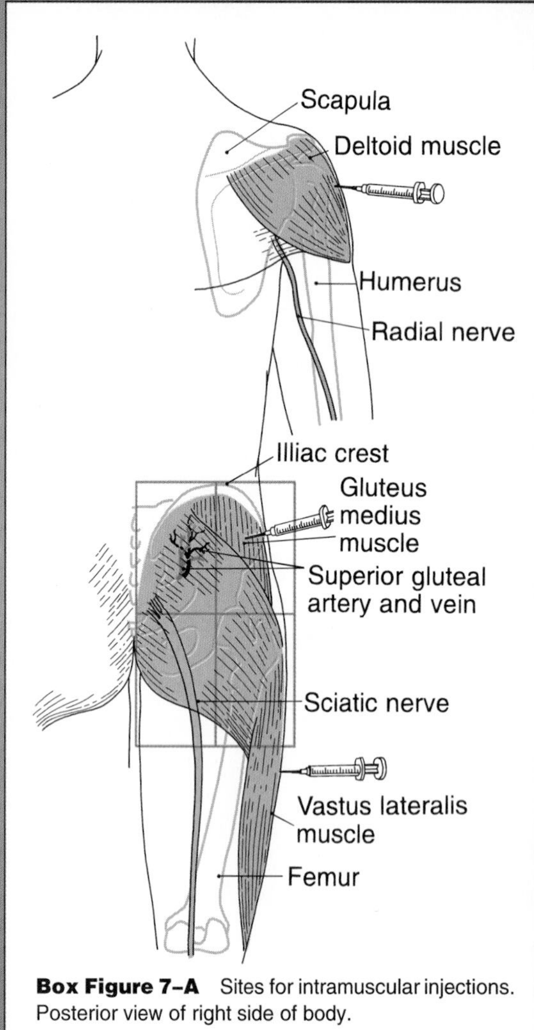

Box Figure 7–A Sites for intramuscular injections. Posterior view of right side of body.

STUDY OUTLINE

Organ Systems Involved in Movement

1. Muscular—moves the bones.
2. Skeletal—bones are moved, at their joints, by muscles.
3. Nervous—transmits impulses to muscles to cause contraction.
4. Respiratory—exchanges O_2 and CO_2 between the air and blood.
5. Circulatory—transports O_2 to muscles and removes CO_2.

Muscle Structure

1. Muscle fibers (cells) are specialized to contract, shorten, and produce movement.
2. A skeletal muscle is made of thousands of muscle fibers. Varying movements require contraction of variable numbers of muscle fibers in a muscle.
3. Tendons attach muscles to bone; the origin is the more stationary bone, the insertion is the more movable bone. A tendon merges with the fascia of a muscle and the periosteum of a bone; all are made of fibrous connective tissue.

Muscle Arrangements

1. Antagonistic muscles have opposite functions. A muscle pulls when it contracts, but exerts no force when it relaxes and cannot push. When one muscle pulls a bone in one direction, another muscle is needed to pull the bone in the other direction (see also Table 7-2 and Fig. 7-1).
2. Synergistic muscles have the same function and alternate as the prime mover depending on the position of the bone to be moved. Synergists also stabilize a joint to make a more precise movement possible.
3. The frontal lobes of the cerebrum generate the impulses necessary for contraction of skeletal muscles. The cerebellum regulates coordination.

Muscle Tone—the state of slight contraction present in muscles

1. Alternate fibers contract to prevent muscle fatigue.
2. Good tone helps maintain posture, produces 25% of body heat (at rest), and improves coordination.
3. Isotonic exercise involves contraction with movement; improves tone and strength and improves cardiovascular and respiratory efficiency (aerobic exercise).
4. Isometric exercise involves contraction without movement; improves tone and strength but is not aerobic.

Muscle Sense—knowing where our muscles are without looking at them

1. Permits us to perform everyday activities without having to concentrate on muscle position.
2. Stretch receptors (proprioceptors) in muscles respond to stretching and generate impulses that the brain interprets as a mental "picture" of where the muscles are. Parietal lobes: conscious muscle sense; cerebellum: unconscious muscle sense used to promote coordination.

Energy Sources for Muscle Contraction

1. ATP is the direct source; the ATP stored in muscles lasts only a few seconds.
2. Creatine phosphate is a secondary energy source; is broken down to creatine + phosphate + energy. The energy is used to synthesize more ATP. Some creatine is converted to creatinine, which must be excreted by the kidneys. Most creatine is used for the resynthesis of creatine phosphate.
3. Glycogen is the most abundant energy source and is first broken down to glucose. Glucose is broken down in cell respiration:

   $$\text{Glucose} + O_2 \rightarrow CO_2 + H_2O + \text{ATP} + \text{heat}$$

   ATP is used for contraction; heat contributes to body temperature; H_2O becomes part of intracellular fluid; CO_2 is eventually exhaled.
4. Oxygen is essential for the completion of cell respiration. Hemoglobin in RBCs carries oxygen to muscles; myoglobin stores oxygen in muscles; both these proteins contain iron, which enables them to bond to oxygen.
5. Oxygen debt: muscle fibers run out of oxygen during strenuous exercise, and glucose is converted to lactic acid, which causes fatigue. Breathing rate remains high after exercise to deliver more oxygen to the liver, which converts lactic acid back to glucose (ATP required).

Muscle Fiber—microscopic structure

1. Neuromuscular junction: axon terminal and sarcolemma; the synapse is the space between. The axon terminal contains acetylcholine (a neuro-

transmitter), and the sarcolemma contains cholinesterase (an inactivator) (see Fig. 7-2).

2. Sarcomeres are the contracting units of a muscle fiber. Myosin and actin filaments are the contracting proteins of sarcomeres. Troponin and tropomyosin are proteins that inhibit the sliding of myosin and actin when the muscle fiber is relaxed (see Fig. 7-3).
3. The sarcoplasmic reticulum surrounds the sarcomeres and is a reservoir for calcium ions.
4. Polarization (resting potential): when the muscle fiber is relaxed, the sarcolemma has a (+) charge outside and a (−) charge inside. Na^+ ions are more abundant outside the cell and K^+ ions are more abundant inside the cell. The Na^+ and K^+ pumps maintain these relative concentrations on either side of the sarcolemma (see Table 7-1 and Fig. 7-4).
5. Depolarization: started by a nerve impulse. Acetylcholine released by the axon terminal makes the sarcolemma very permeable to Na^+ ions, which enter the cell and cause a reversal of charges to (−) outside and (+) inside. The depolarization spreads along the entire sarcolemma and initiates the contraction process.

Mechanism of Contraction—sliding filament theory (see Fig. 7–5)

1. Depolarization stimulates a sequence of events that enables myosin filaments to pull the actin filaments to the center of the sarcomere, which shortens.
2. All the sarcomeres in a muscle fiber contract in response to a nerve impulse; the entire cell contracts.
3. Tetanus—a sustained contraction brought about by continuous nerve impulses; all our movements involve tetanus.
4. Paralysis: muscles that do not receive nerve impulses are unable to contract and will atrophy. Paralysis may be the result of nerve damage, spinal cord damage, or brain damage.

Responses to Exercise—maintaining homeostasis

See section in chapter, and Fig. 7-6.

Major Muscles

See Tables 7-2 through 7-6 and Figs. 7-7 through 7-12.

REVIEW QUESTIONS

1. Name the organ systems directly involved in movement and for each state how they are involved. (p. 128)
2. State the function of tendons. Name the part of a muscle and a bone to which a tendon is attached. (p. 128)
3. State the term for: (pp. 128-129)
   a. Muscles with the same function
   b. Muscles with opposite functions
   c. The muscle that does most of the work in a movement
4. Explain why antagonistic muscle arrangements are necessary. Give two examples. (p. 128)
5. State three reasons why good muscle tone is important. (pp. 129-130)
6. Explain why muscle sense is important. Name the receptors involved and state what they detect. (pp. 130-131)
7. With respect to muscle contraction, state the role of the cerebellum and the frontal lobes of the cerebrum. (p. 129)
8. Name the direct energy source for muscle contraction. Name the two secondary energy sources. Which of these is more abundant? (p. 131)

9. State the simple equation of cell respiration and what happens to each of the products of this reaction. (p. 131)

10. Name the two sources of oxygen for muscle fibers. State what the two proteins have in common. (p. 131)

11. Explain what is meant by oxygen debt. What is needed to correct oxygen debt, and where does it come from? (p. 131)

12. Name these parts of the neuromuscular junction: (pp. 131–132, 134)
    a. The membrane of the muscle fiber
    b. The end of the motor neuron
    c. The space between neuron and muscle cell
    State the locations of acetylcholine and cholinesterase.

13. Name the contracting proteins of sarcomeres, and describe their locations in a sarcomere. Where is the sarcoplasmic reticulum and what does it contain? (p. 134)

14. In terms of ions and charges, describe: (p. 134)
    a. Polarization
    b. Depolarization
    c. Repolarization

15. With respect to the Sliding Filament Theory, explain the function of: (p. 135)
    a. Acetylcholine
    b. Calcium ions
    c. Myosin and actin
    d. Troponin and tropomyosin
    e. Cholinesterase

16. State three of the body's physiological responses to exercise, and explain how each helps maintain homeostasis. (pp. 136, 138)

17. Find the major muscles on yourself, and state a function of each muscle.

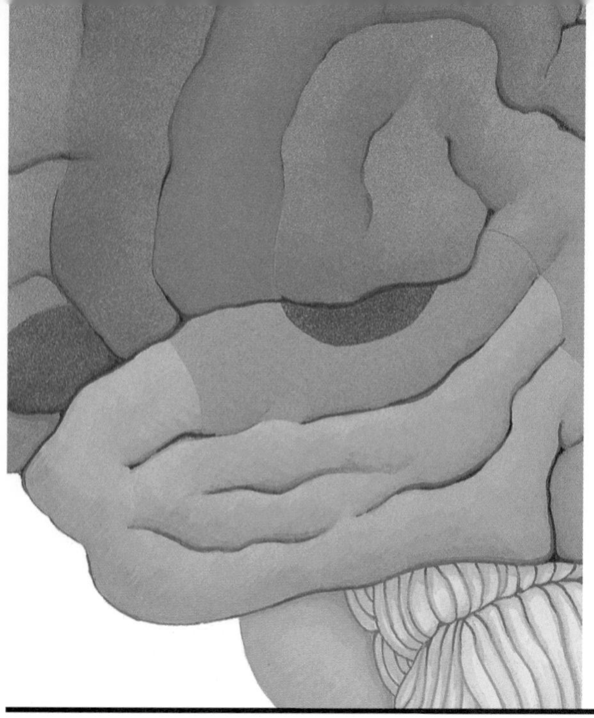

CHAPTER 8

Chapter Outline

Student Objectives

- Name the divisions of the nervous system and the parts of each, and state the general functions of the nervous system.
- Name the parts of a neuron and state the function of each.
- Explain the importance of Schwann cells in the peripheral nervous system and neuroglia in the central nervous system.
- Describe the electrical nerve impulse, and describe impulse transmission at synapses.
- Describe the types of neurons, nerves, and nerve tracts.

The Nervous System

Student Objectives (Continued)

- State the names and numbers of the spinal nerves, and their destinations.
- Explain the importance of stretch reflexes and flexor reflexes.
- State the functions of the parts of the brain; be able to locate each part on a diagram.
- Name the meninges and describe their locations.
- State the locations and functions of cerebrospinal fluid.
- Name the cranial nerves and state their functions.
- Explain how the sympathetic division of the autonomic nervous system enables the body to adapt to a stress situation.
- Explain how the parasympathetic division of the autonomic nervous system promotes normal body functioning in relaxed situations.

New Terminology

Afferent (**AFF**-uh-rent)
Autonomic nervous system (AW-toh-**NOM**-ik)
Cauda equina (**KAW**-dah ee-**KWHY**-nah)
Cerebral cortex (se-**REE**-bruhl **KOR**-teks)
Cerebrospinal fluid (se-**REE**-broh-**SPY**-nuhl)
Choroid plexus (**KOR**-oid **PLEK**-sus)
Corpus callosum (**KOR**-pus kuh-**LOH**-sum)
Cranial nerves (**KRAY**-nee-uhl NERVS)
Efferent (**EFF**-uh-rent)
Gray matter (**GRAY MA**-TUR)
Neuroglia (new-**ROG**-lee-ah)
Neurolemma (NYOO-ro-**LEM**-ah)
Parasympathetic (PAR-uh-SIM-puh-**THET**-ik)
Reflex (**REE**-fleks)
Somatic (soh-**MA**-tik)
Spinal nerves (**SPY**-nuhl NERVS)
Sympathetic (SIM-puh-**THET**-ik)
Ventricles of brain (**VEN**-trick'ls)
Visceral (**VISS**-er-uhl)
White matter (**WIGHT MA**-TUR)

Related Clinical Terminology

Alzheimer's disease (**ALZ**-high-mer's)
Aphasia (ah-**FAY**-zee-ah)
Blood-brain barrier (BLUHD BRAYNE)
Cerebrovascular accident (CVA) (se-**REE**-broh-**VAS**-kyoo-lur)
Lumbar puncture (**LUM**-bar **PUNK**-chur)
Meningitis (MEN-in-**JIGH**-tis)
Multiple sclerosis (MS) (**MULL**-ti-puhl skle-**ROH**-sis)
Neuralgia (new-**RAL**-jee-ah)
Neuritis (new-**RYE**-tis)
Neuropathy (new-**RAH**-puh-thee)
Parkinson's disease (**PAR**-kin-son's)
Remission (ree-**MISH**-uhn)
Spinal shock (**SPY**-nuhl SHAHK)

*Terms that appear in **bold type** in the chapter text are defined in the glossary, which begins on page 528.*

Most of us can probably remember being told, when we were children, not to touch the stove or some other source of potential harm. Because children are curious, such warnings often go unheeded. The result? Touching a hot stove brings about an immediate response of pulling away and a vivid memory of painful fingers. This simple and familiar experience illustrates the functions of the **nervous system**:

1. To detect changes and feel sensations
2. To initiate appropriate responses to changes
3. To organize information for immediate use and store it for future use.

The nervous system is one of the regulating systems (the endocrine system is the other and is discussed in Chapter 10). Electrochemical impulses of the nervous system make it possible to obtain information about the external or internal environment and do whatever is necessary to maintain homeostasis. Some of this activity is conscious, but much of it happens without our awareness.

NERVOUS SYSTEM DIVISIONS

The nervous system has two divisions. The **central nervous system (CNS)** consists of the brain and spinal cord. The **peripheral nervous system (PNS)** consists of cranial nerves and spinal nerves. The PNS includes the autonomic nervous system (ANS).

The peripheral nervous system relays information to and from the central nervous system, and the brain is the center of activity that integrates this information, initiates responses, and makes us the individuals we are.

NERVE TISSUE

Nerve tissue was briefly described in Chapter 4, so we will begin by reviewing what you already know, then adding to it.

Nerve cells are called **neurons**, or **nerve fibers**. Whatever their specific functions, all neurons have the same physical parts. The **cell body** contains the nucleus (Fig. 8-1) and is essential for the continued life of the neuron. As you will see, neuron cell bodies are found in the central nervous system or close to it in the trunk of the body. In these locations, cell bodies are protected by bone. There are no cell bodies in the arms and legs, which are much more subject to injury.

Dendrites are processes (extensions) that transmit impulses toward the cell body. The one **axon** of a neuron transmits impulses away from the cell body. It is the cell membrane of the dendrites, cell body, and axon that carries the electrical nerve impulse.

In the peripheral nervous system, axons and dendrites are "wrapped" in specialized cells called **Schwann cells** (see Fig. 8-1). During embryonic development, Schwann cells grow to surround the neuron processes, enclosing them in several layers of Schwann cell membrane. These layers are the **myelin sheath**; myelin is a phospholipid that electrically insulates neurons from one another. Without the myelin sheath, neurons would short-circuit, just as electrical wires would if they were not insulated (see Box 8-1: Multiple Sclerosis).

The spaces between adjacent Schwann cells, or segments of the myelin sheath, are called nodes of Ranvier (neurofibral nodes). These nodes are the parts of the neuron cell membrane that depolarize when an electrical impulse is transmitted (see "The Nerve Impulse" section, on pages 159-160).

The nuclei and cytoplasm of the Schwann cells are outside the myelin sheath and are called the **neurolemma**, which becomes very important if nerves are damaged. If a peripheral nerve is severed and reattached precisely by microsurgery, the axons and dendrites may regenerate through the tunnels formed by the neurolemmas. The Schwann cells are also believed to produce a chemical growth factor that stimulates regeneration. Although this regeneration may take months, the nerves may eventually reestablish their proper connections, and the person may regain some sensation and movement in the once-severed limb.

In the central nervous system, the myelin sheaths are formed by **oligodendrocytes**, one of the **neuroglia**, the specialized cells found only in the brain and spinal cord. Because no Schwann cells are present, however, there is no neurolemma, and regeneration of neurons is not possible. This is why severing of the spinal cord, for example, results in permanent loss of function (see Table 8-1 for other functions of the neuroglia).

SYNAPSES

Neurons that transmit impulses to other neurons do not actually touch one another. The small gap or space

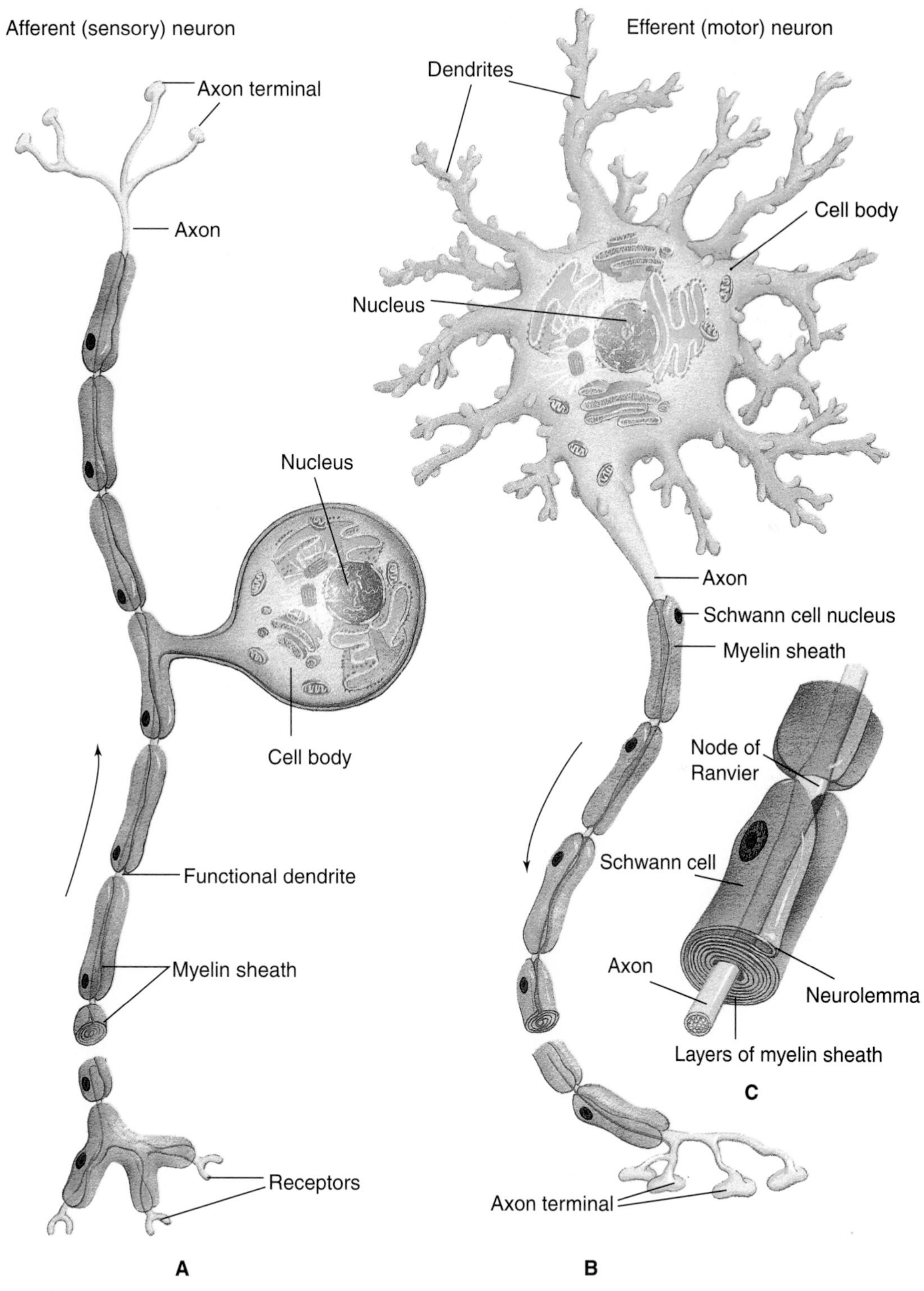

Figure 8–1 Neuron structure. **(A)** A typical sensory neuron. **(B)** A typical motor neuron. The arrows indicate the direction of impulse transmission. **(C)** Details of the myelin sheath and neurolemma formed by Schwann cells.

Box 8–1 MULTIPLE SCLEROSIS

Multiple sclerosis (MS) is a demyelinating disease, that is, it involves deterioration of the myelin sheath of neurons in the central nervous system. Without the myelin sheath, the impulses of these neurons are short-circuited and do not reach their proper destinations.

There is evidence that multiple sclerosis is an **autoimmune** disorder that is triggered by a virus or bacterial infection. Recent research has also uncovered a genetic component to some clusters of MS cases in families. Exactly how such genes would increase a person's susceptibility to an autoimmune disease is not yet known. In MS, the autoantibodies destroy the oligodendrocytes, the myelin-producing neuroglia of the central nervous system, which results in the formation of scleroses, or plaques of scar tissue, that do not provide electrical insulation. Because loss of myelin may occur in many parts of the central nervous system, the symptoms vary, but they usually include muscle weakness or paralysis, numbness or partial loss of sensation, double vision, and loss of spinal cord reflexes, including those for urination and defecation.

The first symptoms usually appear between the ages of 20 and 40 years, and the disease may progress either slowly or rapidly. Some MS patients have **remissions**, periods of time when their symptoms diminish, but remissions and progression of the disease are not predictable. There is still no cure for MS, but new therapies include suppression of the immune response, and interferon, which seems to prolong remissions in some patients.

Table 8–1 NEUROGLIA

Name	Function
Oligodendrocytes	• Produce the myelin sheath to electrically insulate neurons of the CNS.
Microglia	• Capable of movement and phagocytosis of pathogens and damaged tissue.
Astrocytes	• Contribute to the **blood-brain barrier**, which prevents potentially toxic waste products in the blood from diffusing out into brain tissue. A disadvantage of this barrier, however, is that some useful medications cannot cross it; this becomes important during brain infections, inflammation, or other disease or disorder.
Ependyma	• Line the ventricles of the brain; many of the cells have cilia; involved in circulation of cerebrospinal fluid.

between the axon of one neuron and the dendrites or cell body of the next neuron is called the **synapse**. Within the synaptic knob (terminal end) of the presynaptic axon is a chemical **neurotransmitter** that is released into the synapse by the arrival of an electrical nerve impulse (Fig. 8–2). The neurotransmitter diffuses across the synapse, combines with specific receptor sites on the cell membrane of the postsynaptic neuron, and there generates an electrical impulse that in turn is carried by this neuron's axon to the next synapse, and so forth. A chemical **inactivator** at the cell body or dendrite of the postsynaptic neuron quickly inactivates the neurotransmitter. This prevents unwanted, continuous impulses, unless a new impulse from the first neuron releases more neurotransmitter.

Many synapses are termed excitatory, because the postsynaptic neuron depolarizes and transmits an electrical impulse to another neuron, muscle cell, or gland. Some synapses, however, are inhibitory, meaning that the postsynaptic neuron hyperpolarizes (becomes more positive outside) and does *not* transmit an electrical impulse. Such inhibitory synapses are important, for example, in slowing the heart rate.

One important consequence of the presence of synapses is that they ensure one-way transmission of im-

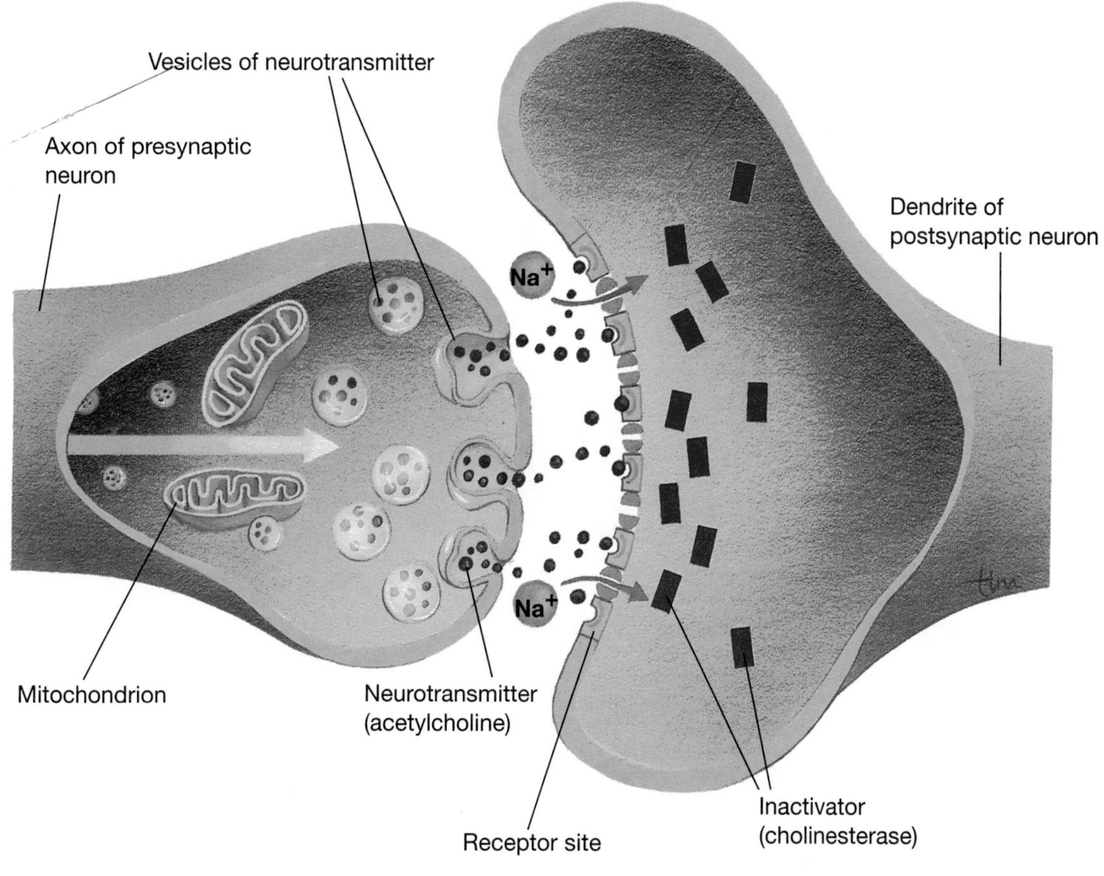

Figure 8–2 Impulse transmission at a synapse. The arrow indicates the direction of the electrical impulse.

pulses in a living person. A nerve impulse cannot go backward across a synapse because there is no neurotransmitter released by the dendrites or cell body. Neurotransmitters can only be released by a neuron's axon, which does not have receptor sites for it, as does the postsynaptic membrane. Keep this in mind when we discuss the types of neurons, below.

An example of a neurotransmitter is **acetylcholine**, which is found in the CNS, at neuromuscular junctions, and in much of the peripheral nervous system. **Cholinesterase** is the inactivator of acetylcholine. There are many other neurotransmitters, especially in the central nervous system. These include dopamine, GABA, norepinephrine, glutamate, and serotonin. Each of these neurotransmitters has its own chemical inactivator.

The complexity and variety of synapses make them frequent targets of medications. For example, drugs that alter mood or behavior often act on specific neurotransmitters in the brain, and antihypertensive drugs affect synapse transmission at the smooth muscle of blood vessels.

TYPES OF NEURONS

Neurons may be classified into three groups: sensory neurons, motor neurons, and interneurons (Fig. 8–3). **Sensory neurons** (or **afferent neurons**) carry impulses from receptors to the central nervous system. **Receptors** detect external or internal changes and

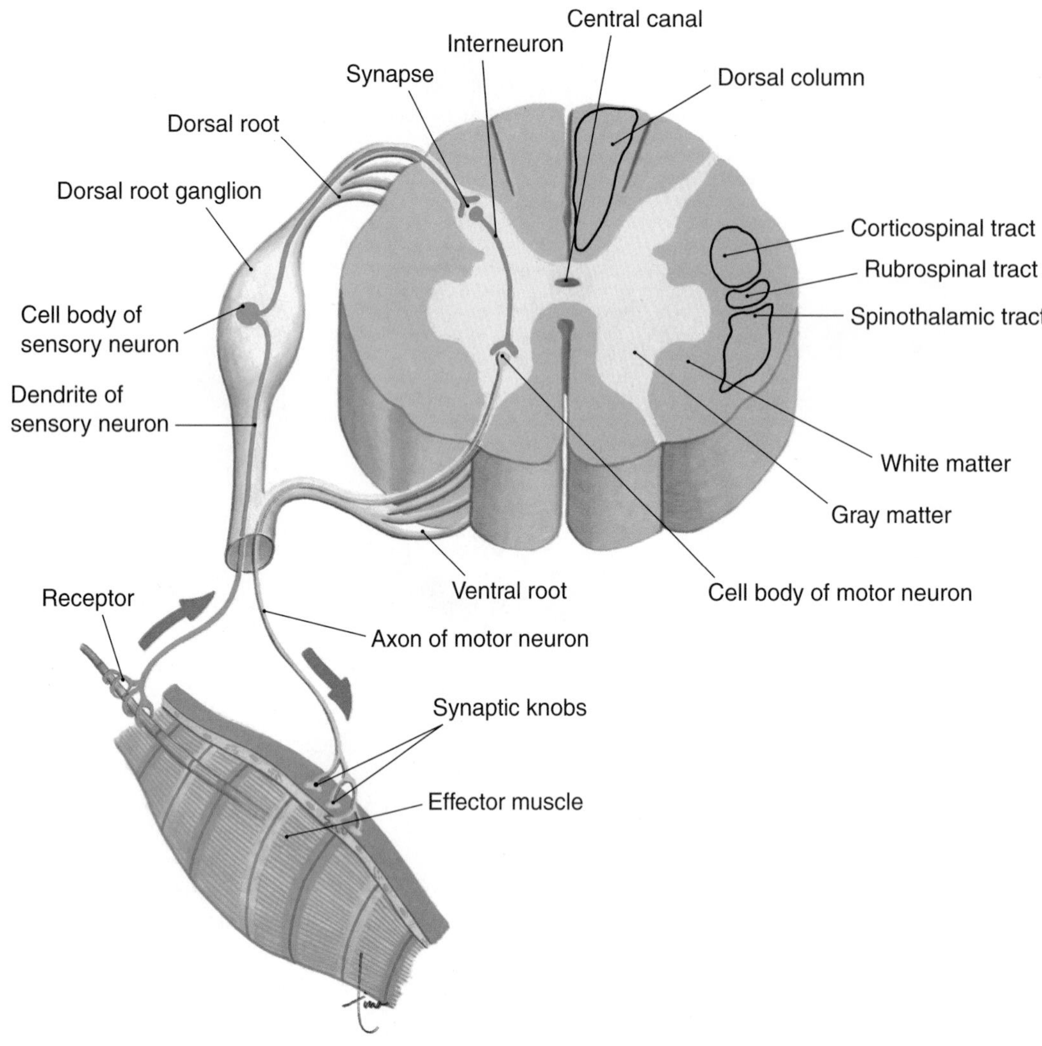

Figure 8–3 Cross-section of the spinal cord. Spinal nerve roots and their neurons are shown on the left side. Spinal nerve tracts are shown in the white matter on the right side. All tracts and nerves are bilateral (both sides).

send the information to the CNS in the form of impulses by way of the afferent neurons. The central nervous system interprets these impulses as a sensation. Sensory neurons from receptors in skin, skeletal muscles, and joints are called **somatic**; those from receptors in internal organs are called **visceral** sensory neurons.

Motor neurons (or **efferent neurons**) carry impulses from the central nervous system to **effectors**. The two types of effectors are muscles and glands. In response to impulses, muscles contract or relax and glands secrete. Motor neurons linked to skeletal muscle are called somatic; those to smooth muscle, cardiac muscle, and glands are called visceral.

Sensory and motor neurons make up the peripheral nervous system. Visceral motor neurons form the autonomic nervous system, a specialized subdivision of the PNS that will be discussed later in this chapter.

Interneurons are found entirely within the central nervous system. They are arranged so as to carry only sensory or motor impulses, or to integrate these functions. Some interneurons in the brain are concerned with thinking, learning, and memory.

A neuron carries impulses in only one direction. This is the result of the neuron's structure and location, as well as its physical arrangement with other neurons and the resulting pattern of synapses. The functioning nervous system, therefore, is an enormous network of "one-way streets," and there is no danger of impulses running into and canceling one another out.

NERVES AND NERVE TRACTS

A **nerve** is a group of axons and/or dendrites of many neurons, with blood vessels and connective tissue. **Sensory nerves** are made only of sensory neurons. The optic nerves for vision are examples of nerves with a purely sensory function. **Motor nerves** are made only of motor neurons; autonomic nerves are motor nerves. A **mixed nerve** contains both sensory and motor neurons. Most of our peripheral nerves, such as the sciatic nerves in the legs, are mixed nerves.

The term **nerve tract** refers to groups of neurons within the central nervous system. All the neurons in a nerve tract are concerned with either sensory or motor activity. These tracts are often referred to as white matter; the myelin sheaths of the neurons give them a white color.

THE NERVE IMPULSE

The events of an electrical nerve impulse are the same as those of the electrical impulse generated in muscle fibers which is discussed in Chapter 7. Stated simply, a neuron not carrying an impulse is in a state of **polarization**, with Na^+ ions more abundant outside the cell, and K^+ ions and negative ions more abundant inside the cell. The neuron has a positive charge on the outside of the cell membrane and a relative negative charge inside. A stimulus (such as a neurotransmitter) makes the membrane very permeable to Na^+ ions, which rush into the cell. This brings about **depolarization**, a reversal of charges on the membrane. The outside now has a negative charge, and the inside has a positive charge.

As soon as depolarization takes place, the neuron membrane becomes very permeable to K^+ ions, which rush out of the cell. This restores the positive charge outside and the negative charge inside, and is called **repolarization**. (The term action potential refers to depolarization followed by repolarization.)

Table 8–2 THE NERVE IMPULSE

State or Event	Description
Polarization (the neuron is not carrying an electrical impulse)	• Neuron membrane has a (+) charge outside and a (−) charge inside. • Na^+ ions are more abundant outside the cell. • K^+ ions and negative ions are more abundant inside the cell. Sodium and potassium pumps maintain these ion concentrations.
Depolarization (generated by a stimulus)	• Neuron membrane becomes very permeable to Na^+ ions, which rush into the cell. • The neuron membrane then has a (−) charge outside and a (+) charge inside.
Propagation of the impulse from point of stimulus	• Depolarization of part of the membrane makes adjacent membrane very permeable to Na^+ ions, and subsequent depolarization, which similarly affects the next part of the membrane, and so on. • The depolarization continues along the membrane of the neuron to the end of the axon.
Repolarization (immediately follows depolarization)	• Neuron membrane becomes very permeable to K^+ ions, which rush out of the cell. This restores the (+) charge outside and (−) charge inside the membrane. • The Na^+ ions are returned outside and the K^+ ions are returned inside by the sodium and potassium pumps. • The neuron is now able to respond to another stimulus and generate another impulse.

Then the sodium and potassium pumps return Na^+ ions outside and K^+ ions inside, and the neuron is ready to respond to another stimulus and transmit another impulse. An action potential in response to a stimulus takes place very rapidly and is measured in milliseconds. An individual neuron is capable of transmitting hundreds of action potentials (impulses) each second. A summary of the events of nerve impulse transmission is given in Table 8-2.

Transmission of electrical impulses is very rapid. The presence of an insulating myelin sheath increases the velocity of impulses, since only the nodes of Ranvier depolarize. This is called **saltatory conduction**. Many of our neurons are capable of transmitting impulses at a speed of many meters per second. Imagine a person 6 feet (about 2 meters) tall who stubs his toe; sensory impulses travel from the toe to the brain in less than a second (crossing a few synapses along the way). You can see how the nervous system can communicate so rapidly with all parts of the body, and why it is such an important regulatory system.

At synapses, nerve impulse transmission changes from electrical to chemical and depends on the release of neurotransmitters. Although diffusion across synapses is slow, the synapses are so small that this does not significantly affect the velocity of impulses in a living person.

THE SPINAL CORD

The **spinal cord** transmits impulses to and from the brain and is the integrating center for the spinal cord reflexes. Although this statement of functions is very brief, the spinal cord is of great importance to the nervous system and to the body as a whole.

Enclosed in the vertebral canal, the spinal cord is well protected from mechanical injury. In length, the spinal cord extends from the foramen magnum of the occipital bone to the disc between the first and second lumbar vertebrae.

A cross-section of the spinal cord is shown in Fig. 8-3; refer to it as you read the following. The internal **gray matter** is shaped like the letter H; gray matter consists of the cell bodies of motor neurons and interneurons. The external **white matter** is made of myelinated axons and dendrites of interneurons. These nerve fibers are grouped into nerve tracts based on their functions. **Ascending tracts** (such as the dorsal columns and spinothalamic tracts) carry sensory impulses to the brain. **Descending tracts** (such as the corticospinal and rubrospinal tracts) carry motor impulses away from the brain. Lastly, find the **central canal**; this contains **cerebrospinal fluid** and is continuous with cavities in the brain called ventricles.

SPINAL NERVES

There are 31 pairs of **spinal nerves**, those that emerge from the spinal cord. The nerves are named according to their respective vertebrae: 8 cervical pairs, 12 thoracic pairs, 5 lumbar pairs, 5 sacral pairs, and 1 very small coccygeal pair. These are shown in Fig. 8-4; notice that each nerve is designated by a letter and a number. The 8th cervical nerve is C8, the 1st thoracic nerve is T1, and so on.

In general, the cervical nerves supply the back of the head, neck, shoulders, arms, and the diaphragm. The first thoracic nerve also contributes to nerves in the arms. The remaining thoracic nerves supply the trunk of the body. The lumbar and sacral nerves supply the hips, pelvic cavity, and legs. Notice that the lumbar and sacral nerves hang below the end of the spinal cord (in order to reach their proper openings to exit from the vertebral canal); this is called the **cauda equina**, literally, the "horse's tail." Some of the important peripheral nerves and their destinations are listed in Table 8-3.

Each spinal nerve has two roots, which are neurons entering or leaving the spinal cord (see Fig. 8-3). The **dorsal root** is made of sensory neurons that carry impulses into the spinal cord. The **dorsal root ganglion** is an enlarged part of the dorsal root that contains the cell bodies of the sensory neurons. The term **ganglion** means a group of cell bodies outside the CNS. These cell bodies are within the vertebral canal and are thereby protected from injury (see Box 8-2: Shingles).

The **ventral root** is the motor root; it is made of motor neurons carrying impulses from the spinal cord to muscles or glands. The cell bodies of these motor neurons, as mentioned above, are in the gray matter of the spinal cord. When the two nerve roots merge, the spinal nerve thus formed is a mixed nerve.

SPINAL CORD REFLEXES

When you hear the term "reflex," you may think of an action that "just happens," and in part this is so. A

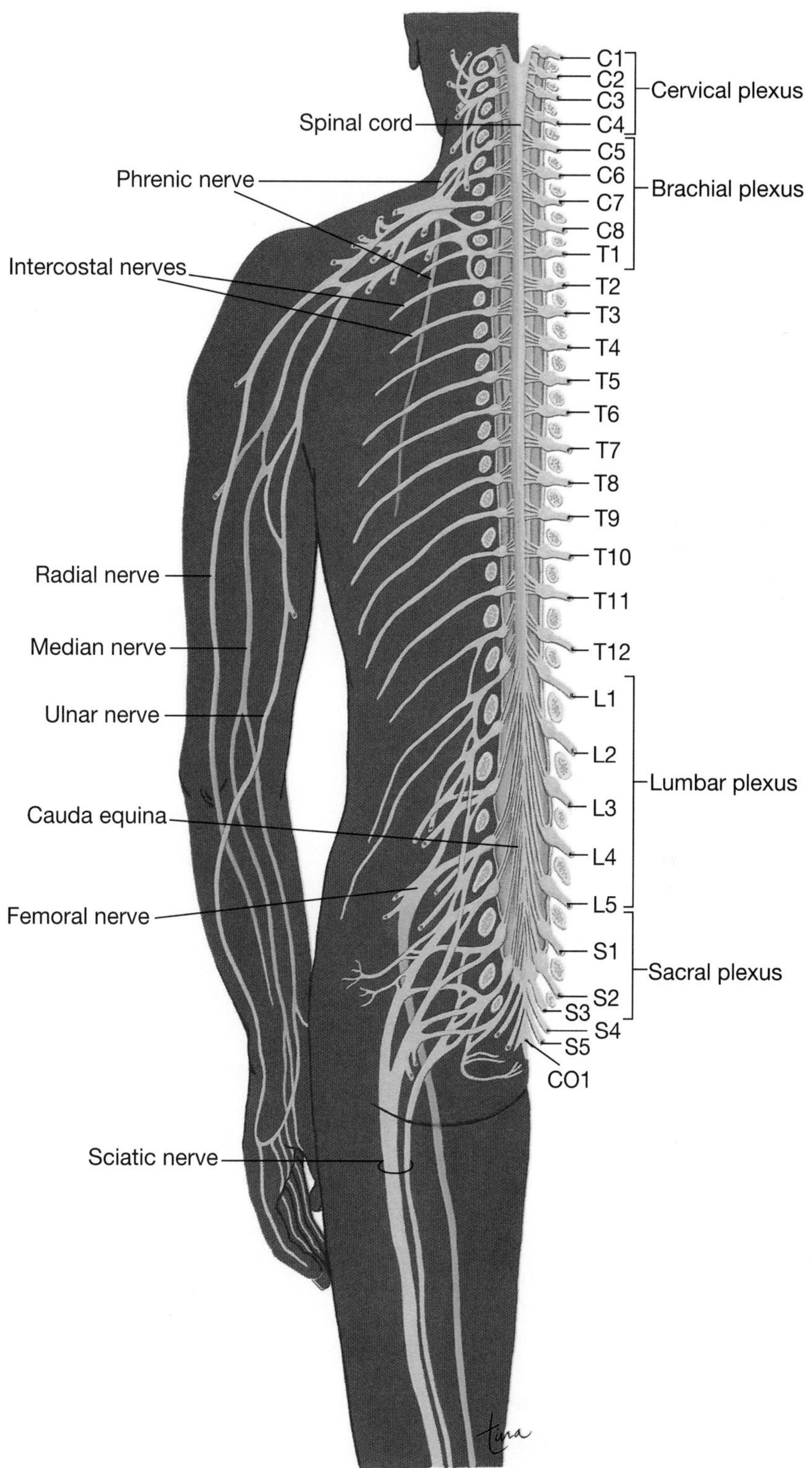

Figure 8–4 The spinal cord and spinal nerves. The distribution of spinal nerves is shown only on the left side. The nerve plexuses are labeled on the right side. A nerve plexus is a network of neurons from several segments of the spinal cord that combine to form nerves to specific parts of the body. For example, the radial and ulnar nerves to the arm emerge from the brachial plexus (see also Table 8–3).

Table 8–3 MAJOR PERIPHERAL NERVES

Nerve	Spinal Nerves That Contribute	Distribution
Phrenic	C3–C5	• Diaphragm
Radial	C5–C8, T1	• Skin and muscles of posterior arm, forearm, and hand; thumb and first 2 fingers
Median	C5–C8, T1	• Skin and muscles of anterior arm, forearm, and hand
Ulnar	C8, T1	• Skin and muscles of medial arm, forearm, and hand; little finger and ring finger
Intercostal	T2–T12	• Intercostal muscles, abdominal muscles; skin of trunk
Femoral	L2–L4	• Skin and muscles of anterior thigh, medial leg, and foot
Sciatic	L4–S3	• Skin and muscles of posterior thigh, leg, and foot

Box 8–2 SHINGLES

Shingles is caused by the same virus that causes chickenpox: the Herpes varicella-zoster virus. Varicella is chickenpox, which most of us probably had as children. When a person recovers from chickenpox, the virus may survive in a dormant (inactive) state in the dorsal root ganglia of some spinal nerves. For most people, the immune system is able to prevent reactivation of the virus. With increasing age, however, the immune system is not as effective, and the virus may become active and cause zoster, or shingles.

The virus is present in sensory neurons, often those of the trunk, but the damage caused by the virus is seen in the skin over the affected nerve. The raised, red lesions of shingles are often very painful and follow the course of the nerve on the skin external to it. Occasionally the virus may affect a cranial nerve and cause facial paralysis called Bell's palsy (7th cranial) or extensive facial lesions, or, rarely, blindness. Although it is not a cure, the antiviral medication acyclovir is now being used to lessen the duration of the illness.

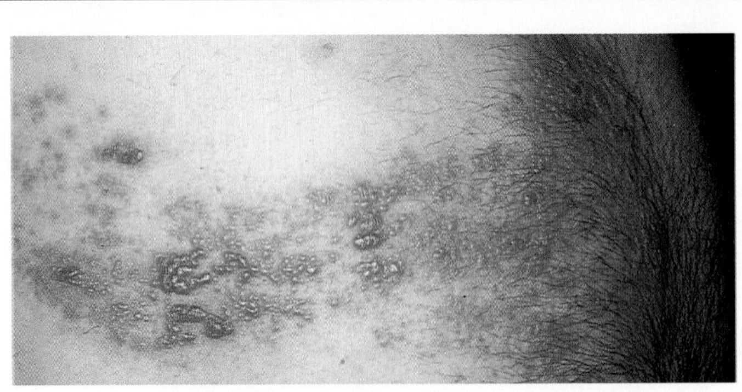

Box Figure 8–A Lesions of shingles on skin of trunk.

reflex is an involuntary response to a stimulus, that is, an automatic action stimulated by a specific change of some kind. **Spinal cord reflexes** are those that do not depend directly on the brain, although the brain may inhibit or enhance them. We do not have to think about these reflexes, which is very important, as you will see.

Reflex Arc

A **reflex arc** is the pathway nerve impulses travel when a reflex is elicited, and there are five essential parts:

1. **Receptors**—detect a change (the stimulus) and generate impulses.
2. **Sensory neurons**—transmit impulses from receptors to the CNS.
3. **Central nervous system**—contains one or more synapses (interneurons may be part of the pathway).
4. **Motor neurons**—transmit impulses from the CNS to the effector.
5. **Effector**—performs its characteristic action.

Let us now look at the reflex arc of a specific reflex, the **patellar** (or kneejerk) **reflex**, with which you are probably familiar. In this reflex, a tap on the patellar tendon just below the kneecap causes extension of the lower leg. This is a **stretch reflex**, which means that a muscle that is stretched will automatically contract. Refer now to Fig. 8–5 as you read the following:

In the quadriceps femoris muscle are (1) stretch receptors that detect the stretching produced by striking the patellar tendon. These receptors generate impulses that are carried along (2) sensory neurons in the femoral nerve to (3) the spinal cord. In the spinal cord, the sensory neurons synapse with (4) motor neurons (this is a two-neuron reflex). The motor neurons in the femoral nerve carry impulses back to (5) the quadriceps femoris, the effector, which contracts and extends the lower leg.

The patellar reflex is one of many that are used clinically to determine whether the nervous system is functioning properly. If the patellar reflex were absent in a patient, the problem could be in the thigh muscle, the femoral nerve, or the spinal cord. Further testing would be needed to determine the precise break in

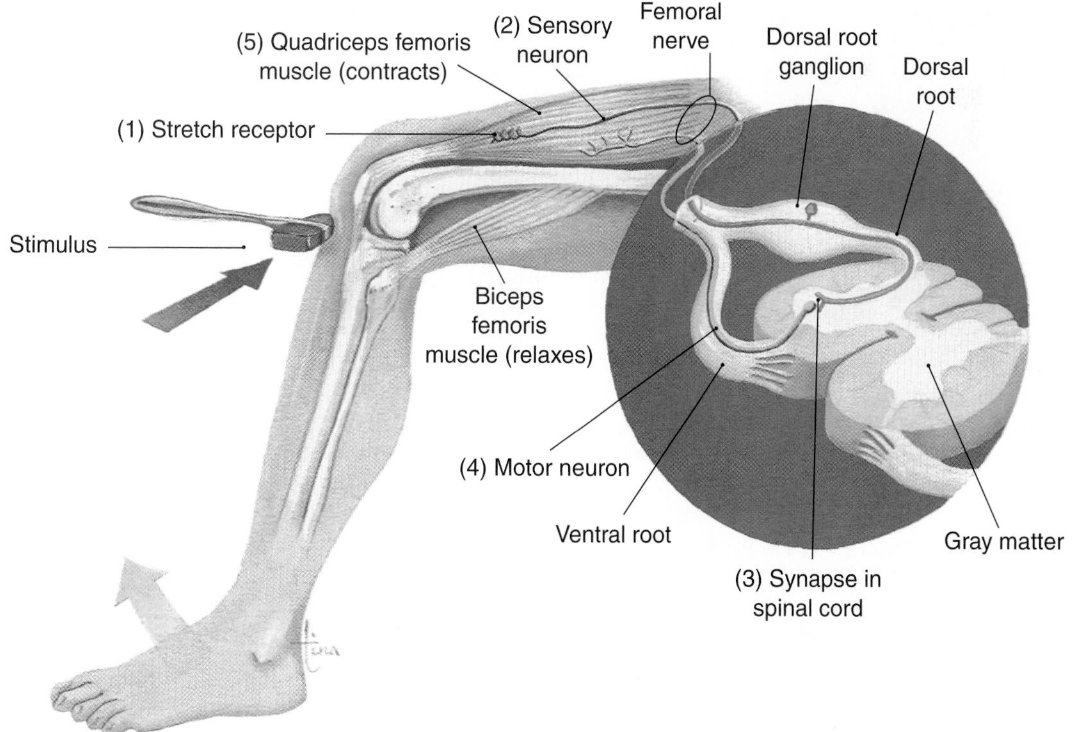

Figure 8–5 Patellar reflex. The reflex arc is shown. See text for description.

the reflex arc. If the reflex is normal, however, that means that all parts of the reflex arc are intact. So the testing of reflexes may be a first step in the clinical assessment of neurological damage.

You may be wondering why we have such reflexes, which are called stretch reflexes. What is their importance in our everyday lives? Imagine a person standing upright—is the body perfectly still? No, it isn't, because gravity exerts a downward pull. However, if the body tilts to the left, the right sides of the leg and trunk are stretched, and these stretched muscles automatically contract and pull the body upright again. This is the purpose of stretch reflexes; they help keep us upright without our having to think about doing so. If the brain had to make a decision every time we swayed a bit, all our concentration would be needed just to remain standing. Since these are spinal cord reflexes, the brain is not directly involved.

Flexor reflexes (or **withdrawal reflexes**) are another type of spinal cord reflex. The stimulus is something painful and potentially harmful, and the response is to pull away from it. If you inadvertently touch a hot stove, you automatically pull your hand away. Flexor reflexes are three-neuron reflexes, because sensory neurons synapse with interneurons in the spinal cord, which in turn synapse with motor neurons. Again, however, the brain does not have to make a decision to protect the body; the flexor reflex does that automatically (see Box 8-3: Spinal Cord Injuries).

THE BRAIN

The **brain** consists of many parts which function as an integrated whole. The major parts are the medulla, pons, and midbrain (collectively called the **brain**

Box 8–3 SPINAL CORD INJURIES

Injuries to the spinal cord are most often caused by auto accidents, falls, and gunshot wounds. The most serious injury is transection, or severing, of the spinal cord. If, for example, the spinal cord is severed at the level of the 8th thoracic segment, there will be paralysis and loss of sensation below that level. Another consequence is spinal shock, the at-least-temporary loss of spinal cord reflexes. In this example, the spinal cord reflexes of the lower trunk and legs will not occur. The stretch reflexes and flexor reflexes of the legs will be at least temporarily abolished, as will the urination and defecation reflexes. Although these reflexes do not depend directly on the brain, spinal cord neurons depend on impulses from the brain to enhance their own ability to generate impulses.

As spinal cord neurons below the injury recover their ability to generate impulses, these reflexes, such as the patellar reflex, often return. Urination and defecation reflexes may also be reestablished, but the person will not have an awareness of the need to urinate or defecate. Nor will voluntary control of these reflexes be possible, because inhibiting impulses from the brain can no longer reach the lower segments of the spinal cord.

Potentially less serious injuries are those in which the spinal cord is crushed rather than severed, and research is providing some promising treatments. Methylprednisolone, a steroid, if given within 8 hours of the injury seems to prevent further damage to the spinal cord by minimizing the inflammation process that occurs in damaged tissue. This helps preserve whatever spinal cord function remains.

A new drug, GM-1 ganglioside, seems to stimulate the production of a nerve growth factor that helps damaged neurons regenerate. Both of these therapies are now undergoing extensive clinical trials in trauma centers and hospitals.

Perhaps the most challenging research is the attempt to stimulate severed spinal cords to regenerate. Partial success has been achieved in rats, with Schwann cells transplanted from their peripheral nerves, and nerve growth factors produced by genetically engineered cells. The researchers caution, however, that it will probably be many years before their procedures will be tested on people.

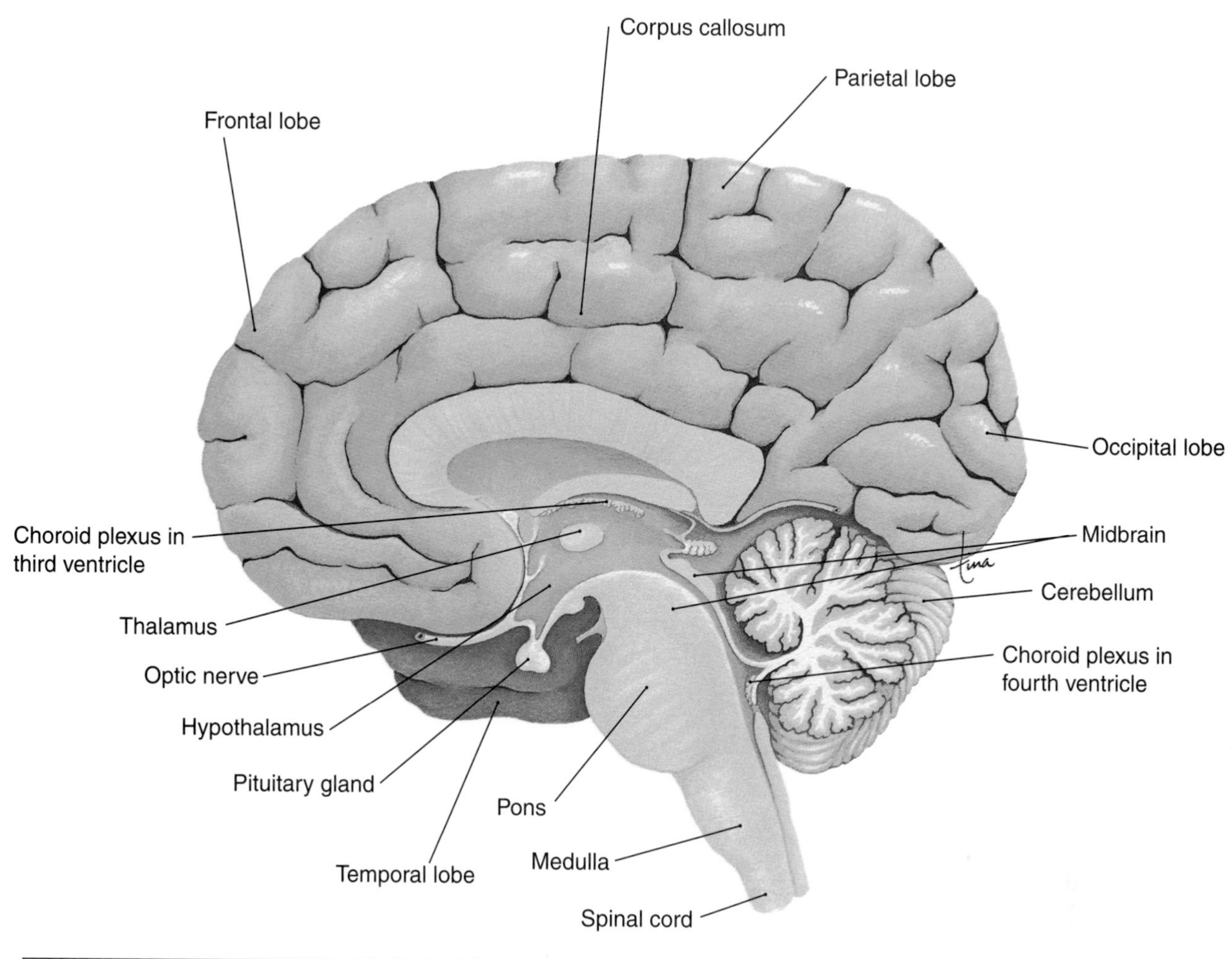

Figure 8–6 Midsagittal section of the brain as seen from the left side. This medial plane shows internal anatomy as well as the lobes of the cerebrum.

stem); the cerebellum, the hypothalamus and thalamus, and the cerebrum. These parts are shown in Fig. 8–6. We will discuss each part separately, but keep in mind that they are all interconnected and work together.

VENTRICLES

The **ventricles** are four cavities within the brain: two lateral ventricles, the third ventricle, and the fourth ventricle (Fig. 8–7). Each ventricle contains a capillary network called a **choroid plexus**, which forms **cerebrospinal fluid** (CSF) from blood plasma. Cerebrospinal fluid is the tissue fluid of the central nervous system; its circulation and functions will be discussed in the section on meninges.

MEDULLA

The **medulla** extends from the spinal cord to the pons and is anterior to the cerebellum. Its functions are those we think of as vital (as in "vital signs"). The medulla contains cardiac centers that regulate heart rate, vasomotor centers that regulate the diameter of blood vessels and, thereby, blood pressure, and respiratory centers that regulate breathing. You can see why a crushing injury to the occipital bone may be rapidly fatal—we cannot survive without the medulla.

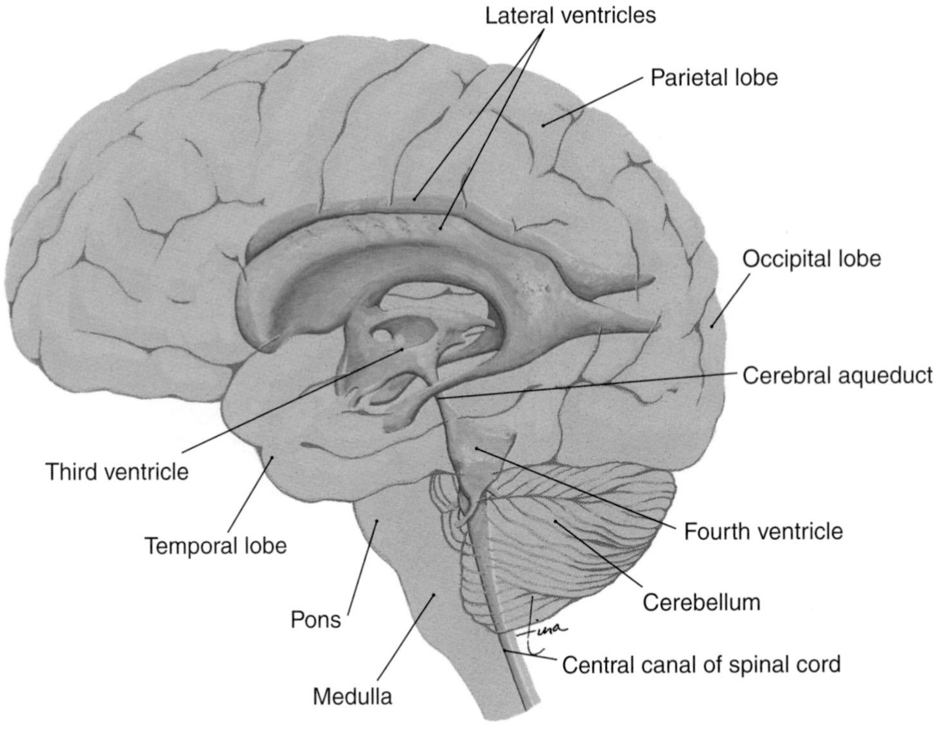

Figure 8–7 Ventricles of the brain as projected into the interior of the brain, which is seen from the left side.

Also in the medulla are reflex centers for coughing, sneezing, swallowing, and vomiting.

PONS

The **pons** bulges anteriorly from the upper part of the medulla. Within the pons are two respiratory centers that work with those in the medulla to produce a normal breathing rhythm. (The function of all the respiratory centers is discussed in Chapter 15.) The many other neurons in the pons connect the medulla with other parts of the brain.

MIDBRAIN

The **midbrain** extends from the pons to the hypothalamus and encloses the **cerebral aqueduct**, a tunnel that connects the third and fourth ventricles. Several different kinds of reflexes are integrated in the midbrain, including visual and auditory reflexes. If you see a wasp flying toward you, you automatically duck or twist away; this is a visual reflex, as is the coordinated movement of the eyeballs. Turning your head (ear) to a sound is an example of an auditory reflex. The midbrain is also concerned with what are called righting reflexes, those that keep the head upright and maintain balance or equilibrium.

CEREBELLUM

The **cerebellum** is separated from the medulla and pons by the fourth ventricle and is inferior to the occipital lobes of the cerebrum. All the functions of the cerebellum are concerned with movement. These include coordination, regulation of muscle tone, the appropriate trajectory and endpoint of movements, and the maintenance of posture and equilibrium. Notice that these are all involuntary, that is, the cerebellum functions below the level of conscious thought. This is important to permit the conscious brain to work without being overburdened. If you decide to pick up a pencil, for example, the impulses for arm movement come from the cerebrum. The cerebellum then modifies these impulses so that your arm and finger movements are coordinated, and you don't reach past the pencil.

In order to regulate equilibrium, the cerebellum (and midbrain) uses information provided by receptors in the inner ears. These receptors are discussed further in Chapter 9.

HYPOTHALAMUS

Located superior to the pituitary gland and inferior to the thalamus, the **hypothalamus** is a small area of the brain with many diverse functions.

1. Production of **antidiuretic hormone** (ADH) and **oxytocin**; these hormones are then stored in the posterior pituitary gland. ADH enables the kidneys to reabsorb water back to the blood and thus helps maintain blood volume. Oxytocin causes contractions of the uterus to bring about labor and delivery.
2. Production of releasing hormones (also called releasing factors) that stimulate the secretion of hormones by the anterior pituitary gland. Because these hormones are covered in Chapter 10, a single example will be given here: the hypothalamus produces **growth hormone releasing hormone** (GHRH), which stimulates the anterior pituitary gland to secrete growth hormone (GH).
3. Regulation of body temperature by promoting responses such as sweating in a warm environment or shivering in a cold environment (see Chapter 17).
4. Regulation of food intake; the hypothalamus is believed to respond to changes in blood nutrient levels or to chemicals secreted by fat cells. When blood nutrient levels are low, we experience a sensation of hunger, and eat. This raises blood nutrient levels and brings about a sensation of satiety, or fullness, and eating ceases.
5. Integration of the functioning of the autonomic nervous system, which in turn regulates the activity of organs such as the heart, blood vessels, and intestines. This will be discussed in more detail later in this chapter.
6. Stimulation of visceral responses during emotional situations. When we are angry, heart rate usually increases. Most of us, when embarrassed, will blush, which is vasodilation in the skin of the face. These responses are brought about by the autonomic nervous system when the hypothalamus perceives a change in emotional state. The neurological basis of our emotions is not well understood, and the visceral responses to emotions are not something most of us can control.
7. Regulation of body rhythms such as secretion of hormones, sleep cycles, changes in mood or mental alertness. This is often referred to as our biological clock, the rhythms as circadian rhythms, meaning "about a day." If you have ever had to stay awake for 24 hours, you know how disorienting it can be, until the hypothalamic biological clock has been reset.

THALAMUS

The **thalamus** is superior to the hypothalamus and inferior to the cerebrum. The third ventricle is a narrow cavity that passes through both the thalamus and hypothalamus. The functions of the thalamus are concerned with sensation. Sensory impulses to the brain (except those for the sense of smell) follow neuron pathways that first enter the thalamus, which groups the impulses before relaying them to the cerebrum, where sensations are felt. For example, holding a cup of hot coffee generates impulses for heat, touch and texture, and the shape of the cup (muscle sense), but we do not experience these as separate sensations. The thalamus integrates the impulses, or puts them together, so that the cerebrum feels the whole and is able to interpret the sensation quickly.

The thalamus may also suppress unimportant sensations. If you are reading an enjoyable book, you may not notice someone coming into the room. By temporarily blocking minor sensations, the thalamus permits the cerebrum to concentrate on important tasks.

CEREBRUM

The largest part of the human brain is the **cerebrum**, which consists of two hemispheres separated by the longitudinal fissure. At the base of this deep groove is the **corpus callosum**, a band of 200 million neurons that connects the right and left hemispheres. Within each hemisphere is a lateral ventricle.

The surface of the cerebrum is gray matter called

the **cerebral cortex**. Gray matter consists of cell bodies of neurons, which carry out the many functions of the cerebrum. Internal to the gray matter is white matter, made of myelinated axons and dendrites that connect the lobes of the cerebrum to one another and to all other parts of the brain.

In the human brain the cerebral cortex is folded extensively. The folds are called **convolutions** or **gyri**, and the grooves between them are **fissures** or **sulci**. This folding permits the presence of millions more neurons in the cerebral cortex. The cerebral cortex of an animal such as a dog or cat does not have this extensive folding. This difference enables us to read, speak, do long division, and so many other "human" things that dogs and cats cannot.

The cerebral cortex is divided into lobes that have the same names as the cranial bones external to them. Therefore, each hemisphere has a frontal lobe, parietal lobe, temporal lobe, and occipital lobe (Fig. 8-8). These lobes have been mapped, that is, certain areas are known to be associated with specific functions. We will discuss the functions of the cerebrum according to these mapped areas.

Frontal Lobes

Within the **frontal lobes** are the **motor areas** that generate the impulses for voluntary movement. The left motor area controls movement on the right side of the body, and the right motor area controls the left side of the body. This is why a patient who has had a cerebrovascular accident, or stroke, in the right frontal lobe will have paralysis of muscles on the left side (see Box 8-4: Cerebrovascular Accidents).

Also in the frontal lobe, usually only the left lobe for most right-handed people, is **Broca's motor speech** area, which controls the movements of the mouth involved in speaking.

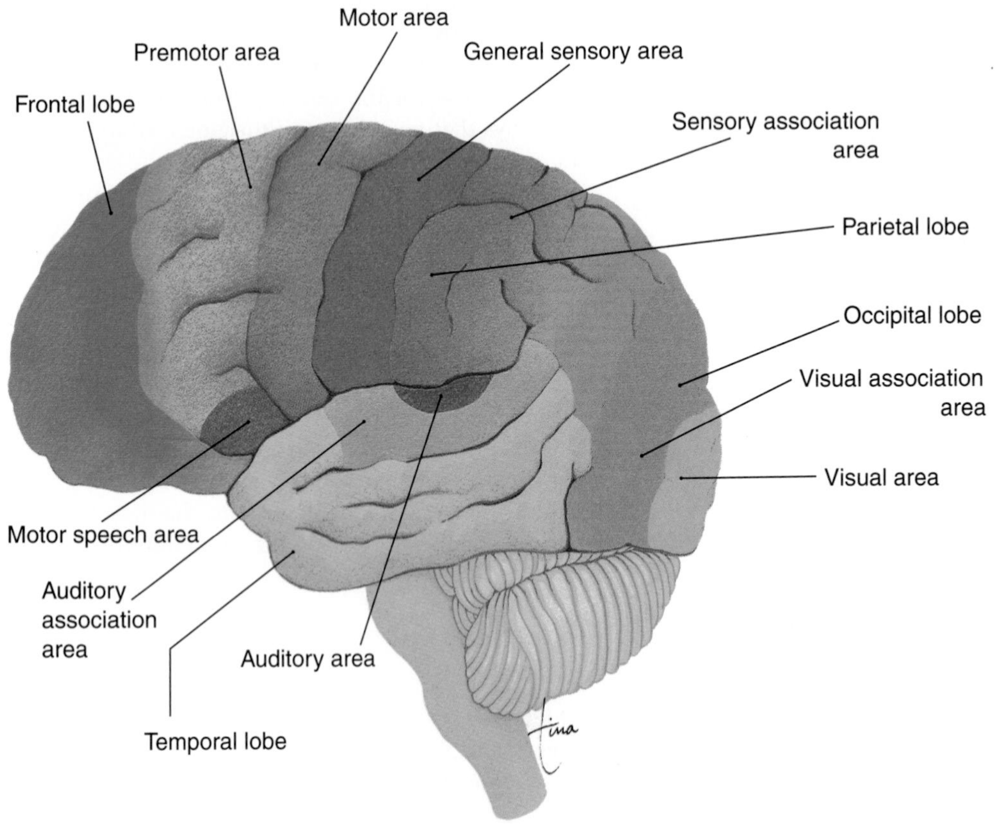

Figure 8-8 Left cerebral hemisphere showing some of the functional areas that have been mapped.

Box 8–4 CEREBROVASCULAR ACCIDENTS

A **cerebrovascular accident (CVA)**, or **stroke**, is damage to a blood vessel in the brain, resulting in lack of oxygen to that part of the brain. Possible types of vessel damage are thrombosis or hemorrhage.

A **thrombus** is a blood clot, which most often is a consequence of atherosclerosis, abnormal lipid deposits in cerebral arteries. The rough surface stimulates clot formation, which obstructs the blood flow to the part of the brain supplied by the artery. The symptoms depend on the part of the brain affected and may be gradual in onset if clot formation is slow. Approximately 80% of CVAs are of this type.

A hemorrhage, the result of arteriosclerosis or **aneurysm** of a cerebral artery, allows blood out into brain tissue, which destroys brain neurons by putting excessive pressure on them as well as depriving them of oxygen. Onset of symptoms in this type of CVA is usually rapid.

If, for example, the CVA is in the left frontal lobe, paralysis of the right side of the body will occur. Speech may also be affected if the speech areas are involved. Some CVAs are fatal because the damage they cause is very widespread or affects vital centers in the medulla or pons.

For CVAs of the thrombus type, a clot-dissolving drug may help reestablish blood flow. To be effective, however, the drug must be administered within 3 hours of symptom onset (see also Box 11–7).

Recovery from a CVA depends on its location and the extent of damage, as well as other factors. One of these is the redundancy of the brain. Redundancy means repetition; the cerebral cortex has many more neurons than we actually use in daily activities. These neurons are available for use, especially in younger people (less than 50 years of age). When a patient recovers from a disabling stroke, what has often happened is that the brain has established new pathways, with previously little-used neurons now carrying impulses "full time." Such recovery is highly individual and may take months. Yet another important factor is that CVA patients be started on rehabilitation therapy as soon as their condition permits.

Parietal Lobes

The **general sensory areas** in the **parietal lobes** receive impulses from receptors in the skin and feel and interpret the cutaneous sensations. The left area is for the right side of the body and vice versa. These areas also receive impulses from stretch receptors in muscles for conscious muscle sense. Impulses from taste buds travel to the **taste areas**, which overlap the parietal and temporal lobes.

Temporal Lobes

The **auditory areas**, as their name suggests, receive impulses from receptors in the inner ear for hearing. The **olfactory areas** receive impulses from receptors in the nasal cavities for the sense of smell.

Also in the temporal and parietal lobes in the left hemisphere (for most of us) are other speech areas concerned with the thought that precedes speech. Each of us can probably recall (and regret) times when we have "spoken without thinking," but in actuality that is not possible. The thinking takes place very rapidly and is essential in order to be able to speak (see Box 8-5: Aphasia).

Occipital Lobes

Impulses from the retinas of the eyes travel along the optic nerves to the **visual areas**. These areas "see" and interpret what is seen. Other parts of the occipital lobes are concerned with spatial relationships; such things as judging distance and seeing in three dimensions.

Association Areas

As you can see in Fig. 8-8, there are many parts of the cerebral cortex not concerned with movement or a

Box 8–5 APHASIA

Our use of language sets us apart from other animals and involves speech, reading, and writing. Language is the use of symbols (words) to designate objects and to express ideas. Damage to the speech areas or interpretation areas of the cerebrum may impair one or more aspects of a person's ability to use language; this is called **aphasia**.

Aphasia may be a consequence of a cerebrovascular accident, or of physical trauma to the skull and brain such as a head injury sustained in an automobile accident. If the motor speech (Broca's) area is damaged, the person is still able to understand written and spoken words and knows what he wants to say, but he cannot say it. Without coordination and impulses from the motor speech area, the muscles used for speech cannot contract to form words properly.

Auditory aphasia is **"word deafness,"** caused by damage to an interpretation area. The person can still hear but cannot comprehend what the words mean. Visual aphasia is **"word blindness"**; the person can still see perfectly well, but cannot make sense of written words (the person retains the ability to understand spoken words). Imagine how you would feel if wms qsbbcljw jmqr rfc yzgjgrw rm pcyb. Frustrating isn't it? You know that those symbols are letters, but you cannot "decode" them right away. Those "words" were formed by shifting the alphabet two letters (A = C, B = D, C = E, etc.), and would normally be read as: "you suddenly lost the ability to read." That may give you a small idea of what word blindness is like.

particular sensation. These may be called **association areas** and perhaps are what truly make us individuals. It is probably these areas that give each of us a personality, a sense of humor, and the ability to reason and use logic. Learning and memory are also functions of these areas.

Although much has been learned about the formation of memories, the processes are still incompletely understood and mostly beyond the scope of this book. Briefly, however, we can say that memories of things such as people or what you did last summer involve the hippocampus (from the Greek for "seahorse," because of its shape), part of the temporal lobe on the floor of the lateral ventricle. The hippocampi seem to collect information from many areas of the cerebral cortex. When you meet a friend, for example, the memory emerges as a whole: "Here's Fred," not in pieces. People whose hippocampi are damaged cannot form new memories that last more than a few seconds.

It is believed that most, if not all, of what we have experienced or learned is stored somewhere in the brain. Sometimes a trigger may bring back memories; a certain scent or a song are possible triggers. Then we find ourselves recalling something from the past and wondering where it came from.

The loss of personality due to destruction of brain neurons is perhaps most dramatically seen in Alzheimer's disease (see Box 8-6: Alzheimer's Disease).

Basal Ganglia

The **basal ganglia** are paired masses of gray matter within the white matter of the cerebral hemispheres. Their functions are certain subconscious aspects of voluntary movement: regulation of muscle tone and accessory movements such as swinging the arms when walking or gesturing while speaking. The most common disorder of the basal ganglia is Parkinson's disease (see Box 8-7: Parkinson's Disease).

Corpus Callosum

As mentioned previously, the **corpus callosum** is a band of nerve fibers that connects the left and right cerebral hemispheres. This enables each hemisphere to know of the activity of the other. This is especially important for people because for most of us, the left hemisphere contains speech areas and the right hemisphere does not. The corpus callosum, therefore, lets the right hemisphere know what the left hemisphere is talking about. The "division of labor" of our cerebral

Box 8–6 ALZHEIMER'S DISEASE

In the United States, Alzheimer's disease, a progressive, incurable form of mental deterioration, affects approximately 5 million people and is the cause of 100,000 deaths each year. The first symptoms, which usually begin after age 65, are memory lapses and slight personality changes. As the disease progresses, there is total loss of memory, reasoning ability, and personality, and those with advanced disease are unable to perform even the simplest tasks or self-care.

Structural changes in the brains of Alzheimer's patients may be seen at autopsy. Neurofibrillary tangles are abnormal fibrous proteins found in the cerebral cortex in areas important for memory and reasoning. Also present are plaques made of another protein called beta-amyloid.

Recently, a defective gene has been found in some patients who have late-onset Alzheimer's disease, the most common type. Yet another gene seems to trigger increased synthesis of beta-amyloid. Current research is focused on the interaction of these genes, much like putting together a puzzle, with the goal of finding drugs to block the overproduction of beta-amyloid.

Some medications help improve memory by delaying the breakdown of acetylcholine in the brain. There is evidence that estrogen and vitamin E may help slow the progression of the disease. It is likely that treatment of Alzheimer's disease will one day mean delaying its onset with a variety of medications, each targeted at a different aspect of this complex disease.

Box 8–7 PARKINSON'S DISEASE

Parkinson's disease is a disorder of the basal ganglia whose cause is unknown, and though there is a genetic component in some families, it is probably not the only factor. The disease usually begins after the age of 60. Neurons in the basal ganglia that produce the neurotransmitter dopamine begin to degenerate and die, and the deficiency of dopamine causes specific kinds of muscular symptoms. Tremor, or involuntary shaking, of the hands is probably the most common symptom. The accessory movements regulated by the basal ganglia gradually diminish, and the affected person walks slowly without swinging the arms. A mask-like face is characteristic of this disease, as the facial muscles become rigid. Eventually all voluntary movements become slower and much more difficult, and balance is seriously impaired.

Dopamine itself cannot be used to treat Parkinson's disease because it does not cross the blood–brain barrier. A substance called L-dopa does cross and can be converted to dopamine by brain neurons. Unfortunately, L-dopa begins to lose its therapeutic effectiveness within a few years.

Recently, another medication called deprenyl has been shown to slow the progression of Parkinson's disease in a small number of patients. The researchers believe that deprenyl may actually prevent the death of basal ganglia neurons, but this has yet to be proven. Studies using larger numbers of patients are now being conducted, and deprenyl has been made available to all those with Parkinson's disease. Also undergoing clinical trials are drugs that mimic dopamine, and transplants of fetal nerve cells.

hemispheres is beyond the scope of this book, but it is a fascinating subject that you may wish to explore further.

MENINGES AND CEREBROSPINAL FLUID

The connective tissue membranes that cover the brain and spinal cord are called **meninges**; the three layers are illustrated in Fig. 8–9. The thick outermost layer, made of fibrous connective tissue, is the **dura mater**, which lines the skull and vertebral canal. The middle **arachnoid membrane** (arachnids are spiders) is made of web-like strands of connective tissue. The innermost **pia mater** is a very thin membrane on the surface of the spinal cord and brain. Between the arachnoid and the pia mater is the **subarachnoid space**, which contains cerebrospinal fluid (CSF), the tissue fluid of the central nervous system.

Recall the ventricles (cavities) of the brain: two lateral ventricles, the third ventricle, and fourth ventricle. Each contains a choroid plexus, a capillary network that forms cerebrospinal fluid from blood plasma. This is a continuous process, and the cerebrospinal fluid then circulates in and around the central nervous system (Fig. 8–10).

From the lateral and third ventricles, cerebrospinal fluid flows through the fourth ventricle, then to the central canal of the spinal cord, and to the cranial and spinal subarachnoid spaces. As more cerebrospinal fluid is formed, you might expect that some must be reabsorbed, and that is just what happens. From the

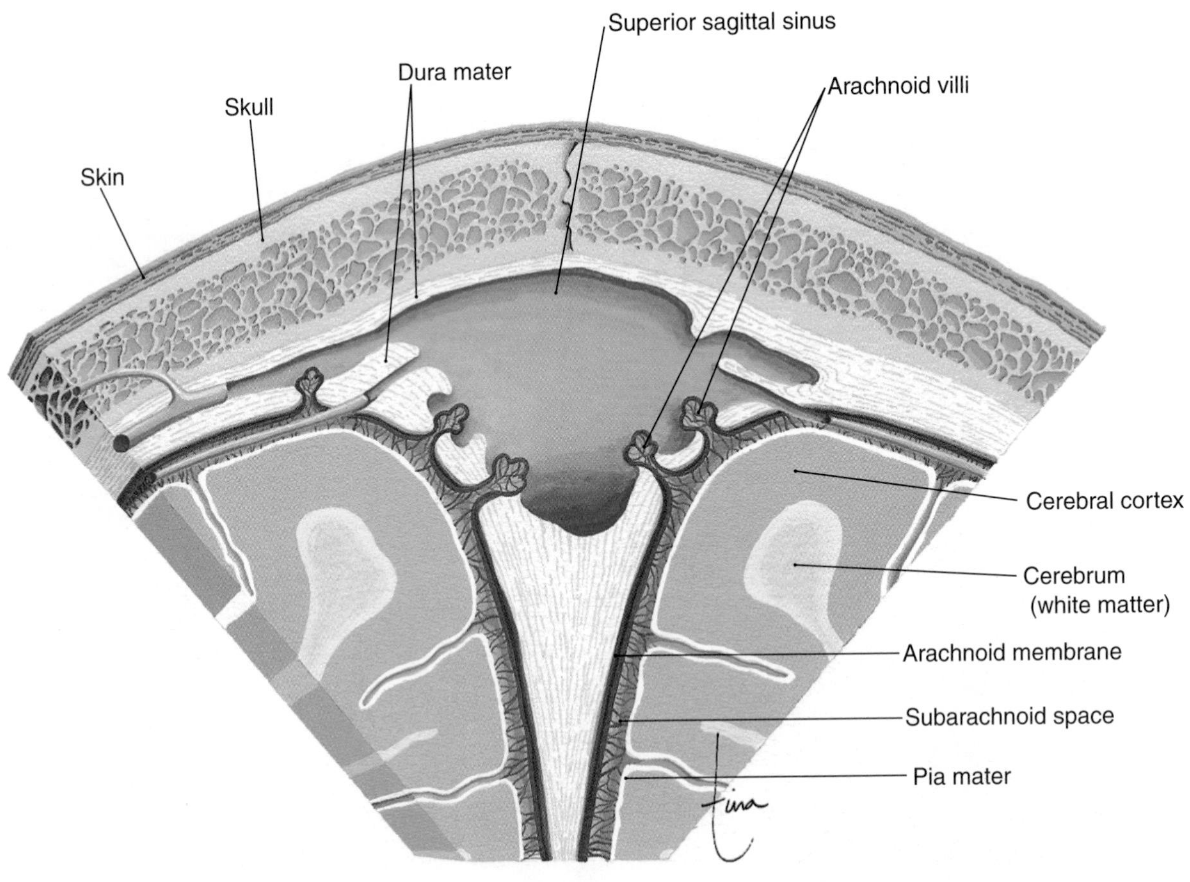

Figure 8–9 Structure of the meninges. Frontal section through the top of the skull showing the double-layered dura mater and one of the cranial venous sinuses.

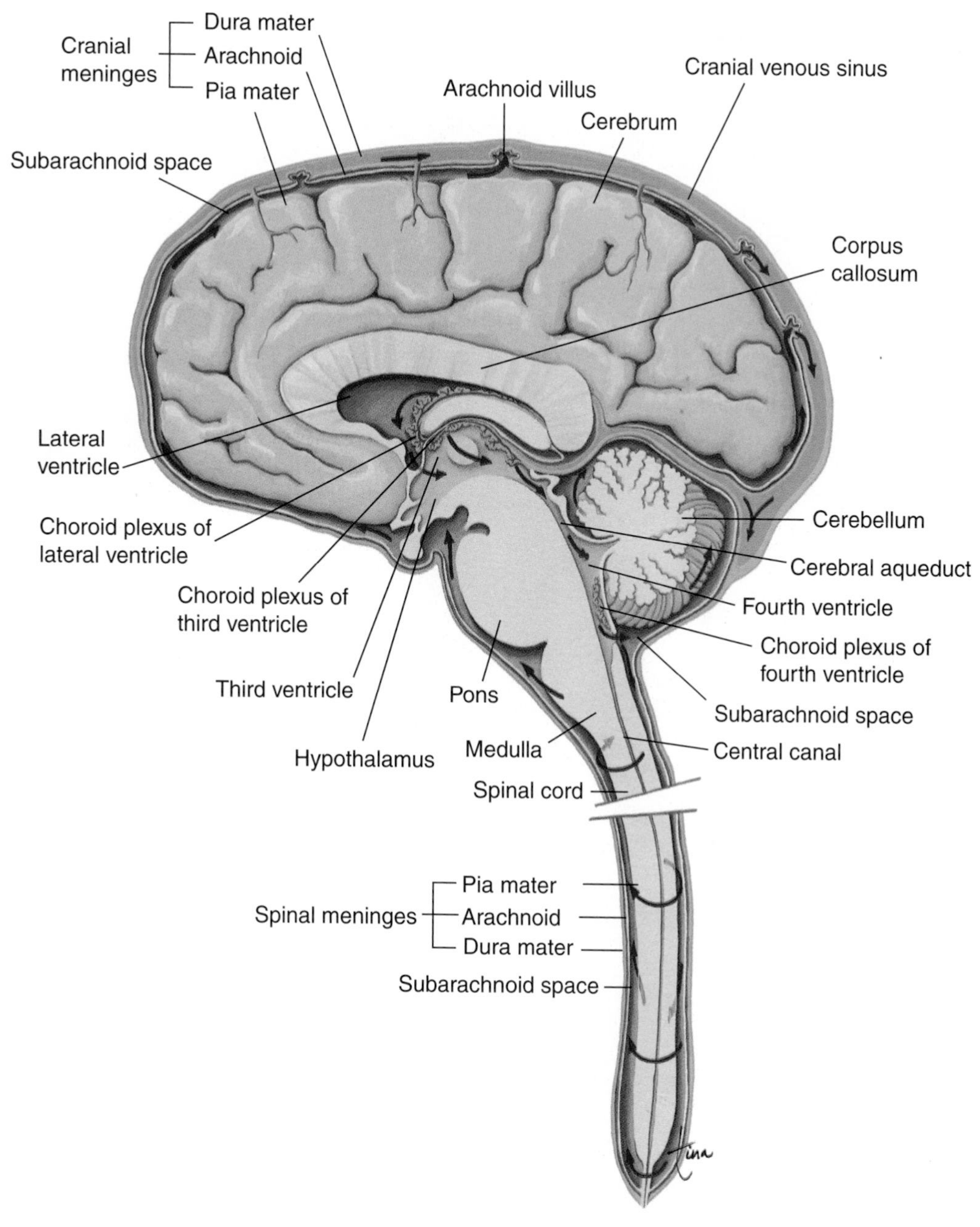

Figure 8–10 Formation, circulation, and reabsorption of cerebrospinal fluid. See text for description.

cranial subarachnoid space, cerebrospinal fluid is reabsorbed through **arachnoid villi** into the blood in **cranial venous sinuses** (large veins within the double-layered cranial dura mater). The cerebrospinal fluid becomes blood plasma again, and the rate of reabsorption normally equals the rate of production.

Since cerebrospinal fluid is tissue fluid, one of its functions is to bring nutrients to CNS neurons and to

Box 8–8 LUMBAR PUNCTURE

A **lumbar puncture** (spinal tap) is a diagnostic procedure that involves the removal of cerebrospinal fluid to determine its pressure and constituents. As the name tells us, the removal, using a syringe, is made in the lumbar area. Because the spinal cord ends between the 1st and 2nd lumbar vertebrae, the needle is usually inserted between the 4th and 5th lumbar vertebrae. The meningeal sac containing cerebrospinal fluid extends to the end of the lumbar vertebrae, permitting access to the cerebrospinal fluid with little chance of damaging the spinal cord.

Cerebrospinal fluid is a circulating fluid and has a normal pressure of 70 to 200 mmH_2O. An abnormal pressure usually indicates an obstruction in circulation, which may be caused by infection, a tumor, or mechanical injury. Other diagnostic tests would be needed to determine the precise cause.

Perhaps the most frequent reason for a lumbar puncture is suspected **meningitis**, which may be caused by several kinds of bacteria. If the patient does have meningitis, the cerebrospinal fluid will be cloudy rather than clear and will be examined for the presence of bacteria and many white blood cells. A few WBCs in CSF is normal, because WBCs are found in all tissue fluid.

Another abnormal constituent of cerebrospinal fluid is red blood cells. Their presence indicates bleeding somewhere in the central nervous system. There may be many causes, and again, further testing would be necessary.

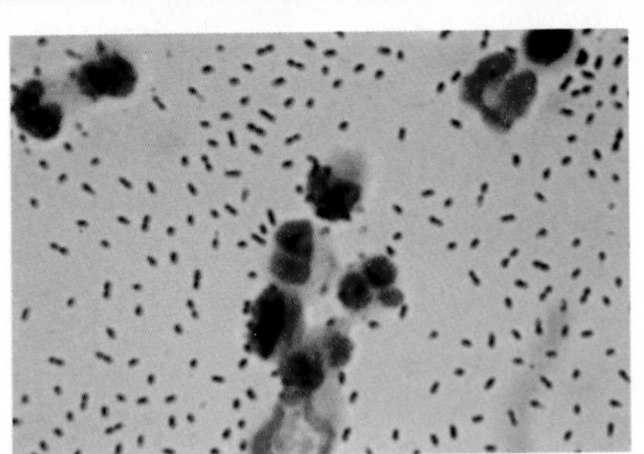

Box Figure 8–B Cerebrospinal fluid from a patient with meningitis. The bacteria are streptococci, found in pairs. The large cells are WBC's. (×500) (From Sacher, RA, and McPherson, RA: Clinical Interpretation of Laboratory Tests, ed. 10. FA Davis, Philadelphia, 1991, Plate 52, with permission.)

remove waste products to the blood as the fluid is reabsorbed. The other function of cerebrospinal fluid is to act as a cushion for the central nervous system. The brain and spinal cord are enclosed in fluid-filled membranes that absorb shock. You can, for example, shake your head vigorously without harming your brain. Naturally, this protection has limits; very sharp or heavy blows to the skull will indeed cause damage to the brain.

Examination of cerebrospinal fluid may be used in the diagnosis of certain diseases (see Box 8–8: Lumbar Puncture).

CRANIAL NERVES

The 12 pairs of **cranial nerves** emerge from the brain stem or other parts of the brain—they are shown in Fig. 8–11. The name "cranial" indicates their origin, and many of them do carry impulses for functions involving the head. Some, however, have more far-reaching destinations.

The impulses for the senses of smell, taste, sight, hearing, and equilibrium are all carried by cranial nerves to their respective sensory areas in the brain.

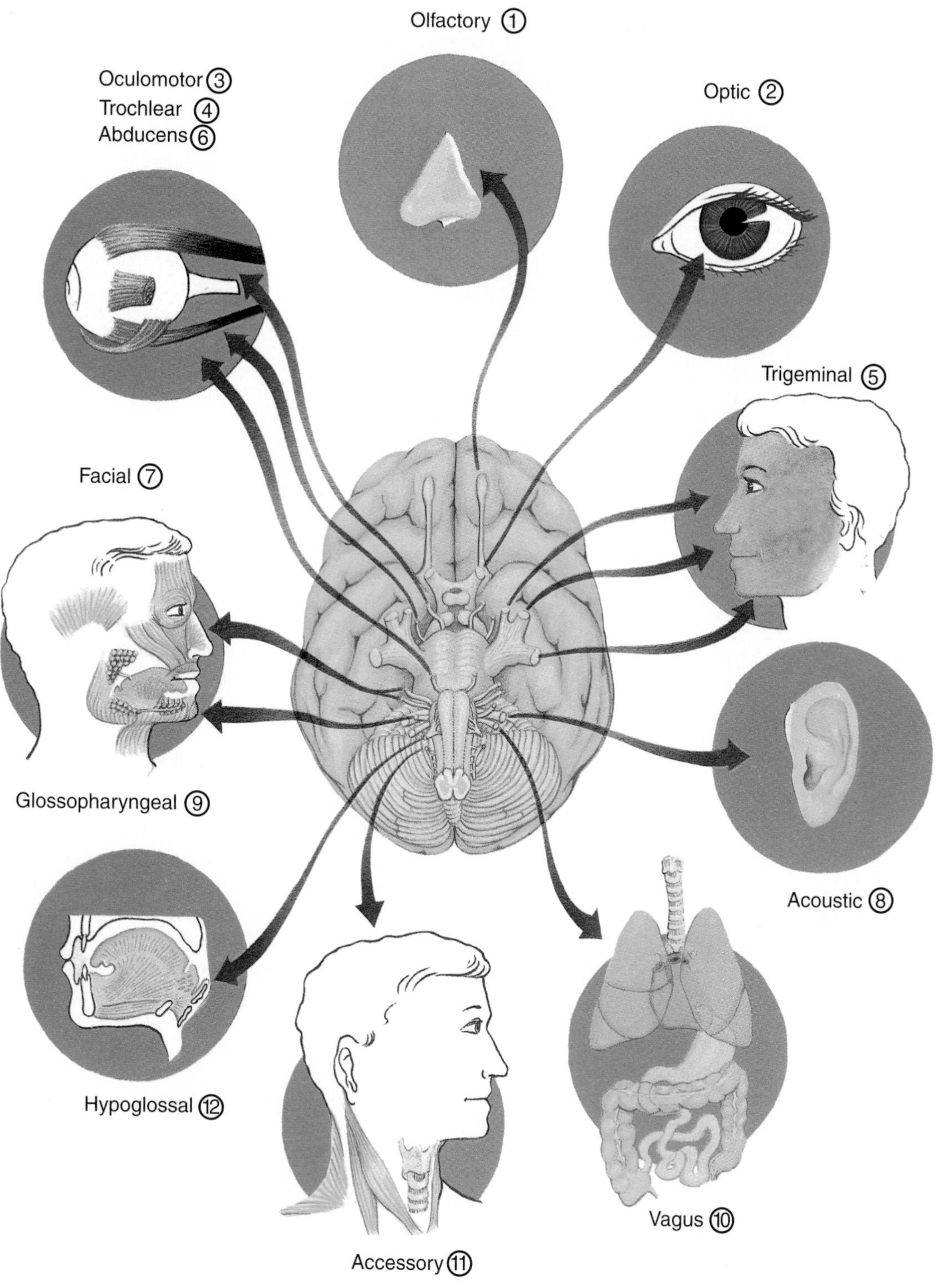

Figure 8–11 Cranial nerves and their distributions. The brain is shown in an inferior view. See Table 8–4 for descriptions.

Table 8–4 CRANIAL NERVES

Number and Name	Function(s)
I Olfactory	• Sense of smell
II Optic	• Sense of sight
III Oculomotor	• Movement of the eyeball; constriction of pupil in bright light or for near vision
IV Trochlear	• Movement of eyeball
V Trigeminal	• Sensation in face, scalp, and teeth; contraction of chewing muscles
VI Abducens	• Movement of the eyeball
VII Facial	• Sense of taste; contraction of facial muscles; secretion of saliva
VIII Acoustic (Vestibulocochlear)	• Sense of hearing; sense of equilibrium
IX Glossopharyngeal	• Sense of taste; sensory for cardiac, respiratory, and blood pressure reflexes; contraction of pharynx; secretion of saliva
X Vagus	• Sensory in cardiac, respiratory, and blood pressure reflexes; sensory and motor to larynx (speaking); decreases heart rate; contraction of alimentary tube (peristalsis); increases digestive secretions
XI Accessory	• Contraction of neck and shoulder muscles; motor to larynx (speaking)
XII Hypoglossal	• Movement of the tongue

Some cranial nerves carry motor impulses to muscles of the face and eyes or to the salivary glands. The vagus nerves ("vagus" means "wanderer") branch extensively to the larynx, heart, stomach and intestines, and the bronchial tubes.

The functions of all the cranial nerves are summarized in Table 8–4.

THE AUTONOMIC NERVOUS SYSTEM

The **autonomic nervous system (ANS)** is actually part of the peripheral nervous system in that it consists of motor portions of some cranial and spinal nerves. Because its functioning is so specialized, however, the autonomic nervous system is usually discussed as a separate entity, as we will do here.

Making up the autonomic nervous system are **visceral motor neurons** to smooth muscle, cardiac muscle, and glands. These are the **visceral effectors**; muscle will either contract or relax, and glands will either increase or decrease their secretions.

The ANS has two divisions: **sympathetic** and **parasympathetic**. Often, they function in opposition to each other, as you will see. The activity of both divisions is integrated by the hypothalamus, which ensures that the visceral effectors will respond appropriately to the situation.

AUTONOMIC PATHWAYS

An autonomic nerve pathway from the central nervous system to a visceral effector consists of two motor neurons that synapse in a ganglion outside the CNS (Fig. 8–12). The first neuron is called the **preganglionic neuron**, from the CNS to the ganglion. The second neuron is called the **postganglionic neuron**, from the ganglion to the visceral effector. The ganglia are actually the cell bodies of the postganglionic neurons.

SYMPATHETIC DIVISION

Another name for the sympathetic division is thoracolumbar division, which tells us where the sympathetic preganglionic neurons originate. Their cell bodies are in the thoracic segments and some of the lumbar segments of the spinal cord. Their axons extend to the sympathetic ganglia, most of which are located in two chains just outside the spinal column (see Fig. 8–12). Within the ganglia are the synapses between preganglionic and postganglionic neurons; the postganglionic axons then go to the visceral effec-

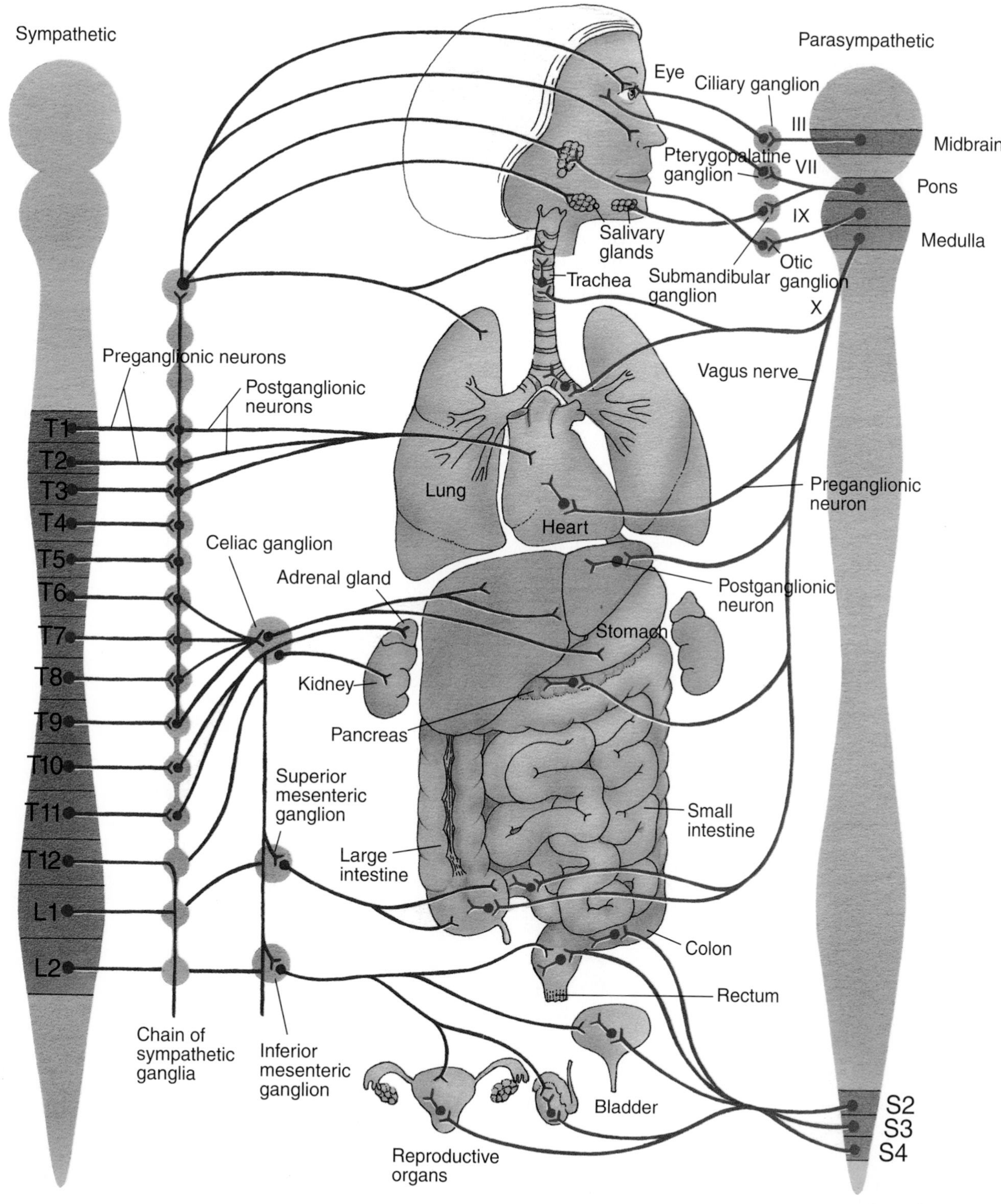

Figure 8–12 The autonomic nervous system. The sympathetic division is shown on the left, and the parasympathetic division is shown on the right (both divisions are bilateral).

tors. One preganglionic neuron often synapses with many postganglionic neurons to many effectors. This anatomic arrangement has physiologic importance: the sympathetic division brings about widespread responses in many organs.

The sympathetic division is dominant in stressful situations, which include anger, fear, or anxiety, as well as exercise. For our prehistoric ancestors, stressful situations often involved the need for intense physical activity—the "fight or flight response." Our nervous systems haven't changed very much in 50,000 years, and if you look at Table 8–5, you will see the kinds of responses the sympathetic division stimulates. The heart rate increases, vasodilation in skeletal muscles supplies them with more oxygen, the bronchioles dilate to take in more air, and the liver changes glycogen to glucose to supply energy. At the same time, digestive secretions decrease and peristalsis slows; these are not important in a stress situation. Vasoconstriction in the skin and viscera shunts blood to more vital organs such as the heart, muscles, and brain. All these responses enabled our ancestors to stay and fight or to get away from potential danger. Even though we may not always be in life-threatening situations during stress (such as figuring out our income taxes), our bodies are prepared for just that.

PARASYMPATHETIC DIVISION

The other name for the parasympathetic division is the craniosacral division. The cell bodies of parasympathetic preganglionic neurons are in the brain stem and the sacral segments of the spinal cord. Their axons are in cranial nerve pairs 3, 7, 9, and 10 and in some sacral nerves and extend to the parasympathetic ganglia. These ganglia are very close to or actually in the visceral effector (see Fig. 8–12), and contain the postganglionic cell bodies, with very short axons to the cells of the effector.

In the parasympathetic division, one preganglionic neuron synapses with just a few postganglionic neurons to only one effector. With this anatomic arrangement, very localized (one organ) responses are possible.

The parasympathetic division dominates in relaxed (non-stress) situations to promote normal functioning of several organ systems. Digestion will be efficient, with increased secretions and peristalsis; defecation

Table 8–5 FUNCTIONS OF THE AUTONOMIC NERVOUS SYSTEM

Organ	Sympathetic Responses	Parasympathetic Response
Heart (cardiac muscle)	• Increase rate	• Decrease rate (to normal)
Bronchioles (smooth muscle)	• Dilate	• Constrict (to normal)
Iris (smooth muscle)	• Pupil dilates	• Pupil constricts (to normal)
Salivary glands	• Decrease secretion	• Increase secretion (to normal)
Stomach and intestines (smooth muscle)	• Decrease peristalsis	• Increase peristalsis for normal digestion
Stomach and intestines (glands)	• Decrease secretion	• Increase secretion for normal digestion
Internal anal sphincter	• Contracts to prevent defecation	• Relaxes to permit defecation
Urinary bladder (smooth muscle)	• Relaxes to prevent urination	• Contracts for normal urination
Internal urethral sphincter	• Contracts to prevent urination	• Relaxes to permit urination
Liver	• Changes glycogen to glucose	• None
Sweat glands	• Increase secretion	• None
Blood vessels in skin and viscera (smooth muscle)	• Constrict	• None
Blood vessels in skeletal muscle (smooth muscle)	• Dilate	• None
Adrenal glands	• Increase secretion of epinephrine and norepinephrine	• None

and urination may occur; and the heart will beat at a normal resting rate. Other functions of this division are listed in Table 8–5.

Notice that when an organ receives both sympathetic and parasympathetic impulses, the responses are opposites. Notice also that some visceral effectors receive only sympathetic impulses. In such cases, the opposite response is brought about by a decrease in sympathetic impulses.

NEUROTRANSMITTERS

Recall that neurotransmitters enable nerve impulses to cross synapses. In autonomic pathways there are two synapses: one between preganglionic and postganglionic neurons, and the second between postganglionic neurons and visceral effectors.

Acetylcholine is the transmitter released by all preganglionic neurons, both sympathetic and parasympathetic; it is inactivated by **cholinesterase** in postganglionic neurons. Parasympathetic postganglionic neurons all release acetylcholine at the synapses with their visceral effectors. Most sympathetic postganglionic neurons release the transmitter **norepinephrine** at the synapses with the effector cells, which is inactivated by **COMT** (Catechol-O-methyl transferase).

AGING AND THE NERVOUS SYSTEM

The aging brain does indeed lose neurons, but this is only a small percentage of the total and not the usual cause of mental impairment in elderly people (far more common causes are depression, malnutrition, hypotension, and the side effects of medications). Some forgetfulness is to be expected, however, as is a decreased ability for *rapid* problem solving, but most memory should remain intact. Voluntary movements become slower, as do reflexes and reaction time. Think of driving a car, an ability most of us take for granted. For elderly people, with their slower perceptions and reaction times, greater *consciousness* of driving is necessary.

As the autonomic nervous system ages, dry eyes and constipation may become problems. Transient hypotension may be the result of decreased sympathetic stimulation of vasoconstriction. In most cases, however, elderly people who are aware of these aspects of aging will be able to work with their physicians or nurses to minimize them.

SUMMARY

The nervous system regulates many of our simplest and our most complex activities. The impulses generated and carried by the nervous system are an example of the chemical level of organization of the body. These nerve impulses then regulate the functioning of tissues, organs, and organ systems, which permits us to perceive and respond to the world around us and the changes within us. The detection of such changes is the function of the sense organs, and they are the subject of our next chapter.

STUDY OUTLINE

Functions of the Nervous System

1. Detect changes and feel sensations.
2. Initiate responses to changes.
3. Organize and store information.

Nervous System Divisions

1. Central Nervous System (CNS)—brain and spinal cord.
2. Peripheral Nervous System (PNS)—cranial nerves and spinal nerves.

Nerve Tissue—neurons (nerve fibers) and specialized cells (Schwann, neuroglia)

1. Neuron cell body contains the nucleus; cell bodies are in the CNS or in the trunk and are protected by bone.
2. Axon carries impulses away from the cell body; dendrites carry impulses toward the cell body.
3. Schwann cells in PNS: layers of cell membrane form the myelin sheath to electrically insulate neurons; nodes of Ranvier are spaces between adjacent

Schwann cells. Nuclei and cytoplasm of Schwann cells form the neurolemma which is essential for regeneration of damaged axons or dendrites.
4. Oligodendrocytes in CNS: form the myelin sheaths (see Table 8-1).
5. Synapse—the space between the axon of one neuron and the dendrites or cell body of the next neuron. A neurotransmitter carries the impulse across a synapse and is then destroyed by a chemical inactivator. Synapses make impulse transmission one-way in the living person.

Types of Neurons—nerve fibers
1. Sensory—carry impulses from receptors to the CNS; may be somatic (from skin, skeletal muscles, joints) or visceral (from internal organs).
2. Motor—carry impulses from the CNS to effectors; may be somatic (to skeletal muscle) or visceral (to smooth muscle, cardiac muscle, or glands). Visceral motor neurons make up the autonomic nervous system.
3. Interneurons—entirely within the CNS.

Nerves and Nerve Tracts
1. Sensory Nerve—made only of sensory neurons.
2. Motor Nerve—made only of motor neurons.
3. Mixed Nerve—made of both sensory and motor neurons.
4. Nerve Tract—a nerve within the CNS; also called white matter.

The Nerve Impulse—see Table 8–2
1. Polarization—neuron membrane has a (+) charge outside and a (−) charge inside.
2. Depolarization—entry of Na^+ ions and reversal of charges on either side of the membrane.
3. Impulse transmission is rapid, often several meters per second.
   - Saltatory conduction—in a myelinated neuron only the nodes of Ranvier depolarize; increases speed of impulses.

The Spinal Cord
1. Functions: transmits impulses to and from the brain, and integrates the spinal cord reflexes.
2. Location: within the vertebral canal; extends from the foramen magnum to the disc between the 1st and 2nd lumbar vertebrae.
3. Cross-section: internal H of gray matter contains cell bodies of motor neurons and interneurons; external white matter is the myelinated axons and dendrites of interneurons.
4. Ascending tracts carry sensory impulses to the brain; descending tracts carry motor impulses away from the brain.
5. Central canal contains cerebrospinal fluid and is continuous with the ventricles of the brain.

Spinal Nerves—see Table 8–3 for major peripheral nerves
1. Eight cervical pairs to head, neck, shoulder, arm, and diaphragm; 12 thoracic pairs to trunk; 5 lumbar pairs and 5 sacral pairs to hip, pelvic cavity and leg; 1 very small coccygeal pair.
2. Cauda Equina—the lumbar and sacral nerves that extend below the end of the spinal cord.
3. Each spinal nerve has two roots: dorsal or sensory root; dorsal root ganglion contains cell bodies of sensory neurons; ventral or motor root; the two roots unite to form a mixed spinal nerve.

Spinal Cord Reflexes—do not depend directly on the brain
1. A reflex is an involuntary response to a stimulus.
2. Reflex Arc—the pathway of nerve impulses during a reflex (1) receptors, (2) sensory neurons, (3) CNS with one or more synapses, (4) motor neurons, (5) effector which responds.
3. Stretch Reflex—a muscle that is stretched will contract; these reflexes help keep us upright against gravity. The patellar reflex is also used clinically to assess neurological functioning, as are many other reflexes (Fig. 8-5).
4. Flexor Reflex—a painful stimulus will cause withdrawal of the body part; these reflexes are protective.

The Brain—many parts that function as an integrated whole; see Figs. 8–6 and 8–8 for locations
1. Ventricles—four cavities: two lateral, 3rd, 4th; each contains a choroid plexus that forms cerebrospinal fluid (Fig. 8-7).
2. Medulla—regulates the vital functions of heart rate, breathing, and blood pressure; regulates reflexes of coughing, sneezing, swallowing, and vomiting.
3. Pons—contains respiratory centers that work with those in the medulla.
4. Midbrain—contains centers for visual reflexes, auditory reflexes, and righting (equilibrium) reflexes.

5. Cerebellum—regulates coordination of voluntary movement, muscle tone, stopping movements, and equilibrium.
6. Hypothalamus—produces antidiuretic hormone (ADH) which increases water reabsorption by the kidneys; produces oxytocin which promotes uterine contractions for labor and delivery; produces releasing hormones that regulate the secretions of the anterior pituitary gland; regulates body temperature; regulates food intake; integrates the functioning of the autonomic nervous system (ANS); promotes visceral responses to emotional situations; acts as a biological clock that regulates body rhythms.
7. Thalamus—groups sensory impulses as to body part before relaying them to the cerebrum; suppresses unimportant sensations to permit concentration.
8. Cerebrum—two hemispheres connected by the corpus callosum, which permits communication between the hemispheres. The cerebral cortex is the surface gray matter, which consists of cell bodies of neurons and is folded extensively into convolutions. The internal white matter consists of nerve tracts that connect the lobes of the cerebrum to one another and to other parts of the brain.
   - Frontal lobes—motor areas initiate voluntary movement; Broca's motor speech area (left hemisphere) regulates the movements involved in speech.
   - Parietal lobes—general sensory area feels and interprets the cutaneous senses and conscious muscle sense; taste area extends into temporal lobe, for sense of taste; speech areas (left hemisphere) for thought before speech.
   - Temporal lobes—auditory areas for hearing; olfactory areas for sense of smell; speech areas for thought before speech.
   - Occipital lobes—visual areas for vision; interpretation areas for spatial relationships.
   - Association areas—in all lobes, for abstract thinking, reasoning, learning, memory, and personality.
   - Basal ganglia—gray matter within the cerebral hemispheres; regulate accessory movements and muscle tone.

Meninges and Cerebrospinal Fluid (CSF) (see Figs. 8–9 and 8–10)

1. Three meningeal layers made of connective tissue: outer—dura mater; middle—arachnoid membrane; inner—pia mater; all three enclose the brain and spinal cord.
2. Subarachnoid space contains CSF, the tissue fluid of the CNS.
3. CSF is formed continuously in the ventricles of the brain by choroid plexuses, from blood plasma.
4. CSF circulates from the ventricles to the central canal of the spinal cord and to the cranial and spinal subarachnoid spaces.
5. CSF is reabsorbed from the cranial subarachnoid space through arachnoid villi into the blood in the cranial venous sinuses. The rate of reabsorption equals the rate of production.
6. As tissue fluid, CSF brings nutrients to CNS neurons and removes waste products. CSF also acts as a shock absorber to cushion the CNS.

Cranial Nerves—12 pairs of nerves that emerge from the brain (see Fig. 8–11)

1. Concerned with vision, hearing and equilibrium, taste and smell, and many other functions.
2. See Table 8-4 for the functions of each pair.

The Autonomic Nervous System (ANS) (see Fig. 8–12 and Table 8–5)

1. Has two divisions: sympathetic and parasympathetic; their functioning is integrated by the hypothalamus.
2. Consists of motor neurons to visceral effectors: smooth muscle, cardiac muscle, and glands.
3. An ANS pathway consists of two neurons that synapse in a ganglion:
   - Preganglionic neurons—from the CNS to the ganglia
   - Postganglionic neurons—from the ganglia to the effectors

   Most sympathetic ganglia are in two chains just outside the vertebral column; parasympathetic ganglia are very near or in the visceral effectors.
4. Neurotransmitters: acetylcholine is released by all preganglionic neurons and by parasympathetic postganglionic neurons; the inactivator is cholinesterase. Norepinephrine is released by most sympathetic postganglionic neurons; the inactivator is COMT.
5. Sympathetic division—dominates the stress situations; responses prepare the body to meet physical demands.
6. Parasympathetic division—dominates in relaxed situations to permit normal functioning.

REVIEW QUESTIONS

1. Name the divisions of the nervous system and state the parts of each. (p. 154)
2. State the function of the following parts of nerve tissue: (pp. 154, 156)
   a. Axon
   b. Dendrites
   c. Myelin sheath
   d. Neurolemma
   e. Microglia
   f. Astrocytes
3. Explain the difference between: (pp. 157-159)
   a. Sensory neurons and motor neurons
   b. Interneurons and nerve tracts
4. Describe an electrical nerve impulse in terms of charges on either side of the neuron membrane. Describe how a nerve impulse crosses a synapse. (pp. 154, 156-157)
5. With respect to the spinal cord: (p. 160)
   a. Describe its location
   b. State what gray matter and white matter are made of
   c. State the function of the dorsal root, ventral root, and dorsal root ganglion
6. State the names and number of pairs of spinal nerves. State the part of the body supplied by the: phrenic nerves, radial nerves, sciatic nerves. (pp. 160-162)
7. Define reflex, and name the five parts of a reflex arc. (p. 163)
8. Define stretch reflexes, and explain their practical importance. Define flexor reflexes, and explain their practical importance. (pp. 163-164)
9. Name the part of the brain concerned with each of the following: (pp. 164-166)
   a. Regulates body temperature
   b. Regulates heart rate
   c. Suppresses unimportant sensations
   d. Regulates respiration (two parts)
   e. Regulates food intake
   f. Regulates coordination of voluntary movement
   g. Regulates secretions of the anterior pituitary gland
   h. Regulates coughing and sneezing
   i. Regulates muscle tone
   j. Regulates visual and auditory reflexes
   k. Regulates blood pressure
10. Name the part of the cerebrum concerned with each of the following: (pp. 167-170)
    a. Feels the cutaneous sensations
    b. Contains the auditory areas
    c. Contains the visual areas
    d. Connects the cerebral hemispheres
    e. Regulates accessory movements
    f. Contains the olfactory areas
    g. Initiates voluntary movement
    h. Contains the speech areas (for most people)

11. Name the three layers of the meninges, beginning with the outermost. (p. 172)
12. State all the locations of cerebrospinal fluid. What is CSF made from? Into what is CSF reabsorbed? State the functions of CSF. (pp. 172–174)
13. State a function of each of the following cranial nerves: (p. 176)
    a. Glossopharyngeal
    b. Olfactory
    c. Trigeminal
    d. Facial
    e. Vagus (three functions)
14. Explain how the sympathetic division of the ANS helps the body adapt to a stress situation; give three specific examples. (p. 178)
15. Explain how the parasympathetic division of the ANS promotes normal body functioning; give three specific examples. (pp. 178–179)

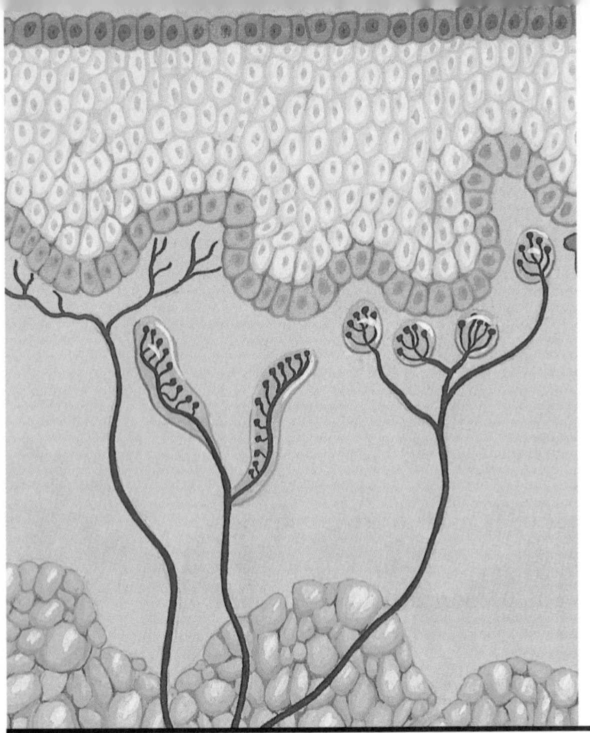

CHAPTER 9

Chapter Outline

Student Objectives

- Explain the general purpose of sensations.
- Name the parts of a sensory pathway, and state the function of each.
- Describe the characteristics of sensations.
- Name the cutaneous senses and explain their purpose.
- Explain referred pain and its importance.
- Explain the importance of muscle sense.
- Describe the pathways for the senses of smell and taste, and explain how these senses are interrelated.
- Name the parts of the eye and their functions.
- Describe the physiology of vision.
- Name the parts of the ear and their functions.
- Describe the physiology of hearing.
- Describe the physiology of equilibrium.
- Explain the importance of the arterial pressoreceptors and chemoreceptors.

The Senses

New Terminology

Adaptation (A-dap-**TAY**-shun)
After-image (**AFF**-ter-im-ije)
Aqueous humor (**AY**-kwee-us **HYOO**-mer)
Cochlea (**KOK**-lee-ah)
Cones (KOHNES)
Conjunctiva (KON-junk-**TIGH**-vah)
Contrast (**KON**-trast)
Cornea (**KOR**-nee-ah)
Eustachian tube (yoo-**STAY**-shee-un TOOB)
Iris (**EYE**-ris)
Lacrimal glands (**LAK**-ri-muhl)
Organ of Corti (**KOR**-tee)
Olfactory receptors (ohl-**FAK**-toh-ree)
Projection (proh-**JEK**-shun)
Referred pain (ree-**FURRD** PAYNE)
Retina (**RET**-i-nah)
Rhodopsin (roh-**DOP**-sin)
Rods (RAHDS)
Sclera (**SKLER**-ah)
Semicircular canals (SEM-ee-**SIR**-kyoo-lur)
Tympanic membrane (tim-**PAN**-ik)
Vitreous humor (**VIT**-ree-us **HYOO**-mer)

Related Clinical Terminology

Astigmatism (un-**STIG**-mah-TIZM)
Cataract (**KAT**-uh-rackt)
Color blindness (**KUHL**-or **BLIND**-ness)
Conjunctivitis (kon-JUNK-ti-**VIGH**-tis)
Deafness (**DEFF**-ness)
Detached retina (dee-**TACHD**)
Glaucoma (glaw-**KOH**-mah)
Hyperopia (HIGH-per-**OH**-pee-ah)
Motion sickness (**MOH**-shun)
Myopia (my-**OH**-pee-ah)
Night blindness (NITE **BLIND**-ness)
Otitis media (oh-**TIGH**-tis **MEE**-dee-ah)
Phantom pain (**FAN**-tum)
Presbyopia (PREZ-bee-**OH**-pee-ah)

Terms that appear in ***bold type*** *in the chapter text are defined in the glossary, which begins on page 528.*

Our **senses** constantly provide us with information about our surroundings: We see, hear, and touch. The senses of taste and smell enable us to enjoy the flavor of our food or warn us that food has spoiled and may be dangerous to eat. Our sense of equilibrium keeps us upright. We also get information from our senses about what is happening inside the body. The pain of a headache, for example, prompts us to do something about it, such as take aspirin. In general, this is the purpose of sensations: to enable the body to respond appropriately to ever-changing situations and maintain homeostasis.

SENSORY PATHWAY

The impulses involved in sensations follow very precise pathways, which all have the following parts:

1. **Receptors**—detect changes **(stimuli)** and generate impulses. Receptors are usually very specific with respect to the kinds of changes they respond to. Those in the retina detect light rays, those in the nasal cavities detect vapors, and so on. Once a specific stimulus has affected receptors, however, they all respond the same way by generating electrical nerve impulses.
2. **Sensory neurons**—transmit impulses from receptors to the central nervous system. These sensory neurons are found in both spinal nerves and cranial nerves, but each carries impulses from only one type of receptor.
3. **Sensory tracts**—white matter in the spinal cord or brain that transmits the impulses to a specific part of the brain.
4. **Sensory area**—most are in the cerebral cortex. These areas feel and interpret the sensations. Learning to interpret sensations begins in infancy, without our awareness of it, and continues throughout life.

CHARACTERISTICS OF SENSATIONS

1. **Projection**—the sensation seems to come from the area where the receptors were stimulated. If you touch this book, the sensation of touch seems to be in your hand but is actually being felt by your cerebral cortex. That it is indeed the brain that feels sensations is demonstrated by patients who feel **phantom pain** after amputation of a limb. After loss of a hand, for example, the person may still feel that the hand is really there. Why does this happen? The receptors in the hand are no longer present, but the severed nerve endings continue to generate impulses. These impulses arrive in the parietal lobe area for the hand, and the brain does what it has always done and creates the projection, the feeling that the hand is still there. For most amputees, phantom pain diminishes as the severed nerves heal, but the person often experiences a phantom "presence" of the missing part. This may be helpful when learning to use an artificial limb.
2. **Intensity**—some sensations are felt more distinctly and to a greater degree than are others. A weak stimulus such as dim light will affect a small number of receptors, but a stronger stimulus, such as bright sunlight, will stimulate many more receptors. When more receptors are stimulated, more impulses will arrive in the sensory area of the brain. The brain "counts" the impulses and projects a more intense sensation.
3. **Contrast**—the effect of a previous or simultaneous sensation on a current sensation, which may then be exaggerated or diminished. Again, this is a function of the brain, which constantly compares sensations. If, on a very hot day, you jump into a swimming pool, the water may feel quite cold at first. The brain compares the new sensation to the previous one, and since there is a significant difference between the two, the water will seem colder than it actually is.
4. **Adaptation**—becoming unaware of a continuing stimulus. Receptors detect changes, but if the stimulus continues it may not be much of a change, and the receptors will generate fewer impulses. Many of us wear a watch and are probably unaware of its presence on the arm most of the time. The cutaneous receptors for touch or pressure adapt very quickly to a continuing stimulus, and if there is no change, there is nothing for the receptors to detect.
5. **After-image**—the sensation remains in the consciousness even after the stimulus has stopped. A familiar example is the bright after-image seen

after watching a flashbulb go off. The very bright light strongly stimulates receptors in the retina, which generate many impulses that are perceived as an intense sensation that lasts longer than the actual stimulus.

CUTANEOUS SENSES

The dermis of the skin contains receptors for the sensations of touch, pressure, heat, cold, and pain. The receptors for pain are **free nerve endings**, which respond to any intense stimulus. Intense cold, for example, may be felt as pain. The receptors for the other cutaneous senses are **encapsulated nerve endings**, meaning that there is a cellular structure around the nerve ending (Fig. 9-1).

The **cutaneous senses** provide us with information about the external environment and also about the skin itself. If you have ever had chickenpox, you may remember the itching sensation of the rash. An itch is actually a mild pain sensation, which may become real pain if not scratched (how scratching relieves the itch has not yet been discovered).

The sensory areas for the skin are in the parietal lobes. The largest parts of this sensory cortex are for the parts of the skin with the most receptors, that is, the hands and face.

REFERRED PAIN

Free nerve endings are also found in internal organs. The smooth muscle of the small intestine, for example, has free nerve endings that are stimulated by excessive stretching or contraction; the resulting pain is called visceral pain. Sometimes pain that originates in an internal organ may be felt in a cutaneous area; this

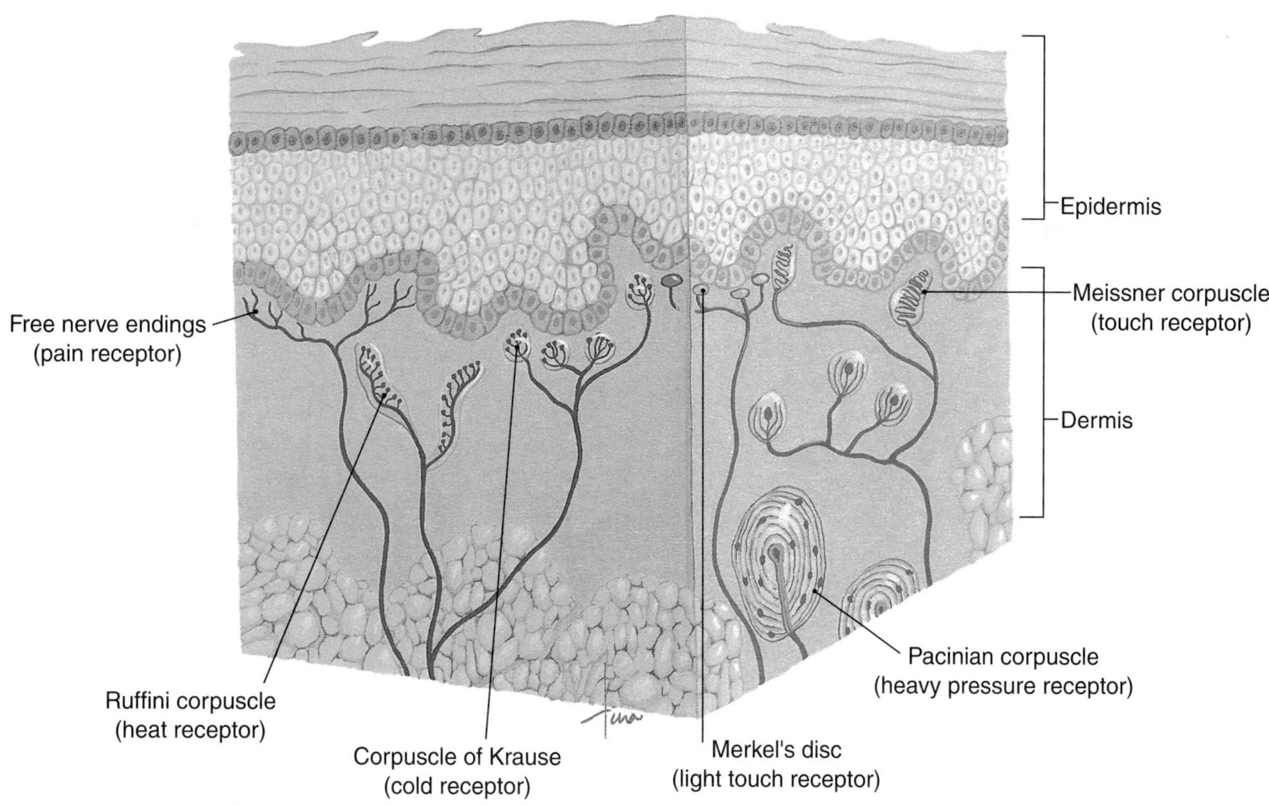

Figure 9–1 Cutaneous receptors in a section of the skin. Free nerve endings and the types of encapsulated nerve endings are shown.

is called **referred pain**. The pain of a heart attack (myocardial infarction) may be felt in the left arm and shoulder, or the pain of gallstones may be felt in the right shoulder.

This referred pain is actually a creation of the brain. Within the spinal cord are sensory tracts that are shared by cutaneous impulses and visceral impulses. Cutaneous impulses are much more frequent, and the brain correctly projects the sensation to the skin. When the impulses have come from an organ such as the heart, however, the brain may still project the sensation to the "usual" cutaneous area. The brain projects sensation based on past experience, and cutaneous pain is far more common than visceral pain. Knowledge of referred pain, as in the examples mentioned earlier, may often be helpful in diagnosis.

MUSCLE SENSE

Muscle sense (also called kinesthetic sense) was discussed in Chapter 7 and will be reviewed only briefly here. Stretch receptors (also called proprioceptors or muscle spindles) detect stretching of muscles and generate impulses, which enable the brain to create a mental picture to know where the muscles are and how they are positioned. Conscious muscle sense is felt by the parietal lobes. Unconscious muscle sense is used by the cerebellum to coordinate voluntary movements. We do not have to see our muscles to be sure that they are performing their intended actions.

SENSE OF TASTE

The receptors for taste are found in **taste buds**, most of which are in papillae on the tongue (Fig. 9–2). These **chemoreceptors** detect chemicals in solution in the mouth. The chemicals are foods and the solvent is saliva (if the mouth is very dry, taste is very indistinct). It is believed that there are four general types of taste receptors: sweet, sour, salty, and bitter. (There is perhaps a fifth, glutamate, which has been described as "savory," like grilled meat.) We experience many different tastes, however, because foods stimulate different combinations of receptors, and the sense of smell also contributes to our perception of food.

Some taste preferences have been found to be genetic. People with more than the average number of taste buds find broccoli very bitter, whereas people with fewer taste buds may like the taste.

The impulses from taste buds are transmitted by the facial and glossopharyngeal (7th and 9th cranial) nerves to the taste areas in the parietal-temporal cortex. The sense of taste is important because it makes eating enjoyable. Some medications may interfere with the sense of taste, and this sense becomes less acute as we get older. These may be contributing factors to poor nutrition in certain patients and in the elderly.

SENSE OF SMELL

The receptors for smell **(olfaction)** are **chemoreceptors** which detect vaporized chemicals that have been sniffed into the upper nasal cavities (see Fig. 9–2). Just as there are basic tastes, there are also believed to be basic scents, and recent research indicates that their number is a thousand or more. When stimulated by vapor molecules, **olfactory receptors** generate impulses carried by the olfactory nerves (1st cranial) through the ethmoid bone to the olfactory bulbs. The pathway for these impulses ends in the olfactory areas of the temporal lobes.

The human sense of smell is very poorly developed compared to those of other animals. Dogs, for example, have a sense of smell at least 200 times more acute than that of people. As mentioned earlier, however, much of what we call taste is actually the smell of food. If you have a cold and your nasal cavities are stuffed up, food just doesn't taste as good as it usually does. Adaptation occurs relatively quickly with odors. Pleasant scents may be sharply distinct at first but rapidly seem to dissipate or fade.

HUNGER AND THIRST

Hunger and thirst may be called **visceral sensations**, in that they are triggered by internal changes. The receptors are thought to be specialized cells in the hypothalamus. Receptors for hunger are believed to detect changes in blood nutrient levels or chemicals released by adipose tissue, and receptors for thirst detect changes in body water content (actually the water–salt proportion).

Naturally we do not feel these sensations in the hy-

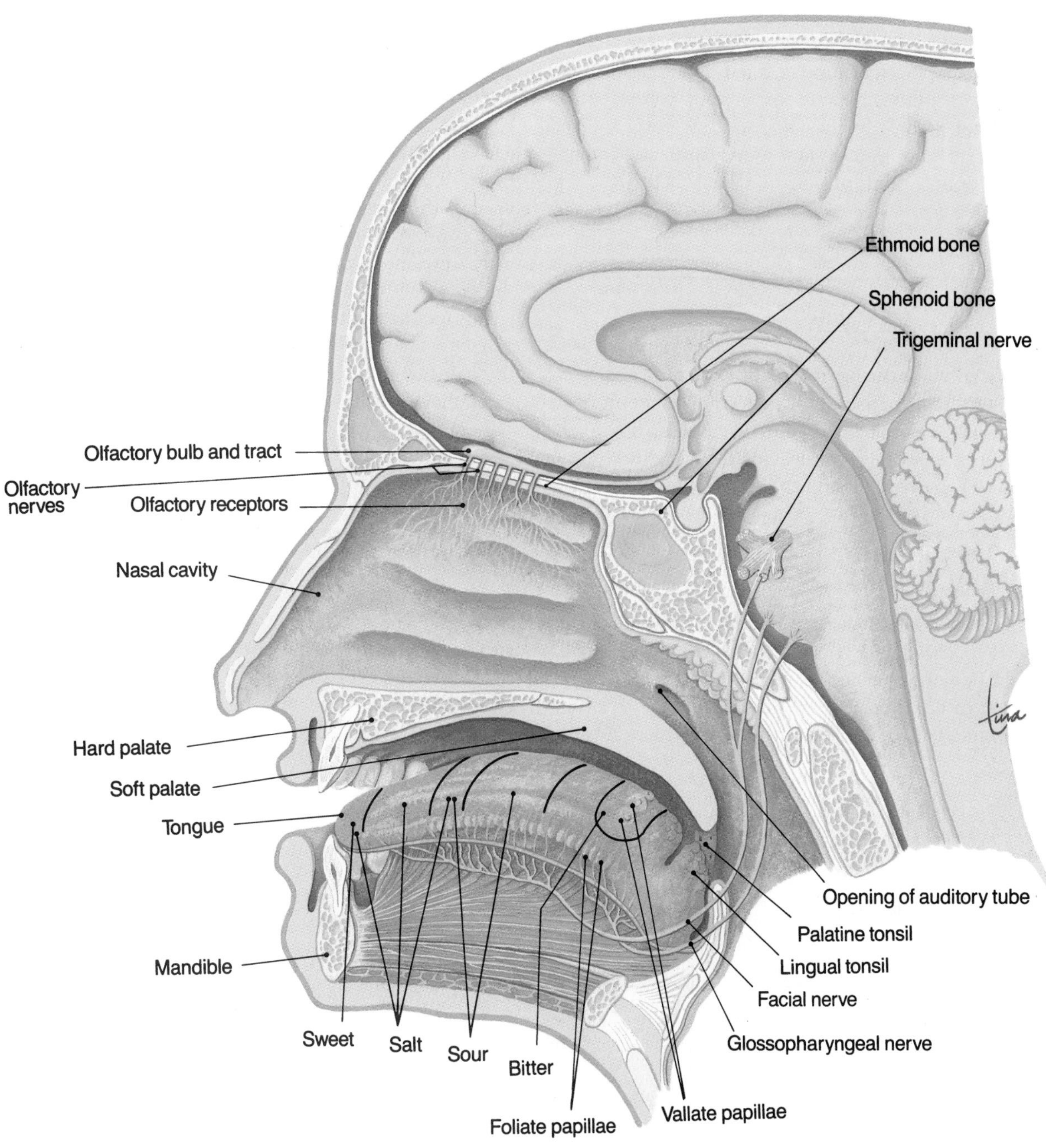

Figure 9–2 Structures concerned with the senses of smell and taste, shown in a midsagittal section of the head.

pothalamus: They are projected. Hunger is projected to the stomach, which contracts. Thirst is projected to the mouth and pharynx, and less saliva is produced.

If not satisfied by eating, the sensation of hunger gradually diminishes, that is, adaptation occurs. The reason is that after blood nutrient levels decrease, they become stable as fat in adipose tissue is used for energy. With no sharp fluctuations, the receptors have few changes to detect, and hunger becomes much less intense.

In contrast, the sensation of thirst, if not satisfied by drinking, continues to worsen. As body water is lost, the amount keeps decreasing and does not stabilize. Therefore, there are constant changes for the receptors to detect, and prolonged thirst may be very painful.

THE EYE

The eye contains the receptors for vision and a refracting system that focuses light rays on the receptors in the retina. We will begin our discussion, however, with the accessory structures of the eye, then later return to the eye itself and the physiology of vision.

EYELIDS AND THE LACRIMAL APPARATUS

The eyelids contain skeletal muscle that enables the eyelids to close and cover the front of the eyeball. Eyelashes along the border of each eyelid help keep dust out of the eyes. The eyelids are lined with a thin membrane called the **conjunctiva**, which is also folded over the white of the eye. Inflammation of this membrane, called **conjunctivitis**, is often caused by allergies and makes the eyes red, itchy, and watery.

Tears are produced by the **lacrimal glands**, located at the upper, outer corner of the eyeball, within the orbit (Fig. 9–3). Small ducts take tears to the anterior of the eyeball, and blinking spreads the tears and washes the surface of the eye. Tears are mostly water and contain **lysozyme**, an enzyme that inhibits the growth of most bacteria on the wet, warm surface of the eye. At the medial corner of the eyelids are two small openings into the superior and inferior lacrimal canals. These ducts take tears to the **lacrimal sac** (in the lacrimal bone), which leads to the **nasolacrimal duct** that empties tears into the nasal cavity. This is why crying often makes the nose run.

EYEBALL

Most of the eyeball is within and protected by the **orbit**, formed by the maxilla, zygomatic, frontal, sphenoid, and ethmoid bones. The six **extrinsic muscles** of the eye are attached to this bony socket and to the surface of the eyeball. There are four rectus muscles that move the eyeball up and down or side to side and two oblique muscles that rotate the eye. These are shown in Fig. 9–4. The cranial nerves that innervate

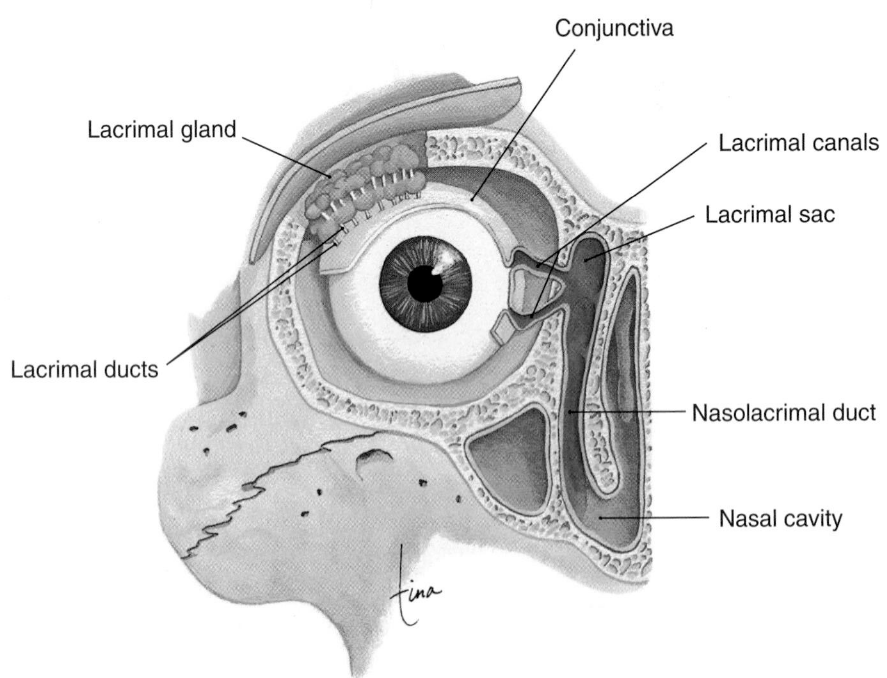

Figure 9–3 Lacrimal apparatus shown in an anterior view of the right eye.

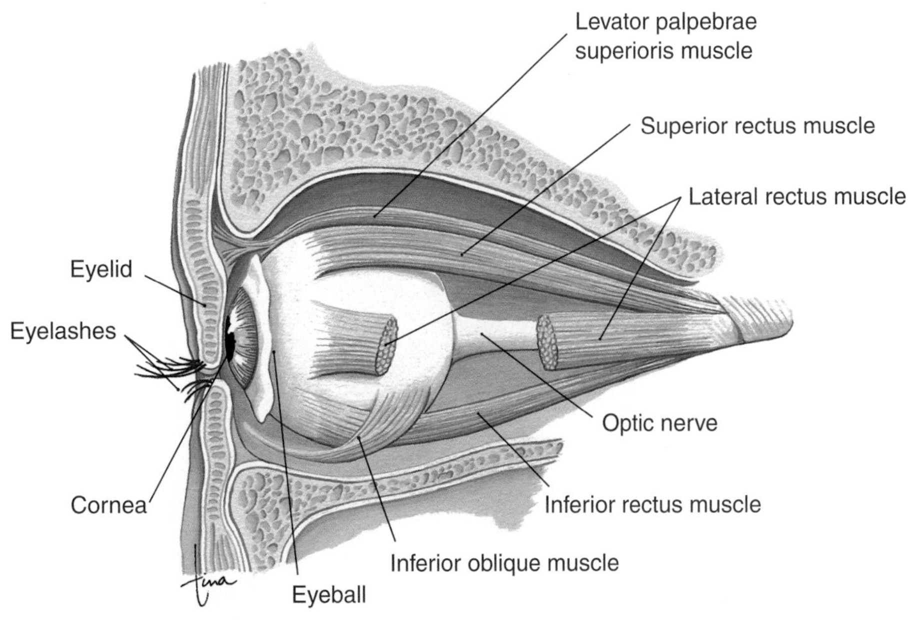

Figure 9–4 Extrinsic muscles of the eye. Lateral view of left eye (the medial rectus and superior oblique are not shown).

these muscles are the oculomotor, trochlear, and abducens (3rd, 4th, and 6th cranial). The complex coordination of these muscles in both eyes is, fortunately, not something we have to think about, and is very important to prevent double vision.

Layers of the Eyeball

In its wall, the eyeball has three layers: the outer sclera, middle choroid layer, and inner retina (Fig. 9–5). The **sclera** is the thickest layer and is made of fibrous connective tissue which is visible as the white of the eye. The most anterior portion is the **cornea**, which differs from the rest of the sclera in that it is transparent and has no capillaries. The cornea is the first part of the eye that **refracts**, or bends, light rays.

The **choroid layer** contains blood vessels and a dark blue pigment that absorbs light within the eyeball and thereby prevents glare. The anterior portion of the choroid is modified into more specialized structures: the ciliary body and the iris. The ciliary body (muscle) is a circular muscle that surrounds the edge of the lens and is connected to the lens by **suspensory ligaments**. The **lens** is made of a transparent, elastic protein, and like the cornea, has no capillaries (see Box 9-1: Cataracts). The shape of the lens is changed by the ciliary muscle, which enables the eye to focus light from objects at varying distances from the eye.

Just in front of the lens is the circular **iris**, the colored part of the eye. Two sets of smooth muscle fibers change the diameter of the **pupil**, the central opening. Contraction of the radial fibers dilates the pupil; this is a sympathetic response. Contraction of the circular fibers constricts the pupil; this is a parasympathetic response (oculomotor nerves). Pupillary constriction is a reflex that protects the retina from intense light or that permits more acute near vision, as when reading.

The **retina** lines the posterior two-thirds of the eyeball and contains the visual receptors, the rods and cones (Fig. 9–6). **Rods** detect only the presence of light, whereas **cones** detect colors which, as you may know from physics, are the different wavelengths of visible light. Cones are most abundant in the center of the retina, especially an area called the **macula lutea** directly behind the center of the lens. The **fovea**, which contains only cones, is a small depression in the macula and is the area for best color vision. Rods are proportionally more abundant toward the periphery, or edge, of the retina. Our best vision in dim light or at night, for which we depend on the rods, is at the sides of our visual fields.

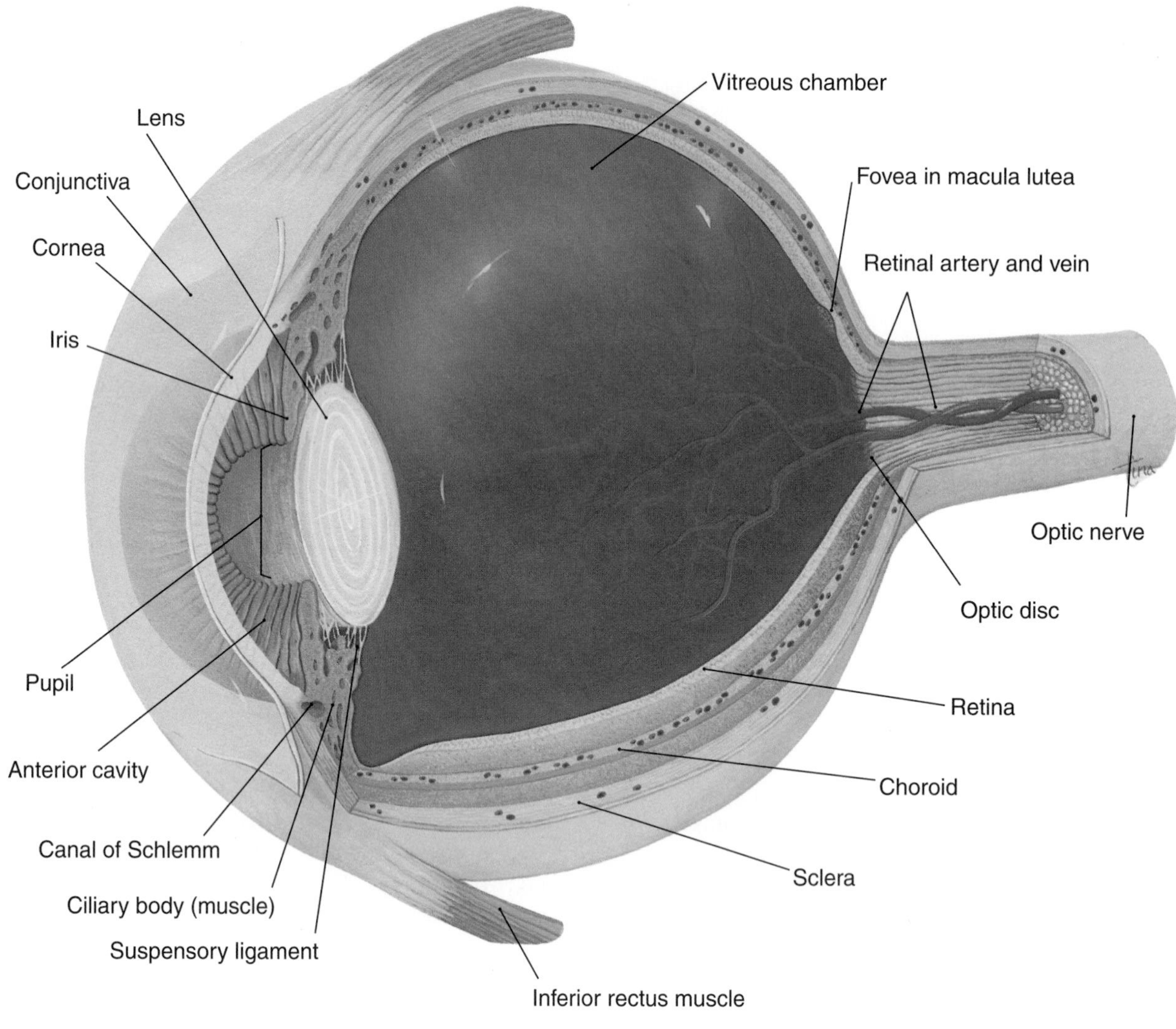

Figure 9–5 Internal anatomy of the eyeball.

Box 9–1 CATARACTS

The lens of the eye is normally transparent but may become opaque; this cloudiness or opacity is called a **cataract**. Cataract formation is most common among elderly people. With age, the proteins of the lens break down and lose their transparency. Long-term exposure to ultraviolet light (sunlight) also seems to be a contributing factor.

The cloudy lens does not refract light properly, and blurred vision is often an early symptom. Treatment of cataracts requires surgical removal of the lens and the use of glasses or contact lenses to replace the refractive function of the lens. The use of very precise laser surgery to destroy small cataracts may permit the lens to retain a portion of its function.

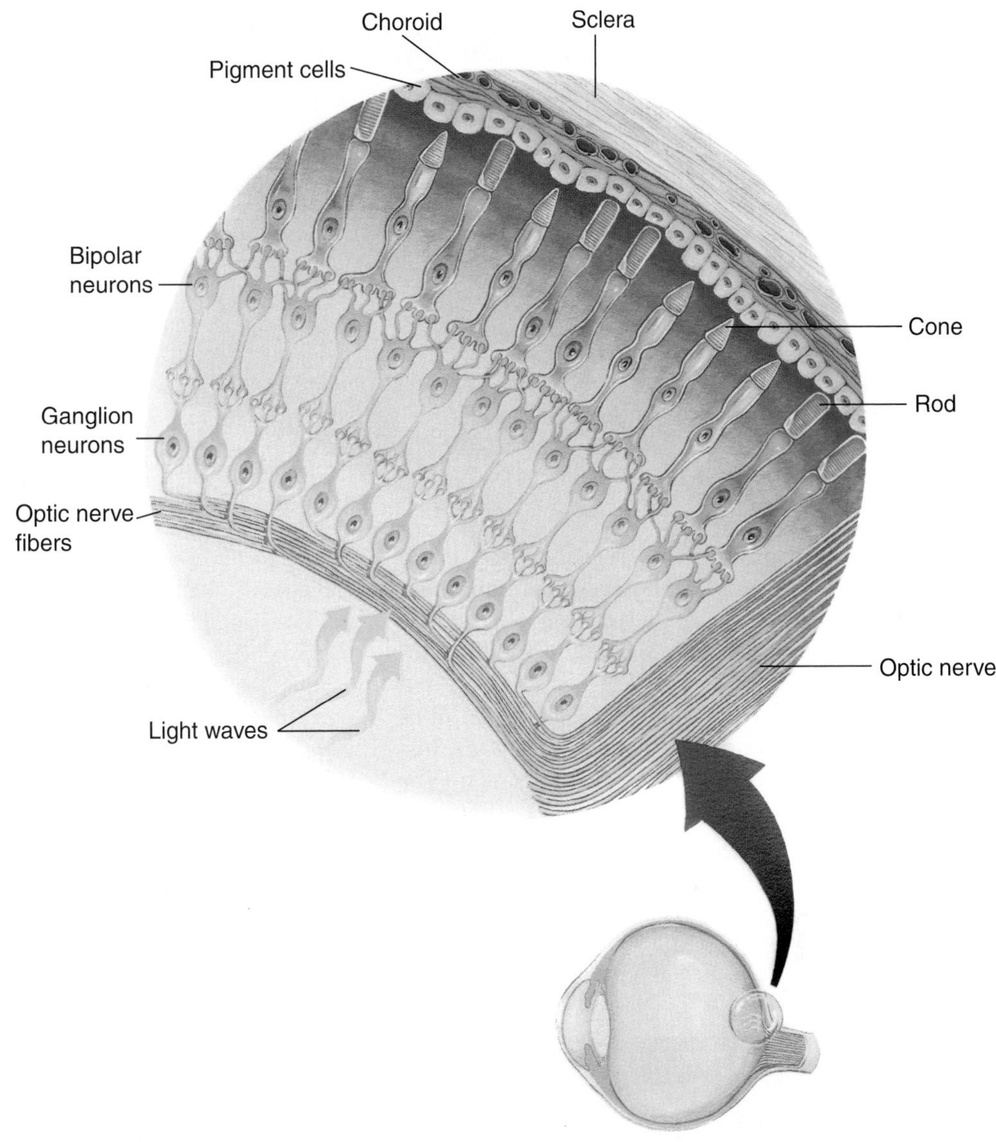

Figure 9–6 Microscopic structure of the retina in the area of the optic disc. See text for description.

Neurons called **ganglion neurons** carry the impulses generated by the rods and cones. These neurons all converge at the **optic disc** (see Figs. 9-5 and 9-6) and pass through the wall of the eyeball as the **optic nerve**. There are no rods or cones in the optic disc, so this part of the retina is sometimes called the "blind spot." We are not aware of a blind spot in our field of vision, however, in part because the eyes are constantly moving, and in part because the brain "fills in" the blank spot to create a "complete" picture.

Cavities of the Eyeball

There are two cavities within the eye: the posterior cavity and the anterior cavity. The larger, **posterior cavity** is found between the lens and retina and con-

Box 9–2 GLAUCOMA

The presence of aqueous humor in the anterior cavity of the eye creates a pressure called intraocular pressure. **Glaucoma** is an increase in intraocular pressure due to decreased reabsorption of aqueous humor into the canal of Schlemm. There are several types of glaucoma, but if untreated, all have the same outcome. Increased pressure in the anterior cavity is transmitted to the lens, vitreous humor, and retina. As pressure on the retina increases, halos may be seen around bright lights. Frequently, however, a person with glaucoma has no symptoms before severe visual impairment or blindness occurs.

Glaucoma may often be controlled with medications that constrict the pupil and flatten the iris, thus opening up access to the canal of Schlemm. If such medications are not effective, surgery may be required to create a larger canal of Schlemm.

Anyone over the age of 40 should have a test for glaucoma; anyone with a family history of glaucoma should have this test annually.

tains **vitreous humor**. This semisolid substance keeps the retina in place. If the eyeball is punctured and vitreous humor is lost, the retina may fall away from the choroid; this is one possible cause of a **detached retina**.

The **anterior cavity** is found between the front of the lens and the cornea and contains **aqueous humor**, the tissue fluid of the eyeball. Aqueous humor is formed by capillaries in the ciliary body, flows anteriorly through the pupil, and is reabsorbed by the **canal of Schlemm** (small veins also called the scleral venous sinus) at the junction of the iris and cornea. Because aqueous humor is tissue fluid, you would expect it to have a nourishing function, and it does. Recall that the lens and cornea have no capillaries; they are nourished by the continuous flow of aqueous humor (see Box 9-2: Glaucoma).

PHYSIOLOGY OF VISION

In order for us to see, light rays must be focused on the retina, and the resulting nerve impulses must be transmitted to the visual areas of the cerebral cortex in the brain.

Refraction of light rays is the deflection or bending of a ray of light as it passes through one object and

Box 9–3 ERRORS OF REFRACTION

Normal visual acuity is referred to as 20/20, that is, the eye should and does clearly see an object 20 feet away. **Nearsightedness (myopia)** means that the eye sees near objects well but not distant ones. If an eye has 20/80 vision, this means that what the normal eye can see at 80 feet, the nearsighted eye can see only if the object is brought to 20 feet away. The nearsighted eye focuses images from distant objects in front of the retina, because the eyeball is too long or the lens too thick (see diagram). These structural characteristics of the eye are hereditary. Correction of nearsightedness requires a concave lens to spread out light rays before they strike the eye.

Farsightedness (hyperopia) means that the eye sees distant objects well. Such an eye may have an acuity of 20/10, that is, it sees at 20 feet what the normal eye can see only at 10 feet. The farsighted eye focuses light from near objects "behind" the retina, because the eyeball is too short or the lens too thin. Correction of the farsighted eye requires a convex lens to converge light rays before they strike the eye.

As we get older, most of us will become more farsighted **(presbyopia)**. As the aging lens loses its elasticity, it is not as able to recoil and thicken for near vision, and glasses for reading are often necessary.

Box 9–3 ERRORS OF REFRACTION (Continued)

Astigmatism is another error of refraction, caused by an irregular curvature of the cornea or lens that scatters light rays and blurs the image on the retina. Correction of astigmatism requires a lens ground specifically for the curvature of the individual eye.

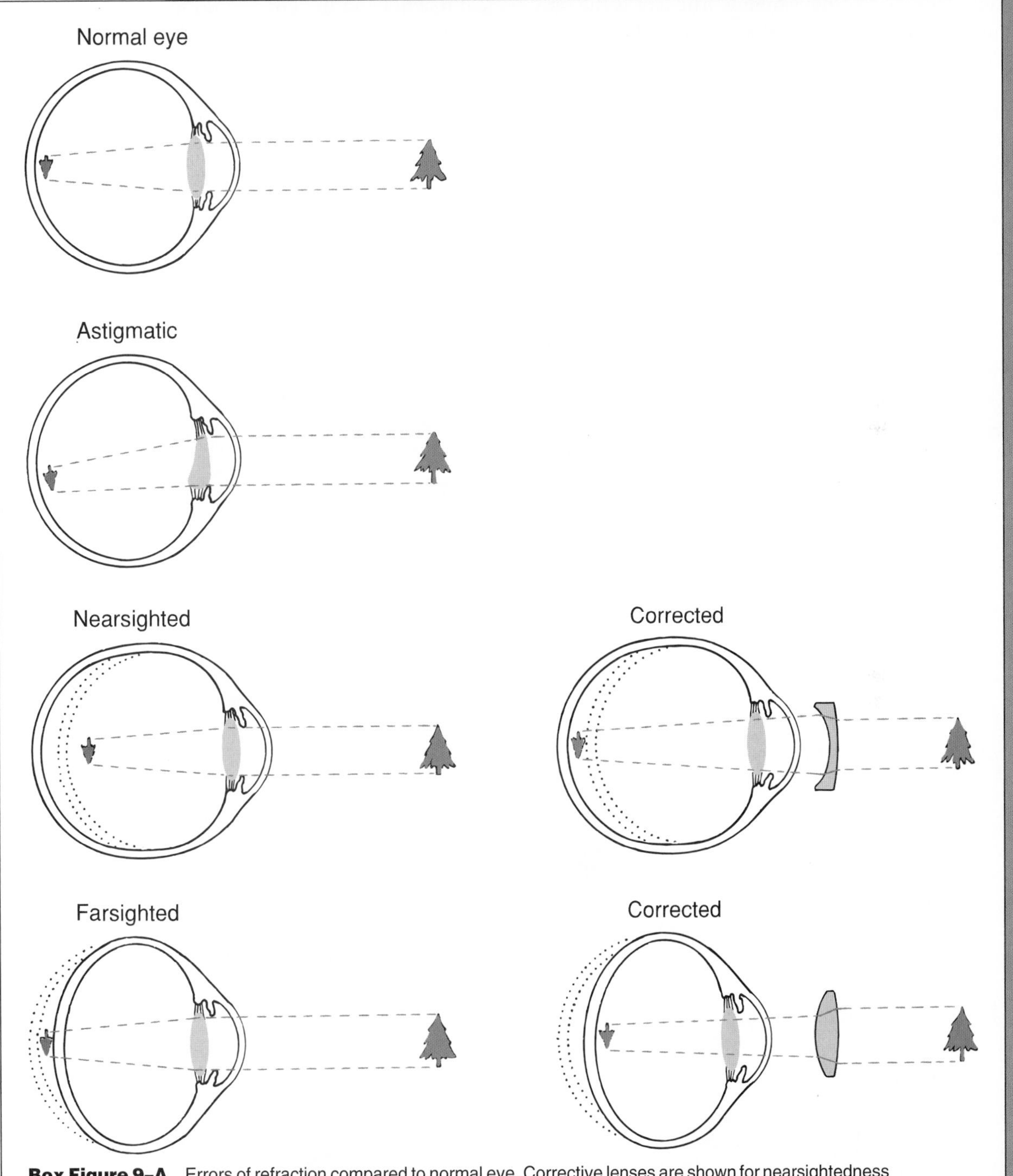

Box Figure 9–A Errors of refraction compared to normal eye. Corrective lenses are shown for nearsightedness and farsightedness.

into another object of greater or lesser density. The refraction of light within the eye takes place in the following pathway of structures: the cornea, aqueous humor, lens, and vitreous humor. The lens is the only adjustable part of the refraction system. When looking at distant objects, the ciliary muscle is relaxed and the lens is elongated and thin. When looking at near objects, the ciliary muscle contracts to form a smaller circle, the elastic lens recoils and bulges in the middle, and has greater refractive power (see Box 9-3: Errors of Refraction).

When light rays strike the retina, they stimulate chemical reactions in the rods and cones. In rods, the chemical **rhodopsin** breaks down to form scotopsin and retinal (a derivative of vitamin A). This chemical reaction generates an electrical impulse, and rhodopsin is then resynthesized in a slower reaction. Chemical reactions in the cones, also involving retinal, are brought about by different wavelengths of light. It is believed that there are three types of cones: red-absorbing, blue-absorbing, and green-absorbing. Each type absorbs wavelengths over about a third of the visible light spectrum, so red cones, for example, absorb light of the red, orange, and yellow wavelengths. The chemical reactions in cones also generate electrical impulses (see Box 9-4: Night Blindness and Color Blindness).

The impulses from the rods and cones are trans-

Box 9–4 NIGHT BLINDNESS AND COLOR BLINDNESS

Night blindness, the inability to see well in dim light or at night is usually caused by a deficiency of vitamin A, although some night blindness may occur with aging. Vitamin A is necessary for the synthesis of rhodopsin in the rods. Without sufficient vitamin A, there is not enough rhodopsin present to respond to low levels of light.

Color blindness is a genetic disorder in which one of the three sets of cones is lacking or nonfunctional. Total color blindness, the inability to see any colors at all, is very rare. The most common form is red-green color blindness, which is the inability to distinguish between these colors. If either the red cones or green cones are missing, the person will still see most of the colors, but will not have the contrast that the missing set of cones would provide. So red and green shades will look somewhat similar, without the definite difference most of us see. (See the accompanying illustration.) This is a sex-linked trait; the recessive gene is on the X chromosome. A woman with one gene for color blindness and a gene for normal color vision on her other X chromosome will not be color blind but may pass the gene for color blindness to her children. A man with a gene for color blindness on his X chromosome has no gene at all for color vision on his Y chromosome and will be color blind.

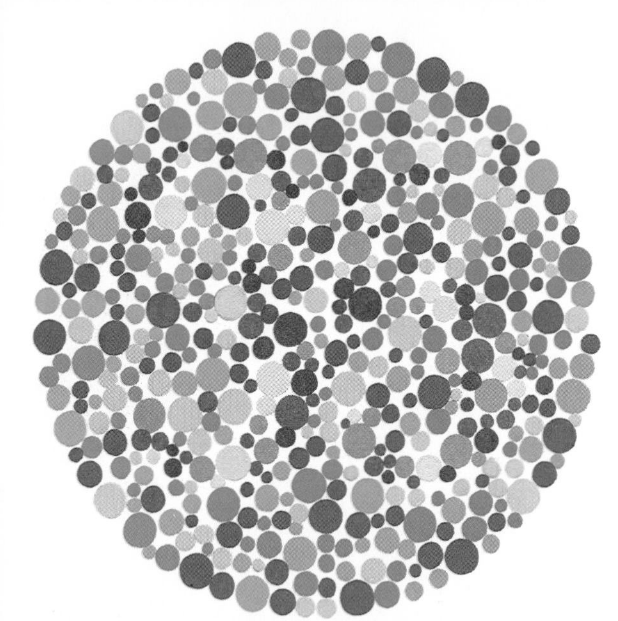

Box Figure 9–B Example of color patterns used to detect color blindness.

mitted to **ganglion neurons** (see Fig. 9-6); these converge at the optic disc and become the **optic nerve**, which passes posteriorly through the wall of the eyeball.

The optic nerves from both eyes converge at the **optic chiasma**, just in front of the pituitary gland. Here, the medial fibers of each optic nerve cross to the other side. This crossing permits each visual area to receive impulses from both eyes, which is important for binocular vision.

The visual areas are in the **occipital lobes** of the cerebral cortex. Although each eye transmits a slightly different picture, the visual areas put them together or integrate them to make a single image. This is what is called **binocular vision**. The visual areas also right the image, because the image on the retina is upside down. The image on film in a camera is also upside down, but we don't even realize that because we look at the pictures right side up. The brain just as automatically ensures that we see our world right side up.

Also for near vision, the pupils constrict to block out peripheral light rays that would otherwise blur the image, and the eyes converge to keep the images on the corresponding parts of both retinas. The importance of pupil constriction can be demonstrated by looking at this page through a pinhole in a piece of paper. You will be able to read with the page much closer to your eye because the paper blocks out light from the sides.

Try looking at your finger placed on the tip of your nose. You can feel your eyes move medially ("cross") in maximum convergence. If the eyes don't converge, the result is double-vision; the brain can't make the very different images into one, and settles for two. This is temporary, however, and the brain will eventually suppress one image.

THE EAR

The ear consists of three areas: the outer ear, the middle ear, and the inner ear (Fig. 9-7). The ear contains the receptors for two senses: hearing and **equilibrium**. These receptors are all found in the inner ear.

OUTER EAR

The **outer ear** consists of the auricle and the ear canal. The **auricle**, or **pinna**, is made of cartilage covered with skin. For animals such as dogs, whose ears are movable, the auricle may act as a funnel for sound waves. For people, however, the stationary auricle is not important. Hearing would not be negatively affected without it, although those of us who wear glasses would have our vision impaired without our auricles. The **ear canal**, also called the **external auditory meatus**, is a tunnel into the temporal bone and curves slightly forward and down.

MIDDLE EAR

The **middle ear** is an air-filled cavity in the temporal bone. The **eardrum**, or **tympanic membrane**, is stretched across the end of the ear canal and vibrates when sound waves strike it. These vibrations are transmitted to the three auditory bones: the **malleus, incus**, and **stapes**. The stapes then transmits vibrations to the fluid-filled inner ear at the **oval window**.

The **eustachian tube** (auditory tube) extends from the middle ear to the nasopharynx and permits air to enter or leave the middle ear cavity. The air pressure in the middle ear must be the same as the external atmospheric pressure in order for the ear drum to vibrate properly. You may have noticed your ears "popping" when in an airplane or when driving to a higher or lower altitude. Swallowing or yawning creates the "pop" by opening the eustachian tubes and equalizing the air pressures.

The eustachian tubes of children are short and nearly horizontal and may permit bacteria to spread from the pharynx to the middle ear. This is why **otitis media** may be a complication of a strep throat.

INNER EAR

Within the temporal bone, the **inner ear** is a cavity called the **bony labyrinth** (maze), which is lined with membrane called the **membranous labyrinth. Perilymph** is the fluid found between bone and membrane, and **endolymph** is the fluid within the membranous structures of the inner ear. These structures are the cochlea, concerned with hearing, and the utricle, saccule, and semicircular canals, all concerned with equilibrium (Fig. 9-8).

Cochlea

The **cochlea** is shaped like a snail shell with two-and-a-half structural turns. Internally, the cochlea is parti-

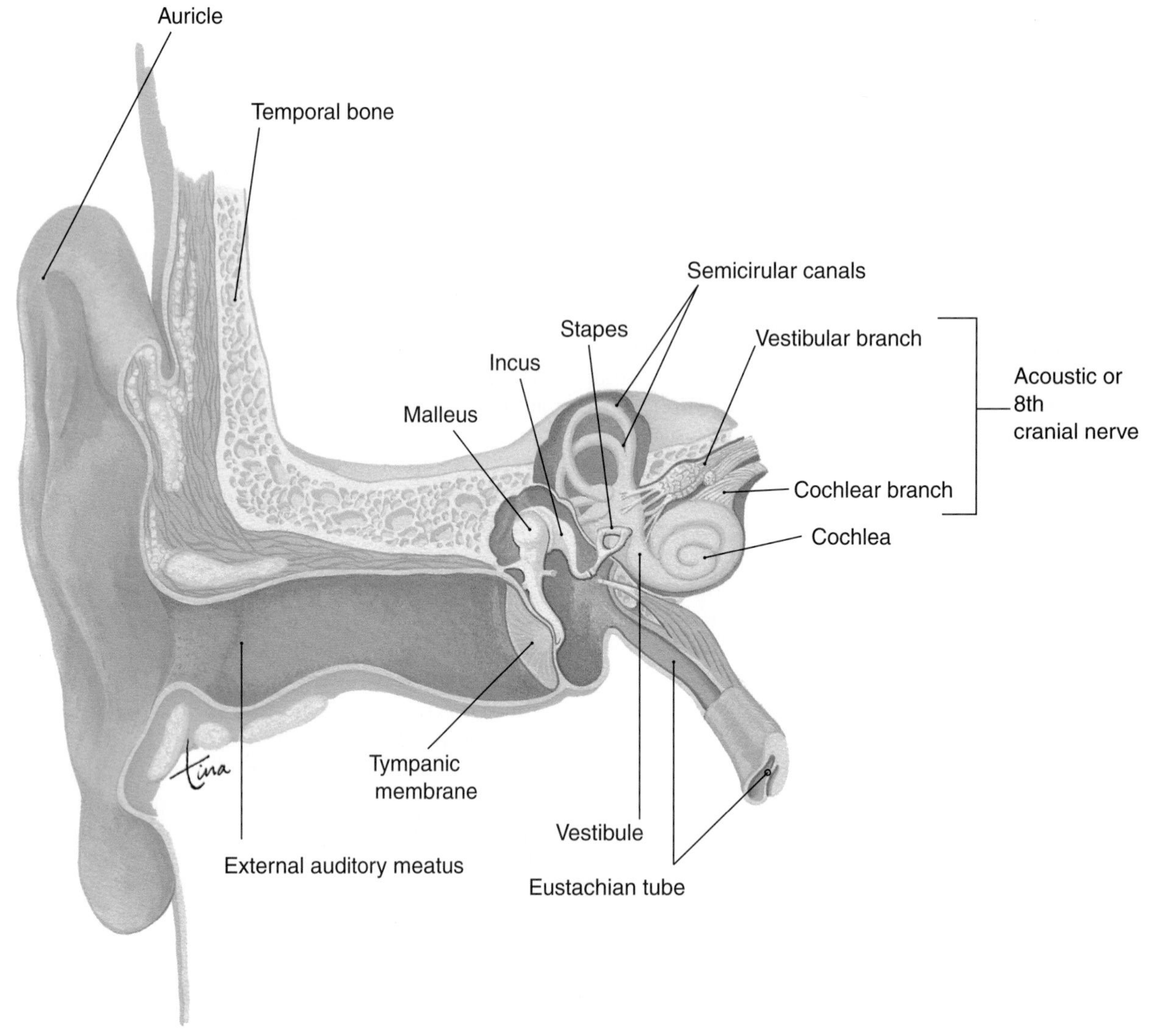

Figure 9–7 Outer, middle, and inner ear structures as shown in a frontal section through the right temporal bone.

tioned into three fluid-filled canals. The medial canal is the cochlear duct, which contains the receptors for hearing in the **organ of Corti (spiral organ)**. The receptors are called hair cells (their projections are not "hair," of course, but rather are specialized microvilli called stereocilia), which contain endings of the cochlear branch of the 8th cranial nerve. Overhanging the hair cells is the tectorial membrane (Fig. 9-9).

Very simply, the process of hearing involves the transmission of vibrations and the generation of nerve impulses. When sound waves enter the ear canal, vibrations are transmitted by the following sequence of structures: ear drum, malleus, incus, stapes, oval window of the inner ear, perilymph and endolymph within the cochlea, and hair cells of the organ of Corti. When the hair cells bend, they generate impulses that are carried by the 8th cranial nerve to the brain. As

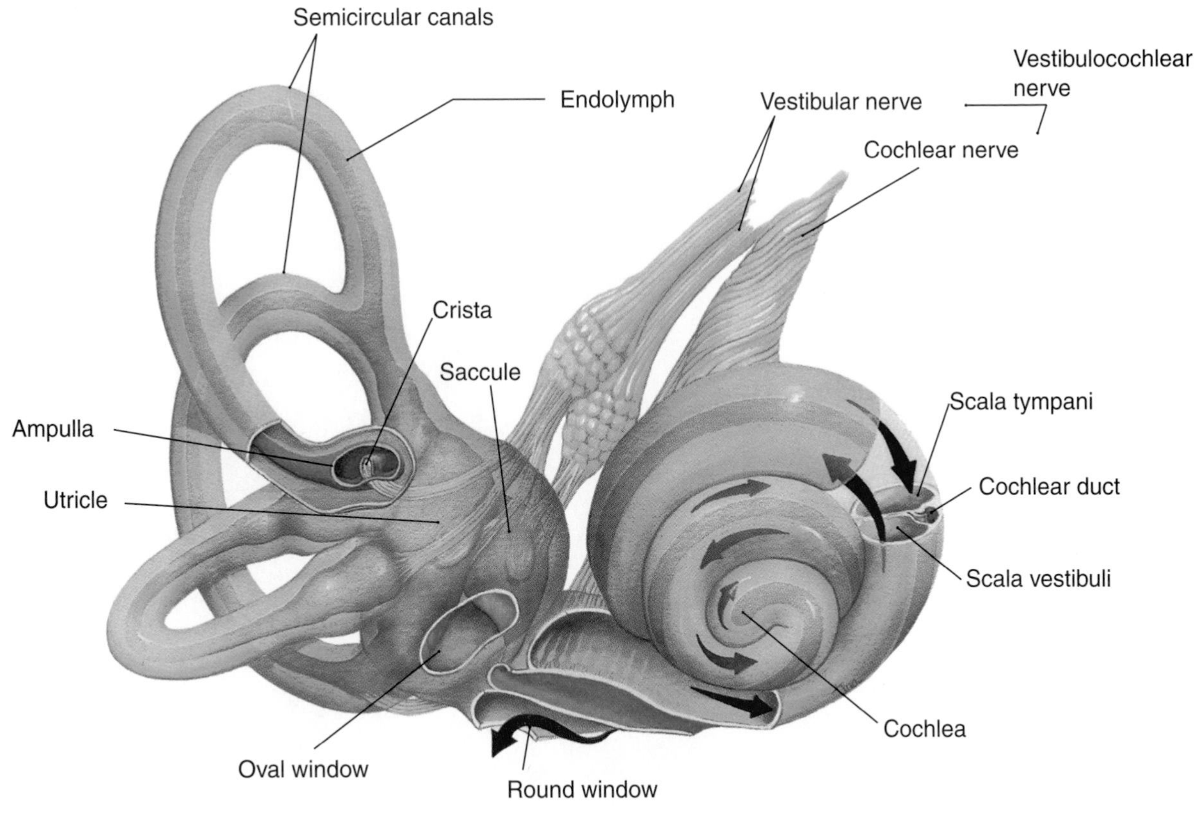

Figure 9–8 Inner ear structures. The arrows show the transmission of vibrations during hearing.

you may recall, the auditory areas are in the **temporal lobes** of the cerebral cortex. It is here that sounds are heard and interpreted (see Box 9-5: Deafness).

The membrane-covered **round window** is important to relieve pressure. When the stapes pushes in the oval window, the round window bulges out, which prevents damage to the hair cells.

Utricle and Saccule

The **utricle** and **saccule** are membranous sacs in an area called the **vestibule**, between the cochlea and semicircular canals. Within the utricle and saccule are hair cells embedded in a gelatinous membrane with tiny crystals of calcium carbonate called **otoliths**. Gravity pulls on the otoliths and bends the hair cells as the position of the head changes (see Fig. 9-10). The impulses generated by these hair cells are carried by the vestibular portion of the 8th cranial nerve to the cerebellum, midbrain, and the temporal lobes of the cerebrum.

The cerebellum and midbrain use this information to maintain equilibrium at a subconscious level. We can, of course, be aware of the position of the head, and it is the cerebrum that provides awareness.

Semicircular Canals

The three **semicircular canals** are fluid-filled membranous ovals oriented in three different planes. At the base of each is an enlarged portion called the ampulla (see Fig. 9-8), which contains hair cells (the crista) that are affected by movement. As the body moves forward, for example, the hair cells are bent backward at first and then straighten (see Fig. 9-10). The bending of the hair cells generates impulses carried by the

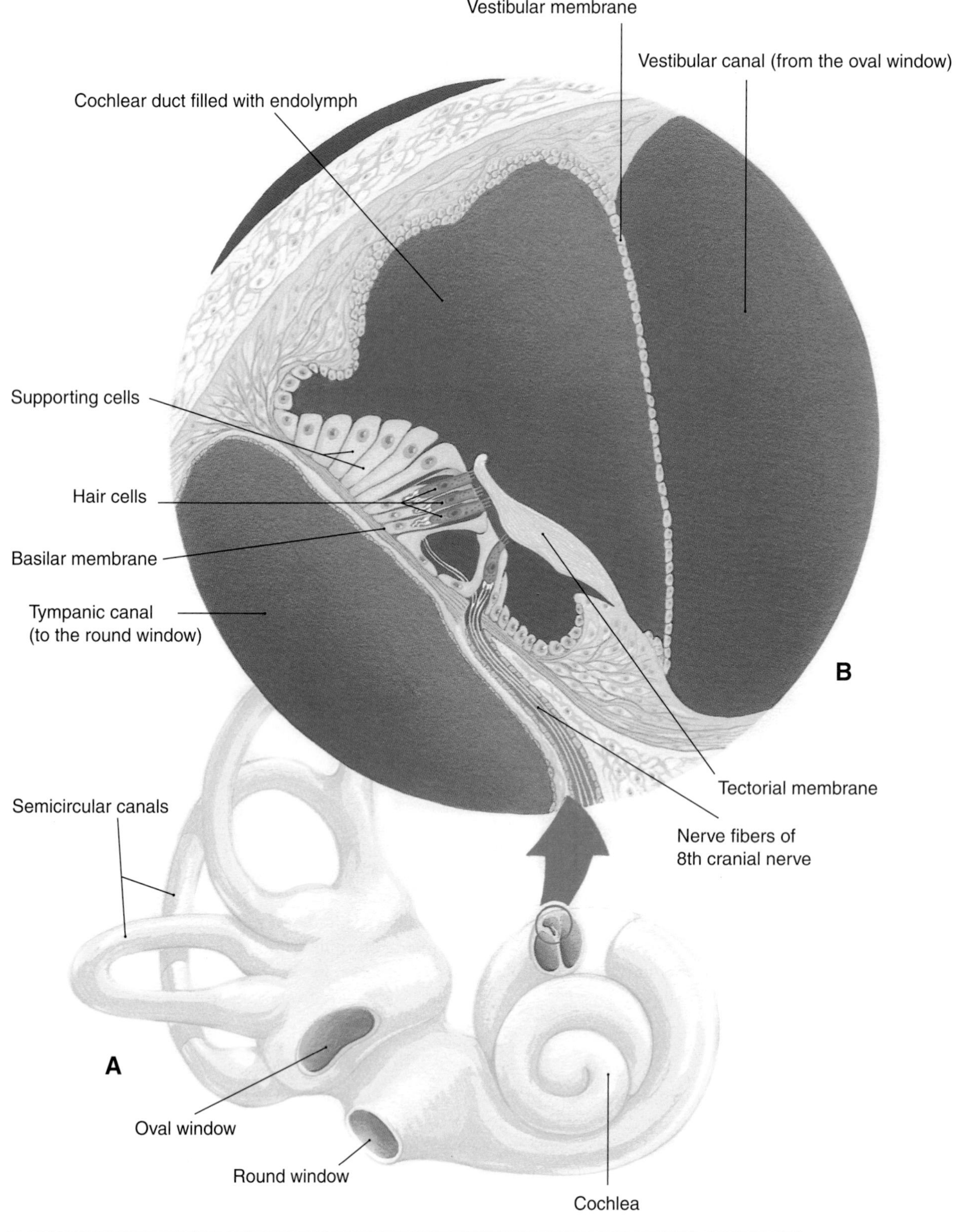

Figure 9–9 Organ of Corti. **(A)** Inner ear structures. **(B)** Magnification of organ of Corti within the cochlea.

Box 9–5 DEAFNESS

Deafness is the inability to hear properly; the types are classified according to the part of the hearing process that is not functioning normally.

Conduction deafness—impairment of one of the structures that transmits vibrations. Examples of this type are a punctured eardrum, arthritis of the auditory bones, or a middle ear infection in which fluid fills the middle ear cavity.

Nerve deafness—impairment of the 8th cranial nerve or the receptors for hearing in the cochlea. The 8th cranial nerve may be damaged by some antibiotics used to treat bacterial infections. Nerve deafness is a rare complication of some viral infections such as mumps or congenital rubella (German measles). Deterioration of the hair cells in the cochlea is a natural consequence of aging, and the acuity of hearing diminishes as we get older. For example, it may be more difficult for an elderly person to distiguish conversation from background noise. Chronic exposure to loud noise accelerates degeneration of the hair cells and onset of this type of deafness.

Central deafness—damage to the auditory areas in the temporal lobes. This type of deafness is rare but may be caused by a brain tumor, meningitis, or a CVA in the temporal lobe.

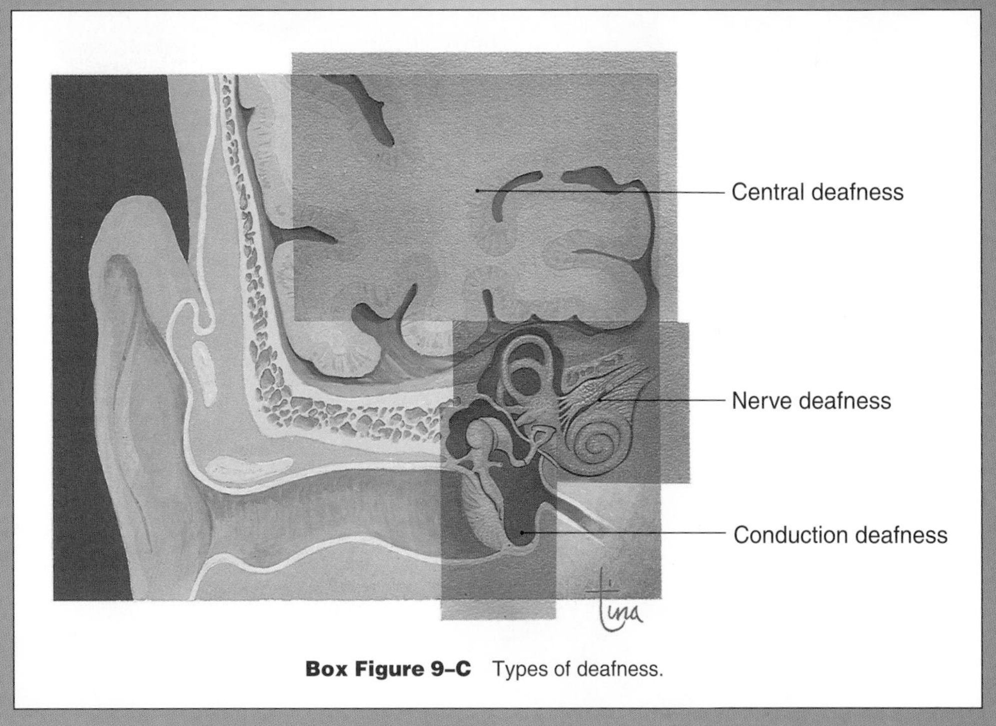

Box Figure 9–C Types of deafness.

vestibular branch of the 8th cranial nerve to the cerebellum, midbrain, and temporal lobes of the cerebrum. These impulses are interpreted as starting or stopping, and accelerating or decelerating, or changing direction, and this information is used to maintain equilibrium while we are moving (see Box 9-6: Motion Sickness).

In summary then, the utricle and saccule provide information about the position of the body at rest, while the semicircular canals provide information

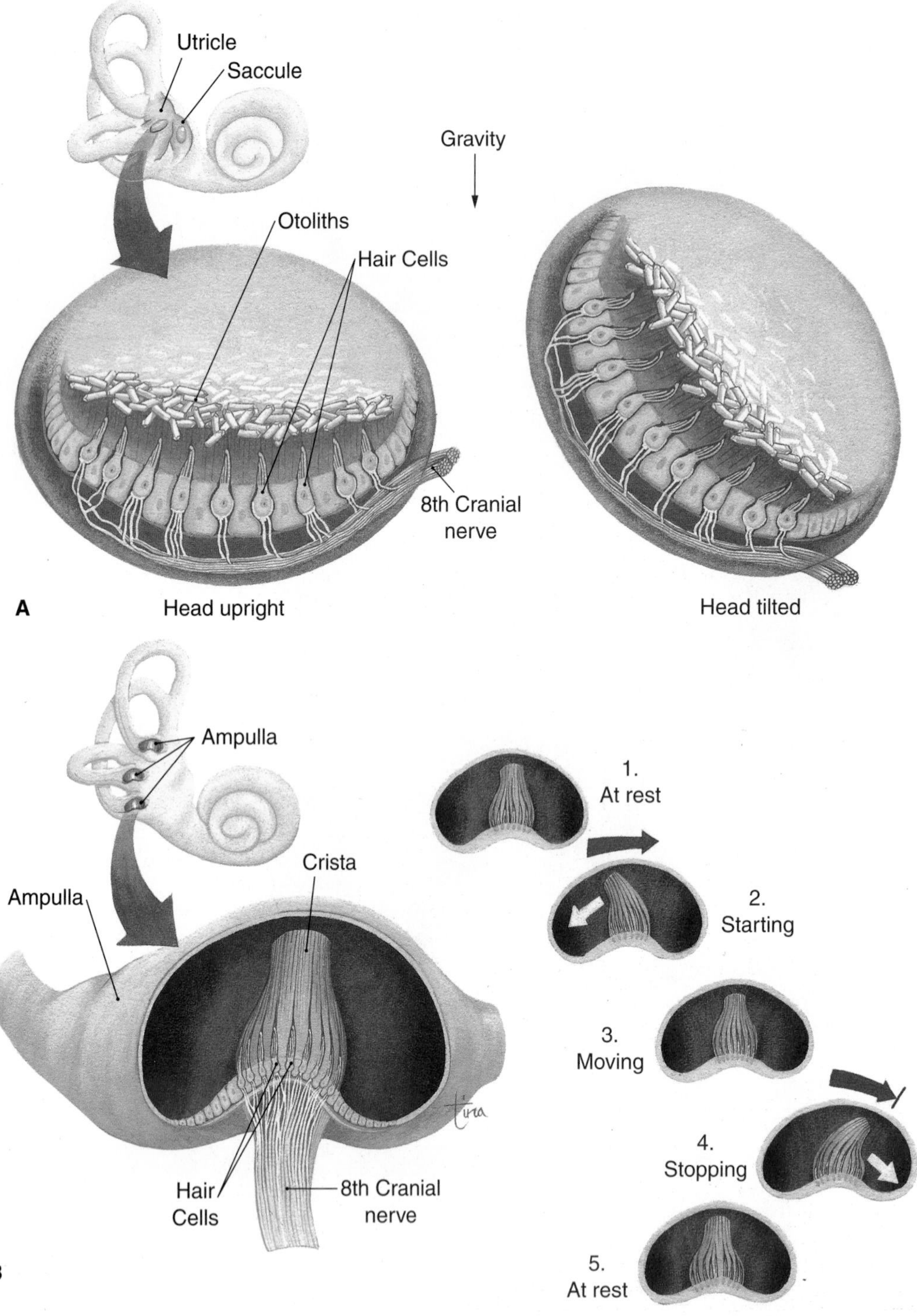

Figure 9–10 Physiology of equilibrium. **(A)** Utricle and saccule. **(B)** Semicircular canals. See text for description.

Box 9–6 MOTION SICKNESS

Motion sickness is characterized by cold sweats, hyperventilation, nausea, and vomiting when the person is exposed to repetitive motion of some kind. Seasickness is a type of motion sickness, as is carsickness (why children are more often carsick than are adults is not known).

Some people are simply not affected by the rolling of a ship or train; for others, the constant stimulation of the receptors for position first becomes uncomfortable, then nauseating. For those who know they are susceptible to motion sickness, medications are available for use before traveling by plane, train, or car.

about the body in motion. Of course there is some overlap, and the brain puts all the information together to create a single sense of body position.

ARTERIAL RECEPTORS

The aorta and carotid arteries contain receptors that detect changes in the blood. The **aortic arch**, which receives blood pumped by the left ventricle of the heart, curves over the top of the heart. The left and right **carotid arteries** are branches of the aortic arch that take blood through the neck on the way to the brain. In each of these vessels are pressoreceptors and chemoreceptors (see Fig. 12–7).

Pressoreceptors in the carotid sinuses and aortic sinus detect changes in blood pressure. **Chemoreceptors** in the carotid bodies and the aortic body detect changes in the oxygen and carbon dioxide content, and the pH, of blood. The impulses generated by these receptors do not give rise to sensations that we feel but rather are information used to make any necessary changes in respiration or circulation. We will return to this in later chapters, so one example will suffice for now.

If the blood level of oxygen decreases significantly, this change (hypoxia) is detected by carotid and aortic chemoreceptors. The sensory impulses are carried by the glossopharyngeal (9th cranial) and vagus (10th cranial) nerves to the medulla. Centers in the medulla may then increase the respiratory rate and the heart rate to obtain and circulate more oxygen. These are the respiratory and cardiac reflexes that were mentioned in Chapter 8 as functions of the glossopharyngeal and vagus nerves. The importance of these reflexes is readily apparent: to maintain normal blood levels of oxygen and carbon dioxide and to maintain normal blood pressure.

AGING AND THE SENSES

All of the senses may be diminished in old age. In the eye, cataracts may make the lens opaque. The lens also loses its elasticity and the eye becomes more far-sighted; a condition called presbyopia. The risk of glaucoma increases, and elderly people should be tested for it because there is treatment that can prevent blindness. Macular degeneration, in which central vision becomes impaired, is a major cause of vision loss for people over 65. In the ear, cumulative damage to the hair cells in the organ of Corti usually becomes apparent some time after the age of 60. Hair cells that have been damaged in a lifetime of noise cannot be replaced. The deafness of old age ranges from slight to profound; very often high-pitched sounds are lost first, while hearing may still be adequate for low-pitched sounds. Both taste and smell become less acute with age, which may contribute to poor nutrition in elderly people.

SUMMARY

Changes take place all around us as well as within us. If the body could not respond appropriately to environmental and internal changes, homeostasis would soon be disrupted, resulting in injury, illness, or even death. In order to respond appropriately to changes, the brain must know what they are. Conveying this information to our brains is the function of our senses. Although we may sometimes take our senses for granted, we could not survive for very long without them.

You have just read about the great variety of internal and external changes that are detected by the sense organs. You are also familiar with the role of the nervous system in the regulation of the body's responses. In the next chapter we will discuss the other regulatory system, the endocrine system. The hormones of the endocrine glands are produced in response to changes, and their regulatory effects all contribute to homeostasis.

STUDY OUTLINE

Purpose of Sensations—to detect changes in the external or internal environment to enable the body to respond appropriately to maintain homeostasis

Sensory Pathway—pathway of impulses for a sensation

1. Receptors—detect a change (usually very specific) and generate impulses.
2. Sensory Neurons—transmit impulses from receptors to the CNS.
3. Sensory Tracts—white matter in the CNS.
4. Sensory Area—most are in the cerebral cortex; feels and interprets the sensation.

Characteristics of Sensations

1. Projection—the sensation seems to come from the area where the receptors were stimulated, even though it is the brain that truly feels the sensation.
2. Intensity—the degree to which a sensation is felt; a strong stimulus affects more receptors, more impulses are sent to the brain and are interpreted as a more intense sensation.
3. Contrast—the effect of a previous or simultaneous sensation on a current sensation as the brain compares them.
4. Adaptation—becoming unaware of a continuing stimulus; if the stimulus remains constant, there is no change for receptors to detect.
5. After-image—the sensation remains in the consciousness after the stimulus has stopped.

Cutaneous Senses—provide information about the external environment and the skin itself

1. In the dermis are free nerve endings for pain and encapsulated nerve endings for touch, pressure, heat, and cold (see Fig. 9-1).
2. Sensory areas are in parietal lobes.
3. Referred pain is visceral pain that is felt as cutaneous pain. Common pathways in the CNS carry both cutaneous and visceral impulses; the brain usually projects sensation to the cutaneous area.

Muscle Sense—knowing where our muscles are without looking at them

1. Stretch receptors in muscles detect stretching.
2. Sensory areas for conscious muscle sense are in parietal lobes.
3. Cerebellum uses unconscious muscle sense to coordinate voluntary movement.

Sense of Taste (see Fig. 9–2)

1. Chemoreceptors are in taste buds on the tongue; detect chemicals (foods) in solution (saliva) in the mouth.
2. Four basic tastes: sweet, sour, salty, bitter; foods stimulate combinations of receptors.
3. Pathway: Facial and glossopharyngeal nerves to taste areas in parietal-temporal lobes.

Sense of Smell (see Fig. 9–2)

1. Chemoreceptors are in upper nasal cavities; detect vaporized chemicals.
2. Pathway: olfactory nerves to olfactory bulbs to olfactory areas in the temporal lobes.
3. Smell contributes greatly to what we call taste.

Hunger and Thirst—visceral (internal) sensations

1. Receptors for Hunger: in hypothalamus, detect changes in nutrient levels in the blood; hunger is projected to the stomach; adaptation does occur.
2. Receptors for Thirst: in hypothalamus, osmoreceptors detect changes in body water (water-salt pro-

portions); thirst is projected to the mouth and pharynx; adaptation does not occur.

The Eye (see Figs. 9–3 through 9–6)

1. Eyelids and eyelashes keep dust out of eyes; conjunctiva line the eyelids and cover white of eye.
2. Lacrimal glands produce tears which flow across eyeball to two lacrimal ducts, to lacrimal sac to nasolacrimal duct to nasal cavity. Tears wash the anterior eyeball and contain lysozyme to inhibit bacterial growth.
3. The eyeball is protected by the bony orbit (socket).
4. The six extrinsic muscles move the eyeball; innervated by the 3rd, 4th, and 6th cranial nerves.
5. Sclera—outermost layer of the eyeball, made of fibrous connective tissue; anterior portion is the transparent cornea, the first light-refracting structure.
6. Choroid Layer—middle layer of eyeball; dark blue pigment absorbs light to prevent glare within the eyeball.
7. Ciliary Body (Muscle) and Suspensory Ligaments—change shape of lens, which is made of a transparent, elastic protein and which refracts light.
8. Iris—two sets of smooth muscle fibers regulate diameter of pupil, that is, how much light strikes the retina.
9. Retina—innermost layer of eyeball; contains rods and cones.
   - Rods—detect light; abundant toward periphery of retina.
   - Cones—detect color; abundant in center of retina. Fovea—in the center of the macula lutea; contains only cones; area of best color vision.
   - Optic Disc—no rods or cones; optic nerve passes through eyeball.
10. Posterior cavity contains vitreous humor (semisolid) that keeps the retina in place.
11. Anterior cavity contains aqueous humor that nourishes the lens and cornea; made by capillaries of the ciliary body, flows through pupil, is reabsorbed to blood at the canal of Schlemm.

Physiology of Vision

1. Refraction (bending and focusing) pathway of light: cornea, aqueous humor, lens, vitreous humor.
2. Lens is adjustable: ciliary muscle relaxes for distant vision, and lens is thin. Ciliary muscle contracts for near vision, and elastic lens thickens and has greater refractive power.
3. Light strikes retina and stimulates chemical reactions in the rods and cones.
4. In rods: rhodopsin breaks down to scotopsin and retinal (from vitamin A), and an electrical impulse is generated. In cones: specific wavelengths of light are absorbed (red, blue, green); chemical reactions generate nerve impulses.
5. Ganglion neurons from the rods and cones form the optic nerve, which passes through the eyeball at the optic disc.
6. Optic Chiasma—site of the crossover of medial fibers of both optic nerves, permitting binocular vision.
7. Visual Areas in Occipital Lobes—each area receives impulses from both eyes; both areas create one image from the two slightly different images of each eye; both areas right the upside down retinal image.

The Ear (see Figs. 9–7 through 9–10)

1. Outer Ear—auricle or pinna has no real function for people; ear canal curves forward and down into temporal bone.
2. Middle Ear—ear drum at end of ear canal vibrates when sound waves strike it. Auditory bones: malleus, incus, stapes; transmit vibrations to inner ear at oval window.
   - Eustachian Tube—extends from middle ear to nasopharynx; allows air in and out of middle ear to permit eardrum to vibrate; air pressure in middle ear should equal atmospheric pressure.
3. Inner Ear—bony labyrinth in temporal bone, lined with membranous labyrinth. Perilymph is fluid between bone and membrane; endolymph is fluid within membrane. Membranous structures are the cochlea, utricle and saccule, and semicircular canals.
4. Cochlea—snail-shell shaped; three internal canals; cochlear duct contains receptors for hearing: hair cells in the organ of Corti; these cells contain endings of the cochlear branch of the 8th cranial nerve.
5. Physiology of Hearing—sound waves stimulate vibration of ear drum, malleus, incus, stapes, oval window of inner ear, perilymph and endolymph of cochlea, and hair cells of organ of Corti. When hair

cells bend, impulses are generated and carried by the 8th cranial nerve to the auditory areas in the temporal lobes. Round window prevents pressure damage to the hair cells.

6. Utricle and Saccule—membranous sacs in the vestibule; each contains hair cells that are affected by gravity. When position of the head changes, otoliths bend the hair cells, which generate impulses along the vestibular branch of the 8th cranial nerve to the cerebellum, midbrain, and cerebrum. Impulses are interpreted as position of the head at rest.
7. Semicircular Canals—three membranous ovals in three planes; enlarged base is the ampulla which contains hair cells (crista) that are affected by movement. As body moves, hair cells bend in opposite direction, generate impulses along vestibular branch of 8th cranial nerve to cerebellum, midbrain, and cerebrum. Impulses are interpreted as movement of the body, changing speed, stopping or starting.

Arterial Receptors—in large arteries; detect changes in blood

1. Aortic Arch—curves over top of heart. Aortic sinus contains pressoreceptors; aortic body contains chemoreceptors; sensory nerve is vagus (10th cranial).
2. Right and left carotid arteries in the neck; carotid sinus contains pressoreceptors; carotid body contains chemoreceptors; sensory nerve is the glossopharyngeal (9th cranial).
3. Pressoreceptors detect changes in blood pressure; chemoreceptors detect changes in pH or oxygen and CO_2 levels in the blood. This information is used by the vital centers in the medulla to change respiration or circulation to maintain normal blood oxygen and CO_2 and normal blood pressure.

REVIEW QUESTIONS

1. State the two general functions of receptors. Explain the purpose of sensory neurons and sensory tracts. (p. 186)

2. Name the receptors for the cutaneous senses, and explain the importance of this information. (p. 187)

3. Name the receptors for muscle sense and the parts of the brain concerned with muscle sense. (p. 188)

4. State what the chemoreceptors for taste and smell detect. Name the cranial nerve(s) for each of these senses and the lobe of the cerebrum where each is felt. (p. 188)

5. Name the part of the eye with each of the following functions: (pp. 190–193)
   a. Change the shape of the lens
   b. Contains the rods and cones
   c. Forms the white of the eye
   d. Form the optic nerve
   e. Keep dust out of eye
   f. Changes the size of the pupil
   g. Produce tears
   h. Absorbs light within the eyeball to prevent glare

6. With respect to vision: (pp. 193–197)
   a. Name the structures and substances that refract light rays (in order).
   b. State what cones detect and what rods detect. What happens within these receptors when light strikes them?
   c. Name the cranial nerve for vision and the lobe of the cerebrum that contains the visual area.

7. With respect to the ear: (pp. 197–203)
   a. Name the parts of the ear that transmit the vibrations of sound waves (in order).
   b. State the location of the receptors for hearing.
   c. State the location of the receptors that respond to gravity.
   d. State the location of the receptors that respond to motion.
   e. State the two functions of the 8th cranial nerve.
   f. Name the lobe of the cerebrum concerned with hearing.
   g. Name the two parts of the brain concerned with maintaining balance and equilibrium.

8. Name the following (p. 203):
   a. The locations of arterial chemoreceptors, and state what they detect
   b. The locations of arterial pressoreceptors, and state what they detect
   c. The cranial nerves involved in respiratory and cardiac reflexes, and state the part of the brain that regulates these vital functions

9. Explain each of the following: adaptation, after-image, projection, contrast. (p. 186)

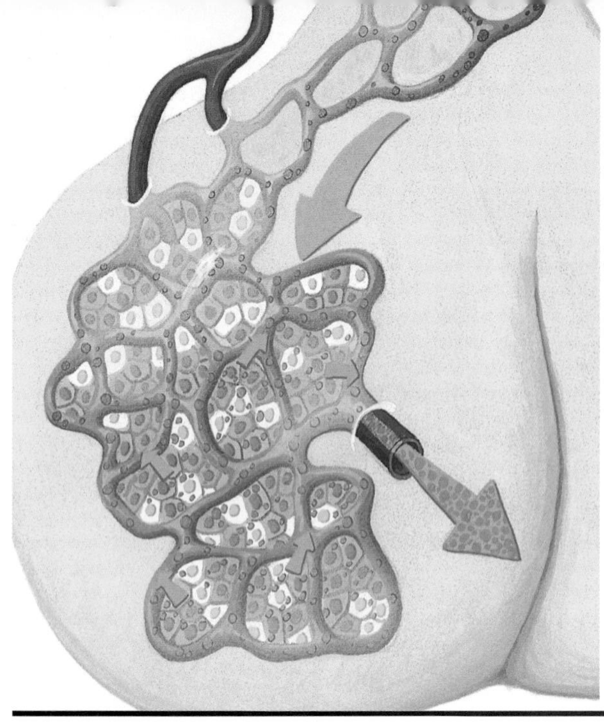

CHAPTER 10

Chapter Outline

Student Objectives

- Name the endocrine glands and the hormones secreted by each.
- Explain how a negative feedback mechanism works.
- Explain how the hypothalamus is involved in the secretion of hormones from the posterior pituitary gland and anterior pituitary gland.
- State the functions of oxytocin and antidiuretic hormone, and explain the stimulus for secretion of each.
- State the functions of the hormones of the anterior pituitary gland, and state the stimulus for secretion of each.

The Endocrine System

Student Objectives (Continued)

- State the functions of thyroxine and T_3, and describe the stimulus for their secretion.
- Explain how parathyroid hormone and calcitonin work as antagonists.
- Explain how insulin and glucagon work as antagonists.
- State the functions of epinephrine and norepinephrine and explain their relationship to the sympathetic division of the autonomic nervous system.
- State the functions of aldosterone and cortisol, and describe the stimulus for secretion of each.
- State the functions of estrogen, progesterone, testosterone, and inhibin and state the stimulus for secretion of each.
- Explain what prostaglandins are made of, and state some of their functions.
- Explain how the protein hormones are believed to exert their effects.
- Explain how the steroid hormones are believed to exert their effects.

New Terminology

Alpha cells (**AL**-fah SELLS)
Beta cells (**BAY**-tah SELLS)
Catecholamines (**KAT**-e-kohl-ah-MEENZ)
Corpus luteum (**KOR**-pus **LOO**-tee-um)
Gluconeogenesis (GLOO-koh-nee-oh-**JEN**-i-sis)
Glycogenesis (GLIGH-koh-**JEN**-i-sis)
Glycogenolysis (GLIGH-ko-jen-**OL**-i-sis)
Hypercalcemia (HIGH-per-kal-**SEE**-mee-ah)
Hyperglycemia (HIGH-per-gligh-**SEE**-mee-ah)
Hypocalcemia (HIGH-poh-kal-**SEE**-mee-ah)
Hypoglycemia (HIGH-poh-gligh-**SEE**-mee-ah)
Hypophysis (high-**POFF**-e-sis)
Islets of Langerhans (**EYE**-lets of **LAHNG**-er-hanz)
Negative feedback mechanism (**NEG**-ah-tiv **FEED**-bak)
Prostaglandins (PRAHS-tah-**GLAND**-ins)
Releasing hormones (ree-**LEE**-sing **HOR**-mohns)
Renin-angiotensin mechanism (**REE**-nin AN-jee-oh-**TEN**-sin)
Sympathomimetic (SIM-pah-tho-mi-**MET**-ik)
Target organ (**TAR**-get **OR**-gan)

Related Clinical Terminology

Acromegaly (AK-roh-**MEG**-ah-lee)
Addison's disease (**ADD**-i-sonz)
Cretinism (**KREE**-tin-izm)
Cushing's syndrome (**KOOSH**-ingz **SIN**-drohm)
Diabetes mellitus (DYE-ah-**BEE**-tis mel-**LYE**-tus)
Giantism (**JIGH**-an-tizm)
Goiter (**GOY**-ter)
Grave's disease (GRAYVES)
Ketoacidosis (KEY-toh-ass-i-**DOH**-sis)
Myxedema (MIK-suh-**DEE**-mah)
Pituitary dwarfism (pi-**TOO**-i-TER-ee **DWORF**-izm)

*Terms that appear in **bold type** in the chapter text are defined in the glossary, which begins on page 528.*

We have already seen how the nervous system regulates body functions by means of nerve impulses and integration of information by the spinal cord and brain. The other regulating system of the body is the **endocrine system**, which consists of endocrine glands that secrete chemicals called **hormones**. These glands are shown in Fig. 10–1.

Endocrine glands are ductless, that is, they do not have ducts to take their secretions to specific sites. Instead, hormones are secreted directly into capillaries and circulate in the blood throughout the body. Each hormone then exerts very specific effects on certain organs, called **target organs** or **target tissues**. Some hormones, such as insulin and thyroxine, have many target organs. Other hormones, such as calcitonin and some pituitary gland hormones, have only one or a few target organs.

In general, the endocrine system and its hormones help regulate growth, the use of foods to produce energy, resistance to stress, the pH of body fluids and fluid balance, and reproduction. In this chapter we will discuss the specific functions of the hormones and how each contributes to homeostasis.

CHEMISTRY OF HORMONES

With respect to their chemical structure, hormones may be classified into three groups: amines, proteins, and steroids.

1. **Amines**—these simple hormones are structural variations of the amino acid tyrosine. This group includes thyroxine from the thyroid gland and epinephrine and norepinephrine from the adrenal medulla.
2. **Proteins**—these hormones are chains of amino acids. Insulin from the pancreas, growth hormone from the anterior pituitary gland, and calcitonin from the thyroid gland are all proteins. Short chains of amino acids may be called **peptides**. Antidiuretic hormone and oxytocin, synthesized by the hypothalamus, are peptide hormones.
3. **Steroids**—cholesterol is the precursor for the steroid hormones, which include cortisol and aldosterone from the adrenal cortex, estrogen and progesterone from the ovaries, and testosterone from the testes.

REGULATION OF HORMONE SECRETION

Hormones are secreted by endocrine glands when there is a need for them, that is, for their effects on their target organs. The cells of endocrine glands respond to changes in the blood, or perhaps to other hormones in the blood. These stimuli are the information they use to increase or decrease secretion of their own hormones. When a hormone brings about its effects, that reverses the stimulus, and secretion of the hormone decreases until the stimulus reoccurs. A specific example will be helpful here; let us use insulin.

Insulin is secreted by the pancreas when the blood glucose level is high, that is, hyperglycemia is the stimulus for secretion of insulin. Once circulating in the blood, insulin enables cells to remove glucose from the blood to use for energy production and enables the liver to store glucose as glycogen. As a result of these actions of insulin, blood glucose level decreases, reversing the stimulus for secretion of insulin. Insulin secretion then decreases until the blood glucose level increases again.

This is an example of a **negative feedback mechanism**, in which information about the effects of the hormone is "fed back" to the gland, which then decreases its secretion of the hormone. This is why the mechanism is called "negative": the effects of the hormone reverse the stimulus and decrease the secretion of the hormone. The secretion of many other hormones is regulated in a similar way.

The hormones of the anterior pituitary gland are secreted in response to **releasing hormones** (also called releasing factors) secreted by the hypothalamus. You may recall this from Chapter 8. Growth hormone, for example, is secreted in response to growth hormone releasing hormone (GHRH) from the hypothalamus. As growth hormone exerts its effects, the secretion of GHRH decreases, which in turn decreases the secretion of growth hormone. This is another type of negative feedback mechanism.

For each of the hormones to be discussed in this chapter, the stimulus for its secretion will also be mentioned. Some hormones function as an **antagonistic pair** to regulate a particular aspect of blood chemistry; these mechanisms will also be covered.

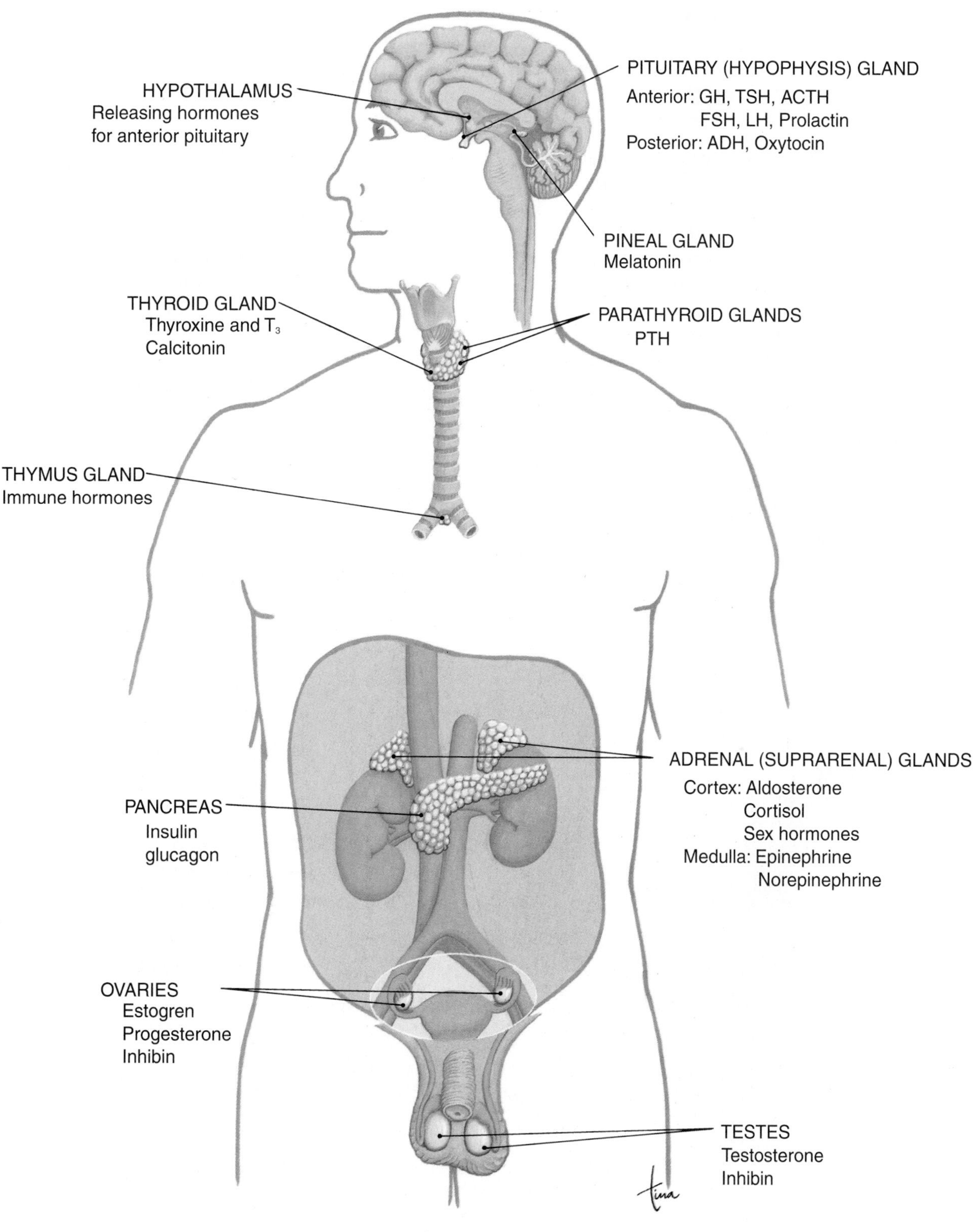

Figure 10–1 The endocrine system. Locations of many endocrine glands. Both male and female gonads (testes and ovaries) are shown.

THE PITUITARY GLAND

The **pituitary gland** (or **hypophysis**) hangs by a short stalk (infundibulum) from the hypothalamus and is enclosed by the sella turcica of the sphenoid bone. Despite its small size, the pituitary gland regulates many body functions. Its two major portions are the posterior pituitary gland **(neurohypophysis)**, which is actually an extension of the nerve tissue of the hypothalamus, and the anterior pituitary gland **(adenohypophysis)** which is separate glandular tissue.

POSTERIOR PITUITARY GLAND

The two hormones of the **posterior pituitary gland** are actually produced by the hypothalamus and simply stored in the posterior pituitary until needed. Their release is stimulated by nerve impulses from the hypothalamus (Fig. 10–2).

Antidiuretic Hormone

Antidiuretic hormone (ADH) increases the reabsorption of water by kidney tubules, which decreases the amount of urine formed. The water is reabsorbed into the blood, so as urinary output is decreased, blood volume is increased, which helps maintain normal blood pressure.

The stimulus for secretion of ADH is decreased water content of the body. If too much water is lost in sweating or diarrhea, for example, **osmoreceptors** in the hypothalamus detect the increased "saltiness" of body fluids. The hypothalamus then transmits impulses to the posterior pituitary to increase the secretion of ADH and decrease the loss of more water in urine.

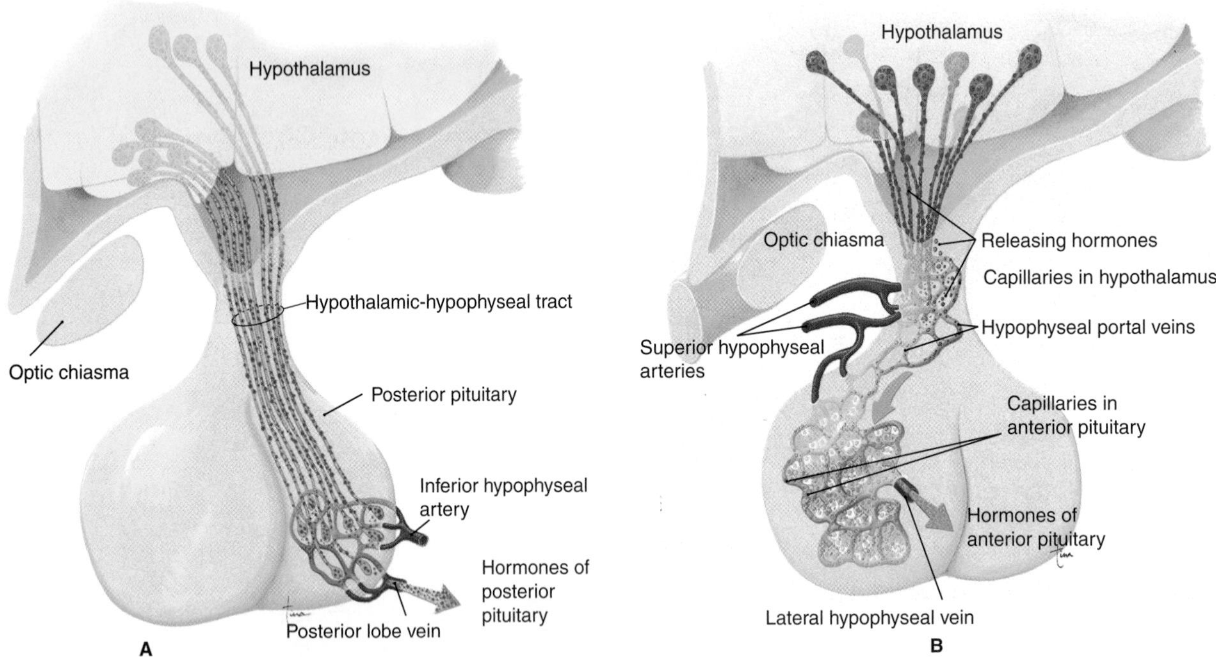

Figure 10–2 Structural relationships of hypothalamus and pituitary gland. **(A)** Posterior pituitary stores hormones produced in the hypothalamus. **(B)** Releasing hormones of the hypothalamus circulate directly to the anterior pituitary and influence its secretions. Notice the two networks of capillaries.

Any type of dehydration stimulates the secretion of ADH to conserve body water. In the case of severe hemorrhage, ADH is released in large amounts and will also cause vasoconstriction, which will contribute to the maintenance of normal blood pressure.

Ingestion of alcohol inhibits the secretion of ADH and increases urinary output. If alcohol intake is excessive and fluid is not replaced, a person will feel thirsty and dizzy the next morning. The thirst is due to the loss of body water, and the dizziness is the result of low blood pressure.

Oxytocin

Oxytocin stimulates contraction of the uterus at the end of pregnancy and stimulates release of milk from the mammary glands.

As labor begins, the cervix of the uterus is stretched, which generates sensory impulses to the hypothalamus, which in turn stimulates the posterior pituitary to release oxytocin. Oxytocin then causes strong contractions of the smooth muscle (myometrium) of the uterus to bring about delivery of the baby and the placenta.

Recently, it has been discovered that the placenta itself secretes oxytocin at the end of gestation and in an amount far higher than that from the posterior pituitary gland. Research is continuing to determine the exact mechanism and precise role of the placenta in labor.

When a baby is breast-fed, the sucking of the baby stimulates sensory impulses from the mother's nipple to the hypothalamus. Nerve impulses from the hypothalamus to the posterior pituitary cause the release of oxytocin, which stimulates contraction of the smooth muscle cells around the mammary ducts. This release of milk is sometimes called the "milk let-down" reflex. The hormones of the posterior pituitary are summarized in Table 10-1.

ANTERIOR PITUITARY GLAND

The hormones of the **anterior pituitary gland** regulate many body functions. They are in turn regulated by **releasing hormones** from the hypothalamus. These releasing hormones are secreted into capillaries in the hypothalamus and pass through the **hypophyseal portal** veins to another capillary network in the anterior pituitary gland. Here, the releasing hormones are absorbed and stimulate secretion of the anterior pituitary hormones. This small but specialized pathway of circulation is shown in Fig. 10-2. This pathway permits the releasing hormones to rapidly stimulate the anterior pituitary, without having to pass through general circulation.

Growth Hormone

Growth hormone (GH) may also be called **somatotropin**, and does indeed stimulate growth. GH increases the transport of amino acids into cells, and increases the rate of protein synthesis. It also stimulates cell division in those tissues capable of mitosis. These functions contribute to the growth of the body during childhood, especially growth of bones and muscles.

You may now be wondering if GH is secreted in adults, and the answer is yes. The use of amino acids for the synthesis of proteins is still necessary, even if the body is not growing in height. GH also stimulates

Table 10–1 HORMONES OF THE POSTERIOR PITUITARY GLAND

Hormone	Function(s)	Regulation of Secretion
Oxytocin	• Promotes contraction of myometrium of uterus (labor) • Promotes release of milk from mammary glands	Nerve impulses from hypothalamus, the result of stretching of cervix or stimulation of nipple. Secretion from placenta at end of gestation—stimulus unknown
Antidiuretic Hormone (ADH)	• Increases water reabsorption by the kidney tubules (water returns to the blood)	Decreased water content in the body (alcohol inhibits secretion)

the release of fat from adipose tissue and the use of fats for energy production. This is important any time we go for extended periods without eating, no matter what our ages.

The secretion of GH is regulated by two releasing hormones from the hypothalamus. Growth hormone releasing hormone (GHRH), which increases the secretion of GH, is produced during hypoglycemia and during exercise. Another stimulus for GHRH is a high blood level of amino acids; the GH then secreted will ensure the conversion of these amino acids into protein. **Somatostatin** may also be called growth hormone inhibiting hormone (GHIH), and as its name tells us, it decreases the secretion of GH. Somatostatin is produced during states of hyperglycemia. Disorders of GH secretion are discussed in Box 10–1.

Thyroid-Stimulating Hormone

Thyroid-stimulating hormone (TSH) may also be called thyrotropin, and its target organ is the thyroid gland. TSH stimulates the normal growth of the thyroid and the secretion of thyroxine (T_4) and triiodothyronine (T_3). The functions of these thyroid hormones will be covered later in this chapter.

The secretion of TSH is stimulated by thyrotropin releasing hormone (TRH) from the hypothalamus. When metabolic rate (energy production) decreases, TRH is produced.

Adrenocorticotropic Hormone

Adrenocorticotropic hormone (ACTH) stimulates the secretion of cortisol and other hormones by the adrenal cortex. Secretion of ACTH is increased by corticotropin releasing hormone (CRH) from the hypothalamus. CRH is produced in any type of physiological stress situation such as injury, disease, exercise, or hypoglycemia (being hungry is stressful).

Prolactin

Prolactin, as its name suggests, is responsible for lactation. More precisely, prolactin initiates and maintains milk production by the mammary glands. The regulation of secretion of prolactin is complex, involving both prolactin releasing hormone (PRH) and prolactin inhibiting hormone (PIH) from the hypothalamus. The mammary glands must first be acted upon by other hormones such as estrogen and progesterone, which are secreted in large amounts by the placenta during pregnancy. Then, after delivery of the

Box 10–1 DISORDERS OF GROWTH HORMONE

A deficiency or excess of growth hormone (GH) during childhood will have marked effects on the growth of a child. Hyposecretion of GH results in **pituitary dwarfism**, in which the person may attain a final height of only 3 to 4 feet but will have normal body proportions. GH can now be produced using genetic engineering and may be used to stimulate growth in children with this disorder. GH will not increase growth of children with the genetic potential for short stature. Reports that GH will reverse the effects of aging are simply not true.

Hypersecretion of GH results in **giantism**, in which the long bones grow excessively and the person may attain a height of 8 feet. Most very tall people, such as basketball players, do *not* have this condition; they are tall as a result of their genetic makeup and good nutrition.

In an adult, hypersecretion of GH is caused by a pituitary tumor, and results in **acromegaly**. The long bones cannot grow because the epiphyseal discs are closed, but the growth of other bones is stimulated. The jaw and other facial bones become disproportionally large, as do the bones of the hands and feet. The skin becomes thicker, and the tongue also grows and may protrude. Other consequences include compression of nerves by abnormally growing bones and growth of the articular cartilages, which then erode and bring on arthritis. Treatment of acromegaly requires surgical removal of the tumor or its destruction by radiation.

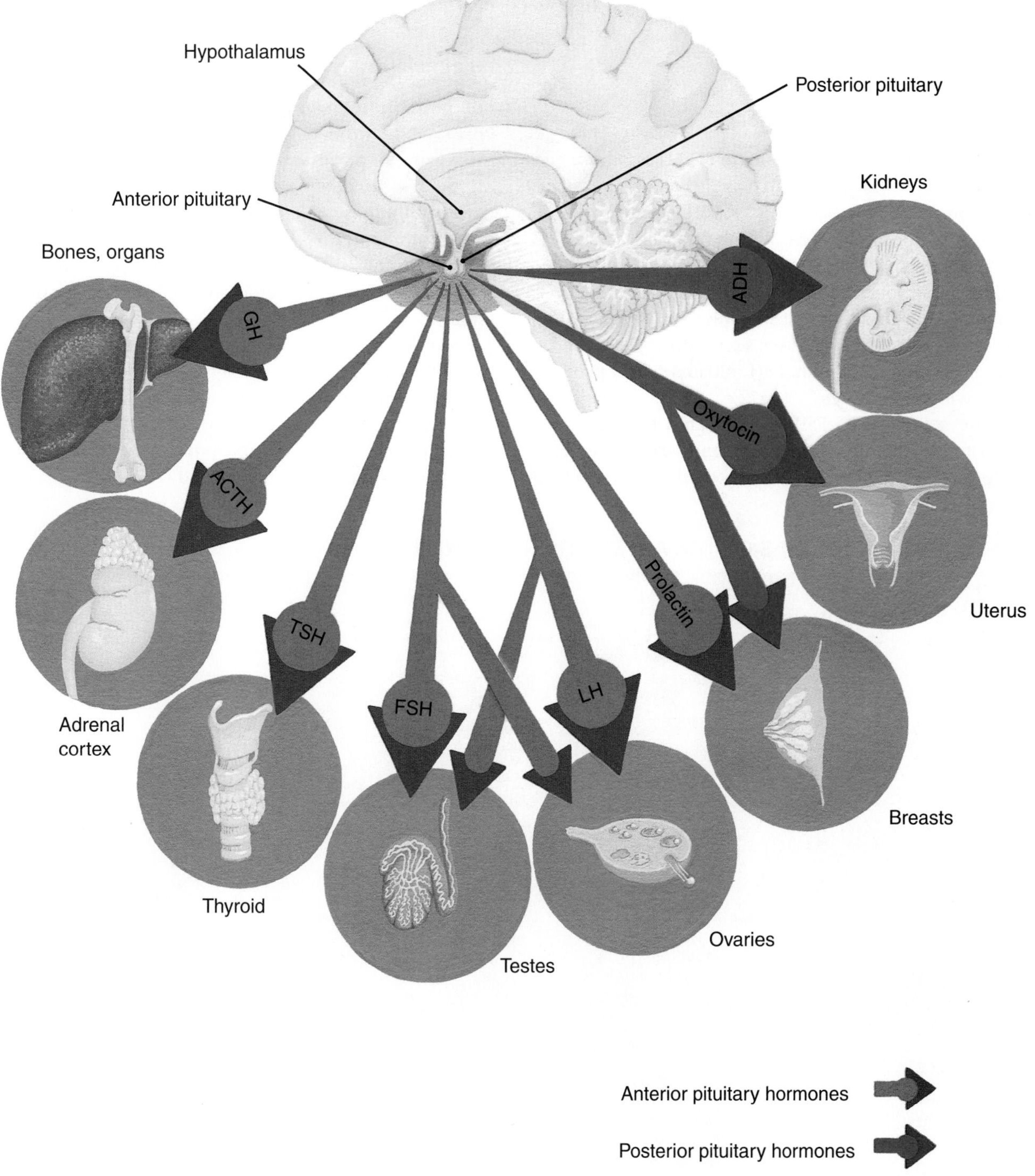

Figure 10–3 Hormones of the pituitary gland and their target organs.

baby, prolactin secretion increases and milk is produced. If the mother continues to breast-feed, prolactin levels remain high.

Follicle-Stimulating Hormone

Follicle-stimulating hormone (FSH) is one of the gonadotropic hormones, that is, it has its effects on the gonads: the ovaries or testes. FSH is named for one of its functions in women. Within the ovaries are ovarian follicles that contain potential ova (egg cells). FSH stimulates the growth of ovarian follicles, that is, it initiates egg development in cycles of approximately 28 days. FSH also stimulates secretion of estrogen by the follicle cells. In men, FSH initiates sperm production within the testes.

The secretion of FSH is stimulated by the hypothalamus, which produces **gonadotropin releasing hormone (GnRH)**. FSH secretion is decreased by **inhibin**, a hormone produced by the ovaries or testes.

Luteinizing Hormone

Luteinizing hormone (LH) is another gonadotropic hormone. In women, LH is responsible for ovulation, the release of a mature ovum from an ovarian follicle. LH then stimulates that follicle to develop into the corpus luteum, which secretes progesterone, also under the influence of LH. In men, LH stimulates the interstitial cells of the testes to secrete testosterone (LH is also called ICSH: interstitial cell stimulating hormone).

Secretion of LH is also regulated by GnRH from the hypothalamus. We will return to FSH and LH, as well as to the sex hormones, in Chapter 20.

All the target organs of the pituitary gland are shown in Fig. 10–3. The hormones of the anterior pituitary are summarized in Table 10–2.

Table 10–2 HORMONES OF THE ANTERIOR PITUITARY GLAND

Hormone	Function(s)	Regulation of Secretion
Growth Hormone (GH)	• Increases rate of mitosis • Increases amino acid transport into cells • Increases rate of protein synthesis • Increases use of fats for energy	• GHRH (hypothalamus) stimulates secretion • GHIH—somatostatin (hypothalamus) inhibits secretion
Thyroid-Stimulating Hormone (TSH)	• Increases secretion of thyroxine and T_3 by thyroid gland	• TRH (hypothalamus)
Adrenocorticotropic Hormone (ACTH)	• Increases secretion of cortisol by the adrenal cortex	• CRH (hypothalamus)
Prolactin	• Stimulates milk production by the mammary glands	• PRH (hypothalamus) stimulates secretion • PIH (hypothalamus) inhibits secretion
Follicle-Stimulating Hormone (FSH)	*In women:* • Initiates growth of ova in ovarian follicles • Increases secretion of estrogen by follicle cells *In men:* • Initiates sperm production in the testes	• GnRH (hypothalamus) stimulates secretion • Inhibin (ovaries or testes) inhibits secretion
Luteinizing Hormone (LH) (ICSH)	*In women:* • Causes ovulation • Causes the ruptured ovarian follicle to become the corpus luteum • Increases secretion of progesterone by the corpus luteum *In men:* • Increases secretion of testosterone by the interstitial cells of the testes	• GnRH (hypothalamus)

THYROID GLAND

The **thyroid gland** is located on the front and sides of the trachea just below the larynx. Its two lobes are connected by a middle piece called the isthmus. The structural units of the thyroid gland are thyroid follicles, which produce **thyroxine (T_4)** and **triiodothyronine (T_3)**. Iodine is necessary for the synthesis of these hormones; thyroxine contains four atoms of iodine, and T_3 contains three atoms of iodine.

The third hormone produced by the thyroid gland is **calcitonin**, which is secreted by parafollicular cells. Its function is very different from those of thyroxine and T_3, which you may recall from Chapter 6.

THYROXINE AND T_3

Thyroxine (T_4) and T_3 have the same functions: regulation of energy production and protein synthesis, which contribute to growth of the body and to normal body functioning throughout life. We will use "thyroxine" to designate both hormones. Thyroxine increases cell respiration of all food types (carbohydrates, fats, and excess amino acids) and thereby increases energy and heat production. Thyroxine also increases the rate of protein synthesis within cells. Normal production of thyroxine is essential for physical growth, normal mental development, and maturation of the reproductive system. Although thyroxine is not a vital hormone, in that it is not crucial to survival, its absence greatly diminishes physical and mental growth and abilities (see Box 10–2: Disorders of Thyroxine).

Secretion of thyroxine and T_3 is stimulated by **thyroid-stimulating hormone (TSH)** from the anterior pituitary gland. When metabolic rate (energy production) decreases, this change is detected by the hypothalamus, which secretes thyrotropin-releasing hormone (TRH). TRH stimulates the anterior pituitary to secrete TSH, which stimulates the thyroid to release thyroxine and T_3, which raise the metabolic rate by increasing energy production. This negative feedback mechanism then shuts off TRH from the hypothalamus until metabolic rate decreases again.

Box 10–2 DISORDERS OF THYROXINE

Iodine is an essential component of thyroxine (and T_3), and a dietary deficiency of iodine causes **goiter**. In an attempt to produce more thyroxine, the thyroid cells become enlarged, and the thyroid gland becomes apparent on the front of the neck. The use of iodized salt has made goiter a rare condition.

Hyposecretion of thyroxine in a newborn has devastating effects on the growth of the child. Without thyroxine, physical growth is diminished, as is mental development. This condition is called **cretinism**, characterized by severe physical and mental retardation. If the thyroxine deficiency is detected shortly after birth, the child may be treated with thyroid hormones to promote normal development.

Hyposecretion of thyroxine in an adult is called **myxedema**. Without thyroxine, the metabolic rate (energy production) decreases, resulting in lethargy, muscular weakness, slow heart rate, a feeling of cold, weight gain, and a characteristic puffiness of the face. The administration of thyroid hormones will return the metabolic rate to normal.

Grave's disease is a hypersecretion of thyroxine that is believed to be an immune disorder. The autoantibodies seem to bind to TSH receptors on the thyroid cells and stimulate secretion of excess thyroxine. The symptoms are those that would be expected when the metabolic rate is abnormally elevated: weight loss accompanied by increased appetite, increased sweating, fast heart rate, feeling of warmth, and fatigue. Also present may be goiter and exophthalmos, which is protrusion of the eyes. Treatment is aimed at decreasing the secretion of thyroxine by the thyroid, and medications or radioactive iodine may be used to accomplish this.

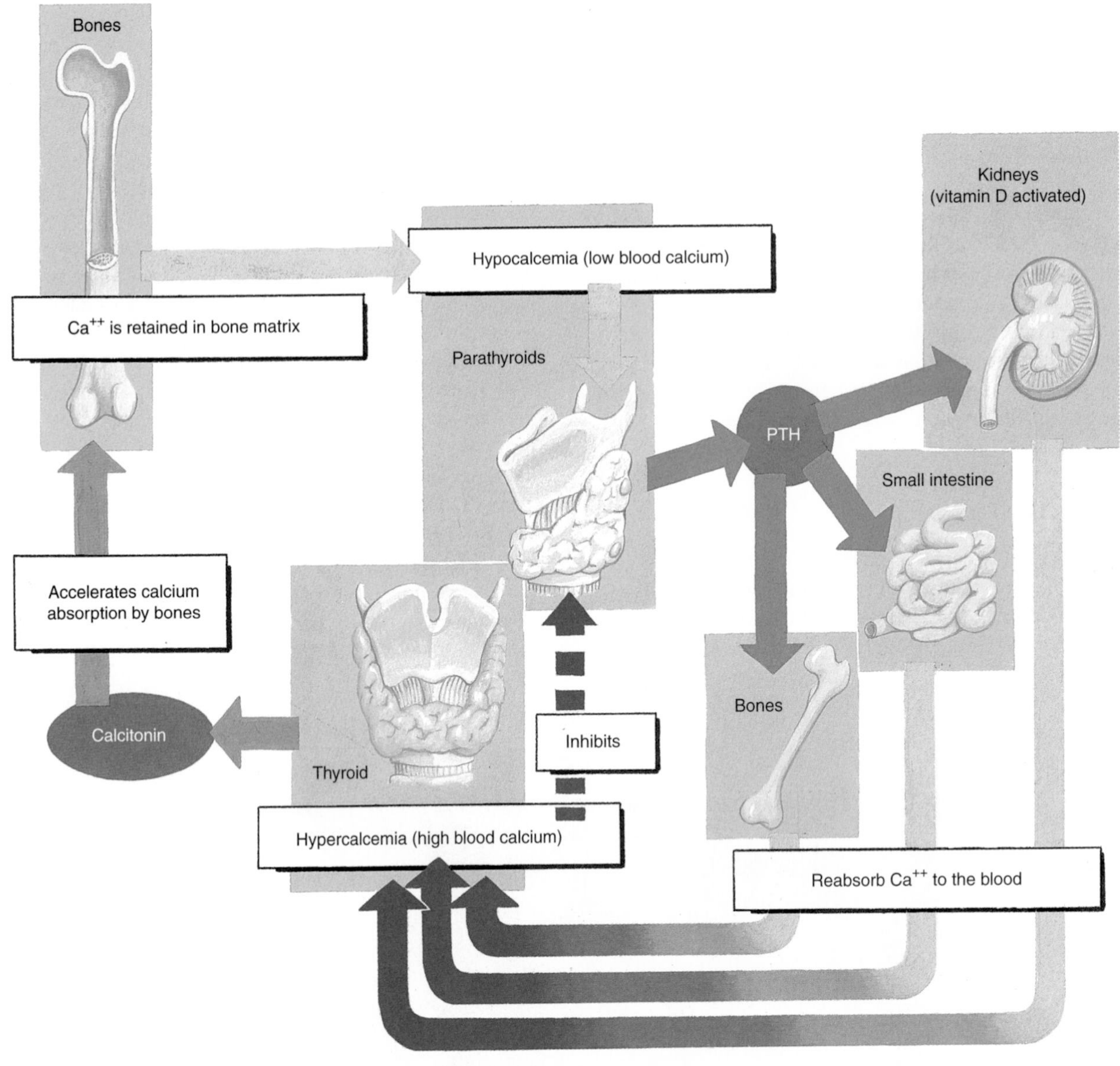

Figure 10–4 Calcitonin and parathyroid hormone (PTH) and their functions related to the maintenance of the blood calcium level.

Table 10–3 HORMONES OF THE THYROID GLAND

Hormone	Function(s)	Regulation of Secretion
Thyroxine (T_4) and Triiodothyronine (T_3)	• Increase energy production from all food types • Increase rate of protein synthesis	TSH (anterior pituitary)
Calcitonin	• Decreases the reabsorption of calcium and phosphate from bones to blood	Hypercalcemia

CALCITONIN

Calcitonin decreases the reabsorption of calcium and phosphate from the bones to the blood, thereby lowering blood levels of these minerals. This function of calcitonin helps maintain normal blood levels of calcium and phosphate and also helps maintain a stable, strong bone matrix. It is believed that calcitonin exerts its most important effects during childhood, when bones are growing.

The stimulus for secretion of calcitonin is **hypercalcemia**, that is, a high blood calcium level. When blood calcium is high, calcitonin ensures that no more will be removed from bones until there is a real need for more calcium in the blood (Fig. 10–4). The hormones of the thyroid gland are summarized in Table 10–3.

PARATHYROID GLANDS

There are four **parathyroid glands**: two on the back of each lobe of the thyroid gland (see Fig. 10–5). The hormone they produce is called parathyroid hormone.

PARATHYROID HORMONE

Parathyroid hormone (PTH) is an antagonist to calcitonin and is important for the maintenance of normal blood levels of calcium and phosphate. The target organs of PTH are the bones, small intestine, and kidneys.

PTH increases the reabsorption of calcium and phosphate from bones to the blood, thereby raising their blood levels. Absorption of calcium and phosphate from food in the small intestine, which also requires vitamin D, is increased by PTH. This too raises blood levels of these minerals. In the kidneys, PTH stimulates the activation of vitamin D and increases the reabsorption of calcium and the excretion of phosphate (more than is obtained from bones). Therefore, the overall effect of PTH is to raise the blood calcium level and lower the blood phosphate level. The functions of PTH are summarized in Table 10–4).

Secretion of PTH is stimulated by **hypocalcemia**, a low blood calcium level, and inhibited by hypercalcemia. The antagonistic effects of PTH and calcitonin are shown in Fig. 10–4. Together, these hormones maintain blood calcium within a normal range. Cal-

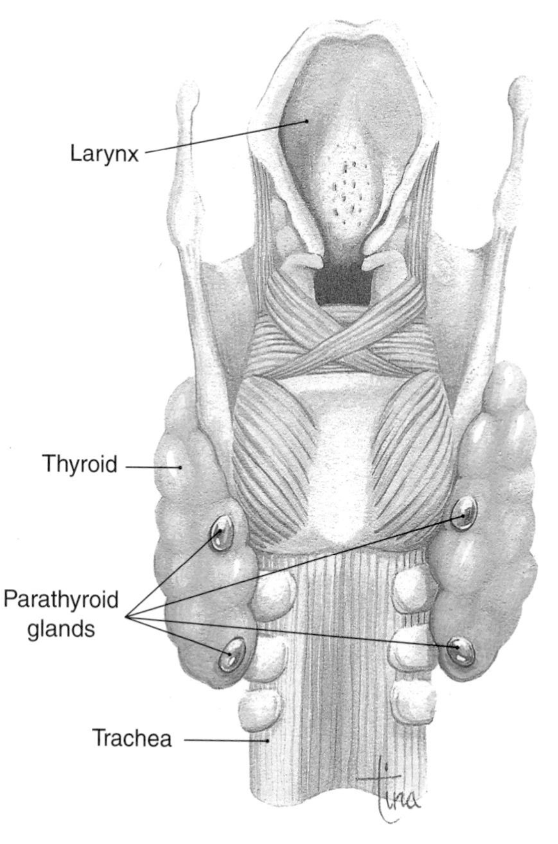

Figure 10–5 Parathyroid glands in posterior view, on lobes of the thyroid gland.

Table 10–4 HORMONE OF THE PARATHYROID GLANDS

Hormone	Functions	Regulation of Secretion
Parathyroid Hormone (PTH)	• Increases the reabsorption of calcium and phosphate from bone to blood • Increases absorption of calcium and phosphate by the small intestine • Increases the reabsorption of calcium and the excretion of phosphate by the kidneys; activates vitamin D	Hypocalcemia stimulates secretion. Hypercalcemia inhibits secretion.

cium in the blood is essential for the process of blood clotting and for normal activity of neurons and muscle cells.

PANCREAS

The **pancreas** is located in the upper left quadrant of the abdominal cavity, extending from the curve of the duodenum to the spleen. Although the pancreas is both an exocrine (digestive) gland as well as an endocrine gland, only its endocrine function will be discussed here. The hormone-producing cells of the pancreas are called **islets of Langerhans** (pancreatic islets); they contain **alpha cells** that produce glucagon and **beta cells** that produce insulin.

GLUCAGON

Glucagon stimulates the liver to change glycogen to glucose (this process is called **glycogenolysis**, which literally means glycogen breakdown) and to increase the use of fats and excess amino acids for energy production. The process of **gluconeogenesis** (literally, making new glucose) is the conversion of excess

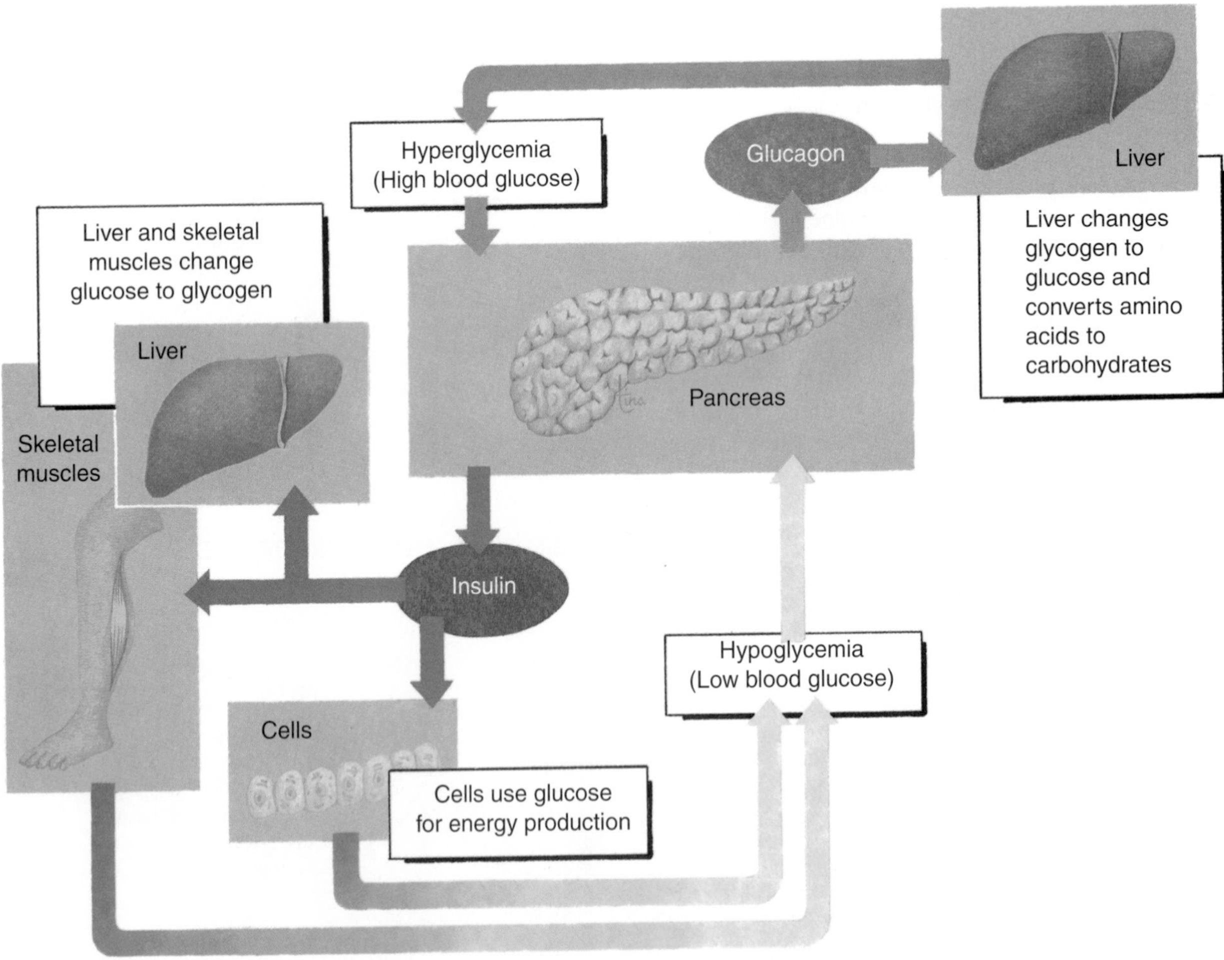

Figure 10–6 Insulin and glucagon and their functions related to the maintenance of the blood glucose level.

Box 10–3 DIABETES MELLITUS

There are two types of **diabetes mellitus**: Type 1 is called insulin-dependent diabetes and its onset is usually in childhood (juvenile onset). Type 2 is called non–insulin-dependent diabetes, and its onset is usually later in life (maturity onset).

Type 1 diabetes is characterized by destruction of the beta cells of the islets of Langerhans and a complete lack of insulin (see below). Destruction of the beta cells is believed to be an autoimmune response, perhaps triggered by a virus; onset of diabetes is usually abrupt. This form of diabetes occurs most often in children, and there may be a genetic predisposition for it (certain HLA types are found more frequently in juvenile-onset dia-

amino acids into simple carbohydrates that may enter the reactions of cell respiration. The overall effect of glucagon, therefore, is to raise the blood glucose level and to make all types of food available for energy production.

The secretion of glucagon is stimulated by **hypoglycemia**, a low blood glucose level. Such a state may occur between meals or during physiological stress situations such as exercise (Fig. 10–6).

INSULIN

Insulin increases the transport of glucose from the blood into cells by increasing the permeability of cell membranes to glucose (brain, liver, and kidney cells, however, are not dependent on insulin for glucose intake). Once inside cells, glucose is used in cell respiration to produce energy. The liver and skeletal muscles also change glucose to glycogen (**glycogenesis**, which literally means glycogen production) to be stored for later use. Insulin also enables cells to take in fatty acids and amino acids to use in the synthesis of lipids and proteins (*not* energy production). With respect to blood glucose, insulin decreases its level by promoting the use of glucose for energy production. The antagonistic functions of insulin and glucagon are shown in Fig. 10–6.

Insulin is a vital hormone; we cannot survive for very long without it. A deficiency of insulin or in its functioning is called **diabetes mellitus**, which is discussed in Box 10–3: Diabetes Mellitus.

Secretion of insulin is stimulated by **hyperglycemia**, a high blood glucose level. This state occurs after eating, especially of meals high in carbohydrates. As glucose is absorbed from the small intestine into the blood, insulin is secreted to enable cells to use the glucose for immediate energy. At the same time, any excess glucose will be stored in the liver and muscles as glycogen. The hormones of the pancreas are summarized in Table 10–5.

Table 10–5 HORMONES OF THE PANCREAS

Hormone	Functions	Regulation of Secretion
Glucagon	• Increases conversion of glycogen to glucose in the liver • Increases the use of excess amino acids and of fats for energy	Hypoglycemia
Insulin	• Increases glucose transport into cells and the use of glucose for energy production • Increases the conversion of excess glucose to glycogen in the liver and muscles • Increases amino acid and fatty acid transport into cells, and their use in synthesis reactions	Hyperglycemia

ADRENAL GLANDS

The two **adrenal glands** are located one on top of each kidney, which gives them their other name of **suprarenal glands**. Each adrenal gland consists of

Box 10–3 DIABETES MELLITUS (Continued)

betics than in other children—see Box 11–5: White Blood Cell Types: HLA). Insulin by injection is essential to control this form of diabetes. Recently, however, an immunosuppressant medication (usually used to prevent rejection of transplanted organs) has been found to slow or stop the progression of type 1 diabetes if given as the disease is developing. By blocking the autoimmune response, this drug seems to permit the survival of some beta cells, which will continue to produce insulin.

In **type 2** diabetes, insulin is produced but cannot exert its effects on cells because of a deficiency of insulin receptors on cell membranes (see below). Onset of type 2 diabetes is usually gradual, and risk factors include a family history of diabetes and being overweight. Control may not require insulin, but rather medications that enable insulin to react with the remaining cell membrane receptors. Some newer medications seem to stimulate the same reactions that insulin does.

Without insulin (or its effects) blood glucose level remains high, and glucose is lost in urine. Since more water is lost as well, symptoms include greater urinary output (polyuria) and thirst (polydipsia).

The long-term effects of hyperglycemia produce distinctive vascular changes. The capillary walls thicken, and exchange of gases and nutrients diminishes. The most damaging effects are seen in the skin (especially of the feet), the retina, and the kidneys. Poorly controlled diabetes may lead to dry gangrene, blindness, and severe kidney damage. In non–insulin-dependent diabetics, atherosclerosis is quite common. It is now possible for diabetics to prevent much of this tissue damage by precise monitoring of blood glucose level and more frequent administration of insulin. The new insulin pumps are able to more closely mimic the natural secretion of insulin.

A very serious potential problem for the insulin-dependent diabetic is **ketoacidosis**. When glucose cannot be used for energy, the body turns to fats and proteins, which are converted by the liver to ketones. Ketones are organic acids (acetone, acetoacetic acid) that can be used in cell respiration, but cells are not able to utilize them rapidly so ketones accumulate in the blood. Since ketones are acids, they lower the pH of the blood as they accumulate. The kidneys excrete excess ketones, but in doing so excrete more water as well, which leads to dehydration and worsens the acidosis. Without administration of insulin to permit the use of glucose for energy, and IV fluids to restore blood volume to normal, ketoacidosis will progress to coma and death.

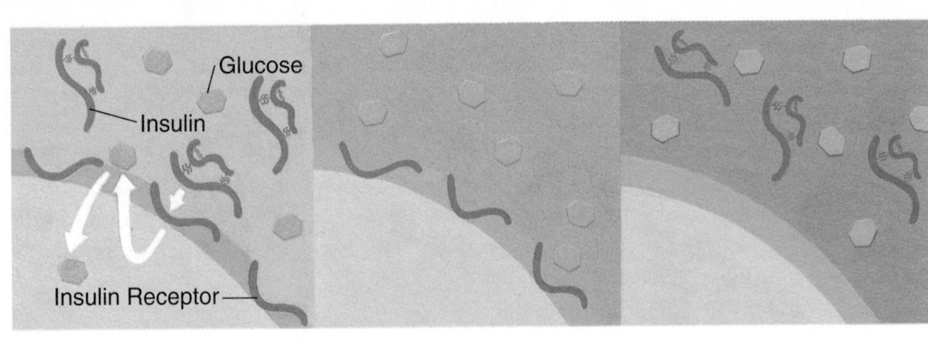

Box Figure 10–A (A) Cell membrane in normal state, with insulin receptors and insulin to regulate glucose intake. **(B)** Cell membrane in type 1 diabetes: Insulin not present, glucose remains outside cell. **(C)** Cell membrane in type 2 diabetes: without insulin receptors, glucose remains outside cell.

two parts: an inner adrenal medulla and an outer adrenal cortex. The hormones produced by each part have very different functions.

ADRENAL MEDULLA

The cells of the **adrenal medulla** secrete epinephrine and norepinephrine, which collectively are called **catecholamines**, and are **sympathomimetic**. The secretion of both hormones is stimulated by sympathetic impulses from the hypothalamus, and their functions duplicate and prolong those of the sympathetic division of the autonomic nervous system ("mimetic" means "to mimic").

Epinephrine and Norepinephrine

Epinephrine (adrenalin) and norepinephrine (noradrenalin) are both secreted in stress situations and help prepare the body for "fight or flight." **Norepinephrine** is secreted in small amounts, and its most significant function is to cause vasoconstriction in the skin, viscera, and skeletal muscles (that is, throughout the body), which raises blood pressure.

Epinephrine, secreted in larger amounts, increases heart rate and force of contraction and stimulates vasoconstriction in skin and viscera and vasodilation in skeletal muscles. It also dilates the bronchioles, decreases peristalsis, stimulates the liver to change glycogen to glucose, increases the use of fats for energy, and increases the rate of cell respiration. Many of these effects do indeed seem to be an echo of sympathetic responses, don't they? Responding to stress is so important that the body is redundant (it repeats itself) and has both a nervous mechanism and a hormonal mechanism. Epinephrine is actually more effective than sympathetic stimulation, however, because the hormone increases energy production and cardiac output to a greater extent. The hormones of the adrenal medulla are summarized in Table 10-6, and their functions are shown in Fig. 10-7.

ADRENAL CORTEX

The **adrenal cortex** secretes three types of steroid hormones: mineralocorticoids, glucocorticoids, and sex hormones. The sex hormones, "female" estrogens and "male" androgens (similar to testosterone), are produced in very small amounts, and their importance is not known with certainty. They may contribute to rapid body growth during early puberty. They may also be important to supply estrogen to women after menopause, and to men throughout life (see section on Estrogen, later in this chapter).

The functions of the other adrenal cortical hormones are well known, however, and these are considered vital hormones.

Aldosterone

Aldosterone is the most abundant of the **mineralocorticoids**, and we will use it as a representative of this group of hormones. The target organs of aldosterone are the kidneys, but there are important secondary effects as well. Aldosterone increases the reabsorption of sodium and the excretion of potassium by

Table 10–6 HORMONES OF THE ADRENAL MEDULLA

Hormone	Function(s)	Regulation of Secretion
Norepinephrine	• Causes vasoconstriction in skin, viscera, and skeletal muscles	
Epinephrine	• Increases heart rate and force of contraction • Dilates bronchioles • Decreases peristalsis • Increases conversion of glycogen to glucose in the liver • Causes vasodilation in skeletal muscles • Causes vasoconstriction in skin and viscera • Increases use of fats for energy • Increases the rate of cell respiration	Sympathetic impulses from the hypothalamus in stress situations

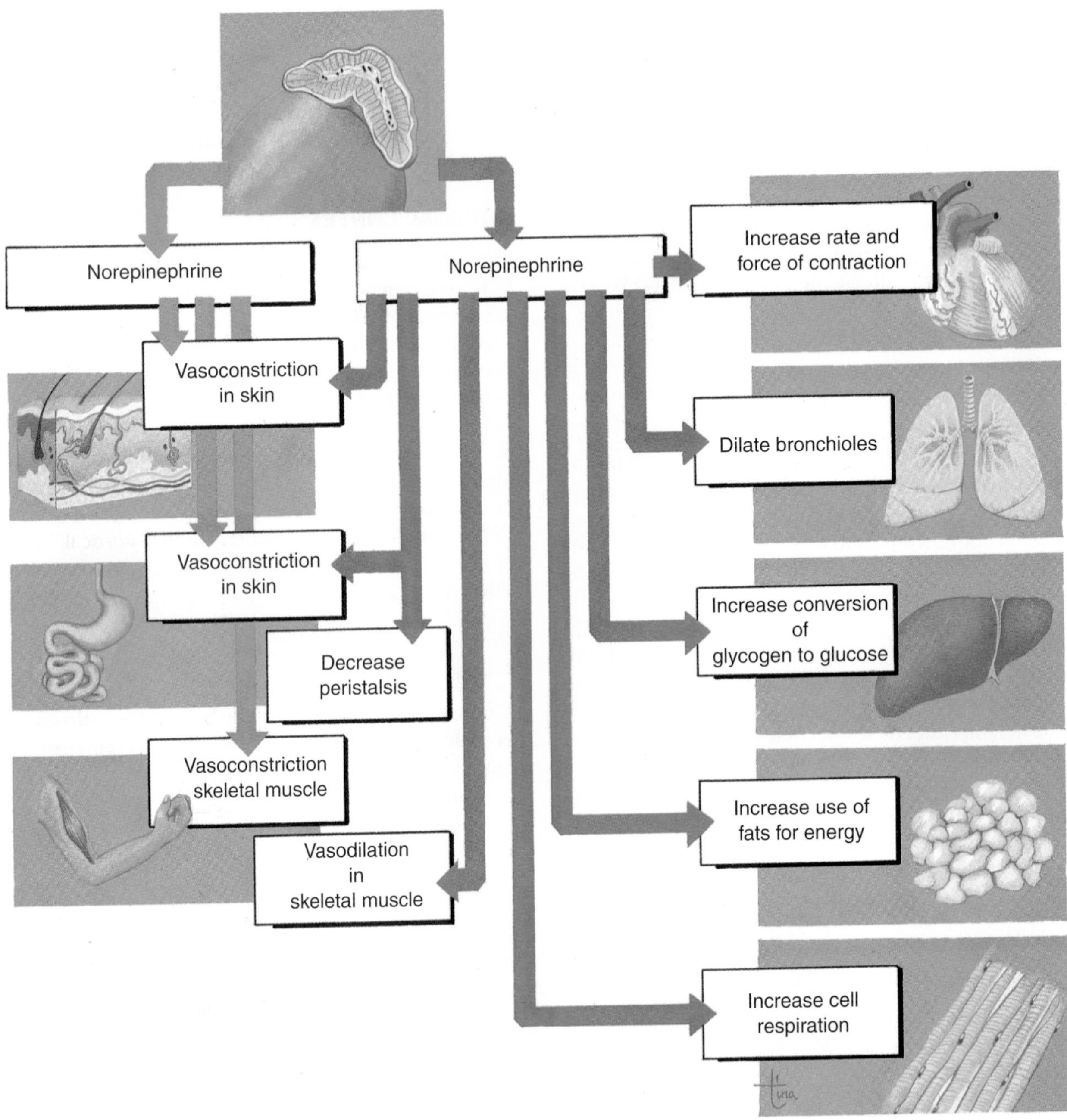

Figure 10–7 Functions of epinephrine and norepinephrine.

the kidney tubules. Sodium ions (Na^+) are returned to the blood, and potassium ions (K^+) are excreted in urine.

As Na^+ ions are reabsorbed, hydrogen ions (H^+) may be excreted in exchange. This is one mechanism to prevent the accumulation of excess H^+ ions which would cause acidosis of body fluids. Also, as Na^+ ions are reabsorbed, negative ions such as chloride (Cl^-) and bicarbonate (HCO_3^-) follow the Na^+ ions back to the blood, and water follows by osmosis. This indirect effect of aldosterone, the reabsorption of water by the kidneys, is very important to maintain normal blood volume and blood pressure. In summary, then, aldosterone maintains normal blood levels of sodium and potassium, and contributes to the maintenance of normal blood pH, blood volume, and blood pressure.

There are a number of factors that stimulate the secretion of aldosterone. These are a deficiency of sodium, loss of blood or dehydration that lowers blood pressure, or an elevated blood level of potassium. Low blood pressure or blood volume activates the **renin-angiotensin mechanism** of the kidneys. This mechanism is discussed in Chapters 13 and 18, so we will say for now that the process culminates in the formation of a chemical called **angiotensin II**. Angiotensin II causes vasoconstriction and stimulates the secretion of aldosterone by the adrenal cortex. Aldosterone then increases sodium and water retention by the kidneys to help restore blood volume and blood pressure to normal.

Cortisol

We will use **cortisol** as a representative of the group of hormones called **glucocorticoids**, since it is responsible for most of the actions of this group. Cortisol increases the use of fats and excess amino acids (gluconeogenesis) for energy, and decreases the use of glucose. This is called the "glucose-sparing effect," and it is important because it conserves glucose for use by the brain. Cortisol is secreted in any type of physiological stress situation: disease, physical injury, hemorrhage, fear or anger, exercise, and hunger. Although most body cells easily use fatty acids and amino acids in cell respiration, brain cells do not and must have glucose. By enabling other cells to use the alternative energy sources, cortisol ensures that whatever glucose is present will be available to the brain.

Cortisol also has an **anti-inflammatory effect**. During inflammation, **histamine** from damaged tissues makes capillaries more permeable, and the lysosomes of damaged cells release their enzymes, which help break down damaged tissue but may also cause destruction of nearby healthy tissue. Cortisol blocks the effects of histamine and stabilizes lysosomal membranes, preventing excessive tissue destruction. Inflammation is a beneficial process up to a point, and is an essential first step if tissue repair is to take place. It may, however, become a vicious cycle of damage, inflammation, more damage, more inflammation, and so on. Normal cortisol secretion seems to limit the inflammation process to what is useful for tissue repair, and to prevent excessive tissue destruction. Too much cortisol, however, decreases the immune response, leaving the body susceptible to infection and significantly slowing the healing of damaged tissue (see Box 10–4: Disorders of the Adrenal Cortex).

The direct stimulus for cortisol secretion is **ACTH** from the anterior pituitary gland, which in turn is stimulated by corticotropin releasing hormone (CRH) from the hypothalamus. CRH is produced in the physiological stress situations mentioned above. Although we often think of epinephrine as a hormone important in stress, cortisol is also important. The hormones of the adrenal cortex are summarized in Table 10–7.

OVARIES

The **ovaries** are located in the pelvic cavity, one on each side of the uterus. The hormones produced by the ovaries are the steroids estrogen and progesterone, and inhibin. Although their functions are an integral part of Chapters 20 and 21, we will briefly discuss some of them here.

ESTROGEN

Estrogen is secreted by the follicle cells of the ovary; secretion is stimulated by **FSH** from the anterior pituitary gland. Estrogen promotes the maturation of the ovum in the ovarian follicle and stimulates the growth of blood vessels in the endometrium (lining) of the uterus in preparation for a possible fertilized egg.

The **secondary sex characteristics** in women also develop in response to estrogen. These include growth of the duct system of the mammary glands,

Box 10–4 DISORDERS OF THE ADRENAL CORTEX

Addison's disease is the result of hyposecretion of the adrenal cortical hormones. Most cases are idiopathic, that is, of unknown cause; atrophy of the adrenal cortex decreases both cortisol and aldosterone secretion.

Deficiency of cortisol is characterized by hypoglycemia, decreased gluconeogenesis, and depletion of glycogen in the liver. Consequences are muscle weakness and the inability to resist physiological stress. Aldosterone deficiency leads to retention of potassium and excretion of sodium and water in urine. The result is severe dehydration, low blood volume, and low blood pressure. Without treatment, circulatory shock and death will follow. Treatment involves administration of hydrocortisone; in high doses this will also compensate for the aldosterone deficiency.

Cushing's syndrome is the result of hypersecretion of the adrenal cortex, primarily cortisol. The cause may be a pituitary tumor that increases ACTH secretion or a tumor of the adrenal cortex itself.

Excessive cortisol promotes fat deposition in the trunk of the body, while the extremities remain thin. The skin becomes thin and fragile, and healing after injury is slow. The bones also become fragile as osteoporosis is accelerated. Also characteristic of this syndrome is the rounded appearance of the face. Treatment is aimed at removal of the cause of the hypersecretion, whether it be a pituitary or adrenal tumor.

Cushing's syndrome may also be seen in people who receive corticosteroids for medical reasons. Transplant recipients or people with rheumatoid arthritis or severe asthma who must take corticosteroids may exhibit any of the above symptoms. In such cases, the disadvantages of this medication must be weighed against the benefits provided.

Table 10–7 HORMONES OF THE ADRENAL CORTEX

Hormone	Functions	Regulation of Secretion
Aldosterone	• Increases reabsorption of Na^+ ions by the kidneys to the blood • Increases excretion of K^+ ions by the kidneys in urine	• Low blood Na^+ level • Low blood volume or blood pressure • High blood K^+ level
Cortisol	• Increases use of fats and excess amino acids for energy • Decreases use of glucose for energy (except for the brain) • Increases conversion of glucose to glycogen in the liver • Anti-inflammatory effect: stabilizes lysosomes and blocks the effects of histamine	• ACTH (anterior pituitary) during physiological stress

growth of the uterus, and the deposition of fat subcutaneously in the hips and thighs. The closure of the epiphyseal discs in long bones is brought about by estrogen, and growth in height stops. Estrogen is also believed to lower blood levels of cholesterol and triglycerides. For women before the age of menopause this is beneficial in that it decreases the risk of atherosclerosis and coronary artery disease.

Recent research suggests that estrogen no longer be considered only a "female" hormone. Estrogen seems to have effects on many organs, including the brain, the heart, and blood vessels. In the brain, testosterone from the testes or the adrenal cortex can be converted to estrogen, which may be important for memory, especially for older people. Estrogen seems to have nonreproductive functions in both men and women, though we cannot yet be as specific as we can be with the reproductive functions in women, mentioned previously.

PROGESTERONE

When a mature ovarian follicle releases an ovum, the follicle becomes the **corpus luteum** and begins to secrete **progesterone** in addition to estrogen. This is stimulated by **LH** from the anterior pituitary gland.

Progesterone promotes the storage of glycogen and the further growth of blood vessels in the endometrium, which thus becomes a potential placenta. The secretory cells of the mammary glands also develop under the influence of progesterone.

Both progesterone and estrogen are secreted by the placenta during pregnancy; these functions are covered in Chapter 21.

INHIBIN

The corpus luteum secretes another hormone, called **inhibin**. Inhibin helps decrease the secretion of FSH by the anterior pituitary gland, and GnRH by the hypothalamus.

TESTES

The **testes** are located in the scrotum, a sac of skin between the upper thighs. Two hormones, testosterone and inhibin, are secreted by the testes.

TESTOSTERONE

Testosterone is a steroid hormone secreted by the interstitial cells of the testes; the stimulus for secretion is LH from the anterior pituitary gland.

Testosterone promotes maturation of sperm in the seminiferous tubules of the testes; this process begins at puberty and continues throughout life. At puberty, testosterone stimulates development of the male **secondary sex characteristics**. These include growth of all the reproductive organs, growth of facial and body hair, growth of the larynx and deepening of the voice, and growth (protein synthesis) of the skeletal muscles. Testosterone also brings about closure of the epiphyses of the long bones.

INHIBIN

The hormone **inhibin** is secreted by the sustentacular cells of the testes; the stimulus for secretion is increased testosterone. The function of inhibin is to decrease the secretion of FSH by the anterior pituitary gland. The interaction of inhibin, testosterone, and the anterior pituitary hormones maintains spermatogenesis at a constant rate.

OTHER HORMONES

Melatonin is a hormone produced by the **pineal gland**, which is located at the back of the third ventricle of the brain. The secretion of melatonin is greatest during darkness and decreases when light enters the eye and the retina signals the hypothalamus. A recent discovery is that the retina also produces melatonin, which seems to indicate that the eyes and pineal gland work with the biological clock of the hypothalamus. In other mammals, melatonin helps regulate seasonal reproductive cycles. For people, melatonin definitely stimulates the onset of sleep and increases its duration. Other claims, such as that melatonin strengthens the immune system or prevents cellular damage and aging, are simply without evidence as yet.

There are other organs that produce hormones that have only one or a few target organs. For example, the stomach and duodenum produce hormones that regulate aspects of digestion. The thymus gland produces hormones necessary for the normal functioning of the immune system, and the kidneys produce a hormone that stimulates red blood cell production. All of these will be discussed in later chapters.

PROSTAGLANDINS

Prostaglandins (PG) are made by virtually all cells from the phospholipids of their cell membranes. They differ from other hormones in that they do not circulate in the blood to target organs, but rather exert their effects locally, where they are produced.

There are many types of prostaglandins, designated by the letters A–I, as in PGA, PGB, and so on. The functions of prostaglandins are also many, and we will list only a few of them here. Prostaglandins are known to be involved in inflammation, pain mechanisms, blood clotting, vasoconstriction and vasodilation, contraction of the uterus, reproduction, secretion of digestive glands, and nutrient metabolism. Current research is directed at determining the normal functioning of prostaglandins in the hope that many of them may eventually be used clinically.

One familiar example may illustrate the widespread activity of prostaglandins. For minor pain such as a headache, many people take aspirin. Aspirin inhibits the synthesis of prostaglandins involved in pain mechanisms, and usually relieves the pain. Some people, however, such as those with rheumatoid arthritis, may take large amounts of aspirin to diminish pain and inflammation. These people may bruise easily because blood clotting has been impaired. This too is an effect of aspirin, which blocks the synthesis of prostaglandins necessary for blood clotting.

MECHANISMS OF HORMONE ACTION

Exactly how hormones exert their effects on their target organs involves a number of complex processes, which will be presented simply here.

A hormone must first bond to a **receptor** for it on or in the target cell. Cells respond to certain hormones and not to others because of the presence of specific receptors, which are proteins. These receptor proteins may be part of the cell membrane or within the cytoplasm or nucleus of the target cells. A hormone will affect only those cells which have its specific receptors. Liver cells, for example, have cell membrane receptors for insulin, glucagon, growth hormone, and epinephrine; bone cells have receptors for growth hormone, PTH, and calcitonin. Cells of the ovaries and testes do not have receptors for PTH and calcitonin, but do have receptors for FSH and LH, which bone cells and liver cells do not have. Once a hormone has bonded to a receptor on or in its target cell, other reactions will take place.

THE TWO-MESSENGER THEORY—PROTEIN HORMONES

The Two-Messenger Theory of hormone action involves "messengers" that make something happen, that is, stimulate specific reactions. **Protein hormones** usually bond to receptors of the cell membrane, and the hormone is called the first messenger. The hormone-receptor bonding activates the enzyme adenyl cyclase on the inner surface of the cell membrane. Adenyl cyclase synthesizes a substance called cyclic adenosine monophosphate **(cyclic AMP)** from ATP, and cyclic AMP is the second messenger.

Cyclic AMP activates specific enzymes within the cell, which bring about the cell's characteristic response to the hormone. These responses include a change in the permeability of the cell membrane to a specific substance, an increase in protein synthesis, activation of other enzymes, or the secretion of a cellular product.

In summary, a cell's response to a hormone is determined by the enzymes within the cell, that is, the reactions of which the cell is capable. These reactions are brought about by the first messenger, the hormone, which stimulates the formation of the second messenger, cyclic AMP. Cyclic AMP then activates the cell's enzymes to elicit a response to the hormone (Fig. 10–8).

ACTION OF STEROID HORMONES

Steroid hormones are soluble in the lipids of the cell membrane, and diffuse easily into a target cell. Once inside the cell, the steroid hormone combines with a protein receptor in the cytoplasm, and this steroid–protein complex enters the nucleus of the cell. Within the nucleus, the steroid–protein complex activates specific genes, which begin the process of **protein synthesis**. The enzymes produced bring about the

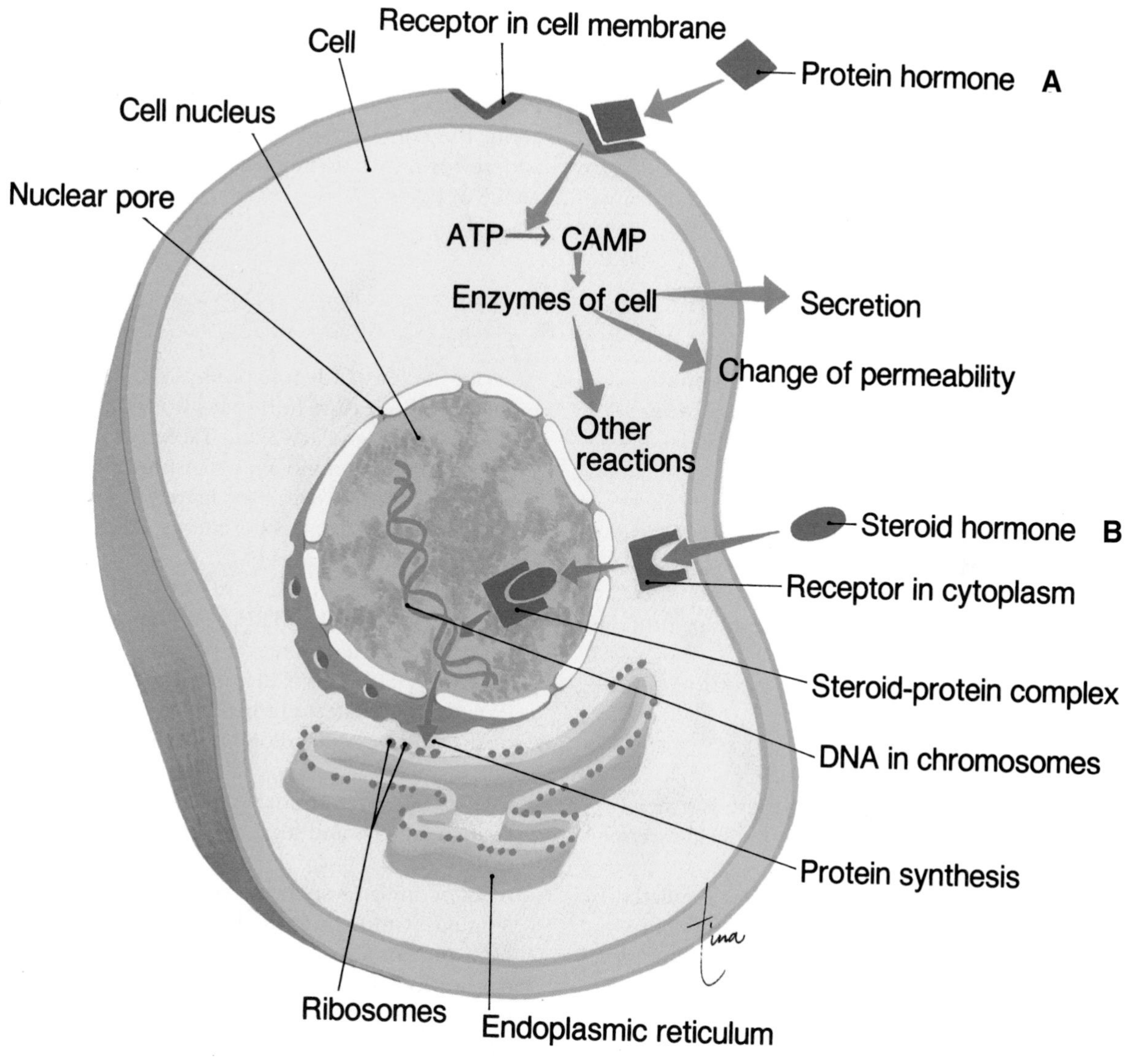

Figure 10–8 Mechanisms of hormone action. **(A)** Two-Messenger model of the action of protein hormones. **(B)** Action of steroid hormones. See text for description.

cell's characteristic response to the hormone (see Fig. 10–8).

AGING AND THE ENDOCRINE SYSTEM

Most of the endocrine glands decrease their secretions with age, but normal aging usually does not lead to serious hormone deficiencies. There are decreases in adrenal cortical hormones, for example, but the levels are usually sufficient to maintain homeostasis of water, electrolytes, and nutrients. The decreased secretion of growth hormone leads to a decrease in muscle mass and an increase in fat storage. A lower basal metabolic rate is common in elderly people as the thyroid slows its secretion of thyroxine. Unless specific pathologies develop, however, the endocrine system usually continues to function adequately in old age.

SUMMARY

The hormones of endocrine glands are involved in virtually all aspects of normal body functioning. The growth and repair of tissues, the utilization of food to produce energy, responses to stress, the maintenance of the proper levels and pH of body fluids, and the continuance of the human species all depend on hormones. Some of these topics will be discussed in later chapters. As you might expect, you will be reading about the contributions of many of these hormones and reviewing their important roles in the maintenance of homeostasis.

STUDY OUTLINE

Endocrine glands are ductless glands that secrete hormones into the blood. Hormones exert their effects on target organs or tissues.

Chemistry of Hormones

1. Amines—structural variations of the amino acid tyrosine; thyroxine, epinephrine.
2. Proteins—chains of amino acids; peptides are short chains. Insulin, GH, glucagon are proteins; ADH and oxytocin are peptides.
3. Steroids—made from cholesterol; cortisol, aldosterone, estrogen, testosterone.

Regulation of Hormone Secretion

1. Hormones are secreted when there is a need for their effects. Each hormone has a specific stimulus for secretion.
2. The secretion of most hormones is regulated by negative feedback mechanisms: as the hormone exerts its effects, the stimulus for secretion is reversed, and secretion of the hormone decreases.

Pituitary Gland (Hypophysis)—hangs from hypothalamus by the infundibulum; enclosed by sella turcica of sphenoid bone (see Fig. 10–1)

Posterior Pituitary (Neurohypophysis)—stores hormones produced by the hypothalamus (Figs. 10-2 and 10-3 and Table 10-1).

- ADH—increases water reabsorption by the kidneys. Result: decreases urinary output and increases blood volume. Stimulus: nerve impulses from hypothalamus when body water decreases.
- Oxytocin—stimulates contraction of myometrium of uterus during labor and release of milk from mammary glands. Stimulus: nerve impulses from hypothalamus as cervix is stretched or as infant sucks on nipple.

Anterior Pituitary (Adenohypophysis)—secretions are regulated by releasing hormones from the hypothalamus (Figs. 10-2 and 10-3 and Table 10-2).

- GH—increases amino acid transport into cells and increases protein synthesis; increases rate of mitosis; increases use of fats for energy. Stimulus: GHRH from the hypothalamus.
- TSH—increases secretion of thyroxine and T_3 by the thyroid. Stimulus: TRH from the hypothalamus.
- ACTH—increases secretion of cortisol by the adrenal cortex. Stimulus: CRH from the hypothalamus.
- Prolactin—initiates and maintains milk production by the mammary glands. Stimulus: PRH from the hypothalamus.
- FSH—*In women:* initiates development of ova in ovarian follicles and secretion of estrogen by follicle cells.
  In men: initiates sperm development in the testes. Stimulus: GnRH from the hypothalamus.
- LH—*In women:* stimulates ovulation, transforms mature follicle into corpus luteum and stimulates secretion of progesterone.
  In men: stimulates secretion of testosterone by the testes. Stimulus: GnRH from the hypothalamus.

Thyroid Gland—on front and sides of trachea below the larynx (see Fig. 10–1 and Table 10–3)

- Thyroxine (T_4) and T_3—produced by thyroid follicles. Increase use of all food types for energy and increase protein synthesis. Necessary for normal physical, mental, and sexual development. Stimulus: TSH from the anterior pituitary.
- Calcitonin—produced by parafollicular cells. Decreases reabsorption of calcium from bones and lowers blood calcium level. Stimulus: hypercalcemia.

Parathyroid Glands—four; two on posterior of each lobe of thyroid (see Figs. 10–4 and 10–5 and Table 10–4)

- PTH—increases reabsorption of calcium and phosphate from bones to the blood; increases absorption of calcium and phosphate by the small intestine; increases reabsorption of calcium and excretion of phosphate by the kidneys, and activates vitamin D. Result: raises blood calcium and lowers blood phosphate levels. Stimulus: hypocalcemia. Inhibitor: hypercalcemia.

Pancreas—extends from curve of duodenum to the spleen. Islets of Langerhans consist of alpha cells and beta cells (see Figs. 10–1 and 10–6 and Table 10–5)

- Glucagon—secreted by alpha cells. Stimulates liver to change glycogen to glucose; increases use of fats and amino acids for energy. Result: raises blood glucose level. Stimulus: hypoglycemia
- Insulin—secreted by beta cells. Increases use of glucose by cells to produce energy; stimulates liver and muscles to change glucose to glycogen; increases cellular intake of fatty acids and amino acids to use for synthesis of lipids and proteins. Result: lowers blood glucose level. Stimulus: hyperglycemia.

Adrenal Glands—one on top of each kidney; each has an inner adrenal medulla and an outer adrenal cortex (see Fig. 10–1)

Adrenal Medulla—produces catecholamines in stress situations (Table 10-6 and Fig. 10-7).

- Norepinephrine—stimulates vasoconstriction and raises blood pressure.
- Epinephrine—increases heart rate and force, causes vasoconstriction in skin and viscera and vasodilation in skeletal muscles; dilates bronchioles; slows peristalsis; causes liver to change glycogen to glucose; increases use of fats for energy; increases rate of cell respiration. Stimulus: sympathetic impulses from the hypothalamus.

Adrenal Cortex—produces mineralocorticoids, glucocorticoids, and very small amounts of sex hormones (function not known with certainty) (Table 10-7).

- Aldosterone—increases reabsorption of sodium and excretion of potassium by the kidneys. Results: hydrogen ions are excreted in exchange for sodium; chloride and bicarbonate ions and water follow sodium back to the blood; maintains normal blood pH, blood volume, and blood pressure. Stimulus: decreased blood sodium or elevated blood potassium; decreased blood volume or blood pressure (activates the renin-angiotensin mechanism of the kidneys).
- Cortisol—increases use of fats and amino acids for energy; decreases use of glucose to conserve glucose for the brain; anti-inflammatory effect: blocks effects of histamine and stabilizes lysosomes to prevent excessive tissue damage. Stimulus: ACTH from hypothalamus during physiological stress.

Ovaries—in pelvic cavity on either side of uterus (see Fig. 10–1)

- Estrogen—produced by follicle cells. Promotes maturation of ovum; stimulates growth of blood vessels in endometrium; stimulates development of secondary sex characteristics: growth of duct system of mammary glands, growth of uterus, fat deposition. Promotes closure of epiphyses of long bones; lowers blood levels of cholesterol and triglycerides.
- Stimulus: FSH from anterior pituitary.
- Progesterone—produced by the corpus luteum. Promotes storage of glycogen and further growth of blood vessels in the endometrium; promotes growth of secretory cells of mammary glands. Stimulus: LH from anterior pituitary.
- Inhibin—inhibits secretion of FSH.

Testes—in scrotum between the upper thighs (see Fig. 10–1)

- Testosterone—produced by interstitial cells. Promotes maturation of sperm in testes; stimulates development of secondary sex characteristics: growth of reproductive organs, facial and body hair, larynx, skeletal muscles; promotes closure of epiphyses of long bones. Stimulus: LH from anterior pituitary.
- Inhibin—produced by sustentacular cells. Inhibits secretion of FSH to maintain a constant rate of sperm production. Stimulus: increased testosterone.

Other Hormones

- Melatonin—secreted by the pineal gland during darkness; brings on sleep.
- Prostaglandins—Synthesized by cells from the phospholipids of their cell membranes; exert their effects locally. Are involved in inflammation and pain, reproduction, nutrient metabolism, changes in blood vessels, blood clotting.

Mechanisms of Hormone Action (see Fig. 10–8)

- A hormone affects cells that have receptors for it. Receptors are proteins that may be part of the cell membrane, or within the cytoplasm or nucleus of the target cell.
- The Two-Messenger Theory: a protein hormone (1st messenger) bonds to a membrane receptor; stimulates formation of cyclic AMP (2nd messenger), which activates the cell's enzymes to bring about the cell's characteristic response to the hormone.
- Steroid hormones diffuse easily through cell membranes and bond to cytoplasmic receptors. Steroid-protein complex enters the nucleus and activates certain genes which initiate protein synthesis.

REVIEW QUESTIONS

1. Use the following to describe a negative feedback mechanism: TSH, TRH, decreased metabolic rate, thyroxine, and T_3. (pp. 210, 218, 219)
2. Name the two hormones stored in the posterior pituitary gland. Where are these hormones produced? State the functions of each of these hormones. (pp. 212–213)
3. Name the two hormones of the anterior pituitary gland that affect the ovaries or testes, and state their functions. (p. 216)
4. Describe the antagonistic effects of PTH and calcitonin on bones and on blood calcium level. State the other functions of PTH. (p. 227)
5. Describe the antagonistic effects of insulin and glucagon on the liver and on blood glucose level. (pp. 220–221)
6. Describe how cortisol affects the use of foods for energy. State the anti-inflammatory effects of cortisol. (p. 226)
7. State the effect of aldosterone on the kidneys. Describe the results of this effect on the composition of the blood. (pp. 223, 226)
8. When are epinephrine and norepinephrine secreted? Describe the effects of these hormones. (p. 223)
9. Name the hormones necessary for development of egg cells in the ovaries. Name the hormones necessary for development of sperm in the testes. (pp. 216, 226–227)
10. State what prostaglandins are made from. State three functions of prostaglandins. (p. 228)
11. Name the hormones that promote the growth of the endometrium of the uterus in preparation for a fertilized egg, and state precisely where each hormone is produced. (p. 226–227)
12. State the functions of thyroxine and T_3. For what aspects of growth are these hormones necessary? (pp. 217)
13. Explain the functions of GH as they are related to normal growth. (pp. 213–214)
14. State the direct stimulus for secretion of each of these hormones: (pp. 213, 216, 218, 219, 221, 225)
    a. Thyroxine
    b. Insulin
    c. Cortisol
    d. PTH
    e. Aldosterone
    f. Calcitonin
    g. GH
    h. Glucagon
    i. Progesterone
    j. ADH

CHAPTER 11

Blood

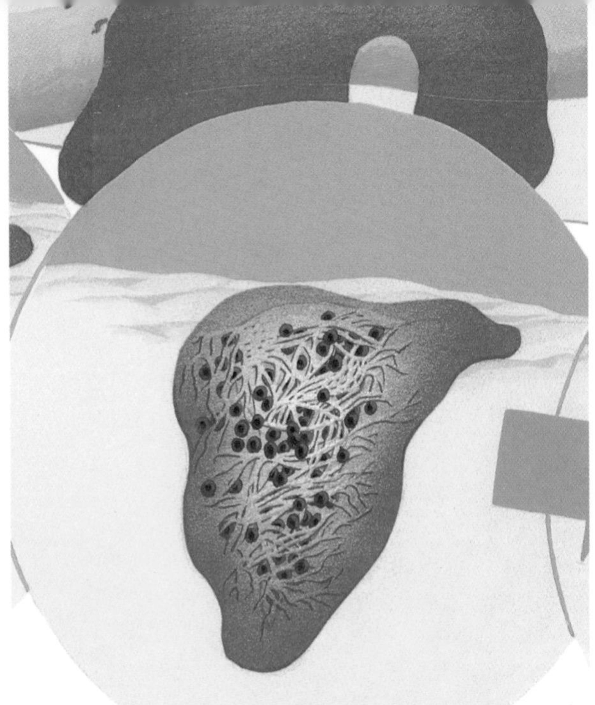

CHAPTER 11

Chapter Outline

Student Objectives

- Describe the composition and explain the functions of blood plasma.
- Name the hemopoietic tissues and the kinds of blood cells each produces.
- State the function of red blood cells, including the protein and the mineral involved.
- Name the nutrients necessary for red blood cell production, and state the function of each.
- Explain how hypoxia may change the rate of red blood cell production.
- Describe what happens to red blood cells that have reached the end of their life span; what happens to the hemoglobin?
- Explain the ABO and Rh blood types.
- Name the five kinds of white blood cells and the function of each.
- State what platelets are, and explain how they are involved in hemostasis.
- Describe the three stages of chemical blood clotting.
- Explain how abnormal clotting is prevented in the vascular system.
- State the normal values in a complete blood count (CBC).

Blood

New Terminology

ABO group (A-B-O GROOP)
Albumin (al-**BYOO**-min)
Bilirubin (**BILL**-ee-roo-bin)
Chemical clotting (**KEM**-i-kuhl **KLAH**-ting)
Embolism (**EM**-boh-lizm)
Erythrocyte (e-**RITH**-roh-sight)
Hemoglobin (**HEE**-muh-GLOW-bin)
Hemostasis (HEE-moh-**STAY**-sis)
Heparin (**HEP**-ar-in)
Immunity (im-**YOO**-ni-tee)
Leukocyte (**LOO**-koh-sight)
Macrophage (**MAK**-roh-fahj)
Normoblast (**NOR**-moh-blast)
Reticulocyte (re-**TIK**-yoo-loh-sight)
Rh factor (R-H **FAK**-ter)
Stem cells (STEM SELLS)
Thrombocyte (**THROM**-boh-sight)
Thrombus (**THROM**-bus)

Related Clinical Terminology

Anemia (uh-**NEE**-mee-yah)
Differential count (**DIFF**-er-EN-shul KOWNT)
Erythroblastosis fetalis (e-RITH-roh-blass-**TOH**-sis fee-**TAL**-is)
Hematocrit (hee-**MAT**-oh-krit)
Hemophilia (HEE-moh-**FILL**-ee-ah)
Jaundice (**JAWN**-diss)
Leukemia (loo-**KEE**-mee-ah)
Leukocytosis (LOO-koh-sigh-**TOH**-sis)
RhoGam (**ROH**-gam)
Tissue-typing (**TISH**-yoo-**TIGH**-ping)
Typing and cross-matching (**TIGH**-ping and **KROSS**-match-ing)

*Terms that appear in **bold type** in the chapter text are defined in the glossary, which begins on page 528.*

One of the simplest and most familiar life-saving medical procedures is a blood transfusion. As you know, however, the blood of one individual is not always compatible with that of another person. The ABO blood types were discovered in the early 1900s by Karl Landsteiner, an Austrian-American. He also contributed to the discovery of the Rh factor in 1940. In the early 1940s, Charles Drew, an African-American, developed techniques for processing and storing blood plasma, which could then be used in transfusions for people with any blood type. When we donate blood today, our blood may be given to a recipient as whole blood, or it may be separated into its component parts, and recipients will then receive only those parts they need, such as red cells, plasma, Factor VIII, or platelets. Each of these parts has a specific function, and all of the functions of blood are essential to our survival.

The general functions of blood are transportation, regulation, and protection. Materials transported by the blood include nutrients, waste products, gases, and hormones. The blood helps regulate fluid-electrolyte balance, acid-base balance, and the body temperature. Protection against pathogens is provided by white blood cells, and the blood clotting mechanism prevents excessive loss of blood after injuries. Each of these functions is covered in more detail in this chapter.

CHARACTERISTICS OF BLOOD

Blood has distinctive physical characteristics:

Amount—a person has 4 to 6 liters of blood, depending on his or her size. Of the total blood volume in the human body, 38% to 48% is composed of the various blood cells, also called "formed elements." The remaining 52% to 62% of the blood volume is plasma, the liquid portion of blood (Fig. 11-1).

Color—you're probably saying to yourself, "of course, it's red!" Mention is made of this obvious fact, however, because the color does vary. Arterial blood is bright red because it contains high levels of oxygen. Venous blood has given up much of its oxygen in tissues, and has a darker, dull red color. This may be important in the assessment of the source of bleeding. If blood is bright red it is probably from a severed artery, and dark red blood is probably venous blood.

pH—the normal pH range of blood is 7.35 to 7.45, which is slightly alkaline. Venous blood normally has a lower pH than does arterial blood because of the presence of more carbon dioxide.

Viscosity—this means thickness or resistance to flow. Blood is about three to five times thicker than water. Viscosity is increased by the presence of blood cells and the plasma proteins, and this thickness contributes to normal blood pressure.

PLASMA

Plasma is the liquid part of blood and is approximately 91% water. The solvent ability of water enables the plasma to transport many types of substances. Nutrients absorbed in the digestive tract are circulated to all body tissues, and waste products of the tissues circulate through the kidneys and are excreted in urine. Hormones produced by endocrine glands are carried in the plasma to their target organs, and antibodies are also transported in plasma. Most of the carbon dioxide produced by cells is carried in the plasma in the form of bicarbonate ions (HCO_3^-). When the blood reaches the lungs, the CO_2 is re-formed, diffuses into the alveoli, and is exhaled.

Also in the plasma are the **plasma proteins**. The clotting factors **prothrombin, fibrinogen**, and others are synthesized by the liver and circulate until activated to form a clot in a ruptured or damaged blood vessel. **Albumin** is the most abundant plasma protein. It too is synthesized by the liver. Albumin contributes to the colloid osmotic pressure of blood, which pulls tissue fluid into capillaries. This is important to maintain normal blood volume and blood pressure. Other plasma proteins are called **globulins**. Alpha and beta globulins are synthesized by the liver and act as carriers for molecules such as fats. The gamma globulins are antibodies produced by lymphocytes. Antibodies initiate the destruction of pathogens and provide us with immunity.

Plasma also carries body heat. Blood is warmed by flowing through active organs such as the liver and muscles. This heat is distributed to cooler parts of the body as blood continues to circulate.

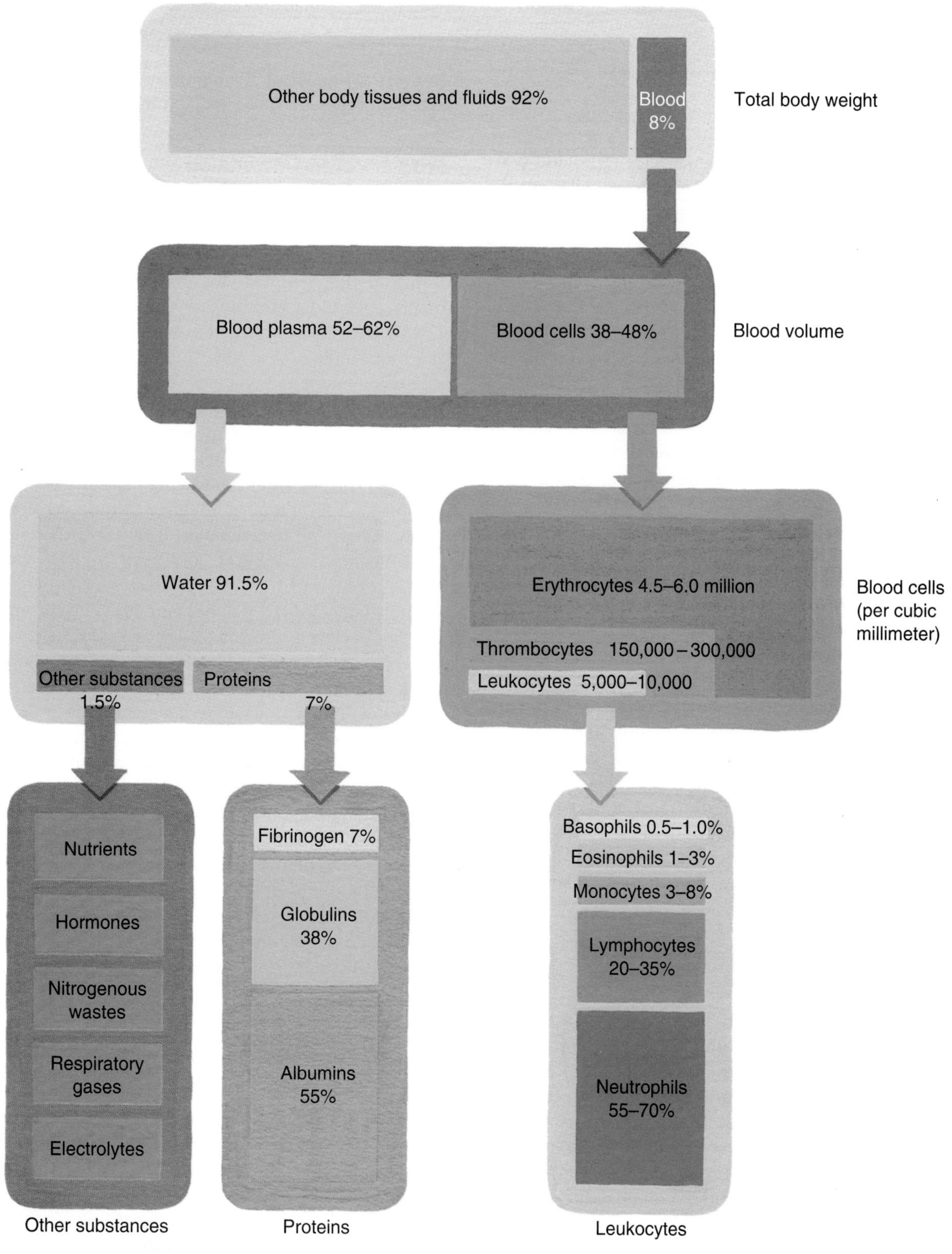

Figure 11–1 Components of blood and the relationship of blood to other body tissues.

BLOOD CELLS

There are three kinds of blood cells: red blood cells, white blood cells, and platelets. Blood cells are produced in **hemopoietic tissues**, of which there are two: **red bone marrow**, found in flat and irregular bones, and **lymphatic tissue**, found in the spleen, lymph nodes, and thymus gland.

RED BLOOD CELLS

Also called **erythrocytes**, red blood cells (RBCs) are biconcave discs, which means their centers are thinner than their edges. You may recall from Chapter 3 that red blood cells are the only human cells without nuclei. Their nuclei disintegrate as the red blood cells mature and are not needed for normal functioning.

A normal RBC count ranges from 4.5 to 6.0 million cells per mm^3 of blood (the volume of a cubic millimeter is approximately that of a very small droplet). RBC counts for men are often toward the high end of this range; those for women are often toward the low end. Another way to measure the amount of RBCs is the **hematocrit**. This test involves drawing blood into a thin glass tube called a capillary tube, and centrifuging the tube to force all the cells to one end. The percentages of cells and plasma can then be determined. Because RBCs are by far the most abundant of the blood cells, a normal hematocrit range is just like that of the total blood cells: 38% to 48%. Both RBC count and hematocrit (Hct) are part of a complete blood count (CBC).

Function

Red blood cells contain the protein **hemoglobin** (Hb), which gives them the ability to carry oxygen. Each red blood cell contains approximately 300 million hemoglobin molecules, each of which can bond to four oxygen molecules. In the pulmonary capillaries, RBCs pick up oxygen and oxyhemoglobin is formed. In the systemic capillaries, hemoglobin gives up much of its oxygen and becomes reduced hemoglobin. A determination of hemoglobin level is also part of a CBC; the normal range is 12 to 18 grams per 100 ml of blood. Essential to the formation of hemoglobin is the mineral iron; there are four atoms of iron in each molecule of hemoglobin. It is the iron that actually bonds to the oxygen and also makes RBCs red.

Production and Maturation

Red blood cells are formed in red bone marrow (RBM) in flat and irregular bones. Within the red bone marrow are precursor cells called **stem cells**, which constantly undergo mitosis to produce all the kinds of blood cells, most of which are RBCs (Figs. 11-2 and 11-3). The rate of production is very rapid (estimated at several million new RBCs per second) and a major regulating factor is oxygen. If the body is in a state of **hypoxia**, or lack of oxygen, the kidneys produce a hormone called **erythropoietin**, which stimulates the red bone marrow to increase the rate of RBC production. This will occur following hemorrhage, or if a person stays for a time at a higher altitude. As a result of the action of erythropoietin, more RBCs will be available to carry oxygen and correct the hypoxic state.

The stem cells that will become RBCs go through a number of developmental stages, only the last two of which we will mention (see Fig. 11-2). The **normoblast** is the last stage with a nucleus, which then disintegrates. The **reticulocyte** has fragments of the endoplasmic reticulum, which are visible when blood smears are stained for microscopic evaluation. These immature cells are usually found in the red bone marrow, although a small number of reticulocytes in the peripheral circulation is considered normal. Large numbers of reticulocytes or normoblasts in the circulating blood mean that the number of mature RBCs is not sufficient to carry the oxygen needed by the body. Such situations include hemorrhage, or when mature RBCs have been destroyed, as in Rh disease of the newborn, and malaria.

The maturation of red blood cells requires many nutrients. Protein and iron are necessary for the synthesis of hemoglobin and become part of hemoglobin molecules. The vitamins folic acid and B_{12} are required for DNA synthesis in the stem cells of the red bone marrow. As these cells undergo mitosis they must continually produce new sets of chromosomes. Vitamin B_{12} is also called the **extrinsic factor** because its source is external, our food. Parietal cells of the stomach lining produce the **intrinsic factor**, a chemical that combines with the vitamin B_{12} in food to prevent its digestion and promote its absorption in the small in-

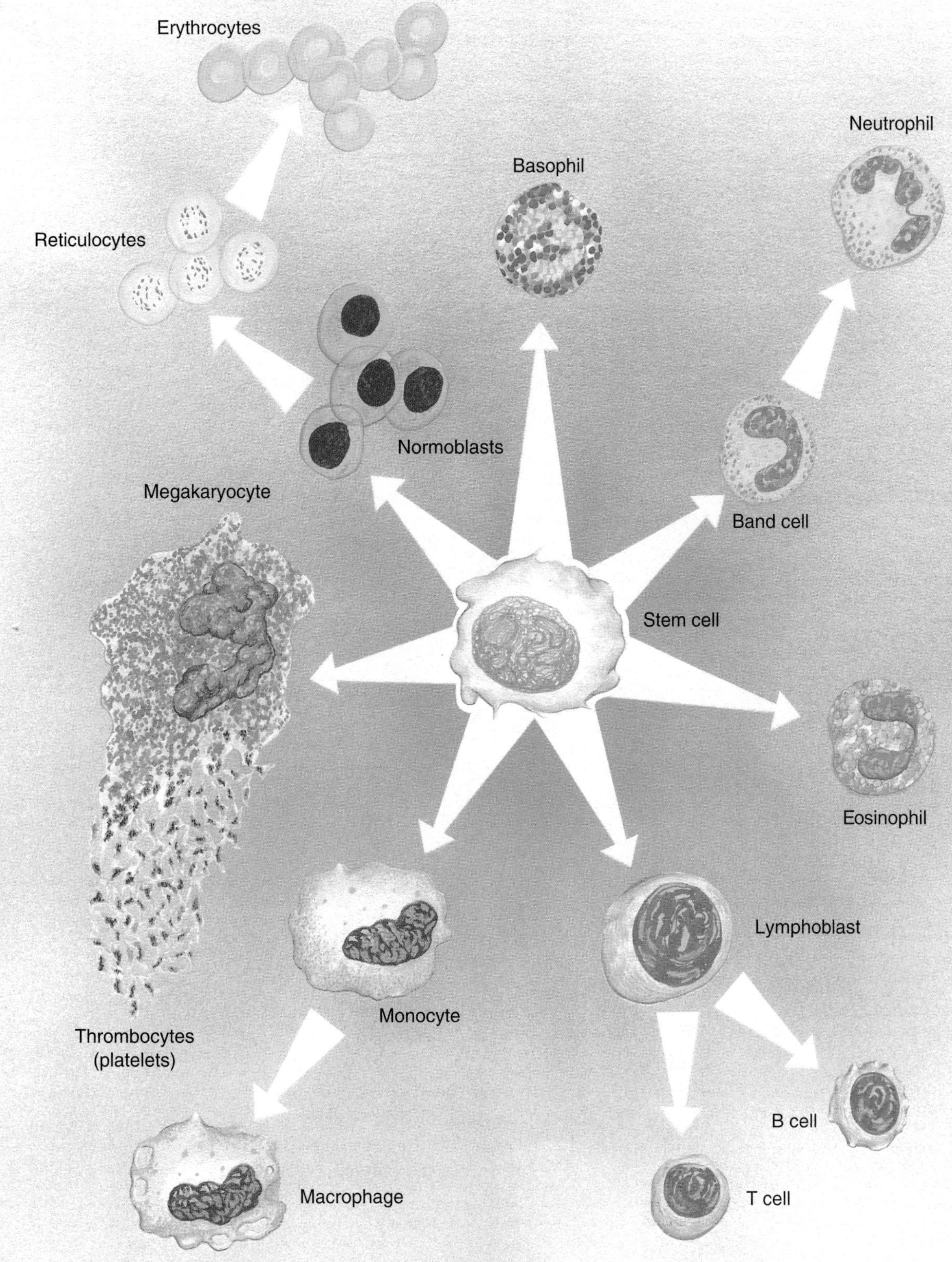

Figure 11–2 Production of blood cells. Stem cells are found in red bone marrow and in lymphatic tissue, and are the precursor cells for all the types of blood cells.

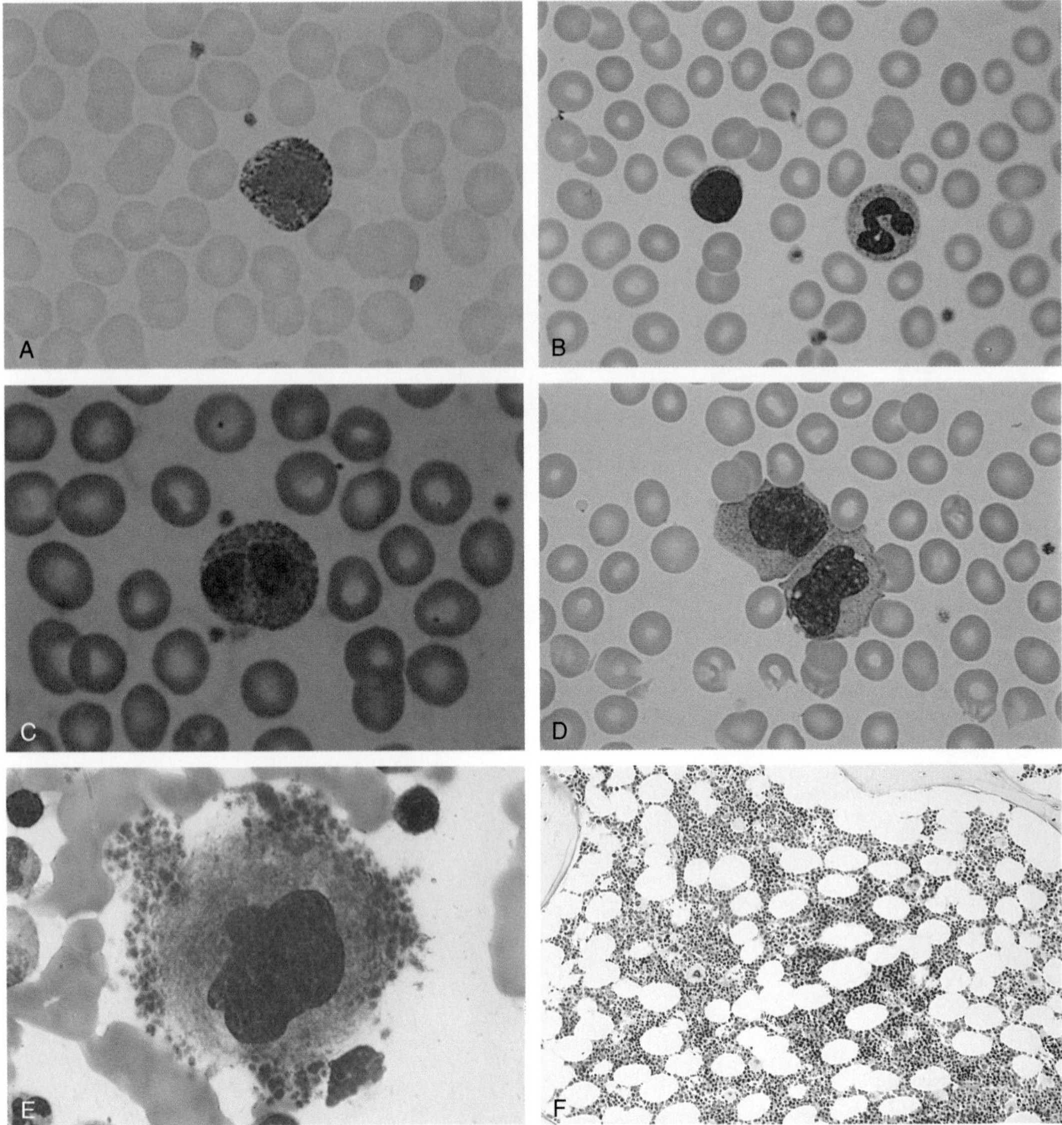

Figure 11–3 Blood cells. **(A)** Red blood cells, platelets, and a basophil. **(B)** Lymphocyte (left) and neutrophil (right). **(C)** Eosinophil. **(D)** Monocytes. **(E)** Megakaryocyte with platelets. (A–E: ×600) **(F)** Normal bone marrow (×200). (From Harmening, DM: Clinical Hematology and Fundamentals of Hemostasis, ed. 3. FA Davis, Philadelphia, 1997, with permission.)

testine. A deficiency of either vitamin B_{12} or the intrinsic factor results in **pernicious anemia** (see Box 11-1: Anemia).

Life Span

Red blood cells live for approximately 120 days. As they reach this age they become fragile and are removed from circulation by cells of the **tissue macrophage system** (formerly called the reticuloendothelial or RE system). The organs that contain macrophages (literally, "big eaters") are the liver, spleen, and red bone marrow. The old RBCs are phagocytized and digested by macrophages, and the iron they contained is put into the blood to be returned to the red bone marrow to be used for the

Box 11–1 ANEMIA

Anemia is a deficiency of red blood cells, or insufficient hemoglobin within the red blood cells. There are many different types of anemia.

Iron-deficiency anemia is caused by a lack of dietary iron, and there is not enough of this mineral to form sufficient hemoglobin. A person with this type of anemia may have a normal RBC count and a normal hematocrit, but the hemoglobin level will be below normal.

A deficiency of vitamin B_{12}, which is found only in animal foods, leads to **pernicious anemia,** in which the RBCs are large, misshapen, and fragile. Another cause of this form of anemia is lack of the intrinsic factor due to autoimmune destruction of the parietal cells of the stomach lining.

Sickle cell anemia has already been discussed in Chapter 3. It is a genetic disorder of hemoglobin, which causes RBCs to sickle, clog capillaries, and rupture.

Aplastic anemia is suppression of the red bone marrow, with decreased production of RBCs, the granular WBCs, and platelets. This is a very serious disorder which may be caused by exposure to radiation, certain chemicals such as benzene, or some medications. There are several antibiotics that must be used with caution since they may have this potentially fatal side effect.

Hemolytic anemia is any disorder that causes rupture of RBCs before the end of their normal life span. Sickle cell anemia and Rh disease of the newborn are examples. Another example is malaria, in which a protozoan parasite reproduces in RBCs and destroys them. Hemolytic anemias are often characterized by jaundice because of the increased production of bilirubin.

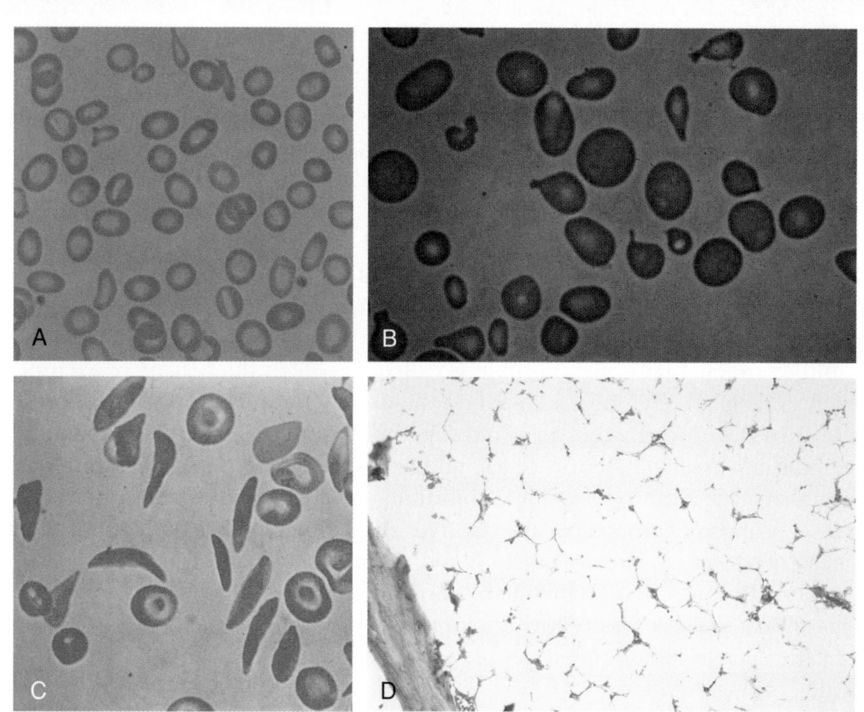

Box Figure 11–A Anemia. **(A)** Iron-deficiency anemia; notice the pale, oval RBCs (×400). **(B)** Pernicious anemia, with large, misshapen RBCs (×400). **(C)** Sickle cell anemia (×400). **(D)** Aplastic anemia, bone marrow (×200). (**A, B, and C** from Listen, Look, and Learn, Vol 3; Coagulation, Hematology. The American Society of Clinical Pathologists Press, Chicago, 1973, with permission. **D** from Harmening, DM: Clinical Hematology and Fundamentals of Hemostasis, ed 3. FA Davis, Philadelphia, 1997, p 49, with permission.)

Box 11–2 JAUNDICE

Jaundice is not a disease, but rather a sign caused by excessive accumulation of bilirubin in the blood. Because one of the liver's many functions is the excretion of bilirubin, jaundice may be a sign of liver disease such as hepatitis or cirrhosis. This may be called **hepatic jaundice,** because the problem is with the liver.

Other types of jaundice are prehepatic jaundice and posthepatic jaundice: The name of each tells us where the problem is. Recall that bilirubin is the waste product formed from the heme portion of the hemoglobin of old RBCs. **Prehepatic jaundice** means that the problem is "before" the liver, that is, hemolysis of RBCs is taking place at a more rapid rate. Rapid hemolysis is characteristic of sickle cell anemia, malaria, and Rh disease of the newborn; these are hemolytic anemias. As excessive numbers of RBCs are destroyed, bilirubin is formed at a faster rate than the liver can excrete it. The bilirubin that the liver cannot excrete remains in the blood and causes jaundice. Another name for this type is **hemolytic jaundice.**

Posthepatic jaundice means that the problem is "after" the liver. The liver excretes bilirubin into bile, which is stored in the gallbladder and then moves to the small intestine. If the bile ducts are obstructed, perhaps by gall stones or inflammation of the gallbladder, bile cannot pass to the small intestine and backs up in the liver. Bilirubin may then be reabsorbed back into the blood and cause jaundice. Another name for this type is **obstructive jaundice.**

synthesis of new hemoglobin. If not needed immediately for this purpose, excess iron is stored in the liver. The iron of RBCs is actually recycled over and over again.

Another part of the hemoglobin molecule is the heme portion, which cannot be recycled and is a waste product. The heme is converted to **bilirubin** by macrophages. The liver removes bilirubin from circulation and excretes it into bile; bilirubin is called a bile pigment. Bile is secreted by the liver into the duodenum and passes through the small intestine and colon, so bilirubin is eliminated in feces, and gives feces their characteristic brown color. If bilirubin is not excreted properly, perhaps because of liver disease such as hepatitis, it remains in the blood. This may cause **jaundice**, a condition in which the whites of the eyes appear yellow. This yellow color may also be seen in the skin of light-skinned people (see Box 11-2: Jaundice).

Blood Types

Our blood types are genetic; that is, we inherit genes from our parents that determine our own types. There are many red blood cell factors or types; we will discuss the two most important ones: the **ABO group** and the **Rh factor**.

The **ABO group** contains four blood types: A, B, AB, and O. The letters A and B represent antigens (protein-oligosaccharides) on the red blood cell membrane. A person with type A blood has the A antigen on the RBCs, and someone with type B blood has the B antigen. Type AB means that both A and B antigens are present, and type O means that neither the A nor the B antigen is present.

In the plasma of each person are natural antibodies for those antigens *not* present on the RBCs. Therefore, a type A person has anti-B antibodies in the plasma; a type B person has anti-A antibodies; a type AB person has neither anti-A nor anti-B antibodies, and a type O person has both anti-A and anti-B antibodies (see Table 11-1 and Fig. 11-4).

Table 11–1 ABO BLOOD TYPES

Type	Antigens Present on RBCs	Antibodies Present in Plasma
A	A	anti-B
B	B	anti-A
AB	both A and B	neither anti-A nor anti-B
O	neither A nor B	both anti-A and anti-B

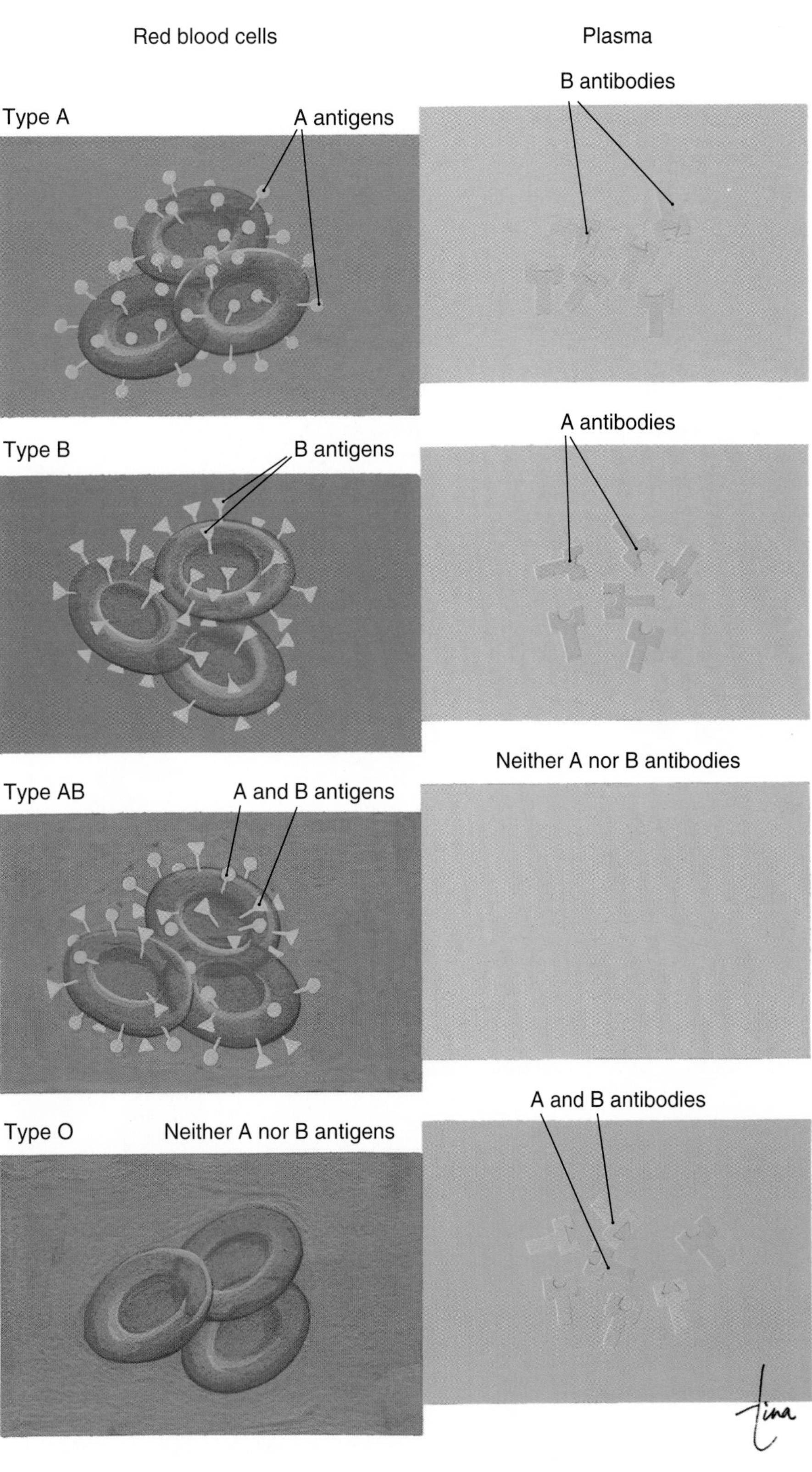

Figure 11–4 **(A)** The ABO blood types. Schematic representation of antigens on the RBCs and antibodies in the plasma.

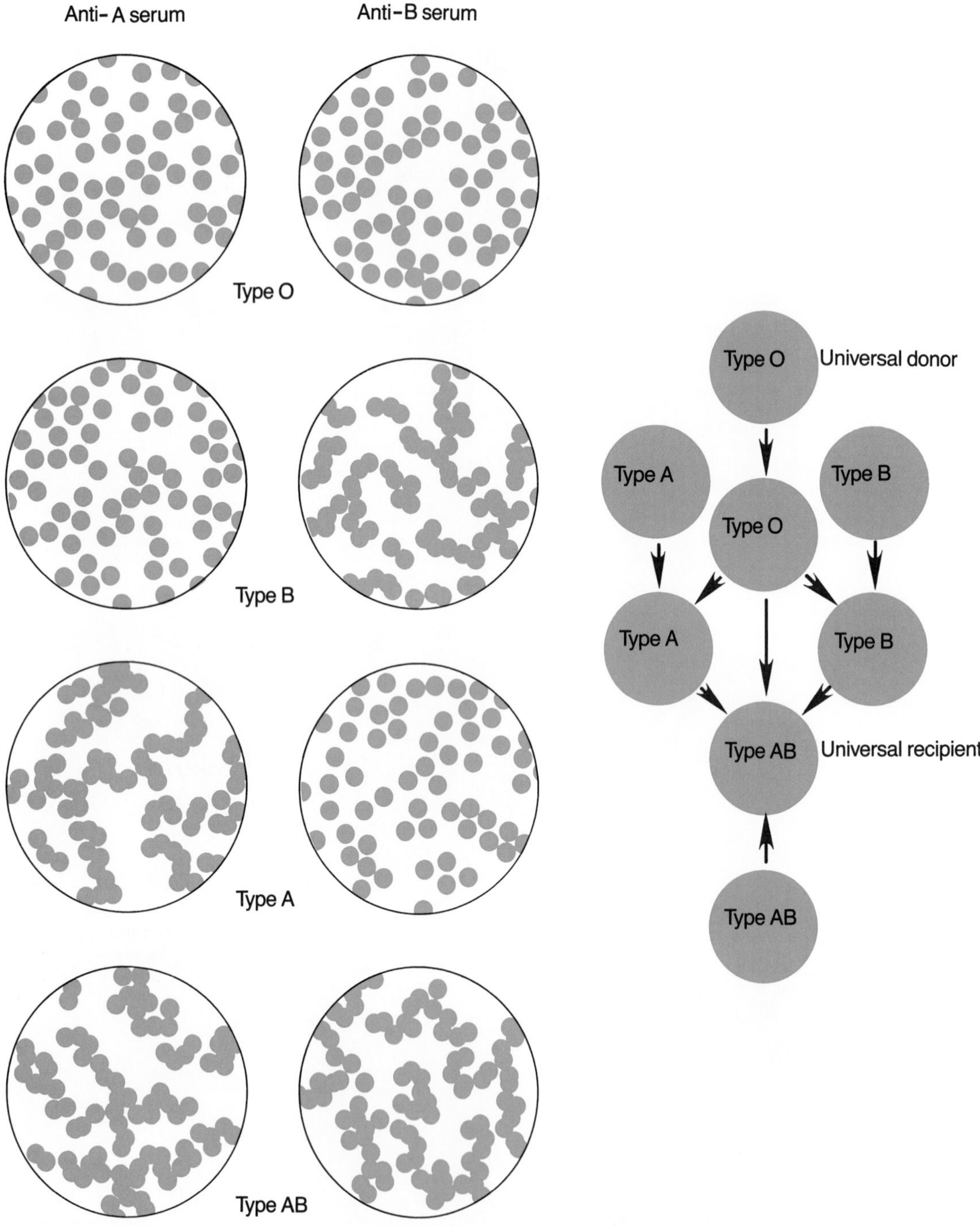

Figure 11-4 **(B)** Typing and cross-matching. The A or B antiserum causes agglutination of RBCs with the matching antigen. Acceptable transfusions are diagrammed on the right and presuppose compatible Rh factors.

These natural antibodies are of great importance for transfusions. If possible, a person should receive blood of his or her own type; only if this type is not available should another type be given. For example, let us say that there is a type A person who needs a transfusion to replace blood lost in hemorrhage. If this person were to receive type B blood, what would happen? The type-A recipient has anti-B antibodies that would bind to the type B antigens of the RBCs of the donated blood. The type-B RBCs would first clump **(agglutination)** then rupture **(hemolysis)**, thus defeating the purpose of the transfusion. An even more serious consequence is that the hemoglobin of the ruptured RBCs, now called free hemoglobin, may clog the capillaries of the kidneys and lead to renal damage or renal failure. You can see why **typing** and **cross-matching** of donor and recipient blood in the hospital laboratory is so important before any transfusion is given (see Fig. 11–4). This procedure helps ensure that donated blood will not bring about a hemolytic transfusion reaction in the recipient.

You may have heard of the concept that type O is the "universal donor." Usually, type O negative blood may be given to people with any other blood type. This is so because type-O RBCs have neither the A nor the B antigens and will not react with whatever antibodies the recipient may have. The term "negative" refers to the Rh factor, which we will now consider.

The **Rh factor** is another antigen (often called D) that may be present on RBCs. People whose RBCs have the Rh antigen are Rh positive; those without the antigen are Rh negative. Rh negative people do not have natural antibodies to the Rh antigen, and for them this antigen is foreign. If an Rh negative person receives Rh positive blood by mistake, antibodies will be formed just as they would be to bacteria or viruses. A first mistaken transfusion often does not cause problems, because antibody production is slow upon the first exposure to Rh positive RBCs. A second transfusion, however, when anti-Rh antibodies are already present, will bring about a transfusion reaction, with hemolysis and possible kidney damage (see also Box 11–3: Rh Disease of the Newborn).

WHITE BLOOD CELLS

White blood cells (WBCs) are also called **leukocytes**. There are five kinds of WBCs; all are larger than RBCs and have nuclei when mature. The nucleus may be in one piece or appear as several lobes. Special staining

Box 11–3 Rh DISEASE OF THE NEWBORN

Rh disease of the newborn may also be called **erythroblastosis fetalis** and is the result of an Rh incompatibility between mother and fetus. During a normal pregnancy, maternal blood and fetal blood do not mix in the placenta. However, during delivery of the placenta (the "afterbirth" that follows the birth of the baby), some fetal blood may enter maternal circulation.

If the woman is Rh negative and her baby is Rh positive, this exposes the woman to Rh positive RBCs. In response, her immune system will now produce anti-Rh antibodies following this first delivery. In a subsequent pregnancy, these maternal antibodies will cross the placenta and enter fetal circulation. If this next fetus is also Rh positive, the maternal antibodies will cause destruction (hemolysis) of the fetal RBCs. In severe cases this may result in the death of the fetus. In less severe cases, the baby will be born anemic and jaundiced from the loss of RBCs. Such an infant may require a gradual exchange transfusion to remove the blood with the maternal antibodies and replace it with Rh negative blood. The baby will continue to produce its own Rh-positive RBCs, which will not be destroyed once the maternal antibodies have been removed.

Much better than treatment, however, is prevention. If an Rh negative woman delivers an Rh positive baby, she should be given **RhoGam** within 72 hours after delivery. RhoGam is an anti-Rh antibody that will destroy any fetal RBCs that have entered the mother's circulation *before* her immune system can respond and produce antibodies. The RhoGam antibodies themselves break down within a few months. The woman's next pregnancy will be like the first, as if she had never been exposed to Rh positive RBCs.

for microscopic examination gives each kind of WBC a distinctive appearance (see Figs. 11–2 and 11–3).

A normal WBC count (part of a CBC) is 5,000 to 10,000 per mm^3. Notice that this number is quite small compared to a normal RBC count. Many of our WBCs are not within blood vessels but are carrying out their functions in tissue fluid.

Classification and Sites of Production

The five kinds of white blood cells may be classified in two groups: granular and agranular. The granular leukocytes are produced in the red bone marrow; these are the **neutrophils, eosinophils**, and **basophils**, which have distinctly colored granules when stained. The agranular leukocytes are **lymphocytes** and **monocytes**, which are produced in the lymphatic tissue of the spleen, lymph nodes, and thymus, as well as in the red bone marrow. A **differential WBC count** (part of a CBC) is the percentage of each kind of leukocyte. Normal ranges are listed in Table 11–2, along with other normal values of a CBC.

Functions

White blood cells all contribute to the same general function, which is to protect the body from infectious disease and to provide **immunity** to certain diseases. Each kind of leukocyte has a role in this very important aspect of homeostasis.

Neutrophils and monocytes are capable of the **phagocytosis** of pathogens. Neutrophils are the more abundant phagocytes, but the monocytes are the more efficient phagocytes, because they differentiate into **macrophages**, which also phagocytize dead or damaged tissue at the site of any injury, helping to make tissue repair possible.

Eosinophils are believed to detoxify foreign proteins. This is especially important in allergic reactions and parasitic infections such as trichinosis (a worm parasite). Basophils contain granules of heparin and histamine. **Heparin** is an anticoagulant that helps prevent abnormal clotting within blood vessels. **Histamine**, you may recall, is released as part of the inflammation process, and it makes capillaries more permeable, allowing tissue fluid, proteins, and white blood cells to accumulate in the damaged area.

There are two major kinds of lymphocytes: T cells and B cells. For now we will say that **T cells** (or T lymphocytes) recognize foreign antigens, may directly destroy some foreign antigens, and stop the immune response when the antigen has been destroyed. **B cells** (or B lymphocytes) become plasma cells that produce antibodies to foreign antigens. These T cell and B cell functions are discussed in the context of the mechanisms of immunity in Chapter 14.

As mentioned earlier, leukocytes function in tissue fluid as well as in the blood. Many WBCs are capable of self-locomotion (ameboid movement) and are able to squeeze between the cells of capillary walls and out into tissue spaces. Macrophages provide a good example of the dual locations of leukocytes. Some macrophages are "fixed," that is, stationary in organs such as the liver, spleen, and red bone marrow (part of the tissue macrophage or RE system) and in the lymph nodes. They phagocytize pathogens that circulate in blood or lymph through these organs (these are the same macrophages that also phagocytize old RBCs). Other "wandering" macrophages move about in tissue fluid, especially in the areolar connective tissue of mucous membranes and below the skin. Pathogens that gain entry into the body through natural openings or through breaks in the skin are usually destroyed by the leukocytes in connective tissue before they can cause serious disease.

A high WBC count, called **leukocytosis**, is often an indication of infection. **Leukopenia** is a low WBC count, which may be present in the early stages of diseases such as tuberculosis. Exposure to radiation or to chemicals such as benzene may destroy WBCs and

Table 11–2 COMPLETE BLOOD COUNT

Measurement	Normal Range*
Red blood cells	• 4.5–6.0 million/mm^3
Hemoglobin	• 12–18 grams/100 ml
Hematocrit	• 38%–48%
Reticulocytes	• 0%–1.5%
White blood cells (total)	• 5000–10,000/mm^3
Neutrophils	• 55%–70%
Eosinophils	• 1%–3%
Basophils	• 0.5%–1%
Lymphocytes	• 20%–35%
Monocytes	• 3%–8%
Platelets	• 150,000–300,000/mm^3

*The values on hospital lab slips may vary somewhat but will be very similar to the normal ranges given here.

Box 11–4 LEUKEMIA

Leukemia is the term for malignancy of the blood-forming tissues, the red bone marrow or lymphatic tissue. There are many types of leukemia, which are classified as acute or chronic, by the types of abnormal cells produced, and by either childhood or adult onset.

In general, leukemia is characterized by an overproduction of immature white blood cells. These immature cells cannot perform their normal functions, and the person becomes very susceptible to infection. As a greater proportion of the body's nutrients are used by malignant cells, the production of other blood cells decreases. Severe anemia is a consequence of decreased red blood cell production, and the tendency to hemorrhage is the result of decreased platelets.

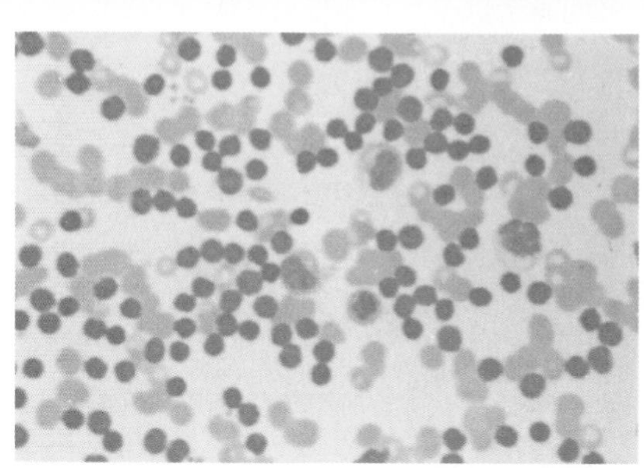

Box Figure 11–B Leukemia. Notice the many darkly staining WBCs (×300). (From Sacher, RA and McPherson, RA: Widmann's Clinical Interpretation of Laboratory Tests, ed 10. FA Davis, Philadelphia, 1991, with permission.)

Chemotherapy may bring about cure or remission for some forms of leukemia, but other forms remain resistant to treatment and may be fatal within a few months of diagnosis. In such cases, the cause of death is often pneumonia or some other serious infection, because the abnormal white blood cells cannot prevent the growth of pathogens within the body.

lower the total count. Such a person is then very susceptible to infection. **Leukemia**, or malignancy of leukocyte-forming tissues, is discussed in Box 11–4: Leukemia.

The white blood cell types (analogous to RBC types such as the ABO group) are called **human leukocyte antigens (HLAs)** and are discussed in Box 11–5: White Blood Cell Types: HLA.

PLATELETS

The more formal name for platelets is **thrombocytes**, which are not whole cells but rather fragments or pieces of cells. A normal platelet count (part of a CBC) is 150,000 to 300,000/mm^3 (the high end of the range may be extended to 500,000). **Thrombocytopenia** is the term for a low platelet count.

Site of Production

Some of the stem cells in the red bone marrow differentiate into large cells called **megakaryocytes** (see Figs. 11–2 and 11–3), which break up into small pieces that enter circulation. These circulating pieces are platelets, which may survive for 5 to 9 days, if not utilized before that.

Function

Platelets are necessary for **hemostasis**, which means prevention of blood loss. There are three mechanisms, and platelets are involved in each. Two of these mechanisms are shown in Fig. 11–5.

1. **Vascular Spasm**—when a large vessel such as an artery or vein is severed, the smooth muscle

Box 11–5 WHITE BLOOD CELL TYPES: HLA

Human leukocyte antigens (HLA) are antigens on WBCs that are representative of the antigens present on all the cells of an individual. These are our "self" antigens that identify cells that belong in the body.

Recall that in the ABO blood group of RBCs, there are only two antigens, A and B, and four possible types: A, B, AB, and O. HLA antigens are also given letter names (A, B, C, D), but there are as many as 40 possible antigens in each of these letter categories. Each individual will have two antigens in each category; the HLA types are inherited, just as RBC types are inherited. Members of the same family may have some of the same HLA types, and identical twins have exactly the same HLA type.

The purpose of the HLA types is to provide a "self" comparison for the immune system to use when pathogens enter the body. The T lymphocytes compare the "self" antigens on macrophages to the antigens on bacteria and viruses. Because these antigens do not match ours, they are recognized as foreign; this is the first step in the destruction of a pathogen.

The surgical transplantation of organs has also focused on the HLA. The most serious problem for the recipient of a transplanted heart or kidney is rejection of the organ and its destruction by the immune system. You may be familiar with the term **"tissue-typing."** This involves determining the HLA types of a donated organ to see if one or several will match the HLA types of the potential recipient. If even one HLA type matches, the chance of rejection is significantly lessened. Although all transplant recipients (except corneal) must receive immunosuppressive medications to prevent rejection, such medications make them more susceptible to infection. The closer the HLA match of the donated organ, the lower the dosage of such medications, and the less chance of serious infections. (The chance of finding a perfect HLA match in the general population is estimated at 1 in 20,000.)

There is yet another aspect of the importance of HLA: People with certain HLA types seem to be more likely to develop certain non-infectious diseases. For example, insulin-dependent diabetes mellitus is often found in people with HLA DR3 or DR4, and a form of arthritis of the spine called ankylosing spondylitis is often found in those with HLA B27. These are *not* genes for these diseases, but may be predisposing factors. What may happen is this: A virus enters the body and stimulates the immune system to produce antibodies. The virus is destroyed, but one of the person's own antigens is so similar to the viral antigen that the immune system continues its activity and begins to destroy this similar part of the body. Another possibility is that a virus damages a self-antigen to the extent that it is now so different that it will be perceived as foreign. These are two theories of how autoimmune diseases are triggered, which is the focus of much research in the field of immunology.

in its wall contracts in response to the damage (called the myogenic response). Platelets in the area of the rupture release serotonin, which also brings about vasoconstriction. The diameter of the vessel is thereby made smaller, and the smaller opening may then be blocked by a blood clot. If the vessel did not constrict first, the clot that forms would quickly be washed out by the force of the blood pressure.

2. **Platelet Plugs**—when capillaries rupture, the damage is too slight to initiate the formation of a blood clot. The rough surface, however, causes platelets to become sticky and stick to the edges of the break and to each other. The platelets form a mechanical barrier or wall to close off the break in the capillary. Capillary ruptures are quite frequent, and platelet plugs, although small, are all that is needed to seal them.

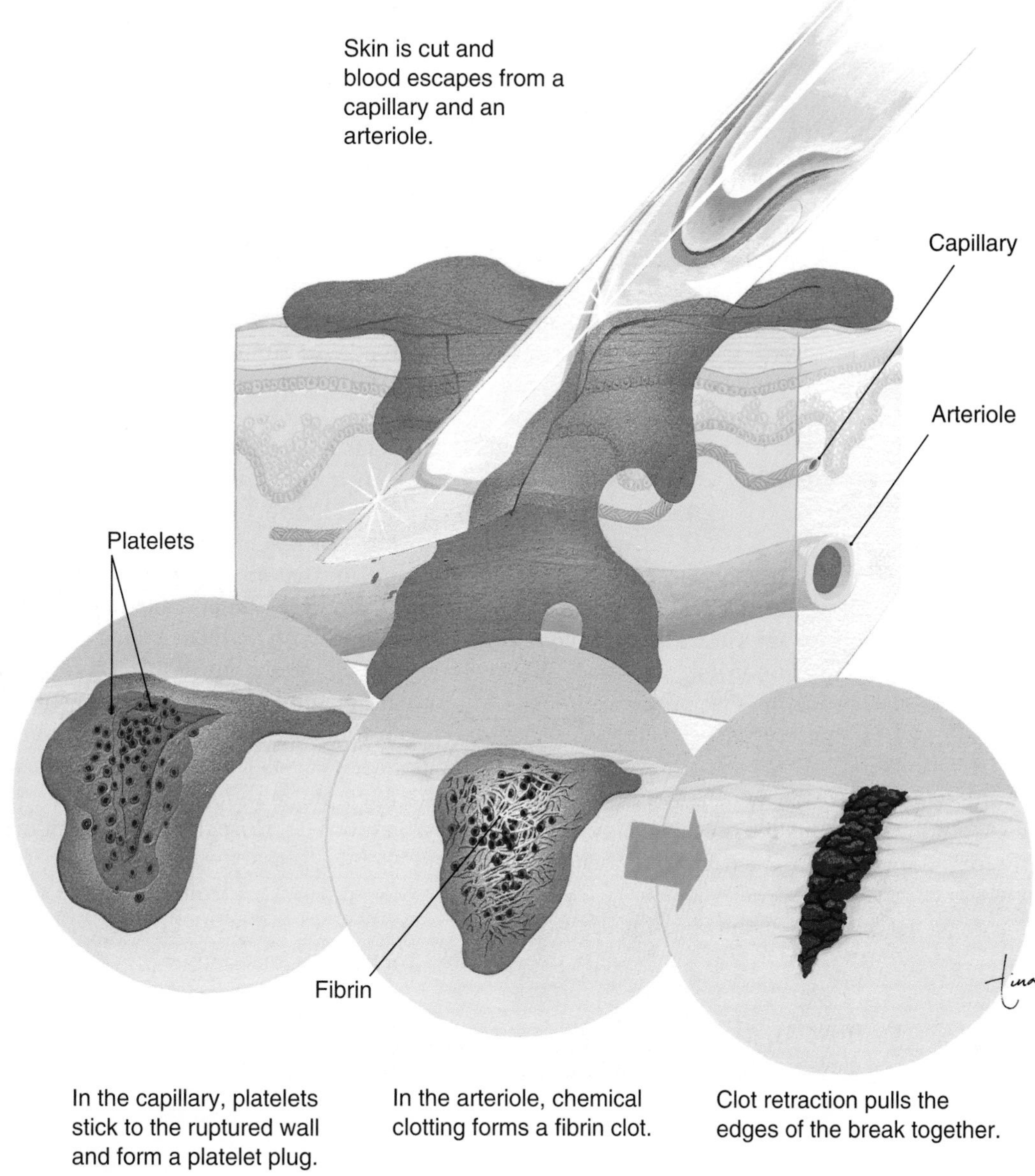
Skin is cut and blood escapes from a capillary and an arteriole.
Capillary
Arteriole
Platelets
Fibrin
In the capillary, platelets stick to the ruptured wall and form a platelet plug.
In the arteriole, chemical clotting forms a fibrin clot.
Clot retraction pulls the edges of the break together.

Figure 11–5 Hemostasis. Platelet plug formation in a capillary and chemical clotting and clot retraction in an arteriole.

Would platelet plugs be effective for breaks in larger vessels? No, they are too small, and would be washed away as fast as they form. Would vascular spasm be effective for capillaries? Again, the answer is no, because capillaries have no smooth muscle and cannot constrict at all.

3. **Chemical Clotting**—The stimulus for clotting is a rough surface within a vessel, or a break in

the vessel, which also creates a rough surface. The more damage there is, the faster clotting begins, usually within 15 to 120 seconds.

The clotting mechanism is a series of reactions involving chemicals that normally circulate in the blood and others that are released when a vessel is damaged.

The chemicals involved in clotting include platelet factors, chemicals released by damaged tissues, calcium ions, and the plasma proteins prothrombin, fibrinogen, Factor 8, and others synthesized by the liver. (These clotting factors are also designated by Roman numerals; Factor 8 would be Factor VIII.) Vitamin K is necessary for the liver to synthesize prothrombin and several other clotting factors (Factors 7, 9, and 10). Most of our vitamin K is produced by the bacteria that live in the colon; the vitamin is absorbed as the colon absorbs water.

Chemical clotting is usually described in three stages, which are shown in Table 11-3 and illustrated in Fig. 11-6. As you follow the pathway, notice that the product of stage 1, prothrombin activator, brings about the stage 2 reaction. The product of stage 2, thrombin, brings about the stage 3 reaction (see Box 11-6: Hemophilia).

The clot itself is made of **fibrin**, the product of stage 3. Fibrin is a thread-like protein. Many strands of fibrin form a mesh that traps RBCs and creates a wall across the break in the vessel.

Once the clot has formed and bleeding has stopped, **clot retraction** and **fibrinolysis** occur. Clot retraction requires platelets, ATP, and Factor 13 and involves folding of the fibrin threads to pull the edges of the rupture in the vessel wall closer together. This will make the area to be repaired smaller. As repair begins, the clot is dissolved, a process called fibrinolysis.

Prevention of Abnormal Clotting

Clotting should take place to stop bleeding, but too much clotting would obstruct vessels and interfere with normal circulation of blood. Clots do not usually form in intact vessels because the simple squamous epithelial lining is very smooth and repels the platelets and clotting factors. If the lining becomes roughened, as happens with the lipid deposits of atherosclerosis, a clot will form.

Heparin, produced by basophils, is a natural anticoagulant that inhibits the clotting process. The liver produces a globulin called **antithrombin**, which combines with and inactivates excess thrombin. This usually limits the fibrin formed to what is needed to create a useful clot but not an obstructive one.

Thrombosis refers to clotting in an intact vessel; the clot itself is called a **thrombus**. Coronary thrombosis, for example, is abnormal clotting in a coronary artery, which will decrease the blood (oxygen) supply to part of the heart muscle. An **embolism** is a clot or other tissue transported from elsewhere that lodges in and obstructs a vessel (see Box 11-7: Dissolving Clots).

Table 11–3 CHEMICAL CLOTTING

Clotting Stage	Factors Needed	Reaction
Stage 1	• Platelet factors • Chemicals from damaged tissue (tissue thromboplastin) • Factors 5,7,8,9,10,11,12 • Calcium ions	Platelet factors + tissue thromboplastin + other clotting factors + calcium ions form prothrombin activator
Stage 2	• Prothrombin activator from stage 1 • Prothrombin • Calcium ions	Prothrombin activator converts prothrombin to thrombin
Stage 3	• Thrombin from stage 2 • Fibrinogen • Calcium ions • Factor 13 (fibrin stabilizing factor)	Thrombin converts fibrinogen to fibrin

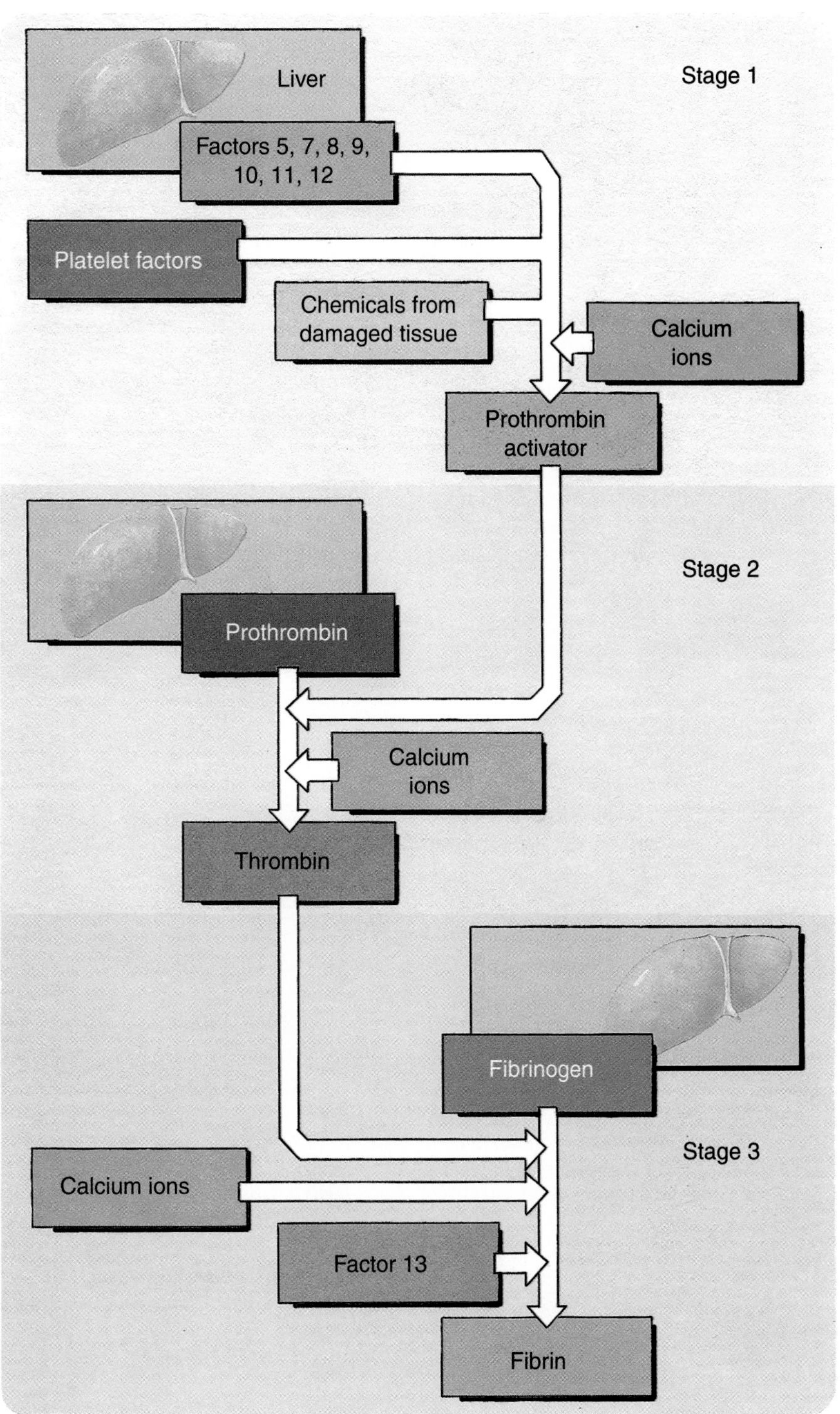

Figure 11–6 Stages of chemical blood clotting.

Box 11–6 HEMOPHILIA

There are three forms of **hemophilia;** all are genetic and are characterized by the inability of the blood to clot properly. Hemophilia A is the most common form and involves a deficiency of clotting Factor 8. The gene for hemophilia A is located on the X chromosome, so this is a **sex-linked trait,** with the same pattern of inheritance as red-green color blindness and Duchenne's muscular dystrophy.

Without Factor 8, the first stage of chemical clotting cannot be completed, and prothrombin activator is not formed. Without treatment, a hemophiliac experiences prolonged bleeding after even minor injuries and extensive internal bleeding, especially in joints subjected to the stresses of weight bearing. In recent years, treatment (but not cure) has become possible with Factor 8 obtained from blood donors. The Factor 8 is extracted from the plasma of donated blood and administered in concentrated form to hemophiliacs, enabling them to live normal lives.

In what is perhaps the most tragic irony of medical progress, many hemophiliacs were inadvertently infected with HIV, the virus that causes AIDS. Prior to 1985, there was no test to detect HIV in donated blood, and the virus was passed to hemophiliacs in the very blood product that was meant to control their disease and prolong their lives. Today, all donated blood and blood products are tested for HIV, and the risk of AIDS transmission to hemophiliacs, or anyone receiving donated blood, is now very small.

Box 11–7 DISSOLVING CLOTS

Abnormal clots may cause serious problems in coronary arteries, pulmonary arteries, cerebral vessels, and even veins in the legs. However, if clots can be dissolved before they cause death of tissue, normal circulation and tissue functioning may be restored.

One of the first substances used to dissolve clots in coronary arteries was **streptokinase,** which is actually a bacterial toxin produced by some members of the genus *Streptococcus.* Streptokinase does indeed dissolve clots, but its use creates the possibility of clot destruction throughout the body, with hemorrhage a potential consequence.

Recently, a natural enzyme has been produced by genetic engineering; this is tissue plasminogen activator, or t–PA. When a blood clot forms, a plasma protein called plasminogen is incorporated into the clot. Plasminogen is converted to plasmin by a substance present in the blood vessel lining and lysosomes, and by thrombin. Plasmin then begins to dissolve the fibrin of which the clot is made. The synthetic t–PA functions in a similar way: it converts the plasminogen in an abnormal clot to plasmin, which dissolves the clot and permits blood to flow through the vessel. In a case of coronary thrombosis, if t–PA can be administered within a few hours, the clot may be dissolved and permanent heart damage prevented. The enzyme t–PA is also used to prevent permanent brain damage after strokes (CVAs) caused by blood clots.

SUMMARY

All the functions of blood described in this chapter contribute to the homeostasis of the body as a whole. However, these functions could not be carried out if the blood did not circulate properly. The circulation of blood throughout the blood vessels is dependent upon the proper functioning of the heart, the pump of the circulatory system.

STUDY OUTLINE

The general functions of blood are transportation, regulation, and protection.

Characteristics of Blood

1. Amount—4 to 6 liters; 38% to 48% is cells; 52% to 62% is plasma (Fig. 11-1).
2. Color—arterial blood has a high oxygen content and is bright red; venous blood has less oxygen and is dark red.
3. pH—7.35 to 7.45; venous blood has more CO_2 and a lower pH than arterial blood.
4. Viscosity—thickness or resistance to flow; due to the presence of cells and plasma proteins; contributes to normal blood pressure.

Plasma—the liquid portion of blood

1. 91% water.
2. Plasma transports nutrients, wastes, hormones, antibodies, CO_2 as HCO_3^-.
3. Plasma proteins: clotting factors are synthesized by the liver; albumin is synthesized by the liver and provides colloid osmotic pressure which pulls tissue fluid into capillaries to maintain normal blood volume and blood pressure; alpha and beta globulins are synthesized by the liver and are carriers for fats and other substances in the blood; gamma globulins are antibodies produced by lymphocytes.

Blood Cells

1. Formed elements are RBCs, WBCs, and platelets (Figs. 11-2 and 11-3).
2. The hemopoietic tissues are red bone marrow (RBM) and lymphatic tissue of the spleen, lymph nodes, and thymus.

Red Blood Cells—erythrocytes (see Table 11–2 for normal values)

1. Biconcave discs; no nuclei when mature.
2. RBCs carry O_2 bonded to the iron in hemoglobin.
3. RBCs are formed in the RBM from stem cells (precursor cells).
4. Hypoxia stimulates the kidneys to produce the hormone erythropoietin, which increases the rate of RBC production in the RBM.
5. Immature RBCs: normoblasts (have nuclei) and reticulocytes (large numbers in peripheral circulation indicate a need for more RBCs to carry oxygen).
6. Vitamin B_{12} is the extrinsic factor, needed for DNA synthesis (mitosis) in stem cells in the RBM. Intrinsic factor is produced by the parietal cells of the stomach lining; it combines with B_{12} to prevent its digestion and promote its absorption.
7. RBCs live for 120 days and are then phagocytized by macrophages in the liver, spleen, and RBM. The iron is returned to the RBM or stored in the liver. The heme of the hemoglobin is converted to bilirubin, which the liver excretes into bile to be eliminated in feces. Jaundice is the accumulation of bilirubin in the blood, perhaps due to liver disease.
8. ABO blood types are hereditary. The type indicates the antigen(s) on the RBCs (see Table 11-1 and Fig. 11-4); antibodies in plasma are for those antigens not present on the RBCs and are important for transfusions.
9. The Rh type is also hereditary. Rh positive means that the D antigen is present on the RBCs; Rh negative means that the D antigen is not present on the RBCs. Rh negative people do not have natural antibodies but will produce them if given Rh positive blood.

White Blood Cells—leukocytes (see Table 11–2 for normal values)

1. Larger than RBCs; have nuclei when mature (Figs. 11-2 and 11-3).
2. Granular WBCs are the neutrophils, eosinophils, and basophils and are produced in the RBM.
3. Agranular WBCs are the lymphocytes and monocytes and are produced in lymphatic tissue, as well as in RBM.

4. Neutrophils and monocytes phagocytize pathogens; monocytes become macrophages which also phagocytize dead tissue.
5. Eosinophils detoxify foreign proteins during allergic reactions and parasitic infections.
6. Basophils contain the anticoagulant heparin and histamine, which contributes to inflammation.
7. Lymphocytes: T cells and B cells. T cells recognize foreign antigens and stop the immune response once the antigen has been destroyed. B cells become plasma cells which produce antibodies to foreign antigens.
8. WBCs carry out their functions in tissue fluid as well as in the blood.

Platelets—thrombocytes (see Table 11–2 for normal values)

1. Platelets are formed in the RBM and are fragments of megakaryocytes.
2. Platelets are involved in all mechanisms of hemostasis (prevention of blood loss) (Fig. 11-5).
3. Vascular Spasm—large vessels constrict when damaged, the myogenic response. Platelets release serotonin, which also causes vasoconstriction. The break in the vessel is made smaller and may be closed with a blood clot.
4. Platelet Plugs—rupture of a capillary creates a rough surface to which platelets stick and form a barrier over the break.
5. Chemical clotting involves platelet factors, chemicals from damaged tissue, prothrombin, fibrinogen and other clotting factors synthesized by the liver, and calcium ions. See Table 11-3 and Fig. 11-6 for the three stages of chemical clotting. The clot is formed of fibrin threads that form a mesh over the break in the vessel.
6. Clot retraction is the folding of the fibrin threads to pull the cut edges of the vessel closer together. Fibrinolysis is the dissolving of the clot once it has served its purpose.
7. Abnormal clotting (thrombosis) is prevented by the very smooth simple squamous epithelium (endothelium) that lines blood vessels; heparin, which inhibits the clotting process; and antithrombin (synthesized by the liver) which inactivates excess thrombin.

REVIEW QUESTIONS

1. Name four different kinds of substances transported in blood plasma. (p. 236)
2. Name the precursor cell of all blood cells. Name the two types of hemopoietic tissue, their locations, and the types of blood cells produced by each. (pp. 238, 246, 247)
3. State the normal values (CBC) for RBCs, WBCs, platelets, hemoglobin, and hematocrit. (p. 246)
4. State the function of RBCs; include the protein and mineral needed. (p. 238)
5. Explain why iron, protein, vitamin B_{12}, and the intrinsic factor are needed for RBC production. (pp. 238, 240)
6. Explain how bilirubin is formed and excreted. (pp. 240, 242)
7. Explain what will happen if a person with type O positive blood receives a transfusion of type A negative blood. (pp. 242, 244)
8. Name the WBC with each of the following functions: (p. 246)
   a. Become macrophages and phagocytize dead tissue
   b. Produce antibodies
   c. Detoxify foreign proteins
   d. Phagocytize pathogens
   e. Contain the anticoagulant heparin
   f. Recognize antigens as foreign
   g. Secrete histamine during inflammation

9. Explain how and why platelet plugs form in ruptured capillaries. (pp. 247–250)
10. Explain how vascular spasm prevents excessive blood loss when a large vessel is severed. (pp. 247–248)
11. With respect to chemical blood clotting: (pp. 249–250)
    a. Name the mineral necessary.
    b. Name the organ that produces many of the clotting factors.
    c. Name the vitamin necessary for prothrombin synthesis.
    d. State what the clot itself is made of.
12. Explain what is meant by clot retraction and fibrinolysis. (p. 250)
13. State two ways abnormal clotting is prevented in the vascular system. (p. 250)
14. Explain what is meant by blood viscosity, the factors that contribute, and why viscosity is important. (p. 236)
15. State the normal pH range of blood. What gas has an effect on blood pH? (p. 236)
16. Define anemia, leukocytosis, and thrombocytopenia. (pp. 241, 246–247)

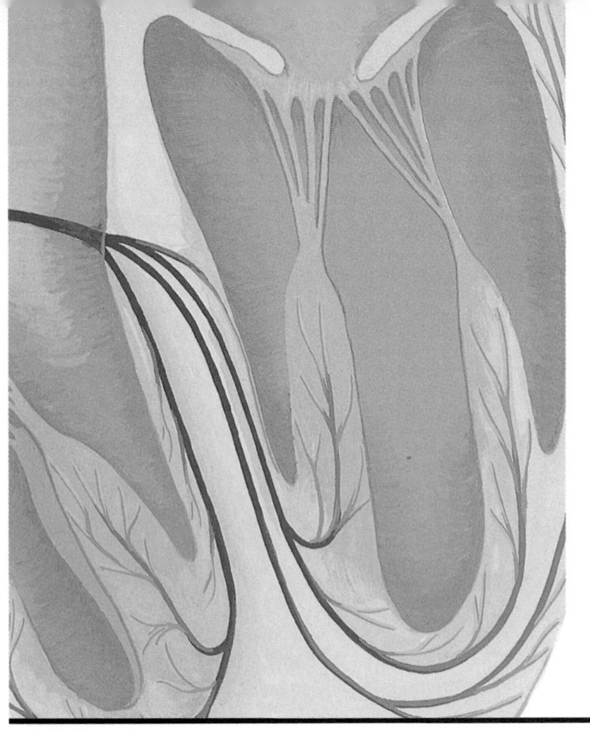

CHAPTER 12

Chapter Outline

Student Objectives

- Describe the location of the heart and the pericardial membranes.
- Name the chambers of the heart and the vessels that enter or leave each.
- Name the valves of the heart, and explain their functions.
- Describe coronary circulation, and explain its purpose.
- Describe the cardiac cycle.
- Explain how heart sounds are created.
- Name the parts of the cardiac conduction pathway, and explain why it is the SA node that initiates each beat.
- Explain stroke volume, cardiac output, and Starling's Law of the Heart.
- Explain how the nervous system regulates heart rate and force of contraction.

The Heart

New Terminology

Aorta (ay-**OR**-tah)
Atrium (**AY**-tree-um)
Cardiac cycle (**KAR**-dee-yak **SIGH**-kuhl)
Cardiac output (**KAR**-dee-yak **OUT**-put)
Coronary arteries (**KOR**-uh-na-ree **AR**-tuh-rees)
Diastole (dye-**AS**-tuh-lee)
Endocardium (EN-doh-**KAR**-dee-um)
Epicardium (EP-ee-**KAR**-dee-um)
Mediastinum (ME-dee-ah-**STYE**-num)
Mitral valve (**MYE**-truhl VALV)
Myocardium (MY-oh-**KAR**-dee-um)
Sinoatrial (SA) node (**SIGH**-noh-AY-tree-al NOHD)
Stroke volume (STROHK **VAHL**-yoom)
Systole (**SIS**-tuh-lee)
Tricuspid valve (try-**KUSS**-pid VALV)
Venous return (**VEE**-nus ree-**TURN**)
Ventricle (**VEN**-tri-kuhl)

Related Clinical Terminology

Arrhythmia (uh-**RITH**-me-yah)
Ectopic focus (ek-**TOP**-ik **FOH**-kus)
Electrocardiogram (ECG) (ee-LEK-troh-**KAR**-dee-oh-GRAM)
Fibrillation (fi-bri-**LAY**-shun)
Heart murmur (HART **MUR**-mur)
Ischemic (iss-**KEY**-mik)
Myocardial infarction (MY-oh-**KAR**-dee-yuhl in-**FARK**-shun)
Pulse (**PULS**)
Stenosis (ste-**NOH**-sis)

*Terms that appear in **bold type** in the chapter text are defined in the glossary, which begins on page 528.*

In the embryo, the heart begins to beat at 4 weeks of age, even before its nerve supply has been established. If a person lives to be 80 years old, his or her heart continues to beat an average of 100,000 times a day, every day for each of those 80 years. Imagine trying to squeeze a tennis ball 70 times a minute. After a few minutes, your arm muscles would begin to tire. Then imagine increasing your squeezing rate to 120 times a minute. Most of us could not keep that up very long, but that is what the heart does during exercise. A healthy heart can increase its rate and force of contraction to meet the body's need for more oxygen, then return to its resting rate and keep on beating as if nothing very extraordinary had happened. In fact, it isn't extraordinary at all; this is the job the heart is meant to do.

The primary function of the heart is to pump blood through the arteries, capillaries, and veins. As you learned in the last chapter, blood transports oxygen and nutrients and has other important functions as well. The heart is the pump that keeps blood circulating properly.

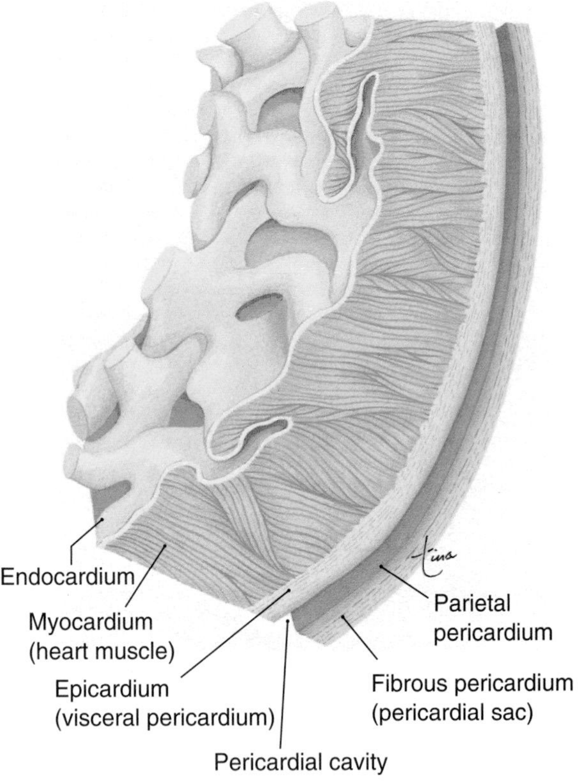

Figure 12–1 Layers of the wall of the heart and the pericardial membranes. The endocardium is the lining of the chambers of the heart. The fibrous pericardium is the outermost layer.

LOCATION AND PERICARDIAL MEMBRANES

The heart is located in the thoracic cavity between the lungs. This area is called the **mediastinum**. The cone-shaped heart has its tip (apex) just above the diaphragm to the left of the midline. This is why we may think of the heart as being on the left side, because the strongest beat can be heard or felt here.

The heart is enclosed in the **pericardial membranes**, of which there are three (Fig. 12–1). The outermost is the **fibrous pericardium**, a loose-fitting sac of fibrous connective tissue that extends inferiorly over the diaphragm and superiorly over the bases of the large vessels that enter and leave the heart. Lining the fibrous pericardium is the **parietal pericardium**, a serous membrane. On the surface of the heart muscle is the **visceral pericardium**, also called the **epicardium**, another serous membrane. Between the parietal and visceral pericardial membranes is **serous fluid**, which prevents friction as the heart beats.

CHAMBERS—VESSELS AND VALVES

The walls of the four chambers of the heart are made of cardiac muscle called the **myocardium**. The chambers are lined with **endocardium**, simple squamous epithelium that also covers the valves of the heart and continues into the vessels as their lining. The important physical characteristic of the endocardium is not its thinness, but rather its smoothness. This very smooth tissue prevents abnormal blood clotting, because clotting would be initiated by contact of blood with a rough surface.

The upper chambers of the heart are the right and left **atria** (singular: **atrium**), which have relatively thin walls and are separated by a common wall of myocardium called the **interatrial septum**. The lower chambers are the right and left **ventricles**, which have thicker walls and are separated by the **interventricular septum** (Fig. 12-2). As you will see, the atria receive blood, either from the body or the lungs, and the ventricles pump blood to either the lungs or the body.

RIGHT ATRIUM

Two large veins return blood from the body to the right atrium (see Fig. 12-2). The **superior vena cava** carries blood from the upper body, and the **inferior vena cava** carries blood from the lower body. From the right atrium, blood will flow through the right **atrioventricular (AV) valve**, or **tricuspid valve**, into the right ventricle.

The tricuspid valve is made of three flaps (or cusps) of endocardium reinforced with connective tissue. The general purpose of all valves in the circulatory system is to prevent backflow of blood. The specific purpose of the tricuspid valve is to prevent backflow of blood from the right ventricle to the right atrium when the right ventricle contracts. As the ventricle contracts, blood is forced behind the three valve flaps, forcing them upward and together to close the valve.

LEFT ATRIUM

The left atrium receives blood from the lungs, by way of four **pulmonary veins**. This blood will then flow into the left ventricle through the left atrioventricular (AV) valve, also called the **mitral valve** or **bicuspid** (two flaps) valve. The mitral valve prevents backflow of blood from the left ventricle to the left atrium when the left ventricle contracts.

A recently discovered function of the atria is the production of a hormone involved in blood pressure maintenance. When the walls of the atria are stretched, as by increased blood volume or blood pressure, the cells produce **atrial natriuretic hormone** (ANH). ANH decreases the reabsorption of sodium ions by the kidneys, so that more sodium ions are excreted in urine, which in turn increases the elimination of water. The loss of water lowers blood volume and blood pressure. You may have noticed that ANH is an antagonist to the hormone aldosterone, which raises blood pressure.

RIGHT VENTRICLE

When the right ventricle contracts, the tricuspid valve closes, and the blood is pumped to the lungs through the pulmonary artery (or trunk). At the junction of this large artery and the right ventricle is the **pulmonary semilunar valve**. Its three flaps are forced open when the right ventricle contracts and pumps blood into the pulmonary artery. When the right ventricle relaxes, blood tends to come back, but this fills the valve flaps and closes the pulmonary semilunar valve to prevent backflow of blood into the right ventricle.

Projecting into the lower part of the right ventricle are columns of myocardium called **papillary muscles** (see Fig. 12-2). Strands of fibrous connective tissue, the **chordae tendineae**, extend from the papillary muscles to the flaps of the tricuspid valve. When the right ventricle contracts, the papillary muscles also contract and pull on the chordae tendineae to prevent inversion of the tricuspid valve. If you have ever had your umbrella blown inside out by a strong wind, you can see what would happen if the flaps of the tricuspid valve were not anchored by the chordae tendineae and papillary muscles.

LEFT VENTRICLE

The walls of the left ventricle are thicker than those of the right ventricle, which enables the left ventricle to contract more forcefully. The left ventricle pumps blood to the body through the **aorta**, the largest artery of the body. At the junction of the aorta and the left ventricle is the **aortic semilunar valve** (see Fig. 12-2). This valve is opened by the force of contraction of the left ventricle, which also closes the mitral valve. The aortic semilunar valve closes when the left ventricle relaxes, to prevent backflow of blood from the aorta to the left ventricle. When the mitral (left AV) valve closes, it prevents backflow of blood to the left atrium; the flaps of the mitral valve are also anchored by chordae tendineae and papillary muscles. All valves are shown in Fig. 12-3.

As you can see from this description of the chambers and their vessels, the heart is really a double, or two-sided, pump. The right side of the heart receives deoxygenated blood from the body and pumps it to the lungs to pick up oxygen and release carbon dioxide. The left side of the heart receives oxygenated blood from the lungs and pumps it to the body. Both

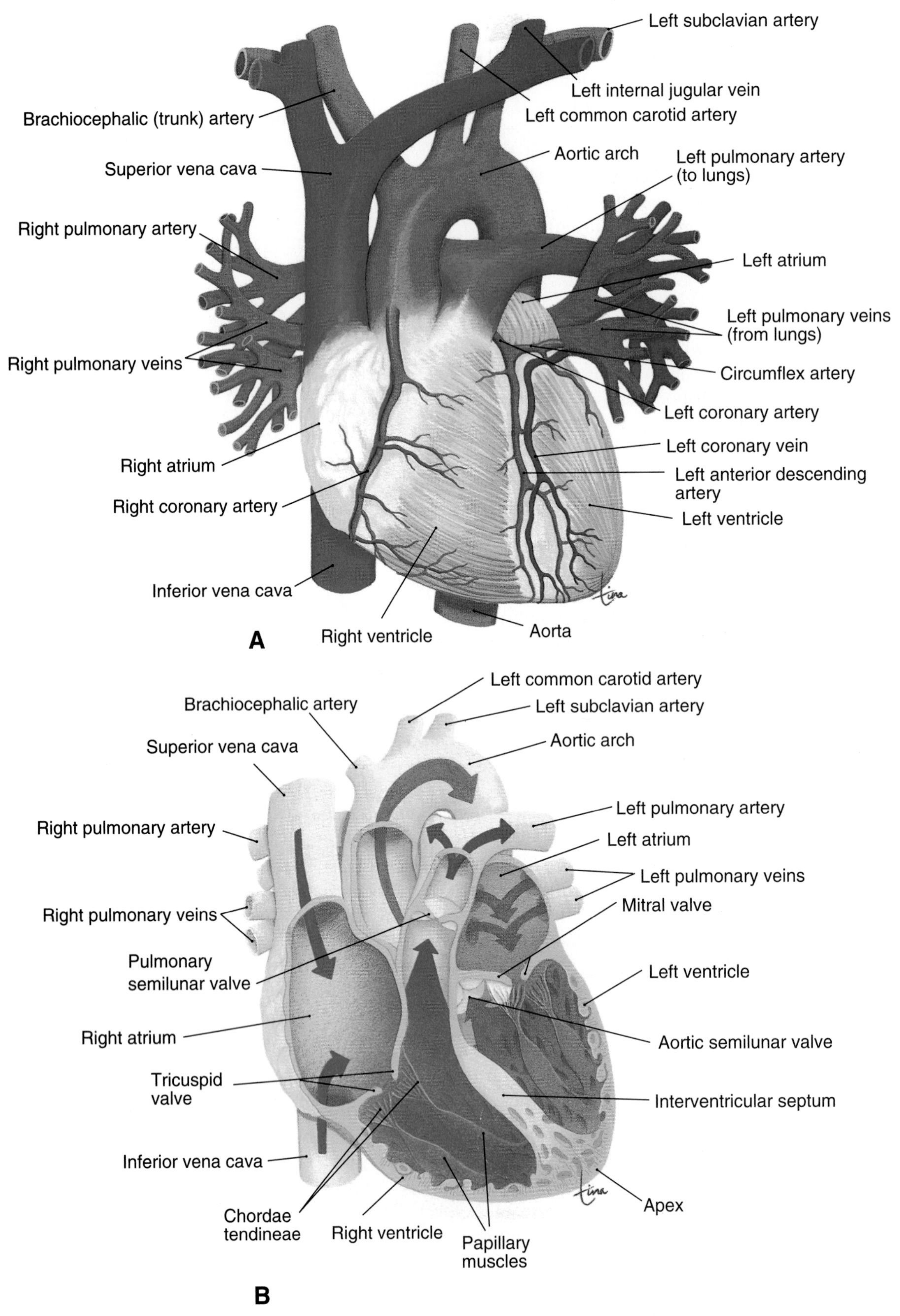

Figure 12–2 **(A)** Anterior view of the heart and major blood vessels. **(B)** Frontal section of the heart in anterior view, showing internal structures.

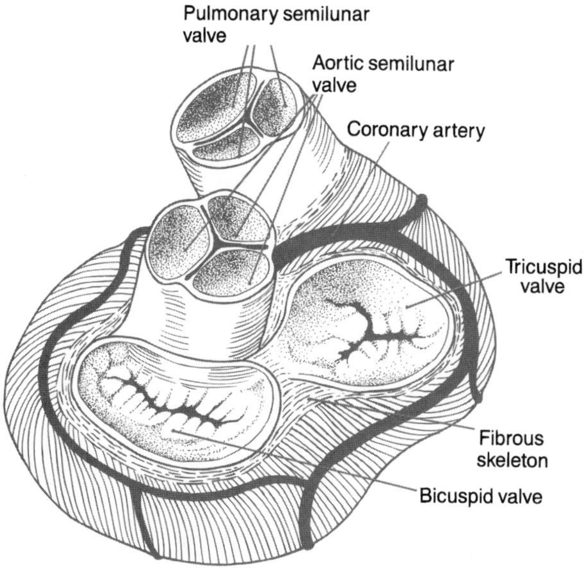

Figure 12–3 Heart valves in superior view. The atria have been removed. The fibrous skeleton of the heart is fibrous connective tissue that anchors the valve flaps and prevents enlargement of the valve openings.

pumps work simultaneously; that is, both atria contract together, followed by the contraction of both ventricles.

CORONARY VESSELS

The right and left **coronary arteries** are the first branches of the ascending aorta, just beyond the aortic semilunar valve (Fig. 12–4). The two arteries branch into smaller arteries and arterioles, then to capillaries. The coronary capillaries merge to form coronary veins, which empty blood into a large coronary sinus that returns blood to the right atrium.

The purpose of the coronary vessels is to supply blood to the myocardium itself, because oxygen is essential for normal myocardial contraction. If a coronary artery becomes obstructed, by a blood clot for example, part of the myocardium becomes **ischemic**, that is, deprived of its blood supply. Prolonged ischemia will create an **infarct**, an area of necrotic (dead) tissue. This is a myocardial infarction, commonly called a heart attack (see also Box 12–1: Risk Factors for Heart Disease).

CARDIAC CYCLE AND HEART SOUNDS

The **cardiac cycle** is the sequence of events in one heartbeat. In its simplest form, the cardiac cycle is the simultaneous contraction of the two atria, followed a fraction of a second later by the simultaneous contraction of the two ventricles. **Systole** is another term for contraction. The term for relaxation is **diastole**. You are probably familiar with these terms as they apply to blood pressure readings. If we apply them to the cardiac cycle, we can say that atrial systole is followed by ventricular systole. There is, however, a significant difference between the movement of blood from the atria to the ventricles and the movement of blood from the ventricles to the arteries. Refer to Fig. 12–5 as you read the following events of the cardiac cycle.

Blood is constantly flowing from the veins into both atria. As more blood accumulates, its pressure forces open the right and left AV valves. Two thirds of the atrial blood flows passively into the ventricles; the atria then contract to pump the remaining blood into the ventricles.

Following their contraction, the atria relax and the ventricles begin to contract. Ventricular contraction forces blood against the flaps of the right and left AV valves and closes them; the force of blood also opens the aortic and pulmonary semilunar valves. As the ventricles continue to contract, they pump blood into the arteries. Notice that blood that enters the arteries must all be pumped. The ventricles then relax, and at the same time blood continues to flow into the atria, and the cycle will begin again.

The important distinction here is that most blood flows passively from atria to ventricles, but *all* blood to the arteries is actively pumped by the ventricles. For this reason, the proper functioning of the ventricles is much more crucial to survival than is atrial functioning.

You may be asking: all this in one heartbeat? The answer is yes. The cardiac cycle is this precise sequence of events that keeps blood moving from the veins, through the heart, and into the arteries.

The cardiac cycle also creates the **heart sounds:** each heartbeat produces two sounds, often called lub-dup, that can be heard with a stethoscope. The first sound, the loudest and longest, is caused by ventricular systole closing the AV valves. The second sound

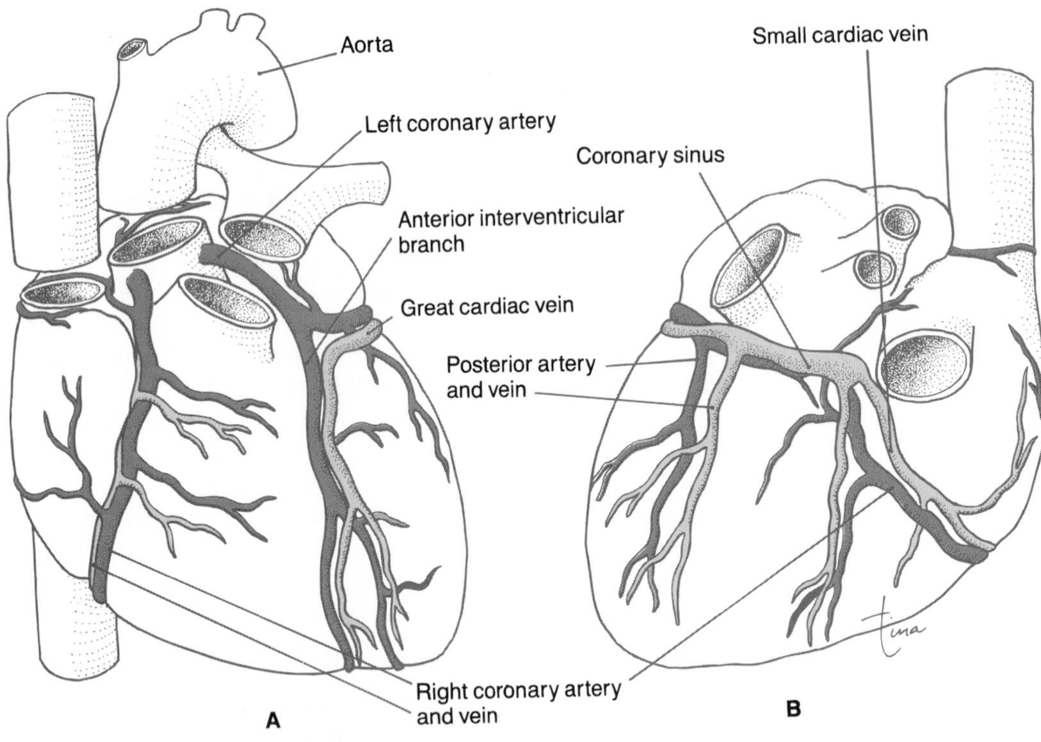

Figure 12–4 **(A)** Coronary vessels in anterior view. The pulmonary artery has been cut to show the left coronary artery emerging from the ascending aorta. **(B)** Coronary vessels in posterior view. The coronary sinus empties blood into the right atrium.

Box 12–1 RISK FACTORS FOR HEART DISEASE

Coronary artery disease results in decreased blood flow to the myocardium. If blood flow is diminished but not completely obstructed, the person may experience angina, which is chest pain caused by lack of oxygen to part of the heart muscle. If blood flow is completely blocked, however, the result is a myocardial infarction (necrosis of cardiac muscle).

The most common cause of coronary artery disease is **atherosclerosis**. Plaques of cholesterol form in the walls of a coronary artery; this narrows the lumen (cavity) and creates a rough surface where a clot (thrombus) may form (see Box Figure 12–A). Predisposing factors for atherosclerosis include cigarette smoking, diabetes mellitus, and high blood pressure. Any one of these may cause damage to the lining of coronary arteries,

Box 12–1 RISK FACTORS FOR HEART DISEASE (Continued)

which is the first step in the abnormal deposition of cholesterol. A diet high in cholesterol and saturated fats and high blood levels of these lipids will increase the rate of cholesterol deposition.

Another predisposing factor is a family history of coronary artery disease. There is no "gene for heart attacks," but we do have genes for the enzymes involved in cholesterol metabolism. Many of these are liver enzymes that regulate the transport of cholesterol in the blood in the form of lipoproteins and regulate the liver's excretion of excess cholesterol in bile. Some people, therefore, have a greater tendency than others to have higher blood levels of cholesterol and certain lipoproteins. In women before menopause, estrogen is believed to exert a protective effect by lowering blood lipid levels. This is why heart attacks in the 30- to 50-year-old age range are far more frequent in men than in women.

What can an individual do to minimize the risk of coronary artery disease? There is no way to change a hereditary predisposition, but other sensible steps can be taken. First, don't smoke. Second, maintain a diet low in cholesterol and saturated fats. Third, maintain normal body weight through proper diet and exercise. Fourth, have regular blood pressure checks and, if hypertension develops, follow the recommendations of a physician to maintain normal blood pressure.

When coronary artery disease may be life-threatening, coronary artery **bypass surgery** may be performed. In this procedure, a synthetic vessel or a vein (such as the saphenous vein of the leg) is grafted around the obstructed coronary vessel to restore blood flow to the myocardium. This is not a cure, for atherosclerosis may occur in a grafted vein or at other sites in the coronary arteries. The person who has had such surgery should follow the guidelines described previously.

Normal artery

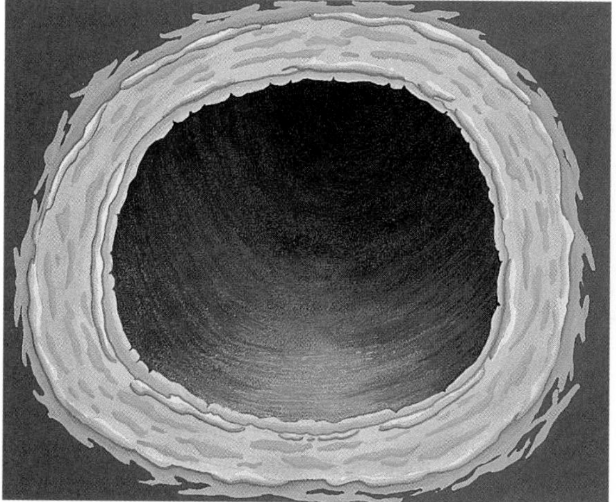

A

Atherosclerotic artery

B

Box Figure 12–A (A) Cross-section of normal coronary artery. **(B)** Coronary artery with atherosclerosis narrowing the lumen.

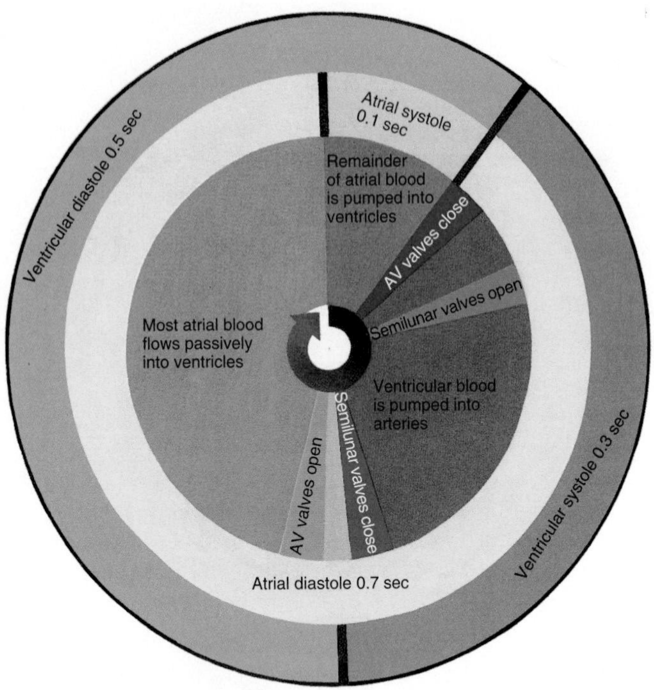

Figure 12–5 The cardiac cycle depicted in one heartbeat (pulse: 75). The outer circle represents the ventricles, the middle circle the atria, and the inner circle the movement of blood and its effect on the heart valves. See text for description.

Box 12–2 HEART MURMUR

A heart murmur is an abnormal or extra heart sound caused by a malfunctioning heart valve. The function of heart valves is to prevent backflow of blood, and when a valve does not close properly, blood will regurgitate (go backward), creating turbulence that may be heard with a stethoscope.

Rheumatic heart disease is a now uncommon complication of a streptococcal infection. In rheumatic fever, the heart valves are damaged by an abnormal response by the immune system. Erosion of the valves makes them "leaky" and inefficient, and a murmur of backflowing blood will be heard. Mitral valve regurgitation, for example, will be heard as a systolic murmur, because this valve is meant to close and prevent backflow during ventricular systole.

Some valve defects involve a narrowing **(stenosis)** and are congenital, that is, the child is born with an abnormally narrow valve. In aortic stenosis, for example, blood cannot easily pass from the left ventricle to the aorta. The ventricle must then work harder to pump blood through the narrow valve to the arteries, and the turbulence created is also heard as a systolic murmur.

Children sometimes have heart murmurs that are called "functional" because no structural cause can be found. These murmurs usually disappear with no adverse effects on the child.

is caused by the closure of the aortic and pulmonary semilunar valves. If any of the valves do not close properly, an extra sound called a **heart murmur** may be heard (see Box 12-2: Heart Murmur).

CARDIAC CONDUCTION PATHWAY

The cardiac cycle is a sequence of mechanical events that is regulated by the electrical activity of the myocardium. Cardiac muscle cells have the ability to contract spontaneously, that is, nerve impulses are not required to cause contraction. The heart generates its own beat, and the electrical impulses follow a very specific route throughout the myocardium. You may find it helpful to refer to Fig. 12-6 as you read the following.

The natural pacemaker of the heart is the **sinoatrial (SA) node**, a specialized group of cardiac muscle cells located in the wall of the right atrium. The SA node is considered specialized because it has the most rapid rate of contraction, that is, it depolarizes more rapidly than any other part of the myocardium. As you may recall, depolarization is the rapid entry of Na^+ ions and the reversal of charges on either side of the cell membrane. The cells of the SA node are more permeable to Na^+ ions than are other cardiac muscle cells. Therefore, they depolarize more rapidly, then contract and initiate each heartbeat.

From the SA node, impulses for contraction travel to the **atrioventricular (AV) node**, located in the lower interatrial septum. The transmission of impulses from the SA node to the AV node and to the rest of the atrial myocardium brings about atrial systole.

Within the upper interventricular septum is the **bundle of His** (AV bundle), which receives impulses from the AV node and transmits them to the right and left **bundle branches**. From the bundle branches, impulses travel along **Purkinje fibers** to the rest of the ventricular myocardium and bring about ventricular systole. The electrical activity of the atria and ventricles is depicted by an electrocardiogram (ECG); this is discussed in Box 12-3: Electrocardiogram.

If the SA node does not function properly, the AV node will initiate the heartbeat, but at a slower rate (50 to 60 beats per minute). The bundle of His is also capable of generating the beat of the ventricles, but at a much slower rate (15 to 40 beats per minute). This may occur in certain kinds of heart disease in which transmission of impulses from the atria to the ventricles is blocked.

Arrhythmias are irregular heartbeats; their effects range from harmless to life-threatening. Nearly everyone experiences heart **palpitations** (becoming aware of an irregular beat) from time to time. These are usually not serious and may be the result of too much caffeine, nicotine, or alcohol. Much more serious is ventricular **fibrillation**, a very rapid and uncoordinated ventricular beat that is totally ineffective for pumping blood (see Box 12-4: Arrhythmias).

HEART RATE

A healthy adult has a resting heart rate **(pulse)** of 60 to 80 beats per minute, which is the rate of depolarization of the SA node. A rate less than 60 (except for athletes) is called **bradycardia**; a rate greater than 100 beats per minute is called **tachycardia**.

A child's normal heart rate may be as high as 100 beats per minute, that of an infant as high as 120, and that of a near-term fetus as high as 140 beats per minute. These higher rates are not related to age, but rather to size: the smaller the individual, the higher the metabolic rate and the faster the heart rate. Parallels may be found among animals of different sizes; the heart rate of a mouse is about 200 beats per minute and that of an elephant about 30 beats per minute.

Let us return to the adult heart rate and consider the person who is in excellent physical condition. As you may know, well-conditioned athletes have low resting pulse rates. Those of basketball players are often around 50 beats per minute, and the pulse of a marathon runner often ranges from 35 to 40 beats per minute. To understand why this is so, remember that the heart is a muscle. When our skeletal muscles are exercised, they become stronger and more efficient. The same is true for the heart; consistent exercise makes it a more efficient pump, as you will see in the next discussion.

CARDIAC OUTPUT

Cardiac output is the amount of blood pumped by a ventricle in 1 minute. A certain level of cardiac output

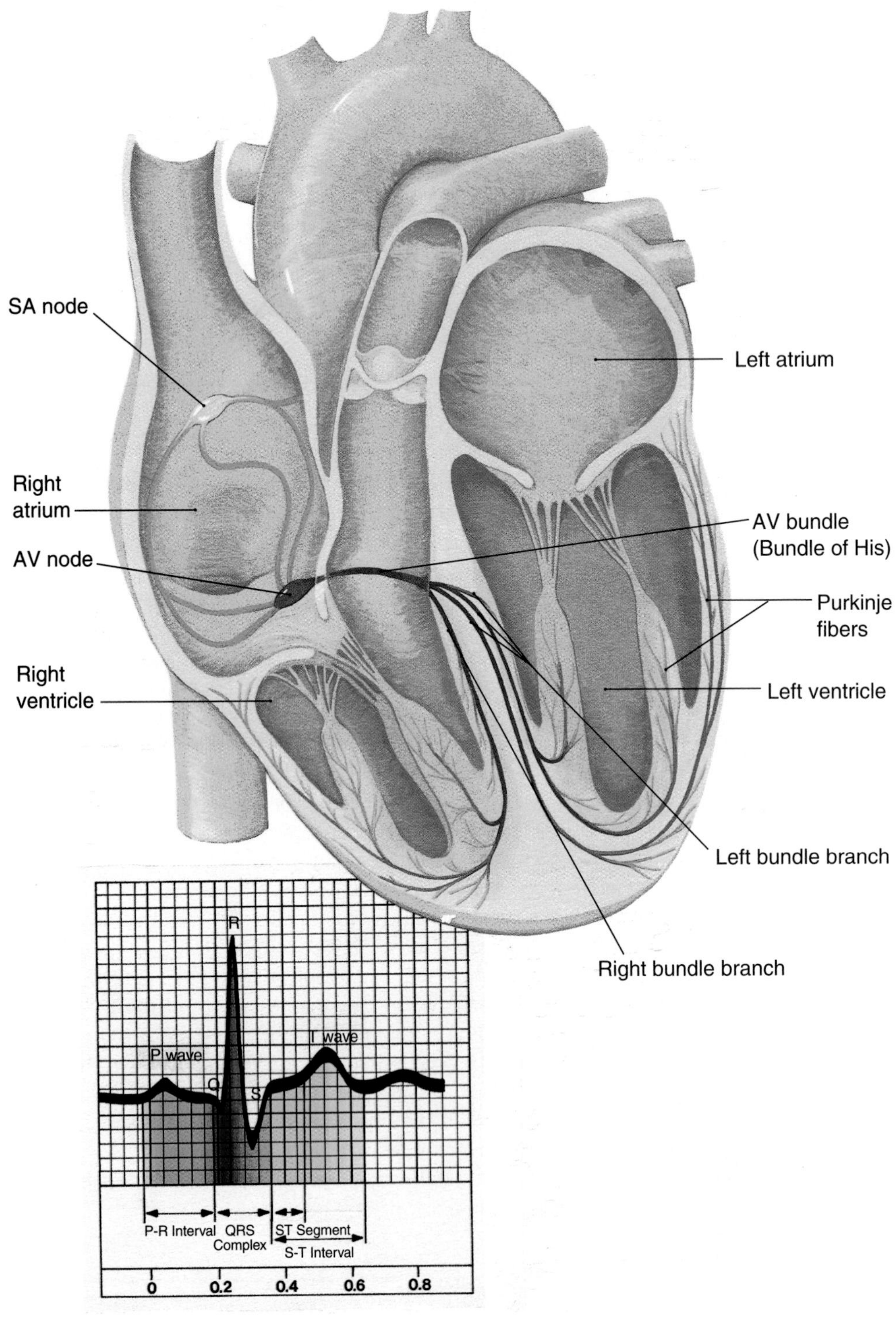

Figure 12–6 Conduction pathway of the heart. Anterior view of the interior of the heart. The electrocardiogram tracing is of one normal heartbeat. See text and Box 12–3 for description.

Box 12–3 ELECTROCARDIOGRAM

A heartbeat is a series of electrical events, and the electrical charges generated by the myocardium can be recorded by placing electrodes on the body surface. Such a recording is called an **electrocardiogram (ECG)** (see Fig. 12–6).

A typical ECG consists of three distinguishable waves or deflections: the P wave, the QRS complex, and the T wave. Each represents a specific electrical event; all are shown in Fig. 12–6 in a normal ECG tracing.

The P wave represents depolarization of the atria, that is, the transmission of electrical impulses from the SA node throughout the atrial myocardium.

The QRS complex represents depolarization of the ventricles as the electrical impulses spread throughout the ventricular myocardium. The T wave represents repolarization of the ventricles (atrial repolarization does not appear as a separate wave because it is masked by the QRS complex).

Detailed interpretation of abnormal ECGs is beyond the scope of this book, but in general, the length of each wave and the time intervals between waves are noted. An ECG may be helpful in the diagnosis of coronary atherosclerosis, which deprives the myocardium of oxygen, or of rheumatic fever or other valve disorders that result in enlargement of a chamber of the heart and prolong a specific wave of an ECG. For example, the enlargement of the left ventricle that is often a consequence of hypertension may be indicated by an abnormal QRS complex.

Box 12–4 ARRHYTHMIAS

Arrhythmias are irregular heartbeats caused by damage to part of the conduction pathway, or by an **ectopic focus**, which is a beat generated in part of the myocardium other than the SA node.

Flutter is a very rapid but fairly regular heartbeat. In atrial flutter, the atria may contract up to 300 times per minute. Because atrial pumping is not crucial, however, blood flow to the ventricles may be maintained for a time, and flutter may not be immediately life-threatening. Ventricular flutter is usually only a brief transition between ventricular tachycardia and fibrillation.

Fibrillation is very rapid and uncoordinated contractions. Ventricular fibrillation is a medical emergency that must be rapidly corrected to prevent death. Normal contraction of the ventricles is necessary to pump blood into the arteries, but fibrillating ventricles are not pumping, and cardiac output decreases sharply.

Ventricular fibrillation may follow a non-fatal heart attack (myocardial infarction). Damaged cardiac muscle cells may not be able to maintain a normal state of polarization, and they depolarize spontaneously and rapidly. From this ectopic focus, impulses spread to other parts of the ventricular myocardium in a rapid and haphazard pattern, and the ventricles quiver rather than contract as a unit.

It is often possible to correct ventricular fibrillation with the use of an electrical defibrillator. This instrument delivers an electric shock to the heart, which causes the entire myocardium to depolarize and contract, then relax. If the first part of the heart to recover is the SA node (which usually has the most rapid rate of contraction), a normal heartbeat may be restored.

is needed at all times to transport oxygen to tissues and to remove waste products. During exercise, cardiac output must increase to meet the body's need for more oxygen. We will return to exercise after first considering resting cardiac output.

In order to calculate cardiac output, we must know the pulse rate and how much blood is pumped per beat. **Stroke volume** is the term for the amount of blood pumped by a ventricle per beat; an average resting stroke volume is 60 to 80 ml per beat. A simple formula then enables us to determine cardiac output:

Cardiac output = stroke volume × pulse (heart rate)

Let us put into this formula an average resting stroke volume, 70 ml, and an average resting pulse, 70 beats per minute (bpm):

Cardiac output = 70 ml × 70 bpm
Cardiac output = 4900 ml per minute (approximately 5 liters)

Naturally, cardiac ouput varies with the size of the person, but the average resting cardiac output is 5 to 6 liters per minute.

If we now reconsider the athlete, you will be able to see precisely why the athlete has a low resting pulse. In our formula, we will use an average resting cardiac output (5 liters) and an athlete's pulse rate (50):

$$\text{Cardiac output} = \text{stroke volume} \times \text{pulse}$$
$$5000 \text{ ml} = \text{stroke volume} \times 50 \text{ bpm}$$
$$\frac{5000}{50} = \text{stroke volume}$$
$$100 \text{ ml} = \text{stroke volume}$$

Notice that the athlete's resting stroke volume is significantly higher than the average. The athlete's more efficient heart pumps more blood with each beat and so can maintain a normal resting cardiac output with fewer beats.

Now let us see how the heart responds to exercise. Heart rate (pulse) increases during exercise, and so does stroke volume. The increase in stroke volume is the result of **Starling's Law of the Heart**, which states that the more the cardiac muscle fibers are stretched, the more forcefully they contract. During exercise, more blood returns to the heart; this is called **venous return**. Increased venous return stretches the myocardium of the ventricles, which contract more forcefully and pump more blood, thereby increasing stroke volume. Therefore, during exercise, our formula might be the following:

Cardiac output = stroke volume × pulse
Cardiac output = 100 ml × 100 bpm
Cardiac output = 10,000 ml (10 liters)

This exercise cardiac output is twice the resting cardiac output we first calculated, which should not be considered unusual. The cardiac output of a healthy young person may increase up to four times the resting level during strenuous exercise. The marathon runner's cardiac output may increase six times or more compared to the resting level; this is the result of the marathoner's extremely efficient heart.

Also related to cardiac output, and another measure of the health of the heart, is the **ejection fraction**. This is the percent of the blood in a ventricle that is pumped during systole. A ventricle doesn't empty completely when it contracts, but should pump out 60% to 70% of the blood within it. A lower percentage would indicate that the ventricle is weakening.

REGULATION OF HEART RATE

Although the heart generates and maintains its own beat, the rate of contraction can be changed to adapt to different situations. The nervous system can and does bring about necessary changes in heart rate as well as in force of contraction.

The **medulla** of the brain contains the two cardiac centers, the **accelerator center** and the **inhibitory center**. These centers send impulses to the heart along autonomic nerves. Recall from Chapter 8 that the autonomic nervous system has two divisions: sympathetic and parasympathetic. Sympathetic impulses from the accelerator center along sympathetic nerves increase heart rate and force of contraction. Parasympathetic impulses from the inhibitory center along the vagus nerves decrease the heart rate.

Our next question might be: What information is received by the medulla to initiate changes? Because the heart pumps blood, it is essential to maintain normal blood pressure. Blood contains oxygen, which all tissues must receive continuously. Therefore, changes in blood pressure and oxygen level of the blood are stimuli for changes in heart rate.

You may also recall from Chapter 9 that pressoreceptors and chemoreceptors are located in the carotid arteries and aortic arch. **Pressoreceptors** in the carotid sinuses and aortic sinus detect changes in blood pressure. **Chemoreceptors** in the carotid bodies and aortic body detect changes in the oxygen content of the blood. The sensory nerves for the carotid receptors are the glossopharyngeal (9th cranial) nerves; the sensory nerves for the aortic arch receptors are the vagus (10th cranial) nerves. If we now put all these facts together in a specific example, you will see that the regulation of heart rate is a reflex. Fig. 12–7 depicts all the structures mentioned above.

A person who stands up suddenly from a lying position may feel light-headed or dizzy for a few moments, because blood pressure to the brain has decreased abruptly. The drop in blood pressure is detected by pressoreceptors in the carotid sinuses—notice that they are "on the way" to the brain, a very strategic location. The drop in blood pressure causes fewer impulses to be generated by the pressoreceptors. These impulses travel along the glossopharyngeal nerves to the medulla, and the decrease in the frequency of impulses stimulates the accelerator center. The accelerator center generates impulses that are carried by sympathetic nerves to the SA node, AV node, and the ventricular myocardium. As heart rate and force increase, blood pressure to the brain is raised to normal, and the sensation of light-headedness passes. When blood pressure to the brain is restored to normal, the heart receives more parasympathetic impulses from the inhibitory center along the vagus nerves to the SA node and AV node. These parasympathetic impulses slow the heart rate to a normal resting pace.

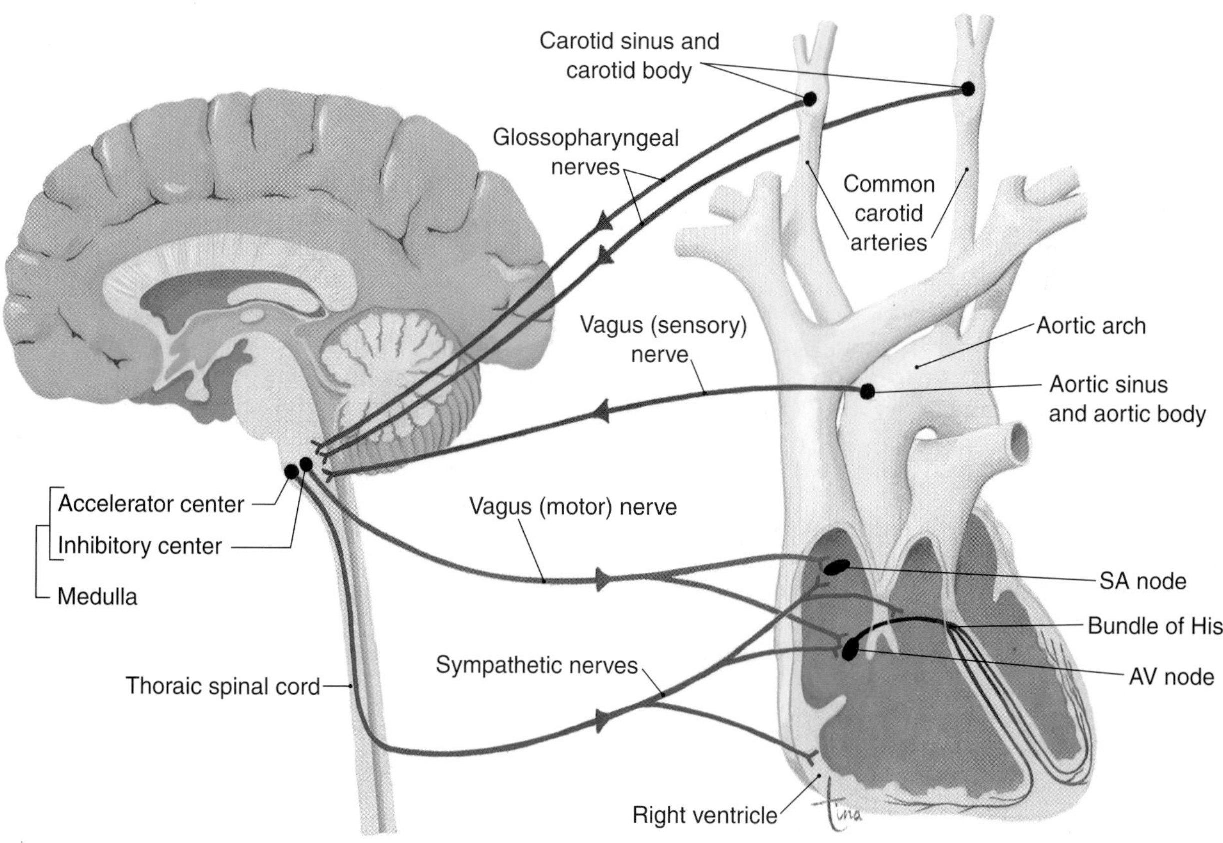

Figure 12–7 Nervous regulation of the heart. The brain and spinal cord are shown on the left. The heart and major blood vessels are shown on the right.

The heart will also be the effector in a reflex stimulated by a decrease in the oxygen content of the blood. The aortic receptors are strategically located so as to detect such an important change as soon as blood leaves the heart. The reflex arc in this situation would be: (1) aortic chemoreceptors, (2) vagus nerves (sensory), (3) accelerator center in the medulla, (4) sympathetic nerves, and (5) the heart, which will increase its rate and force of contraction to circulate more oxygen to correct the hypoxia.

Recall also from Chapter 10 that the hormone epinephrine is secreted by the adrenal medulla in stressful situations. One of the many functions of epinephrine is to increase heart rate and force of contraction. This will help supply more blood to tissues in need of more oxygen.

AGING AND THE HEART

The heart muscle becomes less efficient with age, and there is a decrease in both maximum cardiac output and heart rate, although resting levels may be more than adequate. The health of the myocardium depends on its blood supply, and with age there is greater likelihood that atherosclerosis will narrow the coronary arteries. Atherosclerosis is the deposition of cholesterol on and in the walls of the arteries, which decreases blood flow and forms rough surfaces that may cause intravascular clot formation.

High blood pressure (hypertension) causes the left ventricle to work harder; it may enlarge and outgrow its blood supply, thus becoming weaker. The heart valves may become thickened by fibrosis, leading to heart murmurs. Arrhythmias are also more common with age, as the cells of the conduction pathway become less efficient.

SUMMARY

As you can see, the nervous system regulates the functioning of the heart based on what the heart is supposed to do. The pumping of the heart maintains normal blood pressure and proper oxygenation of tissues, and the nervous system ensures that the heart will be able to meet these demands in different situations.

STUDY OUTLINE

The heart pumps blood, which creates blood pressure, and circulates oxygen, nutrients, and other substances. The heart is located in the mediastinum, the area between the lungs in the thoracic cavity.

Pericardial Membranes—three layers that enclose the heart (see Fig. 12–1)

1. The outer, fibrous pericardium, made of fibrous connective tissue, is a loose-fitting sac that surrounds the heart and extends over the diaphragm and the bases of the great vessels.
2. The parietal pericardium is a serous membrane that lines the fibrous pericardium.
3. The visceral pericardium, or epicardium, is a serous membrane on the surface of the myocardium.
4. Serous fluid between the parietal and visceral pericardial membranes prevents friction as the heart beats.

Chambers of the Heart (see Fig. 12–2)

1. Cardiac muscle tissue, the myocardium, forms the walls of the four chambers of the heart.
2. Endocardium lines the chambers and covers the valves of the heart; is simple squamous epithelium that is very smooth and prevents abnormal clotting.
3. The right and left atria are the upper chambers, separated by the interatrial septum. The atria receive blood from veins.
4. The right and left ventricles are the lower chambers, separated by the interventricular septum. The ventricles pump blood into arteries.

Right Atrium

1. Receives blood from the upper body by way of the superior vena cava and receives blood from the lower body by way of the inferior vena cava.
2. The tricuspid (right AV) valve prevents backflow of blood from the right ventricle to the right atrium when the right ventricle contracts.

Left Atrium

1. Receives blood from the lungs by way of four pulmonary veins.
2. The mitral (left AV or bicuspid) valve prevents backflow of blood from the left ventricle to the left atrium when the left ventricle contracts.
   - The walls of the atria produce atrial natriuretic hormone when stretched by increased blood volume or BP. ANH increases the loss of Na^+ ions and water in urine, which decreases blood volume and BP to normal.

Right Ventricle—has relatively thin walls

1. Pumps blood to the lungs through the pulmonary artery.
2. The pulmonary semilunar valve prevents backflow of blood from the pulmonary artery to the right ventricle when the right ventricle relaxes.
3. Papillary muscles and chordae tendineae prevent inversion of the right AV valve when the right ventricle contracts.

Left Ventricle—has thicker walls than does the right ventricle

1. Pumps blood to the body through the aorta.
2. The aortic semilunar valve prevents backflow of blood from the aorta to the left ventricle when the left ventricle relaxes.
3. Papillary muscles and chordae tendineae prevent inversion of the left AV valve when the left ventricle contracts.
   - The heart is a double pump: The right heart receives deoxygenated blood from the body and pumps it to the lungs; the left heart receives oxygenated blood from the lungs and pumps it to the body. Both sides of the heart work simultaneously.

Coronary Vessels (see Fig. 12–4)

1. Pathway: ascending aorta to right and left coronary arteries, to smaller arteries, to capillaries, to coronary veins, to the coronary sinus, to the right atrium.
2. Coronary circulation supplies oxygenated blood to the myocardium.
3. Obstruction of a coronary artery causes a myocardial infarction: death of an area of myocardium due to lack of oxygen.

Cardiac Cycle—the sequence of events in one heartbeat (see Fig. 12–5)

1. The atria continually receive blood from the veins; as pressure within the atria increases, the AV valves are opened.
2. Two-thirds of the atrial blood flows passively into the ventricles; atrial contraction pumps the remaining blood into the ventricles; the atria then relax.
3. The ventricles contract, which closes the AV valves and opens the aortic and pulmonary semilunar valves.
4. Ventricular contraction pumps all blood into the arteries. The ventricles then relax. Meanwhile, blood is filling the atria, and the cycle begins again.
5. Systole means contraction; diastole means relaxation. In the cardiac cycle, atrial systole is followed by ventricular systole. When the ventricles are in systole, the atria are in diastole.
6. The mechanical events of the cardiac cycle keep blood moving from the veins through the heart and into the arteries.

Heart Sounds—two sounds per heartbeat: lub-dup

1. The first sound is created by closure of the AV valves during ventricular systole.
2. The second sound is created by closure of the aortic and pulmonary semilunar valves.
3. Improper closing of a valve results in a heart murmur.

Cardiac Conduction Pathway—the pathway of impulses during the cardiac cycle (see Fig. 12–6)

1. The SA node in the wall of the right atrium initiates each heartbeat; the cells of the SA node are more permeable to Na^+ ions and depolarize more rapidly than any other part of the myocardium.
2. The AV node is in the lower interatrial septum. Depolarization of the SA node spreads to the AV node and to the atrial myocardium and brings about atrial systole.
3. The bundle of His is in the upper interventricular septum; the first part of the ventricles to depolarize.
4. The right and left bundle branches in the interventricular septum transmit impulses to the Purkinje

fibers in the ventricular myocardium, which complete ventricular systole.
5. An electrocardiogram (ECG) depicts the electrical activity of the heart (see Fig. 12-6).
6. If part of the conduction pathway does not function properly, the next part will initiate contraction, but at a slower rate.
7. Arrhythmias are irregular heartbeats; their effects range from harmless to life-threatening.

Heart Rate

1. Healthy adult: 60 to 80 beats per minute (heart rate equals pulse); children and infants have faster pulses because of their smaller size and higher metabolic rate.
2. A person in excellent physical condition has a slow resting pulse because the heart is a more efficient pump and pumps more blood per beat.

Cardiac Output

1. Cardiac output is the amount of blood pumped by a ventricle in 1 minute.
2. Stroke volume is the amount of blood pumped by a ventricle in one beat; average is 60 to 80 ml.
3. Cardiac output equals stroke volume × pulse; average resting cardiac output is 5 to 6 liters.
4. Starling's Law of the Heart—the more that cardiac muscle fibers are stretched, the more forcefully they contract.
5. During exercise, stroke volume increases as venous return increases and stretches the myocardium of the ventricles (Starling's Law).
6. During exercise, the increase in stroke volume and the increase in pulse result in an increase in cardiac output: two to four times the resting level.
7. The ejection fraction is the percent of its total blood that a ventricle pumps; average is 60% to 70%.

Regulation of Heart Rate (see Fig. 12–7)

1. The heart generates its own beat, but the nervous system brings about changes to adapt to different situations.
2. The medulla contains the cardiac centers: the accelerator center and the inhibitory center.
3. Sympathetic impulses to the heart increase rate and force of contraction; parasympathetic impulses (vagus nerves) to the heart decrease heart rate.
4. Pressoreceptors in the carotid and aortic sinuses detect changes in blood pressure.
5. Chemoreceptors in the carotid and aortic bodies detect changes in the oxygen content of the blood.
6. The glossopharyngeal nerves are sensory for the carotid receptors. The vagus nerves are sensory for the aortic receptors.
7. If blood pressure to the brain decreases, pressoreceptors in the carotid sinuses detect this decrease and send fewer sensory impulses along the glossopharyngeal nerves to the medulla. The accelerator center dominates and sends motor impulses along sympathetic nerves to increase heart rate and force to restore blood pressure to normal.
8. A similar reflex is activated by hypoxia.
9. Epinephrine from the adrenal medulla increases heart rate and force of contraction during stressful situations.

REVIEW QUESTIONS

1. Describe the location of the heart with respect to the lungs and to the diaphragm. (p. 258)
2. Name the three pericardial membranes. Where is serous fluid found and what is its function? (p. 258)
3. Describe the location and explain the function of endocardium. (p. 258)
4. Name the veins that enter the right atrium; name those that enter the left atrium. For each, where does the blood come from? (p. 259)
5. Name the artery that leaves the right ventricle; name the artery that leaves the left ventricle. For each, where is the blood going? (p. 259)
6. Explain the purpose of the right and left AV valves and the purpose of the aortic and pulmonary semilunar valves. (p. 259)

7. Describe the coronary system of vessels and explain the purpose of coronary circulation. (p. 261)
8. Define: systole, diastole, and cardiac cycle. (p. 261)
9. Explain how movement of blood from atria to ventricles differs from movement of blood from ventricles to arteries. (p. 261)
10. Explain why the heart is considered a double pump. Trace the path of blood from the right atrium back to the right atrium, naming the chambers of the heart and their vessels that the blood passes through. (pp. 259, 261)
11. Name the parts, in order, of the cardiac conduction pathway. Explain why it is the SA node that generates each heartbeat. State a normal range of heart rate for a healthy adult. (p. 265)
12. Calculate cardiac output if stroke volume is 75 ml and pulse is 75 bpm. Using the cardiac output you just calculated as a resting normal, what is the stroke volume of a marathoner whose resting pulse is 40 bpm? (pp. 265, 268)
13. Name the two cardiac centers and state their location. Sympathetic impulses to the heart have what effect? Parasympathetic impulses to the heart have what effect? Name the parasympathetic nerves to the heart. (p. 268)
14. State the locations of arterial pressoreceptors and chemoreceptors; what they detect, and their sensory nerves. (p. 269)
15. Describe the reflex arc to increase heart rate and force when blood pressure to the brain decreases. (p. 270)

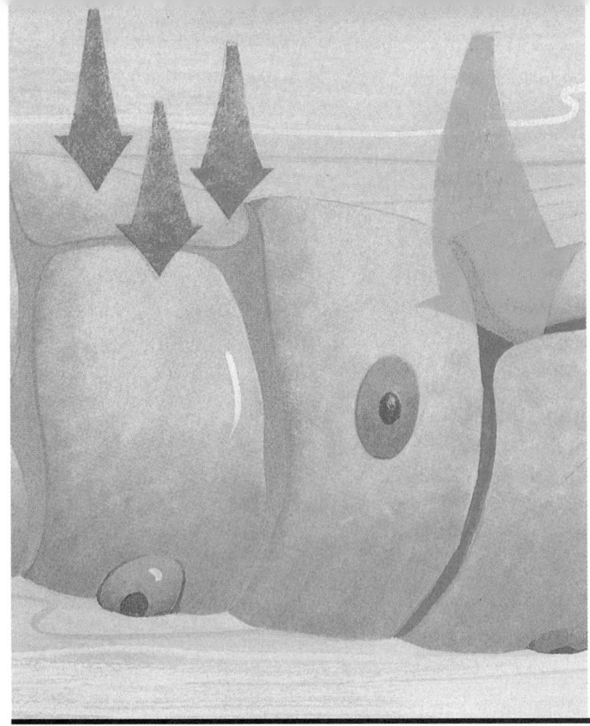

CHAPTER 13

Chapter Outline

Student Objectives

- Describe the structure of arteries and veins, and relate their structure to function.
- Explain the purpose of arterial and venous anastomoses.
- Describe the structure of capillaries, and explain the exchange processes that take place in capillaries.
- Describe the pathway and purpose of pulmonary circulation.
- Name the branches of the aorta and their distributions.
- Name the major systemic veins, and the parts of the body they drain of blood.
- Describe the pathway and purpose of hepatic portal circulation.
- Describe the modifications of fetal circulation, and explain the purpose of each.
- Explain the importance of slow blood flow in capillaries.
- Define blood pressure, and state the normal ranges for systemic and pulmonary blood pressure.
- Explain the factors that maintain systemic blood pressure.
- Explain how the heart and kidneys are involved in the regulation of blood pressure.
- Explain how the medulla and the autonomic nervous system regulate the diameter of blood vessels.

The Vascular System

New Terminology

Anastomosis (a-NAS-ti-**MOH**-sis)
Arteriole (ar-**TIR**-ee-ohl)
Circle of Willis (**SIR**-kuhl of **WILL**-iss)
Ductus arteriosus (**DUK**-tus ar-TIR-ee-**OH**-sis)
Endothelium (EN-doh-**THEEL**-ee-um)
Foramen ovale (for-**RAY**-men oh-**VAHL**-ee)
Hepatic portal (hep-**PAT**-ik **POOR**-tuhl)
Peripheral resistance (puh-**RIFF**-uh-ruhl ree-**ZIS**-tense)
Placenta (pluh-**SEN**-tah)
Precapillary sphincter (pre-**KAP**-i-lar-ee **SFINK**-ter)
Sinusoid (**SIGH**-nuh-soyd)
Umbilical arteries (uhm-**BILL**-i-kull **AR**-tuh-rees)
Umbilical vein (uhm-**BILL**-i-kull VAIN)
Venule (**VEN**-yool)

Related Clinical Terminology

Anaphylactic (AN-uh-fi-**LAK**-tik)
Aneurysm (**AN**-yur-izm)
Arteriosclerosis (ar-TIR-ee-oh-skle-**ROH**-sis)
Hypertension (HIGH-per-**TEN**-shun)
Hypovolemic (HIGH-poh-voh-**LEEM**-ik)
Phlebitis (fle-**BY**-tis)
Pulse deficit (PULS **DEF**-i-sit)
Septic shock (**SEP**-tik SHAHK)
Varicose veins (**VAR**-i-kohs VAINS).

*Terms that appear in **bold type** in the chapter text are defined in the glossary, which begins on page 528.*

The role of blood vessels in the circulation of blood has been known since 1628, when William Harvey, an English anatomist, demonstrated that blood in veins always flowed toward the heart. Before that time, it was believed that blood was static or stationary, some of it within the vessels but the rest sort of in puddles throughout the body. Harvey showed that blood indeed does move, and only in the blood vessels. In the centuries that followed, the active (rather than merely passive) roles of the vascular system were discovered, and all contribute to homeostasis.

The vascular system consists of the arteries, capillaries, and veins through which the heart pumps blood throughout the body. As you will see, the major "business" of the vascular system, which is the exchange of materials between the blood and tissues, takes place in the capillaries. The arteries and veins, however, are just as important, transporting blood between the capillaries and the heart.

Another important topic of this chapter will be blood pressure (BP), which is the force the blood exerts against the walls of the vessels. Normal blood pressure is essential for circulation and for some of the material exchanges that take place in capillaries.

ARTERIES

Arteries carry blood from the heart to capillaries; smaller arteries are called **arterioles**. If we look at an artery in cross-section, we find three layers (or tunics) of tissues, each with different functions (Fig. 13-1).

The innermost layer, the **tunica intima**, is simple squamous epithelium called **endothelium**. This is the same type of tissue that forms the endocardium, the lining of the chambers of the heart. As you might guess, its function is also the same: Its extreme smoothness prevents abnormal blood clotting. The **tunica media**, or middle layer, is made of smooth muscle and elastic connective tissue. Both these tissues are involved in the maintenance of normal blood pressure, especially diastolic blood pressure when the heart is relaxed. Fibrous connective tissue forms the outer layer, the **tunica externa**. This tissue is very strong, which is important to prevent the rupture or bursting of the larger arteries that carry blood under high pressure (see Box 13-1: Disorders of Arteries).

The outer and middle layers of large arteries are quite thick. In the arterioles, only individual smooth muscle cells encircle the tunica intima. The smooth muscle layer enables arteries to constrict or dilate. This is regulated by the medulla and autonomic nervous system and will be discussed in a later section on blood pressure.

VEINS

Veins carry blood from capillaries back to the heart; the smaller veins are called **venules**. The same three tissue layers are present in veins as in the walls of arteries, but there are some differences when compared to the arterial layers. The inner layer of veins is smooth endothelium, but at intervals this lining is folded to form **valves** (see Fig. 13-1). Valves prevent backflow of blood and are most numerous in veins of the legs, where blood must often return to the heart against the force of gravity.

The middle layer of veins is a thin layer of smooth muscle. It is thin because veins do not regulate blood pressure and blood flow into capillaries as arteries do. Veins can constrict extensively, however, and this function becomes very important in certain situations such as severe hemorrhage. The outer layer of veins is also thin; not as much fibrous connective tissue is necessary because blood pressure in veins is very low.

ANASTOMOSES

An **anastomosis** is a connection, or joining, of vessels, that is, artery to artery or vein to vein. The general purpose of these connections is to provide alternate pathways for the flow of blood if one vessel becomes obstructed.

An arterial anastomosis helps ensure that blood will get to the capillaries of an organ to deliver oxygen and nutrients, and to remove waste products. There are arterial anastomoses, for example, between some of the coronary arteries that supply blood to the myocardium.

A venous anastomosis helps ensure that blood will be able to return to the heart in order to be pumped again. Venous anastomoses are most numerous among the veins of the legs, where the possibility of obstruction increases as a person gets older (see Box 13-2: Disorders of Veins).

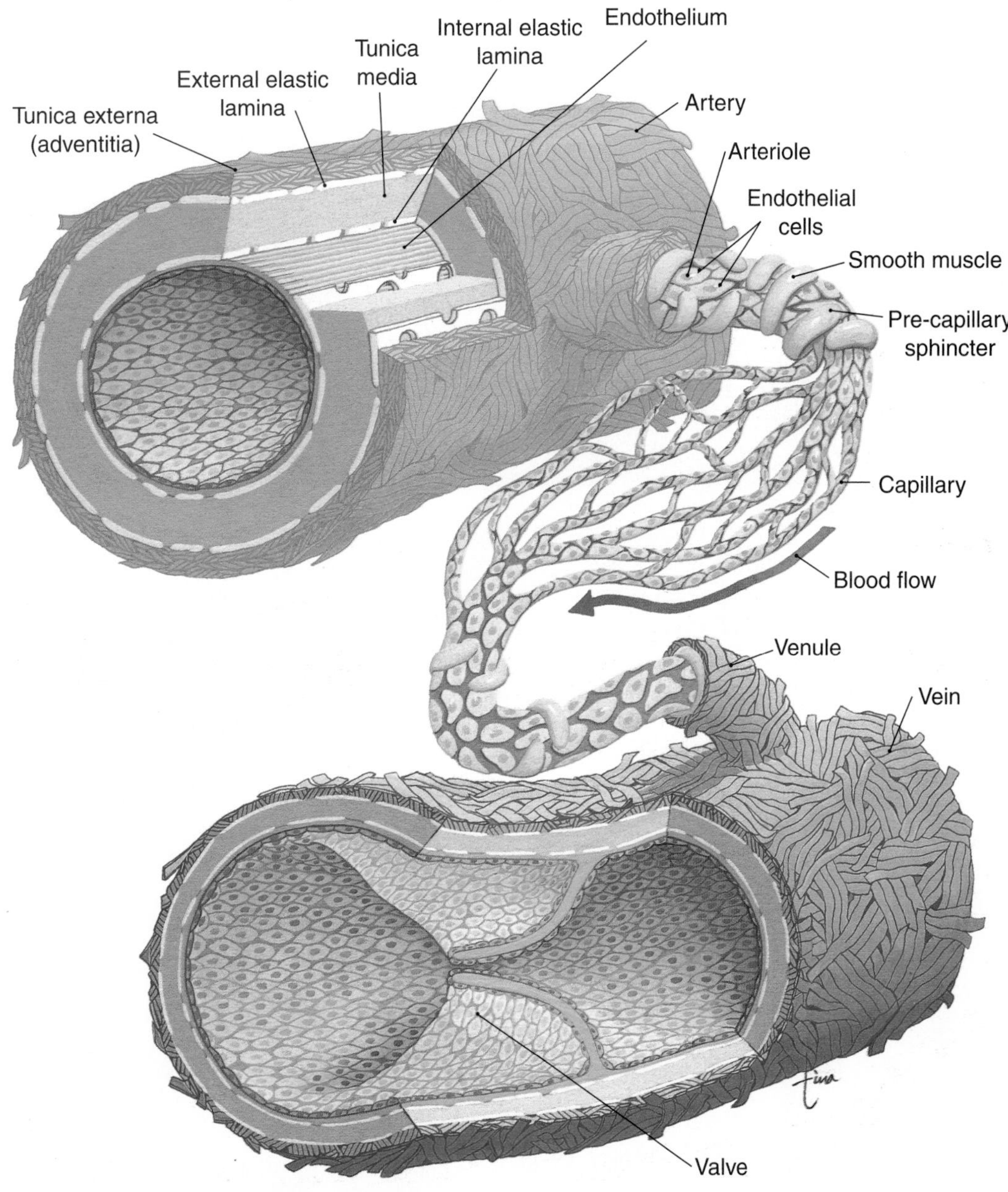

Figure 13–1 Structure of an artery, arteriole, capillary network, venule, and vein. See text for description.

CAPILLARIES

Capillaries carry blood from arterioles to venules. Their walls are only one cell in thickness; capillaries are actually the extension of just the lining of arteries and veins (see Fig. 13–1). Some tissues do not have capillaries; these are the epidermis, cartilage, and the lens and cornea of the eye.

Most tissues, however, have extensive capillary net-

Box 13–1 DISORDERS OF ARTERIES

Arteriosclerosis—although commonly called "hardening of the arteries," arteriosclerosis really means that the arteries lose their elasticity, and their walls become weakened. Arteries carry blood under high pressure, so deterioration of their walls is part of the aging process.

Aneurysm—a weak portion of an arterial wall may bulge out, forming a sac or bubble called an aneurysm. Arteriosclerosis is a possible cause, but some aneurysms are congenital. An aneurysm may be present for many years without any symptoms and may only be discovered during diagnostic procedures for some other purpose.

The most common sites for aneurysm formation are the cerebral arteries and the aorta. Rupture of a cerebral aneurysm is a possible cause of a cerebrovascular accident (CVA). Rupture of an aortic aneurysm is life-threatening and requires immediate corrective surgery. The damaged portion of the artery is removed and replaced with a graft. Such surgery may also be performed when an aneurysm is found before it ruptures.

Atherosclerosis—this condition has been mentioned previously; see Chapter 12.

works. Blood flow into these networks is regulated by smooth muscle cells called **precapillary sphincters**, found at the beginning of each network (see Fig. 13-1). Precapillary sphincters are not regulated by the nervous system but rather constrict or dilate depending on the needs of the tissues. Because there is not enough blood in the body to fill all the capillaries at once, precapillary sphincters are usually slightly constricted. In an active tissue that requires more oxygen, such as exercising muscle, the precapillary sphincters dilate to increase blood flow. These automatic responses ensure that blood, the volume of which is constant, will circulate where it is needed most.

Some organs have another type of capillary called

Box 13–2 DISORDERS OF VEINS

Phlebitis—inflammation of a vein. This condition is most common in the veins of the legs because they are subjected to great pressure as the blood is returned to the heart against the force of gravity. Often no specific cause can be determined, but advancing age, obesity, and blood disorders may be predisposing factors.

If a superficial vein is affected, the area may be tender or painful, but blood flow is usually maintained because there are so many anastomoses among these veins. Deep vein phlebitis is potentially more serious, with the possibility of clot formation (thrombophlebitis) and subsequent dislodging of the clot to form an embolism.

Varicose Veins—swollen and distended veins that occur most often in the superficial veins of the legs. This condition may develop in people who must sit or stand in one place for long periods of time. Without contraction of the leg muscles, blood tends to pool in the leg veins, stretching their walls. If the veins become overly stretched, the valves within them no longer close properly. These incompetent valves no longer prevent backflow of blood, leading to further pooling and even further stretching of the walls of the veins. Varicose veins may cause discomfort and cramping in the legs, or become even more painful. Severe varicosities may be removed surgically.

This condition may also develop during pregnancy, when the enlarged uterus presses against the iliac veins and slows blood flow into the inferior vena cava. Varicose veins of the anal canal are called **hemorrhoids**, which may also be a result of pregnancy or of chronic constipation and straining to defecate. Hemorrhoids that cause discomfort or pain may also be removed surgically. The recent developments in laser surgery have made this a simpler procedure than it was in the past.

sinusoids, which are larger and more permeable than are other capillaries. The permeability of sinusoids permits large substances such as proteins and blood cells to enter or leave the blood. Sinusoids are found in hemopoietic tissues such as the red bone marrow and spleen and in organs such as the liver and pituitary gland, which produce and secrete proteins into the blood.

EXCHANGES IN CAPILLARIES

Capillaries are the sites of exchanges of materials between the blood and the tissue fluid surrounding cells. Some of these substances move from the blood to tissue fluid, and others move from tissue fluid to the blood. The processes by which these substances are exchanged are illustrated in Fig. 13–2.

Gases move by **diffusion**, that is, from their area of greater concentration to their area of lesser concentration. Oxygen, therefore, diffuses from the blood in systemic capillaries to the tissue fluid, and carbon dioxide diffuses from tissue fluid to the blood to be brought to the lungs and exhaled.

Let us now look at the blood pressure as blood enters capillaries from the arterioles. Blood pressure here is about 30 to 35 mmHg, and the pressure of the surrounding tissue fluid is much lower, about 2 mmHg. Because the capillary blood pressure is higher, the process of **filtration** occurs, which forces plasma and dissolved nutrients out of the capillaries and into

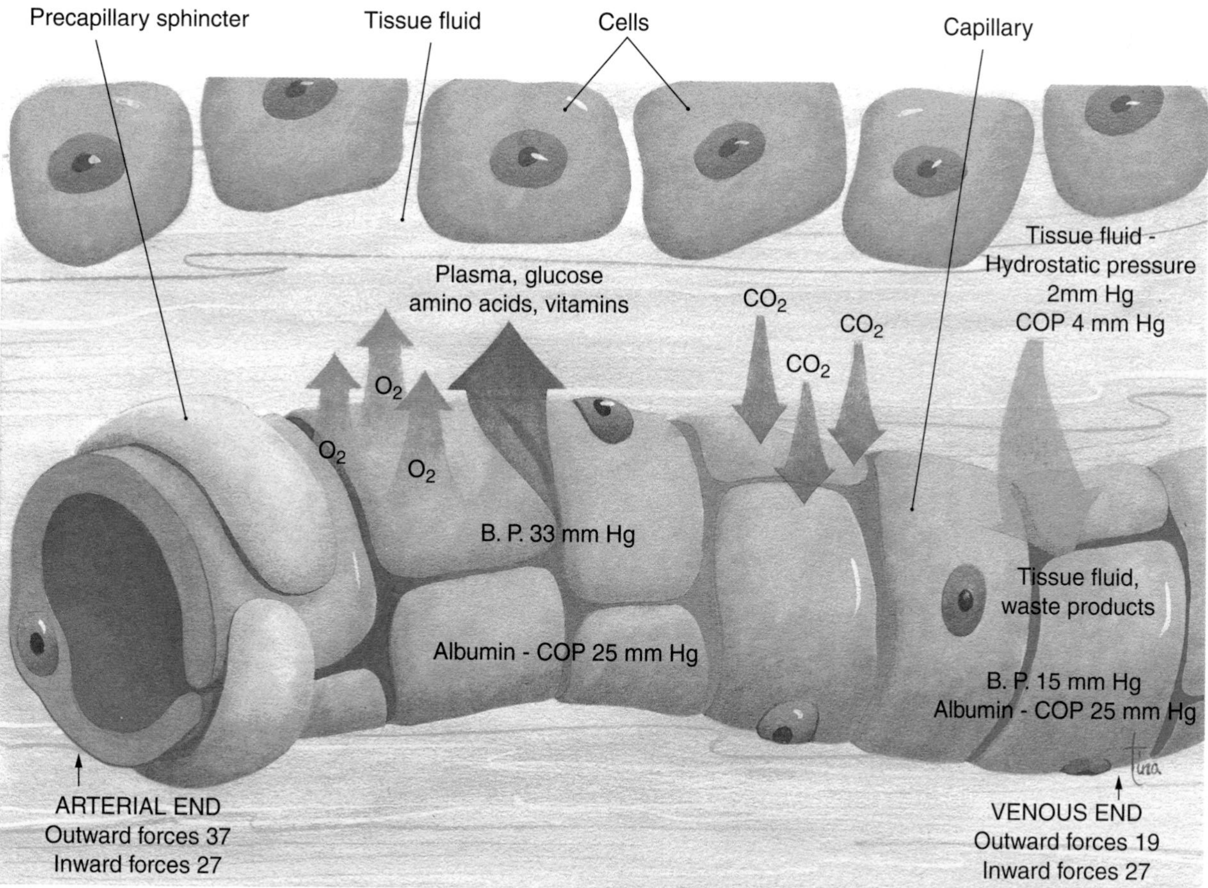

Figure 13–2 Exchanges between blood in a systemic capillary and the surrounding tissue fluid. Arrows depict the direction of movement. Filtration takes place at the arterial end of the capillary. Osmosis takes place at the venous end. Gases are exchanged by diffusion.

tissue fluid. This is how nutrients such as glucose, amino acids, and vitamins are brought to cells.

Blood pressure decreases as blood reaches the venous end of capillaries, but notice that proteins such as albumin have remained in the blood. Albumin contributes to the **colloid osmotic pressure** of blood; this is a "pulling" rather than a "pushing" pressure. At the venous end of capillaries, the presence of albumin in the blood pulls tissue fluid into the capillaries, which also brings into the blood the waste products produced by cells. The tissue fluid that returns to the blood also helps maintain normal blood volume and blood pressure.

The amount of tissue fluid formed is slightly greater than the amount returned to the capillaries. If this were to continue, blood volume would be gradually depleted. The excess tissue fluid, however, enters lymph capillaries. Now called lymph, it will be returned to the blood to be recycled again as plasma, thus maintaining blood volume. This is discussed further in Chapter 14.

PATHWAYS OF CIRCULATION

The two major pathways of circulation are pulmonary and systemic. Pulmonary circulation begins at the right ventricle, and systemic circulation begins at the left ventricle. Hepatic portal circulation is a special segment of systemic circulation that will be covered separately. Fetal circulation involves pathways that are present only before birth and will also be discussed separately.

PULMONARY CIRCULATION

The right ventricle pumps blood into the pulmonary artery (or trunk), which divides into the right and left pulmonary arteries, one to each lung. Within the lungs each artery branches extensively into smaller arteries and arterioles, then to capillaries. The pulmonary capillaries surround the alveoli of the lungs; it is here that exchanges of oxygen and carbon dioxide take place. The capillaries unite to form venules, which merge into veins, and finally into the two pulmonary veins from each lung that return blood to the left atrium. This oxygenated blood will then travel through the systemic circulation. (Notice that the pulmonary veins contain oxygenated blood; these are the only veins that carry blood with a high oxygen content. The blood in systemic veins has a low oxygen content; it is systemic arteries that carry oxygenated blood.)

SYSTEMIC CIRCULATION

The left ventricle pumps blood into the aorta, the largest artery of the body. We will return to the aorta and its branches in a moment, but first we will summarize the rest of systemic circulation. The branches of the aorta take blood into arterioles and capillary networks throughout the body. Capillaries merge to form venules and veins. The veins from the lower body take blood to the inferior vena cava; veins from the upper body take blood to the superior vena cava. These two caval veins return blood to the right atrium. The major arteries and veins are shown in Figs. 13–3 to 13–5, and their functions are listed in Tables 13–1 and 13–2.

The aorta is a continuous vessel, but for the sake of precise description is divided into sections that are named anatomically: ascending aorta, aortic arch, thoracic aorta, and abdominal aorta. The ascending aorta is the first inch that emerges from the top of the left ventricle. The arch of the aorta curves posteriorly over the heart and turns downward. The thoracic aorta continues down through the chest cavity and through the diaphragm. Below the level of the diaphragm, the abdominal aorta continues to the level of the 4th lumbar vertebra, where it divides into the two common iliac arteries. Along its course, the aorta has many branches through which blood travels to specific organs and parts of the body.

The ascending aorta has only two branches: the right and left coronary arteries, which supply blood to the myocardium. This pathway of circulation was described previously in Chapter 12.

The aortic arch has three branches that supply blood to the head and arms: the brachiocephalic artery, left common carotid artery, and left subclavian artery. The brachiocephalic (literally: "arm-head") artery is very short and divides into the right common carotid artery and right subclavian artery. The right and left common carotid arteries extend into the neck, where each divides into an internal carotid artery and external carotid artery, which supply the head. The right and left subclavian arteries are in the shoulders behind the clavicles and continue into the arms. The branches of the carotid and subclavian arteries are di-

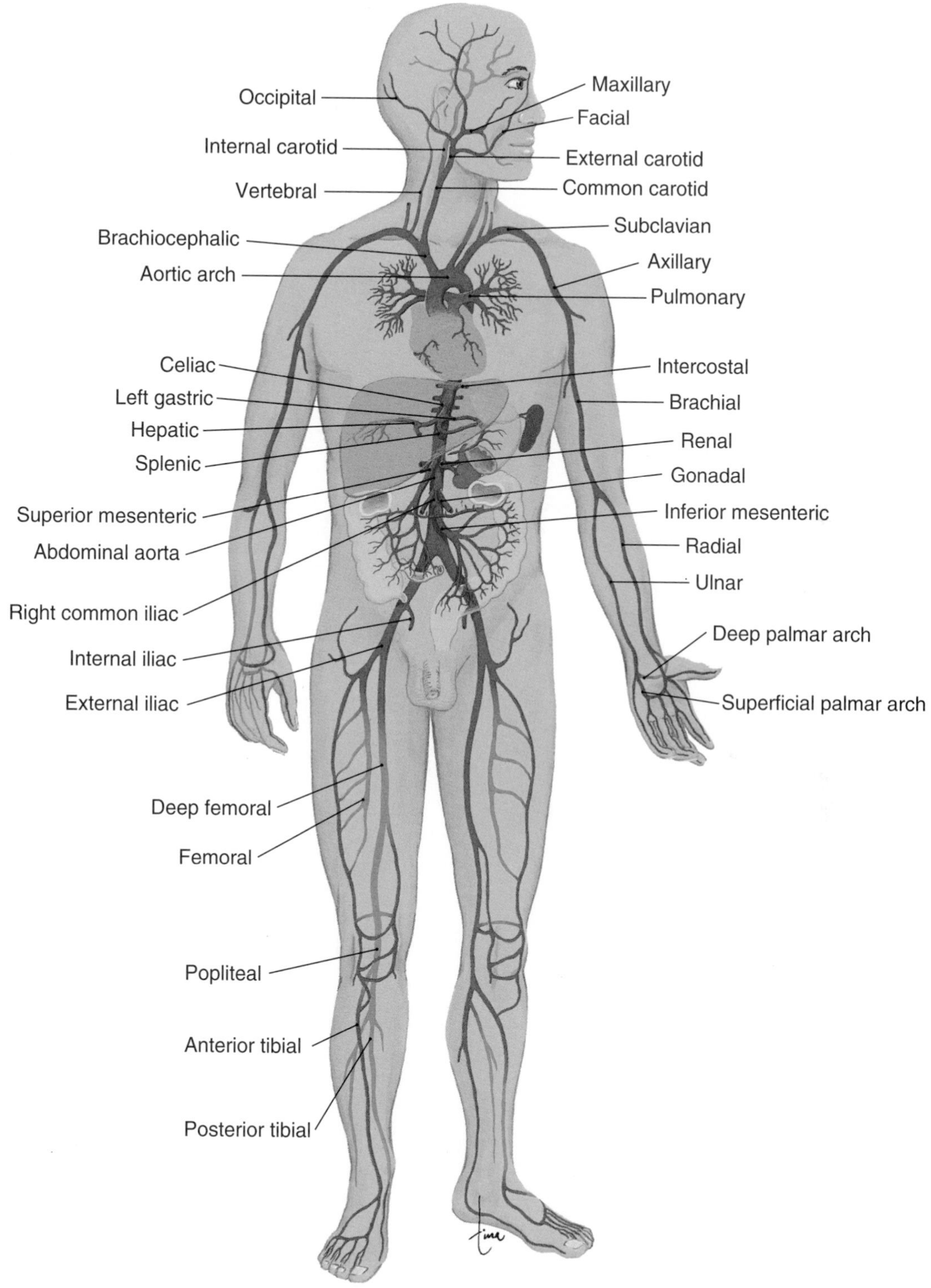

Figure 13–3 Systemic arteries. The aorta and its major branches are shown in anterior view.

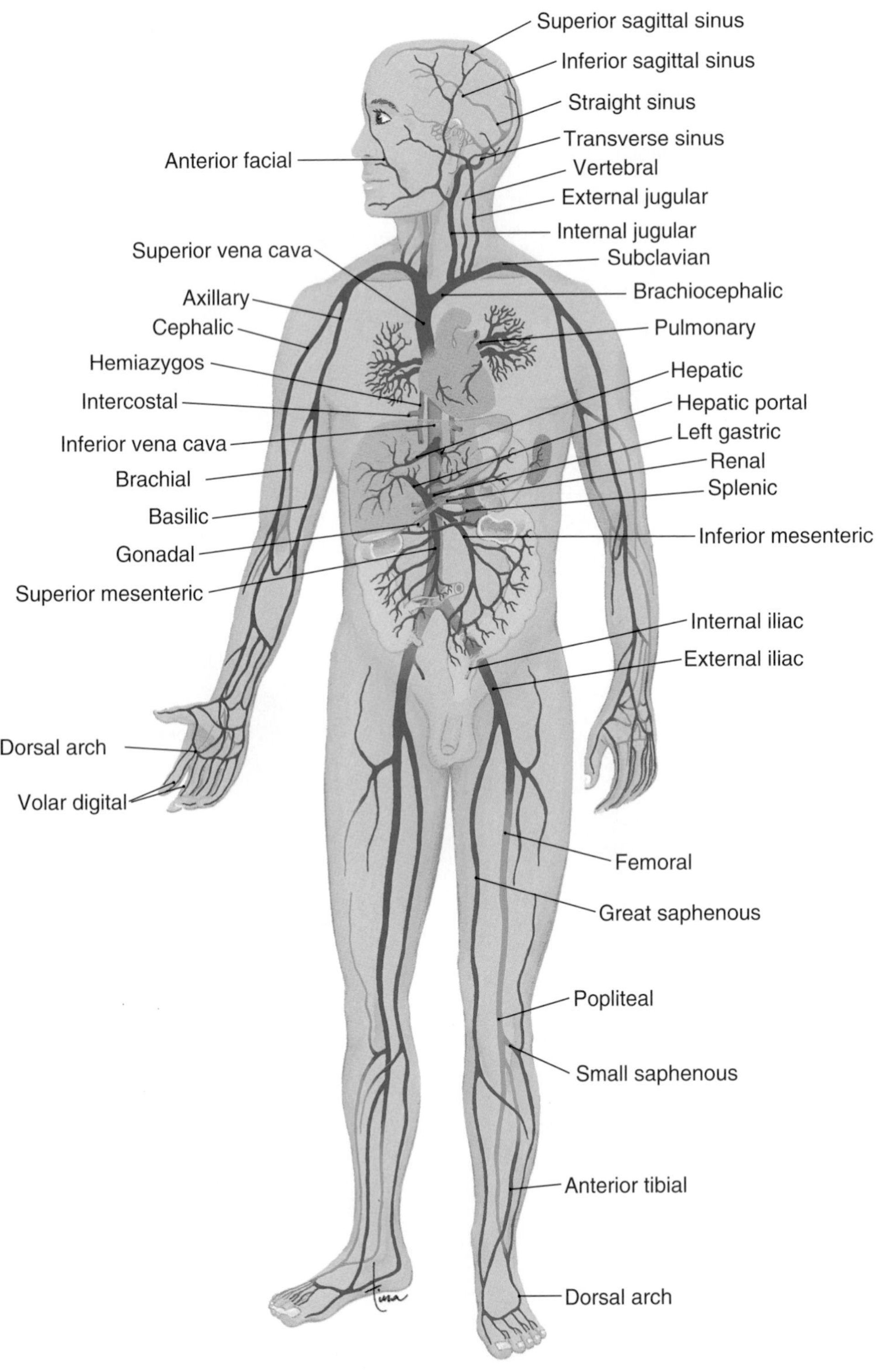

Figure 13–4 Systemic veins shown in anterior view.

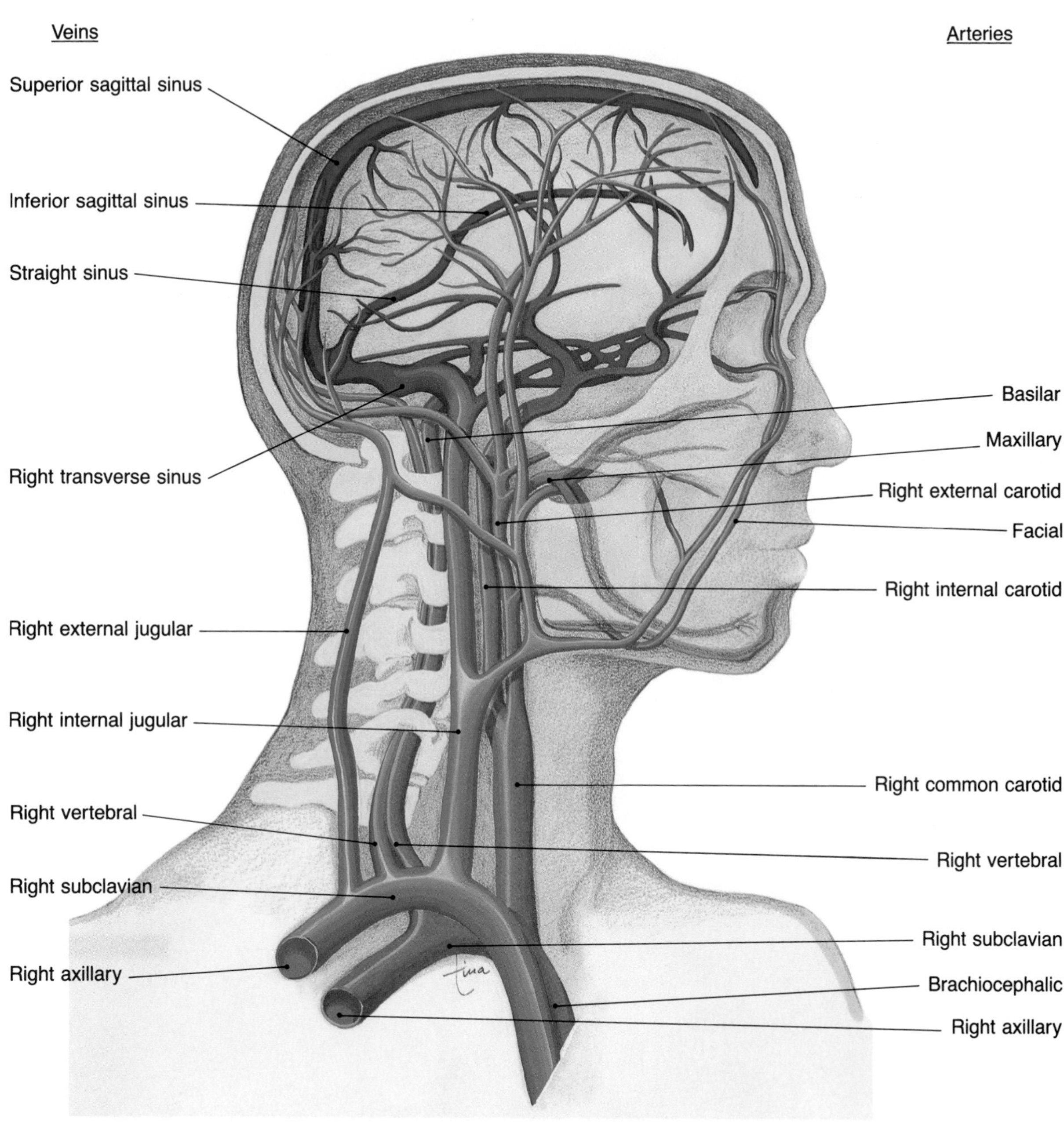

Figure 13–5 Arteries and veins of the head and neck shown in right lateral view. Veins are labeled on the left. Arteries are labeled on the right.

Table 13–1 MAJOR SYSTEMIC ARTERIES

A. Branches of the Ascending Aorta and Aortic Arch

Artery	Branch of	Region Supplied
Coronary a.	Ascending aorta	• Myocardium
Brachiocephalic a.	Aortic arch	• Right arm and head
Right common carotid a.	Brachiocephalic a.	• Right side of head
Right subclavian a.	Brachiocephalic a.	• Right shoulder and arm
Left common carotid a.	Aortic arch	• Left side of head
Left subclavian a.	Aortic arch	• Left shoulder and arm
External carotid a.	Common carotid a.	• Superficial head
Superficial temporal a.	External carotid a.	• Scalp
Internal carotid a.	Common carotid a.	• Brain (circle of Willis)
Ophthalamic a.	Internal carotid a.	• Eye
Vertebral a.	Subclavian a.	• Cervical vertebrae and circle of Willis
Axillary a.	Subclavian a.	• Armpit
Brachial a.	Axillary a.	• Upper arm
Radial a.	Brachial a.	• Forearm
Ulnar a.	Brachial a.	• Forearm
Volar arch	Radial and Ulnar a.	• Hand

B. Branches of the Thoracic Aorta

Artery	Region Supplied
Intercostal a. (9 pairs)	• Skin, muscles, bones of trunk
Superior phrenic a.	• Diaphragm
Pericardial a.	• Pericardium
Esophageal a.	• Esophagus
Bronchial a.	• Bronchioles and connective tissue of the lungs

C. Branches of the Abdominal Aorta

Artery	Region Supplied
Inferior phrenic a.	• Diaphragm
Lumbar a.	• Lumbar area of back
Middle sacral a.	• Sacrum, coccyx, buttocks
Celiac a.	• (see branches)
Hepatic a.	• Liver
Left gastric a.	• Stomach
Splenic a.	• Spleen, pancreas
Superior mesenteric a.	• Small intestine, part of colon
Suprarenal a.	• Adrenal glands
Renal a.	• Kidneys
Inferior mesenteric a.	• Most of colon and rectum
Testicular or ovarian a.	• Testes or ovaries

Table 13–1 MAJOR SYSTEMIC ARTERIES (Continued)

C. Branches of the Abdominal Aorta (Continued)	
Artery	**Region Supplied**
Common iliac a.	• The two large vessels that receive blood from the Abdominal Aorta; each branches as follows below:
Internal iliac a.	• Bladder, rectum, reproductive organs
External iliac a.	• Lower pelvis to leg
Femoral a.	• Thigh
Popliteal a.	• Back of knee
Anterior tibial a.	• Front of lower leg
Dorsalis pedis	• Top of ankle and foot
Plantar arches	• Foot
Posterior tibial a.	• Back of lower leg
Peroneal a.	• Medial lower leg
Plantar arches	• Foot

agrammed in Figs. 13–3 and 13–5. As you look at these diagrams, keep in mind that the name of the vessel often tells us where it is. The ulnar artery, for example, is found in the forearm along the ulna.

Some of these vessels contribute to an important arterial anastomosis, the **circle of Willis** (or cerebral arterial circle), which is a "circle" of arteries around the pituitary gland (Fig. 13–6). The circle of Willis is formed by the right and left internal carotid arteries and the basilar artery, which is the union of the right and left vertebral arteries (branches of the subclavians). The brain must have a constant flow of blood to supply oxygen and remove waste products, and there are four vessels that bring blood to the circle of Willis. From this anastomosis, several paired arteries extend into the brain itself.

The thoracic aorta and its branches supply the chest wall and the organs within the thoracic cavity. These vessels are listed in Table 13–1.

The abdominal aorta gives rise to arteries that supply the abdominal wall and organs and to the common iliac arteries, which continue into the legs. These vessels are also listed in Table 13–1 (see Box 13–3: Pulse Sites).

The systemic veins drain blood from organs or parts of the body and often parallel their corresponding arteries. The most important veins are diagrammed in Fig. 13–4 and listed in Table 13–2.

HEPATIC PORTAL CIRCULATION

Hepatic portal circulation is a subdivision of systemic circulation in which blood from the abdominal digestive organs and spleen circulates through the liver before returning to the heart.

Blood from the capillaries of the stomach, small intestine, colon, pancreas, and spleen flows into two large veins, the superior mesenteric vein and the splenic vein, which unite to form the portal vein (Fig. 13–7). The portal vein takes blood into the liver, where it branches extensively and empties blood into the sinusoids, the capillaries of the liver (see also Fig. 16–6). From the sinusoids, blood flows into hepatic veins, to the inferior vena cava and back to the right atrium. Notice that in this pathway there are two sets of capillaries, and keep in mind that it is in capillaries that exchanges take place. Let us use some specific examples to show the purpose and importance of portal circulation.

Glucose from carbohydrate digestion is absorbed into the capillaries of the small intestine; after a big meal this may greatly increase the blood glucose level. If this blood were to go directly back to the heart and then circulate through the kidneys, some of the glucose might be lost in urine. However, blood from the small intestine passes first through the liver sinusoids, and the liver cells remove the excess glucose and store it as glycogen. The blood that returns to the heart will then have a blood glucose level in the normal range.

Another example: Alcohol is absorbed into the capillaries of the stomach. If it were to circulate directly throughout the body, the alcohol would rapidly impair the functioning of the brain. Portal circulation, however, takes blood from the stomach to the liver, the organ that can detoxify the alcohol and prevent its detrimental effects on the brain. Of course, if al-

Table 13–2 MAJOR SYSTEMIC VEINS

Vein	Vein Joined	Region Drained
Head and Neck		
Cranial venous sinuses	Internal jugular v.	• Brain, including reabsorbed CSF
Internal jugular v.	Brachiocephalic v.	• Face and neck
External jugular v.	Subclavian v.	• Superficial face and neck
Subclavian v.	Brachiocephalic v.	• Shoulder
Brachiocephalic v.	Superior vena cava	• Upper body
Superior vena cava	Right atrium	• Upper body
Arm and Shoulder		
Radial v.	Brachial v.	• Forearm and hand
Ulnar v.	Brachial v.	• Forearm and hand
Cephalic v.	Axillary v.	• Superficial arm and forearm
Basilic v.	Axillary v.	• Superficial upper arm
Brachial v.	Axillary v.	• Upper arm
Axillary v.	Subclavian v.	• Armpit
Subclavian v.	Brachiocephalic v.	• Shoulder
Trunk		
Brachiocephalic v.	Superior vena cava	• Upper body
Azygos v.	Superior vena cava	• Deep structures of chest and abdomen; links inferior vena cava to superior vena cava
Hepatic v.	Inferior vena cava	• Liver
Renal v.	Inferior vena cava	• Kidney
Testicular or Ovarian v.	Inferior vena cava and left renal v.	• Testes or ovaries
Internal iliac v.	Common iliac v.	• Rectum, bladder, reproductive organs
External iliac v.	Common iliac v.	• Leg and abdominal wall
Common iliac v.	Inferior vena cava	• Leg and lower abdomen
Leg		
Anterior and posterior tibial v.	Popliteal v.	• Lower leg and foot
Popliteal v.	Femoral v.	• Knee
Small saphenous v.	Popliteal v.	• Superficial leg and foot
Great saphenous v.	Femoral v.	• Superficial foot, leg, and thigh
Femoral v.	External iliac v.	• Thigh
External iliac v.	Common iliac v.	• Leg and abdominal wall
Common iliac v.	Inferior vena cava	• Leg and lower abdomen
Inferior vena cava	Right atrium	• Lower body

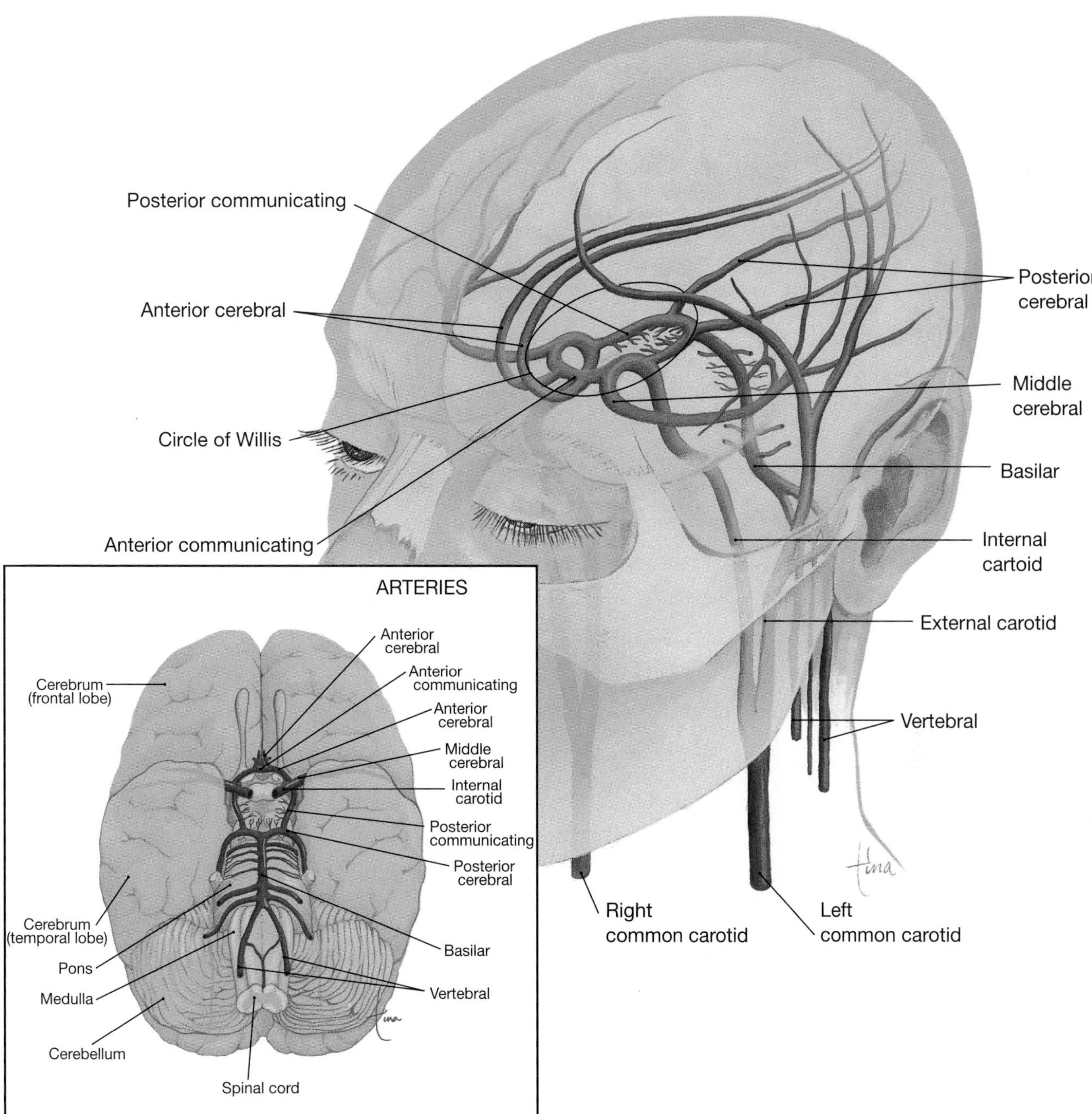

Figure 13–6 Circle of Willis. This anastomosis is formed by the following arteries: internal carotid, anterior communicating, posterior communicating, and basilar. The cerebral arteries extend from the circle of Willis into the brain. The box shows these vessels in an inferior view of the brain.

Box 13–3 PULSE SITES

A pulse is the heartbeat that is felt at an arterial site. What is felt is not actually the force exerted by the blood, but the force of ventricular contraction transmitted through the walls of the arteries. This is why pulses are not felt in veins; they are too far from the heart for the force to be detectable.

The most commonly used pulse sites are:

Radial—the radial artery on the thumb side of the wrist.
Carotid—the carotid artery lateral to the larynx in the neck.
Temporal—the temporal artery just in front of the ear.
Femoral—the femoral artery at the top of the thigh.
Popliteal—the popliteal artery at the back of the knee.
Dorsalis pedis—the dorsalis pedis artery on the top of the foot (commonly called the pedal pulse).

Pulse rate is, of course, the heart rate. However, if the heart is beating weakly, a radial pulse may be lower than an **apical pulse** (listening to the heart itself with a stethoscope). This is called a **pulse deficit** and indicates heart disease of some kind.

When taking a pulse, the careful observer also notes the rhythm and force of the pulse. Abnormal rhythms may reflect cardiac arrhythmias, and the force of the pulse (strong or weak) is helpful in assessing the general condition of the heart and arteries.

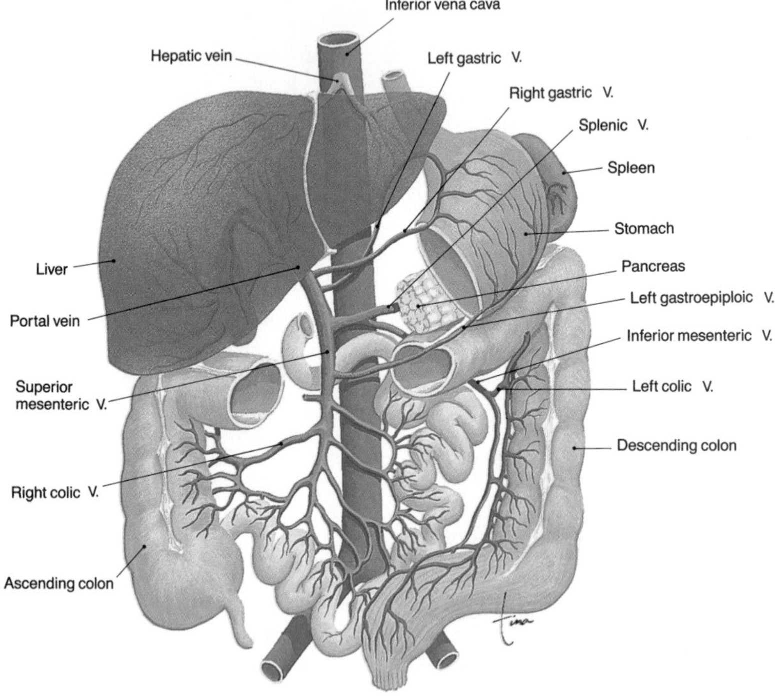

Figure 13–7 Hepatic portal circulation. Portions of some of the digestive organs have been removed to show the veins that unite to form the portal vein.

cohol consumption continues, the blood alcohol level rises faster than the liver's capacity to detoxify, and the well-known signs of alcohol intoxication appear.

As you can see, this portal circulation pathway enables the liver to modify the blood from the digestive organs and spleen. Some nutrients may be stored or changed, bilirubin from the spleen is excreted into bile, and potential poisons are detoxified before the blood returns to the heart and the rest of the body.

FETAL CIRCULATION

The fetus depends upon the mother for oxygen and nutrients and for the removal of carbon dioxide and other waste products. The site of exchange between fetus and mother is the **placenta**, which contains fetal and maternal blood vessels that are very close to one another (see Figs. 13–8 and 21–5). The blood of the fetus does not mix with the blood of the mother; substances are exchanged by diffusion and active transport mechanisms.

The fetus is connected to the placenta by the umbilical cord, which contains two umbilical arteries and one umbilical vein (see Fig. 13–8). The **umbilical arteries** are branches of the fetal internal iliac arteries; they carry blood from the fetus to the placenta. In the placenta, carbon dioxide and waste products in the fetal blood enter maternal circulation, and oxygen and nutrients from the mother's blood enter fetal circulation.

The **umbilical vein** carries this oxygenated blood from the placenta to the fetus. Within the body of the fetus, the umbilical vein branches: One branch takes some blood to the fetal liver, but most of the blood passes through the **ductus venosus** to the inferior vena cava, to the right atrium. After birth, when the umbilical cord is cut, the remnants of these fetal vessels constrict and become nonfunctional.

The other modifications of fetal circulation concern the fetal heart and large arteries (also shown in Fig. 13–8). Because the fetal lungs are deflated and do not provide for gas exchange, blood is shunted away from the lungs and to the body. The **foramen ovale** is an opening in the interatrial septum that permits some blood to flow from the right atrium to the left atrium, not, as usual, to the right ventricle. The blood that does enter the right ventricle is pumped into the pulmonary artery. The **ductus arteriosus** is a short vessel that diverts most of the blood in the pulmonary artery to the aorta, to the body. Both the foramen ovale and the ductus arteriosus permit blood to bypass the fetal lungs.

Just after birth, the baby breathes and expands its lungs, which pulls more blood into the pulmonary circulation. More blood then returns to the left atrium, and a flap on the left side of the foramen ovale is closed. The ductus arteriosus constricts, probably in response to the higher oxygen content of the blood, and pulmonary circulation becomes fully functional within a few days.

VELOCITY OF BLOOD FLOW

The velocity, or speed, with which blood flows differs in the various parts of the vascular system. Velocity is inversely related (meaning as one value goes up, the other goes down) to the cross-sectional area of the particular segment of the vascular system. Refer to Figure 13–9 as you read the following. The aorta receives all the blood from the left ventricle, its cross-sectional area is small, about 3 cm^2 (1 sq. inch), and the blood moves very rapidly, at least 30 cm per second (about 12 inches). Each time the aorta or any artery branches, the total cross-sectional area becomes larger, and the speed of blood flow decreases. Think of a river that begins in a narrow bed and is flowing rapidly. If the river bed widens, the water spreads out to fill it and flows more slowly. If the river were to narrow again, the water would flow faster. This is just what happens in the vascular system.

The capillaries in total have the greatest cross-sectional area, and blood velocity there is slowest, less than 0.1 cm per second. When capillaries unite to form venules, and then veins, the cross-sectional area decreases and blood flow speeds up.

Recall that it is in capillary networks that exchanges of nutrients, wastes, and gases take place between the blood and tissue fluid. The slow rate of blood flow in capillaries permits sufficient time for these essential exchanges. Think of a train slowing down (not actually stopping) at stations to allow people to jump on and off, then speeding up again to get to the next station. The capillaries are the "stations" of the vascular system.

The more rapid blood velocity in other vessels makes circulation time quite short. This is the time it

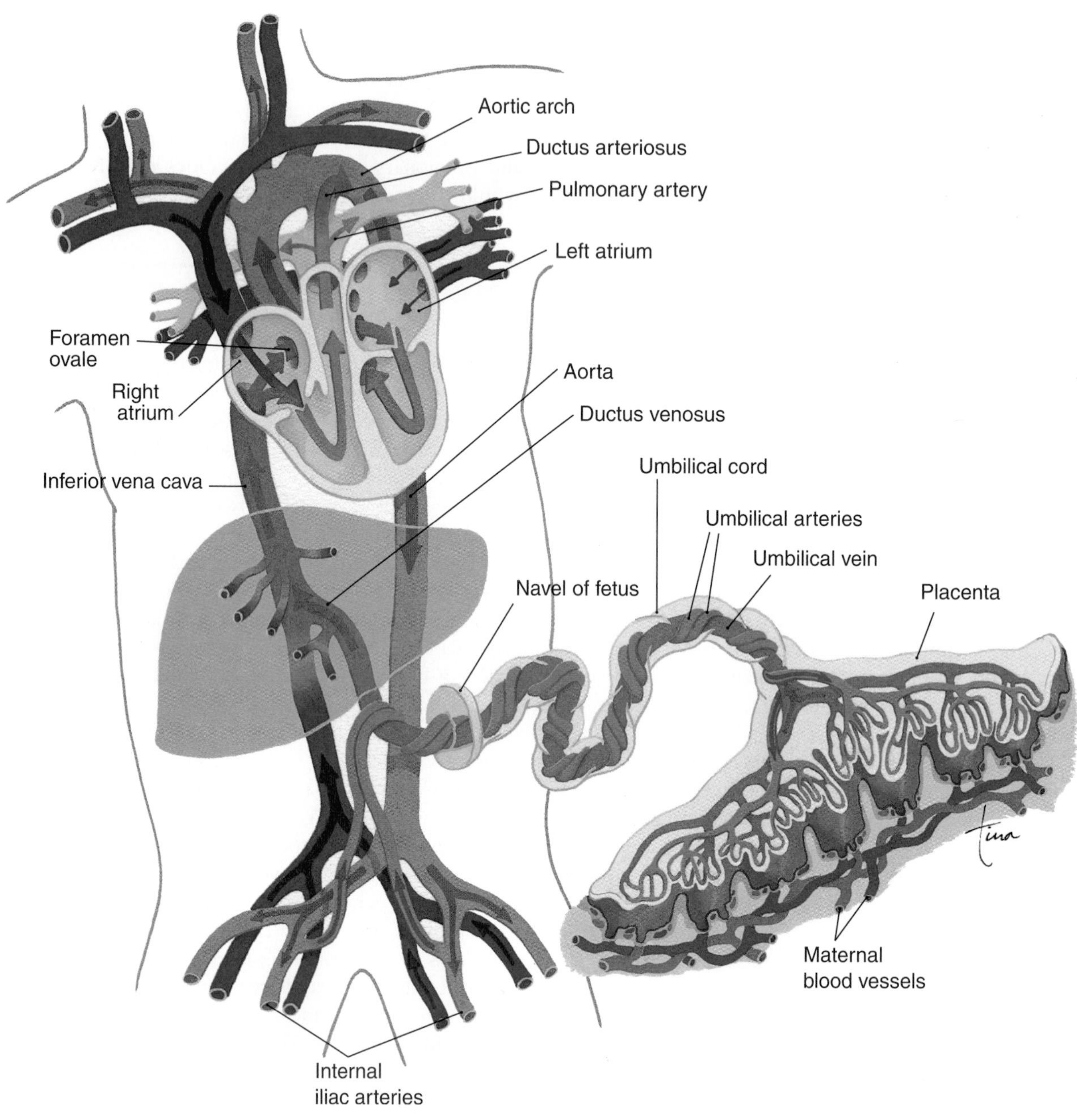

Figure 13–8 Fetal circulation. Fetal heart and blood vessels are shown on the left. Arrows depict the direction of blood flow. The placenta and umbilical blood vessels are shown on the right.

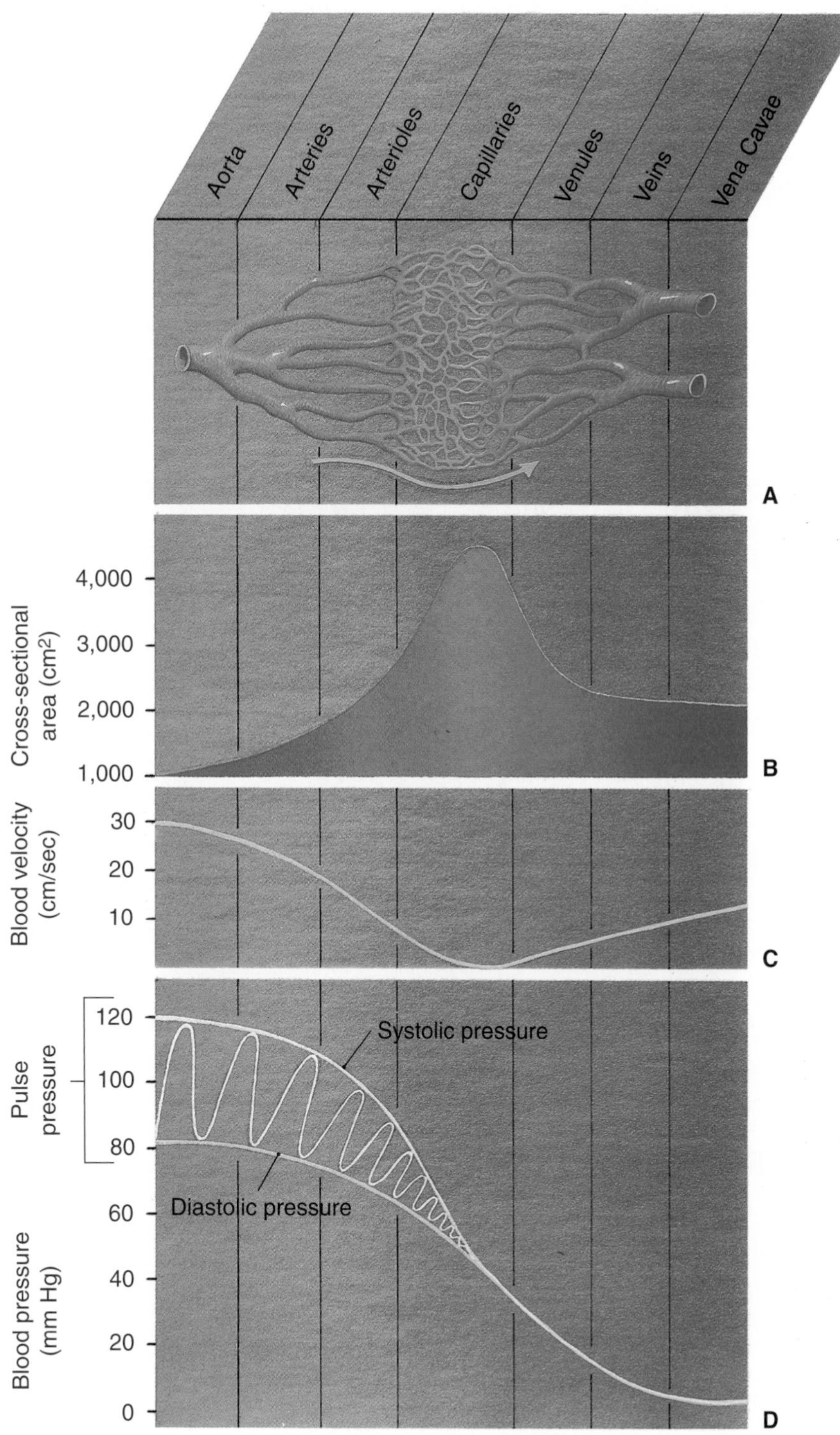

Figure 13–9 Characteristics of the vascular system. **(A)** Schematic of the branching of vessels. **(B)** Cross-sectional area in square centimeters. **(C)** Blood velocity in centimeters per second. **(D)** Systemic blood pressure changes. Notice that systolic and diastolic pressures become one pressure in the capillaries.

takes for blood to go from the right ventricle to the lungs, back to the heart to be pumped by the left ventricle to the body, and return to the heart again. Circulation time is about 1 minute or less, and ensures an adequate exchange of gases.

BLOOD PRESSURE

Blood pressure is the force the blood exerts against the walls of the blood vessels. Filtration in capillaries depends upon blood pressure; filtration brings nutrients to tissues, and as you will see in Chapter 18, is the first step in the formation of urine. Blood pressure is one of the "vital signs" often measured, and indeed a normal blood pressure is essential to life.

The pumping of the ventricles creates blood pressure, which is measured in mmHg (millimeters of mercury). When a systemic blood pressure reading is taken, two numbers are obtained: systolic and diastolic, as in 110/70 mmHg. **Systolic** pressure is always the higher of the two and represents the blood pressure when the left ventricle is contracting. The lower number is the **diastolic** pressure, when the left ventricle is relaxed and does not exert force. Diastolic pressure is maintained by the arteries and arterioles and is discussed in a later section.

Systemic blood pressure is highest in the aorta, which receives all the blood pumped by the left ven-

Box 13–4 HYPERTENSION

Hypertension is high blood pressure, that is, a resting systemic pressure consistently above 140/90. The term "essential hypertension" means that no specific cause can be determined; most cases are in this category. For some people, however, an overproduction of renin by the kidneys is the cause of their hypertension. Excess renin increases the production of angiotensin II, which raises blood pressure. Although hypertension often produces no symptoms, the long-term consequences may be very serious. Chronic hypertension has its greatest effects on the arteries and on the heart.

Although the walls of arteries are strong, hypertension weakens them and contributes to arteriosclerosis. Such weakened arteries may rupture or develop aneurysms, which may in turn lead to a CVA or kidney damage.

Hypertension affects the heart because the left ventricle must now pump blood against the higher arterial pressure. The left ventricle works harder and, like any other muscle, enlarges as more work is demanded; this is called **left ventricular hypertrophy**. This abnormal growth of the myocardium, however, is not accompanied by a corresponding growth in coronary capillaries, and the blood supply of the left ventricle may not be adequate for all situations. Exercise, for example, puts further demands on the heart, and the person may experience angina due to a lack of oxygen or a myocardial infarction if there is a severe oxygen deficiency.

Although there are several different kinds of medications used to treat hypertension, people with moderate hypertension may limit their dependence on medications by following certain guidelines.

1. *Don't smoke,* because nicotine stimulates vasoconstriction, which raises BP. Smoking also damages arteries, contributing to arteriosclerosis.
2. *Lose weight* if overweight. A weight loss of as little as 10 pounds can lower BP. A diet high in fruits and vegetables may, for some people, contribute to lower BP.
3. *Cut salt intake* in half. Although salt consumption may not be the *cause* of hypertension, reducing salt intake may help lower blood pressure by decreasing blood volume.
4. *Exercise* on a regular basis. A moderate amount of aerobic exercise (such as a half-hour walk every day) is beneficial for the entire cardiovascular system and may also contribute to weight loss.

tricle. As blood travels farther away from the heart, blood pressure decreases (see Fig. 13-9). The brachial artery is most often used to take a blood pressure reading; here a normal systolic range is 90 to 135 mmHg, and a normal diastolic range is 60 to 85 mmHg. In the arterioles, blood pressure decreases further, and systolic and diastolic pressures merge into one pressure. At the arterial end of capillary networks, blood pressure is about 30 to 35 mmHg, decreasing to 12 to 15 mmHg at the venous end of capillaries. This is high enough to permit filtration but low enough to prevent rupture of the capillaries. As blood flows through veins, the pressure decreases further, and in the caval veins, blood pressure approaches zero as blood enters the right atrium (see also Box 13-4: Hypertension).

Pulmonary blood pressure is created by the right ventricle, which has relatively thin walls and thus exerts about one-sixth the force of the left ventricle. The result is that pulmonary arterial pressure is always low: 20 to 25/8 to 10 mmHg, and in pulmonary capillaries is lower still. This is important to *prevent* filtration in pulmonary capillaries, to prevent tissue fluid from accumulating in the alveoli of the lungs.

MAINTENANCE OF SYSTEMIC BLOOD PRESSURE

Because blood pressure is so important, there are many factors and physiological processes that interact to keep blood pressure within normal limits.

1. **Venous Return**—the amount of blood that returns to the heart by way of the veins. Venous return is important because the heart can pump only the blood it receives. If venous return decreases, the cardiac muscle fibers will not be stretched, the force of ventricular systole will decrease (Starling's Law), and blood pressure will decrease. This is what might happen following a severe hemorrhage.

   When the body is horizontal, venous return can be maintained fairly easily, but when the body is vertical, gravity must be overcome to return blood from the lower body to the heart. There are three mechanisms that help promote venous return: constriction of veins, the skeletal muscle pump, and the respiratory pump.

   Veins contain smooth muscle, which enables them to constrict and force blood toward the heart; the valves prevent backflow of blood. The second mechanism is the **skeletal muscle pump**, which is especially effective for the deep veins of the legs. These veins are surrounded by skeletal muscles that contract and relax during normal activities such as walking. Contractions of the leg muscles squeeze the veins to force blood toward the heart. The third mechanism is the **respiratory pump**, which affects veins that pass through the chest cavity. The pressure changes of inhalation and exhalation alternately expand and compress the veins, and blood is returned to the heart.
2. **Heart Rate and Force**—in general, if heart rate and force increase, blood pressure increases; this is what happens during exercise. However, if the heart is beating extremely rapidly, the ventricles may not fill completely between beats, and cardiac output and blood pressure will decrease.
3. **Peripheral Resistance**—this term refers to the resistance the vessels offer to the flow of blood. The arteries and veins are usually slightly constricted, which maintains normal diastolic blood pressure. It may be helpful to think of the vessels as the "container" for the blood. If a person's body has 5 liters of blood, the "container" must be smaller in order for the blood to exert a pressure against its walls. This is what normal vasoconstriction does: it makes the container (the vessels) smaller than the volume of blood so that the blood will exert pressure even when the left ventricle is relaxed.

   If more vasoconstriction occurs, blood pressure will increase (the container has become even smaller). This is what happens in a stress situation, when greater vasoconstriction is brought about by sympathetic impulses. If vasodilation occurs, blood pressure will decrease (the container is larger). After eating a large meal, for example, there is extensive vasodilation in the digestive tract to supply more oxygenated blood for all the digestive activities. To keep blood pressure within the normal range, vasoconstriction must, and does, occur elsewhere in the body. This is why strenuous exercise should be avoided right after eating; there is not enough blood to completely supply oxygen to exercising muscles and an active digestive tract at the same time.
4. **Elasticity of the Large Arteries**—when the

left ventricle contracts, the blood that enters the large arteries stretches their walls. The arterial walls are elastic and absorb some of the force. When the left ventricle relaxes, the arterial walls recoil or snap back, which helps keep diastolic pressure within the normal range. Normal elasticity, therefore, lowers systolic pressure, raises diastolic pressure, and maintains a normal pulse pressure. (Pulse pressure is the difference between systolic and diastolic pressure. The usual ratio of systolic to diastolic to pulse pressure is approximately 3:2:1. For example, with a blood pressure of 120/80, the pulse pressure is 40, and the ratio is 120:80:40, or 3:2:1.)

5. **Viscosity of the Blood**—normal blood viscosity depends upon the presence of red blood cells and plasma proteins, especially albumin. Having too many red blood cells is rare but does occur in the disorder called polycythemia vera and in people who are heavy smokers. This will increase blood viscosity and blood pressure.

   Decreased red blood cells, as in severe anemia, or decreased albumin, as may occur in liver disease or kidney disease, will decrease blood viscosity and blood pressure. In these situations, other mechanisms such as vasoconstriction will maintain blood pressure as close to normal as is possible.
6. **Loss of Blood**—a small loss of blood, as when donating a pint of blood, will cause a temporary drop in blood pressure followed by rapid compensation in the form of more rapid heart rate and greater vasoconstriction. After a severe hemorrhage, however, these compensating mechanisms may not be sufficient to maintain normal blood pressure and blood flow to the brain. Although a person may survive blood losses of 50% of the total blood, the possibility of brain damage increases as more blood is lost and not rapidly replaced.
7. **Hormones**—there are several hormones that have effects on blood pressure. You may recall them from Chapters 10 and 12, but let us summarize them here. The adrenal medulla secretes norepinephrine and epinephrine in stress situations. Norepinephrine stimulates vasoconstriction, which raises blood pressure. Epinephrine also causes vasoconstriction, and increases heart rate and force of contraction, which increase blood pressure.

   Antidiuretic hormone (ADH) is secreted by the posterior pituitary gland when the water content of the body decreases. ADH increases the reabsorption of water by the kidneys to prevent further loss of water in urine and a further decrease in blood pressure.

   Aldosterone, a hormone from the adrenal cortex, has a similar effect on blood volume. When blood pressure decreases, secretion of aldosterone stimulates the reabsorption of Na^+ ions by the kidneys. Water follows sodium back to the blood, which maintains blood volume to prevent a further drop in blood pressure.

   Atrial natriuretic hormone (ANH), secreted by the atria of the heart, functions in opposition to aldosterone. ANH increases the excretion of Na^+ ions and water by the kidneys, which decreases blood volume and lowers blood pressure.

REGULATION OF BLOOD PRESSURE

The mechanisms that regulate blood pressure may be divided into two types: intrinsic mechanisms and nervous mechanisms. The nervous mechanisms involve the nervous system, and the intrinsic mechanisms do not require nerve impulses.

INTRINSIC MECHANISMS

The term "intrinsic" means "within." Intrinsic mechanisms work because of the internal characteristics of certain organs. The first such organ is the heart. When

Table 13–3 THE RENIN-ANGIOTENSIN MECHANISM

1. Decreased blood pressure stimulates the kidneys to secrete renin.
2. Renin splits the plasma protein angiotensinogen (synthesized by the liver) to angiotensin I.
3. Angiotensin I is converted to angiotensin II by an enzyme (called converting enzyme) found primarily in lung tissue.
4. Angiotensin II:
   - causes vasoconstriction
   - stimulates the adrenal cortex to secrete aldosterone

venous return increases, cardiac muscle fibers are stretched, and the ventricles pump more forcefully (Starling's Law). Thus, cardiac ouput and blood pressure increase. This is what happens during exercise, when a higher blood pressure is needed. When exercise ends and venous return decreases, the heart pumps less forcefully, which helps return blood pressure to a normal resting level.

The second intrinsic mechanism involves the kidneys. When blood flow through the kidneys decreases, the process of filtration decreases and less urine is formed. This decrease in urinary output preserves blood volume so that it does not decrease further. Following severe hemorrhage or any other type of dehydration, this is very important to maintain blood pressure.

The kidneys are also involved in the **renin-angiotensin mechanism**. When blood pressure decreases, the kidneys secrete the enzyme **renin**, which initiates a series of reactions that result in the formation of **angiotensin II**. These reactions are described in Table 13-3 and depicted in Fig. 13-10. Angiotensin II causes vasoconstriction and stimulates secretion of aldosterone by the adrenal cortex, both of which will increase blood pressure.

NERVOUS MECHANISMS

The medulla and the autonomic nervous system are directly involved in the regulation of blood pressure. The first of these nervous mechanisms concerns the heart; this was described previously, so we will not review it here but refer you to Chapter 12 and Fig. 13-11.

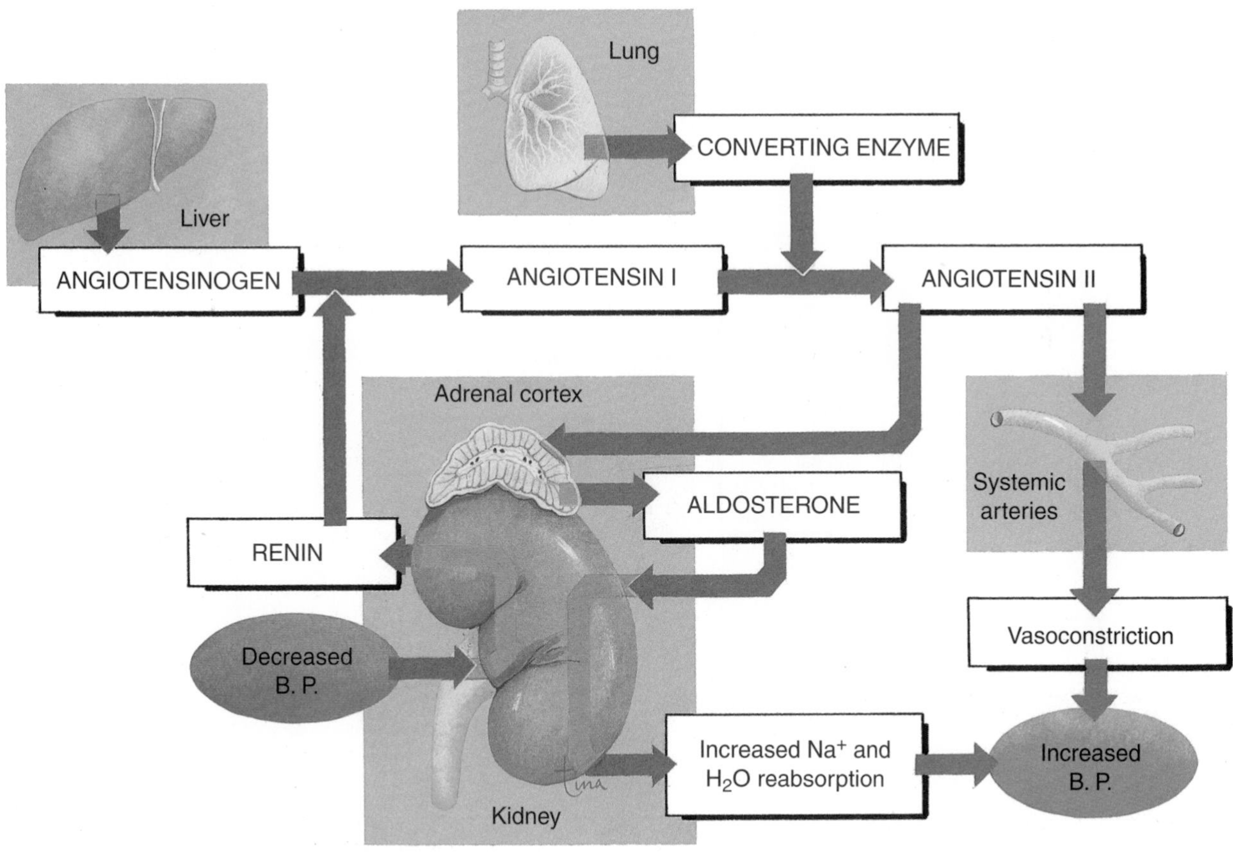

Figure 13–10 The renin-angiotensin mechanism. Begin at "Decreased blood pressure" and see Table 13–3 for numbered steps.

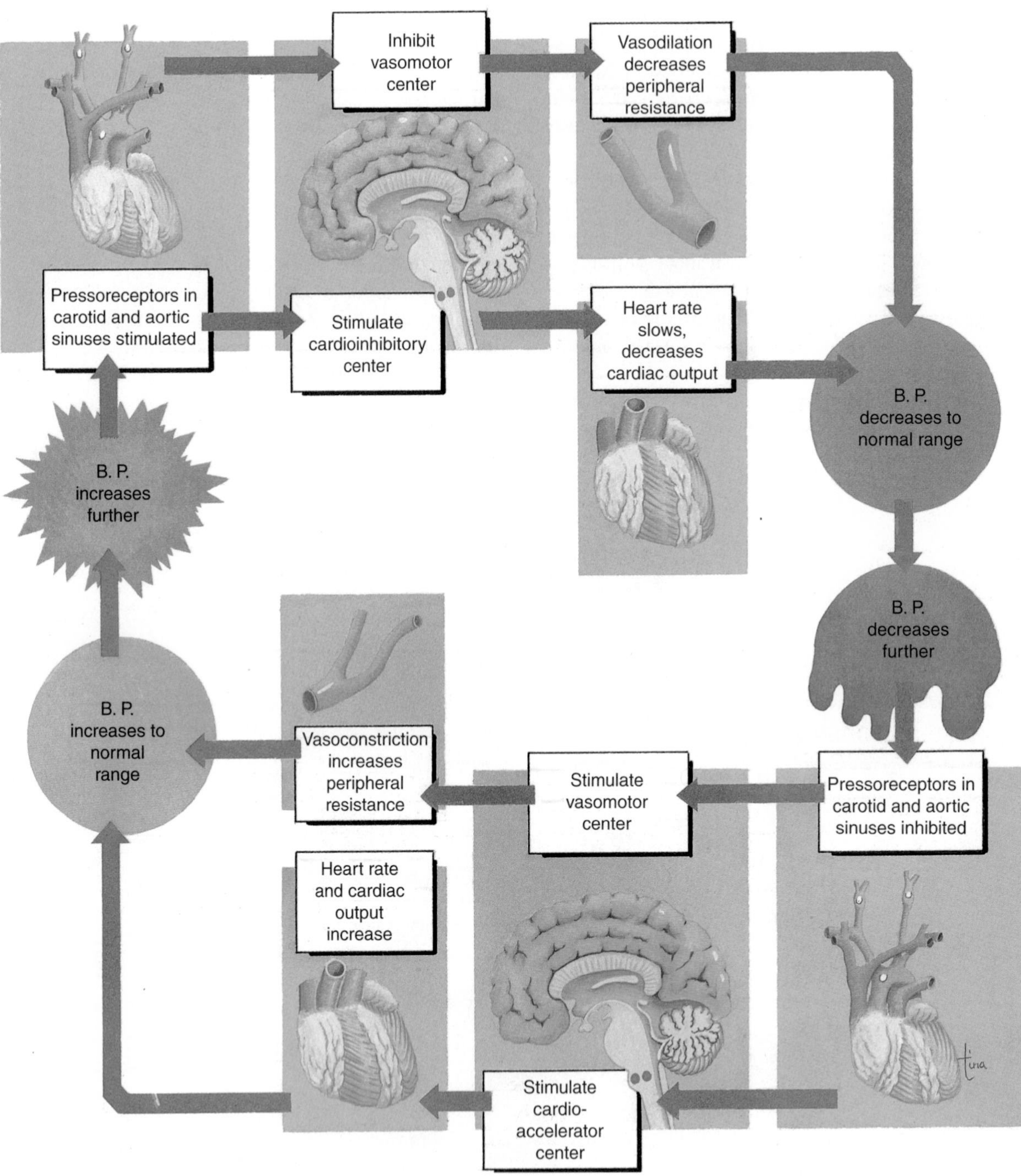

Figure 13–11 Nervous mechanisms that regulate blood pressure. See text for description.

Box 13–5 CIRCULATORY SHOCK

Circulatory shock is any condition in which cardiac output decreases to the extent that tissues are deprived of oxygen and waste products accumulate.

Causes of Shock

Cardiogenic shock occurs most often after a severe myocardial infarction but may also be the result of ventricular fibrillation. In either case, the heart is no longer an efficient pump, and cardiac output decreases.

Hypovolemic shock is the result of decreased blood volume, often due to severe hemorrhage. Other possible causes are extreme sweating (heat stroke) or extreme loss of water through the kidneys (diuresis) or intestines (diarrhea). In these situations, the heart simply does not have enough blood to pump, and cardiac output decreases. Anaphylactic shock, also in this category, is a massive allergic reaction in which great amounts of histamine increase capillary permeability and vasodilation throughout the body. Much plasma is then lost to tissue spaces, which decreases blood volume, blood pressure, and cardiac output.

Septic shock is the result of septicemia, the presence of bacteria in the blood. The bacteria and damaged tissues release inflammatory chemicals that cause vasodilation and extensive loss of plasma into tissue spaces.

Stages of Shock

Compensated shock—the responses by the body maintain cardiac output. Following a small hemorrhage, for example, the heart rate increases, the blood vessels constrict, and the kidneys decrease urinary output to conserve water. These responses help preserve blood volume and maintain blood pressure, cardiac output, and blood flow to tissues.

Progressive shock—the state of shock leads to more shock. Following a severe hemorrhage, cardiac output decreases, and the myocardium itself is deprived of blood. The heart weakens, which further decreases cardiac output. Arteries that are deprived of their blood supply cannot remain constricted. As the arteries dilate, venous return decreases, which in turn decreases cardiac output. Progressive shock is a series of such vicious cycles, and medical intervention is required to restore cardiac output to normal.

Irreversible shock—no amount of medical assistance can restore cardiac output to normal. The usual cause of death is that the heart has been damaged too much to recover. A severe myocardial infarction, massive hemorrhage, or septicemia may all be fatal despite medical treatment.

The second nervous mechanism involves peripheral resistance, that is, the degree of constriction of the arteries and arterioles, and to a lesser extent, the veins (see Fig. 13-11). The medulla contains the **vasomotor center**, which consists of a vasoconstrictor area and a vasodilator area. The vasodilator area may depress the vasoconstrictor area to bring about vasodilation, which will decrease blood pressure. The vasoconstrictor area may bring about more vasoconstriction by way of the sympathetic division of the autonomic nervous system.

Sympathetic vasoconstrictor fibers innervate the smooth muscle of all arteries and veins, and several impulses per second along these fibers maintain normal vasoconstriction. More impulses per second bring about greater vasoconstriction, and fewer impulses per second cause vasodilation. The medulla receives the information to make such changes from the pressoreceptors in the carotid sinuses and the aortic sinus. The inability to maintain normal blood pressure is one aspect of circulatory shock (see Box 13-5: Circulatory Shock).

AGING AND THE VASCULAR SYSTEM

It is believed that the aging of blood vessels, especially arteries, begins in childhood, although the effects are not apparent for decades. The cholesterol deposits of atherosclerosis are to be expected with advancing age, with the most serious consequences in the coronary arteries. A certain degree of arteriosclerosis is to be expected, and average resting blood pressure may increase, which further damages arterial walls. Consequences include stroke and left-sided heart failure.

The veins also deteriorate with age; their thin walls weaken and stretch, making their valves incompetent. This is most likely to occur in the veins of the legs; their walls are subject to great pressure as blood is returned to the heart against the force of gravity. Varicose veins and phlebitis are more likely to occur among elderly people.

SUMMARY

Although the vascular system does form passageways for the blood, you can readily see that the blood vessels are not simply pipes through which the blood flows. The vessels are not passive tubes, but rather active contributors to homeostasis. The arteries and veins help maintain blood pressure, and the capillaries provide sites for the exchanges of materials between the blood and the tissues. Some very important sites of exchange are discussed in the following chapters: the lungs, the digestive tract, and the kidneys.

STUDY OUTLINE

The vascular system consists of the arteries, capillaries, and veins through which blood travels.

Arteries (and arterioles) (see Fig. 13–1)

1. Carry blood from the heart to capillaries; three layers in their walls.
2. Inner layer (tunica intima): simple squamous epithelial tissue (endothelium), very smooth to prevent abnormal blood clotting.
3. Middle layer (tunica media): smooth muscle and elastic connective tissue; contributes to maintenance of diastolic blood pressure (BP).
4. Outer layer (tunica externa): fibrous connective tissue to prevent rupture.
5. Constriction or dilation is regulated by the autonomic nervous system.

Veins (and venules) (see Fig. 13–1)

1. Carry blood from capillaries to the heart; three layers in walls.
2. Inner layer: endothelium folded into valves to prevent the backflow of blood.
3. Middle layer: thin smooth muscle, because veins are not as important in the maintenance of BP.
4. Outer layer: thin fibrous connective tissue because veins do not carry blood under high pressure.

Anastomoses—connections between vessels of the same type

1. Provide alternate pathways for blood flow if one vessel is blocked.
2. Arterial anastomoses provide for blood flow to the capillaries of an organ (e.g., circle of Willis to the brain).
3. Venous anastomoses provide for return of blood to the heart and are most numerous in veins of the legs.

Capillaries (see Fig. 13–1 and Fig. 13–2)

1. Carry blood from arterioles to venules.
2. Walls are one cell thick (simple squamous epithelial tissue) to permit exchanges between blood and tissue fluid.
3. Oxygen and carbon dioxide are exchanged by diffusion.
4. BP in capillaries brings nutrients to tissues and forms tissue fluid in the process of filtration.
5. Albumin in the blood provides colloid osmotic pressure, which pulls waste products and tissue fluid into capillaries. The return of tissue fluid maintains blood volume and BP.
6. Precapillary sphincters regulate blood flow into

capillary networks based on tissue needs; in active tissues they dilate; in less active tissue they constrict.

7. Sinusoids are very permeable capillaries found in the liver, spleen, pituitary gland, and red bone marrow to permit proteins and blood cells to enter or leave the blood.

Pathways of Circulation

1. Pulmonary: Right ventricle → pulmonary artery → pulmonary capillaries (exchange of gases) → pulmonary veins → left atrium.
2. Systemic: left ventricle → aorta → capillaries in body tissues → superior and inferior caval veins → right atrium (see Table 13-1 and Fig. 13-3 for systemic arteries and Table 13-2 and Fig. 13-4 for systemic veins).
3. Hepatic Portal Circulation: blood from the digestive organs and spleen flows through the portal vein to the liver before returning to the heart. Purpose: the liver stores some nutrients or regulates their blood levels and detoxifies potential poisons before blood enters the rest of peripheral circulation (see Fig. 13-7).

Fetal Circulation—the fetus depends on the mother for oxygen and nutrients and for the removal of waste products (see Fig. 13–8)

1. The placenta is the site of exchange between fetal blood and maternal blood.
2. Umbilical arteries (two) carry blood from the fetus to the placenta, where CO_2 and waste products enter maternal circulation.
3. The umbilical vein carries blood with O_2 and nutrients from the placenta to the fetus.
4. The umbilical vein branches: some blood flows through the fetal liver; most blood flows through the ductus venosus to the fetal inferior vena cava.
5. The foramen ovale permits blood to flow from the right atrium to the left atrium to bypass the fetal lungs.
6. The ductus arteriosus permits blood to flow from the pulmonary artery to the aorta to bypass the fetal lungs.
7. These fetal structures become nonfunctional after birth, when the umbilical cord is cut and breathing takes place.

Velocity of Blood Flow (see Fig. 13–9)

1. Velocity is inversely related to the cross-sectional area of a segment of the vascular system.
2. The total capillaries have the greatest cross-sectional area and slowest blood flow.
3. Slow flow is important to permit sufficient time for exchange of gases, nutrients, and wastes.

Blood Pressure (BP)—the force exerted by the blood against the walls of the blood vessels

1. BP is measured in mmHg: systolic/diastolic. Systolic pressure is during ventricular contraction; diastolic pressure is during ventricular relaxation (see Fig. 13-9).
2. Normal range of systemic arterial BP: 90 to 135/60 to 85 mmHg.
3. BP in capillaries is 30 to 35 mmHg at the arterial end and 12 to 15 mmHg at the venous end; high enough to permit filtration but low enough to prevent rupture of the capillaries.
4. BP decreases in the veins and approaches zero in the caval veins.
5. Pulmonary BP is always low (the right ventricle pumps with less force): 20 to 25/8 to 10 mmHg. This low BP prevents filtration and accumulation of tissue fluid in the alveoli.

Maintenance of Systemic BP

1. Venous Return—the amount of blood that returns to the heart. If venous return decreases, the heart contracts less forcefully (Starling's Law) and BP decreases. The mechanisms that maintain venous return when the body is vertical are:
   - constriction of veins with the valves preventing backflow of blood
   - skeletal muscle pump—contraction of skeletal muscles, especially in the legs, squeezes the deep veins
   - respiratory pump—the pressure changes of inhalation and exhalation expand and compress the veins in the chest cavity
2. Heart Rate and Force—if heart rate and force increase, BP increases.
3. Peripheral Resistance—the resistance of the arteries and arterioles to the flow of blood. These vessels are usually slightly constricted to maintain normal diastolic BP. Greater vasoconstriction will increase BP; vasodilation will decrease BP. In the

body, vasodilation in one area requires vasoconstriction in another area to maintain normal BP.

4. Elasticity of the Large Arteries—ventricular systole stretches the walls of large arteries, which recoil during ventricular diastole. Normal elasticity lowers systolic BP, raises diastolic BP, and maintains normal pulse pressure.
5. Viscosity of Blood—depends on RBCs and plasma proteins, especially albumin. Severe anemia tends to decrease BP. Deficiency of albumin as in liver or kidney disease tends to decrease BP. In these cases, compensation such as greater vasoconstriction will keep BP close to normal.
6. Loss of Blood—a small loss will be rapidly compensated for by faster heart rate and greater vasoconstriction. After severe hemorrhage, these mechanisms may not be sufficient to maintain normal BP.
7. Hormones—(a) Norepinephrine stimulates vasoconstriction, which raises BP; (b) epinephrine increases cardiac output and raises BP; (c) ADH increases water reabsorption by the kidneys, which increases blood volume and BP; (d) aldosterone increases reabsorption of Na^+ ions by the kidneys; water follows Na^+ and increases blood volume and BP; (e) ANH increases excretion of Na^+ ions and water by the kidneys, which decreases blood volume and BP.

Regulation of Blood Pressure—intrinsic mechanisms and nervous mechanisms

Intrinsic Mechanisms

1. The Heart—responds to increased venous return by pumping more forcefully (Starling's Law), which increases cardiac output and BP.
2. The Kidneys—decreased blood flow decreases filtration, which decreases urinary output to preserve blood volume. Decreased BP stimulates the kidneys to secrete renin, which initiates the renin-angiotensin mechanism (Table 13-3 and Fig. 13-10) that results in the formation of angiotensin II, which causes vasoconstriction and stimulates secretion of aldosterone.

Nervous Mechanisms (see Fig. 13–11)

1. Heart Rate and Force—see Chapter 12.
2. Peripheral Resistance—the medulla contains the vasomotor center, which consists of a vasoconstrictor area and a vasodilator area. The vasodilator area brings about vasodilation by suppressing the vasoconstrictor area. The vasoconstrictor area maintains normal vasoconstriction by generating several impulses per second along sympathetic vasoconstrictor fibers to all arteries and veins. More impulses per second increase vasoconstriction and raise BP; fewer impulses per second bring about vasodilation and a drop in BP.

REVIEW QUESTIONS

1. Describe the structure of the three layers in the walls of arteries, and state the function of each layer. Describe the structural differences in these layers in veins, and explain the reason for each difference. (p. 276)
2. Describe the structure and purpose of anastomoses, and give a specific example. (p. 276)
3. Describe the structure of capillaries. State the process by which each of the following is exchanged between capillaries and tissue fluid: nutrients, oxygen, waste products, and CO_2. (pp. 277-280)
4. State the part of the body supplied by each of the following arteries: (pp. 284-285)
   a. Bronchial
   b. Femoral
   c. Hepatic
   d. Brachial
   e. Inferior mesenteric
   f. Internal carotid
   g. Subclavian
   h. Intercostal

5. Describe the pathway of blood flow in hepatic portal circulation. Use a specific example to explain the purpose of portal circulation. (pp. 285, 289)

6. Begin at the right ventricle and describe the pathway of pulmonary circulation. Explain the purpose of this pathway. (p. 280)

7. Name the fetal structure with each of the following functions: (p. 289)
   a. Permits blood to flow from the right atrium to the left atrium
   b. Carries blood from the placenta to the fetus
   c. Permits blood to flow from the pulmonary artery to the aorta
   d. Carry blood from the fetus to the placenta
   e. Carries blood from the umbilical vein to the inferior vena cava

8. Describe the three mechanisms that promote venous return when the body is vertical. (pp. 293–294)

9. Explain how the normal elasticity of the large arteries affects both systolic and diastolic blood pressure. (pp. 293–294)

10. Explain how Starling's Law of the Heart is involved in the maintenance of blood pressure. (pp. 294–295)

11. Name two hormones involved in the maintenance of blood pressure, and state the function of each. (p. 294)

12. Describe two different ways the kidneys respond to decreased blood flow and blood pressure. (pp. 294–295)

13. State two compensations that will maintain blood pressure after a small loss of blood. (p. 293)

14. State the location of the vasomotor center and name its two parts. Name the division of the autonomic nervous system that carries impulses to blood vessels. Which blood vessels? Which tissue in these vessels? Explain why normal vasoconstriction is important. Explain how greater vasoconstriction is brought about. Explain how vasodilation is brought about. How will each of these changes affect blood pressure? (pp. 295, 297)

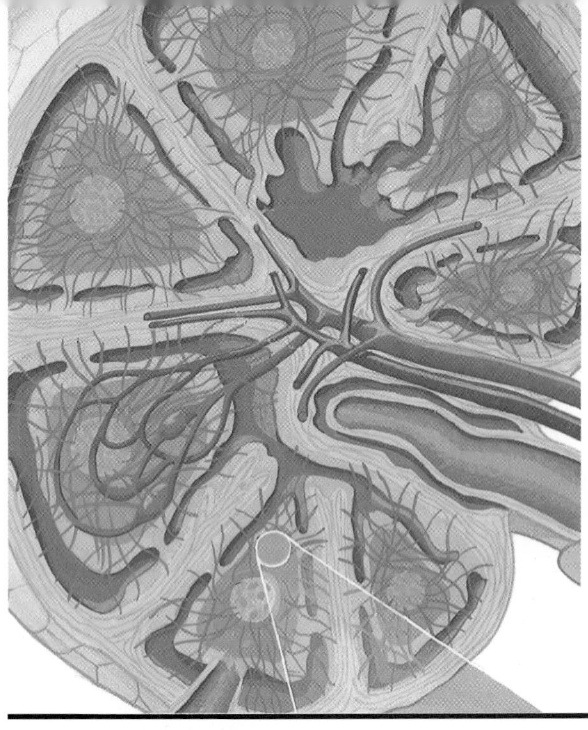

CHAPTER 14

Chapter Outline

Student Objectives

- Describe the functions of the lymphatic system.
- Describe how lymph is formed.
- Describe the system of lymph vessels, and explain how lymph is returned to the blood.
- State the locations and functions of the lymph nodes and nodules.
- State the location and functions of the spleen.
- Explain the role of the thymus in immunity.
- Explain what is meant by immunity.
- Describe humoral immunity and cell-mediated immunity.
- Describe the responses to a first and second exposure to a pathogen.
- Explain the difference between genetic immunity and acquired immunity.
- Explain the difference between passive acquired immunity and active acquired immunity.
- Explain how vaccines work.

The Lymphatic System and Immunity

New Terminology

Acquired immunity (uh-**KWHY**-erd)
Active immunity (**AK**-tiv)
Antibody (**AN**-ti-BAH-dee)
Antigen (**AN**-ti-jen)
B cells (B SELLS)
Cell-mediated immunity (SELL **ME**-dee-ay-ted)
Genetic immunity (je-**NET**-ik)
Humoral immunity (**HYOO**-mohr-uhl)
Lymph (LIMF)
Lymph nodes (LIMF NOHDS)
Lymph nodules (LIMF **NAHD**-yools)
Opsonization (OP-sah-ni-**ZAY**-shun)
Passive immunity (**PASS**-iv)
Plasma cell (**PLAZ**-mah SELL)
Spleen (**SPLEEN**)
T cells (T SELLS)
Thymus (**THIGH**-mus)
Tonsils (**TAHN**-sills)

Related Clinical Terminology

AIDS (AYDS)
Allergy (**AL**-er-jee)
Antibody titer (**AN**-ti-BAH-dee **TIGH**-ter)
Attenuated (uh-**TEN**-yoo-AY-ted)
Complement fixation test (**KOM**-ple-ment fik-**SAY**-shun)
Fluorescent antibody test (floor-**ESS**-ent)
Hodgkin's disease (**HODJ**-kinz)
Interferon (in-ter-**FEER**-on)
Tonsillectomy (TAHN-si-**LEK**-toh-mee)
Toxoid (**TOK**-soyd)
Vaccine (vak-**SEEN**)

Terms that appear in ***bold type*** *in the chapter text are defined in the glossary, which begins on page 528.*

A child falls and scrapes her knee. Is this likely to be a life-threatening injury? Probably not, even though the breaks in the skin have permitted the entry of thousands or even millions of bacteria. Those bacteria, however, will be quickly destroyed by the cells and organs of the lymphatic system.

Although the lymphatic system may be considered part of the circulatory system, we will consider it separately because its functions are so different from those of the heart and blood vessels. Keep in mind, however, that all of these functions are interdependent. The lymphatic system is responsible for returning tissue fluid to the blood and for protecting the body against foreign material. The parts of the lymphatic system are the lymph, the system of lymph vessels, lymph nodes and nodules, the spleen, and the thymus gland.

LYMPH

Lymph is the name for tissue fluid that enters lymph capillaries. As you may recall from Chapter 13, filtration in capillaries creates tissue fluid, most of which returns almost immediately to the blood in the capillaries by osmosis. Some tissue fluid, however, remains in interstitial spaces and must be returned to the blood by way of the lymphatic vessels. Without this return, blood volume and blood pressure would very soon decrease. The relationship of the lymphatic vessels to the cardiovascular system is depicted in Fig. 14–1.

LYMPH VESSELS

The system of lymph vessels begins as dead-end **lymph capillaries** found in most tissue spaces (Fig. 14–2). Lymph capillaries are very permeable and collect tissue fluid and proteins. **Lacteals** are specialized lymph capillaries in the villi of the small intestine; they absorb the fat-soluble end products of digestion, such as fatty acids and vitamin A.

Lymph capillaries unite to form larger lymph vessels, whose structure is very much like that of veins. There is no pump for lymph (as the heart is the pump for blood), but the lymph is kept moving within lymph vessels by the same mechanisms that promote venous return. The smooth muscle layer of the larger lymph vessels constricts, and the one-way valves (just like those of veins) prevent backflow of lymph. Lymph vessels in the extremities are compressed by the skeletal muscles that surround them; this is the **skeletal muscle pump**. The **respiratory pump** alternately expands and compresses the lymph vessels in the chest cavity and keeps the lymph moving.

Where is the lymph going? Back to the blood to become plasma again. Refer to Fig. 14–3 as you read the following. The lymph vessels from the lower body unite in front of the lumbar vertebrae to form a vessel called the **cisterna chyli**, which continues upward in front of the backbone as the **thoracic duct**. Lymph vessels from the upper left quadrant of the body join the thoracic duct, which empties lymph into the left subclavian vein. Lymph vessels from the upper right quadrant of the body unite to form the right lymphatic duct, which empties lymph into the right subclavian vein. Flaps in both subclavian veins permit the entry of lymph but prevent blood from flowing into the lymph vessels.

LYMPH NODES AND NODULES

Lymph nodes and **nodules** are masses of lymphatic tissue. Recall that lymphatic tissue is one of the hemopoietic tissues and that one of its functions is to produce lymphocytes and monocytes. Nodes and nodules differ with respect to size and location. Nodes are usually larger, 10 to 20 mm in length; nodules range from a fraction of a millimeter to several millimeters in length.

Lymph nodes are found in groups along the pathways of lymph vessels, and lymph flows through these nodes on its way to the subclavian veins. Lymph enters a node through several afferent lymph vessels and leaves through one or two efferent vessels (Fig. 14–4). As lymph passes through a lymph node, bacteria and other foreign materials are phagocytized by fixed (stationary) **macrophages**. Fixed **plasma cells** (from lymphocytes) produce antibodies to any pathogens in the lymph; these antibodies, as well as lymphocytes and monocytes, will eventually reach the blood.

There are many groups of lymph nodes along all the lymph vessels throughout the body, but three paired groups deserve mention because of their strategic locations. These are the **cervical, axillary**, and **inguinal** lymph nodes (see Fig. 14–3). Notice that these are at the junctions of the head and extremities with the trunk of the body. Breaks in the skin, with entry of

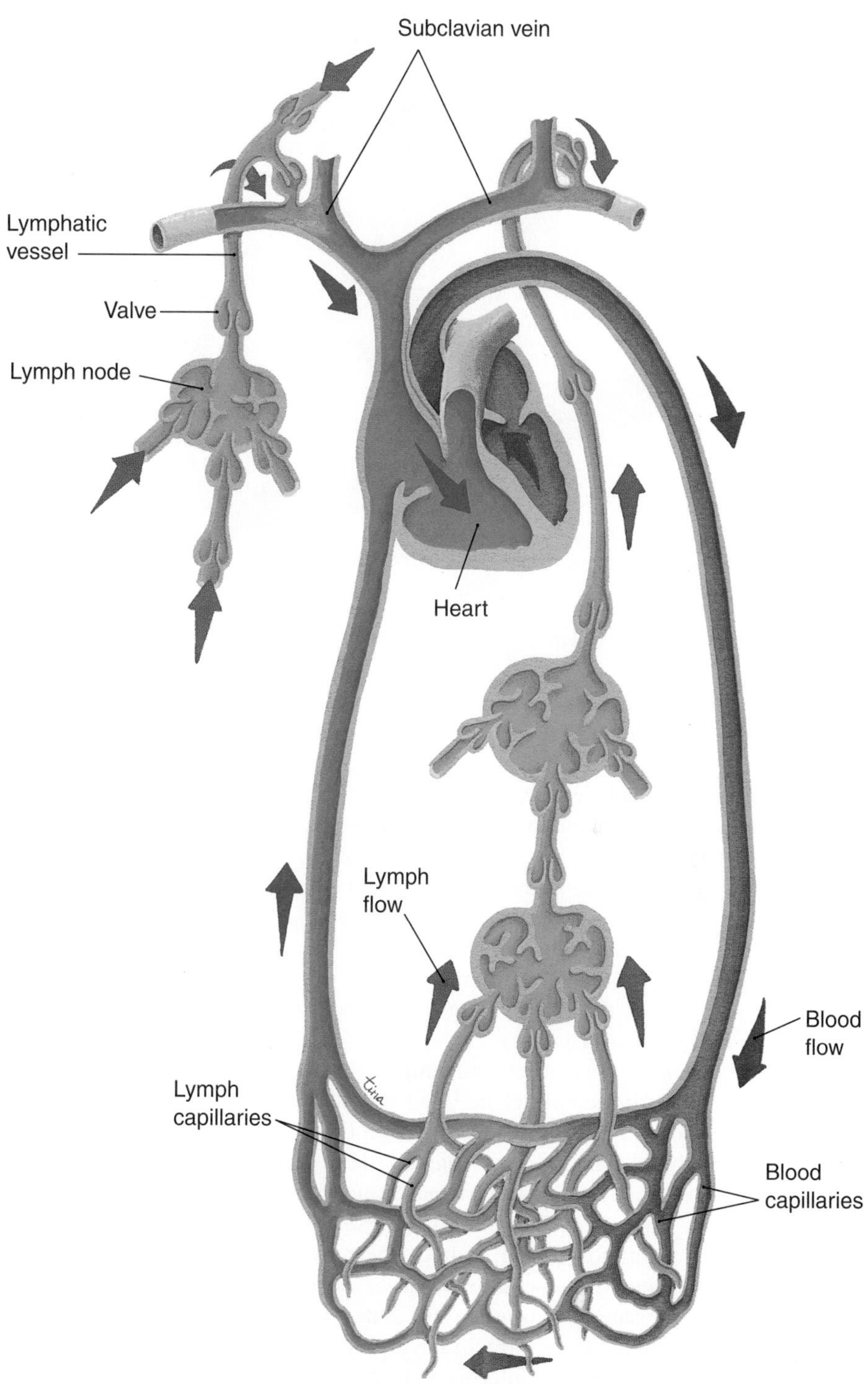

Figure 14–1 Relationship of lymphatic vessels to the cardiovascular system. Lymph capillaries collect tissue fluid, which is returned to the blood. The arrows indicate the direction of flow of the blood and lymph.

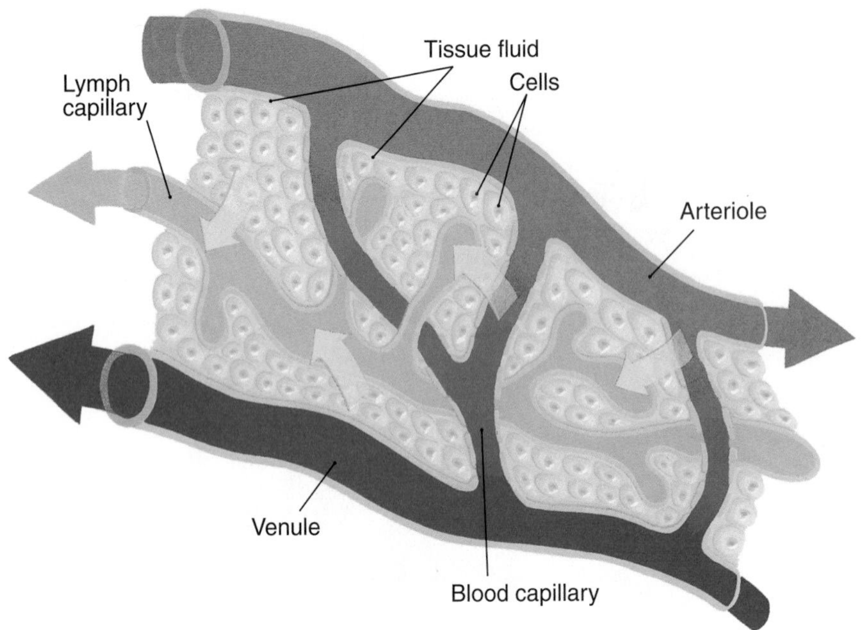

Figure 14–2 Dead-end lymph capillaries found in tissue spaces. Arrows indicate the movement of plasma, lymph, and tissue fluid.

pathogens, are much more likely to occur in the arms or legs or head rather than in the trunk. If these pathogens get to the lymph, they will be destroyed by the lymph nodes before they get to the trunk, before the lymph is returned to the blood in the subclavian veins.

You may be familiar with the expression "swollen glands," as when a child has a strep throat (an inflammation of the pharynx caused by *Streptococcus* bacteria). These "glands" are the cervical lymph nodes that have enlarged as their macrophages attempt to destroy the bacteria in the lymph from the pharynx (see Box 14–1: Hodgkin's Disease).

Lymph nodules are small masses of lymphatic tissue found just beneath the epithelium of all **mucous membranes**. The body systems lined with mucous membranes are those that have openings to the environment: the respiratory, digestive, urinary, and reproductive tracts. You can probably see that these are also strategic locations for lymph nodules, because any natural body opening is a possible portal of entry for pathogens. For example, if bacteria in inhaled air get through the epithelium of the trachea, lymph nodules with their macrophages are in position to destroy these bacteria before they get to the blood.

Some of the lymph nodules have specific names. Those of the small intestine are called **Peyer's patches**, and those of the pharynx are called **tonsils**. The palatine tonsils are on the lateral walls of the pharynx, the adenoid (pharyngeal tonsil) is on the posterior wall, and the lingual tonsils are on the base of the tongue. The tonsils, therefore, form a ring of lymphatic tissue around the pharynx, which is a common pathway for food and air and for the pathogens they contain. A **tonsillectomy** is the surgical removal of the palatine tonsils and the adenoid and may be performed if the tonsils are chronically inflamed and swollen, as may happen in children. As mentioned earlier, the body has redundant structures to help ensure survival if one structure is lost or seriously impaired. Thus, there are many other lymph nodules in the pharynx to take over the function of the surgically removed tonsils.

SPLEEN

The **spleen** is located in the upper left quadrant of the abdominal cavity, just below the diaphragm, behind the stomach. The lower rib cage protects the spleen from physical trauma (see Fig. 14–3).

In the fetus, the spleen produces red blood cells, a function assumed by the red bone marrow after birth.

The functions of the spleen after birth are:

1. Produces lymphocytes and monocytes, which enter the blood.

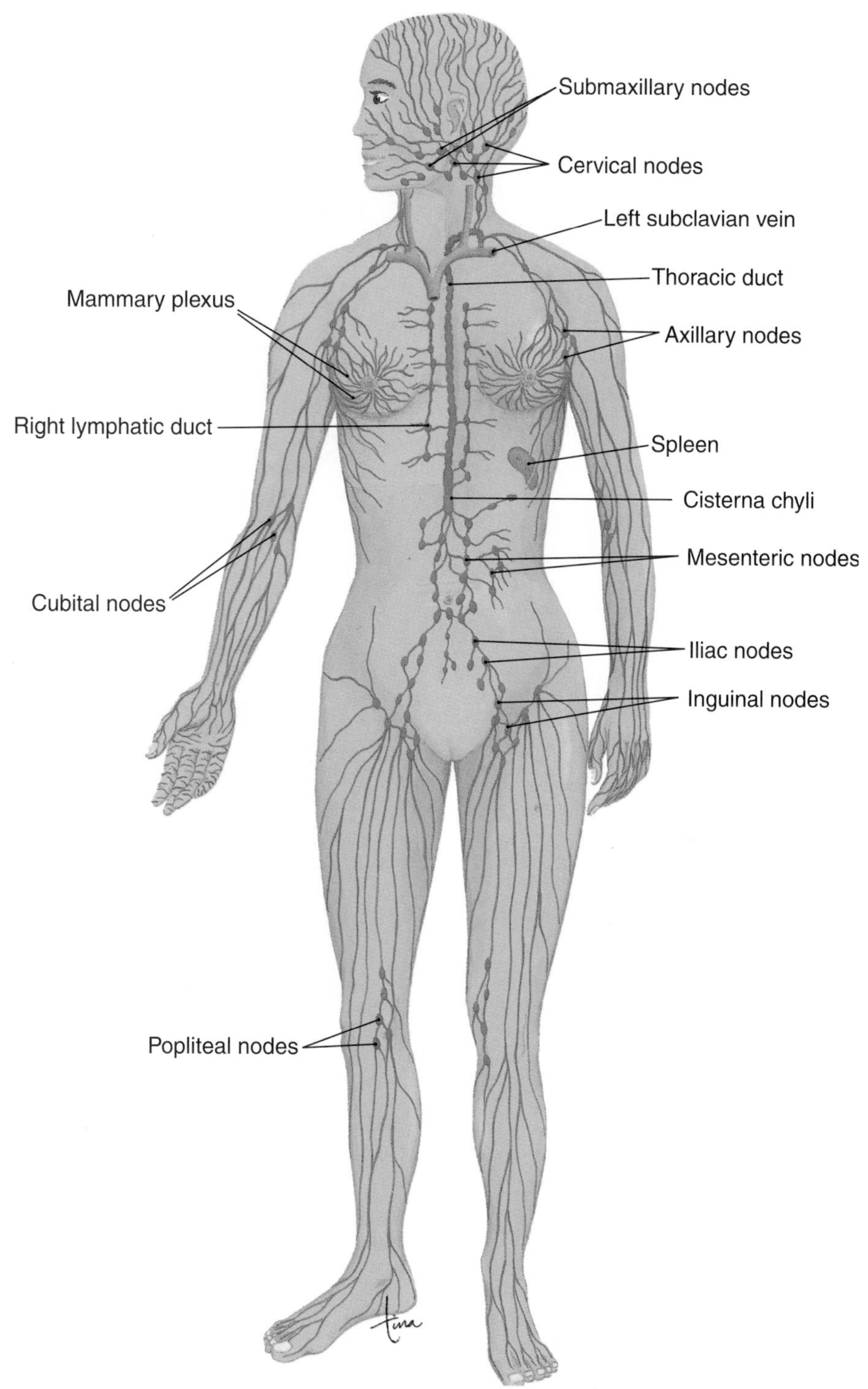

Figure 14–3 System of lymph vessels and the major groups of lymph nodes. Lymph is returned to the blood in the right and left subclavian veins.

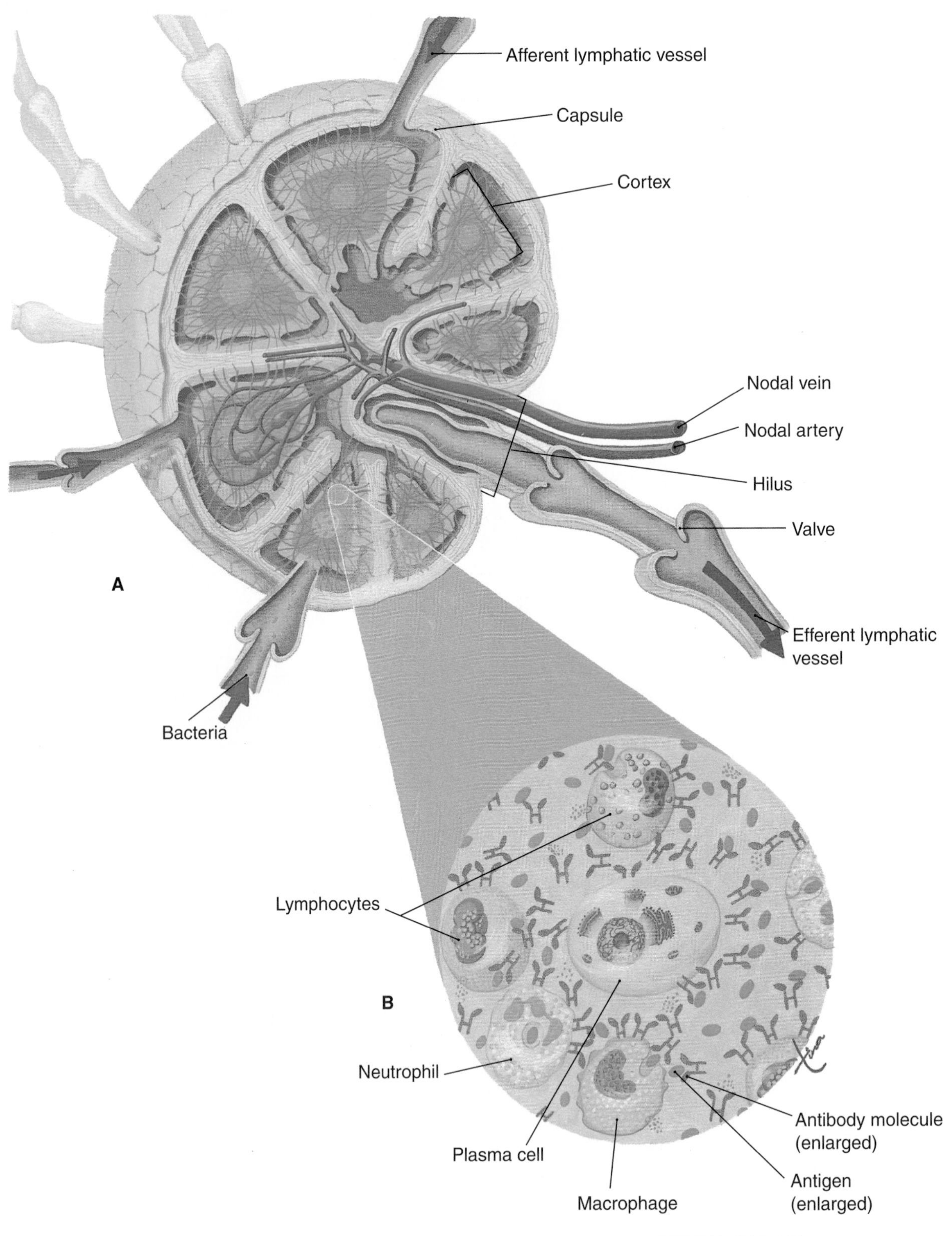

Figure 14–4 Lymph node. **(A)** Section through a lymph node, showing the flow of lymph. **(B)** Microscopic detail of bacteria being destroyed within the lymph node.

Box 14–1 HODGKIN'S DISEASE

Hodgkin's disease is a malignant disorder of the lymph nodes; the cause is not known. The first symptom is usually a swollen but painless lymph node, often in the cervical region. The individual is prompted to seek medical attention by other symptoms: chronic fever, fatigue, and weight loss. The diagnosis involves biopsy of the lymph node and the finding of characteristic cells.

Treatment of Hodgkin's disease requires chemotherapy, radiation, or both. With early diagnosis and proper treatment, this malignancy is very often curable.

2. Contains fixed plasma cells that produce antibodies to foreign antigens.
3. Contains fixed macrophages (RE cells) that phagocytize pathogens or other foreign material in the blood. The macrophages of the spleen also phagocytize old red blood cells and form bilirubin. By way of portal circulation, the bilirubin is sent to the liver for excretion in bile.

The spleen is not considered a vital organ, because other organs compensate for its functions if the spleen must be removed. The liver and red bone marrow will remove old red blood cells from circulation, and the many lymph nodes and nodules will produce lymphocytes and monocytes and phagocytize pathogens (as will the liver). Despite this redundancy, a person without a spleen is somewhat more susceptible to certain bacterial infections such as pneumonia and meningitis.

THYMUS

The **thymus** is located inferior to the thyroid gland. In the fetus and infant, the thymus is large and extends under the sternum (Fig. 14-5). With increasing age, the thymus shrinks, and relatively little thymus tissue is found in adults.

The lymphocytes produced by the thymus are called T lymphocytes or **T cells**; their functions are discussed in the next section. Thymic hormones are necessary for what may be called "immunological

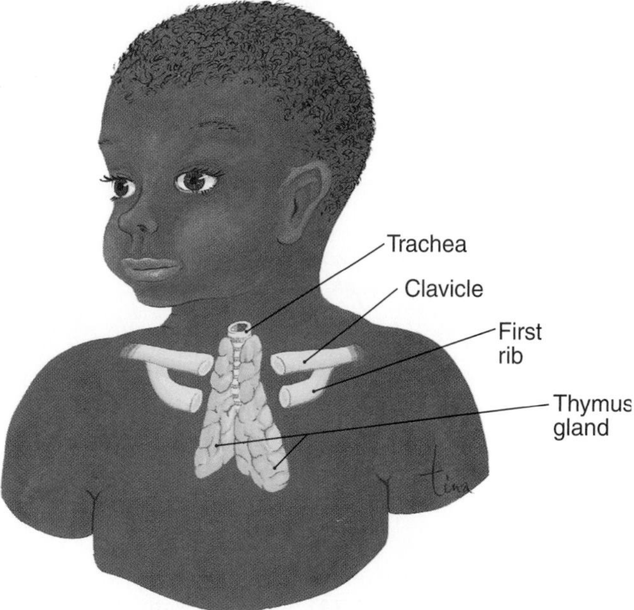

Figure 14–5 Location of the thymus in a young child.

competence." To be competent means to be able to do something well. The thymic hormones enable the T cells to participate in the recognition of foreign antigens and to provide immunity. This capability of T cells is established early in life and then is perpetuated by the lymphocytes themselves. The newborn's immune system is not yet fully mature, and infants are more susceptible to certain infections than are older children and adults. Usually by the age of 2 years, the immune system matures and becomes fully functional. This is why some vaccines, such as the measles vaccine, are not recommended for infants younger than 15 to 18 months of age. Their immune systems are not mature enough to respond strongly to the vaccine, and the protection provided by the vaccine may be incomplete.

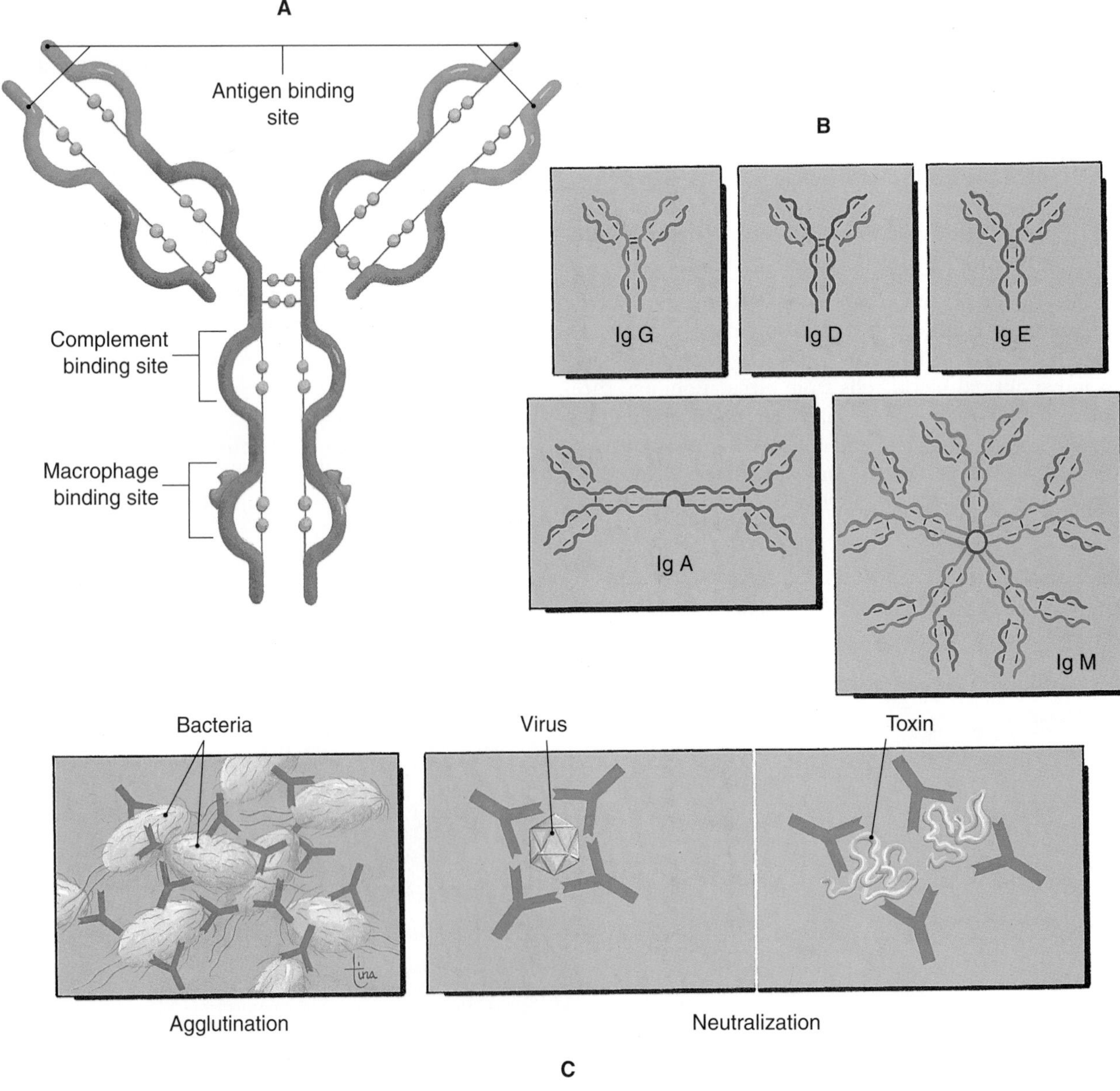

Figure 14–6 Antibodies. **(A)** Structure of one IgC molecule. **(B)** Structure of the five classes of antibodies. **(C)** Antibody activity: agglutination of bacteria and neutralization of viruses or toxins.

IMMUNITY

Immunity may be defined as the ability to destroy pathogens or other foreign material and to prevent further cases of certain infectious diseases. This ability is of vital importance because the body is exposed to pathogens from the moment of birth.

Malignant cells, which may be formed within the body as a result of mutations of normal cells, are also recognized as foreign and are usually destroyed before they can establish themselves and cause cancer. Unfortunately, organ transplants are also foreign tissue, and the immune system may reject (destroy) a transplanted kidney or heart. Sometimes the immune system mistakenly reacts to part of the body itself and causes an autoimmune disease; several of these were mentioned in previous chapters. Most often, however, the immune mechanisms function to protect the body from the microorganisms around us and within us.

LYMPHOCYTES

There are two major types of lymphocytes: T lymphocytes and B lymphocytes, or, more simply, **T cells** and **B cells**. In the embryo, T cells are produced in the bone marrow and thymus. They must pass through the thymus, where the thymic hormones bring about their maturation. The T cells then migrate to the spleen, lymph nodes, and lymph nodules, where they are found after birth.

Produced in the embryo bone marrow, B cells then migrate directly to the spleen and lymph nodes and nodules. When activated during an immune response, some B cells will become plasma cells that produce antibodies to a specific foreign antigen.

ANTIGENS AND ANTIBODIES

Antigens are chemical markers that identify cells. Human cells have their own antigens that identify all the cells in an individual as "self" (recall the HLA types mentioned in Chapter 11). When antigens are foreign, or "non-self," they may be recognized as such and destroyed. Bacteria, viruses, fungi, protozoa, malignant cells, and organ transplants are all foreign antigens that activate immune responses.

Antibodies, also called **immune globulins** or **gamma globulins**, are proteins produced by plasma cells in response to foreign antigens. Antibodies do not themselves destroy foreign antigens, but rather become attached to such antigens to "label" them for destruction. Each antibody produced is specific for only one antigen. Since there are so many different pathogens, you might think that the immune system would have to be capable of producing many different antibodies, and in fact this is so. It is estimated that millions of different antigen-specific antibodies can be produced, should there be a need for them.

The structure of antibodies is shown in Fig. 14–6, and the five classes of antibodies are described in Table 14–1.

Table 14–1 CLASSES OF ANTIBODIES

Name	Location	Functions
IgG	Blood	• Crosses the placenta to provide passive immunity for newborns
	Extracellular fluid	• Provides long-term immunity following recovery or a vaccine
IgA	External secretions (tears, saliva, etc.)	• Present in breast milk to provide passive immunity for breast-fed infants • Found in secretions of all mucous membranes
IgM	Blood	• Produced first by the maturing immune system of infants • Produced first during an infection (IgG production follows)
IgD	B lymphocytes	• Receptors on B lymphocytes
IgE	Mast cells or basophils	• Important in allergic reactions (mast cells release histamine)

MECHANISMS OF IMMUNITY

The first step in the destruction of a pathogen or foreign cell is the recognition of its antigens as foreign. Both T cells and B cells are capable of this, but the immune mechanisms are activated especially well when this recognition is accomplished by macrophages and a specialized group of T lymphocytes called **helper T cells**. The foreign antigen is first phagocytized by a macrophage, and parts of it are "presented" on the macrophage's cell membrane. Also on the macrophage membrane are "self" antigens that are representative of the antigens found on all of the cells of the individual. Therefore, the helper T cell that encounters this macrophage is presented not only with the foreign antigen but also with "self" antigens for comparison. The helper T cell now becomes sensitized to and specific for the foreign antigen, the one that does not belong in the body (see Box 14–2: AIDS).

The recognition of an antigen as foreign initiates one or both of the mechanisms of immunity. These are **cell-mediated immunity**, in which T cells and macrophages participate, and **humoral immunity**, (or antibody-mediated) which involves T cells, B cells, and macrophages.

Cell-Mediated Immunity

This mechanism of immunity does not result in the production of antibodies, but it is effective against intracellular pathogens (such as viruses), fungi, malignant cells, and grafts of foreign tissue. As mentioned above, the first step is the recognition of the foreign antigen by macrophages and helper T cells, which become activated and specific (you may find it helpful to refer to Fig. 14–7 as you read the following).

These activated T cells, which are antigen specific, divide many times, forming **memory T cells** and **cytotoxic (killer) T cells**. The memory T cells will remember the specific foreign antigen and become active if it enters the body again. Cytotoxic T cells are able to chemically destroy foreign antigens by disrupting cell membranes. This is how cytotoxic T cells destroy cells infected with viruses, and prevent the viruses from reproducing. These T cells also produce cytokines, which are chemicals that attract macrophages to the area and activate them to phagocytize the foreign antigen.

Other activated T cells become **suppressor T cells**, which will stop the immune response once the foreign antigen has been destroyed. The memory T cells, however, will quickly initiate the cell-mediated immune response should there be a future exposure to the antigen.

Humoral Immunity

This mechanism of immunity does involve the production of antibodies and is diagrammed in Fig. 14–8. Again, the first step is the recognition of the foreign antigen, this time by B cells as well as by macrophages and helper T cells. The sensitized helper T cell presents the foreign antigen to B cells, which provides a strong stimulus for the activation of B cells specific for this antigen. The activated B cells begin to divide many times, and two types of cells are formed. Some of the new B cells produced are **memory B cells**, which will remember the specific antigen. Other B cells become **plasma cells** that produce antibodies specific for this one foreign antigen.

The antibodies then bond to the antigen, forming an antigen-antibody complex. This complex results in **opsonization**, which means that the antigen is now "labeled" for phagocytosis by macrophages or neutrophils. The antigen-antibody complex also stimulates the process of complement fixation (see Box 14–3: Diagnostic Tests).

Complement is a group of about 20 plasma proteins that circulate in the blood until activated, or fixed, by an antigen-antibody complex. Complement fixation may be complete or partial. If the foreign antigen is cellular, the complement proteins bond to the antigen-antibody complex, then to one another, forming an enzymatic ring that punches a hole in the cell to bring about the death of the cell. This is complete (or entire) complement fixation and is what happens to bacterial cells (it is also the cause of hemolysis in a transfusion reaction).

If the foreign antigen is not a cell, a virus for example, partial complement fixation takes place, in which some of the complement proteins bond to the antigen-antibody complex. This is a chemotaxic factor. Chemotaxis means "chemical movement" and is actually another label that attracts macrophages to engulf and destroy the foreign antigen.

When the foreign antigen has been destroyed, suppressor T cells that have been sensitized to it stop the immune response. This is important to limit antibody production to just what is necessary to eliminate the

Box 14–2 AIDS

In 1981, young homosexual men in New York and California were diagnosed with Kaposi's sarcoma and *Pneumocystis carinii* pneumonia. At that time, Kaposi's sarcoma was known as a rare, slowly growing malignancy in elderly men. Pneumocystis pneumonia was almost unheard of; *P. carinii* is a protozoan that does not cause disease in healthy people. That in itself was a clue. These young men were not healthy; their immune systems were not functioning normally. As the number of patients increased rapidly, the disease was given a name (acquired immune deficiency syndrome—AIDS) and the pathogen was found. Human immunodeficiency virus (HIV) is a retrovirus that infects helper T cells, macrophages, and other human cells. Once infected, the human cells contain HIV genes for the rest of their lives. Without sufficient helper T cells, the immune system is seriously impaired. Foreign antigens are not recognized, B cells are not activated, and killer T cells are not stimulated to proliferate.

The person with AIDS is susceptible to opportunistic infections, that is, those infections caused by fungi and protozoa that would not affect average healthy adults. Some of these infections may be treated with medications and even temporarily cured, but the immune system cannot prevent the next infection, or the next. As of this writing, AIDS is considered an incurable disease, although with proper medical treatment, some people with AIDS may live for many years.

Where did this virus come from? We really do not know. Some researchers believe that a mutation in a previously harmless virus produced HIV, with devastating consequences for those infected with this "new" virus.

The incubation period of AIDS is highly variable, ranging from a few months to several years. An infected person may unknowingly spread HIV to others before any symptoms appear. It should be emphasized that AIDS, although communicable, is not a contagious disease. It is not spread by casual contact as is measles or the common cold. Transmission of AIDS occurs through sexual contact, contact with infected blood, or by placental transmission of the virus from mother to fetus.

In the United States, most of the cases of AIDS during the 1980s were in homosexual men and IV drug users who shared infected syringes contaminated with their blood. By the 1990s, however, it was clear that AIDS was becoming more of a heterosexually transmitted disease, with rapidly increasing case rates among women and teenagers. In much of the rest of the world, especially Africa and Asia, the transmission of AIDS has always been primarily by heterosexual contact, with equal numbers of women and men infected.

How can the spread of HIV be stopped? At present we still have no antiviral medications that will eradicate this virus, although certain combinations of drugs effectively suppress the virus in some people. For these people, AIDS may become a chronic but not fatal disease. Unfortunately, the medications do not work for everyone, and they are very expensive, beyond the means of most of the world's AIDS patients.

Development of an AIDS vaccine is unlikely before the turn of the century, although more than a dozen vaccines are undergoing clinical trials. A vaccine stimulates antibody production to a specific pathogen, but everyone who has died of AIDS had antibodies to HIV, and those antibodies were not protective. Why not? The most likely explanation is that HIV is a mutating virus; it constantly changes itself, making previously produced antibodies ineffective.

If we cannot cure AIDS and we cannot prevent it by vaccination, what recourse is left? Education. Everyone should know how AIDS is spread. The obvious reason is to be able to avoid the high-risk behaviors that make acquiring HIV more likely. Yet another reason, however, is that everyone should know that they need not fear casual contact with people with AIDS. Health-care personnel have a special responsibility, not only to educate themselves, but to provide education about AIDS for their patients and the families of their patients.

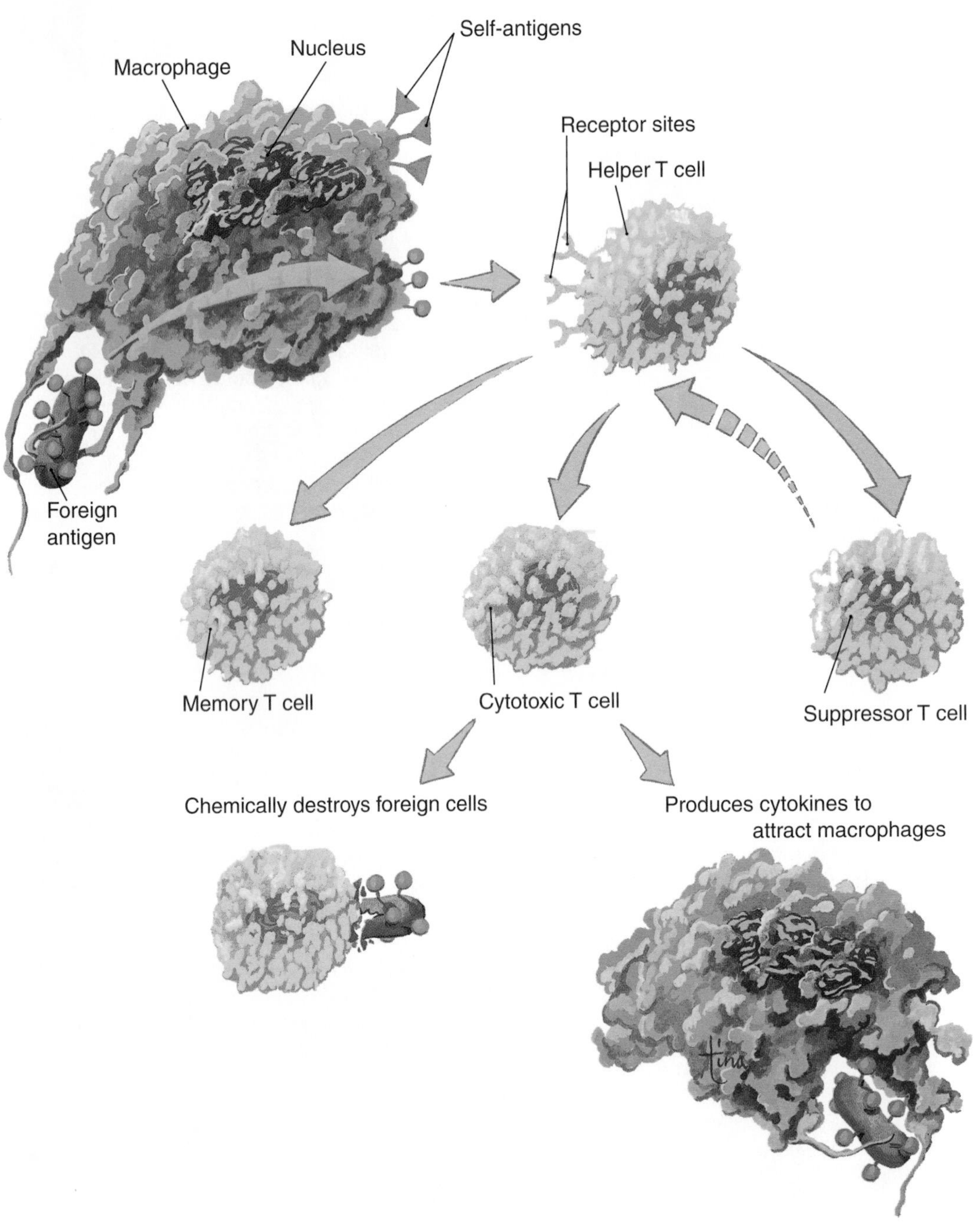

Figure 14–7 Cell-mediated immunity. See text for description of events.

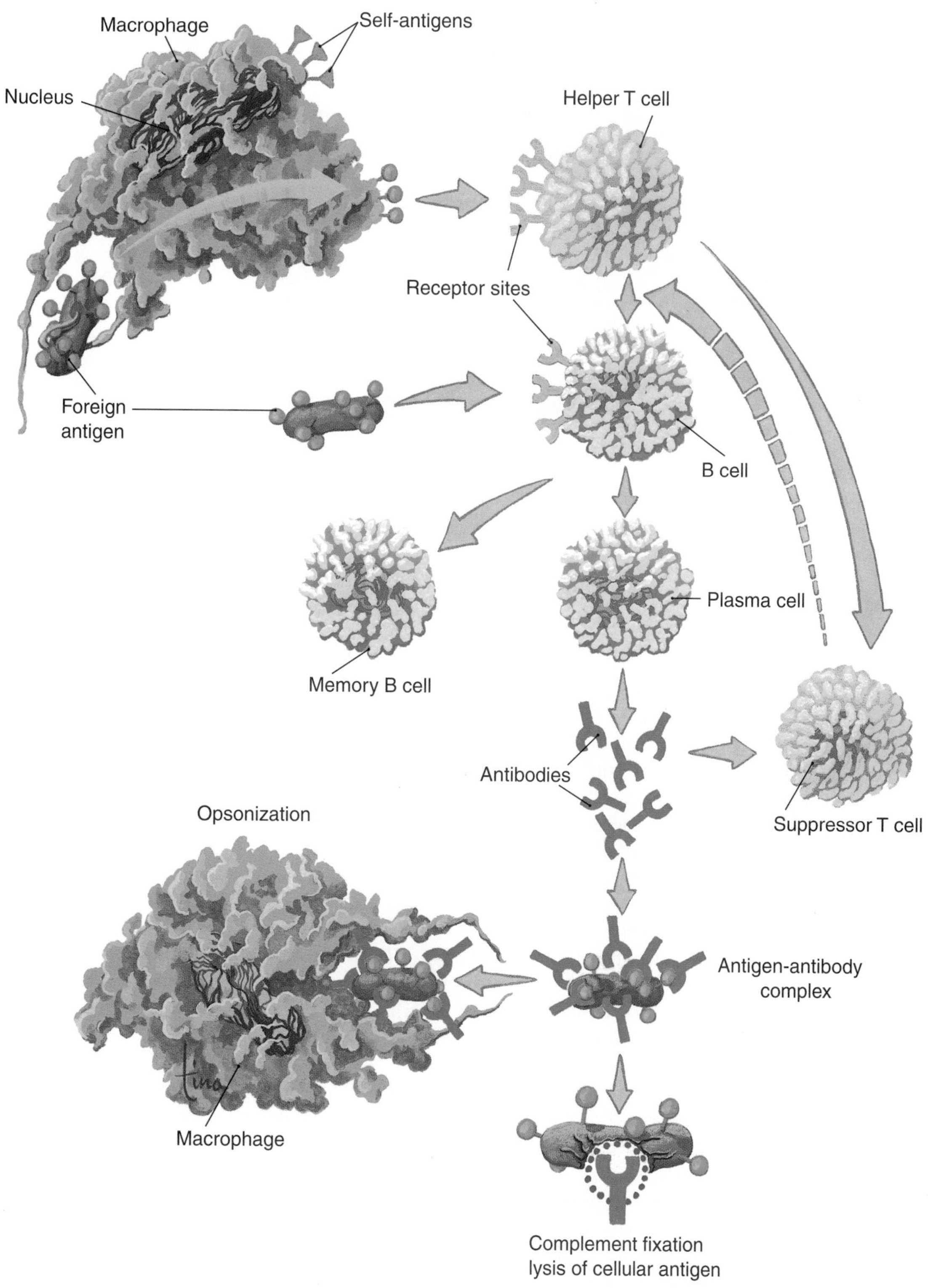

Figure 14–8 Humoral immunity. See text for description of events.

Box 14–3 DIAGNOSTIC TESTS

Several important laboratory tests involve antibodies and may be very useful to confirm a diagnosis.

Complement Fixation Test—determines the presence of a particular antibody in the patient's blood, but does not indicate when the infection occurred.

Antibody Titer—determines the level or amount of a specific antibody in the patient's blood. If another titer is done 1 to several weeks later, an increase in the antibody level shows the infection to be current.

Fluorescent Antibody Test—uses antibodies tagged with fluorescent dyes, which are added to a clinical specimen such as blood, sputum, or a biopsy of tissue. If the suspected pathogen is present, the fluorescent antibodies will bond to it and the antigen-antibody complex will "glow" when examined with a fluorescent microscope.

Tests such as these are used in conjunction with patient history and symptoms to arrive at a diagnosis.

pathogen without triggering an autoimmune response.

ANTIBODY RESPONSES

The first exposure to a foreign antigen does stimulate antibody production, but antibodies are produced slowly and in small amounts (see Fig. 14–9). Let us take as a specific example the measles virus. On a person's first exposure to this virus, antibody production is usually too slow to prevent the disease itself, and the person will have clinical measles. Most people who get measles recover, and upon recovery have antibodies and memory cells that are specific for the measles virus.

On a second exposure to this virus, the memory cells initiate rapid production of large amounts of antibodies, enough to prevent a second case of measles. This is the reason why we develop immunity to certain diseases, and this is also the basis for the protection given by **vaccines** (see Box 14–4: Vaccines).

As mentioned previously, antibodies label pathogens or other foreign antigens for phagocytosis or complement fixation. More specifically, antibodies cause agglutination or neutralization of pathogens before their eventual destruction. **Agglutination** means "clumping," and this is what happens when antibodies bond to bacterial cells. The bacteria that are clumped together by attached antibodies are more easily phagocytized by macrophages (see Fig. 14–6).

The activity of viruses may be neutralized by antibodies. A virus must get inside a living cell in order to reproduce itself. However, a virus with antibodies attached to it is unable to enter a cell, cannot reproduce, and will soon be phagocytized. (Another aspect of antiviral defense is a chemical called **interferon**, which is discussed in Box 14–5.) Bacterial toxins may also be neutralized by attached antibodies. The antibodies change the shape of the toxin, prevent it from exerting its harmful effects, and promote its phagocytosis by macrophages.

Allergies are also the result of antibody activity (see Box 14–6).

TYPES OF IMMUNITY

There are two major categories of immunity: genetic immunity, which is conferred by our DNA, and acquired immunity, which we must develop or acquire by natural or artificial means.

Genetic immunity does not involve antibodies or the immune system; it is the result of our genetic makeup. What it means is that some pathogens cause disease in certain host species but not in others. Dogs and cats, for example, have genetic immunity to the measles virus, which is a pathogen only for people. Plant viruses affect only plants, not people; we have genetic immunity to them. This is not due to antibodies against these plant viruses, but rather that our genetic makeup makes it impossible for such pathogens

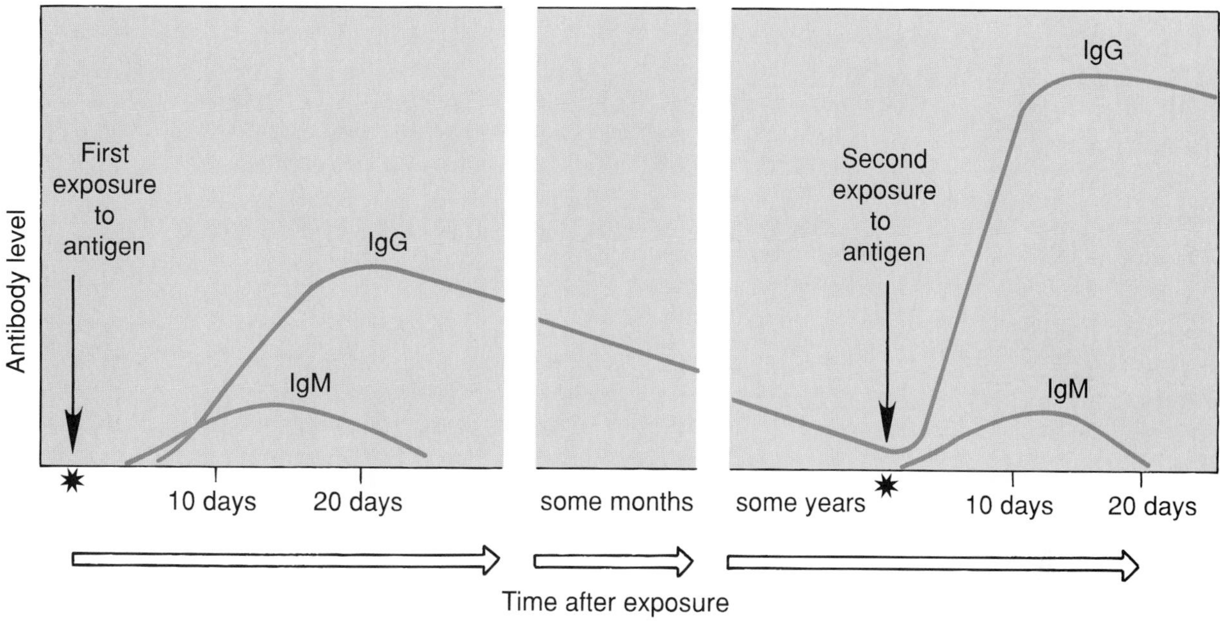

Figure 14–9 Antibody responses to first and subsequent exposures to a pathogen. See text for description.

Box 14–4 VACCINES

The purpose of vaccines is to prevent disease. A vaccine contains an antigen that the immune system will respond to, just as it would to the actual pathogen. The types of vaccine antigens are a killed or weakened **(attenuated)** pathogen, part of a pathogen such as a bacterial capsule, or an inactivated bacterial toxin called a **toxoid**.

Because the vaccine itself does not cause disease (with very rare exceptions), the fact that antibody production to it is slow is not detrimental to the person. The vaccine takes the place of the first exposure to the pathogen and stimulates production of antibodies and memory cells. On exposure to the pathogen itself, the memory cells initiate rapid production of large amounts of antibody, enough to prevent disease.

We now have vaccines for many diseases. The tetanus and diphtheria vaccines contain toxoids, the inactivated toxins of these bacteria. Vaccines for pneumococcal pneumonia and meningitis contain bacterial capsules. These vaccines cannot cause disease because the capsules are nontoxic and nonliving; there is nothing that can reproduce. Influenza and rabies vaccines contain killed viruses. Measles and the oral polio vaccines contain attenuated (weakened) viruses.

Although attenuated pathogens are usually strongly antigenic and stimulate a protective immune response, there is a very small chance that the pathogen may regain its virulence and cause the disease. For example, a few cases each year of polio are caused by the oral polio vaccine itself. The risk is small, however (1 in 500,000), and the vaccine is considered highly effective.

Box 14–5 INTERFERON

Interferon is produced by cells infected with viruses and by T cells. Viruses must be inside a living cell to reproduce, and although interferon cannot prevent the entry of viruses into cells, it does block their reproduction. When viral reproduction is blocked, the viruses cannot infect new cells and cause disease. Interferon is probably a factor in the self-limiting nature of many viral diseases.

There are several types of human interferon (alpha, beta, gamma), which were first produced in amounts sufficient for clinical research in the 1970s. At that time there was hope that interferon would be an effective anti-cancer therapy, but results against the most common forms of cancer were disappointing. Alpha interferon is effective, however, in the treatment of a rare type of leukemia called hairy-cell leukemia and has also been approved for use in cases of genital warts and Kaposi's sarcoma. Most recently, interferon has proved useful in the treatment of hepatitis B and C and multiple sclerosis, although it is not equally effective for all patients.

The research on the interferons is a good example of the way science works. Discovery, research, dead-ends, more research, other possibilities, and so on. Our knowledge is gained piece by piece and often requires many years of careful study. In the future, the interferons will probably become more useful theraputic tools, but there is much to be learned first.

Box 14–6 ALLERGIES

An **allergy** is a hypersensitivity to a particular foreign antigen, called an **allergen**. Allergens include plant pollens, foods, chemicals in cosmetics, antibiotics such as penicillin, dust, and mold spores. Such allergens are not themselves harmful. Most people, for example, can inhale pollen, eat peanuts, or take penicillin with no ill effects.

Hypersensitivity means that the immune system overresponds to the allergen, and produces tissue damage by doing so. Allergic responses are characterized by the production of IgE antibodies, which bond to mast cells. Mast cells are specialized cells that differentiate from basophils, and are numerous in the connective tissue of the skin and mucous membranes. One of several chemicals in mast cells is histamine, which is released by the bonding of IgE antibodies or when tissue damage occurs.

Histamine contributes to the process of inflammation by increasing the permeability of capillaries and venules. When tissue is damaged this promotes greater tissue fluid formation and brings more WBCs to the damaged area.

In an allergic reaction, the effects of histamine and other inflammatory chemicals create symptoms such as watery eyes and runny nose (hay fever) or the more serious wheezing and difficult breathing that characterize asthma. People with seasonal hay fever may take antihistamines to counteract these effects (see Chapter 15 for a description of asthma).

Anaphylactic shock is an extreme allergic response that may be elicited by exposure to penicillin or insect venoms. On the first exposure the person becomes highly sensitized to the foreign antigen. On the second exposure, histamine is released from mast cells throughout the body and causes a drastic decrease in blood volume. The resulting drop in blood pressure may be fatal in only a few minutes. People who know they are allergic to bee stings, for example, may obtain a self-contained syringe of epinephrine to carry with them. Epinephrine can delay the progression of anaphylactic shock long enough for the person to seek medical attention.

to reproduce in our cells and tissues. Because this is a genetic characteristic programmed in DNA, genetic immunity always lasts a lifetime.

Acquired immunity does involve antibodies. **Passive immunity** means that the antibodies are from another source, whereas **active immunity** means that the individual produces his or her own antibodies.

One type of naturally acquired passive immunity is the placental transmission of antibodies (IgG) from maternal blood to fetal circulation. The baby will then be born temporarily immune to the diseases the mother is immune to. Such passive immunity may be prolonged by breast-feeding, because breast milk also contains maternal antibodies (IgA).

Artificially acquired passive immunity is obtained by the injection of immune globulins (gamma globulins or preformed antibodies) after presumed exposure to a particular pathogen. Such immune globulins are available for German measles, hepatitis A and B, tetanus and botulism (anti-toxins), and rabies. These are *not* vaccines; they do not stimulate immune mechanisms, but rather provide immediate antibody protection. Passive immunity is always temporary, lasting a few weeks to a few months, because antibodies from another source eventually break down.

Active immunity is the production of one's own antibodies and may be stimulated by natural or artificial means. Naturally acquired active immunity means that a person has recovered from a disease and now has antibodies and memory cells specific for that pathogen. Artificially acquired active immunity is the result of a vaccine that has stimulated production of antibodies and memory cells (see Box 14–7: Vaccines That Have Changed Our Lives). No general statement can be made about the duration of active immunity. Recovering from plague, for example, confers lifelong immunity, but the plague vaccine does not. Duration of active immunity, therefore, varies with the particular disease or vaccine.

The types of immunity are summarized in Table 14–2.

Box 14–7 VACCINES THAT HAVE CHANGED OUR LIVES

In 1797, Edward Jenner (in England) used the cowpox virus called Vaccinia as the first vaccine for smallpox, a closely related virus. (He was unaware of this, because viruses had not yet been discovered, but he had noticed that milkmaids who got cowpox rarely got smallpox.) In 1980, the World Health Organization declared that smallpox had been eradicated throughout the world. A disease that had killed or disfigured millions of people throughout recorded history is now considered part of history.

In the 19th century in the northern United States, thousands of children died of diphtheria every winter. Today there are fewer than 10 cases of diphtheria each year in the entire country. In the early 1950s, 50,000 cases of paralytic polio were reported in the United States each year. Today, fewer than 10 cases per year are reported.

Smallpox, diphtheria, and polio are no longer the terrible diseases they once were, and this is because of the development and widespread use of vaccines. When people are protected by a vaccine, they are no longer possible reservoirs or sources of the pathogen for others, and the spread of disease may be greatly limited.

Other diseases that have been controlled by the use of vaccines are whooping cough, tetanus, mumps, influenza, measles, and German measles. A new vaccine for hepatitis B has already significantly decreased the number of cases of this disease among health-care workers, and the CDC recommends the vaccine for all children. People who have been exposed to rabies, which is virtually always fatal, can be protected by a new and safe vaccine.

Without such vaccines our lives would be very different. Infant mortality or death in childhood would be much more frequent, and all of us would have to be much more aware of infectious diseases. In many parts of the world this is still true; many of the developing countries in Africa and Asia still cannot afford extensive vaccination programs for their children. Many of the diseases mentioned above, which we may rarely think of, are still a very significant part of the lives of millions of people.

Table 14–2 TYPES OF IMMUNITY

Type	Description
Genetic	• Does not involve antibodies; is programmed in DNA • Some pathogens affect certain host species but not others
Acquired	• Does involve antibodies
Passive	• Antibodies from another source
NATURAL	• Placental transmission of antibodies from mother to fetus • Transmission of antibodies in breast milk
ARTIFICIAL	• Injection of preformed antibodies (gamma globulins or immune globulins) after presumed exposure
Active	• Production of one's own antibodies
NATURAL	• Recovery from a disease, with production of antibodies and memory cells
ARTIFICIAL	• A vaccine stimulates production of antibodies and memory cells

AGING AND THE LYMPHATIC SYSTEM

The aging of the lymphatic system is apparent in the decreased efficiency of immune responses. Elderly people are more susceptible to infections such as influenza and to what are called secondary infections, such as pneumonia following a case of the flu. Vaccines for both of these are available, and elderly people should be encouraged to get them.

Autoimmune disorders are also more common among older people; the immune system mistakenly perceives a body tissue as foreign and initiates its destruction. Rheumatoid arthritis is such an autoimmune disease. The incidence of cancer is also higher. Malignant cells that once might have been quickly destroyed remain alive and proliferate.

SUMMARY

The preceding discussions of immunity will give you a small idea of the complexity of the body's defense system. However, there is still much more to be learned, especially about the effects of the nervous system and endocrine system on immunity. For example, it is known that people under great stress have immune systems that may not function as they did when stress was absent.

At present, there is much research being done in this field. The goal is not to eliminate all disease, for that would not be possible. Rather, the aim is to enable people to live healthier lives by preventing certain diseases.

STUDY OUTLINE

Functions of the Lymphatic System

1. To return tissue fluid to the blood to maintain blood volume (see Fig. 14–1).
2. To protect the body against pathogens and other foreign material.

Parts of the Lymphatic System

1. Lymph
2. Lymph vessels
3. Lymph nodes and nodules
4. Spleen
5. Thymus

Lymph—the tissue fluid that enters lymph capillaries

1. Similar to plasma, but more WBCs are present.
2. Must be returned to the blood to maintain blood volume and blood pressure.

Lymph Vessels

1. Dead-end lymph capillaries are found in most tissue spaces; collect tissue fluid and proteins (see Fig. 14–2).
2. The structure of larger lymph vessels is like that of veins; valves prevent the backflow of lymph.

3. Lymph is kept moving in lymph vessels by:
   - constriction of the lymph vessels
   - the skeletal muscle pump
   - the respiratory pump
4. Lymph from the lower body and upper left quadrant enters the thoracic duct and is returned to the blood in the left subclavian vein (see Fig. 14-3).
5. Lymph from the upper right quadrant enters the right lymphatic duct and is returned to the blood in the right subclavian vein.

Lymph Nodes—masses of lymphatic tissue; produce lymphocytes and monocytes

1. Found in groups along the pathways of lymph vessels.
2. As lymph flows through the nodes:
   - lymphocytes and monocytes enter the lymph
   - foreign materials are phagocytized by fixed macrophages
   - fixed plasma cells produce antibodies to foreign antigens (see Fig. 14-4)
3. The major paired groups of lymph nodes are the cervical, axillary, and inguinal groups. These are at the junctions of the head and extremities with the trunk; remove pathogens from the lymph from the extremities before the lymph is returned to the blood.

Lymph Nodules—small masses of lymphatic tissue; produce lymphocytes and monocytes

1. Found beneath the epithelium of all mucous membranes, that is, the tracts that have natural openings to the environment.
2. Destroy pathogens that penetrate the epithelium of the respiratory, digestive, urinary, or reproductive tracts.
3. Tonsils are the lymph nodules of the pharynx; Peyer's patches are those of the small intestine.

Spleen—located in the upper left abdominal quadrant behind the stomach

1. The fetal spleen produces RBCs.
2. Functions after birth:
   - production of lymphocytes and monocytes
   - fixed plasma cells produce antibodies
   - fixed macrophages (RE cells) phagocytize pathogens and old RBCs; bilirubin is formed and sent to the liver for excretion in bile

Thymus—in the fetus and infant the thymus is large and inferior to the thyroid gland; with age the thymus shrinks (see Fig. 14–5)

1. Produces T lymphocytes (T cells).
2. Produces thymic hormones that make T cells immunologically competent: able to recognize foreign antigens and provide immunity.

Immunity

1. The ability to destroy foreign antigens and prevent future cases of certain infectious diseases.
2. Foreign antigens include bacteria, viruses, fungi, protozoa, and malignant cells.

Lymphocytes

1. T lymphocytes (T cells)—in the embryo are produced in the thymus and RBM; they require the hormones of the thymus for maturation; migrate to the spleen, lymph nodes, and nodules.
2. B lymphocytes (B cells)—in the embryo are produced in the RBM; migrate to the spleen, lymph nodes, and nodules.

Antigens

1. Chemical markers that identify cells.
2. Human cells have "self" antigens—the HLA types.
3. Foreign antigens stimulate antibody production or other immune responses.

Antibodies—immune globulins or gamma globulins (see Table 14–1 and Fig. 14–6)

1. Proteins produced by plasma cells in response to foreign antigens.
2. Each antibody is specific for only one foreign antigen.
3. Bond to the foreign antigen to label it for phagocytosis (opsonization).

Mechanisms of Immunity

1. The antigen must first be recognized as foreign; this is accomplished by B cells or by helper T cells that compare the foreign antigen to "self" antigens present on macrophages.
2. Helper T cells strongly initiate one or both of the immune mechanisms: cell-mediated immunity and humoral immunity.

Cell-Mediated Immunity (see Fig. 14–7)

1. Does not involve antibodies; is effective against intracellular pathogens, malignant cells, and grafts of foreign tissue.
2. Helper T cells recognize the foreign antigen, become antigen specific, and begin to divide to form different groups of T cells.
3. Memory T cells will remember the specific foreign antigen.
4. Cytotoxic (killer) T cells chemically destroy foreign cells and produce cytokines to attract macrophages.
5. Suppressor T cells stop the immune response once the foreign antigen has been destroyed.

Humoral Immunity (see Fig. 14–8)

1. Does involve antibody production; is effective against pathogens and foreign cells.
2. B cells and helper T cells recognize the foreign antigen; the B cells are antigen-specific and begin to divide.
3. Memory B cells will remember the specific foreign antigen.
4. Other B cells become plasma cells that produce antigen-specific antibodies.
5. An antigen-antibody complex is formed, which attracts macrophages (opsonization).
6. Complement fixation is stimulated by antigen-antibody complexes. The complement proteins bind to the antigen-antibody complex and lyse cellular antigens or enhance the phagocytosis of noncellular antigens.
7. Suppressor T cells stop the immune response when the foreign antigen has been destroyed.

Antibody Responses and Functions (see Fig. 14–9)

1. On the first exposure to a foreign antigen, antibodies are produced slowly and in small amounts, and the person may develop clinical disease.
2. On the second exposure, the memory cells initiate rapid production of large amounts of antibodies, and a second case of the disease may be prevented. This is the basis for the protection given by vaccines, which take the place of the first exposure.
3. Antibodies cause agglutination (clumping) of bacterial cells; clumped cells are easier for macrophages to phagocytize (see Fig. 14–6).
4. Antibodies neutralize viruses by bonding to them and preventing their entry into cells.
5. Antibodies neutralize bacterial toxins by bonding to them and changing their shape.

Types of Immunity (see Table 14–2)

REVIEW QUESTIONS

1. Explain the relationships among plasma, tissue fluid, and lymph, in terms of movement of water throughout the body. (p. 304)
2. Describe the system of lymph vessels. Explain how lymph is kept moving in these vessels. Into which veins is lymph emptied? (pp. 304, 307)
3. State the locations of the major groups of lymph nodes, and explain their functions. (pp. 304, 306)
4. State the locations of lymph nodules, and explain their functions. (p. 304)
5. Describe the location of the spleen and explain its functions. If the spleen is removed, what organs will compensate for its functions? (pp. 306, 309)
6. Explain the function of the thymus, and state when (age) this function is important. (pp. 309–310)
7. Name the different kinds of foreign antigens that the immune system responds to. (p. 311)
8. State the functions of helper T cells, cytotoxic T cells, memory T cells, and suppressor T cells. (p. 312)

9. Plasma cells differentiate from which type of lymphocyte? State the function of plasma cells. What other type of cell comes from B lymphocytes? (p. 312)

10. Explain how a foreign antigen is recognized as foreign. Which mechanism of immunity involves antibody production? Explain what opsonization means. (p. 312)

11. What is the stimulus for complement fixation? How does this process destroy cellular antigens and non-cellular antigens? (p. 312)

12. Explain the antibody reactions of agglutination and neutralization. (p. 316)

13. Explain how a vaccine provides protective immunity in terms of first and second exposures to a pathogen. (p. 316)

14. Explain the difference between the following: (pp. 316, 319–320)
    a. Genetic immunity and acquired immunity
    b. Passive acquired immunity and active acquired immunity
    c. Natural and artificial passive acquired immunity
    d. Natural and artificial active acquired immunity

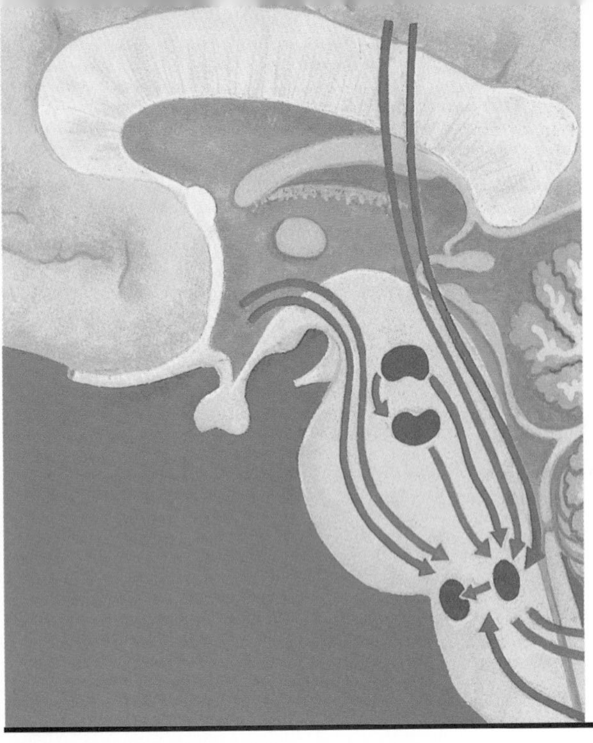

CHAPTER 15

Chapter Outline

Student Objectives

- State the general function of the respiratory system.
- Describe the structure and functions of the nasal cavities and pharynx.
- Describe the structure of the larynx and explain the speaking mechanism.
- Describe the structure and functions of the trachea and bronchial tree.
- State the locations of the pleural membranes, and explain the functions of serous fluid.
- Describe the structure of the alveoli and pulmonary capillaries, and explain the importance of surfactant.
- Name and describe the important air pressures involved in breathing.
- Describe normal inhalation and exhalation and forced exhalation.
- Explain the diffusion of gases in external respiration and internal respiration.
- Describe how oxygen and carbon dioxide are transported in the blood.
- Name the pulmonary volumes and define each.
- Explain the nervous and chemical mechanisms that regulate respiration.
- Explain how respiration affects the pH of body fluids.

The Respiratory System

New Terminology

Alveoli (al-**VEE**-oh-lye)
Bronchial tree (**BRONG**-kee-uhl TREE)
Epiglottis (ep-i-**GLAH**-tis)
Glottis (**GLAH**-tis)
Intrapleural pressure (IN-trah-**PLOOR**-uhl **PRES**-shur)
Intrapulmonic pressure (IN-trah-pull-**MAHN**-ik **PRES**-shur)
Larynx (**LA**-rinks)
Partial pressure (**PAR**-shul **PRES**-shur)
Phrenic nerves (**FREN**-ik NURVZ)
Pulmonary surfactant (**PULL**-muh-ner-ee sir-**FAK**-tent)
Residual air (ree-**ZID**-yoo-al AYRE)
Respiratory acidosis (RES-pi-rah-**TOR**-ee ass-i-**DOH**-sis)
Respiratory alkalosis (RES-pi-rah-**TOR**-ee al-kah-**LOH**-sis)
Soft palate (SAWFT **PAL**-uht)
Tidal volume (**TIGH**-duhl **VAHL**-yoom)
Ventilation (VEN-ti-**LAY**-shun)
Vital capacity (**VY**-tuhl kuh-**PASS**-i-tee)

Related Clinical Terminology

Cyanosis (SIGH-uh-**NOH**-sis)
Dyspnea (**DISP**-nee-ah)
Emphysema (EM-fi-**SEE**-mah)
Heimlich maneuver (**HIGHM**-lik ma-**NEW**-ver)
Hyaline membrane disease (**HIGH**-e-lin **MEM**-brain dis-**EEZ**)
Pneumonia (new-**MOH**-nee-ah)
Pneumothorax (NEW-moh-**THAW**-raks)
Pulmonary edema (**PULL**-muh-ner-ee uh-**DEE**-muh).

Terms that appear in ***bold type*** *in the chapter text are defined in the glossary, which begins on page 528.*

Sometimes a person will describe a habit as being "as natural as breathing." Indeed, what could be more natural? We rarely think about breathing, and it isn't something we look forward to, as we would a good dinner. We just breathe, usually at the rate of 12 to 20 times per minute, and faster when necessary (such as during exercise). You may have heard of trained singers "learning how to breathe," but they are really learning how to make their breathing more efficient.

Most of the **respiratory system** is concerned with what we think of as breathing: moving air into and out of the lungs. The lungs are the site of the exchanges of oxygen and carbon dioxide between the air and the blood. Both of these exchanges are important. All our cells must obtain oxygen to carry out cell respiration to produce ATP. Just as crucial is the elimination of the CO_2 produced as a waste product of cell respiration, and, as you already know, the proper functioning of the circulatory system is essential for the transport of these gases in the blood.

DIVISIONS OF THE RESPIRATORY SYSTEM

The respiratory system may be divided into the upper respiratory tract and the lower respiratory tract. The **upper respiratory tract** consists of the parts outside the chest cavity: the air passages of the nose, nasal cavities, pharynx, larynx, and upper trachea. The **lower respiratory tract** consists of the parts found within the chest cavity: the lower trachea and the lungs themselves, which include the bronchial tubes and alveoli. Also part of the respiratory system are the pleural membranes and the respiratory muscles that form the chest cavity: the diaphragm and intercostal muscles.

Have you recognized some familiar organs and structures thus far? There will be more, for this chapter includes material from Chapters 1 through 9, 11, and 12. Even though we are discussing the body system by system, the respiratory system is an excellent example of the interdependent functioning of all the body systems.

NOSE AND NASAL CAVITIES

Air enters and leaves the respiratory system through the **nose**, which is made of bone and cartilage covered with skin. Just inside the nostrils are hairs, which help block the entry of dust.

The two **nasal cavities** are within the skull, separated by the **nasal septum**, which is a bony plate made of the ethmoid bone and vomer. The **nasal mucosa** (lining) is ciliated epithelium, with goblet cells that produce mucus. The surface area of the nasal mucosa is increased by the conchae, shelf-like bones on the lateral wall of each nasal cavity (Fig. 15–1). As air passes through the nasal cavities it is warmed and humidified, so that air that reaches the lungs is warm and moist. Bacteria and particles of air pollution are trapped on the mucus; the cilia continuously sweep the mucus toward the pharynx. Most of this mucus is eventually swallowed, and any bacteria present will be destroyed by the hydrochloric acid in the gastric juice.

In the upper nasal cavities are the **olfactory receptors**, which detect vaporized chemicals that have been inhaled. The olfactory nerves pass through the ethmoid bone to the brain.

You may also recall the **paranasal sinuses**, air cavities in the maxillae, frontal, sphenoid, and ethmoid bones (see Figs. 15–1 and 6–9). These sinuses are lined with ciliated epithelium, and the mucus produced drains into the nasal cavities. The functions of the paranasal sinuses are to lighten the skull and provide resonance for the voice.

PHARYNX

The **pharynx** is a muscular tube posterior to the nasal and oral cavities and anterior to the cervical vertebrae. For descriptive purposes, the pharynx may be divided into three parts: the nasopharynx, oropharynx, and laryngopharynx (see Fig. 15–1).

The uppermost portion is the **nasopharynx**, which is behind the nasal cavities. The **soft palate** is elevated during swallowing to block the nasopharynx and prevent food or saliva from going up rather than down. The uvula is the part of the soft palate you can see at the back of the throat. On the posterior wall of the nasopharynx is the adenoid or pharyngeal tonsil, a lymph nodule that contains macrophages. Opening into the nasopharynx are the two eustachian tubes, which extend to the middle ear cavities. The purpose of the eustachian tubes is to permit air to enter or leave the middle ears, allowing the eardrums to vibrate properly.

The nasopharynx is a passageway for air only, but

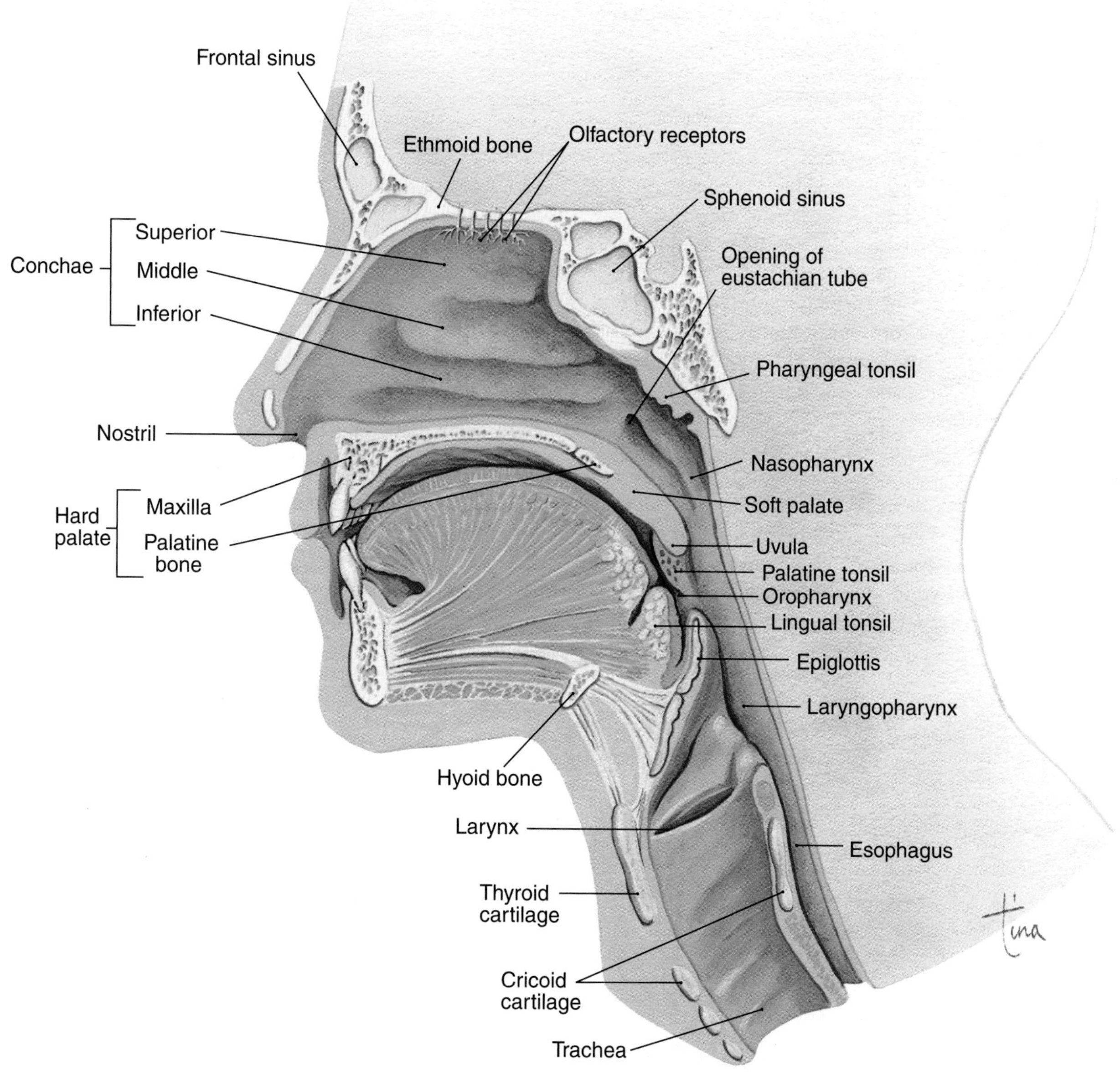

Figure 15–1 Midsagittal section of the head and neck showing the structures of the upper respiratory tract.

the remainder of the pharynx serves as both an air and food passageway, although not for both at the same time. The **oropharynx** is behind the mouth; its mucosa is stratified squamous epithelium, continuous with that of the oral cavity. On its lateral walls are the palatine tonsils, also lymph nodules. Together with the adenoid and the lingual tonsils on the base of the tongue, they form a ring of lymphatic tissue around the pharynx to destroy pathogens that penetrate the mucosa.

The **laryngopharynx** is the most inferior portion of the pharynx. It opens anteriorly into the larynx and posteriorly into the esophagus. Contraction of the muscular wall of the oropharynx and laryngopharynx is part of the swallowing reflex.

LARYNX

The **larynx** is often called the voice box, a name that indicates one of its functions, which is speaking. The other function of the larynx is to be an air passageway between the pharynx and the trachea. Air passages must be kept open at all times, and so the larynx is made of nine pieces of cartilage connected by ligaments. Cartilage is a firm yet flexible tissue that prevents collapse of the larynx. In comparison, the esophagus is a collapsed tube except when food is passing through it.

The largest cartilage of the larynx is the **thyroid cartilage** (Fig. 15–2), which you can feel on the anterior surface of your neck. The **epiglottis** is the uppermost cartilage. During swallowing, the larynx is elevated, and the epiglottis closes over the top to prevent the entry of food into the larynx.

The mucosa of the larynx is ciliated epithelium, except for the vocal cords (stratified squamous epithelium). The cilia of the mucosa sweep upward to remove mucus and trapped dust and microorganisms.

The **vocal cords** (or vocal folds) are on either side of the **glottis**, the opening between them. During breathing, the vocal cords are held at the sides of the glottis, so that air passes freely into and out of the trachea (Fig. 15–3). During speaking, the intrinsic muscles of the larynx pull the vocal cords across the glottis, and exhaled air vibrates the vocal cords to produce sounds which can be turned into speech. It is also physically possible to speak while inhaling, but this is not what we are used to. The cranial nerves that are motor nerves to the larynx for speaking are the vagus and accessory nerves.

TRACHEA AND BRONCHIAL TREE

The **trachea** is about 4 to 5 inches (10 to 13 cm) long and extends from the larynx to the primary bronchi. The wall of the trachea contains 16 to 20 C-shaped pieces of cartilage, which keep the trachea open. The gaps in these incomplete cartilage rings are posterior, to permit the expansion of the esophagus when food is swallowed. The mucosa of the trachea is ciliated

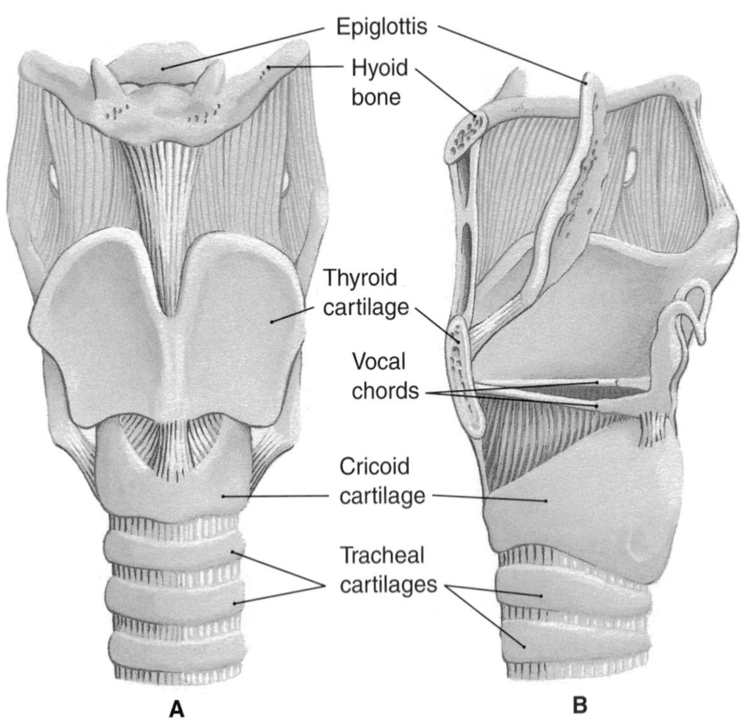

Figure 15–2 Larynx. **(A)** Anterior view. **(B)** Midsagittal section through the larynx, viewed from the left side.

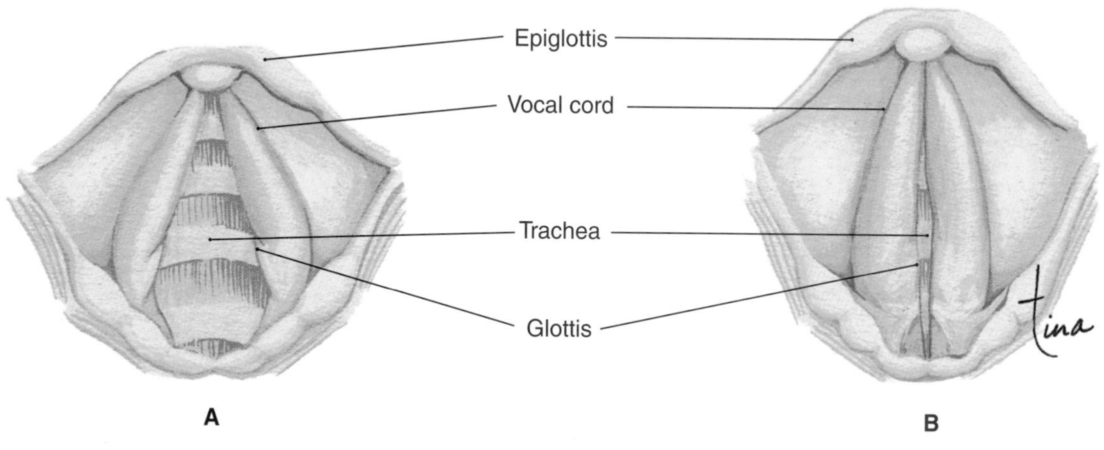

Figure 15–3 Vocal cords and glottis, superior view. **(A)** Position of the vocal cords during breathing. **(B)** Position of the vocal cords during speaking.

epithelium with goblet cells. As in the larynx, the cilia sweep upward toward the pharynx.

The right and left **primary bronchi** (Fig. 15–4) are the branches of the trachea that enter the lungs. Within the lungs, each primary bronchus branches into secondary bronchi leading to the lobes of each lung (three right, two left). The further branching of the bronchial tubes is often called the **bronchial tree**. Imagine the trachea as the trunk of an upside-down tree with extensive branches that become smaller and smaller; these smaller branches are the **bronchioles**. No cartilage is present in the walls of the bronchioles; this becomes clinically important in asthma (see Box 15–1: Asthma). The smallest bronchioles terminate in clusters of alveoli, the air sacs of the lungs.

LUNGS AND PLEURAL MEMBRANES

The **lungs** are located on either side of the heart in the chest cavity and are encircled and protected by the rib cage. The base of each lung rests on the diaphragm below; the apex (superior tip) is at the level of the clavicle. On the medial surface of each lung is an indentation called the **hilus**, where the primary bronchus and the pulmonary artery and veins enter the lung.

The pleural membranes are the serous membranes of the thoracic cavity. The **parietal pleura** lines the chest wall, and the **visceral pleura** is on the surface of the lungs. Between the pleural membranes is serous fluid, which prevents friction and keeps the two membranes together during breathing.

Alveoli

The functional units of the lungs are the air sacs called **alveoli**. The flat alveolar type I cells that form most of the alveolar walls are simple squamous epithelium. In the spaces between clusters of alveoli is elastic connective tissue, which is important for exhalation. Within the alveoli are macrophages that phagocytize pathogens or other foreign material that may not have been swept out by the ciliated epithelium of the bronchial tree. There are millions of alveoli in each lung, and each alveolus is surrounded by a network of pulmonary capillaries (see Fig 15–4). Recall that capillaries are also made of simple squamous epithelium, so there are only two cells between the air in the alveoli and the blood in the pulmonary capillaries, which permits efficient diffusion of gases (Fig. 15–5).

Each alveolus is lined with a thin layer of tissue fluid, which is essential for the diffusion of gases, because a gas must dissolve in a liquid in order to enter or leave a cell (the earthworm principle—an earthworm breathes through its moist skin, and will suffocate if its skin dries out). Although this tissue fluid is necessary, it creates a potential problem in that it would make the walls of an alveolus stick together internally.

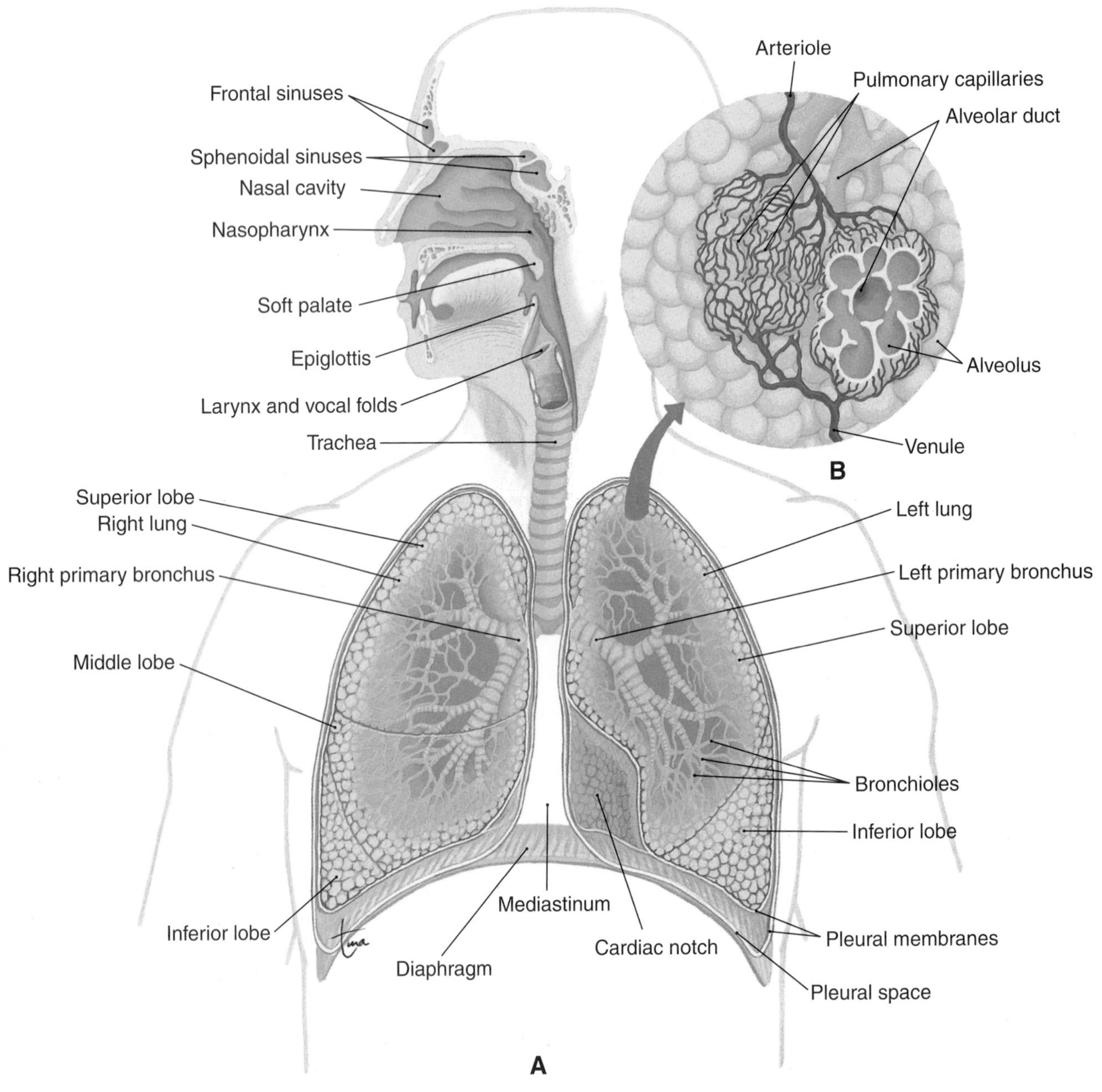

Figure 15–4 Respiratory system. **(A)** Anterior view of the upper and lower respiratory tracts. **(B)** Microscopic view of alveoli and pulmonary capillaries. (The colors represent the vessels, not the oxygen content of the blood within the vessel.)

Box 15–1 ASTHMA

Asthma is usually triggered by an allergic reaction that affects the smooth muscle and glands of the bronchioles. Allergens include foods and inhaled substances such as dust and pollen. Wheezing and dyspnea (difficult breathing) characterize an asthma attack, which may range from mild to fatal.

As part of the allergic response, the smooth muscle of the bronchioles constricts. Because there is no cartilage present in their walls, the bronchioles may close completely. The secretion of mucus increases, perhaps markedly, so the already constricted bronchioles may become clogged or completely obstructed with mucus.

Chronic asthma is a predisposing factor for emphysema. When obstructed bronchioles prevent ventilation of alveoli, the walls of the alveoli begin to deteriorate and break down, leaving large cavities that do not provide much surface area for gas exchange.

Imagine a plastic bag that is wet inside; its walls would stick together because of the surface tension of the water. This is just what would happen in alveoli, and inflation would be very difficult.

This problem is overcome by **pulmonary surfactant**, a lipoprotein secreted by alveolar type II cells, also called septal cells. Surfactant mixes with the tissue fluid within the alveoli and decreases its surface tension, permitting inflation of the alveoli (see Box 15-2: Hyaline Membrane Disease). Normal inflation of the alveoli in turn permits the exchange of gases, but before we discuss this process, we will first see how air gets into and out of the lungs.

MECHANISM OF BREATHING

Ventilation is the term for the movement of air to and from the alveoli. The two aspects of ventilation are inhalation and exhalation, which are brought about by the nervous system and the respiratory muscles. The respiratory centers are located in the medulla and pons. Their specific functions will be covered in a later section, but it is the medulla that generates impulses to the respiratory muscles.

These muscles are the diaphragm and the external and internal intercostal muscles (Fig. 15-6). The **diaphragm** is a dome-shaped muscle below the lungs; when it contracts, the diaphragm flattens and moves downward. The intercostal muscles are found between the ribs. The **external intercostal muscles** pull the ribs upward and outward, and the **internal intercostal muscles** pull the ribs downward and inward. Ventilation is the result of the respiratory muscles producing changes in the pressure within the alveoli and bronchial tree.

With respect to breathing, the important pressures are these three:

1. **Atmospheric Pressure**—the pressure of the air around us. At sea level, atmospheric pressure is 760 mmHg. At higher altitudes, of course, atmospheric pressure is lower.
2. **Intrapleural Pressure**—the pressure within the potential pleural space between the parietal pleura and visceral pleura. This is a potential rather than a real space. A thin layer of serous fluid causes the two pleural membranes to adhere to one another. Intrapleural pressure is always slightly below atmospheric pressure (about 756 mmHg), and is called a "negative" pressure. The elastic lungs are always tending to collapse and pull the visceral pleura away from the parietal pleura. The serous fluid, however, prevents actual separation of the pleural membranes (see Box 15-3: Pneumothorax).
3. **Intrapulmonic Pressure**—the pressure within the bronchial tree and alveoli. This pressure fluctuates below and above atmospheric pressure during each cycle of breathing.

INHALATION

Inhalation, also called **inspiration**, is a precise sequence of events that may be described as follows:

Motor impulses from the medulla travel along the

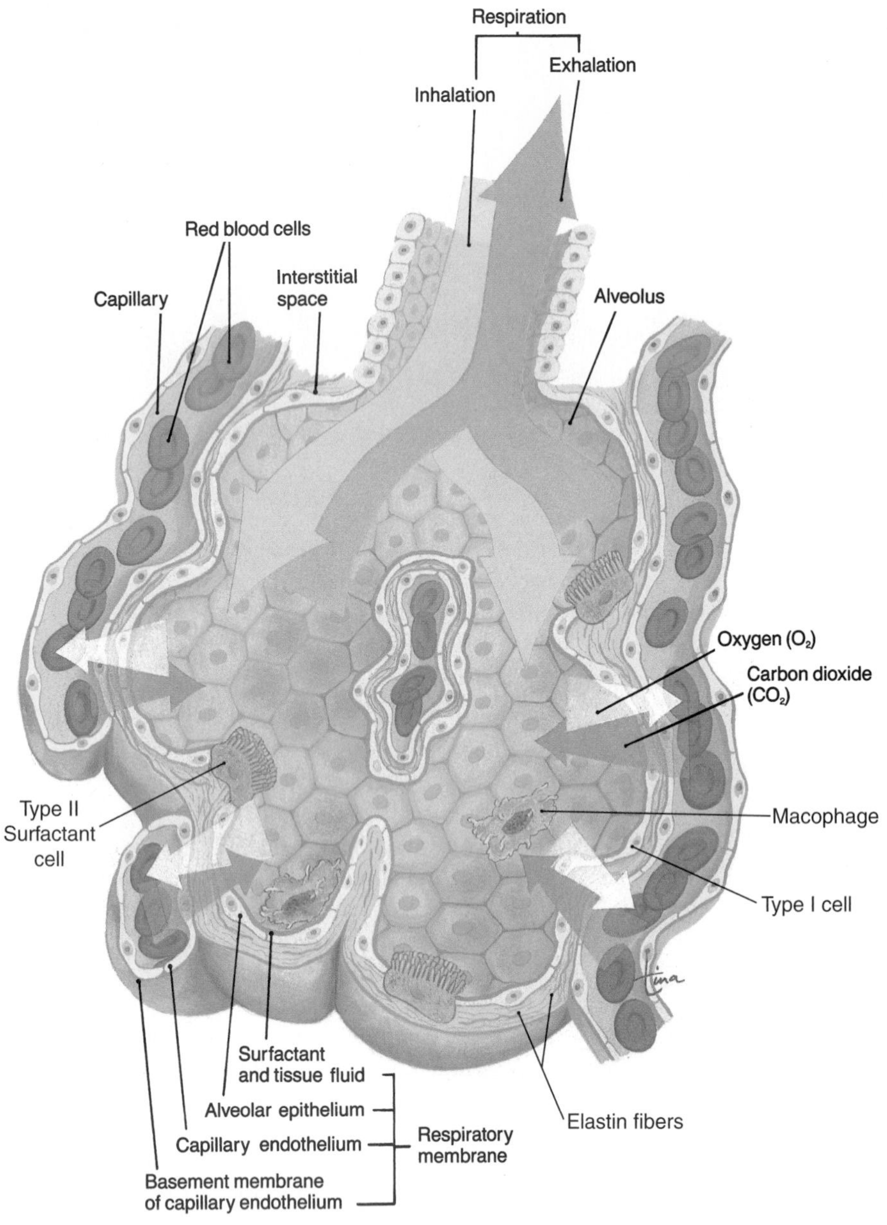

Figure 15–5 Alveolar structure showing type I and type II cells, and alveolar macrophages. The respiratory membrane: the structures and substances through which gases must pass as they diffuse from air to blood (oxygen) or from blood to air (CO_2).

phrenic nerves to the diaphragm and along the **intercostal nerves** to the external intercostal muscles. The diaphragm contracts, moves downward, and expands the chest cavity from top to bottom. The external intercostal muscles pull the ribs up and out, which expands the chest cavity from side to side and front to back.

As the chest cavity is expanded, the parietal pleura

Box 15–2 HYALINE MEMBRANE DISEASE

Hyaline membrane disease is also called respiratory distress syndrome (RDS) of the newborn, and most often affects premature infants whose lungs have not yet produced sufficient quantities of pulmonary surfactant.

The first few breaths of a newborn inflate most of the previously collapsed lungs, and the presence of surfactant permits the alveoli to remain open. The following breaths become much easier, and normal breathing is established.

Without surfactant, the surface tension of the tissue fluid lining the alveoli causes the air sacs to collapse after each breath rather than remain inflated. Each breath, therefore, is difficult, and the newborn must expend a great deal of energy just to breathe.

Premature infants may require respiratory assistance until their lungs are mature enough to produce surfactant. Clinical trials of a synthetic surfactant have shown that some infants are helped significantly, and because they can breathe more normally, their dependence on respirators is minimized. Still undergoing evaluation are the effects of the long-term use of this surfactant in the most premature babies, who may require it for much longer periods of time.

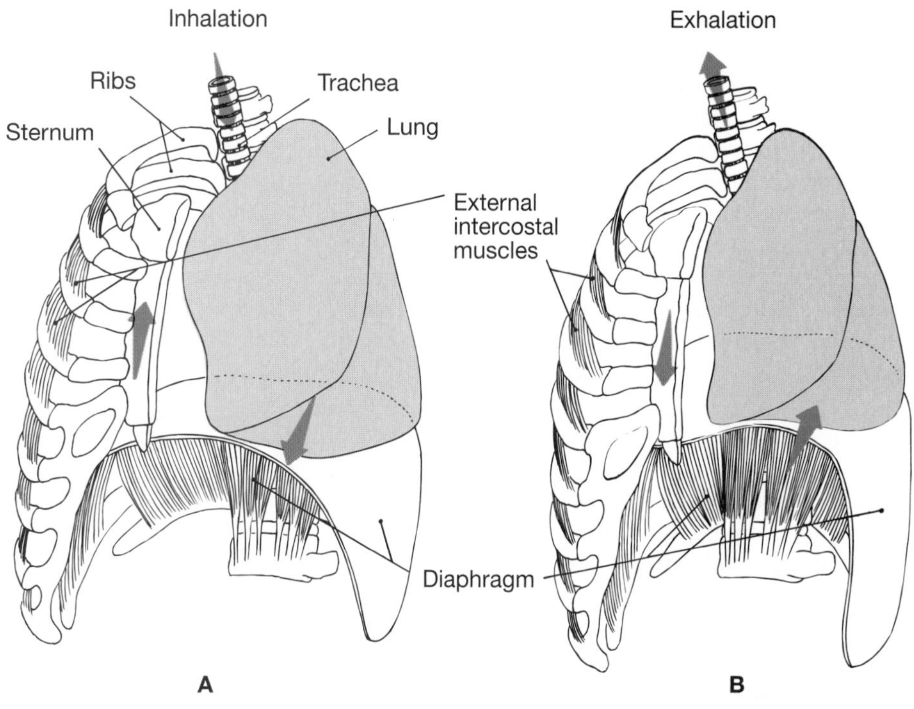

Figure 15–6 Actions of the respiratory muscles. **(A)** Inhalation: diaphragm contracts downward; external intercostal muscles pull rib cage upward and outward; lungs are expanded. **(B)** Normal exhalation: diaphragm relaxes upward; rib cage falls down and in as external intercostal muscles relax; lungs are compressed.

Box 15–3 PNEUMOTHORAX

Pneumothorax is the presence of air in the pleural space, which causes collapse of the lung on that side. Recall that the pleural space is only a potential space because the serous fluid keeps the pleural membranes adhering to one another, and the intrapleural pressure is always slightly below atmospheric pressure. Should air at atmospheric pressure enter the pleural cavity, the suddenly higher pressure outside the lung will contribute to its collapse (the other factor is the normal elasticity of the lungs).

A spontaneous pneumothorax, without apparent trauma, may result from rupture of weakened alveoli on the lung surface. Pulmonary diseases such as emphysema may weaken alveoli.

Puncture wounds of the chest wall also allow air into the pleural space, with resulting collapse of a lung. In severe cases, large amounts of air push the heart, great vessels, trachea, and esophagus toward the opposite side (mediastinal shift), putting pressure on the other lung and making breathing difficult. This is called tension pneumothorax, and requires rapid medical intervention to remove the trapped air.

expands with it. Intrapleural pressure becomes even more negative as a sort of suction is created between the pleural membranes. The adhesion created by the serous fluid, however, permits the visceral pleura to be expanded too, and this expands the lungs as well.

As the lungs expand, intrapulmonic pressure falls below atmospheric pressure, and air enters the nose and travels through the respiratory passages to the alveoli. Entry of air continues until intrapulmonic pressure is equal to atmospheric pressure; this is a normal inhalation. Of course, inhalation can be continued beyond normal, that is, a deep breath. This requires a more forceful contraction of the respiratory muscles to further expand the lungs, permitting the entry of more air.

EXHALATION

Exhalation may also be called **expiration** and begins when motor impulses from the medulla decrease, and the diaphragm and external intercostal muscles relax. As the chest cavity becomes smaller, the lungs are compressed, and their elastic connective tissue, which was stretched during inhalation, recoils and also compresses the alveoli. As intrapulmonic pressure rises above atmospheric pressure, air is forced out of the lungs until the two pressures are again equal.

Notice that inhalation is an active process that requires muscle contraction, but normal exhalation is a passive process, depending to a great extent on the normal elasticity of healthy lungs. In other words, under normal circumstances we must expend energy to inhale but not to exhale (see Box 15–4: Emphysema).

We can, however, go beyond a normal exhalation and expel more air, as when talking, singing, or blowing up a balloon. Such a forced exhalation is an active process that requires contraction of other muscles. Contraction of the internal intercostal muscles pulls the ribs down and in and squeezes even more air out of the lungs. Contraction of abdominal muscles, such as the rectus abdominus, compresses the abdominal organs and pushes the diaphragm upward, which also forces more air out of the lungs (see Box 15–5: The Heimlich Maneuver).

EXCHANGE OF GASES

There are two sites of exchange of oxygen and carbon dioxide: the lungs and the tissues of the body. The exchange of gases between the air in the alveoli and the blood in the pulmonary capillaries is called **external respiration**. This term may be a bit confusing at first, because we often think of "external" as being outside the body. In this case, however, "external" means the exchange that involves air from the external environment. **Internal respiration** is the exchange of gases between the blood in the systemic capillaries and the tissue fluid (cells) of the body.

Box 15–4 EMPHYSEMA

Emphysema, a form of chronic obstructive pulmonary disease (COPD), is a degenerative disease in which the alveoli lose their elasticity and cannot recoil. Perhaps the most common (and avoidable) cause is cigarette smoking; other causes are long-term exposure to severe air pollution or industrial dusts, or chronic asthma. Inhaled irritants damage the alveolar walls and cause deterioration of the elastic connective tissue surrounding the alveoli. Macrophages migrate to the damaged areas, and seem to produce an enzyme that contributes to the destruction of the protein elastin. This is an instance of a useful body response (for cleaning up damaged tissue) becoming damaging when it is excessive. As the alveoli break down, larger air cavities are created that are not efficient in gas exchange (see below).

In progressive emphysema, damaged lung tissue is replaced by fibrous connective tissue (scar tissue), which further limits the diffusion of gases. Blood oxygen level decreases, and blood carbon dioxide level increases. Accumulating carbon dioxide decreases the pH of body fluids; this is a respiratory acidosis.

One of the most characteristic signs of emphysema is that the affected person must make an effort to exhale. The loss of lung elasticity makes normal exhalation an active process, rather than the passive process it usually is. The person must expend energy to exhale in order to make room in the lungs for inhaled air. This extra "work" required for exhalation may be exhausting for the person and contribute to the debilitating nature of emphysema.

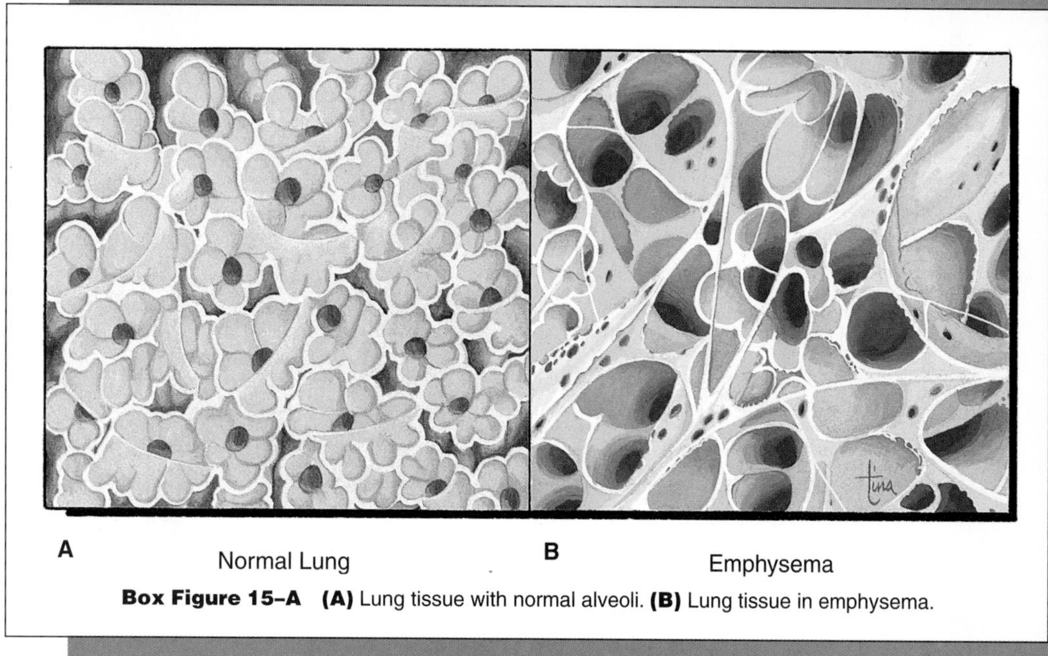

Box Figure 15–A (A) Lung tissue with normal alveoli. **(B)** Lung tissue in emphysema.

The air we inhale (the earth's atmosphere) is approximately 21% oxygen and 0.04% carbon dioxide. Although most (78%) of the atmosphere is nitrogen, this gas is not physiologically available to us, and we simply exhale it. This exhaled air also contains about 16% oxygen and 4.5% carbon dioxide, so it is apparent that some oxygen is retained within the body, and the carbon dioxide produced by cells is exhaled.

Box 15–5 THE HEIMLICH MANEUVER

The **Heimlich maneuver** has received much well-deserved publicity, and indeed it is a life-saving technique.

If a person is choking on a foreign object (such as food) lodged in the pharynx or larynx, the air in the lungs may be utilized to remove the object. The physiology of this technique is illustrated in the accompanying figure.

The person performing the maneuver stands behind the choking victim and puts both arms around the victim's waist. One hand forms a fist that is placed between the victim's navel and rib cage (below the diaphragm), and the other hand covers the fist. It is important to place hands correctly, in order to avoid breaking the victim's ribs. With both hands, a quick, forceful upward thrust is made and repeated if necessary. This forces the diaphragm upward to compress the lungs and force air out. The forcefully expelled air is often sufficient to dislodge the foreign object.

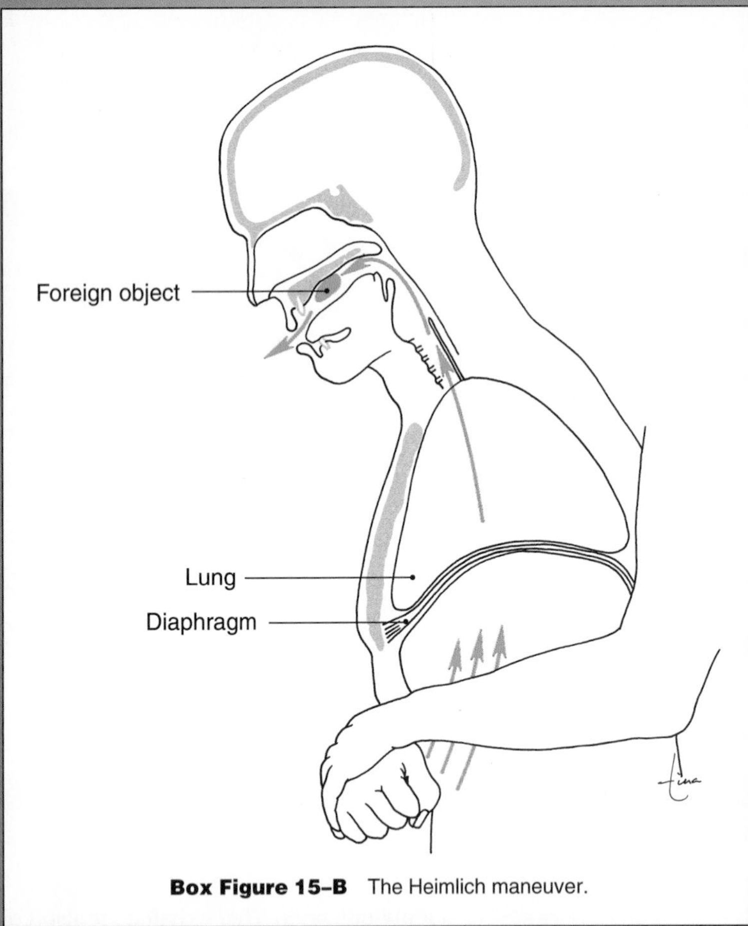

Box Figure 15–B The Heimlich maneuver.

DIFFUSION OF GASES—PARTIAL PRESSURES

Within the body, a gas will diffuse from an area of greater concentration to an area of lesser concentration. The concentration of each gas in a particular site (alveolar air, pulmonary blood, and so on) is expressed in a value called **partial pressure**. The partial pressure of a gas, measured in mmHg, is the pressure it exerts within a mixture of gases, whether the mixture is actually in a gaseous state or is in a liquid such as blood. The partial pressures of oxygen and carbon dioxide in the atmosphere and in the sites of exchange in the body are listed in Table 15-1. The abbreviation for partial pressure is "P," which is used, for example, on hospital lab slips for blood gases and will be used here.

The partial pressures of oxygen and carbon dioxide at the sites of external respiration (lungs) and internal respiration (body) are shown in Fig. 15-7. Because partial pressure reflects concentration, a gas will diffuse from an area of higher partial pressure to an area of lower partial pressure.

The air in the alveoli has a high Po_2 and a low Pco_2. The blood in the pulmonary capillaries, which has just come from the body, has a low Po_2 and a high Pco_2.

Table 15–1 PARTIAL PRESSURES

Site	Po_2 (mmHg)	Pco_2 (mmHg)
Atmosphere	160	0.15
Alveolar air	104	40
Pulmonary blood (venous)	40	45
Systemic blood (arterial)	100	40
Tissue fluid	40	50

Partial pressure is calculated as follows:

% of the gas in the mixture × total pressure = Pgas

Example: **O_2 in the atmosphere**

21% × 760 mmHg = 160 mmHg (Po_2)

Example: **CO_2 in the atmosphere**

0.04% × 760 mmHg = 0.15 mmHg (Pco_2)

Notice that alveolar partial pressures are not exactly those of the atmosphere. Alveolar air contains significant amounts of water vapor and the CO_2 diffusing in from the blood. Oxygen also diffuses readily from the alveoli into the pulmonary capillaries. Therefore, alveolar Po_2 is lower than atmospheric Po_2, and alveolar Pco_2 is significantly higher than atmospheric Pco_2.

Therefore, in external respiration, oxygen diffuses from the air in the alveoli to the blood, and carbon dioxide diffuses from the blood to the air in the alveoli. The blood that returns to the heart now has a high Po_2 and a low Pco_2 and is pumped by the left ventricle into systemic circulation.

The arterial blood that reaches systemic capillaries has a high Po_2 and a low Pco_2. The body cells and tissue fluid have a low Po_2 and a high Pco_2 because cells continuously use oxygen in cell respiration (energy production) and produce carbon dioxide in this process. Therefore, in internal respiration, oxygen diffuses from the blood to tissue fluid (cells), and carbon dioxide diffuses from tissue fluid to the blood. The blood that enters systemic veins to return to the heart now has a low Po_2 and a high Pco_2 and is pumped by the right ventricle to the lungs to participate in external respiration.

Disorders of gas exchange often involve the lungs, that is, external respiration (see Box 15-6: Pulmonary Edema and Box 15-7: Pneumonia).

TRANSPORT OF GASES IN THE BLOOD

As you already know, most oxygen is carried in the blood bonded to the **hemoglobin** in red blood cells (although some oxygen is dissolved in blood plasma, it is not enough to sustain life). The mineral iron is part of hemoglobin and gives this protein its oxygen-carrying ability.

The oxygen-hemoglobin bond is formed in the lungs where Po_2 is high. This bond, however, is relatively unstable, and when blood passes through tissues with a low Po_2, the bond breaks, and oxygen is released to the tissues. The lower the oxygen concentration in a tissue, the more oxygen hemoglobin will release. This ensures that active tissues, such as exercising muscles, receive as much oxygen as possible to continue cell respiration. Other factors that increase the release of oxygen from hemoglobin are a high Pco_2 (actually a lower pH) and a high temperature, both of which are also characteristic of active tissues (see Box 15-8: Carbon Monoxide).

Carbon dioxide transport is a little more complicated. Some carbon dioxide is dissolved in the plasma, and some is carried by hemoglobin (carbaminohemoglobin), but these account for only 10% to 30% of

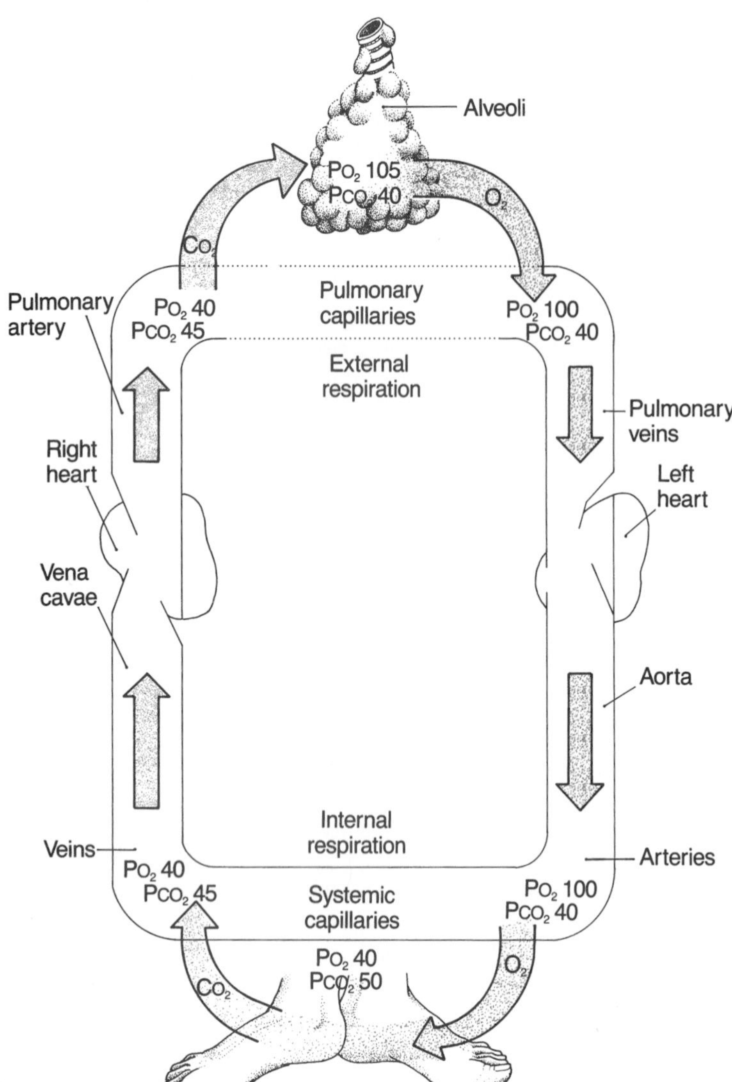

Figure 15–7 External respiration in the lungs and internal respiration in the body. The partial pressures of oxygen and carbon dioxide are shown at each site.

Box 15–6 PULMONARY EDEMA

Pulmonary edema is the accumulation of fluid in the alveoli. This is often a consequence of congestive heart failure in which the left side of the heart (or the entire heart) is not pumping efficiently. If the left ventricle does not pump strongly, the chamber does not empty as it should and cannot receive all the blood flowing in from the left atrium. Blood flow, therefore, is "congested," and blood backs up in the pulmonary veins and then in the pulmonary capillaries. As blood pressure increases in the pulmonary capillaries, filtration creates tissue fluid that collects in the alveoli.

Fluid-filled alveoli are no longer sites of efficient gas exchange, and the resulting hypoxia leads to the symptoms of **dyspnea** and increased respiratory rate. The most effective treatment is that which restores the pumping ability of the heart to normal.

Box 15–7 PNEUMONIA

Pneumonia is a bacterial infection of the lungs. Although there are many bacteria that can cause pneumonia, the most common one is probably *Streptococcus pneumoniae.* This species is estimated to cause at least 500,000 cases of pneumonia every year in the United States, with 50,000 deaths.

S. pneumoniae is a transient inhabitant of the upper respiratory tract, but in otherwise healthy people, the ciliated epithelium and the immune system prevent infection. Most cases of pneumonia occur in elderly people following a primary infection such as influenza.

When the bacteria are able to establish themselves in the alveoli, the alveolar cells secrete fluid that accumulates in the air sacs. Many neutrophils migrate to the site of infection and attempt to phagocytize the bacteria. The alveoli become filled with fluid, bacteria, and neutrophils (this is called consolidation); this decreases the exchange of gases.

Pneumovax is a vaccine for this type of pneumonia. It contains only the capsules of *S. pneumoniae* and cannot cause the disease. This vaccine is recommended for people over the age of 60 years, and for those with chronic pulmonary disorders or any debilitating disease.

total CO_2 transport. Most carbon dioxide is carried in the plasma in the form of bicarbonate ions (HCO_3^-). Let us look at the reactions that transform CO_2 into a bicarbonate ion.

When carbon dioxide enters the blood, most diffuses into red blood cells, which contain the enzyme **carbonic anhydrase**. This enzyme (which contains zinc) catalyzes the reaction of carbon dioxide and water to form carbonic acid:

$$CO_2 + H_2O \rightarrow H_2CO_3$$

The carbonic acid then dissociates:

$$H_2CO_3 \rightarrow H^+ + HCO_3^-$$

The bicarbonate ions diffuse out of the red blood cells into the plasma, leaving the hydrogen ions (H^+) in the red blood cells. The many H^+ ions would tend to make the red blood cells too acidic, but hemoglobin acts as a buffer to prevent acidosis. To maintain an ionic equilibrium, chloride ions (Cl^-) from the plasma enter the red blood cells; this is called the chloride shift. Where

Box 15–8 CARBON MONOXIDE

Carbon monoxide (CO) is a colorless, odorless gas that is produced during the combustion of fuels such as gasoline, coal, oil, and wood. As you know, CO is a poison that may cause death if inhaled in more than very small quantities or for more than a short period of time.

The reason CO is so toxic is that it forms a very strong and stable bond with the hemoglobin in RBCs. Hemoglobin with CO bonded to it cannot bond to and transport oxygen. The effect of CO, therefore, is to drastically decrease the amount of oxygen carried in the blood. As little as 0.1% CO in inhaled air can saturate half the total hemoglobin with CO.

Lack of oxygen is often apparent in people with light skin as **cyanosis**, a bluish cast to the skin, lips, and nailbeds. This is because hemoglobin is dark red unless something (usually oxygen) is bonded to it. When hemoglobin bonds to CO, however, it becomes a bright, cherry red. This color may be seen in light skin and may be very misleading; the person with CO poisoning is in a severely hypoxic state.

Although CO is found in cigarette smoke, it is present in such minute quantities that it is not lethal. Heavy smokers, however, may be in a mild but chronic hypoxic state because much of their hemoglobin is firmly bonded to CO. As a compensation, RBC production may increase, and a heavy smoker may have a hematocrit over 50%.

is the CO_2? In the plasma as part of HCO_3^- ions. When the blood reaches the lungs, an area of lower P_{CO_2}, these reactions are reversed, CO_2 is re-formed and diffuses into the alveoli to be exhaled.

PULMONARY VOLUMES

The capacity of the lungs varies with the size and age of the person. Taller people have larger lungs than do shorter people. Also, as we get older our lung capacity diminishes as lungs lose their elasticity and the respiratory muscles become less efficient. For the following pulmonary volumes, the values given are those for healthy young adults. These are also shown in Fig. 15–8.

1. **Tidal Volume**—the amount of air involved in one normal inhalation and exhalation. The average tidal volume is 500 ml, but many people often have lower tidal volumes due to shallow breathing.
2. **Minute Respiratory Volume** (MRV)—the amount of air inhaled and exhaled in 1 minute. MRV is calculated by multiplying tidal volume by the number of respirations per minute (average range: 12 to 20 per minute). If tidal volume is 500 ml and the respiratory rate is 12 breaths per minute, the MRV is 6000 ml, or 6 liters of air per minute, which is average. Shallow breathing

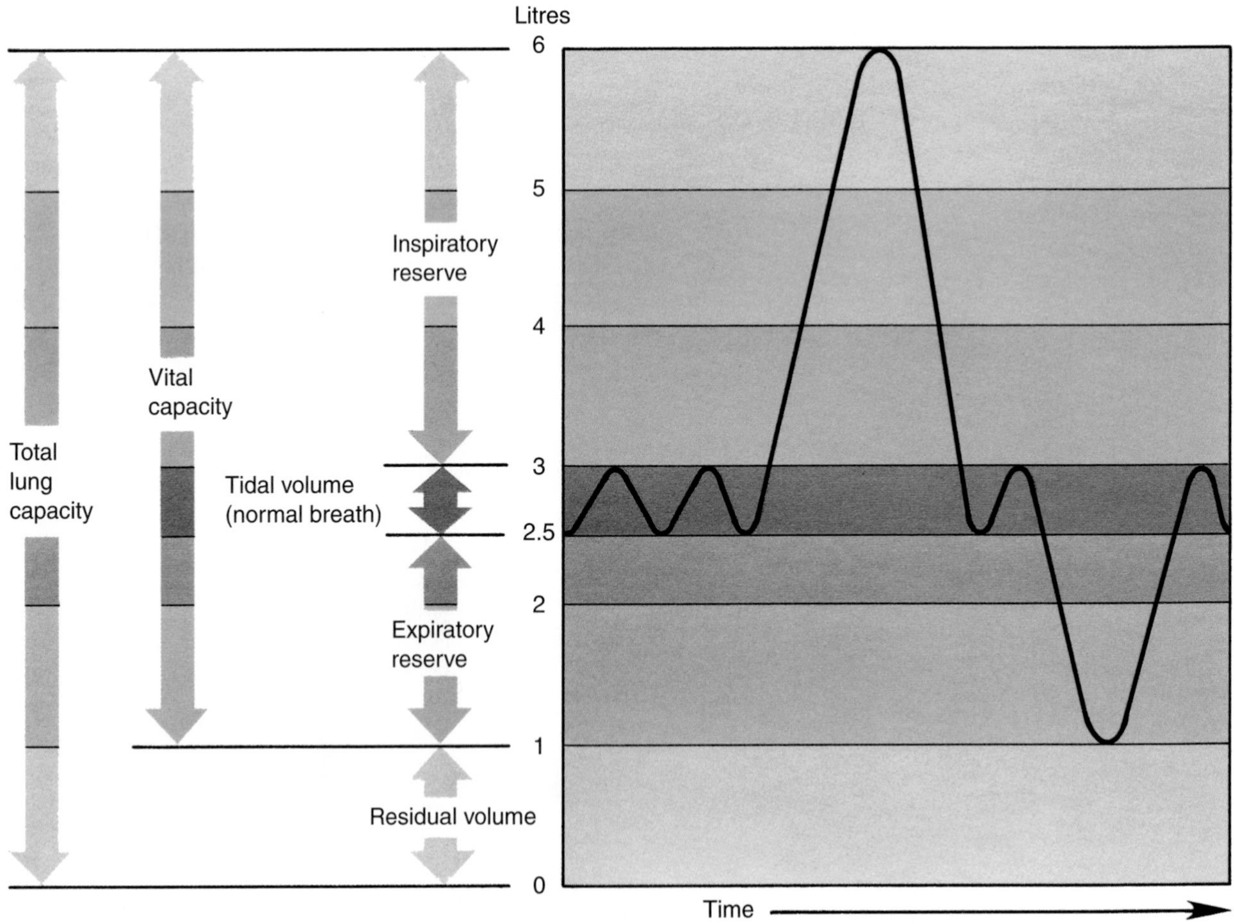

Figure 15–8 Pulmonary volumes. See text for description.

usually indicates a smaller than average tidal volume, and would thus require more respirations per minute to obtain the necessary MRV.

3. **Inspiratory Reserve**—the amount of air, beyond tidal volume, that can be taken in with the deepest possible inhalation. Normal inspiratory reserve ranges from 2000 to 3000 ml.
4. **Expiratory Reserve**—the amount of air, beyond tidal volume, that can be expelled with the most forceful exhalation. Normal expiratory reserve ranges from 1000 to 1500 ml.
5. **Vital Capacity**—the sum of tidal volume, inspiratory reserve, and expiratory reserve. Stated another way, vital capacity is the amount of air involved in the deepest inhalation followed by the most forceful exhalation. Average range of vital capacity is 3500 to 5000 ml.
6. **Residual Air**—the amount of air that remains in the lungs after the most forceful exhalation; the average range is 1000 to 1500 ml. Residual air is important to ensure that there is some air in the lungs at all times, so that exchange of gases is a continuous process, even between breaths.

Some of the volumes described above can be determined with instruments called spirometers, which measure movement of air. Trained singers and musicians who play wind instruments often have vital capacities much larger than would be expected for their height and age, because their respiratory muscles have become more efficient with "practice." The same is true for athletes who exercise regularly. A person with emphysema, however, must "work" to exhale, and vital capacity and expiratory reserve volume are often much lower than average.

REGULATION OF RESPIRATION

There are two types of mechanisms that regulate breathing: nervous mechanisms and chemical mechanisms. Because any changes in the rate or depth of breathing are ultimately brought about by nerve impulses, we will consider nervous mechanisms first.

NERVOUS REGULATION

The respiratory centers are located in the **medulla** and **pons**, which are parts of the brain stem (see Fig. 15-9). Within the medulla are the inspiration center and expiration center.

The **inspiration center** automatically generates impulses in rhythmic spurts. These impulses travel along nerves to the respiratory muscles to stimulate their contraction. The result is inhalation. As the lungs inflate, baroreceptors in lung tissue detect this stretching and generate sensory impulses to the medulla; these impulses begin to depress the inspiration center. This is called the Hering-Breuer inflation reflex, which helps prevent overinflation of the lungs.

As the inspiration center is depressed, the result is a decrease in impulses to the respiratory muscles, which relax to bring about exhalation. Then the inspiration center becomes active again to begin another cycle of breathing. When there is a need for more forceful exhalations, such as during exercise, the inspiration center activates the **expiration center**, which generates impulses to the internal intercostal and abdominal muscles.

The two respiratory centers in the pons work with the inspiration center to produce a normal rhythm of breathing. The **apneustic center** prolongs inhalation, and is then interrupted by impulses from the **pneumotaxic center**, which contributes to exhalation. In normal breathing, inhalation lasts 1 to 2 seconds, followed by a slightly longer (2 to 3 seconds) exhalation, producing the normal range of respiratory rate of 12 to 20 breaths per minute.

What has just been described is normal breathing, but variations are possible and quite common. Emotions often affect respiration; a sudden fright may bring about a gasp or a scream, and anger usually increases the respiratory rate. In these situations, impulses from the **hypothalamus** modify the output from the medulla. The **cerebral cortex** enables us to voluntarily change our breathing rate or rhythm to talk, sing, breathe faster or slower, or even to stop breathing for 1 or 2 minutes. Such changes cannot be continued indefinitely, however, and the medulla will eventually resume control.

Coughing and **sneezing** are reflexes that remove irritants from the respiratory passages; the medulla contains the centers for both of these reflexes. Sneezing is stimulated by an irritation of the nasal mucosa, and coughing is stimulated by irritation of the mucosa of the pharynx, larynx, or trachea. The reflex action is essentially the same for both: An inhalation is followed by exhalation beginning with the glottis closed to build up pressure. Then the glottis opens suddenly,

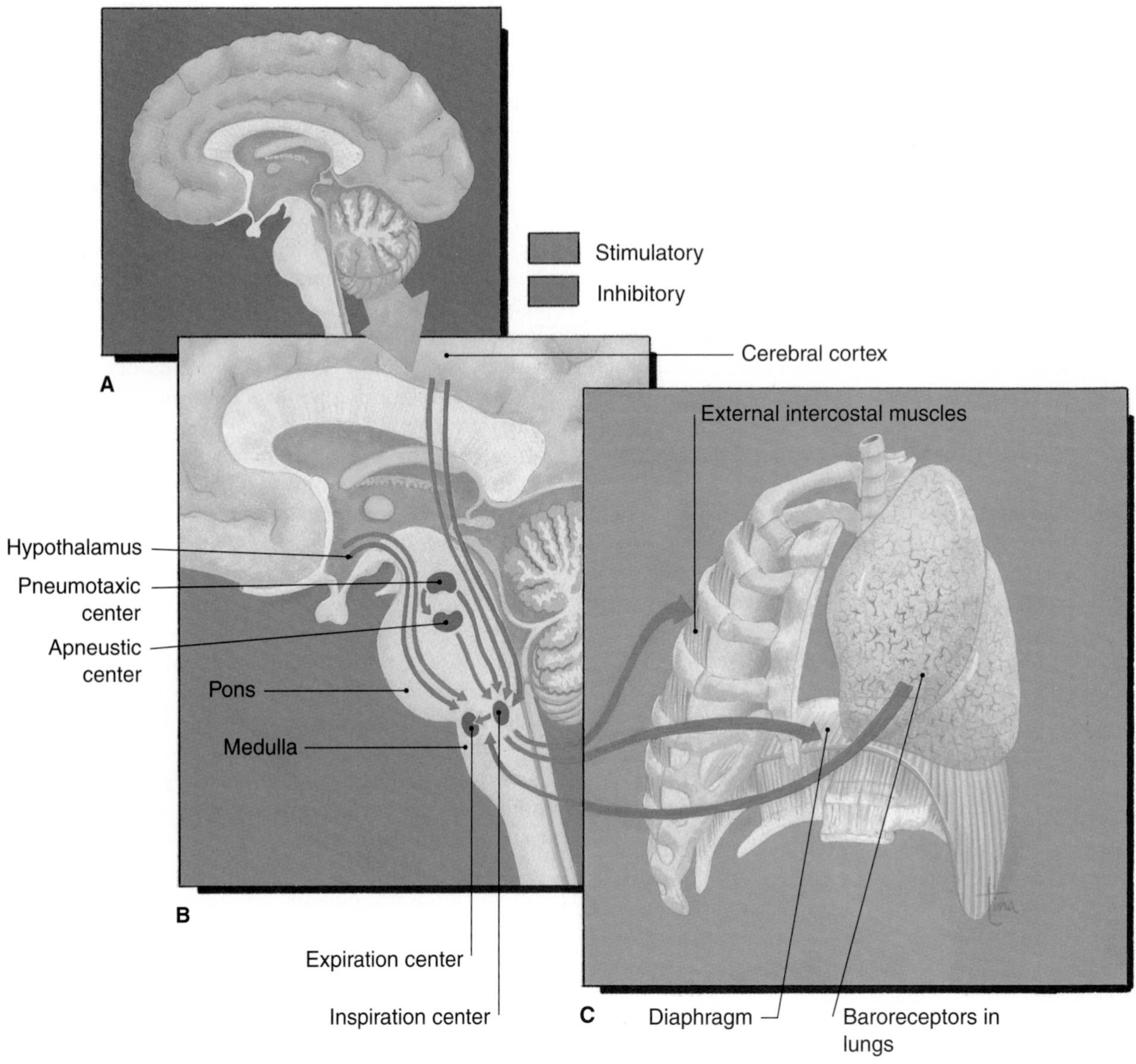

Figure 15–9 Nervous regulation of respiration. **(A)** Midsagittal section of brain. **(B)** Respiratory centers in medulla and pons. **(C)** Respiratory muscles. See text for description.

and the exhalation is explosive. A cough directs the exhalation out the mouth, while a sneeze directs the exhalation out the nose.

Yet another respiratory reflex is yawning. Most of us yawn when we are tired, but the stimulus for and purpose of yawning are not known with certainty. There are several possibilities, such as lack of oxygen or accumulation of carbon dioxide, but we really do not know. Nor do we know why yawning is contagious, but seeing someone yawn is almost sure to elicit a yawn of one's own. You may even have yawned while reading this paragraph about yawning!

CHEMICAL REGULATION

Chemical regulation refers to the effect on breathing of blood pH and blood levels of oxygen and carbon dioxide. This is shown in Fig. 15–10. **Chemorecep-**

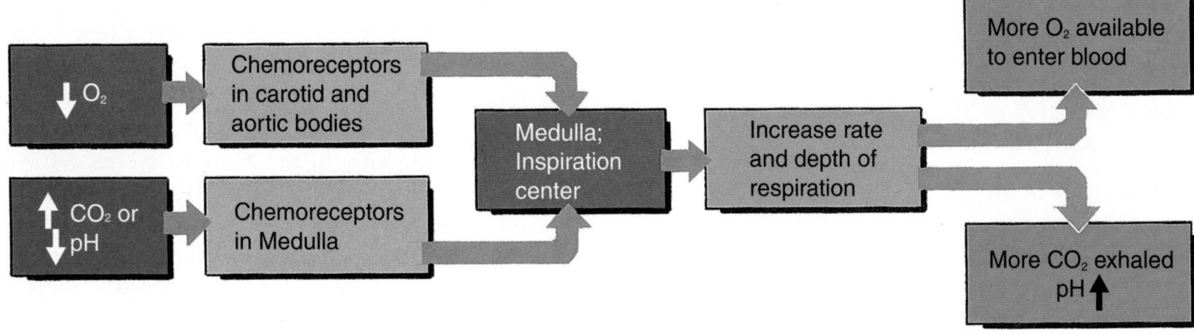

Figure 15–10 Chemical regulation of respiration. See text for description.

tors that detect changes in blood gases and pH are located in the carotid and aortic bodies and in the medulla itself.

A decrease in the blood level of oxygen (hypoxia) is detected by the chemoreceptors in the **carotid** and **aortic bodies**. The sensory impulses generated by these receptors travel along the glossopharyngeal and vagus nerves to the medulla, which responds by increasing respiratory rate or depth (or both). This response will bring more air into the lungs so that more oxygen can diffuse into the blood to correct the hypoxic state.

Carbon dioxide becomes a problem when it is present in excess in the blood, because excess CO_2 lowers the pH when it reacts with water to form carbonic acid (a source of H^+ ions). That is, excess CO_2 makes the blood or other body fluids less alkaline (or more acidic). The **medulla** contains **chemoreceptors** that are very sensitive to changes in pH, especially decreases. If accumulating CO_2 lowers blood pH, the medulla responds by increasing respiration. This is not for the purpose of inhaling, but rather to exhale more CO_2 to raise the pH back to normal.

Of the two respiratory gases, which is the more important as a regulator of respiration? Our guess might be oxygen, because it is essential for energy production in cell respiration. However, the respiratory system can maintain a normal blood level of oxygen even if breathing decreases to half the normal rate or stops for a few moments. Recall that exhaled air is 16% oxygen. This oxygen did not enter the blood but was available to do so if needed. Also, the residual air in the lungs supplies oxygen to the blood even if breathing rate slows.

Therefore, carbon dioxide must be the major regulator of respiration, and the reason is that carbon dioxide affects the pH of the blood. As was just mentioned, an excess of CO_2 causes the blood pH to decrease, a process that must not be allowed to continue. Therefore, any increase in blood CO_2 level is quickly compensated for by increased breathing to exhale more CO_2. If, for example, you hold your breath, what is it that makes you breathe again? Have you run out of oxygen? Probably not, for the reasons mentioned above. What has happened is that accumulating CO_2 has lowered blood pH enough to stimulate the medulla to start the breathing cycle again.

In some situations, oxygen does become the major regulator of respiration. People with severe, chronic pulmonary diseases such as emphysema have decreased exchange of both oxygen and carbon dioxide in the lungs. The decrease in pH caused by accumulating CO_2 is corrected by the kidneys, but the blood oxygen level keeps decreasing. Eventually, the oxygen level may fall so low that it does provide a very strong stimulus to increase the rate and depth of respiration.

RESPIRATION AND ACID–BASE BALANCE

As you have just seen, respiration affects the pH of body fluids because it regulates the amount of carbon dioxide in these fluids. Remember that CO_2 reacts with water to form carbonic acid (H_2CO_3), which ionizes into H^+ ions and HCO_3^- ions. The more hydrogen ions present in a body fluid, the lower the pH, and the fewer hydrogen ions present, the higher the pH.

The respiratory system may be the cause of a pH

imbalance, or it may help correct a pH imbalance created by some other cause.

RESPIRATORY ACIDOSIS AND ALKALOSIS

Respiratory acidosis occurs when the rate or efficiency of respiration decreases, permitting carbon dioxide to accumulate in body fluids. The excess CO_2 results in the formation of more H^+ ions, which decrease the pH. Holding one's breath can bring about a mild respiratory acidosis, which will soon stimulate the medulla to initiate breathing again. More serious causes of respiratory acidosis are pulmonary diseases, such as pneumonia and emphysema, or severe asthma. Each of these impairs gas exchange and allows excess CO_2 to remain in body fluids.

Respiratory alkalosis occurs when the rate of respiration increases, and CO_2 is very rapidly exhaled. Less CO_2 decreases H^+ ion formation, which increases the pH. Breathing faster for a few minutes can bring about a mild state of respiratory alkalosis. Babies who cry for extended periods (crying is a noisy exhalation) experience this condition. In general, however, respiratory alkalosis is not a common occurrence. Severe physical trauma and shock, or certain states of mental and/or emotional anxiety, may be accompanied by hyperventilation and also result in respiratory alkalosis. In addition, traveling to a higher altitude (less oxygen in the atmosphere) may cause a temporary increase in breathing rate before compensation occurs (increased rate of RBC production—see Chapter 11).

RESPIRATORY COMPENSATION

If a pH imbalance is caused by something other than a change in respiration, it is called a metabolic acidosis or alkalosis. In either case, the change in pH stimulates a change in respiration that may help restore the pH of body fluids to normal.

Metabolic acidosis may be caused by untreated diabetes mellitus (ketoacidosis), kidney disease, or severe diarrhea. In such situations, the H^+ ion concentration of body fluids is increased. Respiratory compensation involves an increase in the rate and depth of respiration to exhale more CO_2 to decrease H^+ ion formation, which will raise the pH toward the normal range.

Metabolic alkalosis is not a common occurrence but may be caused by ingestion of excessive amounts of alkaline medications such as those used to relieve gastric disturbances. Another possible cause is vomiting of stomach contents only. In such situations, the H^+ ion concentration of body fluids is decreased. Respiratory compensation involves a decrease in respiration to retain CO_2 in the body to increase H^+ ion formation, which will lower the pH toward the normal range.

Respiratory compensation for an ongoing metabolic pH imbalance cannot be complete, because there are limits to the amounts of CO_2 that may be exhaled or retained. At most, respiratory compensation is only about 75% effective. A complete discussion of acid–base balance is found in Chapter 19.

AGING AND THE RESPIRATORY SYSTEM

Perhaps the most important way to help your respiratory system age gracefully is not to smoke. In the absence of chemical assault, respiratory function does diminish but usually remains adequate. The respiratory muscles, like all skeletal muscles, weaken with age. Lung tissue loses its elasticity and alveoli are lost as their walls deteriorate. All of this results in decreased ventilation and lung capacity, but the remaining capacity is usually sufficient for ordinary activities. The cilia of the respiratory mucosa deteriorate with age, and the alveolar macrophages are not as efficient, which make elderly people more prone to pneumonia, a serious pulmonary infection.

Chronic alveolar hypoxia from diseases such as emphysema or chronic bronchitis may lead to pulmonary hypertension, which in turn overworks the right ventricle of the heart. Systemic hypertension often weakens the left ventricle of the heart, leading to congestive heart failure and pulmonary edema, in which excess tissue fluid collects in the alveoli and decreases gas exchange. Though true at any age, the interdependence of the respiratory and circulatory systems is particularly apparent in elderly people.

SUMMARY

As you have learned, respiration is much more than the simple mechanical actions of breathing. Inhalation provides the body with the oxygen that is necessary for the production of ATP in the process of cell respiration. Exhalation removes the CO_2 that is a product of cell respiration. Breathing also regulates the level of CO_2 within the body, and this contributes to the maintenance of the acid-base balance of body fluids. Although the respiratory gases do not form structural components of the body, their role in the chemical level of organization is essential to the functioning of the body at every level.

STUDY OUTLINE

The respiratory system moves air into and out of the lungs, which are the site of exchange for O_2 and CO_2 between the air and the blood. The functioning of the respiratory system is directly dependent on the proper functioning of the circulatory system.

1. The upper respiratory tract consists of those parts outside the chest cavity.
2. The lower respiratory tract consists of those parts within the chest cavity.

Nose—made of bone and cartilage covered with skin

1. Hairs inside the nostrils block the entry of dust.

Nasal Cavities—within the skull; separated by the nasal septum (see Fig. 15–1)

1. Nasal mucosa is ciliated epithelium with goblet cells; surface area is increased by the conchae.
2. Nasal mucosa warms and moistens the incoming air; dust and microorganisms are trapped on mucus and swept by the cilia to the pharynx.
3. Olfactory receptors respond to vapors in inhaled air.
4. Paranasal sinuses in the maxillae, frontal, sphenoid, and ethmoid bones open into the nasal cavities: functions are to lighten the skull and provide resonance for the voice.

Pharynx—posterior to nasal and oral cavities (see Fig. 15–1)

1. Nasopharynx—above the level of the soft palate, which blocks it during swallowing; a passageway for air only. The eustachian tubes from the middle ears open into it. The adenoid is a lymph nodule on the posterior wall.
2. Oropharynx—behind the mouth; a passageway for both air and food. Palatine tonsils are on the lateral walls.
3. Laryngopharynx—a passageway for both air and food; opens anteriorly into the larynx and posteriorly into the esophagus.

Larynx—the voice box and the airway between the pharynx and trachea (see Fig. 15–2)

1. Made of nine cartilages; the thyroid cartilage is the largest and most anterior.
2. The epiglottis is the uppermost cartilage; covers the larynx during swallowing.
3. The vocal cords are lateral to the glottis, the opening for air (see Fig. 15-3).
4. During speaking, the vocal cords are pulled across the glottis and vibrated by exhaled air, producing sounds which may be turned into speech.
5. The cranial nerves for speaking are the vagus and accessory.

Trachea—extends from the larynx to the primary bronchi (see Fig. 15–4)

1. 16 to 20 C-shaped cartilages in the tracheal wall keep the trachea open.
2. Mucosa is ciliated epithelium with goblet cells; cilia sweep mucus, trapped dust, and microorganisms upward to the pharynx.

Bronchial Tree—extends from the trachea to the alveoli (see Fig. 15–4)

1. The right and left primary bronchi are branches of the trachea; one to each lung.
2. Secondary bronchi: to the lobes of each lung (three right, two left)
3. Bronchioles—no cartilage in their walls.

Pleural Membranes—serous membranes of the thoracic cavity

1. Parietal pleura lines the chest wall.
2. Visceral pleura covers the lungs.
3. Serous fluid between the two layers prevents friction and keeps the membranes together during breathing.

Lungs—on either side of the heart in the chest cavity; extend from the diaphragm below up to the level of the clavicles

1. The rib cage protects the lungs from mechanical injury.
2. Hilus—indentation on the medial side: primary bronchus and pulmonary artery and veins enter (also bronchial vessels).

Alveoli—the sites of gas exchange in the lungs

1. Made of alveolar type I cells, simple squamous epithelium; thin to permit diffusion of gases.
2. Surrounded by pulmonary capillaries, which are also made of simple squamous epithelium (see Fig. 15-4).
3. Elastic connective tissue between alveoli is important for normal exhalation.
4. A thin layer of tissue fluid lines each alveolus; essential to permit diffusion of gases (see Fig. 15-5).
5. Alveolar type II cells produce pulmonary surfactant that mixes with the tissue fluid lining to decrease surface tension to permit inflation of the alveoli.
6. Alveolar macrophages phagocytize foreign material.

Mechanism of Breathing

1. Ventilation is the movement of air into and out of the lungs: inhalation and exhalation.
2. Respiratory centers are in the medulla and pons.
3. Respiratory muscles are the diaphragm and external and internal intercostal muscles (see Fig. 15-6).
   - Atmospheric pressure is air pressure: 760 mmHg at sea level.
   - Intrapleural pressure is within the potential pleural space; always slightly below atmospheric pressure ("negative").
   - Intrapulmonic pressure is within the bronchial tree and alveoli; fluctuates during breathing.

Inhalation (inspiration)

1. Motor impulses from medulla travel along phrenic nerves to diaphragm, which contracts and moves down. Impulses along intercostal nerves to external intercostal muscles, which pull ribs up and out.
2. The chest cavity is expanded and expands the parietal pleura.
3. The visceral pleura adheres to the parietal pleura and is also expanded and in turn expands the lungs.
4. Intrapulmonic pressure decreases, and air rushes into the lungs.

Exhalation (expiration)

1. Motor impulses from the medulla decrease, and the diaphragm and external intercostals relax.
2. The chest cavity becomes smaller and compresses the lungs.
3. The elastic lungs recoil and further compress the alveoli.
4. Intrapulmonic pressure increases, and air is forced out of the lungs. Normal exhalation is passive.
5. Forced exhalation: contraction of the internal intercostal muscles pulls the ribs down and in; contraction of the abdominal muscles forces the diaphragm upward.

Exchange of Gases

1. External respiration is the exchange of gases between the air in the alveoli and the blood in the pulmonary capillaries.
2. Internal respiration is the exchange of gases between blood in the systemic capillaries and tissue fluid (cells).
3. Inhaled air (atmosphere) is 21% O_2 and 0.04% CO_2. Exhaled air is 16% O_2 and 4.5% CO_2.
4. Diffusion of O_2 and CO_2 in the body occurs because of pressure gradients (see Table 15-1). A gas will diffuse from an area of higher partial pressure to an area of lower partial pressure.
5. External respiration: PO_2 in the alveoli is high, and PO_2 in the pulmonary capillaries is low, so O_2 diffuses from the air to the blood. PCO_2 in the alveoli is low, and PCO_2 in the pulmonary capillaries is high, so CO_2 diffuses from the blood to the air and is exhaled (see Fig. 15-7).
6. Internal respiration: PO_2 in the systemic capillaries is high, and PO_2 in the tissue fluid is low, so O_2 diffuses from the blood to the tissue fluid and cells. PCO_2 in the systemic capillaries is low, and PCO_2 in the tissue fluid is high, so CO_2 diffuses from the tissue fluid to the blood (see Fig. 15-7).

Transport of Gases in the Blood

1. Oxygen is carried by the iron of hemoglobin (Hb) in the RBCs. The O_2-Hb bond is formed in the lungs where the PO_2 is high.
2. In tissues, Hb releases much of its O_2; the important factors are low PO_2 in tissues, high PCO_2 in tissues, and a high temperature in tissues.
3. Most CO_2 is carried as HCO_3^- ions in blood plasma. CO_2 enters the RBCs and reacts with H_2O to form carbonic acid (H_2CO_3). Carbonic anhydrase is the enzyme that catalyzes this reaction. H_2CO_3 disso-

ciates to H^+ ions and HCO_3^- ions. The HCO_3^- ions leave the RBCs and enter the plasma; Hb buffers the H^+ ions that remain in the RBCs. Cl^- ions from the plasma enter the RBCs to maintain ionic equilibrium (the chloride shift).

4. When blood reaches the lungs, CO_2 is re-formed, diffuses into the alveoli, and is exhaled.

Pulmonary Volumes (see Fig. 15–8)

1. Tidal Volume—the amount of air in one normal inhalation and exhalation.
2. Minute Respiratory Volume—the amount of air inhaled and exhaled in 1 minute.
3. Inspiratory Reserve—the amount of air beyond tidal in a maximal inhalation.
4. Expiratory Reserve—the amount of air beyond tidal in the most forceful exhalation.
5. Vital Capacity—the sum of tidal volume, inspiratory and expiratory reserves.
6. Residual Volume—the amount of air that remains in the lungs after the most forceful exhalation; provides for continuous exchange of gases.

Nervous Regulation of Respiration (see Fig. 15–9)

1. The medulla contains the inspiration center and expiration center.
2. Impulses from the inspiration center to the respiratory muscles cause their contraction; the chest cavity is expanded.
3. Baroreceptors in lung tissue detect stretching and send impulses to the medulla to depress the inspiration center. This is the Hering-Breuer inflation reflex, which prevents overinflation of the lungs.
4. The expiration center is stimulated by the inspiration center when forceful exhalations are needed.
5. In the pons: the apneustic center prolongs inhalation, and the pneumotaxic center helps bring about exhalation. These centers work with the inspiration center in the medulla to produce a normal breathing rhythm.
6. The hypothalamus influences changes in breathing in emotional situations. The cerebral cortex permits voluntary changes in breathing.
7. Coughing and sneezing remove irritants from the upper respiratory tract; the centers for these reflexes are in the medulla.

Chemical Regulation of Respiration (see Fig. 15–10)

1. Decreased blood O_2 is detected by chemoreceptors in the carotid body and aortic body. Response: increased respiration to take more air into the lungs.
2. Increased blood CO_2 level is detected by chemoreceptors in the medulla. Response: increased respiration to exhale more CO_2.
3. CO_2 is the major regulator of respiration because excess CO_2 decreases the pH of body fluids ($CO_2 + H_2O \rightarrow H_2CO_3 \rightarrow H^+ + HCO_3$). Excess H^+ ions lower pH.
4. Oxygen becomes a major regulator of respiration when blood level is very low, as may occur with severe, chronic pulmonary disease.

Respiration and Acid–Base Balance

1. Respiratory acidosis: a decrease in the rate or efficiency of respiration permits excess CO_2 to accumulate in body fluids, resulting in the formation of excess H^+ ions, which lower pH. Occurs in severe pulmonary disease.
2. Respiratory alkalosis: an increase in the rate of respiration increases the CO_2 exhaled, which decreases the formation of H^+ ions and raises pH. Occurs during hyperventilation or when first at a high altitude.
3. Respiratory compensation for metabolic acidosis: increased respiration to exhale CO_2 to decrease H^+ ion formation to raise pH to normal.
4. Respiratory compensation for metabolic alkalosis: decreased respiration to retain CO_2 to increase H^+ ion formation to lower pH to normal.

REVIEW QUESTIONS

1. State the three functions of the nasal mucosa. (p. 326)
2. Name the three parts of the pharynx; state whether each is an air passage only or an air and food passage. (pp. 326–328)

3. Name the tissue that lines the larynx and trachea, and describe its function. State the function of the cartilage of the larynx and trachea. (pp. 328–329)

4. Name the pleural membranes, state the location of each, and describe the functions of serous fluid. (p. 329)

5. Name the tissue of which the alveoli and pulmonary capillaries are made, and explain the importance of this tissue in these locations. Explain the function of pulmonary surfactant. (pp. 329, 331)

6. Name the respiratory muscles, and describe how they are involved in normal inhalation and exhalation. Define these pressures and relate them to a cycle of breathing: atmospheric pressure, intrapulmonic pressure. (pp. 331–332, 334)

7. Describe external respiration in terms of partial pressures of oxygen and carbon dioxide. (pp. 334–335, 337)

8. Describe internal respiration in terms of partial pressures of oxygen and carbon dioxide. (pp. 334–335, 337)

9. Name the cell, protein, and mineral that transport oxygen in the blood. State the three factors that increase the release of oxygen in tissues. (pp. 337, 339)

10. Most carbon dioxide is transported in what part of the blood, and in what form? Explain the function of hemoglobin with respect to carbon dioxide transport. (pp. 337, 339)

11. Name the respiratory centers in the medulla and pons, and explain how each is involved in a breathing cycle. (p. 341)

12. State the location of chemoreceptors affected by a low blood oxygen level; describe the body's response to hypoxia and its purpose. State the location of chemoreceptors affected by a high blood CO_2 level; describe the body's response and its purpose. (pp. 342–343)

13. For respiratory acidosis and alkalosis: state a cause and explain what happens to the pH of body fluids. (p. 344)

14. Explain how the respiratory system may compensate for metabolic acidosis or alkalosis. For an ongoing pH imbalance, what is the limit of respiratory compensation? (p. 344)

CHAPTER 16

The Digestive System

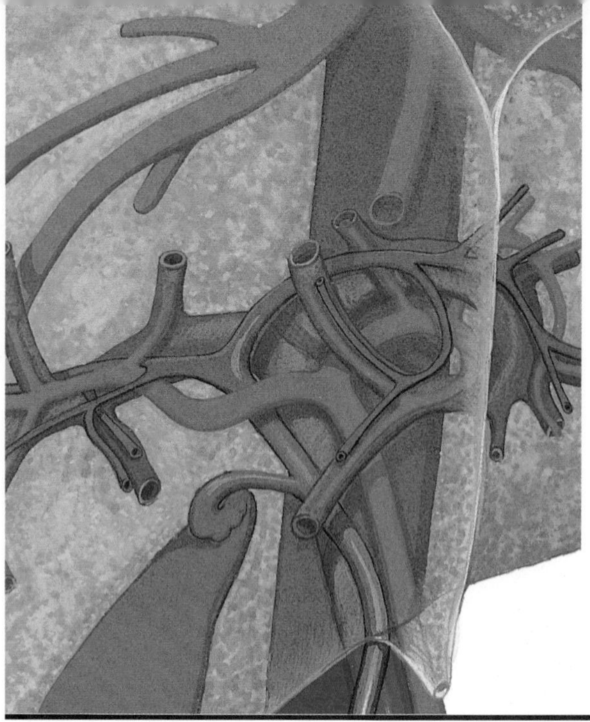

CHAPTER 16

Chapter Outline

Student Objectives

- Describe the general functions of the digestive system, and name its major divisions.
- Explain the difference between mechanical and chemical digestion, and name the end products of digestion.
- Describe the structure and functions of the teeth and tongue.
- Explain the functions of saliva.
- Describe the location and function of the pharynx and esophagus.
- Describe the structure and function of each of the four layers of the alimentary tube.
- Describe the location, structure, and function of the stomach, liver, gallbladder, pancreas, and small intestine.
- Describe absorption in the small intestine.
- Describe the location and functions of the large intestine.
- Explain the functions of the normal flora of the colon.
- Describe the functions of the liver.

The Digestive System

New Terminology

Alimentary tube (AL-i-**MEN**-tah-ree TOOB)
Chemical digestion (**KEM**-i-kuhl dye-**JES**-chun)
Common bile duct (**KOM**-mon BYL DUKT)
Defecation reflex (DEF-e-**KAY**-shun)
Duodenum (dew-**AH**-den-um)
Emulsify (e-**MULL**-si-fye)
Enamel (e-**NAM**-uhl)
Essential amino acids (e-**SEN**-shul ah-**ME**-noh **ASS**-ids)
External anal sphincter (eks-**TER**-nuhl **AY**-nuhl **SFINK**-ter)
Ileocecal valve (ILL-ee-oh-**SEE**-kuhl VALV)
Internal anal sphincter (in-**TER**-nuhl **AY**-nuhl **SFINK**-ter)
Lower esophageal sphincter (e-SOF-uh-**JEE**-uhl **SFINK**-ter)
Mechanical digestion (muh-**KAN**-i-kuhl dye-**JES**-chun)
Non-essential amino acids (NON e-**SEN**-shul ah-**ME**-noh **ASS**-ids)
Normal flora (**NOR**-muhl **FLOOR**-ah)
Periodontal membrane (PER-ee-oh-**DON**-tal)
Pyloric sphincter (pye-**LOR**-ik **SFINK**-ter)
Rugae (**ROO**-gay)
Villi (**VILL**-eye)

Related Clinical Terminology

Appendicitis (uh-PEN-di-**SIGH**-tis)
Diverticulitis (DYE-ver-TIK-yoo-**LYE**-tis)
Gastric ulcer (**GAS**-trik **UL**-ser)
Hepatitis (HEP-uh-**TIGH**-tis)
Lactose intolerance (**LAK**-tohs in-**TAHL**-er-ense)
Lithotripsy (LITH-oh-**TRIP**-see)
Paralytic ileus (**PAR**-uh-LIT-ik **ILL**-ee-us)
Peritonitis (per-i-toh-**NIGH**-tis)
Pyloric stenosis (pye-**LOR**-ik ste-**NOH**-sis)

*Terms that appear in **bold type** in the chapter text are defined in the glossary, which begins on page 528.*

A hurried breakfast when you are late for work or school . . . Thanksgiving dinner . . . going on a diet to lose 5 pounds . . . what do these experiences all have in common? Food. We may take food for granted, celebrate with it, or wish we wouldn't eat quite so much of it. Although food is not as immediate a need for human beings as is oxygen, it is a very important part of our lives. Food provides the raw materials or nutrients that cells use to reproduce and to build new tissue. The energy needed for cell reproduction and tissue building is released from food in the process of cell respiration. In fact, a supply of nutrients from regular food intake is so important, that the body can even store any excess for use later. Those "extra 5 pounds" are often stored fat in adipose tissue.

The food we eat, however, is not in a form that our body cells can use. A turkey sandwich, for example, consists of complex proteins, fats, and carbohydrates. The function of the **digestive system** is to change these complex organic nutrient molecules into simple organic and inorganic molecules that can then be absorbed into the blood or lymph to be transported to cells. In this chapter we will discuss the organs of digestion and the contribution each makes to digestion and absorption.

DIVISIONS OF THE DIGESTIVE SYSTEM

The two divisions of the digestive system are the alimentary tube and the accessory organs (Fig. 16–1). The **alimentary tube** extends from the mouth to the anus. It consists of the oral cavity, pharynx, esophagus, stomach, small intestine, and large intestine. Digestion takes place within the oral cavity, stomach, and small intestine; most absorption of nutrients takes place in the small intestine. Undigestable material, primarily cellulose, is eliminated by the large intestine (also called the colon).

The **accessory organs** of digestion are the teeth, tongue, salivary glands, liver, gallbladder, and pancreas. Digestion does not take place *within* these organs, but each contributes something *to* the digestive process.

TYPES OF DIGESTION

The food we eat is broken down in two complementary processes: mechanical digestion and chemical digestion. **Mechanical digestion** is the physical breaking up of food into smaller pieces. Chewing is an example of this. As food is broken up, more of its surface area is exposed for the action of digestive enzymes. Enzymes are discussed in Chapter 2. The work of the digestive enzymes is the **chemical digestion** of broken up food particles, in which complex chemical molecules are changed into much simpler chemicals that the body can utilize. Such enzymes are specific with respect to the fat, protein, or carbohydrate food molecules each can digest. For example, protein-digesting enzymes work only on proteins, not on carbohydrates or fats. Each enzyme is produced by a particular digestive organ and functions at a specific site. However, the enzyme's site of action may or may not be its site of production. These digestive enzymes and their functions are discussed in later sections.

END PRODUCTS OF DIGESTION

Before we describe the actual organs of digestion, let us see where the process of digestion will take us, or rather, will take our food. The three types of complex organic molecules found in food are carbohydrates, proteins, and fats. Each of these complex molecules is digested to a much more simple substance that the body can then use. Carbohydrates, such as starches and disaccharides, are digested to monosaccharides such as glucose, fructose, and galactose. Proteins are digested to amino acids, and fats are digested to fatty acids and glycerol. Also part of food, and released during digestion, are vitamins, minerals, and water.

We will now return to the beginning of the alimentary tube and consider the digestive organs and the process of digestion.

ORAL CAVITY

Food enters the **oral cavity** (or **buccal cavity**) by way of the mouth. The boundaries of the oral cavity are the hard and soft palates, superiorly; the cheeks, lat-

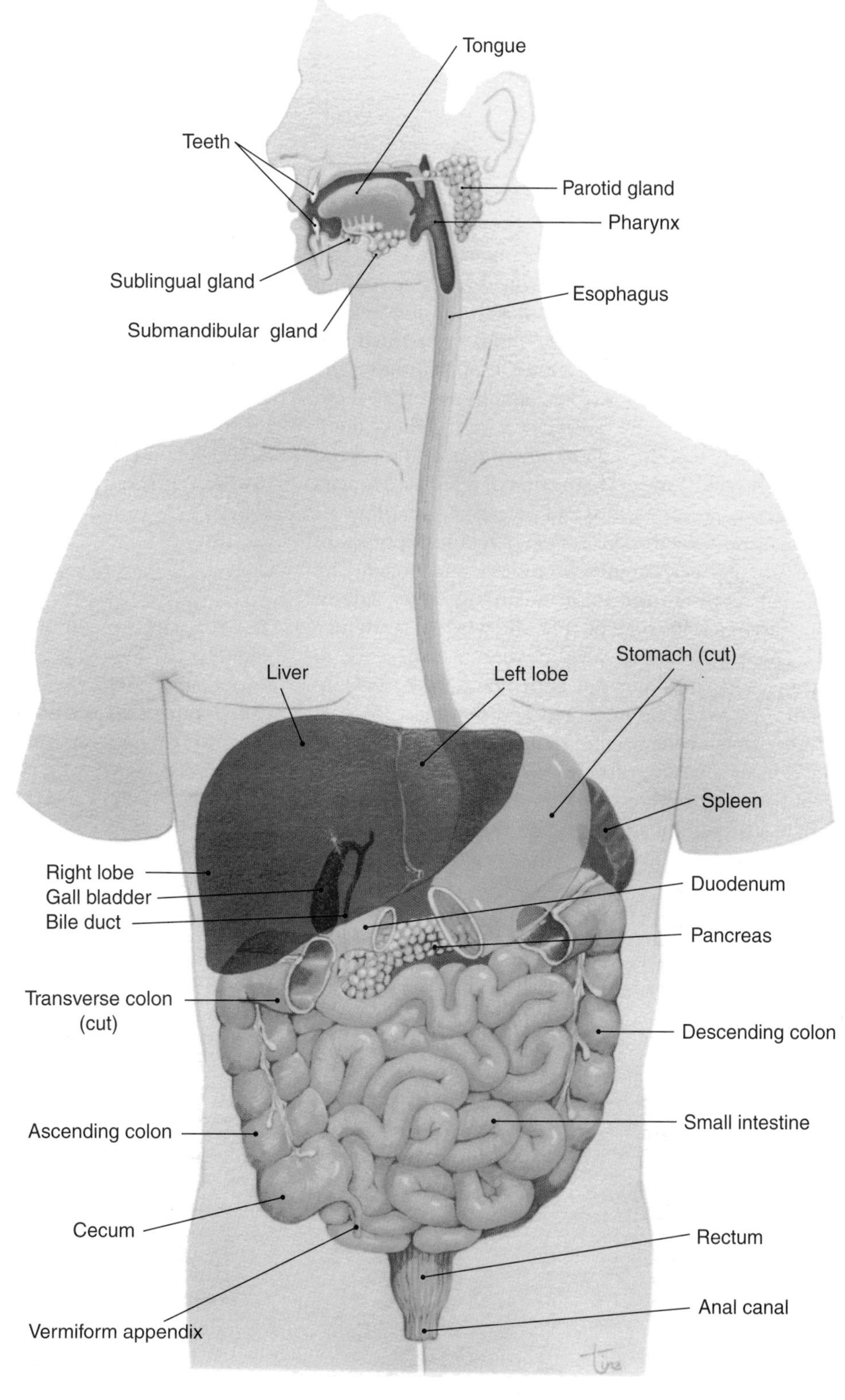

Figure 16–1 The digestive organs shown in anterior view of the trunk and left lateral view of the head. (The spleen is not a digestive organ but is included to show its location relative to the stomach, pancreas, and colon.)

erally; and the floor of the mouth, inferiorly. Within the oral cavity are the teeth and tongue and the openings of the ducts of the salivary glands.

TEETH

The function of the **teeth** is, of course, chewing. This is the process that mechanically breaks food into smaller pieces and mixes it with saliva. An individual develops two sets of teeth: deciduous and permanent. The **deciduous teeth** begin to erupt through the gums at about 6 months of age, and the set of 20 teeth is usually complete by the age of 2 years. These teeth are gradually lost throughout childhood and replaced by the **permanent teeth**, the first of which are molars that emerge around the age of 6 years. A complete set of permanent teeth consists of 32 teeth; the types of teeth are incisors, canines, premolars, and molars. The wisdom teeth are the third molars on either side of each jawbone. In some people, the wisdom teeth may not emerge from the jawbone because there is no room for them along the gum line. These wisdom teeth are said to be impacted and may put pressure on the roots of the second molars. In such cases, extraction of a wisdom tooth may be necessary to prevent damage to other teeth.

The structure of a tooth is shown in Fig. 16–2. The crown is visible above the gum **(gingiva)**. The root is enclosed in a socket in the mandible or maxillae. The **periodontal membrane** lines the socket and produces a bone-like cement that anchors the tooth. The outermost layer of the crown is **enamel**, which is made by cells called ameloblasts. Enamel provides a hard chewing surface and is more resistant to decay than are other parts of the tooth. Within the enamel is **dentin**, which is very similar to bone and is produced by cells called odontoblasts. Dentin also forms the roots of a tooth. The innermost portion of a tooth is the **pulp cavity**, which contains blood vessels and nerve endings of the trigeminal nerve (5th cranial). Erosion of the enamel and dentin layers by bacterial acids (dental caries or cavities) may result in bacterial invasion of the pulp cavity and a very painful toothache.

TONGUE

The **tongue** is made of skeletal muscle that is innervated by the hypoglossal nerves (12th cranial). On the upper surface of the tongue are small projections called **papillae**, many of which contain taste buds (see also Chapter 9). The sensory nerves for taste are also cranial nerves: the facial (7th) and glossopharyngeal (9th). As you know, the sense of taste is important because it makes eating enjoyable, but the tongue has other functions as well.

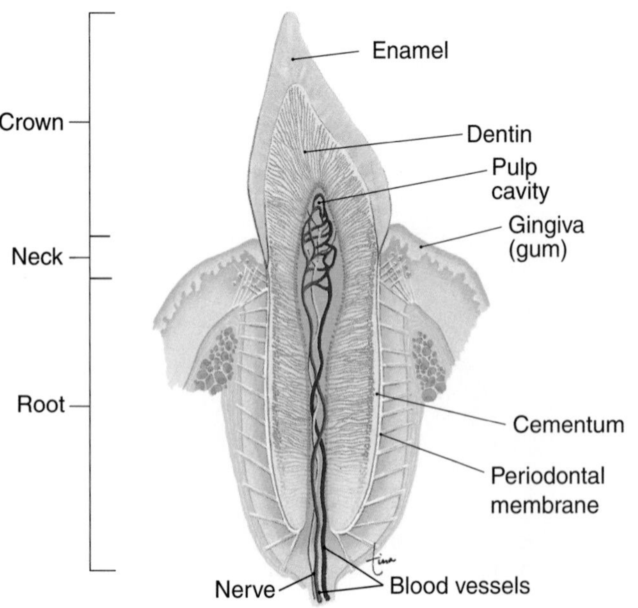

Figure 16–2 Tooth structure. Longitudinal section of a tooth showing internal structure.

Chewing is efficient because of the action of the tongue in keeping the food between the teeth and mixing it with saliva. Elevation of the tongue is the first step in swallowing. This is a voluntary action, in which the tongue contracts and meets the resistance of the hard palate. The mass of food, called a bolus, is thus pushed backward toward the pharynx. The remainder of swallowing is a reflex, which is described in the section on the pharynx.

SALIVARY GLANDS

The digestive secretion in the oral cavity is **saliva**, produced by three pairs of **salivary glands**, which are shown in Fig. 16–3. The **parotid glands** are just below and in front of the ears. The **submandibular** (also called submaxillary) glands are at the posterior corners of the mandible, and the **sublingual** glands are below the floor of the mouth. Each gland has at least one duct that takes saliva to the oral cavity.

Secretion of saliva is continuous, but the amount varies in different situations. The presence of food (or anything else) in the mouth increases saliva secretion. This is a parasympathetic response mediated by the facial and glossopharyngeal nerves. The sight or smell of food also increases secretion of saliva. Sympathetic stimulation in stress situations decreases secretion, making the mouth dry and swallowing difficult.

Saliva is mostly water, which is important to dissolve food for tasting and to moisten food for swallowing. The digestive enzyme in saliva is salivary amylase, which breaks down starch molecules to shorter chains

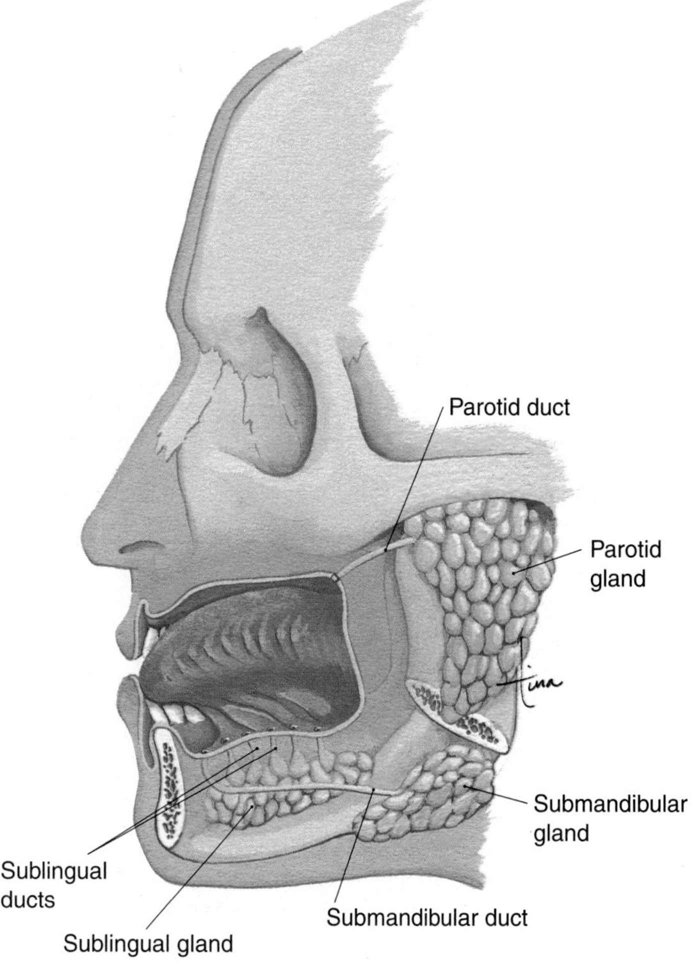

Figure 16–3 The salivary glands shown in left lateral view.

Table 16–1 THE PROCESS OF DIGESTION

Organ	Enzyme or Other Secretion	Function	Site of Action
Salivary glands	• Amylase	• Converts starch to maltose	Oral cavity
Stomach	• Pepsin	• Converts proteins to polypeptides	Stomach
	• HCl	• Changes pepsinogen to pepsin; maintains pH 1–2; destroys pathogens	Stomach
Liver	• Bile salts	• Emulsify fats	Small intestine
Pancreas	• Amylase	• Converts starch to maltose	Small intestine
	• Trypsin	• Converts polypeptides to peptides	Small intestine
	• Lipase	• Converts emulsified fats to fatty acids and glycerol	Small intestine
Small intestine	• Peptidases	• Convert peptides to amino acids	Small intestine
	• Sucrase	• Converts sucrose to glucose and fructose	Small intestine
	• Maltase	• Converts maltose to glucose (2)	Small intestine
	• Lactase	• Converts lactose to glucose and galactose	Small intestine

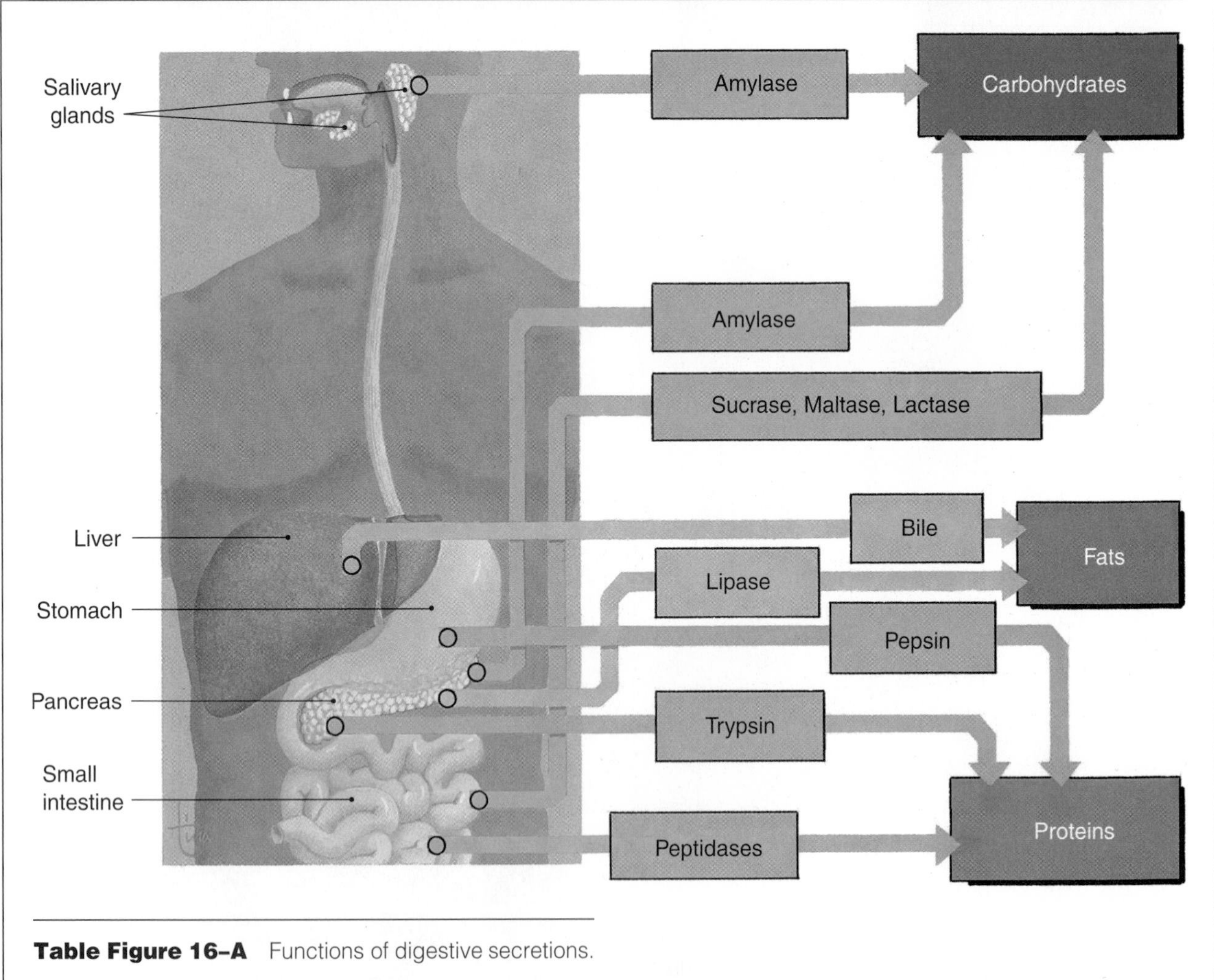

Table Figure 16–A Functions of digestive secretions.

of glucose molecules, or to maltose, a dissacharide. Most of us, however, do not chew our food long enough for the action of salivary amylase to be truly effective. As you will see, another amylase from the pancreas is also available to digest starch. Table 16-1 summarizes the functions of digestive secretions.

PHARYNX

As described in the last chapter, the oropharynx and laryngopharynx are food passageways connecting the oral cavity to the esophagus. No digestion takes place in the pharynx. Its only related function is swallowing, the mechanical movement of food. When the bolus of food is pushed backward by the tongue, the constrictor muscles of the pharynx contract as part of the swallowing reflex. The reflex center for swallowing is in the medulla, which coordinates the many actions that take place: constriction of the pharynx, cessation of breathing, elevation of the soft palate to block the nasopharynx, elevation of the larynx and closure of the epiglottis, and peristalsis of the esophagus. As you can see, swallowing is rather complicated, but because it is a reflex we don't have to think about making it happen correctly. Talking or laughing while eating, however, may interfere with the reflex and cause food to go into the "wrong pipe," the larynx. When that happens, the cough reflex is usually effective in clearing the airway.

ESOPHAGUS

The **esophagus** is a muscular tube that takes food from the pharynx to the stomach; no digestion takes place here. Peristalsis of the esophagus propels food in one direction and ensures that food gets to the stomach even if the body is horizontal or upside down. At the junction with the stomach, the lumen (cavity) of the esophagus is surrounded by the **lower esophageal sphincter** (LES or cardiac sphincter), a circular smooth muscle. The LES relaxes to permit food to enter the stomach, then contracts to prevent the backup of stomach contents. If the LES does not close completely, gastric juice may splash up into the esophagus; this is a painful condition we call heartburn.

STRUCTURAL LAYERS OF THE ALIMENTARY TUBE

Before we continue with our discussion of the organs of digestion, we will first examine the structure of the alimentary tube. When viewed in cross-section, the alimentary tube has four layers (Fig. 16-4): the mucosa, submucosa, external muscle layer, and serosa. Each layer has a specific structure, and its functions contribute to the functioning of the organs of which it is a part.

MUCOSA

The **mucosa**, or lining, of the alimentary tube is made of epithelial tissue. In the esophagus the mucosa is stratified squamous epithelium; the mucosa of the stomach and intestines is simple columnar epithelium. The mucosa secretes mucus, which lubricates the passage of food, and also secretes the digestive enzymes of the stomach and small intestine. Just below the epithelium are lymph nodules which contain macrophages to phagocytize bacteria or other foreign material that get through the epithelium.

SUBMUCOSA

The **submucosa** is made of areolar connective tissue with many blood vessels and lymphatic vessels. Autonomic nerve networks called **Meissner's plexus** (or submucosal plexus) innervate the mucosa to regulate secretions. Parasympathetic impulses increase secretions, whereas sympathetic impulses decrease secretions.

EXTERNAL MUSCLE LAYER

The external muscle layer typically contains two layers of smooth muscle: an inner, circular layer and an outer, longitudinal layer. Variations from the typical do occur, however. In the esophagus, this layer is striated muscle in the upper third, which gradually changes to smooth muscle in the lower portions. The stomach has three layers of smooth muscle, rather than two.

Contractions of this muscle layer help break up food and mix it with digestive juices. The one-way contractions of **peristalsis** move the food toward the anus.

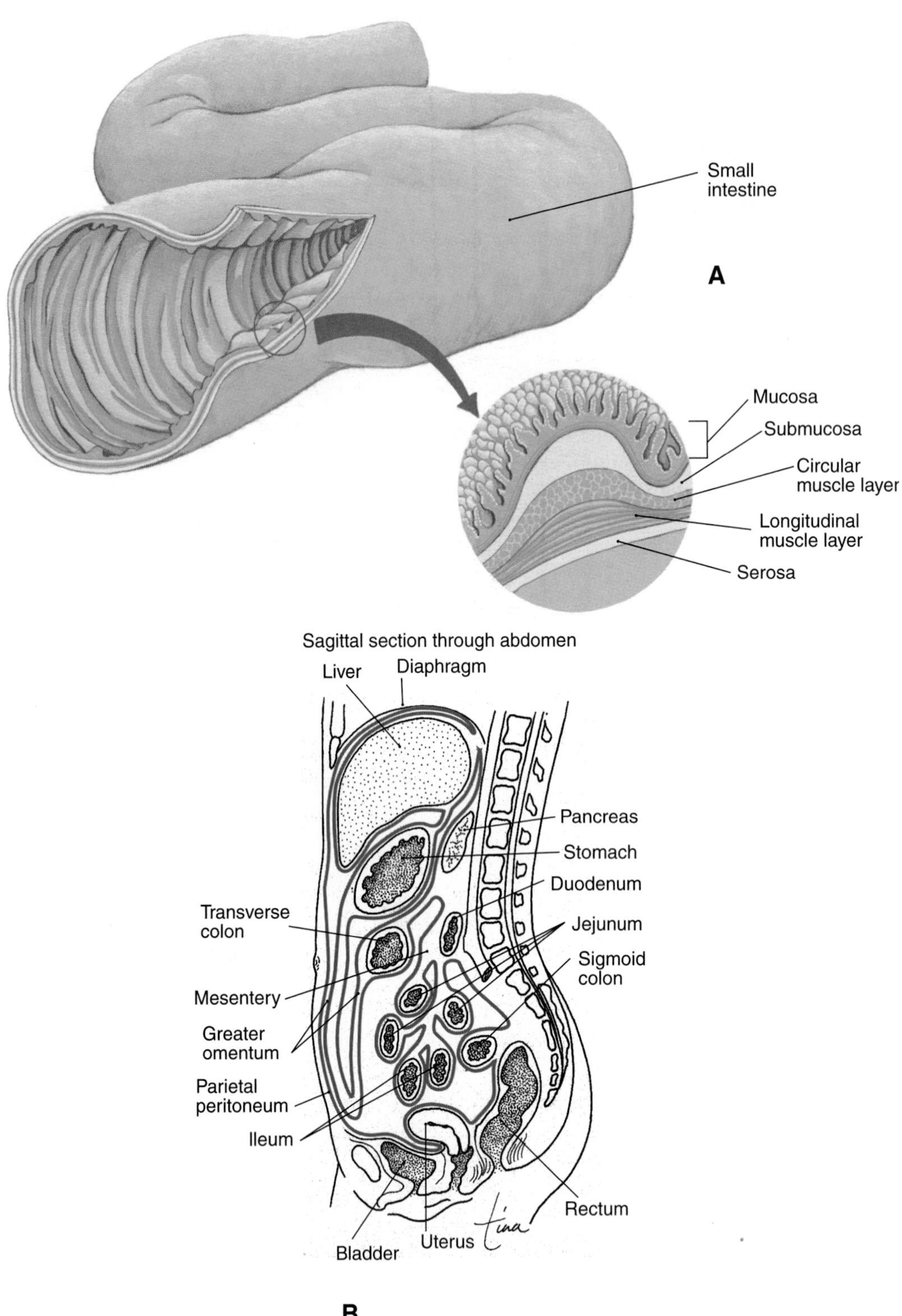

Figure 16–4 **(A)** The four layers of the wall of the alimentary tube. A small part of the wall of the small intestine has been magnified to show the four layers typical of the alimentary tube. **(B)** Sagittal section through the abdomen showing the relationship of the peritoneum and mesentery to the abdominal organs.

Auerbach's plexus (or myenteric plexus) is the autonomic network in this layer: Sympathetic impulses decrease contractions and peristalsis, whereas parasympathetic impulses increase contractions and peristalsis. The parasympathetic nerves are the vagus (10th cranial) nerves; they truly live up to the meaning of "vagus," which is "wanderer."

SEROSA

Above the diaphragm, for the esophagus, the serosa, the outermost layer, is fibrous connective tissue. Below the diaphragm, the serosa is the **mesentery** or visceral peritoneum, a serous membrane. Lining the abdominal cavity is the parietal peritoneum, usually simply called the **peritoneum**. The peritoneum-mesentery is actually one continuous membrane (see Fig. 16–4). The serous fluid between the peritoneum and mesentery prevents friction when the alimentary tube contracts and the organs slide against one another.

The above descriptions are typical of the layers of the alimentary tube. As noted, variations are possible, and any important differences are mentioned in the sections that follow on specific organs.

STOMACH

The **stomach** is located in the upper left quadrant of the abdominal cavity, to the left of the liver and in

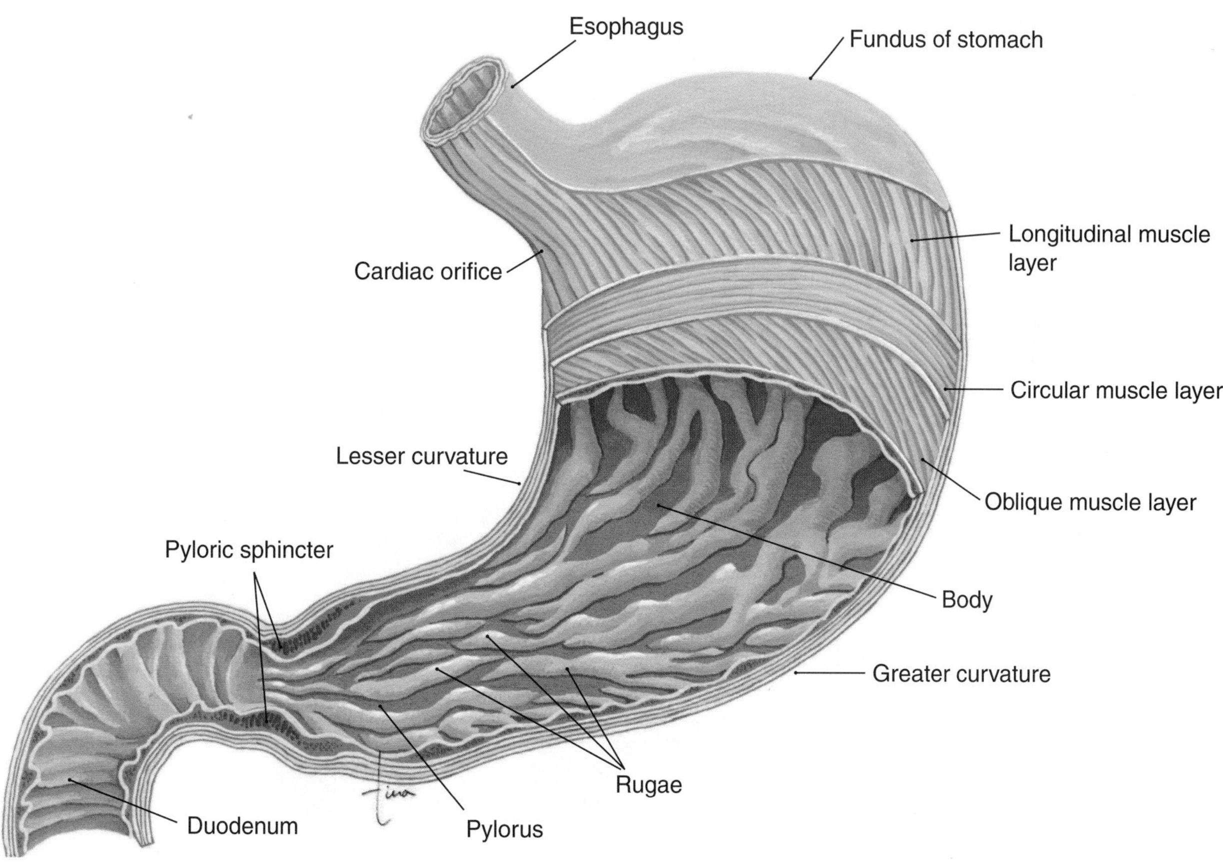

Figure 16–5 The stomach in anterior view. The stomach wall has been sectioned to show the muscle layers and the rugae of the mucosa.

front of the spleen. Although part of the alimentary tube, the stomach is not a tube, but rather a sac that extends from the esophagus to the small intestine. Because it is a sac, the stomach serves as a reservoir for food, so that digestion proceeds gradually and we do not have to eat constantly. Both mechanical and chemical digestion take place in the stomach.

The parts of the stomach are shown in Fig. 16–5. The cardiac orifice is the opening of the esophagus, and the fundus is the portion above the level of this opening. The body of the stomach is the large central portion, bounded laterally by the greater curvature and medially by the lesser curvature. The pylorus is adjacent to the duodenum of the small intestine, and the **pyloric sphincter** surrounds the junction of the two organs. The fundus and body are mainly storage areas, whereas most digestion takes place in the pylorus.

When the stomach is empty, the mucosa appears wrinkled or folded. These folds are called **rugae;** they flatten out as the stomach is filled and permit expansion of the lining without tearing it. The **gastric pits** are the glands of the stomach and consist of several types of cells; their collective secretions are called gastric juice. **Mucous cells** secrete mucus, which coats the stomach lining and helps prevent erosion by the gastric juice. **Chief cells** secrete **pepsinogen**, an inactive form of the enzyme **pepsin. Parietal cells** secrete hydrochloric acid (HCl), which converts pepsinogen to pepsin, which then begins the digestion of proteins to polypeptides. Hydrochloric acid also gives gastric juice its pH of 1 to 2. This very acidic pH is

Box 16–1 DISORDERS OF THE STOMACH

Vomiting is the expulsion of stomach and intestinal contents through the esophagus and mouth. Stimuli include irritation of the stomach, motion sickness, food poisoning, or diseases such as meningitis. The vomiting center is in the medulla, which coordinates the simultaneous contraction of the diaphragm and the abdominal muscles. This squeezes the stomach and upper intestine, expelling their contents. As part of the reflex, the lower esophageal sphincter relaxes, and the glottis closes. If the glottis fails to close, as may happen in alcohol or drug intoxication, aspiration of vomitus may occur and result in fatal obstruction of the respiratory passages.

Pyloric stenosis means that the opening of the pyloric sphincter is narrowed, and emptying of the stomach is impaired. This is most often a congenital disorder caused by hypertrophy of the pyloric sphincter. For reasons unknown, this condition is more common in male infants than in female infants. When the stomach does not empty efficiently, its internal pressure increases. Vomiting relieves the pressure; this is a classic symptom of pyloric stenosis. Correcting this condition requires surgery to widen the opening in the sphincter.

A **gastric ulcer** is an erosion of the mucosa of the stomach. Because the normal stomach lining is adapted to resist the corrosive action of gastric juice, ulcer formation is the result of oversecretion of HCl or undersecretion of mucus. Possible contributing factors include cigarette smoking and ingestion of alcohol or caffeine.

As erosion reaches the submucosa, small blood vessels are ruptured and bleed. If vomiting occurs, the vomitus has a "coffee-ground" appearance due to the presence of blood acted on by gastric juice. A more serious complication is perforation of the stomach wall, with leakage of gastric contents into the abdominal cavity, and **peritonitis**.

Recent clinical studies have confirmed that a bacterium called *Helicobacter pylori* is the cause of most gastric ulcers. For many patients, a few weeks of antibiotic therapy to eradicate this bacterium has produced rapid healing of their ulcers. Other researchers have proposed a link between this bacterium and stomach cancer, and these investigations are continuing.

The medications that decrease the secretion of HCl are useful for ulcer patients not helped by antibiotics.

necessary for pepsin to function and also kills most microorganisms that enter the stomach.

Gastric juice is secreted in small amounts at the sight or smell of food. This is a parasympathetic response that ensures that some gastric juice will be present in the stomach when food arrives. The presence of food in the stomach causes the gastric mucosa to secrete the hormone gastrin, which stimulates the secretion of greater amounts of gastric juice.

The external muscle layer of the stomach consists of three layers of smooth muscle: circular, longitudinal, and oblique layers. These three layers are innervated by the vagus nerves (10th cranial), and provide for very efficient mechanical digestion to change food into a thick liquid called chyme. The pyloric sphincter is usually contracted when the stomach is churning food; it relaxes at intervals to permit small amounts of chyme to pass into the duodenum. This sphincter then contracts again to prevent the backup of intestinal contents into the stomach (see Box 16-1: Disorders of the Stomach).

SMALL INTESTINE

The **small intestine** is about 1 inch (2.5 cm) in diameter and approximately 20 feet (6 m) long and extends from the stomach to the cecum of the large intestine. Within the abdominal cavity, the large intestine encircles the coils of the small intestine (see Fig. 16-1).

The **duodenum** is the first 10 inches (25 cm) of the small intestine. The common bile duct enters the duodenum at the ampulla of Vater (or hepatopancreatic ampulla). The **jejunum** is about 8 feet long, and the **ileum** is about 11 feet in length. In a living person, however, the small intestine is always contracted and is therefore somewhat shorter.

Digestion is completed in the small intestine, and the end products of digestion are absorbed into the blood and lymph. The external muscle layer has the typical circular and longitudinal smooth muscle layers that mix the chyme with digestive secretions and propel the chyme toward the colon. Impulses to these muscle layers are carried by the vagus nerves.

There are three sources of digestive secretions that function within the small intestine: the liver, the pancreas, and the small intestine itself. We will return to the small intestine after considering these other organs.

LIVER

The **liver** consists of two large lobes, right and left, and fills the upper right and center of the abdominal cavity, just below the diaphragm. The cells of the liver have many functions (which are discussed in a later section), but their only digestive function is the production of **bile**. Bile enters the small bile ducts, called bile canaliculi, on the liver cells, which unite to form larger ducts and finally merge to form the **hepatic duct**, which takes bile out of the liver (Fig. 16-6). The hepatic duct unites with the cystic duct of the gallbladder to form the **common bile duct**, which takes bile to the duodenum.

Bile is mostly water and has an excretory function in that it carries bilirubin and excess cholesterol to the intestines for elimination in feces. The digestive function of bile is accomplished by **bile salts**, which **emulsify** fats in the small intestine. Emulsification means that large fat globules are broken into smaller globules. This is mechanical, not chemical, digestion; the fat is still fat but now has more surface area to facilitate chemical digestion.

Production of bile is stimulated by the hormone **secretin**, which is produced by the duodenum when food enters the small intestine. Table 16-2 summarizes the regulation of secretion of all the digestive secretions.

GALLBLADDER

The **gallbladder** is a sac about 3 to 4 inches (7.5 to 10 cm) long located on the undersurface of the right lobe of the liver. Bile in the hepatic duct of the liver flows through the **cystic duct** into the gallbladder (see Fig. 16-6), which stores bile until it is needed in the small intestine. The gallbladder also concentrates bile by absorbing water (see Box 16-2: Gallstones).

When fatty foods enter the duodenum, the duodenal mucosa secretes the hormone **cholecystokinin**. It is this hormone that stimulates contraction of the smooth muscle in the wall of the gallbladder, which

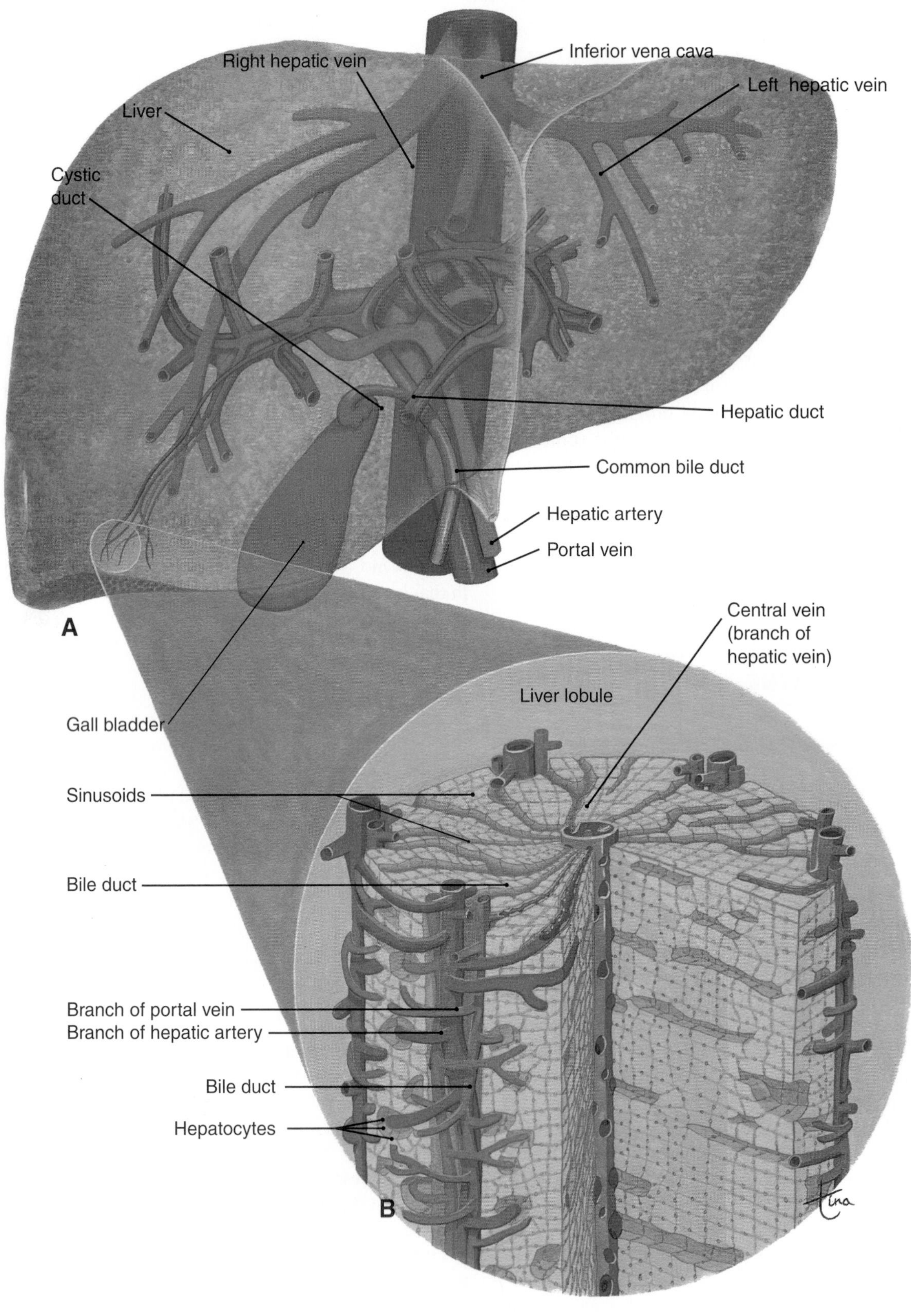

Figure 16–6 **(A)** The liver and gallbladder with blood vessels and bile ducts. **(B)** Magnified view of one liver lobule. See text for description.

Table 16–2 REGULATION OF DIGESTIVE SECRETIONS

Secretion	Nervous Regulation	Chemical Regulation
Saliva	Presence of food in mouth or sight of food; parasympathetic impulses along 7th and 9th cranial nerves	• None
Gastric juice	Sight or smell of food; parasympathetic impulses along 10th cranial nerves	• Gastrin—produced by the gastric mucosa when food is present in the stomach
Bile		
Secretion by the liver	None	• Secretin—produced by the duodenum when chyme enters
Contraction of the gallbladder	None	• Cholecystokinin—produced by the duodenum when chyme enters
Enzyme pancreatic juice	None	• Cholecystokinin—from the duodenum
Bicarbonate pancreatic juice	None	• Secretin—from the duodenum
Intestinal juice	Presence of chyme in the duodenum; parasympathetic impulses along 10th cranial nerves	• None

Box 16–2 GALLSTONES

One of the functions of the gallbladder is to concentrate bile by absorbing water. If the bile contains a high concentration of cholesterol, absorption of water may lead to precipitation and the formation of cholesterol crystals. These crystals are **gallstones**.

If the gallstones are small, they will pass through the cystic duct and common bile duct to the duodenum without causing symptoms. If large, however, the gallstones cannot pass out of the gallbladder, and may cause mild to severe pain that often radiates to the right shoulder. Obstructive jaundice may occur if bile backs up into the liver and bilirubin is reabsorbed into the blood.

There are several possible treatments for gallstones. Medications that dissolve gallstones work slowly, over the course of several months, and are useful if biliary obstruction is not severe. An instrument that generates shock waves (called a lithotripter) may be used to pulverize the stones into smaller pieces that may easily pass into the duodenum; this procedure is called **lithotripsy**. Surgery to remove the gallbladder (cholecystectomy) is required in some cases. The hepatic duct is then connected directly to the common bile duct, and dilute bile flows into the duodenum. Following such surgery, the patient should avoid meals high in fats.

forces bile into the cystic duct, then into the common bile duct, and on into the duodenum.

PANCREAS

The **pancreas** is located in the upper left abdominal quadrant between the curve of the duodenum and the spleen and is about 6 inches (15 cm) in length. The endocrine functions of the pancreas were discussed in Chapter 10, so only the exocrine functions will be considered here. The exocrine glands of the pancreas are called acini. They produce enzymes that are involved in the digestion of all three types of complex food molecules.

The pancreatic enzyme **amylase** digests starch to maltose. You may recall that this is the "backup" enzyme for salivary amylase. **Lipase** converts emulsified fats to fatty acids and glycerol. The emulsifying or fat-separating action of bile salts increases the surface area of fats so that lipase works effectively. Trypsinogen is an inactive enzyme that is changed to active **trypsin** in the duodenum. Trypsin digests polypeptides to shorter chains of amino acids.

The pancreatic enzyme juice is carried by small ducts that unite to form larger ducts, then finally the main **pancreatic duct**. An accessory duct may also be present. The main pancreatic duct emerges from the medial side of the pancreas and joins the common bile duct to the duodenum (Fig. 16–7).

The pancreas also produces a **bicarbonate juice** (containing sodium bicarbonate), which is alkaline. Because the gastric juice that enters the duodenum is very acidic, it must be neutralized to prevent damage to the duodenal mucosa. This neutralizing is accomplished by the sodium bicarbonate in pancreatic juice, and the pH of the duodenal chyme is raised to about 7.5.

Secretion of pancreatic juice is stimulated by the hormones secretin and cholecystokinin, which are produced by the duodenal mucosa when chyme en-

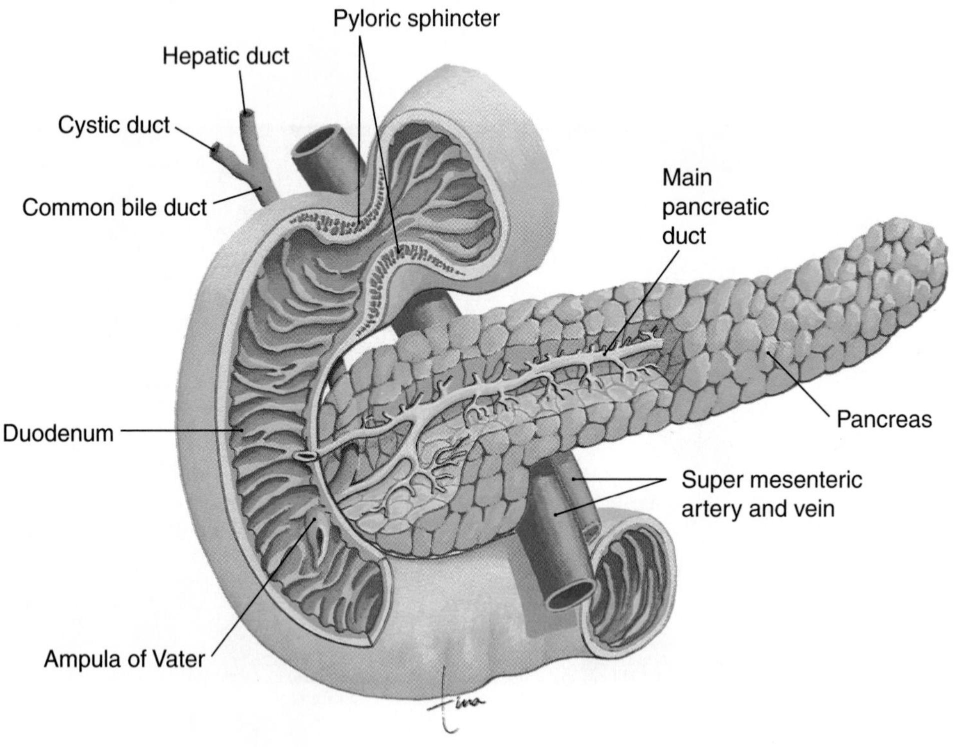

Figure 16–7 The pancreas, sectioned to show the pancreatic ducts. The main pancreatic duct joins the common bile duct.

ters the small intestine. **Secretin** stimulates the production of bicarbonate juice by the pancreas, and **cholecystokinin** stimulates the secretion of the pancreatic enzymes.

COMPLETION OF DIGESTION AND ABSORPTION

SMALL INTESTINE

The secretion of the intestinal glands (or crypts of Lieberkühn) is stimulated by the presence of food in the duodenum. The intestinal enzymes are the peptidases and sucrase, maltase, and lactase. **Peptidases** complete the digestion of protein by breaking down short polypeptide chains to amino acids. **Sucrase, maltase**, and **lactase**, respectively, digest the disaccharides sucrose, maltose, and lactose to monosaccharides.

A summary of the digestive secretions and their functions is found in Table 16-1. Regulation of these secretions is shown in Table 16-2.

ABSORPTION

Most absorption of the end products of digestion takes place in the small intestine (although the stomach does absorb water and alcohol). The process of absorption requires a large surface area, which is provided by several structural modifications of the small intestine; these are shown in Fig. 16-8. **Plica circulares**, or circular folds, are macroscopic folds of the mucosa and submucosa, somewhat like accordion pleats. The mucosa is further folded into projections called **villi**, which give the inner surface of the intestine a velvet-like appearance. Each columnar cell (except the mucus-secreting goblet cells) of the villi also has **microvilli** on its free surface. Microvilli are microscopic folds of the cell membrane. All of these folds greatly increase the surface area of the intestinal lining. It is estimated that if the intestinal mucosa could be flattened out, it would cover more than 2000 square feet (half a basketball court).

The absorption of nutrients takes place from the lumen of the intestine into the vessels within the villi. Refer back to Fig. 16-8 and notice that within each villus is a **capillary network** and a **lacteal**, which is a dead-end lymph capillary. Water-soluble nutrients are absorbed into the blood in the capillary networks. Monosaccharides, amino acids, positive ions, and the water-soluble vitamins (vitamin C and the B vitamins) are absorbed by active transport. Negative ions may be absorbed by either passive or active transport mechanisms. Water is absorbed by osmosis following the absorption of minerals, especially sodium. Certain nutrients have additional special requirements for their absorption: For example, vitamin B_{12} requires the intrinsic factor produced by the gastric mucosa, and the efficient absorption of calcium ions requires parathyroid hormone and vitamin D.

Fat-soluble nutrients are absorbed into the lymph in the lacteals of the villi. Bile salts are necessary for the efficient absorption of fatty acids and the fat-soluble vitamins (A, D, E, K). Once absorbed, fatty acids are recombined with glycerol to form triglycerides. These triglycerides then form globules that include cholesterol and protein; these lipid-protein complexes are called **chylomicrons**. In the form of chylomicrons, most absorbed fat is transported by the lymph and eventually enters the blood in the left subclavian vein.

Blood from the capillary networks in the villi does not return directly to the heart but first travels through the portal vein to the liver. You may recall the importance of portal circulation, discussed in Chapter 13. This pathway enables the liver to regulate the blood levels of glucose and amino acids, store certain vitamins, and remove potential poisons from the blood (see Box 16-3: Disorders of the Intestines).

LARGE INTESTINE

The **large intestine**, also called the **colon**, is approximately 2.5 inches (6.3 cm) in diameter and 5 feet (1.5 m) in length. It extends from the ileum of the small intestine to the anus, the terminal opening. The parts of the colon are shown in Fig. 16-9. The **cecum** is the first portion, and at its junction with the ileum is the **ileocecal valve**, which is not a sphincter but serves the same purpose. After undigested food (which is now mostly cellulose) and water pass from the ileum into the cecum, closure of the ileocecal valve prevents the backflow of fecal material.

Attached to the cecum is the **appendix**, a small, dead-end tube with abundant lymphatic tissue. The appendix seems to be a **vestigial organ**, that is, one whose size and function seem to be reduced. Although there is abundant lymphatic tissue in the wall

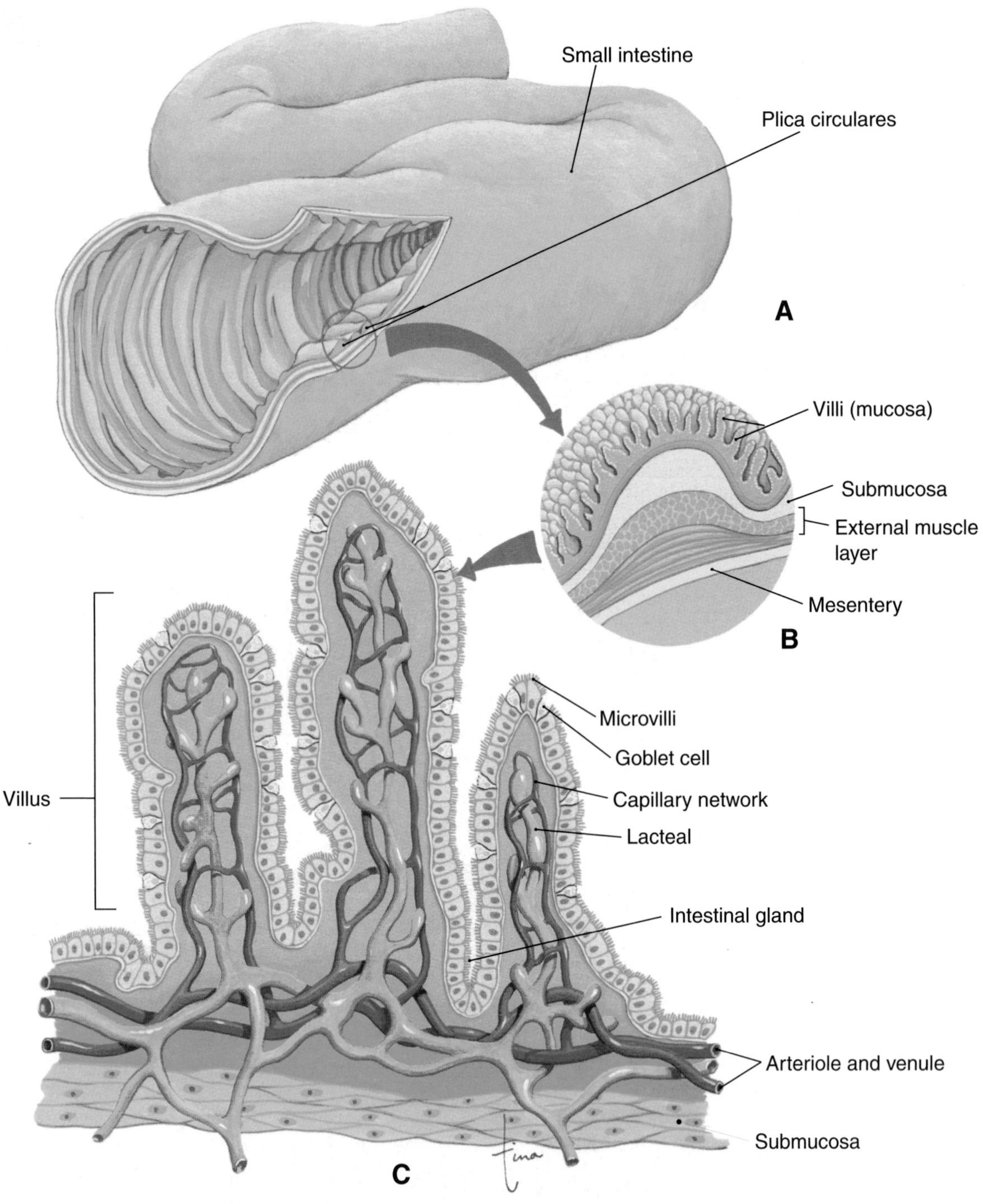

Figure 16–8 The small intestine. **(A)** Section through the small intestine showing plica circulares. **(B)** Magnified view of a section of the intestinal wall showing the villi and the four layers. **(C)** Microscopic view of three villi showing the internal structure.

Box 16–3 DISORDERS OF THE INTESTINES

Duodenal ulcers are erosions of the duodenal wall caused by the gastric juice that enters from the stomach (see Box 16–1).

Paralytic ileus is the cessation of contraction of the smooth muscle layer of the intestine. This is a possible complication of abdominal surgery, but it may also be the result of peritonitis or inflammation elsewhere in the abdominal cavity. In the absence of peristalsis, intestinal obstruction may occur. Bowel movements cease, and vomiting occurs to relieve the pressure within the alimentary tube. Treatment involves suctioning the intestinal contents to eliminate any obstruction and allow the intestine to regain its normal motility.

Lactose intolerance is the inability to digest lactose because of deficiency of the enzyme lactase. Lactase deficiency may be congenital, a consequence of prematurity, or acquired later in life. The delayed form is quite common among people of African or Asian ancestry, and in part is probably genetic. When lactose, or milk sugar, is not digested, it undergoes fermentation in the intestine. Symptoms include diarrhea, abdominal pain, bloating, and flatulence (gas formation).

Salmonella food poisoning is caused by bacteria in the genus *Salmonella.* These are part of the intestinal flora of animals, and animal foods such as meat and eggs may be sources of infection. These bacteria are not normal for people, and they cause the intestines to secrete large amounts of fluid. Symptoms include diarrhea, abdominal cramps, and vomiting and usually last only a few days. For elderly or debilitated people, however, salmonella food poisoning may be very serious or even fatal.

Diverticula are small outpouchings through weakened areas of the intestinal wall. They are more likely to occur in the colon than in the small intestine and may exist for years without causing any symptoms. The presence of diverticula is called **diverticulosis.** Inflammation of diverticula is called **diverticulitis,** which is usually the result of entrapment of feces and bacteria. Symptoms include abdominal pain and tenderness and fever. If uncomplicated, diverticulitis may be treated with antibiotics and modifications in diet. The most serious complication is perforation of diverticula, allowing fecal material into the abdominal cavity, causing peritonitis. A diet high in fiber is believed to be an important aspect of prevention, to provide bulk in the colon and prevent weakening of its wall.

of the appendix, the possibility that the appendix is concerned with immunity is not known with certainty. **Appendicitis** refers to inflammation of the appendix, which may occur if fecal material becomes impacted within it. This usually necessitates an **appendectomy**, the surgical removal of the appendix.

The remainder of the colon consists of the ascending, transverse, and descending colon, which encircle the small intestine; the sigmoid colon, which turns medially and downward; the rectum; and the anal canal. The rectum is about 6 inches long, and the anal canal is the last inch of the colon that surrounds the anus. Clinically, however, the terminal end of the colon is usually referred to as the rectum.

No digestion takes place in the colon. The only secretion of the colonic mucosa is mucus, which lubricates the passage of fecal material. The longitudinal smooth muscle layer of the colon is in three bands called **taeniae coli**. The rest of the colon is "gathered" to fit these bands. This gives the colon a puckered appearance; the puckers or pockets are called **haustra**, which provide for more surface area within the colon.

The functions of the colon are the absorption of water, minerals, and vitamins and the elimination of undigestable material. About 80% of the water that enters the colon is absorbed (400 to 800 ml per day). Positive and negative ions are also absorbed. The vitamins absorbed are those produced by the **normal flora**, the trillions of bacteria that live in the colon. Vitamin K is produced and absorbed in amounts usually sufficient to meet a person's daily need. Other vitamins produced in smaller amounts include riboflavin, thiamin, biotin, and folic acid. Everything

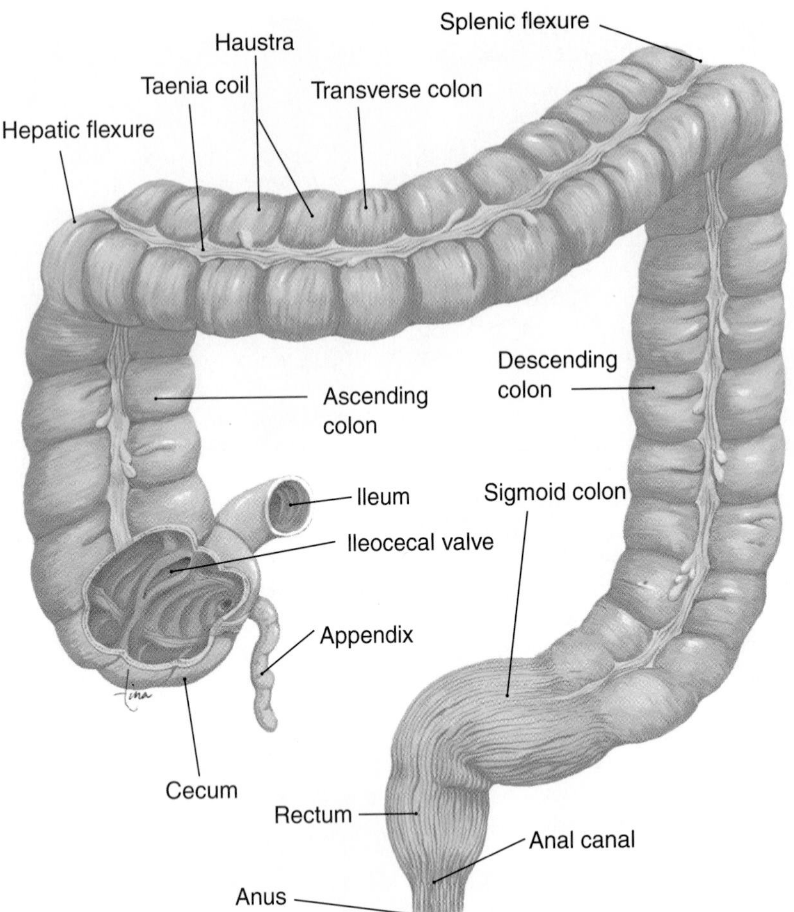

Figure 16–9 The large intestine shown in anterior view. The term *flexure* means a turn or bend.

absorbed by the colon circulates first to the liver by way of portal circulation. Yet another function of the normal colon flora is to inhibit the growth of pathogens (see Box 16–4: Infant Botulism).

ELIMINATION OF FECES

Feces consist of cellulose and other undigestable material, dead and living bacteria, and water. Elimination of feces is accomplished by the **defecation reflex**, a spinal cord reflex that may be controlled voluntarily. The rectum is usually empty until peristalsis of the colon pushes feces into it. These waves of peristalsis tend to occur after eating, especially when food enters the duodenum. The wall of the rectum is stretched by the entry of feces, and this is the stimulus for the defecation reflex.

Stretch receptors in the smooth muscle layer of the rectum generate sensory impulses that travel to the spinal cord. The returning motor impulses cause the smooth muscle of the rectum to contract. Surrounding the anus is the **internal anal sphincter**, which is made of smooth muscle. As part of the reflex this sphincter relaxes, permitting defecation to take place.

The **external anal sphincter** is made of skeletal muscle and surrounds the internal anal sphincter (Fig. 16–10). If defecation must be delayed, the external sphincter may be voluntarily contracted to close the anus. The awareness of the need to defecate passes as the stretch receptors of the rectum adapt. These receptors will be stimulated again when the next wave of peristalsis reaches the rectum (see Box 16–5: Fiber).

Box 16–4 INFANT BOTULISM

Botulism is most often acquired from food. When the spores of the botulism bacteria are in an anaerobic (without oxygen) environment such as a can of food, the spores germinate into active bacteria that produce a neurotoxin. If people ingest food containing this toxin, they will develop the paralysis that is characteristic of botulism.

For infants less than 1 year of age, however, ingestion of just the bacterial spores may be harmful. The infant's stomach does not produce much HCl, so ingested botulism spores may not be destroyed. Of equal importance, the infant's normal colon flora is not yet established. Without the normal population of colon bacteria to provide competition, spores of the botulism bacteria may germinate and produce their toxin.

An affected infant becomes lethargic and weak; paralysis may progress slowly or rapidly. Treatment (antitoxin) is available, but may be delayed if botulism is not suspected. Many cases of infant botulism have been traced to honey that was found to contain botulism spores. Such spores are not harmful to older children and adults, who have a normal colon flora that prevents the botulism bacteria from becoming established.

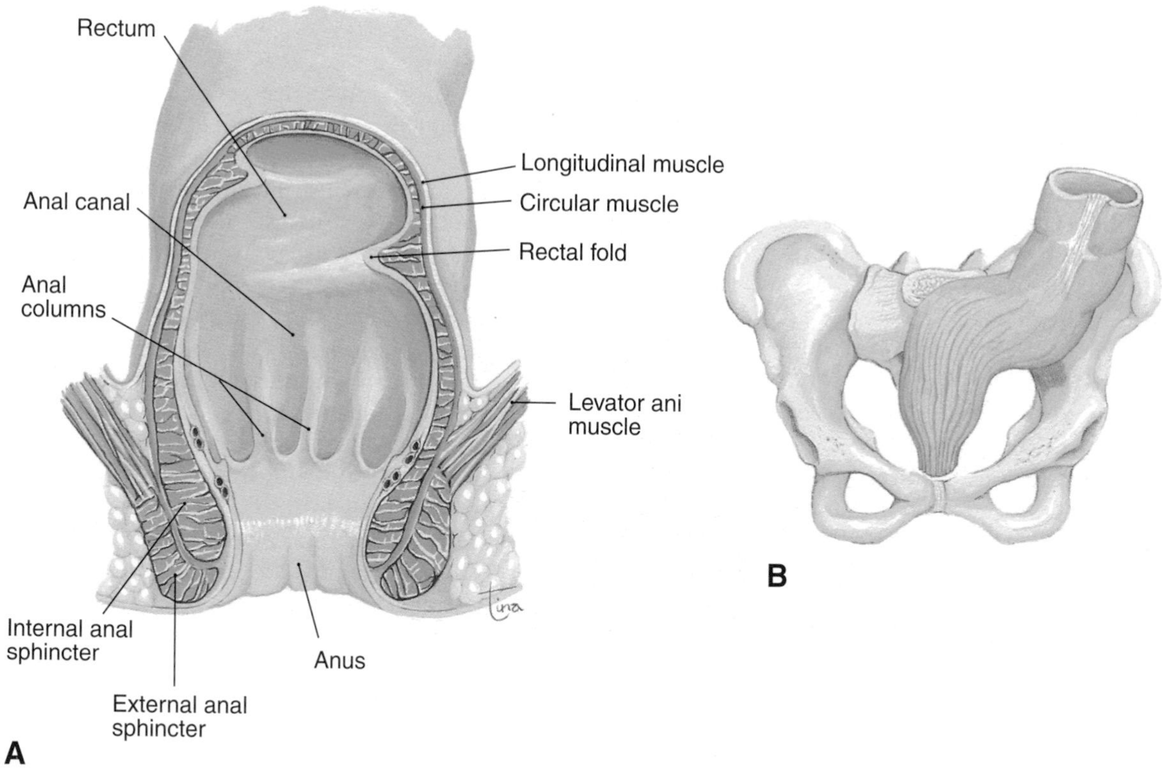

Figure 16–10 **(A)** Internal and external anal sphincters shown in a frontal section through the lower rectum and anal canal. **(B)** Position of rectum and anal canal relative to pelvic bone.

Box 16–5 FIBER

Fiber is a term we use to refer to the organic materials in the cell walls of plants. These are mainly cellulose and pectins. The role of dietary fiber, and possible benefits that a high-fiber diet may provide, are currently the focus of much research. It is important to differentiate what is known from what is, at present, merely speculation.

Many studies have shown that populations (large groups of people) who consume high-fiber diets tend to have a lower frequency of certain diseases. These include diverticulitis, colon cancer, coronary artery disease, diabetes, and hypertension. Such diseases are much more common among populations whose diets are low in vegetables, fruits, and whole grains, and high in meat, dairy products, and processed foods.

Recent claims that high-fiber diets directly lower blood levels of cholesterol and fats are not yet supported by definitive clinical or experimental studies. One possible explanation may be that a person whose diet consists largely of high-fiber foods simply eats less of foods high in cholesterol and fats, and this is the reason for their lower blood levels of fats and cholesterol.

Should people try to make great changes in their diets? Probably not, not if they are careful to limit fat intake and to include significant quantities of vegetables and fruits. Besides the possible benefits of fiber, unprocessed plant foods provide important amounts of vitamins and minerals.

OTHER FUNCTIONS OF THE LIVER

The **liver** is a remarkable organ, and only the brain is capable of a greater variety of functions. The liver cells (hepatocytes) produce many enzymes that catalyze many different chemical reactions. These reactions are the functions of the liver. As blood flows through the sinusoids (capillaries) of the liver (see Fig. 16–6), materials are removed by the liver cells, and the products of the liver cells are secreted into the blood. Some of the liver functions will already be familiar to you. Others are mentioned again and discussed in more detail in the next chapter. Because the liver has such varied effects on so many body systems, we will use the categories below to summarize the liver functions.

1. Carbohydrate Metabolism—As you know, the liver regulates the blood glucose level. Excess glucose is converted to glycogen (glycogenesis) when blood glucose is high; the hormones insulin and cortisol facilitate this process. During hypoglycemia or stress situations, glycogen is converted back to glucose (glycogenolysis) to raise the blood glucose level. Epinephrine and glucagon are the hormones that facilitate this process.

   The liver also changes other monosaccharides to glucose. Fructose and galactose, for example, are end products of the digestion of sucrose and lactose. Because most cells, however, cannot readily use fructose and galactose as energy sources, they are converted by the liver to glucose, which is easily used by cells.

2. Amino Acid Metabolism—The liver regulates blood levels of amino acids based on tissue needs for protein synthesis. Of the 20 different amino acids needed for the production of human proteins, the liver is able to synthesize 12, called the **non-essential amino acids**. The chemical process by which this is done is called **transamination**, the transfer of an amino group (NH_2) from an amino acid present in excess to a free carbon chain that forms a complete, new amino acid molecule. The other eight amino acids, which the liver cannot synthesize, are called the **essential amino acids**. In this case, "essential" means that the amino acids must be supplied by our food, because the liver cannot manufacture them. Similarly, "non-essential" means that the amino acids do not have to be supplied in our food because the liver *can* make them. All 20 amino acids are required in order to make our body proteins.

Excess amino acids, those not needed right away for protein synthesis, cannot be stored. However, they do serve another useful purpose. By the process of **deamination**, which also occurs in the liver, the NH_2 group is removed from an amino acid, and the remaining carbon chain may be converted to a simple carbohydrate molecule or to fat. Thus, excess amino acids are utilized for energy production: either for immediate energy or for the potential energy stored as fat in adipose tissue. The NH_2 groups that were detached from the original amino acids are combined to form urea, a waste product that will be removed from the blood by the kidneys and excreted in urine.

3. Lipid Metabolism—The liver forms lipoproteins, which as their name tells us, are molecules of lipids and proteins, for the transport of fats in the blood to other tissues. The liver also synthesizes cholesterol and excretes excess cholesterol into bile to be eliminated in feces.

   Fatty acids are a potential source of energy, but in order to be used in cell respiration they must be broken down to smaller molecules. In the process of **beta-oxidation**, the long carbon chains of fatty acids are split into two-carbon molecules called acetyl groups, which are simple carbohydrates. These acetyl groups may be used by the liver cells to produce ATP or may be combined to form ketones to be transported in the blood to other cells. These other cells then use the ketones to produce ATP in cell respiration.
4. Synthesis of Plasma Proteins—This is a liver function that you will probably remember from Chapter 11. The liver synthesizes many of the proteins that circulate in the blood. **Albumin**, the most abundant plasma protein, helps maintain blood volume by pulling tissue fluid into capillaries.

   The **clotting factors** are also produced by the liver. These, as you recall, include prothrombin, fibrinogen, and Factor 8, which circulate in the blood until needed in the chemical clotting mechanism. The liver also synthesizes alpha and beta **globulins**, which are proteins that serve as carriers for other molecules, such as fats, in the blood.
5. Formation of Bilirubin—This is another familiar function: The liver contains fixed macrophages that phagocytize old RBCs. Bilirubin is then formed from the heme portion of the hemoglobin. The liver also removes from the blood the bilirubin formed in the spleen and red bone marrow, and excretes it into bile to be eliminated in feces.
6. Phagocytosis by **Kupffer Cells**—The fixed macrophages of the liver are called Kupffer cells (or stellate reticuloendothelial cells). Besides destroying old RBCs, Kupffer cells phagocytize pathogens or other foreign material that circulate through the liver. Many of the bacteria that get to the liver come from the colon. These bacteria are part of the normal flora of the colon but would be very harmful elsewhere in the body. The bacteria that enter the blood with the water absorbed by the colon are carried to the liver by way of portal circulation. The Kupffer cells in the liver phagocytize and destroy these bacteria, removing them from the blood before the blood returns to the heart.
7. Storage—The liver stores the fat-soluble vitamins A, D, E, and K, and the water-soluble vitamin B_{12}. Up to a 6- to 12-month supply of vitamins A and D may be stored, and liver is an excellent dietary source of these vitamins.

   Also stored by the liver are the minerals iron and copper. You already know that iron is needed for hemoglobin and myoglobin and enables these proteins to bond to oxygen. Copper is part of some of the proteins needed for cell respiration.
8. Detoxification—The liver is capable of synthesizing enzymes that will detoxify harmful substances, that is, change them to less harmful ones. Alcohol, for example, is changed to acetate, which is a two-carbon molecule that can be used in cell respiration.

   Medications are all potentially toxic, but the liver produces enzymes that break them down or change them. When given in a proper dosage, a medication exerts its therapeutic effect but is then changed to less active substances that are usually excreted by the kidneys. An overdose of a drug means that there is too much of it for the liver to detoxify in a given time, and the drug will remain in the body with possibly harmful effects. This is why alcohol should never be consumed when taking medication. Such a combination may cause the liver's detoxification ability to be overworked and ineffective, with the result

Box 16–6 HEPATITIS

Hepatitis is inflammation of the liver caused by any of several viruses. The most common of these hepatitis viruses have been designated A, B, and C, although there are others. Symptoms of hepatitis include anorexia, nausea, fatigue, and possibly jaundice. Severity of disease ranges from very mild (even asymptomatic) to fatal. Hundreds of thousands of cases of hepatitis occur in the United States every year, and although liver inflammation is common to all of them, the three hepatitis viruses have different modes of transmission and different consequences for affected people.

Hepatitis A is an intestinal virus that is spread by the fecal-oral route. Food contaminated by the hands of people with mild cases is the usual vehicle of transmission, although shellfish harvested from water contaminated with human sewage are another possible source of this virus. Hepatitis A is most often mild, recovery provides lifelong immunity, and the carrier state is not known to occur. A vaccine is available, but people who have been exposed to hepatitis A may receive gamma globulin by injection to prevent the disease.

Hepatitis B is contracted by exposure to the body fluids of an infected person; these fluids include blood and semen. It is important to mention, however, that neither hepatitis A nor B is spread by blood transfusions, because donated blood is tested for both these viruses. Hepatitis B may be severe or even fatal, and approximately 10% of those who recover become carriers of the virus. Possible consequences of the carrier state are chronic hepatitis progressing to cirrhosis or primary liver cancer. Of equal importance, carriers are sources of the virus for others, especially their sexual partners.

A vaccine is available for hepatitis B, and although it is expensive, health-care workers who have contact with blood, even occasionally, should receive it. Other potential recipients of the vaccine are the sexual partners of carriers. Pediatricians now consider this vaccine one of the standard ones for infants.

The **hepatitis C** virus is also present in body fluids and is spread in the same ways as is hepatitis B. The range of severity of hepatitis C is also very similar to that of hepatitis B, and the carrier state is possible after recovery. Until recently, there was no serological (blood) test to detect this virus in donated blood, and the hepatitis C virus was responsible for several hundred thousand cases of post-transfusion hepatitis every year. The now standard testing done in blood banks has significantly reduced the number of new cases of hepatitis C.

that both the alcohol and the medication will remain toxic for a longer time. Barbiturates taken as sleeping pills after consumption of alcohol have too often proved fatal for just this reason.

Ammonia is a toxic substance produced by the bacteria in the colon. Because it is soluble in water, some ammonia is absorbed into the blood, but it is carried first to the liver by portal circulation. The liver converts ammonia to urea, a less toxic substance, before the ammonia can circulate and damage other organs, especially the brain. The urea formed is excreted by the kidneys (see Box 16–6: Hepatitis).

AGING AND THE DIGESTIVE SYSTEM

Many changes can be expected in the aging digestive system. The sense of taste becomes less acute; less saliva is produced; and there is greater likelihood of periodontal disease and loss of teeth. Secretions are reduced throughout the digestive system, and the effectiveness of peristalsis diminishes. Indigestion may become more frequent, especially if the LES loses its tone, and there is a greater chance of peptic ulcer. In the colon, diverticula may form; these are bubble-like outpouchings of the weakened wall of the colon that may be asymptomatic or become infected. Sluggish peristalsis contributes to constipation, which in turn may contribute to the formation of hemorrhoids. The risk of oral cancer or colon cancer also increases with age.

The liver usually continues to function adequately even well into old age, unless damaged by pathogens such as the hepatitis viruses or by toxins such as alcohol. There is a greater tendency for gallstones to form, perhaps necessitating removal of the gallbladder. In the absence of specific diseases, the pancreas usually functions well, although acute pancreatitis of unknown cause is somewhat more likely in elderly people.

SUMMARY

The processes of the digestion of food and the absorption of nutrients enable the body to use complex food molecules for many purposes. Much of the food we eat literally becomes part of us. The body synthesizes proteins and lipids for the growth and repair of tissues and produces enzymes to catalyze all the reactions that contribute to homeostasis. Some of our food provides the energy required for growth, repair, movement, sensation, and thinking. In the next chapter we will discuss the chemical basis of energy production from food and consider the relationship of energy production to the maintenance of body temperature.

STUDY OUTLINE

Function of the Digestive System—to break down food into simple chemicals that can be absorbed into the blood and lymph and utilized by cells

Divisions of the Digestive System

1. Alimentary Tube—oral cavity, pharynx, esophagus, stomach, small intestine, large intestine. Digestion takes place in the oral cavity, stomach, and small intestine.
2. Accessory Organs—salivary glands, teeth and tongue, liver and gallbladder, pancreas. Each contributes to digestion.

Types of Digestion

1. Mechanical—breaks food into smaller pieces to increase the surface area for the action of enzymes.
2. Chemical—enzymes break down complex organics into simpler organics and inorganics; each enzyme is specific for the food it will digest.

End Products of Digestion

1. Carbohydrates are digested to monosaccharides.
2. Fats are digested to fatty acids and glycerol.
3. Proteins are digested to amino acids.
4. Other end products are vitamins, minerals, and water.

Oral Cavity—food enters by way of the mouth

1. Teeth and tongue break up food and mix it with saliva.
2. Tooth Structure (see Fig. 16-2)—enamel covers the crown and provides a hard chewing surface; dentin is within the enamel and forms the roots; the pulp cavity contains blood vessels and endings of the trigeminal nerve; the periodontal membrane produces cement to anchor the tooth in the jawbone.
3. The tongue is skeletal muscle innervated by the hypoglossal nerves. Papillae on the upper surface contain taste buds (facial and glossopharyngeal nerves). Functions: taste, keeps food between the teeth when chewing, elevates to push food backward for swallowing.
4. Salivary Glands—parotid, submandibular, and sublingual (see Fig. 16-3); ducts take saliva to the oral cavity.
5. Saliva—amylase digests starch to maltose; water dissolves food for tasting and moistens food for swallowing; lysozyme inhibits the growth of bacteria (see Tables 16-1 and 16-2).

Pharynx—food passageway from the oral cavity to the esophagus

1. No digestion takes place.
2. Contraction of pharyngeal muscles is part of swallowing reflex, regulated by the medulla.

Esophagus—food passageway from pharynx to stomach

1. No digestion takes place.
2. Lower esophageal sphincter (LES) at junction with stomach prevents backup of stomach contents.

Structural Layers of the Alimentary Tube (see Fig. 16–4)

1. Mucosa (lining)—made of epithelial tissue that produces the digestive secretions; lymph nodules contain macrophages to phagocytize pathogens that penetrate the mucosa.
2. Submucosa—areolar connective tissue with blood vessels and lymphatic vessels; Meissner's plexus is an autonomic nerve network that innervates the mucosa.
3. External Muscle Layer—typically an inner circular layer and an outer longitudinal layer of smooth muscle; function is mechanical digestion and peristalsis; innervated by Auerbach's plexus: sympathetic impulses decrease motility; parasympathetic impulses increase motility.
4. Serosa—outermost layer; above the diaphragm is fibrous connective tissue; below the diaphragm is the mesentery (serous). The peritoneum (serous) lines the abdominal cavity; serous fluid prevents friction between the serous layers.

Stomach—in upper left abdominal quadrant; a muscular sac that extends from the esophagus to the small intestine (see Fig. 16–5)

1. Reservoir for food; begins the digestion of protein.
2. Gastric juice is secreted by gastric pits (see Tables 16-1 and 16-2).
3. The pyloric sphincter at the junction with the duodenum prevents backup of intestinal contents.

Liver—consists of two lobes in the upper right and center of the abdominal cavity (see Figs. 16–1 and 16–6)

1. The only digestive secretion is bile; the hepatic duct takes bile out of the liver and unites with the cystic duct of the gallbladder to form the common bile duct to the duodenum.
2. Bile salts emulsify fats, a type of mechanical digestion (see Table 16-1).
3. Excess cholesterol and bilirubin are excreted by the liver into bile.

Gallbladder—on undersurface of right lobe of liver (see Fig. 16–6)

1. Stores and concentrates bile until needed in the duodenum (see Table 16-2).
2. The cystic duct joins the hepatic duct to form the common bile duct.

Pancreas—in upper left abdominal quadrant between the duodenum and the spleen (see Fig. 16–1)

1. Pancreatic juice is secreted by acini, carried by pancreatic duct to the common bile duct to the duodenum (see Fig. 16-7).
2. Enzyme pancreatic juice contains enzymes for the digestion of all three food types (see Tables 16-1 and 16-2).
3. Bicarbonate pancreatic juice neutralizes HCl from the stomach in the duodenum.

Small Intestine—coiled within the center of the abdominal cavity (see Fig. 16–1); extends from stomach to colon

1. Duodenum—first 10 inches; the common bile duct brings in bile and pancreatic juice. Jejunum (8 feet) and ileum (11 feet).
2. Enzymes secreted by the intestinal glands complete digestion (see Tables 16-1 and 16-2).
3. Surface area for absorption is increased by plica circulares, villi, and microvilli (see Fig. 16-8).
4. The villi contain capillary networks for the absorption of water-soluble nutrients: monosaccharides, amino acids, vitamin C and B vitamins, minerals, and water. Blood from the small intestine goes to the liver first by way of portal circulation.
5. The villi contain lacteals (lymph capillaries) for the absorption of fat-soluble nutrients: vitamins A, D, E, and K, fatty acids and glycerol, which are combined to form chylomicrons. Lymph from the small intestine is carried back to the blood in the left subclavian vein.

Large Intestine (colon)—extends from the small intestine to the anus

1. Colon—parts (see Fig. 16-9): cecum, ascending colon, transverse colon, descending colon, sigmoid colon, rectum, anal canal.
2. Ileocecal Valve—at the junction of the cecum and ileum; prevents backup of fecal material into the small intestine.
3. Colon—functions: absorption of water, minerals, vitamins; elimination of undigestible material.
4. Normal Flora—the bacteria of the colon; produce vitamins, especially vitamin K, and inhibit the growth of pathogens.
5. Defecation Reflex—stimulus: stretching of the rectum when peristalsis propels feces into it. Sensory

impulses go to the spinal cord, and motor impulses return to the smooth muscle of the rectum, which contracts. The internal anal sphincter relaxes to permit defecation. Voluntary control is provided by the external anal sphincter, made of skeletal muscle (see Fig. 16-10).

Liver—other functions

1. Carbohydrate Metabolism—excess glucose is stored in the form of glycogen and converted back to glucose during hypoglycemia; fructose and galactose are changed to glucose.
2. Amino Acid Metabolism—the non-essential amino acids are synthesized by transamination; excess amino acids are changed to carbohydrates or fats by deamination; the amino groups are converted to urea and excreted by the kidneys.
3. Lipid Metabolism—formation of lipoproteins for transport of fats in the blood; synthesis of cholesterol; excretion of excess cholesterol into bile; beta-oxidation of fatty acids to form two-carbon acetyl groups for energy use.
4. Synthesis of Plasma Proteins—albumin to help maintain blood volume; clotting factors for blood clotting; alpha and beta globulins as carrier molecules.
5. Formation of Bilirubin—old RBCs are phagocytized, and bilirubin is formed from the heme and put into bile to be eliminated in feces.
6. Phagocytosis by Kupffer Cells—fixed macrophages; phagocytize old RBCs and bacteria, especially bacteria absorbed by the colon.
7. Storage—vitamins: B_{12}, A, D, E, K, and the minerals iron and copper.
8. Detoxification—liver enzymes change potential poisons to less harmful substances; examples of toxic substances are alcohol, medications, and ammonia absorbed by the colon.

REVIEW QUESTIONS

1. Name the organs of the alimentary tube, and describe the location of each. Name the accessory digestive organs, and describe the location of each. (pp. 352, 354-355, 359-361, 365)
2. Explain the purpose of mechanical digestion, and give two examples. Explain the purpose of chemical digestion, and give two examples. (pp. 352-356)
3. Name the end products of digestion, and explain how each is absorbed in the small intestine. (pp. 352, 365)
4. Explain the function of teeth and tongue, salivary amylase, enamel of teeth, lysozyme, and water of saliva. (pp. 352, 354-355)
5. Describe the function of the pharynx, esophagus, and lower esophageal sphincter. (p. 357)
6. Name and describe the four layers of the alimentary tube. (pp. 357, 359)
7. State the two general functions of the stomach and the function of the pyloric sphincter. Explain the function of pepsin, HCl, and mucus. (pp. 359-361)
8. Describe the general functions of the small intestine, and name the three parts. Describe the structures that increase the surface area of the small intestine. (pp. 361, 365)
9. Explain how the liver, gallbladder, and pancreas contribute to digestion. (pp. 361, 364)
10. Describe the internal structure of a villus, and explain how structure is related to absorption. (p. 365)
11. Name the parts of the large intestine, and describe the function of the ileocecal valve. (pp. 365, 367)

12. Describe the functions of the colon and of the normal flora of the colon. (p. 367)
13. With respect to the defecation reflex, explain the stimulus, the part of the CNS directly involved, the effector muscle, the function of the internal anal sphincter, and the voluntary control possible. (p. 368)
14. Name the vitamins and minerals stored in the liver. Name the fixed macrophages of the liver, and explain their function. (p. 371)
15. Describe how the liver regulates blood glucose level. Explain the purpose of the processes of deamination and transamination. (pp. 370–371)
16. Name the plasma proteins produced by the liver, and state the function of each. (pp. 370–371)
17. Name the substances excreted by the liver into bile. (p. 371)

CHAPTER 17

Body Temperature and Metabolism

CHAPTER 17

Chapter Outline

Body Temperature
Heat Production
Heat Loss
- Heat loss through the skin
- Heat loss through the respiratory tract
- Heat loss through the urinary and digestive tracts

Regulation of Body Temperature
- Mechanisms to increase heat loss
- Mechanisms to conserve heat

Fever
Metabolism
Cell Respiration
- Glycolysis
- Krebs citric acid cycle
- Cytochrome transport system
- Proteins and fats as energy sources
- Energy available from the three nutrient types

Synthesis Uses of Foods
- Glucose
- Amino acids
- Fatty acids and glycerol

Vitamins and Minerals
Metabolic Rate
Aging and Metabolism

Box 17–1 HEAT-RELATED DISORDERS
Box 17–2 COLD-RELATED DISORDERS
Box 17–3 KETOSIS
Box 17–4 METABOLIC RATE
Box 17–5 WEIGHT LOSS
Box 17–6 LEPTIN AND BODY-MASS INDEX

Student Objectives

- State the normal range of human body temperature.
- Explain how cell respiration produces heat and the factors that affect heat production.
- Describe the pathways of heat loss through the skin and respiratory tract.
- Explain why the hypothalamus is called the thermostat of the body.
- Describe the mechanisms to increase heat loss.
- Describe the mechanisms to conserve heat.
- Explain how a fever is caused and the advantages and disadvantages.
- Define metabolism, anabolism, and catabolism.
- Describe what happens to a glucose molecule during the three stages of cell respiration.
- State what happens to each of the products of cell respiration.
- Explain how amino acids and fats may be used for energy production.

Body Temperature and Metabolism

Student Objectives (Continued)

- Describe the synthesis uses for glucose, amino acids, and fats.
- Explain what is meant by metabolic rate and kilocalories.
- Describe the factors that affect a person's metabolic rate.

New Terminology

Anabolism (an-**AB**-uh-lizm)
Catabolism (kuh-**TAB**-uh-lizm)
Coenzyme (ko-**EN**-zime)
Conduction (kon-**DUK**-shun)
Convection (kon-**VEK**-shun)
Cytochromes (**SIGH**-toh-krohms)
Endogenous pyrogen (en-**DOJ**-en-us **PYE**-roh-jen)
Fever (**FEE**-ver)
Glycolysis (gly-**KAHL**-ah-sis)
Kilocalorie (KILL-oh-**KAL**-oh-ree)
Krebs cycle (KREBS **SIGH**-kuhl)
Metabolism (muh-**TAB**-uh-lizm)
Minerals (**MIN**-er-als)
Pyrogen (**PYE**-roh-jen)
Radiation (RAY-dee-**AY**-shun)
Vitamins (**VY**-tah-mins)

Related Clinical Terminology

Antipyretic (AN-tigh-pye-**RET**-ik)
Basal metabolic rate (**BAY**-zuhl met-ah-**BAHL**-ik RAYT)
Frostbite (**FRAWST**-bite)
Heat exhaustion (HEET eks-**ZAWS**-chun)
Heat stroke (HEET STROHK)
Hypothermia (HIGH-poh-**THER**-mee-ah)

*Terms that appear in **bold type** in the chapter text are defined in the glossary, which begins on page 528.*

During every moment of our lives, our cells are breaking down food molecules to obtain ATP for energy-requiring cellular processes. Naturally, we are not aware of the process of cell respiration, but we may be aware of one of the products, energy in the form of heat. The human body is indeed warm, and its temperature is regulated very precisely, even in a wide range of environmental temperatures.

This chapter discusses the regulation of body temperature and also discusses **metabolism**, which is the total of all the reactions that take place within the body. These reactions include the energy-releasing ones of cell respiration and energy-requiring ones such as protein synthesis, or DNA synthesis for mitosis. As you will see, body temperature and metabolism are inseparable.

BODY TEMPERATURE

The normal range of human body temperature is 96.5 to 99.5°F (36 to 38°C), with an average of 98.6°F (37°C). (A 1992 study suggested a slightly lower average temperature: 98.2°F or 36.8°C. Whether these values will replace the more "traditional" average temperatures remains to be seen.) Within a 24-hour period, an individual's temperature fluctuates 1 to 2°F, with the lowest temperatures occurring during sleep.

At either end of the age spectrum, however, temperature regulation may not be as precise as it is in older children or younger adults. Infants have more surface area (skin) relative to volume and are likely to lose heat more rapidly. In the elderly, the mechanisms that maintain body temperature may not function as efficiently as they once did, and changes in environmental temperature may not be compensated for as quickly or effectively. This is especially important to remember when caring for patients who are very young or very old.

HEAT PRODUCTION

Cell respiration, the process that releases energy from food to produce ATP, also produces heat as one of its energy products. Although cell respiration takes place constantly, there are many factors that influence the rate of this process:

1. The hormone **thyroxine** (and T_3), produced by the thyroid gland, increases the rate of cell respiration and heat production. The secretion of thyroxine is regulated by the body's rate of energy production, the metabolic rate itself (see Chapter 10 for the feedback mechanism involving the hypothalamus and anterior pituitary gland). When the metabolic rate decreases, the thyroid gland is stimulated to secrete more thyroxine. As thyroxine increases the rate of cell respiration, a negative feedback mechanism inhibits further secretion until metabolic rate decreases again. Thus, thyroxine is secreted whenever there is a need for increased cell respiration and is probably the most important regulator of day-to-day energy production.
2. In stress situations, **epinephrine** and norepinephrine are secreted by the adrenal medulla, and the **sympathetic** nervous system becomes more active. Epinephrine increases the rate of cell respiration, especially in organs such as the heart, skeletal muscles, and liver. Sympathetic stimulation also increases the activity of these organs. The increased production of ATP to meet the demands of the stress situation also means that more heat will be produced.
3. Organs that are normally active (producing ATP) are significant sources of heat when the body is at rest. The skeletal muscles, for example, are usually in a state of slight contraction called muscle tone. Because even slight contraction requires ATP, the muscles are also producing heat. This amounts to about 25% of the total body heat at rest and much more during exercise when more ATP is produced.

   The liver is another organ that is continually active, producing ATP to supply energy for its many functions. As a result, the liver produces as much as 20% of the total body heat at rest.

   The heat produced by these active organs is dispersed throughout the body by the blood. As the relatively cooler blood flows through organs such as the muscles and liver, the heat they produce is transferred to the blood, warming it. The warmed blood circulates to other areas of the body, distributing this heat.
4. The intake of food also increases heat production, because the metabolic activity of the digestive tract is increased. Heat is generated as the digestive organs produce ATP for peristalsis and for the synthesis of digestive enzymes.
5. Changes in body temperature also have an effect on metabolic rate and heat production. This be-

Table 17–1 FACTORS THAT AFFECT HEAT PRODUCTION

Factor	Effect
Thyroxine	• The most important regulator of day-to-day metabolism; increases use of foods for ATP production, thereby increasing heat production
Epinephrine and sympathetic stimulation	• Important in stress situations; increases the metabolic activity of many organs; increases ATP and heat production
Skeletal muscles	• Normal muscle tone requires ATP; the heat produced is about 25% of the total body heat at rest
Liver	• Always metabolically active; produces as much as 20% of total body heat at rest
Food intake	• Increases activity of the GI tract; increases ATP and heat production
Higher body temperature	• Increases metabolic rate, which increases heat production, which further increases metabolic rate and heat production. May become detrimental during high fevers.

comes clinically important when a person has a **fever**, an abnormally high body temperature. The higher temperature increases the metabolic rate, which increases heat production and elevates body temperature further. Thus, a high fever may trigger a vicious cycle of ever-increasing heat production. Fever is discussed later in this chapter.

The factors that affect heat production are summarized in Table 17-1.

HEAT LOSS

The pathways of heat loss from the body are the skin, respiratory tract, and to a lesser extent, the urinary and digestive tracts.

Heat Loss through the Skin

Because the skin covers the body, most body heat is lost from the skin to the environment. When the environment is cooler than body temperature (as it usually is), heat loss is unavoidable. The amount of heat that is lost is determined by blood flow through the skin and by the activity of sweat glands.

Blood flow through the skin influences the amount of heat lost by the processes of radiation, conduction, and convection. **Radiation** means that heat from the body is transferred to cooler objects not touching the skin, much as a radiator warms the contents of a room (radiation starts to become less effective when the environmental temperature rises above 88°F). **Conduction** is the loss of heat to cooler air or objects, such as clothing, that touch the skin. **Convection** means that air currents move the warmer air away from the skin surface and facilitate the loss of heat; this is why a fan makes us feel cooler on hot days.

As you may recall from Chapter 5, the temperature of the skin and the subsequent loss of heat are determined by blood flow through the skin. The arterioles in the dermis may constrict or dilate to decrease or increase blood flow. **Vasoconstriction** decreases blood flow through the dermis and thereby decreases heat loss. **Vasodilation** in the dermis increases blood flow to the body surface and loss of heat to the environment.

The other mechanism by which heat is lost from the skin is sweating. The **eccrine sweat glands** secrete sweat (water) onto the skin surface, and excess body heat evaporates the sweat. Think of running water into a hot frying pan; the pan is rapidly cooled as its heat vaporizes the water. Although sweating is not quite as dramatic (no visible formation of steam), the principle is just the same.

Sweating is most efficient when the humidity of the surrounding air is low. Humidity is the percentage of the maximum amount of water vapor the atmosphere can contain. A humidity reading of 90% means that the air is already 90% saturated with water vapor and can hold little more. In such a situation, sweat does not readily evaporate, but rather, remains on the skin even as more sweat is secreted. If the humidity is 40%, however, the air can hold a great deal more water vapor, and sweat evaporates quickly from the skin surface, removing excess body heat. In air that is completely dry, a person may tolerate a temperature of 200°F for nearly 1 hour.

Although sweating is a very effective mechanism of heat loss, it does have a disadvantage in that it requires the loss of water in order to also lose heat. Water loss during sweating may rapidly lead to dehydration, and the water lost must be replaced by drinking fluids (see Box 17-1: Heat-Related Disorders).

Box 17–1 HEAT-RELATED DISORDERS

Heat exhaustion is caused by excessive sweating with loss of water and salts, especially NaCl. The affected person feels very weak, and the skin is usually cool and clammy (moist). Body temperature is normal or slightly below normal, the pulse is often rapid and weak, and blood pressure may be low because of fluid loss. Other symptoms may include dizziness, vomiting, and muscle cramps. Treatment involves rest and consumption of salty fluids or fruit juices (in small amounts at frequent intervals).

Heat stroke is a life-threatening condition that may affect elderly or chronically ill people on hot, humid days, or otherwise healthy people who exercise too strenuously during such weather. High humidity makes sweating an ineffective mechanism of heat loss, but in high heat the sweating process continues. As fluid loss increases, sweating stops to preserve body fluid, and body temperature rises rapidly (over 105°F, possibly as high as 110°F).

The classic symptom of heat stroke is hot, dry skin. The affected person often loses consciousness, reflecting the destructive effect of such a high body temperature on the brain. Treatment should involve hospitalization so that IV fluids may be administered and body temperature lowered under medical supervision. A first-aid measure would be the application of cool (not ice cold) water to as much of the skin as possible. Fluids should never be forced on an unconscious person, because the fluid may be aspirated into the respiratory tract.

Small amounts of heat are also lost in what is called "insensible water loss." Because the skin is not like a plastic bag, but is somewhat permeable to water, a small amount of water diffuses through the skin and is evaporated by body heat. Compared to sweating, however, insensible water loss is a minor source of heat loss.

Heat Loss through the Respiratory Tract

Heat is lost from the respiratory tract as the warmth of the respiratory mucosa evaporates some water from the living epithelial surface. The water vapor formed is exhaled, and a small amount of heat is lost.

Animals such as dogs that do not have numerous sweat glands often pant in warm weather. Panting is the rapid movement of air into and out of the upper respiratory passages, where the warm surfaces evaporate large amounts of water. In this way the animal may lose large amounts of heat.

Heat Loss through the Urinary and Digestive Tracts

When excreted, urine and feces are at body temperature, and their elimination results in a very small amount of heat loss. The pathways of heat loss are summarized in Table 17-2.

REGULATION OF BODY TEMPERATURE

The **hypothalamus** is responsible for the regulation of body temperature and is considered the "thermostat" of the body. As the thermostat, the hypothalamus

Table 17–2 PATHWAYS OF HEAT LOSS

Pathway	Mechanism
Skin (major pathway)	• Radiation and Conduction—heat is lost from the body to cooler air or objects. • Convection—air currents move warm air away from the skin. • Sweating—excess body heat evaporates sweat on the skin surface.
Respiratory tract (secondary pathway)	• Evaporation—body heat evaporates water from the respiratory mucosa, and water vapor is exhaled.
Urinary tract (minor pathway)	• Urination—urine is at body temperature when eliminated.
Digestive tract (minor pathway)	• Defecation—feces are at body temperature when eliminated.

maintains the "setting" of body temperature by balancing heat production and heat loss to keep the body at the set temperature.

In order to do this, the hypothalamus must receive information about the temperature within the body and about the environmental temperature. Specialized neurons of the hypothalamus detect changes in the temperature of the blood that flows through the brain. The temperature receptors in the skin provide information about the external temperature changes the body is exposed to. The hypothalamus then integrates this sensory information and promotes the necessary responses to maintain body temperature within the normal range.

Mechanisms to Increase Heat Loss

In a warm environment or during exercise, the body temperature tends to rise, and greater heat loss is needed. This is accomplished by vasodilation in the dermis and an increase in sweating. Vasodilation brings more warm blood close to the body surface, and heat is lost to the environment. However, if the environmental temperature is close to or higher than body temperature, this mechanism becomes ineffective. The second mechanism is increased sweating, in which excess body heat evaporates the sweat on the skin surface. As mentioned previously, sweating becomes inefficient when the atmospheric humidity is high.

On hot days, heat production may also be decreased by a decrease in muscle tone. This is why we may feel very sluggish on hot days; our muscles are even less slightly contracted than usual and are slower to respond.

Mechanisms to Conserve Heat

In a cold environment, heat loss from the body is unavoidable but may be minimized to some extent. Vasoconstriction in the dermis shunts blood away from the body surface, so that more heat is kept in the core of the body. Sweating decreases, and will stop completely if the temperature of the hypothalamus falls below about 98.6°F.

If these mechanisms are not sufficient to prevent the body temperature from dropping, more heat may be produced by increasing muscle tone. When this greater muscle tone becomes noticeable and rhythmic

Box 17–2 COLD-RELATED DISORDERS

Frostbite is the freezing of part of the body. Fingers, toes, the nose, and ears are most often affected by prolonged exposure to cold, because these areas have little volume in proportion to their surface.

At first the skin tingles, then becomes numb. If body fluids freeze, ice crystals may destroy capillaries and tissues (because water expands when it freezes), and blisters form. In the most severe cases gangrene develops, that is, tissue dies because of lack of oxygen.

Treatment of frostbite includes rewarming the affected area. If skin damage is apparent, it should be treated as if it were a burn injury.

Hypothermia is an abnormally low body temperature (below 95°F) that is most often the result of prolonged exposure to cold. Although the affected person certainly feels cold at first, this sensation may pass and be replaced by confusion, slurred speech, drowsiness, and lack of coordination. At this stage, people often do not realize the seriousness of their condition, and if outdoors (ice skating or skiing) may not seek a warmer environment. In progressive hypothermia, breathing and heart rate slow, and coma and death follow.

Other people at greater risk for hypothermia include the elderly, whose temperature-regulating mechanisms are no longer effective, and quadriplegics, who have no sensation of cold in the body. For both these groups, heat production is or may be low because of inactivity of skeletal muscles.

Artificial hypothermia may be induced during some types of cardiovascular or neurological surgery. This carefully controlled lowering of body temperature decreases the metabolic rate and need for oxygen and makes possible prolonged surgery without causing extensive tissue death in the patient.

it is called shivering and may increase heat production as much as five times the normal.

People also have behavioral responses to cold, and these too are important to prevent heat loss. Such things as putting on a sweater or going indoors reflect our awareness of the discomfort of being cold. For people (we do not have thick fur as do some other mammals), these voluntary activities are of critical importance to the prevention of excessive heat loss when it is very cold (see Box 17–2: Cold-Related Disorders).

FEVER

A fever is an abnormally high body temperature and may accompany infectious diseases, extensive physical trauma, cancer, or damage to the CNS. The substances that may cause a fever are called **pyrogens**. Pyrogens include bacteria, foreign proteins, and chemicals released during inflammation **(endogenous pyrogens)**. It is believed that pyrogens chemically affect the hypothalamus and "raise the setting" of the hypothalamic thermostat. The hypothalamus will then stimulate responses by the body to raise body temperature to this higher setting.

Let us use as a specific example a child who has a strep throat. The bacterial and endogenous pyrogens reset the hypothalamic thermostat upward, to 102°F. At first, the body is "colder" than the setting of the hypothalamus, and the heat conservation and production mechanisms are activated. The child feels cold and begins to shiver (chills). Eventually, sufficient heat

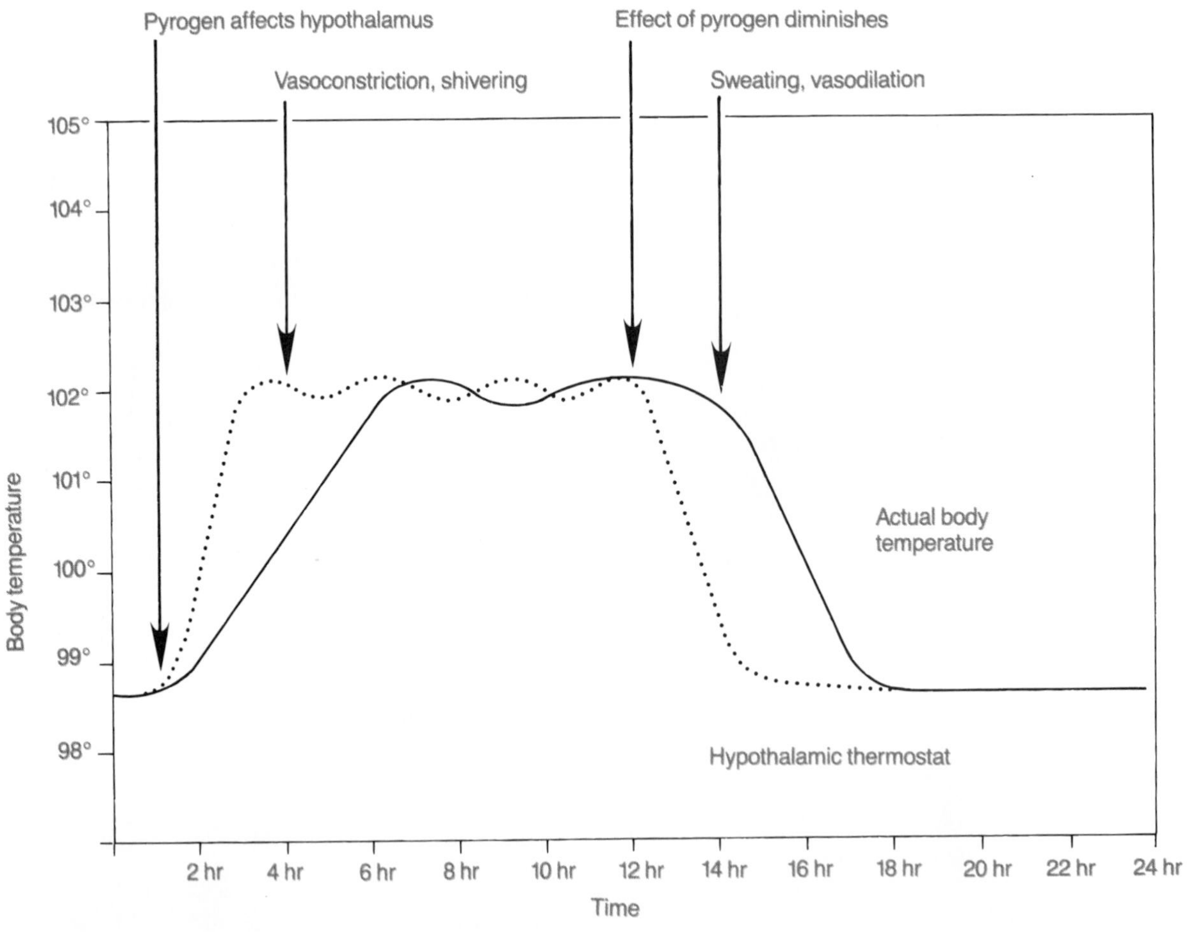

Figure 17–1 Changes in body temperature during an episode of fever. The body temperature *(solid line)* changes lag behind the changes in the hypothalamic thermostat *(dotted line)* but eventually reach whatever the thermostat has called for.

is produced to raise the body temperature to the hypothalamic setting of 102°F. At this time, the child will feel neither too warm nor too cold, because the body temperature is what the hypothalamus wants.

As the effects of the pyrogens diminish, the hypothalamic setting decreases, perhaps close to normal again, 99°F. Now the child will feel warm, and the heat-loss mechanisms will be activated. Vasodilation in the skin and sweating will occur until the body temperature drops to the new hypothalamic setting. This is sometimes referred to as the "crisis," but actually the crisis has passed, because sweating indicates that the body temperature is returning to normal. The sequence of temperature changes during a fever is shown in Fig. 17-1.

You may be wondering if a fever serves a useful purpose. For low fevers that are the result of infection, the answer seems to be yes. White blood cells increase their activity at moderately elevated temperatures, and the metabolism of some pathogens is inhibited. Thus, a fever may be beneficial in that it may shorten the duration of an infection by accelerating the destruction of the pathogen.

High fevers, however, may have serious consequences. When the body temperature rises above 106°F, the hypothalamus begins to lose its ability to regulate temperature. The enzymes of cells are also damaged by such high temperatures. Enzymes become denatured, that is, lose their shape and do not catalyze the reactions necessary within cells. As a result, cells begin to die. This is most serious in the brain, because neurons cannot be replaced, and is the cause of brain damage that may follow a prolonged high fever. The effects of changes in body temperature on the hypothalamus are shown in Fig. 17-2.

A medication such as aspirin is called an **antipyretic** because it lowers a fever, probably by affecting the hypothalamic thermostat. To help lower a very high fever, the body may be cooled by sponging with alcohol or cold water. The excessive body heat will cause these fluids to evaporate, thus reducing temperature.

METABOLISM

The term **metabolism** encompasses all the reactions that take place in the body. Everything that happens within us is part of our metabolism. The reactions of metabolism may be divided into two major categories: anabolism and catabolism.

Anabolism means synthesis or "formation" reactions, the bonding together of smaller molecules to form larger ones. The synthesis of hemoglobin by cells of the red bone marrow, synthesis of glycogen by liver cells, and synthesis of fat to be stored in adipose tissue are all examples of anabolism. Such reactions require energy, usually in the form of ATP.

Catabolism means decomposition, the breaking of bonds of larger molecules to form smaller molecules. Cell respiration is a series of catabolic reactions that break down food molecules to carbon dioxide and water. During catabolism, energy is often released and used to synthesize ATP (the heat energy released was discussed in the previous section). The ATP formed during catabolism is then used for energy-requiring anabolic reactions.

Most of our anabolic and catabolic reactions are catalyzed by enzymes. Enzymes are proteins that enable reactions to take place rapidly at body temperature (see Chapter 2 to review the Active Site Theory of enzyme functioning). The body has thousands of enzymes, and each is specific, that is, will catalyze only one type of reaction. As you read the discussions that follow, keep in mind the essential role of enzymes.

CELL RESPIRATION

You are already familiar with the summary reaction of cell reaction,

$$\underset{\text{(glucose)}}{C_6H_{12}O_6} + O_2 \rightarrow CO_2 + H_2O + ATP + \text{Heat}$$

the purpose of which is to produce ATP. Glucose contains potential energy, and when it is broken down to CO_2 and H_2O, this energy is released in the forms of ATP and heat. The oxygen that is required comes from breathing, and the CO_2 formed is circulated to the lungs to be exhaled. The water formed is called metabolic water, and helps to meet our daily need for water. Energy in the form of heat gives us a body temperature, and the ATP formed is used for energy-requiring reactions. Synthesis of ATP means that energy is used to bond a free phosphate molecule to ADP (adenosine diphosphate). ADP and free phosphates are present in cells after ATP has been broken down for energy-requiring processes.

The breakdown of glucose summarized above is not quite that simple, however, and involves a complex

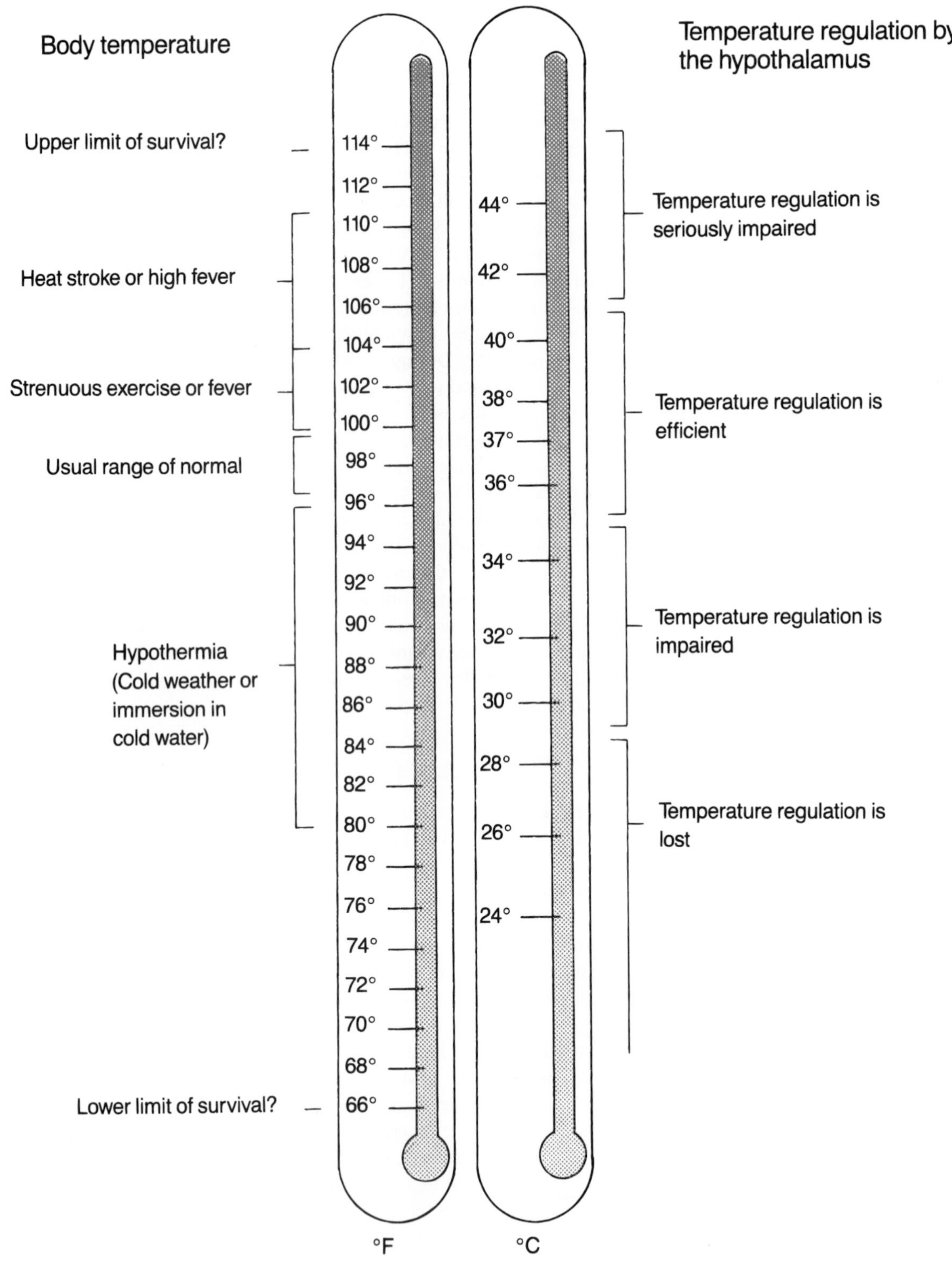

Figure 17–2 Effects of changes in body temperature on the temperature-regulating ability of the hypothalamus. Body temperature is shown in degrees Fahrenheit and degrees Celsius.

series of reactions. Glucose is broken down "piece by piece," with the removal of hydrogens and the splitting of carbon-carbon bonds. This releases the energy of glucose gradually, so that a significant portion (about 40%) is available to synthesize ATP.

Cell respiration of glucose involves three major stages: glycolysis, the Krebs citric acid cycle, and the cytochrome (or electron) transport system. Although all the details of each stage are beyond the scope of this book, we will summarize the most important aspects of each, and then relate to them the use of amino acids and fats for energy.

Glycolysis

The enzymes for the reactions of **glycolysis** are found in the cytoplasm of cells, and oxygen is not required (glycolysis is an anaerobic process). Refer now to Fig. 17-3 as you read the following. In glycolysis, a six-carbon glucose molecule is broken down to two three-carbon molecules of pyruvic acid. Two molecules of ATP are necessary to start the process. The energy they supply is called energy of activation and is necessary to make glucose unstable enough to begin to break down. As a result of these reactions, enough energy is released to synthesize four molecules of ATP, for a net gain of two ATP molecules per glucose molecule. Also during glycolysis, two pairs of hydrogens are removed by NAD, a carrier molecule that contains the vitamin **niacin**. Two NAD molecules thus become $2NADH_2$, and these attached hydrogen pairs will be transported to the cytochrome transport system (stage 3).

If no oxygen is present in the cell, as may happen in muscle cells during exercise, pyruvic acid is converted to lactic acid, which causes muscle fatigue. If oxygen *is* present, however, pyruvic acid continues into the next stage, the Krebs citric acid cycle (or, more simply, the Krebs cycle).

Krebs Citric Acid Cycle

The enzymes for the **Krebs cycle** (or citric acid cycle) are located in the mitochondria of cells. This second stage of cell respiration is aerobic, meaning that oxygen is required. In a series of reactions, a pyruvic acid molecule is "taken apart," and its carbons are converted to CO_2. The first CO_2 molecule is removed by an enzyme that contains the vitamin **thiamine**. This leaves a two-carbon molecule called an acetyl group, which combines with a molecule called coenzyme A to form acetyl coenzyme A (acetyl CoA). As acetyl CoA continues in the Krebs cycle, two more carbons are removed as CO_2, and more pairs of hydrogens are picked up by NAD and FAD (another carrier molecule that contains the vitamin **riboflavin**). $NADH_2$ and $FADH_2$ will carry their hydrogens to the cytochrome transport system.

During the Krebs cycle, a small amount of energy is released, enough to synthesize one molecule of ATP (two per glucose). Notice also that a four-carbon molecule (oxaloacetic acid) is regenerated after the formation of CO_2. This molecule will react with the next acetyl CoA, which is what makes the Krebs cycle truly a self-perpetuating cycle. The results of all the stages of cell respiration are listed in Table 17-3. Before you continue here you may wish to look at that table to see just where the process has gotten thus far.

Cytochrome Transport System

Cytochromes are proteins that contain either **iron** or **copper** and are found in the mitochondria of cells. The pairs of hydrogens that were once part of glucose are brought to the cytochromes by the carrier molecules NAD and FAD. Each hydrogen atom is then split into its proton (H^+ ion) and its electron. The electrons of the hydrogens are passed from one cytochrome to the next, and finally to oxygen. The reactions of the electrons with the cytochromes release most of the energy that was contained in the glucose molecule, enough to synthesize 34 molecules of ATP. As you can see, most of the ATP produced in cell respiration comes from this third stage.

Finally, and very importantly, each oxygen atom that has gained two electrons (from the cytochromes) reacts with two of the H^+ ions (protons) to form water. The formation of metabolic water contributes to the necessary intracellular fluid, and also prevents acidosis. If H^+ ions accumulated, they would rapidly lower the pH of the cell. This does not happen, however, because the H^+ ions react with oxygen to form water, and a decrease in pH is prevented.

The summary of the three stages of cell respiration in Table 17-3 also includes the vitamins and minerals that are essential for this process. An important overall concept is the relationship between eating and breathing. Eating provides us with a potential energy source (often glucose) and with necessary vitamins and minerals. However, to release the energy from

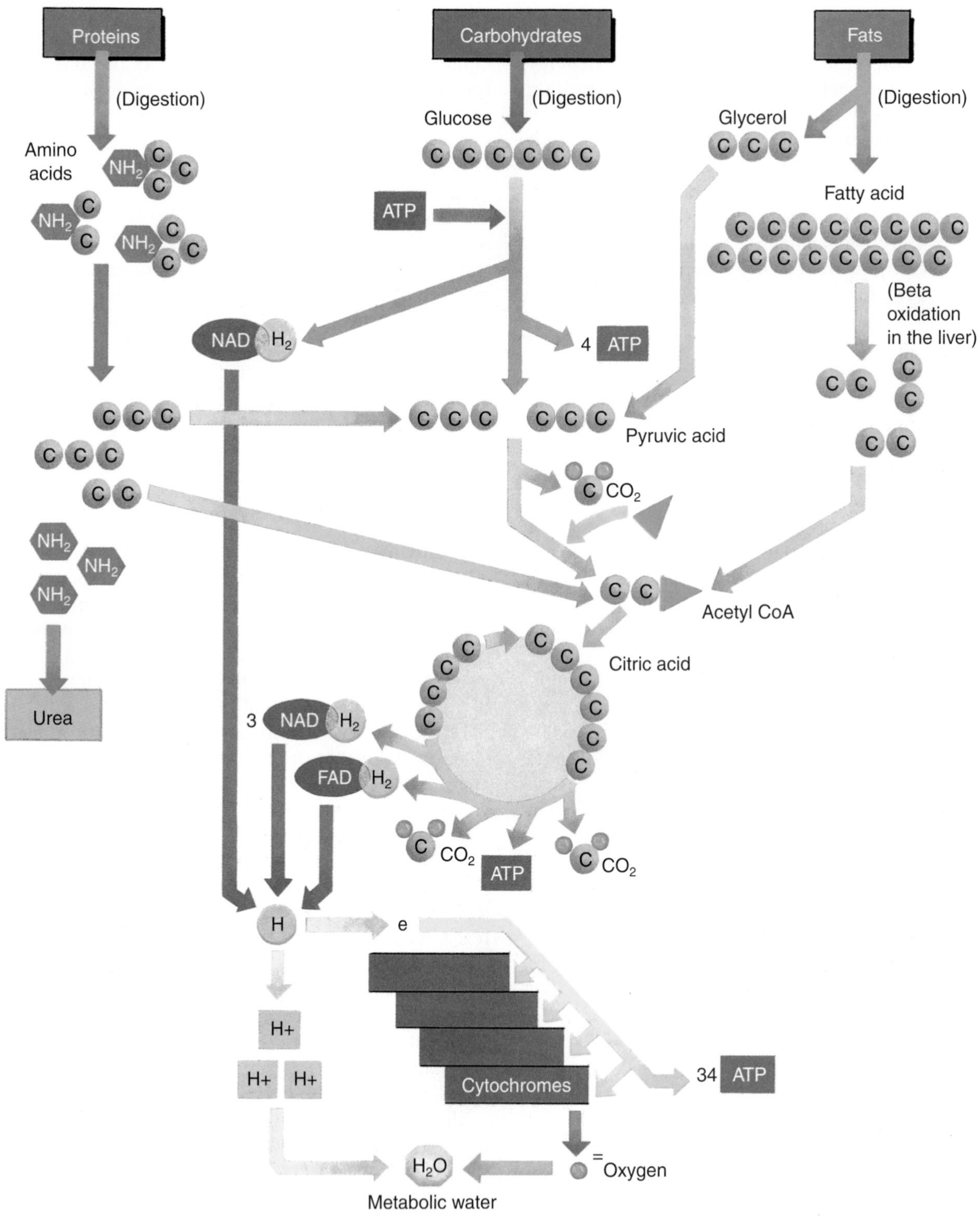

Figure 17–3 Schematic representation of cell respiration. The breakdown of glucose is shown in the center, amino acids on the left, and fatty acids and glycerol on the right. See text for description.

Table 17–3 SUMMARY OF CELL RESPIRATION

Stage	Molecules That Enter the Process	Results	Vitamins or Minerals Needed
Glycolysis (cytoplasm)	Glucose—ATP needed as energy of activation	• 2 ATP (net) • 2 $NADH_2$ (to cytochrome transport system) • 2 pyruvic acid (aerobic: to Krebs cycle; anaerobic: lactic acid formation)	• Niacin (part of NAD)
Krebs citric acid cycle (mitochondria)	Pyruvic acid—from glucose or glycerol or excess amino acids *or* Acetyl CoA—from fatty acids or excess amino acids	• CO_2 (exhaled) • ATP (2 per glucose) • 3 $NADH_2$ and 1 $FADH_2$ (to cytochrome transport system) • A 4-carbon molecule is regenerated for the next cycle	• Thiamine (thiamin) (for removal of CO_2) • Niacin (part of NAD) • Riboflavin (part of FAD) • Pantothenic acid (part of coenzyme A)
Cytochrome transport system (mitochondria)	$NADH_2$ and $FADH_2$—from glycolysis or the Krebs cycle	• 34 ATP • Metabolic water	• Iron and copper (part of some cytochromes)

food, we must breathe. This is *why* we breathe. The oxygen we inhale is essential for the completion of cell respiration, and the CO_2 produced is exhaled.

Proteins and Fats as Energy Sources

Although glucose is the preferred energy source for cells, proteins and fats also contain potential energy and are alternative energy sources in certain situations.

As you know, proteins are made of the smaller molecules called **amino acids**, and the primary use for the amino acids we obtain from food is the synthesis of new proteins. Excess amino acids, however, those not needed immediately for protein synthesis, may be used for energy production. In the liver, excess amino acids are **deaminated**, that is, the amino group (NH_2) is removed. The remaining portion is converted to a molecule that will fit into the Krebs cycle. For example, a deaminated amino acid may be changed to a three-carbon pyruvic acid or to a two-carbon acetyl group. When these molecules enter the Krebs cycle, the results are just the same as if they had come from glucose. This is diagrammed in Fig. 17-3.

Fats are made of glycerol and fatty acids, which are the end products of fat digestion. These molecules may also be changed to ones that will take part in the Krebs cycle, and the reactions that change them usually take place in the liver. Glycerol is a three-carbon molecule that can be converted to the three-carbon pyruvic acid, which enters the Krebs cycle. In the process of beta-oxidation, the long carbon chains of fatty acids are split into two-carbon acetyl groups, which enter a later step in the Krebs cycle (see Fig. 17-3).

Both amino acids and fatty acids may be converted by the liver to **ketones** which are two- or four-carbon molecules such as acetone and acetoacetic acid. Although body cells can use ketones in cell respiration, they do so slowly. In situations in which fats or amino acids have become the primary energy sources, a state called **ketosis** may develop; this is described in Box 17-3: Ketosis. Excess amino acids may also be converted to glucose; this is important to supply the brain when dietary intake of carbohydrates is low. The effects of hormones on the metabolism of food are summarized in Table 17-4.

Energy Available from the Three Nutrient Types

The potential energy in food is measured in units called **Calories** or **kilocalories**. A calorie (lower case "c") is the amount of energy needed to raise the temperature of 1 gram of water 1°C. A kilocalorie or Calorie (capital "C") is 1000 times that amount of energy.

One gram of carbohydrate yields about 4 kilocalo-

Box 17–3 KETOSIS

When fats and amino acids are to be used for energy, they are often converted by the liver to ketones. Ketones are organic molecules such as acetone that may be changed to acetyl CoA and enter the Krebs cycle. Other cells are able to use ketones as an energy source, but they do so slowly. When ketones are produced in small amounts, as they usually are between meals, the blood level does not rise sharply.

A state of **ketosis** exists when fats and proteins become the primary energy sources, and ketones accumulate in the blood faster than cells can utilize them. Because ketones are organic acids, they lower the pH of the blood. As the blood ketone level rises, the kidneys excrete ketones, but they must also excrete more water as a solvent, which leads to dehydration.

Ketosis is clinically important in diabetes mellitus, starvation, and eating disorders such as anorexia nervosa. Diabetics whose disease is poorly controlled may progress to **ketoacidosis**, a form of metabolic acidosis that may lead to confusion, coma, and death. Reversal of this state requires a carbohydrate energy source and the insulin necessary to utilize it.

Table 17–4 HORMONES THAT REGULATE METABOLISM

Hormone (gland)	Effects
Thyroxine (thyroid gland)	• Increases use of all three food types for energy (glucose, fats, amino acids) • Increases protein synthesis
Growth hormone (anterior pituitary)	• Increases amino acid transport into cells • Increases protein synthesis • Increases use of fats for energy
Insulin (pancreas)	• Increases glucose transport into cells and use for energy • Increases conversion of glucose to glycogen in liver and muscles • Increases transport of amino acids and fatty acids into cells to be used for synthesis (*not* energy production)
Glucagon (pancreas)	• Increases conversion of glycogen to glucose • Increases use of amino acids and fats for energy
Cortisol (adrenal cortex)	• Increases conversion of glucose to glycogen in liver • Increases use of amino acids and fats for energy • Decreases protein synthesis except in liver and GI tract
Epinephrine (adrenal medulla)	• Increases conversion of glycogen to glucose • Increases use of fats for energy

ries. A gram of protein also yields about 4 kilocalories. A gram of fat, however, yields 9 kilocalories, and a gram of alcohol yields 7 kilocalories. This is why a diet high in fat is more likely to result in weight gain if the calories are not expended in energy-requiring activities.

You may have noticed that calorie content is part of the nutritional information on food labels. Here, however, the term "calorie" actually means Calorie or kilocalories but is used for the sake of simplicity.

SYNTHESIS USES OF FOODS

Besides being available for energy production, each of the three food types is used in anabolic reactions to synthesize necessary materials for cells and tissues.

Glucose

Glucose is the raw material for the synthesis of another essential monosaccharide, the **pentose sugars** that are part of nucleic acids. Deoxyribose is the five-carbon sugar found in DNA, and ribose is found in RNA. This function of glucose is very important, for without the pentose sugars our cells could neither produce new chromosomes for cell division nor carry out the process of protein synthesis.

Any glucose in excess of immediate energy needs or the need for pentose sugars is converted to **glycogen** in the liver and skeletal muscles. Glycogen is then an energy source during states of hypoglycemia or during exercise.

Amino Acids

As mentioned previously, the primary uses for amino acids are the synthesis of the **non-essential amino acids** by the liver and the synthesis of new **proteins** in all tissues. By way of review, we can mention some proteins with which you are already familiar: keratin and melanin in the epidermis; collagen in the dermis, tendons, and ligaments; myosin, actin, and myoglobin in muscle cells; hemoglobin in RBCs; antibodies produced by WBCs; prothrombin and fibrinogen for clotting; albumin to maintain blood volume; pepsin and amylase for digestion; growth hormone and insulin; and the thousands of enzymes needed to catalyze reactions within the body.

The amino acids we obtain from the proteins in our food are used by our cells to synthesize all of these proteins in the amounts needed by the body. Only when the body's needs for new proteins have been met are amino acids used for energy production.

Fatty Acids and Glycerol

The end products of fat digestion that are not needed immediately for energy production may be stored as fat (triglycerides) in **adipose tissue**. Most adipose tissue is found subcutaneously and is potential energy for times when food intake decreases.

Fatty acids and glycerol are also used for the synthesis of **phospholipids**, which are essential components of all cell membranes. Myelin, for example, is a phospholipid of the membranes of Schwann cells, which form the myelin sheath of peripheral neurons.

When fatty acids are broken down in the process of beta-oxidation, the resulting acetyl groups may also be used for the synthesis of **cholesterol**, a steroid. This takes place primarily in the liver, although all cells are capable of synthesizing cholesterol for their cell membranes. The liver uses cholesterol to synthesize bile salts for the emulsification of fats in digestion. The **steroid hormones** are also synthesized from cholesterol. Cortisol and aldosterone are produced by the adrenal cortex; estrogen and progesterone by the ovaries, and testosterone by the testes.

VITAMINS AND MINERALS

Vitamins are organic molecules needed in very small amounts for normal body functioning. Some vitamins are **coenzymes**, that is, they are necessary for the functioning of certain enzymes. Table 17–5 summarizes some important metabolic and nutritional aspects of the vitamins we need.

Minerals are simple inorganic chemicals and have a variety of functions. Table 17–6 lists some important aspects of minerals. We will return to the minerals as part of fluid-electrolyte balance in Chapter 19.

METABOLIC RATE

Although the term **metabolism** is used to describe all of the chemical reactions that take place within the body, **metabolic rate** is usually expressed as an amount of heat production. This is because many body processes that utilize ATP also produce heat. These processes include the contraction of skeletal muscle, the pumping of the heart, and the normal breakdown of cellular components. Therefore, it is possible to quantify heat production as a measure of metabolic activity.

As mentioned previously, the energy available from food is measured in kilocalories (kcal). Kilocalories are also the units used to measure the energy expended by the body. During sleep, for example, energy expended by a 150-pound person is about 60 to 70 kcal per hour. Getting up and preparing breakfast increases energy expenditure to 80 to 90 kcal per hour. For mothers with several small children, this value may be significantly higher. Clearly, greater activity results in greater energy expenditure.

The energy required for merely living (lying quietly in bed) is the **basal metabolic rate** (BMR). See Box 17–4: Metabolic Rate for a formula to estimate your own metabolic rate. There are a number of factors that affect the metabolic rate of an active person.

1. Exercise—Contraction of skeletal muscle increases energy expenditure and raises metabolic rate (see Box 17–5: Weight Loss).
2. Age—Metabolic rate is highest in young children and decreases with age. The energy requirements for growth and the greater heat loss by a smaller body contribute to the higher rate in children. After growth has stopped, metabolic rate decreases about 2% per decade. If a person becomes less active, the total decrease is almost 5% per decade.
3. Body Configuration of Adults—Tall, thin people usually have higher metabolic rates than do short, stocky people of the same weight. This is so because the tall, thin person has a larger surface area (proportional to weight) through

Table 17–5 VITAMINS

Vitamin	Functions	Food Sources	Comment
Water Soluble			
Thiamine (B_1)	• Conversion of pyruvic acid to acetyl CoA in cell respiration • Synthesis of pentose sugars • Synthesis of acetylcholine	• Meat, eggs, legumes, green leafy vegetables, grains	Rapidly destroyed by heat
Riboflavin (B_2)	• Part of FAD in cell respiration	• Meat, milk, cheese, grains	Small amounts produced by GI bacteria
Niacin (nicotinamide)	• Part of NAD in cell respiration • Metabolism of fat for energy	• Meat, fish, grains, legumes	
Pyridoxine (B_6)	• Part of enzymes needed for amino acid metabolism and protein synthesis, nucleic acid synthesis, synthesis of antibodies	• Meat, fish, grains, yeast, yogurt	Small amounts produced by GI bacteria
B_{12} (cyanocobalamin)	• Synthesis of DNA, especially in RBC production • Metabolism of amino acids for energy	• Liver, meat, fish, eggs, milk, cheese	Contains cobalt; intrinsic factor required for absorption
Biotin	• Synthesis of nucleic acids • Metabolism of fatty acids and amino acids	• Yeast, liver, eggs	Small amounts produced by GI bacteria
Folic acid (folacin)	• Synthesis of DNA, especially in blood cell production	• Liver, grains, legumes, leafy green vegetables	Small amounts produced by GI bacteria
Pantothenic acid	• Part of coenzyme A in cell respiration, use of amino acids and fats for energy	• Meat, fish, grains, legumes, vegetables	Small amounts produced by GI bacteria
Vitamin C (ascorbic acid)	• Synthesis of collagen, especially for wound healing • Metabolism of amino acids • Absorption of iron	• Citrus fruits, tomatoes, potatoes	Rapidly destroyed by heat
Fat Soluble			
Vitamin A	• Synthesis of rhodopsin • Calcification of growing bones • Maintenance of epithelial tissues	• Yellow and green vegetables, liver, milk, eggs	Stored in liver; bile salts required for absorption
Vitamin D	• Absorption of calcium and phosphorus in the small intestine	• Fortified milk, egg yolks, fish liver oils	Produced in skin exposed to UV rays; stored in liver; bile salts required for absorption
Vitamin E	• An antioxidant—prevents destruction of cell membranes • Contributes to wound healing and detoxifying ability of the liver	• Nuts, wheat germ, seed oils	Stored in liver and adipose tissue; bile salts required for absorption
Vitamin K	• Synthesis of prothrombin and other clotting factors	• Liver, spinach, cabbage	Large amounts produced by GI bacteria; bile salts required for absorption; stored in liver

Table 17–6 MINERALS

Mineral	Functions	Food Sources	Comment
Calcium	• Formation of bones and teeth • Neuron and muscle functioning • Blood clotting	• Milk, cheese, yogurt, shellfish, leafy green vegetables	Vitamin D required for absorption; stored in bones
Phosphorus	• Formation of bones and teeth • Part of DNA, RNA, and ATP • Part of phosphate buffer system	• Milk, cheese, fish, meat	Vitamin D required for absorption; stored in bones
Sodium	• Contributes to osmotic pressure of body fluids • Nerve impulse transmission and muscle contraction • Part of bicarbonate buffer system	• Table salt, almost all foods	Most abundant cation (+) in extracellular fluid
Potassium	• Contributes to osmotic pressure of body fluids • Nerve impulse transmission and muscle contraction	• Virtually all foods	Most abundant cation (+) in intracellular fluid
Chlorine	• Contributes to osmotic pressure of body fluids • Part of HCl in gastric juice	• Table salt	Most abundant anion (−) in extracellular fluid
Iron	• Part of hemoglobin and myoglobin • Part of some cytochromes in cell respiration	• Meat, shellfish, dried apricots, legumes, eggs	Stored in liver
Iodine	• Part of thyroxine and T_3	• Iodized salt, seafood	
Sulfur	• Part of some amino acids • Part of thiamine and biotin	• Meat, eggs	Insulin and keratin require sulfur
Magnesium	• Formation of bone • Metabolism of ATP–ADP	• Green vegetables, legumes, seafood, milk	Part of chlorophyll in green plants
Manganese	• Formation of urea • Synthesis of fatty acids and cholesterol	• Legumes, grains, nuts, leafy green vegetables	Some stored in liver
Copper	• Synthesis of hemoglobin • Part of some cytochromes in cell respiration • Synthesis of melanin	• Liver, seafood, grains, nuts, legumes	Stored in liver
Cobalt	• Part of vitamin B_{12}	• Liver, meat, fish	Vitamin B_{12} stored in liver
Zinc	• Part of carbonic anhydrase needed for CO_2 transport • Part of peptidases needed for protein digestion • Necessary for normal taste sensation • Involved in wound healing	• Meat, seafood, grains, legumes	

Box 17–4 METABOLIC RATE

To estimate your own **basal metabolic rate** (BMR), calculate kilocalories (kcal) used per hour as follows:

For women: use the factor of 0.9 kcal per kilogram (kg) of body weight
For men: use the factor of 1.0 kcal per kg of body weight
Then multiply kcal/hour by 24 hours to determine kcal per day.

Example: A **120-pound woman**

1. Change pounds to kilograms:
   120 lb at 2.2 lb/kg = 55 kg
2. Multiply kg weight by the BMR factor:
   55 kg × 0.9 kcal/kg/hr = 49.5 kcal/hr
3. Multiply kcal/hr by 24:
   49.5 kcal/hr × 24 = 1188 kcal/day*
   *This is an approximate BMR, about 1200 kcal/day

Example: **A 160-pound man**

1. 160 lb at 2.2 lb/kg = 73 kg
2. 73 kg × 1.0 kcal/kg/hr = 73 kcal/hr
3. 73 kcal/hr × 24 = 1752 kcal/day

To approximate the amount of energy actually expended during an average day (24 hours), the following percentages may be used:

Sedentary Activity: add 40% to 50% of the BMR to the BMR
Light Activity: add 50% to 65% of the BMR to the BMR
Moderate Activity: add 65% to 75% of the BMR to the BMR
Strenuous Activity: add 75% to 100% of the BMR to the BMR

Using our example of the 120-pound woman with a BMR of 1200 kcal/day:

Sedentary: 1680 to 1800 kcal/day
Light: 1800 to 1980 kcal/day
Moderate: 1980 to 2100 kcal/day
Strenuous: 2100 to 2400 kcal/day

which heat is continuously lost. The metabolic rate, therefore, is slightly higher to compensate for the greater heat loss. The variance of surface to weight ratios for different body configurations is illustrated in Fig. 17–4.

4. Sex Hormones—Testosterone increases metabolic activity to a greater degree than does estrogen, giving men a slightly higher metabolic rate than women. Also, men tend to have more muscle, an active tissue, whereas women tend to have more fat, a relatively inactive tissue.
5. Sympathetic Stimulation—In stress situations, the metabolism of many body cells is increased. Also contributing to this are the hormones epinephrine and norepinephrine. As a result, metabolic rate increases.
6. Decreased Food Intake—If the intake of food decreases for a prolonged period of time, metabolic rate also begins to decrease. It is as if the body's metabolism is "slowing down" to conserve whatever energy sources may still be available. (See also Box 17–6: Leptin and Body-Mass Index.)
7. Climate—People who live in cold climates may have metabolic rates 10% to 20% higher than people who live in tropical regions. This is believed to be due to the variations in the secretion of thyroxine, the hormone most responsible for regulation of metabolic rate. In a cold climate, the necessity for greater heat production brings about an increased secretion of thyroxine and a higher metabolic rate.

Box 17–5 WEIGHT LOSS

Although diet books are often found on the best-seller lists, there is no magic method that will result in weight loss. Losing weight depends on one simple fact: calorie expenditure in activity must exceed calorie intake in food (the term calorie here will be used to mean kilocalorie).

In order to lose 1 pound of body fat, which consists of fat, water, and protein, 3500 calories of energy must be expended. Although any form of exercise requires calories, the more strenuous the exercise, the more calories expended. Some examples are as follows:

Activity	***Calories per 10 minutes*** *(average for a 150-lb person)*	***Activity***	***Calories per 10 minutes*** *(average for a 150-lb person)*
Walking slowly	30	Running (8 mph)	120
Walking briskly	45	Cycling (10 mph)	70
Walking up stairs	170	Cycling (15 mph)	115
Dancing (slow)	40	Swimming	100
Dancing (fast)	65		

Most food packaging contains nutritional information, including the calories per serving of the food. Keeping track of daily caloric intake is an important part of a decision to try to lose weight. It is also important to remember that sustained loss of fat usually does not exceed 1 to 2 pounds per week. In part this is so because as calorie intake decreases, the metabolic rate decreases. There will also be loss of some body protein so that amino acids can be converted to glucose to supply the brain.

A sensible weight-loss diet will include carbohydrate to supply energy needs, will have sufficient protein (40 to 45 grams per day), and will be low in fat. Including vegetables and fruits will supply vitamins, minerals, and fiber.

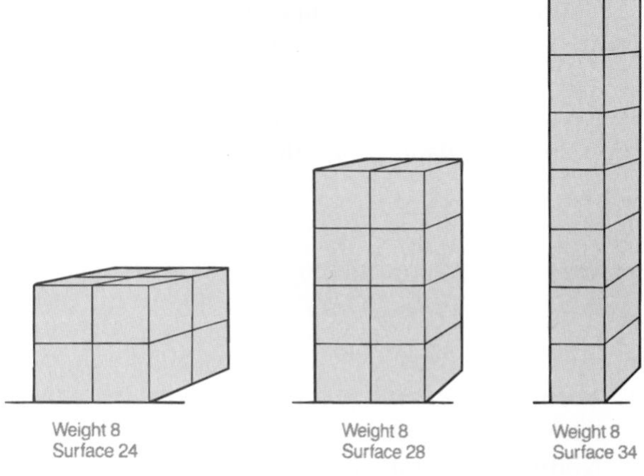

Figure 17–4 Surface-to-weight ratios. Imagine that the three shapes are people who all weigh the same amount. The "tall, thin person" on the right has about 50% more surface area than does the "short, stocky person" on the left. The more surface area (where heat is lost), the higher the metabolic rate.

Box 17–6 LEPTIN AND BODY-MASS INDEX

The 1994 discovery of the hormone leptin was reported to the general public in 1995, along with speculation that leptin could become an anti-obesity medication. Leptin is a protein produced by fat cells, and signals the hypothalamus to release a chemical that acts as an appetite suppressant. It seems to inform the brain of how much stored fat the body has, and is therefore involved in the regulation of body weight.

Another likely role for leptin is as a contributor to the onset of puberty, especially in females. Girls who are very thin, with little body fat, tend to have a later first menstrual period than girls with average body fat, and a certain level of body fat is necessary for continued ovulation. Leptin may be the chemical mediator of this information.

The most recent research indicates that leptin directly decreases fat storage in cells, and improves the efficiency of the pancreatic cells that produce insulin. What was first believed to be a simple chemical signal has proved to be much more complex.

A good measure of leanness or fatness is the **body-mass index**.

To calculate: Multiple weight in pounds by 703.
Divide by height in inches.
Divide again by height in inches = body-mass index

Example: A person five foot six weighing 130 pounds.
130 × 703 = 91,390
91,390 ÷ 66 = 1385
1385 ÷ 66 = 20.98

The optimal body-mass index is considered to be 21. Any index over 28 is considered overweight.

AGING AND METABOLISM

As mentioned in the previous section, metabolic rate decreases with age. Elderly people who remain active, however, can easily maintain a metabolic rate (energy production) adequate for their needs as long as their general health is good.

Sensitivity to external temperature changes may decrease with age, and the regulation of body temperature is no longer as precise. Sweat glands are not as active, and prolonged high environmental temperatures are a real danger for elderly people.

SUMMARY

Food is needed for the synthesis of new cells and tissues, or is utilized to produce the energy required for such synthesis reactions. As a consequence of metabolism, heat energy is released to provide a constant body temperature and permit the continuation of metabolic activity. The metabolic pathways described in this chapter are only a small portion of the body's total metabolism. Even this simple presentation, however, suggests the great chemical complexity of the functioning human being.

STUDY OUTLINE

Body Temperature

1. Normal range is 96.5 to 99.5°F (36 to 38°C), with an average of 98.6°F (37°C).
2. Normal fluctuation in 24 hours is 1 to 2°F.
3. Temperature regulation in infants and the elderly is not as precise as it is at other ages.

Heat Production

Heat is one of the energy products of cell respiration. Many factors affect the total heat actually produced (see Table 17–1).

1. Thyroxine from the Thyroid Gland—the most important regulator of daily heat production. As met-

abolic rate decreases, more thyroxine is secreted to increase the rate of cell respiration.
2. Stress—sympathetic impulses and epinephrine and norepinephrine increase the metabolic activity of many organs, increasing the production of ATP and heat.
3. Active organs continuously produce heat. Skeletal muscle tone produces 25% of the total body heat at rest. The liver provides up to 20% of the resting body heat.
4. Food intake increases the activity of the digestive organs and increases heat production.
5. Changes in body temperature affect metabolic rate. A fever increases the metabolic rate, and more heat is produced; this may become detrimental during very high fevers.

Heat Loss (see Table 17–2)

1. Most heat is lost through the skin.
2. Blood flow through the dermis determines the amount of heat that is lost by radiation, conduction, and convection.
3. Vasodilation in the dermis increases blood flow and heat loss; radiation and conduction are effective only if the environment is cooler than the body.
4. Vasoconstriction in the dermis decreases blood flow and conserves heat in the core of the body.
5. Sweating is a very effective heat loss mechanism; excess body heat evaporates sweat on the skin surface; sweating is most effective when the atmospheric humidity is low.
6. Sweating also has a disadvantage in that water is lost and must be replaced to prevent serious dehydration.
7. Heat is lost from the respiratory tract by the evaporation of water from the warm respiratory mucosa; water vapor is part of exhaled air.
8. A very small amount of heat is lost as urine and feces are excreted at body temperature.

Regulation of Heat Loss

1. The hypothalamus is the thermostat of the body and regulates body temperature by balancing heat production and heat loss.
2. The hypothalamus receives information from its own neurons (blood temperature) and from the temperature receptors in the dermis.
3. Mechanisms to increase heat loss are vasodilation in the dermis and increased sweating. Decreased muscle tone will decrease heat production.
4. Mechanisms to conserve heat are vasoconstriction in the dermis and decreased sweating. Increased muscle tone (shivering) will increase heat production.

Fever—an abnormally elevated body temperature

1. Pyrogens are substances that cause a fever: bacteria, foreign proteins, or chemicals released during inflammation (endogenous pyrogens).
2. Pyrogens raise the setting of the hypothalamic thermostat; the person feels cold and begins to shiver to produce heat.
3. When the pyrogen has been eliminated, the hypothalamic setting returns to normal; the person feels warm, and sweating begins to lose heat to lower the body temperature.
4. A low fever may be beneficial because it increases the activity of WBCs and inhibits the activity of some pathogens.
5. A high fever may be detrimental because enzymes are denatured at high temperatures. This is most critical in the brain, where cells that die cannot be replaced.

Metabolism—all the reactions within the body

1. Anabolism—synthesis reactions that usually require energy in the form of ATP.
2. Catabolism—decomposition reactions that often release energy in the form of ATP.
3. Enzymes catalyze most anabolic and catabolic reactions.

Cell Respiration—the breakdown of food molecules to release their potential energy and synthesize ATP

1. Glucose + oxygen yields CO_2 + H_2O + ATP + heat.
2. The breakdown of glucose involves three stages: glycolysis, Krebs cycle, and the cytochrome transport system (see Table 17-3 and Fig. 17-3).
3. The oxygen necessary comes from breathing.
4. The water formed becomes part of intracellular fluid; CO_2 is exhaled; ATP is used for energy-requiring reactions; heat provides a body temperature.

Proteins and Fats—as energy sources (see Table 17–4 for hormonal regulation)

1. Excess amino acids are deaminated in the liver and converted to pyruvic acid or acetyl groups to enter

the Krebs cycle. Amino acids may also be converted to glucose to supply the brain (Fig. 17-3).
2. Glycerol is converted to pyruvic acid to enter the Krebs cycle.
3. Fatty acids, in the process of beta-oxidation in the liver, are split into acetyl groups to enter the Krebs cycle; ketones are formed for transport to other cells (see Fig. 17-3).

Energy Available from Food
1. Energy is measured in kilocalories (Calories): kcal.
2. There are 4 kcal per gram of carbohydrate, 4 kcal per gram of protein, 9 kcal per gram of fat.

Synthesis Uses of Foods
1. Glucose—used to synthesize the pentose sugars for DNA and RNA; used to synthesize glycogen to store energy in liver and muscles.
2. Amino Acids—used to synthesize new proteins and the non-essential amino acids.
3. Fatty Acids and Glycerol—used to synthesize phospholipids for cell membranes, triglycerides for fat storage in adipose tissue, and cholesterol and other steroids.
4. Vitamins and Minerals—see Tables 17-5 and 17-6.

Metabolic Rate—the heat production by the body; measured in kcal
1. Basal Metabolic Rate (BMR) is the energy required to maintain life (see Box 17-4); several factors influence the metabolic rate of an active person.
2. Age—metabolic rate is highest in young children and decreases with age.
3. Body Configuration—more surface area proportional to weight (tall and thin) means a higher metabolic rate.
4. Sex Hormones—men usually have a higher metabolic rate than do women; men have more muscle proportional to fat than do women.
5. Sympathetic Stimulation—metabolic activity increases in stress situations.
6. Decreased Food Intake—metabolic rate decreases to conserve available energy sources.
7. Climate—people who live in cold climates usually have higher metabolic rates because of a greater need for heat production.

REVIEW QUESTIONS

1. State the normal range of human body temperature in °F and °C. (p. 380)
2. State the summary equation of cell respiration, and state what happens to (or the purpose of) each of the products. (p. 385)
3. Describe the role of each on heat production: thyroxine, skeletal muscles, stress situations, and the liver. (p. 380)
4. Describe the two mechanisms of heat loss through the skin, and explain the role of blood flow. Describe how heat is lost through the respiratory tract. (pp. 381-382)
5. Explain the circumstances when sweating and vasodilation in the dermis are not effective mechanisms of heat loss. (pp. 381-382)
6. Name the part of the brain that regulates body temperature, and explain what is meant by a thermostat. (pp. 382-383)
7. Describe the responses by the body to a warm environment and to a cold environment. (pp. 383-384)
8. Explain how pyrogens are believed to cause a fever, and give two examples of pyrogens. (pp. 384-385)
9. Define metabolism, anabolism, catabolism, kilocalorie, and metabolic rate. (pp. 385, 391)
10. Name the three stages of the cell respiration of glucose and state where in the cell each takes place and whether or not oxygen is required. (pp. 385, 387)

11. For each, state the molecules that enter the process and the results of the process: glycolysis, Krebs cycle, and cytochrome transport system. (pp. 385, 387)
12. Explain how fatty acids, glycerol, and excess amino acids are used for energy production in cell respiration. (pp. 388–389)
13. Describe the synthesis uses for glucose, amino acids, and fatty acids. (pp. 390–391)
14. Describe four factors that affect the metabolic rate of an active person. (pp. 391, 394)
15. If lunch consists of 60 grams of carbohydrate, 15 grams of protein, and 10 grams of fat, how many kilocalories are provided by this meal? (p. 389)

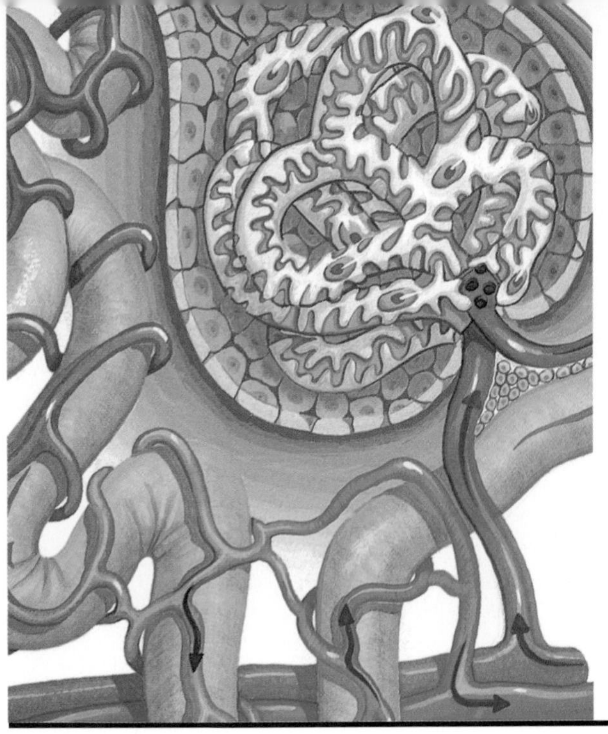

CHAPTER 18

Chapter Outline

Student Objectives

- Describe the location and general function of each organ of the urinary system.
- Name the parts of a nephron and the important blood vessels associated with them.
- Explain how the following are involved in urine formation: glomerular filtration, tubular reabsorption, tubular secretion, and blood flow through the kidney.
- Describe the mechanisms of tubular reabsorption, and explain the importance of tubular secretion.
- Describe how the kidneys help maintain normal blood volume and blood pressure.
- Name and state the functions of the hormones that affect the kidneys.
- Describe how the kidneys help maintain normal pH of blood and tissue fluid.
- Describe the urination reflex, and explain how voluntary control is possible.
- Describe the characteristics of normal urine.

The Urinary System

New Terminology

Bowman's capsule (**BOW**-manz **KAP**-suhl)
Detrusor muscle (de-**TROO**-ser)
External urethral sphincter (yoo-**REE**-thruhl **SFINK**-ter)
Glomerular filtration rate (gloh-**MER**-yoo-ler fill-**TRAY**-shun RAYT)
Glomerulus (gloh-**MER**-yoo-lus)
Internal urethral sphincter (yoo-**REE**-thruhl **SFINK**-ter)
Juxtaglomerular cells (JUKS-tah-gloh-**MER**-yoo-ler SELLS)
Micturition (MIK-tyoo-**RISH**-un)
Nephron (**NEFF**-ron)
Nitrogenous wastes (nigh-**TRAH**-jen-us)
Peritubular capillaries (PER-ee-**TOO**-byoo-ler)
Renal corpuscle (**REE**-nuhl **KOR**-pus'l)
Renal filtrate (**REE**-nuhl **FILL**-trayt)
Renal tubule (**REE**-nuhl **TOO**-byoo'l)
Retroperitoneal (RE-troh-PER-i-toh-**NEE**-uhl)
Specific gravity (spe-**SIF**-ik **GRA**-vi-tee)
Threshold level (**THRESH**-hold **LE**-vuhl)
Trigone (**TRY**-gohn)
Ureter (**YOOR**-uh-ter)
Urethra (yoo-**REE**-thrah)
Urinary bladder (**YOOR**-i-NAR-ee **BLA**-der)

Related Clinical Terminology

Cystitis (sis-**TIGH**-tis)
Dysuria (dis-**YOO**-ree-ah)
Hemodialysis (HEE-moh-dye-**AL**-i-sis)
Nephritis (ne-**FRY**-tis)
Oliguria (AH-li-**GYOO**-ree-ah)
Polyuria (PAH-li-**YOO**-ree-ah)
Renal calculi (**REE**-nuhl **KAL**-kew-lye)
Renal failure (**REE**-nuhl **FAYL**-yer)
Uremia (yoo-**REE**-me-ah)

*Terms that appear in **bold type** in the chapter text are defined in the glossary, which begins on page 528.*

The first successful human organ transplant was a kidney transplant performed in 1953. Because the donor and recipient were identical twins, rejection was not a problem. Thousands of kidney transplants have been performed since then, and the development of immunosuppressive medications has permitted many people to live normal lives with donated kidneys. Although a person usually has two kidneys, it is clear that one kidney can carry out the complex work required to maintain homeostasis of the body fluids.

The urinary system consists of two kidneys, two ureters, the urinary bladder, and the urethra (Fig. 18–1). The formation of urine is the function of the kidneys, and the rest of the system is responsible for eliminating the urine.

Body cells produce waste products such as urea, creatinine, and ammonia, which must be removed from the blood before they accumulate to toxic levels. As the kidneys form urine to excrete these waste products, they also accomplish several other important functions:

1. Regulation of the volume of blood by excretion or conservation of water
2. Regulation of the electrolyte content of the blood by the excretion or conservation of minerals
3. Regulation of the acid–base balance of the blood by excretion or conservation of ions such as H^+ ions or HCO_3^- ions
4. Regulation of all of the above in tissue fluid

The process of urine formation, therefore, helps maintain the normal composition, volume, and pH of both blood and tissue fluid by removing those substances that would upset the normal constancy and balance of these extracellular fluids.

KIDNEYS

The two **kidneys** are located in the upper abdominal cavity on either side of the vertebral column, behind the peritoneum (retroperitoneal). The upper portions of the kidneys rest on the lower surface of the diaphragm and are enclosed and protected by the lower rib cage (see Fig. 18–1). The kidneys are embedded in adipose tissue that acts as a cushion and is in turn covered by a fibrous connective tissue membrane called the **renal fascia**, which helps hold the kidneys in place (see Box 18-1: Floating Kidney).

Each kidney has an indentation called the **hilus** on its medial side. At the hilus, the renal artery enters the kidney, and the renal vein and ureter emerge. The renal artery is a branch of the abdominal aorta, and the renal vein returns blood to the inferior vena cava (see Fig. 18–1). The ureter carries urine from the kidney to the urinary bladder.

INTERNAL STRUCTURE OF THE KIDNEY

In a coronal or frontal section of the kidney, three areas can be distinguished (Fig. 18–2). The outermost area is called the **renal cortex**; it is made of renal corpuscles and convoluted tubules. These are parts of the nephron and are described in the next section. The middle area is the **renal medulla**, which is made of loops of Henle and collecting tubules (also parts of the nephron). The renal medulla consists of wedge-shaped pieces called **renal pyramids**. The tip of each pyramid is its apex or papilla.

The third area is the **renal pelvis**; this is not a layer of tissues, but rather a cavity formed by the expansion of the ureter within the kidney at the hilus. Funnel-shaped extensions of the renal pelvis, called **calyces** (singular, **calyx**), enclose the papillae of the renal pyramids. Urine flows from the renal pyramids into the calyces, then to the renal pelvis and out into the ureter.

THE NEPHRON

The **nephron** is the structural and functional unit of the kidney. Each kidney contains approximately 1 million nephrons. It is in the nephrons, with their associated blood vessels, that urine is formed. Each nephron has two major portions: a renal corpuscle and a renal tubule. Each of these major parts has further subdivisions, which are shown with their blood vessels in Fig. 18–3.

Renal Corpuscle

A **renal corpuscle** consists of a glomerulus surrounded by a Bowman's capsule. The **glomerulus** is a capillary network that arises from an **afferent arteriole** and empties into an **efferent arteriole**. The diameter of the efferent arteriole is smaller than that of the afferent arteriole, which helps maintain a fairly high blood pressure in the glomerulus.

Bowman's capsule (or glomerular capsule) is the expanded end of a renal tubule; it encloses the glo-

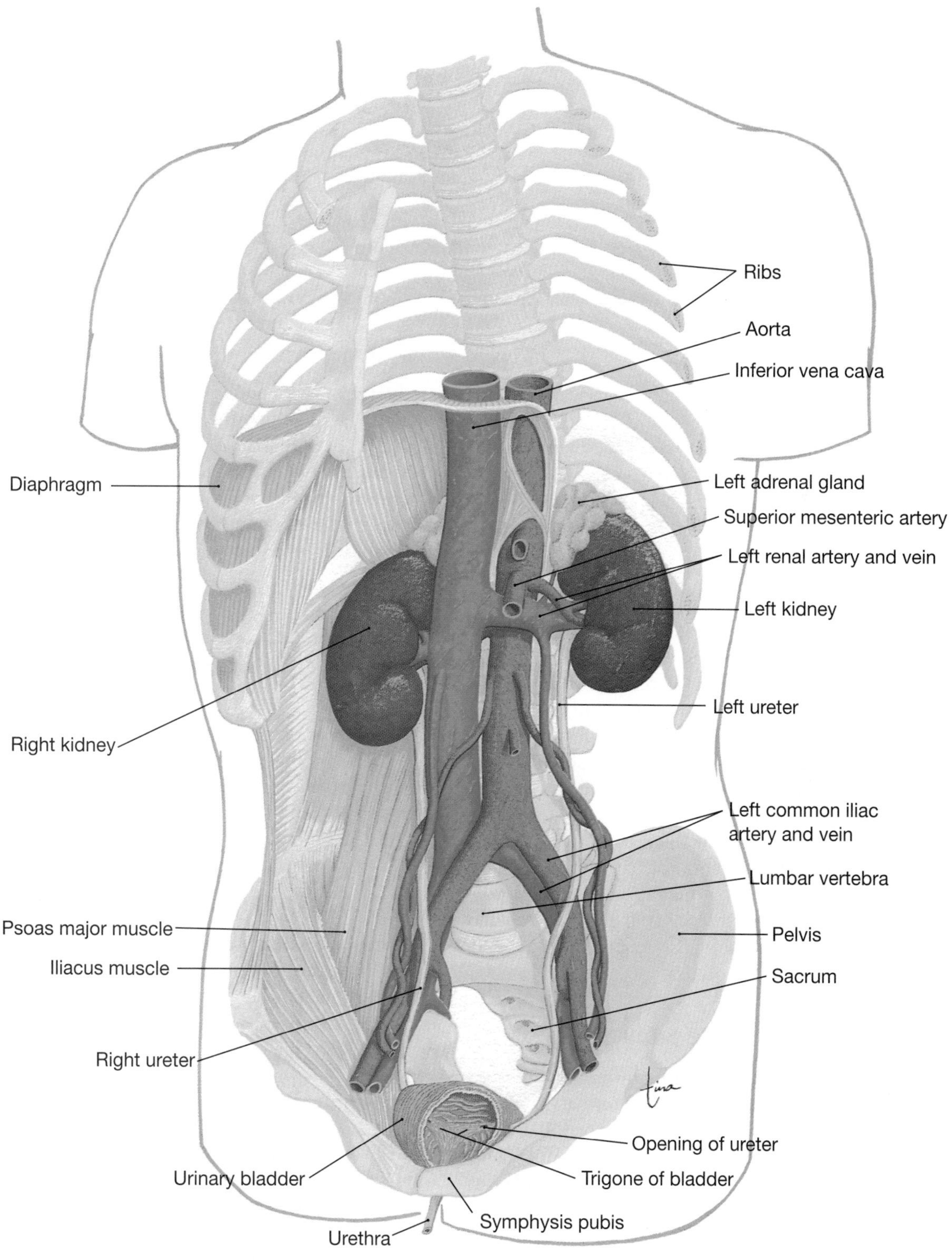

Figure 18–1 The urinary system shown in anterior view.

Box 18–1 FLOATING KIDNEY

A floating kidney is one that has moved out of its normal position. This may happen in very thin people whose renal cushion of adipose tissue is thin, or it may be the result of a sharp blow to the back that dislodges a kidney.

A kidney can function in any position; the problem with a floating kidney is that the ureter may become twisted or kinked. If urine cannot flow through the ureter, the urine backs up and collects in the renal pelvis. Incoming urine from the renal tubules then backs up as well. If the renal filtrate cannot flow out of Bowman's capsules, the pressure within Bowman's capsules increases, opposing the blood pressure in the glomeruli. Glomerular filtration then cannot take place efficiently. If uncorrected, this may lead to permanent kidney damage.

merulus. The inner layer of Bowman's capsule has pores and is very permeable; the outer layer has no pores and is not permeable. The space between the inner and outer layers of Bowman's capsule contains renal filtrate, the fluid that is formed from the blood in the glomerulus and will eventually become urine.

Renal Tubule

The **renal tubule** continues from Bowman's capsule and consists of the following parts: **proximal convoluted tubule** (in the renal cortex), **loop of Henle** (or loop of the nephron, in the renal medulla), and **distal convoluted tubule** (in the renal cortex). The distal convoluted tubules from several nephrons empty into a **collecting tubule**. Several collecting tubules then unite to form a papillary duct that empties urine into a calyx of the renal pelvis.

All the parts of the renal tubule are surrounded by **peritubular capillaries**, which arise from the efferent arteriole. The peritubular capillaries will receive the materials reabsorbed by the renal tubules; this is described in the section on urine formation.

BLOOD VESSELS OF THE KIDNEY

The pathway of blood flow through the kidney is an essential part of the process of urine formation. Blood from the abdominal aorta enters the **renal artery**, which branches extensively within the kidney into smaller arteries (see Fig. 18–2). The smallest arteries give rise to afferent arterioles in the renal cortex (see Fig. 18–3). From the afferent arterioles, blood flows into the glomeruli (capillaries), to efferent arterioles, to peritubular capillaries, to veins within the kidney, to the **renal vein**, and finally to the inferior vena cava. Notice that in this pathway there are two sets of capillaries, and recall that it is in capillaries that exchanges take place between the blood and surrounding tissues. Therefore, in the kidneys there are two sites of exchanges. The exchanges that take place in the capillaries of the kidneys will form urine from blood plasma.

FORMATION OF URINE

The formation of urine involves three major processes. The first is glomerular filtration, which takes place in the renal corpuscles. The second and third are tubular reabsorption and tubular secretion, which take place in the renal tubules.

GLOMERULAR FILTRATION

You may recall that filtration is the process in which blood pressure forces plasma and dissolved material out of capillaries. In **glomerular filtration**, blood pressure forces plasma, dissolved substances and small proteins out of the glomeruli and into Bowman's capsules. This fluid is no longer plasma but is called **renal filtrate**.

The blood pressure in the glomeruli, compared to that in other capillaries, is relatively high, about 60 mmHg. The pressure in Bowman's capsule is very low, and its inner layer is very permeable, so that approximately 20% to 25% of the blood that enters glomeruli becomes renal filtrate in Bowman's capsules. The blood cells and larger proteins are too large to be

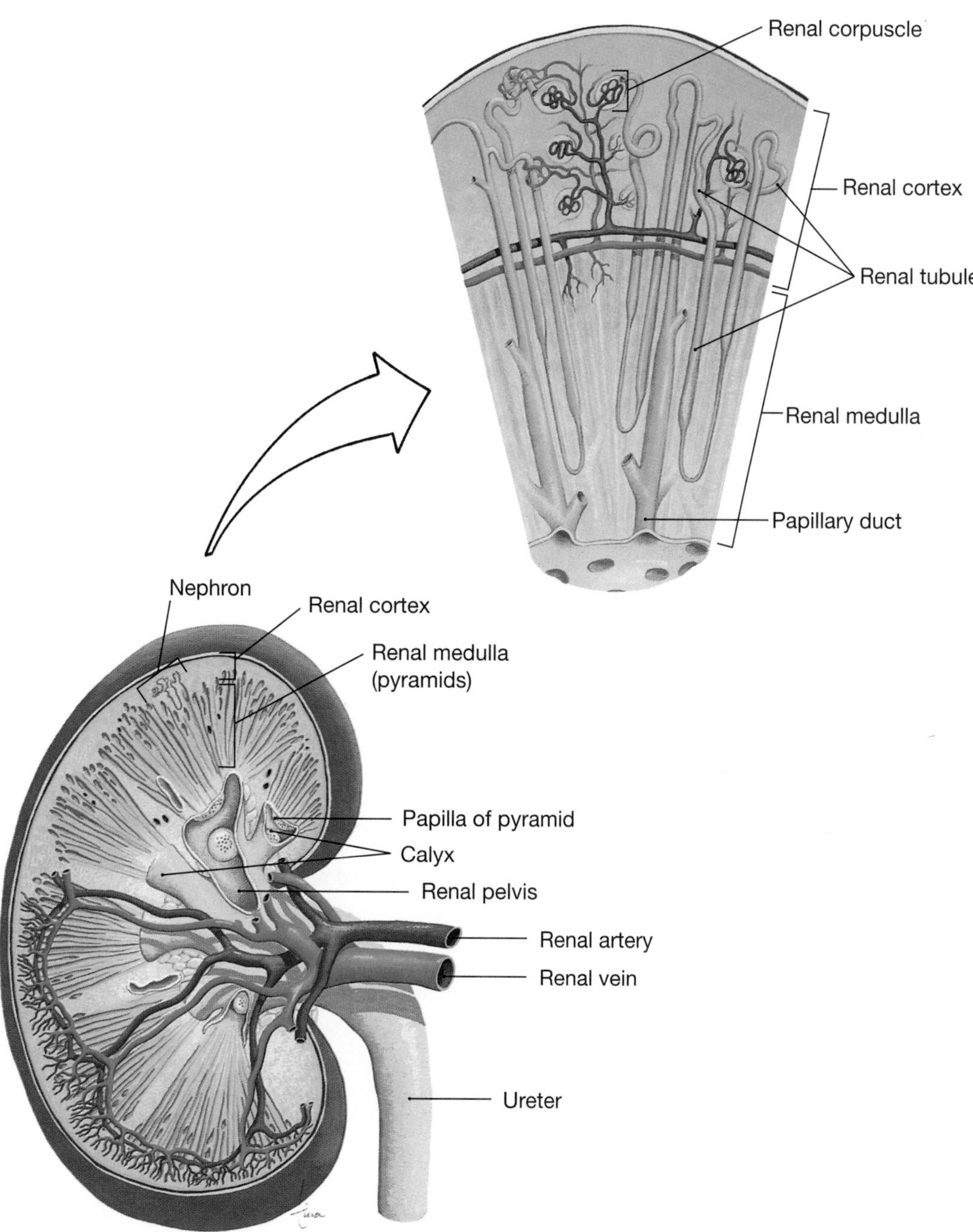

Figure 18–2 Frontal section of the right kidney showing internal structure and blood vessels. The magnified section of the kidney shows several nephrons.

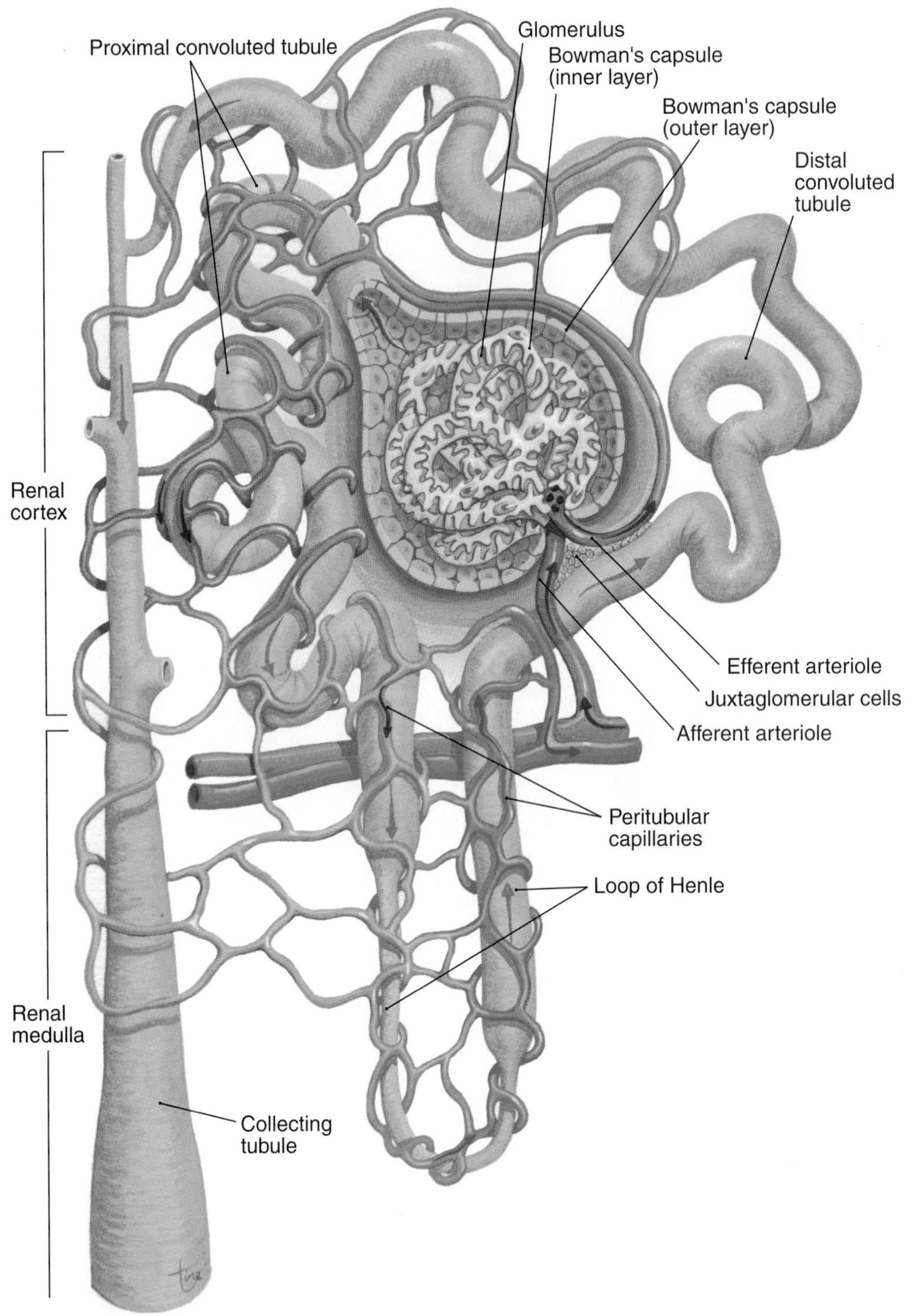

Figure 18–3 A nephron with its associated blood vessels. The arrows indicate the direction of blood flow and flow of renal filtrate. See text for description.

forced out of the glomeruli, so they remain in the blood. Waste products are dissolved in blood plasma, so they pass into the renal filtrate. Useful materials such as nutrients and minerals are also dissolved in plasma and are also present in renal filtrate. Therefore, renal filtrate is very much like blood plasma, except that there is far less protein, and no blood cells are present.

The **glomerular filtration rate** (GFR) is the amount of renal filtrate formed by the kidneys in 1 minute, and averages 100 to 125 ml per minute. GFR may be altered if the rate of blood flow through the kidney changes. If blood flow increases, the GFR increases, and more filtrate is formed. If blood flow decreases (as may happen following a severe hemorrhage), the GFR decreases, less filtrate is formed, and urinary output decreases (see Box 18–2: Renal Failure and Hemodialysis).

TUBULAR REABSORPTION

Tubular reabsorption takes place from the renal tubules into the peritubular capillaries. In a 24-hour period, the kidneys form 150 to 180 liters of filtrate, and normal urinary output in that time is 1 to 2 liters. Therefore, it becomes apparent that most of the renal filtrate does not become urine. Approximately 99% of the filtrate is reabsorbed back into the blood in the peritubular capillaries. Only about 1% of the filtrate will enter the renal pelvis as urine.

Box 18–2 RENAL FAILURE AND HEMODIALYSIS

Renal failure, the inability of the kidneys to function properly, may be the result of three general causes, which may be called: prerenal, intrinsic renal, and postrenal.

Prerenal means that the problem is "before" the kidneys, that is, in the blood flow to the kidneys. Any condition that decreases blood flow to the kidneys may result in renal damage and failure. Examples are severe hemorrhage or very low blood pressure following a heart attack (MI).

Intrinsic renal means that the problem is in the kidneys themselves. Bacterial infections of the kidneys or exposure to chemicals (certain antibiotics) may cause this type of damage. Polycystic kidney disease is a genetic disorder in which the kidney tubules dilate and become nonfunctional. Severe damage may not be apparent until age 40 to 60 years but may then progress to renal failure.

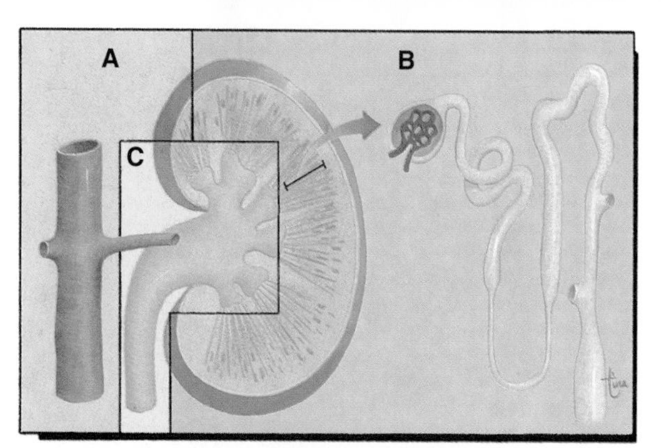

Box Figure 18–A Causes of renal failure. **(A)** Prerenal. **(B)** Intrinsic renal. **(C)** Postrenal.

Postrenal means that the problem is "after" the kidneys, somewhere in the rest of the urinary tract. Obstruction of urine flow may be caused by kidney stones, a twisted ureter, or prostatic hypertrophy.

Treatment of renal failure involves correcting the specific cause, if possible. If not possible, and kidney damage is permanent, the person is said to have chronic renal failure. **Hemodialysis** is the use of an artificial kidney machine to do what the patient's nephrons can no longer do. The patient's blood is passed through minute tubes surrounded by fluid (dialysate) with the same chemical composition as plasma. Waste products and excess minerals diffuse out of the patient's blood into the fluid of the machine.

Although hemodialysis does prolong life for those with chronic renal failure, it does not fully take the place of functioning kidneys. The increasing success rate of kidney transplants, however, does indeed provide the possibility of a normal life for people with chronic renal failure.

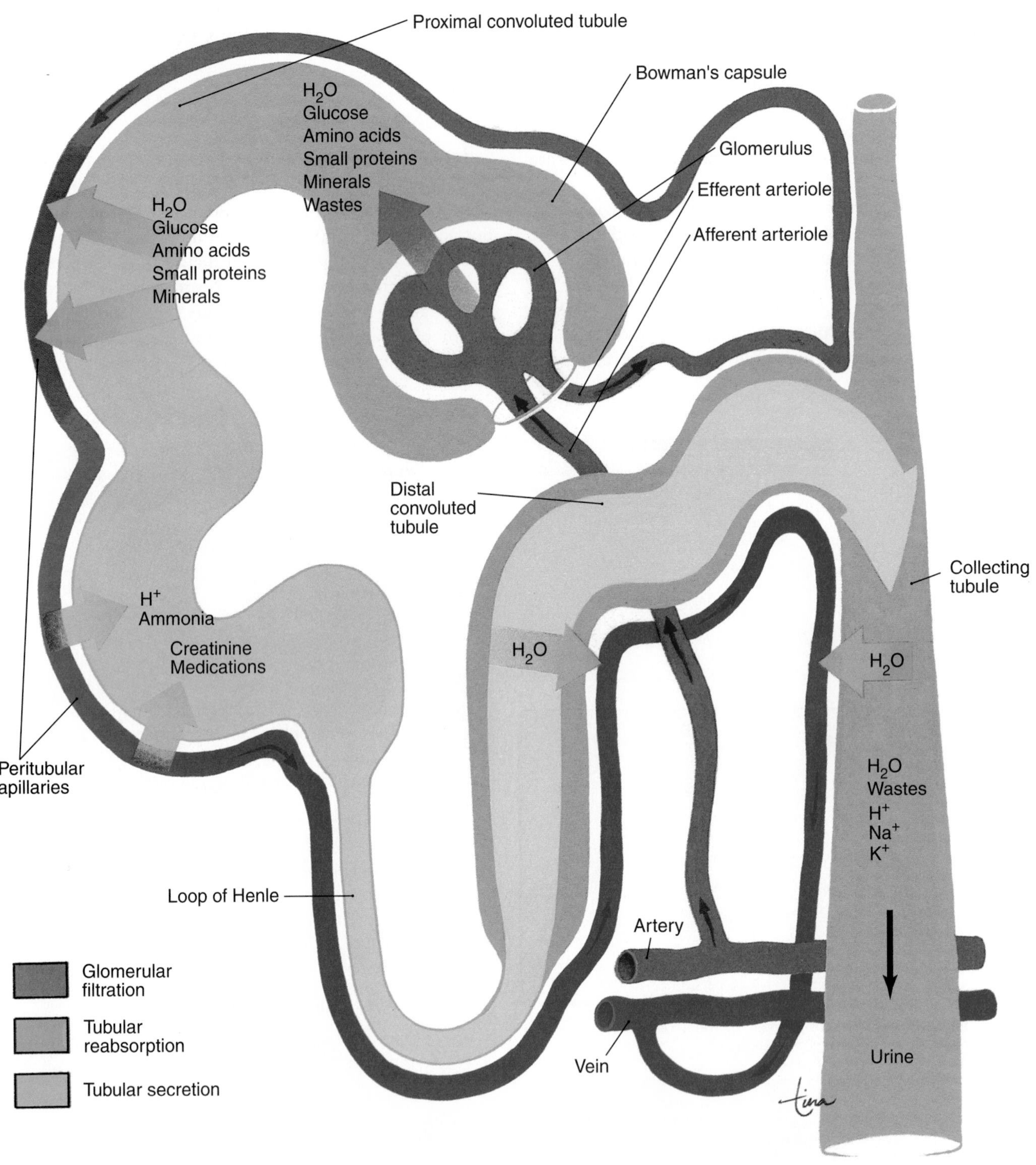

Figure 18–4 Schematic representation of glomerular filtration, tubular reabsorption, and tubular secretion. The renal tubule has been uncoiled, and the peritubular capillaries are shown adjacent to the tubule.

Most reabsorption and secretion (about 65%) take place in the proximal convoluted tubules, whose cells have **microvilli** that greatly increase their surface area. The distal convoluted tubules and collecting tubules are also important sites for the reabsorption of water (Fig. 18-4).

Mechanisms of Reabsorption

1. **Active Transport**—the cells of the renal tubule use ATP to transport most of the useful materials from the filtrate to the blood. These useful materials include glucose, amino acids, vitamins, and positive ions.

   For many of these substances, the renal tubules have a **threshold level** of reabsorption. This means that there is a limit to how much the tubules can remove from the filtrate. For example, if the filtrate level of glucose is normal (reflecting a normal blood glucose level), the tubules will reabsorb all the glucose, and none will be found in the urine. If, however, the blood glucose level is above normal, the amount of glucose in the filtrate will also be above normal and will exceed the threshold level of reabsorption. In this situation, therefore, some glucose will be present in urine.

   The reabsorption of Ca^{+2} ions is increased by parathyroid hormone (PTH). The parathyroid glands secrete PTH when the blood calcium level decreases. The reabsorption of Ca^{+2} ions by the kidneys is one of the mechanisms by which the blood calcium level is raised back to normal.

   The hormone aldosterone, secreted by the adrenal cortex, increases the reabsorption of Na^{+} ions and the excretion of K^{+} ions. Besides regulating the blood levels of sodium and potassium, aldosterone also affects the volume of blood.
2. **Passive Transport**—many of the negative ions that are returned to the blood are reabsorbed following the reabsorption of positive ions, because unlike charges attract.
3. **Osmosis**—the reabsorption of water follows the reabsorption of minerals, especially sodium ions. The hormones that affect reabsorption of water are discussed in the next section.
4. **Pinocytosis**—small proteins are too large to be reabsorbed by active transport. They become adsorbed to the membranes of the cells of the proximal convoluted tubules. The cell membrane then sinks inward and folds around the protein to take it in. Normally all proteins in the filtrate are reabsorbed; none is found in urine.

TUBULAR SECRETION

This mechanism also changes the composition of urine. In **tubular secretion**, substances are actively secreted from the blood in the peritubular capillaries into the filtrate in the renal tubules. Waste products such as ammonia and some creatinine, and the metabolic products of medications may be secreted into the filtrate to be eliminated in urine. Hydrogen ions (H^{+}) may be secreted by the tubule cells to help maintain the normal pH of blood.

HORMONES THAT INFLUENCE REABSORPTION OF WATER

Aldosterone is secreted by the adrenal cortex in response to a high blood potassium level, to a low blood sodium level, or to a decrease in blood pressure. When aldosterone stimulates the reabsorption of Na^{+} ions, water follows from the filtrate back to the blood. This helps maintain normal blood volume and blood pressure.

You may recall that the antagonist to aldosterone is **atrial natriuretic hormone** (ANH), which is secreted by the atria of the heart when the atrial walls are stretched by high blood pressure or greater blood volume. ANH decreases the reabsorption of Na^{+} ions by the kidneys; these remain in the filtrate, as does water, and are excreted. By increasing the elimination of sodium and water, ANH lowers blood volume and blood pressure.

Antidiuretic hormone (ADH) is released by the posterior pituitary gland when the amount of water in the body decreases. Under the influence of ADH, the distal convoluted tubules and collecting tubules are able to reabsorb more water from the renal filtrate. This helps maintain normal blood volume and blood pressure, and also permits the kidneys to produce urine that is more concentrated than body fluids. Producing a concentrated urine is essential to prevent excessive water loss while still excreting all the substances that must be eliminated.

If the amount of water in the body increases, however, the secretion of ADH diminishes, and the kidneys will reabsorb less water. Urine then becomes dilute, and water is eliminated until its concentration in

Table 18–1 EFFECTS OF HORMONES ON THE KIDNEYS

Hormone (gland)	Function
Antidiuretic Hormone (ADH) (posterior pituitary)	• Increases reabsorption of water from the filtrate to the blood.
Parathyroid Hormone (PTH) (parathyroid glands)	• Increases reabsorption of Ca^{+2} ions from filtrate to the blood and excretion of phosphate ions into the filtrate.
Aldosterone (adrenal cortex)	• Increases reabsorption of Na^{+} ions from the filtrate to the blood and excretion of K^{+} ions into the filtrate. Water is reabsorbed following the reabsorption of sodium.
Atrial Natriuretic Hormone (ANH) (atria of heart)	• Decreases reabsorption of Na^{+} ions, which remain in the filtrate. More sodium and water are eliminated in urine.

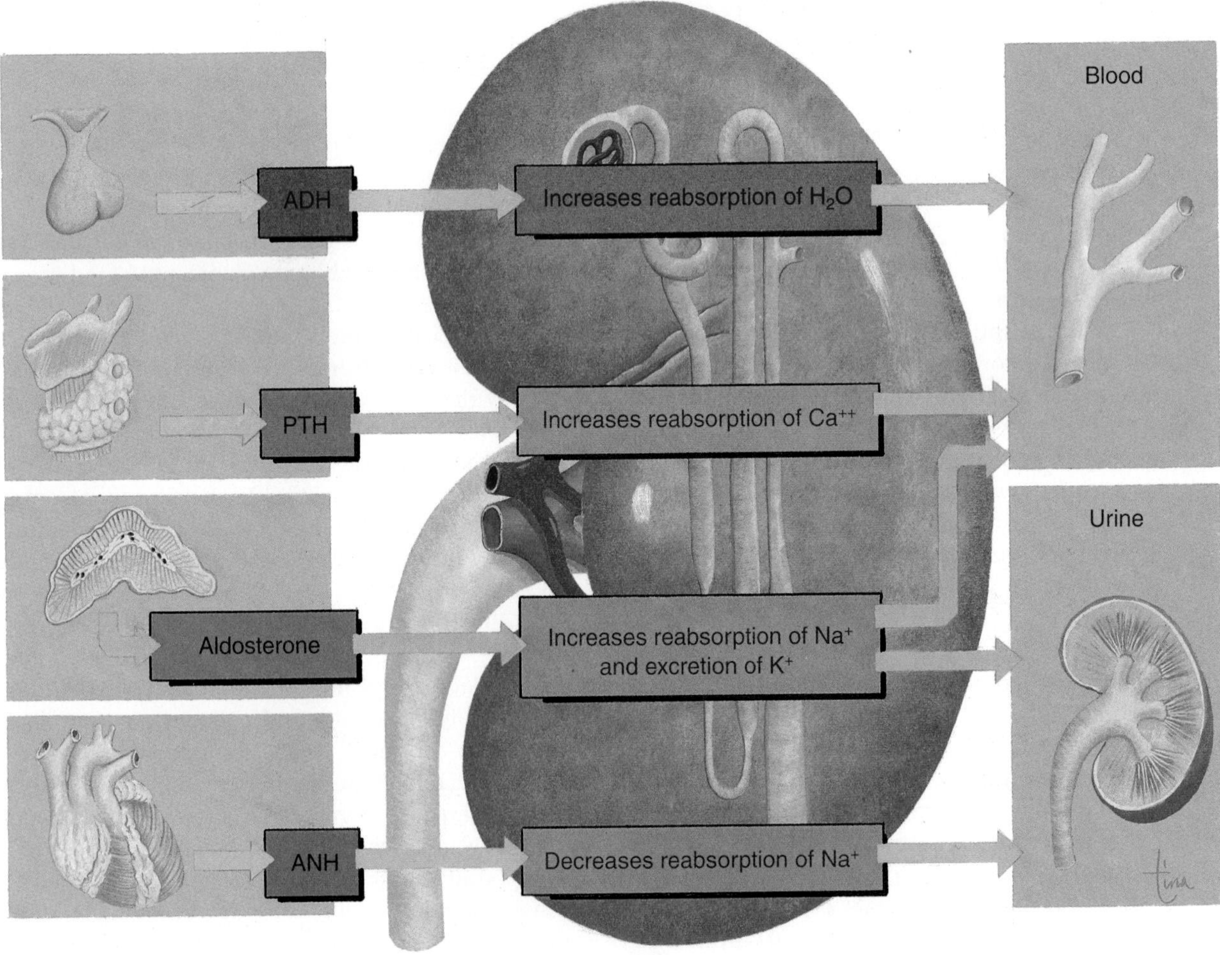

Figure 18–5 Effects of hormones on the kidneys.

the body returns to normal. This may occur following ingestion of excessive quantities of fluids.

SUMMARY OF URINE FORMATION

1. The kidneys form urine from blood plasma. Blood flow through the kidneys is a major factor in determining urinary output.
2. Glomerular filtration is the first step in urine formation. Filtration is not selective in terms of usefulness of materials; it is selective only in terms of size. High blood pressure in the glomeruli forces plasma, dissolved materials, and small proteins into Bowman's capsules; the fluid is now called renal filtrate.
3. Tubular reabsorption is selective in terms of usefulness. Nutrients such as glucose, amino acids, and vitamins are reabsorbed by active transport and may have renal threshold levels. Positive ions are reabsorbed by active transport and negative ions most often by passive transport. Water is reabsorbed by osmosis, and small proteins are reabsorbed by pinocytosis.

   Reabsorption takes place from the filtrate in the renal tubules to the blood in the peritubular capillaries.
4. Tubular secretion takes place from the blood in the peritubular capillaries to the filtrate in the renal tubules and can ensure that wastes such as creatinine or excess H^+ ions are actively put into the filtrate to be excreted.
5. Hormones such as aldosterone, ANH, and ADH influence the reabsorption of water and help maintain normal blood volume and blood pressure. The secretion of ADH determines whether a concentrated or dilute urine will be formed.
6. Waste products remain in the renal filtrate and are excreted in urine. The effects of hormones on the kidneys are summarized in Table 18-1 and illustrated in Fig. 18-5.

THE KIDNEYS AND ACID-BASE BALANCE

The kidneys are the organs most responsible for maintaining the pH of blood and tissue fluid within normal ranges. They have the greatest ability to compensate for the pH changes that are a normal part of body metabolism or the result of disease, and to make the necessary corrections.

This regulatory function of the kidneys is complex, but at its simplest it may be described as follows. If body fluids are becoming too acidic, the kidneys will secrete more H^+ ions into the renal filtrate and will return more HCO_3^- ions to the blood. This will help raise the pH of the blood back to normal. The reactions involved in such a mechanism are shown in Fig. 18-6, and we will return to it later. First, however, let us briefly consider how the kidneys will compensate for body fluids that are becoming too alkaline. You might expect the kidneys to do just the opposite of what is described above, and that is just what happens. The kidneys will return H^+ ions to the blood and excrete HCO_3^- ions in urine. This will help lower the pH of the blood back to normal.

Because the natural tendency is for body fluids to become more acidic, let us look at the pH-raising mechanism in more detail (see Fig. 18-6). The cells of the renal tubules can secrete H^+ ions or ammonia in exchange for Na^+ ions, and by doing so, influence the reabsorption of other ions. Hydrogen ions are obtained from the reaction of CO_2 and water (or other processes). An amine group from an amino acid is combined with a H^+ ion to form ammonia.

The tubule cell secretes the H^+ ion and the ammonia into the renal filtrate, and two Na^+ ions are reabsorbed in exchange. In the filtrate, the H^+ ion and ammonia form NH_4^+ (an ammonium radical), which reacts with a chloride ion (Cl^-) to form NH_4Cl (ammonium chloride) that is excreted in urine.

As the Na^+ ions are returned to the blood in the peritubular capillaries, HCO_3^- ions follow. Notice what has happened: Two H^+ ions have been excreted in urine, and two Na^+ ions and two HCO_3^- ions have been returned to the blood. As reactions like these take place, the body fluids are prevented from becoming too acidic.

Another mechanism used by the cells of the kidney tubules to regulate pH is the phosphate buffer system, which is described in Chapter 19.

OTHER FUNCTIONS OF THE KIDNEYS

In addition to the functions described thus far, the kidneys have other functions, some of which are not

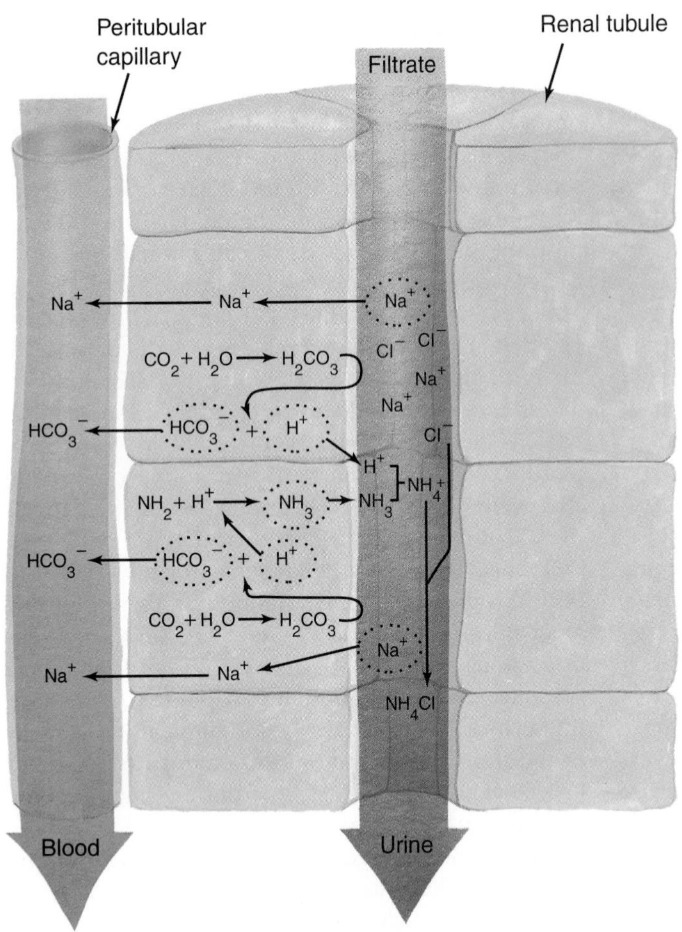

Figure 18–6 Renal regulation of acid–base balance. The cells of the renal tubule secrete H^+ ions and ammonia into the filtrate and return Na^+ ions and HCO_3^- ions to the blood in the peritubular capillaries. See text for further description.

directly related to the formation of urine. These functions are secretion of renin (which does influence urine formation), production of erythropoietin, and activation of vitamin D.

1. Secretion of Renin—When blood pressure decreases, the **juxtaglomerular** (juxta means "next to") cells in the walls of the afferent arterioles secrete the enzyme **renin**. Renin then initiates the renin-angiotensin mechanism to raise blood pressure. This was first described in Chapter 13, and the sequence of events is presented in Table 18–2. The end product of this mechanism is **angiotensin II**, which causes vasoconstriction and increases the secretion of aldosterone, both of which help raise blood pressure.

   A normal blood pressure is essential to normal body functioning. Perhaps the most serious change is a sudden, drastic decrease in blood pressure, as would follow a severe hemorrhage. In response to such a decrease, the kidneys will decrease filtration and urinary output and will

Table 18–2 THE RENIN-ANGIOTENSIN MECHANISM

Sequence
1. Decreased blood pressure stimulates the kidneys to secrete renin.
2. Renin splits the plasma protein angiotensinogen (synthesized by the liver) to angiotensin I.
3. Angiotensin I is converted to angiotensin II by an enzyme found primarily in lung tissue.
4. Angiotensin II causes vasoconstriction and stimulates the adrenal cortex to secrete aldosterone.

Box 18–3 ERYTHROPOIETIN

Anemia is one of the most debilitating consequences of renal failure, one that hemodialysis cannot reverse. Diseased kidneys stop producing erythropoietin, a natural stimulus for RBC production. Recently, however, erythropoietin has been produced by genetic engineering and is available for hemodialysis patients. In the past, their anemia could only be treated with transfusions, which exposed these patients to possible immunologic complications of repeated exposure to donated blood or to viral diseases such as hepatitis C. The synthetic erythropoietin eliminates such risks. Others who benefit from this medication are cancer patients and AIDS patients with severe anemia.

initiate the formation of angiotensin II. In these ways the kidneys help ensure that the heart has enough blood to pump to maintain cardiac output and blood pressure.

2. Secretion of **Erythropoietin**—This hormone is secreted whenever the blood oxygen level decreases (a state of hypoxia). Erythropoietin stimulates the red bone marrow to increase the rate of RBC production. With more RBCs in circulation, the oxygen-carrying capacity of the blood is greater, and the hypoxic state may be corrected (see also Box 18-3: Erythropoietin).
3. Activation of Vitamin D—This vitamin exists in several structural forms that are converted to **calciferol** (D_2) by the kidneys. Calciferol is the most active form of vitamin D, which increases the absorption of calcium and phosphate in the small intestine.

ELIMINATION OF URINE

The ureters, urinary bladder, and urethra do not change the composition or amount of urine, but are responsible for the periodic elimination of urine.

URETERS

Each **ureter** extends from the hilus of a kidney to the lower, posterior side of the urinary bladder (see Fig. 18-1). Like the kidneys, the ureters are retroperitoneal, that is, behind the peritoneum of the dorsal abdominal cavity.

The smooth muscle in the wall of the ureter contracts in peristaltic waves to propel urine toward the urinary bladder. As the bladder fills, it expands and compresses the lower ends of the ureters to prevent backflow of urine.

URINARY BLADDER

The **urinary bladder** is a muscular sac below the peritoneum and behind the pubic bones. In women, the bladder is inferior to the uterus; in men, the bladder is superior to the prostate gland. The bladder is a reservoir for accumulating urine, and it contracts to eliminate urine.

The mucosa of the bladder is **transitional epithelium**, which permits expansion without tearing the lining. When the bladder is empty, the mucosa appears wrinkled; these folds are **rugae**, which also permit expansion. On the floor of the bladder is a triangular area called the **trigone**, which has no rugae and does not expand. The points of the triangle are the openings of the two ureters and that of the urethra (Fig. 18-7).

The smooth muscle layer in the wall of the bladder is called the **detrusor muscle**. It is a muscle in the form of a sphere; when it contracts it becomes a smaller sphere, and its volume diminishes. Around the opening of the urethra the muscle fibers of the detrusor form the **internal urethral sphincter** (or sphincter of the bladder), which is involuntary.

URETHRA

The urethra carries urine from the bladder to the exterior. Within its wall is the **external urethral sphincter**, which is made of skeletal muscle and is under voluntary control.

In women, the urethra is 1 to 1.5 inches (2.5 to 4

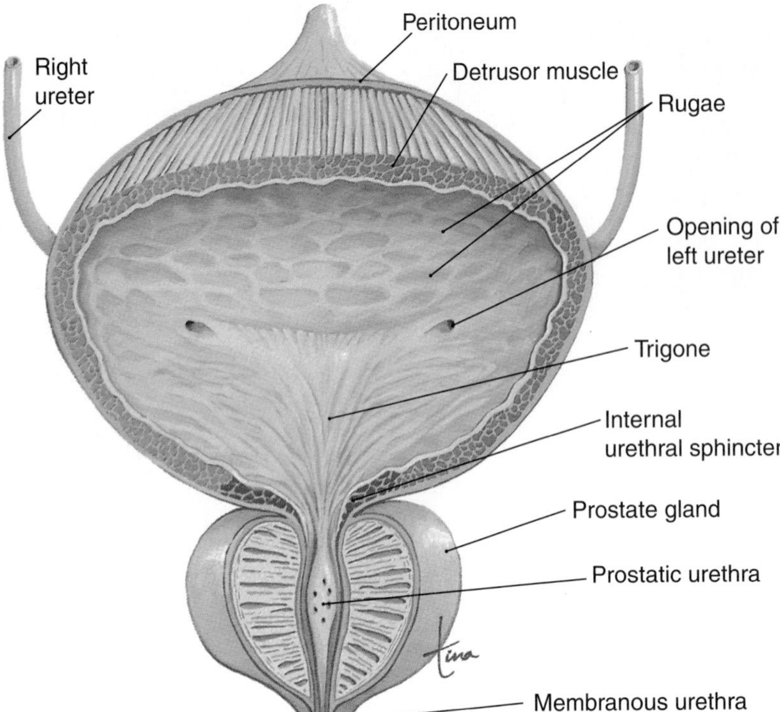

Figure 18–7 The urinary bladder and a portion of the urethra in the male. Anterior view of a frontal section.

cm) long and is anterior to the vagina. In men, the urethra is 7 to 8 inches (17 to 20 cm) long and extends through the prostate gland and penis. The male urethra carries semen as well as urine.

THE URINATION REFLEX

Urination may also be called **micturition** or **voiding**. This reflex is a spinal cord reflex over which voluntary control may be exerted. The stimulus for the reflex is stretching of the detrusor muscle of the bladder. The bladder can hold as much as 800 ml of urine, or even more, but the reflex is activated long before the maximum is reached.

When urine volume reaches 200 to 400 ml, the stretching is sufficient to generate sensory impulses that travel to the sacral spinal cord. Motor impulses return along parasympathetic nerves to the detrusor muscle, causing contraction. At the same time, the internal urethral sphincter relaxes. If the external urethral sphincter is voluntarily relaxed, urine flows into the urethra, and the bladder is emptied.

Urination can be prevented by voluntary contraction of the external urethral sphincter. However, if the bladder continues to fill and be stretched, voluntary control eventually becomes no longer possible.

CHARACTERISTICS OF URINE

The characteristics of urine include the physical and chemical aspects that are often evaluated as part of a urinalysis. Some of these are described in this section, and others are included in Appendix D: Normal Values for Some Commonly Used Urine Tests.

Amount—normal urinary output per 24 hours is 1 to 2 liters. There are many factors that can significantly change output. Excessive sweating or loss of fluid through diarrhea will decrease urinary output (**oliguria**) to conserve body water. Excessive fluid intake will increase urinary output (**polyuria**). Consumption of alcohol will also increase output because alcohol inhibits the secretion of ADH, and the kidneys will reabsorb less water.

Color—the typical yellow color of urine is often referred to as "straw" or "amber." Concentrated

Box 18–4 KIDNEY STONES

Kidney stones, or **renal calculi**, are crystals of the salts that are normally present in urine. A very high concentration of salts in urine may trigger precipitation of the salt and formation of crystals, which can range in size from microscopic to 10 to 20 mm in diameter. The most common type of kidney stone is made of calcium salts; a less common type is made of uric acid.

Kidney stones are most likely to form in the renal pelvis. Predisposing factors include decreased fluid intake or overingestion of minerals (as in mineral supplements), both of which lead to the formation of a very concentrated urine.

The entry of a kidney stone into a ureter may cause intense pain (renal colic) and bleeding. Obstruction of a ureter by a stone may cause backup of urine and possible kidney damage. Treatments include surgery to remove the stone, or lithotripsy, the use of shock waves to crush the stone into pieces small enough to be eliminated without damage to the urinary tract.

urine is a deeper yellow (amber) than is dilute urine. Freshly voided urine is also clear rather than cloudy.

Specific Gravity—the normal range is 1.010 to 1.025; this is a measure of the dissolved materials in urine. The specific gravity of distilled water is 1.000, meaning that there are no solutes present. Therefore, the higher the specific gravity number, the more dissolved material is present. Someone who has been exercising strenuously and has lost body water in sweat will usually produce less urine, which will be more concentrated and have a higher specific gravity.

The specific gravity of the urine is an indicator of the concentrating ability of the kidneys: The kidneys must excrete the waste products that are constantly formed in as little water as possible.

pH—the pH range of urine is between 4.6 and 8.0, with an average value of 6.0. Diet has the greatest influence on urine pH. A vegetarian diet will result in a more alkaline urine, whereas a high-protein diet will result in a more acidic urine.

Constituents—urine is approximately 95% water, which is the solvent for waste products and salts. Salts are not considered true waste products because they may well be utilized by the body when needed, but excess amounts will be excreted in urine (see Box 18-4: Kidney Stones).

Nitrogenous Wastes—as their name indicates, all

Box 18–5 BLOOD TESTS AND KIDNEY FUNCTION

Waste products are normally present in the blood, and the concentration of each varies within a normal range. As part of the standard lab work called Blood Chemistry, the levels of the three nitrogenous waste products are determined (urea, creatinine, and uric acid).

If blood levels of these three substances are within normal ranges, it may be concluded that the kidneys are excreting these wastes at normal rates. If, however, these blood levels are elevated, one possible cause is that kidney function has been impaired. Of the three, the creatinine level is probably the most reliable indicator of kidney functioning. Blood urea nitrogen (BUN) may vary considerably in certain situations not directly related to the kidneys. For example, BUN may be elevated as a consequence of a high-protein diet or of starvation when body protein is being broken down at a faster rate. Uric acid levels may also vary according to diet. However, elevated blood levels of all three nitrogenous wastes usually indicate impaired glomerular filtration.

Table 18–3 CHARACTERISTICS OF NORMAL URINE

Characteristic	Description
Amount	• 1–2 liters per 24 hours; highly variable depending on fluid intake and water loss through the skin and GI tract
Color	• Straw or amber; darker means more concentrated; should be clear, not cloudy
Specific gravity	• 1.010–1.025; a measure of the dissolved material in urine; the lower the value, the more dilute the urine
pH	• Average 6; range 4.6–8.0; diet has the greatest effect on urine pH
Composition	• 95% water; 5% salts and waste products
Nitrogenous wastes	• Urea—from amino acid metabolism • Creatinine—from muscle metabolism • Uric acid—from nucleic acid metabolism

contain nitrogen. Urea is formed by liver cells when excess amino acids are deaminated to be used for energy production. Creatinine comes from the metabolism of creatine phosphate, an energy source in muscles. Uric acid comes from the metabolism of nucleic acids, that is, the breakdown of DNA and RNA. Although these are waste products, there is always a certain amount of each in the blood. Box 18–5: Blood Tests and Kidney Function describes the relationship between blood levels of these waste products and kidney function.

Other non-nitrogenous waste products may include the metabolic products of medications. Table 18–3 summarizes the characteristics of urine.

When a substance not normally found in urine does

Table 18–4 ABNORMAL CONSTITUENTS IN URINE

Characteristic	Reason(s)
Glycosuria (the presence of glucose)	As long as blood glucose levels are within normal limits, filtrate levels will also be normal and will not exceed the threshold level for reabsorption. In an untreated diabetic, for example, blood glucose is too high; therefore the filtrate glucose level is too high. The kidneys reabsorb glucose up to their threshold level, but the excess remains in the filtrate and is excreted in urine.
Proteinuria (the presence of protein)	Most plasma proteins are too large to be forced out of the glomeruli, and the small proteins that enter the filtrate are reabsorbed by pinocytosis. The presence of protein in the urine indicates that the glomeruli have become too permeable, as occurs in some types of kidney disease.
Hematuria (the presence of blood—RBCs)	The presence of RBCs in urine may also indicate that the glomeruli have become too permeable. Another possible cause might be bleeding somewhere in the urinary tract. Pinpointing the site of bleeding would require specific diagnostic tests.
Bacteriuria (the presence of bacteria)	Bacteria give urine a cloudy rather than clear appearance; WBCs may be present also. The presence of bacteria means that there is an infection somewhere in the urinary tract. Further diagnostic tests would be needed to determine the precise location.
Ketonuria (the presence of ketones)	Ketones are formed from fats and proteins that are used for energy production. A trace of ketones in urine is normal. Higher levels of ketones indicate an increased use of fats and proteins for energy. This may be the result of malfunctioning carbohydrate metabolism (as in diabetes mellitus) or simply the result of a high-protein diet.

Box 18–6 URINARY TRACT INFECTIONS

Infections may occur anywhere in the urinary tract and are most often caused by the microbial agents of sexually transmitted diseases (see Chapter 20) or by the bacteria that are part of the normal flora of the colon. In women especially, the urinary and anal openings are in close proximity, and colon bacteria on the skin of the perineum may invade the urinary tract. The use of urinary catheters in hospitalized or bed-ridden patients may also be a factor if sterile technique is not carefully followed.

Cystitis is inflammation of the urinary bladder. Symptoms include frequency of urination, painful voiding, and low back pain. **Nephritis** (or pyelonephritis) is inflammation of the kidneys. Although this may be the result of a systemic bacterial infection, nephritis is a common complication of untreated lower urinary tract infections such as cystitis. Possible symptoms are fever and flank pain (in the area of the kidneys). Untreated nephritis may result in severe damage to nephrons and progress to renal failure.

appear there, there is a reason for it. The reason may be quite specific or more general. Table 18–4 lists some abnormal constituents of urine and possible reasons for each (see Box 18–6: Urinary Tract Infections).

AGING AND THE URINARY SYSTEM

With age, the number of nephrons in the kidneys decreases, often to half the original number by the age of 70 to 80, and the kidneys lose some of their concentrating ability. The glomerular filtration rate also decreases, partly as a consequence of arteriosclerosis and diminished renal blood flow. Despite these changes, excretion of nitrogenous wastes usually remains adequate.

The urinary bladder decreases in size and the tone of the detrusor muscle decreases. These changes may lead to a need to urinate more frequently. Urinary incontinence (the inability to control voiding) is *not* an inevitable consequence of aging, and can be prevented or minimized. Elderly people are, however, more at risk for infections of the urinary tract, especially if voiding leaves residual urine in the bladder.

SUMMARY

The kidneys are the principal regulators of the internal environment of the body. The composition of all body fluids is either directly or indirectly regulated by the kidneys as they form urine from blood plasma. The kidneys are also of great importance in the regulation of the pH of the body fluids. These topics are the subject of the next chapter.

STUDY OUTLINE

The urinary system consists of two kidneys, two ureters, the urinary bladder, and the urethra.

1. The kidneys form urine to excrete waste products and to regulate the volume, electrolytes, and pH of blood and tissue fluid.
2. The other organs of the system are concerned with elimination of urine.

Kidneys (see Fig. 18–1)

1. Retroperitoneal on either side of the backbone in the upper abdominal cavity; partially protected by the lower rib cage.
2. Adipose tissue and the renal fascia cushion the kidneys and help hold them in place.
3. Hilus—an indentation on the medial side; renal artery enters, renal vein and ureter emerge.

Kidney—internal structure (see Fig. 18–2)

1. Renal Cortex—outer area, made of renal corpuscles and convoluted tubules.
2. Renal Medulla (pyramids)—middle area, made of loops of Henle and collecting tubules.
3. Renal Pelvis—a cavity formed by the expanded end of the ureter within the kidney; extensions around the papillae of the pyramids are called calyces, which collect urine.

The Nephron—the functional unit of the kidney (see Fig. 18–3); 1 million per kidney

1. Renal Corpuscle—consists of a glomerulus surrounded by a Bowman's capsule.
   - Glomerulus—a capillary network between an afferent arteriole and an efferent arteriole.
   - Bowman's Capsule—the expanded end of a renal tubule that encloses the glomerulus; inner layer has pores and is very permeable; contains renal filtrate (potential urine).
2. Renal Tubule—consists of the proximal convoluted tubule, loop of Henle, distal convoluted tubule, and collecting tubule. Collecting tubules unite to form papillary ducts that empty urine into the calyces of the renal pelvis.
   - Peritubular Capillaries—arise from the efferent arteriole and surround all parts of the renal tubule.

Blood Vessels of the Kidney (see Figs. 18–1, 18–2, and 18–3)

1. Pathway: abdominal aorta → renal artery → small arteries in the kidney → afferent arterioles → glomeruli → efferent arterioles → peritubular capillaries → small veins in the kidney → renal vein → inferior vena cava.
2. Two sets of capillaries provide for two sites of exchanges between the blood and tissues in the process of urine formation.

Formation of Urine (see Fig. 18–4)

1. Glomerular Filtration—takes place from the glomerulus to Bowman's capsule. High blood pressure (60 mmHg) in the glomerulus forces plasma, dissolved materials and small proteins out of the blood and into Bowman's capsule. The fluid is now called filtrate. Filtration is selective only in terms of size; blood cells and large proteins remain in the blood.
2. GFR is 100 to 125 ml per minute. Increased blood flow to the kidney increases GFR; decreased blood flow decreases GFR.
3. Tubular Reabsorption—takes place from the filtrate in the renal tubule to the blood in the peritubular capillaries; 99% of the filtrate is reabsorbed; only 1% becomes urine.
   - Active Transport—reabsorption of glucose, amino acids, vitamins, and positive ions; threshold level is a limit to the quantity that can be reabsorbed.
   - Passive Transport—most negative ions follow the reabsorption of positive ions.
   - Osmosis—water follows the reabsorption of minerals, especially sodium.
   - Pinocytosis—small proteins are engulfed by proximal tubule cells.
4. Tubular Secretion—takes place from the blood in the peritubular capillaries to the filtrate in the renal tubule; creatinine and other waste products may be secreted into the filtrate to be excreted in urine; secretion of H^+ ions helps maintain pH of blood.
5. Hormones that Affect Reabsorption—aldosterone, atrial natriuretic hormone, antidiuretic hormone, and parathyroid hormone—see Table 18-1 and Fig. 18-5.

The Kidneys and Acid–Base Balance

1. The kidneys have the greatest capacity to compensate for normal and abnormal pH changes.
2. If the body fluids are becoming too acidic, the kidneys excrete H^+ ions and return HCO_3^- ions to the blood (see Fig. 18-6).
3. If the body fluids are becoming too alkaline, the kidneys return H^+ ions to the blood and excrete HCO_3^- ions.

Other Functions of the Kidneys

1. Secretion of renin by juxtaglomerular cells when blood pressure decreases (see Table 18-2). Angiotensin II causes vasoconstriction and increases secretion of aldosterone.
2. Secretion of erythropoietin in response to hypoxia; stimulates RBM to increase rate of RBC production.
3. Activation of vitamin D—conversion of inactive forms to the active form.

Elimination of Urine—the function of the ureters, urinary bladder, and urethra

Ureters (see Figs. 18–1 and 18–7)

1. Each extends from the hilus of a kidney to the lower posterior side of the urinary bladder.
2. Peristalsis of smooth muscle layer propels urine toward bladder.

Urinary Bladder (see Figs. 18–1 and 18–7)

1. A muscular sac below the peritoneum and behind the pubic bones; in women, below the uterus; in men, above the prostate gland.
2. Mucosa—transitional epithelial tissue folded into rugae; permit expansion without tearing.
3. Trigone—triangular area on bladder floor; no rugae, does not expand; bounded by openings of ureters and urethra.
4. Detrusor Muscle—the smooth muscle layer, a spherical muscle; contracts to expel urine (reflex).
5. Internal Urethral Sphincter—involuntary; formed by detrusor muscle fibers around the opening of the urethra.

Urethra—takes urine from the bladder to the exterior

1. In women—1 to 1.5 inches long; anterior to vagina.
2. In men—7 to 8 inches long; passes through the prostate gland and penis.
3. Contains the external urethral sphincter: skeletal muscle (voluntary).

The Urination Reflex—also called micturition or voiding

1. Stimulus: stretching of the detrusor muscle by accumulating urine.
2. Sensory impulses to spinal cord, motor impulses return to detrusor muscle, which contracts; internal urethral sphincter relaxes.
3. Voluntary control is provided by the external urethral sphincter.

Characteristics of Urine (see Table 18–3)

Abnormal Constituents of Urine (see Table 18–4)

REVIEW QUESTIONS

1. Describe the location of the kidneys, ureters, urinary bladder, and urethra. (pp. 402, 413)
2. Name the three areas of the kidney, and state what each consists of. (p. 402)
3. Name the two major parts of a nephron. State the general function of nephrons. (pp. 402, 404)
4. Name the parts of a renal corpuscle. What process takes place here? Name the parts of a renal tubule. What processes take place here? (pp. 402, 404)
5. State the mechanism of tubular reabsorption of each of the following: (p. 409)
   a. Water
   b. Glucose
   c. Small proteins
   d. Positive ions
   e. Negative ions
   f. Amino acids
   g. Vitamins

   Explain what is meant by a threshold level of reabsorption.
6. Explain the importance of tubular secretion. (p. 409)
7. Describe the pathway of blood flow through the kidney from the abdominal aorta to the inferior vena cava. (p. 404)
8. Name the two sets of capillaries in the kidney, and state the processes that take place in each. (pp. 404, 407, 409)
9. Name the hormone that has each of these effects on the kidneys: (p. 409)
   a. Promotes reabsorption of Na^+ ions
   b. Promotes direct reabsorption of water

c. Promotes reabsorption of Ca^{+2} ions
d. Promotes excretion of K^{+} ions
e. Decreases reabsorption of Na^{+} ions

10. In what circumstances will the kidneys excrete H^{+} ions? What ions will be returned to the blood? How will this affect the pH of blood? (p. 411)
11. In what circumstances do the kidneys secrete renin, and what is its purpose? (pp. 411–412)
12. In what circumstances do the kidneys secrete erythropoietin, and what is its purpose? (p. 412)
13. Describe the function of the ureters and that of the urethra. (p. 413)
14. With respect to the urinary bladder, describe the function of rugae and the detrusor muscle. (p. 413)
15. Describe the urination reflex in terms of stimulus, part of the CNS involved, effector muscle, internal urethral sphincter, and voluntary control. (p. 414)
16. Describe the characteristics of normal urine in terms of appearance, amount, pH, specific gravity, and composition. (pp. 414–416)
17. State the source of each of the nitrogenous waste products: creatinine, uric acid, and urea. (pp. 415–416)

CHAPTER 19

Fluid–Electrolyte and Acid–Base Balance

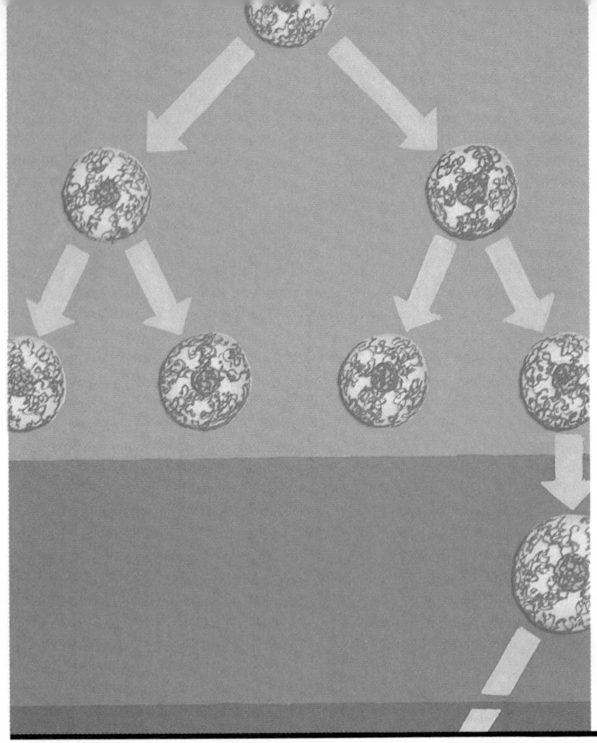

CHAPTER 19

Chapter Outline

Student Objectives

- Describe the water compartments and the name for the water in each.
- Explain how water moves between compartments.
- Explain the regulation of the intake and output of water.
- Name the major electrolytes in body fluids, and state their functions.
- Explain the regulation of the intake and output of electrolytes.
- Describe the three buffer systems in body fluids.
- Explain why the respiratory system has an effect on pH, and describe respiratory compensating mechanisms.
- Explain the renal mechanisms for pH regulation of extracellular fluid.
- Describe the effects of acidosis and alkalosis.

Fluid–Electrolyte and Acid–Base Balance

New Terminology

Amine group (ah-**MEEN**)
Anions (**AN**-eye-ons)
Carboxyl group (kar-**BAHK**-sul)
Cations (**KAT**-eye-ons)
Electrolytes (ee-**LEK**-troh-lites)
Osmolarity (ahs-moh-**LAR**-i-tee)
Osmoreceptors (AHS-moh-re-**SEP**-ters)

Related Clinical Terminology

Edema (uh-**DEE**-muh)
Hypercalcemia (HIGH-per-kal-**SEE**-me-ah)
Hyperkalemia (HIGH-per-kuh-**LEE**-me-ah)
Hypernatremia (HIGH-per-nuh-**TREE**-me-ah)
Hypocalcemia (HIGH-poh-kal-**SEE**-me-ah)
Hypokalemia (HIGH-poh-kuh-**LEE**-me-ah)
Hyponatremia (HIGH-poh-nuh-**TREE**-me-ah)

Terms that appear in ***bold type*** *in the chapter text are defined in the glossary, which begins on page 528.*

The fluid medium of the human body is, of course, water. Water makes up 60% to 75% of the total body weight. Electrolytes are the positive and negative ions present in body fluids. Many of these ions are minerals that are already familiar to you. They each have specific functions in body fluids, and some of them are also involved in the maintenance of the normal pH of the body fluids. In this chapter we will first discuss fluid-electrolyte balance, then review and summarize the mechanisms involved in acid-base balance.

WATER COMPARTMENTS

Most of the water of the body, about two-thirds of the total water volume, is found within individual cells and is called **intracellular fluid** (ICF). The remaining third is called **extracellular fluid** (ECF) and includes blood plasma, lymph, tissue fluid, and the specialized fluids such as cerebrospinal fluid, synovial fluid, aqueous humor, and serous fluid.

Water constantly moves from one fluid site in the body to another by the processes of filtration and osmosis. These fluid sites are called **water compartments** (Fig. 19-1). The chambers of the heart and all of the blood vessels form one compartment, and the water within is called plasma. By the process of filtration in capillaries, some plasma is forced out into tissue spaces (another compartment) and is then called tissue fluid. When tissue fluid enters cells by the process of osmosis, it has moved to still another compartment and is called intracellular fluid. The tissue fluid that enters lymph capillaries is in yet another compartment and is called lymph.

The other process (besides filtration) by which water moves from one compartment to another is osmosis, which, you may recall, is the diffusion of water through a semi-permeable membrane. Water will move through cell membranes from the area of its

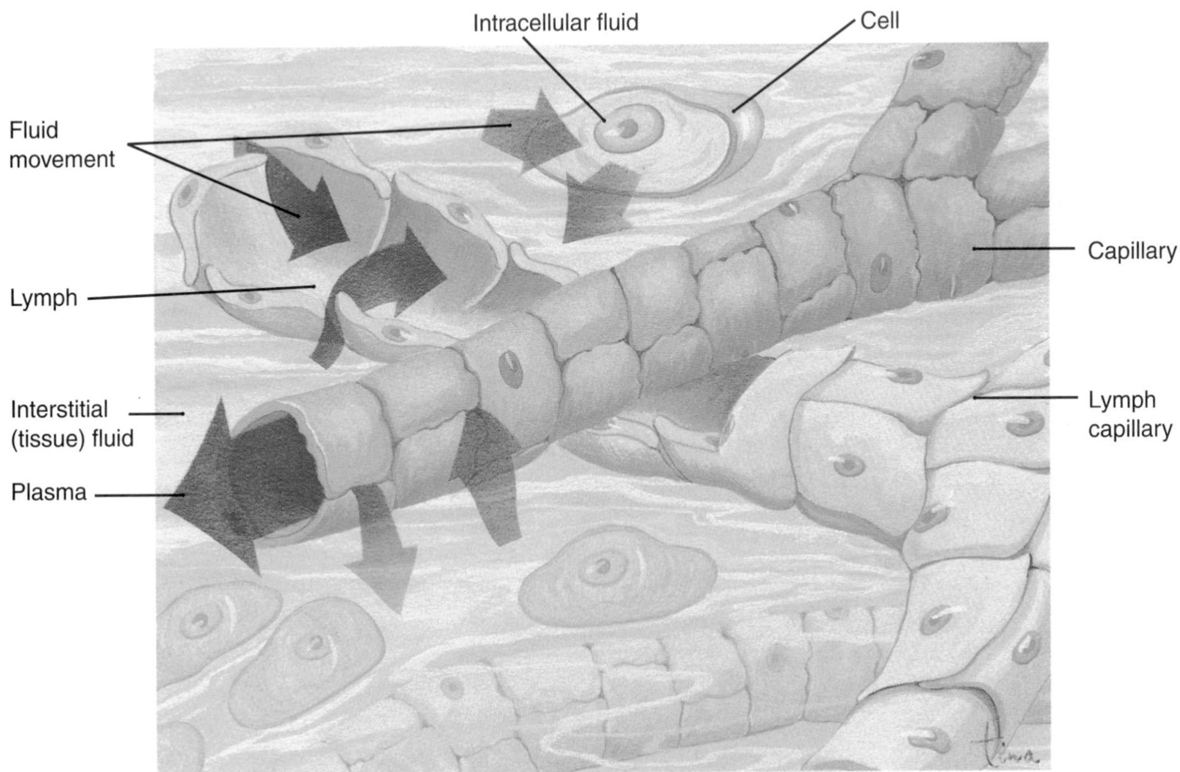

Figure 19–1 Water compartments. The name given to water in each of its locations is indicated.

Box 19–1 EDEMA

Edema is an abnormal increase in the amount of tissue fluid, which may be localized or systemic. Sometimes edema is inapparent, and sometimes it is apparent as swelling.

Localized edema follows injury and inflammation of a body part. Spraining an ankle, for example, damages tissues that then release histamine. Histamine increases the permeability of capillaries, and more tissue fluid is formed. As tissue fluid accumulates, the ankle may become swollen.

Systemic edema is the result of an imbalance between the movement of water out of and into capillaries, that is, between filtration and osmosis. Excessive filtration will occur when capillary pressure rises. This may be caused by venous obstruction due to blood clots or by congestive heart failure. Edema of this type is often apparent in the lower extremities (pulmonary edema was described in Chapter 15).

Systemic bacterial infections may increase capillary permeability, and loss of plasma to tissue spaces is one aspect of septicemia. In this situation, however, the edema is of secondary importance to the hypotension, which may be life threatening.

Insufficient osmosis, the return of tissue fluid into capillaries, is a consequence of a decrease in plasma proteins, especially albumin. This may occur in severe liver diseases such as cirrhosis, kidney disease involving loss of protein in urine, malnutrition, or severe burn injuries.

Because edema is a symptom rather than a disease, treatment is aimed at correcting the specific cause. If that is not possible, the volume of tissue fluid may be diminished by a low-salt diet and the use of diuretics.

greater concentration to the area of its lesser concentration. Another way of expressing this is to say that water will diffuse to an area with a greater concentration of dissolved material. The concentration of electrolytes present in the various water compartments determines just how osmosis will take place. Therefore, if water is in balance in all the compartments, the electrolytes are also in balance. Although water and ions are constantly moving, their relative proportions in the compartments remain constant; this is fluid-electrolyte homeostasis, and its maintenance is essential for life (see Box 19-1: Edema).

WATER INTAKE AND OUTPUT

Most of the water the body requires comes from the ingestion of liquids; this amount averages 1600 ml per day. The food we eat also contains water. Even foods we think of as somewhat dry, such as bread, contain significant amounts of water. The daily water total from food averages 700 ml. The last source of water, about 200 ml per day, is the metabolic water that is a product of cell respiration. The total intake of water per day, therefore, is about 2500 ml, or 2.5 liters.

Most of the water lost from the body is in the form of urine produced by the kidneys; this averages 1500 ml per day. About 500 ml per day is lost in the form of sweat, another 300 ml per day is in the form of water vapor in exhaled air, and another 200 ml per day is lost in feces. The total output of water is thus about 2500 ml per day.

Naturally, any increase in water output must be compensated for by an increase in intake. Someone who exercises strenuously, for example, may lose 1 to 2 liters of water in sweat and must replace that water by drinking more fluids. In a healthy individual, water intake equals water output, even though the amounts of each may vary greatly from the averages used above (Fig. 19-2 and Table 19-1).

REGULATION OF WATER INTAKE AND OUTPUT

The hypothalamus in the brain contains **osmoreceptors** that detect changes in the osmolarity of body fluids. **Osmolarity** is the concentration of dissolved ma-

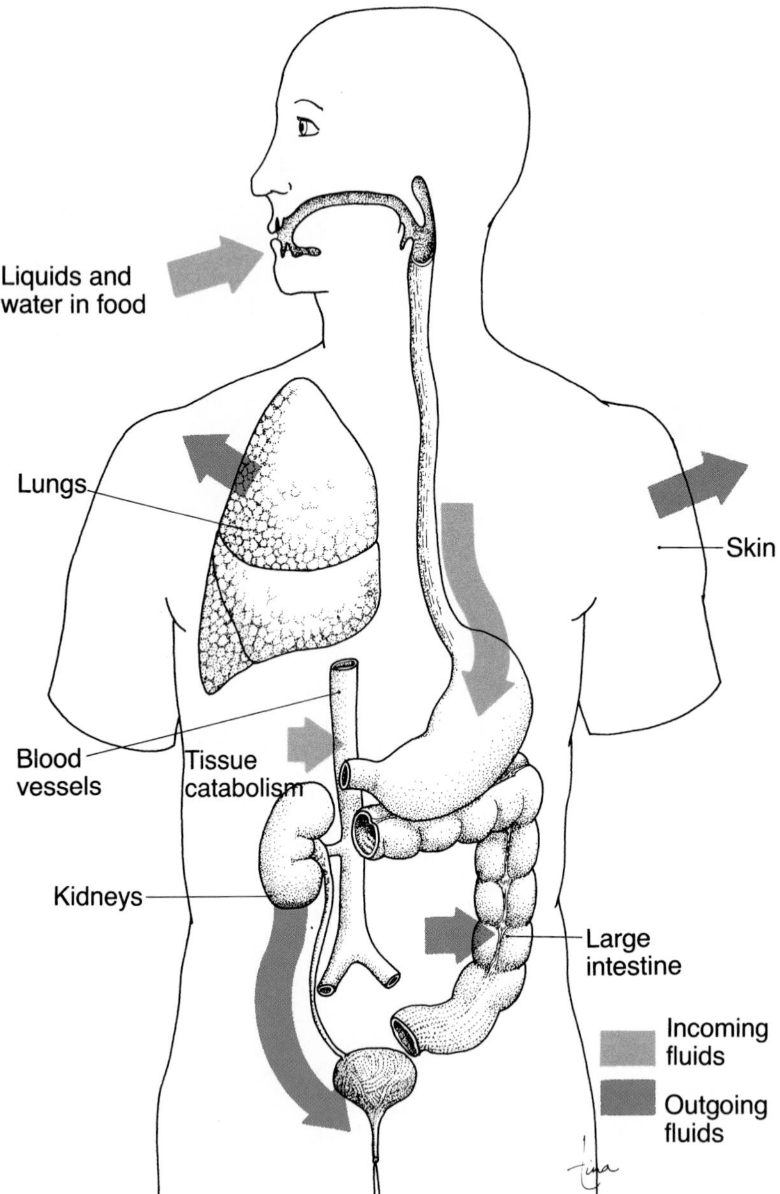

Figure 19–2 Water intake and output. See text for description.

Table 19–1 WATER INTAKE AND OUTPUT

Form	Average Amount per 24 Hours (ml)
Intake	
Liquids	1600
Food	700
Metabolic water	200
Output	
Urine	1500
Sweat (and insensible water loss)	500
Exhaled air (water vapor)	300
Feces	200

terials present in a fluid. Dehydration raises the osmolarity of the blood; that is, there is less water in proportion to the amount of dissolved materials. Another way to express this is to simply say that the blood is now a more concentrated solution. When dehydrated, we experience the sensation of thirst, characterized by dryness of the mouth and throat, as less saliva is produced. Thirst is an uncomfortable sensation, and we drink fluids to relieve it. The water we drink is readily absorbed by the mucosa of the stomach and small intestine and has the effect of decreasing the osmolarity of the blood. In other words, we can say that the water we just drank is causing the blood to become a more dilute solution, and, as the serum osmolarity returns to normal, the sensation of thirst diminishes.

As you may recall, the hypothalamus is also involved in water balance because of its production of antidiuretic hormone (ADH), which is stored in the posterior pituitary gland. In a state of dehydration, the hypothalamus stimulates the release of ADH from the posterior pituitary. Antidiuretic hormone then increases the reabsorption of water by the kidney tubules. Water is returned to the blood to preserve blood volume, and urinary output decreases.

The hormone aldosterone, from the adrenal cortex, also helps regulate water output. Aldosterone increases the reabsorption of Na^+ ions by the kidney tubules, and water from the renal filtrate follows the Na^+ ions back to the blood. Aldosterone is secreted when the Na^+ ion concentration of the blood decreases or whenever there is a significant decrease in blood pressure (the renin-angiotensin mechanism).

Several other factors may also contribute to water loss. These include excessive sweating, hemorrhage, diarrhea or vomiting, severe burns, and fever. In these circumstances, the kidneys will conserve water, but water must also be replaced by increased consumption. Following hemorrhage or during certain disease states, fluids may also be replaced by intravenous administration.

A less common occurrence is that of too much water in the body. This may happen following overconsumption of fluids. The osmolarity of the blood decreases, and there is too much water in proportion to electrolytes (or, the blood is too dilute). Atrial natriuretic hormone (ANH) is secreted by the atria when blood volume or blood pressure increase. ANH then decreases the reabsorption of Na^+ ions by the kidneys, which increases urinary output of sodium and water. Also, secretion of ADH will diminish, which will contribute to a greater urinary output that will return the blood osmolarity to normal.

ELECTROLYTES

Electrolytes are chemicals that dissolve in water and dissociate into their positive and negative ions. Most electrolytes are the inorganic salts, acids, and bases found in all body fluids.

Most organic compounds are non-electrolytes, that is, they do not ionize when in solution. Glucose, for example, dissolves in water but does not ionize; it remains as intact glucose molecules. Some proteins, however, do form ionic bonds and when in solution dissociate into ions.

Positive ions are called **cations**. Examples are Na^+, K^+, Ca^{+2}, Mg^{+2}, Fe^{+2}, and H^+. Negative ions are called **anions**, and examples are Cl^-, HCO_3^-, SO_4^{-2} (sulfate), HPO_4^{-2} (phosphate), and protein anions.

Electrolytes help create the osmolarity of body fluids and, therefore, help regulate the osmosis of water between water compartments. Some electrolytes are involved in acid-base regulatory mechanisms, or they are part of structural components of tissues or part of enzymes.

ELECTROLYTES IN BODY FLUIDS

The three principal fluids in the body are intracellular fluid and the extracellular fluids plasma and tissue

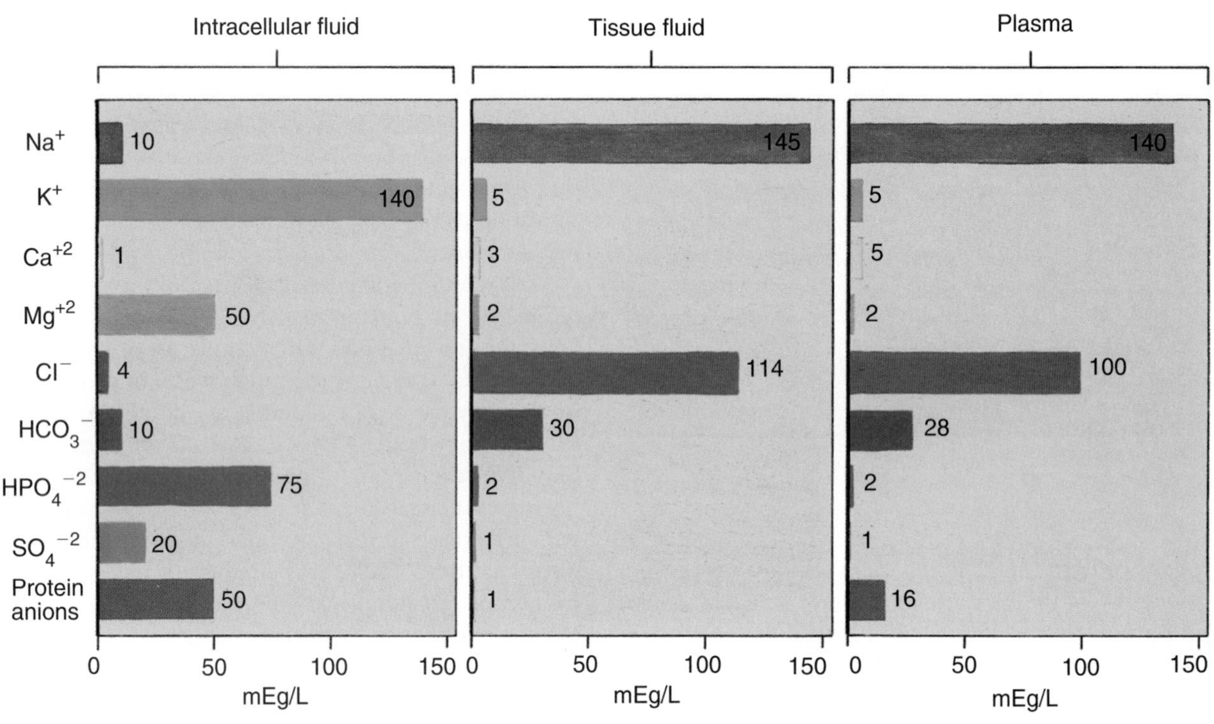

Figure 19–3 Electrolyte concentrations in intracellular fluid, tissue fluid, and plasma. Concentrations are expressed in milliequivalents per liter. See text for summary of major differences among these fluids.

fluid. The relative concentrations of the most important electrolytes in these fluids are depicted in Fig. 19–3. The major differences may be summarized as follows. In intracellular fluid, the most abundant cation is K^+, the most abundant anion is HPO_4^{-2}, and protein anions are also abundant. In both tissue fluid and plasma, the most abundant cation is Na^+, and the most abundant anion is Cl^-. Protein anions form a significant part of plasma but not of tissue fluid. The functions of the major electrolytes are described in Table 19–2.

INTAKE, OUTPUT, AND REGULATION

Electrolytes are part of the food and beverages we consume, are absorbed by the GI tract into the blood, and become part of body fluids. The ECF concentrations of some electrolytes are regulated by hormones. Aldosterone increases the reabsorption of Na^+ ions and the excretion of K^+ ions by the kidneys. The blood sodium level is thereby raised, and the blood potassium level is lowered. Atrial natriuretic hormone (ANH) increases the excretion of Na^+ ions by the kidneys and lowers the blood sodium level. Parathyroid hormone (PTH) and calcitonin regulate the blood levels of calcium and phosphate. PTH increases the reabsorption of these minerals from bones, and increases their absorption from food in the small intestine (vitamin D is also necessary). Calcitonin promotes the removal of calcium and phosphate from the blood to form bone matrix.

Electrolytes are lost in urine, sweat, and feces. Urine contains the electrolytes that are not reabsorbed by the kidney tubules; the major one is Na^+ ions. Other electrolytes are present in urine when their concentrations in the blood exceed the body's need for them.

The most abundant electrolytes in sweat are Na^+ ions and Cl^- ions. Electrolytes lost in feces are those that are not absorbed in either the small intestine or colon.

Some of the major imbalances of electrolyte levels are described in Box 19–2: Electrolyte Imbalances.

Table 19–2 MAJOR ELECTROLYTES

Electrolyte	Plasma Level mEq/L*	ICF Level mEq/L	Functions
Sodium (Na^+)	136–142	10	• Creates much of the osmotic pressure of ECF; the most abundant cation in ECF • Essential for electrical activity of neurons and muscle cells
Potassium (K^+)	3.8–5.0	141	• Creates much of the osmotic pressure in ICF; the most abundant cation in ICF • Essential for electrical activity of neurons and muscle cells
Calcium (Ca^{+2})	4.6–5.5	1	• Most (98%) is found in bones and teeth • Maintains normal excitability of neurons and muscle cells • Essential for blood clotting
Magnesium (Mg^{+2})	1.3–2.1	58	• Most (50%) is found in bone • More abundant in ICF than in ECF • Essential for ATP production and activity of neurons and muscle cells
Chloride (Cl^-)	95–103	4	• Most abundant anion in ECF; diffuses easily into and out of cells; helps regulate osmotic pressure • Part of HCl in gastric juice
Bicarbonate (HCO_3^-)	28	10	• Part of the bicarbonate buffer system
Phosphate (HPO_4^{-2})	1.7–2.6	75	• Most (85%) is found in bones and teeth • Primarily an ICF anion • Part of DNA, RNA, ATP, phospholipids • Part of phosphate buffer system
Sulfate (SO_4^{-2})	1	2	• Part of some amino acids and proteins

*The concentration of an ion is often expressed in milliequivalents per liter, abbreviated mEq/L, which is the number of electrical charges in each liter of solution.

Box 19–2 ELECTROLYTE IMBALANCES

Imbalances of Sodium

Hyponatremia—a consequence of excessive sweating, diarrhea, or vomiting. Characterized by dizziness, confusion, weakness, low BP, shock.

Hypernatremia—a consequence of excessive water loss or sodium ingestion. Characterized by loss of ICF and extreme thirst and agitation.

Imbalances of Potassium

Hypokalemia—a consequence of vomiting or diarrhea or kidney disease. Characterized by fatigue, confusion, possible cardiac failure.

Hyperkalemia—a consequence of acute renal failure or Addison's disease. Characterized by weakness, abnormal sensations, cardiac arrhythmias, and possible cardiac arrest.

Imbalances of Calcium

Hypocalcemia—a consequence of hypoparathyroidism or decreased calcium intake. Characterized by muscle spasms leading to tetany.

Hypercalcemia—a consequence of hyperparathyroidism. Characterized by muscle weakness, bone fragility, and possible kidney stones.

ACID-BASE BALANCE

You have already learned quite a bit about the regulation of the pH of body fluids in the chapters on chemistry, the respiratory system, and the urinary system. In this section, we will put all that information together. You may first wish to review the pH scale, described in Chapter 2.

The normal pH range of blood is 7.35 to 7.45. The pH of tissue fluid is similar but can vary slightly above or below this range. The intracellular fluid has a pH range of 6.8 to 7.0. Notice that these ranges of pH are quite narrow; they must be maintained in order for enzymatic reactions and other processes to proceed normally.

Maintenance of acid-base homeostasis is accomplished by the buffer systems in body fluids, respirations, and the kidneys.

BUFFER SYSTEMS

The purpose of a **buffer system** is to prevent drastic changes in the pH of body fluids by chemically reacting with strong acids or bases that would otherwise greatly change the pH. A buffer system consists of a weak acid and a weak base. These molecules react with strong acids or bases that may be produced and change them to substances that do not have a great effect on pH.

Bicarbonate Buffer System

The two components of this buffer system are carbonic acid (H_2CO_3), a weak acid, and sodium bicarbonate ($NaHCO_3$), a weak base. Each of these molecules participates in a specific type of reaction.

If a potential pH change is created by a strong acid, the following reaction takes place:

$$\underset{\text{(strong acid)}}{HCl} + NaHCO_3 \rightarrow NaCl + \underset{\text{(weak acid)}}{H_2CO_3}$$

The strong acid has reacted with the sodium bicarbonate to produce a salt (NaCl) that has no effect on pH and a weak acid that has little effect on pH.

If a potential pH change is created by a strong base, the following reaction takes place:

$$\underset{\text{(strong base)}}{NaOH} + H_2CO_3 \rightarrow H_2O + \underset{\text{(weak base)}}{NaHCO_3}$$

The strong base has reacted with the carbonic acid to produce water, which has no effect on pH and a weak base that has little effect on pH.

The bicarbonate buffer system is important in both the blood and tissue fluid. During normal metabolism, these fluids tend to become more acidic, so more sodium bicarbonate than carbonic acid is needed. The usual ratio of these molecules to each other is about 20 to 1 ($NaHCO_3$ to H_2CO_3).

Phosphate Buffer System

The two components of this buffer system are sodium dihydrogen phosphate (NaH_2PO_4), a weak acid, and sodium monohydrogen phosphate (Na_2HPO_4), a weak base. Let us use specific reactions to show how this buffer system works.

If a potential pH change is created by a strong acid, the following reaction takes place:

$$\underset{\text{(strong acid)}}{HCl} + Na_2HPO_4 \rightarrow NaCl + \underset{\text{(weak acid)}}{NaH_2PO_4}$$

The strong acid has reacted with the sodium monohydrogen phosphate to produce a salt that has no effect on pH and a weak acid that has little effect on pH.

If a potential pH change is created by a strong base, the following reaction takes place:

$$\underset{\text{(strong base)}}{NaOH} + NaH_2PO_4 \rightarrow H_2O + \underset{\text{(weak base)}}{Na_2HPO_4}$$

The strong base has reacted with the sodium dihydrogen phosphate to form water, which has no effect on pH and a weak base that has little effect on pH.

The phosphate buffer system is important in the regulation of the pH of the blood by the kidneys. The cells of the kidney tubules can remove excess hydrogen ions by forming NaH_2PO_4, which is excreted in urine. The retained Na^+ ions are returned to the blood in the peritubular capillaries, along with bicarbonate ions (Fig. 19–4).

Protein Buffer System

This buffer system is the most important one in the intracellular fluid. The amino acids that make up proteins each have a **carboxyl group** (COOH) and an **amine** (or amino) **group** (NH_2) and may act as either acids or bases.

The carboxyl group may act as an acid because it

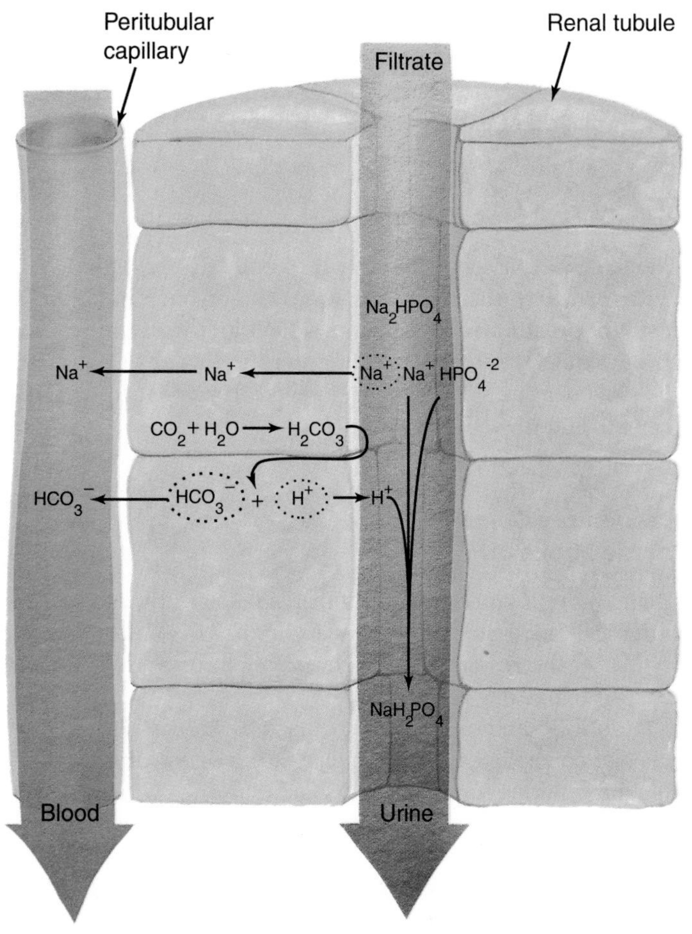

Figure 19–4 The phosphate buffer system. The reactions are shown in a kidney tubule. See text for description.

can donate a hydrogen ion (H^+) to the fluid to counteract increasing alkalinity:

$$NH_2\text{—}\underset{R}{\overset{H}{C}}\text{—}COOH \rightarrow NH_2\text{—}\underset{R}{\overset{H}{C}}\text{—}COO^- + H^+$$

The amine group may act as a base because it can pick up an excess hydrogen ion from the fluid to counteract increasing acidity:

$$COOH\text{—}\underset{R}{\overset{H}{C}}\text{—}NH_2 \rightarrow COOH\text{—}\underset{R}{\overset{H}{C}}\text{—}NH_3^+$$

The buffer systems react within a fraction of a second to prevent drastic pH changes. However, they have the least capacity to prevent great changes in pH because there are a limited number of molecules of these buffers present in body fluids. When an ongoing cause is disrupting the normal pH, the respiratory and renal mechanisms will also be needed.

RESPIRATORY MECHANISMS

The respiratory system affects pH because it regulates the amount of CO_2 present in body fluids. As you know, the respiratory system may be the cause of a pH imbalance or may help correct a pH imbalance from some other cause.

Respiratory Acidosis and Alkalosis

Respiratory acidosis is caused by anything that decreases the rate or efficiency of respiration. Severe pulmonary diseases are possible causes of respiratory acidosis. When CO_2 cannot be exhaled as fast as it is formed during cell respiration, excess CO_2 results in the formation of excess H^+ ions, as shown in this reaction:

$$CO_2 + H_2O \rightarrow H_2CO_3 \rightarrow H^+ + HCO_3^-$$

The excess H^+ ions lower the pH of body fluids.

Respiratory alkalosis is far less common but is the result of breathing more rapidly, which increases the amount of CO_2 exhaled. Because there are fewer CO_2 molecules in the body fluids, fewer H^+ ions are formed, and pH tends to rise.

Respiratory Compensation for Metabolic pH Changes

Changes in pH caused by other than a respiratory disorder are called metabolic acidosis or alkalosis. In either case, the respiratory system may help prevent a drastic change in pH.

Metabolic acidosis may be caused by kidney disease, uncontrolled diabetes mellitus, excessive diarrhea or vomiting, or the use of some diuretics. When excess H^+ ions are present in body fluids, pH begins to decrease, and this stimulates the respiratory centers in the medulla. The response is to increase the rate of breathing to exhale more CO_2 to decrease H^+ ion formation. This helps raise the pH back toward the normal range.

Metabolic alkalosis is not common, but may be caused by the overuse of antacid medications or the vomiting of stomach contents only. As the pH of body fluids begins to rise, breathing slows and decreases the amount of CO_2 exhaled. The CO_2 retained within the body increases the formation of H^+ ions, which will help lower the pH back toward the normal range.

The respiratory system responds quickly to prevent drastic changes in pH, usually within 1 to 3 minutes. For an ongoing metabolic pH imbalance, however, the respiratory mechanism does not have the capacity to fully compensate. In such cases, respiratory compensation is only 50% to 75% effective.

RENAL MECHANISMS

As just discussed in Chapter 18, the kidneys help regulate the pH of extracellular fluid by excreting or conserving H^+ ions and by reabsorbing (or not) Na^+ ions and HCO_3^- ions. One mechanism was depicted in Fig. 18–5, and another, involving the phosphate buffer system is shown in Fig. 19–4.

The kidneys have the greatest capacity to buffer an ongoing pH change. Although the renal mechanisms do not become fully functional for several hours to days, once they do they continue to be effective far longer than respiratory mechanisms. Let us use as an example a patient with untreated diabetes mellitus who is in ketoacidosis, a metabolic acidosis. As acidic ketones accumulate in the blood, the capacity of the ECF buffer systems is quickly exhausted. Breathing rate then increases, and more CO_2 is exhaled to decrease H^+ ion formation and raise the pH of ECF. There is, however, a limit to how much the respiratory rate can increase, but the renal buffering mechanisms will then become effective. At this time it is the kidneys that are keeping the patient alive by preventing

Table 19–3 pH CHANGES

Change	Possible Causes	Compensation
Metabolic acidosis	• Kidney disease, ketosis, diarrhea, or vomiting	• Increased respirations to exhale CO_2
Metabolic alkalosis	• Overingestion of bicarbonate medications, gastric suctioning	• Decreased respirations to retain CO_2
Respiratory acidosis	• Decreased rate or efficiency of respiration: emphysema, asthma, pneumonia, paralysis of respiratory muscles	• Kidneys excrete H^+ ions and reabsorb Na^+ ions and HCO_3^- ions
Respiratory alkalosis	• Increased rate of respiration: anxiety, high altitude	• Kidneys retain H^+ ions and excrete Na^+ ions and HCO_3^- ions

acidosis from reaching a fatal level. Even the kidneys have limits, however, and the cause of the acidosis must be corrected to prevent death.

EFFECTS OF pH CHANGES

A state of **acidosis** is most detrimental to the central nervous system, causing depression of impulse transmission at synapses. A person in acidosis becomes confused and disoriented, then lapses into a coma.

Alkalosis has the opposite effect and affects both the central and peripheral nervous system. Increased synaptic transmission, even without stimuli, is first apparent in irritability and muscle twitches. Progressive alkalosis is characterized by severe muscle spasms and convulsions.

The types of pH changes are summarized in Table 19–3.

AGING AND FLUID AND pH REGULATION

Changes in fluid balance or pH in elderly people are often the result of disease or damage to particular organs. A weak heart (congestive heart failure) that cannot pump efficiently allows blood to back up in circulation. In turn, this may cause edema, an abnormal collection of fluid. Edema may be systemic (often apparent in the lower legs) if the right ventricle is weak, or pulmonary if the left ventricle is failing.

Deficiencies of minerals in elderly people may be the result of poor nutrition or a side effect of some medications, especially those for hypertension that increase urinary output. Disturbances in pH may be caused by chronic pulmonary disease, diabetes, or kidney disease.

STUDY OUTLINE

Fluid–Electrolyte Balance

1. Water makes up 60% to 75% of the total body weight.
2. Electrolytes are the ions found in body fluids; most are minerals.

Water Compartments (see Fig. 19–1)

1. Intracellular Fluid (ICF)—water within cells; about two-thirds of total body water.
2. Extracellular Fluid (ECF)—water outside cells; includes plasma, lymph, tissue fluid, and specialized fluids.
3. Water constantly moves from one compartment to another. Filtration: Plasma becomes tissue fluid. Osmosis: Tissue fluid becomes plasma, or lymph, or ICF.
4. Osmosis is regulated by the concentration of electrolytes in body fluids (osmolarity). Water will diffuse through membranes to areas of greater electrolyte concentration.

Water Intake (see Fig. 19–2)

1. Fluids, Food, Metabolic Water—see Table 19–1.

Water Output (see Fig. 19–2)

1. Urine, Sweat, Exhaled Air, Feces—see Table 19–1.
2. Any variation in output must be compensated for by a change in input.

Regulation of Water Intake and Output

1. Hypothalamus contains osmoreceptors that detect changes in osmolarity of body fluids.
2. Dehydration stimulates the sensation of thirst, and fluids are consumed to relieve it.
3. ADH released from the posterior pituitary increases the reabsorption of water by the kidneys.
4. Aldosterone secreted by the adrenal cortex increases the reabsorption of Na^+ ions by the kidneys; water is then reabsorbed by osmosis.
5. If there is too much water in the body, secretion of ADH decreases, and urinary output increases.
6. If blood volume increases, ANH promotes loss of Na^+ ions and water in urine.

Electrolytes

1. Chemicals that dissolve in water and dissociate into ions; most are inorganic.
2. Cations are positive ions such as Na^+ and K^+.
3. Anions are negative ions such as Cl^- and HCO_3^-.
4. By creating osmotic pressure, electrolytes regulate the osmosis of water between compartments.

Electrolytes in Body Fluids (see Fig. 19–3 and Table 19–2)

1. ICF—principal cation is K^+; principal anion is HPO_4^{-2}; protein anions are also abundant.

2. Plasma—principal cation is Na^+; principal anion is Cl^-; protein anions are significant.
3. Tissue Fluid—same as plasma except that protein anions are insignificant.

Intake, Output, and Regulation

1. Intake—electrolytes are part of food and beverages.
2. Output—urine, sweat, feces.
3. Hormones involved: aldosterone—Na^+ and K^+; ANH—Na^+; PTH and calcitonin—Ca^{+2} and HPO_4^{-2}.

Acid–Base Balance

1. Normal pH Ranges—blood: 7.35 to 7.45; ICF: 6.8 to 7.0; tissue fluid: similar to blood.
2. Normal pH of body fluids is maintained by buffer systems, respirations, and the kidneys.

Buffer Systems

1. Each consists of a weak acid and a weak base; react with strong acids or bases to change them to substances that do not greatly affect pH. React within a fraction of a second, but have the least capacity to prevent pH changes.
2. Bicarbonate Buffer System—see text for reactions; important in both blood and tissue fluid; base to acid ratio is 20:1.
3. Phosphate Buffer System—see Fig. 19–4 and text for reactions; important in ICF and in the kidneys.
4. Protein Buffer System—amino acids may act as either acids or bases. See text for reactions; important in ICF.

Respiratory Mechanisms

1. The respiratory system affects pH because it regulates the amount of CO_2 in body fluids.
2. May be the cause of a pH change or help compensate for a metabolic pH change—see Table 19–3.
3. Respiratory compensation is rapidly effective (within a few minutes), but limited in capacity if the pH imbalance is ongoing.

Renal Mechanisms

1. The kidneys have the greatest capacity to buffer pH changes, but they may take several hours to days to become effective (see Table 19–3).
2. Reactions: see Figs. 18–5 and 19–4.
3. Summary of reactions: in response to acidosis, the kidneys will excrete H^+ ions and retain Na^+ ions and HCO_3^- ions; in response to alkalosis, the kidneys will retain H^+ ions and excrete Na^+ ions and HCO_3^- ions.

Effects of pH Changes

1. Acidosis—depresses synaptic transmission in the CNS; result is confusion, coma, and death.
2. Alkalosis—increases synaptic transmission in the CNS and PNS; result is irritability, muscle spasms, and convulsions.

REVIEW QUESTIONS

1. Name the major water compartments and the name for water in each of them. Name three specialized body fluids and state the location of each. (p. 424)
2. Explain how water moves between compartments; name the processes. (pp. 424–425)
3. Describe the three sources of water for the body and the relative amounts of each. (p. 425)
4. Describe the pathways of water output. Which is the most important? What kinds of variations are possible in water output. (p. 425)
5. Name the hormones that affect fluid volume, and state the function of each. (p. 427)
6. Define electrolyte, cation, anion, osmosis, and osmolarity. (pp. 425, 427)
7. Name the major electrolytes in plasma, tissue fluid, and intracellular fluid, and state their functions. (pp. 427–429)
8. Explain how the bicarbonate buffer system will react to buffer a strong acid. (p. 430)

9. Explain how the phosphate buffer system will react to buffer a strong acid. (p. 430)
10. Explain why an amino acid may act as either an acid or a base. (pp. 430–431)
11. Describe the respiratory compensation for metabolic acidosis and for metabolic alkalosis. (pp. 431–432)
12. If the body fluids are becoming too acidic, what ions will the kidneys excrete? What ions will the kidneys return to the blood? (p. 432)
13. Which of the pH regulatory mechanisms works most rapidly? Most slowly? Which of these mechanisms has the greatest capacity to buffer an ongoing pH change? Which mechanism has the least capacity? (pp. 431–432)
14. Describe the effects of acidosis and alkalosis. (p. 433)

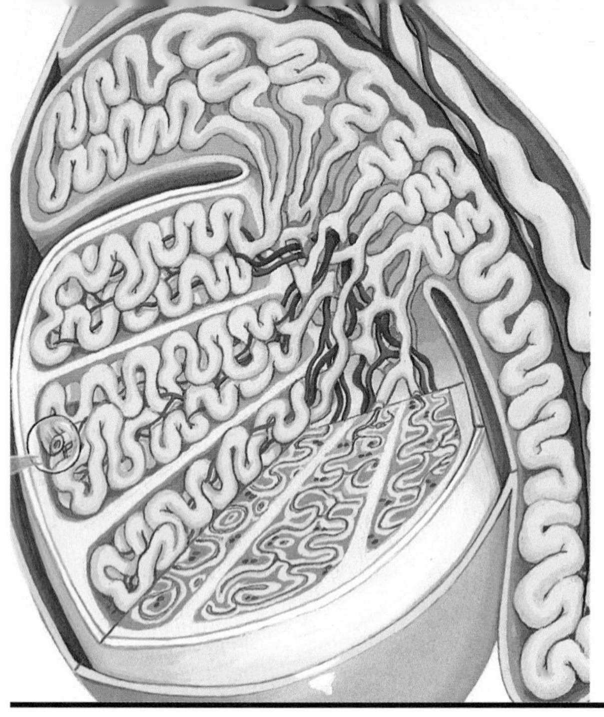

CHAPTER 20

Chapter Outline

Student Objectives

- Describe the process of meiosis. Define diploid and haploid.
- Describe the differences between spermatogenesis and oogenesis.
- Name the hormones necessary for the formation of gametes, and state the function of each.
- Describe the location and functions of the testes.
- Explain the functions of the epididymis, ductus deferens, ejaculatory duct, and urethra.
- Explain the functions of the seminal vesicles, prostate gland, and bulbourethral glands.
- Describe the composition of semen, and explain why its pH must be alkaline.
- Name the parts of a sperm cell, and state the function of each.
- Describe the functions of the ovaries, fallopian tubes, uterus, and vagina.
- Describe the structure and function of the myometrium and endometrium.

The Reproductive Systems

Student Objectives (Continued)

- Describe the structure of the mammary glands and the functions of the hormones involved in lactation.
- Describe the menstrual cycle in terms of the hormones involved and the changes in the ovaries and endometrium.

New Terminology

Cervix (**SIR**-viks)
Ductus deferens (**DUK**-tus **DEF**-er-enz)
Endometrium (EN-doh-**MEE**-tree-uhm)
Fallopian tube (fuh-**LOH**-pee-an TOOB)
Graffian follicle (**GRAF**-ee-uhn **FAH**-li-kuhl)
Inguinal canal (**IN**-gwi-nuhl ka-**NAL**)
Menopause (**MEN**-ah-paws)
Menstrual cycle (**MEN**-stroo-uhl **SIGH**-kuhl)
Myometrium (MY-oh-**MEE**-tree-uhm)
Oogenesis (OH-oh-**JEN**-e-sis)
Prostate gland (**PRAHS**-tayt)
Seminiferous tubules (sem-i-**NIFF**-er-us)
Spermatogenesis (SPER-ma-toh-**JEN**-e-sis)
Vulva (**VUHL**-vah)
Zygote (**ZYE**-goht)

Related Clinical Terminology

Amenorrhea (ay-MEN-uh-**REE**-ah)
Down syndrome (DOWN **SIN**-drohm)
Ectopic pregnancy (ek-**TOP**-ik **PREG**-nun-see)
In vitro fertilization (IN **VEE**-troh FER-ti-li-**ZAY**-shun)
Mammography (mah-**MOG**-rah-fee)
Prostatic hypertrophy (prahs-**TAT**-ik high-**PER**-truh-fee)
Trisomy (**TRY**-suh-mee)
Tubal ligation (**TOO**-buhl lye-**GAY**-shun)
Vasectomy (va-**SEK**-tuh-me)

*Terms that appear in **bold type** in the chapter text are defined in the glossary, which begins on page 528.*

The purpose of the male and female **reproductive systems** is to continue the human species by the production of offspring. How dry and impersonal that sounds, until we remember that each of us is a continuation of our species and that many of us in turn will have our own children. Although some other animals care for their offspring in organized families or societies, the human species is unique, because of cultural influences, in the attention we give to reproduction and to family life.

Yet like other animals, the actual production and growth of offspring is a matter of our anatomy and physiology. The male and female reproductive systems produce **gametes**, that is, sperm and egg cells, and ensure the union of gametes in fertilization following sexual intercourse. In women, the uterus provides the site for the developing embryo/fetus until it is sufficiently developed to survive outside the womb.

This chapter describes the organs of reproduction and the role of each in the creation of new life or the functioning of the reproductive system as a whole. First, however, we will discuss the formation of gametes.

MEIOSIS

The cell division process of **meiosis** produces the gametes, sperm or egg cells. In meiosis, one cell with the diploid number of chromosomes (46 for people) divides twice to form four cells, each with the haploid number of chromosomes. Haploid means half the usual diploid number, so for people the haploid number is 23. Although the process of meiosis is essentially the same in men and women, there are important differences.

SPERMATOGENESIS

Spermatogenesis is the process of meiosis as it takes place in the testes, the site of sperm production. Within each testis are **seminiferous tubules** that contain spermatogonia, or sperm-generating cells. These divide first by mitosis to produce primary spermatocytes (Fig. 20–1). As you may recall from Chapter 10, gamete formation is regulated by hormones. Follicle stimulating hormone (FSH) from the anterior pituitary gland initiates sperm production, and testosterone, secreted by the testes when stimulated by luteinizing hormone (LH) from the anterior pituitary, promotes the maturation of sperm. Inhibin, also produced by the testes, decreases the secretion of FSH. As you can see in Fig. 20–1, for each primary spermatocyte that undergoes meiosis, four functional sperm cells are produced.

Sperm production begins at **puberty** (10 to 14 years of age), and millions of sperm are formed each day in the testes. Although sperm production diminishes with advancing age, there is usually no complete cessation, as there is of egg production in women at menopause.

OOGENESIS

Oogenesis is the process of meiosis for egg cell formation; it begins in the ovaries and is also regulated by hormones. FSH initiates the growth of **ovarian follicles**, each of which contains an oogonium, or egg-generating cell (Fig. 20–2). This hormone also stimulates the follicle cells to secrete estrogen, which promotes the maturation of the ovum. Notice that for each primary oocyte that undergoes meiosis, only one functional egg cell is produced. The other three cells produced are called polar bodies. They have no function, and will simply deteriorate. A mature ovarian follicle actually contains the secondary oocyte; the second meiotic division will take place if and when the egg is fertilized.

The production of ova begins at puberty (10 to 14 years of age) and continues until **menopause** (45 to 55 years of age), when the ovaries atrophy and no longer respond to pituitary hormones. During this 30- to 40-year span, egg production is cyclical, with a mature ovum being produced approximately every 28 days (the menstrual cycle is discussed later in this chapter). Actually, *several* follicles usually begin to develop during each cycle. However, the rupturing (ovulation) of the first follicle to mature stops the growth of the others.

The process of meiosis is like other human processes in that "mistakes" may sometimes occur. One of these, trisomy, is discussed in Box 20–1: Trisomy and Down Syndrome.

The haploid egg and sperm cells produced by meiosis each have 23 chromosomes. When fertilization occurs, the nuclei of the egg and sperm merge, and the fertilized egg **(zygote)** has 46 chromosomes, the

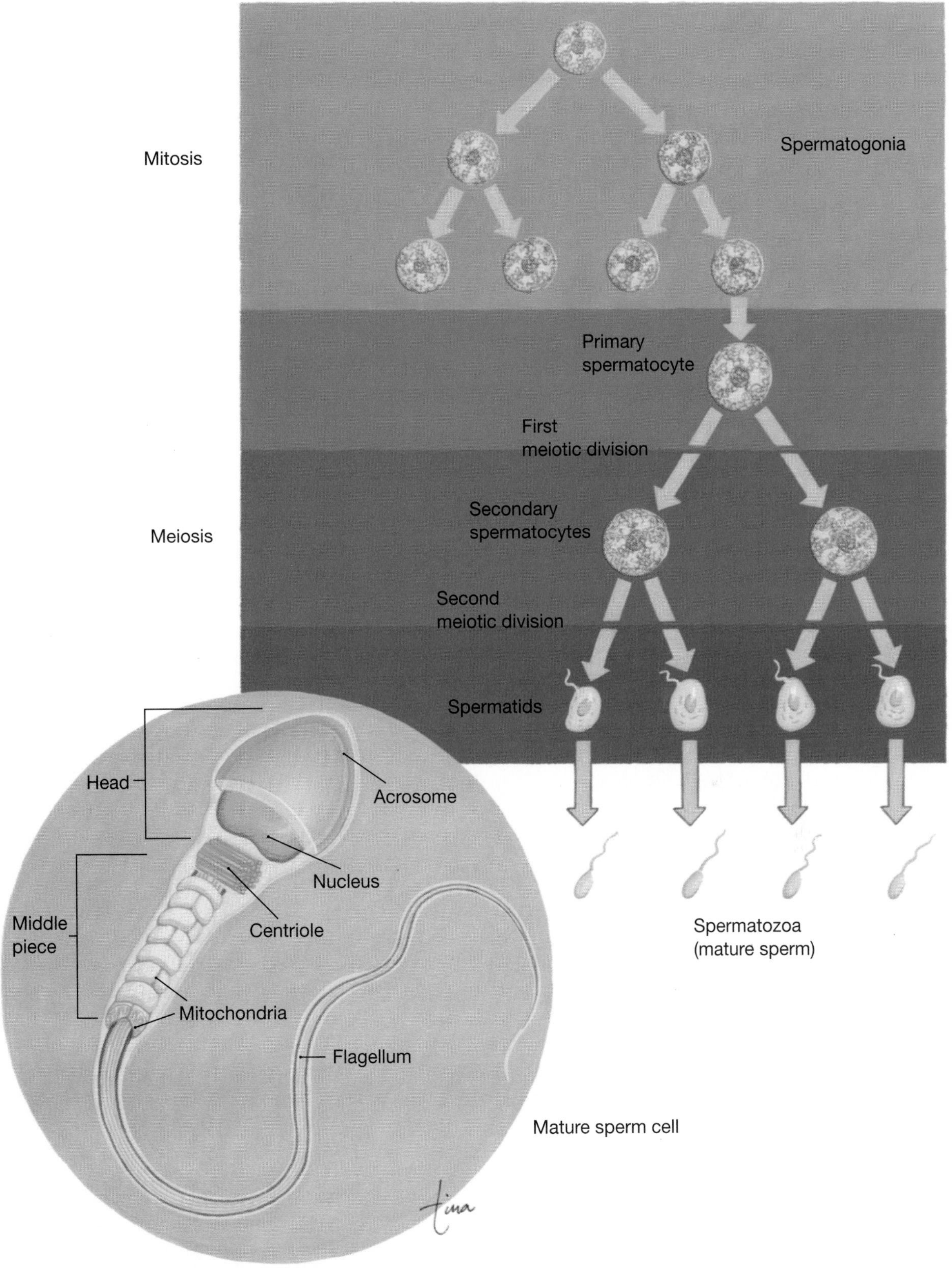

Figure 20–1 Spermatogenesis. The processes of mitosis and meiosis are shown. For each primary spermatocyte that undergoes meiosis, four functional sperm cells are formed. The structure of a mature sperm cell is also shown.

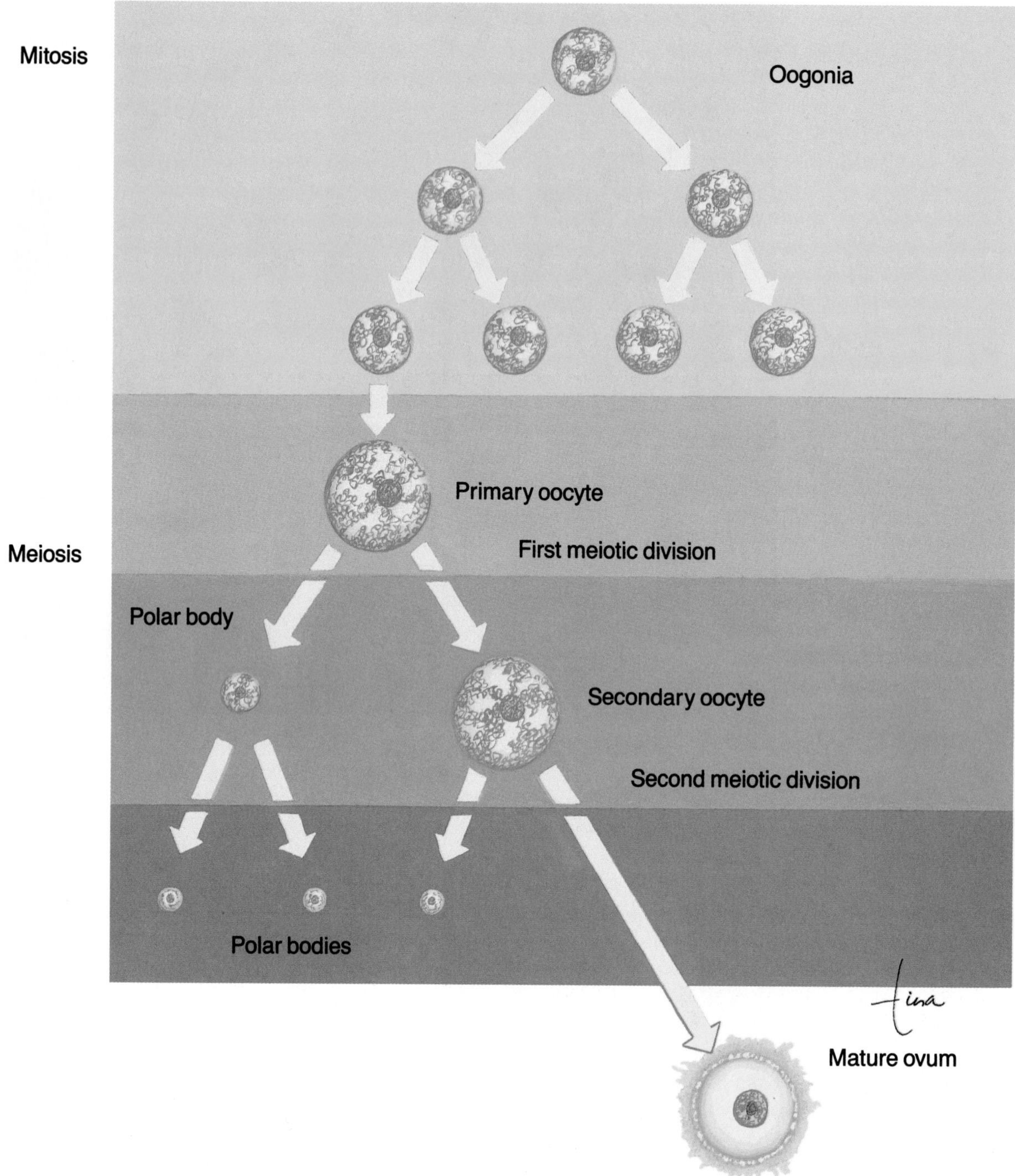

Figure 20–2 Oogenesis. The processes of mitosis and meiosis are shown. For each primary oocyte that undergoes meiosis, only one functional ovum is formed.

Box 20–1 TRISOMY AND DOWN SYNDROME

Trisomy means the presence of three (rather than the normal two) of a particular chromosome in the cells of an individual. This may occur because of non-disjunction (non-separation) of a chromosome pair during the second meiotic division, usually in an egg cell. The egg cell has two of a particular chromosome, and if fertilized by a sperm, will then contain three of that chromosome, and a total of 47 chromosomes.

Most trisomies are probably lethal, that is, the affected embryo will quickly die, even before the woman realizes she is pregnant. When an embryo–fetus survives and a child is born with a trisomy, there are always developmental defects present.

The severity of trisomies may be seen in two of the more rarely occurring ones: Trisomy 13 and Trisomy 18, each of which occurs about once for every 5000 live births. Both of these trisomies are characterized by severe mental and physical retardation, heart defects, deafness, and bone abnormalities. Affected infants usually die within their first year.

Down syndrome (Trisomy 21) is the most common trisomy, with a frequency of about one per 650 live births. Children with Down syndrome are mentally retarded, but there is a great range of mental ability in this group. Physical characteristics include a skin fold above each eye, short stature, poor muscle tone, and heart defects. Again, the degree of severity is highly variable.

Women over the age of 35 are believed to be at greater risk of having a child with Down syndrome. The reason may be that as egg cells age the process of meiosis is more likely to proceed incorrectly.

diploid number. Thus, meiosis maintains the diploid number of the human species by reducing the number of chromosomes by half in the formation of gametes.

MALE REPRODUCTIVE SYSTEM

The male reproductive system consists of the testes and a series of ducts and glands. Sperm are produced in the testes and are transported through the reproductive ducts: epididymis, ductus deferens, ejaculatory duct, and urethra (Fig. 20–3). The reproductive glands produce secretions that become part of semen, the fluid that is ejaculated from the urethra. These glands are the seminal vesicles, prostate gland, and bulbourethral glands.

TESTES

The **testes** are located in the **scrotum**, a sac of skin between the upper thighs. The temperature within the scrotum is about 96°F, slightly lower than body temperature, which is necessary for the production of viable sperm. In the male fetus, the testes develop near the kidneys, then descend into the scrotum just before birth. **Cryptorchidism** is the condition in which the testes fail to descend, and the result is sterility unless the testes are surgically placed in the scrotum.

Each testis is about 1.5 inches long by 1 inch wide (4 cm × 2.5 cm) and is divided internally into lobes (Fig. 20–4). Each lobe contains several **seminiferous tubules**, in which spermatogenesis takes place. Among the spermatogonia of the seminiferous tubules are **sustentacular (Sertoli) cells**, which produce the hormone **inhibin** when stimulated by testosterone. Between the loops of the seminiferous tubules are **interstitial cells**, which produce **testosterone** when stimulated by luteinizing hormone (LH) from the anterior pituitary gland. Besides its role in the maturation of sperm, testosterone is also responsible for the male secondary sex characteristics, which begin to develop at puberty (Table 20–1).

A sperm cell consists of several parts, which are shown in Fig. 20–1. The head contains the 23 chromosomes. On the tip of the head is the **acrosome**, which contains enzymes to digest the membrane of an egg cell. Within the middle piece are mitochondria that produce ATP. The **flagellum** provides motility,

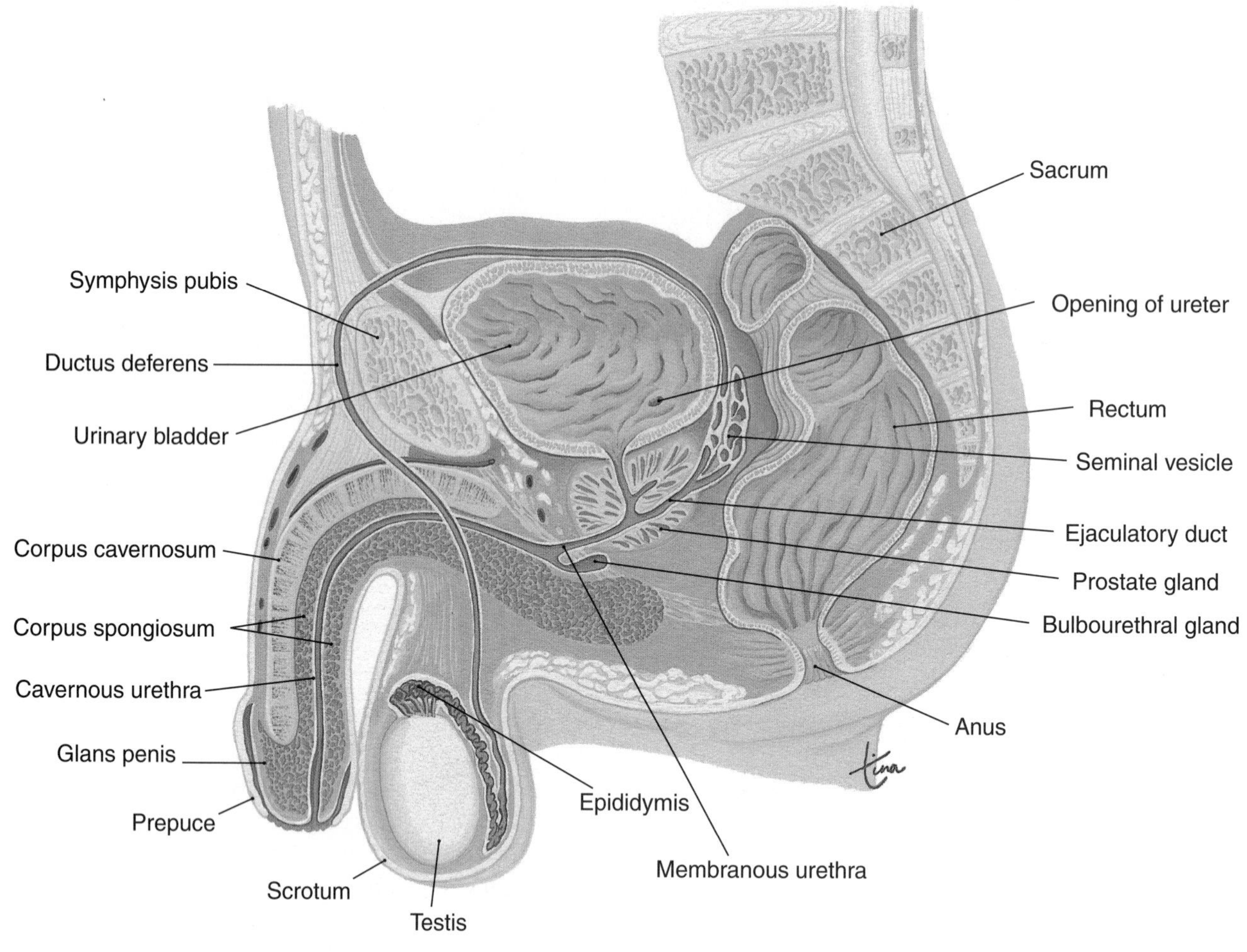

Figure 20–3 Male reproductive system shown in a midsagittal section through the pelvic cavity.

the capability of the sperm cell to move. It is the beating of the flagellum that requires energy from ATP.

Sperm from the seminiferous tubules enter a tubular network called the rete testis, then enter the epididymis, the first of the reproductive ducts.

EPIDIDYMIS

The **epididymis** (plural: epididymides) is a tube about 20 feet (6 m) long that is coiled on the posterior surface of each testis (see Fig. 20–4). Within the epididymis the sperm complete their maturation, and their flagella become functional. Smooth muscle in the wall of the epididymis propels the sperm into the ductus deferens.

DUCTUS DEFERENS

Also called the **vas deferens**, the **ductus deferens** extends from the epididymis in the scrotum on its own side into the abdominal cavity through the **inguinal canal**. This canal is an opening in the abdominal wall for the **spermatic cord**, a connective tissue sheath that contains the ductus deferens, testicular blood vessels, and nerves. Because the inguinal canal is an opening in a muscular wall, it is a natural "weak

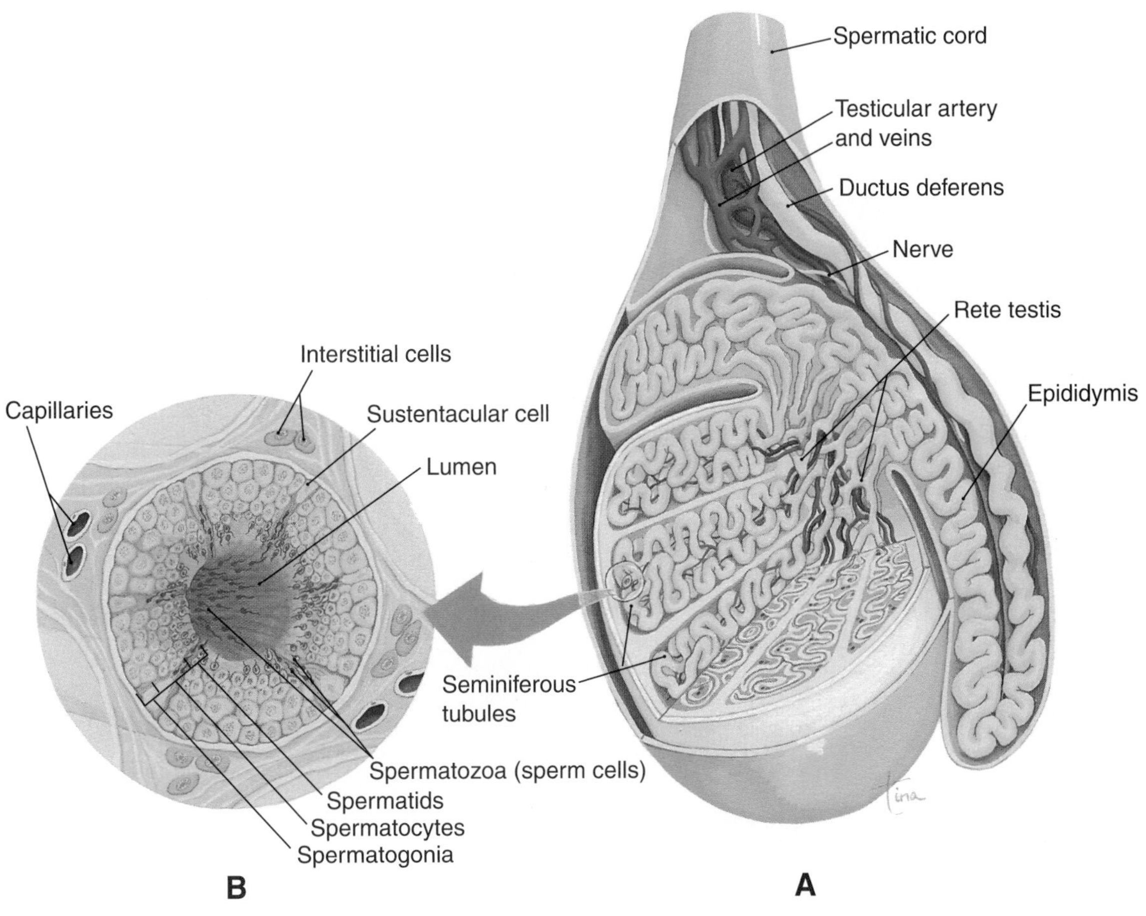

Figure 20–4 **(A)** Midsagittal section of portion of a testis; the epididymis is on the posterior side of the testis. **(B)** Cross-section through a seminiferous tubule showing development of sperm.

Table 20–1 HORMONES OF MALE REPRODUCTION

Hormone	Secreted By	Functions
FSH	Anterior pituitary	• Initiates production of sperm in the testes
LH (ICSH)	Anterior pituitary	• Stimulates secretion of testosterone by the testes
Testosterone*	Testes (interstitial cells)	• Promotes maturation of sperm • Initiates development of the secondary sex characteristics: —growth of the reproductive organs —growth of the larynx —growth of facial and body hair —increased protein synthesis, especially in skeletal muscles
Inhibin	Testes (sustentacular cells)	• Decreases secretion of FSH to maintain constant rate of spermatogenesis

*In both sexes, testosterone (from the adrenal cortex in women) contributes to sex drive and muscle-protein synthesis.

Box 20–2 CONTRACEPTION

There are several methods of contraception, or birth control; some are more effective than others.

Sterilization—Sterilization in men involves a relatively simple procedure called a **vasectomy**. The ductus (vas) deferens is accessible in the scrotum, in which a small incision is made on either side. The ductus is then sutured and cut. Although sperm are still produced in the testes, they cannot pass the break in the ductus, and they simply die and are reabsorbed.

Sterilization in women is usually accomplished by **tubal ligation**, the suturing and severing of the fallopian tubes. Usually this can be done by way of a small incision in the abdominal wall. Ova cannot pass the break in the tube, nor can sperm pass from the uterine side to fertilize an ovum.

When done properly, these forms of surgical sterilization are virtually 100% effective.

Oral Contraceptives ("the pill")—Birth control pills contain progesterone and estrogen in varying proportions. They prevent ovulation by inhibiting the secretion of FSH and LH from the anterior pituitary gland. When taken according to schedule, birth control pills are about 98% effective. Some women report side effects such as headaches, weight gain, and nausea. Women who use this method of contraception should not smoke, for smoking seems to be associated with abnormal clotting and a greater risk of heart attack or stroke.

Barrier Methods—These include the condom, diaphragm, and cervical cap, which prevent sperm from reaching the uterus and fallopian tubes. The use of a spermicide (sperm-killing chemical) increases the effectiveness of these methods. A condom is a latex or rubber sheath that covers the penis and collects and contains ejaculated semen. Leakage is possible, however, and the condom is considered 80% to 90% effective. This is the only contraceptive method that decreases the spread of sexually transmitted diseases.

The diaphragm and cervical cap are plastic structures that are inserted into the vagina to cover the cervix. They are about 80% effective. These methods should not be used, however, by women with vaginal infections or abnormal Pap smears or who have had toxic shock syndrome.

spot," and it is the most common site of hernia formation in men.

Once inside the abdominal cavity, the ductus deferens extends upward over the urinary bladder, then down the posterior side to join the ejaculatory duct on its own side (see Fig. 20–3). The smooth muscle layer of the ductus deferens contracts in waves of peristalsis as part of ejaculation (see Box 20–2: Contraception).

EJACULATORY DUCTS

Each of the two **ejaculatory ducts** receives sperm from the ductus deferens and the secretion of the seminal vesicle on its own side. Both ejaculatory ducts empty into the single urethra (see Fig. 20–3).

SEMINAL VESICLES

The paired **seminal vesicles** are posterior to the urinary bladder (see Fig. 20–3). Their secretion contains fructose to provide an energy source for sperm and is alkaline to enhance sperm motility. The duct of each seminal vesicle joins the ductus deferens on that side to form the ejaculatory duct.

PROSTATE GLAND

A muscular gland just below the urinary bladder, the **prostate gland** surrounds the first inch of the urethra as it emerges from the bladder (see Fig. 20–3). The glandular tissue of the prostate secretes an alkaline fluid that helps maintain sperm motility. The smooth muscle of the prostate gland contracts during **ejacu-**

Box 20–3 PROSTATIC HYPERTROPHY

Prostatic hypertrophy is enlargement of the prostate gland. Benign prostatic hypertrophy is a common occurrence in men over the age of 60 years. The enlarged prostate compresses the urethra within it and may make urination difficult or result in urinary retention. A prostatectomy is a surgical procedure to remove part or all of the prostate. A possible consequence is that ejaculation may be impaired. Many experimental surgical procedures are now undergoing clinical trials to determine if they are more likely to preserve sexual function. Other research involves the use of new medications to shrink enlarged prostate tissue.

Cancer of the prostate is the second most common cancer among men (lung cancer is first). Most cases occur in men over the age of 50 years. Treatment may include surgery to remove the prostate, radiation therapy, or hormone therapy to reduce the patient's level of testosterone.

lation to contribute to the expulsion of semen from the urethra (see Box 20–3: Prostatic Hypertrophy).

BULBOURETHRAL GLANDS

Also called Cowper's glands, the **bulbourethral glands** are located below the prostate gland and empty into the urethra. Their alkaline secretion coats the interior of the urethra just before ejaculation, which will neutralize any acidic urine that might be present.

You have probably noticed that all the secretions of the male reproductive glands are alkaline. This is important because the cavity of the female vagina has an acidic pH due to the normal flora, the natural bacterial population of the vagina. The alkalinity of seminal fluid helps neutralize the acidic vaginal pH and permits sperm motility in what might otherwise be an unfavorable environment.

URETHRA—PENIS

The **urethra** is the last of the ducts through which semen travels, and its longest portion is enclosed within the penis. The **penis** is an external genital organ; its distal end is called the glans penis and is covered with a fold of skin called the prepuce or foreskin. **Circumcision** is the surgical removal of the foreskin. This is a common procedure performed on male infants, although there is considerable medical debate as to whether circumcision truly has a useful purpose.

Within the penis are three masses of cavernous (erectile) tissue (see Fig. 20–3). Each consists of a framework of smooth muscle and connective tissue that contains blood sinuses, which are large, irregular vascular channels.

When blood flow through these sinuses is minimal, the penis is flaccid. During sexual stimulation, the arteries to the penis dilate, the sinuses fill with blood, and the penis becomes erect and firm. The dilation of penile arteries and the resulting erection are brought about by parasympathetic impulses. The erect penis is capable of penetrating the female vagina to deposit sperm. The culmination of sexual stimulation is ejaculation, a sympathetic response which is brought about by peristalsis of all the reproductive ducts, and contraction of the prostate gland and the muscles of the pelvic floor.

SEMEN

Semen consists of sperm and the secretions of the seminal vesicles, prostate gland, and bulbourethral glands; its average pH is about 7.4. During ejaculation, approximately 2 to 4 ml of semen is expelled. Each milliliter of semen contains about 100 million sperm cells.

FEMALE REPRODUCTIVE SYSTEM

The female reproductive system consists of the paired ovaries and fallopian tubes, the single uterus and va-

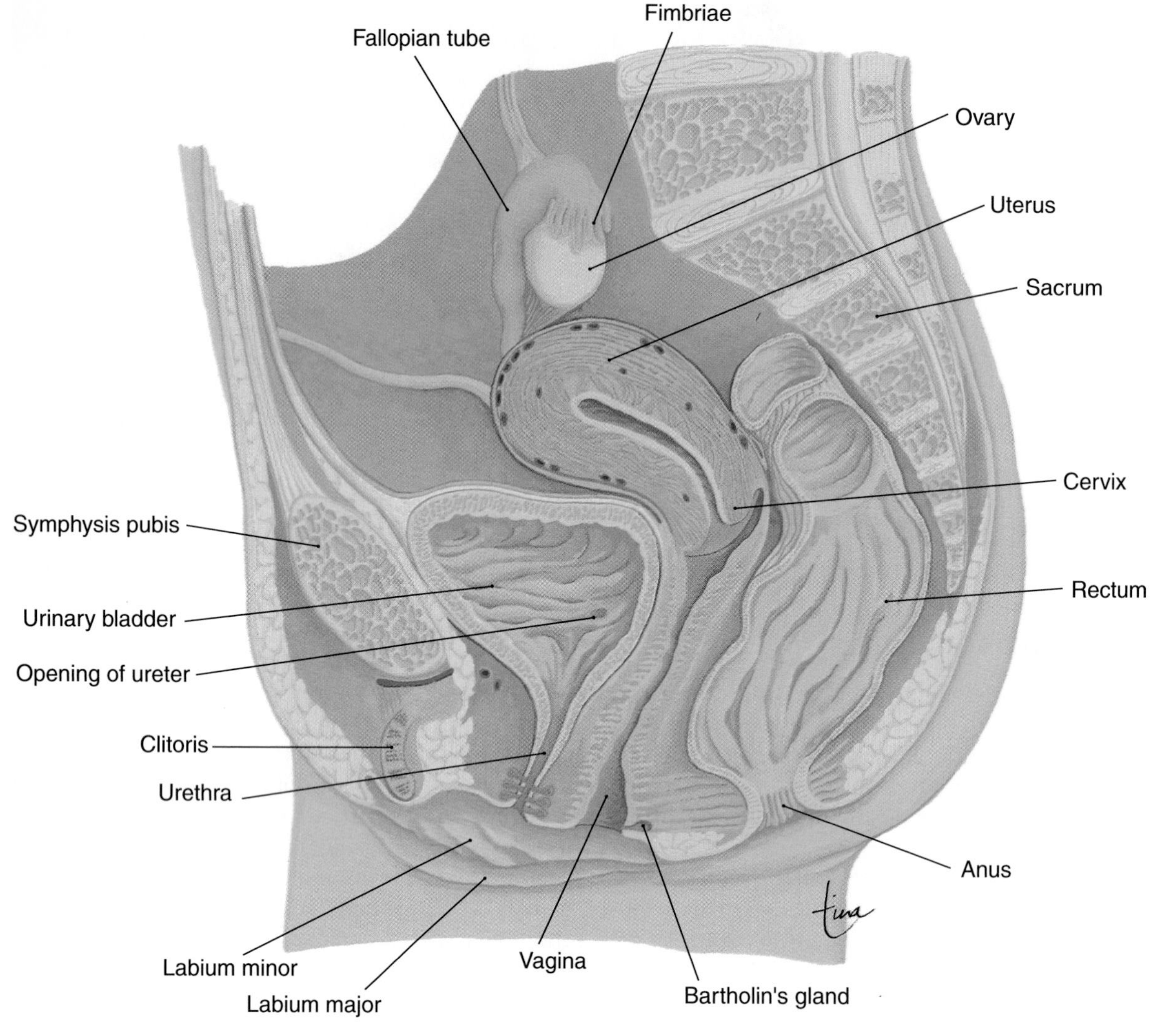

Figure 20–5 Female reproductive system shown in a midsagittal section through the pelvic cavity.

gina, and the external genital structures (Fig. 20–5). Egg cells (ova) are produced in the ovaries and travel through the fallopian tubes to the uterus. The uterus is the site for the growth of the embryo-fetus.

OVARIES

The **ovaries** are a pair of oval structures about 1.5 inches (4 cm) long on either side of the uterus in the pelvic cavity (Fig. 20–6). The ovarian ligament extends from the medial side of an ovary to the uterine wall, and the broad ligament is a fold of the peritoneum that covers the ovaries. These ligaments help keep the ovaries in place.

Within an ovary are several hundred thousand **primary follicles**, which are present at birth. During a woman's childbearing years, only 300 to 400 of these follicles will produce mature ova. As with sperm production in men, the supply of potential gametes far exceeds what is actually needed, but this helps ensure the continuation of the human species.

Each primary ovarian follicle contains an oocyte, a

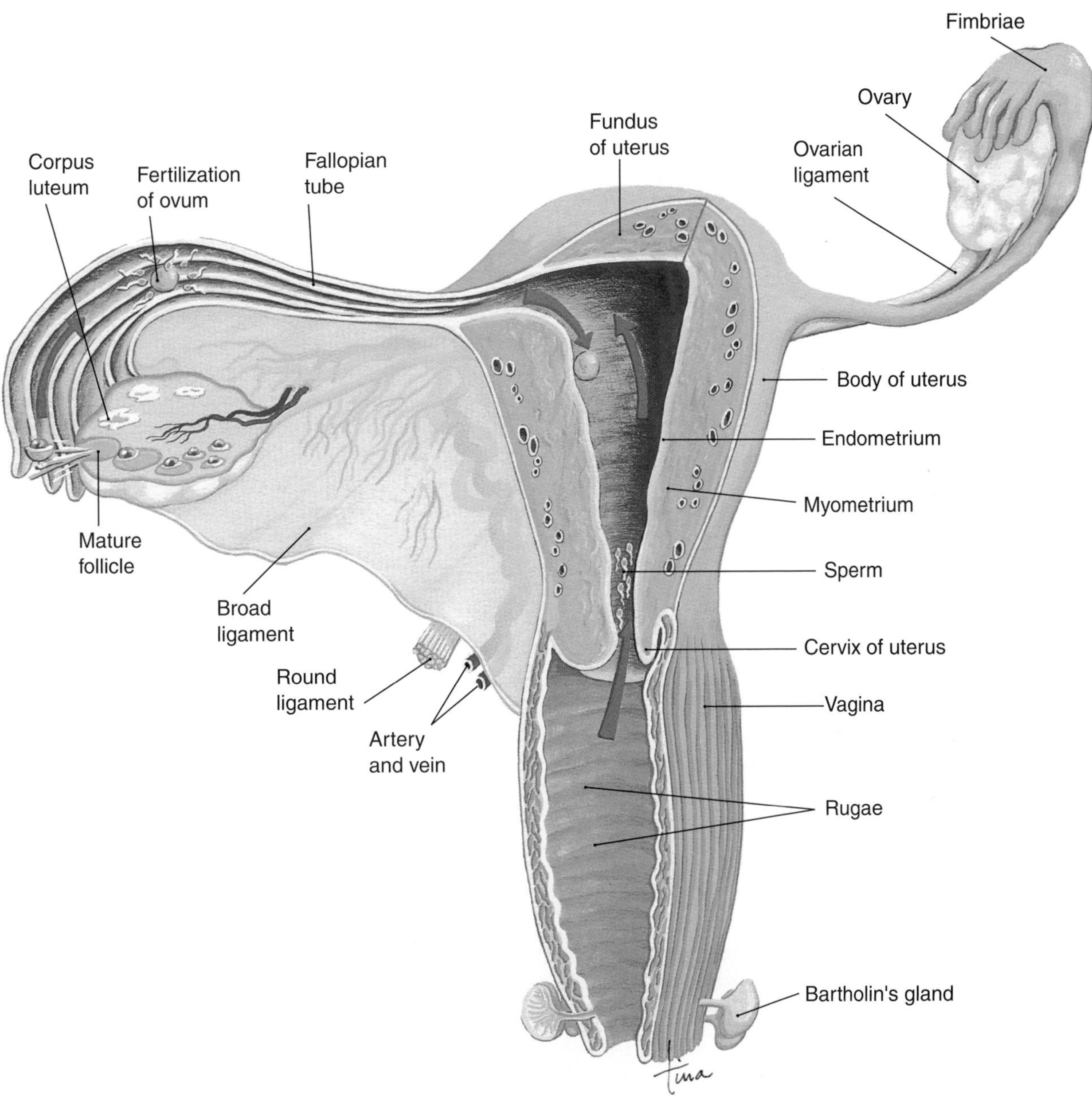

Figure 20–6 Female reproductive system shown in anterior view. The left ovary has been sectioned to show the developing follicles. The left fallopian tube has been sectioned to show fertilization. The uterus and vagina have been sectioned to show internal structures. Arrows indicate the movement of the ovum toward the uterus and the movement of sperm from the vagina toward the fallopian tube.

potential ovum or egg cell. Surrounding the oocyte are the follicle cells, which secrete estrogen. Maturation of a follicle, requiring FSH and estrogen, was described previously in the section on oogenesis. A mature follicle may also be called a **graafian follicle**, and the hormone LH from the anterior pituitary gland causes ovulation, that is, rupture of the mature follicle with release of the ovum. At this time, other developing follicles begin to deteriorate; these are called **atretic follicles** and have no further purpose. Under the influence of LH, the ruptured follicle becomes the **corpus luteum** and begins to secrete progesterone as well as estrogen.

FALLOPIAN TUBES

There are two fallopian tubes (also called uterine tubes or oviducts); each is about 4 inches (10 cm) long. The lateral end of a fallopian tube encloses an ovary, and the medial end opens into the uterus. The end of the tube that encloses the ovary has **fimbriae**, fringe-like projections that create currents in the fluid surrounding the ovary to pull the ovum into the fallopian tube.

Because the ovum has no means of self-locomotion (as do sperm) the structure of the fallopian tube ensures that the ovum will be kept moving toward the uterus. The smooth muscle layer of the tube contracts in peristaltic waves that help propel the ovum (or zygote, as you will see in a moment). The mucosa is extensively folded and is made of ciliated epithelial tissue. The sweeping action of the cilia also moves the ovum toward the uterus.

Fertilization usually takes place in the fallopian tube. If not fertilized, an ovum dies within 24 to 48 hours and disintegrates, either in the tube or the uterus. If fertilized, the ovum becomes a zygote and is swept into the uterus; this takes about 4 to 5 days (see Box 20–4: In Vitro Fertilization).

Sometimes the zygote will not reach the uterus but will still continue to develop. This is called an **ectopic pregnancy**; "ectopic" means in an abnormal site. The developing embryo may become implanted in the fallopian tube, the ovary itself, or even elsewhere in the abdominal cavity. An ectopic pregnancy usually does not progress very long, because these other sites are not specialized to provide a placenta or to expand to accommodate the growth of a fetus, as the uterus is. The spontaneous termination of an ectopic pregnancy is usually the result of bleeding in the mother, and surgery may be necesssary to prevent maternal death from circulatory shock. Occasionally an ectopic pregnancy does go to full term and produces a healthy baby; such an event is a credit to the adaptability of the human body and to the advances of medical science.

UTERUS

The **uterus** is shaped like an upside-down pear, about 3 inches long by 2 inches wide (7.5 × 5 cm), superior to the urinary bladder and between the two ovaries in the pelvic cavity (see Fig. 20–5). The broad ligament also covers the uterus (see Fig. 20–6). During pregnancy the uterus increases greatly in size, contains the placenta to nourish the embryo–fetus, and expels the baby at the end of gestation.

The parts and layers of the uterus are shown in Fig.

Box 20–4 IN VITRO FERTILIZATION

In vitro fertilization is fertilization outside the body, usually in a glass dish. A woman who wishes to conceive by this method is given FSH to stimulate the simultaneous development of several ovarian follicles. LH may then be given to stimulate simultaneous ovulation. The ova are removed by way of a small incision in the abdominal wall and are placed in a solution containing the sperm of the woman's partner (or an anonymous donor). After fertilization and the first mitotic divisions of cleavage, the very early embryo is placed in the woman's uterus.

It is also possible to mix the removed ova with sperm and return them almost immediately to the woman's fallopian tube. Development then proceeds as if the ova had been fertilized naturally.

Since the birth of the first "test tube baby" in 1978, many thousands of babies have been born following in vitro fertilization. The techniques are not always successful, and repeated attempts can be very expensive.

20-6. The **fundus** is the upper portion above the entry of the fallopian tubes, and the **body** is the large central portion. The narrow, lower end of the uterus is the **cervix**, which opens into the vagina.

The outermost layer of the uterus, the serosa or epimetrium, is a fold of the peritoneum. The **myometrium** is the smooth muscle layer; during pregnancy these cells increase in size to accommodate the growing fetus and contract for labor and delivery at the end of pregnancy.

The lining of the uterus is the **endometrium**, which itself consists of two layers. The **basilar layer**, adjacent to the myometrium, is vascular but very thin and is a permanent layer. The **functional layer** is regenerated and lost during each menstrual cycle. Under the influence of estrogen and progesterone from the ovaries, the growth of blood vessels thickens the functional layer in preparation for a possible embryo. If fertilization does not occur, the functional layer sloughs off in menstruation. During pregnancy, the endometrium forms the maternal portion of the placenta.

VAGINA

The **vagina** is a muscular tube about 4 inches (10 cm) long that extends from the cervix to the vaginal orifice in the **perineum** (pelvic floor). It is posterior to the urethra and anterior to the rectum (see Fig. 20-5). The vaginal opening is usually partially covered by a thin membrane called the **hymen**, which is ruptured by the first sexual intercourse or by the use of tampons during the menstrual period.

The functions of the vagina are to receive sperm from the penis during sexual intercourse, to serve as the exit for the menstrual blood flow, and to serve as the birth canal at the end of pregnancy.

The vaginal mucosa after puberty is stratified squa-

Box 20–5 SEXUALLY TRANSMITTED DISEASES (STDs)

Gonorrhea—caused by the bacterium *Neisseria gonorrhoeae.* Infected men have urethritis with painful and frequent urination and pus in the urine. Women are often asymptomatic, and the bacteria may spread from the cervix to other reproductive organs (pelvic inflammatory disease [PID]). The use first of silver nitrate, and more recently antibiotics, in the eyes of all newborns has virtually eliminated neonatal conjunctivitis acquired from an infected mother. Gonorrhea can be treated with antibiotics, but resistant strains of the bacteria complicate treatment. Despite this, the number of reported cases of gonorrhea has been decreasing in recent years.

Syphilis—caused by the bacterium *Treponema pallidum.* Although syphilis can be cured with penicillin, it is a disease that may be ignored by the person who has it because the symptoms may seem minor and often do not last long. If untreated, however, syphilis may cause severe or even fatal damage to the nervous system and heart. In the last few years the number of reported cases of syphilis had increased sharply but is again decreasing.

Genital Herpes—caused by the virus Herpes simplex (usually type 2). Painful lesions in the genital area are the primary symptom. Although the lesions heal within 5 to 9 days, recurrences are possible, perhaps triggered by physiological stresses such as illness. Although herpes is not curable at present, medications have proved useful in suppressing recurrences. It is estimated that 2 million new cases of genital herpes occur every year.

Neonatal herpes is infection of a newborn during passage through the birth canal. The infant's immune system is too immature to control the herpes virus, and this infection may be fatal or cause brain damage. A pregnant woman with a history of genital herpes may choose to have the baby delivered by cesarean section to avoid this possible outcome.

Chlamydial infection—caused by the very simple bacterium *Chlamydia trachomatis.* This is now the most prevalent STD in the United States, with estimates of 4 million new cases yearly. Infected men may have urethritis or epididymitis. Women often have no symptoms at first but may develop PID that increases the risk of ectopic pregnancy. Infants born to infected women may develop conjunctivitis or pneumonia. Chlamydial infection can be successfully treated with antibiotics such as erythromycin or azithromycin.

mous epithelium, which is relatively resistant to pathogens. The normal flora (bacteria) of the vagina creates an acidic pH that helps inhibit the growth of pathogens (see Box 20–5: Sexually Transmitted Diseases [STDs]).

EXTERNAL GENITALS

The female external genital structures may also be called the **vulva** (Fig. 20–7), and include the clitoris, labia majora and minora, and the Bartholin's glands (see Fig. 20–5).

The **clitoris** is a small mass of erectile tissue anterior to the urethral orifice. The only function of the clitoris is sensory, it responds to sexual stimulation, and its vascular sinuses become filled with blood.

The mons pubis is a pad of fat over the pubic symphysis, covered with skin and pubic hair. Extending posteriorly from the mons are the **labia majora** (lateral) and **labia minora** (medial), which are paired folds of skin. The area between the labia minora is called the vestibule and contains the openings of the urethra and vagina. The labia cover these openings and prevent drying of their mucous membranes.

Bartholin's glands, also called vestibular glands (see Figs. 20–5 and 20–6), are within the floor of the vestibule; their ducts open onto the mucosa at the vaginal orifice. The secretion of these glands keeps the mucosa moist and lubricates the vagina during sexual intercourse.

MAMMARY GLANDS

The **mammary glands** are structurally related to the skin but functionally related to the reproductive system because they produce milk for the nourishment of offspring. Enclosed within the breasts, the mammary glands are anterior to the pectoralis major muscles; their structure is shown in Fig. 20–8.

The glandular tissue is surrounded by adipose tissue. The **alveolar glands** produce milk after pregnancy; the milk enters lactiferous ducts which converge at the nipple. The skin around the nipple is a pigmented area called the areola.

The formation of milk is under hormonal control. During pregnancy, high levels of estrogen and progesterone prepare the glands for milk production. **Prolactin** from the anterior pituitary gland causes the

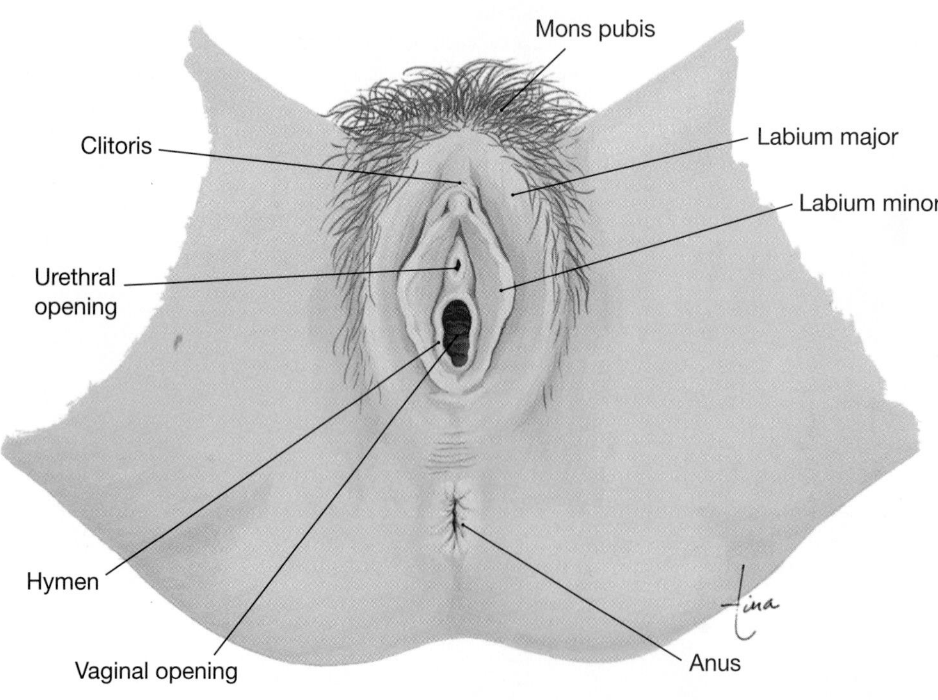

Figure 20–7 Female external genitals (vulva) shown in inferior view of the perineum.

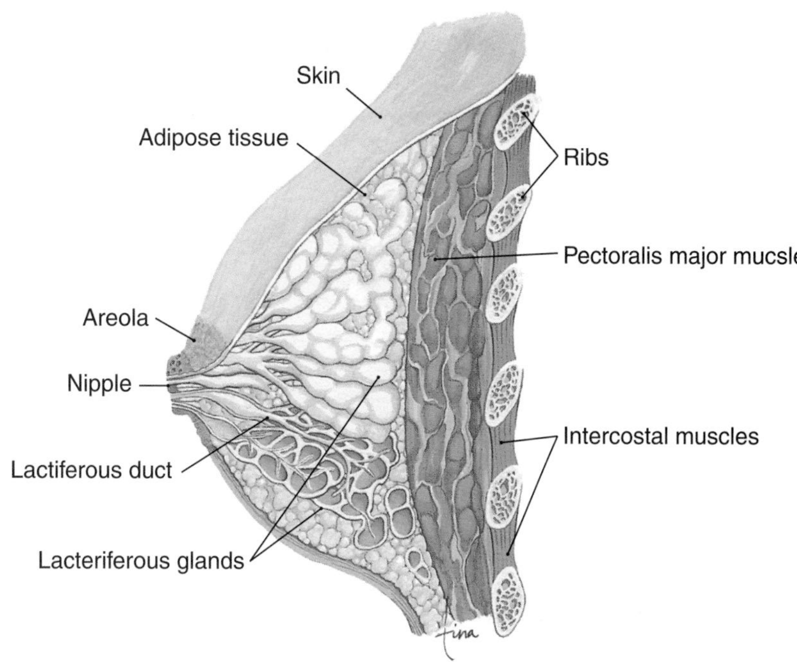

Figure 20–8 Mammary gland shown in midsagittal section.

actual synthesis of milk after pregnancy. The sucking of the infant on the nipple stimulates the hypothalamus to send nerve impulses to the posterior pituitary gland, which secretes **oxytocin** to cause the release of milk. The effects of these hormones on the mammary glands are summarized in Table 20–2 (see also Box 20–6: Mammography).

THE MENSTRUAL CYCLE

The **menstrual cycle** includes the activity of the hormones of the ovaries and anterior pituitary gland and the resultant changes in the ovaries and uterus. These are all incorporated into Fig. 20–9, which may look complicated at first, but refer to it as you read the following.

Notice first the four hormones involved: **FSH** and **LH** from the anterior pituitary gland, **estrogen** from the ovarian follicle, and **progesterone** from the corpus luteum. The fluctuations of these hormones are shown as they would occur in an average 28-day cycle. A cycle may be described in terms of three phases: menstrual phase, follicular phase, and luteal phase.

1. **Menstrual Phase**—The loss of the functional layer of the endometrium is called **menstruation** or the menses. Although this is actually the end of a menstrual cycle, the onset of menstruation is easily pinpointed and is, therefore, a use-

Table 20–2 HORMONE EFFECTS ON THE MAMMARY GLANDS

Hormone	Secreted By	Functions
Estrogen	• Ovary (follicle) • Placenta	• Promotes growth of duct system
Progesterone	• Ovary (corpus luteum) • Placenta	• Promotes growth of secretory cells
Prolactin	• Anterior pituitary	• Promotes production of milk after birth
Oxytocin	• Posterior pituitary (hypothalamus)	• Promotes release of milk

Box 20–6 MAMMOGRAPHY

Mammography is an x-ray technique that is used to evaluate breast tissue for abnormalities. By far the most common usage is to detect breast cancer, which is the most common malignancy in women. If detected early, breast cancer may be cured through a combination of surgery, radiation, and chemotherapy. Women should practice breast self-examination monthly, but mammography can detect lumps that are too small to be felt manually. Women in their 30s may have a mammogram done to serve as a comparison for mammograms later in life.

ful starting point. Menstruation may last 2 to 8 days, with an average of 3 to 6 days. At this time, secretion of FSH is increasing, and several ovarian follicles begin to develop.

2. **Follicular Phase**—FSH stimulates growth of ovarian follicles and secretion of estrogen by the follicle cells. The secretion of LH is also increasing but more slowly. FSH and estrogen promote the growth and maturation of the ovum, and estrogen stimulates the growth of blood vessels in the endometrium to regenerate the functional layer.

   This phase ends with ovulation, when a sharp increase in LH causes rupture of a mature ovarian follicle.
3. **Luteal Phase**—Under the influence of LH, the ruptured follicle becomes the corpus luteum and begins to secrete progesterone. Progesterone stimulates further growth of blood vessels in the functional layer of the endometrium and promotes the storage of nutrients such as glycogen.

   As progesterone secretion increases, LH secretion decreases, and if the ovum is not fertilized, the secretion of progesterone also begins to decrease. Without progesterone, the endometrium cannot be maintained and begins to slough off in menstruation. FSH secretion begins to increase (as estrogen and progesterone decrease), and the cycle begins again.

The 28-day cycle shown in Fig. 20–9 is average. Women may experience cycles of anywhere from 23 to 35 days. Women who engage in strenuous exercise over prolonged periods of time may experience **amenorrhea**, that is, cessation of menses. This seems to be related to reduction of body fat. Apparently the reproductive cycle ceases if a woman does not have sufficient reserves of energy for herself and a developing fetus. The exact mechanism by which this happens is not completely understood at present (see Box 17–6). Amenorrhea may also accompany states of physical or emotional stress, anorexia nervosa, or various endocrine disorders.

The functions of the hormones of female reproduction are summarized in Table 20–3.

AGING AND THE REPRODUCTIVE SYSTEMS

For women there is a definite end to reproductive capability; this is called the menopause and usually occurs between the ages of 45 and 55. Estrogen secretion decreases; ovulation and menstrual cycles become irregular and finally cease. The decrease in estrogen has other effects as well. Loss of bone matrix may lead to osteoporosis and fractures; an increase in blood cholesterol makes women more likely to develop coronary artery disease; drying of the vaginal mucosa increases susceptibility to vaginal infections. Estrogen replacement therapy may delay these consequences of menopause, but there are risks involved, and women should be fully informed of them before starting such therapy. The likelihood of breast cancer also increases with age, and women over age 50 should consider having a mammogram to serve as a baseline, then one at least every other year.

For most men, testosterone secretion continues throughout life, as does sperm production, though both diminish with advancing age. Perhaps the most common reproductive problem for older men is prostatic hypertrophy, enlargement of the prostate gland. As the urethra is compressed, urination becomes difficult, and residual urine in the bladder increases the chance of urinary tract infection. Prostate hypertrophy is usually benign, but cancer of the prostate is one of the more common cancers in elderly men.

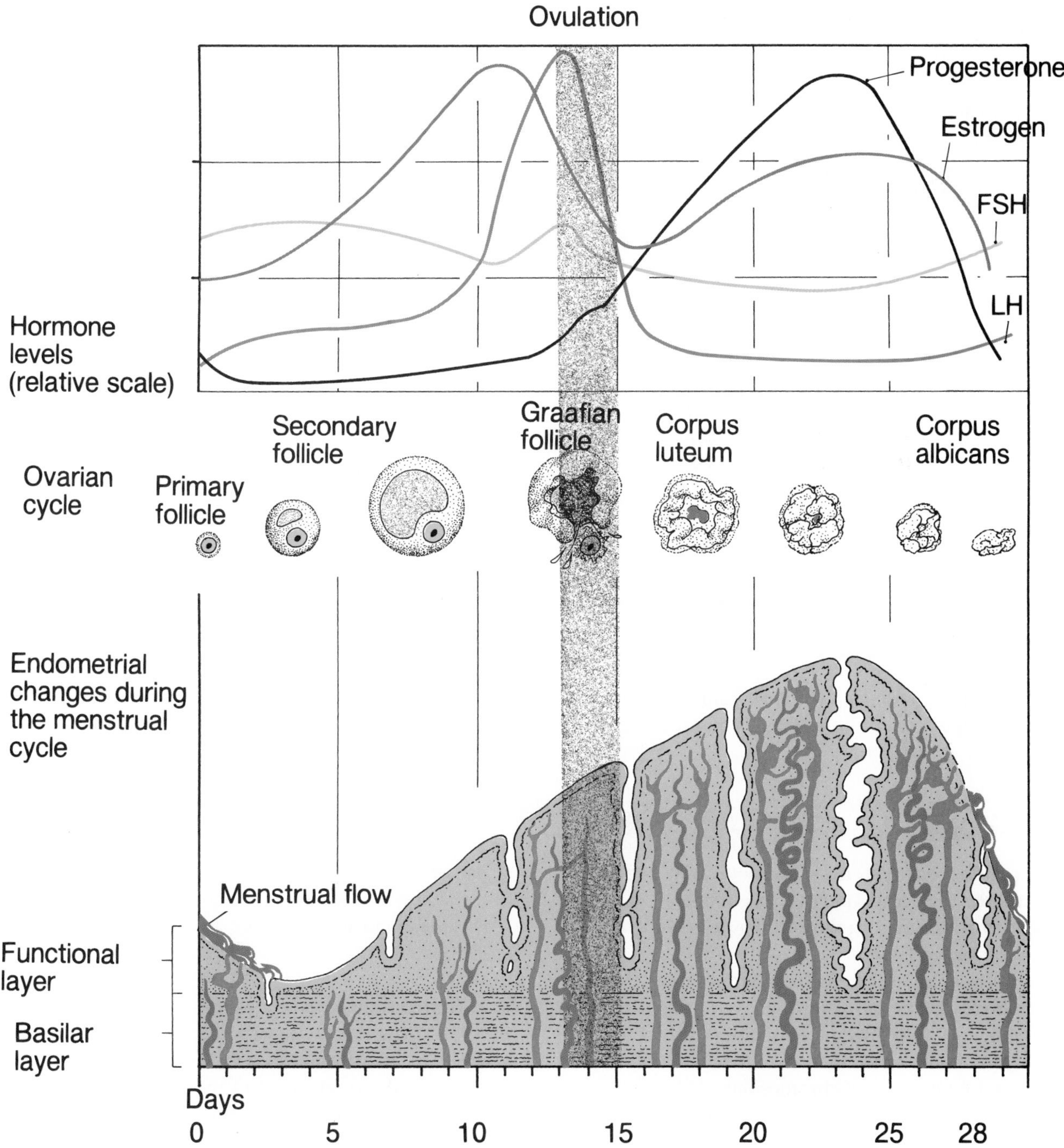

Figure 20–9 The menstrual cycle. The levels of the important hormones are shown relative to one another throughout the cycle. Changes in the ovarian follicle are depicted. The relative thickness of the endometrium is also shown.

Table 20–3 HORMONES OF FEMALE REPRODUCTION

Hormone	Secreted By	Functions
FSH	Anterior pituitary	• Initiates development of ovarian follicles • Stimulates secretion of estrogen by follicle cells
LH	Anterior pituitary	• Causes ovulation • Converts the ruptured ovarian follicle into the corpus luteum • Stimulates secretion of progesterone by the corpus luteum
Estrogen*	Ovary (follicle)	• Promotes maturation of ovarian follicles • Promotes growth of blood vessels in the endometrium • Initiates development of the secondary sex characteristics: —growth of the uterus and other reproductive organs —growth of the mammary ducts and fat deposition in the breasts —broadening of the pelvic bone —subcutaneous fat deposition in hips and thighs
Progesterone	Ovary (corpus luteum)	• Promotes further growth of blood vessels in the endometrium and storage of nutrients • Inhibits contractions of the myometrium
Inhibin	Ovary (corpus luteum)	• Inhibits secretion of FSH

*Estrogen has effects on organs such as bones and blood vessels in both men and women. Estrogen is produced in fat tissue in the breasts and hips. In men, testosterone is converted to estrogen in the brain. Estrogen appears to enhance brain functioning, including memory.

SUMMARY

The production of male or female gametes is a process that is regulated by hormones. When fertilization of an ovum by a sperm cell takes place, the zygote, or fertilized egg, has the potential to become a new human being. The development of the zygote to embryo-fetus to newborn infant is also dependent on hormones and is the subject of our next chapter.

STUDY OUTLINE

Reproductive Systems—purpose is to produce gametes (egg and sperm), to ensure fertilization, and in women to provide a site for the embryo–fetus

Meiosis—the cell division process that produces gametes

1. One cell with the diploid number of chromosomes (46) divides twice to form four cells, each with the haploid number of chromosomes (23).
2. Spermatogenesis takes place in the testes; a continuous process from puberty throughout life; each cell produces four functional sperm (see Fig. 20-1). FSH and testosterone are directly necessary (see Table 20-1).
3. Oogenesis takes place in the ovaries; the process is cyclical (every 28 days) from puberty until menopause; each cell produces one functional ovum and three non-functional polar bodies (see Fig. 20-2). FSH, LH, and estrogen, are necessary (see Table 20-3).

Male Reproductive System—consists of the testes and the ducts and glands that contribute to the formation of semen (see Fig. 20–3)

1. Testes (paired)—located in the scrotum between the upper thighs; temperature in the scrotum is 96°F to permit production of viable sperm. Sperm are produced in seminiferous tubules (see Fig. 20-4 and Table 20-1). A sperm cell consists of the head that contains 23 chromosomes, the middle piece that contains mitochondria, the flagellum for motility, and the acrosome on the tip of the head

to digest the membrane of the egg cell (see Fig. 20-1).

2. Epididymis (paired)—a long coiled tube on the posterior surface of each testis (see Fig. 20-4). Sperm complete their maturation here.
3. Ductus Deferens (paired)—extends from the epididymis into the abdominal cavity through the inguinal canal, over and down behind the urinary bladder to join the ejaculatory duct (see Fig. 20-3). Smooth muscle in the wall contracts in waves of peristalsis.
4. Ejaculatory Ducts (paired)—receive sperm from the ductus deferens and the secretions from the seminal vesicles (see Fig. 20-3); empty into the urethra.
5. Seminal Vesicles (paired)—posterior to urinary bladder; duct of each opens into ejaculatory duct (see Fig. 20-3). Secretion contains fructose to nourish sperm and is alkaline to enhance sperm motility.
6. Prostate Gland (single)—below the urinary bladder, encloses the first inch of the urethra (see Fig. 20-3); secretion is alkaline to maintain sperm motility; smooth muscle contributes to the force required for ejaculation.
7. Bulbourethral Glands (paired)—below the prostate gland; empty into the urethra (see Fig. 20-3); secretion is alkaline to line the urethra prior to ejaculation.
8. Urethra (single)—within the penis; carries semen to exterior (see Fig. 20-3). The penis contains three masses of erectile tissue that have blood sinuses. Sexual stimulation and parasympathetic impulses cause dilation of the penile arteries and an erection. Ejaculation of semen involves peristalsis of all the male ducts and contraction of the prostate gland and pelvic floor.
9. Semen—composed of sperm and the secretions of the seminal vesicles, prostate gland, and bulbourethral glands. The alkaline pH (7.4) neutralizes the acidic pH of the female vagina.

Female Reproductive System—consists of the ovaries, fallopian tubes, uterus, vagina, and external genitals

1. Ovaries (paired)—located on either side of the uterus (see Fig. 20-6). Egg cells are produced in ovarian follicles; each ovum contains 23 chromosomes. Ovulation of a graafian follicle is stimulated by LH (see Table 20-3).
2. Fallopian Tubes (paired)—each extends from an ovary to the uterus (see Fig. 20-6); fimbriae sweep the ovum into the tube; ciliated epithelial tissue and peristalsis of smooth muscle propel the ovum toward the uterus; fertilization usually takes place in the fallopian tube.
3. Uterus (single)—superior to the urinary bladder and between the two ovaries (see Fig. 20-5). Myometrium is the smooth muscle layer that contracts for delivery (see Fig. 20-6). Endometrium is the lining which may become the placenta; basilar layer is permanent; functional layer is lost in menstruation and regenerated. Parts: upper fundus, central body, and lower cervix.
4. Vagina (single)—extends from the cervix to the vaginal orifice (see Figs. 20-5 and 20-6). Receives sperm during intercourse; serves as exit for menstrual blood and as the birth canal during delivery. Normal flora provide an acidic pH that inhibits the growth of pathogens.
5. External Genitals (see Figs. 20-5 and 20-7)—also called the vulva. The clitoris is a small mass of erectile tissue that responds to sexual stimulation; labia majora and minora are paired folds of skin that enclose the vestibule and cover the urethral and vaginal openings; Bartholin's glands open into the vaginal orifice and secrete mucus.

Mammary Glands—anterior to the pectoralis major muscles, surrounded by adipose tissue (see Fig. 20–8)

1. Alveolar glands produce milk; lactiferous ducts converge at the nipple.
2. Hormonal regulation—see Table 20-2.

The Menstrual Cycle—average is 28 days; includes the hormones FSH, LH, estrogen, and progesterone, and changes in the ovaries and endometrium (see Fig. 20–9 and Table 20–3)

1. Menstrual Phase—loss of the endometrium.
2. Follicular Phase—several ovarian follicles develop; ovulation is the rupture of a mature follicle; blood vessels grow in the endometrium.
3. Luteal Phase—the ruptured follicle becomes the corpus luteum; the endometrium continues to develop.
4. If fertilization does not occur, decreased progesterone results in the loss of the endometrium in menstruation.

REVIEW QUESTIONS

1. Describe spermatogenesis and oogenesis in terms of site, number of functional cells produced by each cell that undergoes meiosis, and timing of the process. (p. 438)
2. Describe the functions of FSH, LH, inhibin, and testosterone in spermatogenesis. Describe the functions of FSH and estrogen in oogenesis. (p. 438)
3. Describe the locations of the testes and epididymides, and explain their functions. (pp. 441–442)
4. Name all the ducts, in order, that sperm travel through from the testes to the urethra. (pp. 442, 444–445)
5. Name the male reproductive glands, and state how each contributes to the formation of semen. (pp. 444–445)
6. Explain how the structure of cavernous tissue permits erection of the penis. Name the structures that bring about ejaculation. (pp. 442, 444–445)
7. State the function of each part of a sperm cell: head, middle piece, flagellum, and acrosome. (p. 441)
8. Describe the location of the ovaries, and name the hormones produced by the ovaries. (p. 446)
9. Explain how an ovum or zygote is kept moving through the fallopian tube. (p. 448)
10. Describe the function of myometrium, basilar layer of the endometrium, and functional layer of the endometrium. Name the hormones necessary for growth of the endometrium. (p. 449)
11. State the functions of the vagina, labia majora and minora, and Bartholin's glands. (pp. 449–450)
12. Name the parts of the mammary glands, and state the function of each. (p. 450)
13. Name the hormone that has each of these effects on the mammary glands: (p. 451)
    a. Causes release of milk
    b. Promotes growth of the ducts
    c. Promotes growth of the secretory cells
    d. Stimulates milk production
14. Name the phase of the menstrual cycle in which each takes place: (pp. 451–452)
    a. Rupture of a mature follicle
    b. Loss of the endometrium
    c. Final development of the endometrium
    d. The corpus luteum develops
    e. Several ovarian follicles begin to develop

CHAPTER 21

Human Development and Genetics

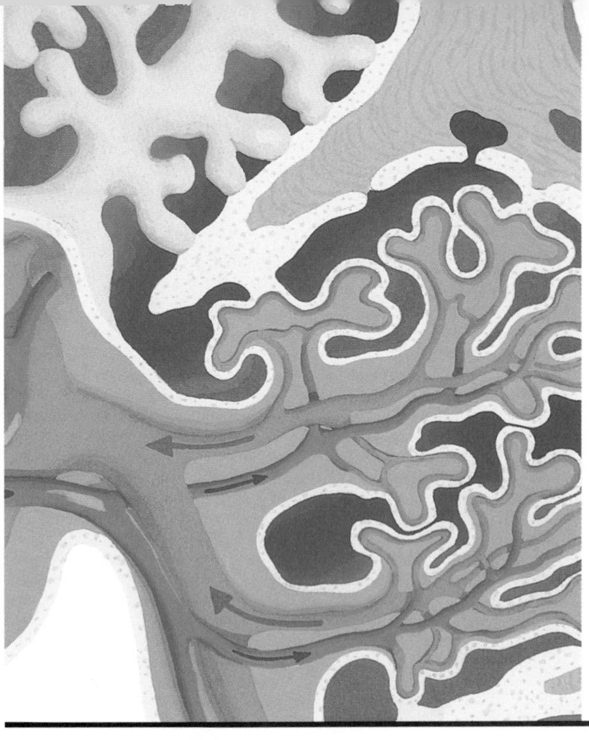

CHAPTER 21

Chapter Outline

Human Development
Fertilization
Implantation
Embryo and Embryonic Membranes
Placenta and Umbilical Cord
Placental hormones
Parturition and Labor
The Infant at Birth
Genetics
Chromosomes and Genes
Genotype and Phenotype
Inheritance: Dominant-Recessive
Inheritance: Multiple Alleles
Inheritance: Sex-Linked Traits

Box 21–1 TWINS
Box 21–2 FETAL DIAGNOSIS
Box 21–3 CONGENITAL FETAL INFECTIONS
Box 21–4 FETAL ALCOHOL SYNDROME
Box 21–5 PREMATURE BIRTH
Box 21–6 APGAR SCORE
Box 21–7 SOLUTION TO GENETICS QUESTION

Student Objectives

- Describe the process of fertilization and cleavage to the blastocyst stage.
- Explain when, where, and how implantation of the embryo occurs.
- Describe the functions of the embryonic membranes.
- Describe the structure and functions of the placenta and umbilical cord.
- Name and explain the functions of the placental hormones.
- State the length of the average gestation period, and describe the stages of labor.
- Describe the major changes in the infant at birth.
- Describe some important maternal changes during pregnancy.
- Explain homologous chromosomes, autosomes, sex chromosomes, and genes.
- Define alleles, genotype, phenotype, homozygous, and heterozygous.
- Explain the following patterns of inheritance: dominant-recessive, multiple alleles, and sex-linked traits.

Human Development and Genetics

New Terminology

Alleles (uh-**LEELZ**)
Amnion (**AM**-nee-on)
Amniotic fluid (**AM**-nee-AH-tik **FLOO**-id)
Autosomes (**AW**-toh-sohms)
Cleavage (**KLEE**-vije)
Embryo (**EM**-bree-oh)
Genotype (**JEE**-noh-type)
Gestation (jes-**TAY**-shun)
Heterozygous (HET-er-oh-**ZYE**-gus)
Homologous pair (hoh-**MAHL**-ah-gus PAYR)
Homozygous (HOH-moh-**ZYE**-gus)
Implantation (IM-plan-**TAY**-shun)
Labor (**LAY**-ber)
Parturition (PAR-tyoo-**RISH**-uhn)
Phenotype (**FEE**-noh-type)
Sex chromosome (SEKS **KROH**-muh-sohm)

Related Clinical Terminology

Amniocentesis (AM-nee-oh-sen-**TEE**-sis)
Apgar score (**APP**-gar SKOR)
Cesarean section (se-**SAR**-ee-an **SEK**-shun)
Chorionic villus sampling (KOR-ee-**ON**-ik **VILL**-us)
Congenital (kon-**JEN**-i-tuhl)
Fetal alcohol syndrome (**FEE**-tuhl **AL**-koh-hol)
Teratogen (te-**RAH**-toh-jen)

*Terms that appear in **bold type** in the chapter text are defined in the glossary, which begins on page 528.*

How often have we heard comments like "She has her mother's eyes" or "That nose is just like his father's"—as people cannot resist comparing a newborn to its parents. Although a child may not resemble either parent, there is a sound basis for such comparisons, because the genetic makeup and many of the traits of a child are the result of the chromosomes inherited from mother and father.

In this chapter, we will cover some of the fundamentals of genetics and inheritance. First, however, we will look at the development of a fertilized egg into a functioning human being.

HUMAN DEVELOPMENT

During the 40 weeks of gestation, the embryo-fetus is protected and nourished in the uterus of the mother. A human being begins life as one cell, a fertilized egg called a zygote, which develops into an individual human being consisting of billions of cells organized into the body systems with whose functions you are now quite familiar.

FERTILIZATION

Although millions of sperm are deposited in the vagina during sexual intercourse, only one will fertilize an ovum. As the sperm swim through the fluid of the uterus and fallopian tube, they undergo a final metabolic change, called **capacitation**. This change involves the **acrosome**, which becomes more fragile and begins to secrete its enzymes. When sperm and egg make contact, these enzymes will digest the layers of cells and membrane around an ovum.

Once a sperm nucleus enters the ovum, changes in the egg cell membrane block the entry of other sperm. The nucleus of the ovum completes the second mei-

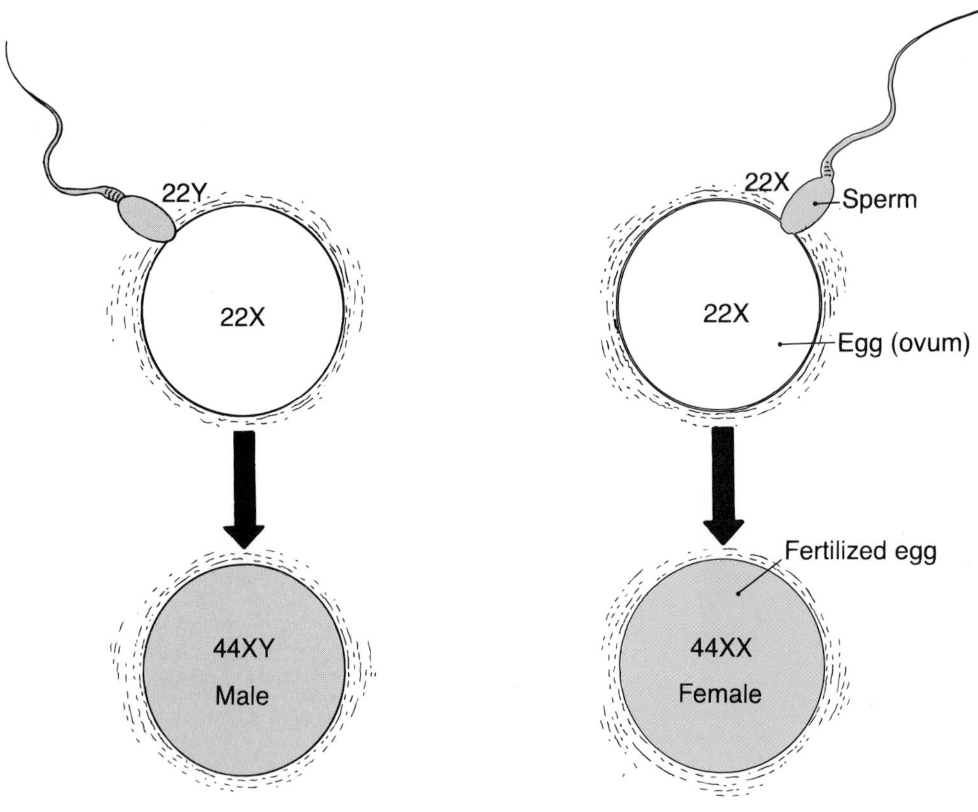

Figure 21–1 Inheritance of gender. Each ovum contains 22 autosomes and an X chromosome. Each sperm contains 22 autosomes and either an X chromosome or a Y chromosome.

otic division, and the nuclei of ovum and sperm fuse, restoring the diploid number of chromosomes in the zygote.

The human diploid number of 46 chromosomes is actually 23 pairs of chromosomes; 23 from the sperm and 23 from the egg. These 23 pairs consist of 22 pairs of **autosomes** (designated by the numerals 1 through 22) and one pair of **sex chromosomes**. Women have the sex chromosomes XX, and men have the sex chromosomes XY. Figure 21-1 shows the inheritance of gender. The Y chromosome has a gene that triggers the development of male gonads in the embryo. In the absence of the Y chromosome, the embryo will develop as a female.

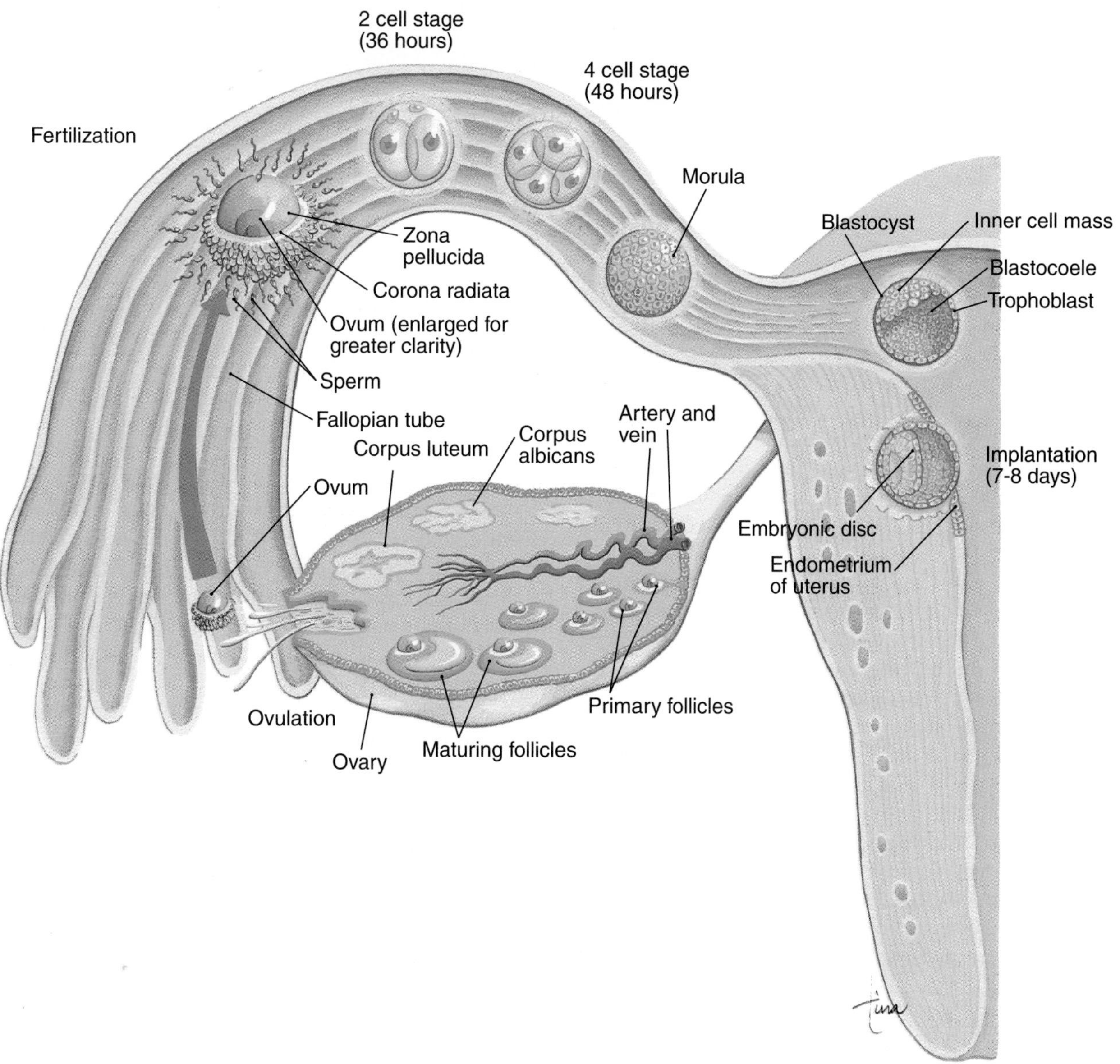

Figure 21-2 Ovulation, fertilization, and early embryonic development. Fertilization takes place in the fallopian tube, and the embryo has reached the blastocyst stage when it becomes implanted in the endometrium of the uterus.

Box 21–1 TWINS

Fraternal twins are the result of two separate ova fertilized by separate sperm. This may occur when two ovarian follicles reach maturity and rupture at the same time. Fraternal twins may be of the same sex or different sexes. Even if of the same sex, however, they are as genetically different as any siblings might be.

Identical twins are the result of the splitting of the very early embryo before the cells start to become specialized. For example, if a 16-cell stage becomes separated into two groups of eight cells each, each group will usually continue to develop in the usual way. Another possible cause is the development of two inner cell masses within the blastocyst. This, too, is before significant specialization has taken place, and each inner cell mass may develop into a complete individual. Twins of this type may be called monozygotic, meaning that they have come from one fertilized egg. Identical twins are always of the same sex, are very much alike in appearance, and in other respects are genetically identical.

IMPLANTATION

Fertilization usually takes place within the fallopian tube, and the zygote begins to divide even as it is being swept toward the uterus. These are mitotic divisions and are called **cleavage**. Refer to Fig. 21-2 as you read the following.

The single-cell zygote divides into a two-cell stage, four-cell stage, eight-cell stage, and so on. Three days after fertilization there are 16 cells, which continue to divide to form a solid sphere of cells called a **morula** (see Box 21-1: Twins). As mitosis proceeds, this sphere becomes hollow and is called a **blastocyst**, which is still about the same size as the original zygote.

A fluid-filled blastocyst consists of an outer layer of cells called the **trophoblast** and an inner cell mass that contains the potential embryo. It is the blastocyst stage that becomes **implanted** in the uterine wall, about 7 to 8 days after fertilization. The trophoblast secretes enzymes to digest the surface of the endometrium, creating a small crater into which the blastocyst sinks. The trophoblast will become the **chorion**, the embryonic membrane that will form the fetal portion of the placenta. Following implantation, the inner cell mass will grow to become the embryo and other membranes.

EMBRYO AND EMBRYONIC MEMBRANES

An **embryo** is the developing human individual from the time of implantation until the eighth week of gestation. Several stages of early embryonic development are shown in Fig. 21-3. At approximately 12 days, the **embryonic disc** (the potential person) is simply a plate of cells within the blastocyst. Very soon thereafter, three primary layers, or germ layers, begin to develop: the **ectoderm, mesoderm**, and **endoderm**. Each primary layer develops into specific organs or parts of organs. "Ecto" means outer; the epidermis is derived from ectoderm. "Meso" means middle; the skeletal muscles develop from mesoderm. "Endo" means inner; the stomach lining is derived from endoderm. Table 21-1 lists some other structures derived from each of the primary germ layers.

At 20 days the **embryonic membranes** can be clearly distinguished from the embryo itself. The **yolk sac** does not contain nutrient yolk, as it does for bird and reptile embryos. It is, however, the site for the formation of the first blood cells and the cells that will become spermatogonia or oogonia. As the embryo grows, the yolk sac membrane is incorporated into the umbilical cord.

The **amnion** is a thin membrane that eventually surrounds the embryo and contains **amniotic fluid**. This fluid provides a cushion for the fetus against mechanical injury as the mother moves. When the fetal kidneys become functional, they excrete urine into the amniotic fluid. Also in this fluid are cells that have sloughed off the fetus; this is clinically important in the procedure called amniocentesis (see Box 21-2: Fetal Diagnosis). The rupture of the amnion (sometimes called the "bag of waters") is usually an indication that labor has begun.

The **chorion** is the name given to the trophoblast

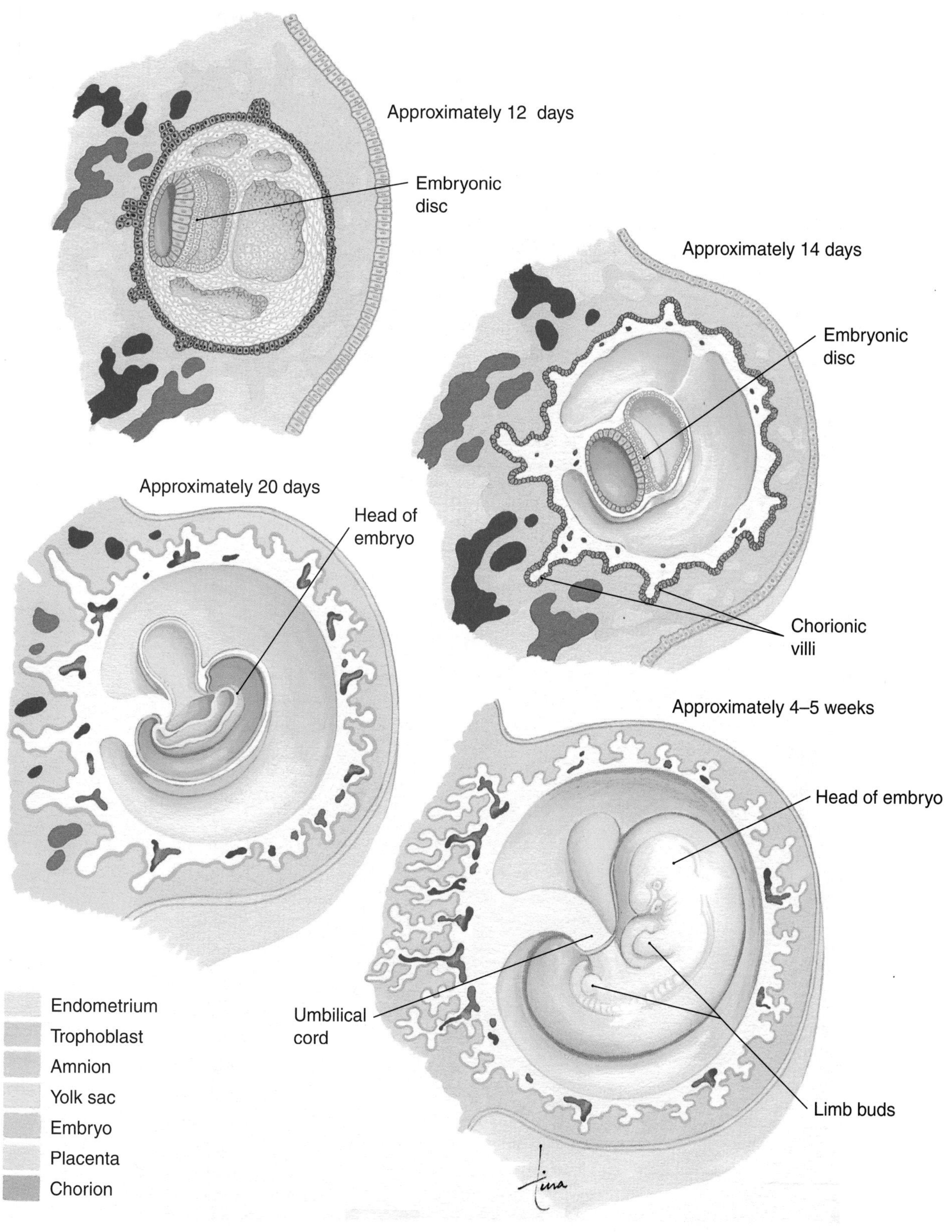

Figure 21–3 Embryonic development at 12 days (after fertilization), 14 days, 20 days, and 4 to 5 weeks. By 5 weeks, the embryo has distinct parts but does not yet look definitely human. See text for description of embryonic membranes.

Table 21–1 STRUCTURES DERIVED FROM THE PRIMARY GERM LAYERS

Layer	Structures Derived*
Ectoderm	• Epidermis; hair and nail follicles; sweat glands • Nervous system; pituitary gland; adrenal medulla • Lens and cornea; internal ear • Mucosa of oral and nasal cavities; salivary glands
Mesoderm	• Dermis; bone and cartilage • Skeletal muscles; cardiac muscle; most smooth muscle • Kidneys and adrenal cortex • Bone marrow and blood; lymphatic tissue; lining of blood vessels
Endoderm	• Mucosa of esophagus, stomach, and intestines • Epithelium of respiratory tract, including lungs • Liver and mucosa of gallbladder • Thyroid gland; pancreas

*These are representative lists, not all-inclusive ones. Keep in mind also that most organs are combinations of tissues from each of the three germ layers. Related structures are grouped together.

Box 21–2 FETAL DIAGNOSIS

Several procedures are currently available to determine certain kinds of abnormalities in a fetus or to monitor development.

Ultrasound (or Fetal Ultrasonography)

This is a non-invasive procedure; high-frequency sound waves are transmitted through the abdominal wall into the uterus. The reflected sound waves are converted to an image called a sonogram. This method is used to confirm multiple pregnancies, fetal age or position, or to detect fetal abnormalities such as heart defects or malformations of other organs.

Amniocentesis

This procedure is usually performed at 16 to 18 weeks of gestation. A hypodermic needle is inserted through the wall of the abdomen into the amniotic sac, and about 10 to 20 ml of amniotic fluid is removed. Within this fluid are fetal cells, which can be cultured so that their chromosomes may be examined. Through such examination and biochemical tests, a number of genetic diseases or chromosome abnormalities may be detected. Because women over the age of 35 years are believed to have a greater chance of having a child with Down syndrome, amniocentesis is often recommended for this age group. A family history of certain genetic diseases is another reason a pregnant woman may wish to have this procedure.

Chorionic Villus Sampling (CVS)

In this procedure, a biopsy catheter is inserted through the vagina and cervix to collect a small portion of the chorionic villi. These cells are derived from the fetus but are not part of the fetus itself. The information obtained is the same as that for amniocentesis, but CVS may be performed earlier in pregnancy, at about 8 weeks. Although there is a risk that the procedure may cause a miscarriage, CVS is considered comparable in safety to amniocentesis. It is important to remember that no invasive procedure is without risks.

Table 21–2 GROWTH OF THE EMBRYO–FETUS

Month of Gestation	Aspects of Development	Approximate Overall Size in Inches
1	• Heart begins to beat; limb buds form; backbone forms; facial features not distinct	.25
2	• Calcification of bones begins; fingers and toes are apparent on limbs; facial features more distinct; body systems are established	1.25–1.5
3	• Facial features distinct but eyes are still closed; nails develop on fingers and toes; ossification of skeleton continues; fetus is distinguishable as male or female	3
4	• Head still quite large in proportion to body, but the arms and legs lengthen; hair appears on head; body systems continue to develop	5–7
5	• Skeletal muscles become active ("quickening" may be felt by the mother); body grows more rapidly than head; body is covered with fine hair (lanugo)	10–12
6	• Eyelashes and eyebrows form; eyelids open; skin is quite wrinkled	11–14
7	• Head and body approach normal infant proportions; deposition of subcutaneous fat makes skin less wrinkled	13–17
8	• Testes of male fetus descend into scrotum; more subcutaneous fat is deposited; production of pulmonary surfactant begins	16–18
9	• Lanugo is shed; nails are fully developed; cranial bones are ossified with fontanels present; lungs are more mature	19–21

as it develops further. Once the embryo has become implanted in the uterus, small projections called **chorionic villi** begin to grow into the endometrium. These will contain the fetal blood vessels that become the fetal portion of the placenta.

At about 4 to 5 weeks of development, the embryo shows definite form. The head is apparent, and limb buds are visible. The period of embryonic growth continues until the eighth week. At this time, all of the organ systems have been established. They will continue to grow and mature until the end of gestation. The period of fetal growth extends from the 9th through the 40th week. Table 21-2 lists some of the major aspects of development in the growth of the embryo-fetus. The fetus at 16 weeks is depicted in Fig. 21-4 (see Box 21-3: Congenital Fetal Infections, and Box 21-4: Fetal Alcohol Syndrome). Maternal changes during pregnancy are summarized in Table 21-3.

PLACENTA AND UMBILICAL CORD

The **placenta** is made of both fetal and maternal tissue. The chorion of the embryo and the endometrium of the uterus contribute, and the placenta is formed by the third month of gestation (12 weeks). The ma-

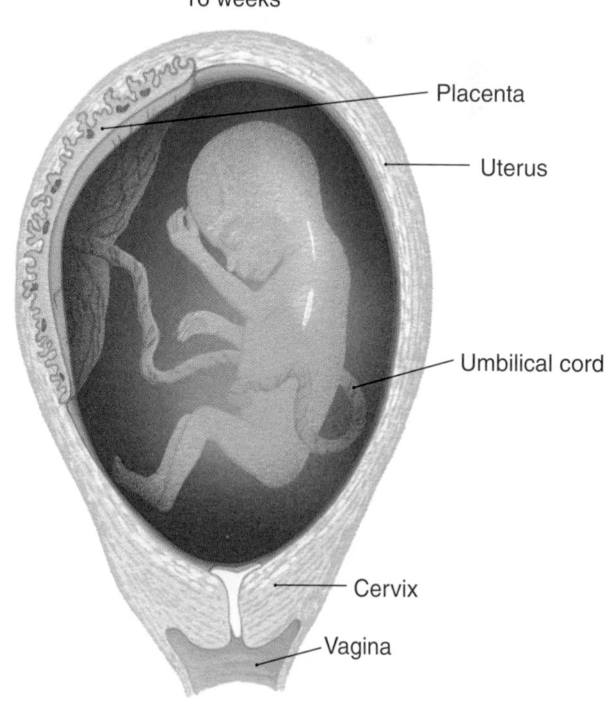

Figure 21–4 Fetal development at 16 weeks.

Box 21–3 CONGENITAL FETAL INFECTIONS

A **teratogen** is anything that may cause developmental abnormalities in an embryo–fetus. Several infectious microorganisms are known to be teratogenic; they may cross the placenta from maternal blood to fetal blood and damage the fetus.

Congenital Rubella Syndrome—the virus of German measles (rubella) is perhaps the best known of the infectious teratogens. If a woman acquires rubella during pregnancy, there is a 20% chance that her fetus will be affected. Consequences for the fetus may be death (stillborn), heart defects, deafness, or mental retardation. Women who have never had rubella should consider receiving the vaccine before they begin to have children. In this way, congenital rubella syndrome is completely prevented.

Congenital Varicella Syndrome—chickenpox *(Herpes varicella)* is a viral disease that most people have during childhood. If a pregnant woman acquires chickenpox, however, the virus may cross the placenta. Consequences for the fetus include malformed limbs, cutaneous scars, blindness, and mental retardation. Although this is not a common congenital syndrome, women who have never had chickenpox should be educated about it.

Congenital Syphilis—syphilis is a sexually transmitted disease caused by a bacterium *(Treponema pallidum).* These bacteria may cross the placenta after the fourth month of gestation. Consequences of early infection of the fetus are death, malformations of bones and teeth, or cataracts. If infection occurs toward the end of gestation, the child may be born with syphilis. This is most often apparent as a rash on the skin and mucous membranes. Syphilis in adults can be cured by penicillin; prevention of congenital syphilis depends upon good prenatal care for women who may be infected.

Congenital Toxoplasmosis—this condition is caused by the protozoan parasite *Toxoplasma gondii.* For most healthy people, toxoplasmosis has no symptoms at all; it is a harmless infection. Because cats and some other mammals are also hosts for this parasite, pregnant women may acquire infection from contact with cat feces or by eating rare beef or lamb. Consequences for the fetus are death, mental retardation, or blindness. Retinal infection may only become apparent during childhood or adolescence. Prevention now depends upon education of pregnant women to avoid potential sources of infection.

Box 21–4 FETAL ALCOHOL SYNDROME

Fetal alcohol syndrome is the term for a group of characteristics present in infants who were exposed to alcohol during their fetal life. Alcohol is a toxin for adults and even more so for fetal tissues that are immature and growing rapidly. Either alcohol or its toxic intermediate product (acetaldehyde) may pass from maternal to fetal circulation and impair fetal development.

Consequences for infants include low birth weight, small head with facial abnormalities, heart defects, malformation of other organs, and irritability and agitation. The infant often grows slowly, both physically and mentally, and mental retardation is also a possible outcome.

Because there is no way to reverse the damage done by intrauterine exposure to alcohol, the best course is prevention. Education is very important; women should be aware of the consequences their consumption of alcohol may have for a fetus.

Table 21–3 MATERNAL CHANGES DURING PREGNANCY

Aspect	Change
Weight	• Gain of 2–3 pounds for each month of gestation
Uterus	• Enlarges considerably and displaces abdominal organs upward
Thyroid gland	• Increases secretion of thyroxine, which increases metabolic rate
Skin	• Appearance of striae (stretch marks) on abdomen
Circulatory system	• Heart rate increases, as do stroke volume and cardiac output; blood volume increases; varicose veins may develop in the legs and anal canal
Digestive system	• Nausea and vomiting may occur in early pregnancy (morning sickness); constipation may occur in later pregnancy
Urinary system	• Kidney activity increases; frequency of urination often increases in later pregnancy (bladder is compressed by uterus)
Respiratory system	• Respiratory rate increases; lung capacity decreases as diaphragm is forced upward by compressed abdominal organs in later pregnancy
Skeletal system	• Lordosis may occur with increased weight at front of abdomen; sacroiliac joints and pubic symphysis become more flexible prior to birth

ture placenta is a flat disc about 7 inches (17 cm) in diameter.

The structure of a small portion of the placenta is shown in Fig. 21-5. Notice that the fetal blood vessels are within maternal blood sinuses, but there is no direct connection between fetal and maternal vessels. Normally, the blood of the fetus does not mix with that of the mother. The placenta has two functions: to serve as the site of exchanges between maternal and fetal blood and to produce hormones to maintain pregnancy. We will consider the exchanges first.

The fetus is dependent upon the mother for oxygen and nutrients and for the removal of waste products. The **umbilical cord** connects the fetus to the placenta. Within the cord are two umbilical arteries that carry blood from the fetus to the placenta and one umbilical vein that returns blood from the placenta to the fetus.

When blood in the umbilical arteries enters the placenta, CO_2 and waste products in the fetal capillaries diffuse into the maternal blood sinuses. Oxygen diffuses from the maternal blood sinuses into the fetal capillaries; nutrients enter the fetal blood by diffusion and active transport mechanisms. This oxygen- and nutrient-rich blood then flows through the umbilical vein back to the fetus. Circulation within the fetus was described in Chapter 13.

When the baby is delivered at the end of gestation, the umbilical cord is cut. The placenta then detaches from the uterine wall and is delivered as the **afterbirth**.

Placental Hormones

The first hormone secreted by the placenta is **human chorionic gonadotropin** (hCG), which is produced by the chorion of the early embryo. The function of hCG is to stimulate the corpus luteum in the maternal ovary, so that it will continue to secrete estrogen and progesterone. The secretion of progesterone is particularly important to prevent contractions of the myometrium, which would otherwise result in miscarriage of the embryo. Once hCG enters maternal circulation, it is excreted in urine, which is the basis for many pregnancy tests. Tests for hCG in maternal blood are even more precise and can determine whether a pregnancy has occurred even before a menstrual period is missed.

The corpus luteum is a small structure, however, and cannot secrete sufficient amounts of estrogen and progesterone to maintain a full-term pregnancy. The placenta itself begins to secrete **estrogen** and **progesterone** within a few weeks, and the levels of these hormones increase until birth. As the placenta takes over, the secretion of hCG decreases, and the corpus luteum becomes non-functional. During pregnancy, estrogen and progesterone inhibit the anterior pituitary secretion of FSH and LH, so no other ovarian fol-

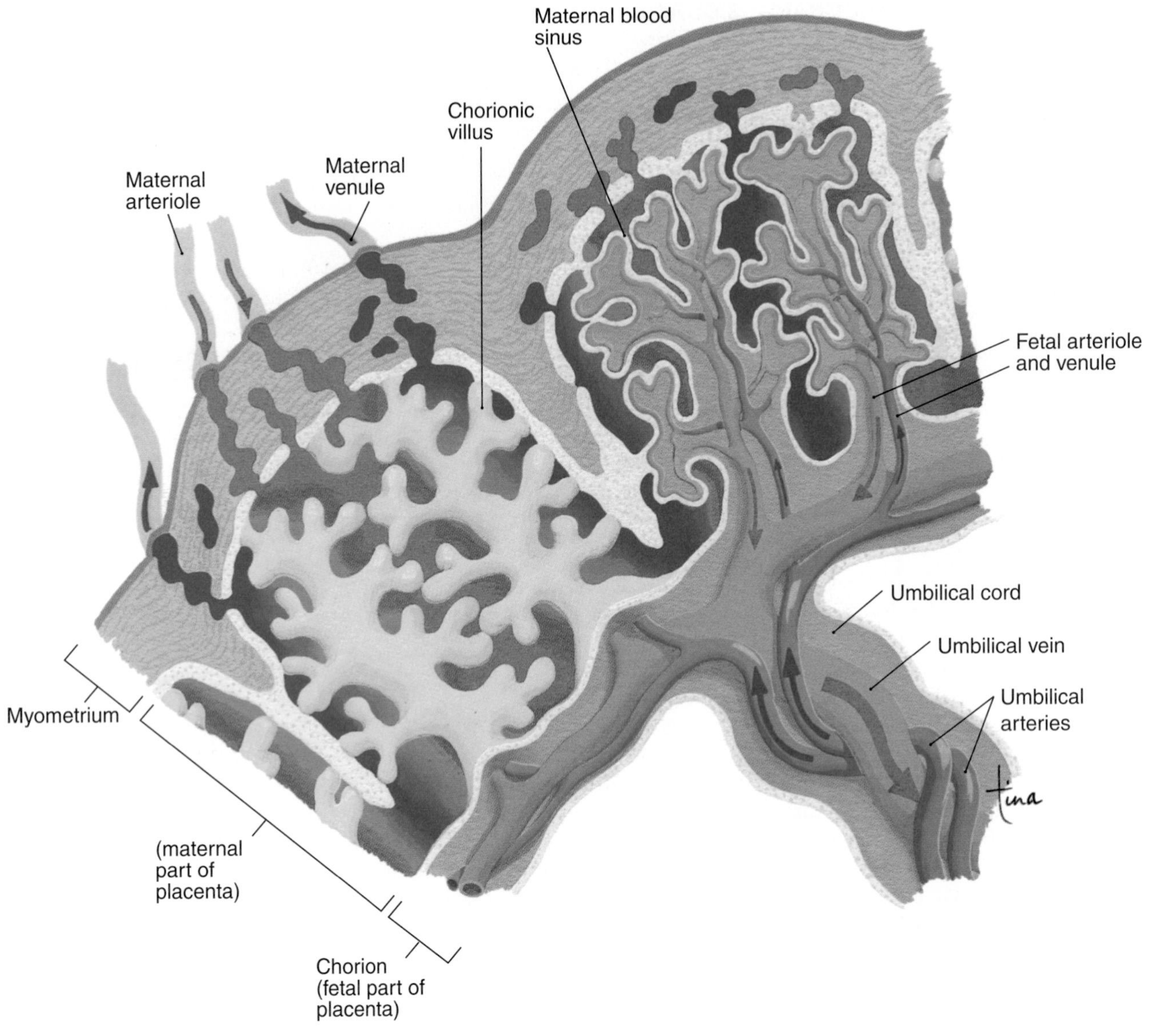

Figure 21–5 Placenta and umbilical cord. The fetal capillaries in chorionic villi are within the maternal blood sinuses. Arrows indicate the direction of blood flow in maternal and fetal vessels.

licles develop. These placental hormones also prepare the mammary glands for lactation.

PARTURITION AND LABOR

Parturition is the rather formal term for birth, and **labor** is the sequence of events that occurs during birth. The average gestation period is 40 weeks (280 days), with a range of 37 to 42 weeks (see Box 21–5: Premature Birth). Toward the end of gestation, the placental secretion of progesterone decreases while estrogen level remains high, and the myometrium begins to contract weakly at irregular intervals. At this time the fetus is often oriented head down within the

Box 21–5 PREMATURE BIRTH

A premature birth is the spontaneous delivery of an infant before the end of the gestation period, or the intentional delivery (for reasons concerning the health of the mother or fetus) before that time. A premature infant is one that weighs less than 2500 grams (5.5 pounds).

In terms of gestation time, an 8-month-old infant has a very good chance of surviving if good medical care is available. Infants of 7 months or less have respiratory distress, because their lungs have not yet produced enough surfactant to permit inflation of the alveoli. This is hyaline membrane disease, which was described in Chapter 15.

Infants as young as 24 weeks have been successfully treated and eventually sent home with their parents. Whether there will be any long-term detrimental effects of their very premature births will only be known in time.

Medical science may have reached its maximum capability in this area. In a fetus of 20 weeks, the cells of the alveoli and the cells of the pulmonary capillaries are just not close enough to each other to permit gas exchange. At present there is no medical intervention that can overcome this anatomical fact.

uterus (Fig. 21-6). Labor itself may be divided into three stages.

First stage—dilation of the cervix. As the uterus contracts, the amniotic sac is forced into the cervix, which dilates (widens) the cervical opening. At the end of this stage, the amniotic sac breaks (rupture of the "bag of waters") and the fluid leaves through the vagina, which may now be called the **birth canal**. This stage lasts an average of 8 to 12 hours but may vary considerably.

Second stage—delivery of the infant. More powerful contractions of the uterus are brought about by **oxytocin** released by the posterior pituitary gland and perhaps by the placenta itself. This stage may be prolonged by several factors. If the fetus is positioned other than head down, delivery may be difficult. This is called a breech birth and may necessitate a **cesarean section** (C-section), which is delivery of the fetus through a surgical incision in the abdominal wall and uterus. For some women, the central opening in the pelvic bone may be too small to permit a vaginal delivery. Fetal distress, as determined by fetal monitoring of heartbeat for example, may also require a cesarean section.

Third stage—delivery of the placenta (afterbirth). Continued contractions of the uterus expel the placenta and membranes, usually within 10 minutes after delivery of the infant. There is some bleeding at this time, but the uterus rapidly decreases in size, and the contractions compress the endometrium to close the ruptured blood vessels at the former site of the placenta. This is important to prevent severe maternal hemorrhage.

THE INFANT AT BIRTH

Immediately after delivery, the umbilical cord is clamped and cut, and the infant's nose and mouth are aspirated to remove any fluid that might interfere with breathing (see Box 21-6: Apgar Score). Now the infant is independent of the mother, and the most rapid changes occur in the respiratory and circulatory systems.

As the level of CO_2 in the baby's blood increases, the respiratory center in the medulla is stimulated and brings about inhalation to expand and inflate the lungs. Full expansion of the lungs may take up to 7 days following birth, and the infant's respiratory rate may be very rapid at this time, as high as 40 respirations per minute.

Breathing promotes greater pulmonary circulation, and the increased amount of blood returning to the left atrium closes the flap of the **foramen ovale**. The **ductus arteriosus** begins to constrict, apparently in response to the higher blood oxygen level. Full closure of the ductus arteriosus may take up to 3 months.

The **ductus venosus** no longer receives blood from the umbilical vein and begins to constrict within a few minutes after birth. Within a few weeks the ductus venosus becomes a non-functional ligament.

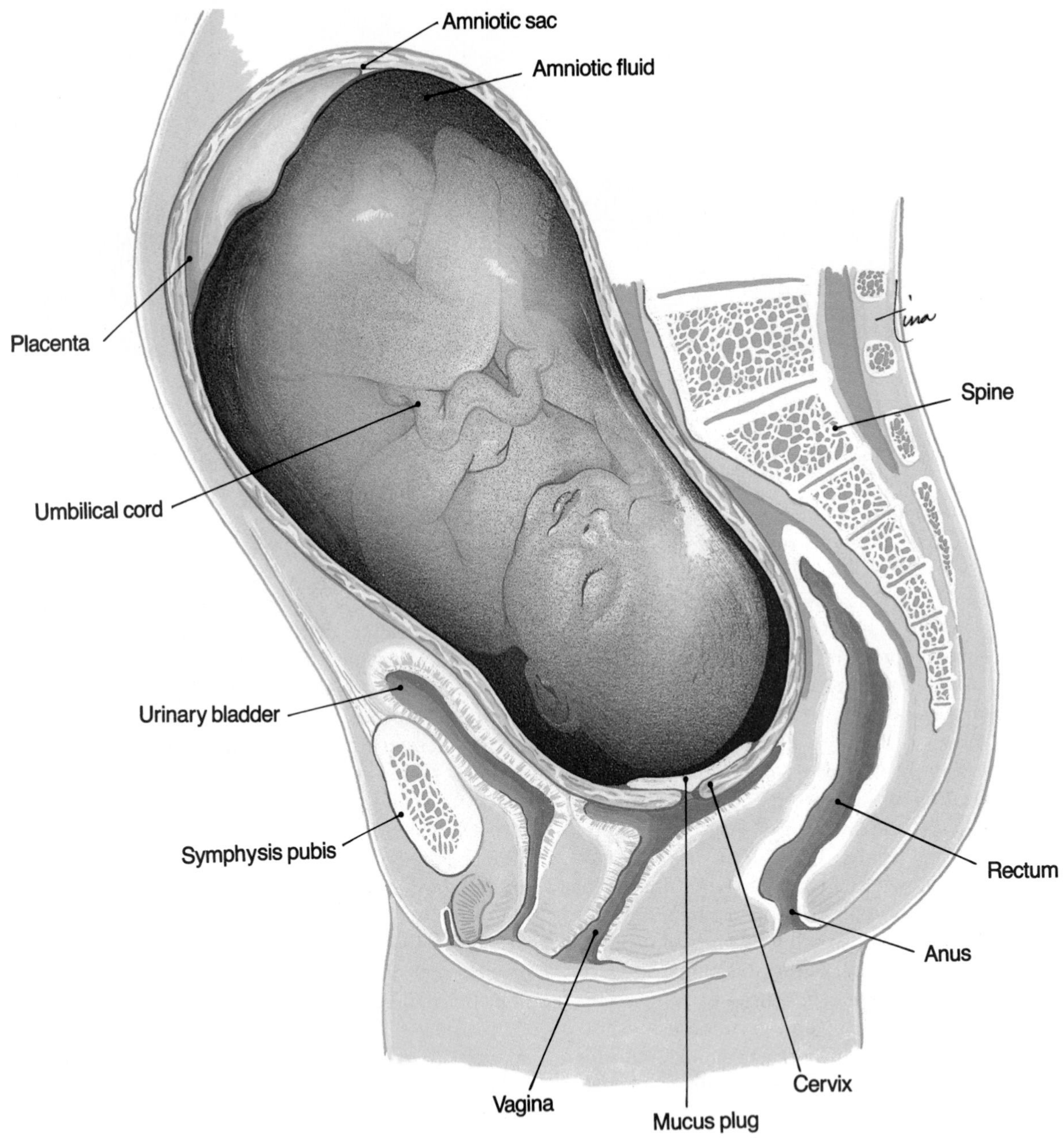

Figure 21–6 Full-term fetus positioned head down within the uterus.

Box 21–6 APGAR SCORE

The Apgar score is an overall assessment of an infant and is usually made 1 minute after birth (may be repeated at 5 minutes if the first score is low). The highest possible score is 10. Infants who score less than 5 require immediate medical attention.

Characteristic	Description	Score
Heartbeat	• Over 100 bpm • Below 100 bpm • No heartbeat	2 1 0
Respiration	• Strong, vigorous cry • Weak cry • No respiratory effort	2 1 0
Muscle Tone	• Spontaneous, active motion • Some motion • No muscle tone	2 1 0
Reflex response to stimulation of sole of the foot	• A cry in response • A grimace in response • No response	2 1 0
Color	• Healthy coloration • Cyanotic extremities • Cyanosis of trunk and extremities	2 1 0

The infant's liver is not fully mature at birth and may be unable to excrete bilirubin efficiently. This may result in jaundice, which may occur in as many as half of all newborns. Such jaundice is not considered serious unless there is another possible cause, such as Rh incompatibility (see Chapter 11).

GENETICS

Genetics is the study of inheritance. Most, if not virtually all, human characteristics are regulated at least partially by genes. We will first look at what genes are, then describe some patterns of inheritance.

CHROMOSOMES AND GENES

Each of the cells of an individual (except mature RBCs and egg and sperm) contains 46 chromosomes, the diploid number. These chromosomes are in 23 pairs called **homologous pairs**. One member of each pair has come from the egg and is called maternal, the other member has come from the sperm and is called paternal. The autosomes are the chromosome pairs designated 1 to 22. The sex chromosomes form the remaining pair. In women these are designated XX and in men XY.

Chromosomes are made of DNA and protein; the DNA is the hereditary material. You may wish to refer to Chapter 3 to review DNA structure. The sequence of bases in the DNA of chromosomes is the genetic code for proteins; structural proteins as well as enzymes. The DNA code for one protein is called a gene. For example, a specific region of the DNA of chromosome 11 is the code for the beta chain of hemoglobin. Because an individual has two of chromosome 11, he or she will have two genes for this protein, a maternal gene inherited from the mother and a paternal gene inherited from the father. This is true for virtually all of the 70,000 genes estimated to be found in our chromosomes. In our genetic makeup, each of us has two genes for each protein.

GENOTYPE AND PHENOTYPE

For each gene of a pair, there may be two or more possibilities for its "expression," that is, its precise nature or how it will affect the individual. These possibilities are called **alleles**. A person, therefore, may be said to have two alleles for each protein or trait; the alleles may be the same or may be different.

If the two alleles are the same, the person is said to be **homozygous** for the trait. If the two alleles are different, the person is said to be **heterozygous** for the trait.

The **genotype** is the actual genetic makeup, that is,

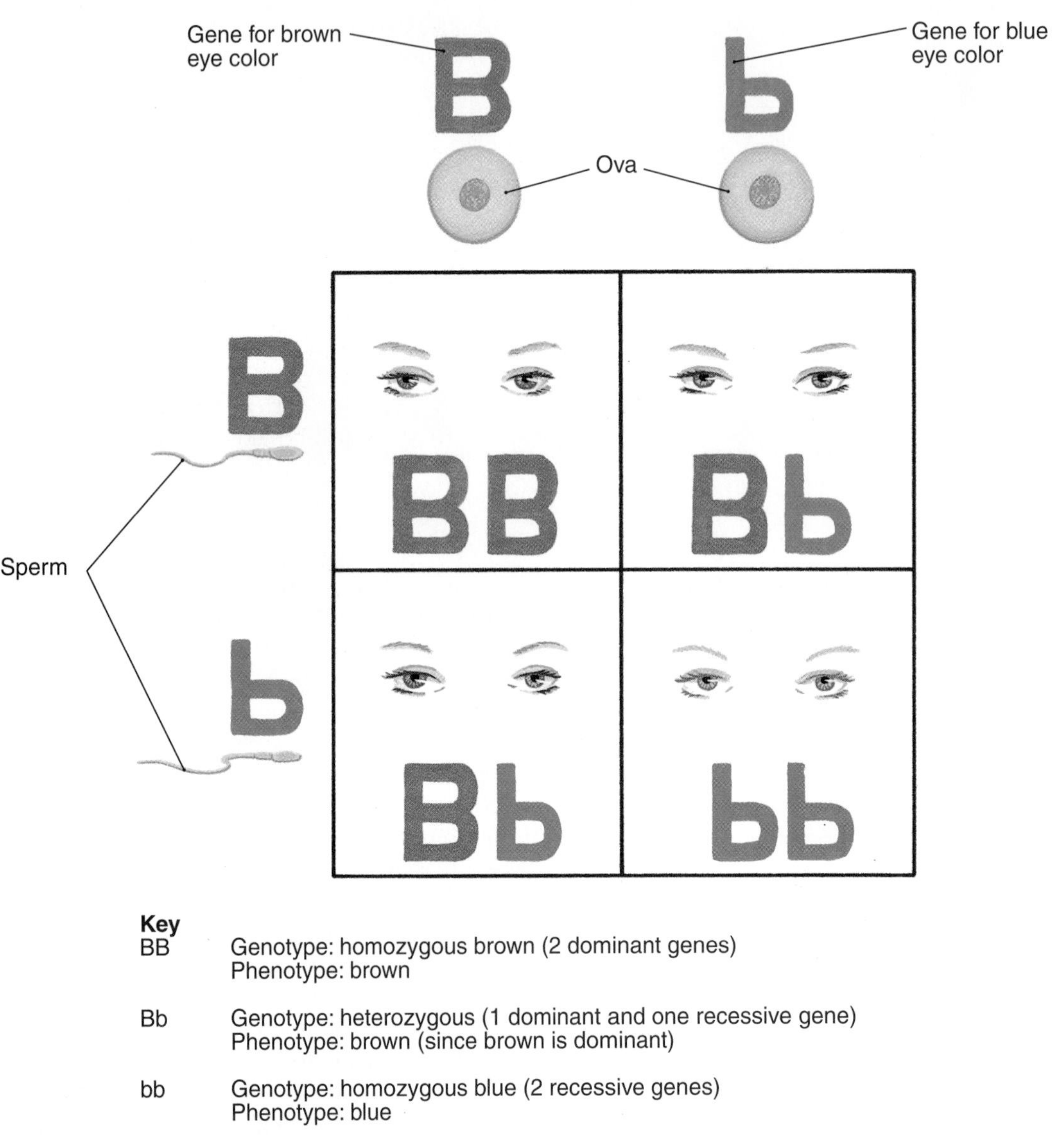

Key

BB Genotype: homozygous brown (2 dominant genes)
Phenotype: brown

Bb Genotype: heterozygous (1 dominant and one recessive gene)
Phenotype: brown (since brown is dominant)

bb Genotype: homozygous blue (2 recessive genes)
Phenotype: blue

Figure 21–7 Inheritance of eye color. Both mother and father are heterozygous for brown eyes. The Punnett square shows the possible combinations of genes for eye color in each child of these parents. See text for further description.

Box 21–7 SOLUTION TO GENETICS QUESTION

Question: Can parents who are both heterozygous for brown eyes have four children with blue eyes? What are the odds of this happening?

Answer: Yes. For each child, the odds of having blue eyes are 1 in 4. To calculate the odds of all four children having blue eyes, multiply the odds for each child separately:

1st child		2nd child		3rd child		4th child		
¼	×	¼	×	¼	×	¼	=	1⁄256

The odds are 256 to 1.

the alleles present. The **phenotype** is the appearance, or how the alleles are expressed. When a gene has two or more alleles, one allele may be **dominant** over the other, which is called **recessive**. For a person who is heterozygous for a trait, the dominant allele (or gene) is the one that will appear in the phenotype. The recessive allele (or gene) is hidden but may be passed to children. For a recessive trait to be expressed in the phenotype, the person must be homozygous recessive, that is, have two recessive alleles (genes) for the trait.

An example will be helpful here to put all this together and is illustrated in Fig. 21–7. When doing genetics problems, a **Punnett square** is used to show the possible combinations of genes in the egg and sperm for a particular set of parents and their children. Remember that an egg or sperm has only 23 chromosomes and, therefore, has only one gene for each trait.

In this example, the inheritance of eye color has been simplified. Although eye color is determined by many pairs of genes, with many possible phenotypes, one pair is considered the principal pair, with brown eyes dominant over blue eyes. A dominant gene is usually represented by a capital letter, and the corresponding recessive gene is represented by the same letter in lower case. The parents in Fig. 21–7 are both heterozygous for eye color. Their genotype consists of a gene for brown eyes and a gene for blue eyes, but their phenotype is brown eyes.

Each egg produced by the mother has a 50% chance of containing the gene for brown eyes, or an equal 50% chance of containing the gene for blue eyes. Similarly, each sperm produced by the father has a 50% chance of containing the gene for brown eyes and a 50% chance of containing the gene for blue eyes.

Now look at the boxes of the Punnett square; these represent the possibilities for the genetic makeup of each child. For eye color there are three possibilities: a 25% (one of four) chance for homozygous brown eyes, a 50% (two of four) chance for heterozygous brown eyes, and a 25% (one of four) chance for homozygous blue eyes. Notice that BB and Bb have the same phenotype (brown eyes) despite their different genotypes, and that the phenotype of blue eyes is only possible with the genotype bb. Can brown-eyed parents have a blue-eyed child? Yes; if each parent is heterozygous for brown eyes, each child has a 25% chance of inheriting blue eyes. Could these parents have four children with blue eyes? What are the odds of this happening? The answers to these questions will be found in Box 21–7: Solution to Genetics Question.

INHERITANCE: DOMINANT–RECESSIVE

The inheritance of eye color just described is an example of a trait determined by a pair of alleles, one of which may dominate the other. Another example is sickle-cell anemia, which was discussed in Chapter 11. The gene for the beta chain of hemoglobin is on chromosome 11; an allele for normal hemoglobin is dominant, and an allele for sickle-cell hemoglobin is recessive. An individual who is heterozygous is said to have sickle-cell trait; an individual who is homozygous recessive will have sickle-cell anemia. The Punnett square in Fig. 21–8 shows that if both parents are heterozygous, each child has a 25% chance of inheriting the two recessive genes. Table 21–4 lists some other human genetic diseases and their patterns of inheritance.

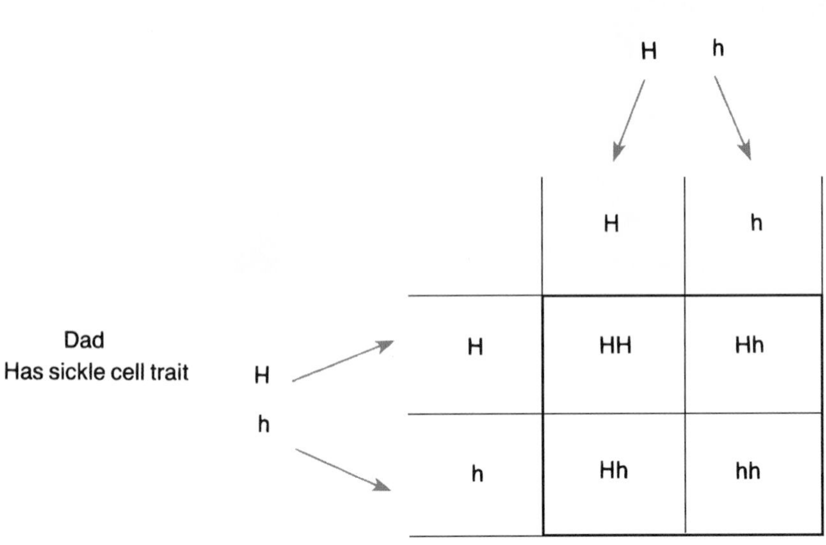

Figure 21–8 Inheritance of sickle-cell anemia (dominant recessive pattern). See text for description.

Table 21–4 HUMAN GENETIC DISEASES

Disease (Pattern of Inheritance)	Description
Sickle-cell anemia (R)	• The most common genetic disease among people of African ancestry. Sickle-cell hemoglobin forms rigid crystals that distort and disrupt RBCs; oxygen-carrying capacity of the blood is diminished.
Cystic fibrosis (R)	• The most common genetic disease among people of European ancestry. Production of thick mucus clogs the bronchial tree and pancreatic ducts. Most severe effects are chronic respiratory infections and pulmonary failure.
Tay-Sachs disease (R)	• The most common genetic disease among people of Jewish ancestry. Degeneration of neurons and the nervous system results in death by the age of 2 years.
Phenylketonuria or PKU (R)	• Lack of an enzyme to metabolize the amino acid phenylalanine leads to severe mental and physical retardation. These effects may be prevented by the use of a diet (beginning at birth) that limits phenylalanine.
Huntington's disease (D)	• Uncontrollable muscle contractions begin between the ages of 30–50 years; followed by loss of memory and personality. There is no treatment that can delay mental deterioration.
Hemophilia (X-linked)	• Lack of Factor 8 impairs chemical clotting; may be controlled with Factor 8 from donated blood.
Duchenne's muscular dystrophy (X-linked)	• Replacement of muscle by adipose or scar tissue, with progressive loss of muscle function; often fatal before age 20 years due to involvement of cardiac muscle.

R = recessive; D = dominant.

Table 21–5 ABO BLOOD TYPES: GENOTYPES

Blood Type	Possible Genotypes
O	OO
AB	AB
A	AA or OA
B	BB or OB

INHERITANCE: MULTIPLE ALLELES

The best example of this pattern of inheritance is human blood type of the ABO group. For each gene of this blood type, there are three possible alleles: A, B, or O. A person will have only two of these alleles, which may be the same or different. O is the recessive allele; A and B are codominant alleles, that is, dominant over O but not over each other.

You already know that in this blood group there are four possible blood types: O, AB, A, and B. Table 21-5 shows the combinations of alleles for each type. Notice that for types O and AB there is only one possible genotype. For types A and B, however, there are two possible genotypes, because both A and B alleles are dominant over an O allele.

Let us now use a problem to illustrate the inheritance of blood type. The Punnett square in Fig. 21-9 shows that Mom has type O blood and Dad has type AB blood. The boxes of the square show the possible blood types for each child. Each child has a 50% chance of having type A blood, and a 50% chance of having type B blood. The genotype, however, will always be heterozygous. Notice that in this example, the blood types of the children will not be the same as those of the parents.

Table 21-6 lists some other human genetic traits, with the dominant and recessive phenotype for each.

INHERITANCE: SEX-LINKED TRAITS

Sex-linked traits may also be called X-linked traits because the genes for them are located only on the X chromosome. The Y does not have corresponding genes for many of the genes on the X chromosome. (The Y chromosome is very small and has only about 20 genes. Some of these genes are active only in the testes and contribute to spermatogenesis. Others are

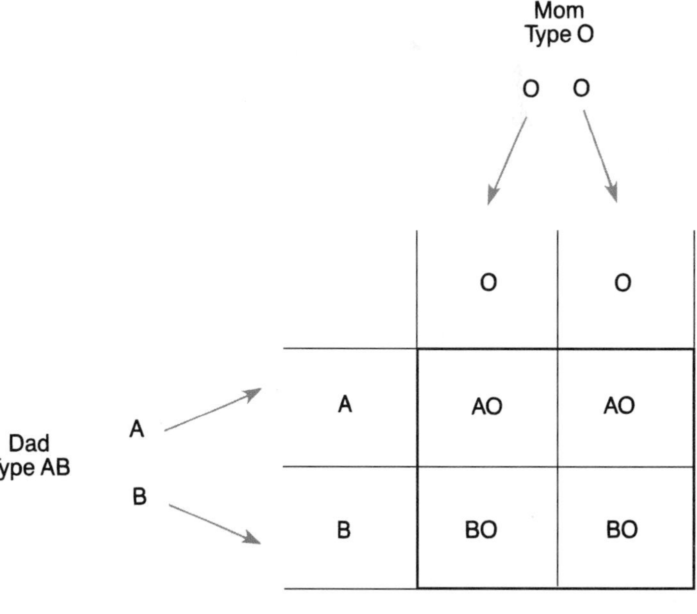

Figure 21-9 Inheritance of blood type (multiple alleles pattern). See text for description.

Table 21–6 HUMAN GENETIC TRAITS

Trait	Dominant Phenotype	Recessive Phenotype
ABO blood type	AB, A, B	O
Rh blood type	Rh positive	Rh negative
Hair color	Dark	Light (blond or red)
Change in hair color	Premature gray	Gray later in life
Hair texture	Curly	Straight
Hairline	Widow's peak	Straight
Eye color	Dark	Light
Color vision	Normal	Color blind
Visual acuity	Nearsighted or farsighted	Normal
Skin color	Dark	Light
Freckles	Abundant	Few
Dimples	Present	Absent
Cleft chin	Present	Absent
Ear lobes	Unattached	Attached
Number of fingers/toes	Polydactyly (more than 5 digits)	5 per hand or foot
Mid-digital hair	Present	Absent
Double-jointed thumb	Present	Absent
Bent little finger	Present	Absent
Ability to roll tongue sides up	Able	Unable

active in many body cells, but their functions are not yet known with certainty.)

The genes for sex-linked traits are recessive, but because there are no corresponding genes on the Y chromosome to mask them, a man needs only one gene to express one of these traits in his phenotype. A woman who has one of these recessive genes on one X chromosome and a dominant gene for normal function on the other X chromosome, will not express this trait. She is called a carrier, however, because the gene is part of her genotype and may be passed to children.

Let us use as an example, red-green color blindness. The Punnett square in Fig. 21–10 shows that Mom is carrier of this trait and that Dad has normal color vision. A Punnett square for a sex-linked trait uses the X and Y chromosomes with a lower case letter on the X to indicate the presence of the recessive gene. The possibilities for each child are divided equally into daughters and sons. In this example, each daughter has a 50% chance of being a carrier and a 50% chance of not being a carrier. In either case, a daughter will have normal color vision. Each son has a 50% chance of being red-green color blind and a 50% chance of having normal color vision. Men can never be carriers of a trait such as this; they either have it or do not have it.

The inheritance of other characteristics is often not as easily depicted as are the examples shown above. Height, for example, is a multiple gene characteristic, meaning that many pairs of genes contribute. Many pairs of genes result in many possible combinations for genotype and many possible phenotypes. In addition, height is a trait that is influenced by environmental factors such as nutrition. These kinds of circumstances or influences are probably important in many other human characteristics.

Another difficulty in predicting genetic outcomes is that we do not know what all our genes are for. Of the estimated 70,000 human genes, about 7000 had been "mapped" as of January 1998. This means that they have been precisely located on a particular chromosome. By the time you read this, many more genes will have been mapped, for this is a project that has been undertaken by many groups of researchers. Their goal is to map all the genes on our 23 pairs of chromosomes by the year 2005.

Someday it will be possible to cure genetic diseases by inserting correct copies of malfunctioning genes into the cells of affected individuals. The first such attempt was undertaken in September 1990, in an effort to supply a missing enzyme in an otherwise fatal disorder of the immune system. The following year, the child, then 5 years old, was able to start school with an immune system that seemed to be completely

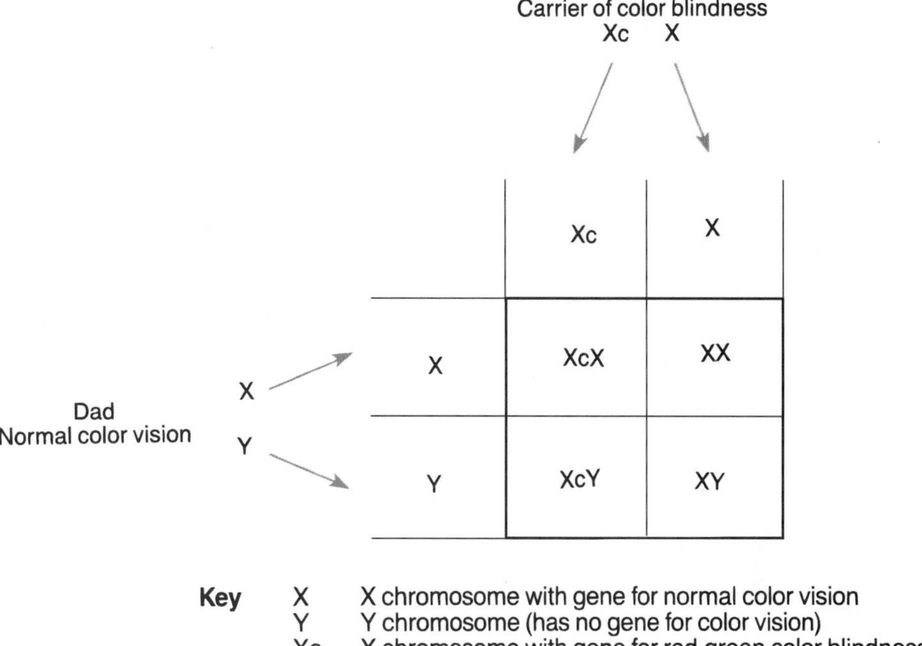

Figure 21–10 Inheritance of red-green color blindness (sex-linked pattern). See text for description.

functional. Despite this success, gene therapy, because of its complexity and high cost, is still limited to federally funded experiments performed on individuals or small groups of people. Also unknown at this time, and yet another reason to proceed slowly, are any possible risks of gene replacement.

Other diseases that may eventually be cured or controlled with gene therapy include cystic fibrosis (for which clinical trials are already underway), Parkinson's disease, diabetes, muscular dystrophy, hemophilia, and sickle-cell anemia. Much more research and experimentation remain to be done before gene replacement becomes the standard treatment available to everyone with these genetic diseases, but the foundation of this remarkable therapy has been established.

STUDY OUTLINE

Human Development—growth of a fertilized egg into a human individual

Fertilization—the union of the nuclei of egg and sperm; usually takes place in the fallopian tube

1. Sperm undergo final maturation (capacitation) within the female reproductive tract; the acrosome secretes enzymes to digest the membrane of the ovum.
2. The 23 chromosomes of the sperm join with the 23 chromosomes of the egg to restore the diploid number of 46 in the zygote.
3. A zygote has 22 pairs of autosomes and one pair of sex chromosomes; XX in females, XY in males (see Fig. 21-1).

Implantation (see Fig. 21–2)

1. Within the fallopian tube, the zygote begins mitotic divisions called cleavage to form two-cell, four-cell, eight-cell stages, and so on.

2. A morula is a solid sphere of cells, which divides further to form a hollow sphere called a blastocyst.
3. A blastocyst consists of an outer layer of cells called the trophoblast and an inner cell mass that contains the potential embryo. The trophoblast secretes enzymes to form a crater in the endometrium into which the blastocyst sinks.

Embryo—weeks 1 through 8 of gestation (see Fig. 21–3)

1. In the embryonic disc, three primary germ layers develop: ectoderm, mesoderm, and endoderm (see Table 21-1).
2. By the eighth week of gestation (end of 2 months), all the organ systems are formed (see Table 21-2).

Embryonic Membranes (see Fig. 21–3)

1. The yolk sac forms the first blood cells and the cells that become spermatogonia or oogonia.
2. The amnion surrounds the fetus and contains amniotic fluid; this fluid absorbs shock around the fetus.
3. The chorion develops chorionic villi that will contain blood vessels that form the fetal portion of the placenta.

Fetus—weeks 9 through 40 of gestation (see Table 21–2)

1. The organ systems grow and mature.
2. The growing fetus brings about structural and functional changes in the mother (see Table 21-3).

Placenta and Umbilical Cord

1. The placenta is formed by the chorion of the embryo and the endometrium of the uterus; the umbilical cord connects the fetus to the placenta.
2. Fetal blood does not mix with maternal blood; fetal capillaries are within maternal blood sinuses (see Fig. 21-5); this is the site of exchanges between maternal and fetal blood.
3. Two umbilical arteries carry blood from the fetus to the placenta; fetal CO_2 and waste products diffuse into maternal blood; oxygen and nutrients enter fetal blood.
4. Umbilical vein returns blood from placenta to fetus.
5. The placenta is delivered after the baby and is called the afterbirth.

Placental Hormones

1. hCG—secreted by the chorion; maintains the corpus luteum so that it secretes estrogen and progesterone during the first few months of gestation. The corpus luteum is too small to maintain a full-term pregnancy.
2. Estrogen and progesterone secretion begins within 4 to 6 weeks and continues until birth in amounts great enough to sustain pregnancy.
3. Estrogen and progesterone inhibit FSH and LH secretion during pregnancy and prepare the mammary glands for lactation.
4. Progesterone inhibits contractions of the myometrium until just before birth, when progesterone secretion begins to decrease.

Parturition and Labor

1. Gestation period ranges from 37 to 42 weeks; the average is 40 weeks.
2. Labor: first stage—dilation of the cervix; uterine contractions force the amniotic sac into the cervix; amniotic sac ruptures and fluid escapes.
3. Labor: second stage—delivery of the infant; oxytocin causes more powerful contractions of the myometrium. If a vaginal delivery is not possible, a cesarean section may be performed.
4. Labor: third stage—delivery of the placenta; the uterus continues to contract to expel the placenta, then contracts further, decreases in size, and compresses endometrial blood vessels.

The Infant at Birth (see Box 21–6)

1. Umbilical cord is clamped and severed; increased CO_2 stimulates breathing, and lungs are inflated.
2. Foramen ovale closes, and ductus arteriosus constricts; ductus venosus constricts; normal circulatory pathways are established.
3. Jaundice may be present if the infant's immature liver cannot rapidly excrete bilirubin.

Genetics—the study of inheritance chromosomes—46 per human cell; in 23 homologous pairs

1. A homologous pair consists of a maternal and a paternal chromosome of the same type (1 or 2, etc.).
2. There are 22 pairs of autosomes and one pair of sex chromosomes (XX or XY).
3. DNA—the hereditary material of chromosomes.
4. Gene—the genetic code for one protein; an indi-

vidual has two genes for each protein or trait, one maternal and one paternal.

5. Alleles—the possibilities for how a gene may be expressed.

Genotype—the alleles present in the genetic makeup

1. Homozygous—having two similar alleles.
2. Heterozygous—having two different alleles.

Phenotype—the appearance, or expression of the alleles present

1. Depends on the dominance or recessiveness of alleles or the particular pattern of inheritance involved.

Inheritance—dominant–recessive

1. A dominant gene will appear in the phenotype of a heterozygous individual (who has only one dominant gene). A recessive gene will appear in the phenotype only if the individual is homozygous, that is, has two recessive genes.
2. See Figs. 21-7 and 21-8 for Punnett squares.

Inheritance—multiple alleles

1. More than two possible alleles for each gene: human ABO blood type.
2. An individual will have only two of the alleles (same or different).
3. See Table 21-5 and Fig. 21-9.

Inheritance—sex-linked traits

1. Genes are recessive and found only on the X chromosome; there are no corresponding genes on the Y chromosome.
2. Women with one gene (and one gene for normal functioning) are called carriers of the trait.
3. Men cannot be carriers; they either have the trait or do not have it.
4. See Fig. 21-10.

REVIEW QUESTIONS

1. Where does fertilization usually take place? How many chromosomes are present in a human zygote? Explain what happens during cleavage, and describe the blastocyst stage. (pp. 460-462)
2. Describe the process of implantation, and state where this takes place. (p. 462)
3. How long is the period of embryonic growth? How long is the period of fetal growth? (pp. 462, 465)
4. Name two body structures derived from ectoderm, mesoderm, and endoderm. (p. 464)
5. Name the embryonic membrane with each of these functions: (p. 462)
   a. Forms the fetal portion of the placenta
   b. Contains fluid to cushion the embryo
   c. Forms the first blood cells for the embryo
6. Explain the function of placenta, umbilical arteries, and umbilical vein. (pp. 465, 467)
7. Explain the functions of the placental hormones: hCG, progesterone, and estrogen and progesterone (together). (p. 467)
8. Describe the three stages of labor, and name the important hormone. (p. 469)
9. Describe the major pulmonary and circulatory changes that occur in the infant after birth. (p. 469)
10. What is the genetic material of chromosomes? Explain what a gene is. Explain why a person has two genes for each protein or trait. (p. 471)

11. Define homologous chromosomes, autosomes, and sex chromosomes. (p. 471)
12. Define allele, homozygous, heterozygous, genotype, and phenotype. (pp. 472–473)
13. Genetics Problem: Mom is heterozygous for brown eyes, and Dad has blue eyes. What is the % chance that a child will have blue eyes? Brown eyes? (pp. 472–473)
14. Genetics Problem: Mom is homozygous for type A blood, and Dad is heterozygous for type B blood. What is the % chance that a child will have type AB blood? Type A? Type B? Type O? (p. 475)
15. Genetics Problem: Mom is red-green color blind, and Dad has normal color vision. What is the % chance that a son will be color blind? That a daughter will be color blind? That a daughter will be a carrier? (pp. 475–477)

CHAPTER 22

An Introduction to Microbiology and Human Disease

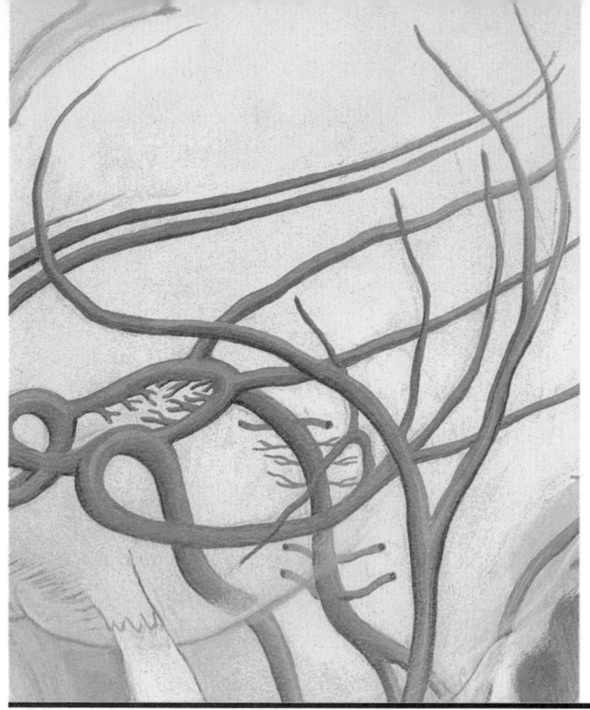

CHAPTER 22

Chapter Outline

Student Objectives

- Explain how microorganisms are classified and named.
- Describe the distribution of and benefits provided by normal flora.
- Explain what is meant by an infectious disease, and describe the different types of infection.
- Describe the ways in which infectious diseases may be spread.
- Explain how the growth of microorganisms may be controlled and the importance of this to public health.
- Describe the general structure of bacteria, viruses, fungi, and protozoa and the diseases they cause.
- Name the common worm infestations, and name the arthropods that are vectors of disease.

An Introduction to Microbiology and Human Disease

New Terminology

Bacillus (buh-**SILL**-us)
Coccus (**KOK**-us)
Communicable (kuhm-**YOO**-ni-kah-b'l)
Contagious (kun-**TAY**-jus)
Epidemiology (EP-i-DEE-mee-**AH**-luh-jee)
Gram stain (**GRAM STAYN**)
Infestation (in-fess-**TAY**-shun)
Mycosis (my-**KOH**-sis)
Non-communicable (NON-kuhm-**YOO**-ni-kah-b'l)
Portal of entry (**POR**-tuhl of **EN**-tree)
Portal of exit (**POR**-tuhl of **EG**-zit)
Reservoir (**REZ**-er-vwor)
Spirillum (spih-**RILL**-uhm)
Spirochaete (**SPY**-roh-keet)
Vector (**VEK**-ter)
Zoonoses (ZOH-oh-**NOH**-seez)

Related Clinical Terminology

Antitoxin (AN-tee-**TAHK**-sin)
Broad spectrum (**BRAWD SPEK**-trum)
Culture and sensitivity testing (**KUL**-chur and SEN-si-**TIV**-i-tee)
Endotoxin (EN-doh-**TAHK**-sin)
Incubation period (IN-kew-**BAY**-shun)
Narrow spectrum (**NAR**-oh **SPEK**-trum)
Nosocomial infection (no-zoh-**KOH**-mee-uhl)
Opportunistic infection (OP-er-too-**NIS**-tick)
Secondary infection (**SEK**-un-dery)
Subclinical infection (sub-**KLIN**-i-kuhl)

*Terms that appear in **bold type** in the chapter text are defined in the glossary, which begins on page 528.*

Microbiology is the study of microorganisms (also called microbes) and of their place or role in their environment. Sometimes, the environment of microbes is us, people, and that will be our focus. Although there are many perspectives in the field of microbiology, we will concentrate on the microorganisms that affect the functioning of the human body. These microbes include our normal flora and those that are capable of causing human disease. A disease is any disruption of normal body functioning. Many diseases are not caused by microorganisms, but those that are may be called infectious diseases. The microorganisms that cause infectious diseases are called pathogens.

Pathogens may also be called parasites, that is, they live on another living organism called a host and cause harm to the host. Some parasites cause diseases that are fatal to the host, but many others live in a more balanced way with their hosts. These cause illnesses from which the host recovers, and the parasite often survives long enough to be spread to other hosts.

Our introduction to microbiology will begin with a brief description of the classification and naming of microorganisms, followed by a discussion of normal flora. We will then consider infectious diseases, types of infection, and the spread and control of infection. Last, we will describe the types of pathogens in more detail and include the methods we have to treat and control the diseases they cause. For a historical perspective on microbiology, see Box 22-1: The Golden Age of Microbiology, and Box 22-2: Microbiology in the 20th Century.

CLASSIFICATION OF MICROORGANISMS

Bacteria are very simple single-celled organisms and are found virtually everywhere. The natural habitats of bacteria include fresh water, salt water, soil, and other living organisms. Most bacteria are not harmful to us,

Box 22–1 THE GOLDEN AGE OF MICROBIOLOGY

Microorganisms were first seen in the 17th century when simple microscopes were developed. The roles of microorganisms, however, especially in relation to disease, were to remain largely unknown for another two centuries. The years between 1875 and 1900 are sometimes called "The Golden Age of Microbiology" because of the number and significance of the discoveries that were made during this time. The following is only a partial list.

1877 Robert Koch proves that anthrax is caused by a bacterium.
1878 Joseph Lister first grows bacteria in pure culture.
The oil immersion lens for microscopes is developed.
1879 The bacterium that causes gonorrhea is discovered.
1880 Louis Pasteur develops a vaccine for chicken cholera.
1881 Discovery that staphylococci cause infections.
1882 The cause of tuberculosis is found to be a bacterium.
1883 Diphtheria is found to be caused by a bacterium.
1884 The bacteria that cause tetanus, typhoid fever, and cholera are each isolated. The first description of phagocytosis by white blood cells is published.
1885 Louis Pasteur first uses his vaccine for rabies.
1890 Development of antitoxins to treat tetanus and diphtheria.
1892 Discovery and demonstration of a virus.
1894 The bacterium that causes plague is discovered.
1897 Discovery of the cause of botulism.
Rat fleas are found to be vectors of plague.
1900 Walter Reed and his associates demonstrate that yellow fever is caused by a virus and that the vector is a mosquito.

Box 22–2 MICROBIOLOGY IN THE 20TH CENTURY

Just a few of the many advances made in microbiology during this century are listed below. Many of them concerned diagnosis, prevention, or treatment of infectious diseases, and some involved using microorganisms and the new genetic technologies.

1928 Discovery of penicillin by Alexander Fleming.
1935 The first of the sulfa drugs is found to be effective against staphylococcal infections.
1938 Penicillin is purified and produced in large quantities.
1941 The fluorescent antibody test is developed.
1943 Streptomycin is discovered and is the first effective drug in the treatment of tuberculosis.
1953 Watson and Crick describe the structure of DNA.
1955 The Salk polio vaccine is first used.
1957 Interferon is discovered.
1959 A vaccine for whooping cough is available.
1962 The Sabin polio vaccine is marketed in the US.
1963 The measles vaccine becomes available.
1969 A vaccine for rubella is developed.
1976 Legionnaire's disease is discovered and described.
1980 Genetically engineered bacteria produce the hormone insulin.
1981 The first cases of AIDS are described.
1982 The first vaccine for hepatitis B is licensed.
1986 Genetically engineered yeast are used to produce a more effective vaccine for hepatitis B.
Genetically engineered bacteria produce human growth hormone.

and within their normal environments, they have the vital role of decomposing dead organic material and recycling their nutrients. However, a number of bacteria cause human bacterial diseases, including strep throat, pneumonia, and meningitis.

Viruses are not cells; they are even smaller and simpler in structure than the bacteria. All viruses are parasites because they can reproduce only within the living cells of a host. Therefore, all viruses cause disease. Common human viral diseases are influenza, the common cold, and chickenpox.

Protozoa are single-celled animals such as amoebas. Most protozoa are freeliving in fresh or salt water, where they consume bacteria, fungi, and one another. Human protozoan parasites include those that cause malaria, amebic dysentery, and pneumocystis pneumonia (common in AIDS patients).

Fungi may be unicellular or multicellular. Molds and mushrooms are familiar fungi. They decompose organic matter in the soil and fresh water and help recycle nutrients. Fungal diseases of people include yeast infections, ringworm, and more serious diseases such as a type of meningitis.

Worms are multicellular animals. Most are freeliving and non-pathogenic; within the soil they consume dead organic matter or smaller living things. Worm infestations of people include trichinosis, hookworm disease, and tapeworms.

Arthropods (the name means "jointed legs") are multicellular animals such as lobsters, shrimp, the insects, ticks, and mites. Some insects (such as mosquitoes and fleas) are **vectors** of disease, that is, they spread pathogens from host to host when they bite to obtain blood. Ticks are also vectors of certain diseases, and some mites may cause infestations of the skin.

BINOMIAL NOMENCLATURE

We refer to bacteria and all other living things using two names (binomial nomenclature), the genus and the species. The genus name is placed first, is always capitalized, and is the larger category. The species name is second, is not capitalized, and is the smaller category. Let us use as examples *Staphylococcus aureus* and *Staphylococcus epidermidis.* These two bacteria are in the same genus, *Staphylococcus,* which tells us that they are related or similar to one another. Yet they are different enough to be given their own species names: *aureus* or *epidermidis.* It may be helpful here to think of our own names. Each of us has a family name, which indicates that we are related to other members of our families, and each of us has a first name indicating that we are individuals in this related group. If we wrote our own names using the method of binomial nomenclature, we would write Smith Mary and Smith John.

In scientific articles and books, for the sake of convenience, the genus name is often abbreviated with its first letter. We might read of *S. aureus* as a cause of a food poisoning outbreak or see *E. coli* (*E.* for *Escherichia*) on a lab report as the cause of a patient's urinary tract infection. Therefore, it is important to learn both genus and species names of important pathogens.

NORMAL FLORA

Each of us has a natural population of microorganisms living on or within us. This is our normal flora. These microbes may be further categorized as residents or

Table 22–1 DISTRIBUTION OF NORMAL FLORA IN THE HUMAN BODY

Body Site	Description of Flora
Skin	Exposed to the environment; therefore has a large bacterial population and small numbers of fungi, especially where the skin is often moist. Flora are kept in check by the continual loss of dead cells from the stratum corneum.
Nasal cavities	Bacteria, mold spores, and viruses constantly enter with inhaled air; the ciliated epithelium limits the microbial population by continuously sweeping mucus and trapped pathogens to the pharynx, where they are swallowed.
Trachea, bronchi, and lungs	The cilia of the trachea and large bronchial tubes sweep mucus and microbes upward toward the pharynx, where they are swallowed. Very few pathogens reach the lungs, and most of these are destroyed by alveolar macrophages.
Oral cavity	Large bacterial population and small numbers of yeasts and protozoa. Kept in check by lysozyme, the enzyme in saliva that inhibits bacterial reproduction. The resident flora help prevent the growth of pathogens.
Esophagus	Contains the microorganisms swallowed with saliva or food.
Stomach	The hydrochloric acid in gastric juice kills most bacteria. This may not be effective if large numbers of a pathogen or bacterial toxins are present in contaminated food.
Small intestine	The ileum, adjacent to the colon, has the largest bacterial population. The duodenum, adjacent to the stomach, has the smallest.
Large intestine	Contains an enormous population of bacteria, which inhibits the growth of pathogens and produces vitamins. The vitamins are absorbed as the colon absorbs water. Vitamin K is obtained in amounts usually sufficient to meet a person's daily need. Smaller amounts of folic acid, riboflavin, and other vitamins are also obtained from colon flora.
Urinary bladder	Is virtually free of bacteria, as is the upper urethra. The lower urethra, especially in women, has a flora similar to that of the skin.
Vagina	A large bacterial population creates an acidic pH that inhibits the growth of pathogens.
Tissue fluid	Small numbers of bacteria and viruses penetrate mucous membranes or get through breaks in the skin. Most are destroyed in lymph nodules or lymph nodes or by wandering macrophages in tissue fluid.
Blood	Should be free of microorganisms.

transients. **Resident flora** are those species that live on or in nearly everyone almost all the time. These residents live in specific sites, and we provide a very favorable environment for them. Some, such as *Staphylococcus epidermidis,* live on the skin. Others, such as *E. coli,* live in the colon and small intestine. When in their natural sites, resident flora do not cause harm to healthy tissue, and some are even beneficial to us. However, residents may become pathogenic if they are introduced into abnormal sites. If *E. coli,* for example, gains access to the urinary bladder, it causes an infection called cystitis. In this situation, *E. coli* is considered an **opportunist**, which is a normally harmless species that has become a pathogen in special circumstances.

Transient flora are those species that are found periodically on or in the body; they are not as well adapted to us as are the residents. *Streptococcus pneumoniae,* for example, is a transient in the upper respiratory tract, where it usually does not cause harm in healthy people. However, transients may become pathogenic when the host's resistance is lowered. In an elderly person with influenza, *S. pneumoniae* may invade the lower respiratory tract and cause a serious or even fatal pneumonia.

The distribution of our normal flora is summarized in Table 22–1. You can see that an important function of normal flora is to inhibit the growth of pathogens in the oral cavity, intestines, and in women, the vagina. The resident bacteria are believed to do this by simply being there and providing competition that makes it difficult for pathogens to establish themselves. An example may be helpful here. Let us use botulism. Typical food-borne botulism is acquired by ingesting the bacterial toxin that has been produced in food. Infants, however, may acquire botulism by ingesting the spores (dormant forms) of the botulism bacteria on foods such as honey or raw vegetables. The infant's colon flora is not yet abundant, and the botulism spores may be able to germinate into active cells that produce toxin in the baby's own intestine. For older children or adults, botulism spores are harmless if ingested, because the normal colon flora prevents the growth of these bacteria.

Resident flora may be diminished by the use of antibiotics to treat bacterial infections. An antibiotic does not distinguish between the pathogen and the resident bacteria. In such circumstances, without the usual competition, yeasts or pathogenic bacteria may be able to overgrow and create new infections. This is most likely to occur on mucous membranes such as those of the oral cavity and vagina.

INFECTIOUS DISEASE

An infectious disease is one that is caused by microorganisms or by the products (toxins) of microorganisms. To cause an infection, a microorganism must enter and establish itself in a host and begin reproducing.

Several factors determine whether a person will develop an infection when exposed to a pathogen. These include the virulence of the pathogen and the resistance of the host. **Virulence** is the ability of the pathogen to cause disease. Host **resistance** is the total of the body's defenses against pathogens. Our defenses include intact skin and mucous membranes, the sweeping of cilia to clear the respiratory tract, adequate nutrition, and the immune responses of our lymphocytes and macrophages (see Chapter 14).

To illustrate these concepts, let us compare the measles virus and rhinoviruses (common cold). The measles virus has at least a 90% infectivity rate, meaning that for every 100 non-immune people exposed, at least 90 will develop clinical measles. Thus, the measles virus is considered highly virulent, even for healthy people. However, people who have recovered from measles or who have received the measles vaccine have developed an active immunity that increases their resistance to measles. Even if exposed many times, such people probably will not develop clinical measles.

In contrast, the rhinoviruses that cause the common cold are not considered virulent pathogens, and healthy people may have them in their upper respiratory tracts without developing illness. However, fatigue, malnutrition, and other physiological stresses may lower a person's resistance and increase the likelihood of developing a cold upon exposure to these viruses.

Once infected, a person may have a **clinical (appparent** or **symptomatic)** infection, in which symptoms appear. **Symptoms** are the observable or measurable changes that indicate illness. For some diseases, **subclinical (inapparent** or **asymptomatic)** infections are possible, in which the person shows no symptoms. Women with the sexually transmitted disease gonorrhea, for example, may have subclinical infections, that is, no symptoms at all. It is important to remember that such people are still **reservoirs** (sources) of the pathogen for others, who may then develop clinical infections.

COURSE OF AN INFECTIOUS DISEASE

When a pathogen establishes itself in a host, there is a period of time before symptoms appear. This is called the **incubation period**. Most infectious diseases have rather specific incubation periods (Table 22-2). Some diseases, however, have more variable incubation periods (see hepatitis in Table 22-2), which may make it difficult to predict the onset of illness after exposure or to trace outbreaks of a disease.

A short time called the prodromal period may follow the incubation period. During this time, vague, non-specific symptoms may begin. These include generalized muscle aches, lethargy and fatigue, or a feeling that "I'm coming down with something."

During the invasion period, the specific symptoms of the illness appear. These might include a high fever, rash, swollen lymph nodes, cough, diarrhea, or such things as the gradual paralysis of botulism. The acme is the height or worst of the disease, and this is followed by recovery or the death of the host.

Some diseases are **self-limiting;** that is, they typically last a certain length of time and are usually followed by recovery. The common cold, chickenpox, and mumps are illnesses that are considered self-limiting.

Table 22–2 INCUBATION PERIODS OF SOME INFECTIOUS DISEASES

Disease	Incubation Period
Chickenpox	14–16 days
Cholera	1–3 days
Diphtheria	2–6 days
Gas gangrene	1–5 days
Gonorrhea	3–5 days
Hepatitis A	2 weeks–2 months
Hepatitis B	6 weeks–6 months
Hepatitis C	2 weeks–6 months
Herpes simplex	4 days
Influenza	1–3 days
Leprosy	3 months–20+ years
Measles	10–12 days
Meningitis (bacterial)	1–7 days
Mumps	2–3 weeks
Pertussis	5 days–3 weeks
Pinworm	2–6 weeks
Plague	2–6 days
Polio	7–14 days
Rabies	2 weeks–2 months (up to 1 year)
Salmonella food poisoning	12–72 hours
Staphyloccus food poisoning	1–8 hours
Syphilis	10 days–3 months
Tetanus	3 days–5 weeks
Tuberculosis	2–10 weeks

TYPES OF INFECTION

The terminology of infection may refer to the location of the pathogens in the body, to the general nature of the disease, or to how or where the pathogen was acquired.

A **localized** infection is one that is confined to one area of the body. Examples are the common cold of the upper respiratory tract, boils of the skin, and salmonella food poisoning that affects the intestines.

In a **systemic** infection, the pathogen is spread throughout the body by way of the lymph or blood. Typhoid fever, for example, begins as an intestinal infection, but the bacteria eventually spread to the liver, gallbladder, kidneys, and other organs. Bubonic plague is an infection that begins in lymph nodes, but again the bacteria are carried throughout the body, and fatal plague is the result of pneumonia.

Bacteremia and **septicemia** are terms that are often used synonymously in clinical practice; they mean that bacteria are present in the blood and are being circulated throughout the body. Septicemia is always serious, for it means that the immune defenses have been completely overwhelmed and are unable to stop the spread of the pathogen.

With respect to timing and duration, some infections may be called acute or chronic. An **acute** infection is one that usually begins abruptly and is severe. In contrast, a **chronic** infection often progresses slowly and may last for a long time.

A **secondary** infection is one that is made possible by a primary infection that has lowered the host's resistance. Influenza in an elderly person, for example, may be followed by bacterial pneumonia. This secondary bacterial infection might not have occurred had not the person first been ill with the flu.

Nosocomial infections are those that are acquired in hospitals or other institutions such as nursing homes. The hospital population includes newborns, the elderly, post-operative patients, people with serious chronic diseases, cancer patients receiving chemotherapy, and others whose resistance to disease is lowered. Some hospital-acquired pathogens, such as *Staphylococcus aureus,* are transmitted from patient to patient by healthy hospital personnel. These staff members are reservoirs for *S. aureus* and carry it on their skin or in their upper respiratory tracts. For this reason, proper handwashing is of critical importance for all hospital staff.

Other nosocomial infections, however, are caused by the patient's own normal flora that has been inadvertently introduced into an abnormal body site. Such infections may be called **endogenous**, which literally means generated from within. Intestinal bacilli such as *E. coli* are now the No. 1 cause of nosocomial infections. Without very careful aseptic technique (and sometimes in spite of it), the patient's own intestinal bacteria may contaminate urinary catheters, decubitus ulcers, surgical incisions, chest tubes, and intravenous lines. Such infections are a significant problem in hospitals, and all those involved in any aspect of patient care should be aware of this.

EPIDEMIOLOGY

Epidemiology is the study of the patterns and spread of disease within a population. As you see, this term is related to **epidemic**, which is an outbreak of disease, that is, more than the usual number of cases in a given time period. An **endemic** disease is one that is present in a population, with an expected or usual number of cases in a given time. Influenza, for example, is endemic in large cities during the winter, and public health personnel expect a certain number of cases. In some winters, however, the number of cases of influenza increases, often markedly, and this is an epidemic.

A **pandemic** is an epidemic that has spread throughout several countries. The bubonic plague pandemic of the 14th century affected nearly all of Europe and killed one-fourth of the population. Just after World War I, in 1918 to 1920, an especially virulent strain of the influenza virus spread around the world and caused 20 million deaths. More recently, an epidemic of cholera began in Peru in January 1991, but soon became a pandemic as cholera spread to neighboring South American countries.

To understand the epidemiology of a disease, we must know several things about the pathogen. These include where it lives in a host, the kinds of hosts it can infect, and whether it can survive outside of hosts.

PORTALS OF ENTRY AND EXIT

The **portal of entry** is the way the pathogen enters a host (Fig. 22–1). Breaks in the skin, even very small ones, are potential portals of entry, as are the natural body openings. Pathogens may be inhaled, consumed with food and water, or acquired during sexual activity. Most pathogens that enter the body by way of these natural routes are destroyed by the white blood cells found in and below the skin and mucous membranes, but some may be able to establish themselves and cause disease.

Insects such as mosquitoes, fleas, and lice, and other arthropods, such as ticks, are vectors of disease. They spread pathogens when they bite to obtain a host's blood. Mosquitoes, for example, are vectors of malaria, yellow fever, and encephalitis. Ticks are vectors of Lyme disease and Rocky Mountain spotted fever.

As mentioned previously, it is important to keep in mind that many hospital procedures may provide portals of entry for pathogens. Any invasive procedure, whether it involves the skin or the mucous membranes, may allow pathogens to enter the body. Thus it is essential that all health-care workers follow aseptic technique for such procedures.

The **portal of exit** (see Fig. 22–1) is the way the pathogen leaves the body or is shed from the host. Skin lesions, such as those of chickenpox, contain pathogens that may be transmitted to others by cutaneous contact. Intestinal pathogens such as the hepatitis A virus and the cholera bacteria are excreted in the host's feces, which may contaminate food or water and be ingested by another host (this is called the fecal-oral route of transmission). Respiratory pathogens such as influenza and measles viruses are shed in respiratory droplets from the mouth and nose and may

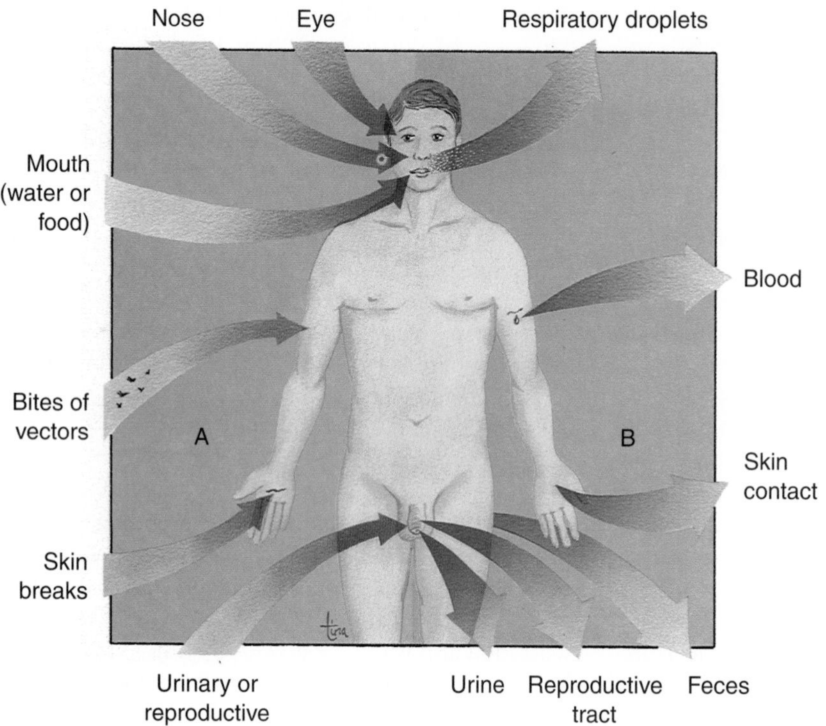

Figure 22–1 **(A)** Portals of entry. **(B)** Portals of exit.

be inhaled by another person. The pathogens of the reproductive tract, such as the bacteria that cause syphilis and gonorrhea, are transmitted to others by sexual contact. Notice that with respect to epidemiology, the pathogen travels from one host's portal of exit to another host's portal of entry.

RESERVOIRS OF INFECTION

Some pathogens cause disease only in people. Measles, whooping cough, syphilis, and bacterial meningitis are strictly human diseases. To acquire such a disease, a person must be exposed to someone who has the illness.

Also of importance is that upon recovery from some diseases, the host may continue to harbor the pathogen and thus be a reservoir of it for others. Such a person is called a **carrier**. Diseases for which the carrier state is possible include typhoid, diphtheria, and hepatitis B.

Many other diseases, however, are really animal diseases that people acquire in certain circumstances. These diseases are called **zoonoses** (singular **zoonosis**) and include plague, Lyme disease, encephalitis, and Rocky Mountain spotted fever, which are spread from animal to person by vectors such as ticks or fleas. Rabies is acquired by contact with infected animal saliva or infected tissue, the virus entering the new host through breaks in the skin. Salmonellosis is a type of food poisoning caused by the intestinal bacilli of animals that contaminate meats such as chicken and turkey. Prevention of such diseases depends upon knowledge of how they are spread. For example, people who live in areas where Lyme disease is endemic should be aware that the disease is acquired by way of a tick bite. If children and pets are examined for ticks after they have been out of doors, the chance of acquiring Lyme disease is greatly diminished.

Some bacteria are pathogenic only by accident, for their natural habitat is soil or water, where they act as decomposers. The bacteria that cause gas gangrene, tetanus, and botulism are normal soil flora and cause disease when they (or their toxins) contaminate a skin wound, or, in the case of botulism, the toxin is present in food.

SPREAD OF INFECTION

Based on our knowledge thus far, we can classify infectious diseases as non-communicable or communicable. A **non-communicable** disease is one in which a resident species causes disease under certain conditions or in which a non-resident species causes disease when it enters the body. Such diseases cannot be transmitted directly or indirectly from host to host. Cystitis caused by *E. coli* in a hospital patient, for example, is not communicable to the nurses who care for that patient. Similarly, a nurse caring for a patient with tetanus or botulism need not worry about acquiring these diseases; both are non-communicable.

A **communicable** disease is one in which the pathogen may be transmitted directly or indirectly from host to host. Direct spread of infection is by way of cutaneous contact (including sexual contact), respiratory droplets, contaminated blood, or placental transmission from mother to fetus. Indirect spread is by way of contaminated food or water, or vectors, or **fomites**, which are inanimate objects that carry the pathogen. Influenza and cold viruses, for example, can survive outside their hosts for a time, so that objects such as eating utensils may be vehicles of transmission for these pathogens.

Some communicable diseases may also be called **contagious**, which means that they are easily spread from person to person by casual cutaneous contact or by respiratory droplets. Chickenpox, measles, and influenza are contagious diseases. In contrast, AIDS is not contagious, because sexual contact, blood contact, or placental transmission is necessary to acquire the virus (HIV). HIV is not spread by cutaneous contact or by respiratory droplets.

METHODS OF CONTROL OF MICROBES

Microorganisms are everywhere in our environment, and although we need not always be aware of their presence, there are times when we must try to diminish or even eliminate them. These situations include the use of chemicals for disinfection, especially in hospitals, and the protection of our food and water supplies.

ANTISEPTICS, DISINFECTANTS, AND STERILIZATION

We are all familiar with the practice of applying iodine, hydrogen peroxide, or alcohol to minor cuts in the skin, and we know the purpose of this: to prevent bacterial infection. The use of such chemicals does indeed destroy many harmful bacteria, although it has no effect on bacterial spores. The chemicals used to prevent infection may be called antiseptics or disinfectants. An **antiseptic** (anti = against; septic = infection) is a chemical that destroys bacteria or inhibits their growth on a living being. The chemicals named above are antiseptics on skin surfaces. A **disinfectant** is a chemical that is used on inanimate objects. Chemicals with antibacterial effects may be further classified as bactericidal or bacteriostatic. **Bactericides** kill bacteria by disrupting important metabolic processes. **Bacteriostatic** chemicals do not destroy bacteria, but rather inhibit their reproduction and slow their growth. Alcohol, for example, is a bactericide that is both an antiseptic and a disinfectant, depending upon the particular surface on which it is used.

Some chemicals are not suitable for use on human skin because they are irritating or damaging, but they may be used on environmental surfaces as disinfectants. Bleach, such as Clorox, and cresols, such as Lysol, may be used in bathrooms, on floors or countertops, and even on dishes and eating utensils (if rinsed thoroughly). These bactericides will also destroy certain viruses, such as those that cause influenza. A dilute (10%) bleach solution will inactivate HIV, the virus that causes AIDS.

In hospitals, environmental surfaces are disinfected, but materials such as surgical instruments, sutures, and dressings must be sterilized. **Sterilization** is a process that destroys all living organisms. Most medical and laboratory products are sterilized by autoclaving. An **autoclave** is a chamber in which steam is generated under pressure. This pressurized steam penetrates the contents of the chamber and kills all microorganisms present, including bacterial spores.

Materials such as disposable plastics that might be damaged by autoclaving are often sterilized by exposure to ionizing radiation. Foods such as meats may also be sterilized by this method. Such food products have a very long shelf life (equivalent to canned food), and this procedure is used for preparing some military field rations.

PUBLIC HEALTH MEASURES

Each of us is rightfully concerned with our own health and the health of our families. People who work in the public health professions, however, consider the health of all of us, that is, the health of a population. Two important aspects of public health are ensuring safe food and safe drinking water.

Food

The safety of our food depends on a number of factors. Most cities have certain standards and practices that must be followed by supermarkets and restaurants, and inspections are conducted on a regular basis.

Food companies prepare their products by using specific methods to prevent the growth of microorganisms. Naturally, it is in the best interests of these companies to do so, for they would soon be out of business if their products made people ill. Also of importance is the willingness of companies to recall products that are only suspected of being contaminated. This is all to the benefit of consumers. For example, since 1925 in the United States, only five fatal cases of botulism have been traced to commercially canned food. If we consider that billions of cans of food have been consumed during this time, we can appreciate the high standards the food industry has maintained.

Milk and milk products provide ideal environments for the growth of bacteria because they contain both protein and sugar (lactose) as food sources. For this reason, milk must be **pasteurized**, that is, heated to 145°F (62.9°C) for 30 minutes. Newer methods of pasteurization use higher temperatures for shorter periods of time, but the result is the same: The pathogens that may be present in milk are killed, although not all bacteria are totally destroyed. Milk products such as cheese and ice cream are also pasteurized, or are made from pasteurized milk.

When a food-related outbreak of disease does take place, public health workers try to trace the outbreak to its source. This stops the immediate spread of disease by preventing access to the contaminated food, and the ensuing publicity on television or in the newspapers may help remind everyone of the need for careful monitoring of food preparation.

Some foods meant to be eaten raw or briefly cooked, such as fruit or rare beef, do carry a small risk. Consumers should realize that food is not sterile, that meat, for example, is contaminated with the animals' intestinal bacteria during slaughtering. Meat should be thoroughly cooked. Fruit and vegetables should be washed or peeled before being eaten raw. (Food-borne diseases, also called food poisoning, are included in the Tables of Diseases at the end of this chapter.)

Finally, the safety of our food may depend on something we often take for granted: our refrigerators. For example, a Thanksgiving turkey that was carved for dinner at 3 P.M. and left on the kitchen counter until midnight probably should not be used for turkey sandwiches the next day. Although we have to rely on others to ensure that commercially prepared food will be safe, once food reaches our homes, all we really need (besides the refrigerator) is our common sense.

Water

When we turn on a faucet to get a glass of water, we usually do not wonder whether the water is safe to drink. It usually is. Having a reliable supply of clean drinking water depends on two things: diverting human sewage away from water supplies and chlorinating water intended for human consumption.

Large cities have sewer systems for the collection of waste water and its subsequent treatment in sewage plants. Once treated, however, the sludge (solid, particulate matter) from these plants must be disposed of. This is becoming more of a problem simply because there is so much sewage sludge (because there are so many of us). Although the sludge is largely free of pathogens, it ought not be put in landfills, and because ocean dumping is being prohibited in many coastal areas, this is a problem that will be with us for a long time.

Drinking water for cities and towns is usually chlorinated. The added chlorine kills virtually all the bacteria that may be present. The importance of chlorination is shown by a 1978 outbreak of enteritis (diarrhea) in a Vermont town of 10,000 people. The chlorination process malfunctioned for 2 days, and 2000 of the town's inhabitants became ill. (The bacterium was *Campylobacter,* a common intestinal inhabitant of animals).

You may now be wondering if all those bottled spring waters are safe to drink. The answer, in general, is yes, because the bottling companies do not wish to make people ill and put themselves out of business. Some bottled waters, however, do have higher mold spore counts than does chlorinated tap water. Usually these molds are not harmful when ingested; they are destroyed by the hydrochloric acid in gastric juice.

In much of North America, nearly everyone has easy access to safe drinking water. We might remind ourselves once in a while that our water will not give us typhoid, polio, or cholera. These diseases are still very common in other parts of the world, where the nearest river or stream is the laundry, the sewer, and the source of drinking water.

THE PATHOGENS

In the sections that follow, each group of pathogens will be described with a summary of important characteristics. Examples of specific pathogens will be given to help you become familiar with them. Tables of important human diseases caused by each group of pathogens are found at the end of this chapter.

BACTERIA (see Table 22–3, pages 502–505)

Bacteria are very simple unicellular organisms. All are microscopic in size, and a magnification of 1000 times is usually necessary to see them clearly. A bacterial cell consists of watery cytoplasm and a single chromosome (made of DNA) surrounded by a cell membrane. Enclosing all of these structures is a cell wall, which is strong and often rigid, giving the bacterium its characteristic shape.

Based on shape, bacteria are classified as one of three groups: coccus, bacillus, or spirillum (Fig. 22–2). A **coccus** (plural: **cocci**) is a sphere; under the microscope cocci appear round. Certain prefixes may be used to describe the arrangement of spheres. **Staphylo** means clusters; **strepto** refers to chains of cells, and **diplo** means pairs of cells.

A **bacillus** (plural: **bacilli**) is a rod-shaped bacterium; rods may vary in length depending on the genus. A **spirillum** (plural: **spirilla**) is a long cell with one or more curves or coils. Some spirilla, such as those that cause syphilis and Lyme disease, are called **spirochaetes**. Many of the bacilli and spirilla are capable of movement because they have **flagella**. These are long, thread-like structures that project from the cell and beat rhythmically.

Bacteria reproduce by the process of **binary fission**, in which the chromosome duplicates itself, and the original cell divides into two identical cells. The presence or absence of oxygen may be important for bacterial reproduction. **Aerobic** bacteria can reproduce only in the presence of oxygen, and **anaerobic** bacteria can reproduce only in the absence of oxygen. **Facultatively anaerobic** bacteria are not inhibited in either situation; they are able to reproduce in either the presence or absence of oxygen. This is obviously an advantage for the bacteria, and many pathogens and potential pathogens are facultative anaerobes.

The Gram Stain

Based on the chemicals in their cell walls, most bacteria can be put into one of two groups, called **gram positive** and **gram negative**. A simple laboratory procedure called the **Gram stain** (Fig. 22–2) shows us the shape of the bacteria and their gram reactions. Gram positive bacteria appear purple or blue, and gram negative bacteria appear pink or red. Some bacteria do not stain with the Gram method, but for those that do, each genus is either gram positive or gram negative. This does not change, just as the characteristic shape of the bacteria does not change. The genus *Streptococcus,* for example, is always a gram positive coccus; the genus *Escherichia* is always a gram negative bacillus. If a Gram stain is done on a sputum specimen from a patient with pneumonia, and a gram positive coccus is found, this eliminates all the gram negative cocci and bacilli that may also cause pneumonia. The Gram stain, therefore, is often an important first step in the indentification of the pathogen that is causing a particular infection. In Table 22–3 (at the end of this chapter), the gram reaction (where applicable) and shape are included for each pathogen.

Special Characteristics

Although bacteria are simple cells, many have special structural or functional characteristics that help them to survive. Some bacilli and cocci have capsules (see Fig. 22–2); a **capsule** is a gelatinous sheath that encloses the entire cell. Capsules are beneficial to the bacteria because they inhibit phagocytosis by the host's white blood cells. This gives the bacteria time to reproduce and possibly establish themselves in the host. This is *not* beneficial from our point of view (remember that we are the hosts), but bacterial capsules are also **antigenic**, which means that they stimulate antibody production by our lymphocytes. This starts the destruction of bacteria by our immune responses. We take advantage of this by using bacterial capsules in some of our vaccines, such as those used to prevent pneumonia and meningitis.

Some bacilli are able to survive unfavorable environments by forming spores. A **spore** is a dormant (inactive) stage that consists of the chromosome and a small amount of cytoplasm surrounded by a thick wall. Spores can survive conditions such as heat (even boiling), freezing, or dehydration, which would kill the vegetative (active) forms of the bacterial cells. Fortunately for us, most pathogens are unable to form spores, but some that do are the causative agents of gas gangrene, botulism, and tetanus. These bacteria are decomposers in the soil environment, and their spore-forming ability enables them to survive the extremes of temperature and lack of water that may occur in the soil.

Many bacteria cause disease because they produce **toxins**, which are chemicals that are harmful to host tissues. Often these toxins are the equivalent of our digestive enzymes, which break down the food we eat. Some bacterial toxins such as hemolysins and pro-

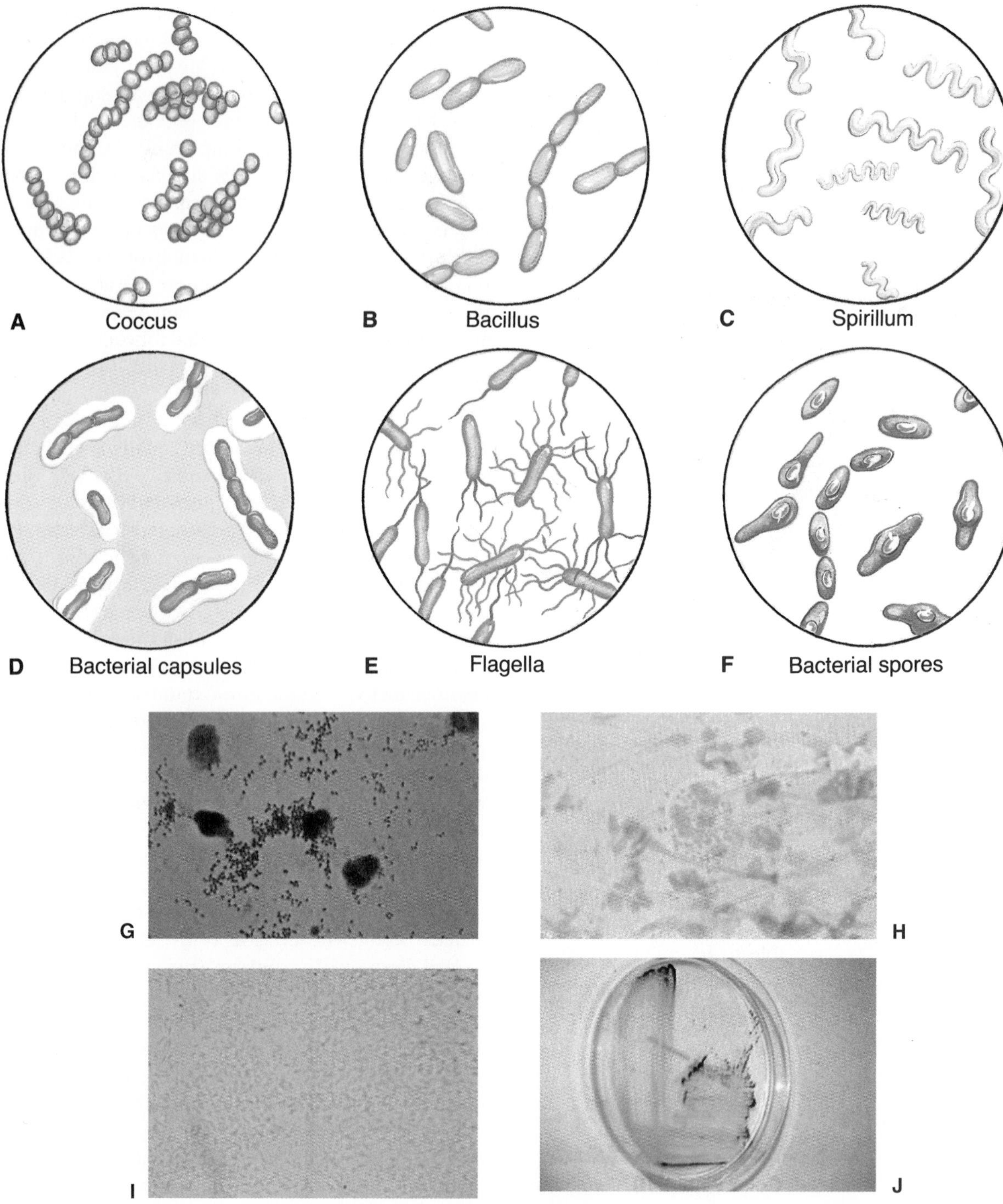

Figure 22–2 **(A–F)** Bacterial shapes and specialized structures (magnification × 2000). **(G)** Gram positive *Staphylococcus aureus* (× 1000). **(H)** Gram negative *Neisseria gonorrhoeae* in WBCs (× 1000). **(I)** Gram negative *Campylobacter jejuni* (× 1000). **(J)** *Salmonella enteritidis* growing on agar in a Petri dish. The black colonies indicate production of hydrogen sulfide gas. (G–J from Sacher, RA, and McPherson, RA: Widmann's Clinical Interpretation of Laboratory Tests, ed 10, FA Davis, Philadelphia, 1991, with permission.)

teases literally digest host tissues such as red blood cells and proteins. The bacteria then absorb the digested nutrients.

The toxins of other bacteria have very specific effects on certain cells of the host. Botulism and tetanus toxins, for example, are **neurotoxins** that disrupt the functioning of nerve cells, leading to the characteristic symptoms of each disease. The diphtheria toxin causes heart failure, the pertussis (whooping cough) toxin immobilizes the cilia of the respiratory tract, and the cholera enterotoxin causes diarrhea (see also Table 22-3).

The cell walls of gram negative bacteria are made of chemicals called endotoxins. **Endotoxins** all have the same effects on the host: They cause fever and circulatory shock (low blood pressure and heart failure). **Endotoxin shock** (also called **gram negative shock**) is a life-threatening condition that may accompany any serious infection with gram negative bacteria.

Rickettsias and Chlamydias

These two groups of bacteria differ from most other bacteria in that they are obligate intracellular parasites. This means that they can reproduce only within the living cells of a host.

The rickettsias are parasites of mammals (including people) and are often spread by arthropod vectors. In the United States, the most common rickettsial disease is Rocky Mountain spotted fever, which is spread by ticks. In other parts of the world, epidemic typhus, which is spread by body lice, is still an important disease. From a historical perspective, until World War I, more people died of epidemic typhus during times of war than were killed by weapons.

The chlamydias cause several human diseases, including ornithosis (parrot fever) and trachoma, which is the leading cause of blindness throughout the world. In the United States, chlamydial infection of the genitourinary tract has become the most prevalent sexually transmitted disease, with estimates of 4 million new cases each year.

Both rickettsial and chlamydial infections can be treated with antibiotics.

Antibiotics

Antibiotics are chemicals that are used to treat bacterial infections. A **broad-spectrum** antibiotic is one that affects many different kinds of bacteria; a **narrow-spectrum** antibiotic affects just a few kinds of bacteria.

The use of antibiotics is based on a very simple principle: Certain chemicals can disrupt or inhibit the chemical reactions that bacteria must carry out to survive. An antibiotic such as penicillin blocks the formation of bacterial cell walls; without their cell walls, bacteria will die. Other antibiotics inhibit DNA synthesis or protein synthesis. These are vital activities for the bacteria, and without them bacteria cannot reproduce and will die.

It is very important to remember that our own cells carry out chemical reactions that are very similar to some of those found in bacteria. For this reason, our own cells may be damaged by antibiotics. This is why some antibiotics have harmful side effects. The most serious side effects are liver and kidney damage or depression of the red bone marrow. The liver is responsible for detoxifying the medication, which may accumulate and damage liver cells. Similar damage may occur in the kidneys, which are responsible for excreting the medication. The red bone marrow is a very active tissue, with constant mitosis and protein synthesis to produce RBCs, WBCs, and platelets. Any antibiotic that interferes with these processes may decrease production of all of these blood cells. Patients who are receiving any of the potentially toxic antibiotics should be monitored with periodic tests of liver and kidney function or with blood counts to assess the state of the red bone marrow.

Another problem with the use of antibiotics is that bacteria may become resistant to them, and so be unaffected. Bacterial **resistance** means that the bacteria are able to produce an enzyme that destroys the antibiotic, rendering it useless. This is a genetic capability on the part of bacteria, and it is, therefore, passed to new generations of bacteria cells. Most strains of *Staphylococcus aureus,* for example, are resistant to penicillin and other antibiotics. Most of the gram negative intestinal bacilli are resistant to a great variety of antibiotics. This is why **culture and sensitivity testing** is so important before an antibiotic is chosen to treat these infections.

To counteract bacterial resistance, new antibiotics are produced that are not inactivated by the destructive bacterial enzymes. Within a few years, however, the usefulness of these new antibiotics will probably diminish as bacteria mutate and develop resistance. This is not a battle that we can ever truly win, because bacteria are living organisms that evolve as their environment changes.

Antibiotics have changed our lives, although we may not always realize that today. A child's strep throat will probably not progress to ear infections and meningitis, and bacterial pneumonia does not have the very high fatality rate that it once did. But, we must keep in mind that antibiotics are not a "cure" for any disease. An infection, especially a serious one, means that the immune system has been overwhelmed by the pathogen. An antibiotic diminishes the number of bacteria to a level with which the immune system can cope. Ultimately, however, the body's own white blood cells must eliminate the very last of the bacteria.

VIRUSES (see Table 22–4, pages 506–507)

Viruses are not cells; their structure is even simpler than that of bacteria, which are the simplest cells. A virus consists of either DNA or RNA surrounded by a protein shell. The protein shell has a shape that is characteristic for each virus (Fig. 22-3). There are no enzymes, cytoplasm, cell membranes, or cell walls in viruses, and they can reproduce only when inside the living cells of a host. Therefore, all viruses are obligate intracellular parasites, and they cause disease when they reproduce inside cells. When a virus enters a host cell, it uses the cell's chromosomes, RNA, and enzymes to make new viruses. Several hundred new viruses may be produced from just one virus. The host cell ruptures and dies, releasing the new viruses, which then enter other cells and reproduce.

The severity of a viral disease depends upon the types of cells infected. If the virus affects skin cells, for example, the disease is usually mild and self-limiting, such as chickenpox. Small numbers of skin cells are not crucial to our survival, and these cells can be replaced by mitosis. If, however, the virus affects nerve cells, the disease is more serious and may be fatal. Rabies is such a disease. Neurons are much more vital to us, and they cannot be replaced once they die.

Some viruses, such as those that cause German measles (rubella) and chickenpox, are able to cross the placenta, that is, pass from maternal circulation to fetal circulation. Although the disease may be very mild for the pregnant woman, the virus may severely damage developing fetal organs and cause congenital birth defects such as blindness, heart malformations, and mental retardation. In the most serious cases, fetal infection may result in miscarriage or stillbirth.

There are some viruses that cause an initial infection, become dormant, then are reactivated, causing another infection months or years later. The herpes viruses that cause cold sores "hide out" in nerves of

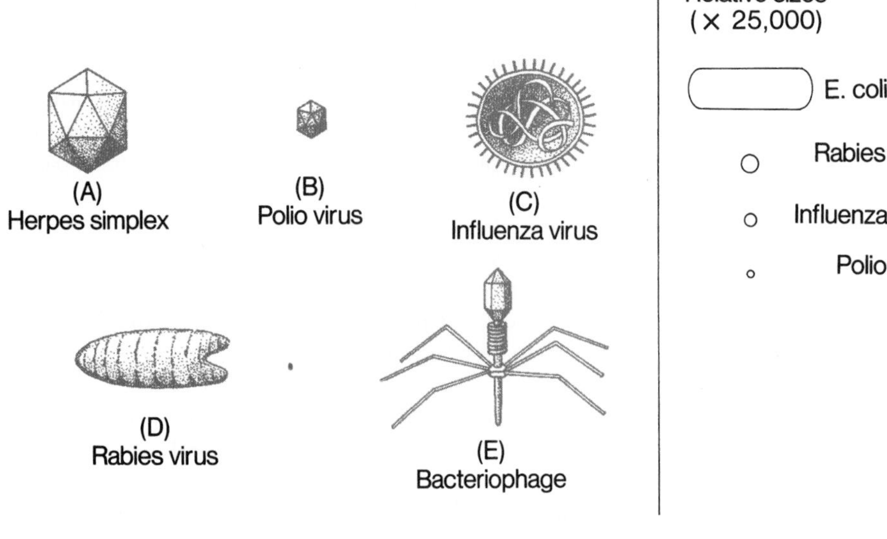

Figure 22–3 Viruses: representative shapes and relative sizes.

the face following the initial skin lesion. At some later time when the host's resistance is lowered, the viruses emerge from the nerves and cause another cold sore. The chickenpox virus, which most of us acquire as children, is a herpes virus that may become dormant in nerves for years, and then be reactivated and cause shingles when we are adults.

A few human viruses are known to be tumor viruses, that is, they cause cells to develop abnormally and form tumors. The Epstein-Barr virus, which causes mononucleosis in North America, is associated with Burkitt's lymphoma in Africa and with nasopharyngeal carcinoma in China. There are environmental factors, as yet unknown, that contribute to the development of these cancers in specific parts of the world. Several of the human papilloma viruses have also been associated with cancers of the mouth or larynx, and three of these viruses are found in 90% of cervical carcinomas in women.

Important viral diseases are described in Table 22-4 at the end of this chapter.

Antiviral Medications

The treatment of viral diseases with chemicals poses some formidable challenges. First, viruses are active (reproducing) only within cells, so the medication must be able to enter infected cells to be effective. Second, viruses are such simple structures that the choice of which of their chemical processes to attempt to disrupt is limited. Third, viruses use the host cell's DNA and enzymes for self-replication, and a medication that interferes with DNA or enzymes may kill the host cell even as it kills the virus.

These problems are illustrated by zidovudine (AZT), the first medication thought to be effective against HIV, the virus that causes AIDS. Zidovudine works by interfering with DNA synthesis, which the virus must carry out to reproduce. The side effects of zidovudine, which are experienced by a significant number of AIDS patients, are caused by the disruption of DNA synthesis in the person's own cells.

Despite these obstacles, a few successful antiviral drugs have been developed. Acyclovir, for example, has proved to be useful in the control (not cure) of herpes viruses. Ribavirin has been quite effective in the treatment of respiratory syncytial virus pneumonia in infants and young children. This is an area of intensive research, and more antiviral medications will undoubtedly be found within the next decade.

FUNGI (see Table 22–5, page 507)

Fungi may be unicellular, such as yeasts, or multicellular, such as the familiar molds and mushrooms. Most fungi are **saprophytes**, that is, they live on dead organic matter and decompose it to recycle the chemicals as nutrients. The pathogenic fungi cause infections that are called **mycoses** (singular: **mycosis**), which may be superficial or systemic.

Yeasts (Fig. 22-4) have been used by people for thousands of years in baking and brewing. In small numbers, yeasts such as *Candida albicans* are part of the resident flora of the skin, mouth, intestines, and vagina. In larger numbers, however, yeasts may cause superficial infections of mucous membranes or the skin, or very serious systemic infections of internal organs. An all-too-common trigger for oral or vaginal yeast infections is the use of an antibiotic to treat a bacterial infection. The antibiotic diminishes the normal bacterial flora, thereby removing competition for the yeasts, which are then able to overgrow. Yeasts may also cause skin infections in diabetics, or in obese people who have skin folds that are always moist. In recent years, *Candida* has become an important cause of nosocomial infections. The resistance of hospital patients is often lowered because of their diseases or treatments, and they are more susceptible to systemic yeast infections in the form of pneumonia or endocarditis.

Another superficial mycosis is ringworm (Tinea), which may be caused by several species of fungi (see Table 22-5 at the end of this chapter). The name *ringworm* is misleading, because there are no worms involved. It is believed to have come from the appearance of the lesions: circular, scaly patches with reddened edges, the center clears as the lesion grows. Athlete's foot, which is probably a bacterial-fungal infection, is perhaps the most common form of ringworm.

The systemic mycoses are more serious diseases that occur when spores of some fungi gain access to the interior of the body. Most of these fungi grow in a mold-like pattern. The molds we sometimes see on stale bread or overripe fruit look fluffy or fuzzy. The fluff is called a mycelium and is made of many threadlike cellular structures called hyphae. The color of a mold is due to the spore cases (sporangia) in which spores are produced (see Fig. 22-4). Each spore may be carried by the air to another site, where it germinates and forms another mycelium.

Because spores of these fungi are common in the

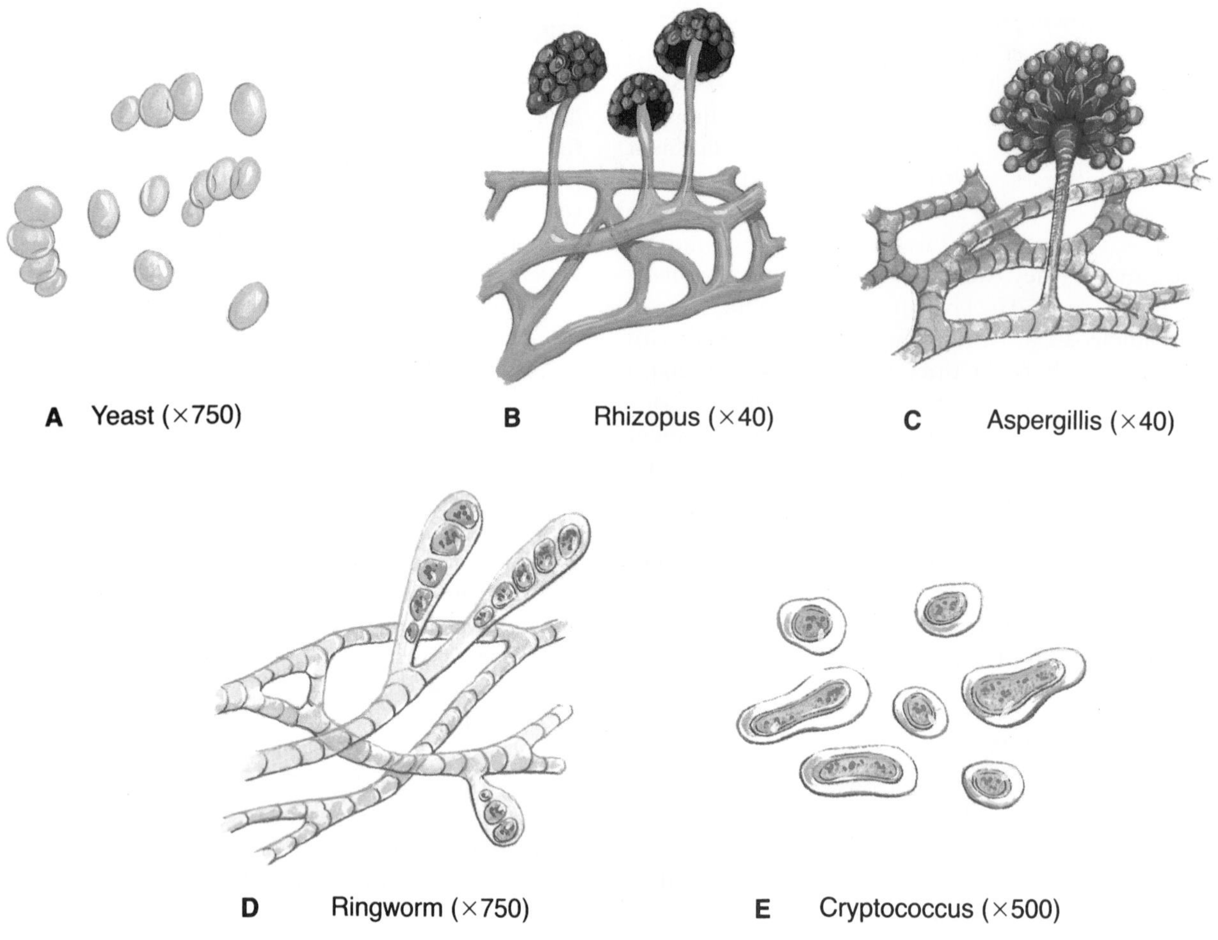

Figure 22–4 Fungi.

environment, they are often inhaled. The immune responses are usually able to prevent infection and healthy people are usually not susceptible to systemic mycoses. Elderly people and those with chronic pulmonary diseases are much more susceptible, however, and they may develop lung infections. The importance of the immune system is clearly evident if we consider people with AIDS. Without the normal immune responses, AIDS patients are very susceptible to invasive fungal diseases, including meningitis caused by *Cryptococcus.*

Antifungal Medications

One of the most effective drugs used to treat serious, systemic mycoses, amphotericin B, has great potential to cause serious side effects. Patients receiving this medication should have periodic tests of liver and kidney function. Newer medications include ketoconazole and fluconazole, which are less toxic to the recipient and may prove to be as effective as amphotericin B.

Superficial mycoses such as ringworm may be treated with drugs such as griseofulvin. Taken orally, the drug is incorporated into living epidermal cells. When these cells die and reach the stratum corneum, they are resistant to the digestive action of the ringworm fungi. An effective topical spray has also been developed.

There are several effective medications for mucosal yeast infections, but it is important that the trigger for the infection (such as antibiotic therapy) be resolved as well. If not, the yeast infection may recur when the medication is stopped.

PROTOZOA (see Table 22–6, page 508)

Protozoa are unicellular animals, single cells that are adapted to life in fresh water (including soil) and salt water. Some are human pathogens and are able to form cysts, which are resistant, dormant cells that are able to survive passage from host to host.

Intestinal protozoan parasites of people include *Entamoeba histolytica,* which causes amebic dysentery, and *Giardia lamblia,* which causes diarrhea called giardiasis (Fig. 22-5 and Table 22-6). People acquire these by ingesting food or water contamifjnated with the cysts of these species. Giardiasis can become a problem in day-care centers if the staff is not careful concerning handwashing and food preparation.

Plasmodium, the genus that causes malaria, affects hundreds of millions of people throughout the world and is probably the most important protozoan parasite. The *Plasmodium* species are becoming increasingly resistant to the standard antimalarial drugs, which are used to prevent disease as well as cure it. Although work is progressing on malaria vaccines, one will probably not be available for several years.

One protozoan of which you have probably heard is *Pneumocystis carinii,* which causes pneumonia in people with AIDS. This species is usually not pathogenic, because the healthy immune system can easily control it. For AIDS patients, however, this form of pneumonia is often the cause of death.

Medications are available that can treat most protozoan infections. Intestinal protozoa, for example, may be treated with metronidazole or furazolidone. The drug pentamidine is used to treat pneumocystis pneumonia, although the underlying cause, AIDS, is not yet curable.

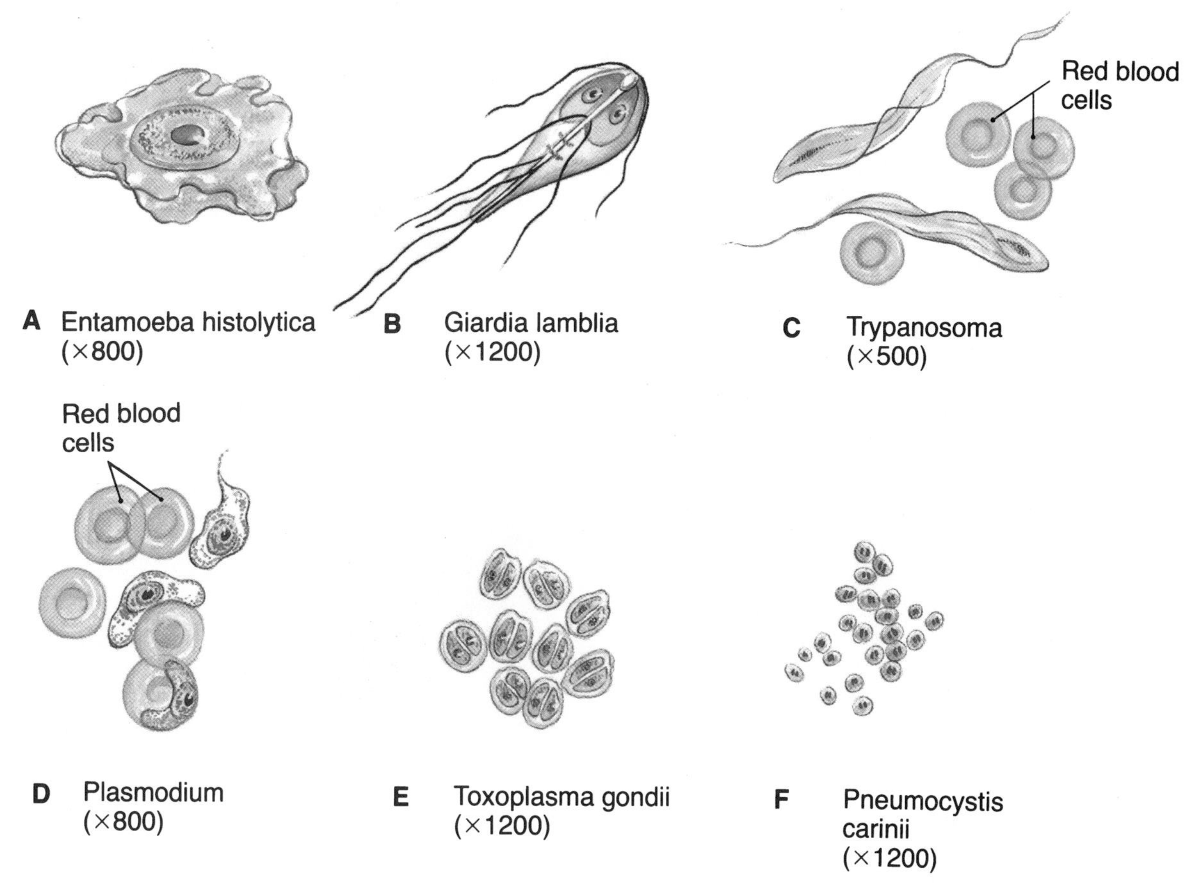

Figure 22–5 Protozoa.

WORMS (HELMINTHS)
(see Table 22–7, page 508)

Most worms are simple multicellular animals. The parasitic worms are even simpler than the familiar earthworm, because they live within hosts and use the host's blood or nutrients as food. Many of the parasitic worms have complex life cycles that involve two or more different host species.

The flukes are flatworms that are rare in most of North America but very common in parts of Africa and Asia. People acquire these species by eating aquatic plants or raw fish in which the larval worms have encysted. Within the person, each species lives in a specific site: the intestine, bile ducts, or even certain veins. Although rarely fatal, these chronic worm infestations are often debilitating, and the host person is a source of the eggs of the fluke, which may then infect others.

Tapeworms are also flatworms (Fig. 22–6). Some are 10 to 15 feet long, and the fish tapeworm can be as long as 60 feet. They are as flat as a ribbon, however, and one could easily be held in the palm of the hand. The tapeworm holds on to the lining of a host's small intestine with the suckers and hooks on its scolex (front end). The segments, called proglottids, are produced continuously in most species and absorb nutrients from the host's digested food. The only function of the proglottids is reproduction: Eggs in one segment are fertilized by sperm from another segment. Mature proglottids containing fertilized eggs break off and are excreted in the host's feces. An intermediate host such as a cow or pig eats food contaminated with human feces, and the eggs hatch within this animal and grow into larval worms that encyst in the animal's muscle tissue. People become infected by eating poorly cooked beef or pork that contains cysts.

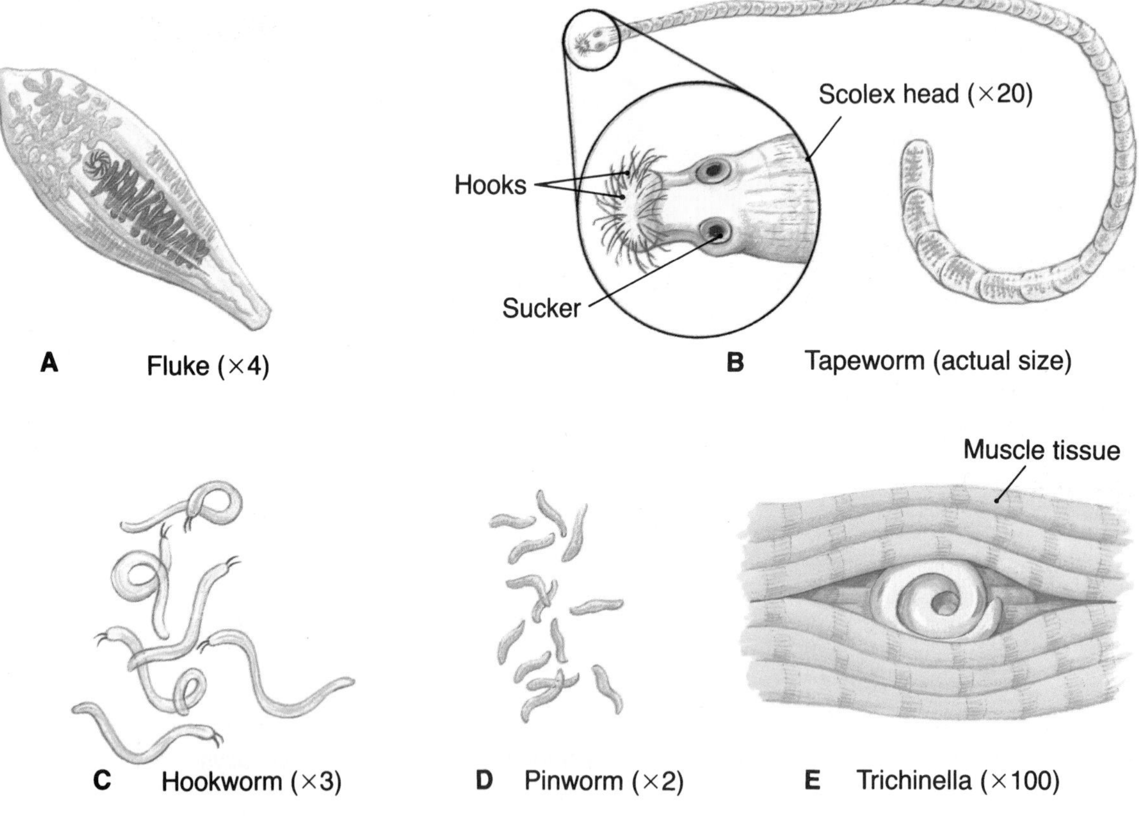

Figure 22–6 Worms.

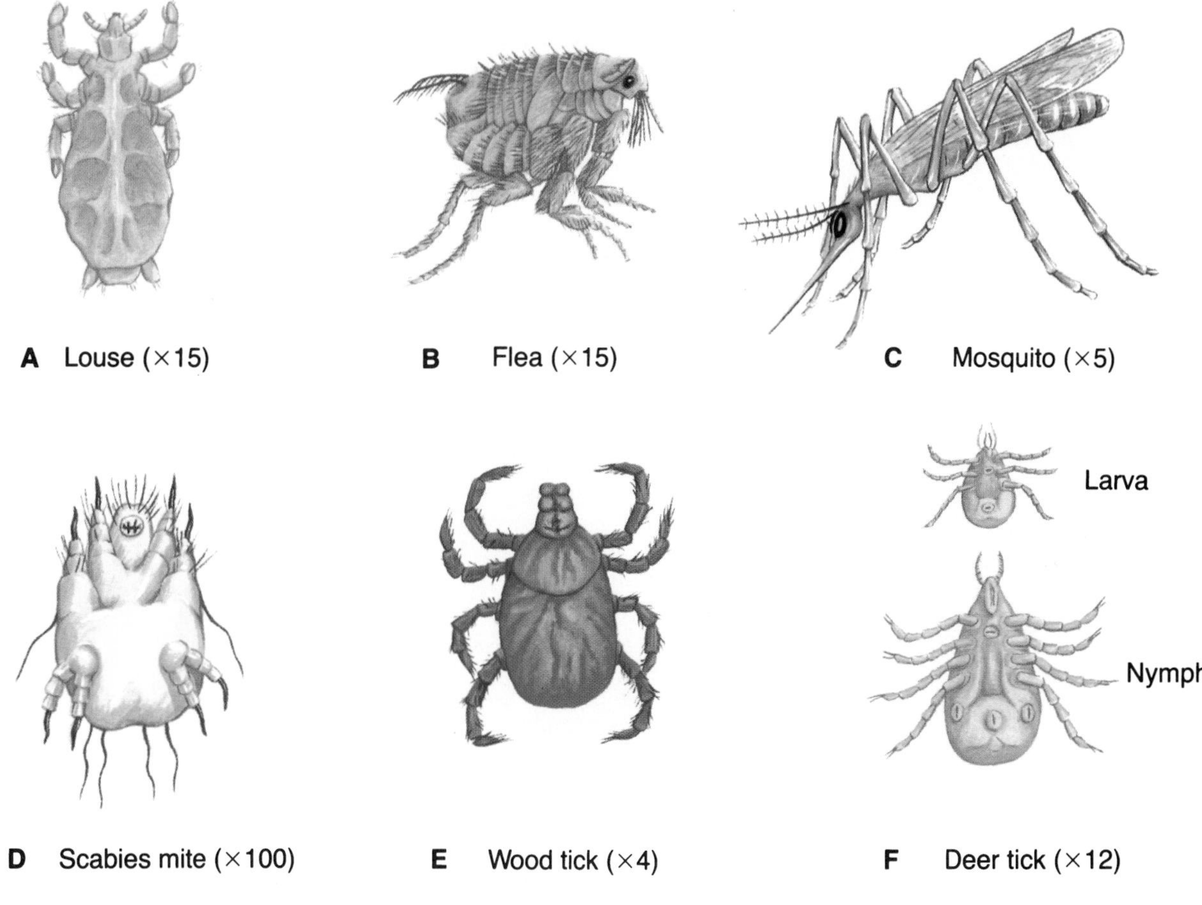

Figure 22–7 Arthropods.

Parasitic roundworms of people include hookworm, pinworm, *Ascaris,* and *Trichinella* (see Table 22-7 at the end of this chapter). Medications are available that can eliminate worm infestations. In endemic areas, however, reinfestation is quite common.

ARTHROPODS (see Table 22–8, page 509)

Arthropods such as the scabies mite and head lice are **ectoparasites** that live on the surface of the body. The infestations they cause are very itchy and uncomfortable but not debilitating or life threatening (Fig. 22-7). Of greater importance are the arthropods that are vectors of disease. These are listed in Table 22-8. Mosquitoes, fleas, lice, and flies are all insects. Ticks are not insects but are more closely related to spiders.

SUMMARY

The preceding discussion is an introduction to microorganisms and human disease, but it is only part of the story. The rest of this story is the remarkable ability of the human body to resist infection. Although we are surrounded and invaded by potential pathogens, most of us remain healthy most of the time. The immune responses that destroy pathogens and enable us to remain healthy are described in Chapter 14. Also in that chapter are discussions of vaccines. The development of vaccines represents the practical application of our knowledge of pathogens and of immunity, and it enables us to prevent many diseases. The availability of specific vaccines is noted in the tables of bacterial and viral diseases that follow.

Table 22–3 DISEASES CAUSED BY BACTERIA

Bacterial Species	Discussion/Disease(s) Caused
Staphylococcus aureus gram (+) coccus	Skin infections such as boils, pneumonia, toxic shock syndrome, osteomyelitis, septicemia. Most strains resistant to penicillin. Second-leading cause of nosocomial infections. Food poisoning characterized by rapid onset (1–8 hours) and vomiting. No vaccine.
Staphylococcus epidermidis gram (+) coccus	Normal skin flora; potential pathogen for those with artificial internal prostheses such as heart valves and joints.
Streptococcus pyogenes gram (+) coccus	Strep throat, otitis media, scarlet fever, endocarditis, puerperal sepsis; possible immunologic complications are rheumatic fever (transient arthritis and permanent damage to heart valves) and glomerulonephritis (transient kidney damage, usually with complete recovery). Rare strains cause necrotizing fasciitis, a potentially fatal hemolytic gangrene. No vaccine.
Streptococcus pneumoniae gram (+) coccus	Pneumonia: accumulation of fluids and white blood cells in the alveoli. The vaccine contains capsules of the most common strains; recommended for the elderly. Possible cause of meningitis in adults with predisposing factors such as sickle-cell anemia, alcoholism, asplenism, or head trauma.
Streptococcus faecalis gram (+) coccus	Normal colon flora. Has become an important cause of nosocomial infections of the urinary tract.
Neisseria gonorrhoeae gram (−) coccus	Gonorrhea: inflammation of the mucous membranes of the reproductive and urinary tracts. May cause scarring of reproductive ducts and subsequent sterility; in women may cause pelvic inflammatory disease. Infants of infected women may acquire the bacteria during birth; this is ophthalmia neonatorum and is prevented by antibiotic eyedrops. No vaccine.
Neisseria meningitidis gram (−) coccus	Meningitis: inflammation and edema of the meninges; pressure on the brain may cause death or permanent brain damage. Most common in older children and young adults. Most cases are sporadic, not part of epidemics. The vaccine is given to military recruits. Post-exposure prophylaxis (prevention) involves antibiotics.
Bacillus anthracis gram (+) bacillus (spore-forming)	Anthrax: spores in soil may be acquired by cattle or sheep. People acquire disease from these animals or from animal products such as wool or leather. Toxin causes death of tissue; may be fatal. Rare in the US, because grazing animals are vaccinated. The vaccine for people is reserved for the military; anthrax is a potential biological weapon.
Clostridium perfringens gram (+) bacillus (spore-forming)	Gas gangrene: normal soil flora may contaminate wounds; spores require anaerobic environment (dead tissue); toxins destroy more tissue, permitting the bacteria to spread; gas produced collects as bubbles in dead tissue. Food poisoning: from contaminated meat, self-limiting diarrhea.
Clostridium tetani gram (+) bacillus (spore-forming)	Tetanus: normal soil flora may contaminate wounds; spores require anaerobic conditions. The toxin prevents muscle relaxation, resulting in muscle spasms. May be fatal if respiratory muscles are affected. The vaccine contains the toxoid (inactivated toxin) and has made this a rare disease in the US. Boosters are strongly recommended for older adults.
Clostridium botulinum gram (+) bacillus (spore-forming)	Botulism: normal soil flora; spores present in anaerobic food containers germinate and produce the toxin, which causes paralysis. Respiratory paralysis may be fatal without assisted ventilation. May cause infant botulism in children less than 2 years of age who have ingested spores. No vaccine. Treatment is antitoxin (antibodies).
Corynebacterium diphtheriae gram (+) bacillus	Diphtheria: toxin causes heart failure and paralysis; a pseudomembrane that grows in the pharynx may cover the larynx and cause suffocation. Vaccination of infants (DTP) has made this a very rare disease in the US. Older adults should receive boosters (combined with tetanus: DT).

Table 22–3 DISEASES CAUSED BY BACTERIA (Continued)

Bacterial Species	Discussion/Disease(s) Caused
Listeria monocytogenes gram (+) bacillus	Listeriosis: septicemia and meningitis in the elderly, infants, and unborn; may cause miscarriage or stillbirth. The bacteria are found in soil and in animals such as cattle. In the US, food poisoning outbreaks traced to contaminated milk or milk products. Sporadic cases traced to undercooked hot dogs or chicken or to cold cuts from delicatessen counters. No vaccine.
Salmonella typhi gram (−) bacillus	Typhoid fever: intestinal infection with erosion and septicemia; subsequent infection of liver, gallbladder, or kidneys. Upon recovery, the carrier state (bacteria in gallbladder) may occur. Rare in the US because of chlorination of drinking water. The vaccine is used in endemic areas of the world such as Asia.
Salmonella enteritidis and other species gram (−) bacillus	Salmonellosis: food poisoning following consumption of contaminated animal products such as poultry or eggs. Diarrhea is usually self-limiting, but it may be fatal for the elderly. In the US estimates are 4 million cases per year. No vaccine.
Shigella species gram (−) bacillus	Bacillary dysentery: mild to severe diarrhea; may be fatal because of dehydration and circulatory shock. Usually transmitted by food prepared by people with mild cases. No vaccine. An important cause of illness in day-care centers.
Escherichia coli, Serratia marcescens, Proteus vulgaris, and other genera of gram (−) bacilli	Normal colon flora; cause opportunistic infections when introduced into any other part of the body. This group is now the most common cause of nosocomial infections (urinary tract, pneumonia, skin infections).
Escherichia coli 0157:H7	Hemorrhagic colitis and hemolytic uremic syndrome, potentially fatal. The bacteria may survive in undercooked meat, especially ground beef. A vaccine is in the research stage.
Pseudomonas aeruginosa gram (−) bacillus	Normal soil and water flora; also transient in human intestines. A serious potential pathogen for patients with severe burns, cystic fibrosis (causes pneumonia), or cancer. May even survive in disinfectant solutions. No vaccine.
Yersinia pestis gram (−) bacillus	Bubonic plague: swollen lymph nodes, septicemia, and hemorrhagic pneumonia; often fatal. Animal reservoirs are prairie dogs, ground squirrels, and other rodents. Rats and people are infected by fleas (the vector). In the US, the few cases each year occur in the Southwest. The vaccine is not reliably protective.
Francisella tularensis gram (−) bacillus	Tularemia (rabbit fever): septicemia and pneumonia; not often fatal but very debilitating. Reservoirs are wild animals and birds. People are infected by vectors (ticks, lice, and biting flies), by ingestion of contaminated animal meat, or by inhalation. No vaccine.
Brucella species gram (−) bacillus	Brucellosis (undulant fever): extreme weakness and fatigue, anorexia, a fever that rises and falls. Reservoirs are cattle, sheep, goats, and pigs; people acquire infection by contact with contaminated animal products. In the US, this is an occupational disease: meat processing workers, vets, farmers. Vaccines are available for animals.
Haemophilis influenzae gram (−) bacillus	Meningitis in children, especially those less than 2 years of age. Older people may have mild upper respiratory infections. The vaccine (Hib) contains the capsules of the bacteria and is recommended for infants beginning at age 2 months. This was the most common cause of meningitis in the US, approximately 20,000 cases per year. The vaccine has reduced the annual number of cases to about 1,000.
Haemophilis aegyptius gram (−) bacillus	Conjunctivitis: painful inflammation of the conjunctiva; spread by direct contact or fomites; may occur in epidemics among groups of children. No vaccine.

Table 22–3 DISEASES CAUSED BY BACTERIA (Continued)

Bacterial Species	Discussion/Disease(s) Caused
Bordetella pertussis gram (−) bacillus	Whooping cough (pertussis): paroxysms of violent coughing that may last for several weeks. Adults may have less severe coughing and go undiagnosed. Pneumonia is a complication that may be fatal, especially for children less than 1 year of age. People are the only host, and the vaccine has eliminated epidemics in the US. Concern about the safety of the original vaccine has prompted the development of newer vaccines, which may further reduce the annual number of cases.
Vibrio cholerae gram (−) bacillus (comma shaped)	Cholera: profuse watery diarrhea; infection ranges from mild to fatal. Spread of infection is usually by way of water contaminated with human feces. Rare in the US; epidemic in Asia, Africa, and South America. The vaccine is not reliably protective.
Vibrio parahaemolyticus gram (−) bacillus (comma shaped)	Enteritis: diarrhea and nausea. Acquired by ingestion of raw or lightly cooked seafood; usually self-limiting.
Vibrio vulnificus gram (−) bacillus (comma shaped)	Gangrene and septicemia: acquired from ocean water that contaminates a wound or by the ingestion of raw shellfish. Illness is often severe and protracted, and it may be fatal, especially for people who are immunosuppressed or who have liver disease. No vaccine.
Campylobacter jejuni gram (−) helical bacillus	Enteritis: diarrhea that is often self-limiting but may be severe in the elderly or very young. Reservoirs are animals such as poultry; people acquire infection from contaminated meat. In the US, estimates are 2 million cases per year. No vaccine.
Helicobacter pylori gram (−) helical bacillus	Gastric ulcers: most are caused by *H. pylori*, which has also been implicated in cancer of the stomach.
Legionella pneumophilia gram (−) bacillus	Legionellosis, which occurs in two forms: Legionnaire's disease is a pneumonia that may be fatal; Pontiac fever is a mild upper respiratory infection that is usually self-limiting. The bacteria are found in natural water, including soil, and may contaminate air conditioning systems or water supplies. Person-to-person transmission does not seem to occur. Has become an important cause of nosocomial pneumonia. No vaccine.
Mycobacterium tuberculosis acid-fast bacillus	Tuberculosis (TB): formation of tubercles containing bacteria and white blood cells, usually in the lung. Lung tissue is destroyed (caseation necrosis) and is removed by macrophages, leaving large cavities. The bacteria are spread by respiratory droplets from people with active cases. Many people acquire a primary infection that becomes dormant and is without symptoms, yet may be triggered later into an active secondary infection. The BCG vaccine is not used in the US, but is in other parts of the world. In the US, TB cases are frequent among homeless people, those with AIDS, and in closed populations such as prisons. Strains of the bacteria resistant to the standard TB medications are becoming much more common and pose a difficult treatment problem.
Mycobacterium species acid-fast bacillus	Atypical mycobacterial infections: clinically similar to TB; usually in the lungs. These bacteria are pathogenic for people with AIDS or other forms of immunosuppression, and for those with chronic pulmonary diseases.
Mycobacterium leprae acid-fast bacillus	Leprosy (Hansen's disease): chronic disease characterized by disfiguring skin lesions and nerve damage that may cause paralysis or loss of sensation. The bacteria are acquired by cutaneous contact or respiratory droplets. The incubation period may be several years; children develop clinical disease more rapidly than do adults. A vaccine is in the testing stage.

Table 22–3 DISEASES CAUSED BY BACTERIA (Continued)

Bacterial Species	Discussion/Disease(s) Caused
Treponema pallidum spirochaete	Syphilis: a sexually transmitted disease that progresses in three stages. Primary syphilis: a painless, hard chancre at the site of entry on skin or mucous membrane. Secondary syphilis: a rash on the skin and mucous membranes (indicates systemic infection). Tertiary syphilis (5 to 40 years later): necrotic lesions (gummas) in the brain, heart valves, aorta, spinal cord, skin, or other organs. No vaccine.
Leptospira interrogans spirochaete	Leptospirosis: a disease of wild or domestic animals that excrete the bacteria in urine. People acquire the bacteria by contact with contaminated water. Disease is usually mild, resembling intestinal virus infection. Weil's disease is the serious form, with hemorrhages in the liver and kidneys. A vaccine is available for dogs.
Borrelia burgdorferi spirochaete	Lyme disease: begins as a flu-like illness, often with a bull's-eye rash at the site of the tick bite. May be followed by cardiac arrhythmias, self-limiting meningitis, or arthritis. Animal reservoirs are deer and field mice; the vector is the deer tick (genus *Ixodes*). Vaccines are in the testing stages.
Borrelia vincenti spirochaete	Trench mouth (Vincent's gingivitis): an ulcerative infection of the gums and pharynx caused by overgrowth of *Borrelia* and other normal oral flora. Triggered by poor oral hygiene or oral infection, which must be corrected to make antibiotic therapy effective.
Rickettsia prowazekii rickettsia	Epidemic typhus: high fever and delirium, hemorrhagic rash; 40% fatality rate. Vector is the human body louse. Very rare in the US but still endemic in other parts of the world.
Rickettsia typhi rickettsia	Endemic typhus: similar to epidemic typhus but milder; 2% fatality rate. Reservoirs are rats and wild rodents; vectors are fleas. The few cases in the US each year usually occur in the Southeast.
Rickettsia rickettsii rickettsia	Rocky Mountain spotted fever (RMSF): high fever, hemorrhagic rash, and pneumonia; 20% fatality rate. Reservoirs are wild rodents and dogs; vectors are ticks. Despite its name, RMSF in the US is most prevalent in the Southeast coastal states (NC, SC) and in Oklahoma.
Chlamydia trachomatis, (serogroups D–K) chlamydia	Genitourinary infection (nongonococcal urethritis): in men, urethritis or epididymitis; in women, cervicitis, although many women are asymptomatic. Complications in women include pelvic inflammatory disease, ectopic pregnancy, and miscarriage. Newborns of infected women may develop conjunctivitis or pneumonia. This is now the most prevalent sexually transmitted disease in the US. No vaccine.
Chlamydia trachomatis, (serogroups A–C) chlamydia	Trachoma: conjunctivitis involving growth of papillae; vascular invasion of the cornea leading to scarring and blindness. Spread by direct contact and fomites. The leading cause of blindness throughout the world, especially in dry, dusty environments.

Table 22–4 DISEASES CAUSED BY VIRUSES

Virus	Discussion/Disease(s) Caused
Herpes simplex	Type 1: fever blisters (cold sores) on the lip or in oral cavity; the virus is dormant in nerves of the face between attacks. Spread in saliva; may cause eye infections (self-inoculation). Type 2: genital herpes; painful lesions in the genital area; a sexually transmitted disease. No vaccine.
Herpes varicella-zoster	Chickenpox: the disease of the first exposure; vesicular rash; pneumonia is a possible complication, especially in adults. The virus then becomes dormant in nerves. Shingles: painful, raised lesions on the skin above the affected nerves following reactivation of the dormant virus. Usually occurs in adults. The vaccine is recommended for children.
Cytomegalovirus (CMV)	Most people have asymptomatic infection; the virus does no harm but remains in the body. Fetal infection may result in mental retardation, blindness, or deafness. CMV is potentially serious for transplant recipients (pneumonia) and AIDS patients (blindness). No vaccine.
Epstein-Barr virus	Mononucleosis: swollen lymph nodes, fatigue, fever, possible spleen or liver enlargement. Spread by saliva. No vaccine.
Adenoviruses	Many different types: some cause acute respiratory disease (ARD) similar to the common cold; others cause pharyngoconjunctival fever and may occur in epidemics related to swimming pools. The ARD vaccine is used only in the military.
Rhinoviruses	Common cold: sore throat, runny nose, low fever; usually self-limiting. No vaccine (there are over 100 types of rhinoviruses).
Influenza viruses	Influenza: muscle aches, fever, fatigue, spread in respiratory droplets. Three types: A, B, and C. Type A is responsible for most epidemics. These are mutating viruses, and new vaccines are needed as the virus changes. The most serious complication is secondary bacterial pneumonia.
Respiratory syncytial virus	Pneumonia: especially at risk are infants and young children, esp. those who must be hospitalized. Decades of work have yet to produce an effective vaccine.
Hantaviruses	Hantavirus pulmonary syndrome: fever, cough, pulmonary edema, hypotension. Mortality rate is 40%–50%. Acquired by contact with rodent feces or urine, or inhalation of virus in rodent-infested areas. No vaccine.
Measles virus	Measles (rubeola): fever, sore throat, Koplik's spots (white) on lining of mouth, rash. Complications are ear infections, pneumonia, and measles encephalitis, which may be fatal. The vaccine is given to infants in combination with mumps and rubella (MMR).
Rubella virus	German measles: mild upper respiratory symptoms; a rash may or may not be present. This virus may cross the placenta and cause congenital rubella syndrome (CRS): blindness, deafness, heart defects, mental retardation, or miscarriage. CRS is most likely to occur if the fetus is infected during the first trimester. The vaccine is given to infants (MMR).
Mumps virus	Mumps: fever, swelling of the parotid salivary glands and perhaps the others (asymptomatic infections do occur); the virus is spread in saliva. Complications are rare in children but include pancreatitis, nerve deafness, and mumps encephalitis. Adult men may develop orchitis, inflammation of the testes. Adult women may develop oophoritis, inflammation of the ovaries. The vaccine is given to infants.
Polio viruses	Polio: most infections are asymptomatic or mild; major infection may result in paralysis. Two vaccines: IPV (Salk) contains a killed virus and cannot cause polio; booster injections are needed. OPV (Sabin) contains an attenuated virus, is given orally and carries a very small risk of causing polio. In the US, polio cases are vaccine related; polio is still endemic in other parts of the world. The WHO has set as a goal the eradication of polio by the year 2000.

Table 22–4 DISEASES CAUSED BY VIRUSES (Continued)

Virus	Discussion/Disease(s) Caused
Rabies virus	Rabies: headache, nausea, fever, spasms of the swallowing muscles; seizures; fatal because of respiratory or heart failure; virtually 100% fatal. Reservoirs are wild animals; the virus is present in their saliva. Post-exposure prevention requires Human Rabies Immune Globulin (antibodies) and the rabies vaccine.
Encephalitis viruses	Encephalitis: most infections are mild; CNS involvement is indicated by confusion, lethargy, or coma. Several types of these viruses occur in the US. Reservoirs are wild birds and small mammals; vectors are mosquitoes. Vaccines are available for horses and for people whose occupations put them at risk.
Rotaviruses	Enteritis: worldwide the leading cause of diarrhea and potentially fatal dehydration in infants and children. A vaccine is in the testing stage.
Norwalk viruses	Gastroenteritis: vomiting; called by many "stomach flu." Acquired from food such as shellfish, or from food handled by people with mild cases. No vaccine.
Ebola viruses	Hemorrhagic fever: high fever, systemic hemorrhages, necrosis of the liver; mortality rate 25% to 90%. The viruses also affect monkeys, which may be reservoirs of infection. No vaccine.
Yellow fever virus	Yellow fever: hemorrhages in the liver, spleen, kidneys, and other organs. The vector is a mosquito. The vaccine is recommended for travelers to endemic areas: Central and South America and Africa.
Hepatitis viruses: types A, B, and C	Hepatitis: anorexia, nausea, fatigue, jaundice (may not be present in mild cases). HAV is spread by the fecal-oral route; contaminated shellfish or food prepared by people with mild cases. No carriers after recovery. HBV is spread by sexual activity or contact with blood or other body fluids. Carrier state is possible; may lead to liver cancer or cirrhosis. HCV transmission is similar to that of HBV. Carrier state is possible. There are vaccines for hepatitis A and B. Other hepatitis viruses are designated D, E, F, and G.
Human immunodeficiency virus (HIV)	AIDS: destruction of helper T cells and suppression of the immune system; opportunistic infections; invariably fatal, often after many years. HIV is spread by sexual activity, contact with blood, or placental transmission. No vaccine.

Table 22–5 DISEASES CAUSED BY FUNGI

Fungus Genus	Discussion/Disease(s) Caused
Microsporum, Trichophyton, Epidermophyton	Ringworm (tinea): scaly red patches on the skin or scalp; loss of hair. Tinea pedis is athlete's foot. May also infect damaged nails. Spores of these fungi are acquired from people or animals.
Candida (*albicans* and other species)	Yeast infections: mucosal infections are called thrush; may be oral or vaginal; yeasts have come from resident flora. Systemic infections include pneumonia and endocarditis. Important nosocomial pathogens.
Cryptococcus	Cryptococcosis: pulmonary infection that may progress to meningitis, especially in AIDS patients. Spores are carried in the air from soil or pigeon droppings.
Histoplasma	Histoplasmosis: pulmonary infection that is often self-limiting. Progressive disease involves ulcerations of the liver, spleen, and lymph nodes; usually fatal. Spores are carried by the air from soil.
Coccidioides	Coccidioidomycosis: pulmonary infection that is often self-limiting. Progressive disease involves the meninges, bones, skin, and other organs; high mortality rate. Spores are carried by the air.

Table 22–6 DISEASES CAUSED BY PROTOZOA

Protozoan	Discussion/Disease(s) Caused
Entamoeba histolytica	Amebic dysentery; ulcerative lesions in the colon, bloody diarrhea; abscesses may form in the liver, lungs, or brain. Spread by the fecal-oral route in water or food.
Naegleria species	Amebic meningoencephalitis: inflammation of the meninges and brain; uncommon in the US but almost always fatal. Amebas in fresh water are acquired when swimmers sniff water into the nasal cavities; the amebas move along the olfactory nerves into the brain.
Balantidium coli	Balantidiasis: abdominal discomfort and diarrhea; often mild. Reservoirs are pigs and other domestic animals; spread by the fecal-oral route.
Giardia lamblia	Giardiasis: fatty diarrhea; may be mild. Reservoirs are wild and domestic animals and people. Spread by the fecal-oral route in water or food prepared by people with mild cases. An important cause of diarrhea in day-care centers.
Trichomonas vaginalis	Trichomoniasis: a sexually transmitted disease. Women: causes cervicitis and vaginitis; men are often asymptomatic.
Plasmodium species	Malaria: the protozoa reproduce in red blood cells, causing hemolysis and anemia. The vector is the *Anopheles* mosquito. No vaccine yet.
Toxoplasma gondii	Toxoplasmosis: asymptomatic infection in healthy people. Congenital infection: miscarriage or mental retardation, blindness. Reservoirs are cats and grazing animals. Pregnant women may acquire cysts from cat feces or from ingestion of rare beef or lamb.
Cryptosporidium species	Diarrhea: ranges from mild to severe; spread by the fecal-oral route. An important cause of diarrhea in day-care centers and in AIDS patients. May also contaminate municipal water supplies and cause extensive epidemics.
Pneumocystis carinii	Pneumonia: only in very debilitated or immunosuppressed persons. A frequent cause of death in AIDS patients.
Trypanosoma species	African sleeping sickness: lethargy progressing to coma and death. Reservoirs are wild and domestic animals; vector is the tse tse fly. No vaccine.

Table 22–7 INFESTATIONS CAUSED BY WORMS

Worm (Genus)	Discussion/Disease(s) Caused
Chinese liver fluke *(Clonorchis)*	Abdominal discomfort; cirrhosis after many years. Adult worms (½ inch) live in bile ducts. Acquired by people from ingestion of raw fish that contains worm cysts.
Tapeworms *(Taenia, Diphyllobothrium)*	Bloating and abdominal discomfort; constipation or diarrhea. People acquire the worms by eating poorly cooked beef, pork, or fish (the alternate hosts) that contain worm cysts.
Pinworm *(Enterobius)*	Adult worms (⅛ inch) live in colon; females lay eggs on perianal skin while host is asleep, causing irritation and itching of skin. Eggs are spread to family members on hands and bed linens. In the US, this is probably the most common worm infestation.
Hookworm *(Necator)*	Adult worms (½ inch) live in the small intestine; their food is blood. Heavy infestations cause anemia and fatigue. Eggs are excreted in feces; larval worms burrow through the skin of a bare foot and migrate to the intestine.
Ascaris	Adults are 10–12 inches long, and live in the small intestine. Large numbers of worms may cause intestinal obstruction. Eggs are excreted in feces and are spread to others on hands or vegetation contaminated by human feces.
Trichinella spiralis	Trichinosis: severe muscle pain as migrating worms form cysts that become calcified. Acquired by eating poorly cooked pork (or wild animals) that contains cysts.

Table 22–8 ARTHROPOD VECTORS

Anthropod	Disease (Type of Pathogen)*
Mosquito	Malaria (protozoan) Encephalitis (virus) Yellow fever (virus)
Flea	Plague (bacterium) Endemic typhus (rickettsia)
Body louse	Epidemic typhus (rickettsia) Tularemia (bacterium) Relapsing fever (bacterium)
Tick	Lyme disease (bacterium) Rocky Mountain spotted fever (rickettsia) Tularemia (bacterium)
Tse tse fly	African sleeping sickness (protozoan)
Deer fly, horse fly	Tularemia (bacterium)

*These diseases are described in previous tables.

STUDY OUTLINE

Classification of Microorganisms

1. Bacteria—unicellular; some are pathogens.
2. Viruses—not cells; all are parasites.
3. Protozoa—unicellular animals; some are pathogens.
4. Fungi—unicellular (yeasts) or multicellular (molds); most are decomposers.
5. Worms—multicellular animals; a few are parasites.
6. Arthropods—insects, ticks, or mites that are vectors of disease or cause infestations.
   - Binomial Nomenclature—the genus and species names.

Normal Flora—see Table 22–1

1. Resident Flora—the microorganisms that live on or in nearly everyone, in specific body sites; cause no harm when in their usual sites.
2. Transient Flora—the microorganisms that periodically inhabit the body and usually cause no harm unless the host's resistance is lowered.

Infectious Disease

1. Caused by microorganisms or their toxins.
2. Clinical infections are characterized by symptoms; in a subclinical infection, the person shows no symptoms.
3. Course of an Infectious Disease: Incubation period—the time between the entry of the pathogen and the onset of symptoms. The acme is the worst stage of the disease, followed by recovery or death. A self-limiting disease typically lasts a certain period of time and is followed by recovery.
4. **Types of Infection**
   - Localized—the pathogen is in one area of the body.
   - Systemic—the pathogen is spread throughout the body by the blood or lymph.
   - Septicemia (Bacteremia)—bacteria in the blood.
   - Acute—usually severe or of abrupt onset.
   - Chronic—progresses slowly or is prolonged.
   - Secondary—made possible by a primary infection that lowered host resistance.
   - Nosocomial—a hospital-acquired infection.
   - Endogenous—caused by the person's own normal flora in an abnormal site.

Epidemiology—see Fig. 22–1

1. The study of the patterns and spread of disease.
2. Portal of Entry—the way a pathogen enters a host.

3. Portal of Exit—the way a pathogen leaves a host.
4. Reservoirs—persons with the disease, carriers after recovery, or animal hosts (for zoonoses).
5. Non-communicable disease—cannot be directly or indirectly transmitted from host to host.
6. Communicable disease—may be transmitted directly from host to host by respiratory droplets, cutaneous or sexual contact, placental transmission, or blood contact. May be transmitted indirectly by food or water, vectors, or fomites.
7. Contagious Disease—easily spread from person to person by casual contact (respiratory droplets).

Methods of Control of Microbes

1. Antiseptics—chemicals that destroy or inhibit bacteria on a living being.
2. Disinfectants—chemicals that destroy or inhibit bacteria on inanimate objects.
3. Sterilization—a process that destroys all living organisms.
4. Public health measures include laws and regulations to ensure safe food and water.

Bacteria—see Fig. 22–2

1. Shapes: coccus, bacillus, and spirillum.
2. Flagella provide motility for some bacilli and spirilla.
3. Aerobes require oxygen, anaerobes are inhibited by oxygen; facultative anaerobes grow in the presence or absence of oxygen.
4. The gram reaction (positive or negative) is based on the chemistry of the cell wall. The Gram stain is a laboratory procedure used in the identification of bacteria.
5. Capsules inhibit phagocytosis by white blood cells. Spores are dormant forms that are resistant to environmental extremes.
6. Toxins are chemicals produced by bacteria that are poisonous to host cells.
7. Rickettsias and Chlamydias differ from other bacteria in that they must be inside living cells to reproduce.
8. Antibiotics are chemicals used in the treatment of bacterial diseases. Broad-spectrum: affects many kinds of bacteria. Narrow-spectrum: affects only a few kinds of bacteria.
9. Bacteria may become resistant to certain antibiotics, which are then of no use in treatment. Culture and sensitivity testing may be necessary before an antibiotic is chosen to treat an infection.
10. Diseases—see Table 22-3.

Viruses—see Fig. 22–3

1. Not cells; a virus consists of either DNA or RNA surrounded by a protein shell.
2. Must be inside living cells to reproduce, which causes death of the host cell.
3. Severity of disease depends on the types of cells infected; some viruses may cross the placenta and infect a fetus.
4. Antiviral medications must interfere with viral reproduction without harming host cells; there are few such chemicals available.
5. Diseases—see Table 22-4.

Fungi—see Fig. 22–4

1. Most are saprophytes, decomposers of dead organic matter. May be unicellular yeasts or multicellular molds.
2. Mycoses may be superficial, involving the skin or mucous membranes, or systemic, involving internal organs such as the lungs or meninges.
3. Effective antifungal medications are available, but some are highly toxic.
4. Diseases—see Table 22-5.

Protozoa—see Fig. 22–5

1. Unicellular animals; some are pathogens.
2. Some are spread by vectors, others by fecal contamination of food or water.
3. Effective medications are available for most diseases.
4. Diseases—see Table 22-6.

Worms—see Fig. 22–6

1. Simple multicellular animals; the parasites are flukes, tapeworms, and some roundworms.
2. May have life cycles that involve other animal hosts as well as people.
3. Effective medications are available for most worm infestations.
4. Diseases—see Table 22-7.

Arthropods—see Fig. 22–7

1. Some cause superficial infestations.
2. Others are vectors of disease—see Table 22-8.

REVIEW QUESTIONS

1. Define resident flora, and explain its importance. (pp. 486-487)
2. State the term described by each statement: (pp. 488-489, 490, 493)
   a. An infection in which the person shows no symptoms
   b. Bacteria that are inhibited by oxygen
   c. A disease that lasts a certain length of time and is followed by recovery
   d. A disease that is usually present in a given population
   e. The presence of bacteria in the blood
   f. An infection made possible by a primary infection that lowers host resistance
   g. A disease of animals that may be acquired by people
   h. Bacteria that are spherical in shape
3. Name these parts of a bacterial cell: (p. 493)
   a. Inhibits phagocytosis by white blood cells
   b. Provides motility
   c. The basis for the gram reaction or Gram stain
   d. A form resistant to heat and drying
   e. Chemicals produced that are poisonous to host cells
4. Explain what is meant by a nosocomial infection, and describe the two general kinds with respect to sources of the pathogen. (p. 489)
5. Name five potential portals of entry for pathogens. (p. 489)
6. Name five potential portals of exit for pathogens. (p. 489)
7. Explain the difference between a communicable disease and a contagious disease. (pp. 490-491)
8. Explain the difference between pasteurization and sterilization. (pp. 491-492)
9. Describe the structure of a virus, and explain how viruses cause disease. (p. 496)
10. Describe the differences between yeasts and molds. (p. 497)
11. Describe the difference between superficial and systemic mycoses. (p. 497)
12. Name some diseases that are spread by vectors, and name the vector for each. (p. 509)

APPENDIX A

Units of Measure

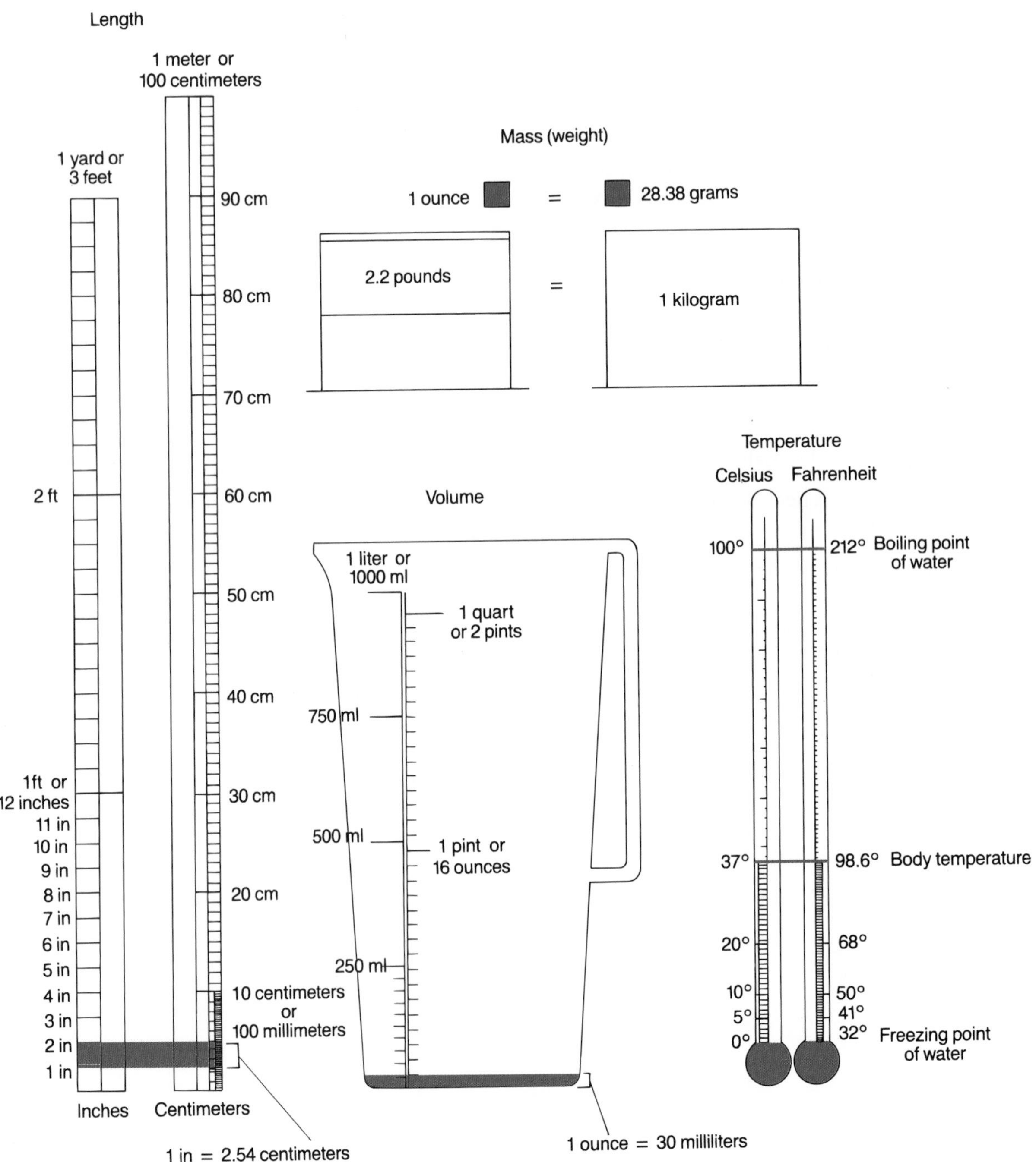

Length
1 meter or
100 centimeters
1 yard or
3 feet
90 cm
80 cm
70 cm
2 ft
60 cm
50 cm
40 cm
1ft or
12 inches
30 cm
11 in
10 in
9 in
8 in
20 cm
7 in
6 in
5 in
4 in
10 centimeters
or
100 millimeters
3 in
2 in
1 in
Inches
Centimeters
1 in = 2.54 centimeters
Mass (weight)
1 ounce
=
28.38 grams
2.2 pounds
=
1 kilogram
Volume
1 liter or
1000 ml
1 quart
or 2 pints
750 ml
500 ml
1 pint or
16 ounces
250 ml
1 ounce = 30 milliliters
Temperature
Celsius
Fahrenheit
100°
212°
Boiling point
of water
37°
98.6°
Body temperature
20°
68°
10°
50°
5°
41°
0°
32°
Freezing point
of water

Units of Length

	mm	cm	in	ft	yd	M
1 millimeter =	1.0	0.1	0.04	0.003	0.001	0.001
1 centimeter =	10.0	1.0	0.39	0.032	0.011	0.01
1 inch =	25.4	2.54	1.0	0.083	0.028	0.025
1 foot =	304.8	30.48	12.0	1.0	0.33	0.305
1 yard =	914.4	91.44	36.0	3.0	1.0	0.914
1 meter =	1000.0	100.0	39.37	3.28	1.09	1.0

1 μ = 1 mu = 1 micrometer (micron) = 0.001 mm = 0.00004 in.
mm = millimeters; **cm** = centimeters; **in** = inches; **ft** = feet; **yd** = yards; **M** = meters.

Units of Weight

	mg	g	oz	lb	kg
1 milligram =	1.0	0.001	0.00004	0.000002	0.000001
1 gram =	1000.0	1.0	0.035	0.002	0.001
1 ounce =	28,380	28.4	1.0	0.06	0.028
1 pound =	454,000	454.0	16	1.0	0.454
1 kilogram =	1,000,000	1000.0	35.2	2.2	1.0

mg = milligrams; **g** = grams; **oz** = ounces; **lb** = pounds; **kg** = kilograms.

Units of Volume

	ml	in^3	oz	qt	l
1 milliliter =	1.0	0.06	0.034	0.001	0.001
1 cubic inch =	16.4	1.0	0.55	0.017	0.016
1 ounce =	29.6	1.8	1.0	0.03	0.029
1 quart =	946.3	57.8	32.0	1.0	0.946
1 liter =	1000.0	61.0	33.8	1.06	1.0

ml = milliliters; **in^3** = cubic inches; **oz** = ounces; **qt** = quarts; **l** = liters.

Temperature Centigrade and Fahrenheit

°C	Is Equivalent to	°F
0°C		32°F
5°C		41°F
10°C		50°F
15°C		59°F
20°C		68°F
25°C		77°F
30°C		86°F
35°C		95°F
40°C		104°F
45°C		113°F
50°C		122°F

APPENDIX B

Abbreviations

The use of abbreviations for medical and scientific terms is timesaving and often standard practice. Some of the most frequently used abbreviations have been listed here. Notice that some have more than one interpretation. It is important to know the abbreviations approved by your institution, use only those, and *never* make up your own abbreviations. If there is any chance of confusion, write out the words.

ABC airway, breathing, circulation
ABG arterial blood gas
ABX antibiotics
ACh acetylcholine
ACTH adrenocorticotropic hormone
AD Alzheimer's disease
ADH antidiuretic hormone
AIDS acquired immunodeficiency syndrome
ALS amyotrophic lateral sclerosis
ANS autonomic nervous system
ARDS acute respiratory distress syndrome
ARF acute renal failure (acute respiratory failure)
ATP adenosine triphosphate
AV atrioventricular
BAL blood alcohol level
BBB blood-brain barrier
BMR basal metabolic rate
BP blood pressure
BPH benign prostatic hypertrophy (hyperplasia)
BPM beats per minute
BS blood sugar (bowel sounds, breath sounds)
BUN blood urea nitrogen
CA cancer
CAD coronary artery disease
CAPD continuous ambulatory peritoneal dialysis
CBC complete blood count
CC creatinine clearance (critical condition, chief complaint)
CCCC closed-chest cardiac compression
CF cystic fibrosis (cardiac failure)
CHD coronary heart disease (congenital heart disease)
CHF congestive heart failure
CI cardiac insufficiency (cerebral infarction)
CNS central nervous system
CO cardiac output; carbon monoxide
COPD chronic obstructive pulmonary disease
CP cerebral palsy
CPR cardiopulmonary resuscitation
CRF chronic renal failure
C-section cesarean section
CSF cerebrospinal fluid
CT (CAT) computed (axial) tomography
CVA cerebrovascular accident
CVP central venous pressure
CVS chorionic villus sampling
D & C dilation and curettage
DM diabetes mellitus (diastolic murmur)
DMD Duchennes muscular dystrophy
DNA deoxyribonucleic acid
DNR do not resuscitate
DOA date of admission (dead on arrival)
DRG diagnosis-related group
Dx diagnosis
EBV Epstein-Barr virus
ECF extracellular fluid (extended care facility)
ECG (EKG) electrocardiogram
EDV end-diastolic volume
EEG electroencephalogram
EFM electronic fetal monitoring
EP ectopic pregnancy
ER endoplasmic reticulum
ERT estrogen replacement therapy
ESR erythrocyte sedimentation rate
ESRD end-stage renal disease
ESV end-systolic volume
FAS fetal alcohol syndrome
FBG fasting blood glucose
FOBT fecal occult blood testing
FSH follicle-stimulating hormone
FUO fever of unknown origin
Fx fracture

GB	gallbladder
GFR	glomerular filtration rate
GH	growth hormone
GI	gastrointestinal
HAV	hepatitis A virus
Hb	hemoglobin
HBV	hepatitis B virus
hCG	human chorionic gonadotropin
Hct	hematocrit
HCV	hepatitis C virus
HDL	high-density lipoprotein
HLA	human leukocyte antigen
HR	heart rate
HRT	hormone replacement therapy
HSV	herpes simplex virus
HTN	hypertension
Hx	history
IBD	inflammatory bowel disease
IBS	irritable bowel syndrome
ICF	intracellular fluid
ICP	intracranial pressure
ICU	intensive care unit
ID	intradermal
IDDM	insulin-dependent diabetes mellitus
Ig	immunoglobulin
IM	intramuscular
IV	intravenous
LA	left atrium
LDL	low-density lipoprotein
LH	luteinizing hormone
LLQ	left lower quadrant
LUQ	left upper quadrant
LV	left ventricule
mEq/L	milliequivalents per liter
MG	myasthenia gravis
MI	myocardial infarction
mm^3	cubic millimeter
mmHg	millimeters of mercury
MRI	magnetic resonance imaging
MS	multiple sclerosis
MSOF	multisystem organ failure
MVP	mitral valve prolapse
NGU	non-gonococcal urethritis
NIDDM	non-insulin-dependent diabetes mellitus
NPN	non-protein nitrogen
OC	oral contraceptive
OTC	over the counter
PE	pulmonary embolism
PET	positron emission tomography
PG	prostaglandin
PID	pelvic inflammatory disease
PKU	phenylketonuria
PMN	polymorphonuclear leukocyte
PMS	premenstrual syndrome
PNS	peripheral nervous system
PT	prothrombin time (patient, patient teaching, physical therapy)
PTH	parathyroid hormone
PTT	partial thromboplastin time
PVC	premature ventricular contraction
RA	right atrium
RBC	red blood cell
RBM	red bone marrow
RDA	recommended daily allowance
RDS	respiratory distress syndrome
REM	rapid eye movement
RES	reticuloendothelial system
Rh	*Rhesus*
RIA	radioimmunoassay
RLQ	right lower quadrant
RNA	ribonucleic acid
RV	right ventricle
RUQ	right upper quadrant
Rx	prescription
SA	sinoatrial
SC	subcutaneous
SCID	severe combined immunodeficiency
SF	synovial fluid
SIDS	sudden infant death syndrome
SLE	systemic lupus erythematosus
SPF	sun protection factor
S/S (sx)	signs and symptoms
STD	sexually transmitted disease
SV	stroke volume
T_3	triiodothyronine
T_4	thyroxine
TIA	transient ischemic attack
TMJ	temporomandibular joint
t-PA	tissue plasminogen activator
TPN	total parenteral nutrition
TSH	thyroid-stimulating hormone
TSS	toxic shock syndrome
Tx	treatment
UA	urinalysis
URI	upper respiratory infection
US	ultrasound
UTI	urinary tract infection
UV	ultraviolet
VD	venereal disease
VPC	ventricular premature contraction
VS	vital signs
WBC	white blood cell
WNL	within normal limits

APPENDIX C

Normal Values for Some Commonly Used Blood Tests

Test	Normal Value	Clinical Significance of Variations
Albumin	3.5—5.5 g/100 ml	• Decreases: kidney disease, severe burns
Bilirubin—Total Direct Indirect	0.3–1.4 mg/100 ml 0.1–0.4 mg/100 ml 0.2–1.0 mg/100 ml	• Increases: liver disease, rapid RBC destruction, biliary obstruction
Calcium	4.3–5.3 mEq/liter	• Increases: hyperparathyroidism • Decreases: hypoparathyroidism, severe diarrhea, malnutrition
Chloride	95–108 mEq/liter	• Decreases: severe diarrhea, severe burns, ketoacidosis
Cholesterol HDL cholesterol LDL cholesterol	150–250 mg/100 ml 29–77 mg/100 ml 62–185 mg/100 ml	• Increases: hypothyroidism, diabetes mellitus
Clotting time	5–10 minutes	• Increases: liver disease
Creatinine	0.6–1.5 mg/100 ml	• Increases: kidney disease
Globulins	2.3–3.5 g/100 ml	• Increases: chronic infections
Glucose	70–110 mg/100 ml	• Increases: diabetes mellitus, liver disease, hyperthyroidism, pregnancy
Hematocrit	38%–48%	• Increases: dehydration, polycythemia • Decreases: anemia, hemorrhage
Hemoglobin	12–18 g/100 ml	• Increases: polycythemia, high altitude, chronic pulmonary disease • Decreases: anemia, hemorrhage
Pco_2	35–45 mmHg	• Increases: pulmonary disease • Decreases: acidosis, diarrhea, kidney disease
pH	7.35–7.45	• Increases: hyperventilation, metabolic alkalosis • Decreases: ketoacidosis, severe diarrhea, hypoventilation
Po_2	75–100 mmHg	• Decreases: anemia, pulmonary disease
Phosphorus	1.8–4.1 mEq/liter	• Increases: kidney disease, hypoparathyroidism • Decreases: hyperparathyroidism
Platelet count	150,000–300,000/mm^3	• Decreases: leukemia, aplastic anemia

Test	Normal Value	Clinical Significance of Variations
Potassium	3.5–5.0 mEq/liter	• Increases: severe cellular destruction • Decreases: diarrhea, kidney disease
Prothrombin time	11–15 seconds	• Increases: liver disease, vitamin K deficiency
Red blood cell count	4.5–6.0 million/mm^3	• Increases: polycythemia, dehydration • Decreases: anemia, hemorrhage, leukemia
Reticulocyte count	0.5%–1.5%	• Increases: anemia, following hemorrhage
Sodium	136–142 mEq/liter	• Increases: dehydration • Decreases: kidney disease, diarrhea, severe burns
Urea nitrogen (BUN)	8–25 mg/100 ml	• Increases: kidney disease, high-protein diet
Uric acid	3.0–7.0 mg/100 ml	• Increases: kidney disease, gout, leukemia
White blood cell count	5000–10,000/mm^3	• Increases: acute infection, leukemia • Decreases: aplastic anemia, radiation sickness

APPENDIX D

Normal Values for Some Commonly Used Urine Tests

Test	Normal Value	Clinical Significance of Variations
Acetone and acetoacetic acid (ketones)	0	• Increases: ketoacidosis, starvation
Albumin	0–trace	• Increases: kidney disease, hypertension
Bilirubin	0	• Increases: biliary obstruction
Calcium	less than 250 mg/24 hrs	• Increases: hyperparathyroidism • Decreases: hypoparathyroidism
Creatinine	1.0–2.0 g/24 hrs	• Increases: infection • Decreases: kidney disease, muscle atrophy
Glucose	0	• Increases: diabetes mellitus
pH	4.5–8.0	• Increases: urinary tract infection, alkalosis, vegetarian diet • Decreases: acidosis, starvation, high-protein diet
Protein	0	• Increases: kidney disease, extensive trauma, hypertension
Specific gravity	1.010–1.025	• Increases: dehydration • Decreases: excessive fluid intake, alcohol intake, severe kidney damage
Urea	25–35 g/24 hrs	• Increases: high-protein diet; excessive tissue breakdown • Decreases: kidney disease
Uric acid	0.4–1.0 g/24 hrs	• Increases: gout, liver disease • Decreases: kidney disease
Urobilinogen	0–4 mg/24 hrs	• Increases: liver disease, hemolytic anemia • Decreases: biliary obstruction

APPENDIX E

Suggestions for Further Reading

Taber's Cyclopedic Medical Dictionary, 18th ed.
Clayton L. Thomas, Editor
F.A. Davis Company, Philadelphia, 1997

Melloni's Illustrated Medical Dictionary
Ida Box, Biagio J. Melloni, Gilbert M. Eisner
The Williams & Wilkins Company, Baltimore, 1979

Tissues and Organs: A Text-Atlas of Scanning Electron Microscopy
Richard G. Kessel and Randy H. Kardon
W.H. Freeman and Company, San Francisco, 1979

Atlas of Normal Histology, 6th ed.
Mariano S.H. diFiore, edited by Victor P. Eroschenko
Lea & Febiger, Philadelphia, 1989

Anatomy & Physiology
Arthur C. Guyton
W.B. Saunders College Publishing, Philadelphia, 1996

Principles of Anatomy & Physiology, 8th ed.
Gerard J. Tortora and Sandra R. Grabowski
Harper & Row, New York, 1997

Introduction to Microbiology for the Health Sciences, 3rd ed.
Marcus M. Jensen and Donald N. Wright
Prentice-Hall, Englewood Cliffs, NJ, 1993

A Manual of Laboratory Diagnostic Tests, 4th ed.
Frances Fischbach
J.B. Lippincott Company, Philadelphia, 1992

Diagnostic Tests—Clinical Pocket Manual
Springhouse Corporation, Springhouse, 1985

Behold Man: A Photographic Journey Inside the Body
Lennart Nilsson
Little, Brown & Co, Boston, 1973

Life, Death, and in Between: Tales of Clinical Neurology
Harold L. Klawans
Paragon House Publishers, New York, 1992

Toscanini's Fumble and Other Tales of Clinical Neurology
Harold L. Klawans
Bantam Books, New York, 1989

New Guinea Tapeworms and Jewish Grandmothers: Tales of Parasites and People
Robert S. Desowitz
W.W. Norton and Company, New York, 1987

The Malaria Capers: Tales of Parasites and People
Robert S. Desowitz
W.W. Norton and Company, New York, 1993

How We Die
Sherwin B. Nuland
Alfred A. Knopf, New York, 1994

APPENDIX F

Prefixes, Combining Word Roots, and Suffixes Used in Medical Terminology

PREFIXES AND COMBINING WORD ROOTS

a-, an- absent, without (amenorrhea: absence of menstruation)
ab- away from (abduct: move away from the midline)
abdomin/o- abdomen (abdominal aorta: the portion of the aorta in the abdomen)
acou- hearing (acoustic nerve: the cranial nerve for hearing)
ad- toward, near, to (adduct: move toward the midline)
aden/o- gland (adenohypophysis: the glandular part of the pituitary gland)
af- to, toward (afferent: toward a center)
alba- white (albino: an animal lacking coloration)
alg- pain (myalgia: muscle pain)
ana- up, back (anabolism: the constructive phase of metabolism)
angi/o- vessel (angiogram: imaging of blood vessels, as in the heart)
ante- before (antenatal: before birth)
anti- against (antiemetic: an agent that prevents vomiting)
arthr/o- joint (arthritis: inflammation of a joint)
atel- imperfect, incomplete (atelectasis: incomplete expansion of a lung)
auto- self (autoimmune disease: a disease in which immune reactions are directed against part of one's own body)

bi- two, twice (biconcave: concave on each side, as a red blood cell)
bio- life (biochemistry: the chemistry of living organisms)
blasto- growth, budding (blastocyst: a rapidly growing embryonic stage)
brachi/o- arm (brachial artery: the artery that passes through the upper arm)
brachy- short (brachydactyly: abnormally short fingers or toes)
brady- slow (bradycardia: slow heart rate)
bronch- air passage (bronchioles: small air passages in the lungs)

carcin/o- cancer (carcinogen: cancer-causing substance)
cardi/o- heart (cardiopathy: heart disease)
carp/o- wrist (carpals: bones of the wrist)
cata- down (catabolism: the breaking down phase of metabolism)
caud- tail (cauda equina: the spinal nerves that hang below the end of the spinal cord and resemble a horse's tail)
celi/o- abdomen (celiac artery: a large artery that supplies abdominal organs)
cephal/o- head (cephaledema: swelling of the head)
cerebr/o- brain (cerebrum: the largest part of the human brain)
cervic- neck (cervical nerves: the spinal nerves from the neck portion of the spinal cord)
chem/o- chemical (chemotherapy: the use of chemicals to treat disease)
chondr/o- cartilage (chondrocyte: cartilage cell)
circum- around (circumoral: around the mouth)
co-, com-, con- with, together (congenital: born with)
contra- opposite, against (contraception: the prevention of conception)
cost/o- ribs (intercostal muscles: muscles between the ribs)
crani/o- skull, head (cranial nerves: the nerves that arise from the brain)
cut- skin (cutaneous: pertaining to the skin)
cyan/o- blue (cyanosis: bluish discoloration of the skin due to lack of oxygen)
cyst- bladder, sac (cystic duct: duct of the gallbladder)
cyt/o- cell (hepatocyte: cell of the liver)

dactyl/o- digits, fingers or toes (polydactyly: more than five fingers or toes)
de- down, from (dehydration: loss of water)
derm- skin (dermatologist: a specialist in diseases of the skin)
di- two, twice (disaccharide: a sugar made of two monosaccharides)
diplo- double (diplopia: double vision)
dis- apart, away from (dissect: to cut apart)

duct- lead, conduct (ductus arteriosus: a fetal artery)
dys- difficult, diseased (dyspnea: difficult breathing)

ecto- outside (ectoparasite: a parasite that lives on the body surface)
edem- swelling (edematous: affected with swelling)
endo- within (endocardium: the innermost layer of the heart wall)
enter/o- intestine (enterotoxin: a toxin that affects the intestine and causes diarrhea)
epi- on, over, upon (epidermis: the outer layer of the skin)
erythr/o- red (erythrocyte: red blood cell)
eu- normal, good (eupnea: normal breathing)
ex- out of (excise: to cut out or remove surgically)
exo- without, outside of (exopthalmia: protrusion of the eyeballs)
extra- outside of, in addition to, beyond (extraembryonic membranes: the membranes that surround the embryo-fetus)

fasci- band (fascia: a fibrous connective tissue membrane)
fore- before, in front (forehead: the front of the head)

gastr/o- stomach (gastric juice: the digestive secretions of the stomach lining)
gluco-, glyco- sugar (glycosuria: glucose in the urine)
gyn/o-, gyne/co- woman, female (gynecology: study of the female reproductive organs)

haplo- single, simple (haploid: a single set, as of chromosomes)
hema-, hemato-, hemo- blood (hemoglobin: the protein of red blood cells)
hemi- half (cerebral hemisphere: the right or left half of the cerebrum)
hepat/o- liver (hepatic duct: the duct that takes bile out of the liver)
hetero- different (heterozygous: having two different genes for a trait)
hist/o- tissue (histology: the study of tissues)
homeo- unchanged (homeostasis: the state of body stability)
homo- same (homozygous: having two similar genes for a trait)
hydr/o- water (hydrophobia: fear of water)
hyper- excessive, above (hyperglycemia: high blood glucose level)
hypo- beneath, under, deficient (hypodermic: below the skin)

idio- distinct, peculiar to the individual (idiopathic: of unknown cause, as a disease)
inter- between, among (interventricular septum: the wall between the ventricles of the heart)
intra- within (intracellular: within cells)
is/o equal, the same (isothermal: having the same temperature)

kinesi/o- movement (kinesthetic sense: muscle sense)

labi- lip (herpes labialis: cold sores of the lips)
lacri- tears (lacrimal glands: tear-producing glands)
lact/o- milk (lactation: milk production)
leuc/o, leuk/o- white (leukocyte: white blood cell)
lip/o- fat (liposuction: removal of fat with a suctioning instrument)

macr/o- large (macromolecule: a large molecule such as a protein)
mal- poor, bad (malnutrition: poor nutrition)
medi- middle (mediastinum: a middle cavity, as in the chest)
mega- large (megacolon: abnormally dilated colon)
meta- next to, beyond (metatarsal: bone of the foot next to the ankle)
micr/o- small (microcephaly: small head)
mon/o- one (monozygotic twins: indentical twins, from one egg)
morph/o- shape, form (amorphous: without definite shape)
multi- many (multicellular: made of many cells)
my/o- muscle (myocardium: heart muscle)

narco- sleep (narcotic: a drug that produces sleep)
nat/a- birth (neonate: a newborn infant)
neo- new (neoplasty: surgical restoration of parts)
nephr/o- kidney (nephrectomy: removal of a kidney)
neur/o- nerve (neuron: nerve cell)
non- not (non-communicable: unable to spread)

ocul/o- eye (oculomotor nerve: a cranial nerve for eye movement)
olig/o- few, scanty (oliguria: diminished amount of urine)
oo- egg (oogenesis: production of an egg cell)
ophthalmo- eye (ophthalmoscope: instrument to examine the eye)
orth/o- straight, normal, correct (orthostatic: related to standing upright)
oste/o- bone (osteocyte: bone cell)
ot/o- ear (otitis media: inflammation of the middle ear)
ovi-, ovo- egg (oviduct: duct for passage of an egg cell, fallopian tube)

path/o- disease (pathology: the study of disease)
ped/ia- child (pediatric: concerning the care of children)
per- through (permeate: to pass through)
peri- around (pericardium: membranes that surround the heart
phag/o- eat (phagocyte: a cell that engulfs other cells)

phleb/o- vein (phlebitis: inflammation of a vein)
pleuro-, pleura- rib (pleurisy: inflammation of the pleural membranes of the chest cavity)
pneumo- lung (pneumonia: lung infection)
pod- foot (pseudopod: false foot, as in ameboid movement)
poly- many (polysaccharide: a carbohydrate made of many monosaccharides)
post- after (postpartum: after delivery of a baby)
pre- before (precancerous: a growth that probably will become malignant)
pro- before, in front of (progeria: premature old age, before its time)
pseudo- false (pseudomembrane: false membrane)
py/o- pus (pyogenic: pus producing)
pyel/o- renal pelvis (pyelogram: an x-ray of the renal pelvis and ureter)

quadr/i- four (quadriceps femoris: a thigh muscle with four parts)

retro- behind, backward (retroperitoneal: located behind the peritoneum)
rhin/o- nose (rhinoviruses that cause the common cold)

salping/o- fallopian tube (salpingitis: inflammation of a fallopian tube)
sarc/o- flesh, muscle (sarcolemma: membrane of a muscle cell)
sclero- hard (sclerosis: hardening of tissue with loss of function)
semi- half (semilunar valve: a valve shaped like a half-moon)
steno- narrow (aortic stenosis: narrowing of the aorta)
sub- below, beheath (subcutaneous: below the skin)
supra- above (suprarenal gland: gland above the kidney, the adrenal gland)
sym- together (symphysis: a joint where two bones meet)
syn- together (synapse: the space between two nerve cells)

tachy- fast (tachycardia: rapid heart rate)
thorac/o- chest (thoracic cavity: chest cavity)
thromb/o- clot (thrombosis: formation of a blood clot)
tox- poison (toxicology: the study of poisons)
trans- across (transmural: across the wall of an organ)
tri- three (trigone: a three-sided area on the floor of the urinary bladder)

ultra- excessive, extreme (ultrasonic: sound waves beyond the normal hearing range)
un/i- one (unicellular: made of one cell)
uria-, uro- urine (urinary calculi: stones in the urine)

vas/o- vessel (vasodilation: dilation of a blood vessel)
viscera-, viscero- organ (visceral pleura: the pleural membrane that covers the lungs)

SUFFIXES

-ac pertaining to (cardiac: pertaining to the heart)
-al pertaining to (intestinal: pertaining to the intestine)
-an, -ian characteristic of, pertaining to, belonging to (ovarian cyst: a cyst of the ovary)
-ar relating to (muscular: relating to muscles)
-ary relating to, connected with (salivary: relating to saliva)
-ase enzyme (sucrase: an enzyme that digests sucrose)
-atresia abnormal closure (biliary atresia: closure or absence of bile ducts)

-blast grow, produce (osteoblast: a bone-producing cell)

-cele swelling, tumor (meningocele: a hernia of the meninges)
-centesis puncture of a cavity (thoracocentesis: puncture of the chest cavity to remove fluid)
-cide kill (bactericide: a chemical that kills bacteria)
-clast destroy, break down (osteoclast: a bone-reabsorbing cell)

-desis binding, stabilizing, fusion (arthrodesis: the surgical immobilization of a joint)
-dipsia thirst (polydipsia: excessive thirst)
-dynia pain (gastrodynia: stomach pain)

-ectasia, -ectasis expansion (atelectasis: without expansion)
-ectomy excision, cutting out (thyroidectomy: removal of the thyroid)
-emia pertaining to blood (hypokalemia: low blood potassium level)

-form structure (spongiform: resembling a sponge)

-gen producing (carcinogen: a substance that produces cancer)
-genesis production of, origin of (spermatogenesis: production of sperm)
-globin protein (myoglobin: a muscle protein)
-gram record, writing (electroencephalogram: a record of the electrical activity of the brain)
-graph an instrument for making records (ultrasonography: the use of ultrasound to produce an image)

-ia condition (pneumonia: condition of inflammation of the lungs)
-iasis diseased condition (cholelithiasis: gall stones)
-ic pertaining to (atomic: pertaining to atoms)
-ile having qualities of (febrile: feverish)
-ism condition, process (alcoholism: condition of being dependent on alcohol)

-ist practitioner, specialist (neurologist: a specialist in diseases of the nervous system)
-itis inflammation (hepatitis: inflammation of the liver)

-lepsy seizure (narcolepsy: a sudden onset of sleep)
-lith stone, crystal (otoliths: stones in the inner ear)
-logy study of (virology: the study of viruses)
-lysis break down (hemolysis: rupture of red blood cells)

-megaly enlargement (splenomegaly: enlargement of the spleen)
-meter a measuring instrument (spirometer: an instrument to measure pulmonary volumes)

-ness state of, quality (illness: state of being ill)

-oid the appearance of (ovoid: resembling an oval or egg)
-ole small, little (arteriole: small artery)
-oma tumor (carcinoma: malignant tumor)
-opia eye (hyperopia: farsightedness)
-ory pertaining to (regulatory: pertaining to regulation)
-ose having qualities of (comatose: having qualities of a coma)
-osis state, condition, action, process (keratosis: abnormal growth of the skin)
-ostomy creation of an opening (colostomy: creation of an opening between the intestine and the abdominal wall)
-otomy cut into (tracheotomy: cut into the trachea)
-ous pertaining to (nervous: pertaining to nerves)

-pathy disease (retinopathy: disease of the retina)
-penia lack of, deficiency (leukopenia: lack of white blood cells)
-philia love of, tendency (hemophilia: a clotting disorder, "love of blood")
-phobia an abnormal fear (acrophobia: fear of heights)
-plasia growth (hyperplasia: excessive growth)
-plasty formation, repair (rhinoplasty: plastic surgery on the nose)
-plegia paralysis (hemiplegia: paralysis of the right or left half of the body)
-poiesis production (erythropoiesis: production of red blood cells)
-ptosis dropping, falling (hysteroptosis: falling of the uterus)

-rrhage burst forth (hemorrhage: loss of blood from blood vessels)
-rrhea discharge, flow (diarrhea: frequent discharge of feces)

-scope instrument to examine (microscope: instrument to examine small objects)
-spasm involuntary contraction (blepharospasm: twitch of the eyelid)
-stasis to be still, control, stop (hemostasis: to stop loss of blood)
-sthenia strength (myasthenia: loss of muscle strength)
-stomy surgical opening (colostomy: a surgical opening in the colon)

-taxia muscle coordination (ataxia: loss of coordination)
-tension pressure (hypertension: high blood pressure)
-tic pertainig to (paralytic: pertaining to paralysis)
-tomy incision, cut into (phlebotomy: incision into a vein)
-tripsy crush (lithotripsy: crushing of stone such as gallstones)
-trophic related to nutrition or growth (autotrophic: capable of making its own food, such as a green plant)
-tropic turning toward (chemotropic: turning toward a chemical)

-ula, -ule small, little (venule: small vein)
-uria urine (hematuria: blood in the urine)

-y condition, process (healthy: condition of health)

APPENDIX G

Eponymous Terms

An eponym is a person for whom something is named, and an eponymous term is a term that uses that name or eponym. For example, fallopian tube is named for Gabriele Fallopio, an Italian anatomist of the 16th century.

In recent years it has been suggested that eponymous terms be avoided because they are not descriptive and that they be replaced with more informative terms. Such changes, however, occur slowly, because the older terms are so familiar to those of us who teach. Some of us may even use them as opportunities to impart a little history, also known as "telling stories."

In this edition, the most familiar eponymous terms have been retained, with the newer term in parentheses after the first usage. The list below is provided to show the extent of reclassification of eponymous terms as related to basic anatomy and physiology.

Eponymous Term	New Term
Achilles reflex	plantar reflex
Achilles tendon	calcaneal tendon
Adam's apple	thyroid cartilage
ampulla of Vater	hepatopancreatic ampulla
aqueduct of Sylvius	cerebral aqueduct
Auerbach's plexus	myenteric plexus
Bartholin's glands	greater vestibular glands
Bowman's capsule	glomerular capsule
Broca's area	Broca's speech area
Brunner's glands	duodenal submucosal glands
bundle of His	atrioventricular bundle
canal of Schlemm	scleral venous sinus
circle of Willis	cerebral arterial circle
Cowper's glands	bulbourethral glands
crypts of Lieberkuhn	intestinal glands
duct of Santorini	accessory pancreatic duct
duct of Wirsung	pancreatic duct
Eustachian tube	auditory tube
Fallopian tube	uterine tube
fissure of Rolando	central sulcus
fissure of Sylvius	lateral cerebral sulcus
Graafian follicle	vesicular ovarian follicle
Grave's disease	hyperthyroidism
Haversian canal	central canal
Haversian system	osteon
Heimlich maneuver	abdominal thrust maneuver
islet of Langerhans	pancreatic islet
Krebs cycle	citric acid cycle
Kupffer cells	stellate reticuloendothelial cells
Langerhans' cell	non-pigmented granular dendrocyte

Eponymous Term	New Term
Leydig cells	interstitial cells
loop of Henle	loop of the nephron
Meissner's corpuscles	tactile corpuscles
Meissner's plexus	submucosal plexus
nodes of Ranvier	neurofibral nodes
organ of Corti	spiral organ
Pacinian corpuscle	lamellated corpuscle
Peyer's patches	aggregated lymph nodules
Purkinje fibers	cardiac conducting myofibers
Schwann cell	neurolemmocyte
Sertoli cells	sustentacular cells
sphincter of Boyden	sphincter of the common bile duct
sphincter of Oddi	sphincter of the hepatopancreatic ampulla
Stensen's ducts	parotid ducts
Volkmann's canal	perforating canal, nutrient canal
Wernicke's area	posterior speech area
Wharton's duct	submandibular duct
Wormian bone	sutural bone

Glossary

PRONUNCIATION GUIDE

This pronunciation guide is intended to help you pronounce the words that appear below in the Glossary. Although it is not a true phonetic key, it does help to suggest the necessary sounds by spelling the sounds of the syllables of frequently encountered words and then using these familiar combinations to "spell out" a pronunciation of the new word being defined in the Glossary.

VOWELS

Long vowel sounds: ay, ee, eye or igh, oh or ow, yoo

The sound spelled as . . .	Is pronounced as it appears in . . .
ay	a as in face
a	a as in atom
aw	au as in cause o as in frost
ah	o as in proper
ee	e as in beat
e	e as in ten
i	i as in it
u	u as in up

CONSONANTS

Consonants are pronounced just as they look, with g pronounced as it appears in gone.

ACCENTS WITHIN WORDS

One accent: boldface capital letters
Two accents: primary accent is in boldface capital letters
secondary accent is in capital letters

—A—

Abdomen (**AB**-doh-men)—Portion of the body between the diaphragm and the pelvis (Chapter 1).

Abdominal cavity (ab-**DAHM**-in-uhl **KAV**-i-tee)—Part of the ventral cavity, inferior to the diaphragm and above the pelvic cavity (Chapter 1).

Abducens nerves (ab-**DEW**-senz)—Cranial nerve pair VI. Motor to an extrinsic muscle of the eye (Chapter 8).

Abduction (ab-**DUK**-shun)—Movement of a body part away from the midline of the body (Chapter 7).

ABO group (A-B-O **GROOP**)—The red blood cell types determined by the presence or absence of A and B antigens on the red blood cell membrane; the four types are A, B, AB, and O (Chapter 11).

Absorption (ab-**ZORB**-shun)—The taking in of materials by cells or tissues (Chapter 4).

Accessory nerves (ak-**SES**-suh-ree)—Cranial nerve pair XI. Motor to the larynx and shoulder muscles (Chapter 8).

Accessory organs (ak-**SES**-suh-ree)—The digestive organs that contribute to the process of digestion, although digestion does not take place within them; consist of the teeth, tongue, salivary glands, liver, gallbladder, and pancreas (Chapter 16).

Acetabulum (ASS-uh-**TAB**-yoo-lum)—The deep socket in the hip bone that articulates with the head of the femur (from the Latin "little vinegar cup") (Chapter 6).

Acetylcholine (as-**SEE**-tull-KOH-leen)—A chemical neurotransmitter released at neuromuscular junctions, as well as by neurons in the central and peripheral nervous systems (Chapter 7).

Acid (**ASS**-id)—A hydrogen ion (H^+) donor; when in solution has a pH less than 7 (Chapter 2).

Acidosis (Ass-i-**DOH**-sis)—The condition in which the pH of the blood falls below 7.35 (Chapter 2).

Acne (**AK**-nee)—Inflammation of the sebaceous glands and hair follicles (Chapter 5).

Acoustic nerves (uh-**KOO**-stik)—Cranial nerve pair VIII. Sensory for hearing and equilibrium (Chapter 8).

Acquired immunity (uh-**KWHY**-erd im-**YOO**-ni-tee)—The immunity obtained upon exposure to a pathogen or a vaccine or upon reception of antibodies for a particular pathogen (Chapter 14).

Acromegaly (AK-roh-**MEG**-ah-lee)—Hypersecretion of growth hormone in an adult, resulting in excessive growth of the bones of the face, hands, and feet (Chapter 10).

Acrosome (**AK**-roh-sohm)—The tip of the head of a sperm cell; contains enzymes to digest the membrane of the ovum (Chapter 20).

Actin (**AK**-tin)—A contractile protein in the sarcomeres of muscle fibers (Chapter 7).

Action potential (**AK**-shun poh-**TEN**-shul)—The changes in electrical charges on either side of a cell membrane in response to a stimulus; depolarization followed by repolarization (Chapter 7).

Active immunity (**AK**-tiv im-**YOO**-ni-tee)—The immunity provided by the production of antibodies after exposure to a foreign antigen; may be natural (recovery from disease) or artificial (reception of a vaccine) (Chapter 14).

Active site theory (**AK**-tiv SITE **THEER**-ree)—The process by which an enzyme catalyzes a specific reaction; depends on the shapes of the enzyme and the substrate molecules (Chapter 2).

Active transport (**AK**-tiv **TRANS**-port)—The process in which there is movement of molecules against a concentration gradient; that is, from an area of lesser concentration to an area of greater concentration. Requires energy (Chapter 3).

Acute (ah-**KEWT**)—1. Characterized by rapid onset 2. Sharp or severe, with respect to symptoms; not chronic. (Chapter 22).

Adaptation (A-dap-**TAY**-shun)—The characteristic of sensations in which awareness of the sensation diminishes despite a continuing stimulus (Chapter 9).

Addison's disease (**ADD**-i-sonz)—Hyposecretion of the hormones of the adrenal cortex, characterized by low blood pressure, dehydration, muscle weakness, and mental lethargy (Chapter 10).

Adduction (ad-**DUK**-shun)—The movement of a body part toward the midline of the body (Chapter 7).

Adenohypophysis (uh-DEN-oh-high-**POFF**-e-sis)—The anterior pituitary gland (Chapter 10).

Adipocyte (**ADD**-i-poh-site)—A cell of adipose tissue, specialized to store fat (Chapter 4).

Adipose tissue (**ADD**-i-pohz **TISH**-yoo)—A connective tissue composed primarily of adipocytes; function is fat storage as a source of potential energy (Chapter 4).

Adrenal cortex (uh-**DREE**-nuhl **KOR**-teks)—The outer layer of the adrenal glands, which secretes cortisol and aldosterone (Chapter 10).

Adrenal glands (uh-**DREE**-nuhl)—The endocrine glands located on the top of the kidneys; each consists of an adrenal cortex, which secretes cortisol and aldosterone, and an adrenal medulla, which secretes epinephrine and norepinephrine (Syn.—suprarenal glands) (Chapter 10).

Adrenal medulla (uh-**DREE**-nuhl muh-**DEW**-lah)—The inner layer of the adrenal glands, which secretes epinephrine and norepinephrine (Chapter 10).

Adrenocorticotropic hormone (ACTH) (uh-DREE-no-KOR-ti-koh-**TROH**-pik)—A hormone produced by the anterior pituitary gland that stimulates the adrenal cortex to secrete cortisol (Chapter 10).

Aerobic (air-**ROH**-bik)—Requiring oxygen (Chapter 3).

Afferent (**AFF**-uh-rent)—To carry toward a center or main part (Chapter 8).

Afferent arteriole (**AFF**-er-ent ar-**TIR**-ee-ohl)—The arteriole that takes blood from the renal artery into a glomerulus; within its wall are juxtaglomerular cells that secrete renin (Chapter 18).

Afterbirth (**AFF**-ter-berth)—The placenta delivered shortly after delivery of the infant (Chapter 21).

After-image (**AFF**-ter-IM-ije)—The characteristic of sensations in which a sensation remains in the consciousness even after the stimulus has stopped (Chapter 9).

Agglutination (uh-GLOO-ti-**NAY**-shun)—Clumping of blood cells or microorganisms; the result of an antigen-antibody reaction (Chapter 11).

AIDS (AYDS)—Acquired immunodeficiency syndrome; caused by a virus (HIV) that infects helper T cells and depresses immune responses (Chapter 14).

Albumin (Al-**BYOO**-min)—A protein synthesized by the liver, which circulates in blood plasma; contributes to the colloid osmotic pressure of the blood (Chapter 11).

Aldosterone (al-**DAH**-ster-ohn)—A hormone (mineralocorticoid) secreted by the adrenal cortex that increases the reabsorption of sodium and the excretion of potassium by the kidneys (Chapter 10).

Alimentary tube (AL-i-**MEN**-tah-ree TOOB)—The series of digestive organs that extends from the mouth to the anus; consists of the oral cavity, pharynx, esophagus, stomach, small intestine, and large intestine (Chapter 16).

Allergen (**AL**-er-jen)—A substance capable of stimulating an allergic response (Chapter 14).

Allergy (**AL**-er-jee)—A hypersensitivity to a foreign antigen that usually does not stimulate an immune response in people; the immune response serves no useful purpose (Chapter 14).

Allele (uh-**LEEL**)—One of two or more different genes for a particular characteristic (Chapter 21).

Alopecia (AL-oh-**PEE**-she-ah)—Loss of hair, especially that of the scalp (Chapter 5).

Alpha cells (**AL**-fah SELLS)—The cells of the islets of Langerhans of the pancreas that secrete the hormone glucagon (Chapter 10).

Alveolar type I cell (al-**VEE**-oh-lar TIGHP WON SELL)—The simple squamous epithelial cell that forms the walls of the alveoli of the lungs (Chapter 15).

Alveolar type II cell (al-**VEE**-oh-lar TIGHP TOO SELL)—The septal cell in the lungs that produces pulmonary surfactant (Chapter 15).

Alveoli (al-**VEE**-oh-lye)—The air sacs of the lungs, made of simple squamous epithelium, in which gas exchange takes place (Chapter 15).

Alzheimer's disease (**ALZ**-high-mers)—A progressive brain disease, of unknown cause, resulting in loss of memory, intellectual ability, speech, and motor control (Chapter 8).

Amenorrhea (ay-MEN-uh-**REE**-ah)—Absence of menstruation (Chapter 20).

Amine/Amino group (ah-**MEEN**/ah-**MEE**-noh)—The NH_2 portion of a molecule such as an amino acid (Chapter 12).

Amino acid (ah-**MEE**-noh **ASS**-id)—An organic compound that contains an amino, or amine, group (NH_2) and a carboxyl group (COOH). Twenty different amino acids are the subunit molecules of which proteins are made (Chapter 2).

Amniocentesis (AM-nee-oh-sen-**TEE**-sis)—A diagnostic procedure in which amniotic fluid is obtained for culture of fetal cells; used to detect genetic diseases or other abnormalities in the fetus (Chapter 21).

Amnion (**AM**-nee-on)—An embryonic membrane that holds the fetus suspended in amniotic fluid; fuses with the chorion by the end of the third month of gestation (Chapter 21).

Amniotic fluid (AM-nee-**AH**-tik **FLOO**-id)—The fluid contained within the amnion; cushions the fetus and absorbs shock (Chapter 21).

Amphiarthrosis (AM-fee-ar-**THROH**-sis)—A slightly movable joint, such as a symphysis (Chapter 6).

Amylase (**AM**-i-lays)—A digestive enzyme that breaks down starch to maltose; secreted by the salivary glands and the pancreas (Chapter 16).

Anabolic steroid (AN-ah-**BAH**-lik **STEER**-oyd)—A chemical similar in structure and action to the male hormone testosterone; increases protein synthesis, especially in muscles (Chapter 7).

Anabolism (an-**AB**-uh-lizm)—Synthesis reactions, in which smaller molecules are bonded together to form larger molecules; require energy (ATP) and are catalyzed by enzymes (Chapter 17).

Anaerobic (AN-air-**ROH**-bik)—1. In the absence of oxygen. 2. Not requiring oxygen (Chapter 7).

Anaphase (**AN**-ah-fayz)—The third stage of mitosis, in which the separate chromatids move toward opposite poles of the cell (Chapter 3).

Anaphylactic shock (AN-uh-fi-**LAK**-tik SHAHK)—A type of circulatory shock that is the result of a massive allergic reaction (from the Greek "unguarded") (Chapter 13).

Anastomosis (a-NAS-ti-**MOH**-sis)—A connection or joining, especially of blood vessels (Chapter 13).

Anatomical position (AN-uh-**TOM**-ik-uhl pa-**ZI**-shun)—The position of the body used in anatomical descriptions: The body is erect and facing forward, the arms are at the sides with the palms facing forward (Chapter 1).

Anatomy (uh-**NAT**-uh-mee)—The study of the structure of the body and the relationships among the parts (Chapter 1).

Anemia (uh-**NEE**-mee-yah)—A deficiency of red blood cells or hemoglobin (Chapter 11).

Aneurysm (**AN**-yur-izm)—A localized sac or bubble that forms in a weak spot in the wall of a blood vessel, usually an artery (Chapter 13).

Angiotensin II (AN-jee-oh-**TEN**-sin 2)—The final product of the renin-angiotensin mechanism; stimulates vasoconstriction and increased secretion of adolsterone, both of which help raise blood pressure (Chapter 13).

Anion (**AN**-eye-on)—An ion with a negative charge (Chapter 2).

Antagonistic muscles (an-**TAG**-on-ISS-tik **MUSS**-uhls)—Muscles that have opposite functions with respect to the movement of a joint (Chapter 7).

Anterior (an-**TEER**-ee-your)—Toward the front (Syn.—ventral) (Chapter 1).

Antibiotic (AN-ti-bye-**AH**-tick)—A chemical medication that stops or inhibits the growth of bacteria or fungi (Chapter 22).

Antibody (**AN**-ti-BAH-dee)—A protein molecule produced by plasma cells that is specific for and will bond to a particular foreign antigen (Syn.—gamma globulin, immune globulin) (Chapter 14).

Antibody titer (**AN**-ti-BAH-dee **TIGH**-ter)—A diagnostic test that determines the level or amount of a particular antibody in blood or serum (Chapter 14).

Anticodon (**AN**-ti-KOH-don)—A triplet of bases on tRNA that matches a codon on mRNA (Chapter 3).

Antidiuretic hormone (ADH) (AN-ti-DYE-yoo-**RET**-ik)—A hormone produced by the hypothalamus and stored in the posterior pituitary gland; increases the reabsorption of water by the kidney tubules (Chapter 8).

Antigen (**AN**-ti-jen)—A chemical marker that identifies cells of a particular species or individual. May be "self" or "foreign." Foreign antigens stimulate immune responses (Chapter 2).

Antigenic (An-ti-**JEN**-ik)—Capable of stimulating antibody production (Chapter 14).

Anti-inflammatory effect (AN-ti-in-**FLAM**-uh-tor-ee)—To lessen the process of inflammation; cortisol is the hormone that has this effect (Chapter 10).

Antipyretic (AN-tigh-pye-**RET**-ik)—A medication, such as aspirin, that lowers a fever (Chapter 17).

Antiseptic (AN-ti-**SEP**-tick)—A chemical that destroys bacteria or inhibits their growth on a living being (Chapter 22).

Antithrombin (**AN**-ti-THROM-bin)—A protein synthesized by the liver that inactivates excess thrombin to prevent abnormal clotting (Chapter 11).

Antitoxin (AN-tee-**TAHK**-sin)—Antibodies specific for a bacterial toxin; used in treatment of diseases such as botulism or tetanus (Chapter 7).

Anus (**AY**-nus)—The terminal opening of the alimentary tube for the elimination of feces; surrounded by the internal and external anal sphincters (Chapter 16).

Aorta (ay-**OR**-tah)—The largest artery of the body; emerges from the left ventricle; has four parts: ascending aorta, aortic arch, thoracic aorta, and abdominal aorta (Chapter 13).

Aortic body (ay-**OR**-tik **BAH**-dee)—The site of chemoreceptors in the aortic arch, which detect changes in blood pH and the blood levels of oxygen and carbon dioxide (Chapter 9).

Aortic semilunar valve (ay-**OR**-tik SEM-ee-**LOO**-nar VALV)—The valve at the junction of the left ventricle and the aorta; prevents backflow of blood from the aorta to the ventricle when the ventricle relaxes (Chapter 12).

Aortic sinus (ay-**OR**-tik **SIGH**-nus)—The location of pressoreceptors in the wall of the aortic arch (Chapter 9).

Apgar score (**APP**-gar SKOR)—A system of evaluating an infant's condition 1 minute after birth; includes heart rate, respiration, muscle tone, response to stimuli, and color (Chapter 21).

Aphasia (ah-**FAY**-zee-ah)—Impairment or absence of the ability to communicate in speech, reading, or writing. May involve word deafness or word blindness (Chapter 8).

Aplastic anemia (ay-**PLAS**-tik un-**NEE**-mee-yah)—Failure of the red bone marrow resulting in decreased numbers of red blood cells, white blood cells, and platelets; may be a side effect of some medications (Chapter 11).

Apneustic center (ap-**NEW**-stik **SEN**-ter)—The respiratory center in the pons that prolongs inhalation (Chapter 15).

Apocrine gland (**AP**-oh-krin)—The type of sweat gland (exocrine) found primarily in the axilla and genital area; actually a modified scent gland (Chapter 5).

Apparent (uh-**PAR**-ent)—1. Readily seen or visible. 2. An infection in which the patient exhibits the symptoms of the disease (Chapter 22).

Appendicitis (uh-PEN-di-**SIGH**-tis)—Inflammation of the appendix (Chapter 16).

Appendicular skeleton (AP-en-**DIK**-yoo-lar)—The portion of the skeleton that consists of the shoulder and pelvic girdles and the bones of the arms and legs (Chapter 6).

Appendix (uh-**PEN**-diks)—A small tubular organ that extends from the cecum; has no known function for people and is considered a vestigial organ (Chapter 16).

Aqueous (**AY**-kwee-us)—Pertaining to water; used especially to refer to solutions (Chapter 2).

Aqueous humor (**AY**-kwee-us **HYOO**-mer)—The tissue fluid of the eye within the anterior cavity of the eyeball; nourishes the lens and cornea (Chapter 9).

Arachnoid membrane (uh-**RAK**-noid)—The middle layer of the meninges, made of web-like connective tissue (from the Greek "spider-like") (Chapter 8).

Arachnoid villi (uh-**RAK**-noid **VILL**-eye)—Projections of the cranial arachnoid membrane into the cranial venous sinuses, through which cerebrospinal fluid is reabsorbed back into the blood (Chapter 8).

Areolar connective tissue (uh-**REE**-oh-lar)—A tissue that consists of tissue fluid, fibroblasts, collagen and elastin fibers, and wandering WBCs; found in all mucous membranes and in subcutaneous tissue (Syn.—loose connective tissue) (Chapter 4).

Arrhythmia (uh-**RITH**-me-yah)—An abnormal or irregular rhythm of the heart (Chapter 12).

Arteriole (ar-**TEER**-ee-ohl)—A small artery (Chapter 5).

Arteriosclerosis (ar-TIR-ee-oh-skle-**ROH**-sis)— Deterioration of arteries with loss of elasticity that is often a consequence of aging or hypertension; a contributing factor to aneurysm or stroke (Chapter 13).

Artery (**AR**-tuh-ree)—A blood vessel that takes blood from the heart toward capillaries (Chapter 13).

Arthropod (**AR**-throw-pod)—Invertebrate animals characterized by an exoskeleton and jointed appendages; includes insects, spiders, ticks, mites, and crustaceans (Chapter 22).

Articular cartilage (ar-**TIK**-yoo-lar **KAR**-ti-lidj)—The cartilage on the joint surfaces of a bone; provides a smooth surface (Chapter 6).

Articulation (ar-TIK-yoo-**LAY**-shun)—A joint (Chapter 6).

Asthma (**AZ**-mah)—A respiratory disorder characterized by constriction of the bronchioles, excessive mucus production, and dyspnea; often caused by allergies (Chapter 15).

Astigmatism (uh-**STIG**-mah-TIZM)—An error of refraction caused by an irregular curvature of the lens or cornea (Chapter 9).

Astrocyte (**ASS**-troh-site)—A type of neuroglia that forms the blood-brain barrier to prevent potentially harmful substances from affecting brain neurons (Chapter 8).

Asymptomatic (AY-simp-toh-**MAT**-ick)—Without symptoms (Chapter 22).

Atelectasis (AT-e-**LEK**-tah-sis)—Collapsed or airless lung, without gas exchange (Chapter 15).

Atherosclerosis (ATH-er-oh-skle-**ROH**-sis)—The abnormal accumulation of lipids and other materials in the walls of arteries; narrows the lumen of the vessel and may stimulate abnormal clot formation (Chapter 2).

Atlas (**AT**-las)—An irregular bone, the first cervical vertebra; supports the skull (Chapter 6).

Atmospheric pressure (AT-mus-**FEER**-ik)—The pressure exerted by the atmosphere on objects on the earth's surface; 760 mmHg at sea level (Chapter 15).

Atom (**A**-tom)—The unit of matter that is the smallest part of an element (Chapter 2).

Atomic number (a-**TOM**-ik)—Number of protons in the nucleus of an atom (Chapter 2).

Atomic weight (a-**TOM**-ik WAYT)—The weight of an atom determined by adding the number of protons and neutrons (Chapter 2).

ATP (Adenosine triphosphate)—A specialized nucleotide that traps and releases biologically useful energy (Chapter 2).

Atrial natriuretic hormone (ANH) (**AY**-tree-uhl NAY-tree-yu-**RET**-ik)—A peptide hormone secreted by the atria of the heart when blood pressure or blood volume increases; increases loss of sodium ions and water by the kidneys (Chapter 12).

Atrioventricular (AV) node (AY-tree-oh-ven-**TRIK**-yoo-lar NOHD)—The part of the cardiac conduction pathway located in the lower interatrial septum (Chapter 12).

Atrium (**AY**-tree-um)—One of the two upper chambers of the heart that receive venous blood from the lungs or the body (Pl.—atria) (Chapter 12).

Atrophy (**AT**-ruh-fee)—Decrease in size of a body part due to lack of use; a wasting (Chapter 7).

Attenuated (uh-**TEN**-yoo-AY-ted)—Weakened, or less harmful; used to describe the microorganisms contained in vaccines, which have been treated to reduce their pathogenicity (Chapter 14).

Auditory bones (**AW**-di-tor-ee)—The malleus, incus, and stapes in the middle ear (Chapter 6).

Auerbach's plexus (**OW**-er-baks **PLEK**-sus)—The autonomic nerve plexus in the external muscle layer of the organs of the alimentary tube; regulates the contractions of the external muscle layer (Chapter 16).

Auricle (**AW**-ri-kuhl)—The portion of the outer ear external to the skull; made of cartilage covered with skin (Syn.—pinna) (Chapter 9).

Autoclave (**AW**-toh-clayve)—A machine that uses steam under pressure for sterilization (Chapter 22).

Autoimmune disease (AW-toh-im-**YOON** di-**ZEEZ**)—A condition in which the immune system produces antibodies to the person's own tissue (Chapter 6).

Autonomic nervous system (AW-toh-**NOM**-ik **NER**-vuhs)—The portion of the peripheral nervous system that consists of visceral motor neurons to smooth muscle, cardiac muscle, and glands (Chapter 8).

Autosomes (**AW**-toh-sohms)—Chromosomes other than the sex chromosomes; for people there are 22 pairs of autosomes in each somatic cell (Chapter 21).

Axial skeleton (**AK**-see-uhl)—The portion of the skeleton that consists of the skull, vertebral column, and rib cage (Chapter 6).

Axis (**AK**-sis)—An irregular bone, the second cervical vertebra; forms a pivot joint with the atlas (Chapter 6).

Axon (**AK**-sahn)—The cellular process of a neuron that carries impulses away from the cell body (Chapter 4).

Axon terminal (**AK**-sahn **TER**-mi-nuhl)—The end of the axon of a motor neuron, part of the neuromuscular junction (Chapter 7).

—B—

B cell (B SELL)—A subgroup of lymphocytes, including memory B cells and plasma cells, both of which are involved in immune responses (Chapter 11).

Bacillus (buh-**SILL**-us) (pl.: bacilli)—A rod-shaped bacterium (Chapter 22).

Bacteremia (back-tah-**REE**-mee-ah)—The presence of bacteria in the blood, which is normally sterile (Chapter 22).

Bacteria (back-**TIR**-ee-yuh) (sing.: bacterium)—The simple unicellular microorganisms of the class Schizomycetes; may be free-living, saprophytic, or parasitic (Chapter 22).

Bactericide (back-**TEER**-i-sigh'd)—A chemical that kills bacteria (Chapter 22).

Bacteriostatic (back-TEE-ree-oh-**STAT**-ick)—Capable of inhibiting the reproduction of bacteria (Chapter 22).

Bacteruria (BAK-tur-**YOO**-ree-ah)—The presence of large numbers of bacteria in urine (Chapter 18).

Ball and socket joint (BAWL and **SOK**-et)—A diarthrosis that permits movement in all planes (Chapter 6).

Bartholin's glands (**BAR**-toh-linz)—The small glands in the wall of the vagina; secrete mucus into the vagina and vestibule (Syn.—vestibular glands) (Chapter 20).

Basal ganglia (**BAY**-zuhl **GANG**-Lee-ah)—Masses of gray matter within the white matter of the cerebral hemispheres; concerned with subconscious aspects of skeletal muscle activity, such as accessory movements (Chapter 8).

Basal metabolic rate (**BAY**-zuhl met-ah-**BAHL**-ik RAYT)—The energy required to maintain the functioning of the body in a resting condition (Chapter 17).

Base (BAYS)—A hydrogen ion (H^+) acceptor, or hydroxyl ion (OH^-) donor; when in solution has a pH greater than 7 (Chapter 2).

Basilar layer (bah-**SILL**-ar **LAY**-er)—The permanent vascular layer of the endometrium that is not lost in menstruation; regenerates the functional layer during each menstrual cycle (Chapter 20).

Basophil (**BAY**-so-fill)—A type of white blood cell (granular); contains heparin and histamine (Chapter 11).

Benign (bee-**NINE**)—Not malignant (Chapter 3).

Beta cells (**BAY**-tah sells)—The cells of the islets of Langerhans of the pancreas that secrete the hormone insulin (Chapter 10).

Beta-oxidation (BAY-tah-AHK-si-**DAY**-shun)—The process by which the long carbon chain of a fatty acid molecule is broken down into two-carbon acetyl groups to be used in cell respiration; takes place in the liver (Chapter 16).

Bile (BYL)—The secretion of the liver that is stored in the gallbladder and passes to the duodenum; contains bile salts to emulsify fats; is the fluid in which bilirubin and excess cholesterol are excreted (Chapter 16).

Bile salts (BYL SAWLTS)—The active component of bile that emulsifies fats in the digestive process (Chapter 16).

Bilirubin (**BILL**-ee-roo-bin)—The bile pigment produced from the heme portion of the hemoglobin of old red blood cells; excreted by the liver in bile (Chapter 11).

Binary fission (**BYE**-na-ree **FISH**-en)—The asexual reproductive process in which one cell divides into two identical new cells (Chapter 22).

Binocular vision (bye-**NOK**-yoo-lur **VI**-zhun)—Normal vision involving the use of both eyes; the ability of the brain to create one image from the slightly different images received from each eye (Chapter 9).

Biopsy (**BYE**-op-see)—Removal of a small piece of living tissue for microscopic examination; a diagnostic procedure (Chapter 5).

Birth canal (BERTH ka-**NAL**)—The vagina during delivery of an infant (Chapter 21).

Blastocyst (**BLAS**-toh-sist)—The early stage of embryonic development that follows the morula; consists of the outer trophoblast and the internal inner cell mass and blastocele (cavity) (Chapter 21).

Blister (**BLISS**-ter)—A collection of fluid below or within the epidermis (Chapter 5).

Blood (**BLUHD**)—The fluid that circulates in the heart and blood vessels; consists of blood cells and plasma (Chapter 4).

Blood–brain barrier (BLUHD BRAYN)—The barrier between the circulating blood and brain tissue, formed by astrocytes and brain capillaries, to prevent harmful substances in the blood from damaging brain neurons (Chapter 8).

Blood pressure (BLUHD **PRE**-shure)—The force exerted by the blood against the walls of the blood vessels; measured in mmHg (Chapter 13).

Body-mass index (**BAH**-dee mass **IN**-deks)—A measure of leanness using height and weight (Chapter 17).

Bond (BAHND)—An attraction or force that holds atoms together in the formation of molecules (Chapter 2).

Bone (**BOWNE**)—1. A connective tissue made of osteocytes in a calcified matrix 2. An organ that is an individual part of the skeleton (Chapter 4).

Botulism (**BOTT**-yoo-lizm)—A disease, characterized by muscle paralysis, caused by the bacterium *Clostridium botulinum* (Chapter 7).

Bowman's capsule (**BOW**-manz **KAP**-suhl)—The expanded end of the renal tubule that encloses a glomerulus; receives filtrate from the glomerulus (Chapter 18).

Bradycardia (BRAY-dee-**KAR**-dee-yah)—An abnormally slow heart rate; less than 60 beats per minute (Chapter 12).

Brain (BRAYN)—The part of the central nervous system within the skull; regulates the activity of the rest of the nervous system (Chapter 8).

Brain stem (**BRAYN** STEM)—The portion of the brain that consists of the medulla, pons, and midbrain (Chapter 8).

Broad-spectrum (**BRAWD SPEK**-trum)—An antibiotic that is effective against a wide variety of bacteria (Chapter 22).

Bronchial tree (**BRONG**-kee-uhl TREE)—The entire system of air passageways formed by the branching of the bronchial tubes within the lungs; the smallest bronchioles terminate in clusters of alveoli (Chapter 15).

Bronchioles (**BRONG**-kee-ohls)—The smallest of the air passageways within the lungs (Chapter 15).

Buffer system (**BUFF**-er **SIS**-tem)—A pair of chemicals that prevents significant changes in the pH of a body fluid (Chapter 2).

Bulbourethral glands (BUHL-boh-yoo-**REE**-thruhl)—The glands on either side of the prostate gland that open into the urethra; secrete an alkaline fluid that becomes part of semen (Syn.—Cowper's glands) (Chapter 20).

Bundle of His (**BUN**-duhl of HISS)—The part of the cardiac conduction pathway located in the upper interventricular septum (Chapter 12).

Burn (BERN)—Damage caused by heat, flames, chemicals, or electricity, especially to the skin; classified as first degree (minor), second degree (blisters), or third degree (extensive damage) (Chapter 5).

Bursa (**BURR**-sah)—A sac of synovial fluid that decreases friction between a tendon and a bone (Chapter 6).

Bursitis (burr-**SIGH**-tiss)—Inflammation of a bursa (Chapter 6).

—C—

Calcaneus (kal-**KAY**-nee-us)—A short bone, the largest of the tarsals; the heel bone (Chapter 6).

Calciferol (kal-**SIF**-er-awl)—A form of vitamin D (Chapter 18).

Calcitonin (KAL-si-**TOH**-nin)—A hormone secreted by the thyroid gland that decreases the reabsorption of calcium from bones (Chapter 10).

Callus (**KAL**-us)—Thickening of an area of epidermis (Chapter 5).

Calorie (**KAL**-oh-ree)—1. Small calorie: the amount of heat energy needed to change the temperature of 1 gram of water 1 degree centigrade; 2. Large calorie or Calorie: a kilocalorie, used to indicate the energy content of foods (Chapter 17).

Calyx (**KAY**-liks)—A funnel-shaped extension of the renal pelvis that encloses the papilla of a renal pyramid and collects urine. (Pl.—calyces) (Chapter 18).

Canal of Schlemm (ka-**NAL** of SHLEM)—Small veins at the junction of the cornea and iris of the eye; the site of reabsorption of aqueous humor into the blood (Chapter 9).

Canaliculi (KAN-a-**LIK**-yoo-lye)—Small channels, such as those in bone matrix, that permit contact between adjacent osteocytes (Chapter 6).

Cancer (**KAN**-ser)—A malignant tumor or growth of cells (Chapter 3).

Capillary (**KAP**-i-lar-ee)—A blood vessel that takes blood from an arteriole to a venule; walls are one cell in thickness to permit exchanges of materials (Chapter 13).

Capsule (**KAP**-suhl)—A gelatinous layer located outside the cell wall of some bacteria; provides resistance to phagocytosis (Chapter 22).

Carbohydrate (KAR-boh-**HIGH**-drayt)—An organic compound that contains carbon, hydrogen, and oxygen; includes sugars and starches (Chapter 2).

Carbonic anhydrase (kar-**BAHN**-ik an-**HIGH**-drays)—The enzyme present in red blood cells and other cells that catalyzes the reaction of carbon dioxide and water to form carbonic acid (Chapter 15).

Carboxyl group (kar-**BAHK**-sul)—The COOH portion of a molecule such as an amino acid (Chapter 19).

Carcinogen (kar-**SIN**-oh-jen)—A substance that increases the risk of developing cancer (Chapter 3).

Carcinoma (KAR-sin-**OH**-mah)—A malignant tumor of epithelial tissue (Chapter 5).

Cardiac cycle (**KAR**-dee-yak **SIGH**-kuhl)—The sequence of events in one heartbeat, in which simultaneous contraction of the atria is followed by simultaneous contraction of the ventricles (Chapter 12).

Cardiac muscle (**KAR**-dee-yak **MUSS**-uhl)—The muscle tissue that forms the walls of the chambers of the heart (Chapter 4).

Cardiac output (**KAR**-dee-yak **OUT**-put)—The amount of blood pumped by a ventricle in 1 minute; the resting average is 5 to 6 liters/min (Chapter 12).

Carotid body (kah-**RAH**-tid **BAH**-dee)—The site of chemoreceptors in the internal carotid artery, which detect changes in blood pH and the levels of oxygen and carbon dioxide in the blood (Chapter 9).

Carotid sinus (kah-**RAH**-tid **SIGH**-nus)—The location of pressoreceptors in the wall of the internal carotid artery, which detect changes in blood pressure (Chapter 9).

Carpals (**KAR**-puhls)—The eight short bones of each wrist (Chapter 6).

Carrier (**KAR**-ree-yur)—A person who recovers from a disease but continues to be a source of the pathogen and may infect others (Chapter 22).

Carrier enzyme (**KA**-ree-er **EN**-zime)—An enzyme that is part of a cell membrane and carries out the process of facilitated diffusion of a specific substance (Chapter 3).

Cartilage (**KAR**-ti-lidj)—A connective tissue made of chondrocytes in a protein matrix (Chapter 4).

Catabolism (kuh-**TAB**-uh-lizm)—Breakdown or degradation reactions, in which larger molecules are broken down to smaller molecules; often release energy (ATP) and catalyzed by enzymes (Chapter 17).

Catalyst (**KAT**-ah-list)—A chemical that affects the speed of a chemical reaction, while remaining itself unchanged (Chapter 2).

Cataract (**KAT**-uh-rackt)—An eye disorder in which the lens becomes opaque and impairs vision (from the Latin "waterfall") (Chapter 9).

Catecholamines (KAT-e-**KOHL**-ah-meens)—Epinephrine and norepinephrine, the hormones secreted by the adrenal medulla (Chapter 10).

Cation (**KAT**-eye-on)—An ion with a positive charge (Chapter 2).

Cauda equina (**KAW**-dah ee-**KWHY**-nah)—The lumbar and sacral spinal nerves that hang below the end of the spinal cord and before they exit from the vertebral canal (Chapter 8).

Cecum (**SEE**-kum)—The first part of the large intestine, the dead-end portion adjacent to the ileum (from the Latin "blindness") (Chapter 16).

Cell (SELL)—The smallest living unit of structure and function of the body (Chapter 1).

Cell body (SELL **BAH**-dee)—The part of a neuron that contains the nucleus (Chapter 4).

Cell mediated immunity (SELL **MEE**-dee-ay-ted im-**YOO**-ni-tee)—The mechanism of immunity that does not involve antibody production, but rather the destruction of foreign antigens by the activities of T cells and macrophages (Chapter 14).

Cell (plasma) membrane (SELL **MEM**-brayn)—The membrane made of phospholipids, protein, and cholesterol that forms the outer boundary of a cell and regulates passage of materials into and out of the cell (Chapter 2).

Cell respiration (SELL RES-pi-**RAY**-shun)—A cellular process in which the energy of nutrients is released in the form of ATP and heat. Oxygen is required, and carbon dioxide and water are produced (Chapter 2).

Central (**SEN**-truhl)—The main part; or in the middle of (Chapter 1).

Central canal (**SEN**-truhl ka-**NAL**)—The hollow center of the spinal cord that contains cerebrospinal fluid (Chapter 8).

Central nervous system (**SEN**-tral **NER**-vuhs)—The part of the nervous system that consists of the brain and spinal cord (Chapter 8).

Centrioles (**SEN**-tree-ohls)—The cell organelles that organize the spindle fibers during cell division (Chapter 3).

Cerebellum (SER-e-**BELL**-uhm)—The part of the brain posterior to the medulla and pons; responsible for many of the subconscious aspects of skeletal muscle functioning, such as coordination and muscle tone (Chapter 7).

Cerebral aqueduct (se-**REE**-bruhl **A**-kwi-dukt)—A tunnel through the midbrain that permits cerebrospinal fluid to flow from the third to the fourth ventricle (Chapter 8).

Cerebral cortex (se-**REE**-bruhl **KOR**-teks)—The gray matter on the surface of the cerebral hemispheres. Includes motor areas, sensory areas, auditory areas, visual areas, taste areas, olfactory areas, speech areas, and association areas (Chapter 8).

Cerebrospinal fluid (se-**REE**-broh-**SPY**-nuhl)—The tissue fluid of the central nervous system; formed by choroid plexuses in the ventricles of the brain, circulates in and around the brain and spinal cord, and is reabsorbed into cranial venous sinuses (Chapter 8).

Cerebrovascular accident (SER-e-broh-**VAS**-kyoo-lur)—A hemorrhagic or ischemic lesion in the brain, often the result of aneurysm, arteriosclerosis, atherosclerosis, or hypertension (Syn.—stroke) (Chapter 8).

Cerebrum (se-**REE**-bruhm)—The largest part of the brain, consisting of the right and left cerebral hemispheres; its many functions include movement, sensation, learning, and memory (Chapter 8).

Cerumen (suh-**ROO**-men)—The waxy secretion of ceruminous glands (Chapter 5).

Ceruminous gland (suh-**ROO**-mi-nus)—An exocrine gland in the dermis of the ear canal that secretes cerumen (ear wax) (Chapter 5).

Cervical (**SIR**-vi-kuhl)—Pertaining to the neck (Chapter 1).

Cervical vertebrae (**SIR**-vi-kuhl **VER**-te-bray)—The seven vertebrae in the neck (Chapter 6).

Cervix (**SIR**-viks)—The most inferior part of the uterus that projects into the vagina (Chapter 20).

Cesarean section (se-**SAR**-ee-an **SEK**-shun)—Removal of the fetus by way of an incision through the abdominal wall and uterus (Chapter 21).

Chemical clotting (**KEM**-i-kuhl **KLAH**-ting)—A series of chemical reactions, stimulated by a rough surface or a break in a blood vessel, that result in the formation of a fibrin clot (Chapter 11).

Chemical digestion (**KEM**-i-kuhl dye-**JES**-chun)—The breakdown of food accomplished by digestive enzymes; complex organic molecules are broken down to simpler organic molecules (Chapter 16).

Chemoreceptors (KEE-moh-re-**SEP**-ters)—1. A sensory receptor that detects a chemical change. 2. Olfactory receptors, taste receptors, and the carotid and aortic chemoreceptors that detect changes in blood gases and blood pH (Chapter 9).

Chemotherapy (KEE-moh-**THER**-uh-pee)—The use of chemicals (medications) to treat disease (Chapter 3).

Chief cells (CHEEF SELLS)—The cells of the gastric pits of the stomach that secrete pepsinogen, the inactive form of the digestive enzyme pepsin (Chapter 16).

Chlamydia (kluh-**MID**-ee-ah)—A group of simple bacteria; *Chlamydia trachomatis* is a sexually transmitted pathogen that may cause conjunctivitis or pneumonia in infants born to infected women (Chapter 20).

Cholecystokinin (KOH-lee-SIS-toh-**KYE**-nin)—A hormone secreted by the duodenum when food enters; stimulates contraction of the gallbladder and secretion of enzyme pancreatic juice (Chapter 16).

Cholesterol (koh-**LESS**-ter-ohl)—A steroid that is synthesized by the liver and is part of cell membranes (Chapter 2).

Cholinesterase (KOH-lin-**ESS**-ter-ays)—The chemical inactivator of acetylcholine (Chapter 7).

Chondrocyte (**KON**-droh-sight)—A cartilage cell (Chapter 4).

Chordae tendineae (**KOR**-day ten-**DIN**-ee-ay)—Strands of connective tissue that connect the flaps of an AV valve to the papillary muscles (Chapter 12).

Chorion (**KOR**-ee-on)—An embryonic membrane that is

formed from the trophoblast of the blastocyst and will develop chorionic villi and become the fetal portion of the placenta (Chapter 21).

Chorionic villi (**KOR**-ee-ON-ik **VILL**-eye)—Projections of the chorion that will develop the fetal blood vessels that will become part of the placenta (Chapter 21).

Chorionic villus sampling (KOR-ee-**ON**-ik **VILL**-us)—A diagnostic procedure in which a biopsy of the chorionic villi is made; used to detect genetic diseases or other abnormalities in the fetus (Chapter 21).

Choroid layer (**KOR**-oyd)—The middle layer of the eyeball, contains a dark pigment to absorb light and prevent glare within the eye (Chapter 9).

Choroid plexus (**KOR**-oyd **PLEK**-sus)—A capillary network in a ventricle of the brain; forms cerebrospinal fluid (Chapter 8).

Chromatin (**KROH**-mah-tin)—The thread-like structure of the genetic material when a cell is not dividing; is not visible as individual chromosomes (Chapter 3).

Chromosomes (**KROH**-muh-sohms)—Structures made of DNA and protein within the nucleus of a cell. A human cell has 46 chromosomes (Chapter 3).

Chronic (**KRAH**-nik)—Characterized by long duration or slow progression (Chapter 22).

Chylomicron (KYE-loh-**MYE**-kron)—A small fat globule formed by the small intestine from absorbed fatty acids and glycerol (Chapter 16).

Cilia (**SILLY**-ah)—Thread-like structures that project through a cell membrane and sweep materials across the cell surface (Chapter 3).

Ciliary body (**SILLY**-air-ee **BAH**-dee)—A circular muscle that surrounds the edge of the lens of the eye and changes the shape of the lens (Chapter 9).

Ciliated epithelium (**SILLY**-ay-ted)—The tissue that has cilia on the free surface of the cells (Chapter 4).

Circle of Willis (**SIR**-kuhl of **WILL**-iss)—An arterial anastomosis that encircles the pituitary gland and supplies the brain with blood; formed by the two internal carotid arteries and the basilar (two vertebral) artery (Chapter 13).

Circulatory shock (**SIR**-kew-lah-TOR-ee SHAHK)—The condition in which decreased cardiac output deprives all tissues of oxygen and permits the accumulation of waste products (Chapter 5).

Cisterna chyli (sis-**TER**-nah **KYE**-lee)—A large lymph vessel formed by the union of lymph vessels from the lower body; continues superiorly as the thoracic duct (Chapter 14).

Clavicle (**KLAV**-i-kuhl)—The flat bone that articulates with the scapula and sternum (from the Latin "little key") (Syn.—collarbone) (Chapter 6).

Cleavage (**KLEE**-vije)—The series of mitotic cell divisions that take place in a fertilized egg; forms the early multicellular embryonic stages (Chapter 21).

Cleft palate (KLEFT **PAL**-uht)—A congenital disorder in which the bones of the hard palate do not fuse, leaving an opening between the oral and nasal cavities (Chapter 6).

Clinical infection (**KLIN**-i-kuhl)—An infection in which the patient exhibits the symptoms of the disease (Chapter 22).

Clitoris (**KLIT**-uh-ris)—An organ that is part of the vulva; a small mass of erectile tissue at the anterior junction of the labia minora; enlarges in response to sexual stimulation (Chapter 20).

Clot retraction (KLAHT ree-**TRAK**-shun)—The shrinking of a blood clot shortly after it forms due to the folding of the fibrin strands; pulls the edges of the ruptured vessel closer together (Chapter 11).

Coccus (**KOK**-us) (pl.: cocci)—A spherical bacterium (Chapter 22).

Coccyx (**KOK**-siks)—The last four to five very small vertebrae; attachment site for some muscles of the pelvic floor (Chapter 6).

Cochlea (**KOK**-lee-ah)—The snail-shell-shaped portion of the inner ear that contains the receptors for hearing in the organ of Corti (Chapter 9).

Codon (**KOH**-don)—The sequence of three bases in DNA or mRNA that is the code for one amino acid; also called a triplet code (Chapter 3).

Coenzyme (ko-**EN**-zime)—A non-protein molecule that combines with an enzyme and is essential for the functioning of the enzyme; some vitamins and minerals are coenzymes (Chapter 17).

Collagen (**KAH**-lah-jen)—A protein that is found in the form of strong fibers in many types of connective tissue (Chapter 4).

Collecting tubule (kah-**LEK**-ting)—The part of a renal tubule that extends from a distal convoluted tubule to a papillary duct (Chapter 18).

Colloid osmotic pressure (**KAH**-loyd ahs-**MAH**-tik)—The force exerted by the presence of protein in a solution; water will move by osmosis to the area of greater protein concentration (Chapter 13).

Colon (**KOH**-lun)—The large intestine (Chapter 16).

Color blindness (**KUHL**-or **BLIND**-ness)—The inability to distinguish certain colors, a hereditary trait (Chapter 9).

Columnar (kuh-**LUM**-nar)—Shaped like a column; height greater than width; used especially in reference to epithelial tissue (Chapter 4).

Common bile duct (**KOM**-mon BYL DUKT)—The duct formed by the union of the hepatic duct from the liver and the cystic duct from the gallbladder, and joined by the main pancreatic duct; carries bile and pancreatic juice to the duodenum (Chapter 16).

Communicable disease (kuhm-**YOO**-ni-kah-b'l)—A disease that may be transmitted from person to person by direct or indirect contact (Chapter 22).

Compact bone (**KOM**-pakt BOWNE)—Bone tissue made

of haversian systems; forms the diaphyses of long bones and covers the spongy bone of other bones (Chapter 6).

Complement (**KOM**-ple-ment)—A group of plasma proteins that are activated by and bond to an antigen-antibody complex; complement fixation results in the lysis of cellular antigens (Chapter 14).

Complement fixation test (**KOM**-ple-ment fik-**SAY**-shun)—A diagnostic test that determines the presence of a particular antibody in blood or serum (Chapter 14).

Computed tomography (CT) scan (kom-**PEW**-ted toh-**MAH**-grah-fee SKAN)—A diagnostic imaging technique that uses x-rays integrated by computer (Chapter 1).

Concentration gradient (KON-sen-**TRAY**-shun **GRAY**-de-ent)—The relative amounts of a substance on either side of a membrane; diffusion occurs with, or along, a concentration gradient, that is, from high concentration to low concentration (Chapter 3).

Conduction (kon-**DUK**-shun)—1. The heat loss process in which heat energy from the skin is transferred to cooler objects touching the skin. 2. The transfer of any energy form from one substance to another; includes nerve and muscle impulses, and the transmission of vibrations in the ear (Chapter 17).

Condyle (**KON**-dyel)—A rounded projection on a bone (Chapter 6).

Condyloid joint (**KON**-di-loyd)—A diarthrosis that permits movement in one plane and some lateral movement (Chapter 6).

Cones (KOHNES)—The sensory receptors in the retina of the eye that detect colors (the different wavelengths of the visible spectrum of light) (Chapter 9).

Congenital (kon-**JEN**-i-tuhl)—Present at birth (Chapter 21).

Conjunctiva (KON-junk-**TIGH**-vah)—The mucous membrane that lines the eyelids and covers the white of the eye (Chapter 9).

Conjunctivitis (kon-JUNK-ti-**VIGH**-tis)—Inflammation of the conjunctiva, most often due to an allergy or bacterial infection (Chapter 9).

Connective tissue (kah-**NEK**-tiv **TISH**-yoo)—Any of the tissues that connects, supports, transports, or stores materials. Consists of cells and matrix (Chapter 4).

Contagious disease (kun-**TAY**-jus)—A disease that is easily transmitted from person to person by casual contact (Chapter 22).

Contrast (**KON**-trast)—The characteristic of sensations in which a previous sensation affects the perception of a current sensation (Chapter 9).

Contusion (kon-**TOO**-zhun)—A bruise; the skin is not broken but may be painful, swollen, and discolored (Chapter 5).

Convection (kon-**VEK**-shun)—The heat loss process in which heat energy is moved away from the skin surface by means of air currents (Chapter 17).

Convolution (kon-voh-**LOO**-shun)—A fold, coil, roll, or twist; the surface folds of the cerebral cortex (Syn.—gyrus) (Chapter 8).

Cornea (**KOR**-nee-ah)—The transparent anterior portion of the sclera of the eye; the first structure that refracts light rays that enter the eye (Chapter 9).

Coronal (frontal) section (koh-**ROH**-nuhl **SEK**-shun)—A plane or cut from side to side, separating front and back parts (Chapter 1).

Coronary vessels (**KOR**-ah-na-ree **VESS**-uhls)—The blood vessels that supply the myocardium with blood; emerge from the ascending aorta and empty into the right atrium (Chapter 12).

Corpus callosum (**KOR**-pus kuh-**LOH**-sum)—The band of white matter that connects the cerebral hemispheres (Chapter 8).

Corpus luteum (**KOR**-pus **LOO**-tee-um)—The temporary endocrine gland formed from an ovarian follicle that has released an ovum; secretes progesterone and estrogen (Chapter 10).

Cortex (**KOR**-teks)—The outer layer of an organ, such as the cerebrum, kidney, or adrenal gland (Chapter 8).

Cortisol (**KOR**-ti-sawl)—A hormone secreted by the adrenal cortex that promotes the efficient use of nutrients in stressful situations and has an anti-inflammatory effect (Chapter 10).

Cough reflex (KAWF)—A reflex integrated by the medulla that expels irritating substances from the pharynx, larynx, or trachea by means of an explosive exhalation (Chapter 15).

Covalent bond (ko-**VAY**-lent)—A chemical bond formed by the sharing of electrons between atoms (Chapter 2).

Cranial cavity (**KRAY**-nee-uhl **KAV**-i-tee)—The cavity formed by the cranial bones; contains the brain; part of the dorsal cavity (Chapter 1).

Cranial nerves (**KRAY**-nee-uhl NERVS)—The 12 pairs of nerves that emerge from the brain (Chapter 8).

Cranial venous sinuses (**KRAY**-nee-uhl **VEE**-nus **SIGH**-nuh-sez)—Large veins between the two layers of the cranial dura mater; the site of reabsorption of the cerebrospinal fluid (Chapter 8).

Cranium (**KRAY**-nee-um)—The cranial bones or bones of the skull that enclose and protect the brain (Chapter 6).

Creatine phosphate (**KREE**-ah-tin **FOSS**-fate)—An energy source in muscle fibers; the energy released is used to synthesize ATP (Chapter 7).

Creatinine (kree-**A**-ti-neen)—A waste product produced when creatine phosphate is used for energy; excreted by the kidneys in urine (Chapter 7).

Cretinism (**KREE**-tin-izm)—Hyposecretion of thyroxine in an infant; if uncorrected, the result is severe mental and physical retardation (Chapter 10).

Cross-section (KRAWS **SEK**-shun)—A plane or cut perpendicular to the long axis of an organ (Chapter 1).

Crypts of Lieberkuhn (KRIPTS of **LEE**-ber-koon)—The digestive glands of the small intestine; secrete digestive enzymes (Chapter 16).

Cuboidal (kew-**BOY**-duhl)—Shaped like a cube; used especially in reference to epithelial tissue (Chapter 4).

Culture and sensitivity testing (**KUL**-chur and SEN-si-**TIV**-i-tee)—A laboratory procedure to determine the best antibiotic with which to treat a bacterial infection (Chapter 22).

Cushing's syndrome (**KOOSH**-ingz **SIN**-drohm)—Hypersecretion of the glucocorticoids of the adrenal cortex, characterized by fragility of skin, poor wound healing, truncal fat deposition, and thin extremities (Chapter 10).

Cutaneous senses (kew-**TAY**-nee-us)—The senses of the skin; the receptors are in the dermis (Chapter 9).

Cyanosis (SIGH-uh-**NOH**-sis)—A blue, gray, or purple discoloration of the skin caused by hypoxia and abnormal amounts of reduced hemoglobin in the blood (Chapter 15).

Cyclic AMP (**SIK**-lik)—A chemical that is the second messenger in the two-messenger theory of hormone action; formed from ATP and stimulates characteristic cellular responses to the hormone (Chapter 10).

Cystic duct (**SIS**-tik DUKT)—The duct that takes bile into and out of the gallbladder; unites with the hepatic duct of the liver to form the common bile duct (Chapter 16).

Cystitis (sis-**TIGH**-tis)—Inflammation of the urinary bladder; most often the result of bacterial infection (Chapter 18).

Cytochrome transport system (**SIGH**-toh-krohm)—The stage of cell respiration in which ATP is formed during reactions of cytochromes with the electrons of the hydrogen atoms that were once part of a food molecule, and metabolic water is formed; aerobic; takes place in the mitochondria of cells (Chapter 17).

Cytokines (**SIGH**-toh-kines)—Chemicals released by activated T cells that attract macrophages (Chapter 14).

Cytokinesis (SIGH-toh-ki-**NEE**-sis)—The division of the cytoplasm of a cell following mitosis (Chapter 3).

Cytoplasm (**SIGH**-toh-plazm)—The cellular material between the nucleus and the cell membrane (Chapter 3).

—D—

Deafness (**DEFF**-ness)—Impairment of normal hearing; may be caused by damage to the vibration conduction pathway (conduction), the acoustic nerve or cochlear receptors (nerve), or the auditory area in the temporal lobe (central) (Chapter 9).

Deamination (DEE-am-i-**NAY**-shun)—The removal of an amino (NH_2) group from an amino acid; takes place in the liver when excess amino acids are used for energy production; the amino groups are converted to urea (Chapter 16).

Decubitus ulcer (dee-**KEW**-bi-tuss **UL**-ser)—The breakdown and death of skin tissue due to prolonged pressure that interrupts blood flow to the area (Chapter 5).

Defecation reflex (DEF-e-**KAY**-shun)—The spinal cord reflex that eliminates feces from the colon (Chapter 16).

Dehydration (DEE-high-**DRAY**-shun)—Excessive loss of water from the body (Chapter 5).

Deltoid (**DELL**-toyd)—1. The shoulder region. 2. The large muscle that covers the shoulder joint (Chapter 1).

Dendrite (**DEN**-dright)—The cellular process of a neuron that carries impulses toward the cell body (Chapter 4).

Dentin (**DEN**-tin)—The bone-like substance that forms the inner crown and the roots of a tooth (Chapter 16).

Depolarization (DE-poh-lahr-i-**ZAY**-shun)—The reversal of electrical charges on either side of a cell membrane in response to a stimulus; negative charge outside and a positive charge inside; brought about by a rapid inflow of sodium ions (Chapter 7).

Dermatology (DER-muh-**TAH**-luh-jee)—The study of the skin and skin diseases (Chapter 5).

Dermis (**DER**-miss)—The inner layer of the skin, made of fibrous connective tissue (Chapter 5).

Detached retina (dee-**TACHD RET**-in-nah)—The separation of the retina from the choroid layer of the eyeball (Chapter 9).

Detrusor muscle (de-**TROO**-ser)—The smooth muscle layer of the wall of the urinary bladder; contracts as part of the urination reflex to eliminate urine (Chapter 18).

Diabetes mellitus (DYE-ah-**BEE**-tis mel-**LYE**-tus)—Hyposecretion of insulin by the pancreas or the inability of insulin to exert its effects; characterized by hyperglycemia, increased urinary output with glycosuria, and thirst (Chapter 10).

Diagnosis (DYE-ag-**NOH**-sis)—The procedures used to identify the cause and nature of a person's illness (Chapter 1).

Diaphragm (**DYE**-uh-fram)—The skeletal muscle that separates the thoracic and abdominal cavities; moves downward when it contracts to enlarge the thoracic cavity to bring about inhalation (Chapter 15).

Diaphysis (dye-**AFF**-i-sis)—The shaft of a long bone; contains a narrow canal filled with yellow bone marrow (Chapter 6).

Diarthrosis (DYE-ar-**THROH**-sis)—A freely movable joint such as hinge, pivot, and ball and socket joints; all are considered synovial joints because synovial membrane is present (Chapter 6).

Diastole (dye-**AS**-tuh-lee)—In the cardiac cycle, the relaxation of the myocardium (Chapter 12).

Differential WBC count (**DIFF**-er-EN-shul KOWNT)—A laboratory test that determines the percentage of each of the five types of white blood cells present in the blood (Chapter 11).

Diffusion (di-**FEW**-zhun)—The process in which there is movement of molecules from an area of greater concentration to an area of lesser concentration; occurs because of the free energy (natural movement) of molecules (Chapter 3).

Digestive system (dye-**JES**-tiv **SIS**-tem)—The organ system that changes food into simpler organic and inorganic molecules that can be absorbed by the blood and lymph and used by cells; consists of the alimentary tube and accessory organs (Chapter 16).

Diploid number (**DIH**-ployd)—The characteristic or usual number of chromosomes found in the somatic (body) cells of a species (human = 46) (Chapter 3).

Disaccharide (dye-**SAK**-ah-ride)—A carbohydrate molecule that consists of two monosaccharides bonded together; includes sucrose, maltose, and lactose (Chapter 2).

Disease (di-**ZEEZ**)—A disorder or disruption of normal body functioning (Chapter 1).

Disinfectant (DIS-in-**FEK**-tent)—A chemical that destroys microorganisms or limits their growth on inanimate objects (Chapter 22).

Dissociation (dih-SEW-see-**AY**-shun)—The separation of an inorganic salt, acid, or base into its ions when dissolved in water (Syn.—ionization) (Chapter 2).

Distal (**DIS**-tuhl)—Furthest from the origin or point of attachment (Chapter 1).

Distal convoluted tubule (**DIS**-tuhl KON-voh-**LOO**-ted)—The part of a renal tubule that extends from a loop of Henle to a collecting tubule (Chapter 18).

Diverticulitis (DYE-ver-tik-yoo-**LYE**-tis)—Inflammation of diverticula in the intestinal tract (Chapter 16).

DNA (Deoxyribonucleic acid)—A nucleic acid in the shape of a double helix. Makes up the chromosomes of cells and is the genetic code for hereditary characteristics (Chapter 2).

DNA replication (REP-li-**KAY**-shun)—The process by which a DNA molecule makes a duplicate of itself. Takes place before mitosis or meiosis, to produce two sets of chromosomes within a cell (Chapter 3).

Dominant (**DAH**-ma-nent)—In genetics, a characteristic that will be expressed even if only one gene for it is present in the homologous pair (Chapter 21).

Dormant (**DOOR**-ment)—Temporarily inactive; a state of little metabolic activity (Chapter 22).

Dorsal (**DOR**-suhl)—Toward the back (Syn.—posterior) (Chapter 1).

Dorsal cavity (DOR-suhl **KAV**-i-tee)—Cavity that consists of the cranial and spinal cavities (Chapter 1).

Dorsal root (**DOR**-suhl ROOT)—The sensory root of a spinal nerve (Chapter 8).

Dorsal root ganglion (**DOR**-suhl ROOT **GANG**-lee-on)—An enlarged area of the dorsal root of a spinal nerve that contains the cell bodies of sensory neurons (Chapter 8).

Down syndrome (DOWN **SIN**-drohm)—A trisomy in which three chromosomes of number 21 are present; characterized by moderate to severe mental retardation and certain physical malformations (Chapter 20).

Duct (DUKT)—A tube or channel, especially one that carries the secretion of a gland (Chapter 4).

Ductus arteriosus (**DUK**-tus ar-TIR-ee-**OH**-sis)—A short fetal blood vessel that takes most blood in the pulmonary artery to the aorta, bypassing the fetal lungs (Chapter 13).

Ductus deferens (**DUK**-tus **DEF**-eer-enz)—The tubular organ that carries sperm from the epididymis to the ejaculatory duct (Syn.—vas deferens) (Chapter 20).

Ductus venosus (**DUK**-tus ve-**NOH**-sus)—A short fetal blood vessel that takes blood from the umbilical vein to the inferior vena cava (Chapter 13).

Duodenum (dew-**AH**-den-um)—The first 10 inches of the small intestine; the common bile duct enters (Chapter 16).

Dura mater (**DEW**-rah **MAH**-ter)—The outermost layer of the meninges, made of fibrous connective tissue (Chapter 8).

Dwarfism (**DWORF**-izm)—The condition of being abnormally small, especially small of stature due to hereditary or endocrine disorder; pituitary dwarfism is caused by a deficiency of growth hormone (Chapter 10).

Dyspnea (**DISP**-nee-ah)—Difficult breathing (Chapter 15).

Dysuria (dis-**YOO**-ree-ah)—Painful or difficult urination (Chapter 18).

—E—

Ear (EER)—The organ that contains the sensory receptors for hearing and equilibrium; consists of the outer ear, middle ear, and inner ear (Chapter 9).

Eccrine gland (**EK**-rin)—The type of sweat gland (exocrine) that produces watery sweat; important in maintenance of normal body temperature (Chapter 5).

Eczema (**EK**-zuh-mah)—An inflammatory condition of the skin that may include the formation of vesicles or pustules (Chapter 5).

Ectoderm (**EK**-toh-derm)—The outer primary germ layer of cells of an embryo; gives rise to epidermis and nervous system (Chapter 21).

Ectoparasite (EK-toh-**PAR**-uh-sight)—A parasite that lives on the surface of the body (Chapter 22).

Ectopic focus (ek-**TOP**-ik **FOH**-kus)—The initiation of a heartbeat by part of the myocardium other than the SA node (Chapter 12).

Ectopic pregnancy (ek-**TOP**-ik **PREG**-nun-see)—Implantation of a fertilized ovum outside the uterus; usually occurs in the fallopian tube but may be in the ovary or abdominal cavity; often results in death of the embryo because a functional placenta cannot be formed in these abnormal sites (Chapter 20).

Edema (uh-**DEE**-muh)—An abnormal accumulation of tissue fluid; may be localized or systemic (Chapter 19).

Effector (e-**FEK**-tur)—An organ such as a muscle or gland that produces a characteristic response after receiving a stimulus (Chapter 8).

Efferent (**EFF**-uh-rent)—To carry away from a center or main part (Chapter 8).

Efferent arteriole (**EFF**-er-ent ar-**TIR**-ee-ohl)—The arteriole that takes blood from a glomerulus to the peritubular capillaries that surround the renal tubule (Chapter 18).

Ejaculation (ee-JAK-yoo-**LAY**-shun)—The ejection of semen from the male urethra (Chapter 20).

Ejaculatory duct (ee-**JAK**-yoo-la-TOR-ee DUKT)—The duct formed by the union of the ductus deferens and the duct of the seminal vesicle; carries sperm to the urethra (Chapter 20).

Ejection fraction (ee-**JEK**-shun **FRAK**-shun)—The percent of blood in a ventricle that is pumped during systole; a measure of the strength of the heart (Chapter 12).

Elastin (eh-**LAS**-tin)—A protein that is found in the form of elastic fibers in several types of connective tissue (Chapter 4).

Electrocardiogram (ECG or EKG) (ee-LEK-troh-**KAR**-dee-oh-GRAM)—A recording of the electrical changes that accompany the cardiac cycle (Chapter 12).

Electrolytes (ee-**LEK**-troh-lites)—Substances that, in solution, dissociate into their component ions; include acids, bases, and salts (Chapter 19).

Electron (e-**LEK**-trahn)—A subatomic particle that has a negative electrical charge; found orbiting the nucleus of an atom (Chapter 2).

Element (**EL**-uh-ment)—A substance that consists of only one type of atom; 92 elements occur in nature (Chapter 2).

Embolism (**EM**-boh-lizm)—Obstruction of a blood vessel by a blood clot or foreign substance that has traveled to and lodged in that vessel (Chapter 11).

Embryo (**EM**-bree-oh)—The developing human individual from the time of implantation until the eighth week of gestation (Chapter 21).

Embryonic disc (EM-bree-**ON**-ik DISK)—The portion of the inner cell mass of the early embryo that will develop into the individual (Chapter 21).

Emphysema (EM-fi-**SEE**-mah)—The deterioration of alveoli and loss of elasticity of the lungs; normal exhalation and gas exchange are impaired (Chapter 15).

Emulsify (e-**MULL**-si-fye)—To physically break up fats into smaller fat globules; the function of bile salts in bile (Chapter 16).

Enamel (en-**AM**-uhl)—The hard substance that covers the crowns of teeth and forms the chewing surface (Chapter 16).

Encapsulated nerve ending (en-**KAP**-sul-LAY-ted NERV **END**-ing)—A sensory nerve ending enclosed in a specialized cellular structure; the cutaneous receptors for touch, pressure, heat, and cold (Chapter 5).

Endemic (en-**DEM**-ik)—A disease that occurs continuously or expectedly in a given population (Chapter 22).

Endocardium (EN-doh-**KAR**-dee-um)—The simple squamous epithelial tissue that lines the chambers of the heart and covers the valves (Chapter 12).

Endocrine gland (**EN**-doh-krin)—A ductless gland that secretes its product (hormone) directly into the blood (Chapter 4).

Endocrine system (**EN**-doh-krin **SIS**-tem)—The organ system that consists of the endocrine glands that secrete hormones into the blood (Chapter 10).

Endoderm (**EN**-doh-derm)—The inner primary germ layer of cells of an embryo; gives rise to respiratory organs and the lining of the digestive tract (Chapter 21).

Endogenous (en-**DOJ**-en-us)—Coming from or produced within the body (Chapter 17).

Endogenous infection (en-**DOJ**-en-us)—An infection caused by a person's own normal flora that have been introduced into an abnormal body site (Chapter 22).

Endolymph (**EN**-doh-limf)—The fluid in the membranous labyrinth of the inner ear (Chapter 9).

Endometrium (EN-doh-**MEE**-tree-um)—The vascular lining of the uterus that forms the maternal portion of the placenta (Chapter 20).

Endoplasmic reticulum (ER) (EN-doh-**PLAZ**-mik re-**TIK**-yoo-lum)—A cell organelle found in the cytoplasm; a network of membranous channels that transport materials within the cell and synthesize lipids (Chapter 3).

Endothelium (EN-doh-**THEE**-lee-um)—The simple squamous epithelial lining of arteries and veins, continuing as the walls of capillaries (Chapter 13).

Endotoxin (EN-doh-**TAHK**-sin)—The toxic portion of the cell walls of gram negative bacteria; causes fever and shock (Chapter 22).

Endotoxin shock (EN-doh-**TAHK**-sin SHAHK)—A state of circulatory shock caused by infection with gram negative bacteria (Chapter 22).

Energy levels (**EN**-er-jee **LEV**-els)—The position of electrons within an atom (Syn.—orbitals or shells) (Chapter 2).

Enzyme (**EN**-zime)—A protein that affects the speed of a chemical reaction. Also called an organic catalyst (Chapter 2).

Eosinophil (EE-oh-**SIN**-oh-fill)—A type of white blood cell (granular); active in allergic reactions (Chapter 11).

Epicardium (EP-ee-**KAR**-dee-um)—The serous membrane on the surface of the myocardium (Syn.—visceral pericardium) (Chapter 12).

Epidemic (EP-i-**DEM**-ik)—A disease that affects many people in a given population in a given time with more than the usual or expected number of cases (Chapter 22).

Epidemiology (EP-i-DEE-mee-**AH**-luh-jee)—The study of the spread of disease and the factors that determine disease frequency and distribution (Chapter 22).

Epidermis (EP-i-**DER**-miss)—The outer layer of the skin, made of stratified squamous epithelium (Chapter 5).

Epididymis (EP-i-**DID**-i-mis)—The tubular organ coiled on the posterior side of a testis; sperm mature here and are carried to the ductus deferens (Pl.—epididymides) (Chapter 20).

Epiglottis (EP-i-**GLAH**-tis)—The uppermost cartilage of the larynx; covers the larynx during swallowing (Chapter 15).

Epinephrine (EP-i-**NEFF**-rin)—A hormone secreted by the adrenal medulla that stimulates many responses to enable the body to react to a stressful situation (Syn.—adrenalin) (Chapter 10).

Epiphyseal disc (e-**PIFF**-i-se-al DISK)—A plate of cartilage at the junction of an epiphysis with the diaphysis of a long bone; the site of growth of a long bone (Chapter 6).

Epiphysis (e-**PIFF**-i-sis)—The end of a long bone (Chapter 6).

Epithelial tissue (EP-i-**THEE**-lee-uhl **TISH**-yoo)—The tissue found on external and internal body surfaces and which forms glands (Chapter 4).

Equilibrium (E-kwe-**LIB**-ree-um)—1. A state of balance. 2. The ability to remain upright and be aware of the position of the body (Chapter 9).

Erythema (ER-i-**THEE**-mah)—Redness of the skin (Chapter 5).

Erythroblastosis fetalis (e-RITH-roh-blass-**TOH**-sis fee-**TAL**-is)—Hemolytic anemia of the newborn, characterized by anemia and jaundice; the result of an Rh incompatibility of fetal blood and maternal blood; also called Rh disease of the newborn (Chapter 11).

Erythrocyte (e-**RITH**-roh-sight)—Red blood cell (Chapter 11).

Erythropoietin (e-RITH-roh-**POY**-e-tin)—A hormone secreted by the kidneys in a state of hypoxia; stimulates the red bone marrow to increase the rate of red blood cell production (Chapter 11).

Esophagus (e-**SOF**-uh-guss)—The organ of the alimentary tube that is a passageway for food from the pharynx to the stomach (Chapter 16).

Essential amino acids (e-**SEN**-shul ah-**ME**-noh **ASS**-ids)—The amino acids that cannot be synthesized by the liver and must be obtained from proteins in the diet (Chapter 16).

Estrogen (**ES**-troh-jen)—The sex hormone secreted by a developing ovarian follicle; contributes to the growth of the female reproductive organs and the secondary sex characteristics (Chapter 10).

Ethmoid bone (**ETH**-moyd)—An irregular cranial bone that forms the upper part of the nasal cavities and a small part of the lower anterior braincase (Chapter 6).

Eustachian tube (yoo-**STAY**-shee-un TOOB)—The air passage between the middle ear cavity and the nasopharynx (Syn.—auditory tube) (Chapter 9).

Exocrine gland (**EK**-so-krin)—A gland that secretes its product into a duct to be taken to a cavity or surface (Chapter 4).

Expiration (EK-spi-**RAY**-shun)—Exhalation; the output of air from the lungs (Chapter 15).

Expiratory reserve (ek-**SPYR**-ah-tor-ee ree-**ZERV**)—The volume of air beyond tidal volume that can be exhaled with the most forceful exhalation; average: 1000–1500 ml (Chapter 15).

Extension (eks-**TEN**-shun)—To increase the angle of a joint (Chapter 7).

External (eks-**TER**-nuhl)—On the outside; toward the surface (Chapter 1).

External anal sphincter (eks-**TER**-nuhl **AY**-nuhl **SFINK**-ter)—The circular skeletal muscle that surrounds the internal anal sphincter and provides voluntary control of defecation (Chapter 16).

External auditory meatus (eks-**TER**-nuhl **AW**-di-TOR-ee me-**AY**-tuss)—The ear canal; the portion of the outer ear that is a tunnel in the temporal bone between the auricle and the eardrum (Chapter 9).

External respiration (eks-**TER**-nuhl RES-pi-**RAY**-shun)—The exchange of gases between the air in the alveoli and the blood in the pulmonary capillaries (Chapter 15).

External urethral sphincter (yoo-**REE**-thruhl **SFINK**-ter)—The skeletal muscle sphincter in the wall of the urethra; provides voluntary control of urination (Chapter 18).

Extracellular fluid (EKS-trah-**SELL**-yoo-ler **FLOO**-id)—The water found outside cells; includes plasma, tissue fluid, lymph, and other fluids (Chapter 2).

Extrinsic factor (eks-**TRIN**-sik **FAK**-ter)—Vitamin B_{12}, obtained from food and necessary for DNA synthesis, especially by stem cells in the red bone marrow (Chapter 11).

Extrinsic muscles (eks-**TRIN**-sik)—The six muscles that move the eyeball (Chapter 9).

—F—

Facial bones (**FAY**-shul)—The 14 irregular bones of the face (Chapter 6).

Facial nerves (**FAY**-shul)—Cranial nerve pair VII; sensory for taste, motor to facial muscles and the salivary glands (Chapter 8).

Facilitated diffusion (fuh-**SILL**-ah-tay-ted di-**FEW**-zhun)—The process in which a substance is transported through a membrane in combination with a carrier molecule (Chapter 3).

Facultative anaerobe (**FAK**-uhl-tay-tive **AN**-air-robe)—A bacterium that is able to reproduce either in the presence or absence of oxygen (Chapter 22).

Fallopian tube (fuh-**LOH**-pee-an TOOB)—The tubular organ that propels an ovum from the ovary to the uterus by means of ciliated epithelium and peristalsis of its smooth muscle layer (Syn.—uterine tube) (Chapter 20).

Fascia (**FASH**-ee-ah)—A fibrous connective tissue membrane that covers individual skeletal muscles and certain organs (Chapter 4).

Fatty acid (**FA**-tee **ASS**-id)—A lipid molecule that consists of an even-numbered carbon chain of 12–24 carbons with hydrogens; may be saturated or unsaturated; an end product of the digestion of fats (Chapter 2).

Femur (**FEE**-mur)—The long bone of the thigh (Chapter 6).

Fertilization (FER-ti-li-**ZAY**-shun)—The union of the nuclei of an ovum and a sperm cell; restores the diploid number (Chapter 3).

Fetal alcohol syndrome (**FEE**-tuhl **AL**-koh-hol)—Birth defects or developmental abnormalities in infants born to women who chronically consumed alcohol during the gestation period (Chapter 21).

Fever (**FEE**-ver)—An abnormally high body temperature, caused by pyrogens; may accompany an infectious disease or severe physical injury (Chapter 17).

Fever blister (**FEE**-ver **BLISS**-ter)—An eruption of the skin caused by the Herpes simplex virus (Syn.—cold sore) (Chapter 5).

Fibrillation (fi-bri-**LAY**-shun)—Very rapid and uncoordinated heart beats; ventricular fibrillation is a life-threatening emergency due to ineffective pumping and decreased cardiac output (Chapter 12).

Fibrin (**FYE**-brin)—A thread-like protein formed by the action of thrombin on fibrinogen; the substance of which a blood clot is made (Chapter 11).

Fibrinogen (fye-**BRIN**-o-jen)—A protein clotting factor produced by the liver; converted to fibrin by thrombin (Chapter 11).

Fibrinolysis (FYE-brin-**AHL**-e-sis)—1. The dissolving of a fibrin clot by natural enzymes, after the clot has served its purpose. 2. The clinical use of clot-dissolving enzymes to dissolve abnormal clots (Chapter 11).

Fibroblast (**FYE**-broh-blast)—A connective tissue cell that produces collagen and elastin fibers (Chapter 4).

Fibrous connective tissue (**FYE**-brus)—The tissue that consists primarily of collagen fibers. Its most important physical characteristic is its strength (Chapter 4).

Fibula (**FIB**-yoo-lah)—A long bone of the lower leg; on the lateral side, thinner than the tibia (Chapter 6).

Filtration (fill-**TRAY**-shun)—The process in which water and dissolved materials move through a membrane from an area of higher pressure to an area of lower pressure (Chapter 3).

Fimbriae (**FIM**-bree-ay)—Finger-like projections at the end of the fallopian tube that enclose the ovary (Chapter 20).

Fissure (**FISH**-er)—A groove or furrow between parts of an organ such as the brain (Syn.—sulcus) (Chapter 8).

Flagellum (flah-**JELL**-um)—A long, thread-like projection through a cell membrane; provides motility for the cell (Chapter 3).

Flexion (**FLEK**-shun)—To decrease the angle of a joint (Chapter 7).

Flexor reflex (**FLEKS**-er **REE**-fleks)—A spinal cord reflex in which a painful stimulus causes withdrawal of a body part (Chapter 8).

Fluorescent antibody test (floor-**ESS**-ent **AN**-ti-BAH-dee)—A diagnostic test that uses fluorescently tagged antibodies to determine the presence of a particular pathogen in the blood or other tissue specimen (Chapter 14).

Flutter (**FLUH**-ter)—A very rapid yet fairly regular heartbeat (Chapter 12).

Follicle-stimulating hormone (FSH) (**FAH**-li-kuhl)—A gonadotropic hormone produced by the anterior pituitary gland, that initiates the production of ova in the ovaries or sperm in the testes (Chapter 10).

Fomites (**FOH**-mights; **FOH**-mi-teez)—Inanimate objects capable of transmitting infectious microorganisms from one host to another (Chapter 22).

Fontanel (FON-tah-**NELL**)—An area of fibrous connective tissue membrane between the cranial bones of an infant's skull, where bone formation is not complete (Chapter 6).

Foramen (for-**RAY**-men)—A hole or opening, as in a bone (from the Latin "ditch") (Chapter 6).

Foramen ovale (for-**RAY**-men oh-**VAHL**-ee)—An opening in the interatrial septum of the fetal heart that permits blood to flow from the right atrium to the left atrium, bypassing the fetal lungs (Chapter 13).

Fossa (**FAH**-sah)—A shallow depression in a bone (from the Latin "ditch") (Chapter 6).

Fovea (**FOH**-vee-ah)—A depression in the retina of the eye directly behind the lens; contains only cones and is the area of best color vision (Chapter 9).

Fracture (**FRAK**-chur)—A break in a bone (Chapter 6).

Free nerve ending (FREE NERV **END**-ing)—The end of a sensory neuron; the receptor for the sense of pain in the skin and viscera (Chapter 5).

Frontal bone (**FRUN**-tuhl)—The flat cranial bone that forms the forehead (Chapter 6).

Frontal lobes (**FRUN**-tuhl LOWBS)—The most anterior parts of the cerebrum; contain the motor areas for voluntary movement and the motor speech area (Chapter 8).

Frontal section (**FRUN**-tuhl **SEK**-shun)—A plane separating the body into front and back portions (Syn.—coronal section) (Chapter 1).

Frostbite (**FRAWST**-bite)—The freezing of part of the body, resulting in tissue damage or death (gangrene) (Chapter 17).

Fructose (**FRUHK**-tohs)—A monosaccharide, a six-carbon sugar that is part of the sucrose in food; converted to glucose by the liver (Chapter 2).

Functional layer (**FUNK**-shun-ul **LAY**-er)—The vascular layer of the endometrium that is lost in menstruation, then regenerated by the basilar layer (Chapter 20).

Fungus (**FUNG**-gus)—Any of the organisms of the kingdom Fungi; they lack chlorophyll; may be unicellular or multicellular; saprophytic or parasitic; include yeasts, molds, and mushrooms (Chapter 22).

—G—

Galactose (guh-**LAK**-tohs)—A monosaccharide, a six-carbon sugar that is part of the lactose in food; converted to glucose by the liver (Chapter 2).

Gallbladder (**GAWL**-bla-der)—An accessory organ of digestion; a sac located on the undersurface of the liver; stores and concentrates bile (Chapter 16).

Gallstones (**GAWL**-stohns)—Crystals formed in the gallbladder or bile ducts; the most common type is made of cholesterol (Chapter 16).

Gametes (**GAM**-eets)—The male or female reproductive cells, sperm cells or ova, each with the haploid number of chromosomes (Chapter 3).

Gamma globulins (**GA**-mah **GLAH**-byoo-lins)—Antibodies (Chapter 14).

Ganglion (**GANG**-lee-on)—A group of neuron cell bodies located outside the CNS (Chapter 8).

Ganglion neurons (**GANG**-lee-on **NYOOR**-onz)—The neurons that form the optic nerve; carry impulses from the retina to the brain (Chapter 9).

Gastric juice (**GAS**-trik JOOSS)—The secretion of the gastric pits of the stomach; contains hydrochloric acid, pepsin, and mucus (Chapter 16).

Gastric pits (**GAS**-trik PITS)—The glands of the mucosa of the stomach; secrete gastric juice (Chapter 16).

Gastric ulcer (**GAS**-trik **UL**-ser)—An erosion of the gastric mucosa and submucosa (Chapter 16).

Gastrin (**GAS**-trin)—A hormone secreted by the gastric mucosa when food enters the stomach; stimulates the secretion of gastric juice (Chapter 16).

Gene (JEEN)—A segment of DNA that is the genetic code for a particular protein and is located in a definite position on a particular chromosome (Chapter 3).

Genetic code (je-**NET**-ik KOHD)—The DNA code for proteins that is shared by all living things; the sequence of bases of the DNA in the chromosomes of cells (Chapter 2).

Genetic disease (je-**NET**-ik di-**ZEEZ**)—A hereditary disorder that is the result of an incorrect sequence of bases in the DNA (gene) of a particular chromosome. May be passed to offspring (Chapter 3).

Genetic immunity (je-**NET**-ik im-**YOO**-ni-tee)—The immunity provided by the genetic makeup of a species; reflects the inability of certain pathogens to cause disease in certain host species (Chapter 14).

Genotype (**JEE**-noh-type)—The genetic makeup of an individual; the genes that are present (Chapter 21).

Gestation (jes-**TAY**-shun)—The length of time from conception to birth; the human gestation period averages 280 days (Chapter 21).

Giantism (**JIGH**-an-tizm)—Excessive growth of the body or its parts; may be the result of hypersecretion of growth hormone in childhood (Chapter 10).

Gingiva (jin-**JIGH**-vah)—The gums; the tissue that covers the upper and lower jaws around the necks of the teeth (Chapter 16).

Gland (GLAND)—A cell or group of epithelial cells that are specialized to secrete a substance (Chapter 4).

Glaucoma (glaw-**KOH**-mah)—An eye disease characterized by increased intraocular pressure due to excessive accumulation of aqueous humor (Chapter 9).

Gliding joint (**GLY**-ding)—A diarthrosis that permits a sliding movement (Chapter 6).

Globulins (**GLAH**-byoo-lins)—Proteins that circulate in blood plasma; alpha and beta globulins are synthesized by the liver; gamma globulins (antibodies) are synthesized by lymphocytes (Chapter 11).

Glomerular filtration (gloh-**MER**-yoo-ler fill-**TRAY**-shun)—The first step in the formation of urine; blood pressure in the glomerulus forces plasma, dissolved materials, and small proteins into Bowman's capsule; this fluid is then called renal filtrate (Chapter 18).

Glomerular filtration rate (gloh-**MER**-yoo-ler fill-**TRAY**-shun RAYT)—The total volume of renal filtrate that the kidneys form in 1 minute; average is 100-125 ml/minute (Chapter 18).

Glomerulus (gloh-**MER**-yoo-lus)—A capillary network that is enclosed by Bowman's capsule; filtration takes place from the glomerulus to Bowman's capsule (from the Latin "little ball") (Chapter 18).

Glossopharyngeal nerves (GLAH-so-fuh-**RIN**-jee-uhl)—Cranial nerve pair IX. Sensory for taste and cardiovascular reflexes. Motor to salivary glands (Chapter 8).

Glottis (**GLAH**-tis)—The opening between the vocal cords; an air passageway (Chapter 15).

Glucagon (**GLOO**-kuh-gahn)—A hormone secreted by the pancreas that increases the blood glucose level (Chapter 10).

Glucocorticoids (GLOO-koh-**KOR**-ti-koids)—The hormones secreted by the adrenal cortex that affect the metabolism of nutrients; cortisol is the major hormone in this group (Chapter 10).

Gluconeogenesis (GLOO-koh-nee-oh-**JEN**-i-sis)—The conversion of excess amino acids to simple carbohydrates or to glucose to be used for energy production (Chapter 10).

Glucose (**GLOO**-kos)—A hexose monosaccharide that is the primary energy source for body cells (Chapter 2).

Glycerol (**GLISS**-er-ol)—A three-carbon molecule that is one of the end products of the digestion of fats (Chapter 2).

Glycogen (**GLY**-ko-jen)—A polysaccharide that is the storage form for excess glucose in the liver and muscles (Chapter 2).

Glycogenesis (GLY-koh-**JEN**-i-sis)—The conversion of

glucose to glycogen to be stored as potential energy (Chapter 10).

Glycogenolysis (GLY-koh-jen-**AHL**-i-sis)—The conversion of stored glycogen to glucose to be used for energy production (Chapter 10).

Glycolysis (gly-**KAHL**-ah-sis)—The first stage of the cell respiration of glucose, in which glucose is broken down to two molecules of pyruvic acid and ATP is formed; anaerobic; takes place in the cytoplasm of cells (Chapter 17).

Glycosuria (GLY-kos-**YOO**-ree-ah)—The presence of glucose in urine; often an indication of diabetes mellitus (Chapter 18).

Goblet cell (**GAHB**-let)—Unicellular glands that secrete mucus; found in the respiratory and GI mucosa (Chapter 4).

Goiter (**GOY**-ter)—An enlargement of the thyroid gland, often due to a lack of dietary iodine (Chapter 10).

Golgi apparatus (**GOHL**-jee)—A cell organelle found in the cytoplasm; synthesizes carbohydrates and packages materials for secretion from the cell (Chapter 3).

Gonadotropic hormone (GAH-nah-doh-**TROH**-pik)—A hormone that has its effects on the ovaries or testes (gonads); FSH and LH (Chapter 10).

Gonorrhea (GAH-nuh-**REE**-ah)—A sexually transmitted disease caused by the bacterium *Neisseria gonorrhoeae;* may also cause conjunctivitis in newborns of infected women (Chapter 20).

Graafian follicle (**GRAFF**-ee-uhn **FAH**-li-kuhl)—A mature ovarian follicle that releases an ovum (Chapter 20).

Gram negative (**GRAM NEG**-uh-tiv)—Bacteria that appear red or pink after gram staining (Chapter 22).

Gram positive (**GRAM PAHS**-uh-tiv)—Bacteria that appear purple or blue after gram staining (Chapter 22).

Gram stain (**GRAM STAYN**)—A staining procedure for bacteria to make them visible microscopically and to determine their gram reaction, which is important in the identification of bacteria (Chapter 22).

Graves' disease (GRAYVES)—Hypersecretion of thyroxine, believed to be an autoimmune disease; symptoms reflect the elevated metabolic rate (Chapter 10).

Gray matter (GRAY)—Nerve tissue within the central nervous system that consists of the cell bodies of neurons (Chapter 8).

Growth hormone (GH) (GROHTH **HOR**-mohn)—A hormone secreted by the anterior pituitary gland that increases the rate of cell division and protein synthesis (Chapter 10).

Gyrus (**JIGH**-rus)—A fold or ridge, as in the cerebral cortex (Syn.—convolution) (Chapter 8).

—H—

Hair (HAIR)—An accessory skin structure produced in a hair follicle (Chapter 5).

Hair cells (HAIR SELLS)—Cells with specialized cilia found in the inner ear; the receptors for hearing (cochlea), static equilibrium (utricle and saccule), and motion equilibrium (semicircular canals) (Chapter 9).

Hair follicle (HAIR **FAH**-li-kuhl)—The structure within the skin in which a hair grows (Chapter 5).

Hair root (HAIR ROOT)—The site of mitosis at the base of a hair follicle; new cells become the hair shaft (Chapter 5).

Haploid number (**HA**-ployd)—Half the usual number of chromosomes found in the cells of a species. Characteristic of the gametes of the species (human = 23) (Chapter 3).

Hard palate (HARD **PAL**-uht)—The anterior portion of the palate formed by the maxillae and the palatine bones (Chapter 6).

Haustra (**HOWS**-trah)—The pouches of the colon (Chapter 16).

Haversian system (ha-**VER**-zhun)—The structural unit of compact bone, consisting of a central haversian canal surrounded by concentric rings of osteocytes within matrix (Chapter 4).

Heart murmur (HART **MUR**-mur)—An abnormal heart sound heard during the cardiac cycle; often caused by a malfunctioning heart valve (Chapter 12).

Heat exhaustion (HEET eks-**ZAWS**-chun)—A state of weakness and dehydration caused by excessive loss of body water and sodium chloride in sweat; the result of exposure to heat or of strenuous exercise (Chapter 17).

Heat stroke (HEET STROHK)—An acute reaction to heat exposure in which there is failure of the temperature-regulating mechanisms; sweating ceases, and body temperature rises sharply (Chapter 17).

Heimlich maneuver (**HIGHM**-lik ma-**NEW**-ver)—A procedure used to remove foreign material lodged in the pharynx, larynx, or trachea (Chapter 15).

Helix (**HEE**-liks)—A coil or spiral. Double helix is the descriptive term used for the shape of a DNA molecule: two strands of nucleotides coiled around each other and resembling a twisted ladder (Chapter 2).

Hematocrit (hee-**MAT**-oh-krit)—A laboratory test that determines the percentage of red blood cells in a given volume of blood; part of a complete blood count (Chapter 11).

Hematuria (HEM-uh-**TYOO**-ree-ah)—The presence of blood (RBCs) in urine (Chapter 18).

Hemodialysis (HEE-moh-dye-**AL**-i-sis)—A technique for providing the function of the kidneys by passing the blood through tubes surrounded by solutions that selectively remove waste products and excess minerals; may be life-saving in cases of renal failure (Chapter 18).

Hemoglobin (**HEE**-muh-GLOW-bin)—The protein in red blood cells that contains iron and transports oxygen in the blood (Chapter 7).

Hemolysis (he-**MAHL**-e-sis)—Lysis or rupture of red blood cells; may be the result of an antigen-antibody reaction or of increased fragility of red blood cells in some types of anemia (Chapter 11).

Hemophilia (HEE-moh-**FILL**-ee-ah)—A hereditary blood disorder characterized by the inability of the blood to clot normally; hemophilia A is caused by a lack of clotting factor 8 (Chapter 11).

Hemopoietic tissue (HEE-moh poy-**ET**-ik)—A blood-forming tissue; the red bone marrow and lymphatic tissue (Chapter 4).

Hemorrhoids (**HEM**-uh-royds)—Varicose veins of the anal canal (Chapter 13).

Hemostasis (HEE-moh-**STAY**-sis)—Prevention of blood loss; the mechanisms are chemical clotting, vascular spasm, and platelet plug formation (Chapter 11).

Heparin (**HEP**-ar-in)—A chemical that inhibits blood clotting, an anticoagulant; produced by basophils. Also used clinically to prevent abnormal clotting, such as following some types of surgery (Chapter 11).

Hepatic duct (hep-**PAT**-ik DUKT)—The duct that takes bile out of the liver; joins the cystic duct of the gallbladder to form the common bile duct (Chapter 16).

Hepatic portal circulation (hep-**PAT**-ik **POOR**-tuhl)—The pathway of systemic circulation in which venous blood from the digestive organs and the spleen circulates through the liver before returning to the heart (Chapter 13).

Hepatitis (HEP-uh-**TIGH**-tis)—Inflammation of the liver, most often caused by the hepatitis viruses A, B, or C (Chapter 16).

Herniated disc (**HER**-nee-ay-ted DISK)—Rupture of an intervertebral disc (Chapter 6).

Herpes simplex (**HER**-peez **SIM**-pleks)—A virus that causes lesions in the skin or mucous membranes of the mouth (usually type 1) or genital area (usually type 2); genital herpes may cause death or mental retardation of infants of infected women (Chapter 20).

Heterozygous (HET-er-oh-**ZYE**-gus)—Having two different alleles for a trait (Chapter 21).

Hexose sugar (**HEKS**-ohs)—A six-carbon sugar, such as glucose, that is an energy source (in the process of cell respiration) (Chapter 2).

Hilus (**HIGH**-lus)—An indentation or depression on the surface of an organ such as a lung or kidney (Chapter 15).

Hinge joint (HINJ)—A diarthrosis that permits movement in one plane (Chapter 6).

Hip bone (HIP BOWNE)—The flat bone that forms half of the pelvic bone; consists of the upper ilium, the lower posterior ischium, and the lower anterior pubis (Chapter 6).

Histamine (**HISS**-tah-meen)—An inflammatory chemical released by damaged tissues; stimulates increased capillary permeability and vasodilation (Chapter 5).

Hives (**HIGHVZ**)—A very itchy eruption of the skin, usually the result of an allergy (Chapter 5).

Hodgkin's disease (**HODJ**-kinz)—A malignancy of the lymphatic tissue; a lymphoma (Chapter 14).

Homeostasis (HOH-mee-oh-**STAY**-sis)—The state in which the internal environment of the body remains relatively stable by responding appropriately to changes (Chapter 1).

Homologous pair (hoh-**MAHL**-ah-gus)—A pair of chromosomes, one maternal and one paternal, that contain genes for the same characteristics (Chapter 21).

Homozygous (HOH-moh-**ZYE**-gus)—Having two similar alleles for a trait (Chapter 21).

Hormone (**HOR**-mohn)—The secretion of an endocrine gland that has specific effects on particular target organs (Chapter 4).

Human leukocyte antigens (HLA) (**HYOO**-man **LOO**-koh-sight **AN**-ti-jens)—The antigens on white blood cells that are representative of the antigens present on all the cells of the individual; the "self" antigens that are controlled by several genes on chromosome number 6; the basis for tissue-typing before an organ transplant is attempted (Chapter 11).

Humerus (**HYOO**-mer-us)—The long bone of the upper arm (Chapter 6).

Humoral immunity (**HYOO**-mohr-uhl im-**YOO**-ni-tee)—The mechanism of immunity that involves antibody production and the destruction of foreign antigens by the activities of B cells, T cells, and macrophages (Chapter 14).

Hyaline membrane disease (**HIGH**-e-lin **MEM**-brayn)—A pulmonary disorder of premature infants whose lungs have not yet produced sufficient pulmonary surfactant to permit inflation of the alveoli (Chapter 15).

Hydrochloric acid (HIGH-droh-**KLOR**-ik **ASS**-id)—An acid secreted by the parietal cells of the gastric pits of the stomach; activates pepsin and maintains a pH of 1-2 in the stomach (Chapter 16).

Hymen (**HIGH**-men)—A thin fold of mucous membrane that partially covers the vaginal orifice (Chapter 20).

Hypercalcemia (HIGH-per-kal-**SEE**-mee-ah)— A high blood calcium level (Chapter 10).

Hyperglycemia (HIGH-per-gligh-**SEE**-mee-ah)— A high blood glucose level (Chapter 10).

Hyperkalemia (HIGH-per-kuh-**LEE**-mee-ah)— A high blood potassium level (Chapter 19).

Hypernatremia (HIGH-per-nuh-**TREE**-mee-ah)— A high blood sodium level (Chapter 19).

Hyperopia (HIGH-per-**OH**-pee-ah)—Farsightedness; an error of refraction in which only distant objects are seen clearly (Chapter 9).

Hypertension (HIGH-per-**TEN**-shun)—An abnormally high blood pressure, consistently above 140/90 mmHg (Chapter 13).

Hypertonic (HIGH-per-**TAHN**-ik)—Having a greater concentration of dissolved materials than the solution used as a comparison (Chapter 3).

Hypertrophy (high-**PER**-troh-fee)—Increase in size of a body part, especially of a muscle following long-term exercise or overuse (Chapter 7).

Hypocalcemia (HIGH-poh-kal-**SEE**-mee-ah)—A low blood calcium level (Chapter 10).

Hypoglossal nerves (HIGH-poh-**GLAH**-suhl)—Cranial nerve pair XII. Motor to the tongue (Chapter 8).

Hypoglycemia (HIGH-poh-gligh-**SEE**-mee-ah)—A low blood glucose level (Chapter 10).

Hypokalemia (HIGH-poh-kuh-**LEE**-mee-ah)—A low blood potassium level (Chapter 19).

Hyponatremia (HIGH-poh-nuh-**TREE**-mee-ah)—A low blood sodium level (Chapter 19).

Hypophyseal portal system (high-POFF-e-**SEE**-al **POR**-tuhl)—The pathway of circulation in which releasing factors from the hypothalamus circulate directly to the anterior pituitary gland (Chapter 10).

Hypophysis (high-**POFF**-e-sis)—The pituitary gland (Chapter 10).

Hypotension (HIGH-poh-**TEN**-shun)—An abnormally low blood pressure, consistently below 90/60 mmHg (Chapter 13).

Hypothalamus (HIGH-poh-**THAL**-uh-muss)—The part of the brain superior to the pituitary gland and inferior to the thalamus; its many functions include regulation of body temperature and regulation of the secretions of the pituitary gland (Chapter 8).

Hypothermia (HIGH-poh-**THER**-mee-ah)—1. The condition in which the body temperature is abnormally low due to excessive exposure to cold. 2. A procedure used during some types of surgery to lower body temperature to reduce the patient's need for oxygen (Chapter 17).

Hypotonic (HIGH-po-**TAHN**-ik)—Having a lower concentration of dissolved materials than the solution used as a comparison (Chapter 3).

Hypovolemic shock (HIGH-poh-voh-**LEEM**-ik SHAHK)—A type of circulatory shock caused by a decrease in blood volume (Chapter 13).

Hypoxia (high-**PAHK**-see-ah)—A deficiency or lack of oxygen (Chapter 2).

—I—

Idiopathic (ID-ee-oh-**PATH**-ik)—A disease or disorder of unknown cause (Chapter 11).

Ileocecal valve (ILL-ee-oh-**SEE**-kuhl VALV)—The tissue of the ileum that extends into the cecum and acts as a sphincter; prevents the backup of fecal material into the small intestine (Chapter 16).

Ileum (**ILL**-ee-um)—The third and last portion of the small intestine, about 11 feet long (Chapter 16).

Ilium (**ILL**-ee-um)—The upper, flared portion of the hip bone (Chapter 6).

Immunity (im-**YOO**-ni-tee)—The state of being protected from an infectious disease, usually by having been exposed to the infectious agent or a vaccine (Chapter 11).

Impetigo (IM-pe-**TYE**-go)—A bacterial infection of the skin that occurs most often in children (Chapter 5).

Implantation (IM-plan-**TAY**-shun)—Embedding of the embryonic blastocyst in the endometrium of the uterus 6 to 8 days after fertilization (Chapter 21).

Inactivator (in-**AK**-ti-vay-tur)—A chemical that inactivates a neurotransmitter to prevent continuous impulses (Chapter 8).

Inapparent infection (**IN**-uh-PAR-ent)—An infection without symptoms (Chapter 22).

Incubation period (IN-kew-**BAY**-shun)—In the course of an infectious disease, the time between the entry of the pathogen and the onset of symptoms (Chapter 22).

Incus (**ING**-kuss)—The second of the three auditory bones in the middle ear; transmits vibrations from the malleus to the stapes (Chapter 9).

Infarct (**IN**-farkt)—An area of tissue that has died due to lack of a blood supply (Chapter 12).

Infection (in-**FEK**-shun)—A disease process caused by the invasion and multiplication of a microorganism (Chapter 22).

Inferior (in-**FEER**-ee-your)—Below or lower (Chapter 1).

Inferior vena cava (**VEE**-nah **KAY**-vah)—The vein that returns blood from the lower body to the right atrium (Chapter 12).

Infestation (in-fess-**TAY**-shun)—The harboring of parasites, especially worms or arthropods (Chapter 22).

Inflammation (in-fluh-**MAY**-shun)—The reactions of tissue to injury (Chapter 5).

Inguinal canal (**IN**-gwi-nuhl ka-**NAL**)—The opening in the lower abdominal wall that contains a spermatic cord in men and the round ligament of the uterus in women; a natural weak spot that may be the site of hernia formation (Chapter 20).

Inhibin (in-**HIB**-in)—A protein hormone secreted by the sustentacular cells of the testes and by the ovaries; inhibits secretion of FSH (Chapter 10).

Inorganic (**IN**-or-GAN-ik)—A chemical compound that does not contain carbon-hydrogen covalent bonds; includes water, salts, and oxygen (Chapter 1).

Insertion (in-**SIR**-shun)—The more movable attachment point of a muscle to a bone (Chapter 7).

Inspiration (in-spi-**RAY**-shun)—Inhalation; the intake of air to the lungs (Chapter 15).

Inspiratory reserve (in-**SPYR**-ah-tor-ee ree-**ZERV**)—The volume of air beyond tidal volume that can be inhaled with the deepest inhalation; average: 2000–3000 ml (Chapter 15).

Insulin (**IN**-syoo-lin)—A hormone secreted by the pancreas that decreases the blood glucose level by increasing storage of glycogen and use of glucose by cells for energy production (Chapter 10).

Integumentary system (in-TEG-yoo-**MEN**-tah-ree)—The organ system that consists of the skin and its accessory structures and the subcutaneous tissue (Chapter 5).

Intensity (in-**TEN**-si-tee)—The degree to which a sensation is felt (Chapter 9).

Intercostal muscles (IN-ter-**KAHS**-tuhl **MUSS**-uhls)—The skeletal muscles between the ribs; the external intercostals pull the ribs up and out for inhalation; the internal intercostals pull the ribs down and in for a forced exhalation (Syn.—spareribs) (Chapter 15).

Intercostal nerves (IN-ter-**KAHS**-tuhl NERVS)—The pairs of peripheral nerves that are motor to the intercostal muscles (Chapter 15).

Interferon (in-ter-**FEER**-on)—A chemical produced by T cells or by cells infected with viruses; prevents the reproduction of viruses (Chapter 14).

Internal (in-**TER**-nuhl)—On the inside, or away from the surface (Chapter 1).

Internal anal sphincter (in-**TER**-nuhl **AY**-nuhl **SFINK**-ter)—The circular smooth muscle that surrounds the anus; relaxes as part of the defecation reflex to permit defecation (Chapter 16).

Internal respiration (in-**TER**-nuhl RES-pi-**RAY**-shun)—The exchange of gases between the blood in the systemic capillaries and the surrounding tissue fluid and cells (Chapter 15).

Internal urethral sphincter (yoo-**REE**-thruhl **SFINK**-ter)—The smooth muscle sphincter at the junction of the urinary bladder and the urethra; relaxes as part of the urination reflex to permit urination (Chapter 18).

Interneuron (**IN**-ter-NYOOR-on)—A nerve cell entirely within the central nervous system (Chapter 8).

Interphase (**IN**-ter-fayz)—The period of time between mitotic divisions during which DNA replication takes place (Chapter 3).

Interstitial cells (in-ter-**STISH**-uhl SELLS)—The cells in the testes that secrete testosterone when stimulated by LH (Chapter 20).

Intestinal glands (in-**TESS**-tin-uhl)—1. The glands of the small intestine that secrete digestive enzymes. 2. The glands of the large intestine that secrete mucus (Chapter 16).

Intracellular fluid (IN-trah-**SELL**-yoo-ler **FLOO**-id)—The water found within cells (Chapter 2).

Intramuscular injection (in-trah-**MUSS**-kew-ler in-**JEK**-shun)—An injection of a medication into a muscle (Chapter 7).

Intrapleural pressure (in-trah-**PLOOR**-uhl)—The pressure within the potential pleural space; always slightly below atmospheric pressure, about 756 mmHg (Chapter 15).

Intrapulmonic pressure (in-trah-pull-**MAHN**-ik)—The air pressure within the bronchial tree and alveoli; fluctuates below and above atmospheric pressure during breathing (Chapter 15).

Intrinsic factor (in-**TRIN**-sik **FAK**-ter)—A chemical produced by the parietal cells of the gastric mucosa; necessary for the absorption of vitamin B_{12} (Chapter 11).

In vitro fertilization (IN **VEE**-troh FER-ti-li-**ZAY**-shun)—Fertilization outside the body, in which sperm and ova are mixed in laboratory glassware; early embryos may then be introduced into the uterus for implantation (Chapter 20).

Involuntary muscle (in-**VAHL**-un-tary **MUSS**-uhl)—Another name for smooth muscle tissue (Chapter 4).

Ion (**EYE**-on)—An atom or group of atoms with an electrical charge (Chapter 2).

Ionic bond (eye-**ON**-ik)—A chemical bond formed by the loss and gain of electrons between atoms (Chapter 2).

Iris (**EYE**-ris)—The colored part of the eye, between the cornea and lens; made of two sets of smooth muscle fibers that regulate the size of the pupil, the opening in the center of the iris (from the Latin "rainbow") (Chapter 9).

Ischemic (iss-**KEY**-mik)—Lack of blood to a body part, often due to an obstruction in circulation (Chapter 12).

Ischium (**ISH**-ee-um)—The lower posterior part of the hip bone (Chapter 6).

Islets of Langerhans (**EYE**-lets of **LAHNG**-er-hanz)—The endocrine portions of the pancreas that secrete insulin and glucagon (Chapter 10).

Isometric exercise (EYE-so-**MEH**-trik)—Contraction of muscles without movement of a body part (Chapter 7).

Isotonic (EYE-so-**TAHN**-ik)—Having the same concentration of dissolved materials as the solution used as a comparison (Chapter 3).

Isotonic exercise (EYE-so-**TAHN**-ik)—Contraction of muscles with movement of a body part (Chapter 7).

—J—

Jaundice (**JAWN**-diss)—A condition characterized by a yellow color in the whites of the eyes and the skin; caused by an elevated blood level of bilirubin. May be hepatic, pre-hepatic, or post-hepatic in origin (Chapter 11).

Jejunum (je-**JOO**-num)—The second portion of the small intestine, about 8 feet long (Chapter 16).

Joint capsule (JOYNT **KAP**-suhl)—The fibrous connective tissue sheath that encloses a joint (Chapter 6).

Juxtaglomerular cells (JUKS-tah-gloh-**MER**-yoo-ler SELLS)—Cells in the wall of the afferent arteriole that secrete renin when blood pressure decreases (Chapter 18)

—K—

Keratin (**KER**-uh-tin)—A protein produced by epidermal cells; found in the epidermis, hair, and nails (from the Greek "horn") (Chapter 5).

Keratinocyte (KER-un-**TIN**-oh-sight)—A cell of the epidermis, produces keratin before dying (Chapter 5).

Ketoacidosis (KEY-toh-ass-i-**DOH**-sis)—A metabolic acidosis that results from the accumulation of ketones in the blood when fats and proteins are used for energy production (Chapter 10).

Ketones (**KEY**-tohns)—Organic acid molecules that are formed from fats or amino acids when these nutrients are used for energy production; include acetone and acetoacetic acid (Chapter 10).

Ketonuria (KEY-ton-**YOO**-ree-ah)—The presence of ketones in urine (Chapter 18).

Kidneys (**KID**-nees)—The two organs on either side of the vertebral column in the upper abdomen that produce urine to eliminate waste products and to regulate the volume, pH, and fluid-electrolyte balance of the blood (Chapter 6).

Kilocalorie (KILL-oh-**KAL**-oh-ree)—One thousand calories; used to indicate the energy content of foods or the energy expended in activity (Chapter 17).

Kinesthetic sense (KIN-ess-**THET**-ik)—Muscle sense (Chapter 9).

Krebs cycle (KREBS **SIGH**-kuhl)—The stage of cell respiration comprised of a series of reactions in which pyruvic acid or acetyl CoA is broken down to carbon dioxide and ATP is formed; aerobic; takes place in the mitochondria of cells (Syn.—citric acid cycle) (Chapter 17).

Kupffer cells (**KUP**-fer SELLS)—The macrophages of the liver; phagocytize pathogens and old red blood cells (Chapter 16).

Kyphosis (kye-**FOH**-sis)—An exaggerated thoracic curvature of the vertebral column (Chapter 6).

—L—

Labia majora (**LAY**-bee-uh muh-**JOR**-ah)—The outer folds of skin of the vulva; enclose the labia minora and the vestibule (Chapter 20).

Labia minora (**LAY**-bee-uh min-**OR**-ah)—The inner folds of the vulva; enclose the vestibule (Chapter 20).

Labor (**LAY**-ber)—The process by which a fetus is expelled from the uterus through the vagina to the exterior of the body (Chapter 21).

Labyrinth (**LAB**-i-rinth)—1. A maze; an interconnected series of passageways. 2. In the inner ear: the bony labyrinth is a series of tunnels in the temporal bone lined with membrane called the membranous labyrinth (Chapter 9).

Lacrimal glands (**LAK**-ri-muhl)—The glands that secrete tears, located at the upper, outer corner of each eyeball (Chapter 9).

Lactase (**LAK**-tays)—A digestive enzyme that breaks down lactose to glucose and galactose; secreted by the small intestine (Chapter 16).

Lacteals (lak-**TEELS**)—The lymph capillaries in the villi of the small intestine, which absorb the fat-soluble end products of digestion (Chapter 14).

Lactic acid (**LAK**-tik **ASS**-id)—The chemical end product of anaerobic cell respiration; contributes to fatigue in muscle cells (Chapter 7).

Lactose (**LAK**-tohs)—A disaccharide made of one glucose and one galactose molecule (Syn.—milk sugar) (Chapter 2).

Lactose intolerance (**LAK**-tohs in-**TAHL**-er-ense)—The inability to digest lactose due to a deficiency of the enzyme lactase; may be congenital or acquired (Chapter 16).

Langerhans cell (**LAHNG**-er-hanz SELL)—A phagocytic cell of the epidermis (Chapter 5).

Large intestine (LARJ in-**TESS**-tin)—The organ of the alimentary tube that extends from the small intestine to the anus; absorbs water, minerals, and vitamins and eliminates undigested materials (Syn.—colon) (Chapter 16).

Laryngopharynx (la-RIN-goh-**FA**-rinks)—The lower portion of the pharynx that opens into the larynx and the esophagus; a passageway for both air and food (Chapter 15).

Larynx (**LA**-rinks)—The organ between the pharynx and the trachea; contains the vocal cords for speech (Syn.—voice box) (Chapter 15).

Lateral (**LAT**-er-uhl)—Away from the midline, or at the side (Chapter 1).

Lens (LENZ)—The oval structure of the eye posterior to the pupil, made of transparent protein; the only adjustable portion of the refraction pathway for the focusing of light rays (Chapter 9).

Lesion (**LEE**-zhun)—An area of pathologically altered tissue; an injury or wound (Chapter 5).

Leukemia (loo-**KEE**-mee-ah)—Malignancy of blood-forming tissues, in which large numbers of immature and nonfunctional white blood cells are produced (Chapter 11).

Leukocyte (**LOO**-koh-sight)—White blood cell; the five kinds are neutrophils, eosinophils, basophils, lymphocytes, and monocytes (Chapter 11).

Leukocytosis (LOO-koh-sigh-**TOH**-sis)—An elevated white blood cell count, often an indication of infection (Chapter 11).

Leukopenia (LOO-koh-**PEE**-nee-ah)—An abnormally low white blood cell count; may be the result of aplastic anemia, or a side effect of some medications (Chapter 11).

Ligament (**LIG**-uh-ment)—A fibrous connective tissue structure that connects bone to bone (Chapter 6).

Lipase (**LYE**-pays)—A digestive enzyme that breaks down emulsified fats to fatty acids and glycerol; secreted by the pancreas (Chapter 16).

Lipid (**LIP**-id)—An organic chemical insoluble in water; includes true fats, phospholipids, and steroids (Chapter 2).

Lipoprotein (li-poh-**PRO**-teen)—A large molecule that is a combination of proteins, triglycerides, and cholesterol;

formed by the liver to circulate lipids in the blood (Chapter 16).

Lithotripsy (LITH-oh-**TRIP**-see)—Crushing of gallstones or renal calculi by an instrument that uses ultrasonic waves applied to the exterior of the body (Chapter 16).

Liver (**LIV**-er)—The organ in the upper right and center of the abdominal cavity; secretes bile for the emulsification of fats in digestion; has many other functions related to the metabolism of nutrients and the composition of blood (Chapter 16).

Localized infection (**LOH**-kuhl-IZ'D)—An infection confined to one body organ or site (Chapter 22).

Longitudinal section (LAWNJ-i-**TOO**-din-uhl **SEK**-shun)—A plane or cut along the long axis of an organ or the body (Chapter 1).

Loop of Henle (LOOP of **HEN**-lee)—The part of a renal tubule that extends from the proximal convoluted tubule to the distal convoluted tubule (Chapter 18).

Lordosis (lor-**DOH**-sis)—An exaggerated lumbar curvature of the vertebral column (Chapter 6).

Lower esophageal sphincter (e-SOF-uh-**JEE**-uhl **SFINK**-ter)—The circular smooth muscle at the lower end of the esophagus; prevents backup of stomach contents (Syn.—cardiac sphincter) (Chapter 16).

Lower respiratory tract (**LOH**-er **RES**-pi-rah-TOR-ee TRAKT)—The respiratory organs located within the chest cavity (Chapter 15).

Lumbar puncture (**LUM**-bar **PUNK**-chur)—A diagnostic procedure that involves removal of cerebrospinal fluid from the lumbar meningeal sac to assess the pressure and constituents of cerebrospinal fluid (Chapter 8).

Lumbar vertebrae (**LUM**-bar **VER**-te-bray)—The five large vertebrae in the small of the back (Chapter 6).

Lungs (LUHNGS)—The paired organs in the thoracic cavity in which gas exchange takes place between the air in the alveoli and the blood in the pulmonary capillaries (Chapter 15).

Luteinizing hormone (LH or ICSH) (LOO-tee-in-**EYE**-zing)—A gonadotropic hormone produced by the anterior pituitary gland that, in men, stimulates secretion of testosterone by the testes or, in women, stimulates ovulation and secretion of progesterone by the corpus luteum in the ovary (Chapter 10).

Lymph (LIMF)—The water found within lymphatic vessels (Chapter 2).

Lymph node (LIMF NOHD)—A small mass of lymphatic tissue located along the pathway of a lymph vessel; produces lymphocytes and monocytes and destroys pathogens in the lymph (Chapter 14).

Lymph nodule (LIMF **NAHD**-yool)—A small mass of lymphatic tissue located in a mucous membrane; produces lymphocytes and monocytes and destroys pathogens that penetrate mucous membranes (Chapter 14).

Lymphatic tissue (lim-**FAT**-ik **TISH**-yoo)—A hemopoietic tissue that produces lymphocytes and monocytes; found in the spleen and lymph nodes and nodules (Chapter 11).

Lymphocyte (**LIM**-foh-sight)—A type of white blood cell (agranular); the two kinds are T cells and B cells, both of which are involved in immune responses (Chapter 11).

Lysosome (**LYE**-soh-zome)—A cell organelle found in the cytoplasm; contains enzymes that digest damaged cell parts or material ingested by the cell (Chapter 3).

Lysozyme (**LYE**-soh-zyme)—An enzyme in tears and saliva that inhibits the growth of bacteria in these fluids (Chapter 9).

—M—

Macrophage (**MAK**-roh-fahj)—A phagocytic cell derived from monocytes; capable of phagocytosis of pathogens, dead or damaged cells, and old red blood cells (Chapter 11).

Macula lutea (**MAK**-yoo-lah **LOO**-tee-ah)—A spot in the center of the retina; contains the fovea (Chapter 9).

Magnetic resonance imaging (MRI) (mag-**NET**-ik **REZ**-ah-nanse **IM**-ah-jing)—A diagnostic imaging technique that uses a magnetic field and a computer to integrate the images (Chapter 1).

Malignant (muh-**LIG**-nunt)—Tending to spread and become worse; used especially with reference to cancer (Chapter 3).

Malleus (**MAL**-ee-us)—The first of the three auditory bones in the middle ear; transmits vibrations from the eardrum to the incus (Chapter 9).

Maltase (**MAWL**-tays)—A digestive enzyme that breaks down maltose to glucose; secreted by the small intestine (Chapter 16).

Maltose (**MAWL**-tohs)—A disaccharide made of two glucose molecules (Chapter 2).

Mammary glands (**MAM**-uh-ree)—The glands of the female breasts that secrete milk; secretion and release of milk are under hormonal control (Chapter 20).

Mammography (mah-**MOG**-rah-fee)—A diagnostic procedure that uses radiography to detect breast cancer (Chapter 20).

Mandible (**MAN**-di-buhl)—The lower jaw bone (Chapter 6).

Manubrium (muh-**NOO**-bree-um)—The upper part of the sternum (Chapter 6).

Marrow canal (**MA**-roh ka-**NAL**)—The cavity within the diaphysis of a long bone; contains yellow bone marrow (Chapter 4).

Mastoid sinus (**MASS**-toyd)—An air cavity within the mastoid process of the temporal bone (Chapter 6).

Matrix (**MAY**-triks)—The non-living intercellular material that is part of connective tissues (Chapter 4).

Matter (**MAT**-ter)—Anything that occupies space; may be solid, liquid, or gas; may be living or non-living (Chapter 2).

Maxilla (mak-**SILL**-ah)—The upper jaw bone (Chapter 6).

Mechanical digestion (muh-**KAN**-i-kuhl dye-**JES**-chun)—The physical breakdown of food into smaller pieces, which increases the surface area for the action of digestive enzymes (Chapter 16).

Medial (**MEE**-dee-uhl)—Toward the midline, or in the middle (Chapter 1).

Mediastinum (ME-dee-ah-**STYE**-num)—The area or space between the lungs; contains the heart and great vessels (Chapter 12).

Medulla (muh-**DEW**-lah) (muh-**DULL**-ah)—1. The part of the brain superior to the spinal cord; regulates vital functions such as heart rate, respiration, and blood pressure. 2. The inner part of an organ, such as the renal medulla or the adrenal medulla (Chapter 8).

Megakaryocyte (MEH-ga-**KA**-ree-oh-sight)—A cell in the red bone marrow that breaks up into small fragments called platelets, which then circulate in peripheral blood (Chapter 11).

Meiosis (my-**OH**-sis)—The process of cell division in which one cell with the diploid number of chromosomes divides twice to form four cells, each with the haploid number of chromosomes (Chapter 3).

Meissner's plexus (**MIZE**-ners **PLEK**-sus)—The autonomic nerve plexus in the submucosa of the organs of the alimentary tube; regulates secretions of the glands in the mucosa of these organs (Chapter 16).

Melanin (**MEL**-uh-nin)—A protein pigment produced by melanocytes. Absorbs ultraviolet light; gives color to the skin, hair, iris, and choroid layer of the eye (Chapter 5).

Melanocyte (muh-**LAN**-o-sight)—A cell in the lower epidermis that synthesizes the pigment melanin (Chapter 5).

Melanoma (MEL-ah-**NOH**-mah)—Malignant pigmented mole or nevus (Chapter 5).

Melatonin (mel-ah-**TOH**-nin)—A hormone produced by the pineal gland; influences sleep cycles (Chapter 10).

Membrane (**MEM**-brayn)—A sheet of tissue; may be made of epithelial tissue or connective tissue (Chapter 4).

Meninges (me-**NIN**-jeez)—The connective tissue membranes that line the dorsal cavity and cover the brain and spinal cord (Chapter 1).

Meningitis (MEN-in-**JIGH**-tis)—Inflammation of the meninges, most often the result of bacterial infection (Chapter 8).

Menopause (**MEN**-ah-paws)—The period during life in which menstrual activity ceases; usually occurs between the ages of 45 and 55 years (Chapter 20).

Menstrual cycle (**MEN**-stroo-uhl **SIGH**-kuhl)—The periodic series of changes that occur in the female reproductive tract; the average cycle is 28 days (Chapter 20).

Menstruation (MEN-stroo-**AY**-shun)—The periodic discharge of a bloody fluid from the uterus that occurs at regular intervals from puberty to menopause (Chapter 20).

Mesentery (**MEZ**-en-TER-ee)—The visceral peritoneum (serous) that covers the abdominal organs; a large fold attaches the small intestine to the posterior abdominal wall (Chapter 1).

Mesoderm (**MEZ**-oh-derm)—The middle primary germ layer of cells of an embryo; gives rise to muscles, bones, and connective tissues (Chapter 21).

Metabolic acidosis (MET-uh-**BAH**-lik ass-i-**DOH**-sis)—A condition in which the blood pH is lower than normal, caused by any disorder that increases the number of acidic molecules in the body or increases the loss of alkaline molecules (Chapter 15).

Metabolic alkalosis (MET-uh-**BAH**-lik al-kah-**LOH**-sis)—A condition in which the blood pH is higher than normal, caused by any disorder that decreases the number of acidic molecules in the body or increases the number of alkaline molecules (Chapter 15).

Metabolism (muh-**TAB**-uh-lizm)—All the reactions that take place within the body; includes anabolism and catabolism (Chapter 17).

Metacarpals (MET-uh-**KAR**-puhls)—The five long bones in the palm of the hand (Chapter 6).

Metaphase (**MET**-ah-fayz)—The second stage of mitosis, in which the pairs of chromatids line up on the equator of the cell (Chapter 3).

Metastasis (muh-**TASS**-tuh-sis)—The spread of disease from one part of the body to another (Chapter 3).

Metatarsals (MET-uh-**TAR**-suhls)—The five long bones in the arch of the foot (Chapter 6).

Microglia (my-kroh-**GLEE**-ah)—A type of neuroglia capable of movement and phagocytosis of pathogens (Chapter 8).

Micron (**MY**-kron)—A unit of linear measure equal to 0.001 millimeter (Syn.—micrometer) (Chapter 3).

Microvilli (MY-kro-**VILL**-eye)—Folds of the cell membrane on the free surface of an epithelial cell; increase the surface area for absorption (Chapter 4).

Micturition (MIK-tyoo-**RISH**-un)—Urination; the voiding or elimination of urine from the urinary bladder (Chapter 18).

Midbrain (**MID**-brayn)—The part of the brain between the pons and hypothalamus; regulates visual, auditory, and righting reflexes (Chapter 8).

Mineral (**MIN**-er-al)—An inorganic element or compound; many are needed by the body for normal metabolism and growth (Chapter 17).

Mineralocorticoids (MIN-er-al-oh-**KOR**-ti-koidz)—The hormones secreted by the adrenal cortex that affect fluid-electrolyte balance; aldosterone is the major hormone in this group (Chapter 10).

Minute respiratory volume (**MIN**-uht RES-pi-rah-**TOR**-ee **VAHL**-yoom)—The volume of air inhaled and exhaled in 1 minute; calculated by multiplying tidal volume by number of respirations per minute (Chapter 15).

Mitochondria (MY-toh-**KAHN**-dree-ah)—The cell organelles in which aerobic cell respiration takes place and energy (ATP) is produced; found in the cytoplasm of a cell (Chapter 3).

Mitosis (my-**TOH**-sis)—The process of cell division in which one cell with the diploid number of chromosomes divides once to form two identical cells, each with the diploid number of chromosomes (Chapter 3).

Mitral valve (**MY**-truhl VALV)—The left AV valve (bicuspid valve), which prevents backflow of blood from the left ventricle to the left atrium when the ventricle contracts (Chapter 12).

Mixed nerve (MIKSD NERV)—A nerve that contains both sensory and motor neurons (Chapter 8).

Molecule (**MAHL**-e-kuhl)—A chemical combination of two or more atoms (Chapter 2).

Monocyte (**MAH**-no-sight)—A type of white blood cell (agranular); differentiates into a macrophage, which is capable of phagocytosis of pathogens and dead or damaged cells (Chapter 11).

Monosaccharide (MAH-noh-**SAK**-ah-ride)—A carbohydrate molecule that is a single sugar; includes the hexose and pentose sugars (Chapter 2).

Morula (**MOR**-yoo-lah)—An early stage of embryonic development, a solid mass of cells (Chapter 21).

Motility (moh-**TILL**-e-tee)—The ability to move (Chapter 3).

Motor neuron (**MOH**-ter **NYOOR**-on)—A nerve cell that carries impulses from the central nervous system to an effector (Syn.—efferent neuron) (Chapter 8).

Mucosa (mew-**KOH**-suh)—A mucous membrane, the epithelial lining of a body cavity that opens to the environment (Chapter 4).

Mucous membrane (**MEW**-kuss **MEM**-brayn)—The epithelial tissue lining of a body tract that opens to the environment (Chapter 4).

Mucus (**MEW**-kuss)—The thick fluid secreted by mucous membranes or mucous glands (Chapter 4).

Multicellular (MULL-tee-**SELL**-yoo-lar)—Consisting of more than one cell; made of many cells (Chapter 4).

Multiple sclerosis (**MULL**-ti-puhl skle-**ROH**-sis)—A progressive nervous system disorder, possibly an autoimmune disease, characterized by the degeneration of the myelin sheaths of CNS neurons (Chapter 8).

Muscle fatigue (**MUSS**-uhl fah-**TEEG**)—The state in which muscle fibers cannot contract efficiently, due to a lack of oxygen and the accumulation of lactic acid (Chapter 7).

Muscle fiber (**MUSS**-uhl **FYE**-ber)—A muscle cell (Chapter 7).

Muscle sense (**MUSS**-uhl SENSE)—The conscious or unconscious awareness of where the muscles are, and their degree of contraction, without having to look at them (Chapter 7).

Muscle tissue (**MUSS**-uhl **TISH**-yoo)—The tissue specialized for contraction and movement of parts of the body (Chapter 4).

Muscle tone (**MUSS**-uhl TONE)—The state of slight contraction present in healthy muscles (Chapter 7).

Muscular dystrophy (**MUSS**-kyoo-ler **DIS**-truh-fee)—A genetic disease characterized by the replacement of muscle tissue by fibrous connective tissue or adipose tissue, with progressive loss of muscle functioning; the most common form is Duchenne's muscular dystrophy (Chapter 7).

Muscular system (**MUSS**-kew-ler)—The organ system that consists of the skeletal muscles and tendons; its functions are to move the skeleton and produce body heat (Chapter 7).

Mutation (mew-**TAY**-shun)—A change in DNA; a genetic change that may be passed to offspring (Chapter 3).

Myalgia (my-**AL**-jee-ah)—Pain or tenderness in a muscle (Chapter 7).

Myasthenia gravis (MY-ass-**THEE**-nee-yuh **GRAH**-viss)—An autoimmune disease characterized by extreme muscle weakness and fatigue following minimal exertion (Chapter 7).

Mycosis (my-**KOH**-sis) (pl.: mycoses)—An infection caused by a pathogenic fungus (Chapter 22).

Myelin (**MY**-uh-lin)—A phospholipid produced by Schwann cells and oligodendrocytes that forms the myelin sheath of axons and dendrites (Chapter 2).

Myelin sheath (**MY**-uh-lin SHEETH)—The white, segmented, phospholipid sheath of most axons and dendrites; provides electrical insulation and increases the speed of impulse transmission (Chapter 4).

Myocardial infarction (MI) (MY-oh-**KAR**-dee-yuhl in-**FARK**-shun)—Death of part of the heart muscle due to lack of oxygen; often the result of an obstruction in a coronary artery (Syn.—heart attack) (Chapter 12).

Myocardium (MY-oh-**KAR**-dee-um)—The cardiac muscle tissue that forms the walls of the chambers of the heart (Chapter 4).

Myofibril (MY-oh-**FYE**-bril)—A linear arrangement of sarcomeres within a muscle fiber (Chapter 7).

Myoglobin (**MY**-oh-GLOW-bin)—The protein in muscle fibers that contains iron and stores oxygen in muscle fibers (Chapter 7).

Myometrium (MY-oh-**MEE**-tree-uhm)—The smooth muscle layer of the uterus; contracts for labor and delivery of an infant (Chapter 20).

Myopathy (my-**AH**-puh-thee)—A disease or abnormal condition of skeletal muscles (Chapter 7).

Myopia (my-**OH**-pee-ah)—Nearsightedness; an error of refraction in which only near objects are seen clearly (Chapter 9).

Myosin (**MY**-oh-sin)—A contractile protein in the sarcomeres of muscle fibers (Chapter 7).

Myxedema (MIK-suh-**DEE**-mah)—Hyposecretion of thyroxine in an adult; decreased metabolic rate results in physical and mental lethargy (Chapter 10).

—N—

Nail follicle (NAYL **FAH**-li-kuhl)—The structure within the skin of a finger or toe in which a nail grows; mitosis takes place in the nail root (Chapter 5).

Narrow-spectrum (**NAR**-oh **SPEK**-trum)—An antibiotic that is effective against only a few kinds of bacteria (Chapter 22).

Nasal cavities (**NAY**-zuhl **KAV**-i-tees)—The two air cavities within the skull through which air passes from the nostrils to the nasopharynx; separated by the nasal septum (Chapter 15).

Nasal mucosa (NAY-zuhl mew-**KOH**-sah)—The lining of the nasal cavities; made of ciliated epithelium that warms and moistens the incoming air and sweeps mucus, dust, and pathogens toward the nasopharynx (Chapter 15).

Nasal septum (**NAY**-zuhl **SEP**-tum)—The verticle plate made of bone and cartilage that separates the two nasal cavities (Chapter 15).

Nasolacrimal duct (NAY-zo-**LAK**-ri-muhl)—A duct that carries tears from the lacrimal sac to the nasal cavity (Chapter 9).

Nasopharynx (NAY-zo-**FA**-rinks)—The upper portion of the pharynx above the level of the soft palate; an air passageway (Chapter 15).

Negative feedback mechanism (**NEG**-ah-tiv **FEED**-bak)—A control system in which a stimulus initiates a response that reverses or reduces the stimulus, thereby stopping the response until the stimulus occurs again (Chapter 1).

Nephritis (ne-**FRY**-tis)—Inflammation of the kidney; may be caused by bacterial infection or toxic chemicals (Chapter 18).

Nephron (**NEFF**-ron)—The structural and functional unit of the kidney that forms urine; consists of a renal corpuscle and a renal tubule (Chapter 18).

Nerve (NERV)—A group of neurons, together with blood vessels and connective tissue (Chapter 8).

Nerve tissue (NERV **TISH**-yoo)—The tissue specialized to generate and transmit electrochemical impulses that have many functions in the maintenance of homeostasis (Chapter 4).

Nerve tract (NERV TRAKT)—A group of neurons that share a common function within the central nervous system; a tract may be ascending (sensory) or descending (motor) (Chapter 8).

Nervous system (**NERV**-us **SIS**-tem)—The organ system that regulates body functions by means of electrochemical impulses; consists of the brain, spinal cord, cranial nerves, and spinal nerves (Chapter 8).

Neuralgia (new-**RAL**-jee-ah)—Sharp, severe pain along the course of a nerve (Chapter 8).

Neuritis (new-**RYE**-tis)—Inflammation of a nerve (Chapter 8).

Neuroglia (new-**ROG**-lee-ah)—The non-neuronal cells of the central nervous system (Chapter 4).

Neurohypophysis (NEW-roo-high-**POFF**-e-sis)—The posterior pituitary gland (Chapter 10).

Neurolemma (NEW-roh-**LEM**-ah)—The sheath around peripheral axons and dendrites, formed by the cytoplasm and nuclei of Schwann cells; is essential for the regeneration of damaged peripheral neurons (Chapter 8).

Neuromuscular junction (NYOOR-oh-**MUSS**-kew-lar **JUNK**-shun)—The termination of a motor neuron on the sarcolemma of a muscle fiber; the synapse is the microscopic space between the two structures (Chapter 7).

Neuron (**NYOOR**-on)—A nerve cell; consists of a cell body, an axon, and dendrites (Chapter 4).

Neuropathy (new-**RAH**-puh-thee)—Any disease or disorder of the nerves (Chapter 8).

Neurotoxin (NEW-roh-**TOK**-sin)—A chemical that disrupts an aspect of the functioning of the nervous system (Chapter 7).

Neurotransmitter (NYOOR-oh-**TRANS**-mih-ter)—A chemical released by the axon of a neuron, which crosses a synapse and affects the electrical activity of the post-synaptic membrane (neuron or muscle cell or gland) (Chapter 4).

Neutron (**NEW**-trahn)—A subatomic particle that has no electrical charge; found in the nucleus of an atom (Chapter 2).

Neutrophil (**NEW**-troh-fill)—A type of white blood cell (granular); capable of phagocytosis of pathogens (Chapter 11).

Nevus (**NEE**-vus)—A pigmented area of the skin; a mole (Chapter 5).

Night blindness (NITE **BLIND**-ness)—The inability to see well in dim light or at night; may result from a vitamin A deficiency (Chapter 9).

Nine areas (NYNE)—The subdivision of the abdomen into nine equal areas to facilitate the description of locations (Chapter 1).

Non-communicable disease (NON-kuhm-**YOO**-ni-kah-b'l)—A disease that cannot be directly or indirectly transmitted from host to host (Chapter 22).

Non-essential amino acids (NON-e-**SEN**-shul ah-**ME**-noh **ASS**-ids)—The amino acids that can be synthesized by the liver (Chapter 16).

Norephinephrine (NOR-ep-i-**NEFF**-rin)—A hormone secreted by the adrenal medulla that causes vasoconstriction throughout the body, which raises blood pressure in stressful situations (Chapter 10).

Normal flora (**NOR**-muhl **FLOOR**-uh)—1. The population of microorganisms that is usually present in certain parts of the body. 2. In the colon, the bacteria that produce vitamins and inhibit the growth of pathogens (Chapter 16).

Normoblast (**NOR**-mow-blast)—A red blood cell with a nucleus, an immature stage in red blood cell formation; usually found in the red bone marrow and not in peripheral circulation (Chapter 11).

Nosocomial infection (no-zoh-**KOH**-mee-uhl)—An infection acquired in a hospital or other health-care institution (Chapter 22).

Nuclear membrane (**NEW**-klee-er **MEM**-brain)—The double-layer membrane that encloses the nucleus of a cell (Chapter 3).

Nucleic acid (new-**KLEE**-ik **ASS**-id)—An organic chemical that is made of nucleotide subunits. Examples are DNA and RNA (Chapter 2).

Nucleolus (new-**KLEE**-oh-lus)—A small structure made of DNA, RNA, and protein. Found in the nucleus of a cell; produces ribosomal RNA (Chapter 3).

Nucleotide (**NEW**-klee-oh-tide)—An organic compound that consists of a pentose sugar, a phosphate group, and one of five nitrogenous bases (adenine, guanine, cytosine, thymine, or uracil); the subunits of DNA and RNA (Chapter 2).

Nucleus (**NEW**-klee-us)—1. The membrane bound part of a cell that contains the hereditary material in chromosomes. 2. The central part of an atom containing protons and neutrons (Chapters 2, 3).

—O—

Occipital bone (ok-**SIP**-i-tuhl)—The flat bone that forms the back of the skull (Chapter 6).

Occipital lobes (ok-**SIP**-i-tuhl LOWBS)—The most posterior part of the cerebrum; contain the visual areas (Chapter 8).

Oculomotor nerves (OK-yoo-loh-**MOH**-tur)—Cranial nerve pair III. Motor to the extrinsic muscles of the eye, the ciliary body, and the iris (Chapter 8).

Olfactory nerves (ohl-**FAK**-tuh-ree)—Cranial nerve pair I. Sensory for smell (Chapter 8).

Olfactory receptors (ohl-**FAK**-tuh-ree ree-**SEP**-ters)—The sensory receptors in the upper nasal cavities that detect vaporized chemicals, providing a sense of smell (Chapter 9).

Oligodendrocyte (ah-li-goh-**DEN**-droh-sight)—A type of neuroglia that produces the myelin sheath around neurons of the central nervous system (Chapter 8).

Oligosaccharide (ah-lig-oh-**SAK**-ah-ride)—A carbohydrate molecule that consists of from 3-20 monosaccharides bonded together; form "self" antigens on cell membranes (Chapter 2).

Oliguria (AH-li-**GYOO**-ree-ah)—Decreased urine formation and output (Chapter 18).

Oogenesis (Oh-oh-**JEN**-e-sis)—The process of meiosis in the ovary to produce an ovum (Chapter 3).

Opportunistic infection (OP-er-too-**NIS**-tik)—An infection caused by a microorganism that is usually a saprophyte but may become a parasite under certain conditions, such as lowered host resistance (Chapter 22).

Opsonization (OP-sah-ni-**ZAY**-shun)—The action of antibodies or complement that upon binding to a foreign antigen attracts macrophages and facilitates phagocytosis (from the Greek "to purchase food") (Chapter 14).

Optic chiasma (**OP**-tik kye-**AS**-muh)—The site of the crossing of the medial fibers of each optic nerve, anterior to the pituitary gland; important for binocular vision (Chapter 9).

Optic disc (**OP**-tik DISK)—The portion of the retina where the optic nerve passes through; no rods or cones are present (Syn.—blind spot) (Chapter 9).

Optic nerves (**OP**-tik)—Cranial nerve pair II. Sensory for vision (Chapter 8).

Oral cavity (**OR**-uhl **KAV**-i-tee)—The cavity in the skull bounded by the hard palate, cheeks, and tongue (Chapter 16).

Orbit (**OR**-bit)—The cavity in the skull that contains the eyeball (Syn.—eyesocket) (Chapter 9).

Organ (**OR**-gan)—A structure with specific functions; made of two or more tissues (Chapter 1).

Organ of Corti (**KOR**-tee) (spiral organ)—The structure in the cochlea of the inner ear that contains the receptors for hearing (Chapter 9).

Organ system (**OR**-gan **SIS**-tem)—A group of related organs that work together to perform specific functions (Chapter 1).

Organelle (OR-gan-**ELL**)—An intracellular structure that has a specific function (Chapter 3).

Organic (or-**GAN**-ik)—A chemical compound that contains carbon-hydrogen covalent bonds; includes carbohydrates, lipids, proteins, and nucleic acids (Chapter 1).

Origin (**AHR**-i-jin)—1. The more stationary attachment point of a muscle to a bone. 2. The beginning (Chapter 7).

Oropharynx (OR-oh-**FA**-rinks)—The middle portion of the pharynx behind the oral cavity; a passageway for both air and food (Chapter 15).

Osmolarity (ahs-moh-**LAR**-i-tee)—The concentration of osmotically active particles in a solution (Chapter 19).

Osmoreceptors (AHS-moh-re-**SEP**-ters)—Specialized cells in the hypothalamus that detect changes in the water content of the body (Chapter 10).

Osmosis (ahs-**MOH**-sis)—The diffusion of water through a selectively permeable membrane (Chapter 3).

Osmotic pressure (ahs-**MAH**-tik)—Pressure that develops when two solutions of different concentration are separated by a selectively permeable membrane. A hypertonic solution that would cause cells to shrivel has a higher osmotic pressure. A hypotonic solution that would cause cells to swell has a lower osmotic pressure (Chapter 3).

Ossification (AHS-i-fi-**KAY**-shun)—The process of bone formation; bone matrix is produced by osteoblasts during the growth or repair of bones (Chapter 6).

Osteoarthritis (AHS-tee-oh-ar-**THRY**-tiss)—The inflammation of a joint, especially a weight-bearing joint, that is most often a consequence of aging (Chapter 6).

Osteoblast (**AHS**-tee-oh-BLAST)—A bone-producing cell; produces bone matrix for the growth or repair of bones (Chapter 6).

Osteoclast (**AHS**-tee-oh-KLAST)—A bone-destroying cell; reabsorbs bone matrix as part of the growth or repair of bones (Chapter 6).

Osteocyte (**AHS**-tee-oh-sight)—A bone cell (Chapter 4).

Osteomyelitis (AHS-tee-oh-my-uh-**LYE**-tiss)—Inflammation of a bone caused by a pathogenic microorganism (Chapter 6).

Osteoporosis (AHS-tee-oh-por-**OH**-sis)—A condition in which bone matrix is lost and not replaced, resulting in weakened bones that are then more likely to fracture (Chapter 6).

Otitis media (oh-**TIGH**-tis **MEE**-dee-ah)—Inflammation of the middle ear (Chapter 9).

Otoliths (**OH**-toh-liths)—Microscopic crystals of calcium carbonate in the utricle and saccule of the inner ear; are pulled by gravity (Chapter 9).

Oval window (**OH**-vul **WIN**-doh)—The membrane-covered opening through which the stapes transmit vibrations to the fluid in the inner ear (Chapter 9).

Ovary (**OH**-vuh-ree)—The female gonad that produces ova; also an endocrine gland that produces the hormones estrogen and progesterone (Chapter 10).

Ovum (**OH**-vuhm)—An egg cell, produced by an ovary (Pl.—ova) (Chapter 20).

Oxygen debt (**OKS**-ah-jen DET)—The state in which there is not enough oxygen to complete the process of cell respiration; lactic acid is formed, which contributes to muscle fatigue (Chapter 7).

Oxytocin (OK-si-**TOH**-sin)—A hormone produced by the hypothalamus and stored in the posterior pituitary gland; stimulates contraction of the myometrium and release of milk by the mammary glands (Chapter 8).

—P—

Palate (**PAL**-uht)—The roof of the mouth, which separates the oral cavity from the nasal cavities (Chapter 16).

Palpitation (pal-pi-**TAY**-shun)—An irregular heart beat of which the person is aware (Chapter 12).

Pancreas (**PAN**-kree-us)—1. An endocrine gland located between the curve of the duodenum and the spleen; secretes insulin and glucagon. 2. An exocrine gland that secretes digestive enzymes for the digestion of starch, fats, and proteins (Chapter 10).

Pancreatic duct (PAN-kree-**AT**-ik DUKT)—The duct that takes pancreatic juices to the common bile duct (Chapter 16).

Pandemic (pan-**DEM**-ik)—An epidemic that affects several countries at the same time (Chapter 22).

Papillae (pah-**PILL**-ay)—1. Elevated, pointed projections. 2. On the tongue, the projections that contain taste buds (Chapter 16).

Papillary layer (**PAP**-i-lar-ee **LAY**-er)—The uppermost layer of the dermis; contains capillaries to nourish the epidermis (Chapter 5).

Papillary muscles (**PAP**-i-lar-ee **MUSS**-uhls)—Columns of myocardium that project from the floor of a ventricle and anchor the flaps of the AV valve by way of the chordae tendineae (Chapter 12).

Paralysis (pah-**RAL**-i-sis)—Complete or partial loss of function, especially of a muscle (Chapter 7).

Paralytic ileus (**PAR**-uh-LIT-ik **ILL**-ee-us)—Paralysis of the intestines that may occur following abdominal surgery (Chapter 16).

Paraplegia (PAR-ah-**PLEE**-gee-ah)—Paralysis of the legs (Chapter 8).

Paranasal sinus (PAR-uh-**NAY**-zuhl **SIGH**-nus)—An air cavity in the frontal, maxilla, sphenoid, or ethmoid bones; opens into the nasal cavities (Chapter 6).

Parasite (**PAR**-uh-sight)—An organism that lives on or in another living organism, called a host, to which it causes harm (from the Greek for "to eat at another's table") (Chapter 22).

Parasympathetic (PAR-ah-SIM-puh-**THET**-ik)—The division of the autonomic nervous system that dominates during non-stressful situations (Chapter 8).

Parathyroid glands (PAR-ah-**THIGH**-royd)—The four endocrine glands located on the posterior side of the thyroid gland; secrete parathyroid hormone (Chapter 10).

Parathyroid hormone (PTH) (PAR-ah-**THIGH**-royd)—A hormone secreted by the parathyroid glands; increases the reabsorption of calcium from bones and the absorption of calcium by the small intestine and kidneys (Chapter 10).

Parietal (puh-**RYE**-uh-tuhl)—1. Pertaining to the walls of a body cavity (Chapter 1). 2. The flat bone that forms the crown of the cranial cavity (from the Latin "wall") (Chapter 6).

Parietal cells (puh-**RYE**-uh-tuhl SELLS)—The cells of the gastric pits of the stomach that secrete hydrochloric acid and the intrinsic factor (Chapter 16).

Parietal lobes (puh-**RYE**-uh-tuhl LOWBS)—The parts of

the cerebrum posterior to the frontal lobes; contain the sensory areas for cutaneous sensation and conscious muscle sense (Chapter 8).

Parkinson's disease (**PAR**-kin-sonz)—A progressive disorder of the basal ganglia, characterized by tremor, muscle weakness and rigidity, and a peculiar gait (Chapter 8).

Parotid glands (pah-**RAH**-tid)—The pair of salivary glands located just below and in front of the ears (Chapter 16).

Partial pressure (**PAR**-shul **PRES**-shur)—1. The pressure exerted by a gas in a mixture of gases. 2. The value used to measure oxygen and carbon dioxide concentrations in the blood or other body fluid (Chapter 15).

Parturition (PAR-tyoo-**RISH**-uhn)—The act of giving birth (Chapter 21).

Passive immunity (**PASS**-iv im-**YOO**-ni-tee)—The immunity provided by the reception of antibodies from another source; may be natural (placental, breast milk) or artificial (injection of gamma globulins) (Chapter 14).

Pasteurization (PAS-tyoor-i-**ZAY**-shun)—The process of heating a fluid to moderate temperatures in order to destroy pathogenic bacteria (Chapter 22).

Patella (puh-**TELL**-ah)—The kneecap, a short bone (from the Latin "flat dish") (Chapter 6).

Patellar reflex (puh-**TELL**-ar **REE**-fleks)—A stretch reflex integrated in the spinal cord, in which a tap on the patellar tendon causes extension of the lower leg (Syn.—kneejerk reflex) (Chapter 8).

Pathogen (**PATH**-oh-jen)—A microorganism capable of producing disease; includes bacteria, viruses, fungi, protozoa, and worms (Chapter 14).

Pathophysiology (PATH-oh-FIZZ-ee-**AH**-luh-jee)—The study of diseases as they are related to functioning (Chapter 1).

Pelvic cavity (**PELL**-vik **KAV**-i-tee)—Inferior portion of the ventral cavity, below the abdominal cavity (Chapter 1).

Penis (**PEE**-nis)—The male organ of copulation when the urethra serves as a passage for semen; an organ of elimination when the urethra serves as a passage for urine (Chapter 20).

Pentose sugar (**PEN**-tohs)—A five-carbon sugar (monosaccharide) that is a structural part of the nucleic acids DNA and RNA (Chapter 2).

Pepsin (**PEP**-sin)—The enzyme found in gastric juice that begins protein digestion; secreted by chief cells (Chapter 16).

Peptidases (**PEP**-ti-day-ses)—Digestive enzymes that break down polypeptides to amino acids; secreted by the small intestine (Chapter 16).

Peptide bond (**PEP**-tyde)—A chemical bond that links two amino acids in a protein molecule (Chapter 2).

Pericardium (PER-ee-**KAR**-dee-um)—The three membranes that enclose the heart, consisting of an outer fibrous layer and two serous layers (Chapter 12).

Perichondrium (PER-ee-**KON**-dree-um)—The fibrous connective tissue membrane that covers cartilage (Chapter 4).

Perilymph (**PER**-i-limf)—The fluid in the bony labyrinth of the inner ear (Chapter 9).

Periodontal membrane (PER-ee-oh-**DON**-tal)—The membrane that lines the tooth sockets in the upper and lower jaws; produces a bone-like cement to anchor the teeth (Chapter 16).

Periosteum (PER-ee-**AHS**-tee-um)—The fibrous connective tissue membrane that covers bone; contains osteoblasts for bone growth or repair (Chapter 4).

Peripheral (puh-**RIFF**-uh-ruhl)—Extending from a main part; closer to the surface (Chapter 1).

Peripheral nervous system (puh-**RIFF**-uh-ruhl **NERV**-vuhs)—The part of the nervous system that consists of the cranial nerves and spinal nerves (Chapter 8).

Peripheral resistance (puh-**RIFF**-uh-ruhl ree-**ZIS**-tense)—The resistance of the blood vessels to the flow of blood; changes in the diameter of arteries have effects on blood pressure (Chapter 13).

Peristalsis (per-i-**STALL**-sis)—Waves of muscular contraction (one-way) that propel the contents through a hollow organ (Chapter 2).

Peritoneum (PER-i-toh-**NEE**-um)—The serous membrane that lines the abdominal cavity (Chapter 1).

Peritonitis (per-i-toh-**NIGH**-tis)—Inflammation of the peritoneum (Chapter 16).

Peritubular capillaries (PER-ee-**TOO**-byoo-ler)—The capillaries that surround the renal tubule and receive the useful materials reabsorbed from the renal filtrate; carry blood from the efferent arteriole to the renal vein (Chapter 18).

Pernicious anemia (per-**NISH**-us uh-**NEE**-mee-yah)—An anemia that is the result of a deficiency of vitamin B_{12} or the intrinsic factor (Chapter 11).

Peyer's patches (**PYE**-erz)—The lymph nodules in the mucosa of the small intestine, especially in the ileum (Chapter 14).

pH —A symbol of the measure of the concentration of hydrogen ions in a solution. The pH scale extends from 0-14, with a value of 7 being neutral. Values lower than 7 are acidic, values higher than 7 are alkaline (basic) (Chapter 2).

Phagocytosis (FAG-oh-sigh-**TOH**-sis)—The process by which a cell engulfs a particle; especially, the ingestion of microorganisms by white blood cells (Chapter 3).

Phalanges (fuh-**LAN**-jees)—The long bones of the fingers and toes. There are 14 in each hand or foot (from the Latin "line of soldiers") (Chapter 6).

Phantom pain (**FAN**-tum PAYN)—Pain following amputation of a limb that seems to come from the missing limb (Chapter 9).

Pharynx (**FA**-rinks)—A muscular tube located behind the nasal and oral cavities; a passageway for air and food (Chapter 15).

Phenotype (**FEE**-noh-type)—The appearance of the individual as related to genotype; the expression of the genes that are present (Chapter 21).

Phlebitis (fle-**BY**-tis)—Inflammation of a vein (Chapter 13).

Phospholipid (**FOSS**-foh-LIP-id)—An organic compound in the lipid group that is made of one glycerol, two fatty acids, and a phosphate molecule (Chapter 2).

Phrenic nerves (**FREN**-ik NERVZ)—The pair of peripheral nerves that are motor to the diaphragm (Chapter 15).

Physiology (FIZZ-ee-**AH**-luh-jee)—The study of the functioning of the body and its parts (Chapter 1).

Pia mater (**PEE**-ah **MAH**-ter)—The innermost layer of the meninges, made of thin connective tissue on the surface of the brain and spinal cord (Chapter 8).

Pineal gland (**PIN**-ee-uhl)—An endocrine gland on the posterior wall of the third ventricle of the brain; secretes melatonin (Chapter 10).

Pilomotor muscle (**PYE**-loh-MOH-ter)—A smooth muscle attached to a hair follicle; contraction pulls the follicle upright, resulting in "goose bumps" (Syn.—arrector pili muscle) (Chapter 5).

Pinocytosis (PIN-oh-sigh-**TOH**-sis)—The process by which a stationary cell ingests very small particles or a liquid (Chapter 3).

Pituitary gland (pi-**TOO**-i-TER-ee)—An endocrine gland located below the hypothalamus, consisting of anterior and posterior lobes (Syn.—hypophysis) (Chapter 10).

Pivot joint (**PI**-vot)—A diarthrosis that permits rotation (Chapter 6).

Placenta (pluh-**SEN**-tah)—The organ formed in the uterus during pregnancy, made of both fetal and maternal tissue; the site of exchanges of materials between fetal blood and maternal blood (Chapter 13).

Plane (PLAYN)—An imaginary flat surface that divides the body in a specific way (Chapter 1).

Plasma (**PLAZ**-mah)—The water found within the blood vessels. Plasma comprises 52%–62% of the total blood (Chapter 2).

Plasma cell (**PLAZ**-mah SELL)—A cell derived from an activated B cell that produces antibodies to a specific antigen (Chapter 14).

Plasma proteins (**PLAZ**-mah **PRO**-teenz)—The proteins that circulate in the liquid portion of the blood; include albumin, globulins, and clotting factors (Chapter 11).

Platelets (**PLAYT**-lets)—Blood cells that are fragments of larger cells (megakaryocytes) of the red bone marrow; involved in blood clotting and other mechanisms of hemostasis (Syn.—thrombocytes) (Chapter 4).

Pleural membranes (**PLOOR**-uhl **MEM**-braynz)—The serous membranes of the thoracic cavity (Chapter 1).

Plica circulares (**PLEE**-ka SIR-kew-**LAR**-es)—The circular folds of the mucosa and submucosa of the small intestine; increase the surface area for absorption (Chapter 16).

Pneumonia (new-**MOH**-nee-ah)—Inflammation of the lungs caused by bacteria, viruses, or chemicals (Chapter 15).

Pneumotaxic center (NEW-moh-**TAK**-sik **SEN**-ter)—The respiratory center in the pons that helps bring about exhalation (Chapter 15).

Pneumothorax (NEW-moh-**THAW**-raks)—The accumulation of air in the potential pleural space, which increases intrapleural pressure and causes collapse of a lung (Chapter 15).

Polarization (POH-lahr-i-**ZAY**-shun)—The distribution of ions on either side of a membrane; in a resting neuron or muscle cell, sodium ions are more abundant outside the cell, and potassium and negative ions are more abundant inside the cell, giving the membrane a positive charge outside and a relative negative charge inside (Chapter 7).

Polypeptide (PAH-lee-**PEP**-tyde)—A short chain of amino acids, not yet a specific protein (Chapter 2).

Polysaccharide (PAH-lee-**SAK**-ah-ryde)—A carbohydrate molecule that consists of many monosaccharides (usually glucose) bonded together; includes glycogen, starch, and cellulose (Chapter 2).

Polyuria (PAH-li-**YOO**-ree-ah)—Increased urine formation and output (Chapter 18).

Pons (PONZ)—The part of the brain anterior and superior to the medulla; contributes to the regulation of respiration (from the Latin "bridge") (Chapter 8).

Pore (POR)—An opening on a surface to permit the passage of materials (Chapter 3).

Portal of entry (**POR**-tuhl of **EN**-tree)—The way a pathogen enters the body, such as natural body openings or breaks in the skin (Chapter 22).

Portal of exit (**POR**-tuhl of **EG**-zit)—The way a pathogen leaves a host, such as in respiratory droplets, feces, or reproductive secretions (Chaper 22).

Positron emission tomography (PET) (**PAHS**-i-tron eh-**MISH**-shun toh-**MAH**-grah-fee)—An imaging technique that depicts rate of metabolism or blood flow in organs or tissues (Chapter 1).

Posterior (poh-**STEER**-ee-your)—Toward the back (Syn.—dorsal) (Chapter 1).

Postganglionic neuron (POST-gang-lee-**ON**-ik)—In the autonomic nervous system, a neuron that extends from a ganglion to a visceral effector (Chapter 8).

Precapillary sphincter (pree-**KAP**-i-lar-ee **SFINK**-ter)—A smooth muscle cell at the beginning of a capillary network that regulates the flow of blood through the network (Chapter 13).

Preganglionic neuron (PRE-gang-lee-**ON**-ik)—In the autonomic nervous system, a neuron that extends from the CNS to a ganglion and synapses with a postganglionic neuron (Chapter 8).

Presbyopia (PREZ-bee-**OH**-pee-ah)—Farsightedness that is a consequence of aging and the loss of elasticity of the lens (Chapter 9).

Pressoreceptors (**PRESS**-oh-ree-SEP-ters)—The sensory receptors in the carotid sinuses and aortic sinus that detect changes in blood pressure (Chapter 9).

Primary bronchi (**PRY**-ma-ree **BRONG**-kye)—The two branches of the lower end of the trachea; air passageways to the right and left lungs (Chapter 15).

Prime mover (PRYME **MOO**-ver)—The muscle responsible for the main action when a joint is moved (Chapter 7).

Progesterone (proh-**JESS**-tuh-rohn)—The sex hormone secreted by the corpus luteum of the ovary; contributes to the growth of the endometrium and the maintenance of pregnancy (Chapter 10).

Projection (proh-**JEK**-shun)—The characteristic of sensations in which the sensation is felt in the area where the receptors were stimulated (Chapter 9).

Prolactin (proh-**LAK**-tin)—A hormone produced by the anterior pituitary gland, that stimulates milk production by the mammary glands (Chapter 10).

Pronation (pro-**NAY**-shun)—Turning the palm downward, or lying face down (Chapter 7).

Prophase (**PROH**-fayz)—The first stage of mitosis, in which the pairs on chromatids become visible (Chapter 3).

Proprioceptor (**PROH**-pree-oh-SEP-ter)—A sensory receptor in a muscle that detects stretching of the muscle (Syn.—stretch receptor) (Chapter 7).

Prostaglandins (PRAHS-tah-**GLAND**-ins)—Locally acting hormone-like substances produced by virtually all cells from the phospholipids of their cell membranes; the many types have many varied functions (Chapter 10).

Prostate gland (**PRAHS**-tayt)—A muscular gland that surrounds the first inch of the male urethra; secretes an alkaline fluid that becomes part of semen; its smooth muscle contributes to ejaculation (Chapter 20).

Prostatic hypertrophy (prahs-**TAT**-ik high-**PER**-truh-fee)—Enlargement of the prostate gland; may be benign or malignant (Chapter 20).

Protein (**PRO**-teen)—An organic compound made of amino acids linked by peptide bonds (Chapter 2).

Proteinuria (PRO-teen-**YOO**-ree-ah)—The presence of protein in urine (Chapter 18).

Prothrombin (proh-**THROM**-bin)—A clotting factor synthesized by the liver and released into the blood; converted to thrombin in the process of chemical clotting (Chapter 11).

Proton (**PRO**-tahn)—A subatomic particle that has a positive electrical charge; found in the nucleus of an atom (Chapter 2).

Protozoa (PROH-tuh-**ZOH**-ah) (Sing.: Protozoan)—The simplest animal-like microorganisms in the kingdom Protista; usually unicellular, some are colonial; may be free-living or parasitic (Chapter 22).

Proximal (**PROK**-si-muhl)—Closest to the origin or point of attachment (Chapter 1).

Proximal convoluted tubule (**PROK**-si-muhl KON-voh-**LOO**-ted)—The part of a renal tubule that extends from Bowman's capsule to the loop of Henle (Chapter 18).

Pruritus (proo-**RYE**-tus)—Severe itching (Chapter 5).

Puberty (**PEW**-ber-tee)—The period during life in which members of both sexes become sexually mature and capable of reproduction; usually occurs between the ages of 10 and 14 years (Chapter 20).

Pubic symphysis (**PEW**-bik **SIM**-fi-sis)—The joint between the right and left pubic bones, in which a disc of cartilage separates the two bones (Chapter 6).

Pubis (**PEW**-biss)—The lower anterior part of the hip bone (Syn.—pubic bone) (Chapter 6).

Pulmonary artery (**PULL**-muh-NER-ee **AR**-tuh-ree)—The artery that takes blood from the right ventricle to the lungs (Chapter 12).

Pulmonary edema (**PULL**-muh-NER-ee uh-**DEE**-muh)—Accumulation of tissue fluid in the alveoli of the lungs (Chapter 15).

Pulmonary semilunar valve (**PULL**-muh-NER-ee SEM-ee-**LOO**-nar VALV)—The valve at the junction of the right ventricle and the pulmonary artery; prevents backflow of blood from the artery to the ventricle when the ventricle relaxes (Chapter 12).

Pulmonary surfactant (**PULL**-muh-NER-ee sir-**FAK**-tent)—A lipid substance secreted by the alveoli in the lungs; reduces the surface tension within alveoli to permit inflation (Chapter 15).

Pulmonary veins (**PULL**-muh-NER-ee VAYNS)—The four veins that return blood from the lungs to the left atrium (Chapter 12).

Pulp cavity (PUHLP)—The innermost portion of a tooth that contains blood vessels and nerve endings (Chapter 16).

Pulse (PULS)—The force of the heartbeat detected at an arterial site such as the radial artery (Chapter 12).

Pulse deficit (PULS **DEF**-i-sit)—The condition in which the radial pulse count is lower than the rate of the heartbeat heard with a stethoscope; may occur in some types of heart disease in which the heartbeat is weak (Chapter 13).

Pulse pressure (PULS **PRES**-shur)—The difference between systolic and diastolic blood pressure; averages about 40 mmHg (Chapter 13).

Punnett square (**PUHN**-net SKWAIR)—A diagram used to determine the possible combinations of genes in the offspring of a particular set of parents (Chapter 21).

Pupil (**PYOO**-pil)—The opening in the center of the iris; light rays pass through the aqueous humor in the pupil (Chapter 9).

Purkinje fibers (purr-**KIN**-jee)—Specialized cardiac muscle fibers that are part of the cardiac conduction pathway (Chapter 12).

Pyloric sphincter (pye-**LOR**-ik **SFINK**-ter)—The circular smooth muscle at the junction of the stomach and the duodenum; prevents backup of intestinal contents into the stomach (from the Greek "gatekeeper") (Chapter 16).

Pyloric stenosis (pye-**LOR**-ik ste-**NOH**-sis)—Narrowing of the opening between the stomach and duodenum caused by hypertrophy of the pyloric sphincter; a congenital disorder (Chapter 16).

Pyrogen (**PYE**-roh-jen)—Any microorganism or substance that causes a fever; includes bacteria, viruses, or chemicals released during inflammation (called endogenous pyrogens); activates the heat production and conservation mechanisms regulated by the hypothalamus (Chapter 17).

—Q—

QRS wave —The portion of an ECG that depicts depolarization of the ventricles (Chapter 12).

Quadrants (**KWAH**-drants)—A division into four parts, used especially to divide the abdomen into four areas to facilitate description of locations (Chapter 1).

Quadriplegia (KWA-dri-**PLEE**-jee-ah)—Paralysis of all four limbs (Chapter 17).

—R—

Radiation (RAY-dee-**AY**-shun)—1. The heat loss process in which heat energy from the skin is emitted to the cooler surroundings. 2. The emissions of certain radioactive elements; may be used for diagnostic or therapeutic purposes (Chapter 17).

Radius (**RAY**-dee-us)—The long bone of the forearm on the thumb side (from the Latin "a spoke") (Chapter 6).

Range of motion exercises (RANJE of **MOH**-shun)—Movements of joints through their full range of motion; used to preserve mobility or to regain mobility following an injury (Chapter 7).

Receptor (ree-**SEP**-tur)—A specialized cell or nerve ending that responds to a particular change such as light, sound, heat, touch, or pressure (Chapter 5).

Receptor site (ree-**SEP**-ter SIGHT)—An arrangement of molecules, often part of the cell membrane, that will accept only molecules with a complementary shape (Chapter 3).

Recessive (ree-**SESS**-iv)—In genetics, a characteristic that will be expressed only if two genes for it are present in the homologous pair (Chapter 21).

Red blood cells (RED BLUHD SELLS)—The most numerous cells in the blood; carry oxygen bonded to the hemoglobin within them (Syn.—erythrocytes) (Chapter 4).

Red bone marrow (RED BOWN **MAR**-row)—A hemopoietic tissue found in flat and irregular bones; produces all the types of blood cells (Chapter 6).

Reduced hemoglobin (re-**DOOSD HEE**-muh-GLOW-bin)—Hemoglobin that has released its oxygen in the systemic capillaries (Chapter 11).

Referred pain (ree-**FURD** PAYNE)—Visceral pain that is projected and felt as cutaneous pain (Chapter 9).

Reflex (**REE**-fleks)—An involuntary response to a stimulus (Chapter 8).

Reflex arc (**REE**-fleks ARK)—The pathway nerve impulses follow when a reflex is stimulated (Chapter 8).

Refraction (ree-**FRAK**-shun)—The bending of light rays as they pass through the eyeball; normal refraction focuses an image on the retina (Chapter 9).

Releasing hormones (ree-**LEE**-sing **HOR**-mohns)—Hormones released by the hypothalamus that stimulate secretion of hormones by the anterior pituitary gland (SYN-releasing factors) (Chapter 10).

Remission (ree-**MISH**-uhn)—Lessening of severity of symptoms (Chapter 8).

Renal artery (**REE**-nuhl **AR**-te-ree)—The branch of the abdominal aorta that takes blood into a kidney (Chapter 18).

Renal calculi (**REE**-nuhl **KAL**-kew-lye)—Kidney stones; made of precipitated minerals in the form of crystals (Chapter 18).

Renal corpuscle (**REE**-nuhl **KOR**-pusl)—The part of a nephron that consists of a glomerulus enclosed by a Bowman's capsule; the site of glomerular filtration (Chapter 18).

Renal cortex (**REE**-nuhl **KOR**-teks)—The outermost area of the kidney; consists of renal corpuscles and convoluted tubules (Chapter 18).

Renal failure (**REE**-nuhl **FAYL**-yer)—The inability of the kidneys to function properly and form urine; causes include severe hemorrhage, toxins, and obstruction of the urinary tract (Chapter 18).

Renal fascia (**REE**-nuhl **FASH**-ee-ah)—The fibrous connective tissue membrane that covers the kidneys and the surrounding adipose tissue and helps keep the kidneys in place (Chapter 18).

Renal filtrate (**REE**-nuhl **FILL**-trayt)—The fluid formed from blood plasma by the process of filtration in the renal corpuscles; flows from Bowman's capsules through the renal tubules where most is reabsorbed; the filtrate that enters the renal pelvis is called urine (Chapter 18).

Renal medulla (**REE**-nuhl muh-**DEW**-lah)—The middle area of the kidney; consists of loops of Henle and collecting tubules; the triangular segments of the renal medulla are called renal pyramids (Chapter 18).

Renal pelvis (**REE**-nuhl **PELL**-vis)—The innermost area of the kidney; a cavity formed by the expanded end of the ureter within the medial side of the kidney (Chapter 18).

Renal pyramids (**REE**-nuhl **PEER**-ah-mids)—The triangular segments of the renal medulla; the papillae of the pyramids empty urine into the calyces of the renal pelvis (Chapter 18).

Renal tubule (**REE**-nuhl **TOO**-byool)—The part of a nephron that consists of a proximal convoluted tubule, loop of Henle, distal convoluted tubule, and collecting tubule; the site of tubular reabsorption and tubular secretion (Chapter 18).

Renal vein (**REE**-nuhl VAYN)—The vein that returns blood from a kidney to the inferior vena cava (Chapter 18).

Renin-angiotensin mechanism (**REE**-nin AN-jee-oh-**TEN**-sin)—A series of chemical reactions initiated by a decrese in blood pressure that stimulates the kidneys to secrete the enzyme renin; culminates in the formation of angiotensin II (Chapter 10).

Repolarization (RE-pol-lahr-i-**ZAY**-shun)—The restoration of electrical charges on either side of a cell membrane following depolarization; positive charge outside and a negative charge inside brought about by a rapid outflow of potassium ions (Chapter 7).

Reproductive system (REE-proh-**DUK**-tive **SIS**-tem)—The male or female organ system that produces gametes, ensures fertilization, and in women, provides a site for the developing embryo-fetus (Chapter 20).

Reservoir (**REZ**-er-vwor)—A person or animal who harbors a pathogen and is a source of the pathogen for others (Chapter 22).

Resident flora (**REZ**-i-dent **FLOOR**-uh)—Part of normal flora; those microorganisms on or in nearly everyone in specific body sites nearly all the time (Chapter 22).

Residual air (ree-**ZID**-yoo-al)—The volume of air that remains in the lungs after the most forceful exhalation; important to provide for continuous gas exchange; average: 1000–1500 ml (Chapter 15).

Resistance (re-**ZIS**-tenss)—The total of all of the body's defenses against pathogens; includes non-specific barriers such as unbroken skin and specific mechanisms such as antibody production (Chapter 22).

Respiratory acidosis (RES-pi-rah-**TOR**-ee ass-i-**DOH**-sis)—A condition in which the blood pH is lower than normal, caused by disorders that decrease the rate or efficiency of respiration and permit the accumulation of carbon dioxide (Chapter 15).

Respiratory alkalosis (RES-pi-rah-**TOR**-ee al-kah-**LOH**-sis)—A condition in which the blood pH is higher than normal, caused by disorders that increase the rate of respiration and decrease the level of carbon dioxide in the blood (Chapter 15).

Respiratory pump (**RES**-pi-rah-**TOR**-ee)—A mechanism that increases venous return; pressure changes during breathing compress the veins that pass through the thoracic cavity (Chapter 13).

Respiratory system (**RES**-pi-rah-TOR-ee **SIS**-tem)—The organ system that moves air into and out of the lungs so that oxygen and carbon dioxide may be exchanged between the air and the blood (Chapter 15).

Resting potential (**RES**-ting poh-**TEN**-shul)—The difference in electrical charges on either side of a cell membrane not transmitting an impulse; positive charge outside and a negative charge inside (Chapter 7).

Reticulocyte (re-**TIK**-yoo-loh-sight)—A red blood cell that contains remnants of the ER, an immature stage in red blood cell formation; makes up about 1% of the red blood cells in peripheral circulation (Chapter 11).

Reticuloendothelial system (re-TIK-yoo-loh-en-doh-**THEE**-lee-al)—Former name for the tissue macrophage system, the organs or tissues that contain macrophages which phagocytize old red blood cells; the liver, spleen, and red bone marrow (Chapter 11).

Retina (**RET**-i-nah)—The innermost layer of the eyeball that contains the photoreceptors, the rods and cones (Chapter 9).

Retroperitoneal (RE-troh-PER-i-toh-**NEE**-uhl)—Located behind the peritoneum (Chapter 18).

Rh factor (R-H **FAK**-ter)—The red blood cell types determined by the presence or absence of the Rh (D) antigen on the red blood cell membranes; the two types are Rh positive and Rh negative (Chapter 11).

Rheumatoid arthritis (**ROO**-muh-toyd ar-**THRY**-tiss)—Inflammation of a joint; believed to be an autoimmune disease. The joint damage may progress to fusion and immobility of the joint (Chapter 6).

Rhodopsin (roh-**DOP**-sin)—The chemical in the rods of the retina that breaks down when light waves strike it; this chemical change initiates a nerve impulse (Chapter 9).

RhoGam (**ROH**-gam)—The trade name for the Rh (D) antibody administered to an Rh negative woman who has delivered an Rh positive infant; it will destroy any fetal red blood cells that may have entered maternal circulation (Chapter 11).

Ribosome (**RYE**-boh-sohme)—A cell organelle found in the cytoplasm; the site of protein synthesis (Chapter 3).

Ribs (RIBZ)—The 24 flat bones that, together with the sternum, form the rib cage. The first seven pairs are true ribs, the next three pairs are false ribs, and the last two pairs are floating ribs (Chapter 6).

Rickets (**RIK**-ets)—A deficiency of vitamin D in children, resulting in poor and abnormal bone growth (Chapter 6).

RNA —Ribonucleic acid; a nucleic acid that is a single strand of nucleotides; essential for protein synthesis within cells; messenger RNA (mRNA) is a copy of the genetic code of DNA; transfer RNA (tRNA) aligns amino acids in the proper sequence on the mRNA (Chapter 2).

Rods (RAHDZ)—The sensory receptors in the retina of the eye that detect the presence of light (Chapter 9).

Round window (ROWND **WIN**-doh)—The membrane-covered opening of the inner ear that bulges to prevent pressure damage to the organ of Corti (Chapter 9).

Rugae (**ROO**-gay)—Folds of the mucosa of organs such as the stomach, urinary bladder, and vagina; permit expansion of these organs (Chapter 16).

—S—

Saccule (**SAK**-yool)—A membranous sac in the vestibule of the inner ear that contains receptors for static equilibrium (Chapter 9).

Sacrum (**SAY**-krum)—The five fused sacral vertebrae at the base of the spine (Chapter 6).

Sacroiliac joint (SAY-kroh-**ILL**-ee-ak)—The slightly movable joint between the sacrum and the ilium (Chapter 6).

Saddle joint (**SA**-duhl)—The carpometacarpal joint of the thumb, a diarthrosis (Chapter 6).

Sagittal section (**SAJ**-i-tuhl **SEK**-shun)—A plane or cut from front to back, separating right and left parts (from the Latin "arrow") (Chapter 1).

Saliva (sah-**LYE**-vah)—The secretion of the salivary glands; mostly water and containing the enzyme amylase (Chapter 16).

Salivary glands (**SAL**-i-va-ree)—The three pairs of exocrine glands that secrete saliva into the oral cavity; parotid, submandibular, and sublingual pairs (Chapter 16).

Salt (SAWLT)—A chemical compound that consists of a positive ion other than hydrogen and a negative ion other than hydroxyl (Chapter 2).

Saltatory conduction (**SAWL**-tah-taw-ree kon-**DUK**-shun)—The rapid transmission of a nerve impulse from one node of Ranvier to the next; characteristic of myelineated neurons (Chapter 8).

Saprophyte (**SAP**-roh-fight)—An organism that lives on dead organic matter; a decomposer (Chapter 22).

Sarcolemma (SAR-koh-**LEM**-ah)—The cell membrane of a muscle fiber (Chapter 7).

Sarcomere (**SAR**-koh-meer)—The unit of contraction in a skeletal muscle fiber; a precise arrangement of myosin and actin filaments between two Z lines (Chapter 7).

Sarcoplasmic reticulum (SAR-koh-**PLAZ**-mik re-**TIK**-yoo-lum)—The endoplasmic reticulum of a muscle fiber; is a reservoir for calcium ions (Chapter 7).

Saturated fat (**SAT**-uhr-ay-ted)—A true fat that is often solid at room temperature and of animal origin (Chapter 2).

Scapula (**SKAP**-yoo-luh)—The flat bone of the shoulder that articulates with the humerus (Syn.—shoulderblade) (Chapter 6).

Schwann cell (SHWAHN SELL)—A cell of the peripheral nervous system that forms the myelin sheath and neurolemma of peripheral axons and dendrites (Chapter 4).

Sclera (**SKLER**-ah)—The outermost layer of the eyeball, made of fibrous connective tissue; the anterior portion is the transparent cornea (Chapter 9).

Scoliosis (SKOH-lee-**OH**-sis)—A lateral curvature of the vertebral column (Chapter 6).

Scrotum (**SKROH**-tum)—The sac of skin between the upper thighs in males; contains the testes, epididymides, and part of the ductus deferens (Chapter 20).

Sebaceous gland (suh-**BAY**-shus)—An exocrine gland in the dermis that produces sebum (Chapter 5).

Sebum (**SEE**-bum)—The lipid (oil) secretion of sebaceous glands (Chapter 5).

Secondary infection (**SEK**-un-DAR-ee)—An infection made possible by a primary infection that has lowered the host's resistance (Chapter 22).

Secondary sex characteristics (**SEK**-un-DAR-ee SEKS)—The features that develop at puberty in males or females; they are under the influence of the sex hormones but are not directly involved in reproduction. Examples are growth of facial or body hair and growth of muscles.

Secretin (se-**KREE**-tin)—A hormone secreted by the duodenum when food enters; stimulates secretion of bile by the liver and secretion of bicarbonate pancreatic juice (Chapter 16).

Secretion (see-**KREE**-shun)—The production and release of a cellular product with a useful purpose (Chapter 4).

Section (**SEK**-shun)—The cutting of an organ or the body to make internal structures visible (Chapter 1).

Selectively permeable (se-**LEK**-tiv-lee **PER**-me-uh-buhl)—A characteristic of cell membranes; permits the passage of some materials but not of others (Chapter 3).

Self-limiting disease (sellf-**LIM**-i-ting)—A disease that typically lasts a certain period of time and is followed by recovery (Chapter 22).

Semen (**SEE**-men)—The thick, alkaline fluid that contains sperm and the secretions of the seminal vesicles, prostate gland, and bulbourethral glands (Chapter 20).

Semicircular canals (SEM-eye-**SIR**-kyoo-lur)—Three oval canals in the inner ear that contain the receptors that detect motion (Chapter 9).

Seminal vesicles (**SEM**-i-nuhl **VESS**-i-kulls)—The glands located posterior to the prostate gland and inferior to the urinary bladder; secrete an alkaline fluid that enters the ejaculatory ducts and becomes part of semen (Chapter 20).

Seminiferous tubules (sem-i-**NIFF**-er-us)—The site of spermatogenesis in the testes (Chapter 20).

Sensation (sen-**SAY**-shun)—A feeling or awareness of conditions outside or inside the body, resulting from the stimulation of sensory receptors (Chapter 9).

Sensory neuron (**SEN**-suh-ree **NYOOR**-on)—A nerve cell that carries impulses from a receptor to the central nervous system. (Syn.—afferent neuron) (Chapter 8).

Septic shock (**SEP**-tik SHAHK)—A type of circulatory shock that is a consequence of a bacterial infection (Chapter 13)

Septicemia (SEP-tih-**SEE**-mee-ah)—The presence of bacteria in the blood (Chapter 5).

Septum (**SEP**-tum)—A wall that separates two cavities, such as the nasal septum between the nasal cavities or the interventricular septum between the two ventricles of the heart (Chapter 12).

Serous fluid (**SEER**-us **FLOO**-id)—A fluid that prevents friction between the two layers of a serous membrane (Chapter 4).

Serous membrance (**SEER**-us **MEM**-brayn)—An epithelial membrane that lines a closed body cavity and covers the organs in that cavity (Chapter 4).

Sex chromosomes (**SEKS** KROH-muh-sohms)—The pair of chromosomes that determines the gender of an individual; designated XX in females and XY in males (Chapter 21).

Sex-linked trait (**SEKS** LINKED TRAYT)—A genetic characteristic in which the gene is located on the X chromosome (Chapter 7).

Simple (**SIM**-puhl)—Having only one layer, used especially to describe certain types of epithelial tissue (Chapter 4).

Sinoatrial (SA) node (**SIGH**-noh-AY-tree-al NOHD)—The first part of the cardiac conduction pathway, located in the wall of the right atrium; initiates each heartbeat (Chapter 12).

Sinusoid (SIGH-nuh-soyd)—A large, very permeable capillary; permits proteins or blood cells to enter or leave the blood (Chapter 13).

Skeletal muscle pump (**SKEL**-e-tuhl **MUSS**-uhl)—A mechanism that increases venous return; contractions of the skeletal muscles compress the deep veins, especially those of the legs (Chapter 13).

Skeletal system (**SKEL**-e-tuhl)—The organ system that consists of the bones, ligaments, and cartilage; supports the body and is a framework for muscle attachment (Chapter 6).

Skin (SKIN)—An organ that is part of the integumentary system; consists of the outer epidermis and the inner dermis (Chapter 5).

Sliding filament theory (**SLY**-ding **FILL**-ah-ment)—The sequence of events that occurs within sarcomeres when a muscle fiber contracts (Chapter 7).

Small intestine (SMAWL in-**TESS**-tin)—The organ of the alimentary tube between the stomach and the large intestine; secretes enyzmes that complete the digestive process and absorbs the end products of digestion (Chapter 16).

Smooth muscle (SMOOTH **MUSS**-uhl)—The muscle tissue that forms the walls of hollow internal organs. Also called visceral or involuntary muscle (Chapter 4).

Sneeze reflex (SNEEZ)—A reflex integrated by the medulla that expels irritating substances from the nasal cavities by means of an explosive exhalation (Chapter 15).

Sodium-potassium pumps (**SEW**-dee-um pa-**TASS**-ee-um)—The active transport mechanisms that maintain a high sodium ion concentration outside the cell and a high potassium ion concentration inside the cell (Chapter 7).

Soft palate (SAWFT **PAL**-uht)—The posterior portion of the palate that is elevated during swallowing to block the nasopharynx (Chapter 15).

Solute (**SAH**-loot)—The substance that is dissolved in a solution (Chapter 3).

Solution (suh-**LOO**-shun)—The dispersion of one or more compounds (solutes) in a liquid (solvent) (Chapter 2).

Solvent (**SAHL**-vent)—A liquid in which substances (solutes) will dissolve (Chapter 2).

Somatic (soh-**MA**-tik)—Pertaining to structures of the body wall, such as skeletal muscles and the skin (Chapter 8).

Somatostatin (GHIH) (SOH-mat-oh-**STAT**-in)—Growth hormone inhibiting hormone; produced by the hypothalamus (Chapter 10).

Somatotropin (SOH-mat-oh-**TROH**-pin)—Growth hormone (Chapter 10).

Specialized fluids (**SPEH**-shul-eyezd **FLUIDS**)—Specific compartments of extracellular fluid (ECF) which include cerebrospinal fluid, synovial fluid, aqueous humor in the eye, and others.

Spermatic cord (sper-**MAT**-ik KORD)—The cord that suspends the testis; composed of the ductus deferens, blood vessels, and nerves (Chapter 20).

Spermatogenesis (SPER-ma-toh-**JEN**-e-sis)—The process of meiosis in the testes to produce sperm cells (Chapter 3).

Spermatozoa (sper-MAT-oh-**ZOH**-ah)—Sperm cells; produced by the testes (Sing.—spermatozoon) (Chapter 20).

Sphenoid bone (**SFEE**-noyd)—The flat bone that forms part of the anterior floor of the cranial cavity and encloses the pituitary gland (Chapter 6).

Sphincter (**SFINK**-ter)—A circular muscle that regulates the size of an opening (Chapter 7).

Spinal cavity (**SPY**-nuhl **KAV**-i-tee)—The cavity within the vertebral column that contains the spinal cord; part of the dorsal cavity (Syn.—vertebral canal or cavity) (Chapter 1).

Spinal cord (**SPY**-nuhl KORD)—The part of the central nervous system within the vertebral canal; transmits impulses to and from the brain (Chapter 8).

Spinal cord reflex (**SPY**-nuhl KORD **REE**-fleks)—A reflex integrated in the spinal cord, in which the brain is not directly involved (Chapter 8).

Spinal nerves (**SPY**-nuhl NERVS)—The 31 pairs of nerves that emerge from the spinal cord (Chapter 8).

Spinal shock (**SPY**-nuhl SHAHK)—The temporary or permanent loss of spinal cord reflexes following injury to the spinal cord (Chapter 8).

Spirillum (spih-**RILL**-uhm) (Pl.: spirilla)—A bacterium with a spiral shape (Chapter 22).

Spirochaete (**SPY**-roh-keet)—Spiral bacteria of the order Spirochaetales (Chapter 22).

Spleen (SPLEEN)—An organ located in the upper left abdominal quadrant behind the stomach; consists of lymphatic tissue that produces lymphocytes and monocytes; also contains macrophages that phagocytize old red blood cells (Chapter 14).

Spongy bone (**SPUN**-jee BOWNE)—Bone tissue not organized into Haversian systems; forms most of short, flat, and irregular bones and forms epiphyses of long bones (Chapter 6).

Spontaneous fracture (spahn-**TAY**-nee-us)—A fracture that occurs without apparent trauma; often a consequence of osteoporosis (Chapter 6).

Spore (**SPOOR**)—1. A bacterial form that is dormant and highly resistant to environmental extremes such as heat. 2. A unicellular fungal reproductive form (Chapter 22).

Squamous (**SKWAY**-mus)—Flat or scale-like; used especially in reference to epithelial tissue (Chapter 4).

Stapes (**STAY**-peez)—The third of the auditory bones in the middle ear; transmits vibrations from the incus to the oval window of the inner ear (Chapter 9).

Starling's Law of the Heart (**STAR**-lingz LAW)—The force of contraction of cardiac muscle fibers is determined by the length of the fibers; the more cardiac muscle fibers are stretched, the more forcefully they contract (Chapter 12).

Stem cell (STEM SELL)—The immature cell found in red bone marrow and lymphatic tissue that is the precursor cell for all the types of blood cells (Chapter 11).

Stenosis (ste-**NOH**-sis)—An abnormal constriction or narrowing of an opening or duct (Chapter 12).

Sterilization (STIR-ill-i-**ZAY**-shun)—The process of completely destroying all of the microorganisms on or in a substance or object (Chapter 22).

Sternum (**STIR**-num)—The flat bone that forms part of the anterior rib cage; consists of the manubrium, body, and xiphoid process (Syn.—breastbone) (Chapter 6).

Steroid (**STEER**-oyd)—An organic compound in the lipid group; includes cholesterol and certain hormones (Chapter 2).

Stimulus (**STIM**-yoo-lus)—A change, especially one that affects a sensory receptor or which brings about a response in a living organism (Chapter 9).

Stomach (**STUM**-uk)—The sac-like organ of the alimentary tube between the esophagus and the small intestine; is a reservoir for food and secretes gastric juice to begin protein digestion (Chapter 16).

Stratified (**STRA**-ti-fyed)—Having two or more layers (Chapter 4).

Stratum corneum (**STRA**-tum **KOR**-nee-um)—The outermost layer of the epidermis, made of many layers of dead, keratinized cells (Chapter 5).

Stratum germinativum (**STRA**-tum JER-min-ah-**TEE**-vum)—The innermost layer of the epidermis; the cells undergo mitosis to produce new epidermis (Chapter 5).

Streptokinase (STREP-toh-**KYE**-nase)—An enzyme produced by bacteria of the genus *Streptococcus* that is used clinically to dissolve abnormal clots, such as those in coronary arteries (Chapter 11).

Stretch receptor (STRETCH ree-**SEP**-ter)—A sensory receptor in a muscle that detects stretching of the muscle (Syn.—proprioceptor) (Chapter 7).

Stretch reflex (STRETCH **REE**-fleks)—A spinal cord reflex in which a muscle that is stretched will contract (Chapter 8).

Striated muscle (**STRY**-ay-ted **MUSS**-uhl)—The muscle tissue that forms the skeletal muscles that move bones (Chapter 4).

Stroke volume (STROHK **VAHL**-yoom)—The amount of blood pumped by a ventricle in one beat; the resting average is 60–80 ml/beat (Chapter 12).

Subarachnoid space (SUB-uh-**RAK**-noid)—The space between the arachnoid membrane and the pia mater; contains cerebrospinal fluid (Chapter 8).

Subclinical infection (sub-**KLIN**-i-kuhl)—An infection in which the person shows no symptoms (Chapter 22.)

Subcutaneous (SUB-kew-**TAY**-nee-us)—Below the skin; the tissues between the dermis and the muscles (Chapter 5).

Sublingual glands (sub-**LING**-gwal)—The pair of salivary glands located below the floor of the mouth (Chapter 16).

Submandibular glands (SUB-man-**DIB**-yoo-lar)—The pair of salivary glands located at the posterior corners of the mandible (Chapter 16).

Submucosa (SUB-mew-**KOH**-sah)—The layer of connective tissue and blood vessels located below the mucosa (lining) of a mucous membrane (Chapter 16).

Substrates (**SUB**-strayts)—The substances acted upon, as by enzymes (Chapter 2).

Sucrase (**SOO**-krays)—A digestive enzyme that breaks down sucrose to glucose and fructose; secreted by the small intestine (Chapter 16).

Sucrose (**SOO**-krohs)—A disaccharide made of one glucose and one fructose molecule (Syn.—cane sugar, table sugar) (Chapter 2).

Sulcus (**SUHL**-kus)—A furrow or groove, as between the gyri of the cerebrum (Syn.—fissure) (Chapter 8).

Superficial (soo-per-**FISH**-uhl)—Toward the surface (Chapter 1).

Superficial fascia (soo-per-**FISH**-uhl **FASH**-ee-ah)—The subcutaneous tissue, between the dermis and the muscles. Consists of areolar connective tissue and adipose tissue (Chapter 4).

Superior (soo-**PEER**-ee-your)—Above, or higher (Chapter 1).

Superior vena cava (**VEE**-nah **KAY**-vah)—The vein that returns blood from the upper body to the right atrium (Chapter 12).

Suspensory ligaments (suh-**SPEN**-suh-ree **LIG**-uh-ments)—The strands of connective tissue that connect the ciliary body to the lens of the eye (Chapter 9).

Suture (**SOO**-cher)—A synarthrosis, an immovable joint between cranial bones or facial bones (from the Latin "seam") (Chapter 6).

Supination (SOO-pi-**NAY**-shun)—Turning the palm upward, or lying face up (Chapter 7).

Sympathetic (SIM-puh-**THET**-ik)—The division of the autonomic nervous system that dominates during stressful situations (Chapter 8).

Sympathomimetic (SIM-pah-tho-mi-**MET**-ik)—Having the same effects as sympathetic impulses, as has epinephrine, a hormone of the adrenal medulla (Chapter 10).

Symphysis (**SIM**-fi-sis)—An amphiarthrosis in which a disc of cartilage is found between two bones, as in the vertebral column (Chapter 6).

Symptomatic infection (**SIMP**-toh-MAT-ik)—An infection in which the patient exhibits the symptoms of the disease (Chapter 22).

Synapse (**SIN**-aps)—The space between the axon of one neuron and the cell body or dendrite of the next neuron or between the end of a motor neuron and an effector cell (Chapter 4).

Synaptic knob (si-**NAP**-tik NAHB)—The end of an axon of a neuron that releases a neurotransmitter (Chapter 8).

Synarthrosis (SIN-ar-**THROH**-sis)—An immovable joint, such as a suture (Chapter 6).

Synergistic muscles (SIN-er-**JIS**-tik **MUSS**-uhls)—Muscles that have the same function, or a stabilizing function, with respect to the movement of a joint (Chapter 7).

Synovial fluid (sin-**OH**-vee-uhl **FLOO**-id)—A thick slippery fluid that prevents friction within joint cavities (Chapter 6).

Synovial membrane (sin-**OH**-ve-uhl **MEM**-brayn)—The connective tissue membrane that lines joint cavities and secretes synovial fluid (Chapter 4).

Synthesis (**SIN**-the-siss)—The process of forming complex molecules or compounds from simpler compounds or elements (Chapter 2).

Syphilis (**SIFF**-i-lis)—A sexually transmitted disease caused by the bacterium *Treponema pallidum;* may also cause congenital syphilis in newborns of infected women (Chapter 20).

Systemic infection (sis-**TEM**-ik)—An infection that has spread throughout the body from an initial site (Chapter 22).

Systole (**SIS**-tuh-lee)—In the cardiac cycle, the contraction of the myocardium; ventricular systole pumps blood into the arteries (Chapter 12).

—T—

T cell (T SELL)—A subgroup of lymphocytes; include helper T cells, cytotoxic T cells, and suppressor T cells, all of which are involved in immune responses (Chapter 11).

Tachycardia (TAK-ee-**KAR**-dee-yah)—An abnormally rapid heart rate; more than 100 beats per minute (Chapter 12).

Taenia coli (**TAY**-nee-uh **KOH**-lye)—The longitudinal muscle layer of the colon; three bands of smooth muscle fibers that extend from the cecum to the sigmoid colon (Chapter 16).

Talus (**TAL**-us)—One of the tarsals; articulates with the tibia (Chapter 6).

Target organ (**TAR**-get **OR**-gan)—The organ (or tissue) in which a hormone exerts its specific effects. (Chapter 10).

Tarsals (**TAR**-suhls)—The seven short bones in each ankle (Chapter 6).

Taste buds (TAYST BUDS)—Structures on the papillae of the tongue that contain the chemoreceptors for the detection of chemicals (food) dissolved in saliva (Chapter 9).

Tears (TEERS)—The watery secretion of the lacrimal glands; wash the anterior surface of the eyeball and keep it moist (Chapter 9).

Teeth (TEETH)—Bony projections in the upper and lower jaws that function in chewing (Chapter 16).

Telophase (**TELL**-ah-fayz)—The fourth stage of mitosis, in which two nuclei are reformed (Chapter 3).

Temporal bone (**TEM**-puh-ruhl)—The flat bone that forms the side of the cranial cavity and contains middle and inner ear structures (Chapter 6).

Temporal lobes (**TEM**-puh-ruhl LOWBS)—The lateral parts of the cerebrum; contain the auditory, olfactory, and taste areas (Chapter 8).

Tendon (**TEN**-dun)—A fibrous connective tissue structure that connects muscle to bone (Chapter 7).

Teratogen (te-**RAH**-toh-jen)—Anything that causes developmental abnormalities in an embryo; may be a chemical or microorganism to which an embryo is exposed by way of the mother (Chapter 21).

Testes (**TES**-teez)—The male gonads that produce sperm cells; also endocrine glands that secrete the hormone testosterone (Sing.—testis) (Chapter 10).

Testosterone (tes-**TAHS**-ter-ohn)—The sex hormone secreted by the interstitial cells of the testes; responsible for the growth of the male reproductive organs and the secondary sex characteristics (Chapter 10).

Tetanus (**TET**-uh-nus)—1. A sustained contraction of a muscle fiber in response to rapid nerve impulses. 2. A disease, characterized by severe muscle spasms, caused by the bacterium *Clostridium tetani* (Chapter 7).

Thalamus (**THAL**-uh-muss)—The part of the brain superior to the hypothalamus; regulates subconscious aspects of sensation (Chapter 8).

Theory (**THEER**-ree)—A statement that is the best explanation of all the available evidence on a particular action or mechanism. A theory is *not* a guess (Chapter 3).

Thoracic cavity (thaw-**RASS**-ik **KAV**-i-tee)—Part of the ventral cavity, superior to the diaphragm (Chapter 1).

Thoracic duct (thaw-**RASS**-ik DUKT)—The lymph vessel that empties lymph from the lower half and upper left quadrant of the body into the left subclavian vein (Chapter 14).

Thoracic vertebrae (thaw-**RASS**-ik **VER**-te-bray)—The 12 vertebrae that articulate with the ribs (Chapter 6).

Threshold level–renal (**THRESH**-hold **LE**-vuhl)—The concentration at which a substance in the blood *not* normally excreted by the kidneys begins to appear in the urine; for several substances, such as glucose, in the renal filtrate, there is a limit to how much the renal tubules can reabsorb (Chapter 18).

Thrombocyte (**THROM**-boh-sight)—Platelet, a fragment of a megakaryocyte (Chapter 11).

Thrombocytopenia (THROM-boh-SIGH-toh-**PEE**-nee-ah)—An abnormally low platelet count (Chapter 11).

Thrombus (**THROM**-bus)—A blood clot that obstructs blood flow through a blood vessel (Chapter 11).

Thymus (**THIGH**-mus)—An organ made of lymphatic tissue located inferior to the thyroid gland; large in the fetus and child, and shrinks with age; produces T cells and hormones necessary for the maturation of the immune system (Chapter 14).

Thyroid cartilage (**THIGH**-roid **KAR**-ti-ledj)—The largest and most anterior cartilage of the larynx; may be felt in the front of the neck (Chapter 15).

Thyroid gland (**THIGH**-roid)—An endocrine gland on the anterior side of the trachea below the larynx; secretes thyroxine, triiodothyronine, and calcitonin (Chapter 10).

Thyroid-stimulating hormone (TSH) —A hormone secreted by the anterior pituitary gland that causes the thyroid gland to secrete triiodothyronine, and T_3 (Chapter 10).

Thyroxine (T_4) (thigh-**ROK**-sin)—A hormone secreted by the thyroid gland that increases energy production and protein synthesis (Chapter 10).

Tibia (**TIB**-ee-yuh)—The larger long bone of the lower leg (Syn.—shinbone) (Chapter 6).

Tidal volume (**TIGH**-duhl **VAHL**-yoom)—The volume of air in one normal inhalation and exhalation; average: 400-600 ml (Chapter 15).

Tissue (**TISH**-yoo)—A group of cells with similar structure and function (Chapter 1).

Tissue fluid (**TISH**-yoo **FLOO**-id)—The water found in intercellular spaces. Also called interstitial fluid (Chapter 2).

Tissue macrophage system (**TISH**-yoo **MAK**-roh-fayj)—The organs or tissues that contain macrophages which phagocytize old red blood cells: the liver, spleen, and red bone marrow (Chapter 11).

Tissue typing (**TISH**-yoo **TIGH**-ping)—A laboratory procedure that determines the HLA types of a donated organ, prior to an organ transplant (Chapter 11).

Titin (**TIGH**-tin)—The protein in sarcomeres that anchors myosin filaments to the Z lines (Chapter 7).

Tongue (TUHNG)—A muscular organ on the floor of the oral cavity; contributes to chewing and swallowing and contains taste buds (Chapter 16).

Tonsillectomy (TAHN-si-**LEK**-toh-mee)—The surgical removal of the palatine tonsils and/or adenoid (Chapter 14).

Tonsils (**TAHN**-sills)—The lymph nodules in the mucosa of the pharynx, the palatine tonsils, and the adenoid; also the lingual tonsils on the base of the tongue (Chapter 14).

Toxin (**TAHK**-sin)—A chemical that is poisonous to cells (Chapter 22).

Toxoid (**TAHK**-soyd)—An inactivated bacterial toxin that is no longer harmful yet is still antigenic; used as a vaccine (Chapter 14).

Trace element (TRAYS **EL**-uh-ment)—Those elements needed in very small amounts by the body for normal functioning (Chapter 2).

Trachea (**TRAY**-kee-ah)—The organ that is the air passageway between the larynx and the primary bronchi (Syn.—windpipe) (Chapter 15).

Transamination (TRANS-am-i-**NAY**-shun)—The transfer of an amino (NH_2) group from an amino acid to a carbon chain to form a non-essential amino acid; takes place in the liver (Chapter 16).

Transient flora (**TRAN**-zee-ent **FLOOR**-uh)—Part of normal flora; those microorganisms that may inhabit specific sites in the body for short periods of time (Chapter 22).

Transitional (trans-**ZI**-shun-uhl)—Changing from one form to another (Chapter 4).

Transitional epithelium (tran-**ZI**-shun-uhl)—A type of epithelium in which the surface cells change from rounded to flat as the organ changes shape (Chapter 4).

Transverse section (trans-**VERS SEK**-shun)—A plane or cut from front to back, separating upper and lower parts (Chapter 1).

Tricuspid valve (try-**KUSS**-pid VALV)—The right AV valve, which prevents backflow of blood from the right ventricle to the right atrium when the ventricle contracts (Chapter 12).

Trigeminal nerves (try-**JEM**-in-uhl)—Cranial nerve pair V; sensory for the face and teeth; motor to chewing muscles (Chapter 8).

Trigone (**TRY**-gohn)—Triangular area on the floor of the urinary bladder bounded by the openings of the two ureters and the urethra (Chapter 18).

Triglyceride (try-**GLI**-si-ryde)—An organic compound, a true fat, that is made of one glycerol and three fatty acids (Chapter 2).

Triiodothyronine (T_3) (TRY-eye-oh-doh-**THIGH**-roh-neen)—A hormone secreted by the thyroid gland that increases energy production and protein synthesis (Chapter 10).

Trisomy (**TRY**-suh-mee)—In genetics, having three homologous chromosomes instead of the usual two (Chapter 20).

Trochlear nerves (**TROK**-lee-ur)—Cranial nerve pair IV; motor to an extrinsic muscle of the eye (Chapter 8).

Trophoblast (**TROH**-foh-blast)—The outermost layer of the embryonic blastocyst; will become the chorion, one of the embryonic membranes (Chapter 21).

Tropomyosin (TROH-poh-**MYE**-oh-sin)—A protein that inhibits the contraction of sarcomeres in a muscle fiber (Chapter 7).

Troponin (**TROH**-poh-nin)—A protein that inhibits the contraction of the sarcomeres in a muscle fiber (Chapter 7).

True fat (TROO FAT)—An organic compound in the lipid group that is made of glycerol and fatty acids (Chapter 2).

Trypsin (**TRIP**-sin)—A digestive enzyme that breaks down proteins into polypeptides; secreted by the pancreas (Chapter 16).

Tubal ligation (**TOO**-buhl lye-**GAY**-shun)—A surgical procedure to remove or sever the fallopian tubes; usually done as a method of contraception in women (Chapter 20).

Tubular reabsorption (**TOO**-byoo-ler REE-ab-**SORP**-shun)—The processes by which useful substances in the renal filtrate are returned to the blood in the peritubular capillaries (Chapter 18).

Tubular secretion (**TOO**-byoo-ler se-**KREE**-shun)—The processes by which cells of the renal tubules secrete substances into the renal filtrate to be excreted in urine (Chapter 18).

Tunica (**TOO**-ni-kah)—A layer or coat, as in the wall of an artery (Chapter 13).

Tympanic membrane (tim-**PAN**-ik)—The eardrum, the membrane that is stretched across the end of the ear canal; vibrates when sound waves strike it (Chapter 9).

Typing and cross matching (**TIGH**-ping and **KROSS**-match-ing)—A laboratory test that determines whether or not donated blood is compatible, with respect to the red blood cell types.

—U—

Ulna (**UHL**-nuh)—The long bone of the forearm on the little finger side (Chapter 6).

Ultrasound (**UHL**-tra-sownd)—1. Inaudible sound. 2. A technique used in diagnosis in which ultrasound waves provide outlines of the shapes of organs or tissues (Chapter 21).

Umbilical arteries (uhm-**BILL**-i-kull **AR**-tuh-rees)—The fetal blood vessels contained in the umbilical cord that carry deoxygenated blood from the fetus to the placenta (Chapter 13).

Umbilical cord (um-**BILL**-i-kull KORD)—The structure that connects the fetus to the placenta; contains two umbilical arteries and one umbilical vein (Chapter 13).

Umbilical vein (uhm-**BILL**-i-kull VAYN)—The fetal blood vessel contained in the umbilical cord that carries oxygenated blood from the placenta to the fetus (Chapter 13).

Unicellular (YOO-nee-**SELL**-yoo-lar)—Composed of one cell (Chapter 4).

Unsaturated fat (un-**SAT**-uhr-ay-ted)—A true fat that is often liquid at room temperature; of plant origin (Chapter 2).

Upper respiratory tract (**UH**-per **RES**-pi-rah-TOR-ee TRAKT)—The respiratory organs located outside the chest cavity (Chapter 15).

Urea (yoo-**REE**-ah)—A nitrogenous waste product formed in the liver from the deamination of amino acids or from ammonia (Chapter 5).

Uremia (yoo-**REE**-me-ah)—The condition in which blood levels of nitrogenous waste products are elevated; caused by renal insufficiency or failure (Chapter 18).

Ureter (**YOOR**-uh-ter)—The tubular organ that carries urine from the renal pelvis (kidney) to the urinary bladder (Chapter 18).

Urethra (yoo-**REE**-thrah)—The tubular organ that carries urine from the urinary bladder to the exterior of the body (Chapter 18).

Urinary bladder (**YOOR**-i-NAR-ee **BLA**-der)—The organ that stores urine temporarily and contracts to eliminate urine by way of the urethra (Chapter 18).

Urinary system (**YOOR**-i-NAR-ee **SIS**-tem)—The organ system that produces and eliminates urine; consists of the kidneys, ureters, urinary bladder, and urethra (Chapter 18).

Urine (**YOOR**-in)—The fluid formed by the kidneys from blood plasma (Chapter 18).

Uterus (**YOO**-ter-us)—The organ of the female reproductive system in which the placenta is formed to nourish a developing embryo-fetus (Chapter 20).

Utricle (**YOO**-tri-kuhl)—A membranous sac in the vestibule of the inner ear that contains receptors for static equilibrium (Chapter 9).

—V—

Vaccine (vak-**SEEN**)—A preparation of a foreign antigen that is administered by injection or other means in order to stimulate an antibody response to provide immunity to a particular pathogen (Chapter 14).

Vagina (vuh-**JIGH**-nah)—The muscular tube that extends from the cervix of the uterus to the vaginal orifice; serves as the birth canal (Chapter 20).

Vagus nerves (**VAY**-gus)—Cranial nerve pair X; sensory for cardiovascular and respiratory reflexes; motor to larynx, bronchioles, stomach, and intestines (Chapter 8).

Valence (**VAY**-lens)—The combining power of an atom when compared to a hydrogen atom; expressed as a positive or negative number (Chapter 2).

Varicose vein (**VAR**-i-kohs VAYN)—An enlarged, abnormally dilated vein; most often occurs in the legs (Chapter 13).

Vasectomy (va-**SEK**-tuh-me)—A surgical procedure to remove or sever the ductus deferens; usually done as a method of contraception in men (Chapter 20).

Vasoconstriction (VAY-zoh-kon-**STRIK**-shun)—A decrease in the diameter of a blood vessel caused by contraction of the smooth muscle in the wall of the vessel (Chapter 5).

Vasodilation (VAY-zoh-dye-**LAY**-shun)—An increase in the diameter of a blood vessel caused by relaxation of the smooth muscle in the wall of the vessel (Chapter 5).

Vector (**VEK**-ter)—An arthropod that transmits pathogens from host to host, usually when it bites to obtain blood (Chapter 22).

Vein (VAYN)—A blood vessel that takes blood from capillaries back to the heart (Chapter 13).

Venous return (**VEE**-nus ree-**TURN**)—The amount of blood returned by the veins to the heart; is directly related to cardiac output, which depends on adequate venous return (Chapter 12).

Ventilation (**VEN**-ti-**LAY**-shun)—The movement of air into and out of the lungs (Chapter 15).

Ventral (**VEN**-truhl)—Toward the front (Syn.—anterior) (Chapter 1).

Ventral cavity (**VEN**-truhl **KAV**-i-tee)—Cavity that consists of the thoracic, abdominal, and pelvic cavities (Chapter 1).

Ventral root (**VEN**-truhl ROOT)—The motor root of a spinal nerve (Chapter 8).

Ventricle (**VEN**-tri-kul)—1. A cavity, such as the four ventricles of the brain that contain cerebrospinal fluid. 2. One of the two lower chambers of the heart that pump blood to the body or to the lungs (Chapter 8).

Venule (**VEN**-yool)—A small vein (Chapter 13).

Vertebra (**VER**-te-brah)—One of the bones of the spine or backbone (Chapter 6).

Vertebral canal (**VER**-te-brahl ka-**NAL**)—The spinal cavity that contains and protects the spinal cord (Chapter 6).

Vertebral column (**VER**-te-brahl **KAH**-luhm)—The spine or backbone (Chapter 6).

Vestibule (**VES**-ti-byool)—1. The bony chamber of the inner ear that contains the utricle and saccule (Chapter 9). 2. The female external genital area between the labia minor that contains the openings of the urethra, vagina, and Bartholin's glands (Chapter 20).

Vestigial organ (ves-**TIJ**-ee-uhl)—An organ that is reduced in size and function when compared with that of evolutionary ancestors; includes the appendix, ear muscles that move the auricle, and wisdom teeth (Chapter 16).

Villi (**VILL**-eye)—1. Folds of the mucosa of the small intestine that increase the surface area for absorption; each villus contains a capillary network and a lacteal (Chapter 16). 2. Projections of the chorion, an embryonic membrane that forms the fetal portion of the placenta (Chapter 21).

Virulence (**VIR**-yoo-lents)—The ability of a microorganism to cause disease; the degree of pathogenicity (Chapter 22).

Virus (**VIGH**-rus)—The simplest type of microorganism, consisting of either DNA or RNA within a protein shell; all are obligate intracellular parasites (Chapter 14).

Visceral (**VISS**-er-uhl)—Pertaining to organs within a body cavity, especially thoracic and abdominal organs (Chapter 8).

Visceral effectors (**VISS**-er-uhl e-**FEK**-turs)—Smooth muscle, cardiac muscle, and glands; receive motor nerve fibers of the autonomic nervous system; responses are involuntary (Chapter 8).

Visceral muscle (**VIS**-ser-uhl **MUSS**-uhl)—Another name for smooth muscle tissue (Chapter 4).

Vital capacity (**VY**-tuhl kuh-**PASS**-i-tee)—The volume of air involved in the deepest inhalation followed by the most forceful exhalation; average: 3500-5000 ml (Chapter 15).

Vitamin (**VY**-tah-min)—An organic molecule needed in small amounts by the body for normal metabolism or growth (Chapter 17).

Vitreous humor (**VIT**-ree-us **HYOO**-mer)—The semisolid, gelatinous substance in the posterior cavity of the eyeball; helps keep the retina in place (Chapter 9).

Vocal cords (**VOH**-kul KORDS)—The pair of folds within the larynx that are vibrated by the passage of air, producing sounds that may be turned into speech (Chapter 15).

Voluntary muscle (**VAHL**-un-tary **MUSS**-uhl)—Another name for striated or skeletal muscle tissue (Chapter 4).

Vomiting (**VAH**-mi-ting)—Ejection through the mouth of stomach and intestinal contents (Chapter 16).

Vulva (**VUHL**-vah)—The female external genital organs (Chapter 20).

—W-X-Y-Z—

Wart (WART)—An elevated, benign skin lesion caused by a virus (Chapter 5).

White blood cells (WIGHT BLUHD SELLS)—The cells that destroy pathogens that enter the body and provide immunity to some diseases; the five types are neutrophils, eosinophils, basophils, lymphocytes, and monocytes (Syn.—leukocytes) (Chapter 4).

White matter (WIGHT)—Nerve tissue within the central nervous system that consists of myelinated axons and dendrites of interneurons (Chapter 8).

Worm (WURM)—An elongated invertebrate; parasitic worms include tapeworms and hookworm (Chapter 22).

Xiphoid process (**ZYE**-foyd)—The most inferior part of the sternum (from the Greek "sword-like") (Chapter 6).

Yellow bone marrow (**YELL**-oh BOWN **MAR**-roh)—Primarily adipose tissue, found in the marrow cavities of the diaphyses of long bones and in the spongy bone of the epiphyses of adult bones (Chapter 6).

Yolk sac (YOHK SAK)—An embryonic membrane that forms the first blood cells for the developing embryo (Chapter 21).

Z line (ZEE LYEN)—The lateral boundary of a sarcomere in muscle tissue (Chapter 7).

Zoonoses (ZOH-oh-**NOH**-seez) (Sing.: zoonosis)—Diseases of animals that may be transmitted to people under certain conditions (Chapter 22).

Zygote (**ZYE**-goht)—A fertilized egg, formed by the union of the nuclei of egg and sperm; the diploid number of chromosomes (46 for people) is restored (Chapter 20).

Index

Page numbers followed by a "t" or "f" indicate tables or figures, respectively. Page numbers followed by "b" indicate boxed material.

"This book helped me solve a Windows networking problem that I had been struggling with for a long time. This book is a must-have for anyone who is trying to implement and maintain a Microsoft file and print server environment—that even includes environments without Samba!"

—*Brad Jones, Senior Network/Systems Architect*

"The book is outstanding. The sections on printing are THE BEST I have ever seen—totally excellent. This book tells network admins the how-and-why of subjects that no one else can cover like [the authors]. In plain language, the authors tell you how to properly configure the various sections to get it all working. [The authors] really, really understand what an admin is going through and have combined that with their own global experience to make a smashingly great book. A must-have manual to use and learn from."

—*Ismet Kursunoglu, MD, FCCP, Medical Director, Orbitonix*

"This book is for anyone looking to implement Samba for the first time, upgrade from an existing version, or even streamline an existing installation—in any type of environment. It is a much-needed, comprehensive reference. I would feel confident going into any environment to implement a Samba solution armed with this book. Written by the authors of the Samba software, the book's first-hand perspective gives it an edge in dealing with the new features present in the updated release."

—*Steve Elgersma, Systems Administrator, Princeton University, Department of Computer Science*

"[This book contains] some absolutely outstanding documentation. It is broken into self-contained chapters that start by laying out how a certain task or protocol works in general, and then how to configure Samba to take part in it. Considering that Samba can perform so many different roles, the mix-and-match method is a lot more sensible. Even if you don't use Samba, consider [this book] as a reference for troubleshooting Windows problems—I've found [this book offers] a far more complete and focused discussion of Windows technologies for the sysadmin than any Microsoft book or webpage."

—*Kristopher Magnusson, Open Source Software Advocate*

"When I showed some Microsoft network admins I have been working with *The Official Samba-3 HOWTO and Reference Guide*, they were more than just impressed. Good work on good docs."

—*C. Lee Taylor, National IT/IS Manager for Scania South Africa (Pty.) Ltd.*

Bruce Perens' Open Source Series

- *Managing Linux Systems with Webmin: System Administration and Module Development*
  Jamie Cameron
- *Understanding the Linux Virtual Memory Manager*
  Mel Gorman
- *Implementing CIFS: The Common Internet File System*
  Christopher R. Hertel
- *Embedded Software Development with eCos*
  Anthony J. Massa
- *Rapid Application Development with Mozilla*
  Nigel McFarlane
- *The Linux Development Platform: Configuring, Using, and Maintaining a Complete Programming Environment*
  Rafeeq Ur Rehman, Christopher Paul
- *Intrusion Detection Systems with Snort: Advanced IDS Techniques with Snort, Apache, MySQL, PHP, and ACID*
  Rafeeq Ur Rehman
- *The Official Samba-3 HOWTO and Reference Guide*
  John H. Terpstra, Jelmer R. Vernooij, Editors

The Official Samba-3 HOWTO and Reference Guide

John H. Terpstra
and
Jelmer R. Vernooij,
Editors

Prentice Hall PTR
Upper Saddle River, New Jersey 07458
www.phptr.com

Library of Congress Cataloging-in-Publication Data

A CIP catalog record for this book can be obtained from the Library of Congress.

Editorial/production supervision: *Mary Sudul*
Cover design director: *Jerry Votta*
Cover design: *DesignSource*
Manufacturing manager: *Maura Zaldivar*
Acquisitions editor: *Jill Harry*
Editorial assistant: *Brenda Mulligan*
Marketing manager: *Dan DePasquale*

Published by Prentice Hall PTR
Upper Saddle River, New Jersey 07458

Prentice Hall books are widely used by corporations and government agencies for training, marketing, and resale.
The publisher offers discounts on this book when ordered in bulk quantities. For more information, contact Corporate Sales Department, Phone: 800-382-3419; FAX: 201-236-7141;
E-mail: corpsales@prenhall.com
Or write: Prentice Hall PTR, Corporate Sales Dept., One Lake Street, Upper Saddle River, NJ 07458.

Printed in the United States of America

1st Printing

ISBN 0-13-145355-6

Pearson Education LTD.
Pearson Education Australia PTY, Limited
Pearson Education Singapore, Pte. Ltd.
Pearson Education North Asia Ltd.
Pearson Education Canada, Ltd.
Pearson Educación de Mexico, S.A. de C.V.
Pearson Education — Japan
Pearson Education Malaysia, Pte. Ltd.

ATTRIBUTION

How to Install and Test SAMBA

- Andrew Tridgell <tridge@samba.org>
- Jelmer R. Vernooij <jelmer@samba.org>
- John H. Terpstra <jht@samba.org>
- Karl Auer <kauer@biplane.com.au>
- Dan Shearer <dan@samba.org>

Fast Start: Cure for Impatience

- John H. Terpstra <jht@samba.org>

Server Types and Security Modes

- Andrew Tridgell <tridge@samba.org>
- Jelmer R. Vernooij <jelmer@samba.org>
- John H. Terpstra <jht@samba.org>

Domain Control

- John H. Terpstra <jht@samba.org>
- Gerald (Jerry) Carter <jerry@samba.org>
- David Bannon <dbannon@samba.org>
- Guenther Deschner <gd@suse.de> (LDAP updates)

Backup Domain Control

- John H. Terpstra <jht@samba.org>
- Volker Lendecke <Volker.Lendecke@SerNet.DE>
- Guenther Deschner <gd@suse.de> (LDAP updates)

Domain Membership

- John H. Terpstra <jht@samba.org>

- Jeremy Allison <jra@samba.org>
- Gerald (Jerry) Carter <jerry@samba.org>
- Andrew Tridgell <tridge@samba.org>
- Jelmer R. Vernooij <jelmer@samba.org>
- Guenther Deschner <gd@suse.de> (LDAP updates)

Stand-alone Servers

- John H. Terpstra <jht@samba.org>

MS Windows Network Configuration Guide

- John H. Terpstra <jht@samba.org>

Network Browsing

- John H. Terpstra <jht@samba.org>
- Jelmer R. Vernooij <jelmer@samba.org>

Account Information Databases

- Jelmer R. Vernooij <jelmer@samba.org>
- John H. Terpstra <jht@samba.org>
- Gerald (Jerry) Carter <jerry@samba.org>
- Jeremy Allison <jra@samba.org>
- Guenther Deschner <gd@suse.de> (LDAP updates)
- Olivier (lem) Lemaire <olem@IDEALX.org>

Group Mapping MS Windows and UNIX

- John H. Terpstra <jht@samba.org>
- Jean François Micouleau
- Gerald (Jerry) Carter <jerry@samba.org>

File, Directory and Share Access Controls

- John H. Terpstra <jht@samba.org>
- Jeremy Allison <jra@samba.org>
- Jelmer R. Vernooij <jelmer@samba.org> (drawing)

File and Record Locking

- Jeremy Allison <jra@samba.org>
- Jelmer R. Vernooij <jelmer@samba.org>
- John H. Terpstra <jht@samba.org>
- Eric Roseme <eric.roseme@hp.com>

Securing Samba

- Andrew Tridgell <tridge@samba.org>
- John H. Terpstra <jht@samba.org>

Interdomain Trust Relationships

- John H. Terpstra <jht@samba.org>
- Rafal Szczesniak <mimir@samba.org>
- Jelmer R. Vernooij <jelmer@samba.org> (drawing)
- Stephen Langasek <vorlon@netexpress.net>

Hosting a Microsoft Distributed File System tree

- Shirish Kalele <samba@samba.org>
- John H. Terpstra <jht@samba.org>

Classical Printing Support

- Kurt Pfeifle <kpfeifle@danka.de>
- Gerald (Jerry) Carter <jerry@samba.org>
- John H. Terpstra <jht@samba.org>

CUPS Printing Support

- Kurt Pfeifle <kpfeifle@danka.de>
- Ciprian Vizitiu <CVizitiu@gbif.org> (drawings)
- Jelmer R. Vernooij <jelmer@samba.org> (drawings)

Stackable VFS modules

- Jelmer R. Vernooij <jelmer@samba.org>
- John H. Terpstra <jht@samba.org>
- Tim Potter <tpot@samba.org>
- Simo Sorce (original vfs_skel README)
- Alexander Bokovoy (original vfs_netatalk docs)
- Stefan Metzmacher (Update for multiple modules)

Winbind: Use of Domain Accounts

- Tim Potter <tpot@linuxcare.com.au>
- Andrew Tridgell <tridge@samba.org>
- Naag Mummaneni <getnag@rediffmail.com> (Notes for Solaris)
- John Trostel <jtrostel@snapserver.com>
- Jelmer R. Vernooij <jelmer@samba.org>

- John H. Terpstra <jht@samba.org>

Advanced Network Management

- John H. Terpstra <jht@samba.org>

System and Account Policies

- John H. Terpstra <jht@samba.org>

Desktop Profile Management

- John H. Terpstra <jht@samba.org>

PAM-Based Distributed Authentication

- John H. Terpstra <jht@samba.org>
- Stephen Langasek <vorlon@netexpress.net>

Integrating MS Windows Networks with Samba

- John H. Terpstra <jht@samba.org>

Unicode/Charsets

- Jelmer R. Vernooij <jelmer@samba.org>
- John H. Terpstra <jht@samba.org>
- TAKAHASHI Motonobu <monyo@home.monyo.com>

Backup Techniques

- John H. Terpstra <jht@samba.org>

High Availability

- John H. Terpstra <jht@samba.org>
- Jeremy Allison <jra@samba.org>

Upgrading from Samba-2.x to Samba-3.0.0

- Jelmer R. Vernooij <jelmer@samba.org>
- John H. Terpstra <jht@samba.org>
- Gerald (Jerry) Carter <jerry@samba.org>

Migration from NT4 PDC to Samba-3 PDC

- John H. Terpstra <jht@samba.org>

SWAT The Samba Web Administration Tool

- John H. Terpstra <jht@samba.org>

The Samba Checklist

- Andrew Tridgell <tridge@samba.org>
- Jelmer R. Vernooij <jelmer@samba.org>

- Dan Shearer <dan@samba.org>

Analyzing and Solving Samba Problems

- Gerald (Jerry) Carter <jerry@samba.org>
- Jelmer R. Vernooij <jelmer@samba.org>
- David Bannon <dbannon@samba.org>
- Dan Shearer <dan@samba.org>

Reporting Bugs

- John H. Terpstra <jht@samba.org>
- Jelmer R. Vernooij <jelmer@samba.org>
- Andrew Tridgell <tridge@samba.org>

How to Compile Samba

- Jelmer R. Vernooij <jelmer@samba.org>
- John H. Terpstra <jht@samba.org>
- Andrew Tridgell <tridge@samba.org>

Portability

- Jelmer R. Vernooij <jelmer@samba.org>
- John H. Terpstra <jht@samba.org>

Samba and Other CIFS Clients

- Jelmer R. Vernooij <jelmer@samba.org>
- John H. Terpstra <jht@samba.org>
- Dan Shearer <dan@samba.org>
- Jim McDonough <jmcd@us.ibm.com> (OS/2)

Samba Performance Tuning

- Paul Cochrane <paulc@dth.scot.nhs.uk>
- Jelmer R. Vernooij <jelmer@samba.org>
- John H. Terpstra <jht@samba.org>

DNS and DHCP Configuration Guide

- John H. Terpstra <jht@samba.org>

ABSTRACT

The editors wish to thank you for your decision to purchase this book. The Official Samba-3 HOWTO and Reference Guide is the result of many years of accumulation of information, feedback, tips, hints, and happy solutions.

Please note that this book is a living document, the contents of which are constantly being updated. We encourage you to contribute your tips, techniques, helpful hints, and your special insight into the Windows networking world to help make the next generation of this book even more valuable to Samba users.

We have made a concerted effort to document more comprehensively than has been done previously the information that may help you to better deploy Samba and to gain more contented network users.

This book provides example configurations, it documents key aspects of Microsoft Windows networking, provides in-depth insight into the important configuration of Samba-3, and helps to put all of these into a useful framework.

The most recent electronic versions of this document can be found at `http://www.samba.org/` on the "*Documentation*" page.

Updates, patches and corrections are most welcome. Please email your contributions to any one of the following:

Jelmer Vernooij (jelmer@samba.org)
John H. Terpstra (jht@samba.org)
Gerald (Jerry) Carter (jerry@samba.org)

We wish to advise that only original and unencumbered material can be published. Please do not submit content that is not your own work unless proof of consent from the copyright holder accompanies your submission.

FOREWORD

Over the last few years, the Samba project has undergone a major tranformation. From a small project used only by people who dream in machine code, Samba has grown to be an integral part of the IT infrastructure of many businesses. Along with the growth in the popularity of Samba there has been a corresponding growth in the ways that it can be used, and a similar growth in the number of configuration options and the interactions between them.

To address this increasing complexity a wealth of documentation has been written on Samba, including numerous HOWTOs, diagnostic tips, manual pages, and explanations of important pieces of technology that Samba relies on. While it has been gratifying to see so much documentation being written, the sheer volume of different types of documentation has proved difficult to navigate, thus reducing its value to system administrators trying to cope with the complexity.

This book gathers together that wealth of information into a much more accessible form, to allow system administrators to quickly find what they need. The breadth of technical information provided ensures that even the most demanding administrators will find something helpful.

I am delighted that the Samba documentation has now developed to the extent that it can be presented usefully as a book, and I am grateful for the efforts of the many people who have contributed so much toward this result. Enjoy!

Andrew Tridgell
President, Samba Team
July 2003

CONTENTS

List of Figures

List of Tables

Part I

General Installation

PREFACE AND INTRODUCTION

"A man's gift makes room for him before great men. Gifts are like hooks that can catch hold of the mind taking it beyond the reach of forces that otherwise might constrain it." — Anon.

This is a book about Samba. It is a tool, a derived work of the labors of many and of the diligence and goodwill of more than a few. This book contains material that has been contributed in a persistent belief that each of us can add value to our neighbors as well as to those who will follow us.

This book is designed to meet the needs of the Microsoft network administrator. UNIX administrators will benefit from this book also, though they may complain that it is hard to find the information they think they need. So if you are a Microsoft certified specialist, this book should meet your needs rather well. If you are a UNIX or Linux administrator, there is no need to feel badly — you should have no difficulty finding answers to your current concerns also.

What Is Samba?

Samba is a big, complex project. The Samba project is ambitious and exciting. The team behind Samba is a group of some thirty individuals who are spread the world over and come from an interesting range of backgrounds. This team includes scientists, engineers, programmers, business people, and students.

Team members were drawn into active participation through the desire to help deliver an exciting level of transparent interoperability between Microsoft Windows and the non-Microsoft information technology world.

The slogan that unites the efforts behind the Samba project says: *Samba, Opening Windows to a Wider World!* The goal behind the project is one of removing barriers to interoperability.

Samba provides file and print services for Microsoft Windows clients. These services may be hosted off any TCP/IP-enabled platform. The original deployment platforms were UNIX and Linux, though today it is in common use across a broad variety of systems.

The Samba project includes not only an impressive feature set in file and print serving capabilities, but has been extended to include client functionality, utilities to ease migration to Samba, tools to aid interoperability with Microsoft Windows, and adminstration tools.

The real people behind Samba are users like you. You have inspired the developers (the Samba Team) to do more than any of them imagined could or should be done. User feedback drives Samba development. Samba-3 in particular incorporates a huge amount of work done as a result of user requests, suggestions and direct code contributions.

Why This Book?

There is admittedly a large number of Samba books on the market today and each book has its place. Despite the apparent plethora of books, Samba as a project continues to receive much criticism for failing to provide sufficient documentation. Samba is also criticized for being too complex and too difficult to configure. In many ways this is evidence of the success of Samba as there would be no complaints if it was not successful.

The Samba Team members work predominantly with UNIX and Linux, so it is hardly surprising that existing Samba documentation should reflect that orientation. The original HOWTO text documents were intended to provide some tips, a few golden nuggets, and if they helped anyone then that was just wonderful. But the HOWTOs lacked structure and context. They were isolated snapshots of information that were written to pass information on to someone else who might benefit. They reflected a need to transmit more information that could be conveniently put into manual pages.

The original HOWTO documents were written by different authors. Most HOWTO documents are the result of feedback and contributions from numerous authors. In this book we took care to preserve as much original content as possible. As you read this book you will note that chapters were written by multiple authors, each of whom has his own style. This demonstrates the nature of the Open Source software development process.

Out of the original HOWTO documents sprang a collection of unoffical HOWTO documents that are spread over the Internet. It is sincerely intended that this work will *not* replace the valuable unofficial HOWTO work that continues to flourish. If you are involved in unofficial HOWTOs then please continue your work!

Those of you who have dedicated your labors to the production of unofficial HOWTOs, to Web page information regarding Samba, or to answering questions on the mailing lists or elsewhere, may be aware that this is a labor of love. We would like to know about your contribution and willingly receive the precious perls of wisdom you have collected. Please email your contribution to John H. Terpstra (jht@samba.org). As a service to other users we will gladly adopt material that is technically accurate.

Existing Samba books are largely addressed to the UNIX administrator. From the perspective of this target group the existing books serve an adequate purpose, with one exception — now that Samba-3 is out they need to be updated!

This book, the *Official Samba-3 HOWTO and Reference Guide*, includes the Samba-HOWTO-Collection.pdf that ships with Samba. These documents have been written with a new design intent and purpose.

Over the past two years many Microsoft network administrators have adopted Samba and have become interested in its deployment. Their information needs are very different from that of the UNIX administrator. This book has been arranged and the information presented from the perspective of someone with previous Microsoft Windows network administrative training and experience.

Book Structure and Layout

This book is presented in six parts:

General Installation — Designed to help you get Samba-3 running quickly. The Fast Start chapter is a direct response to requests from Microsoft network administrators for some sample configurations that *just work.*

Server Configuration Basics — The purpose of this section is to aid the transition from existing Microsoft Windows network knowledge to Samba terminology and norms. The chapters in this part each cover the installation of one type of Samba server.

Advanced Configuration — The mechanics of network browsing have long been the Achilles heel of all Microsoft Windows users. Samba-3 introduces new user and machine account management facilities, a new way to map UNIX groups and Windows groups, Interdomain trusts, new loadable file system drivers (VFS), and more. New with this document is expanded printing documentation, as well as a wealth of information regarding desktop and user policy handling, use of desktop profiles, and techniques for enhanced network integration. This section makes up the core of the book. Read and enjoy.

Migration and Updating — A much requested addition to the book is information on how to migrate from Microsoft Windows NT4 to Samba-3, as well as an overview of what the issues are when moving from Samba-2.x to Samba-3.

Troubleshooting — This short section should help you when all else fails.

Appendix — Here you will find a collection of things that are either too peripheral for most users, or are a little left of field to be included in the main body of information.

Welcome to Samba-3 and the first published document to help you and your users to enjoy a whole new world of interoperability between Microsoft Windows and the rest of the world.

Chapter 1

HOW TO INSTALL AND TEST SAMBA

1.1 Obtaining and Installing Samba

Binary packages of Samba are included in almost any Linux or UNIX distribution. There are also some packages available at the Samba homepage[1]. Refer to the manual of your operating system for details on installing packages for your specific operating system.

If you need to compile Samba from source, check Chapter 35, *How to Compile Samba.*

1.2 Configuring Samba (smb.conf)

Samba's configuration is stored in the `smb.conf` file, which usually resides in `/etc/samba/smb.conf` or `/usr/local/samba/lib/smb.conf`. You can either edit this file yourself or do it using one of the many graphical tools that are available, such as the Web-based interface SWAT, that is included with Samba.

1.2.1 Configuration file syntax

The `smb.conf` file uses the same syntax as the various old .ini files in Windows 3.1: Each file consists of various sections, which are started by putting the section name between brackets ([]) on a new line. Each contains zero or more key/value-pairs seperated by an equality sign (=). The file is just a plain-text file, so you can open and edit it with your favorite editing tool.

Each section in the `smb.conf` file represents a share on the Samba server. The section "*global*" is special, since it contains settings that apply to the whole Samba server and not to one share in particular.

Example 1.1 contains a very minimal `smb.conf`.

[1] http://samba.org/

Example 1.1. A minimal smb.conf

```
[global]
workgroup = WKG
netbios name = MYNAME

[share1]
path = /tmp

[share2]
path = /my_shared_folder
comment = Some random files
```

1.2.2 Example Configuration

There are sample configuration files in the examples subdirectory in the distribution. It is suggested you read them carefully so you can see how the options go together in practice. See the man page for all the options. It might be worthwhile to start out with the smb.conf.default configuration file and adapt it to your needs. It contains plenty of comments.

The simplest useful configuration file would contain something like shown in Example 1.2.

Example 1.2. Another simple smb.conf File

```
[global]
workgroup = MIDEARTH

[homes]
guest ok = no
read only = no
```

This will allow connections by anyone with an account on the server, using either their login name or *homes* as the service name. (Note: The workgroup that Samba should appear in must also be set. The default workgroup name is WORKGROUP.)

Make sure you put the `smb.conf` file in the correct place.

For more information about security settings for the *[homes]* share please refer to Chapter 14, *Securing Samba*.

1.2.2.1 Test Your Config File with testparm

It's important to validate the contents of the `smb.conf` file using the testparm program. If testparm runs correctly, it will list the loaded services. If not, it will give an error message. Make sure it runs correctly and that the services look reasonable before proceeding. Enter the command:

```
root#  testparm /etc/samba/smb.conf
```

Testparm will parse your configuration file and report any unknown parameters or incorrect syntax.

Always run testparm again whenever the `smb.conf` file is changed!

1.2.3 SWAT

SWAT is a Web-based interface that can be used to facilitate the configuration of Samba. SWAT might not be available in the Samba package that shipped with your platform, but in a separate package. Please read the SWAT manpage on compiling, installing and configuring SWAT from source.

To launch SWAT, just run your favorite Web browser and point it to `http://localhost:901/`. Replace *`localhost`* with the name of the computer on which Samba is running if that is a different computer than your browser.

SWAT can be used from a browser on any IP-connected machine, but be aware that connecting from a remote machine leaves your connection open to password sniffing as passwords will be sent over the wire in the clear.

More information about SWAT can be found in Chapter 31, *SWAT The Samba Web Administration Tool.*

1.3 List Shares Available on the Server

To list shares that are available from the configured Samba server execute the following command:

```
$ smbclient -L yourhostname
```

You should see a list of shares available on your server. If you do not, then something is incorrectly configured. This method can also be used to see what shares are available on other SMB servers, such as Windows 2000.

If you choose user-level security you may find that Samba requests a password before it will list the shares. See the **smbclient** man page for details. You can force it to list the shares without a password by adding the option `-N` to the command line.

1.4 Connect with a UNIX Client

Enter the following command:

```
$ smbclient  //yourhostname/aservice
```

Typically *yourhostname* is the name of the host on which smbd has been installed. The *aservice* is any service that has been defined in the `smb.conf` file. Try your user name if you just have a *[homes]* section in the `smb.conf` file.

Example: If the UNIX host is called *bambi* and a valid login name is *fred*, you would type:

```
$ smbclient //bambi/fred
```

1.5 Connect from a Remote SMB Client

Now that Samba is working correctly locally, you can try to access it from other clients. Within a few minutes, the Samba host should be listed in the Network Neighborhood on all Windows clients of its subnet. Try browsing the server from another client or 'mounting' it.

Mounting disks from a DOS, Windows or OS/2 client can be done by running a command such as:

```
C:\> net use d: \\servername\service
```

Try printing, e.g.

```
C:\> net use lpt1:   \\servername\spoolservice
```

```
C:\> print filename
```

1.6 What If Things Don't Work?

You might want to read Chapter 32, *The Samba Checklist.* If you are still stuck, refer to Chapter 33, *Analyzing and Solving Samba Problems.* Samba has been successfully installed at thousands of sites worldwide. It is unlikely that your particular problem is unique, so it might be productive to perform an Internet search to see if someone else has encountered your problem and has found a way to overcome it.

1.7 Common Errors

The following questions and issues are raised repeatedly on the Samba mailing list.

1.7.1 Large Number of smbd Processes

Samba consists of three core programs: nmbd, smbd, and winbindd. nmbd is the name server message daemon, smbd is the server message daemon, and winbindd is the daemon that handles communication with Domain Controllers.

If Samba is *not* running as a WINS server, then there will be one single instance of nmbd running on your system. If it is running as a WINS server then there will be two instances — one to handle the WINS requests.

smbd handles all connection requests. It spawns a new process for each client connection made. That is why you may see so many of them, one per client connection.

winbindd will run as one or two daemons, depending on whether or not it is being run in *split mode* (in which case there will be two instances).

1.7.2 Error Message: open_oplock_ipc

An error message is observed in the log files when smbd is started: "*open_oplock_ipc: Failed to get local UDP socket for address 100007f. Error was Cannot assign requested.*"

Your loopback device isn't working correctly. Make sure it is configured correctly. The loopback device is an internal (virtual) network device with the IP address *127.0.0.1*. Read your OS documentation for details on how to configure the loopback on your system.

1.7.3 "*The network name cannot be found*"

This error can be caused by one of these misconfigurations:

- You specified an nonexisting path for the share in `smb.conf`.
- The user you are trying to access the share with does not have sufficient permissions to access the path for the share. Both read (r) and access (x) should be possible.
- The share you are trying to access does not exist.

Chapter 2

FAST START: CURE FOR IMPATIENCE

When we first asked for suggestions for inclusion in the Samba HOWTO documentation, someone wrote asking for example configurations — and lots of them. That is remarkably difficult to do, without losing a lot of value that can be derived from presenting many extracts from working systems. That is what the rest of this document does. It does so with extensive descriptions of the configuration possibilites within the context of the chapter that covers it. We hope that this chapter is the medicine that has been requested.

2.1 Features and Benefits

Samba needs very little configuration to create a basic working system. In this chapter we progress from the simple to the complex, for each providing all steps and configuration file changes needed to make each work. Please note that a comprehensively configured system will likely employ additional smart features. The additional features are covered in the remainder of this document.

The examples used here have been obtained from a number of people who made requests for example configurations. All identities have been obscured to protect the guilty and any resemblance to unreal non-existent sites is deliberate.

2.2 Description of Example Sites

In the first set of configuration examples we consider the case of exceptionally simple system requirements. There is a real temptation to make something that should require little effort much too complex.

Section 2.3.1.1 documents the type of server that might be sufficient to serve CD-ROM images, or reference document files for network client use. This configuration is also discussed in Chapter 7, *Stand-alone Servers*, Section 7.3.1. The purpose for this configuration is to provide a shared volume that is read-only that anyone, even guests, can access.

The second example shows a minimal configuration for a print server that anyone can print to as long as they have the correct printer drivers installed on their computer. This is a mirror of the system described in Chapter 7, *Stand-alone Servers*, Section 7.3.2.

The next example is of a secure office file and print server that will be accessible only to users who have an account on the system. This server is meant to closely resemble a Workgroup file and print server, but has to be more secure than an anonymous access machine. This type of system will typically suit the needs of a small office. The server does not provide network logon facilities, offers no Domain Control, instead it is just a network attached storage (NAS) device and a printserver.

Finally, we start looking at more complex systems that will either integrate into existing Microsoft Windows networks, or replace them entirely. The examples provided covers domain member servers as well as Samba Domain Control (PDC/BDC) and finally describes in detail a large distributed network with branch offices in remote locations.

2.3 Worked Examples

The configuration examples are designed to cover everything necessary to get Samba running. They do not cover basic operating system platform configuration, which is clearly beyond the scope of this text.

It is also assumed that Samba has been correctly installed, either by way of installation of the packages that are provided by the operating system vendor, or through other means.

2.3.1 Stand-alone Server

A Stand-alone Server implies no more than the fact that it is not a Domain Controller and it does not participate in Domain Control. It can be a simple workgroup-like server, or it may be a complex server that is a member of a domain security context.

2.3.1.1 Annonymous Read-Only Document Server

The purpose of this type of server is to make available to any user any documents or files that are placed on the shared resource. The shared resource could be a CD-ROM drive, a CD-ROM image, or a file storage area.

As the examples are developed, every attempt is made to progress the system toward greater capability, just as one might expect would happen in a real business office as that office grows in size and its needs change.

The configuration file is:

- The file system share point will be `/export`.
- All files will be owned by a user called Jack Baumbach. Jack's login name will be *jackb*. His password will be *m0r3pa1n* — of course, that's just the example we are using; do not use this in a production environment because all readers of this document will know it.

INSTALLATION PROCEDURE — READ-ONLY SERVER

1. Add user to system (with creation of the users' home directory):

Example 2.1. Anonymous Read-Only Server Configuration

```
# Global parameters

[global]
workgroup = MIDEARTH
netbios name = HOBBIT
security = share

[data]
comment = Data
path = /export
read only = No
guest only = Yes
```

```
root# useradd -c "Jack Baumbach" -m -g users -p m0r3pa1n jackb
```

2. Create directory, and set permissions and ownership:

```
root# mkdir /export
root# chmod u+rwx,g+rx,o+rx /export
root# chown jackb.users /export
```

3. Copy the files that should be shared to the `/export` directory.
4. Install the Samba configuration file (`/etc/samba/smb.conf`) as shown.
5. Test the configuration file:

```
root# testparm
```

Note any error messages that might be produced. Do not procede until you obtain error-free output. An example of the output with the following file will list the file.

```
Load smb config files from /etc/samba/smb.conf
Processing section "[data]"
Loaded services file OK.
Server role: ROLE_STANDALONE
Press enter to see a dump of your service definitions
[Press enter]

# Global parameters
[global]
   workgroup = MIDEARTH
```

```
    netbios name = HOBBIT
    security = share

[data]
    comment = Data
    path = /export
    read only = No
    guest only = Yes
```

6. Start Samba using the method applicable to your operating system platform.
7. Configure your Microsoft Windows client for workgroup *MIDEARTH*, set the machine name to ROBBINS, reboot, wait a few (2 - 5) minutes, then open Windows Explorer and visit the network neighborhood. The machine HOBBIT should be visible. When you click this machine icon, it should open up to reveal the *data* share. After clicking the share it, should open up to revel the files previously placed in the `/export` directory.

The information above (following # Global parameters) provides the complete contents of the `/etc/samba/smb.conf` file.

2.3.1.2 Anonymous Read-Write Document Server

We should view this configuration as a progression from the previous example. The difference is that shared access is now forced to the user identity of jackb and to the primary group jackb belongs to. One other refinement we can make is to add the user *jackb* to the `smbpasswd` file. To do this execute:

```
root# smbpasswd -a jackb
New SMB password: m0r3pa1n
Retype new SMB password: m0r3pa1n
Added user jackb.
```

Addition of this user to the `smbpasswd` file allows all files to be displayed in the Explorer Properties boxes as belonging to *jackb* instead of to *User Unknown*.

The complete, modified `smb.conf` file is as shown in Example 2.2.

2.3.1.3 Anonymous Print Server

An anonymous print server serves two purposes:

- It allows printing to all printers from a single location.
- It reduces network traffic congestion due to many users trying to access a limited number of printers.

In the simplest of anonymous print servers, it is common to require the installation of the correct printer drivers on the Windows workstation. In this case the print server will be

Example 2.2. Modified Anonymous Read-Write smb.conf

```
# Global parameters

[global]
workgroup = MIDEARTH
netbios name = HOBBIT
security = SHARE

[data]
comment = Data
path = /export
force user = jackb
force group = users
read only = No
guest ok = Yes
```

designed to just pass print jobs through to the spooler, and the spooler should be configured to do raw passthrough to the printer. In other words, the print spooler should not filter or process the data stream being passed to the printer.

In this configuration it is undesirable to present the Add Printer Wizard and we do not want to have automatic driver download, so we will disable it in the following configuration. Example 2.3 is the resulting `smb.conf` file.

Example 2.3. Anonymous Print Server smb.conf

```
# Global parameters

[global]
workgroup = MIDEARTH
netbios name = LUTHIEN
security = share
printcap name = cups
disable spoolss = Yes
show add printer wizard = No
printing = cups

[printers]
comment = All Printers
path = /var/spool/samba
guest ok = Yes
printable = Yes
use client driver = Yes
browseable = No
```

The above configuration is not ideal. It uses no smart features, and it deliberately presents a less than elegant solution. But it is basic, and it does print.

NOTE

Windows users will need to install a local printer and then change the print to device after installation of the drivers. The print to device can then be set to the network printer on this machine.

Make sure that the directory `/var/spool/samba` is capable of being used as intended. The following steps must be taken to achieve this:

- The directory must be owned by the superuser (root) user and group:

```
root# chown root.root /var/spool/samba
```

- Directory permissions should be set for public read-write with the sticky-bit set as shown:

```
root# chmod a+rwtx /var/spool/samba
```

NOTE

On CUPS enabled systems there is a facility to pass raw data directly to the printer without intermediate processing via CUPS print filters. Where use of this mode of operation is desired it is necessary to configure a raw printing device. It is also necessary to enable the raw mime handler in the `/etc/mime.conv` and `/etc/mime.types` files. Refer to Section 18.3.4.

2.3.1.4 Secure Read-Write File and Print Server

We progress now from simple systems to a server that is slightly more complex.

Our new server will require a public data storage area in which only authenticated users (i.e., those with a local account) can store files, as well as a home directory. There will be one printer that should be available for everyone to use.

In this hypothetical environment (no espionage was conducted to obtain this data), the site is demanding a simple environment that is *secure enough* but not too difficult to use.

Site users will be: Jack Baumbach, Mary Orville and Amed Sehkah. Each will have a password (not shown in further examples). Mary will be the printer administrator and will own all files in the public share.

This configuration will be based on *User Level Security* that is the default, and for which the default is to store Microsoft Windows-compatible encrypted passwords in a file called `/etc/samba/smbpasswd`. The default `smb.conf` entry that makes this happen is: ***passdb backend*** = smbpasswd, guest. Since this is the default it is not necessary to enter it into the configuration file. Note that guest backend is added to the list of active passdb backends not matter was it specified directly in Samba configuration file or not.

Installing the Secure Office Server

1. Add all users to the Operating System:

```
root# useradd -c "Jack Baumbach" -m -g users -p m0r3pa1n jackb
root# useradd -c "Mary Orville" -m -g users -p secret maryo
root# useradd -c "Amed Sehkah" -m -g users -p secret ameds
```

2. Configure the Samba `smb.conf` file as shown in Example 2.4.
3. Initialize the Microsoft Windows password database with the new users:

```
root# smbpasswd -a root
New SMB password: bigsecret
Reenter smb password: bigsecret
Added user root.

root# smbpasswd -a jackb
New SMB password: m0r3pa1n
Retype new SMB password: m0r3pa1n
Added user jackb.

root# smbpasswd -a maryo
New SMB password: secret
Reenter smb password: secret
Added user maryo.

root# smbpasswd -a ameds
New SMB password: mysecret
Reenter smb password: mysecret
Added user ameds.
```

4. Install printer using the CUPS Web interface. Make certain that all printers that will be shared with Microsoft Windows clients are installed as raw printing devices.
5. Start Samba using the operating system administrative interface. Alternately, this can be done manually by running:

```
root# nmbd; smbd;
```

Example 2.4. Secure Office Server smb.conf

```
# Global parameters

[global]
workgroup = MIDEARTH
netbios name = OLORIN
printcap name = cups
disable spoolss = Yes
show add printer wizard = No
printing = cups

[homes]
comment = Home Directories
valid users = %S
read only = No
browseable = No

[public]
comment = Data
path = /export
force user = maryo
force group = users
guest ok = Yes

[printers]
comment = All Printers
path = /var/spool/samba
printer admin = root, maryo
create mask = 0600
guest ok = Yes
printable = Yes
use client driver = Yes
browseable = No
```

6. Configure the /export directory:

```
root# mkdir /export
root# chown maryo.users /export
root# chmod u=rwx,g=rwx,o-rwx /export
```

7. Check that Samba is running correctly:

```
root# smbclient -L localhost -U%
Domain=[MIDEARTH] OS=[UNIX] Server=[Samba-3.0.0]
```

```
Sharename      Type      Comment
---------      ----      -------
public         Disk      Data
IPC$           IPC       IPC Service (Samba-3.0.0)
ADMIN$         IPC       IPC Service (Samba-3.0.0)
hplj4          Printer   hplj4

Server              Comment
---------           -------
OLORIN              Samba-3.0.0

Workgroup           Master
---------           -------
MIDEARTH            OLORIN
```

8. Connect to OLORIN as maryo:

```
root# smbclient //olorin/maryo -Umaryo%secret
OS=[UNIX] Server=[Samba-3.0.0]
smb: \> dir
.                               D        0  Sat Jun 21 10:58:16 2003
..                              D        0  Sat Jun 21 10:54:32 2003
Documents                        D        0  Fri Apr 25 13:23:58 2003
DOCWORK                          D        0  Sat Jun 14 15:40:34 2003
OpenOffice.org                   D        0  Fri Apr 25 13:55:16 2003
.bashrc                          H     1286  Fri Apr 25 13:23:58 2003
.netscape6                      DH        0  Fri Apr 25 13:55:13 2003
.mozilla                        DH        0  Wed Mar  5 11:50:50 2003
.kermrc                          H      164  Fri Apr 25 13:23:58 2003
.acrobat                        DH        0  Fri Apr 25 15:41:02 2003

      55817 blocks of size 524288. 34725 blocks available
smb: \> q
```

By now you should be getting the hang of configuration basics. Clearly, it is time to explore slightly more complex examples. For the remainder of this chapter we will abbreviate instructions since there are previous examples.

2.3.2 Domain Member Server

In this instance we will consider the simplest server configuration we can get away with to make an accounting department happy. Let's be warned, the users are accountants and they do have some nasty demands. There is a budget for only one server for this department.

The network is managed by an internal Information Services Group (ISG), to which we belong. Internal politics are typical of a medium-sized organization; Human Resources is of the opinion that they run the ISG because they are always adding and disabling users. Also, departmental managers have to fight tooth and nail to gain basic network resources access for their staff. Accounting is different though, they get exactly what they want. So this should set the scene.

We will use the users from the last example. The accounting department has a general printer that all departmental users may. There is also a check printer that may be used only by the person who has authority to print checks. The Chief Financial Officer (CFO) wants that printer to be completely restricted and for it to be located in the private storage area in her office. It therefore must be a network printer.

Accounting department uses an accounting application called *SpytFull* that must be run from a central application server. The software is licensed to run only off one server, there are no workstation components, and it is run off a mapped share. The data store is in a UNIX-based SQL backend. The UNIX gurus look after that, so is not our problem.

The accounting department manager (maryo) wants a general filing system as well as a separate file storage area for form letters (nastygrams). The form letter area should be read-only to all accounting staff except the manager. The general filing system has to have a structured layout with a general area for all staff to store general documents, as well as a separate file area for each member of her team that is private to that person, but she wants full access to all areas. Users must have a private home share for personal work-related files and for materials not related to departmental operations.

2.3.2.1 Example Configuration

The server *valinor* will be a member server of the company domain. Accounting will have only a local server. User accounts will be on the Domain Controllers as will desktop profiles and all network policy files.

1. Do not add users to the UNIX/Linux server; all of this will run off the central domain.
2. Configure `smb.conf` according to Example 2.5 and Example 2.6.
3. Join the domain. Note: Do not start Samba until this step has been completed!

```
root# net rpc join -Uroot%'bigsecret'
Joined domain MIDEARTH.
```

4. Make absolutely certain that you disable (shut down) the **nscd** daemon on any system on which **winbind** is configured to run.
5. Start Samba following the normal method for your operating system platform. If you wish to this manually execute as root:

```
root# nmbd; smbd; winbindd;
```

6. Configure the name service switch control file on your system to resolve user and group names via winbind. Edit the following lines in `/etc/nsswitch.conf`:

Example 2.5. Member server smb.conf (globals)

```
# Global parameters

[global]
workgroup = MIDEARTH
netbios name = VALINOR
security = DOMAIN
printcap name = cups
disable spoolss = Yes
show add printer wizard = No
idmap uid = 15000-20000
idmap gid = 15000-20000
winbind separator = +
winbind use default domain = Yes
use sendfile = Yes
printing = cups
```

```
passwd: files winbind
group:  files winbind
hosts:  files dns winbind
```

7. Set the password for **wbinfo** to use:

```
root# wbinfo --set-auth-user=root%'bigsecret'
```

8. Validate that domain user and group credentials can be correctly resolved by executing:

```
root# wbinfo -u
MIDEARTH+maryo
MIDEARTH+jackb
MIDEARTH+ameds
...
MIDEARTH+root

root# wbinfo -g
MIDEARTH+Domain Users
MIDEARTH+Domain Admins
MIDEARTH+Domain Guests
...
MIDEARTH+Accounts
```

9. Check that **winbind** is working. The following demonstrates correct username resolution via the **getent** system utility:

Example 2.6. Member server smb.conf (shares and services)

```
[homes]
comment = Home Directories
valid users = %S
read only = No
browseable = No

[spytfull]
comment = Accounting Application Only
path = /export/spytfull
valid users = @Accounts
admin users = maryo
read only = Yes

[public]
comment = Data
path = /export/public
read only = No

[printers]
comment = All Printers
path = /var/spool/samba
printer admin = root, maryo
create mask = 0600
guest ok = Yes
printable = Yes
use client driver = Yes
browseable = No
```

```
root# getent passwd maryo
maryo:x:15000:15003:Mary Orville:/home/MIDEARTH/maryo:/bin/false
```

10. A final test that we have this under control might be reassuring:

```
root# touch /export/a_file
root# chown maryo /export/a_file
root# ls -al /export/a_file
...
-rw-r--r--    1 maryo    users        11234 Jun 21 15:32 a_file
...

root# rm /export/a_file
```

11. Configuration is now mostly complete, so this is an opportune time to configure the directory structure for this site:

```
root# mkdir -p /export/{spytfull,public}
root# chmod ug=rwxS,o=x /export/{spytfull,public}
root# chown maryo.Accounts /export/{spytfull,public}
```

2.3.3 Domain Controller

For the remainder of this chapter the focus is on the configuration of Domain Control. The examples that follow are for two implementation strategies. Remember, our objective is to create a simple but working solution. The remainder of this book should help to highlight opportunity for greater functionality and the complexity that goes with it.

A Domain Controller configuration can be achieved with a simple configuration using the new tdbsam password backend. This type of configuration is good for small offices, but has limited scalabilty (cannot be replicated) and performance can be expected to fall as the size and complexity of the domain increases.

The use of tdbsam is best limited to sites that do not need more than a primary Domain Controller (PDC). As the size of a domain grows the need for additional Domain Controllers becomes apparent. Do not attempt to under-resource a Microsoft Windows network environment; Domain Controllers provide essential authentication services. The following are symptoms of an under-resourced Domain Control environment:

- Domain logons intermittently fail.
- File access on a Domain Member server intermittently fails, giving a permission denied error message.

A more scalable Domain Control authentication backend option might use Microsoft Active Directory, or an LDAP-based backend. Samba-3 provides for both options as a Domain Member server. As a PDC Samba-3 is not able to provide an exact alternative to the functionality that is available with Active Directory. Samba-3 can provide a scalable LDAP-based PDC/BDC solution.

The tdbsam authentication backend provides no facility to replicate the contents of the database, except by external means. (i.e., there is no self-contained protocol in Samba-3 for Security Account Manager database [SAM] replication.)

> NOTE
>
> 
>
> If you need more than one Domain Controller, do not use a tdbsam authentication backend.

2.3.3.1 Example: Engineering Office

The engineering office network server we present here is designed to demonstrate use of the new tdbsam password backend. The tdbsam facility is new to Samba-3. It is designed to provide many user and machine account controls that are possible with Microsoft Windows NT4. It is safe to use this in smaller networks.

1. A working PDC configuration using the tdbsam password backend can be found in Example 2.7 together with Example 2.8:

Example 2.7. Engineering Office smb.conf (globals)

```
[global]
workgroup = MIDEARTH
netbios name = FRODO
passdb backend = tdbsam
printcap name = cups
add user script = /usr/sbin/useradd -m %u
delete user script = /usr/sbin/userdel -r %u
add group script = /usr/sbin/groupadd %g
delete group script = /usr/sbin/groupdel %g
add user to group script = /usr/sbin/usermod -G %g %u
add machine script = /usr/sbin/useradd -s /bin/false -d /dev/null %u
# Note: The following specifies the default logon script.
# Per user logon scripts can be specified in the user account using pdbedit
logon script = scripts\logon.bat
# This sets the default profile path. Set per user paths with pdbedit
logon path = \\%L\Profiles\%U
logon drive = H:
logon home = \\%L\%U
domain logons = Yes
os level = 35
preferred master = Yes
domain master = Yes
idmap uid = 15000-20000
idmap gid = 15000-20000
printing = cups
```

2. Create UNIX group accounts as needed using a suitable operating system tool:

```
root# groupadd ntadmins
root# groupadd designers
root# groupadd engineers
root# groupadd qateam
```

3. Create user accounts on the system using the appropriate tool provided with the operating system. Make sure all user home directories are created also. Add users to

Example 2.8. Eningeering Office smb.conf (shares and services)

```
[homes]
comment = Home Directories
valid users = %S
read only = No
browseable = No
# Printing auto-share (makes printers available thru CUPS)

[printers]
comment = All Printers
path = /var/spool/samba
printer admin = root, maryo
create mask = 0600
guest ok = Yes
printable = Yes
browseable = No

[print$]
comment = Printer Drivers Share
path = /var/lib/samba/drivers
write list = maryo, root
printer admin = maryo, root
# Needed to support domain logons

[netlogon]
comment = Network Logon Service
path = /var/lib/samba/netlogon
admin users = root, maryo
guest ok = Yes
browseable = No
# For profiles to work, create a user directory under the path
# shown. i.e., mkdir -p /var/lib/samba/profiles/maryo

[Profiles]
comment = Roaming Profile Share
path = /var/lib/samba/profiles
read only = No
profile acls = Yes
# Other resource (share/printer) definitions would follow below.
...
```

groups as required for access control on files, directories, printers, and as required for use in the Samba environment.

4. Assign each of the UNIX groups to NT groups: (It may be useful to copy this text to a shell script called `initGroups.sh`.)

```
#!/bin/bash
#### Keep this as a shell script for future re-use

# First assign well known groups
net groupmap modify ntgroup="Domain Admins" unixgroup=ntadmins rid=512
net groupmap modify ntgroup="Domain Users"  unixgroup=users    rid=513
net groupmap modify ntgroup="Domain Guests" unixgroup=nobody   rid=514

# Now for our added Domain Groups
net groupmap add ntgroup="Designers" unixgroup=designers type=d rid=1112
net groupmap add ntgroup="Engineers" unixgroup=engineers type=d rid=1113
net groupmap add ntgroup="QA Team"   unixgroup=qateam    type=d rid=1114
```

5. Create the `scripts` directory for use in the *[NETLOGON]* share:

```
root# mkdir -p /var/lib/samba/netlogon/scripts
```

   Place the logon scripts that will be used (batch or cmd scripts) in this directory.

The above configuration provides a functional Primary Domain Control (PDC) system to which must be added file shares and printers as required.

2.3.3.2 A Big Organization

In this section we finally get to review in brief a Samba-3 configuration that uses a Light Weight Directory Access (LDAP)-based authentication backend. The main reasons for this choice are to provide the ability to host primary and Backup Domain Control (BDC), as well as to enable a higher degree of scalability to meet the needs of a very distributed environment.

The Primary Domain Controller This is an example of a minimal configuration to run a Samba-3 PDC using an LDAP authentication backend. It is assumed that the operating system has been correctly configured.

1. Obtain from the Samba sources `~/examples/LDAP/samba.schema` and copy it to the `/etc/openldap/schema/` directory.
2. Set up the LDAP server. This example is suitable for OpenLDAP 2.1.x. The `/etc/openldap/slapd.conf` file:

```
# Note commented out lines have been removed
include         /etc/openldap/schema/core.schema
include         /etc/openldap/schema/cosine.schema
include         /etc/openldap/schema/inetorgperson.schema
include         /etc/openldap/schema/nis.schema
include         /etc/openldap/schema/samba.schema
```

```
pidfile          /var/run/slapd/slapd.pid
argsfile         /var/run/slapd/slapd.args

database         bdb
suffix           "dc=quenya,dc=org"
rootdn           "cn=Manager,dc=quenya,dc=org"
rootpw           {SSHA}06qDkonA8hk6W6SSnRzWj0/pBcU3m0/P
# The password for the above is 'nastyon3'

directory      /var/lib/ldap

index   objectClass     eq
index cn                      pres,sub,eq
index sn                      pres,sub,eq
index uid                     pres,sub,eq
index displayName             pres,sub,eq
index uidNumber               eq
index gidNumber               eq
index memberUid               eq
index   sambaSID              eq
index   sambaPrimaryGroupSID  eq
index   sambaDomainName       eq
index   default               sub
```

3. Create the following file samba-ldap-init.ldif:

```
# Organization for SambaXP Demo
dn: dc=quenya,dc=org
objectclass: dcObject
objectclass: organization
dc: quenya
o: SambaXP Demo
description: The SambaXP Demo LDAP Tree

# Organizational Role for Directory Management
dn: cn=Manager,dc=quenya,dc=org
objectclass: organizationalRole
cn: Manager
description: Directory Manager

# Setting up the container for users
dn: ou=People, dc=quenya, dc=org
objectclass: top
objectclass: organizationalUnit
ou: People
```

```
# Set up an admin handle for People OU
dn: cn=admin, ou=People, dc=quenya, dc=org
cn: admin
objectclass: top
objectclass: organizationalRole
objectclass: simpleSecurityObject
userPassword: {SSHA}0jBHgQ1vp4EDX2rEMMfIudvRMJoGwjVb
# The password for above is 'mordonL8'
```

4. Load the initial data above into the LDAP database:

```
root# slapadd -v -l initdb.ldif
```

5. Start the LDAP server using the appropriate tool or method for the operating system platform on which it is installed.
6. The `smb.conf` file that drives this backend can be found in example Example 2.9.
7. Add the LDAP password to the `secrets.tdc` file so Samba can update the LDAP database:

```
root# smbpasswd -w mordonL8
```

8. Add users and groups as required. Users and groups added using Samba tools will automatically be added to both the LDAP backend as well as to the operating system as required.

Backup Domain Controller Example 2.10 shows the example configuration for the BDC.

1. Decide if the BDC should have its own LDAP server or not. If the BDC is to be the LDAP server change the following `smb.conf` as indicated. The default configuration in Example 2.10 uses a central LDAP server.
2. Configure the NETLOGON and PROFILES directory as for the PDC in Example 2.10.

Example 2.9. LDAP backend smb.conf for PDC

```
# Global parameters

[global]
workgroup = MIDEARTH
netbios name = FRODO
passdb backend = ldapsam:ldap://localhost
username map = /etc/samba/smbusers
printcap name = cups
add user script = /usr/sbin/useradd -m %u
delete user script = /usr/sbin/userdel -r %u
add group script = /usr/sbin/groupadd %g
delete group script = /usr/sbin/groupdel %g
add user to group script = /usr/sbin/usermod -G %g %u
add machine script = /usr/sbin/useradd -s /bin/false -d /dev/null \
-g machines %u
logon script = scripts\logon.bat
logon path = \\%L\Profiles\%U
logon drive = H:
logon home = \\%L\%U
domain logons = Yes
os level = 35
preferred master = Yes
domain master = Yes
ldap suffix = dc=quenya,dc=org
ldap machine suffix = ou=People
ldap user suffix = ou=People
ldap group suffix = ou=People
ldap idmap suffix = ou=People
ldap admin dn = cn=Manager
ldap ssl = no
ldap passwd sync = Yes
idmap uid = 15000-20000
idmap gid = 15000-20000
winbind separator = +
printing = cups
...
```

Example 2.10. Remote LDAP BDC smb.conf

```
# Global parameters

[global]
workgroup = MIDEARTH
netbios name = GANDALF
passdb backend = ldapsam:ldap://frodo.quenya.org
username map = /etc/samba/smbusers
printcap name = cups
add user script = /usr/sbin/useradd -m %u
delete user script = /usr/sbin/userdel -r %u
add group script = /usr/sbin/groupadd %g
delete group script = /usr/sbin/groupdel %g
add user to group script = /usr/sbin/usermod -G %g %u
add machine script = /usr/sbin/useradd -s /bin/false -d /dev/null \
-g machines %u
logon script = scripts\logon.bat
logon path = \\%L\Profiles\%U
logon drive = H:
logon home = \\%L\%U
domain logons = Yes
os level = 33
preferred master = Yes
domain master = No
ldap suffix = dc=quenya,dc=org
ldap machine suffix = ou=People
ldap user suffix = ou=People
ldap group suffix = ou=People
ldap idmap suffix = ou=People
ldap admin dn = cn=Manager
ldap ssl = no
ldap passwd sync = Yes
idmap uid = 15000-20000
idmap gid = 15000-20000
winbind separator = +
printing = cups
...
```

Part II

Server Configuration Basics

Chapter 3

SERVER TYPES AND SECURITY MODES

This chapter provides information regarding the types of server that Samba may be configured to be. A Microsoft network administrator who wishes to migrate to or use Samba will want to know the meaning, within a Samba context, of terms familiar to MS Windows administrator. This means that it is essential also to define how critical security modes function before we get into the details of how to configure the server itself.

The chapter provides an overview of the security modes of which Samba is capable and how they relate to MS Windows servers and clients.

A question often asked is, "*Why would I want to use Samba?*" Most chapters contain a section that highlights features and benefits. We hope that the information provided will help to answer this question. Be warned though, we want to be fair and reasonable, so not all features are positive towards Samba. The benefit may be on the side of our competition.

3.1 Features and Benefits

Two men were walking down a dusty road, when one suddenly kicked up a small red stone. It hurt his toe and lodged in his sandal. He took the stone out and cursed it with a passion and fury befitting his anguish. The other looked at the stone and said, "*This is a garnet. I can turn that into a precious gem and some day it will make a princess very happy!*"

The moral of this tale: Two men, two very different perspectives regarding the same stone. Like it or not, Samba is like that stone. Treat it the right way and it can bring great pleasure, but if you are forced to use it and have no time for its secrets, then it can be a source of discomfort.

Samba started out as a project that sought to provide interoperability for MS Windows 3.x clients with a UNIX server. It has grown up a lot since its humble beginnings and now provides features and functionality fit for large scale deployment. It also has some warts. In sections like this one we tell of both.

So, what are the benefits of features mentioned in this chapter?

- Samba-3 can replace an MS Windows NT4 Domain Controller.

- Samba-3 offers excellent interoperability with MS Windows NT4-style domains as well as natively with Microsoft Active Directory domains.
- Samba-3 permits full NT4-style Interdomain Trusts.
- Samba has security modes that permit more flexible authentication than is possible with MS Windows NT4 Domain Controllers.
- Samba-3 permits use of multiple account database backends.
- The account (password) database backends can be distributed and replicated using multiple methods. This gives Samba-3 greater flexibility than MS Windows NT4 and in many cases a significantly higher utility than Active Directory domains with MS Windows 200x.

3.2 Server Types

Administrators of Microsoft networks often refer to three different type of servers:

- Domain Controller
  - Primary Domain Controller
  - Backup Domain Controller
  - ADS Domain Controller
- Domain Member Server
  - Active Directory Domain Server
  - NT4 Style Domain Domain Server
- Stand-alone Server

The chapters covering Domain Control, Backup Domain Control and Domain Membership provide pertinent information regarding Samba configuration for each of these server roles. The reader is strongly encouraged to become intimately familiar with the information presented.

3.3 Samba Security Modes

In this section the function and purpose of Samba's security modes are described. An accurate understanding of how Samba implements each security mode as well as how to configure MS Windows clients for each mode will significantly reduce user complaints and administrator heartache.

In the SMB/CIFS networking world, there are only two types of security: *User Level* and *Share Level*. We refer to these collectively as *security levels*. In implementing these two security levels, Samba provides flexibilities that are not available with Microsoft Windows NT4/200x servers. In actual fact, Samba implements *Share Level* security only one way, but has four ways of implementing *User Level* security. Collectively, we call the Samba implementations *Security Modes*. They are known as: *SHARE*, *USER*, *DOMAIN*, *ADS*, and *SERVER* modes. They are documented in this chapter.

An SMB server tells the client at startup what security level it is running. There are two options: Share Level and User Level. Which of these two the client receives affects the way the client then tries to authenticate itself. It does not directly affect (to any great extent) the way the Samba server does security. This may sound strange, but it fits in with the client/server approach of SMB. In SMB everything is initiated and controlled by the client, and the server can only tell the client what is available and whether an action is allowed.

3.3.1 User Level Security

We will describe User Level Security first, as its simpler. In User Level Security, the client will send a session setup request directly following protocol negotiation. This request provides a username and password. The server can either accept or reject that username/password combination. At this stage the server has no idea what share the client will eventually try to connect to, so it can't base the *accept/reject* on anything other than:

1. the username/password.
2. the name of the client machine.

If the server accepts the username/password then the client expects to be able to mount shares (using a *tree connection*) without specifying a password. It expects that all access rights will be as the username/password specified in the *session setup*.

It is also possible for a client to send multiple *session setup* requests. When the server responds, it gives the client a *uid* to use as an authentication tag for that username/password. The client can maintain multiple authentication contexts in this way (WinDD is an example of an application that does this).

3.3.1.1 Example Configuration

The `smb.conf` parameter that sets user level security is:

```
security = user
```

This is the default setting since Samba-2.2.x.

3.3.2 Share Level Security

In Share Level security, the client authenticates itself separately for each share. It sends a password along with each tree connection (share mount). It does not explicitly send a username with this operation. The client expects a password to be associated with each share, independent of the user. This means that Samba has to work out what username the client probably wants to use. It is never explicitly sent the username. Some commercial SMB servers such as NT actually associate passwords directly with shares in Share Level security, but Samba always uses the UNIX authentication scheme where it is a username/password pair that is authenticated, not a share/password pair.

To understand the MS Windows networking parallels, one should think in terms of MS Windows 9x/Me where one can create a shared folder that provides read-only or full access, with or without a password.

Many clients send a session setup even if the server is in Share Level security. They normally send a valid username but no password. Samba records this username in a list of possible usernames. When the client then does a tree connection it also adds to this list the name of the share they try to connect to (useful for home directories) and any users listed in the *user* parameter in the `smb.conf` file. The password is then checked in turn against these possible usernames. If a match is found then the client is authenticated as that user.

3.3.2.1 Example Configuration

The `smb.conf` parameter that sets Share Level security is:

```
security = share
```

There are reports that recent MS Windows clients do not like to work with share mode security servers. You are strongly discouraged from using Share Level security.

3.3.3 Domain Security Mode (User Level Security)

When Samba is operating in *security* = domain mode, the Samba server has a domain security trust account (a machine account) and causes all authentication requests to be passed through to the Domain Controllers. In other words, this configuration makes the Samba server a Domain Member server.

3.3.3.1 Example Configuration

Samba as a Domain Member Server

This method involves addition of the following parameters in the `smb.conf` file:

```
security = domain
workgroup = MIDEARTH
```

In order for this method to work, the Samba server needs to join the MS Windows NT security domain. This is done as follows:

1. On the MS Windows NT Domain Controller, using the Server Manager, add a machine account for the Samba server.

2. On the UNIX/Linux system execute:

```
root# net rpc join -U administrator%password
```

> **NOTE**
>
> 
>
> Samba-2.2.4 and later can auto-join a Windows NT4-style Domain just by executing:
>
> ```
> root# smbpasswd -j DOMAIN_NAME -r PDC_NAME \
>     -U Administrator%password
> ```
>
> Samba-3 can do the same by executing:
>
> ```
> root# net rpc join -U Administrator%password
> ```
>
> It is not necessary with Samba-3 to specify the *DOMAIN_NAME* or the *PDC_NAME* as it figures this out from the `smb.conf` file settings.

Use of this mode of authentication does require there to be a standard UNIX account for each user in order to assign a UID once the account has been authenticated by the remote Windows DC. This account can be blocked to prevent logons by clients other than MS Windows through means such as setting an invalid shell in the `/etc/passwd` entry.

An alternative to assigning UIDs to Windows users on a Samba member server is presented in Chapter 20, *Winbind: Use of Domain Accounts*.

For more information regarding Domain Membership, see Chapter 6, *Domain Membership*.

3.3.4 ADS Security Mode (User Level Security)

Both Samba-2.2, and Samba-3 can join an Active Directory domain. This is possible if the domain is run in native mode. Active Directory in native mode perfectly allows NT4-style Domain Members. This is contrary to popular belief. Active Directory in native mode prohibits only the use of Backup Domain Controllers running MS Windows NT4.

If you are using Active Directory, starting with Samba-3 you can join as a native AD member. Why would you want to do that? Your security policy might prohibit the use of NT-compatible authentication protocols. All your machines are running Windows 2000 and above and all use Kerberos. In this case Samba as an NT4-style domain would still require NT-compatible authentication data. Samba in AD-member mode can accept Kerberos tickets.

3.3.4.1 Example Configuration

```
realm = your.kerberos.REALM
security = ADS
```

The following parameter may be required:

```
password server = your.kerberos.server
```

Please refer to Chapter 6, *Domain Membership* and Section 6.4 for more information regarding this configuration option.

3.3.5 Server Security (User Level Security)

Server Security Mode is left over from the time when Samba was not capable of acting as a Domain Member server. It is highly recommended not to use this feature. Server security mode has many drawbacks that include:

- Potential Account Lockout on MS Windows NT4/200x password servers.
- Lack of assurance that the password server is the one specified.
- Does not work with Winbind, which is particularly needed when storing profiles remotely.
- This mode may open connections to the password server, and keep them open for extended periods.
- Security on the Samba server breaks badly when the remote password server suddenly shuts down.
- With this mode there is NO security account in the domain that the password server belongs to for the Samba server.

In Server Security Mode the Samba server reports to the client that it is in User Level security. The client then does a session setup as described earlier. The Samba server takes the username/password that the client sends and attempts to login to the *password server* by sending exactly the same username/password that it got from the client. If that server is in User Level Security and accepts the password, then Samba accepts the client's connection. This allows the Samba server to use another SMB server as the *password server*.

You should also note that at the start of all this where the server tells the client what security level it is in, it also tells the client if it supports encryption. If it does, it supplies the client with a random cryptkey. The client will then send all passwords in encrypted form. Samba supports this type of encryption by default.

The parameter *security* = server means that Samba reports to clients that it is running in *user mode* but actually passes off all authentication requests to another *user mode* server. This requires an additional parameter *password server* that points to the real authentication server. The real authentication server can be another Samba server, or it can be a Windows NT server, the latter being natively capable of encrypted password support.

NOTE

When Samba is running in *Server Security Mode* it is essential that the parameter *password server* is set to the precise NetBIOS machine name of the target authentication server. Samba cannot determine this from NetBIOS name lookups because the choice of the target authentication server is arbitrary and cannot be determined from a domain name. In essence, a Samba server that is in *Server Security Mode* is operating in what used to be known as workgroup mode.

3.3.5.1 Example Configuration

Using MS Windows NT as an Authentication Server

This method involves the additions of the following parameters in the `smb.conf` file:

```
encrypt passwords = Yes
security = server
password server = "NetBIOS_name_of_a_DC"
```

There are two ways of identifying whether or not a username and password pair is valid. One uses the reply information provided as part of the authentication messaging process, the other uses just an error code.

The downside of this mode of configuration is the fact that for security reasons Samba will send the password server a bogus username and a bogus password and if the remote server fails to reject the username and password pair then an alternative mode of identification of validation is used. Where a site uses password lock out after a certain number of failed authentication attempts this will result in user lockouts.

Use of this mode of authentication requires a standard UNIX account for the user. This account can be blocked to prevent logons by non-SMB/CIFS clients.

3.4 Password Checking

MS Windows clients may use encrypted passwords as part of a challenge/response authentication model (a.k.a. NTLMv1 and NTLMv2) or alone, or cleartext strings for simple password-based authentication. It should be realized that with the SMB protocol, the password is passed over the network either in plain-text or encrypted, but not both in the same authentication request.

When encrypted passwords are used, a password that has been entered by the user is encrypted in two ways:

- An MD4 hash of the unicode of the password string. This is known as the NT hash.
- The password is converted to upper case, and then padded or truncated to 14 bytes. This string is then appended with 5 bytes of NULL characters and split to form two

56-bit DES keys to encrypt a "*magic*" 8-byte value. The resulting 16 bytes form the LanMan hash.

MS Windows 95 pre-service pack 1, MS Windows NT versions 3.x and version 4.0 pre-service pack 3 will use either mode of password authentication. All versions of MS Windows that follow these versions no longer support plain text passwords by default.

MS Windows clients have a habit of dropping network mappings that have been idle for 10 minutes or longer. When the user attempts to use the mapped drive connection that has been dropped, the client re-establishes the connection using a cached copy of the password.

When Microsoft changed the default password mode, support was dropped for caching of the plain-text password. This means that when the registry parameter is changed to re-enable use of plain-text passwords it appears to work, but when a dropped service connection mapping attempts to revalidate, this will fail if the remote authentication server does not support encrypted passwords. It is definitely not a good idea to re-enable plain-text password support in such clients.

The following parameters can be used to work around the issue of Windows 9x/Me clients upper-casing usernames and passwords before transmitting them to the SMB server when using cleartext authentication:

```
password level = integer
username level = integer
```

By default Samba will convert to lower case the username before attempting to lookup the user in the database of local system accounts. Because UNIX usernames conventionally only contain lower-case characters, the ***username level*** parameter is rarely needed.

However, passwords on UNIX systems often make use of mixed-case characters. This means that in order for a user on a Windows 9x/Me client to connect to a Samba server using cleartext authentication, the ***password level*** must be set to the maximum number of upper case letters that *could* appear in a password. Note that if the server OS uses the traditional DES version of crypt(), a ***password level*** of 8 will result in case insensitive passwords as seen from Windows users. This will also result in longer login times as Samba has to compute the permutations of the password string and try them one by one until a match is located (or all combinations fail).

The best option to adopt is to enable support for encrypted passwords wherever Samba is used. Most attempts to apply the registry change to re-enable plain-text passwords will eventually lead to user complaints and unhappiness.

3.5 Common Errors

We all make mistakes. It is okay to make mistakes, as long as they are made in the right places and at the right time. A mistake that causes lost productivity is seldom tolerated, however a mistake made in a developmental test lab is expected.

Here we look at common mistakes and misapprehensions that have been the subject of discussions on the Samba mailing lists. Many of these are avoidable by doing your homework before attempting a Samba implementation. Some are the result of a misunderstanding of the English language. The English language, which has many phrases that are potentially vague and may be highly confusing to those for whom English is not their native tongue.

3.5.1 What Makes Samba a Server?

To some the nature of the Samba *security* mode is obvious, but entirely wrong all the same. It is assumed that *security* = server means that Samba will act as a server. Not so! This setting means that Samba will *try* to use another SMB server as its source for user authentication alone.

3.5.2 What Makes Samba a Domain Controller?

The smb.conf parameter *security* = domain does not really make Samba behave as a Domain Controller. This setting means we want Samba to be a Domain Member.

3.5.3 What Makes Samba a Domain Member?

Guess! So many others do. But whatever you do, do not think that *security* = user makes Samba act as a Domain Member. Read the manufacturer's manual before the warranty expires. See Chapter 6, *Domain Membership* for more information.

3.5.4 Constantly Losing Connections to Password Server

"*Why does server_validate() simply give up rather than re-establish its connection to the password server? Though I am not fluent in the SMB protocol, perhaps the cluster server process passes along to its client workstation the session key it receives from the password server, which means the password hashes submitted by the client would not work on a subsequent connection whose session key would be different. So server_validate() must give up.*"

Indeed. That's why *security* = server is at best a nasty hack. Please use *security* = domain; *security* = server mode is also known as pass-through authentication.

Chapter 4

DOMAIN CONTROL

There are many who approach MS Windows networking with incredible misconceptions. That's okay, because it gives the rest of us plenty of opportunity to be of assistance. Those who really want help would be well advised to become familiar with information that is already available.

The reader is advised not to tackle this section without having first understood and mastered some basics. MS Windows networking is not particularly forgiving of misconfiguration. Users of MS Windows networking are likely to complain of persistent niggles that may be caused by a broken network configuration. To a great many people, however, MS Windows networking starts with a Domain Controller that in some magical way is expected to solve all network operational ills.

The diagram in Figure 4.1 shows a typical MS Windows Domain Security network environment. Workstations A, B and C are representative of many physical MS Windows network clients.

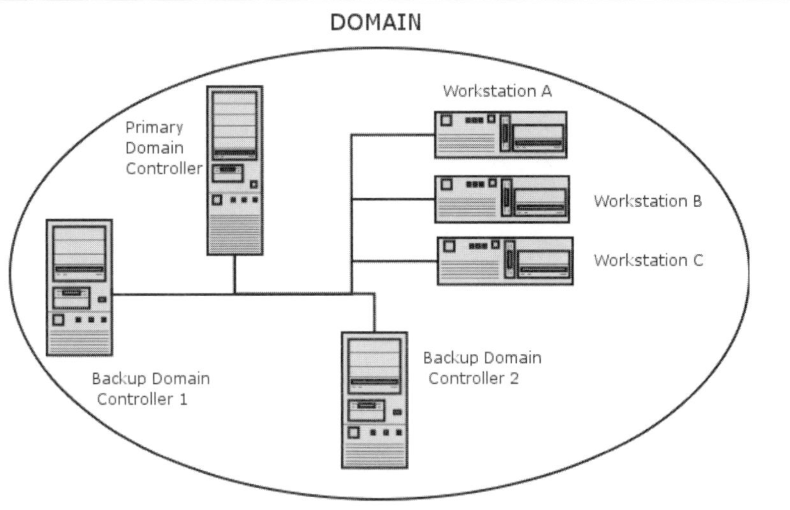

Figure 4.1. An Example Domain.

From the Samba mailing list one can readily identify many common networking issues. If you are not clear on the following subjects, then it will do much good to read the sections of this HOWTO that deal with it. These are the most common causes of MS Windows networking problems:

- Basic TCP/IP configuration.
- NetBIOS name resolution.
- Authentication configuration.
- User and group configuration.
- Basic file and directory permission control in UNIX/Linux.
- Understanding how MS Windows clients interoperate in a network environment.

Do not be put off; on the surface of it MS Windows networking seems so simple that anyone can do it. In fact, it is not a good idea to set up an MS Windows network with inadequate training and preparation. But let's get our first indelible principle out of the way: *It is perfectly okay to make mistakes!* In the right place and at the right time, mistakes are the essence of learning. It is very much not okay to make mistakes that cause loss of productivity and impose an avoidable financial burden on an organization.

Where is the right place to make mistakes? Only out of harm's way. If you are going to make mistakes, then please do it on a test network, away from users and in such a way as to not inflict pain on others. Do your learning on a test network.

4.1 Features and Benefits

What is the key benefit of Microsoft Domain Security?

In a word, *Single Sign On*, or SSO for short. To many, this is the Holy Grail of MS Windows NT and beyond networking. SSO allows users in a well-designed network to log onto any workstation that is a member of the domain that their user account is in (or in a domain that has an appropriate trust relationship with the domain they are visiting) and they will be able to log onto the network and access resources (shares, files and printers) as if they are sitting at their home (personal) workstation. This is a feature of the Domain Security protocols.

The benefits of Domain Security are available to those sites that deploy a Samba PDC. A Domain provides a unique network security identifier (SID). Domain user and group security identifiers are comprised of the network SID plus a relative identifier (RID) that is unique to the account. User and Group SIDs (the network SID plus the RID) can be used to create Access Control Lists (ACLs) attached to network resources to provide organizational access control. UNIX systems recognize only local security identifiers.

Note

Network clients of an MS Windows Domain Security Environment must be Domain Members to be able to gain access to the advanced features provided. Domain Membership involves more than just setting the workgroup name to the Domain name. It requires the creation of a Domain trust account for the workstation (called a machine account). Refer to Chapter 6, *Domain Membership* for more information.

The following functionalities are new to the Samba-3 release:

- Windows NT4 domain trusts.
- Adding users via the User Manager for Domains. This can be done on any MS Windows client using the `Nexus.exe` toolkit that is available from Microsoft's Web site. Samba-3 supports the use of the Microsoft Management Console for user management.
- Introduces replaceable and multiple user account (authentication) backends. In the case where the backend is placed in an LDAP database, Samba-3 confers the benefits of a backend that can be distributed, replicated and is highly scalable.
- Implements full Unicode support. This simplifies cross locale internationalization support. It also opens up the use of protocols that Samba-2.2.x had but could not use due to the need to fully support Unicode.

The following functionalities are not provided by Samba-3:

- SAM replication with Windows NT4 Domain Controllers (i.e., a Samba PDC and a Windows NT BDC or vice versa). This means Samba cannot operate as a BDC when the PDC is Microsoft-based or replicate account data to Windows BDCs.
- Acting as a Windows 2000 Domain Controller (i.e., Kerberos and Active Directory). In point of fact, Samba-3 does have some Active Directory Domain Control ability that is at this time purely experimental that is certain to change as it becomes a fully supported feature some time during the Samba-3 (or later) life cycle. However, Active Directory is more then just SMB — it's also LDAP, Kerberos, DHCP, and other protocols (with proprietary extensions, of course).
- The Windows 200x/XP MMC (Computer Management) Console can not be used to manage a Samba-3 server. For this you can use only the MS Windows NT4 Domain Server manager and the MS Windows NT4 Domain User Manager. Both are part of the SVRTOOLS.EXE package mentioned later.

Windows 9x/Me/XP Home clients are not true members of a domain for reasons outlined in this chapter. The protocol for support of Windows 9x/Me style network (domain) logons is completely different from NT4/Windows 200x type domain logons and has been officially supported for some time. These clients use the old LanMan Network Logon facilities that are supported in Samba since approximately the Samba-1.9.15 series.

Samba-3 implements group mapping between Windows NT groups and UNIX groups (this is really quite complicated to explain in a short space). This is discussed more fully in

Chapter 11, *Group Mapping MS Windows and UNIX*.

Samba-3, like an MS Windows NT4 PDC or a Windows 200x Active Directory, needs to store user and Machine Trust Account information in a suitable backend datastore. Refer to Section 6.2. With Samba-3 there can be multiple backends for this. A complete discussion of account database backends can be found in Chapter 10, *Account Information Databases*.

4.2 Basics of Domain Control

Over the years, public perceptions of what Domain Control really is has taken on an almost mystical nature. Before we branch into a brief overview of Domain Control, there are three basic types of Domain Controllers.

4.2.1 Domain Controller Types

- Primary Domain Controller
- Backup Domain Controller
- ADS Domain Controller

The *Primary Domain Controller* or PDC plays an important role in MS Windows NT4. In Windows 200x Domain Control architecture, this role is held by Domain Controllers. Folklore dictates that because of its role in the MS Windows network, the Domain Controller should be the most powerful and most capable machine in the network. As strange as it may seem to say this here, good overall network performance dictates that the entire infrastructure needs to be balanced. It is advisable to invest more in Stand-alone (Domain Member) servers than in the Domain Controllers.

In the case of MS Windows NT4-style domains, it is the PDC that initiates a new Domain Control database. This forms a part of the Windows registry called the Security Account Manager (SAM). It plays a key part in NT4-type domain user authentication and in synchronization of the domain authentication database with Backup Domain Controllers.

With MS Windows 200x Server-based Active Directory domains, one Domain Controller initiates a potential hierarchy of Domain Controllers, each with their own area of delegated control. The master domain controller has the ability to override any downstream controller, but a downline controller has control only over its downline. With Samba-3, this functionality can be implemented using an LDAP-based user and machine account backend.

New to Samba-3 is the ability to use a backend database that holds the same type of data as the NT4-style SAM database (one of the registry files)[1].

The *Backup Domain Controller* or BDC plays a key role in servicing network authentication requests. The BDC is biased to answer logon requests in preference to the PDC. On a network segment that has a BDC and a PDC, the BDC will most likely service network logon requests. The PDC will answer network logon requests when the BDC is too busy (high load). A BDC can be promoted to a PDC. If the PDC is online at the time that a BDC is promoted to PDC, the previous PDC is automatically demoted to a BDC. With Samba-3,

[1]See also Chapter 10, *Account Information Databases*.

this is not an automatic operation; the PDC and BDC must be manually configured and changes also need to be made.

With MS Windows NT4, a decision is made at installation to determine what type of machine the server will be. It is possible to promote a BDC to a PDC and vice versa. The only way to convert a Domain Controller to a Domain Member server or a Stand-alone Server is to reinstall it. The install time choices offered are:

- *Primary Domain Controller* — the one that seeds the domain SAM.
- *Backup Domain Controller* — one that obtains a copy of the domain SAM.
- *Domain Member Server* — one that has no copy of the domain SAM, rather it obtains authentication from a Domain Controller for all access controls.
- *Stand-alone Server* — one that plays no part is SAM synchronization, has its own authentication database and plays no role in Domain Security.

With MS Windows 2000, the configuration of Domain Control is done after the server has been installed. Samba-3 is capable of acting fully as a native member of a Windows 200x server Active Directory domain.

New to Samba-3 is the ability to function fully as an MS Windows NT4-style Domain Controller, excluding the SAM replication components. However, please be aware that Samba-3 also supports the MS Windows 200x Domain Control protocols.

At this time any appearance that Samba-3 is capable of acting as an *Domain Controller* in native ADS mode is limited and experimental in nature. This functionality should not be used until the Samba Team offers formal support for it. At such a time, the documentation will be revised to duly reflect all configuration and management requirements. Samba can act as a NT4-style DC in a Windows 2000/XP environment. However, there are certain compromises:

- No machine policy files.
- No Group Policy Objects.
- No synchronously executed AD logon scripts.
- Can't use Active Directory management tools to manage users and machines.
- Registry changes tattoo the main registry, while with AD they do not leave permanent changes in effect.
- Without AD you cannot perform the function of exporting specific applications to specific users or groups.

4.2.2 Preparing for Domain Control

There are two ways that MS Windows machines may interact with each other, with other servers and with Domain Controllers: either as *Stand-alone* systems, more commonly called *Workgroup* members, or as full participants in a security system, more commonly called *Domain* members.

It should be noted that *Workgroup* membership involves no special configuration other than the machine being configured so the network configuration has a commonly used name for its workgroup entry. It is not uncommon for the name WORKGROUP to be used for this. With this mode of configurationi, there are no Machine Trust Accounts and any concept of membership as such is limited to the fact that all machines appear in the network neighborhood to be logically grouped together. Again, just to be clear: *workgroup mode does not involve security machine accounts.*

Domain Member machines have a machine account in the Domain accounts database. A special procedure must be followed on each machine to effect Domain Membership. This procedure, which can be done only by the local machine Administrator account, will create the Domain machine account (if it does not exist), and then initializes that account. When the client first logs onto the Domain it triggers a machine password change.

NOTE

When Samba is configured as a Domain Controller, secure network operation demands that all MS Windows NT4/200x/XP Professional clients should be configured as Domain Members. If a machine is not made a member of the Domain, then it will operate like a workgroup (Stand-alone) machine. Please refer to Chapter 6, *Domain Membership* for information regarding Domain Membership.

The following are necessary for configuring Samba-3 as an MS Windows NT4-style PDC for MS Windows NT4/200x/XP clients:

- Configuration of basic TCP/IP and MS Windows networking.
- Correct designation of the Server Role (*security* = user).
- Consistent configuration of Name Resolution[2].
- Domain logons for Windows NT4/200x/XP Professional clients.
- Configuration of Roaming Profiles or explicit configuration to force local profile usage.
- Configuration of network/system policies.
- Adding and managing domain user accounts.
- Configuring MS Windows client machines to become Domain Members.

The following provisions are required to serve MS Windows 9x/Me clients:

- Configuration of basic TCP/IP and MS Windows networking.
- Correct designation of the server role (*security* = user).
- Network Logon Configuration (since Windows 9x/Me/XP Home are not technically domain members, they do not really participate in the security aspects of Domain logons as such).

[2]See Chapter 9, *Network Browsing*, and Chapter 25, *Integrating MS Windows Networks with Samba.*

- Roaming Profile Configuration.
- Configuration of System Policy handling.
- Installation of the network driver "*Client for MS Windows Networks*" and configuration to log onto the domain.
- Placing Windows 9x/Me clients in User Level Security — if it is desired to allow all client share access to be controlled according to domain user/group identities.
- Adding and managing domain user accounts.

NOTE

Roaming Profiles and System/Network policies are advanced network administration topics that are covered in the Chapter 23, *Desktop Profile Management* and Chapter 22, *System and Account Policies* chapters of this document. However, these are not necessarily specific to a Samba PDC as much as they are related to Windows NT networking concepts.

A Domain Controller is an SMB/CIFS server that:

- Registers and advertises itself as a Domain Controller (through NetBIOS broadcasts as well as by way of name registrations either by Mailslot Broadcasts over UDP broadcast, to a WINS server over UDP unicast, or via DNS and Active Directory).
- Provides the NETLOGON service. (This is actually a collection of services that runs over mulitple protocols. These include the LanMan Logon service, the Netlogon service, the Local Security Account service, and variations of them.)
- Provides a share called NETLOGON.

It is rather easy to configure Samba to provide these. Each Samba Domain Controller must provide the NETLOGON service that Samba calls the *domain logons* functionality (after the name of the parameter in the `smb.conf` file). Additionally, one server in a Samba-3 Domain must advertise itself as the Domain Master Browser[3]. This causes the Primary Domain Controller to claim a domain-specific NetBIOS name that identifies it as a Domain Master Browser for its given domain or workgroup. Local master browsers in the same domain or workgroup on broadcast-isolated subnets then ask for a complete copy of the browse list for the whole wide area network. Browser clients will then contact their Local Master Browser, and will receive the domain-wide browse list, instead of just the list for their broadcast-isolated subnet.

4.3 Domain Control — Example Configuration

The first step in creating a working Samba PDC is to understand the parameters necessary in `smb.conf`. An example `smb.conf` for acting as a PDC can be found in Example 4.1.

[3]See Chapter 9, *Network Browsing.*

Example 4.1. smb.conf for being a PDC

```
[global]
netbios name = BELERIAND
workgroup = MIDEARTH
passdb backend = tdbsam
os level = 33
preferred master = yes
domain master = yes
local master = yes
security = user
domain logons = yes
logon path = \\%N\profiles\%u
logon drive = H:
logon home = \\homeserver\%u\winprofile
logon script = logon.cmd

[netlogon]
path = /var/lib/samba/netlogon
read only = yes
write list = ntadmin

[profiles]
path = /var/lib/samba/profiles
read only = no
create mask = 0600
directory mask = 0700
```

The basic options shown in Example 4.1 are explained as follows:

passdb backend — This contains all the user and group account information. Acceptable values for a PDC are: *smbpasswd, tdbsam, and ldapsam.* The "*guest*" entry provides default accounts and is included by default, there is no need to add it explicitly.

Where use of backup Domain Controllers (BDCs) is intended, the only logical choice is to use LDAP so the passdb backend can be distributed. The tdbsam and smbpasswd files cannot effectively be distributed and therefore should not be used.

Domain Control Parameters — The parameters *os level, preferred master, domain master, security, encrypt passwords, and domain logons* play a central role in assuring domain control and network logon support.

The *os level* must be set at or above a value of 32. A Domain Controller must be the Domain Master Browser, must be set in *user* mode security, must support Microsoft-compatible encrypted passwords, and must provide the network logon service (domain logons). Encrypted passwords must be enabled. For more details on how to do this, refer to Chapter 10, *Account Information Databases.*

Environment Parameters — The parameters *logon path, logon home, logon drive, and logon script* are environment support settings that help to facilitate client logon operations and that help to provide automated control facilities to ease network management overheads. Please refer to the man page information for these parameters.

NETLOGON Share — The NETLOGON share plays a central role in domain logon and Domain Membership support. This share is provided on all Microsoft Domain Controllers. It is used to provide logon scripts, to store Group Policy files (NTConfig.POL), as well as to locate other common tools that may be needed for logon processing. This is an essential share on a Domain Controller.

PROFILE Share — This share is used to store user desktop profiles. Each user must have a directory at the root of this share. This directory must be write-enabled for the user and must be globally read-enabled. Samba-3 has a VFS module called "*fake_permissions*" that may be installed on this share. This will allow a Samba administrator to make the directory read-only to everyone. Of course this is useful only after the profile has been properly created.

NOTE

The above parameters make for a full set of parameters that may define the server's mode of operation. The following `smb.conf` parameters are the essentials alone:

```
netbios name = BELERIAND
workgroup = MIDEARTH
domain logons = Yes
domain master = Yes
security = User
```

The additional parameters shown in the longer listing above just makes for a more complete explanation.

4.4 Samba ADS Domain Control

Samba-3 is not, and cannot act as, an Active Directory Server. It cannot truly function as an Active Directory Primary Domain Controller. The protocols for some of the functionality of Active Directory Domain Controllers has been partially implemented on an experimental only basis. Please do not expect Samba-3 to support these protocols. Do not depend on any such functionality either now or in the future. The Samba Team may remove these experimental features or may change their behavior. This is mentioned for the benefit of those who have discovered secret capabilities in Samba-3 and who have asked when this functionality will be completed. The answer is maybe or maybe never!

To be sure, Samba-3 is designed to provide most of the functionality that Microsoft Windows NT4-style Domain Controllers have. Samba-3 does not have all the capabilities of Windows NT4, but it does have a number of features that Windows NT4 domain contollers do not have. In short, Samba-3 is not NT4 and it is not Windows Server 200x, it is not an Active Directory server. We hope this is plain and simple enough for all to understand.

4.5 Domain and Network Logon Configuration

The subject of Network or Domain Logons is discussed here because it forms an integral part of the essential functionality that is provided by a Domain Controller.

4.5.1 Domain Network Logon Service

All Domain Controllers must run the netlogon service (*domain logons* in Samba). One Domain Controller must be configured with *domain master* = Yes (the Primary Domain Controller); on all Backup Domain Controllers *domain master* = No must be set.

4.5.1.1 Example Configuration

Example 4.2. smb.conf for being a PDC

```
[global]
domain logons = Yes
domain master = (Yes on PDC, No on BDCs)

[netlogon]
comment = Network Logon Service
path = /var/lib/samba/netlogon
guest ok = Yes
browseable = No
```

4.5.1.2 The Special Case of MS Windows XP Home Edition

To be completely clear: If you want MS Windows XP Home Edition to integrate with your MS Windows NT4 or Active Directory Domain Security, understand it cannot be done. The only option is to purchase the upgrade from MS Windows XP Home Edition to MS Windows XP Professional.

> NOTE
>
> 
>
> MS Windows XP Home Edition does not have the ability to join any type of Domain Security facility. Unlike MS Windows 9x/Me, MS Windows XP Home Edition also completely lacks the ability to log onto a network.

Now that this has been said, please do not ask the mailing list or email any of the Samba Team members with your questions asking how to make this work. It can't be done. If it can be done, then to do so would violate your software license agreement with Microsoft, and we recommend that you do not do that.

4.5.1.3 The Special Case of Windows 9x/Me

A domain and a workgroup are exactly the same in terms of network browsing. The difference is that a distributable authentication database is associated with a domain, for secure login access to a network. Also, different access rights can be granted to users if they successfully authenticate against a domain logon server. Samba-3 does this now in the same way as MS Windows NT/200x.

The SMB client logging on to a domain has an expectation that every other server in the domain should accept the same authentication information. Network browsing functionality of domains and workgroups is identical and is explained in this documentation under the browsing discussions. It should be noted that browsing is totally orthogonal to logon support.

Issues related to the single-logon network model are discussed in this section. Samba supports domain logons, network logon scripts and user profiles for MS Windows for workgroups and MS Windows 9X/ME clients, which are the focus of this section.

When an SMB client in a domain wishes to logon, it broadcasts requests for a logon server. The first one to reply gets the job, and validates its password using whatever mechanism the Samba administrator has installed. It is possible (but ill advised) to create a domain where the user database is not shared between servers, i.e., they are effectively workgroup servers advertising themselves as participating in a domain. This demonstrates how authentication is quite different from but closely involved with domains.

Using these features you can make your clients verify their logon via the Samba server; make clients run a batch file when they logon to the network and download their preferences, desktop and start menu.

MS Windows XP Home edition is not able to join a domain and does not permit the use of domain logons.

Before launching into the configuration instructions, it is worthwhile to look at how a Windows 9x/Me client performs a logon:

1. The client broadcasts (to the IP broadcast address of the subnet it is in) a NetLogon request. This is sent to the NetBIOS name DOMAIN<#1c> at the NetBIOS layer.

The client chooses the first response it receives, which contains the NetBIOS name of the logon server to use in the format of `\\SERVER`.

2. The client connects to that server, logs on (does an SMBsessetupX) and then connects to the IPC$ share (using an SMBtconX).

3. The client does a NetWkstaUserLogon request, which retrieves the name of the user's logon script.

4. The client then connects to the NetLogon share and searches for said script. If it is found and can be read, it is retrieved and executed by the client. After this, the client disconnects from the NetLogon share.

5. The client sends a NetUserGetInfo request to the server to retrieve the user's home share, which is used to search for profiles. Since the response to the NetUserGetInfo request does not contain much more than the user's home share, profiles for Windows 9x clients must reside in the user home directory.

6. The client connects to the user's home share and searches for the user's profile. As it turns out, you can specify the user's home share as a sharename and path. For example, `\\server\fred\.winprofile`. If the profiles are found, they are implemented.

7. The client then disconnects from the user's home share and reconnects to the NetLogon share and looks for `CONFIG.POL`, the policies file. If this is found, it is read and implemented.

The main difference between a PDC and a Windows 9x/Me logon server configuration is:

- Password encryption is not required for a Windows 9x/Me logon server. But note that beginning with MS Windows 98 the default setting is that plain-text password support is disabled. It can be re-enabled with the registry changes that are documented in Chapter 22, *System and Account Policies*.
- Windows 9x/Me clients do not require and do not use Machine Trust Accounts.

A Samba PDC will act as a Windows 9x/Me logon server; after all, it does provide the network logon services that MS Windows 9x/Me expect to find.

NOTE

Use of plain-text passwords is strongly discouraged. Where used they are easily detected using a sniffer tool to examine network traffic.

4.5.2 Security Mode and Master Browsers

There are a few comments to make in order to tie up some loose ends. There has been much debate over the issue of whether it is okay to configure Samba as a Domain Controller in security modes other than user. The only security mode that will not work due to technical reasons is share-mode security. Domain and server mode security are really just a variation on SMB User Level Security.

Actually, this issue is also closely tied to the debate on whether Samba must be the Domain Master Browser for its workgroup when operating as a DC. While it may technically be possible to configure a server as such (after all, browsing and domain logons are two distinctly different functions), it is not a good idea to do so. You should remember that the DC must register the DOMAIN<#1b> NetBIOS name. This is the name used by Windows clients to locate the DC. Windows clients do not distinguish between the DC and the DMB. A DMB is a Domain Master Browser — see Section 9.4.1. For this reason, it is wise to configure the Samba DC as the DMB.

Now back to the issue of configuring a Samba DC to use a mode other than *security* = user. If a Samba host is configured to use another SMB server or DC in order to validate user connection requests, it is a fact that some other machine on the network (the *password server*) knows more about the user than the Samba host. About 99% of the time, this other host is a Domain Controller. Now to operate in domain mode security, the *workgroup* parameter must be set to the name of the Windows NT domain (which already has a Domain Controller). If the domain does not already have a Domain Controller, you do not yet have a Domain.

Configuring a Samba box as a DC for a domain that already by definition has a PDC is asking for trouble. Therefore, you should always configure the Samba DC to be the DMB for its domain and set *security* = user. This is the only officially supported mode of operation.

4.6 Common Errors

4.6.1 "$" Cannot Be Included in Machine Name

A machine account, typically stored in `/etc/passwd`, takes the form of the machine name with a "$" appended. FreeBSD (and other BSD systems) will not create a user with a "$" in the name.

The problem is only in the program used to make the entry. Once made, it works perfectly. Create a user without the "$". Then use **vipw** to edit the entry, adding the "$". Or create the whole entry with vipw if you like; make sure you use a unique user login ID.

NOTE

The machine account must have the exact name that the workstation has.

NOTE

The UNIX tool **vipw** is a common tool for directly editing the /etc/passwd file.

4.6.2 Joining Domain Fails Because of Existing Machine Account

"I get told, 'You already have a connection to the Domain....' or 'Cannot join domain, the credentials supplied conflict with an existing set...' when creating a Machine Trust Account."

This happens if you try to create a Machine Trust Account from the machine itself and already have a connection (e.g., mapped drive) to a share (or IPC$) on the Samba PDC. The following command will remove all network drive connections:

```
C:\> net use * /d
```

Further, if the machine is already a "*member of a workgroup*" that is the same name as the domain you are joining (bad idea) you will get this message. Change the workgroup name to something else, it does not matter what, reboot, and try again.

4.6.3 The System Cannot Log You On (C000019B)

"I joined the domain successfully but after upgrading to a newer version of the Samba code I get the message, 'The system cannot log you on (C000019B), Please try again or consult your system administrator when attempting to logon.'"

This occurs when the domain SID stored in the secrets.tdb database is changed. The most common cause of a change in domain SID is when the domain name and/or the server name (NetBIOS name) is changed. The only way to correct the problem is to restore the original domain SID or remove the domain client from the domain and rejoin. The domain SID may be reset using either the net or rpcclient utilities.

To reset or change the domain SID you can use the net command as follows:

```
root# net getlocalsid 'OLDNAME'
root# net setlocalsid 'SID'
```

Workstation Machine Trust Accounts work only with the Domain (or network) SID. If this SID changes Domain Members (workstations) will not be able to log onto the domain. The original Domain SID can be recovered from the secrets.tdb file. The alternative is to visit each workstation to re-join it to the domain.

4.6.4 The Machine Trust Account Is Not Accessible

"When I try to join the domain I get the message, 'The machine account for this computer either does not exist or is not accessible'. What's wrong?"

This problem is caused by the PDC not having a suitable Machine Trust Account. If you are using the ***add machine script*** method to create accounts then this would indicate that it has not worked. Ensure the domain admin user system is working.

Alternately, if you are creating account entries manually then they have not been created correctly. Make sure that you have the entry correct for the Machine Trust Account in `smbpasswd` file on the Samba PDC. If you added the account using an editor rather than using the smbpasswd utility, make sure that the account name is the machine NetBIOS name with a "$" appended to it (i.e., computer_name$). There must be an entry in both /etc/passwd and the smbpasswd file.

Some people have also reported that inconsistent subnet masks between the Samba server and the NT client can cause this problem. Make sure that these are consistent for both client and server.

4.6.5 Account Disabled

"When I attempt to login to a Samba Domain from a NT4/W200x workstation, I get a message about my account being disabled."

Enable the user accounts with `smbpasswd -e` ***`username`***. This is normally done as an account is created.

4.6.6 Domain Controller Unavailable

"Until a few minutes after Samba has started, clients get the error 'Domain Controller Unavailable'"

A Domain Controller has to announce its role on the network. This usually takes a while. Be patient for up to fifteen minutes, then try again.

4.6.7 Cannot Log onto Domain Member Workstation After Joining Domain

After successfully joining the domain, user logons fail with one of two messages: one to the effect that the Domain Controller cannot be found; the other claims that the account does not exist in the domain or that the password is incorrect. This may be due to incompatible settings between the Windows client and the Samba-3 server for *schannel* (secure channel) settings or *smb signing* settings. Check your Samba settings for *client schannel, server schannel, client signing, server signing* by executing:

```
testparm -v | more and looking for the value of these parameters.
```

Also use the Microsoft Management Console — Local Security Settings. This tool is available from the Control Panel. The Policy settings are found in the Local Policies/Securty Options area and are prefixed by *Secure Channel: ..., and Digitally sign .....*

It is important that these be set consistently with the Samba-3 server settings.

Chapter 5

BACKUP DOMAIN CONTROL

Before you continue reading this section, please make sure that you are comfortable with configuring a Samba Domain Controller as described in Chapter 4, *Domain Control*.

5.1 Features and Benefits

This is one of the most difficult chapters to summarize. It does not matter what we say here for someone will still draw conclusions and/or approach the Samba Team with expectations that are either not yet capable of being delivered, or that can be achieved far more effectively using a totally different approach. In the event that you should have a persistent concern that is not addressed in this book, please email John H. Terpstra[1] clearly setting out your requirements and/or question and we will do our best to provide a solution.

Samba-3 is capable of acting as a Backup Domain Controller (BDC) to another Samba Primary Domain Controller (PDC). A Samba-3 PDC can operate with an LDAP Account backend. The LDAP backend can be either a common master LDAP server, or a slave server. The use of a slave LDAP server has the benefit that when the master is down, clients may still be able to log onto the network. This effectively gives Samba a high degree of scalability and is an effective solution for large organizations. Do not use an LDAP slave server for a PDC, this may cause serious stability and operational problems.

While it is possible to run a Samba-3 BDC with non-LDAP backend, the administrator will need to figure out precisely what is the best way to replicate (copy/distribute) the user and machine accounts' backend.

The use of a non-LDAP backend SAM database is particularly problematic because Domain Member servers and workstations periodically change the Machine Trust Account password. The new password is then stored only locally. This means that in the absence of a centrally stored accounts database (such as that provided with an LDAP-based solution) if Samba-3 is running as a BDC, the BDC instance of the Domain Member trust account password will not reach the PDC (master) copy of the SAM. If the PDC SAM is then replicated to BDCs, this results in overwriting the SAM that contains the updated (changed) trust account password with resulting breakage of the domain trust.

[1] mailto:jht@samba.org

Considering the number of comments and questions raised concerning how to configure a BDC, let's consider each possible option and look at the pros and cons for each possible solution. Table 5.1 lists possible design configurations for a PDC/BDC infrastructure.

Table 5.1. Domain Backend Account Distribution Options

PDC Backend	BDC Backend	Notes/Discussion
Master LDAP Server	Slave LDAP Server	The optimal solution that provides high integrity. The SAM will be replicated to a common master LDAP server.
Single Central LDAP Server	Single Central LDAP Server	A workable solution without fail-over ability. This is a useable solution, but not optimal.
tdbsam	tdbsam + **net rpc vampire**	Does not work with Samba-3.0.0; may be implemented in a later release. The downside of this solution is that an external process will control account database integrity. This solution may appeal to sites that wish to avoid the complexity of LDAP. The **net rpc vampire** is used to synchronize domain accounts from the PDC to the BDC.
tdbsam	tdbsam + **rsync**	Do not use this configuration. Does not work because the TDB files are live and data may not have been flushed to disk. Use **rsync** to synchronize the TDB database files from the PDC to the BDC.
smbpasswd file	smbpasswd file	Do not use this configuration. Not an elegant solution due to the delays in synchronization. Use **rsync** to synchronize the TDB database files from the PDC to the BDC. Can be made to work using a **cron** job to synchronize data from the PDC to the BDC.

5.2 Essential Background Information

A Domain Controller is a machine that is able to answer logon requests from network workstations. Microsoft LanManager and IBM LanServer were two early products that provided this capability. The technology has become known as the LanMan Netlogon service.

When MS Windows NT3.10 was first released, it supported a new style of Domain Control and with it a new form of the network logon service that has extended functionality. This service became known as the NT NetLogon Service. The nature of this service has changed with the evolution of MS Windows NT and today provides a complex array of services that are implemented over an intricate spectrum of technologies.

5.2.1 MS Windows NT4-style Domain Control

Whenever a user logs into a Windows NT4/200x/XP Professional Workstation, the workstation connects to a Domain Controller (authentication server) to validate that the username and password the user entered are valid. If the information entered does not match account information that has been stored in the Domain Control database (the SAM, or Security Account Manager database), a set of error codes is returned to the workstation that has made the authentication request.

When the username/password pair has been validated, the Domain Controller (authentication server) will respond with full enumeration of the account information that has been stored regarding that user in the User and Machine Accounts database for that Domain. This information contains a complete network access profile for the user but excludes any information that is particular to the user's desktop profile, or for that matter it excludes all desktop profiles for groups that the user may belong to. It does include password time limits, password uniqueness controls, network access time limits, account validity information, machine names from which the user may access the network, and much more. All this information was stored in the SAM in all versions of MS Windows NT (3.10, 3.50, 3.51, 4.0).

The account information (user and machine) on Domain Controllers is stored in two files, one containing the Security information and the other the SAM. These are stored in files by the same name in the `C:\Windows NT\System32\config` directory. These are the files that are involved in replication of the SAM database where Backup Domain Controllers are present on the network.

There are two situations in which it is desirable to install Backup Domain Controllers:

- On the local network that the Primary Domain Controller is on, if there are many workstations and/or where the PDC is generally very busy. In this case the BDCs will pick up network logon requests and help to add robustness to network services.
- At each remote site, to reduce wide area network traffic and to add stability to remote network operations. The design of the network, the strategic placement of Backup Domain Controllers, together with an implementation that localizes as much of network to client interchange as possible will help to minimize wide area network bandwidth needs (and thus costs).

The inter-operation of a PDC and its BDCs in a true Windows NT4 environemt is worth mentioning here. The PDC contains the master copy of the SAM. In the event that an administrator makes a change to the user account database while physically present on the local network that has the PDC, the change will likely be made directly to the PDC instance of the master copy of the SAM. In the event that this update may be performed in a branch office, the change will likely be stored in a delta file on the local BDC. The BDC will then send a trigger to the PDC to commence the process of SAM synchronization. The PDC will then request the delta from the BDC and apply it to the master SAM. The PDC will then contact all the BDCs in the Domain and trigger them to obtain the update and then apply that to their own copy of the SAM.

Samba-3 can not participate in true SAM replication and is therefore not able to employ precisely the same protocols used by MS Windows NT4. A Samba-3 BDC will not create

SAM update delta files. It will not inter-operate with a PDC (NT4 or Samba) to synchronize the SAM from delta files that are held by BDCs.

Samba-3 cannot function as a BDC to an MS Windows NT4 PDC, and Samba-3 can not function correctly as a PDC to an MS Windows NT4 BDC. Both Samba-3 and MS Windows NT4 can function as a BDC to its own type of PDC.

The BDC is said to hold a *read-only* of the SAM from which it is able to process network logon requests and authenticate users. The BDC can continue to provide this service, particularly while, for example, the wide area network link to the PDC is down. A BDC plays a very important role in both the maintenance of Domain Security as well as in network integrity.

In the event that the NT4 PDC should need to be taken out of service, or if it dies, one of the NT4 BDCs can be promoted to a PDC. If this happens while the original NT4 PDC is on line, it is automatically demoted to an NT4 BDC. This is an important aspect of Domain Controller management. The tool that is used to effect a promotion or a demotion is the Server Manager for Domains. It should be noted that Samba-3 BDCs can not be promoted in this manner because reconfiguration of Samba requires changes to the `smb.conf` file.

5.2.1.1 Example PDC Configuration

Beginning with Version 2.2, Samba officially supports domain logons for all current Windows clients, including Windows NT4, 2003 and XP Professional. For Samba to be enabled as a PDC, some parameters in the *[global]*-section of the `smb.conf` have to be set. Refer to Example 5.1 for an example of the minimum required settings.

Example 5.1. Minimal smb.conf for a PDC in Use With a BDC LDAP Server on PDC.

```
workgroup = MIDEARTH
passdb backend = ldapsam://localhost:389
domain master = yes
domain logons = yes
```

Several other things like a *[homes]* and a *[netlogon]* share also need to be set along with settings for the profile path, the user's home drive, and so on. This is not covered in this chapter; for more information please refer to Chapter 4, *Domain Control*.

5.2.2 LDAP Configuration Notes

When configuring a master and a slave LDAP server, it is advisable to use the master LDAP server for the PDC and slave LDAP servers for the BDCs. It is not essential to use slave LDAP servers, however, many administrators will want to do so in order to provide redundant services. Of course, one or more BDCs may use any slave LDAP server. Then again, it is entirely possible to use a single LDAP server for the entire network.

When configuring a master LDAP server that will have slave LDAP servers, do not forget to configure this in the `/etc/openldap/slapd.conf` file. It must be noted that the DN of a server certificate must use the CN attribute to name the server, and the CN must carry the servers' fully qualified domain name. Additional alias names and wildcards may be present in the subjectAltName certificate extension. More details on server certificate names are in RFC2830.

It does not really fit within the scope of this document, but a working LDAP installation is basic to LDAP enabled Samba operation. When using an OpenLdap server with Transport Layer Security (TLS), the machine name in `/etc/ssl/certs/slapd.pem` must be the same as in `/etc/openldap/sldap.conf`. The Red Hat Linux startup script creates the `slapd.pem` file with hostname *"localhost.localdomain."* It is impossible to access this LDAP server from a slave LDAP server (i.e., a Samba BDC) unless the certificate is recreated with a correct hostname.

Do not install a Samba PDC on a OpenLDAP slave server. Joining client machines to the domain will fail in this configuration because the change to the machine account in the LDAP tree must take place on the master LDAP server. This is not replicated rapidly enough to the slave server that the PDC queries. It therfore gives an error message on the client machine about not being able to set up account credentials. The machine account is created on the LDAP server but the password fields will be empty.

Possible PDC/BDC plus LDAP configurations include:

- PDC+BDC -> One Central LDAP Server.
- PDC -> LDAP master server, BDC -> LDAP slave server.
- PDC -> LDAP master, with secondary slave LDAP server.

  BDC -> LDAP master, with secondary slave LDAP server.
- PDC -> LDAP master, with secondary slave LDAP server.

  BDC -> LDAP slave server, with secondary master LDAP server.

In order to have a fall-back configuration (secondary) LDAP server one would specify the secondary LDAP server in the `smb.conf` file as shown in Example 5.2.

Example 5.2. Multiple LDAP Servers in smb.conf

```
...
passdb backend = ldapsam:ldap://master.quenya.org
ldapsam:ldap://slave.quenya.org
...
```

5.2.3 Active Directory Domain Control

As of the release of MS Windows 2000 and Active Directory, this information is now stored in a directory that can be replicated and for which partial or full administrative control can be delegated. Samba-3 is not able to be a Domain Controller within an Active Directory tree, and it cannot be an Active Directory server. This means that Samba-3 also cannot act as a Backup Domain Controller to an Active Directory Domain Controller.

5.2.4 What Qualifies a Domain Controller on the Network?

Every machine that is a Domain Controller for the domain MIDEARTH has to register the NetBIOS group name MIDEARTH<#1c> with the WINS server and/or by broadcast on the local network. The PDC also registers the unique NetBIOS name MIDEARTH<#1b> with the WINS server. The name type <#1b> name is normally reserved for the Domain Master Browser, a role that has nothing to do with anything related to authentication, but the Microsoft Domain implementation requires the Domain Master Browser to be on the same machine as the PDC.

Where a WINS server is not used, broadcast name registrations alone must suffice. Refer to Section 9.3 for more information regarding TCP/IP network protocols and how SMB/CIFS names are handled.

5.2.5 How does a Workstation find its Domain Controller?

There are two different mechanisms to locate a domain controller, one method is used when NetBIOS over TCP/IP is enabled and the other when it has been disabled in the TCP/IP network configuration.

Where NetBIOS over TCP/IP is disabled, all name resolution involves the use of DNS, broadcast messaging over UDP, as well as Active Directory communication technologies. In this type of environment all machines require appropriate DNS entries. More information may be found in Section 9.3.3.

5.2.5.1 NetBIOS Over TCP/IP Enabled

An MS Windows NT4/200x/XP Professional workstation in the domain MIDEARTH that wants a local user to be authenticated has to find the Domain Controller for MIDEARTH. It does this by doing a NetBIOS name query for the group name MIDEARTH<#1c>. It assumes that each of the machines it gets back from the queries is a Domain Controller and can answer logon requests. To not open security holes, both the workstation and the selected Domain Controller authenticate each other. After that the workstation sends the user's credentials (name and password) to the local Domain Controller for validation.

5.2.5.2 NetBIOS Over TCP/IP Disabled

An MS Windows NT4/200x/XP Professional workstation in the realm `quenya.org` that has a need to affect user logon authentication will locate the Domain Controller by requerying DNS servers for the `_ldap._tcp.pdc.ms-dcs.quenya.org` record. More information regarding this subject may be found in Section 9.3.3.

5.3 Backup Domain Controller Configuration

The creation of a BDC requires some steps to prepare the Samba server before smbd is executed for the first time. These steps are outlines as follows:

- The domain SID has to be the same on the PDC and the BDC. In Samba versions pre-2.2.5, the domain SID was stored in the file `private/MACHINE.SID`. The domain SID is now stored in the file `private/secrets.tdb`. This file is unique to each server and can not be copied from a PDC to a BDC, the BDC will generate a new SID at start-up. It will over-write the PDC domain SID with the newly created BDC SID. There is a procedure that will allow the BDC to aquire the Domain SID. This is described here.

  To retrieve the domain SID from the PDC or an existing BDC and store it in the `secrets.tdb`, execute:

  ```
  root# net rpc getsid
  ```

- Specification of the *ldap admin dn* is obligatory. This also requires the LDAP administration password to be set in the `secrets.tdb` using the **smbpasswd -w** *mysecret*.
- Either *ldap suffix* or *ldap idmap suffix* must be specified in the `smb.conf` file.
- The UNIX user database has to be synchronized from the PDC to the BDC. This means that both the `/etc/passwd` and `/etc/group` have to be replicated from the PDC to the BDC. This can be done manually whenever changes are made. Alternately, the PDC is set up as an NIS master server and the BDC as an NIS slave server. To set up the BDC as a mere NIS client would not be enough, as the BDC would not be able to access its user database in case of a PDC failure. NIS is by no means the only method to synchronize passwords. An LDAP solution would also work.
- The Samba password database must be replicated from the PDC to the BDC. Although it is possible to synchronize the `smbpasswd` file with **rsync** and **ssh**, this method is broken and flawed, and is therefore not recommended. A better solution is to set up slave LDAP servers for each BDC and a master LDAP server for the PDC.
- The netlogon share has to be replicated from the PDC to the BDC. This can be done manually whenever login scripts are changed, or it can be done automatically using a **cron** job that will replicate the directory structure in this share using a tool like **rsync**.

5.3.1 Example Configuration

Finally, the BDC has to be found by the workstations. This can be done by setting Samba as shown in Example 5.3.

Example 5.3. Minimal setup for being a BDC

```
workgroup = MIDEARTH
passdb backend = ldapsam:ldap://slave-ldap.quenya.org
domain master = no
domain logons = yes
idmap backend = ldapsam:ldap://slave-ldap.quenya.org
```

In the *[global]*-section of the `smb.conf` of the BDC. This makes the BDC only register the name SAMBA<#1c> with the WINS server. This is no problem as the name SAMBA<#1c> is a NetBIOS group name that is meant to be registered by more than one machine. The parameter *domain master* = no forces the BDC not to register SAMBA<#1b> which as a unique NetBIOS name is reserved for the Primary Domain Controller.

The *idmap backend* will redirect the **winbindd** utility to use the LDAP database to resolve all UIDs and GIDs for UNIX accounts.

> NOTE
>
> 
>
> Samba-3 has introduced a new ID mapping facility. One of the features of this facility is that it allows greater flexibility in how user and group IDs are handled in respect to NT Domain User and Group SIDs. One of the new facilities provides for explicitly ensuring that UNIX/Linux UID and GID values will be consistent on the PDC, all BDCs and all Domain Member servers. The parameter that controls this is called *idmap backend*. Please refer to the man page for `smb.conf` for more information regarding its behavior.

The use of the *idmap backend* = ldap://master.quenya/org option on a BDC only make sense where ldapsam is used on a PDC. The purpose for an LDAP based idmap backend is also to allow a domain-member (without its own passdb backend) to use winbindd to resolve Windows network users and groups to common UID/GIDs. In other words, this option is generally intended for use on BDCs and on Domain Member servers.

5.4 Common Errors

As this is a rather new area for Samba, there are not many examples that we may refer to. Updates will be published as they become available and may be found in later Samba releases or from the Samba web site.[2]

5.4.1 Machine Accounts Keep Expiring

This problem will occur when the passdb (SAM) files are copied from a central server but the local Backup Domain Controller is acting as a PDC. This results in the application of Local Machine Trust Account password updates to the local SAM. Such updates are not copied back to the central server. The newer machine account password is then over written when the SAM is re-copied from the PDC. The result is that the Domain Member machine on start up will find that its passwords do not match the one now in the database and since the startup security check will now fail, this machine will not allow logon attempts to proceed and the account expiry error will be reported.

[2] http://samba.org

The solution is to use a more robust passdb backend, such as the ldapsam backend, setting up a slave LDAP server for each BDC, and a master LDAP server for the PDC.

5.4.2 Can Samba Be a Backup Domain Controller to an NT4 PDC?

No. The native NT4 SAM replication protocols have not yet been fully implemented.

Can I get the benefits of a BDC with Samba? Yes, but only to a Samba PDC.The main reason for implementing a BDC is availability. If the PDC is a Samba machine, a second Samba machine can be set up to service logon requests whenever the PDC is down.

5.4.3 How Do I Replicate the smbpasswd File?

Replication of the smbpasswd file is sensitive. It has to be done whenever changes to the SAM are made. Every user's password change is done in the smbpasswd file and has to be replicated to the BDC. So replicating the smbpasswd file very often is necessary.

As the smbpasswd file contains plain text password equivalents, it must not be sent unencrypted over the wire. The best way to set up smbpasswd replication from the PDC to the BDC is to use the utility rsync. rsync can use ssh as a transport. **ssh** itself can be set up to accept *only* **rsync** transfer without requiring the user to type a password.

As said a few times before, use of this method is broken and flawed. Machine trust accounts will go out of sync, resulting in a broken domain. This method is *not* recommended. Try using LDAP instead.

5.4.4 Can I Do This All with LDAP?

The simple answer is yes. Samba's pdb_ldap code supports binding to a replica LDAP server, and will also follow referrals and rebind to the master if it ever needs to make a modification to the database. (Normally BDCs are read only, so this will not occur often).

Chapter 6

DOMAIN MEMBERSHIP

Domain Membership is a subject of vital concern. Samba must be able to participate as a member server in a Microsoft Domain Security context, and Samba must be capable of providing Domain machine member trust accounts, otherwise it would not be able to offer a viable option for many users.

This chapter covers background information pertaining to Domain Membership, the Samba configuration for it, and MS Windows client procedures for joining a domain. Why is this necessary? Because both are areas in which there exists within the current MS Windows networking world and particularly in the UNIX/Linux networking and administration world, a considerable level of misinformation, incorrect understanding and a lack of knowledge. Hopefully this chapter will fill the voids.

6.1 Features and Benefits

MS Windows workstations and servers that want to participate in Domain Security need to be made Domain Members. Participating in Domain Security is often called *Single Sign On* or SSO for short. This chapter describes the process that must be followed to make a workstation (or another server — be it an MS Windows NT4 / 200x server) or a Samba server a member of an MS Windows Domain Security context.

Samba-3 can join an MS Windows NT4-style domain as a native member server, an MS Windows Active Directory Domain as a native member server, or a Samba Domain Control network. Domain Membership has many advantages:

- MS Windows workstation users get the benefit of SSO.
- Domain user access rights and file ownership/access controls can be set from the single Domain Security Account Manager (SAM) database (works with Domain Member servers as well as with MS Windows workstations that are Domain Members).
- Only MS Windows NT4/200x/XP Professional workstations that are Domain Members can use network logon facilities.
- Domain Member workstations can be better controlled through the use of Policy files (`NTConfig.POL`) and Desktop Profiles.

- Through the use of logon scripts, users can be given transparent access to network applications that run off application servers.
- Network administrators gain better application and user access management abilities because there is no need to maintain user accounts on any network client or server, other than the central Domain database (either NT4/Samba SAM style Domain, NT4 Domain that is backended with an LDAP directory, or via an Active Directory infrastructure).

6.2 MS Windows Workstation/Server Machine Trust Accounts

A Machine Trust Account is an account that is used to authenticate a client machine (rather than a user) to the Domain Controller server. In Windows terminology, this is known as a "*Computer Account.*" The purpose of the machine account is to prevent a rogue user and Domain Controller from colluding to gain access to a domain member workstation.

The password of a Machine Trust Account acts as the shared secret for secure communication with the Domain Controller. This is a security feature to prevent an unauthorized machine with the same NetBIOS name from joining the domain and gaining access to domain user/group accounts. Windows NT/200x/XP Professional clients use machine trust accounts, but Windows 9x/Me/XP Home clients do not. Hence, a Windows 9x/Me/XP Home client is never a true member of a Domain because it does not possess a Machine Trust Account, and, thus, has no shared secret with the Domain Controller.

A Windows NT4 PDC stores each Machine Trust Account in the Windows Registry. The introduction of MS Windows 2000 saw the introduction of Active Directory, the new repository for Machine Trust Accounts. A Samba PDC, however, stores each Machine Trust Account in two parts, as follows:

- A Domain Security Account (stored in the *`passdb backend`* that has been configured in the `smb.conf` file. The precise nature of the account information that is stored depends on the type of backend database that has been chosen.

  The older format of this data is the `smbpasswd` database that contains the UNIX login ID, the UNIX user identifier (UID), and the LanMan and NT encrypted passwords. There is also some other information in this file that we do not need to concern ourselves with here.

  The two newer database types are called ldapsam, and tdbsam. Both store considerably more data than the older `smbpasswd` file did. The extra information enables new user account controls to be implemented.

- A corresponding UNIX account, typically stored in `/etc/passwd`. Work is in progress to allow a simplified mode of operation that does not require UNIX user accounts, but this may not be a feature of the early releases of Samba-3.

There are three ways to create Machine Trust Accounts:

- Manual creation from the UNIX/Linux command line. Here, both the Samba and corresponding UNIX account are created by hand.

- Using the MS Windows NT4 Server Manager, either from an NT4 Domain Member server, or using the Nexus toolkit available from the Microsoft Web site. This tool can be run from any MS Windows machine as long as the user is logged on as the administrator account.
- "*On-the-fly*" creation. The Samba Machine Trust Account is automatically created by Samba at the time the client is joined to the domain. (For security, this is the recommended method.) The corresponding UNIX account may be created automatically or manually.

6.2.1 Manual Creation of Machine Trust Accounts

The first step in manually creating a Machine Trust Account is to manually create the corresponding UNIX account in `/etc/passwd`. This can be done using **vipw** or another "*add user*" command that is normally used to create new UNIX accounts. The following is an example for a Linux-based Samba server:

```
root# /usr/sbin/useradd -g machines -d /dev/null -c "machine nickname" \
   -s /bin/false machine_name$

root# passwd -l machine_name$
```

In the above example above there is an existing system group "*machines*" which is used as the primary group for all machine accounts. In the following examples the "*machines*" group has numeric GID equal 100.

On *BSD systems, this can be done using the **chpass** utility:

```
root# chpass -a \
'machine_name$:*:101:100::0:0:Windows machine_name:/dev/null:/sbin/nologin'
```

The `/etc/passwd` entry will list the machine name with a "*$*" appended, will not have a password, will have a null shell and no home directory. For example, a machine named "*doppy*" would have an `/etc/passwd` entry like this:

```
doppy$:x:505:100:machine_nickname:/dev/null:/bin/false
```

Above, *machine_nickname* can be any descriptive name for the client, i.e., BasementComputer. *machine_name* absolutely must be the NetBIOS name of the client to be joined to the domain. The "*$*" must be appended to the NetBIOS name of the client or Samba will not recognize this as a Machine Trust Account.

Now that the corresponding UNIX account has been created, the next step is to create the Samba account for the client containing the well-known initial Machine Trust Account password. This can be done using the **smbpasswd** command as shown here:

```
root# smbpasswd -a -m machine_name
```

where *machine_name* is the machine's NetBIOS name. The RID of the new machine account is generated from the UID of the corresponding UNIX account.

JOIN THE CLIENT TO THE DOMAIN IMMEDIATELY

Manually creating a Machine Trust Account using this method is the equivalent of creating a Machine Trust Account on a Windows NT PDC using the Server Manager. From the time at which the account is created to the time the client joins the domain and changes the password, your domain is vulnerable to an intruder joining your domain using a machine with the same NetBIOS name. A PDC inherently trusts members of the domain and will serve out a large degree of user information to such clients. You have been warned!

6.2.2 Managing Domain Machine Accounts using NT4 Server Manager

A working *add machine script* script is essential for machine trust accounts to be automatically created. This applies no matter whether one uses automatic account creation, or if one wishes to use the NT4 Domain Server Manager.

If the machine from which you are trying to manage the domain is an MS Windows NT4 workstation or MS Windows 200x/XP Professional, the tool of choice is the package called **SRVTOOLS.EXE**. When executed in the target directory it will unpack **SrvMgr.exe** and **UsrMgr.exe** (both are domain management tools for MS Windows NT4 workstation).

If your workstation is a Microsoft Windows 9x/Me family product you should download the **Nexus.exe** package from the Microsoft web site. When executed from the target directory this will unpack the same tools but for use on this platform.

Further information about these tools may be obtained from the following locations:

```
http://support.microsoft.com/default.aspx?scid=kb;en-us;173673
http://support.microsoft.com/default.aspx?scid=kb;en-us;172540
```

Launch the **srvmgr.exe** (Server Manager for Domains) and follow these steps:

SERVER MANAGER ACCOUNT MACHINE ACCOUNT MANAGEMENT

1. From the menu select **Computer**.
2. Click **Select Domain**.
3. Click the name of the domain you wish to administer in the **Select Domain** panel and then click **OK**.
4. Again from the menu select **Computer**.

5. Select **Add to Domain**.
6. In the dialog box, click the radio button to **Add NT Workstation of Server**, then enter the machine name in the field provided, and click the **Add** button.

6.2.3 On-the-Fly Creation of Machine Trust Accounts

The second (and recommended) way of creating Machine Trust Accounts is simply to allow the Samba server to create them as needed when the client is joined to the domain.

Since each Samba Machine Trust Account requires a corresponding UNIX account, a method for automatically creating the UNIX account is usually supplied; this requires configuration of the add machine script option in `smb.conf`. This method is not required, however, corresponding UNIX accounts may also be created manually.

Here is an example for a Red Hat Linux system.

```
add machine script = /usr/sbin/useradd -d /dev/null -g 100 \
-s /bin/false -M %u
```

6.2.4 Making an MS Windows Workstation or Server a Domain Member

The procedure for making an MS Windows workstation or server a member of the domain varies with the version of Windows.

6.2.4.1 Windows 200x/XP Professional Client

When the user elects to make the client a Domain Member, Windows 200x prompts for an account and password that has privileges to create machine accounts in the domain. A Samba Administrator Account (i.e., a Samba account that has `root` privileges on the Samba server) must be entered here; the operation will fail if an ordinary user account is given.

For security reasons, the password for this Administrator Account should be set to a password that is other than that used for the root user in `/etc/passwd`.

The name of the account that is used to create Domain Member machine accounts can be anything the network administrator may choose. If it is other than `root` then this is easily mapped to `root` in the file named in the `smb.conf` parameter *username map* = /etc/samba/smbusers.

The session key of the Samba Administrator Account acts as an encryption key for setting the password of the machine trust account. The Machine Trust Account will be created on-the-fly, or updated if it already exists.

6.2.4.2 Windows NT4 Client

If the Machine Trust Account was created manually, on the Identification Changes menu enter the domain name, but do not check the box **Create a Computer Account in the Domain**. In this case, the existing Machine Trust Account is used to join the machine to the domain.

If the Machine Trust Account is to be created on-the-fly, on the Identification Changes menu enter the domain name and check the box **Create a Computer Account in the Domain**. In this case, joining the domain proceeds as above for Windows 2000 (i.e., you must supply a Samba Administrator Account when prompted).

6.2.4.3 Samba Client

Joining a Samba client to a domain is documented in Section 6.3.

6.3 Domain Member Server

This mode of server operation involves the Samba machine being made a member of a domain security context. This means by definition that all user authentication will be done from a centrally defined authentication regime. The authentication regime may come from an NT3/4-style (old domain technology) server, or it may be provided from an Active Directory server (ADS) running on MS Windows 2000 or later.

Of course it should be clear that the authentication backend itself could be from any distributed directory architecture server that is supported by Samba. This can be LDAP (from OpenLDAP), or Sun's iPlanet, or NetWare Directory Server, and so on.

NOTE

When Samba is configured to use an LDAP, or other identity management and/or directory service, it is Samba that continues to perform user and machine authentication. It should be noted that the LDAP server does not perform authentication handling in place of what Samba is designed to do.

Please refer to Chapter 4, *Domain Control*, for more information regarding how to create a domain machine account for a Domain Member server as well as for information on how to enable the Samba Domain Member machine to join the domain and be fully trusted by it.

6.3.1 Joining an NT4-type Domain with Samba-3

Table 6.1 lists names that have been used in the remainder of this chapter.

Table 6.1. Assumptions

NetBIOS name:	SERV1
Windows 200x/NT domain name:	MIDEARTH
Domain's PDC NetBIOS name:	DOMPDC
Domain's BDC NetBIOS names:	DOMBDC1 and DOMBDC2

First, you must edit your `smb.conf` file to tell Samba it should now use domain security.

Change (or add) your *security* line in the [global] section of your `smb.conf` to read:

security = domain

Next change the *workgroup* line in the *[global]* section to read:

workgroup = MIDEARTH

This is the name of the domain we are joining.

You must also have the parameter *encrypt passwords* set to `yes` in order for your users to authenticate to the NT PDC. This is the defaulty setting if this parameter is not specified. There is no need to specify this parameter, but if it is specified in the `smb.conf` file, it must be set to `Yes`.

Finally, add (or modify) a *password server* line in the [global] section to read:

password server = DOMPDC DOMBDC1 DOMBDC2

These are the primary and backup Domain Controllers Samba will attempt to contact in order to authenticate users. Samba will try to contact each of these servers in order, so you may want to rearrange this list in order to spread out the authentication load among Domain Controllers.

Alternately, if you want smbd to automatically determine the list of Domain Controllers to use for authentication, you may set this line to be:

*password server = **

This method allows Samba to use exactly the same mechanism that NT does. The method either uses broadcast-based name resolution, performs a WINS database lookup in order to find a Domain Controller against which to authenticate, or locates the Domain Controller using DNS name resolution.

To join the domain, run this command:

```
root# net join -S DOMPDC -UAdministrator%password
```

If the `-S DOMPDC` argument is not given, the domain name will be obtained from `smb.conf`.

The machine is joining the domain DOM, and the PDC for that domain (the only machine that has write access to the domain SAM database) is DOMPDC, therefore use the `-S` option. The *Administrator%password* is the login name and password for an account that has the necessary privilege to add machines to the domain. If this is successful, you will see the message in your terminal window the text shown below. Where the older NT4 style domain architecture is used:

```
Joined domain DOM.
```

Where Active Directory is used:

```
Joined SERV1 to realm MYREALM.
```

Refer to the **net** man page for further information.

This process joins the server to the domain without having to create the machine trust account on the PDC beforehand.

This command goes through the machine account password change protocol, then writes the new (random) machine account password for this Samba server into a file in the same directory in which a smbpasswd file would be normally stored:

```
/usr/local/samba/private/secrets.tdb
or
/etc/samba/secrets.tdb.
```

This file is created and owned by root and is not readable by any other user. It is the key to the Domain-level security for your system, and should be treated as carefully as a shadow password file.

Finally, restart your Samba daemons and get ready for clients to begin using domain security. The way you can restart your Samba daemons depends on your distribution, but in most cases the following will suffice:

```
root# /etc/init.d/samba restart
```

6.3.2 Why Is This Better Than security = server?

Currently, domain security in Samba does not free you from having to create local UNIX users to represent the users attaching to your server. This means that if Domain user `DOM\fred` attaches to your Domain Security Samba server, there needs to be a local UNIX user fred to represent that user in the UNIX file system. This is similar to the older Samba security mode *security* = server, where Samba would pass through the authentication request to a Windows NT server in the same way as a Windows 95 or Windows 98 server would.

Please refer to Chapter 20, *Winbind: Use of Domain Accounts*, for information on a system to automatically assign UNIX UIDs and GIDs to Windows NT Domain users and groups.

The advantage to Domain-level security is that the authentication in Domain-level security is passed down the authenticated RPC channel in exactly the same way that an NT server would do it. This means Samba servers now participate in domain trust relationships in exactly the same way NT servers do (i.e., you can add Samba servers into a resource domain and have the authentication passed on from a resource domain PDC to an account domain PDC).

In addition, with *security* = server, every Samba daemon on a server has to keep a connection open to the authenticating server for as long as that daemon lasts. This can drain the connection resources on a Microsoft NT server and cause it to run out of available connections. With *security* = domain, however, the Samba daemons connect to the PDC/BDC only for as long as is necessary to authenticate the user and then drop the connection, thus conserving PDC connection resources.

And finally, acting in the same manner as an NT server authenticating to a PDC means that as part of the authentication reply, the Samba server gets the user identification information such as the user SID, the list of NT groups the user belongs to, and so on.

NOTE

Much of the text of this document was first published in the Web magazine LinuxWorld[a] as the article `http://www.linuxworld.com/linuxworld/lw-1998-10/lw-10-samba.html`[b] *Doing the NIS/NT Samba.*

[a] http://www.linuxworld.com
[b] http://www.linuxworld.com/linuxworld/lw-1998-10/lw-10-samba.html

6.4 Samba ADS Domain Membership

This is a rough guide to setting up Samba-3 with Kerberos authentication against a Windows 200x KDC. A familiarity with Kerberos is assumed.

6.4.1 Configure smb.conf

You must use at least the following three options in `smb.conf`:

```
realm = your.kerberos.REALM
security = ADS
encrypt passwords = yes
```

In case samba cannot correctly identify the appropriate ADS server using the realm name, use the *password server* option in `smb.conf`:

```
password server = your.kerberos.server
```

NOTE

You do *not* need a smbpasswd file, and older clients will be authenticated as if *security* = domain, although it will not do any harm and allows you to have local users not in the domain.

6.4.2 Configure /etc/krb5.conf

With both MIT and Heimdal Kerberos, this is unnecessary, and may be detrimental. All ADS domains will automatically create SRV records in the DNS zone *_kerberos.REALM.NAME* for each KDC in the realm. MIT's, as well as Heimdal's, KRB5 libraries default to checking for these records, so they will automatically find the KDCs. In addition, `krb5.conf` only allows specifying a single KDC, even there if there is more than one. Using the DNS lookup allows the KRB5 libraries to use whichever KDCs are available.

When manually configuring `krb5.conf`, the minimal configuration is:

```
[libdefaults]
   default_realm = YOUR.KERBEROS.REALM

   [realms]
   YOUR.KERBEROS.REALM = {
   kdc = your.kerberos.server
       }
```

When using Heimdal versions before 0.6 use the following configuration settings:

```
[libdefaults]
   default_realm        = YOUR.KERBEROS.REALM
   default_etypes       = des-cbc-crc des-cbc-md5
   default_etypes_des = des-cbc-crc des-cbc-md5

        [realms]
        YOUR.KERBEROS.REALM = {
        kdc = your.kerberos.server
            }
```

Test your config by doing a `kinit` *USERNAME@REALM* and making sure that your password is accepted by the Win2000 KDC.

With Heimdal versions earlier than 0.6.x you only can use newly created accounts in ADS or accounts that have had the password changed once after migration, or in case of `Administrator` after installation. At the moment, a Windows 2003 KDC can only be used with a Heimdal releases later than 0.6 (and no default etypes in krb5.conf). Unfortunatly this whole area is still in a state of flux.

NOTE

The realm must be in uppercase or you will get "*Cannot find KDC for requested realm while getting initial credentials*" error (Kerberos is case-sensitive!).

NOTE

Time between the two servers must be synchronized. You will get a "*kinit(v5): Clock skew too great while getting initial credentials*" if the time difference is more than five minutes.

Clock skew limits are configurable in the Kerberos protocols. The default setting is five minutes.

You also must ensure that you can do a reverse DNS lookup on the IP address of your KDC. Also, the name that this reverse lookup maps to must either be the NetBIOS name of the KDC (i.e., the hostname with no domain attached) or it can alternately be the NetBIOS name followed by the realm.

The easiest way to ensure you get this right is to add a `/etc/hosts` entry mapping the IP address of your KDC to its NetBIOS name. If you do not get this correct then you will get a local error when you try to join the realm.

If all you want is Kerberos support in smbclient then you can skip directly to Section 6.4.5 now. Section 6.4.3 and Section 6.4.4 are needed only if you want Kerberos support for smbd and winbindd.

6.4.3 Create the Computer Account

As a user who has write permission on the Samba private directory (usually root), run:

```
root#  net ads join -U Administrator%password
```

When making a Windows client a member of an ADS domain within a complex organization, you may want to create the machine account within a particular organizational unit. Samba-3 permits this to be done using the following syntax:

```
root#   kinit Administrator@your.kerberos.REALM
root#  net ads join organizational_unit
```

For example, you may want to create the machine account in a container called "*Servers*" under the organizational directory "*Computers\BusinessUnit\Department*" like this:

```
root#  net ads join "Computers\BusinessUnit\Department\Servers"
```

6.4.3.1 Possible Errors

ADS support not compiled in — Samba must be reconfigured (remove config.cache) and recompiled (make clean all install) after the Kerberos libiraries and headers files are installed.

net ads join prompts for user name — You need to login to the domain using `kinit` *`USERNAME@REALM`*. *`USERNAME`* must be a user who has rights to add a machine to the domain.

Unsupported encryption/or checksum types — Make sure that the `/etc/krb5.conf` is correctly configured for the type and version of Kerberos installed on the system.

6.4.4 Testing Server Setup

If the join was successful, you will see a new computer account with the NetBIOS name of your Samba server in Active Directory (in the "*Computers*" folder under Users and Computers.

On a Windows 2000 client, try `net use * \\server\share`. You should be logged in with Kerberos without needing to know a password. If this fails then run `klist tickets`. Did you get a ticket for the server? Does it have an encryption type of DES-CBC-MD5?

> NOTE
>
> 
>
> Samba can use both DES-CBC-MD5 encryption as well as ARCFOUR-HMAC-MD5 encoding.

6.4.5 Testing with smbclient

On your Samba server try to login to a Win2000 server or your Samba server using smbclient and Kerberos. Use smbclient as usual, but specify the `-k` option to choose Kerberos authentication.

6.4.6 Notes

You must change administrator password at least once after DC install, to create the right encryption types.

Windows 200x does not seem to create the *`_kerberos._udp`* and *`_ldap._tcp`* in the default DNS setup. Perhaps this will be fixed later in service packs.

6.5 Sharing User ID Mappings between Samba Domain Members

Samba maps UNIX users and groups (identified by UIDs and GIDs) to Windows users and groups (identified by SIDs). These mappings are done by the *idmap* subsystem of Samba.

In some cases it is useful to share these mappings between Samba Domain Members, so *name->id* mapping is identical on all machines. This may be needed in particular when sharing files over both CIFS and NFS.

To use the *LDAP ldap idmap suffix*, set:

```
ldap idmap suffix = ou=Idmap,dc=quenya,dc=org
```

See the `smb.conf` man page entry for the *ldap idmap suffix* parameter for further information.

Do not forget to specify also the *ldap admin dn* and to make certain to set the LDAP administrative password into the `secrets.tdb` using:

```
root#  smbpasswd -w ldap-admin-password
```

6.6 Common Errors

In the process of adding/deleting/re-adding Domain Member machine accounts, there are many traps for the unwary player and many "*little*" things that can go wrong. It is particularly interesting how often subscribers on the Samba mailing list have concluded after repeated failed attempts to add a machine account that it is necessary to "*re-install*" MS Windows on the machine. In truth, it is seldom necessary to reinstall because of this type of problem. The real solution is often quite simple and with an understanding of how MS Windows networking functions, it is easy to overcome.

6.6.1 Cannot Add Machine Back to Domain

"*A Windows workstation was re-installed. The original domain machine account was deleted and added immediately. The workstation will not join the domain if I use the same machine name. Attempts to add the machine fail with a message that the machine already exists on the network — I know it does not. Why is this failing?*"

The original name is still in the NetBIOS name cache and must expire after machine account deletion before adding that same name as a Domain Member again. The best advice is to delete the old account and then add the machine with a new name.

6.6.2 Adding Machine to Domain Fails

"*Adding a Windows 200x or XP Professional machine to the Samba PDC Domain fails with a message that, 'The machine could not be added at this time, there is a network problem. Please try again later.' Why?*"

You should check that there is an *add machine script* in your `smb.conf` file. If there is not, please add one that is appropriate for your OS platform. If a script has been defined, you will need to debug its operation. Increase the *log level* in the `smb.conf` file to level 10, then try to rejoin the domain. Check the logs to see which operation is failing.

Possible causes include:

- The script does not actually exist, or could not be located in the path specified.

  Corrective action: Fix it. Make sure when run manually that the script will add both the UNIX system account and the Samba SAM account.

- The machine could not be added to the UNIX system accounts file `/etc/passwd`.

  Corrective action: Check that the machine name is a legal UNIX system account name. If the UNIX utility **useradd** is called, then make sure that the machine name you are trying to add can be added using this tool. **Useradd** on some systems will not allow any upper case characters nor will it allow spaces in the name.

The *add machine script* does not create the machine account in the Samba backend database, it is there only to create a UNIX system account to which the Samba backend database account can be mapped.

6.6.3 I Can't Join a Windows 2003 PDC

Windows 2003 requires SMB signing. Client side SMB signing has been implemented in Samba-3.0. Set *client use spnego* = yes when communicating with a Windows 2003 server.

Chapter 7

STAND-ALONE SERVERS

Stand-alone Servers are independent of Domain Controllers on the network. They are not Domain Members and function more like workgroup servers. In many cases a Stand-alone Server is configured with a minimum of security control with the intent that all data served will be readily accessible to all users.

7.1 Features and Benefits

Stand-alone Servers can be as secure or as insecure as needs dictate. They can have simple or complex configurations. Above all, despite the hoopla about Domain Security they remain a common installation.

If all that is needed is a server for read-only files, or for printers alone, it may not make sense to effect a complex installation. For example: A drafting office needs to store old drawings and reference standards. Noone can write files to the server as it is legislatively important that all documents remain unaltered. A share mode read-only Stand-alone Server is an ideal solution.

Another situation that warrants simplicity is an office that has many printers that are queued off a single central server. Everyone needs to be able to print to the printers, there is no need to effect any access controls and no files will be served from the print server. Again, a share mode Stand-alone Server makes a great solution.

7.2 Background

The term *Stand-alone Server* means that it will provide local authentication and access control for all resources that are available from it. In general this means that there will be a local user database. In more technical terms, it means resources on the machine will be made available in either SHARE mode or in USER mode.

No special action is needed other than to create user accounts. Stand-alone servers do not provide network logon services. This means that machines that use this server do not perform a domain logon to it. Whatever logon facility the workstations are subject to is independent of this machine. It is, however, necessary to accommodate any network user so the logon name they use will be translated (mapped) locally on the Stand-alone Server to a locally known user name. There are several ways this can be done.

Samba tends to blur the distinction a little in respect of what is a Stand-alone Server. This is because the authentication database may be local or on a remote server, even if from the SMB protocol perspective the Samba server is not a member of a domain security context.

Through the use of Pluggable Authentication Modules (PAM) and the name service switcher (NSSWITCH), which maintains the UNIX-user database) the source of authentication may reside on another server. We would be inclined to call this the authentication server. This means that the Samba server may use the local UNIX/Linux system password database (`/etc/passwd` or `/etc/shadow`), may use a local smbpasswd file, or may use an LDAP backend, or even via PAM and Winbind another CIFS/SMB server for authentication.

7.3 Example Configuration

The examples, Example 7.1, and link linkend="SimplePrintServer"/>, are designed to inspire simplicity. It is too easy to attempt a high level of creativity and to introduce too much complexity in server and network design.

7.3.1 Reference Documentation Server

Configuration of a read-only data server that everyone can access is very simple. Example 7.1 is the `smb.conf` file that will do this. Assume that all the reference documents are stored in the directory `/export`, and the documents are owned by a user other than nobody. No home directories are shared, and there are no users in the `/etc/passwd` UNIX system database. This is a simple system to administer.

Example 7.1. smb.conf for Reference Documentation Server

```
# Global parameters

[global]
workgroup = MIDEARTH
netbios name = GANDALF
security = SHARE
passdb backend = guest
wins server = 192.168.1.1

[data]
comment = Data
path = /export
guest only = Yes
```

In Example 7.1 above, the machine name is set to GANDALF, the workgroup is set to the name of the local workgroup (MIDEARTH) so the machine will appear together with systems with which users are familiar. The only password backend required is the "*guest*" backend to allow default unprivileged account names to be used. As there is a WINS server on this networki, we of obviously make use of it.

7.3.2 Central Print Serving

Configuration of a simple print server is easy if you have all the right tools on your system.

Assumptions:

1. The print server must require no administration.
2. The print spooling and processing system on our print server will be CUPS. (Please refer to Chapter 18, *CUPS Printing Support* for more information).
3. The print server will service only network printers. The network administrator will correctly configure the CUPS environment to support the printers.
4. All workstations will use only postscript drivers. The printer driver of choice is the one shipped with the Windows OS for the Apple Color LaserWriter.

In this example our print server will spool all incoming print jobs to `/var/spool/samba` until the job is ready to be submitted by Samba to the CUPS print processor. Since all incoming connections will be as the anonymous (guest) user, two things will be required:

Enabling Anonymous Printing

- The UNIX/Linux system must have a **guest** account. The default for this is usually the account **nobody**. To find the correct name to use for your version of Samba, do the following:

  ```
  $ testparm -s -v | grep "guest account"
  ```

  Make sure that this account exists in your system password database (`/etc/passwd`).

- The directory into which Samba will spool the file must have write access for the guest account. The following commands will ensure that this directory is available for use:

  ```
  root# mkdir /var/spool/samba
  root# chown nobody.nobody /var/spool/samba
  root# chmod a+rwt /var/spool/samba
  ```

The contents of the `smb.conf` file is shown in Example 7.2.

Example 7.2. smb.conf for Anonymous Printing

```
# Global parameters

[global]
workgroup = MIDEARTH
netbios name = GANDALF
security = SHARE
passdb backend = guest
printing = cups
printcap name = cups

[printers]
comment = All Printers
path = /var/spool/samba
printer admin = root
guest ok = Yes
printable = Yes
use client driver = Yes
browseable = No
```

NOTE

On CUPS-enabled systems there is a facility to pass raw data directly to the printer without intermediate processing via CUPS print filters. Where use of this mode of operation is desired, it is necessary to configure a raw printing device. It is also necessary to enable the raw mime handler in the `/etc/mime.conv` and `/etc/mime.types` files. Refer to Section 18.3.4.

7.4 Common Errors

The greatest mistake so often made is to make a network configuration too complex. It pays to use the simplest solution that will meet the needs of the moment.

Chapter 8

MS WINDOWS NETWORK CONFIGURATION GUIDE

8.1 Features and Benefits

Occasionally network administrators will report difficulty getting Microsoft Windows clients to interoperate correctly with Samba servers. It would appear that some folks just can not accept the fact that the right way to configure MS Windows network client is precisely as one would do when using Microsoft Windows NT4 or 200x servers. Yet there is repetitious need to provide detailed Windows client configuration instructions.

The purpose of this chapter is to graphically illustrate MS Windows client configuration for the most common critical aspects of such configuration. An experienced network administrator will not be interested in the details of this chapter.

8.2 Technical Details

This chapter discusses TCP/IP protocol configuration as well as network membership for the platforms that are in common use today. These are:

- Microsoft Windows XP Professional.
- Windows 2000 Professional.
- Windows Millenium edition (Me).

8.2.1 TCP/IP Configuration

The builder of a house must ensure that all construction takes place on a firm foundation. The same is true of TCP/IP-based networking. Fundamental network configuration problems will plague all network users until they are resolved.

Microsoft Windows workstations and servers can be configured either with fixed IP addresses or via DHCP. The examples that follow demonstrate the use of DHCP and make only passing reference to those situations where fixed IP configuration settings can be effected.

It is posible to use shortcuts or abreviated keystrokes to arrive at a particular configuration screen. The decision was made to base all examples in this chapter on use of the **Start** button.

8.2.1.1 MS Windows XP Professional

There are two paths to the Windows XP TCP/IP configuration panel. Choose the access method that you prefer:

Click **Start** -> **Control Panel** -> **Network Connections**

Alternately, click **Start** ->, and right click **My Network Places** then select **Properties**

The following procedure steps through the Windows XP Professional TCP/IP configuration process:

1. On some installations the interface will be called **Local Area Connection** and on others it will be called **Network Bridge**. On our system it is called **Network Bridge**. Right click on **Network Bridge** -> **Properties**. See Figure 8.1.

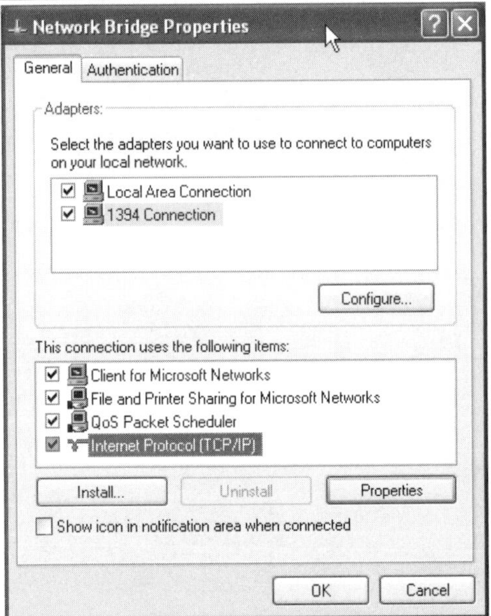

Figure 8.1. Network Bridge Configuration.

2. The Network Bridge Configuration, or Local Area Connection, panel is used to set TCP/IP protocol settings. In **This connection uses the following items:** box, click on **Internet Protocol (TCP/IP)**, then click the on **Properties**. The default setting is DHCP enbled operation. (i.e., "*Obtain an IP address automatically*"). See Figure 8.2.

   Many network administrators will want to use DHCP to configure all client TCP/IP protocol stack settings. (For information on how to configure the ISC DHCP server for Microsoft Windows client support see, Section 39.2.2. If it is necessary to provide

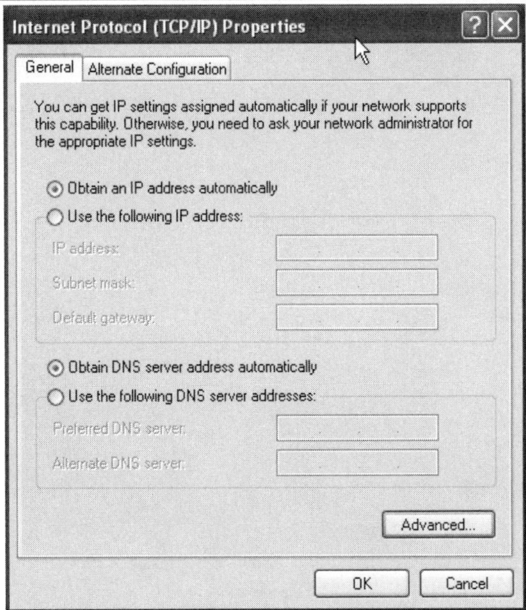

Figure 8.2. Internet Protocol (TCP/IP) Properties.

a fixed IP address, click on "*Use the following IP address*" and procede to enter the IP Address, the subnet mask, and the default gateway address in the boxes provided.

3. Click the **Advanced** button to procede with TCP/IP configuration. This opens a panel in which it is possible to create additional IP Addresses for this interface. The technical name for the additional addresses is *IP Aliases*, and additionally this panel permits the setting of more default gateways (routers). In most cases where DHCP is used, it will not be necessary to create additional settings. See Figure 8.3 to see the appearance of this panel.

   Fixed settings may be required for DNS and WINS if these settings are not provided automatically via DHCP.

4. Click the **DNS** tab to add DNS server settings. The example system uses manually configured DNS settings. When finished making changes, click the **OK** to commit the settings. See Figure 8.4.

5. Click the **WINS** tab to add manual WINS server entries. This step demonstrates an example system that uses manually configured WINS settings. When finished making, changes click the **OK** to commit the settings. See Figure 8.5.

8.2.1.2 MS Windows 2000

There are two paths to the Windows 2000 Professional TCP/IP configuration panel. Choose the access method that you prefer:

Click **Start** -> **Control Panel** -> **Network and Dial-up Connections**

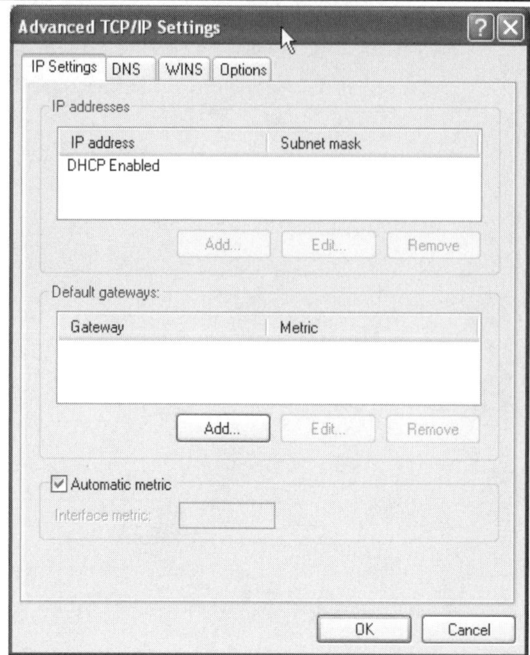

Figure 8.3. Advanced Network Settings

Alternately, click on **Start**, then right click **My Network Places** and select **Properties**.

The following procedure steps through the Windows XP Professional TCP/IP configuration process:

1. Right click on **Local Area Connection**, now click the **Properties**. See Figure 8.6.

2. The Local Area Connection Properties is used to set TCP/IP protocol settings. Click on **Internet Protocol (TCP/IP)** in the **Components checked are used by this connection:** box, then click the **Properties** button.

3. The default setting is DHCP enbled operation. (i.e., "*Obtain an IP address automatically*"). See Figure 8.7.

   Many network administrators will want to use DHCP to configure all client TCP/IP protocol stack settings. (For information on how to configure the ISC DHCP server for Microsoft Windows client support, see Section 39.2.2. If it is necessary to provide a fixed IP address, click on "*Use the following IP address*" and procede to enter the IP Address, the subnet mask, and the default gateway address in the boxes provided. For this example we are assuming that all network clients will be configured using DHCP.

4. Click the **Advanced** button to procede with TCP/IP configuration. Refer to Figure 8.8.

   Fixed settings may be required for DNS and WINS if these settings are not provided automatically via DHCP.

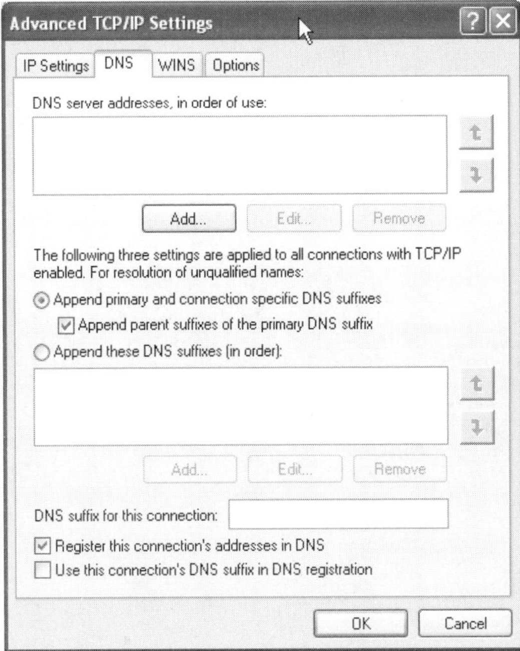

Figure 8.4. DNS Configuration.

5. Click the **DNS** tab to add DNS server settings. The example system uses manually configured DNS settings. When finished making changes, click on **OK** to commit the settings. See Figure 8.9.

6. Click the **WINS** tab to add manual WINS server entries. This step demonstrates an example system that uses manually configured WINS settings. When finished making changes, click on **OK** to commit the settings. See Figure 8.10.

8.2.1.3 MS Windows Me

There are two paths to the Windows Millenium edition (Me) TCP/IP configuration panel. Choose the access method that you prefer:

Click **Start** -> **Control Panel** -> **Network Connections**

Alternately, click on **Start** ->, and right click on **My Network Places** then select **Properties**.

The following procedure steps through the Windows Me TCP/IP configuration process:

1. In the box labelled **The following network components are installed:**, click on **Internet Protocol TCP/IP**, now click on the **Properties** button. See Figure 8.11.

2. Many network administrators will want to use DHCP to configure all client TCP/IP protocol stack settings. (For information on how to configure the ISC DHCP server for Microsoft Windows client support see, Section 39.2.2. The default setting on

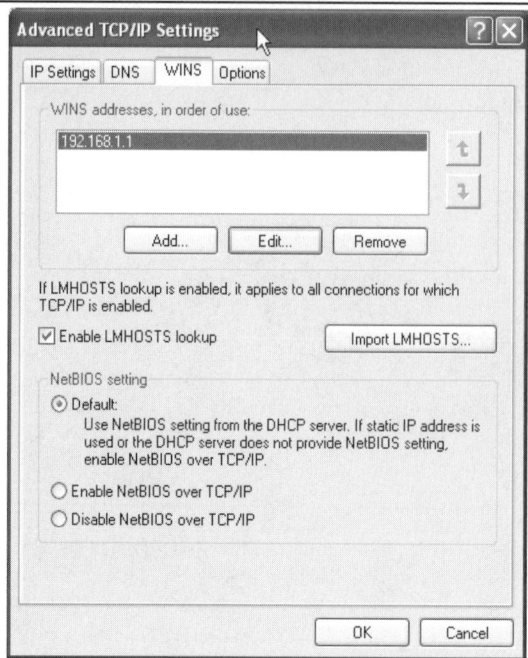

Figure 8.5. WINS Configuration

Microsoft Windows Me workstations is for DHCP enbled operation, i.e., **Obtain IP address automatically** is enabled. See Figure 8.12.

If it is necessary to provide a fixed IP address, click on **Specify an IP address** and procede to enter the IP Address and the subnet mask in the boxes provided. For this example we are assuming that all network clients will be configured using DHCP.

3. Fixed settings may be required for DNS and WINS if these settings are not provided automatically via DHCP.

4. If necessary, click the **DNS Configuration** tab to add DNS server settings. Click the **WINS Configuration** tab to add WINS server settings. The **Gateway** tab allows additional gateways (router addresses) to be added to the network interface settings. In most cases where DHCP is used, it will not be necessary to create these manual settings.

5. The following example uses manually configured WINS settings. See Figure 8.13. When finished making changes, click on **OK** to commit the settings.

   This is an example of a system that uses manually configured WINS settings. One situation where this might apply is on a network that has a single DHCP server that provides settings for multiple Windows workgroups or domains. See Figure 8.14.

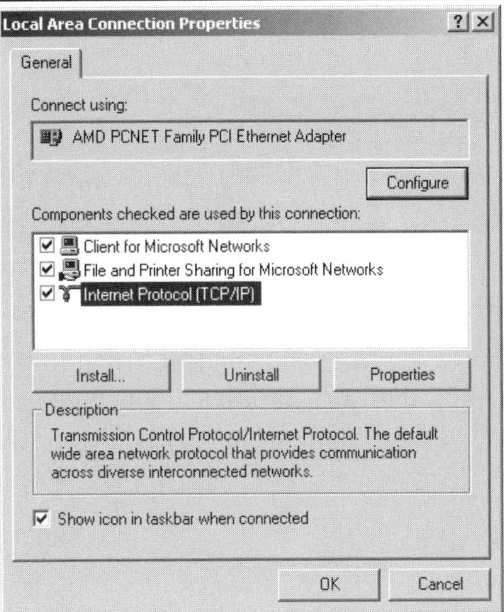

Figure 8.6. Local Area Connection Properties.

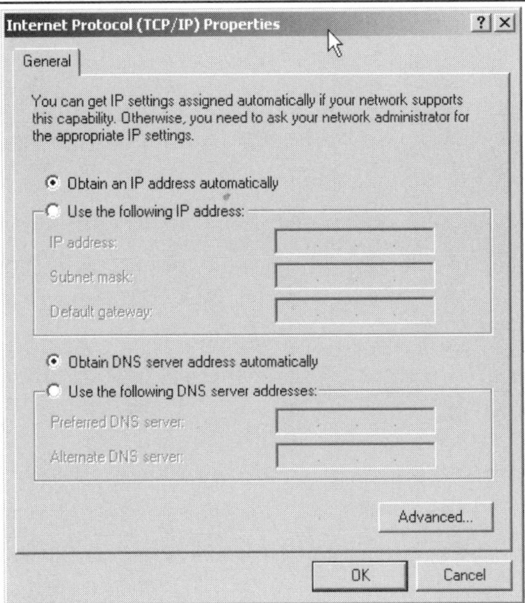

Figure 8.7. Internet Protocol (TCP/IP) Properties.

8.2.2 Joining a Domain: Windows 2000/XP Professional

Microsoft Windows NT/200x/XP Professional platforms can participate in Domain Security. This section steps through the process for making a Windows 200x/XP Professional machine

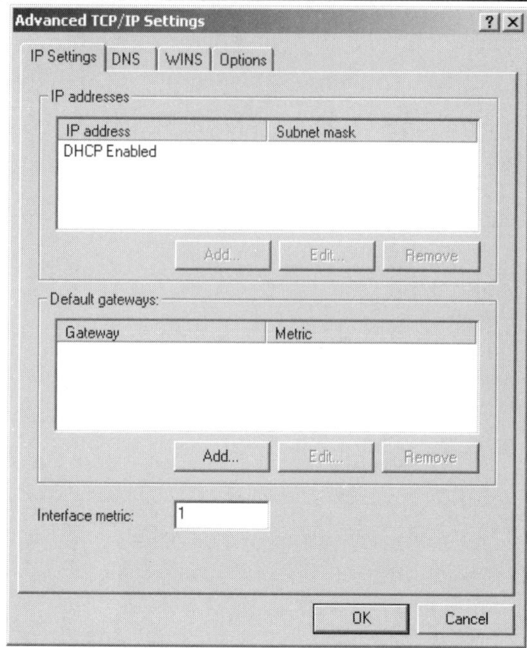

Figure 8.8. Advanced Network Settings.

a member of a Domain Security environment. It should be noted that this process is identical when joining a domain that is controlled by Windows NT4/200x as well as a Samba PDC.

1. Click **Start**.

2. Right click **My Computer**, then select **Properties**.

3. The opening panel is the same one that can be reached by clicking **System** on the Control Panel. See Figure 8.15.

4. Click the **Computer Name** tab. This panel shows the **Computer Description**, the **Full computer name**, and the **Workgroup** or **Domain name**. Clicking the **Network ID** button will launch the configuration wizard. Do not use this with Samba-3. If you wish to change the computer name, join or leave the domain, click the **Change** button. See Figure 8.16.

5. Click on **Change**. This panel shows that our example machine (TEMPTATION) is in a workgroup called WORKGROUP. We will join the domain called MIDEARTH. See Figure 8.17.

6. Enter the name **MIDEARTH** in the field below the Domain radio button. This panel shows that our example machine (TEMPTATION) is set to join the domain called MIDEARTH. See Figure 8.18.

7. Now click the **OK** button. A dialog box should appear to allow you to provide the credentials (username and password) of a Domain administrative account that has the rights to add machines to the Domain. Enter the name "*root*" and the root password

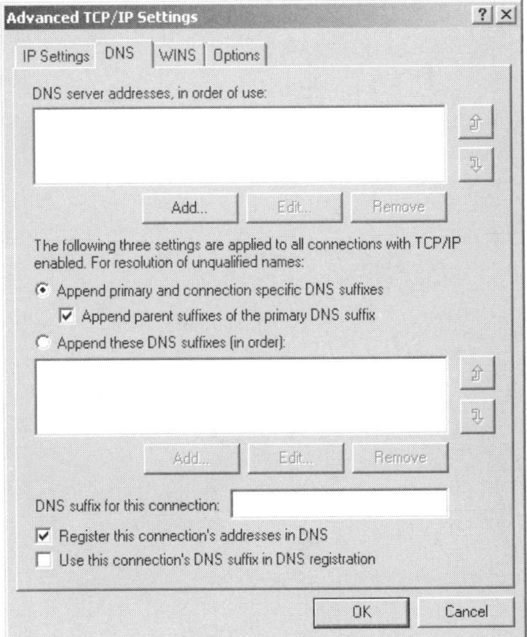

Figure 8.9. DNS Configuration.

from your Samba-3 server. See Figure 8.19.

8. Click on **OK**. The "*Welcome to the MIDEARTH domain.*" dialog box should appear. At this point the machine must be rebooted. Joining the domain is now complete.

8.2.3 Domain Logon Configuration: Windows 9x/Me

We follow the convention used by most in saying that Windows 9x/Me machines can participate in Domain logons. The truth is that these platforms can use only the LanManager network logon protocols.

> NOTE
>
> 
>
> Windows XP Home edition cannot participate in Domain or LanManager network logons.

1. Right click on the **Network Neighborhood** icon.
2. The Network Configuration Panel allows all common network settings to be changed. See Figure 8.20.

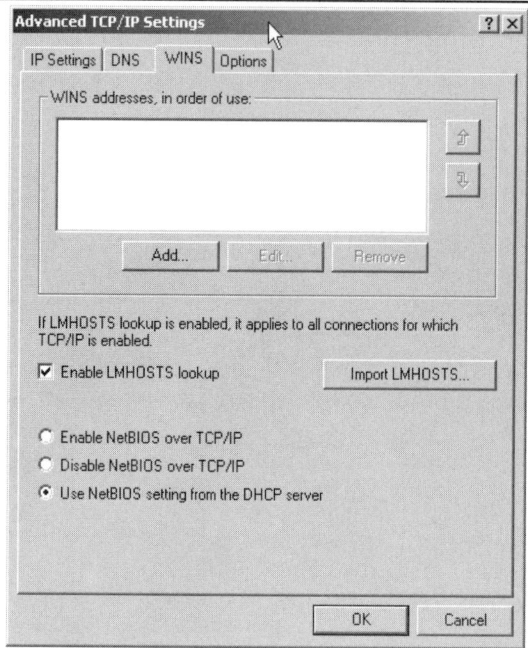

Figure 8.10. WINS Configuration.

Make sure that the **Client for Microsoft Networks** driver is installed as shown. Click on the **Client for Microsoft Networks** entry in **The following network components are installed:** box. Then click the **Properties** button.

3. The Client for Microsoft Networks Properties panel is the correct location to configure network logon settings. See Figure 8.21.

   Enter the Windows NT domain name, check the **Log on to Windows NT domain** box, click **OK**.

4. Click on the **Identification** button. This is the location at which the workgroup (domain) name and the machine name (computer name) need to be set. See Figure 8.22.

5. Now click the **Access Control** button. If you want to be able to assign share access permissions using domain user and group accounts, it is necessary to enable **User-level access control** as shown in this panel. See Figure 8.23.

8.3 Common Errors

The most common errors that can afflict Windows networking systems include:

- Incorrect IP address.
- Incorrect or inconsistent netmasks.

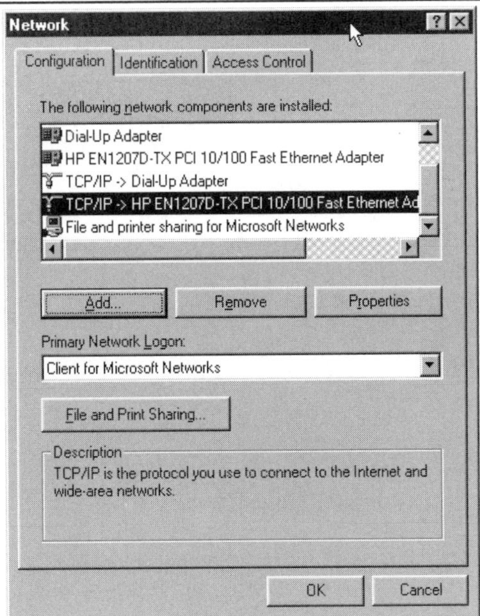

Figure 8.11. The Windows Me Network Configuration Panel.

- Incorrect router address.
- Incorrect DNS server address.
- Incorrect WINS server address.
- Use of a Network Scope setting — watch out for this one!

The most common reasons for which a Windows NT/200x/XP Professional client cannot join the Samba controlled domain are:

- `smb.conf` does not have correct ***add machine script*** settings.
- "*root*" account is not in password backend database.
- Attempt to use a user account instead of the "*root*" account to join a machine to the domain.
- Open connections from the workstation to the server.
- Firewall or filter configurations in place on either the client or on the Samba server.

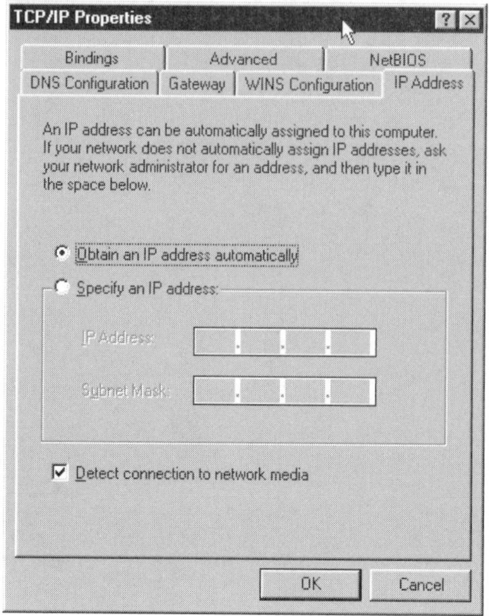

Figure 8.12. IP Address.

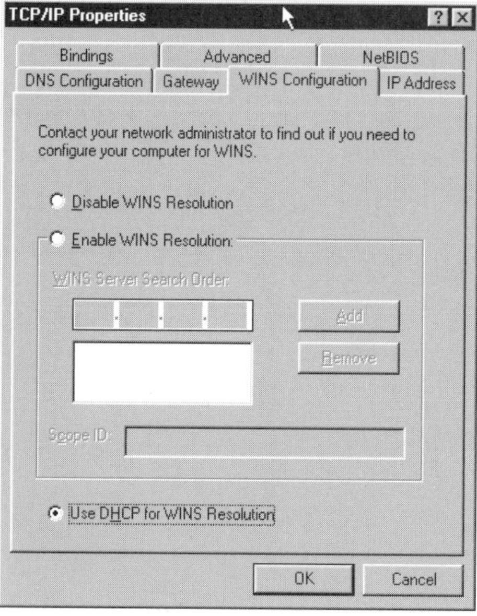

Figure 8.13. DNS Configuration.

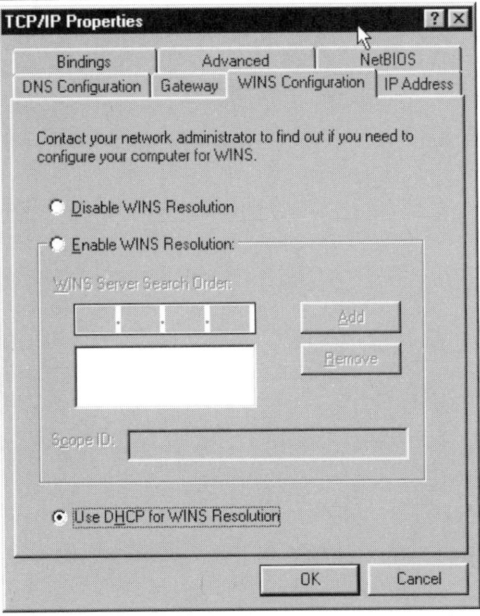

Figure 8.14. WINS Configuration.

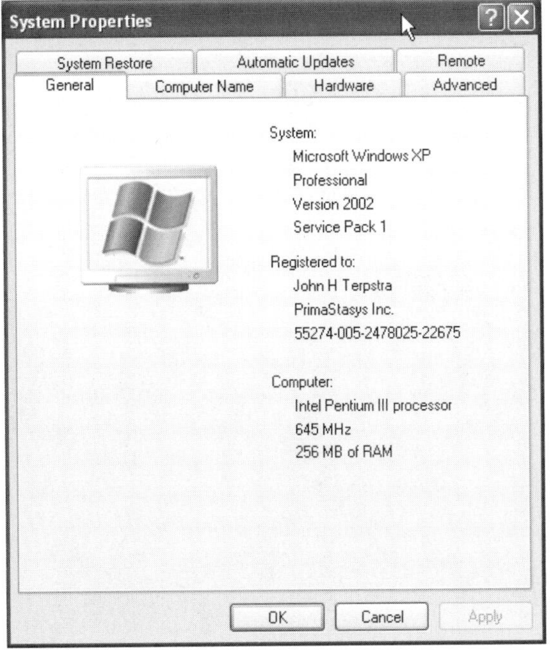

Figure 8.15. The General Panel.

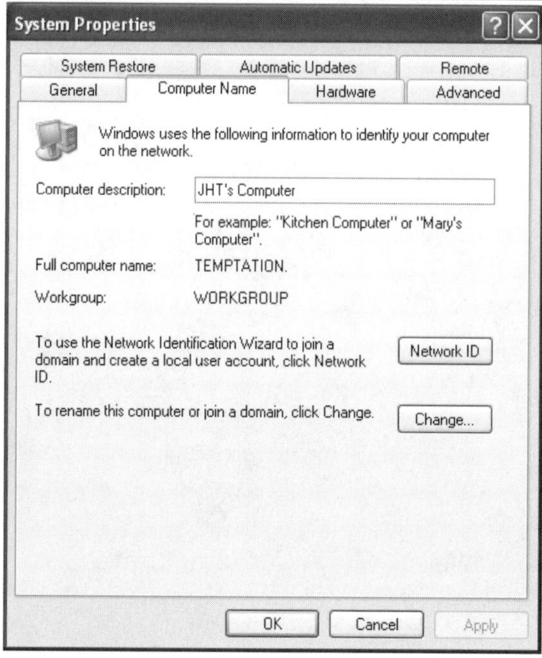

Figure 8.16. The Computer Name Panel.

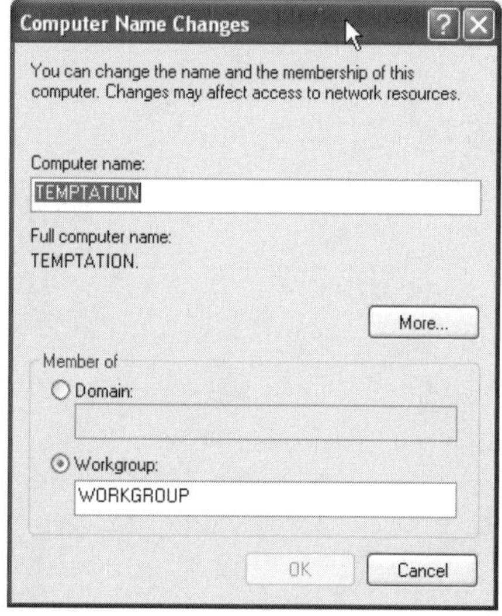

Figure 8.17. The Computer Name Changes Panel.

Figure 8.18. The Computer Name Changes Panel Domain MIDEARTH.

Figure 8.19. Computer Name Changes User name and Password Panel.

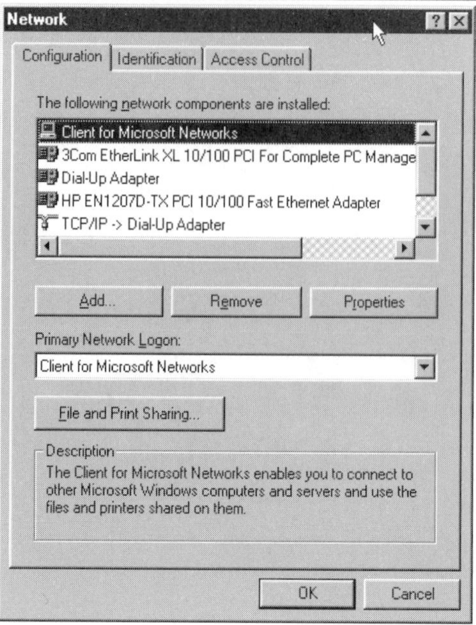

Figure 8.20. The Network Panel.

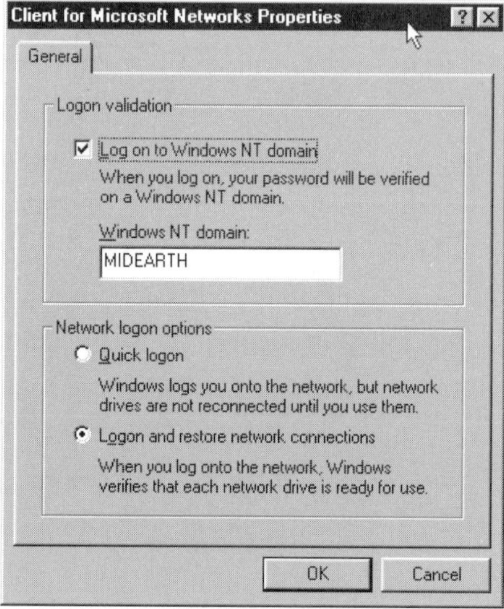

Figure 8.21. Client for Microsoft Networks Properties Panel.

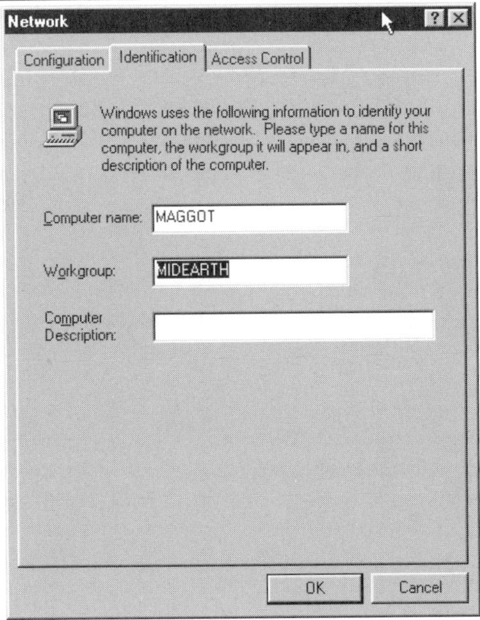

Figure 8.22. Identification Panel.

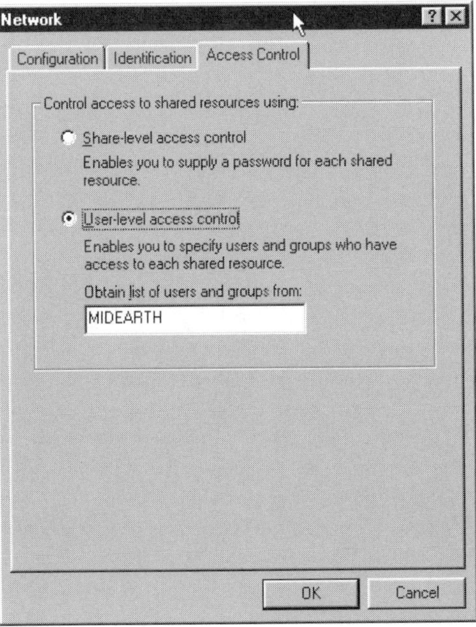

Figure 8.23. Identification Panel.

Part III

Advanced Configuration

Chapter 9

NETWORK BROWSING

This document contains detailed information as well as a fast track guide to implementing browsing across subnets and/or across workgroups (or domains). WINS is the best tool for resolution of NetBIOS names to IP addresses. WINS is not involved in browse list handling except by way of name to address resolution.

NOTE

MS Windows 2000 and later versions can be configured to operate with no NetBIOS over TCP/IP. Samba-3 and later versions also support this mode of operation. When the use of NetBIOS over TCP/IP has been disabled, the primary means for resolution of MS Windows machine names is via DNS and Active Directory. The following information assumes that your site is running NetBIOS over TCP/IP.

9.1 Features and Benefits

Someone once referred to the past in these words "It was the best of times, it was the worst of times." The more we look back, the more we long for what was and hope it never returns.

For many MS Windows network administrators, that statement sums up their feelings about NetBIOS networking precisely. For those who mastered NetBIOS networking, its fickle nature was just par for the course. For those who never quite managed to tame its lusty features, NetBIOS is like Paterson's Curse.

For those not familiar with botanical problems in Australia, Paterson's Curse, *Echium plantagineum*, was introduced to Australia from Europe during the mid-nineteenth century. Since then it has spread rapidly. The high seed production, with densities of thousands of seeds per square meter, a seed longevity of more than seven years, and an ability to germinate at any time of year, given the right conditions, are some of the features which make it such a persistent weed.

In this chapter we explore vital aspects of Server Message Block (SMB) networking with a particular focus on SMB as implemented through running NetBIOS (Network Basic Input/Output System) over TCP/IP. Since Samba does not implement SMB or NetBIOS over any other protocols, we need to know how to configure our network environment and simply remember to use nothing but TCP/IP on all our MS Windows network clients.

Samba provides the ability to implement a WINS (Windows Internetworking Name Server) and implements extensions to Microsoft's implementation of WINS. These extensions help Samba to effect stable WINS operations beyond the normal scope of MS WINS.

WINS is exclusively a service that applies only to those systems that run NetBIOS over TCP/IP. MS Windows 200x/XP have the capacity to operate with support for NetBIOS disabled, in which case WINS is of no relevance. Samba supports this also.

For those networks on which NetBIOS has been disabled (i.e., WINS is not required) the use of DNS is necessary for host name resolution.

9.2 What Is Browsing?

To most people browsing means they can see the MS Windows and Samba servers in the Network Neighborhood, and when the computer icon for a particular server is clicked, it opens up and shows the shares and printers available on the target server.

What seems so simple is in fact a complex interaction of different technologies. The technologies (or methods) employed in making all of this work include:

- MS Windows machines register their presence to the network.
- Machines announce themselves to other machines on the network.
- One or more machine on the network collates the local announcements.
- The client machine finds the machine that has the collated list of machines.
- The client machine is able to resolve the machine names to IP addresses.
- The client machine is able to connect to a target machine.

The Samba application that controls browse list management and name resolution is called `nmbd`. The configuration parameters involved in nmbd's operation are:

Browsing options: *os level* (*), *lm announce*, *lm interval*, *preferred master* (*), *local master* (*), *domain master* (*), *browse list*, *enhanced browsing*.

Name Resolution Method: *name resolve order* (*).

WINS options: *dns proxy*, *wins proxy*, *wins server* (*), *wins support* (*), *wins hook*.

For Samba, the WINS Server and WINS Support are mutually exclusive options. Those marked with an (*) are the only options that commonly may need to be modified. Even if none of these parameters is set, `nmbd` will still do its job.

9.3 Discussion

All MS Windows networking uses SMB-based messaging. SMB messaging may be implemented with or without NetBIOS. MS Windows 200x supports NetBIOS over TCP/IP for backwards compatibility. Microsoft appears intent on phasing out NetBIOS support.

9.3.1 NetBIOS over TCP/IP

Samba implements NetBIOS, as does MS Windows NT/200x/XP, by encapsulating it over TCP/IP. MS Windows products can do likewise. NetBIOS-based networking uses broadcast messaging to effect browse list management. When running NetBIOS over TCP/IP, this uses UDP-based messaging. UDP messages can be broadcast or unicast.

Normally, only unicast UDP messaging can be forwarded by routers. The *remote announce* parameter to smb.conf helps to project browse announcements to remote network segments via unicast UDP. Similarly, the *remote browse sync* parameter of `smb.conf` implements browse list collation using unicast UDP.

Secondly, in those networks where Samba is the only SMB server technology, wherever possible `nmbd` should be configured on one machine as the WINS server. This makes it easy to manage the browsing environment. If each network segment is configured with its own Samba WINS server, then the only way to get cross-segment browsing to work is by using the *remote announce* and the *remote browse sync* parameters to your `smb.conf` file.

If only one WINS server is used for an entire multi-segment network, then the use of the *remote announce* and the *remote browse sync* parameters should not be necessary.

As of Samba-3 WINS replication is being worked on. The bulk of the code has been committed, but it still needs maturation. This is not a supported feature of the Samba-3.0.0 release. Hopefully, this will become a supported feature of one of the Samba-3 release series.

Right now Samba WINS does not support MS-WINS replication. This means that when setting up Samba as a WINS server, there must only be one `nmbd` configured as a WINS server on the network. Some sites have used multiple Samba WINS servers for redundancy (one server per subnet) and then used *remote browse sync* and *remote announce* to effect browse list collation across all segments. Note that this means clients will only resolve local names, and must be configured to use DNS to resolve names on other subnets in order to resolve the IP addresses of the servers they can see on other subnets. This setup is not recommended, but is mentioned as a practical consideration (i.e., an "*if all else fails*" scenario).

Lastly, take note that browse lists are a collection of unreliable broadcast messages that are repeated at intervals of not more than 15 minutes. This means that it will take time to establish a browse list and it can take up to 45 minutes to stabilize, particularly across network segments.

9.3.2 TCP/IP without NetBIOS

All TCP/IP-enabled systems use various forms of host name resolution. The primary methods for TCP/IP hostname resolution involve either a static file (`/etc/hosts`) or the Domain Name System (DNS). DNS is the technology that makes the Internet usable. DNS-based host name resolution is supported by nearly all TCP/IP-enabled systems. Only a few embedded TCP/IP systems do not support DNS.

When an MS Windows 200x/XP system attempts to resolve a host name to an IP address it follows a defined path:

1. Checks the `hosts` file. It is located in `C:\Windows NT\System32\Drivers\etc`.
2. Does a DNS lookup.
3. Checks the NetBIOS name cache.
4. Queries the WINS server.
5. Does a broadcast name lookup over UDP.
6. Looks up entries in LMHOSTS. It is located in `C:\Windows NT\System32\Drivers\etc`.

Windows 200x/XP can register its host name with a Dynamic DNS server. You can force register with a Dynamic DNS server in Windows 200x/XP using: **ipconfig /registerdns**.

With Active Directory (ADS), a correctly functioning DNS server is absolutely essential. In the absence of a working DNS server that has been correctly configured, MS Windows clients and servers will be unable to locate each other, so consequently network services will be severely impaired.

The use of Dynamic DNS is highly recommended with Active Directory, in which case the use of BIND9 is preferred for its ability to adequately support the SRV (service) records that are needed for Active Directory.

9.3.3 DNS and Active Directory

Occasionally we hear from UNIX network administrators who want to use a UNIX-based Dynamic DNS server in place of the Microsoft DNS server. While this might be desirable to some, the MS Windows 200x DNS server is auto-configured to work with Active Directory. It is possible to use BIND version 8 or 9, but it will almost certainly be necessary to create service records so MS Active Directory clients can resolve host names to locate essential network services. The following are some of the default service records that Active Directory requires:

_ldap._tcp.pdc.ms-dcs.*Domain* — This provides the address of the Windows NT PDC for the Domain.

_ldap._tcp.pdc.ms-dcs.*DomainTree* — Resolves the addresses of Global Catalog servers in the domain.

_ldap._tcp.*site*.sites.writable.ms-dcs.*Domain* — Provides list of Domain Controllers based on sites.

_ldap._tcp.writable.ms-dcs.*Domain* — Enumerates list of Domain Controllers that have the writable copies of the Active Directory datastore.

_ldap._tcp.*GUID*.domains.ms-dcs.*DomainTree* — Entry used by MS Windows clients to locate machines using the Global Unique Identifier.

_ldap._tcp.*Site*.gc.ms-dcs.*DomainTree* — Used by MS Windows clients to locate site configuration dependent Global Catalog server.

9.4 How Browsing Functions

MS Windows machines register their NetBIOS names (i.e., the machine name for each service type in operation) on start-up. The exact method by which this name registration takes place is determined by whether or not the MS Windows client/server has been given a WINS server address, whether or not LMHOSTS lookup is enabled, or if DNS for NetBIOS name resolution is enabled, etc.

In the case where there is no WINS server, all name registrations as well as name lookups are done by UDP broadcast. This isolates name resolution to the local subnet, unless LMHOSTS is used to list all names and IP addresses. In such situations, Samba provides a means by which the Samba server name may be forcibly injected into the browse list of a remote MS Windows network (using the *remote announce* parameter).

Where a WINS server is used, the MS Windows client will use UDP unicast to register with the WINS server. Such packets can be routed and thus WINS allows name resolution to function across routed networks.

During the startup process an election will take place to create a Local Master Browser if one does not already exist. On each NetBIOS network one machine will be elected to function as the Domain Master Browser. This domain browsing has nothing to do with MS security Domain Control. Instead, the Domain Master Browser serves the role of contacting each local master browser (found by asking WINS or from LMHOSTS) and exchanging browse list contents. This way every master browser will eventually obtain a complete list of all machines that are on the network. Every 11 to 15 minutes an election is held to determine which machine will be the master browser. By the nature of the election criteria used, the machine with the highest uptime, or the most senior protocol version or other criteria, will win the election as Domain Master Browser.

Clients wishing to browse the network make use of this list, but also depend on the availability of correct name resolution to the respective IP address/addresses.

Any configuration that breaks name resolution and/or browsing intrinsics will annoy users because they will have to put up with protracted inability to use the network services.

Samba supports a feature that allows forced synchronization of browse lists across routed networks using the *remote browse sync* parameter in the `smb.conf` file. This causes Samba to contact the local master browser on a remote network and to request browse list synchronization. This effectively bridges two networks that are separated by routers. The two remote networks may use either broadcast-based name resolution or WINS-based name resolution, but it should be noted that the *remote browse sync* parameter provides browse list synchronization — and that is distinct from name to address resolution. In other words, for cross-subnet browsing to function correctly it is essential that a name-to-address resolution mechanism be provided. This mechanism could be via DNS, `/etc/hosts`, and so on.

9.4.1 Configuring WORKGROUP Browsing

To configure cross-subnet browsing on a network containing machines in a WORKGROUP, not an NT Domain, you need to set up one Samba server to be the Domain Master Browser (note that this is not the same as a Primary Domain Controller, although in an NT Domain the same machine plays both roles). The role of a Domain Master Browser is to collate the browse lists from Local Master Browsers on all the subnets that have a machine participating in the workgroup. Without one machine configured as a Domain Master Browser, each subnet would be an isolated workgroup unable to see any machines on another subnet. It is the presence of a Domain Master Browser that makes cross-subnet browsing possible for a workgroup.

In a WORKGROUP environment the Domain Master Browser must be a Samba server, and there must only be one Domain Master Browser per workgroup name. To set up a Samba server as a Domain Master Browser, set the following option in the *[global]* section of the `smb.conf` file:

```
domain master = yes
```

The Domain Master Browser should preferably be the local master browser for its own subnet. In order to achieve this, set the following options in the *[global]* section of the `smb.conf` file as shown in Example 9.1.

Example 9.1. Domain Master Browser smb.conf

```
[global]
domain master = yes
local master = yes
preferred master = yes
os level = 65
```

The Domain Master Browser may be the same machine as the WINS server, if necessary.

Next, you should ensure that each of the subnets contains a machine that can act as a Local Master Browser for the workgroup. Any MS Windows NT/200x/XP machine should be able to do this, as will Windows 9x/Me machines (although these tend to get rebooted more often, so it is not such a good idea to use these). To make a Samba server a Local Master Browser set the following options in the *[global]* section of the `smb.conf` file as shown in Example 9.2:

Do not do this for more than one Samba server on each subnet, or they will war with each other over which is to be the Local Master Browser.

The *local master* parameter allows Samba to act as a Local Master Browser. The *preferred master* causes **nmbd** to force a browser election on startup and the *os level* parameter sets Samba high enough so it should win any browser elections.

Example 9.2. Local master browser smb.conf

```
[global]
domain master = no
local master = yes
preferred master = yes
os level = 65
```

If you have an NT machine on the subnet that you wish to be the Local Master Browser, you can disable Samba from becoming a Local Master Browser by setting the following options in the *[global]* section of the `smb.conf` file as shown in Example 9.3:

Example 9.3. smb.conf for not being a Master Browser

```
[global]
domain master = no
local master = no
preferred master = no
os level = 0
```

9.4.2 DOMAIN Browsing Configuration

If you are adding Samba servers to a Windows NT Domain, then you must not set up a Samba server as a Domain Master Browser. By default, a Windows NT Primary Domain Controller for a domain is also the Domain Master Browser for that domain. Network browsing may break if a Samba server registers the domain master browser NetBIOS name (*DOMAIN*<1B>) with WINS instead of the PDC.

For subnets other than the one containing the Windows NT PDC, you may set up Samba servers as Local Master Browsers as described. To make a Samba server a Local Master Browser, set the following options in the *[global]* section of the `smb.conf` file as shown in Example 9.4:

Example 9.4. Local Master Browser smb.conf

```
[global]
domain master = no
local master = yes
preferred master = yes
os level = 65
```

If you wish to have a Samba server fight the election with machines on the same subnet you may set the *os level* parameter to lower levels. By doing this you can tune the order of machines that will become Local Master Browsers if they are running. For more details on this refer to Section 9.4.3.

If you have Windows NT machines that are members of the domain on all subnets and you are sure they will always be running, you can disable Samba from taking part in browser elections and ever becoming a Local Master Browser by setting the following options in the *[global]* section of the `smb.conf` file as shown in Example 9.5:

Example 9.5. smb.conf for not being a master browser

```
[global]
domain master = no
local master = no
preferred master = no
os level = 0
```

9.4.3 Forcing Samba to Be the Master

Who becomes the master browser is determined by an election process using broadcasts. Each election packet contains a number of parameters that determine what precedence (bias) a host should have in the election. By default Samba uses a low precedence and thus loses elections to just about every Windows network server or client.

If you want Samba to win elections, set the *os level* global option in `smb.conf` to a higher number. It defaults to zero. Using 34 would make it win all elections every other system (except other samba systems).

An *os level* of two would make it beat Windows for Workgroups and Windows 9x/Me, but not MS Windows NT/200x Server. An MS Windows NT/200x Server Domain Controller uses level 32. The maximum os level is 255.

If you want Samba to force an election on startup, set the *preferred master* global option in `smb.conf` to `yes`. Samba will then have a slight advantage over other potential master browsers that are not Perferred Master Browsers. Use this parameter with care, as if you have two hosts (whether they are Windows 9x/Me or NT/200x/XP or Samba) on the same local subnet both set with *preferred master* to `yes`, then periodically and continually they will force an election in order to become the Local Master Browser.

If you want Samba to be a *Domain Master Browser*, then it is recommended that you also set *preferred master* to `yes`, because Samba will not become a Domain Master Browser for the whole of your LAN or WAN if it is not also a Local Master Browser on its own broadcast isolated subnet.

It is possible to configure two Samba servers to attempt to become the Domain Master Browser for a domain. The first server that comes up will be the Domain Master Browser. All other Samba servers will attempt to become the Domain Master Browser every five minutes. They will find that another Samba server is already the domain master browser and will fail. This provides automatic redundancy, should the current Domain Master Browser fail.

9.4.4 Making Samba the Domain Master

The domain master is responsible for collating the browse lists of multiple subnets so browsing can occur between subnets. You can make Samba act as the Domain Master by setting *domain master* = yes in `smb.conf`. By default it will not be a Domain Master.

Do not set Samba to be the Domain Master for a workgroup that has the same name as an NT/200x Domain. If Samba is configured to be the Domain Master for a workgroup that is present on the same network as a Windows NT/200x domain that has the same name, network browsing problems will certainly be experienced.

When Samba is the Domain Master and the Master Browser, it will listen for master announcements (made roughly every twelve minutes) from Local Master Browsers on other subnets and then contact them to synchronize browse lists.

If you want Samba to be the domain master, you should also set the *os level* high enough to make sure it wins elections, and set *preferred master* to `yes`, to get Samba to force an election on startup.

All servers (including Samba) and clients should be using a WINS server to resolve NetBIOS names. If your clients are only using broadcasting to resolve NetBIOS names, then two things will occur:

1. Local Master Browsers will be unable to find a Domain Master Browser, as they will be looking only on the local subnet.
2. If a client happens to get hold of a domain-wide browse list and a user attempts to access a host in that list, it will be unable to resolve the NetBIOS name of that host.

If, however, both Samba and your clients are using a WINS server, then:

1. Local master browsers will contact the WINS server and, as long as Samba has registered that it is a Domain Master Browser with the WINS server, the Local Master Browser will receive Samba's IP address as its Domain Master Browser.
2. When a client receives a domain-wide browse list and a user attempts to access a host in that list, it will contact the WINS server to resolve the NetBIOS name of that host. As long as that host has registered its NetBIOS name with the same WINS server, the user will be able to see that host.

9.4.5 Note about Broadcast Addresses

If your network uses a 0 based broadcast address (for example, if it ends in a 0) then you will strike problems. Windows for Workgroups does not seem to support a zeros broadcast and you will probably find that browsing and name lookups will not work.

9.4.6 Multiple Interfaces

Samba supports machines with multiple network interfaces. If you have multiple interfaces, you will need to use the *interfaces* option in `smb.conf` to configure them.

9.4.7 Use of the Remote Announce Parameter

The *remote announce* parameter of `smb.conf` can be used to forcibly ensure that all the NetBIOS names on a network get announced to a remote network. The syntax of the *remote announce* parameter is:

```
remote announce = a.b.c.d [e.f.g.h] ...
```

or

```
remote announce = a.b.c.d/WORKGROUP [e.f.g.h/WORKGROUP] ...
```

where:

a.b.c.d **and** *e.f.g.h* — is either the LMB (Local Master Browser) IP address or the broadcast address of the remote network. i.e., the LMB is at 192.168.1.10, or the address could be given as 192.168.1.255 where the netmask is assumed to be 24 bits (255.255.255.0). When the remote announcement is made to the broadcast address of the remote network, every host will receive our announcements. This is noisy and therefore undesirable but may be necessary if we do not know the IP address of the remote LMB.

WORKGROUP — is optional and can be either our own workgroup or that of the remote network. If you use the workgroup name of the remote network, our NetBIOS machine names will end up looking like they belong to that workgroup. This may cause name resolution problems and should be avoided.

9.4.8 Use of the Remote Browse Sync Parameter

The *remote browse sync* parameter of `smb.conf` is used to announce to another LMB that it must synchronize its NetBIOS name list with our Samba LMB. This works only if the Samba server that has this option is simultaneously the LMB on its network segment.

The syntax of the *remote browse sync* parameter is:

```
remote browse sync = a.b.c.d
```

where *a.b.c.d* is either the IP address of the remote LMB or else is the network broadcast address of the remote segment.

9.5 WINS — The Windows Internetworking Name Server

Use of WINS (either Samba WINS or MS Windows NT Server WINS) is highly recommended. Every NetBIOS machine registers its name together with a name_type value for each of several types of service it has available. It registers its name directly as a unique (the type 0x03) name. It also registers its name if it is running the LanManager compatible server service (used to make shares and printers available to other users) by registering the server (the type 0x20) name.

All NetBIOS names are up to 15 characters in length. The name_type variable is added to the end of the name, thus creating a 16 character name. Any name that is shorter than 15 characters is padded with spaces to the 15th character. Thus, all NetBIOS names are 16 characters long (including the name_type information).

WINS can store these 16-character names as they get registered. A client that wants to log onto the network can ask the WINS server for a list of all names that have registered the NetLogon service name_type. This saves broadcast traffic and greatly expedites logon processing. Since broadcast name resolution cannot be used across network segments this type of information can only be provided via WINS or via a statically configured `lmhosts` file that must reside on all clients in the absence of WINS.

WINS also serves the purpose of forcing browse list synchronization by all LMBs. LMBs must synchronize their browse list with the DMB (Domain Master Browser) and WINS helps the LMB to identify its DMB. By definition this will work only within a single workgroup. Note that the Domain Master Browser has nothing to do with what is referred to as an MS Windows NT Domain. The later is a reference to a security environment while the DMB refers to the master controller for browse list information only.

WINS will work correctly only if every client TCP/IP protocol stack has been configured to use the WINS servers. Any client that has not been configured to use the WINS server will continue to use only broadcast-based name registration so WINS may never get to know about it. In any case, machines that have not registered with a WINS server will fail name to address lookup attempts by other clients and will therefore cause workstation access errors.

To configure Samba as a WINS server just add *wins support* = yes to the `smb.conf` file [global] section.

To configure Samba to register with a WINS server just add *wins server* = a.b.c.d to your `smb.conf` file *[global]* section.

IMPORTANT

Never use both *wins support* = yes together with *wins server* = a.b.c.d particularly not using its own IP address. Specifying both will cause nmbd to refuse to start!

9.5.1 WINS Server Configuration

Either a Samba Server or a Windows NT Server machine may be set up as a WINS server. To configure a Samba Server to be a WINS server you must add to the `smb.conf` file on the selected Server the following line to the *[global]* section:

```
wins support = yes
```

Versions of Samba prior to 1.9.17 had this parameter default to yes. If you have any older versions of Samba on your network it is strongly suggested you upgrade to a recent version, or at the very least set the parameter to "*no*" on all these machines.

Machines configured with *wins support* = yes will keep a list of all NetBIOS names registered with them, acting as a DNS for NetBIOS names.

It is strongly recommended to set up only one WINS server. Do not set the *wins support* = yes option on more than one Samba server.

To configure Windows NT/200x Server as a WINS server, install and configure the WINS service. See the Windows NT/200x documentation for details. Windows NT/200x WINS servers can replicate to each other, allowing more than one to be set up in a complex subnet environment. As Microsoft refuses to document the replication protocols, Samba cannot currently participate in these replications. It is possible in the future that a Samba-to-Samba WINS replication protocol may be defined, in which case more than one Samba machine could be set up as a WINS server. Currently only one Samba server should have the *wins support* = yes parameter set.

After the WINS server has been configured, you must ensure that all machines participating on the network are configured with the address of this WINS server. If your WINS server is a Samba machine, fill in the Samba machine IP address in the **Primary WINS Server** field of the **Control Panel->Network->Protocols->TCP->WINS Server** dialogs in Windows 9x/Me or Windows NT/200x. To tell a Samba server the IP address of the WINS server, add the following line to the *[global]* section of all `smb.conf` files:

```
wins server = <name or IP address>
```

where <name or IP address> is either the DNS name of the WINS server machine or its IP address.

This line must not be set in the `smb.conf` file of the Samba server acting as the WINS server itself. If you set both the *wins support* = yes option and the *wins server* = <name> option then **nmbd** will fail to start.

There are two possible scenarios for setting up cross-subnet browsing. The first details setting up cross-subnet browsing on a network containing Windows 9x/Me, Samba and Windows NT/200x machines that are not configured as part of a Windows NT Domain. The second details setting up cross-subnet browsing on networks that contain NT Domains.

9.5.2 WINS Replication

Samba-3 permits WINS replication through the use of the `wrepld` utility. This tool is not currently capable of being used as it is still in active development. As soon as this tool becomes moderately functional, we will prepare man pages and enhance this section of the documentation to provide usage and technical details.

9.5.3 Static WINS Entries

Adding static entries to your Samba WINS server is actually fairly easy. All you have to do is add a line to `wins.dat`, typically located in `/usr/local/samba/var/locks`.

Entries in `wins.dat` take the form of:

```
"NAME#TYPE" TTL ADDRESS+ FLAGS
```

where NAME is the NetBIOS name, TYPE is the NetBIOS type, TTL is the time-to-live as an absolute time in seconds, ADDRESS+ is one or more addresses corresponding to the registration and FLAGS are the NetBIOS flags for the registration.

A typical dynamic entry looks like this:

```
"MADMAN#03" 1055298378 192.168.1.2 66R
```

To make it static, all that has to be done is set the TTL to 0, like this:

```
"MADMAN#03" 0 192.168.1.2 66R
```

Though this method works with early Samba-3 versions, there is a possibility that it may change in future versions if WINS replication is added.

9.6 Helpful Hints

The following hints should be carefully considered as they are stumbling points for many new network administrators.

9.6.1 Windows Networking Protocols

WARNING

Do not use more than one protocol on MS Windows machines.

A common cause of browsing problems results from installing more than one protocol on an MS Windows machine.

Every NetBIOS machine takes part in a process of electing the LMB (and DMB) every 15 minutes. A set of election criteria is used to determine the order of precedence for winning this election process. A machine running Samba or Windows NT will be biased so the most suitable machine will predictably win and thus retain its role.

The election process is "*fought out*" so to speak over every NetBIOS network interface. In the case of a Windows 9x/Me machine that has both TCP/IP and IPX installed and has NetBIOS enabled over both protocols, the election will be decided over both protocols. As often happens, if the Windows 9x/Me machine is the only one with both protocols then the LMB may be won on the NetBIOS interface over the IPX protocol. Samba will then lose the LMB role as Windows 9x/Me will insist it knows who the LMB is. Samba will then cease to function as an LMB and thus browse list operation on all TCP/IP-only machines will fail.

Windows 95, 98, 98se, and Me are referred to generically as Windows 9x/Me. The Windows NT4, 200x, and XP use common protocols. These are roughly referred to as the Windows NT family, but it should be recognized that 2000 and XP/2003 introduce new protocol extensions that cause them to behave differently from MS Windows NT4. Generally, where a server does not support the newer or extended protocol, these will fall back to the NT4 protocols.

The safest rule of all to follow is: use only one protocol!

9.6.2 Name Resolution Order

Resolution of NetBIOS names to IP addresses can take place using a number of methods. The only ones that can provide NetBIOS name_type information are:

- WINS — the best tool.
- LMHOSTS — static and hard to maintain.
- Broadcast — uses UDP and cannot resolve names across remote segments.

Alternative means of name resolution include:

- Static `/etc/hosts`— hard to maintain, and lacks name_type info.
- DNS — is a good choice but lacks essential name_type info.

Many sites want to restrict DNS lookups and avoid broadcast name resolution traffic. The *name resolve order* parameter is of great help here. The syntax of the *name resolve order* parameter is:

```
name resolve order = wins lmhosts bcast host
```

or

```
name resolve order = wins lmhosts (eliminates bcast and host)
```

The default is:

```
name resolve order = host lmhost wins bcast
```

where "*host*" refers to the native methods used by the UNIX system to implement the gethostbyname() function call. This is normally controlled by `/etc/host.conf`, `/etc/nsswitch.conf` and `/etc/resolv.conf`.

9.7 Technical Overview of Browsing

SMB networking provides a mechanism by which clients can access a list of machines in a network, a so-called *browse list*. This list contains machines that are ready to offer file and/or print services to other machines within the network. Thus it does not include machines that aren't currently able to do server tasks. The browse list is heavily used by all

SMB clients. Configuration of SMB browsing has been problematic for some Samba users, hence this document.

MS Windows 2000 and later versions, as with Samba-3 and later versions, can be configured to not use NetBIOS over TCP/IP. When configured this way, it is imperative that name resolution (using DNS/LDAP/ADS) be correctly configured and operative. Browsing will not work if name resolution from SMB machine names to IP addresses does not function correctly.

Where NetBIOS over TCP/IP is enabled, use of a WINS server is highly recommended to aid the resolution of NetBIOS (SMB) names to IP addresses. WINS allows remote segment clients to obtain NetBIOS name_type information that cannot be provided by any other means of name resolution.

9.7.1 Browsing Support in Samba

Samba facilitates browsing. The browsing is supported by nmbd and is also controlled by options in the `smb.conf` file. Samba can act as a local browse master for a workgroup and the ability to support domain logons and scripts is now available.

Samba can also act as a Domain Master Browser for a workgroup. This means that it will collate lists from Local Master Browsers into a wide area network server list. In order for browse clients to resolve the names they may find in this list, it is recommended that both Samba and your clients use a WINS server.

Do not set Samba to be the Domain Master for a workgroup that has the same name as an NT Domain. On each wide area network, you must only ever have one Domain Master Browser per workgroup, regardless of whether it is NT, Samba or any other type of domain master that is providing this service.

NOTE

nmbd can be configured as a WINS server, but it is not necessary to specifically use Samba as your WINS server. MS Windows NT4, Server or Advanced Server 200x can be configured as your WINS server. In a mixed NT/200x server and Samba environment on a Wide Area Network, it is recommended that you use the Microsoft WINS server capabilities. In a Samba-only environment, it is recommended that you use one and only one Samba server as the WINS server.

To get browsing to work you need to run nmbd as usual, but will need to use the *`workgroup`* option in `smb.conf` to control what workgroup Samba becomes a part of.

Samba also has a useful option for a Samba server to offer itself for browsing on another subnet. It is recommended that this option is only used for "*unusual*" purposes: announcements over the Internet, for example. See *`remote announce`* in the `smb.conf` man page.

9.7.2 Problem Resolution

If something does not work, the `log.nmbd` file will help to track down the problem. Try a *log level* of 2 or 3 for finding problems. Also note that the current browse list usually gets stored in text form in a file called `browse.dat`.

If it does not work, you should still be able to type the server name as `\\SERVER` in **filemanager**, then press enter and **filemanager** should display the list of available shares.

Some people find browsing fails because they do not have the global *guest account* set to a valid account. Remember that the IPC$ connection that lists the shares is done as guest and, thus, you must have a valid guest account.

MS Windows 2000 and later (as with Samba) can be configured to disallow anonymous (i.e., guest account) access to the IPC$ share. In that case, the MS Windows 2000/XP/2003 machine acting as an SMB/CIFS client will use the name of the currently logged-in user to query the IPC$ share. MS Windows 9x/Me clients are not able to do this and thus will not be able to browse server resources.

The other big problem people have is that their broadcast address, netmask or IP address is wrong (specified with the *interfaces* option in `smb.conf`)

9.7.3 Cross-Subnet Browsing

Since the release of Samba 1.9.17 (alpha1), Samba has supported the replication of browse lists across subnet boundaries. This section describes how to set this feature up in different settings.

To see browse lists that span TCP/IP subnets (i.e., networks separated by routers that do not pass broadcast traffic), you must set up at least one WINS server. The WINS server acts as a DNS for NetBIOS names. This will allow NetBIOS name-to-IP address translation to be completed by a direct query of the WINS server. This is done via a directed UDP packet on port 137 to the WINS server machine. The WINS server avoids the necessity of default NetBIOS name-to-IP address translation, which is done using UDP broadcasts from the querying machine. This means that machines on one subnet will not be able to resolve the names of machines on another subnet without using a WINS server.

Remember, for browsing across subnets to work correctly, all machines, be they Windows 95, Windows NT or Samba servers, must have the IP address of a WINS server given to them by a DHCP server, or by manual configuration (for Windows 9x/Me and Windows NT/200x/XP, this is in the TCP/IP Properties, under Network settings); for Samba, this is in the `smb.conf` file.

9.7.3.1 Behavior of Cross-Subnet Browsing

Cross-subnet Browsing is a complicated dance, containing multiple moving parts. It has taken Microsoft several years to get the code that achieves this correct, and Samba lags behind in some areas. Samba is capable of cross-subnet browsing when configured correctly.

Consider a network set up as Figure 9.1.

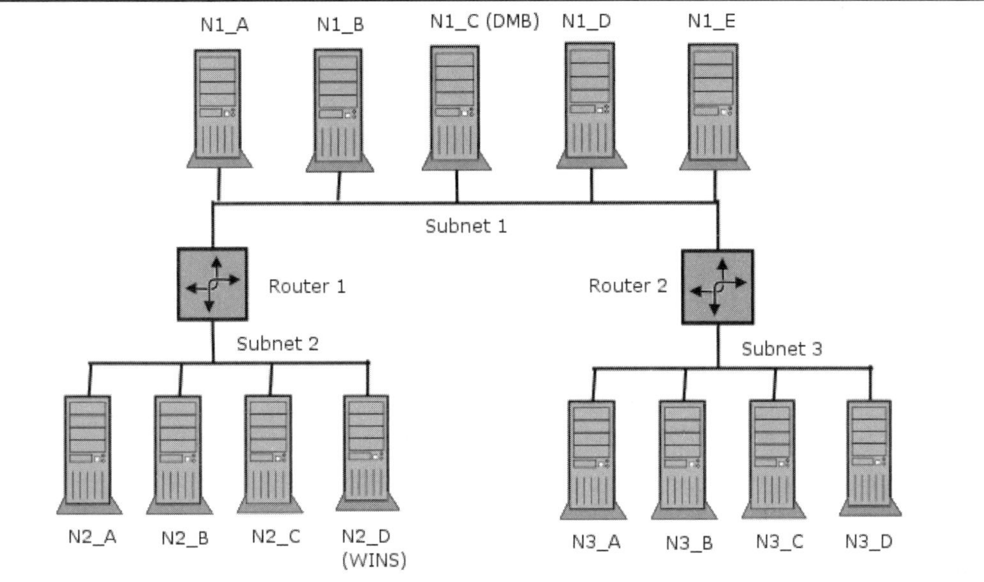

Figure 9.1. Cross-Subnet Browsing Example.

This consists of 3 subnets (1, 2, 3) connected by two routers (R1, R2) which do not pass broadcasts. Subnet 1 has five machines on it, subnet 2 has four machines, subnet 3 has four machines. Assume for the moment that all machines are configured to be in the same workgroup (for simplicity's sake). Machine N1_C on subnet 1 is configured as Domain Master Browser (i.e., it will collate the browse lists for the workgroup). Machine N2_D is configured as WINS server and all the other machines are configured to register their NetBIOS names with it.

As these machines are booted up, elections for master browsers will take place on each of the three subnets. Assume that machine N1_C wins on subnet 1, N2_B wins on subnet 2, and N3_D wins on subnet 3. These machines are known as Local Master Browsers for their particular subnet. N1_C has an advantage in winning as the Local Master Browser on subnet 1 as it is set up as Domain Master Browser.

On each of the three networks, machines that are configured to offer sharing services will broadcast that they are offering these services. The Local Master Browser on each subnet will receive these broadcasts and keep a record of the fact that the machine is offering a service. This list of records is the basis of the browse list. For this case, assume that all the machines are configured to offer services, so all machines will be on the browse list.

For each network, the Local Master Browser on that network is considered "*authoritative*" for all the names it receives via local broadcast. This is because a machine seen by the Local Master Browser via a local broadcast must be on the same network as the Local Master Browser and thus is a "*trusted*" and "*verifiable*" resource. Machines on other networks that the Local Master Browsers learn about when collating their browse lists have not been directly seen. These records are called "*non-authoritative.*"

At this point the browse lists appear as shown in Table 9.1 (these are the machines you would see in your network neighborhood if you looked in it on a particular network right now).

Table 9.1. Browse Subnet Example 1

Subnet	Browse Master	List
Subnet1	N1_C	N1_A, N1_B, N1_C, N1_D, N1_E
Subnet2	N2_B	N2_A, N2_B, N2_C, N2_D
Subnet3	N3_D	N3_A, N3_B, N3_C, N3_D

At this point all the subnets are separate, and no machine is seen across any of the subnets.

Now examine subnet 2. As soon as N2_B has become the Local Master Browser it looks for a Domain Master Browser with which to synchronize its browse list. It does this by querying the WINS server (N2_D) for the IP address associated with the NetBIOS name WORKGROUP<1B>. This name was registered by the Domain Master Browser (N1_C) with the WINS server as soon as it was started.

Once N2_B knows the address of the Domain Master Browser, it tells it that is the Local Master Browser for subnet 2 by sending a *MasterAnnouncement* packet as a UDP port 138 packet. It then synchronizes with it by doing a *NetServerEnum2* call. This tells the Domain Master Browser to send it all the server names it knows about. Once the Domain Master Browser receives the *MasterAnnouncement* packet, it schedules a synchronization request to the sender of that packet. After both synchronizations are complete the browse lists look as shown in Table 9.2:

Table 9.2. Browse Subnet Example 2

Subnet	Browse Master	List
Subnet1	N1_C	N1_A, N1_B, N1_C, N1_D, N1_E, N2_A(*), N2_B(*), N2_C(*), N2_D(*)
Subnet2	N2_B	N2_A, N2_B, N2_C, N2_D, N1_A(*), N1_B(*), N1_C(*), N1_D(*), N1_E(*)
Subnet3	N3_D	N3_A, N3_B, N3_C, N3_D

Servers with an (*) after them are non-authoritative names.

At this point users looking in their network neighborhood on subnets 1 or 2 will see all the servers on both, users on subnet 3 will still only see the servers on their own subnet.

The same sequence of events that occurred for N2_B now occurs for the Local Master Browser on subnet 3 (N3_D). When it synchronizes browse lists with the Domain Master Browser (N1_A) it gets both the server entries on subnet 1, and those on subnet 2. After N3_D has synchronized with N1_C and vica versa, the browse lists will appear as shown in Table 9.3.

Servers with an (*) after them are non-authoritative names.

At this point, users looking in their network neighborhood on subnets 1 or 3 will see all the servers on all subnets, while users on subnet 2 will still only see the servers on subnets 1 and 2, but not 3.

Finally, the Local Master Browser for subnet 2 (N2_B) will sync again with the Domain Master Browser (N1_C) and will receive the missing server entries. Finally, as when a

Table 9.3. Browse Subnet Example 3

Subnet	Browse Master	List
Subnet1	N1_C	N1_A, N1_B, N1_C, N1_D, N1_E, N2_A(*), N2_B(*), N2_C(*), N2_D(*), N3_A(*), N3_B(*), N3_C(*), N3_D(*)
Subnet2	N2_B	N2_A, N2_B, N2_C, N2_D, N1_A(*), N1_B(*), N1_C(*), N1_D(*), N1_E(*)
Subnet3	N3_D	N3_A, N3_B, N3_C, N3_D, N1_A(*), N1_B(*), N1_C(*), N1_D(*), N1_E(*), N2_A(*), N2_B(*), N2_C(*), N2_D(*)

steady state (if no machines are removed or shut off) has been achieved, the browse lists will appear as shown in Table 9.4.

Table 9.4. Browse Subnet Example 4

Subnet	Browse Master	List
Subnet1	N1_C	N1_A, N1_B, N1_C, N1_D, N1_E, N2_A(*), N2_B(*), N2_C(*), N2_D(*), N3_A(*), N3_B(*), N3_C(*), N3_D(*)
Subnet2	N2_B	N2_A, N2_B, N2_C, N2_D, N1_A(*), N1_B(*), N1_C(*), N1_D(*), N1_E(*), N3_A(*), N3_B(*), N3_C(*), N3_D(*)
Subnet3	N3_D	N3_A, N3_B, N3_C, N3_D, N1_A(*), N1_B(*), N1_C(*), N1_D(*), N1_E(*), N2_A(*), N2_B(*), N2_C(*), N2_D(*)

Servers with an (*) after them are non-authoritative names.

Synchronizations between the Domain Master Browser and Local Master Browsers will continue to occur, but this should remain a steady state operation.

If either router R1 or R2 fails, the following will occur:

1. Names of computers on each side of the inaccessible network fragments will be maintained for as long as 36 minutes in the network neighborhood lists.

2. Attempts to connect to these inaccessible computers will fail, but the names will not be removed from the network neighborhood lists.

3. If one of the fragments is cut off from the WINS server, it will only be able to access servers on its local subnet using subnet-isolated broadcast NetBIOS name resolution. The effects are similar to that of losing access to a DNS server.

9.8 Common Errors

Many questions are asked on the mailing lists regarding browsing. The majority of browsing problems originate from incorrect configuration of NetBIOS name resolution. Some are of particular note.

9.8.1 How Can One Flush the Samba NetBIOS Name Cache without Restarting Samba?

Samba's **nmbd** process controls all browse list handling. Under normal circumstances it is safe to restart **nmbd**. This will effectively flush the Samba NetBIOS name cache and cause it to be rebuilt. This does not make certain that a rogue machine name will not re-appear in the browse list. When **nmbd** is taken out of service, another machine on the network will become the Browse Master. This new list may still have the rogue entry in it. If you really want to clear a rogue machine from the list, every machine on the network will need to be shut down and restarted after all machines are down. Failing a complete restart, the only other thing you can do is wait until the entry times out and is then flushed from the list. This may take a long time on some networks (perhaps months).

9.8.2 Server Resources Can Not Be Listed

"*My Client Reports* 'This server is not configured to list shared resources"

Your guest account is probably invalid for some reason. Samba uses the guest account for browsing in **smbd**. Check that your guest account is valid.

Also see *`guest account`* in the `smb.conf` man page.

9.8.3 I get an 'Unable to browse the network' error

This error can have multiple causes:

- There is no Local Master Browser. Configure nmbd or any other machine to serve as Local Master Browser.
- You cannot log onto the machine that is the local master browser. Can you logon to it as a guest user?
- There is no IP connectivity to the Local Master Browser. Can you reach it by broadcast?

9.8.4 Browsing of Shares and Directories is Very Slow

" *There are only two machines on a test network. One a Samba server, the other a Windows XP machine. Authentication and logons work perfectly, but when I try to explore shares on the Samba server, the Windows XP client becomes unrespsonsive. Sometimes it does not respond for some minutes. Eventually, Windows Explorer will respond and displays files and directories without problem. display file and directory.*"

"*But, the share is immediately available from a command shell (**cmd**, followed by exploration with dos command. Is this a Samba problem or is it a Windows problem? How can I solve this?*"

Here are a few possibilities:

Bad Networking Hardware — Most common defective hardware problems center around low cost or defective HUBs, routers, Network Interface Controllers (NICs) and bad wiring. If one piece of hardware is defective the whole network may suffer. Bad networking hardware can cause data corruption. Most bad networking hardware problems are accompanied by an increase in apparent network traffic, but not all.

The Windows XP WebClient — A number of sites have reported similar slow network browsing problems and found that when the WebClient service is turned off, the problem dissapears. This is certainly something that should be explored as it is a simple solution — if it works.

Inconsistent WINS Configuration — This type of problem is common when one client is configured to use a WINS server (that is a TCP/IP configuration setting) and there is no WINS server on the network. Alternately, this will happen is there is a WINS server and Samba is not configured to use it. The use of WINS is highly recommended if the network is using NetBIOS over TCP/IP protocols. If use of NetBIOS over TCP/IP is disabled on all clients, Samba should not be configured as a WINS server neither should it be configured to use one.

Incorrect DNS Configuration — If use of NetBIOS over TCP/IP is disabled, Active Directory is in use and the DNS server has been incorrectly configured. Refer Section 9.3.3 for more information.

Chapter 10

ACCOUNT INFORMATION DATABASES

Samba-3 implements a new capability to work concurrently with multiple account backends. The possible new combinations of password backends allows Samba-3 a degree of flexibility and scalability that previously could be achieved only with MS Windows Active Directory. This chapter describes the new functionality and how to get the most out of it.

In the development of Samba-3, a number of requests were received to provide the ability to migrate MS Windows NT4 SAM accounts to Samba-3 without the need to provide matching UNIX/Linux accounts. We called this the *Non-UNIX Accounts (NUA)* capability. The intent was that an administrator could decide to use the *tdbsam* backend and by simply specifying *passdb backend* = tdbsam_nua, this would allow Samba-3 to implement a solution that did not use UNIX accounts per se. Late in the development cycle, the team doing this work hit upon some obstacles that prevents this solution from being used. Given the delays with the Samba-3 release, a decision was made to not deliver this functionality until a better method of recognizing NT Group SIDs from NT User SIDs could be found. This feature may return during the life cycle for the Samba-3 series.

> NOTE
>
> 
>
> Samba-3 does not support Non-UNIX Account (NUA) operation for user accounts. Samba-3 does support NUA operation for machine accounts.

10.1 Features and Benefits

Samba-3 provides for complete backward compatibility with Samba-2.2.x functionality as follows:

10.1.1 Backward Compatibility Backends

Plain Text — This option uses nothing but the UNIX/Linux `/etc/passwd` style backend. On systems that have Pluggable Authentication Modules (PAM) support, all PAM modules are supported. The behavior is just as it was with Samba-2.2.x, and the protocol limitations imposed by MS Windows clients apply likewise. Please refer to Section 10.2 for more information regarding the limitations of Plain Text password usage.

smbpasswd — This option allows continued use of the `smbpasswd` file that maintains a plain ASCII (text) layout that includes the MS Windows LanMan and NT encrypted passwords as well as a field that stores some account information. This form of password backend does not store any of the MS Windows NT/200x SAM (Security Account Manager) information required to provide the extended controls that are needed for more comprehensive interoperation with MS Windows NT4/200x servers.

This backend should be used only for backward compatibility with older versions of Samba. It may be deprecated in future releases.

ldapsam_compat (Samba-2.2 LDAP Compatibility) — There is a password backend option that allows continued operation with an existing OpenLDAP backend that uses the Samba-2.2.x LDAP schema extension. This option is provided primarily as a migration tool, although there is no reason to force migration at this time. This tool will eventually be deprecated.

Samba-3 introduces a number of new password backend capabilities.

10.1.2 New Backends

tdbsam — This backend provides a rich database backend for local servers. This backend is not suitable for multiple Domain Controllers (i.e., PDC + one or more BDC) installations.

The *tdbsam* password backend stores the old *smbpasswd* information plus the extended MS Windows NT / 200x SAM information into a binary format TDB (trivial database) file. The inclusion of the extended information makes it possible for Samba-3 to implement the same account and system access controls that are possible with MS Windows NT4/200x-based systems.

The inclusion of the *tdbsam* capability is a direct response to user requests to allow simple site operation without the overhead of the complexities of running OpenLDAP. It is recommended to use this only for sites that have fewer than 250 users. For larger sites or implementations, the use of OpenLDAP or of Active Directory integration is strongly recommended.

ldapsam — This provides a rich directory backend for distributed account installation.

Samba-3 has a new and extended LDAP implementation that requires configuration of OpenLDAP with a new format Samba schema. The new format schema file is included in the `examples/LDAP` directory of the Samba distribution.

The new LDAP implementation significantly expands the control abilities that were possible with prior versions of Samba. It is now possible to specify *"per user"* profile settings, home directories, account access controls, and much more. Corporate sites will see that the Samba Team has listened to their requests both for capability and to allow greater scalability.

mysqlsam (MySQL based backend) — It is expected that the MySQL-based SAM will be very popular in some corners. This database backend will be of considerable interest to sites that want to leverage existing MySQL technology.

xmlsam (XML based datafile) — Allows the account and password data to be stored in an XML format data file. This backend cannot be used for normal operation, it can only be used in conjunction with **pdbedit**'s pdb2pdb functionality. The DTD that is used might be subject to changes in the future.

The *xmlsam* option can be useful for account migration between database backends or backups. Use of this tool will allow the data to be edited before migration into another backend format.

10.2 Technical Information

Old Windows clients send plain text passwords over the wire. Samba can check these passwords by encrypting them and comparing them to the hash stored in the UNIX user database.

Newer Windows clients send encrypted passwords (so-called Lanman and NT hashes) over the wire, instead of plain text passwords. The newest clients will send only encrypted passwords and refuse to send plain text passwords, unless their registry is tweaked.

These passwords can't be converted to UNIX-style encrypted passwords. Because of that, you can't use the standard UNIX user database, and you have to store the Lanman and NT hashes somewhere else.

In addition to differently encrypted passwords, Windows also stores certain data for each user that is not stored in a UNIX user database. For example, workstations the user may logon from, the location where the user's profile is stored, and so on. Samba retrieves and stores this information using a *passdb backend*. Commonly available backends are LDAP, plain text file, and MySQL. For more information, see the man page for `smb.conf` regarding the *passdb backend* parameter.

The resolution of SIDs to UIDs is fundamental to correct operation of Samba. In both cases shown, if winbindd is not running, or cannot be contacted, then only local SID/UID resolution is possible. See Figure 10.1 and Figure 10.2.

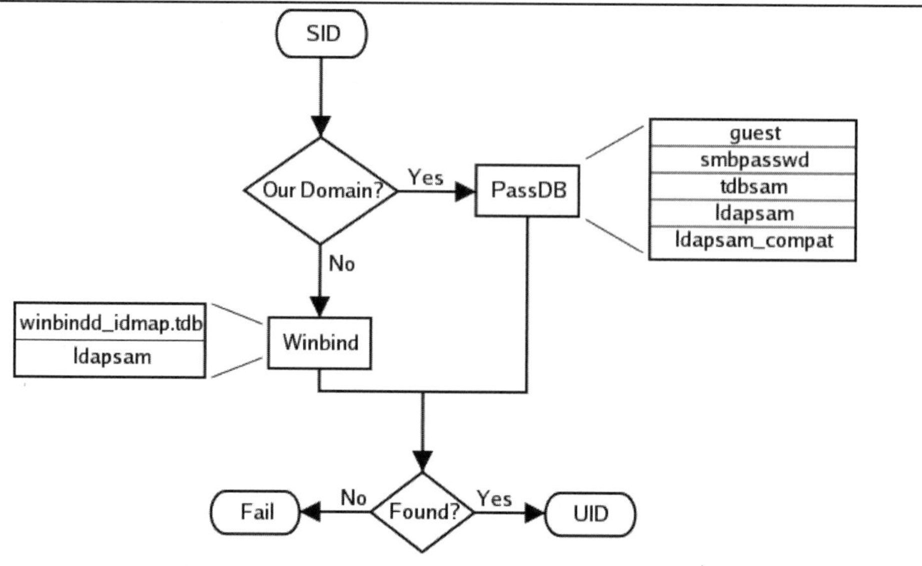

Figure 10.1. IDMAP: Resolution of SIDs to UIDs.

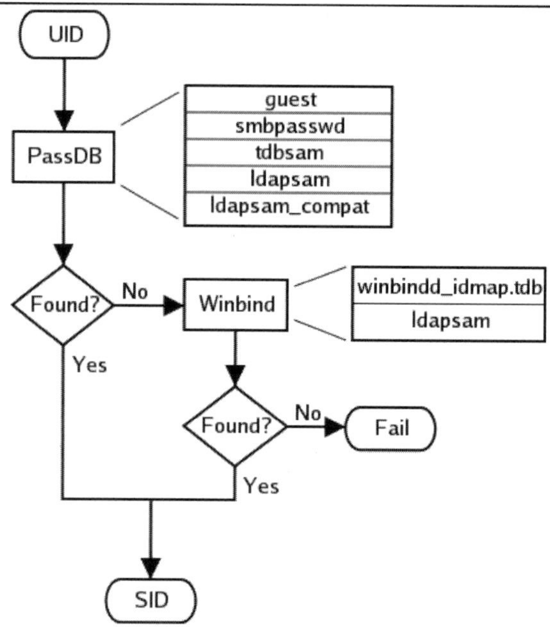

Figure 10.2. IDMAP: Resolution of UIDs to SIDs.

10.2.1 Important Notes About Security

The UNIX and SMB password encryption techniques seem similar on the surface. This similarity is, however, only skin deep. The UNIX scheme typically sends cleartext passwords over the network when logging in. This is bad. The SMB encryption scheme never sends the cleartext password over the network but it does store the 16 byte hashed values on disk. This is also bad. Why? Because the 16 byte hashed values are a *"password equivalent."* You cannot derive the user's password from them, but they could potentially be used in a modified client to gain access to a server. This would require considerable technical knowledge on behalf of the attacker but is perfectly possible. You should thus treat the datastored in whatever passdb backend you use (smbpasswd file, LDAP, MYSQL) as though it contained the cleartext passwords of all your users. Its contents must be kept secret and the file should be protected accordingly.

Ideally, we would like a password scheme that involves neither plain text passwords on the network nor on disk. Unfortunately, this is not available as Samba is stuck with having to be compatible with other SMB systems (Windows NT, Windows for Workgroups, Windows 9x/Me).

Windows NT 4.0 Service Pack 3 changed the default setting so plaintext passwords are disabled from being sent over the wire. This mandates either the use of encrypted password support or editing the Windows NT registry to re-enable plaintext passwords.

The following versions of Microsoft Windows do not support full domain security protocols, although they may log onto a domain environment:

- MS DOS Network client 3.0 with the basic network redirector installed.
- Windows 95 with the network redirector update installed.
- Windows 98 [Second Edition].
- Windows Me.

NOTE

MS Windows XP Home does not have facilities to become a Domain Member and it cannot participate in domain logons.

The following versions of MS Windows fully support domain security protocols.

- Windows NT 3.5x.
- Windows NT 4.0.
- Windows 2000 Professional.
- Windows 200x Server/Advanced Server.
- Windows XP Professional.

All current releases of Microsoft SMB/CIFS clients support authentication via the SMB Challenge/Response mechanism described here. Enabling cleartext authentication does not disable the ability of the client to participate in encrypted authentication. Instead, it allows the client to negotiate either plain text or encrypted password handling.

MS Windows clients will cache the encrypted password alone. Where plain text passwords are re-enabled through the appropriate registry change, the plain text password is never cached. This means that in the event that a network connections should become disconnected (broken), only the cached (encrypted) password will be sent to the resource server to effect an auto-reconnect. If the resource server does not support encrypted passwords the auto-reconnect will fail. Use of encrypted passwords is strongly advised.

10.2.1.1 Advantages of Encrypted Passwords

- Plaintext passwords are not passed across the network. Someone using a network sniffer cannot just record passwords going to the SMB server.
- Plaintext passwords are not stored anywhere in memory or on disk.
- Windows NT does not like talking to a server that does not support encrypted passwords. It will refuse to browse the server if the server is also in User Level security mode. It will insist on prompting the user for the password on each connection, which is very annoying. The only things you can do to stop this is to use SMB encryption.
- Encrypted password support allows automatic share (resource) reconnects.
- Encrypted passwords are essential for PDC/BDC operation.

10.2.1.2 Advantages of Non-Encrypted Passwords

- Plaintext passwords are not kept on disk, and are not cached in memory.
- Uses same password file as other UNIX services such as Login and FTP.
- Use of other services (such as Telnet and FTP) that send plain text passwords over the network, so sending them for SMB is not such a big deal.

10.2.2 Mapping User Identifiers between MS Windows and UNIX

Every operation in UNIX/Linux requires a user identifier (UID), just as in MS Windows NT4/200x this requires a Security Identifier (SID). Samba provides two means for mapping an MS Windows user to a UNIX/Linux UID.

First, all Samba SAM (Security Account Manager database) accounts require a UNIX/Linux UID that the account will map to. As users are added to the account information database, Samba will call the *add user script* interface to add the account to the Samba host OS. In essence all accounts in the local SAM require a local user account.

The second way to effect Windows SID to UNIX UID mapping is via the *idmap uid* and *idmap gid* parameters in `smb.conf`. Please refer to the man page for information about these parameters. These parameters are essential when mapping users from a remote SAM server.

10.2.3 Mapping Common UIDs/GIDs on Distributed Machines

Samba-3 has a special facility that makes it possible to maintain identical UIDs and GIDs on all servers in a distributed network. A distributed network is one where there exists a PDC, one or more BDCs and/or one or more Domain Member servers. Why is this important? This is important if files are being shared over more than one protocol (e.g., NFS) and where users are copying files across UNIX/Linux systems using tools such as **rsync**.

The special facility is enabled using a parameter called *idmap backend*. The default setting for this parameter is an empty string. Technically it is possible to use an LDAP based idmap backend for UIDs and GIDs, but it makes most sense when this is done for network configurations that also use LDAP for the SAM backend. A sample use is shown in Example 10.1.

Example 10.1. Example configuration with the LDAP idmap backend

```
[global]
idmap backend = ldapsam:ldap://ldap-server.quenya.org:636
idmap backend = ldapsam:ldaps://ldap-server.quenya.org
```

A network administrator who wants to make significant use of LDAP backends will sooner or later be exposed to the excellent work done by PADL Software. PADL `http://www.padl.com`[1] have produced and released to open source an array of tools that might be of interest. These tools include:

- *nss_ldap:* An LDAP Name Service Switch module to provide native name service support for AIX, Linux, Solaris, and other operating systems. This tool can be used for centralized storage and retrieval of UIDs/GIDs.
- *pam_ldap:* A PAM module that provides LDAP integration for UNIX/Linux system access authentication.
- *idmap_ad:* An IDMAP backend that supports the Microsoft Services for UNIX RFC 2307 schema available from their web site[2].

10.3 Account Management Tools

Samba provides two tools for management of user and machine accounts. These tools are called **smbpasswd** and **pdbedit**. A third tool is under development but is not expected to ship in time for Samba-3.0.0. The new tool will be a TCL/TK GUI tool that looks much like the MS Windows NT4 Domain User Manager. Hopefully this will be announced in time for the Samba-3.0.1 release.

[1] http://www.padl.com

[2] http://www.padl.com/download/xad_oss_plugins.tar.gz

10.3.1 The *smbpasswd* Command

The smbpasswd utility is similar to the **passwd** or **yppasswd** programs. It maintains the two 32 byte password fields in the passdb backend.

smbpasswd works in a client-server mode where it contacts the local smbd to change the user's password on its behalf. This has enormous benefits.

smbpasswd has the capability to change passwords on Windows NT servers (this only works when the request is sent to the NT Primary Domain Controller if changing an NT Domain user's password).

smbpasswd can be used to:

- *add* user or machine accounts.
- *delete* user or machine accounts.
- *enable* user or machine accounts.
- *disable* user or machine accounts.
- *set to NULL* user passwords.
- *manage interdomain trust accounts.*

To run smbpasswd as a normal user just type:

```
$ smbpasswd
Old SMB password: secret
```

For `secret`, type old value here or press return if there is no old password.

```
New SMB Password: new secret
Repeat New SMB Password: new secret
```

If the old value does not match the current value stored for that user, or the two new values do not match each other, then the password will not be changed.

When invoked by an ordinary user, the command will only allow the user to change his or her own SMB password.

When run by root, **smbpasswd** may take an optional argument specifying the user name whose SMB password you wish to change. When run as root, **smbpasswd** does not prompt for or check the old password value, thus allowing root to set passwords for users who have forgotten their passwords.

smbpasswd is designed to work in the way familiar to UNIX users who use the **passwd** or **yppasswd** commands. While designed for administrative use, this tool provides essential User Level password change capabilities.

For more details on using **smbpasswd**, refer to the man page (the definitive reference).

10.3.2 The *pdbedit* Command

pdbedit is a tool that can be used only by root. It is used to manage the passdb backend. **pdbedit** can be used to:

- add, remove or modify user accounts.
- list user accounts.
- migrate user accounts.

The **pdbedit** tool is the only one that can manage the account security and policy settings. It is capable of all operations that smbpasswd can do as well as a super set of them.

One particularly important purpose of the **pdbedit** is to allow the migration of account information from one passdb backend to another. See the XML password backend section of this chapter.

The following is an example of the user account information that is stored in a tdbsam password backend. This listing was produced by running:

```
$ pdbedit -Lv met
UNIX username:        met
NT username:
Account Flags:        [UX         ]
User SID:             S-1-5-21-1449123459-1407424037-3116680435-2004
Primary Group SID:    S-1-5-21-1449123459-1407424037-3116680435-1201
Full Name:            Melissa E Terpstra
Home Directory:       \\frodo\met\Win9Profile
HomeDir Drive:        H:
Logon Script:         scripts\logon.bat
Profile Path:         \\frodo\Profiles\met
Domain:               MIDEARTH
Account desc:
Workstations:         melbelle
Munged dial:
Logon time:           0
Logoff time:          Mon, 18 Jan 2038 20:14:07 GMT
Kickoff time:         Mon, 18 Jan 2038 20:14:07 GMT
Password last set:    Sat, 14 Dec 2002 14:37:03 GMT
Password can change:  Sat, 14 Dec 2002 14:37:03 GMT
Password must change: Mon, 18 Jan 2038 20:14:07 GMT
```

The **pdbedit** tool allows migration of authentication (account) databases from one backend to another. For example: To migrate accounts from an old `smbpasswd` database to a *`tdbsam`* backend:

1. Set the *`passdb backend`* = tdbsam, smbpasswd.
2. Execute:

```
root# pdbedit -i smbpassed -e tdbsam
```

3. Now remove the *smbpasswd* from the passdb backend configuration in `smb.conf`.

10.4 Password Backends

Samba offers the greatest flexibility in backend account database design of any SMB/CIFS server technology available today. The flexibility is immediately obvious as one begins to explore this capability.

It is possible to specify not only multiple different password backends, but even multiple backends of the same type. For example, to use two different tdbsam databases:

passdb backend = tdbsam:/etc/samba/passdb.tdb \ tdbsam:/etc/samba/old-passdb.tdb

10.4.1 Plaintext

Older versions of Samba retrieved user information from the UNIX user database and eventually some other fields from the file `/etc/samba/smbpasswd` or `/etc/smbpasswd`. When password encryption is disabled, no SMB specific data is stored at all. Instead all operations are conducted via the way that the Samba host OS will access its `/etc/passwd` database. Linux systems For example, all operations are done via PAM.

10.4.2 smbpasswd — Encrypted Password Database

Traditionally, when configuring *encrypt passwords* = yes in Samba's `smb.conf` file, user account information such as username, LM/NT password hashes, password change times, and account flags have been stored in the `smbpasswd(5)` file. There are several disadvantages to this approach for sites with large numbers of users (counted in the thousands).

- The first problem is that all lookups must be performed sequentially. Given that there are approximately two lookups per domain logon (one for a normal session connection such as when mapping a network drive or printer), this is a performance bottleneck for large sites. What is needed is an indexed approach such as used in databases.
- The second problem is that administrators who desire to replicate a smbpasswd file to more than one Samba server were left to use external tools such as **rsync(1)** and **ssh(1)** and wrote custom, in-house scripts.
- Finally, the amount of information that is stored in an smbpasswd entry leaves no room for additional attributes such as a home directory, password expiration time, or even a Relative Identifier (RID).

As a result of these deficiencies, a more robust means of storing user attributes used by smbd was developed. The API which defines access to user accounts is commonly referred to as the samdb interface (previously this was called the passdb API, and is still so named in the Samba CVS trees).

Samba provides an enhanced set of passdb backends that overcome the deficiencies of the smbpasswd plain text database. These are tdbsam, ldapsam and xmlsam. Of these, ldapsam will be of most interest to large corporate or enterprise sites.

10.4.3 tdbsam

Samba can store user and machine account data in a "*TDB*" (Trivial Database). Using this backend does not require any additional configuration. This backend is recommended for new installations that do not require LDAP.

As a general guide, the Samba Team does not recommend using the tdbsam backend for sites that have 250 or more users. Additionally, tdbsam is not capable of scaling for use in sites that require PDB/BDC implementations that require replication of the account database. Clearly, for reason of scalability, the use of ldapsam should be encouraged.

The recommendation of a 250 user limit is purely based on the notion that this would generally involve a site that has routed networks, possibly spread across more than one physical location. The Samba Team has not at this time established the performance based scalability limits of the tdbsam architecture.

10.4.4 ldapsam

There are a few points to stress that the ldapsam does not provide. The LDAP support referred to in this documentation does not include:

- A means of retrieving user account information from an Windows 200x Active Directory server.
- A means of replacing /etc/passwd.

The second item can be accomplished by using LDAP NSS and PAM modules. LGPL versions of these libraries can be obtained from PADL Software[3]. More information about the configuration of these packages may be found at *LDAP, System Administration*; Gerald Carter by O'Reilly; Chapter 6: Replacing NIS."[4]

This document describes how to use an LDAP directory for storing Samba user account information traditionally stored in the smbpasswd(5) file. It is assumed that the reader already has a basic understanding of LDAP concepts and has a working directory server already installed. For more information on LDAP architectures and directories, please refer to the following sites:

- OpenLDAP[5]
- Sun iPlanet Directory Server[6]

Two additional Samba resources which may prove to be helpful are:

- The Samba-PDC-LDAP-HOWTO[7] maintained by Ignacio Coupeau.

[3]http://www.padl.com/
[4]http://safari.oreilly.com/?XmlId=1-56592-491-6
[5]http://www.openldap.org/
[6]http://iplanet.netscape.com/directory
[7]http://www.unav.es/cti/ldap-smb/ldap-smb-3-howto.html

- The NT migration scripts from IDEALX[8] that are geared to manage users and group in such a Samba-LDAP Domain Controller configuration.

10.4.4.1 Supported LDAP Servers

The LDAP ldapsam code has been developed and tested using the OpenLDAP 2.0 and 2.1 server and client libraries. The same code should work with Netscape's Directory Server and client SDK. However, there are bound to be compile errors and bugs. These should not be hard to fix. Please submit fixes via the process outlined in Chapter 34, *Reporting Bugs*.

10.4.4.2 Schema and Relationship to the RFC 2307 posixAccount

Samba-3.0 includes the necessary schema file for OpenLDAP 2.0 in `examples/LDAP/samba.schema`. The sambaSamAccount objectclass is given here:

```
objectclass (1.3.6.1.4.1.7165.2.2.6 NAME 'sambaSamAccount' SUP top AUXILIARY
    DESC 'Samba-3.0 Auxiliary SAM Account'
    MUST ( uid $ sambaSID )
    MAY  ( cn $ sambaLMPassword $ sambaNTPassword $ sambaPwdLastSet $
          sambaLogonTime $ sambaLogoffTime $ sambaKickoffTime $
          sambaPwdCanChange $ sambaPwdMustChange $ sambaAcctFlags $
          displayName $ sambaHomePath $ sambaHomeDrive $ sambaLogonScript $
          sambaProfilePath $ description $ sambaUserWorkstations $
          sambaPrimaryGroupSID $ sambaDomainName ))
```

The `samba.schema` file has been formatted for OpenLDAP 2.0/2.1. The Samba Team owns the OID space used by the above schema and recommends its use. If you translate the schema to be used with Netscape DS, please submit the modified schema file as a patch to jerry@samba.org[9].

Just as the smbpasswd file is meant to store information that provides information additional to a user's `/etc/passwd` entry, so is the sambaSamAccount object meant to supplement the UNIX user account information. A sambaSamAccount is a `AUXILIARY` objectclass so it can be used to augment existing user account information in the LDAP directory, thus providing information needed for Samba account handling. However, there are several fields (e.g., uid) that overlap with the posixAccount objectclass outlined in RFC2307. This is by design.

In order to store all user account information (UNIX and Samba) in the directory, it is necessary to use the sambaSamAccount and posixAccount objectclasses in combination. However, smbd will still obtain the user's UNIX account information via the standard C library calls (e.g., getpwnam(), et al). This means that the Samba server must also have the LDAP NSS library installed and functioning correctly. This division of information makes it possible to store all Samba account information in LDAP, but still maintain UNIX account information in NIS while the network is transitioning to a full LDAP infrastructure.

[8] http://samba.idealx.org/
[9] mailto:jerry@samba.org

10.4.4.3 OpenLDAP Configuration

To include support for the sambaSamAccount object in an OpenLDAP directory server, first copy the samba.schema file to slapd's configuration directory. The samba.schema file can be found in the directory `examples/LDAP` in the Samba source distribution.

```
root# cp samba.schema /etc/openldap/schema/
```

Next, include the `samba.schema` file in `slapd.conf`. The sambaSamAccount object contains two attributes that depend on other schema files. The ***uid*** attribute is defined in `cosine.schema` and the ***displayName*** attribute is defined in the `inetorgperson.schema` file. Both of these must be included before the `samba.schema` file.

```
## /etc/openldap/slapd.conf

## schema files (core.schema is required by default)
include             /etc/openldap/schema/core.schema

## needed for sambaSamAccount
include            /etc/openldap/schema/cosine.schema
include            /etc/openldap/schema/inetorgperson.schema
include            /etc/openldap/schema/samba.schema
include            /etc/openldap/schema/nis.schema
....
```

It is recommended that you maintain some indices on some of the most useful attributes, as in the following example, to speed up searches made on sambaSamAccount objectclasses (and possibly posixAccount and posixGroup as well):

```
# Indices to maintain
## required by OpenLDAP
index objectclass             eq

index cn                      pres,sub,eq
index sn                      pres,sub,eq
## required to support pdb_getsampwnam
index uid                     pres,sub,eq
## required to support pdb_getsambapwrid()
index displayName             pres,sub,eq

## uncomment these if you are storing posixAccount and
## posixGroup entries in the directory as well
##index uidNumber               eq
##index gidNumber               eq
##index memberUid               eq
```

```
index   sambaSID               eq
index   sambaPrimaryGroupSID   eq
index   sambaDomainName        eq
index   default                sub
```

Create the new index by executing:

```
root# ./sbin/slapindex -f slapd.conf
```

Remember to restart slapd after making these changes:

```
root# /etc/init.d/slapd restart
```

10.4.4.4 Initialize the LDAP Database

Before you can add accounts to the LDAP database you must create the account containers that they will be stored in. The following LDIF file should be modified to match your needs (DNS entries, and so on):

```
# Organization for Samba Base
dn: dc=quenya,dc=org
objectclass: dcObject
objectclass: organization
dc: quenya
o: Quenya Org Network
description: The Samba-3 Network LDAP Example

# Organizational Role for Directory Management
dn: cn=Manager,dc=quenya,dc=org
objectclass: organizationalRole
cn: Manager
description: Directory Manager

# Setting up container for users
dn: ou=People,dc=quenya,dc=org
objectclass: top
objectclass: organizationalUnit
ou: People

# Setting up admin handle for People OU
dn: cn=admin,ou=People,dc=quenya,dc=org
cn: admin
objectclass: top
```

```
objectclass: organizationalRole
objectclass: simpleSecurityObject
userPassword: {SSHA}c3ZM9tBaBo9autm1dL3waDS21+JSfQVz

# Setting up container for groups
dn: ou=Groups,dc=quenya,dc=org
objectclass: top
objectclass: organizationalUnit
ou: People

# Setting up admin handle for Groups OU
dn: cn=admin,ou=Groups,dc=quenya,dc=org
cn: admin
objectclass: top
objectclass: organizationalRole
objectclass: simpleSecurityObject
userPassword: {SSHA}c3ZM9tBaBo9autm1dL3waDS21+JSfQVz

# Setting up container for computers
dn: ou=Computers,dc=quenya,dc=org
objectclass: top
objectclass: organizationalUnit
ou: People

# Setting up admin handle for Computers OU
dn: cn=admin,ou=Computers,dc=quenya,dc=org
cn: admin
objectclass: top
objectclass: organizationalRole
objectclass: simpleSecurityObject
userPassword: {SSHA}c3ZM9tBaBo9autm1dL3waDS21+JSfQVz
```

The userPassword shown above should be generated using **slappasswd**.

The following command will then load the contents of the LDIF file into the LDAP database.

```
$ slapadd -v -l initldap.dif
```

Do not forget to secure your LDAP server with an adequate access control list as well as an admin password.

NOTE

Before Samba can access the LDAP server you need to store the LDAP admin password into the Samba-3 `secrets.tdb` database by:

```
root# smbpasswd -w secret
```

10.4.4.5 Configuring Samba

The following parameters are available in smb.conf only if your version of Samba was built with LDAP support. Samba automatically builds with LDAP support if the LDAP libraries are found.

LDAP related smb.conf options: *passdb backend* = ldapsam:url, *ldap admin dn*, *ldap delete dn*, *ldap filter*, *ldap group suffix*, *ldap idmap suffix*, *ldap machine suffix*, *ldap passwd sync*, *ldap ssl*, *ldap suffix*, *ldap user suffix*,

These are described in the `smb.conf` man page and so will not be repeated here. However, a sample `smb.conf` file for use with an LDAP directory could appear as shown in Example 10.2.

10.4.4.6 Accounts and Groups Management

As user accounts are managed through the sambaSamAccount objectclass, you should modify your existing administration tools to deal with sambaSamAccount attributes.

Machine accounts are managed with the sambaSamAccount objectclass, just like users accounts. However, it is up to you to store those accounts in a different tree of your LDAP namespace. You should use "*ou=Groups,dc=quenya,dc=org*" to store groups and "*ou=People,dc=quenya,dc=org*" to store users. Just configure your NSS and PAM accordingly (usually, in the `/etc/openldap/sldap.conf` configuration file).

In Samba-3, the group management system is based on POSIX groups. This means that Samba makes use of the posixGroup objectclass. For now, there is no NT-like group system management (global and local groups). Samba-3 knows only about `Domain Groups` and, unlike MS Windows 2000 and Active Directory, Samba-3 does not support nested groups.

10.4.4.7 Security and sambaSamAccount

There are two important points to remember when discussing the security of sambaSamAccount entries in the directory.

- *Never* retrieve the lmPassword or ntPassword attribute values over an unencrypted LDAP session.
- *Never* allow non-admin users to view the lmPassword or ntPassword attribute values.

Example 10.2. Configuration with LDAP

```
[global]
security = user
encrypt passwords = yes
netbios name = MORIA
workgroup = NOLDOR
# ldap related parameters
# define the DN to use when binding to the directory servers
# The password for this DN is not stored in smb.conf. Rather it
# must be set by using 'smbpasswd -w secretpw' to store the
# passphrase in the secrets.tdb file. If the "ldap admin dn" values
# change, this password will need to be reset.
ldap admin dn = "cn=Manager,ou=People,dc=quenya,dc=org"
# Define the SSL option when connecting to the directory
# ('off', 'start tls', or 'on' (default))
ldap ssl = start tls
# syntax: passdb backend = ldapsam:ldap://server-name[:port]
passdb backend = ldapsam:ldap://frodo.quenya.org
# smbpasswd -x delete the entire dn-entry
ldap delete dn = no
# the machine and user suffix added to the base suffix
# wrote WITHOUT quotes. NULL suffixes by default
ldap user suffix = ou=People
ldap group suffix = ou=Groups
ldap machine suffix = ou=Computers
# Trust UNIX account information in LDAP
# (see the smb.conf manpage for details)
# specify the base DN to use when searching the directory
ldap suffix = ou=People,dc=quenya,dc=org
# generally the default ldap search filter is ok
ldap filter = (&(uid=%u)(objectclass=sambaSamAccount))
```

These password hashes are cleartext equivalents and can be used to impersonate the user without deriving the original cleartext strings. For more information on the details of LM/NT password hashes, refer to the Account Information Database section of this chapter.

To remedy the first security issue, the *ldap ssl* `smb.conf` parameter defaults to require an encrypted session (*ldap ssl* = on) using the default port of `636` when contacting the directory server. When using an OpenLDAP server, it is possible to use the use the Start-TLS LDAP extended operation in the place of LDAPS. In either case, you are strongly discouraged to disable this security (*ldap ssl* = off).

Note that the LDAPS protocol is deprecated in favor of the LDAPv3 StartTLS extended operation. However, the OpenLDAP library still provides support for the older method of securing communication between clients and servers.

The second security precaution is to prevent non-administrative users from harvesting password hashes from the directory. This can be done using the following ACL in `slapd.conf`:

```
## allow the "ldap admin dn" access, but deny everyone else
access to attrs=lmPassword,ntPassword
     by dn="cn=Samba Admin,ou=People,dc=quenya,dc=org" write
     by * none
```

10.4.4.8 LDAP Special Attributes for sambaSamAccounts

The sambaSamAccount objectclass is composed of the attributes shown in Table 10.1, and Table 10.2.

The majority of these parameters are only used when Samba is acting as a PDC of a domain (refer to Chapter 4, *Domain Control*, for details on how to configure Samba as a Primary Domain Controller). The following four attributes are only stored with the sambaSamAccount entry if the values are non-default values:

- sambaHomePath
- sambaLogonScript
- sambaProfilePath
- sambaHomeDrive

These attributes are only stored with the sambaSamAccount entry if the values are non-default values. For example, assume MORIA has now been configured as a PDC and that *logon home* = \\%L\%u was defined in its `smb.conf` file. When a user named "*becky*" logons to the domain, the *logon home* string is expanded to \\MORIA\becky. If the smbHome attribute exists in the entry "*uid=becky,ou=People,dc=samba,dc=org*", this value is used. However, if this attribute does not exist, then the value of the *logon home* parameter is used in its place. Samba will only write the attribute value to the directory entry if the value is something other than the default (e.g., `\\MOBY\becky`).

10.4.4.9 Example LDIF Entries for a sambaSamAccount

The following is a working LDIF that demonstrates the use of the SambaSamAccount objectclass:

```
dn: uid=guest2, ou=People,dc=quenya,dc=org
sambaLMPassword: 878D8014606CDA29677A44EFA1353FC7
sambaPwdMustChange: 2147483647
sambaPrimaryGroupSID: S-1-5-21-2447931902-1787058256-3961074038-513
sambaNTPassword: 552902031BEDE9EFAAD3B435B51404EE
sambaPwdLastSet: 1010179124
sambaLogonTime: 0
objectClass: sambaSamAccount
```

```
uid: guest2
sambaKickoffTime: 2147483647
sambaAcctFlags: [UX         ]
sambaLogoffTime: 2147483647
sambaSID: S-1-5-21-2447931902-1787058256-3961074038-5006
sambaPwdCanChange: 0
```

The following is an LDIF entry for using both the sambaSamAccount and posixAccount objectclasses:

```
dn: uid=gcarter, ou=People,dc=quenya,dc=org
sambaLogonTime: 0
displayName: Gerald Carter
sambaLMPassword: 552902031BEDE9EFAAD3B435B51404EE
sambaPrimaryGroupSID: S-1-5-21-2447931902-1787058256-3961074038-1201
objectClass: posixAccount
objectClass: sambaSamAccount
sambaAcctFlags: [UX         ]
userPassword: {crypt}BpM2ej8Rkzogo
uid: gcarter
uidNumber: 9000
cn: Gerald Carter
loginShell: /bin/bash
logoffTime: 2147483647
gidNumber: 100
sambaKickoffTime: 2147483647
sambaPwdLastSet: 1010179230
sambaSID: S-1-5-21-2447931902-1787058256-3961074038-5004
homeDirectory: /home/moria/gcarter
sambaPwdCanChange: 0
sambaPwdMustChange: 2147483647
sambaNTPassword: 878D8014606CDA29677A44EFA1353FC7
```

10.4.4.10 Password Synchronization

Samba-3 and later can update the non-samba (LDAP) password stored with an account. When using pam_ldap, this allows changing both UNIX and Windows passwords at once.

The *ldap passwd sync* options can have the values shown in Table 10.3.

More information can be found in the `smb.conf` manpage.

10.4.5 MySQL

Every so often someone will come along with a great new idea. Storing user accounts in a SQL backend is one of them. Those who want to do this are in the best position to know what the specific benefits are to them. This may sound like a cop-out, but in truth we cannot attempt to document every little detail why certain things of marginal utility to the bulk of Samba users might make sense to the rest. In any case, the following instructions should help the determined SQL user to implement a working system.

10.4.5.1 Creating the Database

You can set up your own table and specify the field names to pdb_mysql (see below for the column names) or use the default table. The file `examples/pdb/mysql/mysql.dump` contains the correct queries to create the required tables. Use the command:

```
$ mysql -uusername -hhostname -ppassword \
   databasename < /path/to/samba/examples/pdb/mysql/mysql.dump
```

10.4.5.2 Configuring

This plugin lacks some good documentation, but here is some brief infoormation. Add the following to the ***passdb backend*** variable in your `smb.conf`:

passdb backend = [other-plugins] mysql:identifier [other-plugins]

The identifier can be any string you like, as long as it does not collide with the identifiers of other plugins or other instances of pdb_mysql. If you specify multiple pdb_mysql.so entries in ***passdb backend***, you also need to use different identifiers.

Additional options can be given through the `smb.conf` file in the *[global]* section. Refer to Table 10.4.

WARNING

Since the password for the MySQL user is stored in the `smb.conf` file, you should make the `smb.conf` file readable only to the user who runs Samba. This is considered a security bug and will soon be fixed.

Names of the columns are given in Table 10.5. The default column names can be found in the example table dump.

You can put a colon (:) after the name of each column, which should specify the column to update when updating the table. You can also specify nothing behind the colon. Then the field data will not be updated. Setting a column name to *NULL* means the field should not be used.

An example configuration can be found in Example 10.3.

Example 10.3. Example configuration for the MySQL passdb backend

```
[global]
passdb backend = mysql:foo
foo:mysql user = samba
foo:mysql password = abmas
foo:mysql database = samba
# domain name is static and can't be changed
foo:domain column = 'MYWORKGROUP':
# The fullname column comes from several other columns
foo:fullname column = CONCAT(firstname,' ',surname):
# Samba should never write to the password columns
foo:lanman pass column = lm_pass:
foo:nt pass column = nt_pass:
# The unknown 3 column is not stored
foo:unknown 3 column = NULL
```

10.4.5.3 Using Plaintext Passwords or Encrypted Password

I strongly discourage the use of plaintext passwords, however, you can use them.

If you would like to use plaintext passwords, set 'identifier:lanman pass column' and 'identifier:nt pass column' to 'NULL' (without the quotes) and 'identifier:plain pass column' to the name of the column containing the plaintext passwords.

If you use encrypted passwords, set the 'identifier:plain pass column' to 'NULL' (without the quotes). This is the default.

10.4.5.4 Getting Non-Column Data from the Table

It is possible to have not all data in the database by making some 'constant'.

For example, you can set 'identifier:fullname column' to something like **CONCAT(Firstname,' ',Surname)**

Or, set 'identifier:workstations column' to: **NULL**

See the MySQL documentation for more language constructs.

10.4.6 XML

This module requires libxml2 to be installed.

The usage of pdb_xml is fairly straightforward. To export data, use:

`$ pdbedit -e xml:filename`

(where filename is the name of the file to put the data in)

To import data, use: `$ pdbedit -i xml:filename`

10.5 Common Errors

10.5.1 Users Cannot Logon

"I've installed Samba, but now I can't log on with my UNIX account!"

Make sure your user has been added to the current Samba ***passdb backend***. Read the section Section 10.3 for details.

10.5.2 Users Being Added to the Wrong Backend Database

A few complaints have been received from users that just moved to Samba-3. The following `smb.conf` file entries were causing problems, new accounts were being added to the old smbpasswd file, not to the tdbsam passdb.tdb file:

```
...
passdb backend = smbpasswd, tdbsam
...
```

Samba will add new accounts to the first entry in the *passdb backend* parameter entry. If you want to update to the tdbsam, then change the entry to:

```
passdb backend = tdbsam, smbpasswd
```

10.5.3 Configuration of auth methods

When explicitly setting an ***auth methods*** parameter, ***guest*** must be specified as the first entry on the line, for example, ***auth methods*** = guest sam.

This is the exact opposite of the requirement for the ***passdb backend*** option, where it must be the *LAST* parameter on the line.

Table 10.1. Attributes in the sambaSamAccount objectclass (LDAP) — Part A

`sambaLMPassword`	The LANMAN password 16-byte hash stored as a character representation of a hexadecimal string.
`sambaNTPassword`	The NT password hash 16-byte stored as a character representation of a hexadecimal string.
`sambaPwdLastSet`	The integer time in seconds since 1970 when the `sambaLMPassword` and `sambaNTPassword` attributes were last set.
`sambaAcctFlags`	String of 11 characters surrounded by square brackets [] representing account flags such as U (user), W (workstation), X (no password expiration), I (Domain trust account), H (Home dir required), S (Server trust account), and D (disabled).
`sambaLogonTime`	Integer value currently unused
`sambaLogoffTime`	Integer value currently unused
`sambaKickoffTime`	Specifies the time (UNIX time format) when the user will be locked down and cannot login any longer. If this attribute is ommited, then the account will never expire. If you use this attribute together with 'shadowExpire' of the 'shadowAccount' objectClass, will enable accounts to expire completly on an exact date.
`sambaPwdCanChange`	Specifies the time (UNIX time format) from which on the user is allowed to change his password. If attribute is not set, the user will be free to change his password whenever he wants.
`sambaPwdMustChange`	Specifies the time (UNIX time format) since when the user is forced to change his password. If this value is set to '0', the user will have to change his password at first login. If this attribute is not set, then the password will never expire.
`sambaHomeDrive`	Specifies the drive letter to which to map the UNC path specified by sambaHomePath. The drive letter must be specified in the form "*X:*" where X is the letter of the drive to map. Refer to the "*logon drive*" parameter in the smb.conf(5) man page for more information.
`sambaLogonScript`	The sambaLogonScript property specifies the path of the user's logon script, .CMD, .EXE, or .BAT file. The string can be null. The path is relative to the netlogon share. Refer to the *logon script* parameter in the `smb.conf` man page for more information.
`sambaProfilePath`	Specifies a path to the user's profile. This value can be a null string, a local absolute path, or a UNC path. Refer to the *logon path* parameter in the `smb.conf` man page for more information.
`sambaHomePath`	The sambaHomePath property specifies the path of the home directory for the user. The string can be null. If sambaHomeDrive is set and specifies a drive letter, sambaHomePath should be a UNC path. The path must be a network UNC path of the form `\\server\share\directory`. This value can be a null string. Refer to the **logon home** parameter in the `smb.conf` man page for more information.

Table 10.2. Attributes in the sambaSamAccount objectclass (LDAP) — Part B

`sambaUserWorkstations`	Here you can give a comma-seperated list of machines on which the user is allowed to login. You may observe problems when you try to connect to an Samba Domain Member. Bacause Domain Members are not in this list, the Domain Controllers will reject them. Where this attribute is ommited, the default implies no restrictions.
`sambaSID`	The security identifier(SID) of the user. The Windows equivalent of UNIX UIDs.
`sambaPrimaryGroupSID`	The Security IDentifier (SID) of the primary group of the user.
`sambaDomainName`	Domain the user is part of.

Table 10.3. Possible *ldap passwd sync* values

Value	Description
yes	When the user changes his password, update `ntPassword`, `lmPassword` and the `password` fields.
no	Only update `ntPassword` and `lmPassword`.
only	Only update the LDAP password and let the LDAP server worry about the other fields. This option is only available on some LDAP servers. Only when the LDAP server supports LDAP_EXOP_X_MODIFY_PASSWD.

Table 10.4. Basic smb.conf options for MySQL passdb backend

Field	Contents
mysql host	Host name, defaults to 'localhost'
mysql password	
mysql user	Defaults to 'samba'
mysql database	Defaults to 'samba'
mysql port	Defaults to 3306
table	Name of the table containing the users

Table 10.5. MySQL field names for MySQL passdb backend

Field	Type	Contents
logon time column	int(9)	UNIX time stamp of last logon of user
logoff time column	int(9)	UNIX time stamp of last logoff of user
kickoff time column	int(9)	UNIX time stamp of moment user should be kicked off workstation (not enforced)
pass last set time column	int(9)	UNIX time stamp of moment password was last set
pass can change time column	int(9)	UNIX time stamp of moment from which password can be changed
pass must change time column	int(9)	UNIX time stamp of moment on which password must be changed
username column	varchar(255)	UNIX username
domain column	varchar(255)	NT domain user belongs to
nt username column	varchar(255)	NT username
fullname column	varchar(255)	Full name of user
home dir column	varchar(255)	UNIX homedir path
dir drive column	varchar(2)	Directory drive path (e.g., "*H:*")
logon script column	varchar(255)	Batch file to run on client side when logging on
profile path column	varchar(255)	Path of profile
acct desc column	varchar(255)	Some ASCII NT user data
workstations column	varchar(255)	Workstations user can logon to (or NULL for all)
unknown string column	varchar(255)	Unknown string
munged dial column	varchar(255)	Unknown
user sid column	varchar(255)	NT user SID
group sid column	varchar(255)	NT group SID
lanman pass column	varchar(255)	Encrypted lanman password
nt pass column	varchar(255)	Encrypted nt passwd
plain pass column	varchar(255)	Plaintext password
acct ctrl column	int(9)	NT user data
unknown 3 column	int(9)	Unknown
logon divs column	int(9)	Unknown
hours len column	int(9)	Unknown
bad password count column	int(5)	Number of failed password tries before disabling an account
logon count column	int(5)	Number of logon attempts
unknown 6 column	int(9)	Unknown

Chapter 11

GROUP MAPPING — MS WINDOWS AND UNIX

Starting with Samba-3, new group mapping functionality is available to create associations between Windows group SIDs and UNIX groups. The **groupmap** subcommand included with the net tool can be used to manage these associations.

The new facility for mapping NT Groups to UNIX system groups allows the administrator to decide which NT Domain Groups are to be exposed to MS Windows clients. Only those NT Groups that map to a UNIX group that has a value other than the default (`-1`) will be exposed in group selection lists in tools that access domain users and groups.

WARNING

The *domain admin group* parameter has been removed in Samba-3 and should no longer be specified in `smb.conf`. This parameter was used to give the listed users membership in the `Domain Admins` Windows group which gave local admin rights on their workstations (in default configurations).

11.1 Features and Benefits

Samba allows the administrator to create MS Windows NT4/200x group accounts and to arbitrarily associate them with UNIX/Linux group accounts.

Group accounts can be managed using the MS Windows NT4 or MS Windows 200x/XP Professional MMC tools. Appropriate interface scripts should be provided in `smb.conf` if it is desired that UNIX/Linux system accounts should be automatically created when these tools are used. In the absence of these scripts, and so long as **winbindd** is running, Samba group accounts that are created using these tools will be allocated UNIX UIDs/GIDs from the ID range specified by the *idmap uid*/*idmap gid* parameters in the `smb.conf` file.

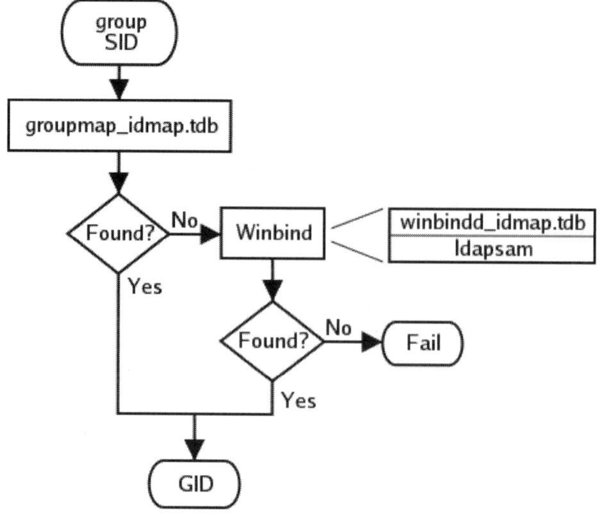

Figure 11.1. IDMAP: group SID to GID resolution.

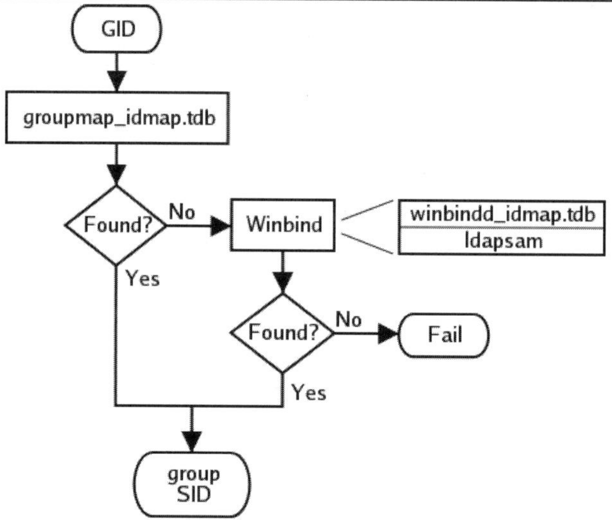

Figure 11.2. IDMAP: GID resolution to matching SID.

In both cases, when winbindd is not running, only locally resolvable groups can be recognized. Please refer to Figure 11.1 and Figure 11.2. The **net groupmap** is used to establish UNIX group to NT SID mappings as shown in Figure 11.3.

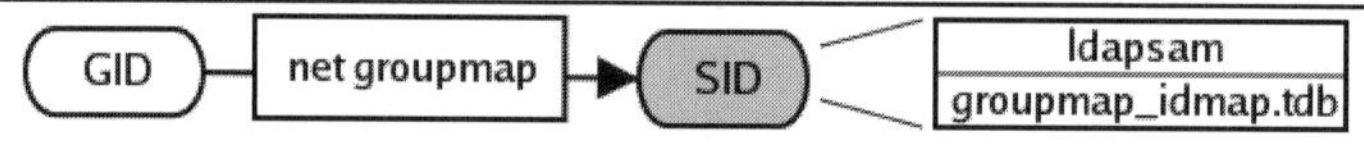

Figure 11.3. IDMAP storing group mappings.

Administrators should be aware that where `smb.conf` group interface scripts make direct calls to the UNIX/Linux system tools (the shadow utilities, **groupadd**, **groupdel**, and **groupmod**), the resulting UNIX/Linux group names will be subject to any limits imposed by these tools. If the tool does not allow upper case characters or space characters, then the creation of an MS Windows NT4/200x style group of *Engineering Managers* will attempt to create an identically named UNIX/Linux group, an attempt that will of course fail.

There are several possible work-arounds for the operating system tools limitation. One method is to use a script that generates a name for the UNIX/Linux system group that fits the operating system limits, and that then just passes the UNIX/Linux group ID (GID) back to the calling Samba interface. This will provide a dynamic work-around solution.

Another work-around is to manually create a UNIX/Linux group, then manually create the MS Windows NT4/200x group on the Samba server and then use the **net groupmap** tool to connect the two to each other.

11.2 Discussion

When installing MS Windows NT4/200x on a computer, the installation program creates default users and groups, notably the `Administrators` group, and gives that group privileges necessary privileges to perform essential system tasks, such as the ability to change the date and time or to kill (or close) any process running on the local machine.

The `Administrator` user is a member of the `Administrators` group, and thus inherits `Administrators` group privileges. If a `joe` user is created to be a member of the `Administrators` group, `joe` has exactly the same rights as the user, `Administrator`.

When an MS Windows NT4/200x/XP machine is made a Domain Member, the "*Domain Admins*" group of the PDC is added to the local `Administrators` group of the workstation. Every member of the `Domain Administrators` group inherits the rights of the local `Administrators` group when logging on the workstation.

The following steps describe how to make Samba PDC users members of the `Domain Admins` group?

1. Create a UNIX group (usually in `/etc/group`), let's call it `domadm`.
2. Add to this group the users that must be "*Administrators*". For example, if you want `joe, john` and `mary` to be administrators, your entry in `/etc/group` will look like this:

   ```
   domadm:x:502:joe,john,mary
   ```

3. Map this domadm group to the "*Domain Admins*" group by running the command:

   ```
   root# net groupmap add ntgroup=Domain Admins UNIXgroup=domadm
   ```

   The quotes around "*Domain Admins*" are necessary due to the space in the group name. Also make sure to leave no white-space surrounding the equal character (=).

Now joe, john and mary are domain administrators.

It is possible to map any arbitrary UNIX group to any Windows NT4/200x group as well as making any UNIX group a Windows domain group. For example, if you wanted to include a UNIX group (e.g., acct) in an ACL on a local file or printer on a Domain Member machine, you would flag that group as a domain group by running the following on the Samba PDC:

```
root# net groupmap add rid=1000 ntgroup="Accounting" UNIXgroup=acct
```

Be aware that the RID parameter is a unsigned 32-bit integer that should normally start at 1000. However, this RID must not overlap with any RID assigned to a user. Verification for this is done differently depending on the passdb backend you are using. Future versions of the tools may perform the verification automatically, but for now the burden is on you.

11.2.1 Default Users, Groups and Relative Identifiers

When first installed, Microsoft Windows NT4/200x/XP are preconfigured with certain User, Group, and Alias entities. Each has a well-known Relative Identifier (RID). These must be preserved for continued integrity of operation. Samba must be provisioned with certain essential Domain Groups that require the appropriate RID value. When Samba-3 is configured to use tdbsam the essential Domain Groups are automatically created. It is the LDAP administrators' responsibility to create (provision) the default NT Groups.

Each essential Domain Group must be assigned its respective well-kown RID. The default Users, Groups, Aliases, and RIDs are shown in Table 11.1.

NOTE

When the *passdb backend* uses LDAP (ldapsam) it is the admininstrators' responsibility to create the essential Domain Groups, and to assign each its default RID.

It is permissible to create any Domain Group that may be necessary, just make certain that the essential Domain Groups (well known) have been created and assigned its default RID. Other groups you create may be assigned any arbitrary RID you care to use.

Be sure to map each Domain Group to a UNIX system group. That is the only way to ensure that the group will be available for use as an NT Domain Group.

11.2.2 Example Configuration

You can list the various groups in the mapping database by executing **net groupmap list**. Here is an example:

```
root#  net groupmap list
```

Table 11.1. Well-Known User Default RIDs

Well-Known Entity	RID	Type	Essential
Domain Administrator	500	User	No
Domain Guest	501	User	No
Domain KRBTGT	502	User	No
Domain Admins	512	Group	Yes
Domain Users	513	Group	Yes
Domain Guests	514	Group	Yes
Domain Computers	515	Group	No
Domain Controllers	516	Group	No
Domain Certificate Admins	517	Group	No
Domain Schema Admins	518	Group	No
Domain Enterprise Admins	519	Group	No
Domain Policy Admins	520	Group	No
Builtin Admins	544	Alias	No
Builtin users	545	Alias	No
Builtin Guests	546	Alias	No
Builtin Power Users	547	Alias	No
Builtin Account Operators	548	Alias	No
Builtin System Operators	549	Alias	No
Builtin Print Operators	550	Alias	No
Builtin Backup Operators	551	Alias	No
Builtin Replicator	552	Alias	No
Builtin RAS Servers	553	Alias	No

```
Domain Admins (S-1-5-21-2547222302-1596225915-2414751004-512) -> domadmin
Domain Users (S-1-5-21-2547222302-1596225915-2414751004-513) -> domuser
Domain Guests (S-1-5-21-2547222302-1596225915-2414751004-514) -> domguest
```

For complete details on **net groupmap**, refer to the net(8) man page.

11.3 Configuration Scripts

Everyone needs tools. Some of us like to create our own, others prefer to use canned tools (i.e., prepared by someone else for general use).

11.3.1 Sample smb.conf Add Group Script

A script to create complying group names for use by the Samba group interfaces is provided in Example 11.1.

The `smb.conf` entry for the above script would be something like that in Example 11.2.

Example 11.1. smbgrpadd.sh

```
#!/bin/bash

# Add the group using normal system groupadd tool.
groupadd smbtmpgrp00

thegid=`cat /etc/group | grep smbtmpgrp00 | cut -d ":" -f3`

# Now change the name to what we want for the MS Windows networking end
cp /etc/group /etc/group.bak
cat /etc/group.bak | sed s/smbtmpgrp00/$1/g > /etc/group

# Now return the GID as would normally happen.
echo $thegid
exit 0
```

Example 11.2. Configuration of smb.conf for the add group script.

```
[global]
...
add group script = /path_to_tool/smbgrpadd.sh %g
...
```

11.3.2 Script to Configure Group Mapping

In our example we have created a UNIX/Linux group called *ntadmin*. Our script will create the additional groups *Orks*, *Elves*, and *Gnomes*. It is a good idea to save this shell script for later re-use just in case you ever need to rebuild your mapping database. For the sake of concenience we elect to save this script as a file called `initGroups.sh`. This script is given in Example 11.3.

Of course it is expected that the administrator will modify this to suit local needs. For information regarding the use of the **net groupmap** tool please refer to the man page.

11.4 Common Errors

At this time there are many little surprises for the unwary administrator. In a real sense it is imperative that every step of automated control scripts must be carefully tested manually before putting them into active service.

Example 11.3. Script to Set Group Mapping

```
#!/bin/bash

net groupmap modify ntgroup="Domain Admins" unixgroup=ntadmin
net groupmap modify ntgroup="Domain Users" unixgroup=users
net groupmap modify ntgroup="Domain Guests" unixgroup=nobody

groupadd Orks
groupadd Elves
groupadd Gnomes

net groupmap add ntgroup="Orks"   unixgroup=Orks   type=d
net groupmap add ntgroup="Elves"  unixgroup=Elves  type=d
net groupmap add ntgroup="Gnomes" unixgroup=Gnomes type=d
```

11.4.1 Adding Groups Fails

This is a common problem when the **groupadd** is called directly by the Samba interface script for the *add group script* in the `smb.conf` file.

The most common cause of failure is an attempt to add an MS Windows group account that has either an upper case character and/or a space character in it.

There are three possible work-arounds. First, use only group names that comply with the limitations of the UNIX/Linux **groupadd** system tool. Second, it involves the use of the script mentioned earlier in this chapter, and third is the option is to manually create a UNIX/Linux group account that can substitute for the MS Windows group name, then use the procedure listed above to map that group to the MS Windows group.

11.4.2 Adding MS Windows Groups to MS Windows Groups Fails

Samba-3 does not support nested groups from the MS Windows control environment.

11.4.3 Adding *Domain Users* to the *Power Users* Group

"*What must I do to add Domain Users to the Power Users group?*" The Power Users group is a group that is local to each Windows 200x/XP Professional workstation. You cannot add the Domain Users group to the Power Users group automatically, it must be done on each workstation by logging in as the local workstation *administrator* and then using the following procedure:

1. Click **Start** -> **Control Panel** -> **Users and Passwords**.
2. Click the **Advanced** tab.
3. Click the **Advanced** button.

4. Click `Groups`.
5. Double click `Power Users`. This will launch the panel to add users or groups to the local machine `Power Uses` group.
6. Click the **Add** button.
7. Select the domain from which the `Domain Users` group is to be added.
8. Double click the `Domain Users` group.
9. Click the **Ok** button. If a logon box is presented during this process please remember to enter the connect as `DOMAIN\UserName`. i.e., For the domain `MIDEARTH` and the user `root` enter `MIDEARTH\root`.

Chapter 12

FILE, DIRECTORY AND SHARE ACCESS CONTROLS

Advanced MS Windows users are frequently perplexed when file, directory and share manipulation of resources shared via Samba do not behave in the manner they might expect. MS Windows network administrators are often confused regarding network access controls and how to provide users with the access they need while protecting resources from unauthorized access.

Many UNIX administrators are unfamiliar with the MS Windows environment and in particular have difficulty in visualizing what the MS Windows user wishes to achieve in attempts to set file and directory access permissions.

The problem lies in the differences in how file and directory permissions and controls work between the two environments. This difference is one that Samba cannot completely hide, even though it does try to bridge the chasm to a degree.

POSIX Access Control List technology has been available (along with Extended Attributes) for UNIX for many years, yet there is little evidence today of any significant use. This explains to some extent the slow adoption of ACLs into commercial Linux products. MS Windows administrators are astounded at this, given that ACLs were a foundational capability of the now decade-old MS Windows NT operating system.

The purpose of this chapter is to present each of the points of control that are possible with Samba-3 in the hope that this will help the network administrator to find the optimum method for delivering the best environment for MS Windows desktop users.

This is an opportune point to mention that Samba was created to provide a means of interoperability and interchange of data between differing operating environments. Samba has no intent to change UNIX/Linux into a platform like MS Windows. Instead the purpose was and is to provide a sufficient level of exchange of data between the two environments. What is available today extends well beyond early plans and expectations, yet the gap continues to shrink.

12.1 Features and Benefits

Samba offers a lot of flexibility in file system access management. These are the key access control facilities present in Samba today:

Samba Access Control Facilities

- *UNIX File and Directory Permissions*

  Samba honors and implements UNIX file system access controls. Users who access a Samba server will do so as a particular MS Windows user. This information is passed to the Samba server as part of the logon or connection setup process. Samba uses this user identity to validate whether or not the user should be given access to file system resources (files and directories). This chapter provides an overview for those to whom the UNIX permissions and controls are a little strange or unknown.

- *Samba Share Definitions*

  In configuring share settings and controls in the `smb.conf` file, the network administrator can exercise overrides to native file system permissions and behaviors. This can be handy and convenient to effect behavior that is more like what MS Windows NT users expect but it is seldom the *best* way to achieve this. The basic options and techniques are described herein.

- *Samba Share ACLs*

  Just like it is possible in MS Windows NT to set ACLs on shares themselves, so it is possible to do this in Samba. Few people make use of this facility, yet it remains on of the easiest ways to affect access controls (restrictions) and can often do so with minimum invasiveness compared with other methods.

- *MS Windows ACLs through UNIX POSIX ACLs*

  The use of POSIX ACLs on UNIX/Linux is possible only if the underlying operating system supports them. If not, then this option will not be available to you. Current UNIX technology platforms have native support for POSIX ACLs. There are patches for the Linux kernel that also provide this. Sadly, few Linux platforms ship today with native ACLs and Extended Attributes enabled. This chapter has pertinent information for users of platforms that support them.

12.2 File System Access Controls

Perhaps the most important recognition to be made is the simple fact that MS Windows NT4/200x/XP implement a totally divergent file system technology from what is provided in the UNIX operating system environment. First we consider what the most significant differences are, then we look at how Samba helps to bridge the differences.

12.2.1 MS Windows NTFS Comparison with UNIX File Systems

Samba operates on top of the UNIX file system. This means it is subject to UNIX file system conventions and permissions. It also means that if the MS Windows networking environment requires file system behavior that differs from UNIX file system behavior then somehow Samba is responsible for emulating that in a transparent and consistent manner.

It is good news that Samba does this to a large extent and on top of that provides a high degree of optional configuration to override the default behavior. We look at some of these over-rides, but for the greater part we will stay within the bounds of default behavior. Those wishing to explore the depths of control ability should review the `smb.conf` man page.

The following compares file system features for UNIX with those of Microsoft Windows NT/200x:

Name Space — MS Windows NT4/200x/XP files names may be up to 254 characters long, and UNIX file names may be 1023 characters long. In MS Windows, file extensions indicate particular file types, in UNIX this is not so rigorously observed as all names are considered arbitrary.

What MS Windows calls a folder, UNIX calls a directory.

Case Sensitivity — MS Windows file names are generally upper case if made up of 8.3 (8 character file name and 3 character extension. File names that are longer than 8.3 are case preserving and case insensitive.

UNIX file and directory names are case sensitive and case preserving. Samba implements the MS Windows file name behavior, but it does so as a user application. The UNIX file system provides no mechanism to perform case insensitive file name lookups. MS Windows does this by default. This means that Samba has to carry the processing overhead to provide features that are not native to the UNIX operating system environment.

Consider the following. All are unique UNIX names but one single MS Windows file name:

```
MYFILE.TXT
MyFile.txt
myfile.txt
```

So clearly, in an MS Windows file name space these three files cannot co-exist, but in UNIX they can.

So what should Samba do if all three are present? That which is lexically first will be accessible to MS Windows users, the others are invisible and unaccessible — any other solution would be suicidal.

Directory Separators — MS Windows and DOS uses the backslash \ as a directory delimiter, and UNIX uses the forward-slash / as its directory delimiter. This is handled transparently by Samba.

Drive Identification — MS Windows products support a notion of drive letters, like **C:** to represent disk partitions. UNIX has no concept of separate identifiers for file partitions, each such file system is mounted to become part of the overall directory tree. The UNIX directory tree begins at / just like the root of a DOS drive is specified as `C:\`.

File Naming Conventions — MS Windows generally never experiences file names that begin with a dot (`.`) while in UNIX these are commonly found in a user's home directory. Files that begin with a dot (`.`) are typically either start-up files for various UNIX applications, or they may be files that contain start-up configuration data.

Links and Short-Cuts — MS Windows make use of "*links and short-cuts*" that are actually special types of files that will redirect an attempt to execute the file to the real location of the file. UNIX knows of file and directory links, but they are entirely different from what MS Windows users are used to.

Symbolic links are files in UNIX that contain the actual location of the data (file or directory). An operation (like read or write) will operate directly on the file referenced. Symbolic links are also referred to as "*soft links.*" A hard link is something that MS Windows is not familiar with. It allows one physical file to be known simultaneously by more than one file name.

There are many other subtle differences that may cause the MS Windows administrator some temporary discomfort in the process of becoming familiar with UNIX/Linux. These are best left for a text that is dedicated to the purpose of UNIX/Linux training and education.

12.2.2 Managing Directories

There are three basic operations for managing directories: **create, delete, rename**.

Table 12.1. Managing Directories with UNIX and Windows

Action	MS Windows Command	UNIX Command
create	md folder	mkdir folder
delete	rd folder	rmdir folder
rename	rename oldname newname	mv oldname newname

12.2.3 File and Directory Access Control

The network administrator is strongly advised to read foundational training manuals and reference materials regarding file and directory permissions maintenance. Much can be achieved with the basic UNIX permissions without having to resort to more complex facilities like POSIX Access Control Lists (ACLs) or Extended Attributes (EAs).

UNIX/Linux file and directory access permissions involves setting three primary sets of data and one control set. A UNIX file listing looks as follows:

```
$ ls -la
total 632
drwxr-xr-x   13 maryo   gnomes      816 2003-05-12 22:56 .
drwxrwxr-x   37 maryo   gnomes     3800 2003-05-12 22:29 ..
dr-xr-xr-x    2 maryo   gnomes       48 2003-05-12 22:29 muchado02
drwxrwxrwx    2 maryo   gnomes       48 2003-05-12 22:29 muchado03
drw-rw-rw-    2 maryo   gnomes       48 2003-05-12 22:29 muchado04
d-w--w--w-    2 maryo   gnomes       48 2003-05-12 22:29 muchado05
dr--r--r--    2 maryo   gnomes       48 2003-05-12 22:29 muchado06
drwsrwsrwx    2 maryo   gnomes       48 2003-05-12 22:29 muchado08
----------    1 maryo   gnomes     1242 2003-05-12 22:31 mydata00.lst
--w--w--w-    1 maryo   gnomes     7754 2003-05-12 22:33 mydata02.lst
-r--r--r--    1 maryo   gnomes    21017 2003-05-12 22:32 mydata04.lst
-rw-rw-rw-    1 maryo   gnomes    41105 2003-05-12 22:32 mydata06.lst
$
```

The columns above represent (from left to right): permissions, number of hard links to file, owner, group, size (bytes), access date, access time, file name.

An overview of the permissions field can be found in Figure 12.1.

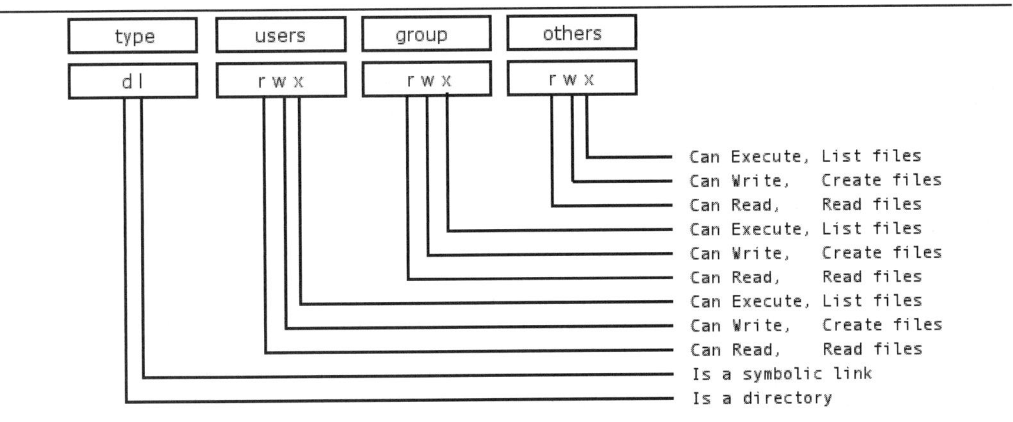

Figure 12.1. Overview of UNIX permissions field.

Any bit flag may be unset. An unset bit flag is the equivalent of "*cannot*" and is represented as a "-" character.

Example 12.1. Example File

```
-rwxr-x---   Means: The owner (user) can read, write, execute
                    the group can read and execute
                    everyone else cannot do anything with it.
```

Additional possibilities in the [type] field are: c = character device, b = block device, p = pipe device, s = UNIX Domain Socket.

The letters `rwxXst` set permissions for the user, group and others as: read (r), write (w), execute (or access for directories) (x), execute only if the file is a directory or already has execute permission for some user (X), set user or group ID on execution (s), sticky (t).

When the sticky bit is set on a directory, files in that directory may be unlinked (deleted) or renamed only by root or their owner. Without the sticky bit, anyone able to write to the directory can delete or rename files. The sticky bit is commonly found on directories, such as `/tmp`, that are world-writable.

When the set user or group ID bit (s) is set on a directory, then all files created within it will be owned by the user and/or group whose 'set user or group' bit is set. This can be helpful in setting up directories for which it is desired that all users who are in a group should be able to write to and read from a file, particularly when it is undesirable for that file to be exclusively owned by a user whose primary group is not the group that all such users belong to.

When a directory is set `drw-r-----` this means that the owner can read and create (write) files in it, but because the (x) execute flags are not set, files cannot be listed (seen) in the directory by anyone. The group can read files in the directory but cannot create new files. If files in the directory are set to be readable and writable for the group, then group members will be able to write to (or delete) them.

12.3 Share Definition Access Controls

The following parameters in the `smb.conf` file sections define a share control or effect access controls. Before using any of the following options, please refer to the man page for `smb.conf`.

12.3.1 User and Group-Based Controls

User and group-based controls can prove quite useful. In some situations it is distinctly desirable to affect all file system operations as if a single user were doing so. The use of the *force user* and *force group* behavior will achieve this. In other situations it may be necessary to effect a paranoia level of control to ensure that only particular authorized persons will be able to access a share or its contents. Here the use of the *valid users* or the *invalid users* may be most useful.

As always, it is highly advisable to use the least difficult to maintain and the least ambiguous method for controlling access. Remember, when you leave the scene someone else will need to provide assistance and if he finds too great a mess or does not understand what you have done, there is risk of Samba being removed and an alternative solution being adopted.

Table 12.2 enumerates these controls.

Table 12.2. User and Group Based Controls

Control Parameter	Description - Action - Notes
admin users	List of users who will be granted administrative privileges on the share. They will do all file operations as the super-user (root). Any user in this list will be able to do anything they like on the share, irrespective of file permissions.
force group	Specifies a UNIX group name that will be assigned as the default primary group for all users connecting to this service.
force user	Specifies a UNIX user name that will be assigned as the default user for all users connecting to this service. This is useful for sharing files. Incorrect use can cause security problems.
guest ok	If this parameter is set for a service, then no password is required to connect to the service. Privileges will be those of the guest account.
invalid users	List of users that should not be allowed to login to this service.
only user	Controls whether connections with usernames not in the user list will be allowed.
read list	List of users that are given read-only access to a service. Users in this list will not be given write access, no matter what the read only option is set to.
username	Refer to the `smb.conf` man page for more information – this is a complex and potentially misused parameter.
valid users	List of users that should be allowed to login to this service.
write list	List of users that are given read-write access to a service.

12.3.2 File and Directory Permissions-Based Controls

The following file and directory permission-based controls, if misused, can result in considerable difficulty to diagnose causes of misconfiguration. Use them sparingly and carefully. By gradually introducing each one by one, undesirable side effects may be detected. In the event of a problem, always comment all of them out and then gradually reintroduce them in a controlled way.

Refer to Table 12.3 for information regarding the parameters that may be used to affect file and directory permission-based access controls.

12.3.3 Miscellaneous Controls

The following are documented because of the prevalence of administrators creating inadvertent barriers to file access by not understanding the full implications of `smb.conf` file settings. See Table 12.4.

12.4 Access Controls on Shares

This section deals with how to configure Samba per share access control restrictions. By default, Samba sets no restrictions on the share itself. Restrictions on the share itself can

Table 12.3. File and Directory Permission Based Controls

Control Parameter	Description - Action - Notes
create mask	Refer to the `smb.conf` man page.
directory mask	The octal modes used when converting DOS modes to UNIX modes when creating UNIX directories. See also: directory security mask.
dos filemode	Enabling this parameter allows a user who has write access to the file to modify the permissions on it.
force create mode	This parameter specifies a set of UNIX mode bit permissions that will always be set on a file created by Samba.
force directory mode	This parameter specifies a set of UNIX mode bit permissions that will always be set on a directory created by Samba.
force directory security mode	Controls UNIX permission bits modified when a Windows NT client is manipulating UNIX permissions on a directory.
force security mode	Controls UNIX permission bits modified when a Windows NT client manipulates UNIX permissions.
hide unreadable	Prevents clients from seeing the existence of files that cannot be read.
hide unwriteable files	Prevents clients from seeing the existence of files that cannot be written to. Unwriteable directories are shown as usual.
nt acl support	This parameter controls whether smbd will attempt to map UNIX permissions into Windows NT access control lists.
security mask	Controls UNIX permission bits modified when a Windows NT client is manipulating the UNIX permissions on a file.

be set on MS Windows NT4/200x/XP shares. This can be an effective way to limit who can connect to a share. In the absence of specific restrictions the default setting is to allow the global user `Everyone - Full Control` (full control, change and read).

At this time Samba does not provide a tool for configuring access control setting on the share itself. Samba does have the capacity to store and act on access control settings, but the only way to create those settings is to use either the NT4 Server Manager or the Windows 200x MMC for Computer Management.

Samba stores the per share access control settings in a file called `share_info.tdb`. The location of this file on your system will depend on how Samba was compiled. The default location for Samba's tdb files is under `/usr/local/samba/var`. If the `tdbdump` utility has been compiled and installed on your system, then you can examine the contents of this file by executing: **tdbdump share_info.tdb** in the directory containing the tdb files.

Table 12.4. Other Controls

Control Parameter	Description - Action - Notes
case sensitive, default case, short preserve case	This means that all file name lookup will be done in a case sensitive manner. Files will be created with the precise file name Samba received from the MS Windows client.
csc policy	Client Side Caching Policy - parallels MS Windows client side file caching capabilities.
dont descend	Allows specifying a comma-delimited list of directories that the server should always show as empty.
dos filetime resolution	This option is mainly used as a compatibility option for Visual C++ when used against Samba shares.
dos filetimes	DOS and Windows allow users to change file time stamps if they can write to the file. POSIX semantics prevent this. This option allows DOS and Windows behavior.
fake oplocks	Oplocks are the way that SMB clients get permission from a server to locally cache file operations. If a server grants an oplock, the client is free to assume that it is the only one accessing the file and it will aggressively cache file data.
hide dot files, hide files, veto files	Note: MS Windows Explorer allows override of files marked as hidden so they will still be visible.
read only	If this parameter is yes, then users of a service may not create or modify files in the service's directory.
veto files	List of files and directories that are neither visible nor accessible.

12.4.1 Share Permissions Management

The best tool for the task is platform dependant. Choose the best tool for your environment.

12.4.1.1 Windows NT4 Workstation/Server

The tool you need to use to manage share permissions on a Samba server is the NT Server Manager. Server Manager is shipped with Windows NT4 Server products but not with Windows NT4 Workstation. You can obtain the NT Server Manager for MS Windows NT4 Workstation from Microsoft — see details below.

Instructions

1. Launch the NT4 Server Manager, click on the Samba server you want to administer. From the menu select **Computer**, then click on **Shared Directories**.

2. Click on the share that you wish to manage, then click the **Properties** tab. then click the **Permissions** tab. Now you can add or change access control settings as you wish.

12.4.1.2 Windows 200x/XP

On MS Windows NT4/200x/XP system access control lists on the share itself are set using native tools, usually from File Manager. For example, in Windows 200x, right click on the shared folder, then select **Sharing**, then click on **Permissions**. The default Windows NT4/200x permission allows "*Everyone*" full control on the share.

MS Windows 200x and later versions come with a tool called the Computer Management snap-in for the Microsoft Management Console (MMC). This tool is located by clicking on **Control Panel** -> **Administrative Tools** -> **Computer Management**.

Instructions

1. After launching the MMC with the Computer Management snap-in, click the menu item **Action**, and select **Connect to another computer**. If you are not logged onto a domain you will be prompted to enter a domain login user identifier and a password. This will authenticate you to the domain. If you are already logged in with administrative privilege, this step is not offered.

2. If the Samba server is not shown in the **Select Computer** box, type in the name of the target Samba server in the field **Name:**. Now click the on **[+]** next to **System Tools**, then on the **[+]** next to **Shared Folders** in the left panel.

3. In the right panel, double-click on the share on which you wish to set access control permissions. Then click the tab **Share Permissions**. It is now possible to add access control entities to the shared folder. Remember to set what type of access (full control, change, read) you wish to assign for each entry.

Warning

Be careful. If you take away all permissions from the `Everyone` user without removing this user, effectively no user will be able to access the share. This is a result of what is known as ACL precedence. Everyone with *no access* means that `MaryK` who is part of the group `Everyone` will have no access even if she is given explicit full control access.

12.5 MS Windows Access Control Lists and UNIX Interoperability

12.5.1 Managing UNIX Permissions Using NT Security Dialogs

Windows NT clients can use their native security settings dialog box to view and modify the underlying UNIX permissions.

This ability is careful not to compromise the security of the UNIX host on which Samba is running, and still obeys all the file permission rules that a Samba administrator can set.

Samba does not attempt to go beyond POSIX ACLs, so the various finer-grained access control options provided in Windows are actually ignored.

> NOTE
>
> 
>
> All access to UNIX/Linux system files via Samba is controlled by the operating system file access controls. When trying to figure out file access problems, it is vitally important to find the identity of the Windows user as it is presented by Samba at the point of file access. This can best be determined from the Samba log files.

12.5.2 Viewing File Security on a Samba Share

From an NT4/2000/XP client, right click on any file or directory in a Samba-mounted drive letter or UNC path. When the menu pops up, click on the **Properties** entry at the bottom of the menu. This brings up the file `Properties` dialog box. Click on the **Security** tab and you will see three buttons: **Permissions**, **Auditing**, and **Ownership**. The **Auditing** button will cause either an error message 'A requested privilege is not held by the client' to appear if the user is not the NT Administrator, or a dialog which is intended to allow an Administrator to add auditing requirements to a file if the user is logged on as the NT Administrator. This dialog is non-functional with a Samba share at this time, as the only useful button, the **Add** button, will not currently allow a list of users to be seen.

12.5.3 Viewing File Ownership

Clicking on the **Ownership** button brings up a dialog box telling you who owns the given file. The owner name will be displayed like this:

"***SERVER**user (Long name)***"

SERVER is the NetBIOS name of the Samba server, *user* is the user name of the UNIX user who owns the file, and *(Long name)* is the descriptive string identifying the user (normally found in the GECOS field of the UNIX password database). Click on the **Close** button to remove this dialog.

If the parameter *nt acl support* is set to `false`, the file owner will be shown as the NT user *Everyone*.

The **Take Ownership** button will not allow you to change the ownership of this file to yourself (clicking it will display a dialog box complaining that the user you are currently logged onto the NT client cannot be found). The reason for this is that changing the ownership of a file is a privileged operation in UNIX, available only to the *root* user. As clicking on this button causes NT to attempt to change the ownership of a file to the current user logged into the NT clienti, this will not work with Samba at this time.

There is an NT **chown** command that will work with Samba and allow a user with Administrator privilege connected to a Samba server as root to change the ownership of files on both a local NTFS filesystem or remote mounted NTFS or Samba drive. This is available as part of the Seclib NT security library written by Jeremy Allison of the Samba Team, and is available from the main Samba FTP site.

12.5.4 Viewing File or Directory Permissions

The third button is the **Permissions** button. Clicking on this brings up a dialog box that shows both the permissions and the UNIX owner of the file or directory. The owner is displayed like this:

SERVER\\`user (Long name)`

Where ***SERVER*** is the NetBIOS name of the Samba server, ***user*** is the user name of the UNIX user who owns the file, and *(Long name)* is the descriptive string identifying the user (normally found in the GECOS field of the UNIX password database).

If the parameter ***nt acl support*** is set to `false`, the file owner will be shown as the NT user `Everyone` and the permissions will be shown as NT "*Full Control*".

The permissions field is displayed differently for files and directories, so I'll describe the way file permissions are displayed first.

12.5.4.1 File Permissions

The standard UNIX user/group/world triplet and the corresponding `read, write, execute` permissions triplets are mapped by Samba into a three element NT ACL with the "*r*", "*w*" and "*x*" bits mapped into the corresponding NT permissions. The UNIX world permissions are mapped into the global NT group `Everyone`, followed by the list of permissions allowed for UNIX world. The UNIX owner and group permissions are displayed as an NT **user** icon and an NT **local group** icon, respectively, followed by the list of permissions allowed for the UNIX user and group.

Because many UNIX permission sets do not map into common NT names such as `read`, `change` or `full control`, usually the permissions will be prefixed by the words `Special Access` in the NT display list.

But what happens if the file has no permissions allowed for a particular UNIX user group or world component? In order to allow "*no permissions*" to be seen and modified Samba then overloads the NT `Take Ownership` ACL attribute (which has no meaning in UNIX) and reports a component with no permissions as having the NT **O** bit set. This was chosen, of course, to make it look like a zero, meaning zero permissions. More details on the decision behind this is given below.

12.5.4.2 Directory Permissions

Directories on an NT NTFS file system have two different sets of permissions. The first set is the ACL set on the directory itself, which is usually displayed in the first set of parentheses in the normal `RW` NT style. This first set of permissions is created by Samba in exactly the same way as normal file permissions are, described above, and is displayed in the same way.

The second set of directory permissions has no real meaning in the UNIX permissions world and represents the `inherited` permissions that any file created within this directory would inherit.

Samba synthesises these inherited permissions for NT by returning as an NT ACL the UNIX permission mode that a new file created by Samba on this share would receive.

12.5.5 Modifying File or Directory Permissions

Modifying file and directory permissions is as simple as changing the displayed permissions in the dialog box, and clicking on **OK**. However, there are limitations that a user needs to be aware of, and also interactions with the standard Samba permission masks and mapping of DOS attributes that need to also be taken into account.

If the parameter ***nt acl support*** is set to `false`, any attempt to set security permissions will fail with an 'Access Denied' message.

The first thing to note is that the **Add** button will not return a list of users in Samba (it will give an error message saying 'The remote procedure call failed and did not execute'). This means that you can only manipulate the current user/group/world permissions listed in the dialog box. This actually works quite well as these are the only permissions that UNIX actually has.

If a permission triplet (either user, group, or world) is removed from the list of permissions in the NT dialog box, then when the **OK** button is pressed it will be applied as "*no permissions*" on the UNIX side. If you then view the permissions again, the "*no permissions*" entry will appear as the NT **O** flag, as described above. This allows you to add permissions back to a file or directory once you have removed them from a triplet component.

As UNIX supports only the "*r*", "*w*" and "*x*" bits of an NT ACL, if other NT security attributes such as `Delete Access` are selected they will be ignored when applied on the Samba server.

When setting permissions on a directory, the second set of permissions (in the second set of parentheses) is by default applied to all files within that directory. If this is not what you want, you must uncheck the **Replace permissions on existing files** checkbox in the NT dialog before clicking on **OK**.

If you wish to remove all permissions from a user/group/world component, you may either highlight the component and click on the **Remove** button, or set the component to only have the special `Take Ownership` permission (displayed as **O**) highlighted.

12.5.6 Interaction with the Standard Samba "*create mask*" Parameters

There are four parameters that control interaction with the standard Samba *create mask* parameters. These are:

- *security mask*
- *force security mode*
- *directory security mask*
- *force directory security mode*

Once a user clicks on **OK** to apply the permissions, Samba maps the given permissions into a user/group/world r/w/x triplet set, and then checks the changed permissions for a file against the bits set in the *security mask* parameter. Any bits that were changed that are not set to "*1*" in this parameter are left alone in the file permissions.

Essentially, zero bits in the *security mask* may be treated as a set of bits the user is *not* allowed to change, and one bits are those the user is allowed to change.

If not explicitly set, this parameter defaults to the same value as the *create mask* parameter. To allow a user to modify all the user/group/world permissions on a file, set this parameter to 0777.

Next Samba checks the changed permissions for a file against the bits set in the *force security mode* parameter. Any bits that were changed that correspond to bits set to "*1*" in this parameter are forced to be set.

Essentially, bits set in the *force security mode* parameter may be treated as a set of bits that, when modifying security on a file, the user has always set to be "*on*".

If not explicitly set, this parameter defaults to the same value as the *force create mode* parameter. To allow a user to modify all the user/group/world permissions on a file with no restrictions set this parameter to 000. The *security mask* and *force security mode* parameters are applied to the change request in that order.

For a directory, Samba will perform the same operations as described above for a file except it uses the parameter *directory security mask* instead of *security mask*, and *force directory security mode* parameter instead of *force security mode*.

The *directory security mask* parameter by default is set to the same value as the *directory mask* parameter and the *force directory security mode* parameter by default is set to the same value as the *force directory mode* parameter. In this way Samba enforces the permission restrictions that an administrator can set on a Samba share, while still allowing users to modify the permission bits within that restriction.

If you want to set up a share that allows users full control in modifying the permission bits on their files and directories and does not force any particular bits to be set "*on*", then set the following parameters in the `smb.conf` file in that share-specific section:

```
security mask = 0777
force security mode = 0
directory security mask = 0777
force directory security mode = 0
```

12.5.7 Interaction with the Standard Samba File Attribute Mapping

NOTE

Samba maps some of the DOS attribute bits (such as "*read only*") into the UNIX permissions of a file. This means there can be a conflict between the permission bits set via the security dialog and the permission bits set by the file attribute mapping.

If a file has no UNIX read access for the owner, it will show up as "*read only*" in the standard file attributes tabbed dialog. Unfortunately, this dialog is the same one that contains the security information in another tab.

What this can mean is that if the owner changes the permissions to allow himself read access using the security dialog, clicks on **OK** to get back to the standard attributes tab dialog, and clicks on **OK** on that dialog, then NT will set the file permissions back to read-only (as that is what the attributes still say in the dialog). This means that after setting permissions and clicking on **OK** to get back to the attributes dialog, you should always press **Cancel** rather than **OK** to ensure that your changes are not overridden.

12.6 Common Errors

File, directory and share access problems are common on the mailing list. The following are examples taken from the mailing list in recent times.

12.6.1 Users Cannot Write to a Public Share

"*We are facing some troubles with file/directory permissions. I can log on the domain as admin user(root), and there's a public share on which everyone needs to have permission to create/modify files, but only root can change the file, no one else can. We need to constantly go to the server to* `chgrp -R users *` *and* `chown -R nobody *` *to allow others users to change the file.*"

There are many ways to solve this problem and here are a few hints:

1. Go to the top of the directory that is shared.

2. Set the ownership to what ever public owner and group you want

```
$ find 'directory_name' -type d -exec chown user.group {}\;
$ find 'directory_name' -type d -exec chmod 6775 'directory_name'
$ find 'directory_name' -type f -exec chmod 0775 {} \;
$ find 'directory_name' -type f -exec chown user.group {}\;
```

> **NOTE**
>
> 
>
> The above will set the `sticky bit` on all directories. Read your UNIX/Linux man page on what that does. It causes the OS to assign to all files created in the directories the ownership of the directory.

3. Directory is: */foodbar*

```
$ chown jack.engr /foodbar
```

> **NOTE**
>
> 
>
> This is the same as doing:
>
> ```
> $ chown jack /foodbar
> $ chgrp engr /foodbar
> ```

4. Now type:

```
$ chmod 6775 /foodbar
$ ls -al /foodbar/..
```

   You should see:

```
drwsrwsr-x  2 jack  engr    48 2003-02-04 09:55 foodbar
```

5. Now type:

```
$ su - jill
$ cd /foodbar
$ touch Afile
$ ls -al
```

   You should see that the file `Afile` created by Jill will have ownership and permissions of Jack, as follows:

```
-rw-r--r--  1 jack  engr     0 2003-02-04 09:57 Afile
```

6. Now in your `smb.conf` for the share add:

   force create mode = 0775
   force direcrtory mode = 6775

> **NOTE**
>
> 
>
> These procedures are needed only if your users are not members of the group you have used. That is if within the OS do not have write permission on the directory.

An alternative is to set in the `smb.conf` entry for the share:

force user = jack
force group = engr

12.6.2 File Operations Done as *root* with *force user* Set

When you have a user in ***admin users***, Samba will always do file operations for this user as *root*, even if ***force user*** has been set.

12.6.3 MS Word with Samba Changes Owner of File

Question: "When user B saves a word document that is owned by user A the updated file is now owned by user B. Why is Samba doing this? How do I fix this?"

Answer: Word does the following when you modify/change a Word document: MS Word creates a NEW document with a temporary name, Word then closes the old document and deletes it, Word then renames the new document to the original document name. There is no mechanism by which Samba can in any way know that the new document really should be owned by the owners of the original file. Samba has no way of knowing that the file will be renamed by MS Word. As far as Samba is able to tell, the file that gets created is a NEW file, not one that the application (Word) is updating.

There is a work-around to solve the permissions problem. That work-around involves understanding how you can manage file system behavior from within the `smb.conf` file, as well as understanding how UNIX file systems work. Set on the directory in which you are changing Word documents: **chmod g+s 'directory_name'** This ensures that all files will be created with the group that owns the directory. In `smb.conf` share declaration section set:

force create mode = 0660
force directory mode = 0770

These two settings will ensure that all directories and files that get created in the share will be read/writable by the owner and group set on the directory itself.

Chapter 13

FILE AND RECORD LOCKING

One area that causes trouble for many network administrators is locking. The extent of the problem is readily evident from searches over the Internet.

13.1 Features and Benefits

Samba provides all the same locking semantics that MS Windows clients expect and that MS Windows NT4/200x servers also provide.

The term *locking* has exceptionally broad meaning and covers a range of functions that are all categorized under this one term.

Opportunistic locking is a desirable feature when it can enhance the perceived performance of applications on a networked client. However, the opportunistic locking protocol is not robust and, therefore, can encounter problems when invoked beyond a simplistic configuration or on extended slow or faulty networks. In these cases, operating system management of opportunistic locking and/or recovering from repetitive errors can offset the perceived performance advantage that it is intended to provide.

The MS Windows network administrator needs to be aware that file and record locking semantics (behavior) can be controlled either in Samba or by way of registry settings on the MS Windows client.

NOTE

Sometimes it is necessary to disable locking control settings on both the Samba server as well as on each MS Windows client!

13.2 Discussion

There are two types of locking that need to be performed by an SMB server. The first is *record locking* that allows a client to lock a range of bytes in a open file. The second is the *deny modes* that are specified when a file is open.

Record locking semantics under UNIX are very different from record locking under Windows. Versions of Samba before 2.2 have tried to use the native fcntl() UNIX system call to implement proper record locking between different Samba clients. This cannot be fully correct for several reasons. The simplest is the fact that a Windows client is allowed to lock a byte range up to 2^32 or 2^64, depending on the client OS. The UNIX locking only supports byte ranges up to 2^31. So it is not possible to correctly satisfy a lock request above 2^31. There are many more differences, too many to be listed here.

Samba 2.2 and above implements record locking completely independent of the underlying UNIX system. If a byte range lock that the client requests happens to fall into the range of 0-2^31, Samba hands this request down to the UNIX system. All other locks cannot be seen by UNIX, anyway.

Strictly speaking, an SMB server should check for locks before every read and write call on a file. Unfortunately with the way fcntl() works, this can be slow and may overstress the **rpc.lockd**. This is almost always unnecessary as clients are supposed to independently make locking calls before reads and writes if locking is important to them. By default, Samba only makes locking calls when explicitly asked to by a client, but if you set ***strict locking*** = yes, it will make lock checking calls on *every* read and write call.

You can also disable byte range locking completely by using ***locking*** = no. This is useful for those shares that do not support locking or do not need it (such as CDROMs). In this case, Samba fakes the return codes of locking calls to tell clients that everything is okay.

The second class of locking is the *deny modes*. These are set by an application when it opens a file to determine what types of access should be allowed simultaneously with its open. A client may ask for `DENY_NONE`, `DENY_READ`, `DENY_WRITE`, or `DENY_ALL`. There are also special compatibility modes called `DENY_FCB` and `DENY_DOS`.

13.2.1 Opportunistic Locking Overview

Opportunistic locking (Oplocks) is invoked by the Windows file system (as opposed to an API) via registry entries (on the server and the client) for the purpose of enhancing network performance when accessing a file residing on a server. Performance is enhanced by caching the file locally on the client that allows:

Read-ahead: — The client reads the local copy of the file, eliminating network latency.

Write caching: — The client writes to the local copy of the file, eliminating network latency.

Lock caching: — The client caches application locks locally, eliminating network latency.

The performance enhancement of oplocks is due to the opportunity of exclusive access to the file — even if it is opened with deny-none — because Windows monitors the file's status for concurrent access from other processes.

WINDOWS DEFINES 4 KINDS OF OPLOCKS:

Level1 Oplock — The redirector sees that the file was opened with deny none (allowing concurrent access), verifies that no other process is accessing the file, checks that oplocks are enabled, then grants deny-all/read-write/exclusive access to the file. The client now performs operations on the cached local file.

If a second process attempts to open the file, the open is deferred while the redirector "*breaks*" the original oplock. The oplock break signals the caching client to write the local file back to the server, flush the local locks and discard read-ahead data. The break is then complete, the deferred open is granted, and the multiple processes can enjoy concurrent file access as dictated by mandatory or byte-range locking options. However, if the original opening process opened the file with a share mode other than deny-none, then the second process is granted limited or no access, despite the oplock break.

Level2 Oplock — Performs like a Level1 oplock, except caching is only operative for reads. All other operations are performed on the server disk copy of the file.

Filter Oplock — Does not allow write or delete file access.

Batch Oplock — Manipulates file openings and closings and allows caching of file attributes.

An important detail is that oplocks are invoked by the file system, not an application API. Therefore, an application can close an oplocked file, but the file system does not relinquish the oplock. When the oplock break is issued, the file system then simply closes the file in preparation for the subsequent open by the second process.

Opportunistic locking is actually an improper name for this feature. The true benefit of this feature is client-side data caching, and oplocks is merely a notification mechanism for writing data back to the networked storage disk. The limitation of opportunistic locking is the reliability of the mechanism to process an oplock break (notification) between the server and the caching client. If this exchange is faulty (usually due to timing out for any number of reasons), then the client-side caching benefit is negated.

The actual decision that a user or administrator should consider is whether it is sensible to share among multiple users data that will be cached locally on a client. In many cases the answer is no. Deciding when to cache or not cache data is the real question, and thus "*opportunistic locking*" should be treated as a toggle for client-side caching. Turn it "*on*" when client-side caching is desirable and reliable. Turn it "*off*" when client-side caching is redundant, unreliable or counter-productive.

Opportunistic locking is by default set to "*on*" by Samba on all configured shares, so careful attention should be given to each case to determine if the potential benefit is worth the potential for delays. The following recommendations will help to characterize the environment where opportunistic locking may be effectively configured.

Windows opportunistic locking is a lightweight performance-enhancing feature. It is not a robust and reliable protocol. Every implementation of opportunistic locking should be evaluated as a tradeoff between perceived performance and reliability. Reliability decreases as each successive rule above is not enforced. Consider a share with oplocks enabled, over a wide area network, to a client on a South Pacific atoll, on a high-availability server, serving a mission-critical multi-user corporate database during a tropical storm. This configuration will likely encounter problems with oplocks.

Oplocks can be beneficial to perceived client performance when treated as a configuration toggle for client-side data caching. If the data caching is likely to be interrupted, then oplock usage should be reviewed. Samba enables opportunistic locking by default on all shares. Careful attention should be given to the client usage of shared data on the server, the server network reliability and the opportunistic locking configuration of each share. In mission critical high availability environments, data integrity is often a priority. Complex and expensive configurations are implemented to ensure that if a client loses connectivity with a file server, a failover replacement will be available immediately to provide continuous data availability.

Windows client failover behavior is more at risk of application interruption than other platforms because it is dependent upon an established TCP transport connection. If the connection is interrupted — as in a file server failover — a new session must be established. It is rare for Windows client applications to be coded to recover correctly from a transport connection loss, therefore, most applications will experience some sort of interruption — at worst, abort and require restarting.

If a client session has been caching writes and reads locally due to opportunistic locking, it is likely that the data will be lost when the application restarts or recovers from the TCP interrupt. When the TCP connection drops, the client state is lost. When the file server recovers, an oplock break is not sent to the client. In this case, the work from the prior session is lost. Observing this scenario with oplocks disabled and with the client writing data to the file server real-time, the failover will provide the data on disk as it existed at the time of the disconnect.

In mission-critical high-availability environments, careful attention should be given to opportunistic locking. Ideally, comprehensive testing should be done with all affected applications with oplocks enabled and disabled.

13.2.1.1 Exclusively Accessed Shares

Opportunistic locking is most effective when it is confined to shares that are exclusively accessed by a single user, or by only one user at a time. Because the true value of opportunistic locking is the local client caching of data, any operation that interrupts the caching mechanism will cause a delay.

Home directories are the most obvious examples of where the performance benefit of opportunistic locking can be safely realized.

13.2.1.2 Multiple-Accessed Shares or Files

As each additional user accesses a file in a share with opportunistic locking enabled, the potential for delays and resulting perceived poor performance increases. When multiple users are accessing a file on a share that has oplocks enabled, the management impact of sending and receiving oplock breaks and the resulting latency while other clients wait for the caching client to flush data offset the performance gains of the caching user.

As each additional client attempts to access a file with oplocks set, the potential performance improvement is negated and eventually results in a performance bottleneck.

13.2.1.3 UNIX or NFS Client-Accessed Files

Local UNIX and NFS clients access files without a mandatory file-locking mechanism. Thus, these client platforms are incapable of initiating an oplock break request from the server to a Windows client that has a file cached. Local UNIX or NFS file access can therefore write to a file that has been cached by a Windows client, which exposes the file to likely data corruption.

If files are shared between Windows clients, and either local UNIX or NFS users, turn opportunistic locking off.

13.2.1.4 Slow and/or Unreliable Networks

The biggest potential performance improvement for opportunistic locking occurs when the client-side caching of reads and writes delivers the most differential over sending those reads and writes over the wire. This is most likely to occur when the network is extremely slow, congested, or distributed (as in a WAN). However, network latency also has a high impact on the reliability of the oplock break mechanism, and thus increases the likelihood of encountering oplock problems that more than offset the potential perceived performance gain. Of course, if an oplock break never has to be sent, then this is the most advantageous scenario to utilize opportunistic locking.

If the network is slow, unreliable, or a WAN, then do not configure opportunistic locking if there is any chance of multiple users regularly opening the same file.

13.2.1.5 Multi-User Databases

Multi-user databases clearly pose a risk due to their very nature — they are typically heavily accessed by numerous users at random intervals. Placing a multi-user database on a share with opportunistic locking enabled will likely result in a locking management bottleneck on the Samba server. Whether the database application is developed in-house or a commercially available product, ensure that the share has opportunistic locking disabled.

13.2.1.6 PDM Data Shares

Process Data Management (PDM) applications such as IMAN, Enovia and Clearcase are increasing in usage with Windows client platforms, and therefore SMB datastores. PDM applications manage multi-user environments for critical data security and access. The typical PDM environment is usually associated with sophisticated client design applications that will load data locally as demanded. In addition, the PDM application will usually monitor the data-state of each client. In this case, client-side data caching is best left to the local application and PDM server to negotiate and maintain. It is appropriate to eliminate the client OS from any caching tasks, and the server from any oplock management, by disabling opportunistic locking on the share.

13.2.1.7 Beware of Force User

Samba includes an `smb.conf` parameter called *force user* that changes the user accessing a share from the incoming user to whatever user is defined by the smb.conf variable. If opportunistic locking is enabled on a share, the change in user access causes an oplock break to be sent to the client, even if the user has not explicitly loaded a file. In cases where the network is slow or unreliable, an oplock break can become lost without the user even accessing a file. This can cause apparent performance degradation as the client continually reconnects to overcome the lost oplock break.

Avoid the combination of the following:

- *force user* in the `smb.conf` share configuration.
- Slow or unreliable networks
- Opportunistic locking enabled

13.2.1.8 Advanced Samba Opportunistic Locking Parameters

Samba provides opportunistic locking parameters that allow the administrator to adjust various properties of the oplock mechanism to account for timing and usage levels. These parameters provide good versatility for implementing oplocks in environments where they would likely cause problems. The parameters are: ***oplock break wait time***, ***oplock contention limit***.

For most users, administrators and environments, if these parameters are required, then the better option is to simply turn oplocks off. The Samba SWAT help text for both parameters reads: "*Do not change this parameter unless you have read and understood the Samba oplock code.*" This is good advice.

13.2.1.9 Mission-Critical High-Availability

In mission-critical high-availability environments, data integrity is often a priority. Complex and expensive configurations are implemented to ensure that if a client loses connectivity with a file server, a failover replacement will be available immediately to provide continuous data availability.

Windows client failover behavior is more at risk of application interruption than other platforms because it is dependant upon an established TCP transport connection. If the connection is interrupted — as in a file server failover — a new session must be established. It is rare for Windows client applications to be coded to recover correctly from a transport connection loss, therefore, most applications will experience some sort of interruption — at worst, abort and require restarting.

If a client session has been caching writes and reads locally due to opportunistic locking, it is likely that the data will be lost when the application restarts, or recovers from the TCP interrupt. When the TCP connection drops, the client state is lost. When the file server recovers, an oplock break is not sent to the client. In this case, the work from the prior session is lost. Observing this scenario with oplocks disabled, and the client was writing data to the file server real-time, then the failover will provide the data on disk as it existed at the time of the disconnect.

In mission-critical high-availability environments, careful attention should be given to opportunistic locking. Ideally, comprehensive testing should be done with all effected applications with oplocks enabled and disabled.

13.3 Samba Opportunistic Locking Control

Opportunistic locking is a unique Windows file locking feature. It is not really file locking, but is included in most discussions of Windows file locking, so is considered a de facto locking feature. Opportunistic locking is actually part of the Windows client file caching mechanism. It is not a particularly robust or reliable feature when implemented on the variety of customized networks that exist in enterprise computing.

Like Windows, Samba implements opportunistic locking as a server-side component of the client caching mechanism. Because of the lightweight nature of the Windows feature design, effective configuration of opportunistic locking requires a good understanding of its limitations, and then applying that understanding when configuring data access for each particular customized network and client usage state.

Opportunistic locking essentially means that the client is allowed to download and cache a file on their hard drive while making changes; if a second client wants to access the file, the first client receives a break and must synchronize the file back to the server. This can give significant performance gains in some cases; some programs insist on synchronizing the contents of the entire file back to the server for a single change.

Level1 Oplocks (also known as just plain "*oplocks*") is another term for opportunistic locking.

Level2 Oplocks provides opportunistic locking for a file that will be treated as *read only*. Typically this is used on files that are read-only or on files that the client has no initial intention to write to at time of opening the file.

Kernel Oplocks are essentially a method that allows the Linux kernel to co-exist with Samba's oplocked files, although this has provided better integration of MS Windows network file locking with the underlying OS, SGI IRIX and Linux are the only two OSs that are oplock-aware at this time.

Unless your system supports kernel oplocks, you should disable oplocks if you are accessing the same files from both UNIX/Linux and SMB clients. Regardless, oplocks should always be disabled if you are sharing a database file (e.g., Microsoft Access) between multiple clients, as any break the first client receives will affect synchronization of the entire file (not just the single record), which will result in a noticeable performance impairment and, more likely, problems accessing the database in the first place. Notably, Microsoft Outlook's personal folders (*.pst) react quite badly to oplocks. If in doubt, disable oplocks and tune your system from that point.

If client-side caching is desirable and reliable on your network, you will benefit from turning on oplocks. If your network is slow and/or unreliable, or you are sharing your files among other file sharing mechanisms (e.g., NFS) or across a WAN, or multiple people will be accessing the same files frequently, you probably will not benefit from the overhead of your client sending oplock breaks and will instead want to disable oplocks for the share.

Another factor to consider is the perceived performance of file access. If oplocks provide no measurable speed benefit on your network, it might not be worth the hassle of dealing with them.

13.3.1 Example Configuration

In the following section we examine two distinct aspects of Samba locking controls.

13.3.1.1 Disabling Oplocks

You can disable oplocks on a per-share basis with the following:

```
oplocks = False
level2 oplocks = False
```

The default oplock type is Level1. Level2 oplocks are enabled on a per-share basis in the `smb.conf` file.

Alternately, you could disable oplocks on a per-file basis within the share:

```
veto oplock files = /*.mdb/*.MDB/*.dbf/*.DBF/
```

If you are experiencing problems with oplocks as apparent from Samba's log entries, you may want to play it safe and disable oplocks and Level2 oplocks.

13.3.1.2 Disabling Kernel Oplocks

Kernel oplocks is an `smb.conf` parameter that notifies Samba (if the UNIX kernel has the capability to send a Windows client an oplock break) when a UNIX process is attempting to open the file that is cached. This parameter addresses sharing files between UNIX and Windows with oplocks enabled on the Samba server: the UNIX process can open the file that is Oplocked (cached) by the Windows client and the smbd process will not send an oplock break, which exposes the file to the risk of data corruption. If the UNIX kernel has the ability to send an oplock break, then the kernel oplocks parameter enables Samba to send the oplock break. Kernel oplocks are enabled on a per-server basis in the `smb.conf` file.

```
kernel oplocks = yes
```
The default is no.

Veto opLocks is an `smb.conf` parameter that identifies specific files for which oplocks are disabled. When a Windows client opens a file that has been configured for veto oplocks, the client will not be granted the oplock, and all operations will be executed on the original file on disk instead of a client-cached file copy. By explicitly identifying files that are shared with UNIX processes and disabling oplocks for those files, the server-wide Oplock configuration can be enabled to allow Windows clients to utilize the performance benefit of file caching without the risk of data corruption. Veto Oplocks can be enabled on a per-share basis, or globally for the entire server, in the `smb.conf` file as shown in Example 13.1.

Example 13.1. Share with some files oplocked

```
[global]
veto oplock files = /filename.htm/*.txt/

[share_name]
veto oplock files = /*.exe/filename.ext/
```

oplock break wait time is an `smb.conf` parameter that adjusts the time interval for Samba to reply to an oplock break request. Samba recommends: "*Do not change this parameter unless you have read and understood the Samba oplock code.*" Oplock break Wait Time can only be configured globally in the `smb.conf` file as shown below.

```
oplock break wait time = 0 (default)
```

Oplock break contention limit is an `smb.conf` parameter that limits the response of the Samba server to grant an oplock if the configured number of contending clients reaches the limit specified by the parameter. Samba recommends "*Do not change this parameter unless you have read and understood the Samba oplock code.*" Oplock break Contention Limit can be enable on a per-share basis, or globally for the entire server, in the `smb.conf` file as shown in Example 13.2.

Example 13.2. Configuration with oplock break contention limit

```
[global]
oplock break contention limit = 2 (default)

[share_name]
oplock break contention limit = 2 (default)
```

13.4 MS Windows Opportunistic Locking and Caching Controls

There is a known issue when running applications (like Norton Anti-Virus) on a Windows 2000/ XP workstation computer that can affect any application attempting to access shared database files across a network. This is a result of a default setting configured in the Windows 2000/XP operating system known as *opportunistic locking*. When a workstation attempts to access shared data files located on another Windows 2000/XP computer, the Windows 2000/XP operating system will attempt to increase performance by locking the files and caching information locally. When this occurs, the application is unable to properly function, which results in an "*Access Denied*" error message being displayed during network operations.

All Windows operating systems in the NT family that act as database servers for data files (meaning that data files are stored there and accessed by other Windows PCs) may need to have opportunistic locking disabled in order to minimize the risk of data file corruption. This includes Windows 9x/Me, Windows NT, Windows 200x, and Windows XP.

If you are using a Windows NT family workstation in place of a server, you must also disable opportunistic locking (oplocks) on that workstation. For example, if you use a PC with the Windows NT Workstation operating system instead of Windows NT Server, and you have data files located on it that are accessed from other Windows PCs, you may need to disable oplocks on that system.

The major difference is the location in the Windows registry where the values for disabling oplocks are entered. Instead of the LanManServer location, the LanManWorkstation location may be used.

You can verify (change or add, if necessary) this registry value using the Windows Registry Editor. When you change this registry value, you will have to reboot the PC to ensure that the new setting goes into effect.

The location of the client registry entry for opportunistic locking has changed in Windows 2000 from the earlier location in Microsoft Windows NT.

> **NOTE**
>
> 
>
> Windows 2000 will still respect the EnableOplocks registry value used to disable oplocks in earlier versions of Windows.

You can also deny the granting of opportunistic locks by changing the following registry entries:

```
HKEY_LOCAL_MACHINE\System\
   CurrentControlSet\Services\MRXSmb\Parameters\

   OplocksDisabled REG_DWORD 0 or 1
   Default: 0 (not disabled)
```

> **NOTE**
>
> 
>
> The OplocksDisabled registry value configures Windows clients to either request or not request opportunistic locks on a remote file. To disable oplocks, the value of OplocksDisabled must be set to 1.

```
HKEY_LOCAL_MACHINE\System\
CurrentControlSet\Services\LanmanServer\Parameters

EnableOplocks REG_DWORD 0 or 1
Default: 1 (Enabled by Default)

EnableOpLockForceClose REG_DWORD 0 or 1
Default: 0 (Disabled by Default)
```

NOTE

The EnableOplocks value configures Windows-based servers (including Workstations sharing files) to allow or deny opportunistic locks on local files.

To force closure of open oplocks on close or program exit, EnableOpLockForceClose must be set to 1.

An illustration of how Level2 oplocks work:

- Station 1 opens the file requesting oplock.
- Since no other station has the file open, the server grants station 1 exclusive oplock.
- Station 2 opens the file requesting oplock.
- Since station 1 has not yet written to the file, the server asks station 1 to break to Level2 oplock.
- Station 1 complies by flushing locally buffered lock information to the server.
- Station 1 informs the server that it has Broken to Level2 Oplock (alternately, station 1 could have closed the file).
- The server responds to station 2's open request, granting it Level2 oplock. Other stations can likewise open the file and obtain Level2 oplock.
- Station 2 (or any station that has the file open) sends a write request SMB. The server returns the write response.
- The server asks all stations that have the file open to break to none, meaning no station holds any oplock on the file. Because the workstations can have no cached writes or locks at this point, they need not respond to the break-to-none advisory; all they need do is invalidate locally cashed read-ahead data.

13.4.1 Workstation Service Entries

```
\HKEY_LOCAL_MACHINE\System\
   CurrentControlSet\Services\LanmanWorkstation\Parameters
```

```
UseOpportunisticLocking   REG_DWORD   0 or 1
Default: 1 (true)
```

This indicates whether the redirector should use opportunistic-locking (oplock) performance enhancement. This parameter should be disabled only to isolate problems.

13.4.2 Server Service Entries

```
\HKEY_LOCAL_MACHINE\System\
   CurrentControlSet\Services\LanmanServer\Parameters
```

```
EnableOplocks   REG_DWORD   0 or 1
Default: 1 (true)
```

This specifies whether the server allows clients to use oplocks on files. Oplocks are a significant performance enhancement, but have the potential to cause lost cached data on some networks, particularly wide area networks.

```
MinLinkThroughput   REG_DWORD   0 to infinite bytes per second
Default: 0
```

This specifies the minimum link throughput allowed by the server before it disables raw and opportunistic locks for this connection.

```
MaxLinkDelay   REG_DWORD   0 to 100,000 seconds
Default: 60
```

This specifies the maximum time allowed for a link delay. If delays exceed this number, the server disables raw I/O and opportunistic locking for this connection.

```
OplockBreakWait   REG_DWORD   10 to 180 seconds
Default: 35
```

This specifies the time that the server waits for a client to respond to an oplock break request. Smaller values can allow detection of crashed clients more quickly but can potentially cause loss of cached data.

13.5 Persistent Data Corruption

If you have applied all of the settings discussed in this chapter but data corruption problems and other symptoms persist, here are some additional things to check out.

We have credible reports from developers that faulty network hardware, such as a single faulty network card, can cause symptoms similar to read caching and data corruption. If you see persistent data corruption even after repeated reindexing, you may have to rebuild the data files in question. This involves creating a new data file with the same definition as the file to be rebuilt and transferring the data from the old file to the new one. There are several known methods for doing this that can be found in our Knowledge Base.

13.6 Common Errors

In some sites, locking problems surface as soon as a server is installed; in other sites locking problems may not surface for a long time. Almost without exception, when a locking problem does surface it will cause embarrassment and potential data corruption.

Over the past few years there have been a number of complaints on the Samba mailing lists that have claimed that Samba caused data corruption. Three causes have been identified so far:

- Incorrect configuration of opportunistic locking (incompatible with the application being used. This is a common problem even where MS Windows NT4 or MS Windows 200x-based servers were in use. It is imperative that the software application vendors' instructions for configuration of file locking should be followed. If in doubt, disable oplocks on both the server and the client. Disabling of all forms of file caching on the MS Windows client may be necessary also.
- Defective network cards, cables, or HUBs/Switched. This is generally a more prevalent factor with low cost networking hardware, although occasionally there have also been problems with incompatibilities in more up-market hardware.
- There have been some random reports of Samba log files being written over data files. This has been reported by very few sites (about five in the past three years) and all attempts to reproduce the problem have failed. The Samba Team has been unable to catch this happening and thus has not been able to isolate any particular cause. Considering the millions of systems that use Samba, for the sites that have been affected by this as well as for the Samba Team this is a frustrating and a vexing challenge. If you see this type of thing happening, please create a bug report on Samba Bugzilla[1] without delay. Make sure that you give as much information as you possibly can help isolate the cause and to allow replication of the problem (an essential step in problem isolation and correction).

13.6.1 locking.tdb Error Messages

"*We are seeing lots of errors in the Samba logs, like:*

```
tdb(/usr/local/samba_2.2.7/var/locks/locking.tdb): rec_read bad magic
 0x4d6f4b61 at offset=36116
```

[1] https://bugzilla.samba.org

What do these mean?"

This error indicated a corrupted tdb. Stop all instances of smbd, delete locking.tdb, and restart smbd.

13.6.2 Problems Saving Files in MS Office on Windows XP

This is a bug in Windows XP. More information can be found in Microsoft Knowledge Base article 812937.[2]

13.6.3 Long Delays Deleting Files Over Network with XP SP1

"It sometimes takes approximately 35 seconds to delete files over the network after XP SP1 has been applied."

This is a bug in Windows XP. More information can be found in Microsoft Knowledge Base article 811492.[3]

13.7 Additional Reading

You may want to check for an updated version of this white paper on our Web site from time to time. Many of our white papers are updated as information changes. For those papers, the last edited date is always at the top of the paper.

Section of the Microsoft MSDN Library on opportunistic locking:

Opportunistic Locks, Microsoft Developer Network (MSDN), Windows Development > Windows Base Services > Files and I/O > SDK Documentation > File Storage > File Systems > About File Systems > Opportunistic Locks, Microsoft Corporation. `http://msdn.microsoft.com/library/en-us/fileio/storage_5yk3.asp`

Microsoft Knowledge Base Article Q224992 "*Maintaining Transactional Integrity with OPLOCKS*", Microsoft Corporation, April 1999, `http://support.microsoft.com/default.aspx?scid=kb;en-us;Q224992`.

Microsoft Knowledge Base Article Q296264 "*Configuring Opportunistic Locking in Windows 2000*", Microsoft Corporation, April 2001, `http://support.microsoft.com/default.aspx?scid=kb;en-us;Q296264`.

Microsoft Knowledge Base Article Q129202 "*PC Ext: Explanation of Opportunistic Locking on Windows NT*", Microsoft Corporation, April 1995, `http://support.microsoft.com/default.aspx?scid=kb;en-us;Q129202`.

[2] http://support.microsoft.com/?id=812937
[3] http://support.microsoft.com/?id=811492

Chapter 14

SECURING SAMBA

14.1 Introduction

This note was attached to the Samba 2.2.8 release notes as it contained an important security fix. The information contained here applies to Samba installations in general.

> A new apprentice reported for duty to the chief engineer of a boiler house. He said, "*Here I am, if you will show me the boiler I'll start working on it.*" Then engineer replied, "*You're leaning on it!*"

Security concerns are just like that. You need to know a little about the subject to appreciate how obvious most of it really is. The challenge for most of us is to discover that first morsel of knowledge with which we may unlock the secrets of the masters.

14.2 Features and Benefits

There are three levels at which security principals must be observed in order to render a site at least moderately secure. They are the perimeter firewall, the configuration of the host server that is running Samba and Samba itself.

Samba permits a most flexible approach to network security. As far as possible Samba implements the latest protocols to permit more secure MS Windows file and print operations.

Samba may be secured from connections that originate from outside the local network. This may be done using *host-based protection* (using samba's implementation of a technology known as "*tcpwrappers,*" or it may be done be using *interface-based exclusion* so smbd will bind only to specifically permitted interfaces. It is also possible to set specific share or resource-based exclusions, for example on the *[IPC$]* auto-share. The *[IPC$]* share is used for browsing purposes as well as to establish TCP/IP connections.

Another method by which Samba may be secured is by setting Access Control Entries (ACEs) in an Access Control List (ACL) on the shares themselves. This is discussed in Chapter 12, *File, Directory and Share Access Controls.*

14.3 Technical Discussion of Protective Measures and Issues

The key challenge of security is the fact that protective measures suffice at best only to close the door on known exploits and breach techniques. Never assume that because you have followed these few measures that the Samba server is now an impenetrable fortress! Given the history of information systems so far, it is only a matter of time before someone will find yet another vulnerability.

14.3.1 Using Host-Based Protection

In many installations of Samba, the greatest threat comes from outside your immediate network. By default, Samba will accept connections from any host, which means that if you run an insecure version of Samba on a host that is directly connected to the Internet you can be especially vulnerable.

One of the simplest fixes in this case is to use the *hosts allow* and *hosts deny* options in the Samba `smb.conf` configuration file to only allow access to your server from a specific range of hosts. An example might be:

```
hosts allow = 127.0.0.1 192.168.2.0/24 192.168.3.0/24
hosts deny = 0.0.0.0/0
```

The above will only allow SMB connections from `localhost` (your own computer) and from the two private networks 192.168.2 and 192.168.3. All other connections will be refused as soon as the client sends its first packet. The refusal will be marked as not listening on called name error.

14.3.2 User-Based Protection

If you want to restrict access to your server to valid users only, then the following method may be of use. In the `smb.conf` *[global]* section put:

```
valid users = @smbusers, jacko
```

This restricts all server access to either the user *jacko* or to members of the system group *smbusers*.

14.3.3 Using Interface Protection

By default, Samba will accept connections on any network interface that it finds on your system. That means if you have a ISDN line or a PPP connection to the Internet then Samba will accept connections on those links. This may not be what you want.

You can change this behavior using options like this:

```
interfaces = eth* lo
bind interfaces only = yes
```

This tells Samba to only listen for connections on interfaces with a name starting with `eth` such as `eth0`, `eth1` plus on the loopback interface called `lo`. The name you will need to use depends on what OS you are using. In the above, I used the common name for Ethernet adapters on Linux.

If you use the above and someone tries to make an SMB connection to your host over a PPP interface called `ppp0`, then they will get a TCP connection refused reply. In that case, no Samba code is run at all as the operating system has been told not to pass connections from that interface to any Samba process.

14.3.4 Using a Firewall

Many people use a firewall to deny access to services they do not want exposed outside their network. This can be a good idea, although I recommend using it in conjunction with the above methods so you are protected even if your firewall is not active for some reason.

If you are setting up a firewall, you need to know what TCP and UDP ports to allow and block. Samba uses the following:

UDP/137 - used by nmbd
UDP/138 - used by nmbd
TCP/139 - used by smbd
TCP/445 - used by smbd

The last one is important as many older firewall setups may not be aware of it, given that this port was only added to the protocol in recent years.

14.3.5 Using IPC$ Share-Based Denials

If the above methods are not suitable, then you could also place a more specific deny on the IPC$ share that is used in the recently discovered security hole. This allows you to offer access to other shares while denying access to IPC$ from potentially untrustworthy hosts.

To do this you could use:

```
hosts allow = 192.168.115.0/24 127.0.0.1
hosts deny = 0.0.0.0/0
```

This instructs Samba that IPC$ connections are not allowed from anywhere except from the two listed network addresses (localhost and the 192.168.115 subnet). Connections to other shares are still allowed. As the IPC$ share is the only share that is always accessible anonymously, this provides some level of protection against attackers that do not know a valid username/password for your host.

If you use this method, then clients will be given an 'access denied' reply when they try to access the IPC$ share. Those clients will not be able to browse shares, and may also be unable to access some other resources. This is not recommended unless you cannot use one of the other methods listed above for some reason.

14.3.6 NTLMv2 Security

To configure NTLMv2 authentication, the following registry keys are worth knowing about:

```
[HKEY_LOCAL_MACHINE\SYSTEM\CurrentControlSet\Control\Lsa]
"lmcompatibilitylevel"=dword:00000003
```

The value 0x00000003 means send NTLMv2 response only. Clients will use NTLMv2 authentication, use NTLMv2 session security if the server supports it. Domain Controllers accept LM, NTLM and NTLMv2 authentication.

```
[HKEY_LOCAL_MACHINE\SYSTEM\CurrentControlSet\Control\Lsa\MSV1_0]
"NtlmMinClientSec"=dword:00080000
```

The value 0x00080000 means permit only NTLMv2 session security. If either NtlmMinClientSec or NtlmMinServerSec is set to 0x00080000, the connection will fail if NTLMv2 session security is not negotiated.

14.4 Upgrading Samba

Please check regularly on `http://www.samba.org/` for updates and important announcements. Occasionally security releases are made and it is highly recommended to upgrade Samba when a security vulnerability is discovered. Check with your OS vendor for OS specific upgrades.

14.5 Common Errors

If all of Samba and host platform configuration were really as intuitive as one might like them to be, this section would not be necessary. Security issues are often vexing for a support person to resolve, not because of the complexity of the problem, but for the reason that most administrators who post what turns out to be a security problem request are totally convinced that the problem is with Samba.

14.5.1 Smbclient Works on Localhost, but the Network Is Dead

This is a common problem. Red Hat Linux (and others) installs a default firewall. With the default firewall in place, only traffic on the loopback adapter (IP address 127.0.0.1) is allowed through the firewall.

The solution is either to remove the firewall (stop it) or modify the firewall script to allow SMB networking traffic through. See section above in this chapter.

14.5.2 Why Can Users Access Home Directories of Other Users?

"*We are unable to keep individual users from mapping to any other user's home directory once they have supplied a valid password! They only need to enter their own password. I have not found any method to configure Samba so that users may map only their own home directory."*

"User xyzzy can map his home directory. Once mapped user xyzzy can also map anyone else's home directory."

This is not a security flaw, it is by design. Samba allows users to have exactly the same access to the UNIX file system as when they were logged onto the UNIX box, except that it only allows such views onto the file system as are allowed by the defined shares.

If your UNIX home directories are set up so that one user can happily **cd** into another users directory and execute **ls**, the UNIX security solution is to change file permissions on the user's home directories such that the **cd** and **ls** are denied.

Samba tries very hard not to second guess the UNIX administrators security policies, and trusts the UNIX admin to set the policies and permissions he or she desires.

Samba allows the behavior you require. Simply put the *only user* = %S option in the *[homes]* share definition.

The *only user* works in conjunction with the *users* = list, so to get the behavior you require, add the line :

```
users = %S
```

this is equivalent to adding

```
valid users = %S
```

to the definition of the *[homes]* share, as recommended in the `smb.conf` man page.

Chapter 15

INTERDOMAIN TRUST RELATIONSHIPS

Samba-3 supports NT4-style domain trust relationships. This is a feature that many sites will want to use if they migrate to Samba-3 from an NT4-style domain and do not want to adopt Active Directory or an LDAP-based authentication backend. This section explains some background information regarding trust relationships and how to create them. It is now possible for Samba-3 to trust NT4 (and vice versa), as well as to create Samba-to-Samba trusts.

15.1 Features and Benefits

Samba-3 can participate in Samba-to-Samba as well as in Samba-to-MS Windows NT4-style trust relationships. This imparts to Samba similar scalability as with MS Windows NT4.

Given that Samba-3 has the capability to function with a scalable backend authentication database such as LDAP, and given its ability to run in Primary as well as Backup Domain Control modes, the administrator would be well advised to consider alternatives to the use of Interdomain trusts simply because by the very nature of how this works it is fragile. That was, after all, a key reason for the development and adoption of Microsoft Active Directory.

15.2 Trust Relationship Background

MS Windows NT3/4 type security domains employ a non-hierarchical security structure. The limitations of this architecture as it effects the scalability of MS Windows networking in large organizations is well known. Additionally, the flat namespace that results from this design significantly impacts the delegation of administrative responsibilities in large and diverse organizations.

Microsoft developed Active Directory Service (ADS), based on Kerberos and LDAP, as a means of circumventing the limitations of the older technologies. Not every organization is ready or willing to embrace ADS. For small companies the older NT4-style domain security paradigm is quite adequate, there remains an entrenched user base for whom there is no direct desire to go through a disruptive change to adopt ADS.

With MS Windows NT, Microsoft introduced the ability to allow differing security domains to effect a mechanism so users from one domain may be given access rights and privileges in another domain. The language that describes this capability is couched in terms of *Trusts*. Specifically, one domain will *trust* the users from another domain. The domain from which users are available to another security domain is said to be a trusted domain. The domain in which those users have assigned rights and privileges is the trusting domain. With NT3.x/4.0 all trust relationships are always in one direction only, thus if users in both domains are to have privileges and rights in each others' domain, then it is necessary to establish two relationships, one in each direction.

In an NT4-style MS security domain, all trusts are non-transitive. This means that if there are three domains (let's call them RED, WHITE and BLUE) where RED and WHITE have a trust relationship, and WHITE and BLUE have a trust relationship, then it holds that there is no implied trust between the RED and BLUE domains. Relationships are explicit and not transitive.

New to MS Windows 2000 ADS security contexts is the fact that trust relationships are two-way by default. Also, all inter-ADS domain trusts are transitive. In the case of the RED, WHITE and BLUE domains above, with Windows 2000 and ADS the RED and BLUE domains can trust each other. This is an inherent feature of ADS domains. Samba-3 implements MS Windows NT4-style Interdomain trusts and interoperates with MS Windows 200x ADS security domains in similar manner to MS Windows NT4-style domains.

15.3 Native MS Windows NT4 Trusts Configuration

There are two steps to creating an interdomain trust relationship. To effect a two-way trust relationship, it is necessary for each domain administrator to create a trust account for the other domain to use in verifying security credentials.

15.3.1 Creating an NT4 Domain Trust

For MS Windows NT4, all domain trust relationships are configured using the Domain User Manager. This is done from the Domain User Manager Policies entry on the menu bar. From the **Policy** menu, select **Trust Relationships**. Next to the lower box labeled **Permitted to Trust this Domain** are two buttons, **Add** and **Remove**. The **Add** button will open a panel in which to enter the name of the remote domain that will be able to assign access rights to users in your domain. You will also need to enter a password for this trust relationship, which the trusting domain will use when authenticating users from the trusted domain. The password needs to be typed twice (for standard confirmation).

15.3.2 Completing an NT4 Domain Trust

A trust relationship will work only when the other (trusting) domain makes the appropriate connections with the trusted domain. To consummate the trust relationship, the administrator will launch the Domain User Manager from the menu select **Policies**, then select **Trust Relationships**, click on the **Add** button next to the box that is labeled **Trusted Domains**.

A panel will open in which must be entered the name of the remote domain as well as the password assigned to that trust.

15.3.3 Inter-Domain Trust Facilities

A two-way trust relationship is created when two one-way trusts are created, one in each direction. Where a one-way trust has been established between two MS Windows NT4 domains (let's call them DomA and DomB), the following facilities are created:

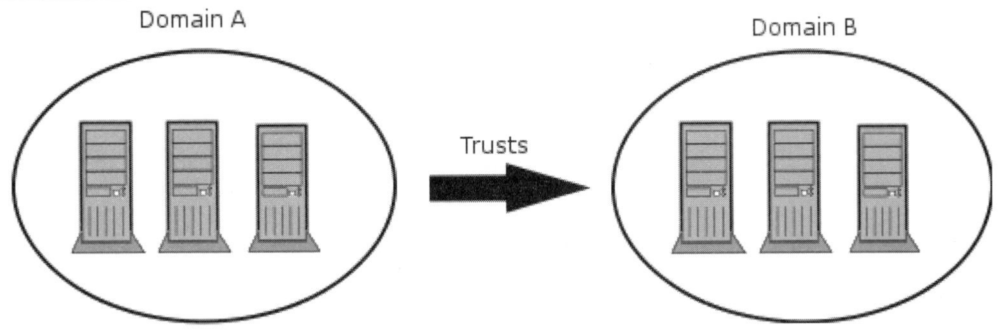

Figure 15.1. Trusts overview.

- DomA (completes the trust connection) ***Trusts*** DomB.
- DomA is the ***Trusting*** domain.
- DomB is the ***Trusted*** domain (originates the trust account).
- Users in DomB can access resources in DomA.
- Users in DomA cannot access resources in DomB.
- Global groups from DomB can be used in DomA.
- Global groups from DomA cannot be used in DomB.
- DomB does appear in the logon dialog box on client workstations in DomA.
- DomA does not appear in the logon dialog box on client workstations in DomB.
- Users/Groups in a trusting domain cannot be granted rights, permissions or access to a trusted domain.
- The trusting domain can access and use accounts (Users/Global Groups) in the trusted domain.
- Administrators of the trusted domain can be granted admininstrative rights in the trusting domain.
- Users in a trusted domain can be given rights and privileges in the trusting domain.
- Trusted domain Global Groups can be given rights and permissions in the trusting domain.

- Global Groups from the trusted domain can be made members in Local Groups on MS Windows Domain Member machines.

15.4 Configuring Samba NT-Style Domain Trusts

This description is meant to be a fairly short introduction about how to set up a Samba server so that it can participate in interdomain trust relationships. Trust relationship support in Samba is at an early stage, so do not be surprised if something does not function as it should.

Each of the procedures described below assumes the peer domain in the trust relationship is controlled by a Windows NT4 server. However, the remote end could just as well be another Samba-3 domain. It can be clearly seen, after reading this document, that combining Samba-specific parts of what's written below leads to trust between domains in a purely Samba environment.

15.4.1 Samba as the Trusted Domain

In order to set the Samba PDC to be the trusted party of the relationship, you first need to create a special account for the domain that will be the trusting party. To do that, you can use the **smbpasswd** utility. Creating the trusted domain account is similar to creating a trusted machine account. Suppose, your domain is called SAMBA, and the remote domain is called RUMBA. The first step will be to issue this command from your favorite shell:

```
root#  smbpasswd -a -i rumba
New SMB password: XXXXXXXX
Retype SMB password: XXXXXXXX
Added user rumba$
```

where `-a` means to add a new account into the passdb database and `-i` means: "*create this account with the InterDomain trust flag*".

The account name will be "*rumba$*" (the name of the remote domain).

After issuing this command, you will be asked to enter the password for the account. You can use any password you want, but be aware that Windows NT will not change this password until seven days following account creation. After the command returns successfully, you can look at the entry for the new account (in the standard way as appropriate for your configuration) and see that account's name is really RUMBA$ and it has the "*I*" flag set in the flags field. Now you are ready to confirm the trust by establishing it from Windows NT Server.

Open User Manager for Domains and from the **Policies** menu, select **Trust Relationships...**. Beside the **Trusted domains** list box click the **Add...** button. You will be prompted for the trusted domain name and the relationship password. Type in SAMBA, as this is the name of the remote domain and the password used at the time of account creation. Click on **OK** and, if everything went without incident, you will see the `Trusted domain relationship successfully established` message.

15.4.2 Samba as the Trusting Domain

This time activities are somewhat reversed. Again, we'll assume that your domain controlled by the Samba PDC is called SAMBA and the NT-controlled domain is called RUMBA.

The very first step is to add an account for the SAMBA domain on RUMBA's PDC.

Launch the Domain User Manager, then from the menu select **Policies**, **Trust Relationships**. Now, next to the **Trusted Domains** box press the **Add** button and type in the name of the trusted domain (SAMBA) and the password to use in securing the relationship.

The password can be arbitrarily chosen. It is easy to change the password from the Samba server whenever you want. After confirming the password your account is ready for use. Now its Samba's turn.

Using your favorite shell while being logged in as root, issue this command:

```
root#net rpc trustdom establish rumba
```

You will be prompted for the password you just typed on your Windows NT4 Server box. An error message 'NT_STATUS_NOLOGON_INTERDOMAIN_TRUST_ACCOUNT' that may be reported periodically is of no concern and may safely be ignored. It means the password you gave is correct and the NT4 Server says the account is ready for interdomain connection and not for ordinary connection. After that, be patient; it can take a while (especially in large networks), but eventually you should see the `Success` message. Congratulations! Your trust relationship has just been established.

NOTE

You have to run this command as root because you must have write access to the `secrets.tdb` file.

15.5 NT4-Style Domain Trusts with Windows 2000

Although Domain User Manager is not present in Windows 2000, it is also possible to establish an NT4-style trust relationship with a Windows 2000 domain controller running in mixed mode as the trusting server. It should also be possible for Samba to trust a Windows 2000 server, however, more testing is still needed in this area.

After creating the interdomain trust account on the Samba server as described above, open Active Directory Domains and Trusts on the AD controller of the domain whose resources you wish Samba users to have access to. Remember that since NT4-style trusts are not transitive, if you want your users to have access to multiple mixed-mode domains in your AD forest, you will need to repeat this process for each of those domains. With Active Directory Domains and Trusts open, right-click on the name of the Active Directory domain that will trust our Samba domain and choose **Properties**, then click on the **Trusts** tab. In the upper part of the panel, you will see a list box labeled **Domains trusted by this domain:**, and an

Add... button next to it. Press this button and just as with NT4, you will be prompted for the trusted domain name and the relationship password. Press OK and after a moment, Active Directory will respond with `The trusted domain has been added and the trust has been verified.` Your Samba users can now be granted acess to resources in the AD domain.

15.6 Common Errors

Interdomain trust relationships should not be attempted on networks that are unstable or that suffer regular outages. Network stability and integrity are key concerns with distributed trusted domains.

Chapter 16

HOSTING A MICROSOFT DISTRIBUTED FILE SYSTEM TREE

16.1 Features and Benefits

The Distributed File System (DFS) provides a means of separating the logical view of files and directories that users see from the actual physical locations of these resources on the network. It allows for higher availability, smoother storage expansion, load balancing, and so on.

For information about DFS, refer to the Microsoft documentation[1]. This document explains how to host a DFS tree on a UNIX machine (for DFS-aware clients to browse) using Samba.

To enable SMB-based DFS for Samba, configure it with the `--with-msdfs` option. Once built, a Samba server can be made a DFS server by setting the global Boolean ***host msdfs*** parameter in the `smb.conf` file. You designate a share as a DFS root using the Share Level Boolean *msdfs root* parameter. A DFS root directory on Samba hosts DFS links in the form of symbolic links that point to other servers. For example, a symbolic link `junction->msdfs:storage1\share1` in the share directory acts as the DFS junction. When DFS-aware clients attempt to access the junction link, they are redirected to the storage location (in this case, *storage1**share1*).

DFS trees on Samba work with all DFS-aware clients ranging from Windows 95 to 200x. Example 16.1 shows how to setup a DFS tree on a Samba server. In the `/export/dfsroot` directory, you set up your DFS links to other servers on the network.

```
root# cd /export/dfsroot
root# chown root /export/dfsroot
root# chmod 755 /export/dfsroot
root# ln -s msdfs:storageA\\shareA linka
root# ln -s msdfs:serverB\\share,serverC\\share linkb
```

[1] http://www.microsoft.com/NTServer/nts/downloads/winfeatures/NTSDistrFile/AdminGuide.asp

Example 16.1. smb.conf with DFS configured

```
[global]
netbios name = GANDALF
host msdfs = yes

[dfs]
path = /export/dfsroot
msdfs root = yes
```

You should set up the permissions and ownership of the directory acting as the DFS root so that only designated users can create, delete or modify the msdfs links. Also note that symlink names should be all lowercase. This limitation exists to have Samba avoid trying all the case combinations to get at the link name. Finally, set up the symbolic links to point to the network shares you want and start Samba.

Users on DFS-aware clients can now browse the DFS tree on the Samba server at \\samba\dfs. Accessing links linka or linkb (which appear as directories to the client) takes users directly to the appropriate shares on the network.

16.2 Common Errors

- Windows clients need to be rebooted if a previously mounted non-DFS share is made a DFS root or vice versa. A better way is to introduce a new share and make it the DFS root.
- Currently, there's a restriction that msdfs symlink names should all be lowercase.
- For security purposes, the directory acting as the root of the DFS tree should have ownership and permissions set so only designated users can modify the symbolic links in the directory.

16.2.1 MSDFS UNIX Path Is Case-Critical

A network administrator sent advice to the Samba mailing list after a long sessions trying to determine why DFS was not working. His advice is worth noting.

"I spent some time trying to figure out why my particular dfs root wasn't working. I noted in the documenation that the symlink should be in all lowercase. It should be amended that the entire path to the symlink should all be in lowercase as well."

For example, I had a share defined as such:

```
[pub]
   path = /export/home/Shares/public_share
   msdfs root = yes
```

and I could not make my Windows 9x/Me (with the dfs client installed) follow this symlink:

```
damage1 -> msdfs:damage\test-share
```

Running a debug level of 10 reveals:

```
[2003/08/20 11:40:33, 5] msdfs/msdfs.c:is_msdfs_link(176)
  is_msdfs_link: /export/home/shares/public_share/* does not exist.
```

Curious. So I changed the directory name from .../Shares/... to .../shares/... (along with my service definition) and it worked!

Chapter 17

CLASSICAL PRINTING SUPPORT

17.1 Features and Benefits

Printing is often a mission-critical service for the users. Samba can provide this service reliably and seamlessly for a client network consisting of Windows workstations.

A Samba print service may be run on a Stand-alone or Domain Member server, side by side with file serving functions, or on a dedicated print server. It can be made as tight or as loosely secured as needs dictate. Configurations may be simple or complex. Available authentication schemes are essentially the same as described for file services in previous chapters. Overall, Samba's printing support is now able to replace an NT or Windows 2000 print server full-square, with additional benefits in many cases. Clients may download and install drivers and printers through their familiar "*Point'n'Print*" mechanism. Printer installations executed by "*Logon Scripts*" are no problem. Administrators can upload and manage drivers to be used by clients through the familiar "*Add Printer Wizard*". As an additional benefit, driver and printer management may be run from the command line or through scripts, making it more efficient in case of large numbers of printers. If a central accounting of print jobs (tracking every single page and supplying the raw data for all sorts of statistical reports) is required, this function is best supported by the newer Common UNIX Printing System (CUPS) as the print subsystem underneath the Samba hood.

This chapter deals with the foundations of Samba printing as they are implemented by the more traditional UNIX (BSD- and System V-style) printing systems. Many things covered in this chapter apply also to CUPS. If you use CUPS, you may be tempted to jump to the next chapter but you will certainly miss a few things if you do. It is recommended that you read this chapter as well as Chapter 18, *CUPS Printing Support*.

NOTE

Most of the following examples have been verified on Windows XP Professional clients. Where this document describes the responses to commands given, bear in mind that Windows 200x/XP clients are quite similar, but may differ in minor details. Windows NT is somewhat different again.

17.2 Technical Introduction

Samba's printing support always relies on the installed print subsystem of the UNIX OS it runs on. Samba is a *"middleman."* It takes print files from Windows (or other SMB) clients and passes them to the real printing system for further processing, therefore, it needs to communicate with both sides: the Windows print clients and the UNIX printing system. Hence, we must differentiate between the various client OS types, each of which behave differently, as well as the various UNIX print subsystems, which themselves have different features and are accessed differently.

This deals with the traditional way of UNIX printing. The next chapter covers in great detail the more modern *Common UNIX Printing System* (CUPS).

IMPORTANT

CUPS users, be warned: do not just jump on to the next chapter. You might miss important information only found here!

It is apparent from postings on the Samba mailing list that print configuration is one of the most problematic aspects of Samba administration today. Many new Samba administrators have the impression that Samba performs some sort of print processing. Rest assured, Samba does not peform any type of print processing. It does not do any form of print filtering.

Samba obtains from its clients a data stream (print job) that it spools to a local spool area. When the entire print job has been received, Samba invokes a local UNIX/Linux print command and passes the spooled file to it. It is up to the local system printing subsystems to correctly process the print job and to submit it to the printer.

17.2.1 Client to Samba Print Job Processing

Successful printing from a Windows client via a Samba print server to a UNIX printer involves six (potentially seven) stages:

1. Windows opens a connection to the printer share.
2. Samba must authenticate the user.

3. Windows sends a copy of the print file over the network into Samba's spooling area.
4. Windows closes the connection.
5. Samba invokes the print command to hand the file over to the UNIX print subsystem's spooling area.
6. The UNIX print subsystem processes the print job.
7. The print file may need to be explicitly deleted from the Samba spooling area. This item depends on your print spooler configuration settings.

17.2.2 Printing Related Configuration Parameters

There are a number of configuration parameters to control Samba's printing behavior. Please refer to the man page for `smb.conf` for an overview of these. As with other parameters, there are Global Level (tagged with a *G* in the listings) and Service Level (*S*) parameters.

Global Parameters — These *may not* go into individual share definitions. If they go in by error, the **testparm** utility can discover this (if you run it) and tell you so.

Service Level Parameters — These may be specified in the *[global]* section of `smb.conf`. In this case they define the default behavior of all individual or service level shares (provided they do not have a different setting defined for the same parameter, thus overriding the global default).

17.3 Simple Print Configuration

Example 17.1 shows a simple printing configuration. If you compare this with your own, you may find additional parameters that have been pre-configured by your OS vendor. Below is a discussion and explanation of the parameters. This example does not use many parameters. However, in many environments these are enough to provide a valid `smb.conf` file that enables all clients to print.

Example 17.1. Simple configuration with BSD printing

```
[global]
printing = bsd
load printers = yes

[printers]
path = /var/spool/samba
printable = yes
public = yes
writable = no
```

This is only an example configuration. Samba assigns default values to all configuration parameters. The defaults are conservative and sensible. When a parameter is specified in the `smb.conf` file, this overwrites the default value. The **testparm** utility when run as root is capable of reporting all setting, both default as well as `smb.conf` file settings. **Testparm** gives warnings for all misconfigured settings. The complete output is easily 340 lines and more, so you may want to pipe it through a pager program.

The syntax for the configuration file is easy to grasp. You should know that is not very picky about its syntax. As has been explained elsewhere in this document, Samba tolerates some spelling errors (such as *browsable* instead of *browseable*), and spelling is case-insensitive. It is permissible to use *Yes/No* or *True/False* for Boolean settings. Lists of names may be separated by commas, spaces or tabs.

17.3.1 Verifing Configuration with testparm

To see all (or at least most) printing-related settings in Samba, including the implicitly used ones, try the command outlined below. This command greps for all occurrences of `lp`, `print`, `spool`, `driver`, `ports` and `[` in testparms output. This provides a convenient overview of the running **smbd** print configuration. This command does not show individually created printer shares or the spooling paths they may use. Here is the output of my Samba setup, with settings shown in Example 17.1:

```
root# testparm -s -v | egrep "(lp|print|spool|driver|ports|\[)"
 Load smb config files from /etc/samba/smb.conf
 Processing section "[homes]"
 Processing section "[printers]"

 [global]
        smb ports = 445 139
        lpq cache time = 10
        total print jobs = 0
        load printers = Yes
        printcap name = /etc/printcap
        disable spoolss = No
        enumports command =
        addprinter command =
        deleteprinter command =
        show add printer wizard = Yes
        os2 driver map =
        printer admin =
        min print space = 0
        max print jobs = 1000
        printable = No
        printing = bsd
        print command = lpr -r -P'%p' %s
        lpq command = lpq -P'%p'
        lprm command = lprm -P'%p' %j
```

```
        lppause command =
        lpresume command =
        printer name =
        use client driver = No

[homes]

[printers]
        path = /var/spool/samba
        printable = Yes
```

You can easily verify which settings were implicitly added by Samba's default behavior. *Remember: it may be important in your future dealings with Samba.*

NOTE

testparm in Samba-3 behaves differently from that in 2.2.x: used without the "-v" switch it only shows you the settings actually written into! To see the complete configuration used, add the "-v" parameter to testparm.

17.3.2 Rapid Configuration Validation

Should you need to troubleshoot at any stage, please always come back to this point first and verify if **testparm** shows the parameters you expect. To give you a warning from personal experience, try to just comment out the *load printers* parameter. If your 2.2.x system behaves like mine, you'll see this:

```
root# grep "load printers" /etc/samba/smb.conf
        #  load printers = Yes
        # This setting is commented out!!

root# testparm -v /etc/samba/smb.conf | egrep "(load printers)"
        load printers = Yes
```

I assumed that commenting out of this setting should prevent Samba from publishing my printers, but it still did. It took some time to figure out the reason. But I am no longer fooled ... at least not by this.

```
root# grep -A1 "load printers" /etc/samba/smb.conf
        load printers = No
        # The above setting is what I want!
        #  load printers = Yes
```

```
        # This setting is commented out!

root# testparm -s -v smb.conf.simpleprinting | egrep "(load printers)"
        load printers = No
```

Only when the parameter is explicitly set to *load printers* = No would Samba conform with my intentions. So, my strong advice is:

- Never rely on commented out parameters.
- Always set parameters explicitly as you intend them to behave.
- Use **testparm** to uncover hidden settings that might not reflect your intentions.

The following is the most minimal configuration file:

```
root# cat /etc/samba/smb.conf-minimal
        [printers]
```

This example should show that you can use testparm to test any Samba configuration file. Actually, we encourage you *not* to change your working system (unless you know exactly what you are doing). Don't rely on the assumption that changes will only take effect after you re-start smbd! This is not the case. Samba re-reads it every 60 seconds and on each new client connection. You might have to face changes for your production clients that you didn't intend to apply. You will now note a few more interesting things; **testparm** is useful to identify what the Samba print configuration would be if you used this minimalistic configuration. Here is what you can expect to find:

```
root# testparm -v smb.conf-minimal | egrep "(print|lpq|spool|driver|ports|[)"
 Processing section "[printers]"
 WARNING: [printers] service MUST be printable!
 No path in service printers - using /tmp

        lpq cache time = 10
        total print jobs = 0
        load printers = Yes
        printcap name = /etc/printcap
        disable spoolss = No
        enumports command =
        addprinter command =
        deleteprinter command =
        show add printer wizard = Yes
        os2 driver map =
        printer admin =
        min print space = 0
        max print jobs = 1000
        printable = No
```

```
        printing = bsd
        print command = lpr -r -P%p %s
        lpq command = lpq -P%p
        printer name =
        use client driver = No

 [printers]
        printable = Yes
```

testparm issued two warnings:

- We did not specify the *[printers]* section as printable.
- We did not tell Samba which spool directory to use.

However, this was not fatal and Samba will default to values that will work. Please, do not rely on this and do not use this example. This was included to encourage you to be careful to design and specify your setup to do precisely what you require. The outcome on your system may vary for some parameters given, since Samba may have been built with different compile-time options. *Warning:* do not put a comment sign *at the end* of a valid line. It will cause the parameter to be ignored (just as if you had put the comment sign at the front). At first I regarded this as a bug in my Samba versions. But the man page clearly says: "*Internal whitespace in a parameter value is retained verbatim.*" This means that a line consisting of, for example:

printing = lprng

will regard the whole of the string after the "=" sign as the value you want to define. This is an invalid value that will be ignored and a default value will be used in its place.

17.4 Extended Printing Configuration

In Example 17.2 we show a more verbose example configuration for print-related settings in a BSD-style printing environment. What follows is a discussion and explanation of the various parameters. We chose to use BSD-style printing here because it is still the most commonly used system on legacy UNIX/Linux installations. New installations predominantly use CUPS, which is discussed in a separate chapter. Example 17.2 explicitly names many parameters that do not need to be specified because they are set by default. You could use a much leaner `smb.conf` file. Alternately, you can use **testparm** or **SWAT** to optimize the `smb.conf` file to remove all parameters that are set at default.

This is an example configuration. You may not find all the settings that are in the confioguration file that was provided by the OS vendor. Samba configuration parameters, if not explicitly set default to a sensible value. To see all settings, as `root` use the **testparm** utility. **testparm** gives warnings for misconfigured settings.

Example 17.2. Extended BSD Printing Configuration

```
[global]
printing = bsd
load printers = yes
show add printer wizard = yes
printcap name = /etc/printcap
printer admin = @ntadmin, root
total print jobs = 100
lpq cache time = 20
use client driver = no

[printers]
comment = All Printers
printable = yes
path = /var/spool/samba
browseable = no
guest ok = yes
public = yes
read only = yes
writable = no

[my_printer_name]
comment = Printer with Restricted Access
path = /var/spool/samba_my_printer
printer admin = kurt
browseable = yes
printable = yes
writeable = no
hosts allow = 0.0.0.0
hosts deny = turbo_xp, 10.160.50.23, 10.160.51.60
guest ok = no
```

17.4.1 Detailed Explanation Settings

The following is a discussion of the settings from above shown example.

17.4.1.1 The [global] Section

The *[global]* section is one of four special sections (along with [*[homes]*, *[printers]* and *[print$]*...). The *[global]* contains all parameters which apply to the server as a whole. It is the place for parameters that have only a global meaning. It may also contain service level parameters that then define default settings for all other sections and shares. This way you can simplify the configuration and avoid setting the same value repeatedly. (Within each individual section or share you may, however, override these globally set share settings and specify other values).

printing = **bsd** — Causes Samba to use default print commands applicable for the BSD (also known as RFC 1179 style or LPR/LPD) printing system. In general, the *printing* parameter informs Samba about the print subsystem it should expect. Samba supports CUPS, LPD, LPRNG, SYSV, HPUX, AIX, QNX, and PLP. Each of these systems defaults to a different *print command* (and other queue control commands).

CAUTION

The *printing* parameter is normally a service level parameter. Since it is included here in the *[global]* section, it will take effect for all printer shares that are not defined differently. Samba-3 no longer supports the SOFTQ printing system.

load printers = **yes** — Tells Samba to create automatically all available printer shares. Available printer shares are discovered by scanning the printcap file. All created printer shares are also loaded for browsing. If you use this parameter, you do not need to specify separate shares for each printer. Each automatically created printer share will clone the configuration options found in the *[printers]* section. (The *load printers = no* setting will allow you to specify each UNIX printer you want to share separately, leaving out some you do not want to be publicly visible and available).

show add printer wizard = **yes** — Setting is normally enabled by default (even if the parameter is not specified in `smb.conf`). It causes the **Add Printer Wizard** icon to appear in the **Printers** folder of the Samba host's share listing (as shown in **Network Neighborhood** or by the **net view** command). To disable it, you need to explicitly set it to `no` (commenting it out will not suffice). The *Add Printer Wizard* lets you upload printer drivers to the *[print$]* share and associate it with a printer (if the respective queue exists before the action), or exchange a printer's driver against any other previously uploaded driver.

total print jobs = **100** — Sets the upper limit to 100 print jobs being active on the Samba server at any one time. Should a client submit a job that exceeds this number, a "*no more space available on server*" type of error message will be returned by Samba to the client. A setting of zero (the default) means there is *no* limit at all.

printcap name = **/etc/printcap** — Tells Samba where to look for a list of available printer names. Where CUPS is used, make sure that a printcap file is written. This is controlled by the `Printcap` directive in the `cupsd.conf` file.

printer admin = **@ntadmin** — Members of the ntadmin group should be able to add drivers and set printer properties (`ntadmin` is only an example name, it needs to be a valid UNIX group name); root is implicitly always a *printer admin*. The @ sign precedes group names in the `/etc/group`. A printer admin can do anything to printers via the remote administration interfaces offered by MS-RPC (see below). In larger installations, the *printer admin* parameter is normally a per-share parameter. This permits different groups to administer each printer share.

lpq cache time = **20** — Controls the cache time for the results of the lpq command. It prevents the lpq command being called too often and reduces the load on a heavily used print server.

use client driver = **no** — If set to `yes`, only takes effect for Windows NT/200x/XP clients (and not for Win 95/98/ME). Its default value is `No` (or `False`). It must *not* be enabled on print shares (with a `yes` or `true` setting) that have valid drivers installed on the Samba server. For more detailed explanations see the `smb.conf` man page.

17.4.1.2 The [printers] Section

This is the second special section. If a section with this name appears in the `smb.conf`, users are able to connect to any printer specified in the Samba host's printcap file, because Samba on startup then creates a printer share for every printername it finds in the printcap file. You could regard this section as a general convenience shortcut to share all printers with minimal configuration. It is also a container for settings that should apply as default to all printers. (For more details see the `smb.conf` man page.) Settings inside this container must be Share Level parameters.

comment = **All printers** — The *comment* is shown next to the share if a client queries the server, either via **Network Neighborhood** or with the **net view** command to list available shares.

printable = **yes** — The *[printers]* service *must* be declared as printable. If you specify otherwise, smbd will refuse to load at startup. This parameter allows connected clients to open, write to and submit spool files into the directory specified with the *path* parameter for this service. It is used by Samba to differentiate printer shares from file shares.

path = **/var/spool/samba** — Must point to a directory used by Samba to spool incoming print files. *It must not be the same as the spool directory specified in the configuration of your UNIX print subsystem!* The path typically points to a directory that is world writeable, with the "*sticky*" bit set to it.

browseable = **no** — Is always set to `no` if *printable* = yes. It makes the *[printer]* share itself invisible in the list of available shares in a **net view** command or in the Explorer browse list. (You will of course see the individual printers).

guest ok = **yes** — If this parameter is set to **yes**, no password is required to connect to the printer's service. Access will be granted with the privileges of the ***guest account***. On many systems the guest account will map to a user named "*nobody*". This user will usually be found in the UNIX passwd file with an empty password, but with no valid UNIX login. (On some systems the guest account might not have the privilege to be able to print. Test this by logging in as your guest user using **su - guest** and run a system print command like:

```
lpr -P printername /etc/motd
```

public = **yes** — Is a synonym for ***guest ok*** = yes. Since we have ***guest ok*** = yes, it really does not need to be here. (This leads to the interesting question: "*What if I by accident have two contradictory settings for the same share?*" The answer is the last one encountered by Samba wins. Testparm does not complain about different settings of the same parameter for the same share. You can test this by setting up multiple lines for the ***guest account*** parameter with different usernames, and then run testparm to see which one is actually used by Samba.)

read only = **yes** — Normally (for other types of shares) prevents users from creating or modifying files in the service's directory. However, in a "*printable*" service, it is *always* allowed to write to the directory (if user privileges allow the connection), but only via print spooling operations. Normal write operations are not permitted.

writeable = **no** — Is a synonym for ***read only*** = yes.

17.4.1.3 Any [my_printer_name] Section

If a section appears in the `smb.conf` file, which when given the parameter ***printable*** = yes causes Samba to configure it as a printer share. Windows 9x/Me clients may have problems with connecting or loading printer drivers if the share name has more than eight characters. Do not name a printer share with a name that may conflict with an existing user or file share name. On Client connection requests, Samba always tries to find file shares with that name first. If it finds one, it will connect to this and will not connect to a printer with the same name!

comment = **Printer with Restricted Access** — The comment says it all.

path = **/var/spool/samba_my_printer** — Sets the spooling area for this printer to a directory other than the default. It is not necessary to set it differently, but the option is available.

printer admin = **kurt** — The printer admin definition is different for this explicitly defined printer share from the general *[printers]* share. It is not a requirement; we did it to show that it is possible.

browseable = **yes** — This makes the printer browseable so the clients may conveniently find it when browsing the **Network Neighborhood**.

printable = **yes** — See Section 17.4.1.2.

writeable = **no** — See Section 17.4.1.2.

hosts allow = **10.160.50.,10.160.51.** — Here we exercise a certain degree of access control by using the *hosts allow* and *hosts deny* parameters. This is not by any means a safe bet. It is not a way to secure your printers. This line accepts all clients from a certain subnet in a first evaluation of access control.

hosts deny = **turbo_xp,10.160.50.23,10.160.51.60** — All listed hosts are not allowed here (even if they belong to the allowed subnets). As you can see, you could name IP addresses as well as NetBIOS hostnames here.

guest ok = **no** — This printer is not open for the guest account.

17.4.1.4 Print Commands

In each section defining a printer (or in the *[printers]* section), a *print command* parameter may be defined. It sets a command to process the files that have been placed into the Samba print spool directory for that printer. (That spool directory was, if you remember, set up with the *path* parameter). Typically, this command will submit the spool file to the Samba host's print subsystem, using the suitable system print command. But there is no requirement that this needs to be the case. For debugging or some other reason, you may want to do something completely different than print the file. An example is a command that just copies the print file to a temporary location for further investigation when you need to debug printing. If you craft your own print commands (or even develop print command shell scripts), make sure you pay attention to the need to remove the files from the Samba spool directory. Otherwise, your hard disk may soon suffer from shortage of free space.

17.4.1.5 Default UNIX System Printing Commands

You learned earlier on that Samba, in most cases, uses its built-in settings for many parameters if it cannot find an explicitly stated one in its configuration file. The same is true for the *print command*. The default print command varies depending on the *printing* parameter setting. In the commands listed below, you will notice some parameters of the form *%X* where *X* is *p, s, J*, and so on. These letters stand for printer name, spoolfile and job ID, respectively. They are explained in more detail further below. Table 17.1 presents an overview of key printing options but excludes the special case of CUPS that is discussed in Chapter 18, *CUPS Printing Support*.

Table 17.1. Default Printing Settings

Setting	Default Printing Commands
printing = bsd|aix|lprng|plp	print command is **lpr -r -P%p %s**
printing = sysv|hpux	print command is **lp -c -P%p %s; rm %s**
printing = qnx	print command is **lp -r -P%p -s %s**
printing = bsd|aix|lprng|plp	lpq command is **lpq -P%p**
printing = sysv|hpux	lpq command is **lpstat -o%p**
printing = qnx	lpq command is **lpq -P%p**
printing = bsd|aix|lprng|plp	lprm command is **lprm -P%p %j**
printing = sysv|hpux	lprm command is **cancel %p-%j**
printing = qnx	lprm command is **cancel %p-%j**
printing = bsd|aix|lprng|plp	lppause command is **lp -i %p-%j -H hold**
printing = sysv|hpux	lppause command (...is empty)
printing = qnx	lppause command (...is empty)
printing = bsd|aix|lprng|plp	lpresume command is **lp -i %p-%j -H resume**
printing = sysv|hpux	lpresume command (...is empty)
printing = qnx	lpresume command (...is empty)

We excluded the special case of CUPS here, because it is discussed in the next chapter. For *printing = CUPS*, if Samba is compiled against libcups, it uses the CUPS API to submit jobs. (It is a good idea also to set *printcap* = cups in case your `cupsd.conf` is set to write its autogenerated printcap file to an unusual place). Otherwise, Samba maps to the System V printing commands with the -oraw option for printing, i.e., it uses **lp -c -d%p -oraw; rm %s**. With *printing = cups*, and if Samba is compiled against libcups, any manually set print command will be ignored!

17.4.1.6 Custom Print Commands

After a print job has finished spooling to a service, the *print command* will be used by Samba via a *system()* call to process the spool file. Usually the command specified will submit the spool file to the host's printing subsystem. But there is no requirement at all that this must be the case. The print subsystem may not remove the spool file on its own. So whatever command you specify, you should ensure that the spool file is deleted after it has been processed.

There is no difficulty with using your own customized print commands with the traditional printing systems. However, if you do not wish to roll your own, you should be well informed about the default built-in commands that Samba uses for each printing subsystem (see Table 17.1). In all the commands listed in the last paragraphs, you see parameters of the form *%X*. These are *macros*, or shortcuts, used as placeholders for the names of real objects. At the time of running a command with such a placeholder, Samba will insert the appropriate value automatically. Print commands can handle all Samba macro substitutions. In regard to printing, the following ones do have special relevance:

- *%s, %f* — the path to the spool file name.
- *%p* — the appropriate printer name.

- *%J* — the job name as transmitted by the client.
- *%c* — the number of printed pages of the spooled job (if known).
- *%z* — the size of the spooled print job (in bytes).

The print command must contain at least one occurrence of *%s* or the *%f*. The *%p* is optional. If no printer name is supplied, the *%p* will be silently removed from the print command. In this case, the job is sent to the default printer.

If specified in the *[global]* section, the print command given will be used for any printable service that does not have its own print command specified. If there is neither a specified print command for a printable service nor a global print command, spool files will be created but not processed! Most importantly, print files will not be removed, so they will consume disk space.

Printing may fail on some UNIX systems when using the "*nobody*" account. If this happens, create an alternative guest account and give it the privilege to print. Set up this guest account in the *[global]* section with the ***guest account*** parameter.

You can form quite complex print commands. You need to realize that print commands are just passed to a UNIX shell. The shell is able to expand the included environment variables as usual. (The syntax to include a UNIX environment variable *$variable* in the Samba print command is *%$variable*.) To give you a working ***print command*** example, the following will log a print job to `/tmp/print.log`, print the file, then remove it. The semicolon (";" is the usual separator for commands in shell scripts:

```
print command = echo Printing %s >> /tmp/print.log; lpr -P %p %s; rm %s
```

You may have to vary your own command considerably from this example depending on how you normally print files on your system. The default for the ***print command*** parameter varies depending on the setting of the ***printing*** parameter. Another example is:

```
print command = /usr/local/samba/bin/myprintscript %p %s
```

17.5 Printing Developments Since Samba-2.2

Prior to Samba-2.2.x, print server support for Windows clients was limited to *LanMan* printing calls. This is the same protocol level as Windows 9x/Me PCs offer when they share printers. Beginning with the 2.2.0 release, Samba started to support the native Windows NT printing mechanisms. These are implemented via *MS-RPC* (RPC = *Remote Procedure Calls*). MS-RPCs use the *SPOOLSS* named pipe for all printing.

The additional functionality provided by the new SPOOLSS support includes:

- Support for downloading printer driver files to Windows 95/98/NT/2000 clients upon demand (*Point'n'Print*).
- Uploading of printer drivers via the Windows NT *Add Printer Wizard* (APW) or the Imprints[1] tool set.

[1] http://imprints.sourceforge.net/

- Support for the native MS-RPC printing calls such as StartDocPrinter, EnumJobs(), and so on. (See the MSDN documentation[2] for more information on the Win32 printing API).
- Support for NT *Access Control Lists* (ACL) on printer objects.
- Improved support for printer queue manipulation through the use of internal databases for spooled job information (implemented by various `*.tdb` files).

A benefit of updating is that Samba-3 is able to publish its printers to Active Directory (or LDAP).

A fundamental difference exists between MS Windows NT print servers and Samba operation. Windows NT permits the installation of local printers that are not shared. This is an artifact of the fact that any Windows NT machine (server or client) may be used by a user as a workstation. Samba will publish all printers that are made available, either by default or by specific declaration via printer-specific shares.

Windows NT/200x/XP Professional clients do not have to use the standard SMB printer share; they can print directly to any printer on another Windows NT host using MS-RPC. This, of course, assumes that the client has the necessary privileges on the remote host that serves the printer resource. The default permissions assigned by Windows NT to a printer gives the Print permissions to the well-known *Everyone* group. (The older clients of type Windows 9x/Me can only print to shared printers).

17.5.1 Point'n'Print Client Drivers on Samba Servers

There is much confusion about what all this means. The question is often asked, "*Is it or is it not necessary for printer drivers to be installed on a Samba host in order to support printing from Windows clients?*" The answer to this is no, it is not necessary.

Windows NT/2000 clients can, of course, also run their APW to install drivers *locally* (which then connect to a Samba-served print queue). This is the same method used by Windows 9x/Me clients. (However, a *bug* existed in Samba 2.2.0 that made Windows NT/2000 clients require that the Samba server possess a valid driver for the printer. This was fixed in Samba 2.2.1).

But it is a new capability to install the printer drivers into the *[print$]* share of the Samba server, and a big convenience, too. Then *all* clients (including 95/98/ME) get the driver installed when they first connect to this printer share. The *uploading* or *depositing* of the driver into this *[print$]* share and the following binding of this driver to an existing Samba printer share can be achieved by different means:

- Running the *APW* on an NT/200x/XP Professional client (this does not work from 95/98/ME clients).
- Using the *Imprints* toolset.
- Using the *smbclient* and *rpcclient* commandline tools.
- Using *cupsaddsmb* (only works for the CUPS printing system, not for LPR/LPD, LPRng, and so on).

[2]http://msdn.microsoft.com/

Samba does not use these uploaded drivers in any way to process spooled files. These drivers are utilized entirely by the clients who download and install them via the *"Point'n'Print"* mechanism supported by Samba. The clients use these drivers to generate print files in the format the printer (or the UNIX print system) requires. Print files received by Samba are handed over to the UNIX printing system, which is responsible for all further processing, as needed.

17.5.2 The Obsoleted [printer$] Section

Versions of Samba prior to 2.2 made it possible to use a share named *[printer$]*. This name was taken from the same named service created by Windows 9x/Me clients when a printer was shared by them. Windows 9x/Me printer servers always have a *[printer$]* service that provides read-only access (with no password required) to support printer driver downloads. However, Samba's initial implementation allowed for a parameter named ***printer driver location*** to be used on a per share basis. This specified the location of the driver files associated with that printer. Another parameter named ***printer driver*** provided a means of defining the printer driver name to be sent to the client.

These parameters, including the ***printer driver file*** parameter, are now removed and cannot be used in installations of Samba-3. The share name *[print$]* is now used for the location of downloadable printer drivers. It is taken from the *[print$]* service created by Windows NT PCs when a printer is shared by them. Windows NT print servers always have a *[print$]* service that provides read-write access (in the context of its ACLs) to support printer driver downloads and uploads. This does not mean Windows 9x/Me clients are now thrown aside. They can use Samba's *[print$]* share support just fine.

17.5.3 Creating the [print$] Share

In order to support the uploading and downloading of printer driver files, you must first configure a file share named *[print$]*. The public name of this share is hard coded in the MS Windows clients. It cannot be renamed since Windows clients are programmed to search for a service of exactly this name if they want to retrieve printer driver files.

You should modify the server's file to add the global parameters and create the *[print$]* file share (of course, some of the parameter values, such as ***path*** are arbitrary and should be replaced with appropriate values for your site). See Example 17.3.

Of course, you also need to ensure that the directory named by the ***path*** parameter exists on the UNIX file system.

17.5.4 [print$] Section Parameters

The *[print$]* is a special section in `smb.conf`. It contains settings relevant to potential printer driver download and is used by windows clients for local print driver installation. The following parameters are frequently needed in this share section:

Example 17.3. [print$] example

```
[global]
# members of the ntadmin group should be able to add drivers and set
# printer properties. root is implicitly always a 'printer admin'.
printer admin = @ntadmin
...

[printers]
...

[print$]
comment = Printer Driver Download Area
path = /etc/samba/drivers
browseable = yes
guest ok = yes
read only = yes
write list = @ntadmin, root
```

comment = **Printer Driver Download Area** — The comment appears next to the share name if it is listed in a share list (usually Windows clients will not see it, but it will also appear up in a **smbclient -L sambaserver** output).

path = **/etc/samba/printers** — Is the path to the location of the Windows driver file deposit from the UNIX point of view.

browseable = **no** — Makes the *[print$]* share invisible to clients from the **Network Neighborhood**. However, you can still mount it from any client using the **net use g:\\sambaserver\print$** command in a DOS-box or the **Connect network drive menu**> from Windows Explorer.

guest ok = **yes** — Gives read-only access to this share for all guest users. Access may be granted to download and install printer drivers on clients. The requirement for *guest ok = yes* depends on how your site is configured. If users will be guaranteed to have an account on the Samba host, then this is a non-issue.

NOTE

If all your Windows NT users are guaranteed to be authenticated by the Samba server (for example, if Samba authenticates via an NT domain server and the user has already been validated by the Domain Controller in order to logon to the Windows NT session), then guest access is not necessary. Of course, in a workgroup environment where you just want to print without worrying about silly accounts and security, then configure the share for guest access. You should consider adding *map to guest* = Bad User in the *[global]* section as well. Make sure you understand what this parameter does before using it.

read only = **yes** — Because we do not want everybody to upload driver files (or even change driver settings), we tagged this share as not writeable.

write list = **@ntadmin, root** — The *[print$]* was made read-only by the previous setting so we should create a *write list* entry also. UNIX groups (denoted with a leading "@" character). Users listed here are allowed write-access (as an exception to the general public's read-only access), which they need to update files on the share. Normally, you will want to only name administrative-level user account in this setting. Check the file system permissions to make sure these accounts can copy files to the share. If this is a non-root account, then the account should also be mentioned in the global *printer admin* parameter. See the `smb.conf` man page for more information on configuring file shares.

17.5.5 The [print$] Share Directory

In order for a Windows NT print server to support the downloading of driver files by multiple client architectures, you must create several subdirectories within the *[print$]* service (i.e., the UNIX directory named by the *path* parameter). These correspond to each of the supported client architectures. Samba follows this model as well. Just like the name of the *[print$]* share itself, the subdirectories must be exactly the names listed below (you may leave out the subdirectories of architectures you do not need to support).

Therefore, create a directory tree below the *[print$]* share for each architecture you wish to support like this:

```
[print$]--+
          |--W32X86           # serves drivers to Windows NT x86
          |--WIN40            # serves drivers to Windows 95/98
          |--W32ALPHA         # serves drivers to Windows NT Alpha_AXP
          |--W32MIPS          # serves drivers to Windows NT R4000
          |--W32PPC           # serves drivers to Windows NT PowerPC
```

REQUIRED PERMISSIONS

In order to add a new driver to your Samba host, one of two conditions must hold true:

- The account used to connect to the Samba host must have a UID of 0 (i.e., a root account).
- The account used to connect to the Samba host must be named in the *printer admin*list.

Of course, the connected account must still have write access to add files to the subdirectories beneath *[print$]* . Remember that all file shares are set to *"read-only"* by default.

Once you have created the required *[print$]* service and associated subdirectories, go to a Windows NT 4.0/200x/XP client workstation. Open **Network Neighborhood** or **My Network Places** and browse for the Samba host. Once you have located the server, navigate to its **Printers and Faxes** folder. You should see an initial listing of printers that matches the printer shares defined on your Samba host.

17.6 Installing Drivers into [print$]

Have you successfully created the *[print$]* share in `smb.conf`, and have your forced Samba to re-read its `smb.conf` file? Good. But you are not yet ready to use the new facility. The client driver files need to be installed into this share. So far it is still an empty share. Unfortunately, it is not enough to just copy the driver files over. They need to be correctly installed so that appropriate records for each driver will exist in the Samba internal databases so it can provide the correct drivers as they are requested from MS Windows clients. And that is a bit tricky, to say the least. We now discuss two alternative ways to install the drivers into *[print$]*:

- Using the Samba commandline utility **rpcclient** with its various subcommands (here: **adddriver** and **setdriver**) from any UNIX workstation.
- Running a GUI (**Printer Properties** and **Add Printer Wizard**) from any Windows NT/200x/XP client workstation.

The latter option is probably the easier one (even if the process may seem a little bit weird at first).

17.6.1 Add Printer Wizard Driver Installation

The initial listing of printers in the Samba host's **Printers** folder accessed from a client's Explorer will have no real printer driver assigned to them. By default this driver name is set to a null string. This must be changed now. The local **Add Printer Wizard** (APW), run from NT/2000/XP clients, will help us in this task.

Installation of a valid printer driver is not straightforward. You must attempt to view the printer properties for the printer to which you want the driver assigned. Open the Windows Explorer, open **Network Neighborhood**, browse to the Samba host, open Samba's **Printers** folder, right-click on the printer icon and select **Properties...**. You are now trying to view printer and driver properties for a queue that has this default `NULL` driver assigned. This will result in the following error message:

Device settings cannot be displayed. The driver for the specified printer is not installed, only spooler properties will be displayed. Do you want to install the driver now?

Do not click on **Yes**! Instead, click on **No** in the error dialog. Only now you will be presented with the printer properties window. From here, the way to assign a driver to a printer is open to us. You now have the choice of:

- Select a driver from the pop-up list of installed drivers. Initially this list will be empty.
- Click on **New Driver** to install a new printer driver (which will start up the APW).

Once the APW is started, the procedure is exactly the same as the one you are familiar with in Windows (we assume here that you are familiar with the printer driver installations procedure on Windows NT). Make sure your connection is, in fact, setup as a user with *printer admin* privileges (if in doubt, use **smbstatus** to check for this). If you wish to install printer drivers for client operating systems other than Windows NT x86, you will need to use the **Sharing** tab of the printer properties dialog.

Assuming you have connected with an administrative (or root) account (as named by the *printer admin* parameter), you will also be able to modify other printer properties such as ACLs and default device settings using this dialog. For the default device settings, please consider the advice given further in Section 17.6.2.

17.6.2 Installing Print Drivers Using rpcclient

The second way to install printer drivers into *[print$]* and set them up in a valid way is to do it from the UNIX command line. This involves four distinct steps:

1. Gather info about required driver files and collect the files.
2. Deposit the driver files into the *[print$]* share's correct subdirectories (possibly by using **smbclient**).
3. Run the **rpcclient** command line utility once with the **adddriver** subcommand.
4. Run **rpcclient** a second time with the **setdriver** subcommand.

We provide detailed hints for each of these steps in the paragraphs that follow.

17.6.2.1 Identifying Driver Files

To find out about the driver files, you have two options. You could check the contents of the driver CDROM that came with your printer. Study the `*.inf` files lcoated on the CDROM. This may not be possible, since the `*.inf` file might be missing. Unfortunately, vendors have now started to use their own installation programs. These installations packages are often in some Windows platform archive format. Additionally, the files may be re-named during the installation process. This makes it extremely difficult to identify the driver files required.

Then you only have the second option. Install the driver locally on a Windows client and investigate which file names and paths it uses after they are installed. (You need to repeat this procedure for every client platform you want to support. We show it here for the W32X86 platform only, a name used by Microsoft for all Windows NT/200x/XP clients.)

A good method to recognize the driver files is to print the test page from the driver's **Properties** dialog (**General** tab). Then look at the list of driver files named on the printout. You'll need to recognize what Windows (and Samba) are calling the **Driver File**, **Data File**, **Config File**, **Help File** and (optionally) the **Dependent Driver Files** (this may vary slightly for Windows NT). You need to take a note of all file names for the next steps.

Another method to quickly test the driver filenames and related paths is provided by the **rpcclient** utility. Run it with **enumdrivers** or with the **getdriver** subcommand, each at the 3 info level. In the following example, *TURBO_XP* is the name of the Windows PC (in this case it was a Windows XP Professional laptop). I installed the driver locally to TURBO_XP, from a Samba server called `KDE-BITSHOP`. We could run an interactive **rpcclient** session; then we would get an **rpcclient />** prompt and would type the subcommands at this prompt. This is left as a good exercise to the reader. For now, we use **rpcclient** with the `-c` parameter to execute a single subcommand line and exit again. This is the method you would use if you want to create scripts to automate the procedure for a large number of printers and drivers. Note the different quotes used to overcome the different spaces in between words:

```
root# rpcclient -U'Danka%xxxx' -c \
   'getdriver "Heidelberg Digimaster 9110 (PS)" 3' TURBO_XP
cmd = getdriver "Heidelberg Digimaster 9110 (PS)" 3

[Windows NT x86]
Printer Driver Info 3:
  Version: [2]
  Driver Name: [Heidelberg Digimaster 9110 (PS)]
  Architecture: [Windows NT x86]
  Driver Path: [C:\WINNT\System32\spool\DRIVERS\W32X86\2\HDNIS01_de.DLL]
  Datafile: [C:\WINNT\System32\spool\DRIVERS\W32X86\2\Hddm91c1_de.ppd]
  Configfile: [C:\WINNT\System32\spool\DRIVERS\W32X86\2\HDNIS01U_de.DLL]
  Helpfile: [C:\WINNT\System32\spool\DRIVERS\W32X86\2\HDNIS01U_de.HLP]

  Dependentfiles: [C:\WINNT\System32\spool\DRIVERS\W32X86\2\Hddm91c1_de.DLL]
  Dependentfiles: [C:\WINNT\System32\spool\DRIVERS\W32X86\2\Hddm91c1_de.INI]
  Dependentfiles: [C:\WINNT\System32\spool\DRIVERS\W32X86\2\Hddm91c1_de.dat]
  Dependentfiles: [C:\WINNT\System32\spool\DRIVERS\W32X86\2\Hddm91c1_de.cat]
  Dependentfiles: [C:\WINNT\System32\spool\DRIVERS\W32X86\2\Hddm91c1_de.def]
  Dependentfiles: [C:\WINNT\System32\spool\DRIVERS\W32X86\2\Hddm91c1_de.hre]
  Dependentfiles: [C:\WINNT\System32\spool\DRIVERS\W32X86\2\Hddm91c1_de.vnd]
  Dependentfiles: [C:\WINNT\System32\spool\DRIVERS\W32X86\2\Hddm91c1_de.hlp]
  Dependentfiles: [C:\WINNT\System32\spool\DRIVERS\W32X86\2\HDNIS01Aux.dll]
```

```
Dependentfiles: [C:\WINNT\System32\spool\DRIVERS\W32X86\2\HDNIS01_de.NTF]

Monitorname: []
Defaultdatatype: []
```

You may notice that this driver has quite a large number of **Dependent files** (there are worse cases, however). Also, strangely, the **Driver File** is tagged here **Driver Path**. We do not yet have support for the so-called WIN40 architecture installed. This name is used by Microsoft for the Windows 9x/Me platforms. If we want to support these, we need to install the Windows 9x/Me driver files in addition to those for W32X86 (i.e., the Windows NT72000/XP clients) onto a Windows PC. This PC can also host the Windows 9x/Me drivers, even if it runs on Windows NT, 2000 or XP.

Since the *[print$]* share is usually accessible through the **Network Neighborhood**, you can also use the UNC notation from Windows Explorer to poke at it. The Windows 9x/Me driver files will end up in subdirectory `0` of the `WIN40` directory. The full path to access them will be `\\WINDOWSHOST\print$\WIN40\0\`.

NOTE

More recent drivers on Windows 2000 and Windows XP are installed into the "*3*" subdirectory instead of the "*2*". The version 2 of drivers, as used in Windows NT, were running in Kernel Mode. Windows 2000 changed this. While it still can use the Kernel Mode drivers (if this is enabled by the Admin), its native mode for printer drivers is User Mode execution. This requires drivers designed for this. These types of drivers install into the "*3*" subdirectory.

17.6.2.2 Obtaining Driver Files from Windows Client [print$] Shares

Now we need to collect all the driver files we identified in our previous step. Where do we get them from? Well, why not retrieve them from the very PC and the same *[print$]* share that we investigated in our last step to identify the files? We can use **smbclient** to do this. We will use the paths and names that were leaked to us by **getdriver**. The listing is edited to include linebreaks for readability:

```
root# smbclient //TURBO_XP/print\$ -U'Danka%xxxx' \
   -c 'cd W32X86/2;mget HD*_de.* hd*ppd Hd*_de.* Hddm*dll HDN*Aux.DLL'

added interface ip=10.160.51.60 bcast=10.160.51.255 nmask=255.255.252.0
Got a positive name query response from 10.160.50.8 ( 10.160.50.8 )
Domain=[DEVELOPMENT] OS=[Windows 5.1] Server=[Windows 2000 LAN Manager]
Get file Hddm91c1_de.ABD? n
Get file Hddm91c1_de.def? y
```

```
getting file \W32X86\2\Hddm91c1_de.def of size 428 as Hddm91c1_de.def
Get file Hddm91c1_de.DLL? y
getting file \W32X86\2\Hddm91c1_de.DLL of size 876544 as Hddm91c1_de.DLL
[...]
```

After this command is complete, the files are in our current local directory. You probably have noticed that this time we passed several commands to the `-c` parameter, separated by semi-colons. This effects that all commands are executed in sequence on the remote Windows server before smbclient exits again.

Remember to repeat the procedure for the WIN40 architecture should you need to support Windows 9x/Me/XP clients. Remember too, the files for these architectures are in the `WIN40/0/` subdirectory. Once this is complete, we can run **smbclient ... put** to store the collected files on the Samba server's *[print$]* share.

17.6.2.3 Installing Driver Files into [print$]

We are now going to locate the driver files into the *[print$]* share. Remember, the UNIX path to this share has been defined previously in your words missing here. You also have created subdirectories for the different Windows client types you want to support. Supposing your *[print$]* share maps to the UNIX path `/etc/samba/drivers/`, your driver files should now go here:

- For all Windows NT, 2000 and XP clients into `/etc/samba/drivers/W32X86/` but not (yet) into the `2` subdirectory.
- For all Windows 95, 98 and ME clients into `/etc/samba/drivers/WIN40/` but not (yet) into the `0` subdirectory.

We again use smbclient to transfer the driver files across the network. We specify the same files and paths as were leaked to us by running **getdriver** against the original *Windows* install. However, now we are going to store the files into a *Samba/UNIX* print server's *[print$]* share.

```
root# smbclient //SAMBA-CUPS/print\$ -U'root%xxxx' -c \
  'cd W32X86; put HDNIS01_de.DLL; \
   put Hddm91c1_de.ppd; put HDNIS01U_de.DLL;          \
   put HDNIS01U_de.HLP; put Hddm91c1_de.DLL;          \
   put Hddm91c1_de.INI; put Hddm91c1KMMin.DLL;        \
   put Hddm91c1_de.dat; put Hddm91c1_de.dat;          \
   put Hddm91c1_de.def; put Hddm91c1_de.hre;          \
   put Hddm91c1_de.vnd; put Hddm91c1_de.hlp;          \
   put Hddm91c1_de_reg.HLP; put HDNIS01Aux.dll;       \
   put HDNIS01_de.NTF'

added interface ip=10.160.51.60 bcast=10.160.51.255 nmask=255.255.252.0
Got a positive name query response from 10.160.51.162 ( 10.160.51.162 )
Domain=[CUPS-PRINT] OS=[UNIX] Server=[Samba 2.2.7a]
```

```
putting file HDNIS01_de.DLL as \W32X86\HDNIS01_de.DLL
putting file Hddm91c1_de.ppd as \W32X86\Hddm91c1_de.ppd
putting file HDNIS01U_de.DLL as \W32X86\HDNIS01U_de.DLL
putting file HDNIS01U_de.HLP as \W32X86\HDNIS01U_de.HLP
putting file Hddm91c1_de.DLL as \W32X86\Hddm91c1_de.DLL
putting file Hddm91c1_de.INI as \W32X86\Hddm91c1_de.INI
putting file Hddm91c1KMMin.DLL as \W32X86\Hddm91c1KMMin.DLL
putting file Hddm91c1_de.dat as \W32X86\Hddm91c1_de.dat
putting file Hddm91c1_de.dat as \W32X86\Hddm91c1_de.dat
putting file Hddm91c1_de.def as \W32X86\Hddm91c1_de.def
putting file Hddm91c1_de.hre as \W32X86\Hddm91c1_de.hre
putting file Hddm91c1_de.vnd as \W32X86\Hddm91c1_de.vnd
putting file Hddm91c1_de.hlp as \W32X86\Hddm91c1_de.hlp
putting file Hddm91c1_de_reg.HLP as \W32X86\Hddm91c1_de_reg.HLP
putting file HDNIS01Aux.dll as \W32X86\HDNIS01Aux.dll
putting file HDNIS01_de.NTF as \W32X86\HDNIS01_de.NTF
```

Whew — that was a lot of typing! Most drivers are a lot smaller — many only having three generic PostScript driver files plus one PPD. While we did retrieve the files from the 2 subdirectory of the `W32X86` directory from the Windows box, we do not put them (for now) in this same subdirectory of the Samba box. This relocation will automatically be done by the **adddriver** command, which we will run shortly (and do not forget to also put the files for the Windows 9x/Me architecture into the `WIN40/` subdirectory should you need them).

17.6.2.4 smbclient to Confirm Driver Installation

For now we verify that our files are there. This can be done with **smbclient**, too (but, of course, you can log in via SSH also and do this through a standard UNIX shell access):

```
root# smbclient //SAMBA-CUPS/print\$ -U 'root%xxxx' \
   -c 'cd W32X86; pwd; dir; cd 2; pwd; dir'
 added interface ip=10.160.51.60 bcast=10.160.51.255 nmask=255.255.252.0
Got a positive name query response from 10.160.51.162 ( 10.160.51.162 )
Domain=[CUPS-PRINT] OS=[UNIX] Server=[Samba 2.2.8a]

Current directory is \\SAMBA-CUPS\print$\W32X86\
.                                  D        0  Sun May  4 03:56:35 2003
..                                 D        0  Thu Apr 10 23:47:40 2003
2                                  D        0  Sun May  4 03:56:18 2003
HDNIS01Aux.dll                     A    15356  Sun May  4 03:58:59 2003
Hddm91c1KMMin.DLL                  A    46966  Sun May  4 03:58:59 2003
HDNIS01_de.DLL                     A   434400  Sun May  4 03:58:59 2003
HDNIS01_de.NTF                     A   790404  Sun May  4 03:56:35 2003
Hddm91c1_de.DLL                    A   876544  Sun May  4 03:58:59 2003
Hddm91c1_de.INI                    A      101  Sun May  4 03:58:59 2003
Hddm91c1_de.dat                    A     5044  Sun May  4 03:58:59 2003
```

```
Hddm91c1_de.def                     A      428  Sun May  4 03:58:59 2003
Hddm91c1_de.hlp                     A    37699  Sun May  4 03:58:59 2003
Hddm91c1_de.hre                     A   323584  Sun May  4 03:58:59 2003
Hddm91c1_de.ppd                     A    26373  Sun May  4 03:58:59 2003
Hddm91c1_de.vnd                     A    45056  Sun May  4 03:58:59 2003
HDNIS01U_de.DLL                     A   165888  Sun May  4 03:58:59 2003
HDNIS01U_de.HLP                     A    19770  Sun May  4 03:58:59 2003
Hddm91c1_de_reg.HLP                 A   228417  Sun May  4 03:58:59 2003
             40976 blocks of size 262144. 709 blocks available

Current directory is \\SAMBA-CUPS\print$\W32X86\2\
.                                  D        0  Sun May  4 03:56:18 2003
..                                 D        0  Sun May  4 03:56:35 2003
ADOBEPS5.DLL                        A   434400  Sat May  3 23:18:45 2003
laserjet4.ppd                       A     9639  Thu Apr 24 01:05:32 2003
ADOBEPSU.DLL                        A   109568  Sat May  3 23:18:45 2003
ADOBEPSU.HLP                        A    18082  Sat May  3 23:18:45 2003
PDFcreator2.PPD                     A    15746  Sun Apr 20 22:24:07 2003
             40976 blocks of size 262144. 709 blocks available
```

Notice that there are already driver files present in the 2 subdirectory (probably from a previous installation). Once the files for the new driver are there too, you are still a few steps away from being able to use them on the clients. The only thing you could do now is to retrieve them from a client just like you retrieve ordinary files from a file share, by opening print$ in Windows Explorer. But that wouldn't install them per Point'n'Print. The reason is: Samba does not yet know that these files are something special, namely *printer driver files* and it does not know to which print queue(s) these driver files belong.

17.6.2.5 Running rpcclient with adddriver

Next, you must tell Samba about the special category of the files you just uploaded into the *[print$]* share. This is done by the **adddriver** command. It will prompt Samba to register the driver files into its internal TDB database files. The following command and its output has been edited, again, for readability:

```
root# rpcclient -Uroot%xxxx -c 'adddriver "Windows NT x86" \
  "dm9110:HDNIS01_de.DLL: \
  Hddm91c1_de.ppd:HDNIS01U_de.DLL:HDNIS01U_de.HLP:   \
  NULL:RAW:Hddm91c1_de.DLL,Hddm91c1_de.INI,          \
  Hddm91c1_de.dat,Hddm91c1_de.def,Hddm91c1_de.hre,   \
  Hddm91c1_de.vnd,Hddm91c1_de.hlp,Hddm91c1KMMin.DLL, \
  HDNIS01Aux.dll,HDNIS01_de.NTF,                     \
  Hddm91c1_de_reg.HLP' SAMBA-CUPS

cmd = adddriver "Windows NT x86" \
  "dm9110:HDNIS01_de.DLL:Hddm91c1_de.ppd:HDNIS01U_de.DLL:   \
```

```
  HDNIS01U_de.HLP:NULL:RAW:Hddm91c1_de.DLL,Hddm91c1_de.INI, \
  Hddm91c1_de.dat,Hddm91c1_de.def,Hddm91c1_de.hre,          \
  Hddm91c1_de.vnd,Hddm91c1_de.hlp,Hddm91c1KMMin.DLL,        \
  HDNIS01Aux.dll,HDNIS01_de.NTF,Hddm91c1_de_reg.HLP"

Printer Driver dm9110 successfully installed.
```

After this step, the driver should be recognized by Samba on the print server. You need to be very careful when typing the command. Don't exchange the order of the fields. Some changes would lead to an NT_STATUS_UNSUCCESSFUL error message. These become obvious. Other changes might install the driver files successfully, but render the driver unworkable. So take care! Hints about the syntax of the adddriver command are in the man page. The CUPS printing chapter provides a more detailed description, should you need it.

17.6.2.6 Checking adddriver Completion

One indication for Samba's recognition of the files as driver files is the `successfully installed` message. Another one is the fact that our files have been moved by the **adddriver** command into the 2 subdirectory. You can check this again with **smbclient**:

```
root# smbclient //SAMBA-CUPS/print\$ -Uroot%xx \
   -c 'cd W32X86;dir;pwd;cd 2;dir;pwd'
 added interface ip=10.160.51.162 bcast=10.160.51.255 nmask=255.255.252.0
 Domain=[CUPS-PRINT] OS=[UNIX] Server=[Samba 2.2.7a]

  Current directory is \\SAMBA-CUPS\print$\W32X86\
  .                                   D        0  Sun May  4 04:32:48 2003
  ..                                  D        0  Thu Apr 10 23:47:40 2003
  2                                   D        0  Sun May  4 04:32:48 2003
              40976 blocks of size 262144. 731 blocks available

  Current directory is \\SAMBA-CUPS\print$\W32X86\2\
  .                                   D        0  Sun May  4 04:32:48 2003
  ..                                  D        0  Sun May  4 04:32:48 2003
  DigiMaster.PPD                      A   148336  Thu Apr 24 01:07:00 2003
  ADOBEPS5.DLL                        A   434400  Sat May  3 23:18:45 2003
  laserjet4.ppd                       A     9639  Thu Apr 24 01:05:32 2003
  ADOBEPSU.DLL                        A   109568  Sat May  3 23:18:45 2003
  ADOBEPSU.HLP                        A    18082  Sat May  3 23:18:45 2003
  PDFcreator2.PPD                     A    15746  Sun Apr 20 22:24:07 2003
  HDNIS01Aux.dll                      A    15356  Sun May  4 04:32:18 2003
  Hddm91c1KMMin.DLL                   A    46966  Sun May  4 04:32:18 2003
  HDNIS01_de.DLL                      A   434400  Sun May  4 04:32:18 2003
  HDNIS01_de.NTF                      A   790404  Sun May  4 04:32:18 2003
  Hddm91c1_de.DLL                     A   876544  Sun May  4 04:32:18 2003
  Hddm91c1_de.INI                     A      101  Sun May  4 04:32:18 2003
```

```
  Hddm91c1_de.dat                       A      5044  Sun May  4 04:32:18 2003
  Hddm91c1_de.def                       A       428  Sun May  4 04:32:18 2003
  Hddm91c1_de.hlp                       A     37699  Sun May  4 04:32:18 2003
  Hddm91c1_de.hre                       A    323584  Sun May  4 04:32:18 2003
  Hddm91c1_de.ppd                       A     26373  Sun May  4 04:32:18 2003
  Hddm91c1_de.vnd                       A     45056  Sun May  4 04:32:18 2003
  HDNIS01U_de.DLL                       A    165888  Sun May  4 04:32:18 2003
  HDNIS01U_de.HLP                       A     19770  Sun May  4 04:32:18 2003
  Hddm91c1_de_reg.HLP                   A    228417  Sun May  4 04:32:18 2003
                40976 blocks of size 262144. 731 blocks available
```

Another verification is that the timestamp of the printing TDB files is now updated (and possibly their file size has increased).

17.6.2.7 Check Samba for Driver Recognition

Now the driver should be registered with Samba. We can easily verify this, and will do so in a moment. However, this driver is not yet associated with a particular printer. We may check the driver status of the files by at least three methods:

- From any Windows client browse Network Neighborhood, find the Samba host and open the Samba **Printers and Faxes** folder. Select any printer icon, right-click and select the printer **Properties**. Click the **Advanced** tab. Here is a field indicating the driver for that printer. A drop-down menu allows you to change that driver (be careful not to do this unwittingly). You can use this list to view all drivers known to Samba. Your new one should be among them. (Each type of client will only see his own architecture's list. If you do not have every driver installed for each platform, the list will differ if you look at it from Windows95/98/ME or WindowsNT/2000/XP.)
- From a Windows 200x/XP client (not Windows NT) browse **Network Neighborhood**, search for the Samba server and open the server's **Printers** folder, right-click on the white background (with no printer highlighted). Select **Server Properties**. On the **Drivers** tab you will see the new driver listed. This view enables you to also inspect the list of files belonging to that driver (this does not work on Windows NT, but only on Windows 2000 and Windows XP; Windows NT does not provide the **Drivers** tab). An alternative and much quicker method for Windows 2000/XP to start this dialog is by typing into a DOS box (you must of course adapt the name to your Samba server instead of *SAMBA-CUPS*):

  ```
  rundll32 printui.dll,PrintUIEntry /s /t2 /n\\SAMBA-CUPS
  ```

- From a UNIX prompt, run this command (or a variant thereof) where *SAMBA-CUPS* is the name of the Samba host and xxxx represents the actual Samba password assigned to root:

  ```
  rpcclient -U'root%xxxx' -c 'enumdrivers' SAMBA-CUPS
  ```

  You will see a listing of all drivers Samba knows about. Your new one should be among them. But it is only listed under the *[Windows NT x86]* heading, not under *[Windows 4.0]*, since you didn't install that part. Or did you? You will see a listing of all drivers Samba knows about. Your new one should be among them. In our example it is named dm9110. Note that the third column shows the other installed drivers twice, one time for each supported architecture. Our new driver only shows up for Windows NT 4.0 or 2000. To have it present for Windows 95, 98 and ME, you'll have to repeat the whole procedure with the WIN40 architecture and subdirectory.

17.6.2.8 Specific Driver Name Flexibility

You can name the driver as you like. If you repeat the **adddriver** step with the same files as before but with a different driver name, it will work the same:

```
root# rpcclient -Uroot%xxxx          \
  -c 'adddriver "Windows NT x86"                          \
  "mydrivername:HDNIS01_de.DLL:                   \
  Hddm91c1_de.ppd:HDNIS01U_de.DLL:HDNIS01U_de.HLP:   \
  NULL:RAW:Hddm91c1_de.DLL,Hddm91c1_de.INI,          \
  Hddm91c1_de.dat,Hddm91c1_de.def,Hddm91c1_de.hre,   \
  Hddm91c1_de.vnd,Hddm91c1_de.hlp,Hddm91c1KMMin.DLL, \
  HDNIS01Aux.dll,HDNIS01_de.NTF,Hddm91c1_de_reg.HLP' SAMBA-CUPS


cmd = adddriver "Windows NT x86" \
 "mydrivername:HDNIS01_de.DLL:Hddm91c1_de.ppd:HDNIS01U_de.DLL:\
  HDNIS01U_de.HLP:NULL:RAW:Hddm91c1_de.DLL,Hddm91c1_de.INI,              \
  Hddm91c1_de.dat,Hddm91c1_de.def,Hddm91c1_de.hre,                       \
  Hddm91c1_de.vnd,Hddm91c1_de.hlp,Hddm91c1KMMin.DLL,                     \
  HDNIS01Aux.dll,HDNIS01_de.NTF,Hddm91c1_de_reg.HLP"

Printer Driver mydrivername successfully installed.
```

You will be able to bind that driver to any print queue (however, you are responsible that you associate drivers to queues that make sense with respect to target printers). You cannot run the **rpcclient adddriver** command repeatedly. Each run consumes the files you had put into the *[print$]* share by moving them into the respective subdirectories. So you must execute an **smbclient ... put** command before each **rpcclient ... adddriver** command.

17.6.2.9 Running rpcclient with the setdriver

Samba needs to know which printer owns which driver. Create a mapping of the driver to a printer, and store this info in Samba's memory, the TDB files. The **rpcclient setdriver** command achieves exactly this:

```
root# rpcclient -U'root%xxxx' -c 'setdriver dm9110 mydrivername' SAMBA-CUPS
 cmd = setdriver dm9110 mydrivername
```

```
Successfully set dm9110 to driver mydrivername.
```

Ah, no, I did not want to do that. Repeat, this time with the name I intended:

```
root# rpcclient -U'root%xxxx' -c 'setdriver dm9110 dm9110' SAMBA-CUPS
 cmd = setdriver dm9110 dm9110
Successfully set dm9110 to driver dm9110.
```

The syntax of the command is:

```
rpcclient -U'root%sambapassword' -c 'setdriver printername \
 drivername' SAMBA-Hostname.
```

Now we have done most of the work, but not all of it.

NOTE

The **setdriver** command will only succeed if the printer is already known to Samba. A bug in 2.2.x prevented Samba from recognizing freshly installed printers. You had to restart Samba, or at least send an HUP signal to all running smbd processes to work around this: `kill -HUP 'pidof smbd'`.

17.7 Client Driver Installation Procedure

As Don Quixote said: "*The proof of the pudding is in the eating.*" The proof for our setup lies in the printing. So let's install the printer driver onto the client PCs. This is not as straightforward as it may seem. Read on.

17.7.1 First Client Driver Installation

Especially important is the installation onto the first client PC (for each architectural platform separately). Once this is done correctly, all further clients are easy to setup and shouldn't need further attention. What follows is a description for the recommended first procedure. You work now from a client workstation. You should guarantee that your connection is not unwittingly mapped to *bad user* nobody. In a DOS box type:

`net use \\`*`SAMBA-SERVER`*`\print$ /user:root`

Replace root, if needed, by another valid *`printer admin`* user as given in the definition. Should you already be connected as a different user, you will get an error message. There is no easy way to get rid of that connection, because Windows does not seem to know a concept of logging off from a share connection (do not confuse this with logging off from the local workstation; that is a different matter). You can try to close all Windows file explorer and Internet Explorer for Windows. As a last resort, you may have to reboot. Make sure there is no automatic reconnection set up. It may be easier to go to a different workstation and try from there. After you have made sure you are connected as a printer admin user (you can check this with the **smbstatus** command on Samba), do this from the Windows workstation:

1. Open **Network Neighborhood**.
2. Browse to Samba server.
3. Open its **Printers and Faxes** folder.
4. Highlight and right-click on the printer.
5. Select **Connect** (for Windows NT4/200x it is possibly **Install**).

A new printer (named *printername* on Samba-server) should now have appeared in your *local* Printer folder (check **Start** – **Settings** – **Control Panel** – **Printers and Faxes**).

Most likely you are now tempted to try to print a test page. After all, you now can open the printer properties, and on the **General** tab there is a button offering to do just that. But chances are that you get an error message saying Unable to print Test Page. The reason might be that there is not yet a valid Device Mode set for the driver, or that the "*Printer Driver Data*" set is still incomplete.

You must make sure that a valid *Device Mode* is set for the driver. We now explain what that means.

17.7.2 Setting Device Modes on New Printers

For a printer to be truly usable by a Windows NT/200x/XP client, it must possess:

- A valid *Device Mode* generated by the driver for the printer (defining things like paper size, orientation and duplex settings).
- A complete set of *Printer Driver Data* generated by the driver.

If either of these is incomplete, the clients can produce less than optimal output at best. In the worst cases, unreadable garbage or nothing at all comes from the printer or it produces a harvest of error messages when attempting to print. Samba stores the named values and all printing related information in its internal TDB database files (`ntprinters.tdb`, `ntdrivers.tdb`, `printing.tdb` and `ntforms.tdb`).

What do these two words stand for? Basically, the Device Mode and the set of Printer Driver Data is a collection of settings for all print queue properties, initialized in a sensible way. Device Modes and Printer Driver Data should initially be set on the print server (the Samba host) to healthy values so the clients can start to use them immediately. How do we set these initial healthy values? This can be achieved by accessing the drivers remotely from an NT (or 200x/XP) client, as is discussed in the following paragraphs.

Be aware that a valid Device Mode can only be initiated by a *printer admin*, or root (the reason should be obvious). Device Modes can only be correctly set by executing the printer driver program itself. Since Samba cannot execute this Win32 platform driver code, it sets this field initially to NULL (which is not a valid setting for clients to use). Fortunately, most drivers automatically generate the Printer Driver Data that is needed when they are uploaded to the *[print$]* share with the help of the APW or rpcclient.

The generation and setting of a first valid Device Mode, however, requires some tickling from a client, to set it on the Samba server. The easiest means of doing so is to simply change the page orientation on the server's printer. This executes enough of the printer driver program on the client for the desired effect to happen, and feeds back the new Device Mode to our Samba server. You can use the native Windows NT/200x/XP printer properties page from a Window client for this:

1. Browse the **Network Neighborhood.**
2. Find the Samba server.
3. Open the Samba server's **Printers and Faxes** folder.
4. Highlight the shared printer in question.
5. Right-click on the printer (you may already be here, if you followed the last section's description).
6. At the bottom of the context menu select **Properties** (if the menu still offers the **Connect** entry further above, you need to click on that one first to achieve the driver installation as shown in the last section).
7. Go to the **Advanced** tab; click on **Printing Defaults**.
8. Change the **Portrait** page setting to **Landscape** (and back).
9. Make sure to apply changes between swapping the page orientation to cause the change to actually take effect.
10. While you are at it, you may also want to set the desired printing defaults here, which then apply to all future client driver installations on the remaining from now on.

This procedure has executed the printer driver program on the client platform and fed back the correct Device Mode to Samba, which now stored it in its TDB files. Once the driver is installed on the client, you can follow the analogous steps by accessing the *local* **Printers** folder, too, if you are a Samba printer admin user. From now on, printing should work as expected.

Samba includes a service level parameter name ***default devmode*** for generating a default Device Mode for a printer. Some drivers will function well with Samba's default set of properties. Others may crash the client's spooler service. So use this parameter with caution. It is always better to have the client generate a valid device mode for the printer and store it on the server for you.

17.7.3 Additional Client Driver Installation

Every additional driver may be installed, along the lines described above. Browse network, open the **Printers** folder on Samba server, right-click on **Printer** and choose **Connect...**. Once this completes (should be not more than a few seconds, but could also take a minute, depending on network conditions), you should find the new printer in your client workstation local **Printers and Faxes** folder.

You can also open your local **Printers and Faxes** folder by using this command on Windows 200x/XP Professional workstations:

```
rundll32 shell32.dll,SHHelpShortcuts_RunDLL PrintersFolder
```

or this command on Windows NT 4.0 workstations:

```
rundll32 shell32.dll,Control_RunDLL MAIN.CPL @2
```

You can enter the commands either inside a **DOS box** window or in the **Run command...** field from the **Start** menu.

17.7.4 Always Make First Client Connection as root or "*printer admin*"

After you installed the driver on the Samba server (in its *[print$]* share, you should always make sure that your first client installation completes correctly. Make it a habit for yourself to build the very first connection from a client as *printer admin*. This is to make sure that:

- A first valid *Device Mode* is really initialized (see above for more explanation details).
- The default print settings of your printer for all further client installations are as you want them.

Do this by changing the orientation to landscape, click on **Apply**, and then change it back again. Next, modify the other settings (for example, you do not want the default media size set to **Letter** when you are all using **A4**, right? You may want to set the printer for **duplex** as the default, and so on).

To connect as root to a Samba printer, try this command from a Windows 200x/XP DOS box command prompt:

```
C:\> runas /netonly /user:root "rundll32 printui.dll,PrintUIEntry /p /t3 /n
   \\SAMBA-SERVER\printername"
```

You will be prompted for root's Samba-password; type it, wait a few seconds, click on **Printing Defaults**, and proceed to set the job options that should be used as defaults by all clients. Alternately, instead of root you can name one other member of the *printer admin* from the setting.

Now all the other users downloading and installing the driver the same way (called "*Point'n'Print*") will have the same defaults set for them. If you miss this step you'll get a lot of Help Desk calls from your users, but maybe you like to talk to people.

17.8 Other Gotchas

Your driver is installed. It is now ready for Point'n'Print installation by the clients. You may have tried to download and use it onto your first client machine, but wait. Let's make sure you are acquainted first with a few tips and tricks you may find useful. For example, suppose you did not set the defaults on the printer, as advised in the preceding paragraphs. Your users complain about various issues (such as, "*We need to set the paper size for each job from Letter to A4 and it will not store it.*")

17.8.1 Setting Default Print Options for Client Drivers

The last sentence might be viewed with mixed feelings by some users and admins. They have struggled for hours and could not arrive at a point where their settings seemed to be saved. It is not their fault. The confusing thing is that in the multi-tabbed dialog that pops up when you right-click on the printer name and select **Properties**, you can arrive at two dialogs that appear identical, each claiming that they help you to set printer options in three different ways. Here is the definite answer to the Samba default driver setting FAQ:

How are you doing it? I bet the wrong way. (It is not easy to find out, though). There are three different ways to bring you to a dialog that seems to set everything. All three dialogs look the same, but only one of them does what you intend. You need to be Administrator or Print Administrator to do this for all users. Here is how I reproduce it in an XP Professional:

A The first "*wrong*" way:

  1 Open the **Printers** folder.

  2 Right-click on the printer (*remoteprinter on cupshost*) and select in context menu **Printing Preferences...**

  3 Look at this dialog closely and remember what it looks like.

B The second "*wrong*" way:

  1 Open the **Printers** folder.

  2 Right-click on the printer (*remoteprinter on cupshost*) and select in the context menu **Properties**

  3 Click on the **General** tab

  4 Click on the **Printing Preferences...**

  5 A new dialog opens. Keep this dialog open and go back to the parent dialog.

C The third and correct way: (should you do this from the beginning, just carry out steps 1 and 2 from the second method above).

  1 Click on the **Advanced** tab. (If everything is "*grayed out,*" then you are not logged in as a user with enough privileges).

  2 Click on the **Printing Defaults** button.

  3 On any of the two new tabs, click on the **Advanced** button.

  4 A new dialog opens. Compare this one to the other. Are they identical looking comparing one from "*B.5*" and one from A.3".

Do you see any difference in the two settings dialogs? I do not either. However, only the last one, which you arrived at with steps C.1 through 6 will permanently save any settings which will then become the defaults for new users. If you want all clients to have the same defaults, you need to conduct these steps as administrator (*printer admin* in) before a client downloads the driver (the clients can later set their own per-user defaults by following procedures A or B above). Windows 200x/XP allow per-user default settings and the ones the administrator gives them, before they set up their own. The parents of the identically-looking dialogs have a slight difference in their window names; one is called `Default Print Values for Printer Foo on Server Bar"` (which is the one you need) and the other is called "*Print Settings for Printer Foo on Server Bar*". The last one is the one you arrive at when you right-click on the printer and select **Print Settings....** This is the one that you were taught to use back in the days of Windows NT, so it is only natural to try the same way with Windows 200x/XP. You would not dream that there is now a different path to arrive at an identically looking, but functionally different, dialog to set defaults for all users.

TIP

Try (on Windows 200x/XP) to run this command (as a user with the right privileges):

`rundll32 printui.dll,PrintUIEntry /p /t3 /n\\SAMBA-SERVER\printersharename`

To see the tab with the **Printing Defaults** button (the one you need),also run this command:

`rundll32 printui.dll,PrintUIEntry /p /t0 /n\\SAMBA-SERVER\printersharename`

To see the tab with the **Printing Preferences** button (the one which does not set system-wide defaults), you can start the commands from inside a DOS box" or from **Start** -> **Run**.

17.8.2 Supporting Large Numbers of Printers

One issue that has arisen during the recent development phase of Samba is the need to support driver downloads for hunderds of printers. Using Windows NT APW here is somewhat awkward (to say the least). If you do not want to acquire RSS pains from the printer installation clicking orgy alone, you need to think about a non-interactive script.

If more than one printer is using the same driver, the **rpcclient setdriver** command can be used to set the driver associated with an installed queue. If the driver is uploaded to *[print$]* once and registered with the printing TDBs, it can be used by multiple print queues. In this case, you just need to repeat the **setprinter** subcommand of **rpcclient** for every queue (without the need to conduct the **adddriver** repeatedly). The following is an example of how this could be accomplished:

```
root# rpcclient SAMBA-CUPS -U root%secret -c 'enumdrivers'
```

```
 cmd = enumdrivers

 [Windows NT x86]
 Printer Driver Info 1:
   Driver Name: [infotec  IS 2075 PCL 6]

 Printer Driver Info 1:
   Driver Name: [DANKA InfoStream]

 Printer Driver Info 1:
   Driver Name: [Heidelberg Digimaster 9110 (PS)]

 Printer Driver Info 1:
   Driver Name: [dm9110]

 Printer Driver Info 1:
   Driver Name: [mydrivername]

 [....]
```

```
root# rpcclient SAMBA-CUPS -U root%secret -c 'enumprinters'
 cmd = enumprinters
   flags:[0x800000]
   name:[\\SAMBA-CUPS\dm9110]
   description:[\\SAMBA-CUPS\dm9110,,110ppm HiVolume DANKA Stuttgart]
   comment:[110 ppm HiVolume DANKA Stuttgart]
 [....]
```

```
root# rpcclient SAMBA-CUPS -U root%secret -c \
  'setdriver dm9110 "Heidelberg Digimaster 9110 (PS)"'
 cmd = setdriver dm9110 Heidelberg Digimaster 9110 (PPD)
 Successfully set dm9110 to driver Heidelberg Digimaster 9110 (PS).
```

```
root# rpcclient SAMBA-CUPS -U root%secret -c 'enumprinters'
 cmd = enumprinters
   flags:[0x800000]
   name:[\\SAMBA-CUPS\dm9110]
   description:[\\SAMBA-CUPS\dm9110,Heidelberg Digimaster 9110 (PS),\
     110ppm HiVolume DANKA Stuttgart]
   comment:[110ppm HiVolume DANKA Stuttgart]
 [....]
```

```
root# rpcclient SAMBA-CUPS -U root%secret -c 'setdriver dm9110 mydrivername'
 cmd = setdriver dm9110 mydrivername
 Successfully set dm9110 to mydrivername.
```

```
root# rpcclient SAMBA-CUPS -U root%secret -c 'enumprinters'
 cmd = enumprinters
   flags:[0x800000]
   name:[\\SAMBA-CUPS\dm9110]
   description:[\\SAMBA-CUPS\dm9110,mydrivername,\
     110ppm HiVolume DANKA Stuttgart]
   comment:[110ppm HiVolume DANKA Stuttgart]
 [....]
```

It may not be easy to recognize that the first call to **enumprinters** showed the "*dm9110*" printer with an empty string where the driver should have been listed (between the 2 commas in the description field). After the **setdriver** command succeeded, all is well.

17.8.3 Adding New Printers with the Windows NT APW

By default, Samba exhibits all printer shares defined in `smb.conf` in the **Printers** folder. Also located in this folder is the Windows NT Add Printer Wizard icon. The APW will be shown only if:

- The connected user is able to successfully execute an **OpenPrinterEx(\\server)** with administrative privileges (i.e., root or *printer admin*).

> TIP
>
> 
>
> Try this from a Windows 200x/XP DOS box command prompt:
>
> `runas /netonly /user:root rundll32 printui.dll,PrintUIEntry /p /t0 /n \\SAMBA-SERVER\printersharename`
>
> Click on **Printing Preferences**.

- ... contains the setting *show add printer wizard* = yes (the default).

The APW can do various things:

- Upload a new driver to the Samba *[print$]* share.
- Associate an uploaded driver with an existing (but still driverless) print queue.
- Exchange the currently used driver for an existing print queue with one that has been uploaded before.

- Add an entirely new printer to the Samba host (only in conjunction with a working *add printer command*. A corresponding *delete printer command* for removing entries from the **Printers** folder may also be provided).

The last one (add a new printer) requires more effort than the previous ones. To use the APW to successfully add a printer to a Samba server, the *add printer command* must have a defined value. The program hook must successfully add the printer to the UNIX print system (i.e., to `/etc/printcap`, `/etc/cups/printers.conf` or other appropriate files) and to `smb.conf` if necessary.

When using the APW from a client, if the named printer share does not exist, smbd will execute the *add printer command* and reparse to the to attempt to locate the new printer share. If the share is still not defined, an error of Access Denied is returned to the client. The *add printer command* is executed under the context of the connected user, not necessarily a root account. A *map to guest* = bad user may have connected you unwittingly under the wrong privilege. You should check it by using the **smbstatus** command.

17.8.4 Error Message: "*Cannot connect under a different Name*"

Once you are connected with the wrong credentials, there is no means to reverse the situation other than to close all Explorer Windows, and perhaps reboot.

- The **net use \\SAMBA-SERVER\sharename /user:root** gives you an error message: "*Multiple connections to a server or a shared resource by the same user utilizing the several user names are not allowed. Disconnect all previous connections to the server, resp. the shared resource, and try again.*"
- Every attempt to "*connect a network drive*" to `\\SAMBASERVER\\print$` to `z:` is countered by the pertinacious message: "*This network folder is currently connected under different credentials (username and password). Disconnect first any existing connection to this network share in order to connect again under a different username and password*".

So you close all connections. You try again. You get the same message. You check from the Samba side, using **smbstatus**. Yes, there are more connections. You kill them all. The client still gives you the same error message. You watch the smbd.log file on a high debug level and try reconnect. Same error message, but not a single line in the log. You start to wonder if there was a connection attempt at all. You run ethereal and tcpdump while you try to connect. Result: not a single byte goes on the wire. Windows still gives the error message. You close all Explorer windows and start it again. You try to connect — and this times it works! Windows seems to cache connection informtion somewhere and does not keep it up-to-date (if you are unlucky you might need to reboot to get rid of the error message).

17.8.5 Take Care When Assembling Driver Files

You need to be extremely careful when you take notes about the files and belonging to a particular driver. Don't confuse the files for driver version "*0*" (for Windows 9x/Me, going into `[print$]/WIN/0/`), driver version 2 (Kernel Mode driver for Windows NT, going into `[print$]/W32X86/2/` may be used on Windows 200x/XP also), and driver version "*3*" (non-Kernel Mode driver going into `[print$]/W32X86/3/` cannot be used on Windows NT). Quite often these different driver versions contain files that have the same name but actually are very different. If you look at them from the Windows Explorer (they reside in `%WINDOWS%\system32\spool\drivers\W32X86\`), you will probably see names in capital letters, while an **enumdrivers** command from Samba would show mixed or lower case letters. So it is easy to confuse them. If you install them manually using **rpcclient** and subcommands, you may even succeed without an error message. Only later, when you try install on a client, you will encounter error messages like `This server has no appropriate driver for the printer`.

Here is an example. You are invited to look closely at the various files, compare their names and their spelling, and discover the differences in the composition of the version 2 and 3 sets. Note: the version 0 set contained 40 ***Dependentfiles***, so I left it out for space reasons:

```
root# rpcclient -U 'Administrator%secret' -c 'enumdrivers 3' 10.160.50.8

 Printer Driver Info 3:
        Version: [3]
        Driver Name: [Canon iR8500 PS3]
        Architecture: [Windows NT x86]
        Driver Path: [\\10.160.50.8\print$\W32X86\3\cns3g.dll]
        Datafile: [\\10.160.50.8\print$\W32X86\3\iR8500sg.xpd]
        Configfile: [\\10.160.50.8\print$\W32X86\3\cns3gui.dll]
        Helpfile: [\\10.160.50.8\print$\W32X86\3\cns3g.hlp]

        Dependentfiles: [\\10.160.50.8\print$\W32X86\3\aucplmNT.dll]
        Dependentfiles: [\\10.160.50.8\print$\W32X86\3\ucs32p.dll]
        Dependentfiles: [\\10.160.50.8\print$\W32X86\3\tnl32.dll]
        Dependentfiles: [\\10.160.50.8\print$\W32X86\3\aussdrv.dll]
        Dependentfiles: [\\10.160.50.8\print$\W32X86\3\cnspdc.dll]
        Dependentfiles: [\\10.160.50.8\print$\W32X86\3\aussapi.dat]
        Dependentfiles: [\\10.160.50.8\print$\W32X86\3\cns3407.dll]
        Dependentfiles: [\\10.160.50.8\print$\W32X86\3\CnS3G.cnt]
        Dependentfiles: [\\10.160.50.8\print$\W32X86\3\NBAPI.DLL]
        Dependentfiles: [\\10.160.50.8\print$\W32X86\3\NBIPC.DLL]
        Dependentfiles: [\\10.160.50.8\print$\W32X86\3\cpcview.exe]
        Dependentfiles: [\\10.160.50.8\print$\W32X86\3\cpcdspl.exe]
        Dependentfiles: [\\10.160.50.8\print$\W32X86\3\cpcedit.dll]
        Dependentfiles: [\\10.160.50.8\print$\W32X86\3\cpcqm.exe]
        Dependentfiles: [\\10.160.50.8\print$\W32X86\3\cpcspl.dll]
        Dependentfiles: [\\10.160.50.8\print$\W32X86\3\cfine32.dll]
        Dependentfiles: [\\10.160.50.8\print$\W32X86\3\cpcr407.dll]
        Dependentfiles: [\\10.160.50.8\print$\W32X86\3\Cpcqm407.hlp]
        Dependentfiles: [\\10.160.50.8\print$\W32X86\3\cpcqm407.cnt]
        Dependentfiles: [\\10.160.50.8\print$\W32X86\3\cns3ggr.dll]

        Monitorname: []
        Defaultdatatype: []
```

```
Printer Driver Info 3:
        Version: [2]
        Driver Name: [Canon iR5000-6000 PS3]
        Architecture: [Windows NT x86]
        Driver Path: [\\10.160.50.8\print$\W32X86\2\cns3g.dll]
        Datafile: [\\10.160.50.8\print$\W32X86\2\IR5000sg.xpd]
        Configfile: [\\10.160.50.8\print$\W32X86\2\cns3gui.dll]
        Helpfile: [\\10.160.50.8\print$\W32X86\2\cns3g.hlp]

        Dependentfiles: [\\10.160.50.8\print$\W32X86\2\AUCPLMNT.DLL]
        Dependentfiles: [\\10.160.50.8\print$\W32X86\2\aussdrv.dll]
        Dependentfiles: [\\10.160.50.8\print$\W32X86\2\cnspdc.dll]
        Dependentfiles: [\\10.160.50.8\print$\W32X86\2\aussapi.dat]
        Dependentfiles: [\\10.160.50.8\print$\W32X86\2\cns3407.dll]
        Dependentfiles: [\\10.160.50.8\print$\W32X86\2\CnS3G.cnt]
        Dependentfiles: [\\10.160.50.8\print$\W32X86\2\NBAPI.DLL]
        Dependentfiles: [\\10.160.50.8\print$\W32X86\2\NBIPC.DLL]
        Dependentfiles: [\\10.160.50.8\print$\W32X86\2\cns3gum.dll]

        Monitorname: [CPCA Language Monitor2]
        Defaultdatatype: []
```

If we write the "*version 2*" files and the "*version 3*" files into different text files and compare the result, we see this picture:

```
root# sdiff 2-files 3-files
```

```
 cns3g.dll                           cns3g.dll
 iR8500sg.xpd                        iR8500sg.xpd
 cns3gui.dll                         cns3gui.dll
 cns3g.hlp                           cns3g.hlp
 AUCPLMNT.DLL                      | aucplmNT.dll
                                   > ucs32p.dll
                                   > tnl32.dll
 aussdrv.dll                         aussdrv.dll
 cnspdc.dll                          cnspdc.dll
 aussapi.dat                         aussapi.dat
 cns3407.dll                         cns3407.dll
 CnS3G.cnt                           CnS3G.cnt
 NBAPI.DLL                           NBAPI.DLL
 NBIPC.DLL                           NBIPC.DLL
 cns3gum.dll                       | cpcview.exe
                                   > cpcdspl.exe
                                   > cpcqm.exe
```

```
> cpcspl.dll
> cfine32.dll
> cpcr407.dll
> Cpcqm407.hlp
> cpcqm407.cnt
> cns3ggr.dll
```

Do not be fooled! Driver files for each version with identical names may be different in their content, as you can see from this size comparison:

```
root# for i in cns3g.hlp cns3gui.dll cns3g.dll; do                  \
           smbclient //10.160.50.8/print\$ -U 'Administrator%xxxx' \
           -c "cd W32X86/3; dir $i; cd .. ; cd 2; dir $i";        \
         done

  CNS3G.HLP                  A   122981  Thu May 30 02:31:00 2002
  CNS3G.HLP                  A    99948  Thu May 30 02:31:00 2002

  CNS3GUI.DLL                A  1805824  Thu May 30 02:31:00 2002
  CNS3GUI.DLL                A  1785344  Thu May 30 02:31:00 2002

  CNS3G.DLL                  A  1145088  Thu May 30 02:31:00 2002
  CNS3G.DLL                  A    15872  Thu May 30 02:31:00 2002
```

In my example were even more differences than shown here. Conclusion: you must be careful to select the correct driver files for each driver version. Don't rely on the names alone and don't interchange files belonging to different driver versions.

17.8.6 Samba and Printer Ports

Windows NT/2000 print servers associate a port with each printer. These normally take the form of `LPT1:`, `COM1:`, `FILE:`, and so on. Samba must also support the concept of ports associated with a printer. By default, only one printer port, named "*Samba Printer Port*", exists on a system. Samba does not really need such a "*port*" in order to print; rather it is a requirement of Windows clients. They insist on being told about an available port when they request this information, otherwise they throw an error message at you. So Samba fakes the port information to keep the Windows clients happy.

Samba does not support the concept of `Printer Pooling` internally either. Printer Pooling assigns a logical printer to multiple ports as a form of load balancing or fail over.

If you require multiple ports be defined for some reason or another (my users and my boss should not know that they are working with Samba), configure *`enumports command`* which can be used to define an external program that generates a listing of ports on a system.

17.8.7 Avoiding Common Client Driver Misconfiguration

So now the printing works, but there are still problems. Most jobs print well, some do not print at all. Some jobs have problems with fonts, which do not look good. Some jobs print fast and some are dead-slow. We cannot cover it all, but we want to encourage you to read the brief paragraph about "*Avoiding the Wrong PostScript Driver Settings*" in the CUPS Printing part of this document.

17.9 The Imprints Toolset

The Imprints tool set provides a UNIX equivalent of the Windows NT Add Printer Wizard. For complete information, please refer to the Imprints Web site at `http://imprints.sourceforge.net/` as well as the documentation included with the imprints source distribution. This section only provides a brief introduction to the features of Imprints.

Unfortunately, the Imprints toolset is no longer maintained. As of December 2000, the project is in need of a new maintainer. The most important skill to have is Perl coding and an interest in MS-RPC-based printing used in Samba. If you wish to volunteer, please coordinate your efforts on the Samba technical mailing list. The toolset is still in usable form, but only for a series of older printer models where there are prepared packages to use. Packages for more up-to-date print devices are needed if Imprints should have a future.

17.9.1 What is Imprints?

Imprints is a collection of tools for supporting these goals:

- Providing a central repository of information regarding Windows NT and 95/98 printer driver packages.
- Providing the tools necessary for creating the Imprints printer driver packages.
- Providing an installation client that will obtain printer drivers from a central Internet (or intranet) Imprints Server repository and install them on remote Samba and Windows NT4 print servers.

17.9.2 Creating Printer Driver Packages

The process of creating printer driver packages is beyond the scope of this document (refer to Imprints.txt also included with the Samba distribution for more information). In short, an Imprints driver package is a gzipped tarball containing the driver files, related INF files, and a control file needed by the installation client.

17.9.3 The Imprints Server

The Imprints server is really a database server that may be queried via standard HTTP mechanisms. Each printer entry in the database has an associated URL for the actual downloading of the package. Each package is digitally signed via GnuPG which can be used to verify that the package downloaded is actually the one referred in the Imprints database. It is strongly recommended that this security check not be disabled.

17.9.4 The Installation Client

More information regarding the Imprints installation client is available from the the documentation file `Imprints-Client-HOWTO.ps` that is included with the Imprints source package. The Imprints installation client comes in two forms:

- A set of command line Perl scripts.
- A GTK+ based graphical interface to the command line Perl scripts.

The installation client (in both forms) provides a means of querying the Imprints database server for a matching list of known printer model names as well as a means to download and install the drivers on remote Samba and Windows NT print servers.

The basic installation process is in four steps and Perl code is wrapped around smbclient and rpcclient.

- For each supported architecture for a given driver:
    1. rpcclient: Get the appropriate upload directory on the remote server.
    2. smbclient: Upload the driver files.
    3. rpcclient: Issues an AddPrinterDriver() MS-RPC.
- rpcclient: Issue an AddPrinterEx() MS-RPC to actually create the printer.

One of the problems encountered when implementing the Imprints tool set was the name space issues between various supported client architectures. For example, Windows NT includes a driver named "*Apple LaserWriter II NTX v51.8*" and Windows 95 calls its version of this driver "*Apple LaserWriter II NTX*".

The problem is how to know what client drivers have been uploaded for a printer. An astute reader will remember that the Windows NT Printer Properties dialog only includes space for one printer driver name. A quick look in the Windows NT 4.0 system registry at:

`HKLM\System\CurrentControlSet\Control\Print\Environment`

will reveal that Windows NT always uses the NT driver name. This is okay as Windows NT always requires that at least the Windows NT version of the printer driver is present. Samba does not have the requirement internally, therefore, "*How can you use the NT driver name if it has not already been installed?*"

The way of sidestepping this limitation is to require that all Imprints printer driver packages include both the Intel Windows NT and 95/98 printer drivers and that the NT driver is installed first.

17.10 Adding Network Printers without User Interaction

The following MS Knowledge Base article may be of some help if you need to handle Windows 2000 clients: *How to Add Printers with No User Interaction in Windows 2000,* (`http://support.microsoft.com/default.aspx?scid=kb;en-us;189105`[3]). It also applies to Windows XP Professional clients. The ideas sketched out in this section are inspired by this article, which describes a commandline method that can be applied to install network and local printers and their drivers. This is most useful if integrated in Logon Scripts. You can see what options are available by typing in the command prompt (**DOS box**):

```
rundll32 printui.dll,PrintUIEntry /?
```

A window pops up that shows you all of the commandline switches available. An extensive list of examples is also provided. This is only for Win 200x/XP, it does not work on Windows NT. Windows NT probably has some other tools in the respective Resource Kit. Here is a suggestion about what a client logon script might contain, with a short explanation of what the lines actually do (it works if 200x/XP Windows clients access printers via Samba, and works for Windows-based print servers too):

```
rundll32 printui.dll,PrintUIEntry /dn /n "\\cupsserver\infotec2105-IPDS" /q
rundll32 printui.dll,PrintUIEntry /in /n "\\cupsserver\infotec2105-PS"
rundll32 printui.dll,PrintUIEntry /y /n "\\cupsserver\infotec2105-PS"
```

Here is a list of the used commandline parameters:

/dn — deletes a network printer

/q — quiet modus

/n — names a printer

/in — adds a network printer connection

/y — sets printer as default printer

- Line 1 deletes a possibly existing previous network printer *infotec2105-IPDS* (which had used native Windows drivers with LPRng that were removed from the server that was converted to CUPS). The **/q** at the end eliminates Confirm or error dialog boxes from popping up. They should not be presented to the user logging on.
- Line 2 adds the new printer *infotec2105-PS* (which actually is the same physical device but is now run by the new CUPS printing system and associated with the CUPS/Adobe PS drivers). The printer and its driver must have been added to Samba prior to the user logging in (e.g., by a procedure as discussed earlier in this chapter, or by running **cupsaddsmb**). The driver is now auto-downloaded to the client PC where the user is about to log in.
- Line 3 sets the default printer to this new network printer (there might be several other printers installed with this same method and some may be local as well, so we

[3] http://support.microsoft.com/default.aspx?scid=kb;en-us;189105

decide for a default printer). The default printer selection may, of course, be different for different users.

The second line only works if the printer *infotec2105-PS* has an already working print queue on the `cupsserver`, and if the printer drivers have been successfully uploaded (via the **APW**, **smbclient/rpcclient**, or **cupsaddsmb**) into the *[print$]* driver repository of Samba. Some Samba versions prior to version 3.0 required a re-start of smbd after the printer install and the driver upload, otherwise the script (or any other client driver download) would fail.

Since there no easy way to test for the existence of an installed network printer from the logon script, do not bother checking, just allow the deinstallation/reinstallation to occur every time a user logs in; it's really quick anyway (1 to 2 seconds).

The additional benefits for this are:

- It puts in place any printer default setup changes automatically at every user logon.
- It allows for "*roaming*" users' login into the domain from different workstations.

Since network printers are installed per user, this much simplifies the process of keeping the installation up-to-date. The few extra seconds at logon time will not really be noticeable. Printers can be centrally added, changed and deleted at will on the server with no user intervention required from the clients (you just need to keep the logon scripts up-to-date).

17.11 The addprinter Command

The **addprinter** command can be configured to be a shell script or program executed by Samba. It is triggered by running the APW from a client against the Samba print server. The APW asks the user to fill in several fields (such as printer name, driver to be used, comment, port monitor, and so on). These parameters are passed on to Samba by the APW. If the addprinter command is designed in a way that it can create a new printer (through writing correct printcap entries on legacy systems, or execute the **lpadmin** command on more modern systems) and create the associated share in, then the APW will in effect really create a new printer on Samba and the UNIX print subsystem!

17.12 Migration of Classical Printing to Samba

The basic NT-style printer driver management has not changed considerably in 3.0 over the 2.2.x releases (apart from many small improvements). Here migration should be quite easy, especially if you followed previous advice to stop using deprecated parameters in your setup. For migrations from an existing 2.0.x setup, or if you continued Windows 9x/Me-style printing in your Samba 2.2 installations, it is more of an effort. Please read the appropriate release notes and the HOWTO Collection for Samba-2.2.x. You can follow several paths. Here are possible scenarios for migration:

- You need to study and apply the new Windows NT printer and driver support. Previously used parameters *printer driver file*, *printer driver* and *printer driver location* are no longer supported.

- If you want to take advantage of Windows NT printer driver support, you also need to migrate the Windows 9x/Me drivers to the new setup.

- An existing `printers.def` file (the one specified in the now removed parameter *printer driver file*) will no longer work with Samba-3. In 3.0, smbd attempts to locate a Windows 9x/Me driver files for the printer in *[print$]* and additional settings in the TDB and only there; if it fails, it will *not* (as 2.2.x used to do) drop down to using a `printers.def` (and all associated parameters). The make_printerdef tool is removed and there is no backward compatibility for this.

- You need to install a Windows 9x/Me driver into the *[print$]* share for a printer on your Samba host. The driver files will be stored in the "*WIN40/0*" subdirectory of *[print$]*, and some other settings and information go into the printing-related TDBs.

- If you want to migrate an existing `printers.def` file into the new setup, the only current solution is to use the Windows NT APW to install the NT drivers and the 9x/Me drivers. This can be scripted using smbclient and rpcclient. See the Imprints installation client at:

  `http://imprints.sourceforge.net/`

  for an example. See also the discussion of rpcclient usage in the "*CUPS Printing*" section.

17.13 Publishing Printer Information in Active Directory or LDAP

This will be addressed in a later update of this document. If you wish to volunteer your services to help document this, please contact John H Terpstra.[4]

17.14 Common Errors

17.14.1 I Give My Root Password but I Do Not Get Access

Do not confuse the root password which is valid for the UNIX system (and in most cases stored in the form of a one-way hash in a file named `/etc/shadow`), with the password used to authenticate against Samba. Samba does not know the UNIX password. Root access to Samba resources requires that a Samba account for root must first be created. This is done with the **smbpasswd** command as follows:

```
root#  smbpasswd -a root
New SMB password: secret
Retype new SMB password: secret
```

[4] mail://jht@samba.org

17.14.2 My Print Jobs Get Spooled into the Spooling Directory, but Then Get Lost

Do not use the existing UNIX print system spool directory for the Samba spool directory. It may seem convenient and a savings of space, but it only leads to problems. The two must be separate.

Chapter 18

CUPS PRINTING SUPPORT

18.1 Introduction

18.1.1 Features and Benefits

The Common UNIX Print System (CUPS[1]) has become quite popular. All major Linux distributions now ship it as their default printing system. To many, it is still a mystical tool. Mostly, it just works. People tend to regard it as a "*black box*" that they do not want to look into as long as it works. But once there is a little problem, they are in trouble to find out where to start debugging it. Refer to the chapter "*Classical Printing*" that contains a lot of information that is relevant for CUPS.

CUPS sports quite a few unique and powerful features. While their basic functions may be grasped quite easily, they are also new. Because they are different from other, more traditional printing systems, it is best not to try and apply any prior knowledge about printing to this new system. Rather, try to understand CUPS from the beginning. This documentation will lead you to a complete understanding of CUPS. Let's start with the most basic things first.

18.1.2 Overview

CUPS is more than just a print spooling system. It is a complete printer management system that complies with the new Internet Printing Protocol (IPP). IPP is an industry and Internet Engineering Task Force (IETF) standard for network printing. Many of its functions can be managed remotely (or locally) via a Web browser (giving you a platform-independent access to the CUPS print server). Additionally, it has the traditional command line and several more modern GUI interfaces (GUI interfaces developed by third parties, like KDE's overwhelming KDEPrint[2]).

CUPS allows creation of "*raw*" printers (i.e., no print file format translation) as well as "*smart*" printers (i.e., CUPS does file format conversion as required for the printer). In many ways this gives CUPS similar capabilities to the MS Windows print monitoring system. Of course, if you are a CUPS advocate, you would argue that CUPS is better! In any case, let

[1] http://www.cups.org/

[2] http://printing.kde.org/

us now move on to explore how one may configure CUPS for interfacing with MS Windows print clients via Samba.

18.2 Basic CUPS Support Configuration

Printing with CUPS in the most basic `smb.conf` setup in Samba-3.0 (as was true for 2.2.x) only needs two settings: *printing* = cups and *printcap* = cups. CUPS does not need a printcap file. However, the **`cupsd.conf`** configuration file knows of two related directives that control how such a file will be automatically created and maintained by CUPS for the convenience of third-party applications (example: *Printcap /etc/printcap* and *PrintcapFormat BSD*). Legacy programs often require the existence of a printcap file containing printer names or they will refuse to print. Make sure CUPS is set to generate and maintain a printcap file. For details, see **man cupsd.conf** and other CUPS-related documentation, like the wealth of documents on your CUPS server itself: `http://localhost:631/documentation.html`.

18.2.1 Linking smbd with libcups.so

Samba has a special relationship to CUPS. Samba can be compiled with CUPS library support. Most recent installations have this support enabled. Per default, CUPS linking is compiled into smbd and other Samba binaries. Of course, you can use CUPS even if Samba is not linked against `libcups.so` — but there are some differences in required or supported configuration.

When Samba is compiled against `libcups`, *printcap* = cups uses the CUPS API to list printers, submit jobs, query queues, and so on. Otherwise it maps to the System V commands with an additional **-oraw** option for printing. On a Linux system, you can use the **ldd** utility to find out details (ldd may not be present on other OS platforms, or its function may be embodied by a different command):

```
root# ldd 'which smbd'
libssl.so.0.9.6 => /usr/lib/libssl.so.0.9.6 (0x4002d000)
libcrypto.so.0.9.6 => /usr/lib/libcrypto.so.0.9.6 (0x4005a000)
libcups.so.2 => /usr/lib/libcups.so.2 (0x40123000)
[....]
```

The line `libcups.so.2 => /usr/lib/libcups.so.2 (0x40123000)` shows there is CUPS support compiled into this version of Samba. If this is the case, and printing = cups is set, then *any otherwise manually set print command in `smb.conf` is ignored.* This is an important point to remember!

TIP

Should it be necessary, for any reason, to set your own print commands, you can do this by setting *printing* = sysv. However, you will loose all the benefits of tight CUPS/Samba integration. When you do this you must manually configure the printing system commands (most important: *print command*; other commands are *lppause command*, *lpresume command*, *lpq command*, *lprm command*, *queuepause command* and *queue resume command*).

18.2.2 Simple smb.conf Settings for CUPS

To summarize, Example 18.1 shows simplest printing-related setup for `smb.conf` to enable basic CUPS support:

Example 18.1. Simplest printing-related smb.conf

```
[global]
load printers = yes
printing = cups
printcap name = cups

[printers]
comment = All Printers
path = /var/spool/samba
browseable = no
public = yes
guest ok = yes
writable = no
printable = yes
printer admin = root, @ntadmins
```

This is all you need for basic printing setup for CUPS. It will print all graphic, text, PDF, and PostScript files submitted from Windows clients. However, most of your Windows users would not know how to send these kinds of files to print without opening a GUI application. Windows clients tend to have local printer drivers installed, and the GUI application's print buttons start a printer driver. Your users also rarely send files from the command line. Unlike UNIX clients, they hardly submit graphic, text or PDF formatted files directly to the spooler. They nearly exclusively print from GUI applications with a "*printer driver*" hooked in between the application's native format and the print-data-stream. If the backend printer is not a PostScript device, the print data stream is "*binary,*" sensible only for the target printer. Read on to learn which problem this may cause and how to avoid it.

18.2.3 More Complex CUPS smb.conf Settings

Example 18.2 is a slightly more complex printing-related setup for `smb.conf`. It enables general CUPS printing support for all printers, but defines one printer share, which is set up differently.

Example 18.2. Overriding global CUPS settings for one printer

```
[global]
printing = cups
printcap name = cups
load printers = yes

[printers]
comment = All Printers
path = /var/spool/samba
public = yes
guest ok = yes
writable = no
printable = yes
printer admin = root, @ntadmins

[special_printer]
comment = A special printer with his own settings
path = /var/spool/samba-special
printing = sysv
printcap = lpstat
print command = echo "NEW: `date`: printfile %f" >> /tmp/smbprn.log ; \
echo " `date`: p-%p s-%s f-%f" >> /tmp/smbprn.log ; \
echo " `date`: j-%j J-%J z-%z c-%c" >> /tmp/smbprn.log : rm %f
public = no
guest ok = no
writeable = no
printable = yes
printer admin = kurt
hosts deny = 0.0.0.0
hosts allow = turbo_xp, 10.160.50.23, 10.160.51.60
```

This special share is only there for testing purposes. It does not write the print job to a file. It just logs the job parameters known to Samba into the `/tmp/smbprn.log` file and deletes the jobfile. Moreover, the ***printer admin*** of this share is "*kurt*" (not the "*@ntadmins*" group), guest access is not allowed, the share isn't published to the Network Neighborhood (so you need to know it is there), and it only allows access from only three hosts. To prevent CUPS kicking in and taking over the print jobs for that share, we need to set ***printing*** = sysv and ***printcap*** = lpstat.

18.3 Advanced Configuration

Before we delve into all the configuration options, let us clarify a few points. *Network printing needs to be organized and setup correctly.* This frequently doesn't happen. Legacy systems or small business LAN environments often lack design and good housekeeping.

18.3.1 Central Spooling vs. "*Peer-to-Peer*" Printing

Many small office or home networks, as well as badly organized larger environments, allow each client a direct access to available network printers. This is generally a bad idea. It often blocks one client's access to the printer when another client's job is printing. It might freeze the first client's application while it is waiting to get rid of the job. Also, there are frequent complaints about various jobs being printed with their pages mixed with each other. A better concept is the usage of a print server: it routes all jobs through one central system, which responds immediately, takes jobs from multiple concurrent clients at the same time, and in turn transfers them to the printer(s) in the correct order.

18.3.2 Raw Print Serving — Vendor Drivers on Windows Clients

Most traditionally configured UNIX print servers acting on behalf of Samba's Windows clients represented a really simple setup. Their only task was to manage the "*raw*" spooling of all jobs handed to them by Samba. This approach meant that the Windows clients were expected to prepare the print job file that its ready to be sent to the printing device. Here is a native (vendor-supplied) Windows printer driver for the target device needed to be installed on each and every client.

It is possible to configure CUPS, Samba and your Windows clients in the same traditional and simple way. When CUPS printers are configured for RAW print-through mode operation, it is the responsibility of the Samba client to fully render the print job (file). The file must be sent in a format that is suitable for direct delivery to the printer. Clients need to run the vendor-provided drivers to do this. In this case, CUPS will not do any print file format conversion work.

18.3.3 Installation of Windows Client Drivers

The printer drivers on the Windows clients may be installed in two functionally different ways:

- Manually install the drivers locally on each client, one by one; this yields the old *LanMan* style printing and uses a `\\sambaserver\printershare` type of connection.
- Deposit and prepare the drivers (for later download) on the print server (Samba); this enables the clients to use "*Point'n'Print*" to get drivers semi-automatically installed the first time they access the printer; with this method NT/200x/XP clients use the *SPOOLSS/MS-RPC* type printing calls.

The second method is recommended for use over the first.

18.3.4 Explicitly Enable *"raw"* Printing for *application/octet-stream*

If you use the first option (drivers are installed on the client side), there is one setting to take care of: CUPS needs to be told that it should allow *"raw"* printing of deliberate (binary) file formats. The CUPS files that need to be correctly set for RAW mode printers to work are:

- `/etc/cups/mime.types`
- `/etc/cups/mime.convs`

Both contain entries (at the end of the respective files) which must be uncommented to allow RAW mode operation. In `/etc/cups/mime.types`, make sure this line is present:

```
application/octet-stream
```

In `/etc/cups/mime.convs`, have this line:

```
application/octet-stream   application/vnd.cups-raw   0   -
```

If these two files are not set up correctly for raw Windows client printing, you may encounter the dreaded `Unable to convert file 0` in your CUPS error_log file.

> NOTE
>
> 
>
> Editing the `mime.convs` and the `mime.types` file does not *enforce* *"raw"* printing, it only *allows* it.

CUPS being a more security-aware printing system than traditional ones does not by default allow a user to send deliberate (possibly binary) data to printing devices. This could be easily abused to launch a *"Denial of Service"* attack on your printer(s), causing at least the loss of a lot of paper and ink. *"Unknown"* data are tagged by CUPS as *`MIME type: application/octet-stream`* and not allowed to go to the printer. By default, you can only send other (known) MIME types *"raw"*. Sending data *"raw"* means that CUPS does not try to convert them and passes them to the printer untouched (see the next chapter for even more background explanations).

This is all you need to know to get the CUPS/Samba combo printing *"raw"* files prepared by Windows clients, which have vendor drivers locally installed. If you are not interested in background information about more advanced CUPS/Samba printing, simply skip the remaining sections of this chapter.

18.3.5 Driver Upload Methods

This section describes three familiar methods, plus one new one, by which printer drivers may be uploaded.

If you want to use the MS-RPC type printing, you must upload the drivers onto the Samba server first (*[print$]* share). For a discussion on how to deposit printer drivers on the Samba host (so the Windows clients can download and use them via "*Point'n'Print*"), please refer to the previous chapter of this HOWTO Collection. There you will find a description or reference to three methods of preparing the client drivers on the Samba server:

- The GUI, "*Add Printer Wizard*" *upload-from-a-Windows-client* method.
- The command line, "*smbclient/rpcclient*" upload-from-a-UNIX-workstation method.
- The Imprints Toolset method.

These three methods apply to CUPS all the same. A new and more convenient way to load the Windows drivers into Samba is provided if you use CUPS:

- the *cupsaddsmb* utility.

cupsaddsmb is discussed in much detail further below. But we first explore the CUPS filtering system and compare the Windows and UNIX printing architectures.

18.4 Advanced Intelligent Printing with PostScript Driver Download

We now know how to set up a "*dump*" printserver, that is, a server which is spooling printjobs "*raw*", leaving the print data untouched.

Possibly you need to setup CUPS in a smarter way. The reasons could be manifold:

- Maybe your boss wants to get monthly statistics: Which printer did how many pages? What was the average data size of a job? What was the average print run per day? What are the typical hourly peaks in printing? Which department prints how much?
- Maybe you are asked to setup a print quota system: Users should not be able to print more jobs, once they have surpassed a given limit per period.
- Maybe your previous network printing setup is a mess and must be re-organized from a clean beginning.
- Maybe you have experiencing too many "*blue screens*" originating from poorly debugged printer drivers running in NT "*kernel mode*"?

These goals cannot be achieved by a raw print server. To build a server meeting these requirements, you'll first need to learn about how CUPS works and how you can enable its features.

What follows is the comparison of some fundamental concepts for Windows and UNIX printing; then follows a description of the CUPS filtering system, how it works and how you can tweak it.

18.4.1 GDI on Windows – PostScript on UNIX

Network printing is one of the most complicated and error-prone day-to-day tasks any user or administrator may encounter. This is true for all OS platforms. And there are reasons for this.

You can't expect most file formats to just throw them toward printers and they get printed. There needs to be a file format conversion in between. The problem is that there is no common standard for print file formats across all manufacturers and printer types. While PostScript (trademark held by Adobe) and, to an extent, PCL (trademark held by HP) have developed into semi-official "*standards*" by being the most widely used PDLs Page Description Languages (PDLs), there are still many manufacturers who "*roll their own*" (their reasons may be unacceptable license fees for using printer-embedded PostScript interpreters, and so on).

18.4.2 Windows Drivers, GDI and EMF

In Windows OS, the format conversion job is done by the printer drivers. On MS Windows OS platforms all application programmers have at their disposal a built-in API, the Graphical Device Interface (GDI), as part and parcel of the OS itself to base themselves on. This GDI core is used as one common unified ground for all Windows programs to draw pictures, fonts and documents *on screen* as well as *on paper* (print). Therefore, printer driver developers can standardize on a well-defined GDI output for their own driver input. Achieving WYSIWYG ("*What You See Is What You Get*") is relatively easy, because the on-screen graphic primitives, as well as the on-paper drawn objects, come from one common source. This source, the GDI, often produces a file format called Enhanced MetaFile (EMF). The EMF is processed by the printer driver and converted to the printer-specific file format.

NOTE

To the GDI foundation in MS Windows, Apple has chosen to put paper and screen output on a common foundation for their (BSD-UNIX-based, did you know?) Mac OS X and Darwin Operating Systems. Their *Core Graphic Engine* uses a *PDF* derivative for all display work.

18.4.3 UNIX Printfile Conversion and GUI Basics

In UNIX and Linux, there is no comparable layer built into the OS kernel(s) or the X (screen display) server. Every application is responsible for itself to create its print output. Fortunately, most use PostScript and that at least gives some common ground. Unfortunately, there are many different levels of quality for this PostScript. And worse, there is a huge difference (and no common root) in the way the same document is displayed on screen and how it is presented on paper. WYSIWYG is more difficult to achieve. This goes back to the time, decades ago, when the predecessors of X.org, designing the UNIX foundations and protocols for Graphical User Interfaces, refused to take responsibility for "*paper output*"

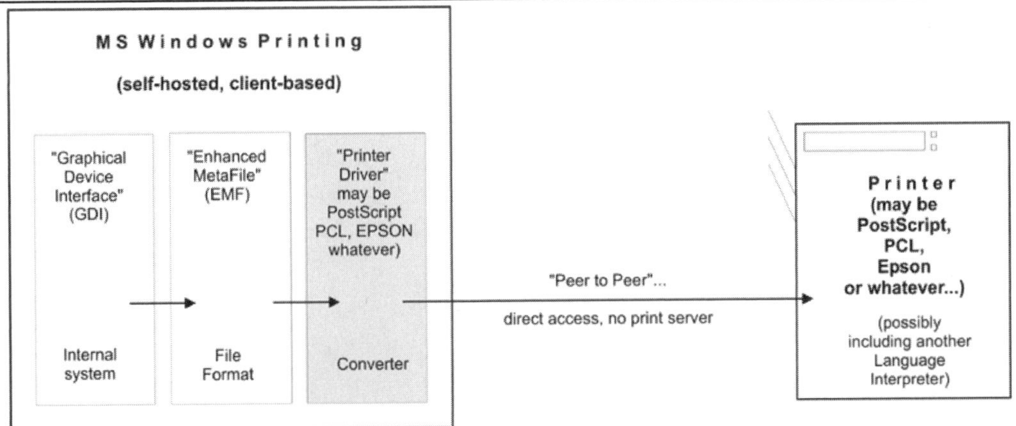

Figure 18.1. Windows printing to a local printer.

also, as some had demanded at the time, and restricted itself to "*on-screen only.*" (For some years now, the "*Xprint*" project has been under development, attempting to build printing support into the X framework, including a PostScript and a PCL driver, but it is not yet ready for prime time.) You can see this unfavorable inheritance up to the present day by looking into the various "*font*" directories on your system; there are separate ones for fonts used for X display and fonts to be used on paper.

The PostScript programming language is an "*invention*" by Adobe Inc., but its specifications have been published to the full. Its strength lies in its powerful abilities to describe graphical objects (fonts, shapes, patterns, lines, curves, and dots), their attributes (color, linewidth) and the way to manipulate (scale, distort, rotate, shift) them. Because of its open specification, anybody with the skill can start writing his own implementation of a PostScript interpreter and use it to display PostScript files on screen or on paper. Most graphical output devices are based on the concept of "*raster images*" or "*pixels*" (one notable exception is pen plotters). Of course, you can look at a PostScript file in its textual form and you will be reading its PostScript code, the language instructions which need to be interpreted by a rasterizer. Rasterizers produce pixel images, which may be displayed on screen by a viewer program or on paper by a printer.

18.4.4 PostScript and Ghostscript

So, UNIX is lacking a common ground for printing on paper and displaying on screen. Despite this unfavorable legacy for UNIX, basic printing is fairly easy if you have PostScript printers at your disposal. The reason is these devices have a built-in PostScript language "*interpreter,*" also called a Raster Image Processor (RIP) (which makes them more expensive than other types of printers); throw PostScript toward them, and they will spit out your printed pages. Their RIP is doing all the hard work of converting the PostScript drawing commands into a bitmap picture as you see it on paper, in a resolution as done by your printer. This is no different to PostScript printing a file from a Windows origin.

NOTE

Traditional UNIX programs and printing systems — while using PostScript — are largely not PPD-aware. PPDs are "*PostScript Printer Description*" files. They enable you to specify and control all options a printer supports: duplexing, stapling and punching. Therefore, UNIX users for a long time couldn't choose many of the supported device and job options, unlike Windows or Apple users. But now there is CUPS.

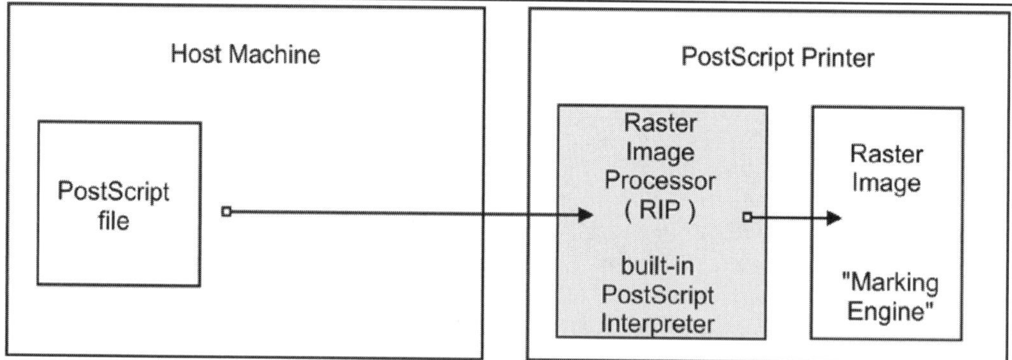

Figure 18.2. Printing to a PostScript printer.

However, there are other types of printers out there. These do not know how to print PostScript. They use their own Page Description Language (PDL, often proprietary). To print to them is much more demanding. Since your UNIX applications mostly produce PostScript, and since these devices do not understand PostScript, you need to convert the printfiles to a format suitable for your printer on the host before you can send it away.

18.4.5 Ghostscript — the Software RIP for Non-PostScript Printers

Here is where Ghostscript kicks in. Ghostscript is the traditional (and quite powerful) PostScript interpreter used on UNIX platforms. It is a RIP in software, capable of doing a *lot* of file format conversions for a very broad spectrum of hardware devices as well as software file formats. Ghostscript technology and drivers are what enable PostScript printing to non-PostScript hardware.

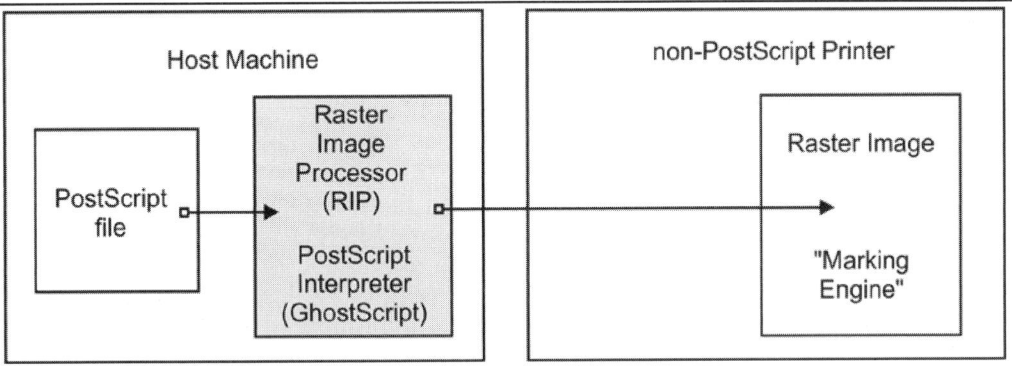

Figure 18.3. Ghostscript as a RIP for non-postscript printers.

TIP

Use the "*gs -h*" command to check for all built-in "*devices*" of your Ghostscript version. If you specify a parameter of *-sDEVICE=png256* on your Ghostscript command line, you are asking Ghostscript to convert the input into a PNG file. Naming a "*device*" on the command line is the most important single parameter to tell Ghostscript exactly how it should render the input. New Ghostscript versions are released at fairly regular intervals, now by artofcode LLC. They are initially put under the "*AFPL*" license, but re-released under the GNU GPL as soon as the next AFPL version appears. GNU Ghostscript is probably the version installed on most Samba systems. But it has some deficiencies. Therefore, ESP Ghostscript was developed as an enhancement over GNU Ghostscript, with lots of bug-fixes, additional devices and improvements. It is jointly maintained by developers from CUPS, Gimp-Print, MandrakeSoft, SuSE, RedHat, and Debian. It includes the "*cups*" device (essential to print to non-PS printers from CUPS).

18.4.6 PostScript Printer Description (PPD) Specification

While PostScript in essence is a Page Description Language (PDL) to represent the page layout in a device-independent way, real-world print jobs are always ending up being output on hardware with device-specific features. To take care of all the differences in hardware and to allow for innovations, Adobe has specified a syntax and file format for PostScript Printer Description (PPD) files. Every PostScript printer ships with one of these files.

PPDs contain all the information about general and special features of the given printer model: Which different resolutions can it handle? Does it have a Duplexing Unit? How many paper trays are there? What media types and sizes does it take? For each item, it also names the special command string to be sent to the printer (mostly inside the PostScript file) in order to enable it.

Information from these PPDs is meant to be taken into account by the printer drivers. Therefore, installed as part of the Windows PostScript driver for a given printer is the printer's PPD. Where it makes sense, the PPD features are presented in the drivers' UI dialogs to display to the user a choice of print options. In the end, the user selections are somehow written (in the form of special PostScript, PJL, JCL or vendor-dependent commands) into the PostScript file created by the driver.

WARNING

A PostScript file that was created to contain device-specific commands for achieving a certain print job output (e.g., duplexed, stapled and punched) on a specific target machine, may not print as expected, or may not be printable at all on other models; it also may not be fit for further processing by software (e.g., by a PDF distilling program).

18.4.7 Using Windows-Formatted Vendor PPDs

CUPS can handle all spec-compliant PPDs as supplied by the manufacturers for their PostScript models. Even if a vendor might not have mentioned our favorite OS in his manuals and brochures, you can safely trust this: *If you get the Windows NT version of the PPD, you can use it unchanged in CUPS* and thus access the full power of your printer just like a Windows NT user could!

TIP

To check the spec compliance of any PPD online, go to `http://www.cups.org/testppd.php` and upload your PPD. You will see the results displayed immediately. CUPS in all versions after 1.1.19 has a much more strict internal PPD parsing and checking code enabled; in case of printing trouble, this online resource should be one of your first pitstops.

WARNING

For real PostScript printers, *do not* use the *Foomatic* or *cupsomatic* PPDs from Linuxprinting.org. With these devices, the original vendor-provided PPDs are always the first choice!

TIP

If you are looking for an original vendor-provided PPD of a specific device, and you know that an NT4 box (or any other Windows box) on your LAN has the PostScript driver installed, just use **smbclient //NT4-box/print\$ -U username** to access the Windows directory where all printer driver files are stored. First look in the `W32X86/2` subdir for the PPD you are seeking.

18.4.8 CUPS Also Uses PPDs for Non-PostScript Printers

CUPS also uses specially crafted PPDs to handle non-PostScript printers. These PPDs are usually not available from the vendors (and no, you can't just take the PPD of a PostScript printer with the same model name and hope it works for the non-PostScript version too). To understand how these PPDs work for non-PS printers, we first need to dive deeply into the CUPS filtering and file format conversion architecture. Stay tuned.

18.5 The CUPS Filtering Architecture

The core of the CUPS filtering system is based on Ghostscript. In addition to Ghostscript, CUPS uses some other filters of its own. You (or your OS vendor) may have plugged in even more filters. CUPS handles all data file formats under the label of various MIME types. Every incoming printfile is subjected to an initial auto-typing. The auto-typing determines its given MIME type. A given MIME type implies zero or more possible filtering chains relevant to the selected target printer. This section discusses how MIME types recognition and conversion rules interact. They are used by CUPS to automatically setup a working filtering chain for any given input data format.

If CUPS rasterizes a PostScript file natively to a bitmap, this is done in two stages:

- The first stage uses a Ghostscript device named "*cups*" (this is since version 1.1.15) and produces a generic raster format called "*CUPS raster*".
- The second stage uses a "*raster driver*" that converts the generic CUPS raster to a device-specific raster.

Make sure your Ghostscript version has the "*cups*" device compiled in (check with **gs -h | grep cups**). Otherwise you may encounter the dreaded `Unable to convert file 0` in your CUPS error_log file. To have "*cups*" as a device in your Ghostscript, you either need to patch GNU Ghostscript and re-compile, or use ESP Ghostscript[3]. The superior alternative is ESP Ghostscript. It supports not just CUPS, but 300 other devices too (while GNU Ghostscript supports only about 180). Because of this broad output device support, ESP Ghostscript is the first choice for non-CUPS spoolers, too. It is now recommended by Linuxprinting.org for all spoolers.

[3] http://www.cups.org/ghostscript.php

CUPS printers may be setup to use external rendering paths. One of the most common is provided by the Foomatic/cupsomatic concept from Linuxprinting.org.[4] This uses the classical Ghostscript approach, doing everything in one step. It does not use the "*cups*" device, but one of the many others. However, even for Foomatic/cupsomatic usage, best results and broadest printer model support is provided by ESP Ghostscript (more about cupsomatic/Foomatic, particularly the new version called now *foomatic-rip*, follows below).

18.5.1 MIME Types and CUPS Filters

CUPS reads the file `/etc/cups/mime.types` (and all other files carrying a `*.types` suffix in the same directory) upon startup. These files contain the MIME type recognition rules that are applied when CUPS runs its auto-typing routines. The rule syntax is explained in the man page for `mime.types` and in the comments section of the `mime.types` file itself. A simple rule reads like this:

```
application/pdf         pdf string(0,%PDF)
```

This means if a filename has either a `.pdf` suffix or if the magic string *%PDF* is right at the beginning of the file itself (offset 0 from the start), then it is a PDF file (*application/pdf*). Another rule is this:

```
application/postscript  ai eps ps string(0,%!) string(0,<04>%!)
```

If the filename has one of the suffixes `.ai`, `.eps`, `.ps` or if the file itself starts with one of the strings *%!* or *<04>%!*, it is a generic PostScript file (*application/postscript*).

WARNING

Don't confuse the other mime.types files your system might be using with the one in the `/etc/cups/` directory.

[4] http://www.linuxprinting.org/

NOTE

There is an important difference between two similar MIME types in CUPS: one is *application/postscript*, the other is *application/vnd.cups-postscript*. While *application/postscript* is meant to be device independent (job options for the file are still outside the PS file content, embedded in command line or environment variables by CUPS), *application/vnd.cups-postscript* may have the job options inserted into the PostScript data itself (where applicable). The transformation of the generic PostScript (*application/postscript*) to the device-specific version (*application/vnd.cups-postscript*) is the responsibility of the CUPS *pstops* filter. pstops uses information contained in the PPD to do the transformation.

CUPS can handle ASCII text, HP-GL, PDF, PostScript, DVI, and many image formats (GIF. PNG, TIFF, JPEG, Photo-CD, SUN-Raster, PNM, PBM, SGI-RGB, and more) and their associated MIME types with its filters.

18.5.2 MIME Type Conversion Rules

CUPS reads the file `/etc/cups/mime.convs` (and all other files named with a `*.convs` suffix in the same directory) upon startup. These files contain lines naming an input MIME type, an output MIME type, a format conversion filter that can produce the output from the input type and virtual costs associated with this conversion. One example line reads like this:

```
application/pdf         application/postscript   33   pdftops
```

This means that the ***pdftops*** filter will take ***application/pdf*** as input and produce ***application/postscript*** as output; the virtual cost of this operation is 33 CUPS-$. The next filter is more expensive, costing 66 CUPS-$:

```
application/vnd.hp-HPGL application/postscript   66   hpgltops
```

This is the ***hpgltops***, which processes HP-GL plotter files to PostScript.

```
application/octet-stream
```

Here are two more examples:

```
application/x-shell     application/postscript   33    texttops
```

```
text/plain                 application/postscript   33    texttops
```

The last two examples name the *texttops* filter to work on *text/plain* as well as on *application/x-shell*. (Hint: This differentiation is needed for the syntax highlighting feature of *texttops*).

18.5.3 Filtering Overview

There are many more combinations named in `mime.convs`. However, you are not limited to use the ones pre-defined there. You can plug in any filter you like into the CUPS framework. It must meet, or must be made to meet, some minimal requirements. If you find (or write) a cool conversion filter of some kind, make sure it complies to what CUPS needs and put in the right lines in `mime.types` and `mime.convs`, then it will work seamlessly inside CUPS.

18.5.3.1 Filter requirements

The mentioned "*CUPS requirements*" for filters are simple. Take filenames or `stdin` as input and write to `stdout`. They should take these 5 or 6 arguments: *printer job user title copies options [filename]*

Printer — The name of the printer queue (normally this is the name of the filter being run).

job — The numeric job ID for the job being printed.

user — The string from the originating-user-name attribute.

title — The string from the job-name attribute.

copies — The numeric value from the number-copies attribute.

options — The job options.

filename — (Optionally) The print request file (if missing, filters expected data fed through `stdin`). In most cases, it is easy to write a simple wrapper script around existing filters to make them work with CUPS.

18.5.4 Prefilters

As previously stated, PostScript is the central file format to any UNIX-based printing system. From PostScript, CUPS generates raster data to feed non-PostScript printers.

But what happens if you send one of the supported non-PS formats to print? Then CUPS runs "*pre-filters*" on these input formats to generate PostScript first. There are pre-filters to create PS from ASCII text, PDF, DVI, or HP-GL. The outcome of these filters is always of MIME type ***application/postscript*** (meaning that any device-specific print options are not yet embedded into the PostScript by CUPS, and that the next filter to be called is pstops). Another pre-filter is running on all supported image formats, the ***imagetops*** filter. Its outcome is always of MIME type ***application/vnd.cups-postscript*** (not application/postscript), meaning it has the print options already embedded into the file.

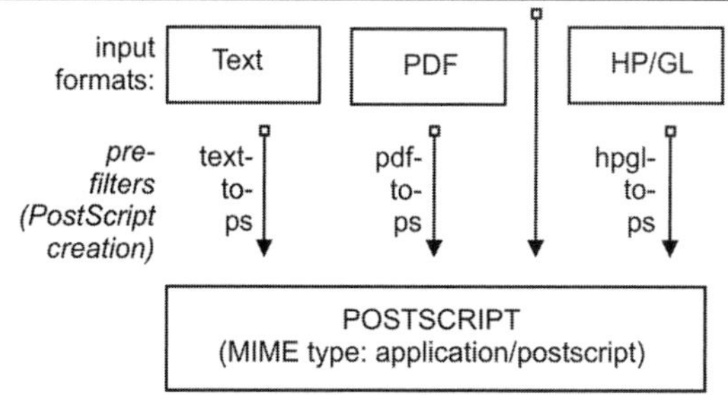

Figure 18.4. Pre-filtering in CUPS to form PostScript.

18.5.5 pstops

pstops is the filter to convert ***application/postscript*** to ***application/vnd.cups-postscript***. It was said above that this filter inserts all device-specific print options (commands to the printer to ask for the duplexing of output, or stapling and punching it, and so on) into the PostScript file.

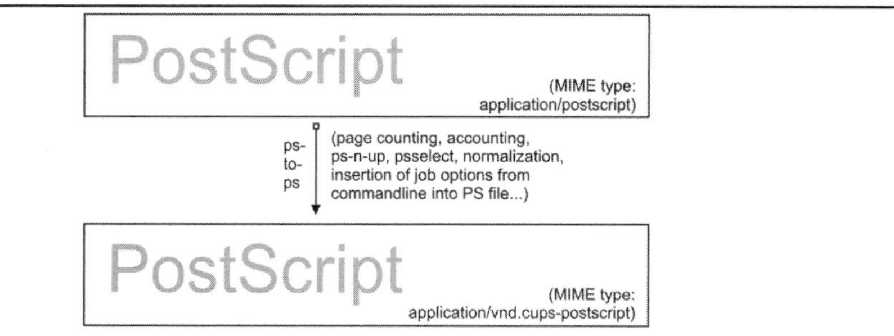

Figure 18.5. Adding device-specific print options.

This is not all. Other tasks performed by it are:

- Selecting the range of pages to be printed (if you choose to print only pages "*3, 6, 8-11, 16, 19-21*", or only the odd numbered ones).
- Putting 2 or more logical pages on one sheet of paper (the so-called "*number-up*" function).

- Counting the pages of the job to insert the accounting information into the `/var/log/cups/page_log`.

18.5.6 pstoraster

pstoraster is at the core of the CUPS filtering system. It is responsible for the first stage of the rasterization process. Its input is of MIME type application/vnd.cups-postscript; its output is application/vnd.cups-raster. This output format is not yet meant to be printable. Its aim is to serve as a general purpose input format for more specialized *raster drivers* that are able to generate device-specific printer data.

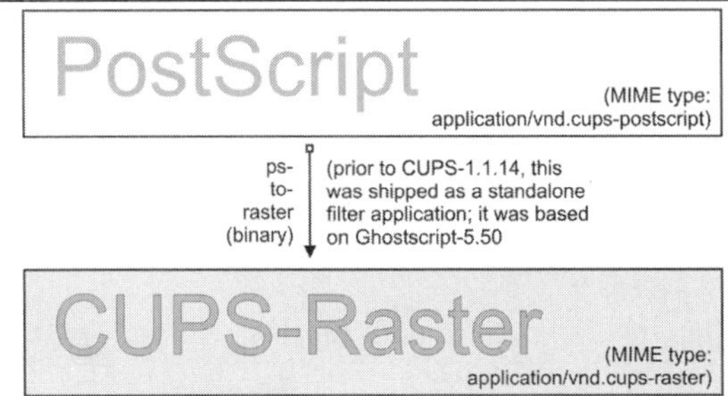

Figure 18.6. PostScript to intermediate raster format.

CUPS raster is a generic raster format with powerful features. It is able to include per-page information, color profiles, and more, to be used by the following downstream raster drivers. Its MIME type is registered with IANA and its specification is, of course, completely open. It is designed to make it quite easy and inexpensive for manufacturers to develop Linux and UNIX raster drivers for their printer models, should they choose to do so. CUPS always takes care for the first stage of rasterization so these vendors do not need to care about Ghostscript complications (in fact, there is currently more than one vendor financing the development of CUPS raster drivers).

CUPS versions before version 1.1.15 were shipping a binary (or source code) standalone filter, named *pstoraster*. *pstoraster* was derived from GNU Ghostscript 5.50, and could be installed besides and in addition to any GNU or AFPL Ghostscript package without conflicting.

>From version 1.1.15, this has changed. The functions for this have been integrated back into Ghostscript (now based on GNU Ghostscript version 7.05). The *pstoraster* filter is now a simple shell script calling **gs** with the **-sDEVICE=cups** parameter. If your Ghostscript does not show a success on asking for **gs -h |grep cups**, you might not be able to print. Update your Ghostscript.

18.5.7 imagetops and imagetoraster

In the section about pre-filters, we mentioned the pre-filter that generates PostScript from image formats. The *imagetoraster* filter is used to convert directly from image to raster,

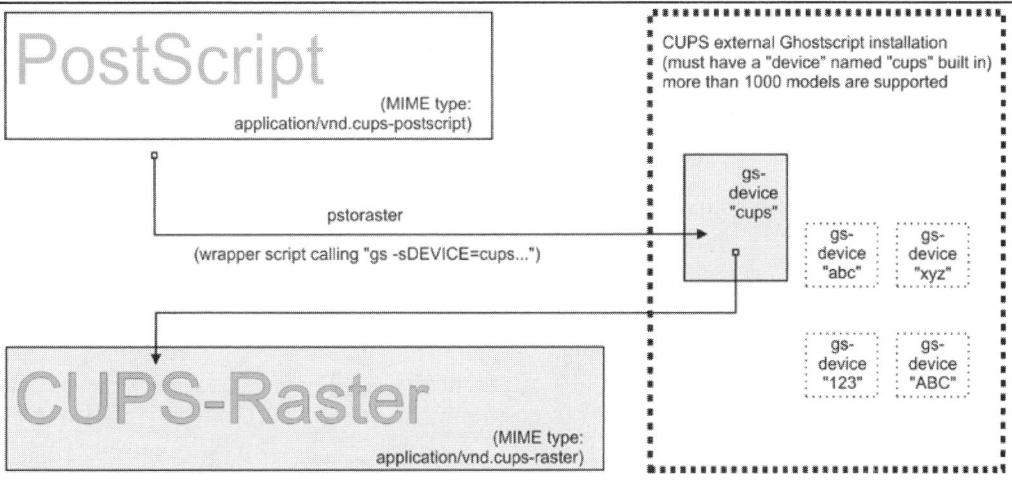

Figure 18.7. CUPS-raster production using Ghostscript.

without the intermediate PostScript stage. It is used more often than the above mentioned pre-filters. A summarizing flowchart of image file filtering is shown in Figure 18.8.

18.5.8 rasterto [printers specific]

CUPS ships with quite different raster drivers processing CUPS raster. On my system I find in /usr/lib/cups/filter/ these: *rastertoalps*, *rastertobj*, *rastertoepson*, *rastertoescp*, *rastertopcl*, *rastertoturboprint*, *rastertoapdk*, *rastertodymo*, *rastertoescp*, *rastertohp*, and *rastertoprinter*. Don't worry if you have less than this; some of these are installed by commercial add-ons to CUPS (like *rastertoturboprint*), others (like *rastertoprinter*) by third-party driver development projects (such as Gimp-Print) wanting to cooperate as closely as possible with CUPS.

18.5.9 CUPS Backends

The last part of any CUPS filtering chain is a backend. Backends are special programs that send the print-ready file to the final device. There is a separate backend program for any transfer protocol of sending printjobs over the network, or for every local interface. Every CUPS print queue needs to have a CUPS "*device-URI*" associated with it. The device URI is the way to encode the backend used to send the job to its destination. Network device-URIs are using two slashes in their syntax, local device URIs only one, as you can see from the following list. Keep in mind that local interface names may vary much from my examples, if your OS is not Linux:

usb — This backend sends printfiles to USB-connected printers. An example for the CUPS device-URI to use is: `usb:/dev/usb/lp0`.

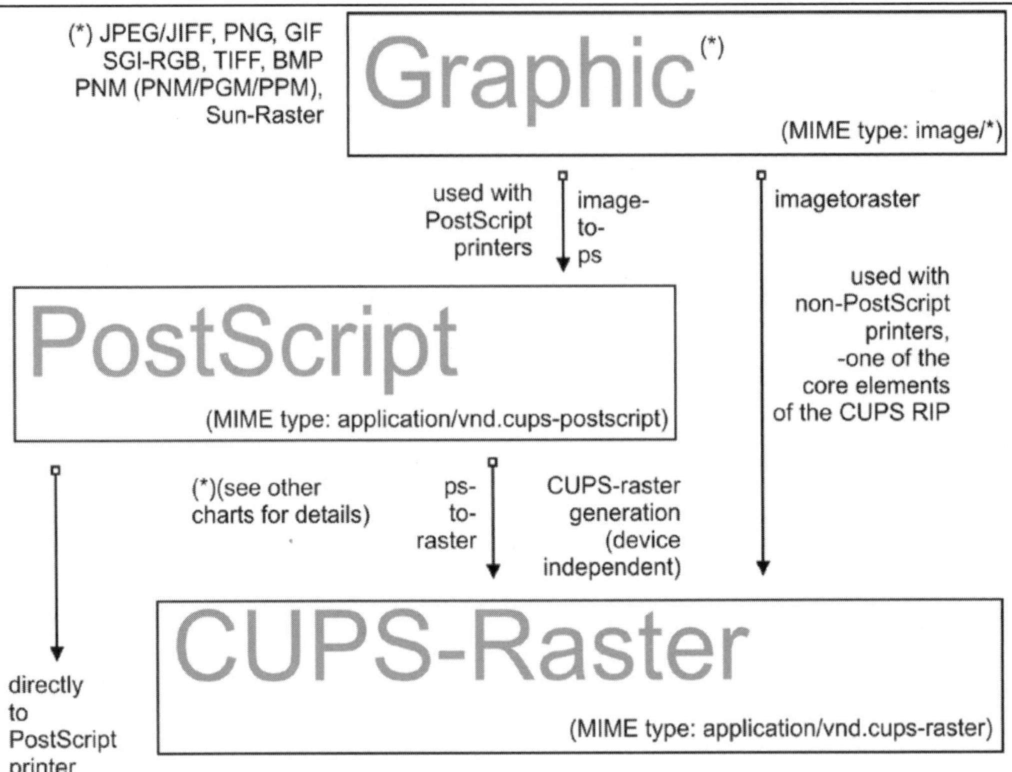

Figure 18.8. Image format to CUPS-raster format conversion.

serial — This backend sends printfiles to serially connected printers. An example for the CUPS device-URI to use is: `serial:/dev/ttyS0?baud=11500`.

parallel — This backend sends printfiles to printers connected to the parallel port. An example for the CUPS device-URI to use is: `parallel:/dev/lp0`.

scsi — This backend sends printfiles to printers attached to the SCSI interface. An example for the CUPS device-URI to use is: `scsi:/dev/sr1`.

lpd — This backend sends printfiles to LPR/LPD connected network printers. An example for the CUPS device-URI to use is: `lpd://remote_host_name/remote_queue_name`.

AppSocket/HP JetDirect — This backend sends printfiles to AppSocket (a.k.a. "HP JetDirect") connected network printers. An example for the CUPS device-URI to use is: `socket://10.11.12.13:9100`.

ipp — This backend sends printfiles to IPP connected network printers (or to other CUPS

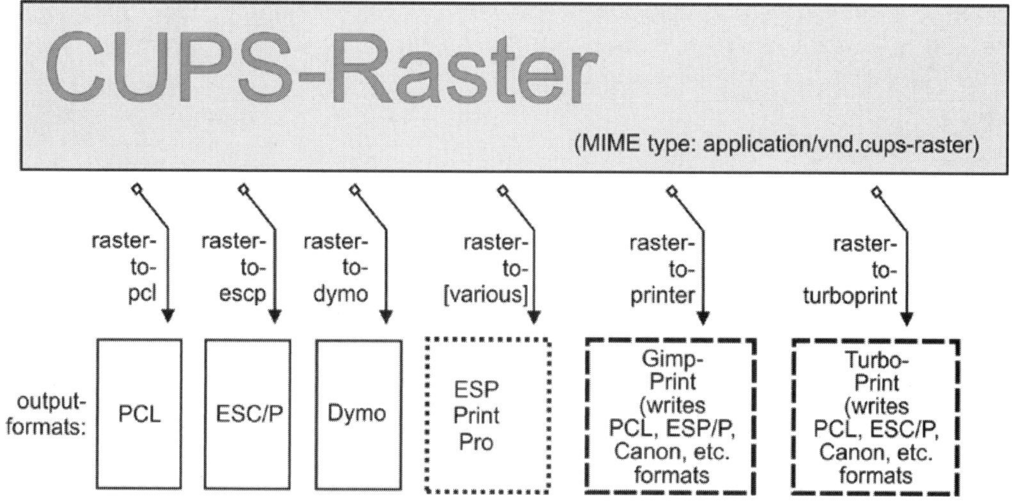

Figure 18.9. Raster to printer-specific formats.

servers). Examples for CUPS device-URIs to use are: `ipp::/192.193.194.195/ipp` (for many HP printers) or `ipp://remote_cups_server/printers/remote_printer_name`.

http — This backend sends printfiles to HTTP connected printers. (The http:// CUPS backend is only a symlink to the ipp:// backend.) Examples for the CUPS device-URIs to use are: `http:://192.193.194.195:631/ipp` (for many HP printers) or `http://remote_cups_server:631/printers/remote_printer_name`.

smb — This backend sends printfiles to printers shared by a Windows host. An example for CUPS device-URIs that may be used includes:

```
smb://workgroup/server/printersharename
smb://server/printersharename
smb://username:password@workgroup/server/printersharename
smb://username:password@server/printersharename
```

The smb:// backend is a symlink to the Samba utility ***smbspool*** (does not ship with CUPS). If the symlink is not present in your CUPS backend directory, have your root user create it: **ln -s 'which smbspool' /usr/lib/cups/backend/smb**.

It is easy to write your own backends as shell or Perl scripts, if you need any modification or extension to the CUPS print system. One reason could be that you want to create "*special*" printers that send the printjobs as email (through a "*mailto:/*" backend), convert them to PDF (through a "*pdfgen:/*" backend) or dump them to "*/dev/null*". (In fact I have the system-wide default printer set up to be connected to a devnull:/ backend: there are just too many people sending jobs without specifying a printer, or scripts and programs which do not name a printer. The system-wide default deletes the job and sends a polite email back to the $USER asking him to always specify the correct printer name.)

Not all of the mentioned backends may be present on your system or usable (depending on your hardware configuration). One test for all available CUPS backends is provided by the *lpinfo* utility. Used with the `-v` parameter, it lists all available backends:

```
$ lpinfo -v
```

18.5.10 The Role of cupsomatic/foomatic

cupsomatic filters may be the most widely used on CUPS installations. You must be clear about the fact that these were not developed by the CUPS people. They are a third party add-on to CUPS. They utilize the traditional Ghostscript devices to render jobs for CUPS. When troubleshooting, you should know about the difference. Here the whole rendering process is done in one stage, inside Ghostscript, using an appropriate device for the target printer. *cupsomatic* uses PPDs that are generated from the Foomatic Printer & Driver Database at Linuxprinting.org.

You can recognize these PPDs from the line calling the *cupsomatic* filter:

```
*cupsFilter: "application/vnd.cups-postscript  0  cupsomatic"
```

You may find this line among the first 40 or so lines of the PPD file. If you have such a PPD installed, the printer shows up in the CUPS Web interface with a *foomatic* namepart for the driver description. *cupsomatic* is a Perl script that runs Ghostscript with all the complicated command line options auto-constructed from the selected PPD and command line options give to the printjob.

However, *cupsomatic* is now deprecated. Its PPDs (especially the first generation of them, still in heavy use out there) are not meeting the Adobe specifications. You might also suffer difficulties when you try to download them with *"Point'n'Print"* to Windows clients. A better and more powerful successor is now in a stable beta-version: it is called *foomatic-rip*. To use *foomatic-rip* as a filter with CUPS, you need the new-type PPDs. These have a similar but different line:

```
*cupsFilter: "application/vnd.cups-postscript  0  foomatic-rip"
```

The PPD generating engine at Linuxprinting.org has been revamped. The new PPDs comply to the Adobe spec. On top, they also provide a new way to specify different quality levels (hi-res photo, normal color, grayscale, and draft) with a single click, whereas before you could have required five or more different selections (media type, resolution, inktype and dithering algorithm). There is support for custom-size media built in. There is support to switch print-options from page to page in the middle of a job. And the best thing is the new foomatic-rip now works seamlessly with all legacy spoolers too (like LPRng, BSD-LPD, PDQ, PPR and so on), providing for them access to use PPDs for their printing.

18.5.11 The Complete Picture

If you want to see an overview of all the filters and how they relate to each other, the complete picture of the puzzle is at the end of this document.

18.5.12 mime.convs

CUPS auto-constructs all possible filtering chain paths for any given MIME type, and every printer installed. But how does it decide in favor or against a specific alternative? (There may often be cases where there is a choice of two or more possible filtering chains for the same target printer.) Simple. You may have noticed the figures in the third column of the mime.convs file. They represent virtual costs assigned to this filter. Every possible filtering chain will sum up to a total *"filter cost."* CUPS decides for the most *"inexpensive"* route.

> TIP
>
> 
>
> The setting of *`FilterLimit 1000`* in `cupsd.conf` will not allow more filters to run concurrently than will consume a total of 1000 virtual filter cost. This is an efficient way to limit the load of any CUPS server by setting an appropriate *"FilterLimit"* value. A FilterLimit of 200 allows roughly one job at a time, while a FilterLimit of 1000 allows approximately five jobs maximum at a time.

18.5.13 "*Raw*" Printing

You can tell CUPS to print (nearly) any file *"raw"*. *"Raw"* means it will not be filtered. CUPS will send the file to the printer *"as is"* without bothering if the printer is able to digest it. Users need to take care themselves that they send sensible data formats only. Raw printing can happen on any queue if the *"`-o raw`"* option is specified on the command line. You can also set up raw-only queues by simply not associating any PPD with it. This command:

```
$ lpadmin -P rawprinter -v socket://11.12.13.14:9100 -E
```

sets up a queue named *"rawprinter"*, connected via the *"socket"* protocol (a.k.a. *"HP JetDirect"*) to the device at IP address 11.12.1.3.14, using port 9100. (If you had added a PPD with **-P /path/to/PPD** to this command line, you would have installed a *"normal"* print queue.

CUPS will automatically treat each job sent to a queue as a *"raw"* one, if it can't find a PPD associated with the queue. However, CUPS will only send known MIME types (as defined in its own mime.types file) and refuse others.

18.5.14 application/octet-stream Printing

Any MIME type with no rule in the `/etc/cups/mime.types` file is regarded as unknown or *`application/octet-stream`* and will not be sent. Because CUPS refuses to print unknown MIME types per default, you will probably have experienced the fact that print jobs originating from Windows clients were not printed. You may have found an error message in your CUPS logs like:

```
Unable to convert file 0 to printable format for job
```

To enable the printing of *`application/octet-stream`* files, edit these two files:

- `/etc/cups/mime.convs`
- `/etc/cups/mime.types`

Both contain entries (at the end of the respective files) which must be uncommented to allow RAW mode operation for *`application/octet-stream`*. In `/etc/cups/mime.types` make sure this line is present:

```
application/octet-stream
```

This line (with no specific auto-typing rule set) makes all files not otherwise auto-typed a member of *`application/octet-stream`*. In `/etc/cups/mime.convs`, have this line:

```
application/octet-stream   application/vnd.cups-raw   0   -
```

This line tells CUPS to use the *Null Filter* (denoted as "-", doing nothing at all) on *`application/octet-stream`*, and tag the result as *`application/vnd.cups-raw`*. This last one is always a green light to the CUPS scheduler to now hand the file over to the backend connecting to the printer and sending it over.

> NOTE
>
> 
>
> Editing the `mime.convs` and the `mime.types` file does not *enforce* "*raw*" printing, it only *allows* it.

CUPS being a more security-aware printing system than traditional ones does not by default allow one to send deliberate (possibly binary) data to printing devices. (This could be easily abused to launch a Denial of Service attack on your printer(s), causing at least the loss of a lot of paper and ink...) "*Unknown*" data are regarded by CUPS as *MIME type application/octet-stream*. While you *can* send data "*raw*", the MIME type for these must be one that is known to CUPS and an allowed one. The file `/etc/cups/mime.types` defines the "*rules*" of how CUPS recognizes MIME types. The file `/etc/cups/mime.convs` decides which file conversion filter(s) may be applied to which MIME types.

18.5.15 PostScript Printer Descriptions (PPDs) for Non-PS Printers

Originally PPDs were meant to be used for PostScript printers only. Here, they help to send device-specific commands and settings to the RIP which processes the jobfile. CUPS has extended this scope for PPDs to cover non-PostScript printers too. This was not difficult, because it is a standardized file format. In a way it was logical too: CUPS handles PostScript and uses a PostScript RIP (Ghostscript) to process the jobfiles. The only difference is: a PostScript printer has the RIP built-in, for other types of printers the Ghostscript RIP runs on the host computer.

PPDs for a non-PS printer have a few lines that are unique to CUPS. The most important one looks similar to this:

```
*cupsFilter: application/vnd.cups-raster  66   rastertoprinter
```

It is the last piece in the CUPS filtering puzzle. This line tells the CUPS daemon to use as a last filter *rastertoprinter*. This filter should be served as input an *application/vnd.cups-raster* MIME type file. Therefore, CUPS should auto-construct a filtering chain, which delivers as its last output the specified MIME type. This is then taken as input to the specified *rastertoprinter* filter. After this the last filter has done its work (*rastertoprinter* is a Gimp-Print filter), the file should go to the backend, which sends it to the output device.

CUPS by default ships only a few generic PPDs, but they are good for several hundred printer models. You may not be able to control different paper trays, or you may get larger margins than your specific model supports. See Table 18.1 for summary information.

Table 18.1. PPDs shipped with CUPS

PPD file	Printer type
deskjet.ppd	older HP inkjet printers and compatible
deskjet2.ppd	newer HP inkjet printers and compatible
dymo.ppd	label printers
epson9.ppd	Epson 24pin impact printers and compatible
epson24.ppd	Epson 24pin impact printers and compatible
okidata9.ppd	Okidata 9pin impact printers and compatible
okidat24.ppd	Okidata 24pin impact printers and compatible
stcolor.ppd	older Epson Stylus Color printers
stcolor2.ppd	newer Epson Stylus Color printers
stphoto.ppd	older Epson Stylus Photo printers
stphoto2.ppd	newer Epson Stylus Photo printers
laserjet.ppd	all PCL printers. Further below is a discussion of several other driver/PPD-packages suitable for use with CUPS.

18.5.16 *cupsomatic/foomatic-rip* Versus *native CUPS* Printing

Native CUPS rasterization works in two steps:

- First is the *pstoraster* step. It uses the special CUPS device from ESP Ghostscript 7.05.x as its tool.
- Second comes the *rasterdriver* step. It uses various device-specific filters; there are several vendors who provide good quality filters for this step. Some are free software, some are shareware/non-free and some are proprietary.

Often this produces better quality (and has several more advantages) than other methods.

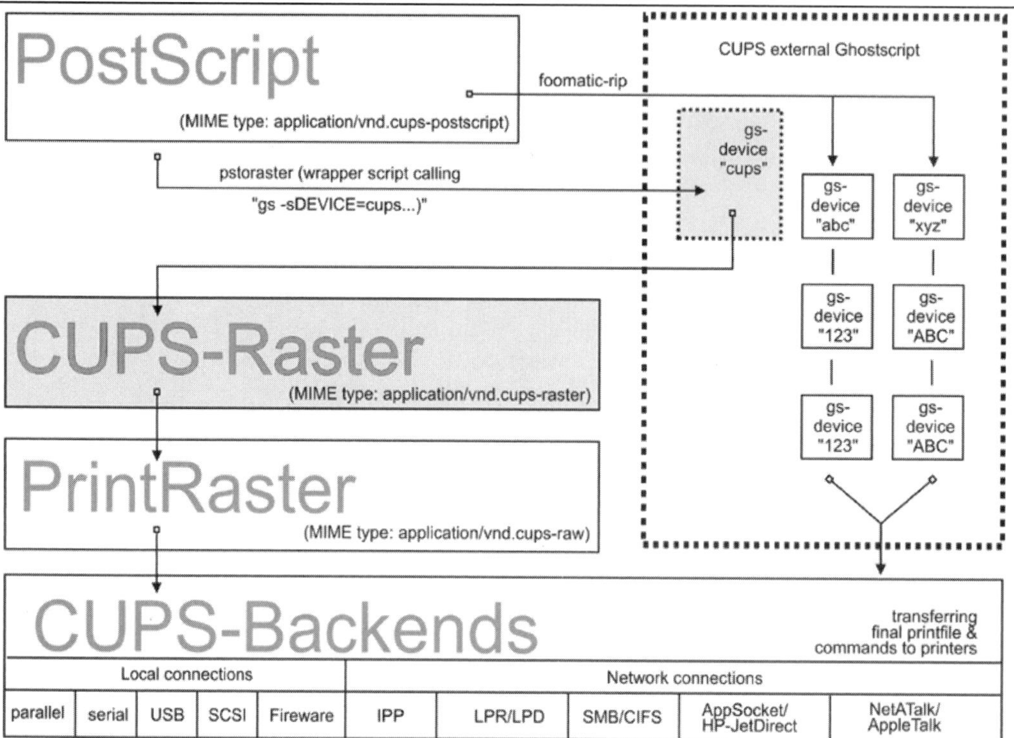

Figure 18.10. cupsomatic/foomatic Processing versus Native CUPS.

One other method is the *cupsomatic/foomatic-rip* way. Note that *cupsomatic* is *not* made by the CUPS developers. It is an independent contribution to printing development, made by people from Linuxprinting.org [5]. *cupsomatic* is no longer developed and maintained and is no longer supported. It has now been replaced by *foomatic-rip*. *foomatic-rip* is a complete re-write of the old *cupsomatic* idea, but very much improved and generalized to other (non-CUPS) spoolers. An upgrade to foomatic-rip is strongly advised, especially if you are upgrading to a recent version of CUPS, too.

Both the *cupsomatic* (old) and the *foomatic-rip* (new) methods from Linuxprinting.org use the traditional Ghostscript print file processing, doing everything in a single step. It therefore relies on all the other devices built into Ghostscript. The quality is as good (or bad) as Ghostscript rendering is in other spoolers. The advantage is that this method supports many printer models not supported (yet) by the more modern CUPS method.

[5] see also `http://www.cups.org/cups-help.html`

Of course, you can use both methods side by side on one system (and even for one printer, if you set up different queues) and find out which works best for you.

cupsomatic kidnaps the printfile after the *application/vnd.cups-postscript* stage and deviates it through the CUPS-external, system-wide Ghostscript installation. Therefore the printfile bypasses the *pstoraster* filter (and also bypasses the CUPS-raster-drivers *rastertosomething*). After Ghostscript finished its rasterization, *cupsomatic* hands the rendered file directly to the CUPS backend. The flowchart in Figure 18.10 illustrates the difference between native CUPS rendering and the *Foomatic/cupsomatic* method.

18.5.17 Examples for Filtering Chains

Here are a few examples of commonly occurring filtering chains to illustrate the workings of CUPS.

Assume you want to print a PDF file to an HP JetDirect-connected PostScript printer, but you want to print the pages 3-5, 7, 11-13 only, and you want to print them "*two-up*" and "*duplex*":

- Your print options (page selection as required, two-up, duplex) are passed to CUPS on the command line.
- The (complete) PDF file is sent to CUPS and autotyped as *application/pdf*.
- The file therefore must first pass the *pdftops* pre-filter, which produces PostScript MIME type *application/postscript* (a preview here would still show all pages of the original PDF).
- The file then passes the *pstops* filter that applies the command line options: it selects the pages 2-5, 7 and 11-13, creates an imposed layout "*2 pages on 1 sheet*" and inserts the correct "*duplex*" command (as defined in the printer's PPD) into the new PostScript file; the file is now of PostScript MIME type *application/vnd.cups-postscript*.
- The file goes to the *socket* backend, which transfers the job to the printers.

The resulting filter chain, therefore, is as drawn in Figure 18.11.

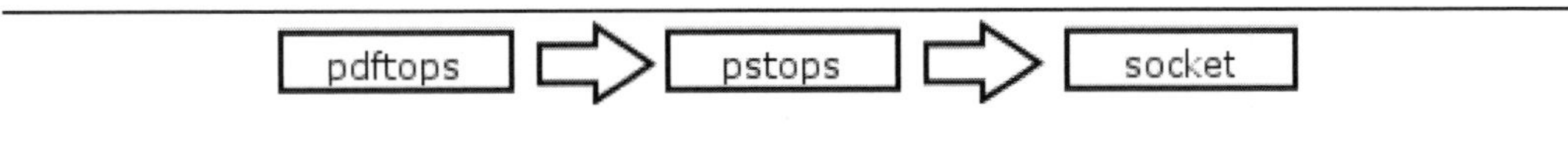

Figure 18.11. PDF to socket chain.

Assume your want to print the same filter to an USB-connected Epson Stylus Photo printer installed with the CUPS `stphoto2.ppd`. The first few filtering stages are nearly the same:

- Your print options (page selection as required, two-up, duplex) are passed to CUPS on the commandline.
- The (complete) PDF file is sent to CUPS and autotyped as *application/pdf*.

- The file must first pass the ***pdftops*** pre-filter, which produces PostScript MIME type ***application/postscript*** (a preview here would still show all pages of the original PDF).

- The file then passes the "*pstops*" filter that applies the commandline options: it selects the pages 2-5, 7 and 11-13, creates an imposed layout "*two pages on one sheet*" and inserts the correct "*duplex*" command... (Oops — this printer and PPD do not support duplex printing at all — so this option will be ignored) into the new PostScript file; the file is now of PostScript MIME type ***application/vnd.cups-postscript***.

- The file then passes the ***pstoraster*** stage and becomes MIME type ***application/cups-raster***.

- Finally, the ***rastertoepson*** filter does its work (as indicated in the printer's PPD), creating the rinter-specific raster data and embedding any user-selected print-options into the print data stream.

- The file goes to the ***usb*** backend, which transfers the job to the printers.

The resulting filter chain therefore is as drawn in Figure 18.12.

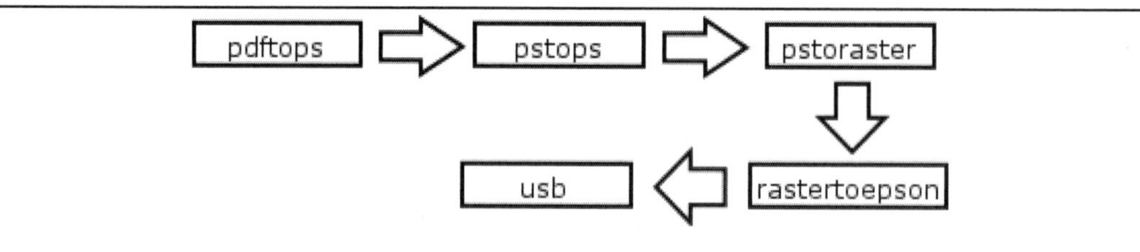

Figure 18.12. PDF to USB chain.

18.5.18 Sources of CUPS Drivers/PPDs

On the Internet you can now find many thousands of CUPS-PPD files (with their companion filters), in many national languages supporting more than thousand non-PostScript models.

- ESP PrintPro[6] (commercial, non-free) is packaged with more than three thousand PPDs, ready for successful use "*out of the box*" on Linux, Mac OS X, IBM-AIX, HP-UX, Sun-Solaris, SGI-IRIX, Compaq Tru64, Digital UNIX, and some more commercial Unices (it is written by the CUPS developers themselves and its sales help finance the further development of CUPS, as they feed their creators).

- The Gimp-Print-Project[7] (GPL, free software) provides around 140 PPDs (supporting nearly 400 printers, many driven to photo quality output), to be used alongside the Gimp-Print CUPS filters.

- TurboPrint[8] (shareware, non-free) supports roughly the same amount of printers in excellent quality.

[6] http://wwwl.easysw.com/printpro/
[7] http://gimp-print.sourceforge.net/
[8] http://www.turboprint.com/

- OMNI[9] (LPGL, free) is a package made by IBM, now containing support for more than 400 printers, stemming from the inheritance of IBM OS/2 Know-How ported over to Linux (CUPS support is in a beta-stage at present).
- HPIJS[10] (BSD-style licenses, free) supports around 150 of HP's own printers and is also providing excellent print quality now (currently available only via the Foomatic path).
- Foomatic/cupsomatic[11] (LPGL, free) from Linuxprinting.org are providing PPDs for practically every Ghostscript filter known to the world (including Omni, Gimp-Print and HPIJS).

18.5.19 Printing with Interface Scripts

CUPS also supports the usage of "*interface scripts*" as known from System V AT&T printing systems. These are often used for PCL printers, from applications that generate PCL print jobs. Interface scripts are specific to printer models. They have a similar role as PPDs for PostScript printers. Interface scripts may inject the Escape sequences as required into the print data stream, if the user has chosen to select a certain paper tray, or print landscape, or use A3 paper, etc. Interfaces scripts are practically unknown in the Linux realm. On HP-UX platforms they are more often used. You can use any working interface script on CUPS too. Just install the printer with the **-i** option:

```
root# lpadmin -p pclprinter -v socket://11.12.13.14:9100 \
  -i /path/to/interface-script
```

Interface scripts might be the "*unknown animal*" to many. However, with CUPS they provide the easiest way to plug in your own custom-written filtering script or program into one specific print queue (some information about the traditional usage of interface scripts is to be found at `http://playground.sun.com/printing/documentation/interface.html`).

18.6 Network Printing (Purely Windows)

Network printing covers a lot of ground. To understand what exactly goes on with Samba when it is printing on behalf of its Windows clients, let's first look at a "*purely Windows*" setup: Windows clients with a Windows NT print server.

18.6.1 From Windows Clients to an NT Print Server

Windows clients printing to an NT-based print server have two options. They may:

- Execute the driver locally and render the GDI output (EMF) into the printer-specific format on their own.

[9]http://www-124.ibm.com/developerworks/oss/linux/projects/omni/
[10]http://hpinkjet.sourceforge.net/
[11]http://www.linuxprinting.org/

- Send the GDI output (EMF) to the server, where the driver is executed to render the printer specific output.

Both print paths are shown in the flowcharts in Figure 18.13 and Figure 18.14.

18.6.2 Driver Execution on the Client

In the first case the print server must spool the file as raw, meaning it shouldn't touch the jobfile and try to convert it in any way. This is what a traditional UNIX-based print server can do too, and at a better performance and more reliably than an NT print server. This is what most Samba administrators probably are familiar with. One advantage of this setup is that this "*spooling-only*" print server may be used even if no driver(s) for UNIX are available it is sufficient to have the Windows client drivers available; and installed on the clients.

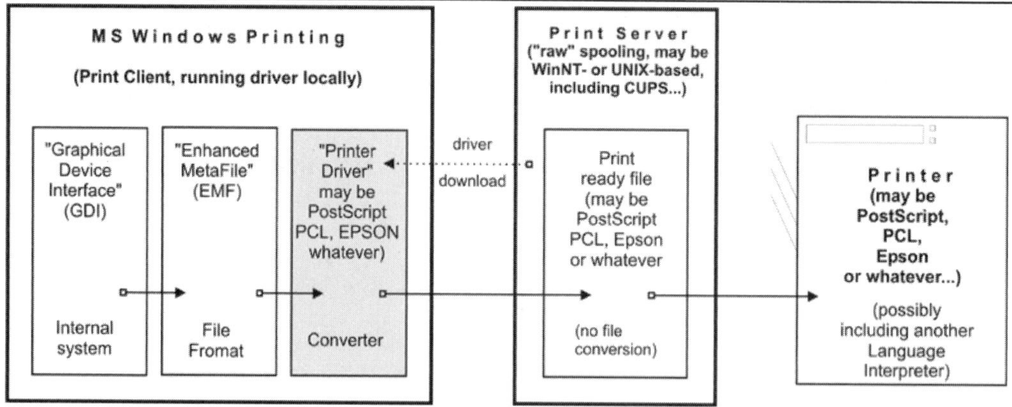

Figure 18.13. Print driver execution on the client.

18.6.3 Driver Execution on the Server

The other path executes the printer driver on the server. The client transfers print files in EMF format to the server. The server uses the PostScript, PCL, ESC/P or other driver to convert the EMF file into the printer-specific language. It is not possible for UNIX to do the same. Currently, there is no program or method to convert a Windows client's GDI output on a UNIX server into something a printer could understand.

However, there is something similar possible with CUPS. Read on.

18.7 Network Printing (Windows Clients — UNIX/Samba Print Servers)

Since UNIX print servers *cannot* execute the Win32 program code on their platform, the picture is somewhat different. However, this does not limit your options all that much. On the contrary, you may have a way here to implement printing features that are not possible otherwise.

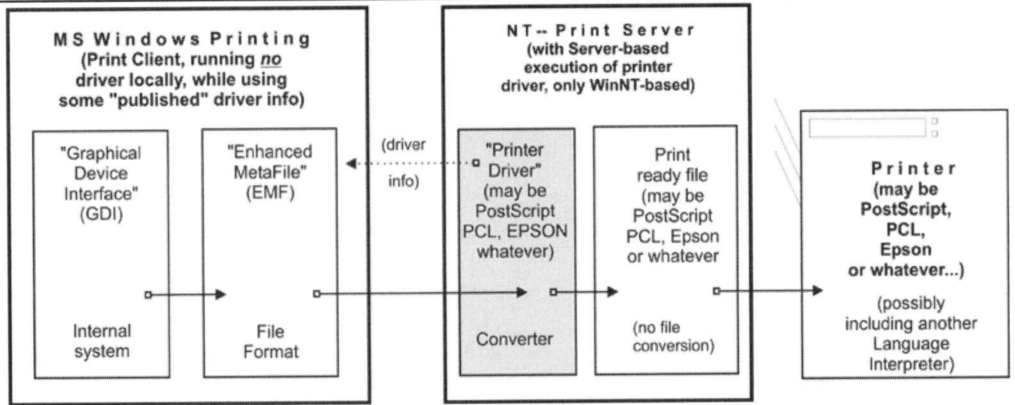

Figure 18.14. Print driver execution on the server.

18.7.1 From Windows Clients to a CUPS/Samba Print Server

Here is a simple recipe showing how you can take advantage of CUPS' powerful features for the benefit of your Windows network printing clients:

- Let the Windows clients send PostScript to the CUPS server.
- Let the CUPS server render the PostScript into device-specific raster format.

This requires the clients to use a PostScript driver (even if the printer is a non-PostScript model. It also requires that you have a driver on the CUPS server.

First, to enable CUPS-based rinting through Samba the following options should be set in your `smb.conf` file [global] section:

```
printing = cups
printcap = cups
```

When these parameters are specified, all manually set print directives (like *print command*, or *lppause command*) in `smb.conf` (as well as in Samba itself) will be ignored. Instead, Samba will directly interface with CUPS through its application program interface (API), as long as Samba has been compiled with CUPS library (libcups) support. If Samba has not been compiled with CUPS support, and if no other print commands are set up, then printing will use the *System V* AT&T command set, with the -oraw option automatically passing through (if you want your own defined print commands to work with a Samba that has CUPS support compiled in, simply use *printing* = sysv).

18.7.2 Samba Receiving Jobfiles and Passing Them to CUPS

Samba *must* use its own spool directory (it is set by a line similar to *path* = /var/spool/samba, in the *[printers]* or *[printername]* section of `smb.conf`). Samba receives the job in its own spool space and passes it into the spool directory of CUPS (the CUPS spooling directory is set by the *RequestRoot* directive, in a line that defaults to *RequestRoot /var/spool/cups*). CUPS checks the access rights of its spool dir and resets it to healthy

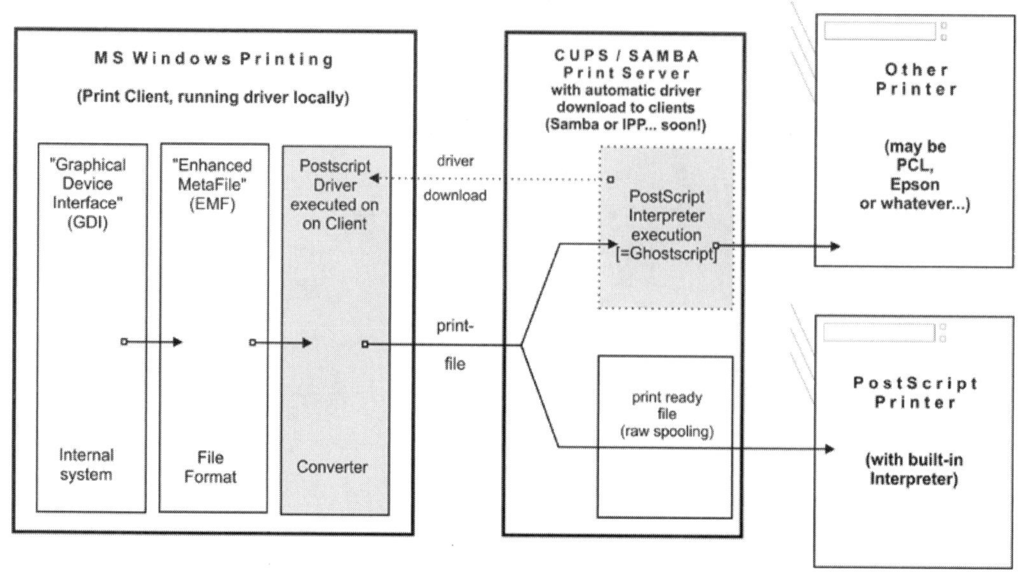

Figure 18.15. Printing via CUPS/Samba server.

values with every restart. We have seen quite a few people who had used a common spooling space for Samba and CUPS, and were struggling for weeks with this "*problem*."

A Windows user authenticates only to Samba (by whatever means is configured). If Samba runs on the same host as CUPS, you only need to allow "*localhost*" to print. If they run on different machines, you need to make sure the Samba host gets access to printing on CUPS.

18.8 Network PostScript RIP

This section discusses the use of CUPS filters on the server — configuration where clients make use of a PostScript driver with CUPS-PPDs.

PPDs can control all print device options. They are usually provided by the manufacturer, if you own a PostScript printer, that is. PPD files (PostScript Printer Descriptions) are always a component of PostScript printer drivers on MS Windows or Apple Mac OS systems. They are ASCII files containing user-selectable print options, mapped to appropriate PostScript, PCL or PJL commands for the target printer. Printer driver GUI dialogs translate these options "*on-the-fly*" into buttons and drop-down lists for the user to select.

CUPS can load, without any conversions, the PPD file from any Windows (NT is recommended) PostScript driver and handle the options. There is a Web browser interface to the print options (select `http://localhost:631/printers/` and click on one **Configure Printer** button to see it), or a command line interface (see **man lpoptions** or see if you have **lphelp** on your system). There are also some different GUI frontends on Linux/UNIX, which can present PPD options to users. PPD options are normally meant to be evaluated by the PostScript RIP on the real PostScript printer.

18.8.1 PPDs for Non-PS Printers on UNIX

CUPS does not limit itself to "*real*" PostScript printers in its usage of PPDs. The CUPS developers have extended the scope of the PPD concept to also describe available device and driver options for non-PostScript printers through CUPS-PPDs.

This is logical, as CUPS includes a fully featured PostScript interpreter (RIP). This RIP is based on Ghostscript. It can process all received PostScript (and additionally many other file formats) from clients. All CUPS-PPDs geared to non-PostScript printers contain an additional line, starting with the keyword `*cupsFilter`. This line tells the CUPS print system which printer-specific filter to use for the interpretation of the supplied PostScript. Thus CUPS lets all its printers appear as PostScript devices to its clients, because it can act as a PostScript RIP for those printers, processing the received PostScript code into a proper raster print format.

18.8.2 PPDs for Non-PS Printers on Windows

CUPS-PPDs can also be used on Windows-Clients, on top of a "*core*" PostScript driver (now recommended is the "CUPS PostScript Driver for WindowsNT/200x/XP"; you can also use the Adobe one, with limitations). This feature enables CUPS to do a few tricks no other spooler can do:

- Act as a networked PostScript RIP (Raster Image Processor), handling printfiles from all client platforms in a uniform way.
- Act as a central accounting and billing server, since all files are passed through the pstops filter and are, therefore, logged in the CUPS `page_log` file. *Note:* this cannot happen with "*raw*" print jobs, which always remain unfiltered per definition.
- Enable clients to consolidate on a single PostScript driver, even for many different target printers.

Using CUPS PPDs on Windows clients enables these to control all print job settings just as a UNIX client can do.

18.9 Windows Terminal Servers (WTS) as CUPS Clients

This setup may be of special interest to people experiencing major problems in WTS environments. WTS often need a multitude of non-PostScript drivers installed to run their clients' variety of different printer models. This often imposes the price of much increased instability.

18.9.1 Printer Drivers Running in "*Kernel Mode*" Cause Many Problems

In Windows NT printer drivers which run in "*Kernel Mode*", introduces a high risk for the stability of the system if the driver is not really stable and well-tested. And there are a lot of bad drivers out there! Especially notorious is the example of the PCL printer driver that had an additional sound module running, to notify users via soundcard of their finished jobs. Do I need to say that this one was also reliably causing "*blue screens of death*" on a regular basis?

PostScript drivers are generally well tested. They are not known to cause any problems, even though they also run in kernel mode. This might be because there have been so far only two different PostScript drivers: the ones from Adobe and the one from Microsoft. Both are well tested and are as stable as you can imagine on Windows. The CUPS driver is derived from the Microsoft one.

18.9.2 Workarounds Impose Heavy Limitations

In many cases, in an attempt to work around this problem, site administrators have resorted to restricting the allowed drivers installed on their WTS to one generic PCL and one PostScript driver. This, however, restricts the clients in the number of printer options available for them. Often they can't get out more than simplex prints from one standard paper tray, while their devices could do much better, if driven by a different driver!

18.9.3 CUPS: A "*Magical Stone*"?

Using a PostScript driver, enabled with a CUPS-PPD, seems to be a very elegant way to overcome all these shortcomings. There are, depending on the version of Windows OS you use, up to three different PostScript drivers available: Adobe, Microsoft and CUPS PostScript drivers. None of them is known to cause major stability problems on WTS (even if used with many different PPDs). The clients will be able to (again) chose paper trays, duplex printing and other settings. However, there is a certain price for this too: a CUPS server acting as a PostScript RIP for its clients requires more CPU and RAM than when just acting as a "*raw spooling*" device. Plus, this setup is not yet widely tested, although the first feedbacks look very promising.

18.9.4 PostScript Drivers with No Major Problems — Even in Kernel Mode

More recent printer drivers on W200x and XP no longer run in kernel mode (unlike Windows NT). However, both operating systems can still use the NT drivers, running in kernel mode (you can roughly tell which is which as the drivers in subdirectory "*2*" of "*W32X86*" are "*old*" ones). As was said before, the Adobe as well as the Microsoft PostScript drivers are not known to cause any stability problems. The CUPS driver is derived from the Microsoft one. There is a simple reason for this: The MS DDK (Device Development Kit) for Windows NT (which used to be available at no cost to licensees of Visual Studio) includes the source code of the Microsoft driver, and licensees of Visual Studio are allowed to use and modify it for their own driver development efforts. This is what the CUPS people have done. The license does not allow them to publish the whole of the source code. However, they have released the "*diff*" under the GPL, and if you are the owner of an "*MS DDK for Windows NT,*" you can check the driver yourself.

18.10 Configuring CUPS for Driver Download

As we have said before, all previously known methods to prepare client printer drivers on the Samba server for download and Point'n'Print convenience of Windows workstations are working with CUPS, too. These methods were described in the previous chapter. In reality, this is a pure Samba business and only relates to the Samba/Windows client relationship.

18.10.1 *cupsaddsmb*: The Unknown Utility

The **cupsaddsmb** utility (shipped with all current CUPS versions) is an alternate method to transfer printer drivers into the Samba *[print$]* share. Remember, this share is where clients expect drivers deposited and setup for download and installation. It makes the sharing of any (or all) installed CUPS printers quite easy. **cupsaddsmb** can use the Adobe PostScript driver as well as the newly developed CUPS PostScript Driver for Windows NT/200x/XP. *cupsaddsmb* does *not* work with arbitrary vendor printer drivers, but only with the *exact* driver files that are named in its man page.

The CUPS printer driver is available from the CUPS download site. Its package name is `cups-samba-[version].tar.gz` . It is preferred over the Adobe drivers since it has a number of advantages:

- It supports a much more accurate page accounting.
- It supports banner pages, and page labels on all printers.
- It supports the setting of a number of job IPP attributes (such as job-priority, page-label and job-billing).

However, currently only Windows NT, 2000 and XP are supported by the CUPS drivers. You will also need to get the respective part of Adobe driver if you need to support Windows 95, 98 and ME clients.

18.10.2 Prepare Your smb.conf for cupsaddsmb

Prior to running **cupsaddsmb**, you need the settings in `smb.conf` as shown in Example 18.3:

18.10.3 CUPS "*PostScript Driver for Windows NT/200x/XP*"

CUPS users may get the exact same packages from `http://www.cups.org/software.html`. It is a separate package from the CUPS base software files, tagged as CUPS 1.1.x Windows NT/200x/XP Printer Driver for Samba (tar.gz, 192k). The filename to download is `cups-samba-1.1.x.tar.gz`. Upon untar and unzipping, it will reveal these files:

```
root# tar xvzf cups-samba-1.1.19.tar.gz
cups-samba.install
cups-samba.license
cups-samba.readme
cups-samba.remove
```

Example 18.3. smb.conf for cupsaddsmb usage

```
[global]
load printers = yes
printing = cups
printcap name = cups

[printers]
comment = All Printers
path = /var/spool/samba
browseable = no
public = yes
# setting depends on your requirements
guest ok = yes
writable = no
printable = yes
printer admin = root

[print$]
comment = Printer Drivers
path = /etc/samba/drivers
browseable = yes
guest ok = no
read only = yes
write list = root
```

```
cups-samba.ss
```

These have been packaged with the ESP meta packager software EPM. The `*.install` and `*.remove` files are simple shell scripts, which untars the `*.ss` (the `*.ss` is nothing else but a tar-archive, which can be untarred by "*tar*" too). Then it puts the content into `/usr/share/cups/drivers/`. This content includes three files:

```
root# tar tv cups-samba.ss
cupsdrvr.dll
cupsui.dll
cups.hlp
```

The *cups-samba.install* shell scripts are easy to handle:

```
root# ./cups-samba.install
[....]
Installing software...
Updating file permissions...
```

```
Running post-install commands...
Installation is complete.
```

The script should automatically put the driver files into the /usr/share/cups/drivers/ directory.

WARNING

Due to a bug, one recent CUPS release puts the cups.hlp driver file into/usr/share/drivers/ instead of /usr/share/cups/drivers/. To work around this, copy/move the file (after running the **./cups-samba.install** script) manually to the correct place.

```
root# cp /usr/share/drivers/cups.hlp /usr/share/cups/drivers/
```

This new CUPS PostScript driver is currently binary-only, but free of charge. No complete source code is provided (yet). The reason is that it has been developed with the help of the Microsoft Driver Developer Kit (DDK) and compiled with Microsoft Visual Studio 6. Driver developers are not allowed to distribute the whole of the source code as free software. However, CUPS developers released the "*diff*" in source code under the GPL, so anybody with a license of Visual Studio and a DDK will be able to compile for him/herself.

18.10.4 Recognizing Different Driver Files

The CUPS drivers do not support the older Windows 95/98/Me, but only the Windows NT/2000/XP client.

Windows NT, 2000 and XP are supported by:

- cups.hlp
- cupsdrvr.dll
- cupsui.dll

Adobe drivers are available for the older Windows 95/98/Me as well as the Windows NT/2000/XP clients. The set of files is different from the different platforms.

Windows 95, 98 and ME are supported by:

- ADFONTS.MFM
- ADOBEPS4.DRV
- ADOBEPS4.HLP
- DEFPRTR2.PPD
- ICONLIB.DLL

- PSMON.DLL

Windows NT, 2000 and XP are supported by:

- ADOBEPS5.DLL
- ADOBEPSU.DLL
- ADOBEPSU.HLP

NOTE

If both the Adobe driver files and the CUPS driver files for the support of Windows NT/200x/XP are present in FIXME, the Adobe ones will be ignored and the CUPS ones will be used. If you prefer — for whatever reason — to use Adobe-only drivers, move away the three CUPS driver files. The Windows 9x/Me clients use the Adobe drivers in any case.

18.10.5 Acquiring the Adobe Driver Files

Acquiring the Adobe driver files seems to be unexpectedly difficult for many users. They are not available on the Adobe Web site as single files and the self-extracting and/or self-installing Windows-.exe is not easy to locate either. Probably you need to use the included native installer and run the installation process on one client once. This will install the drivers (and one Generic PostScript printer) locally on the client. When they are installed, share the Generic PostScript printer. After this, the client's *[print$]* share holds the Adobe files, from where you can get them with smbclient from the CUPS host.

18.10.6 ESP Print Pro PostScript Driver for Windows NT/200x/XP

Users of the ESP Print Pro software are able to install their Samba drivers package for this purpose with no problem. Retrieve the driver files from the normal download area of the ESP Print Pro software at `http://www.easysw.com/software.html`. You need to locate the link labelled "*SAMBA*" among the **Download Printer Drivers for ESP Print Pro 4.x** area and download the package. Once installed, you can prepare any driver by simply highlighting the printer in the Printer Manager GUI and select **Export Driver...** from the menu. Of course you need to have prepared Samba beforehand to handle the driver files; i.e., setup the *[print$]* share, and so on. The ESP Print Pro package includes the CUPS driver files as well as a (licensed) set of Adobe drivers for the Windows 95/98/Me client family.

18.10.7 Caveats to be Considered

Once you have run the install script (and possibly manually moved the `cups.hlp` file to `/usr/share/cups/drivers/`), the driver is ready to be put into Samba's *[print$]* share (which often maps to `/etc/samba/drivers/` and contains a subdirectory tree with *WIN40* and *W32X86* branches). You do this by running **cupsaddsmb** (see also **man cupsaddsmb** for CUPS since release 1.1.16).

TIP

You may need to put root into the smbpasswd file by running **smbpasswd**; this is especially important if you should run this whole procedure for the first time, and are not working in an environment where everything is configured for *single sign on* to a Windows Domain Controller.

Once the driver files are in the *[print$]* share and are initialized, they are ready to be downloaded and installed by the Windows NT/200x/XP clients.

NOTE

Win 9x/Me clients will not work with the CUPS PostScript driver. For these you still need to use the `ADOBE*.*` drivers as previously stated.

NOTE

It is not harmful if you still have the `ADOBE*.*` driver files from previous installations in the `/usr/share/cups/drivers/` directory. The new **cupsaddsmb** (from 1.1.16) will automatically prefer its own drivers if it finds both.

NOTE

Should your Windows clients have had the old `ADOBE*.*` files for the Adobe PostScript driver installed, the download and installation of the new CUPS PostScript driver for Windows NT/200x/XP will fail at first. You need to wipe the old driver from the clients first. It is not enough to "*delete*" the printer, as the driver files will still be kept by the clients and re-used if you try to re-install the printer. To really get rid of the Adobe driver files on the clients, open the **Printers** folder (possibly via **Start** > **Settings** > **Control Panel** > **Printers**), right-click on the folder background and select **Server Properties**. When the new dialog opens, select the **Drivers** tab. On the list select the driver you want to delete and click the **Delete** button. This will only work if there is not one single printer left that uses that particular driver. You need to "*delete*" all printers using this driver in the **Printers** folder first. You will need Administrator privileges to do this.

NOTE

Once you have successfully downloaded the CUPS PostScript driver to a client, you can easily switch all printers to this one by proceeding as described in Chapter 17, *Classical Printing Support*. Either change a driver for an existing printer by running the **Printer Properties** dialog, or use **rpcclient** with the **setdriver** subcommand.

18.10.8 Windows CUPS PostScript Driver Versus Adobe Driver

Are you interested in a comparison between the CUPS and the Adobe PostScript drivers? For our purposes these are the most important items that weigh in favor of the CUPS ones:

- No hassle with the Adobe EULA.
- No hassle with the question "*Where do I get the ADOBE*.* driver files from?*"
- The Adobe drivers (on request of the printer PPD associated with them) often put a PJL header in front of the main PostScript part of the print file. Thus, the printfile starts with *<1B >%-12345X* or *<escape>%-12345X* instead of *%!PS*). This leads to the CUPS daemon auto-typing the incoming file as a print-ready file, not initiating a pass through the *pstops* filter (to speak more technically, it is not regarded as the generic MIME-type *application/postscript*, but as the more special MIME type *application/cups.vnd-postscript*), which therefore also leads to the page accounting in */var/log/cups/page_log* not receiving the exact number of pages; instead the dummy page number of "*1*" is logged in a standard setup).

- The Adobe driver has more options to misconfigure the PostScript generated by it (like setting it inadvertently to **Optimize for Speed**, instead of **Optimize for Portability**, which could lead to CUPS being unable to process it).
- The CUPS PostScript driver output sent by Windows clients to the CUPS server is guaranteed to auto-type as the generic MIME type *application/postscript*, thus passing through the CUPS *pstops* filter and logging the correct number of pages in the `page_log` for accounting and quota purposes.
- The CUPS PostScript driver supports the sending of additional standard (IPP) print options by Windows NT/200x/XP clients. Such additional print options are: naming the CUPS standard *banner pages* (or the custom ones, should they be installed at the time of driver download), using the CUPS page-label option, setting a job-priority, and setting the scheduled time of printing (with the option to support additional useful IPP job attributes in the future).
- The CUPS PostScript driver supports the inclusion of the new **cupsJobTicket* comments at the beginning of the PostScript file (which could be used in the future for all sort of beneficial extensions on the CUPS side, but which will not disturb any other applications as they will regard it as a comment and simply ignore it).
- The CUPS PostScript driver will be the heart of the fully fledged CUPS IPP client for Windows NT/200x/XP to be released soon (probably alongside the first beta release for CUPS 1.2).

18.10.9 Run cupsaddsmb (Quiet Mode)

The **cupsaddsmb** command copies the needed files into your *[print$]* share. Additionally, the PPD associated with this printer is copied from `/etc/cups/ppd/` to *[print$]*. There the files wait for convenient Windows client installations via Point'n'Print. Before we can run the command successfully, we need to be sure that we can authenticate toward Samba. If you have a small network, you are probably using user-level security (*security* = user).

Here is an example of a successfully run **cupsaddsmb** command:

```
root# cupsaddsmb -U root infotec_IS2027
Password for root required to access localhost via Samba: ['secret']
```

To share *all* printers and drivers, use the `-a` parameter instead of a printer name. Since **cupsaddsmb** "*exports*" the printer drivers to Samba, it should be obvious that it only works for queues with a CUPS driver associated.

18.10.10 Run cupsaddsmb with Verbose Output

Probably you want to see what's going on. Use the `-v` parameter to get a more verbose output. The output below was edited for better readability: all "\" at the end of a line indicate that I inserted an artificial line break plus some indentation here:

WARNING

You will see the root password for the Samba account printed on screen.

```
root# cupsaddsmb -U root -v infotec_2105
Password for root required to access localhost via GANDALF:
Running command: smbclient //localhost/print\$ -N -U'root%secret' \
    -c 'mkdir W32X86; \
    put /var/spool/cups/tmp/3e98bf2d333b5 W32X86/infotec_2105.ppd; \
   put /usr/share/cups/drivers/cupsdrvr.dll W32X86/cupsdrvr.dll; \
    put /usr/share/cups/drivers/cupsui.dll W32X86/cupsui.dll; \
    put /usr/share/cups/drivers/cups.hlp W32X86/cups.hlp'
added interface ip=10.160.51.60 bcast=10.160.51.255 nmask=255.255.252.0
Domain=[CUPS-PRINT] OS=[UNIX] Server=[Samba 2.2.7a]
NT_STATUS_OBJECT_NAME_COLLISION making remote directory \W32X86
putting file /var/spool/cups/tmp/3e98bf2d333b5 as \W32X86/infotec_2105.ppd
putting file /usr/share/cups/drivers/cupsdrvr.dll as \W32X86/cupsdrvr.dll
putting file /usr/share/cups/drivers/cupsui.dll as \W32X86/cupsui.dll
putting file /usr/share/cups/drivers/cups.hlp as \W32X86/cups.hlp

Running command: rpcclient localhost -N -U'root%secret'
   -c 'adddriver "Windows NT x86"   \
   "infotec_2105:cupsdrvr.dll:infotec_2105.ppd:cupsui.dll:cups.hlp:NULL: \
    RAW:NULL"'
cmd = adddriver "Windows NT x86" \
   "infotec_2105:cupsdrvr.dll:infotec_2105.ppd:cupsui.dll:cups.hlp:NULL: \
   RAW:NULL"
Printer Driver infotec_2105 successfully installed.

Running command: smbclient //localhost/print\$ -N -U'root%secret' \
-c 'mkdir WIN40; \
    put /var/spool/cups/tmp/3e98bf2d333b5 WIN40/infotec_2105.PPD; \
   put /usr/share/cups/drivers/ADFONTS.MFM WIN40/ADFONTS.MFM;    \
    put /usr/share/cups/drivers/ADOBEPS4.DRV WIN40/ADOBEPS4.DRV; \
    put /usr/share/cups/drivers/ADOBEPS4.HLP WIN40/ADOBEPS4.HLP; \
    put /usr/share/cups/drivers/DEFPRTR2.PPD WIN40/DEFPRTR2.PPD; \
   put /usr/share/cups/drivers/ICONLIB.DLL WIN40/ICONLIB.DLL; \
   put /usr/share/cups/drivers/PSMON.DLL WIN40/PSMON.DLL;'
  added interface ip=10.160.51.60 bcast=10.160.51.255 nmask=255.255.252.0
  Domain=[CUPS-PRINT] OS=[UNIX] Server=[Samba 2.2.7a]
  NT_STATUS_OBJECT_NAME_COLLISION making remote directory \WIN40
  putting file /var/spool/cups/tmp/3e98bf2d333b5 as \WIN40/infotec_2105.PPD
  putting file /usr/share/cups/drivers/ADFONTS.MFM as \WIN40/ADFONTS.MFM
```

```
putting file /usr/share/cups/drivers/ADOBEPS4.DRV as \WIN40/ADOBEPS4.DRV
putting file /usr/share/cups/drivers/ADOBEPS4.HLP as \WIN40/ADOBEPS4.HLP
putting file /usr/share/cups/drivers/DEFPRTR2.PPD as \WIN40/DEFPRTR2.PPD
putting file /usr/share/cups/drivers/ICONLIB.DLL as \WIN40/ICONLIB.DLL
putting file /usr/share/cups/drivers/PSMON.DLL as \WIN40/PSMON.DLL

Running command: rpcclient localhost -N -U'root%secret' \
 -c 'adddriver "Windows 4.0"      \
 "infotec_2105:ADOBEPS4.DRV:infotec_2105.PPD:NULL:ADOBEPS4.HLP: \
 PSMON.DLL:RAW:ADOBEPS4.DRV,infotec_2105.PPD,ADOBEPS4.HLP,PSMON.DLL, \
  ADFONTS.MFM,DEFPRTR2.PPD,ICONLIB.DLL"'
 cmd = adddriver "Windows 4.0" "infotec_2105:ADOBEPS4.DRV:\
 infotec_2105.PPD:NULL:ADOBEPS4.HLP:PSMON.DLL:RAW:ADOBEPS4.DRV,\
 infotec_2105.PPD,ADOBEPS4.HLP,PSMON.DLL,ADFONTS.MFM,DEFPRTR2.PPD,\
 ICONLIB.DLL"
Printer Driver infotec_2105 successfully installed.

Running command: rpcclient localhost -N -U'root%secret'  \
 -c 'setdriver infotec_2105 infotec_2105'
cmd = setdriver infotec_2105 infotec_2105
Successfully set infotec_2105 to driver infotec_2105.
```

If you look closely, you'll discover your root password was transferred unencrypted over the wire, so beware! Also, if you look further, you'll discover error messages like NT_STATUS_OBJECT_NAME_COLLISION in between. They occur, because the directories WIN40 and W32X86 already existed in the *[print$]* driver download share (from a previous driver installation). They are harmless here.

18.10.11 Understanding cupsaddsmb

What has happened? What did **cupsaddsmb** do? There are five stages of the procedure:

1. Call the CUPS server via IPP and request the driver files and the PPD file for the named printer.
2. Store the files temporarily in the local TEMPDIR (as defined in `cupsd.conf`).
3. Connect via smbclient to the Samba server's *[print$]* share and put the files into the share's WIN40 (for Windows 9x/Me) and W32X86/ (for Windows NT/200x/XP) subdirectories.
4. Connect via rpcclient to the Samba server and execute the **adddriver** command with the correct parameters.
5. Connect via rpcclient to the Samba server a second time and execute the **setdriver** command.

Note

You can run the **cupsaddsmb** utility with parameters to specify one remote host as Samba host and a second remote host as CUPS host. Especially if you want to get a deeper understanding, it is a good idea to try it and see more clearly what is going on (though in real life most people will have their CUPS and Samba servers run on the same host):

```
root# cupsaddsmb -H sambaserver -h cupsserver -v printer
```

18.10.12 How to Recognize If cupsaddsmb Completed Successfully

You *must* always check if the utility completed successfully in all fields. You need as a minimum these three messages among the output:

1. *Printer Driver infotec_2105 successfully installed.* # (for the W32X86 == Windows NT/200x/XP architecture).
2. *Printer Driver infotec_2105 successfully installed.* # (for the WIN40 == Windows 9x/Me architecture).
3. *Successfully set [printerXPZ] to driver [printerXYZ].*

These messages are probably not easily recognized in the general output. If you run **cupsaddsmb** with the `-a` parameter (which tries to prepare *all* active CUPS printer drivers for download), you might miss if individual printers drivers had problems installing properly. Here a redirection of the output will help you analyze the results in retrospective.

Note

It is impossible to see any diagnostic output if you do not run **cupsaddsmb** in verbose mode. Therefore, we strongly recommend to not use the default quiet mode. It will hide any problems from you that might occur.

18.10.13 cupsaddsmb with a Samba PDC

Can't get the standard **cupsaddsmb** command to run on a Samba PDC? Are you asked for the password credential all over again and again and the command just will not take off at all? Try one of these variations:

```
root# cupsaddsmb -U MIDEARTH\\root -v printername
```

```
root# cupsaddsmb -H SAURON -U MIDEARTH\\root -v printername
root# cupsaddsmb -H SAURON -U MIDEARTH\\root -h cups-server -v printername
```

(Note the two backslashes: the first one is required to "*escape*" the second one).

18.10.14 cupsaddsmb Flowchart

Figure 18.16 shows a chart about the procedures, commandflows and dataflows of the **cupaddsmb** command. Note again: cupsaddsmb is not intended to, and does not work with, raw queues!

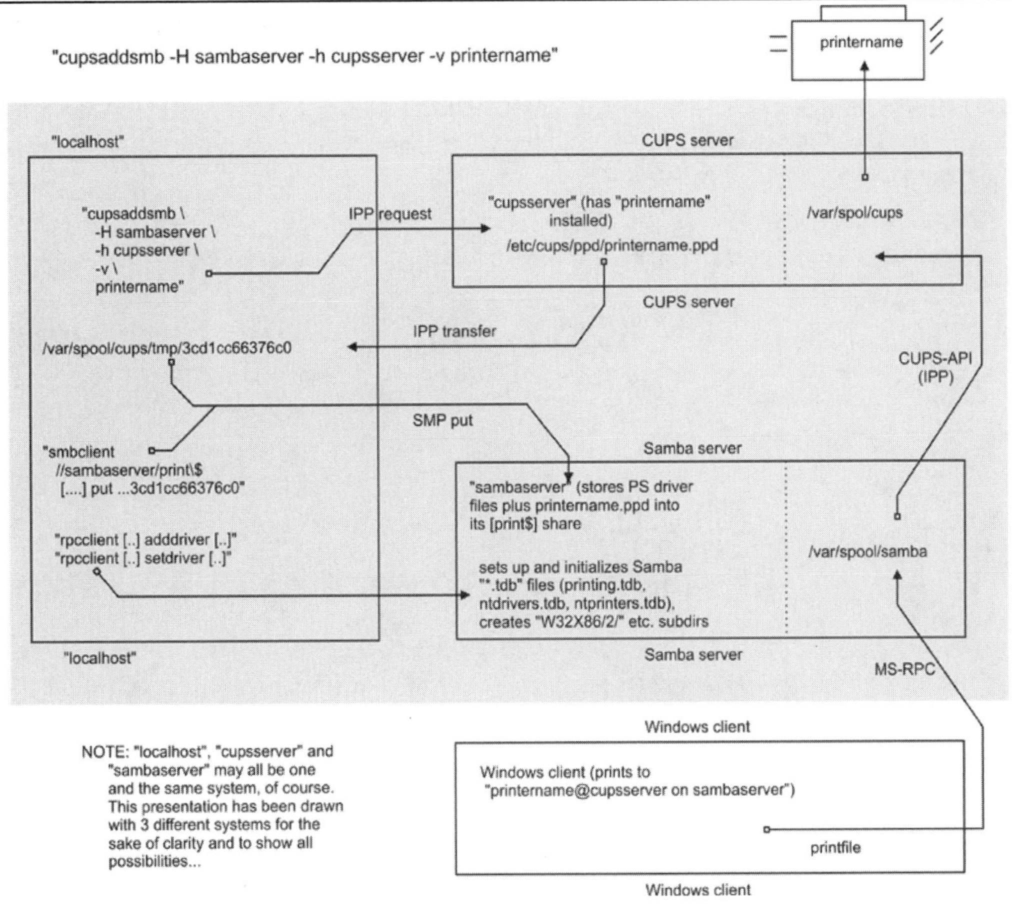

Figure 18.16. cupsaddsmb flowchart.

18.10.15 Installing the PostScript Driver on a Client

After **cupsaddsmb** is completed, your driver is prepared for the clients to use. Here are the steps you must perform to download and install it via Point'n'Print. From a Windows client, browse to the CUPS/Samba server:

- Open the **Printers** share of Samba in Network Neighborhood.
- Right-click on the printer in question.
- From the opening context-menu select **Install...** or **Connect...** (depending on the Windows version you use).

After a few seconds, there should be a new printer in your client's *local* **Printers** folder. On Windows XP it will follow a naming convention of *PrinterName on SambaServer*. (In my current case it is "infotec_2105 on kde-bitshop"). If you want to test it and send your first job from an application like Winword, the new printer appears in a `\\SambaServer\PrinterName` entry in the drop-down list of available printers.

cupsaddsmb will only reliably work with CUPS version 1.1.15 or higher and Samba from 2.2.4. If it does not work, or if the automatic printer driver download to the clients does not succeed, you can still manually install the CUPS printer PPD on top of the Adobe PostScript driver on clients. Then point the client's printer queue to the Samba printer share for a UNC type of connection:

```
C:\> net use lpt1: \\sambaserver\printershare /user:ntadmin
```

should you desire to use the CUPS networked PostScript RIP functions. (Note that user "*ntadmin*" needs to be a valid Samba user with the required privileges to access the printershare.) This sets up the printer connection in the traditional *LanMan* way (not using MS-RPC).

18.10.16 Avoiding Critical PostScript Driver Settings on the Client

Printing works, but there are still problems. Most jobs print well, some do not print at all. Some jobs have problems with fonts, which do not look very good. Some jobs print fast and some are dead-slow. Many of these problems can be greatly reduced or even completely eliminated if you follow a few guidelines. Remember, if your print device is not PostScript-enabled, you are treating your Ghostscript installation on your CUPS host with the output your client driver settings produce. Treat it well:

- Avoid the PostScript Output Option: Optimize for Speed setting. Use the Optimize for Portability instead (Adobe PostScript driver).
- Don't use the Page Independence: NO setting. Instead, use Page Independence YES (CUPS PostScript Driver).
- Recommended is the True Type Font Downloading Option: Native True Type over Automatic and Outline; you should by all means avoid Bitmap (Adobe PostScript Driver).
- Choose True Type Font: Download as Softfont into Printer over the default Replace by Device Font (for exotic fonts, you may need to change it back to get a printout at all) (Adobe).

- Sometimes you can choose PostScript Language Level: In case of problems try 2 instead of 3 (the latest ESP Ghostscript package handles Level 3 PostScript very well) (Adobe).
- Say Yes to PostScript Error Handler (Adobe).

18.11 Installing PostScript Driver Files Manually Using rpcclient

Of course, you can run all the commands that are embedded into the cupsaddsmb convenience utility yourself, one by one, and hereby upload and prepare the driver files for future client downloads.

1. Prepare Samba (A CUPS print queue with the name of the printer should be there. We are providing the driver now).
2. Copy all files to *[print$]*.
3. Run **rpcclient adddriver** (for each client architecture you want to support).
4. Run **rpcclient setdriver.**

We are going to do this now. First, read the man page on *rpcclient* to get a first idea. Look at all the printing related subcommands. **enumprinters**, **enumdrivers**, **enumports**, **adddriver**, **setdriver** are among the most interesting ones. *rpcclient* implements an important part of the MS-RPC protocol. You can use it to query (and command) a Windows NT (or 200x/XP) PC, too. MS-RPC is used by Windows clients, among other things, to benefit from the Point'n'Print features. Samba can now mimic this as well.

18.11.1 A Check of the rpcclient man Page

First let's check the *rpcclient* man page. Here are two relevant passages:

adddriver <arch> <config> Execute an **AddPrinterDriver()** RPC to install the printer driver information on the server. The driver files should already exist in the directory returned by **getdriverdir**. Possible values for *arch* are the same as those for the **getdriverdir** command. The *config* parameter is defined as follows:

```
Long Printer Name:\
Driver File Name:\
Data File Name:\
Config File Name:\
Help File Name:\
Language Monitor Name:\
Default Data Type:\
Comma Separated list of Files
```

Any empty fields should be enter as the string "*NULL*".

Samba does not need to support the concept of Print Monitors since these only apply to local printers whose driver can make use of a bi-directional link for communication. This field should be "*NULL*". On a remote NT print server, the Print Monitor for a driver must already be installed prior to adding the driver or else the RPC will fail.

setdriver <printername> <drivername> Execute a **SetPrinter()** command to update the printer driver associated with an installed printer. The printer driver must already be correctly installed on the print server.

See also the **enumprinters** and **enumdrivers** commands for obtaining a list of installed printers and drivers.

18.11.2 Understanding the rpcclient man Page

The *exact* format isn't made too clear by the man page, since you have to deal with some parameters containing spaces. Here is a better description for it. We have line-broken the command and indicated the breaks with "\". Usually you would type the command in one line without the linebreaks:

```
adddriver "Architecture" \
          "LongPrinterName:DriverFile:DataFile:ConfigFile:HelpFile:\
          LanguageMonitorFile:DataType:ListOfFiles,Comma-separated"
```

What the man pages denote as a simple <*config*> keyword, in reality consists of eight colon-separated fields. The last field may take multiple (in some very insane cases, even 20 different additional) files. This might sound confusing at first. What the man pages names the "*LongPrinterName*" in reality should be called the "*Driver Name*". You can name it anything you want, as long as you use this name later in the **rpcclient ... setdriver** command. For practical reasons, many name the driver the same as the printer.

It isn't simple at all. I hear you asking: "*How do I know which files are "Driver File"*, "*Data File*", "*Config File*", "*Help File*" and "*Language Monitor File" in each case?*" — For an answer, you may want to have a look at how a Windows NT box with a shared printer presents the files to us. Remember, that this whole procedure has to be developed by the Samba team by overhearing the traffic caused by Windows computers on the wire. We may as well turn to a Windows box now and access it from a UNIX workstation. We will query it with **rpcclient** to see what it tells us and try to understand the man page more clearly that we've read just now.

18.11.3 Producing an Example by Querying a Windows Box

We could run **rpcclient** with a **getdriver** or a **getprinter** subcommand (in level 3 verbosity) against it. Just sit down at a UNIX or Linux workstation with the Samba utilities installed, then type the following command:

```
root# rpcclient -U'user%secret' NT-SERVER -c 'getdriver printername 3'
```

From the result it should become clear which is which. Here is an example from my installation:

```
root# rpcclient -U'Danka%xxxx' W200xSERVER \
   -c'getdriver "DANKA InfoStream Virtual Printer" 3'
 cmd = getdriver "DANKA InfoStream Virtual Printer" 3

 [Windows NT x86]
 Printer Driver Info 3:
         Version: [2]
         Driver Name: [DANKA InfoStream]
         Architecture: [Windows NT x86]
         Driver Path: [C:\WINNT\System32\spool\DRIVERS\W32X86\2\PSCRIPT.DLL]
         Datafile: [C:\WINNT\System32\spool\DRIVERS\W32X86\2\INFOSTRM.PPD]
         Configfile: [C:\WINNT\System32\spool\DRIVERS\W32X86\2\PSCRPTUI.DLL]
         Helpfile: [C:\WINNT\System32\spool\DRIVERS\W32X86\2\PSCRIPT.HLP]

         Dependentfiles: []
         Dependentfiles: []
         Dependentfiles: []
         Dependentfiles: []
         Dependentfiles: []
         Dependentfiles: []
         Dependentfiles: []

         Monitorname: []
         Defaultdatatype: []
```

Some printer drivers list additional files under the label *Dependentfiles* and these would go into the last field *ListOfFiles,Comma-separated*. For the CUPS PostScript drivers, we do not need any (nor would we for the Adobe PostScript driver), therefore, the field will get a "*NULL*" entry.

18.11.4 Requirements for adddriver and setdriver to Succeed

>From the man page (and from the quoted output of **cupsaddsmb** above) it becomes clear that you need to have certain conditions in order to make the manual uploading and initializing of the driver files succeed. The two **rpcclient** subcommands (**adddriver** and **setdriver**) need to encounter the following preconditions to complete successfully:

- You are connected as *printer admin* or root (this is *not* the "*Printer Operators*" group in NT, but the *printer admin* group as defined in the *[global]* section of `smb.conf`).
- Copy all required driver files to `\\SAMBA\print$\w32x86` and `\\SAMBA\print$\win40` as appropriate. They will end up in the "*0*" respective "*2*" subdirectories later. For now, *do not* put them there, they'll be automatically used by the **adddriver** subcommand. (If you use **smbclient** to put the driver files into the share, note that you need to escape the "*$*": **smbclient //sambaserver/print\$ -U root.**)

- The user you're connecting as must be able to write to the *[print$]* share and create subdirectories.
- The printer you are going to setup for the Windows clients needs to be installed in CUPS already.
- The CUPS printer must be known to Samba, otherwise the **setdriver** subcommand fails with an NT_STATUS_UNSUCCESSFUL error. To check if the printer is known by Samba, you may use the **enumprinters** subcommand to **rpcclient**. A long-standing bug prevented a proper update of the printer list until every smbd process had received a SIGHUP or was restarted. Remember this in case you've created the CUPS printer just recently and encounter problems: try restarting Samba.

18.11.5 Manual Driver Installation in 15 Steps

We are going to install a printer driver now by manually executing all required commands. As this may seem a rather complicated process at first, we go through the procedure step by step, explaining every single action item as it comes up.

Manual Driver Installation

1. **Install the printer on CUPS.**

```
root# lpadmin -p mysmbtstprn -v socket://10.160.51.131:9100 -E \
        -P canonIR85.ppd
```

   This installs a printer with the name *mysmbtstprn* to the CUPS system. The printer is accessed via a socket (a.k.a. JetDirect or Direct TCP/IP) connection. You need to be root for this step.

2. **(Optional) Check if the printer is recognized by Samba.**

```
    root# rpcclient -Uroot%xxxx -c 'enumprinters' localhost \
   | grep -C2 mysmbtstprn
flags:[0x800000]
name:[\\kde-bitshop\mysmbtstprn]
description:[\\kde-bitshop\mysmbtstprn,,mysmbtstprn]
comment:[mysmbtstprn]
```

   This should show the printer in the list. If not, stop and restart the Samba daemon (smbd), or send a HUP signal:

```
root# kill -HUP `pidof smbd`
```

   Check again. Troubleshoot and repeat until successful. Note the "*empty*" field between the two commas in the "*description*" line. The driver name would appear here if there was one already. You need to know root's Samba password (as set by the

smbpasswd command) for this step and most of the following steps. Alternately, you can authenticate as one of the users from the "*write list*" as defined in `smb.conf` for *[print$]*.

3. **(Optional) Check if Samba knows a driver for the printer.**

```
root# rpcclient -Uroot%xxxx -c 'getprinter mysmbtstprn 2' localhost \
         | grep driver
drivername:[]

root# rpcclient -Uroot%xxxx -c 'getprinter mysmbtstprn 2' localhost \
   | grep -C4 driv
servername:[\\kde-bitshop]
printername:[\\kde-bitshop\mysmbtstprn]
sharename:[mysmbtstprn]
portname:[Samba Printer Port]
drivername:[]
comment:[mysmbtstprn]
location:[]
sepfile:[]
printprocessor:[winprint]

root# rpcclient -U root%xxxx -c 'getdriver mysmbtstprn' localhost
 result was WERR_UNKNOWN_PRINTER_DRIVER
```

None of the three commands shown above should show a driver. This step was done for the purpose of demonstrating this condition. An attempt to connect to the printer at this stage will prompt the message along the lines of: "*The server does not have the required printer driver installed.*"

4. **Put all required driver files into Samba's [print$].**

```
root# smbclient //localhost/print\$ -U 'root%xxxx' \
   -c 'cd W32X86; \
   put /etc/cups/ppd/mysmbtstprn.ppd mysmbtstprn.PPD; \
   put /usr/share/cups/drivers/cupsui.dll cupsui.dll; \
   put /usr/share/cups/drivers/cupsdrvr.dll cupsdrvr.dll; \
   put /usr/share/cups/drivers/cups.hlp cups.hlp'
```

(This command should be entered in one long single line. Line-breaks and the line-end indicated by "\" have been inserted for readability reasons.) This step is *required* for the next one to succeed. It makes the driver files physically present in the *[print$]* share. However, clients would still not be able to install them, because Samba does not yet treat them as driver files. A client asking for the driver would still be presented with a "*not installed here*" message.

5. **Verify where the driver files are now.**

```
root# ls -l /etc/samba/drivers/W32X86/
total 669
drwxr-sr-x    2 root     ntadmin        532 May 25 23:08 2
drwxr-sr-x    2 root     ntadmin        670 May 16 03:15 3
-rwxr--r--    1 root     ntadmin      14234 May 25 23:21 cups.hlp
-rwxr--r--    1 root     ntadmin     278380 May 25 23:21 cupsdrvr.dll
-rwxr--r--    1 root     ntadmin     215848 May 25 23:21 cupsui.dll
-rwxr--r--    1 root     ntadmin     169458 May 25 23:21 mysmbtstprn.PPD
```

The driver files now are in the W32X86 architecture "*root*" of *[print$]*.

6. **Tell Samba that these are driver files (adddriver).**

```
root# rpcclient -Uroot%xxxx -c 'adddriver "Windows NT x86" \
   "mydrivername:cupsdrvr.dll:mysmbtstprn.PPD: \
  cupsui.dll:cups.hlp:NULL:RAW:NULL" \
  localhost
Printer Driver mydrivername successfully installed.
```

You cannot repeat this step if it fails. It could fail even as a result of a simple typo. It will most likely have moved a part of the driver files into the "*2*" subdirectory. If this step fails, you need to go back to the fourth step and repeat it before you can try this one again. In this step, you need to choose a name for your driver. It is normally a good idea to use the same name as is used for the printer name; however, in big installations you may use this driver for a number of printers that obviously have different names, so the name of the driver is not fixed.

7. **Verify where the driver files are now.**

```
root# ls -l /etc/samba/drivers/W32X86/
total 1
drwxr-sr-x    2 root     ntadmin        532 May 25 23:22 2
drwxr-sr-x    2 root     ntadmin        670 May 16 03:15 3

root# ls -l /etc/samba/drivers/W32X86/2
total 5039
[....]
-rwxr--r--    1 root     ntadmin      14234 May 25 23:21 cups.hlp
-rwxr--r--    1 root     ntadmin     278380 May 13 13:53 cupsdrvr.dll
-rwxr--r--    1 root     ntadmin     215848 May 13 13:53 cupsui.dll
-rwxr--r--    1 root     ntadmin     169458 May 25 23:21 mysmbtstprn.PPD
```

Notice how step 6 also moved the driver files to the appropriate subdirectory. Compare this with the situation after step 5.

8. **(Optional) Verify if Samba now recognizes the driver.**

```
root# rpcclient -Uroot%xxxx -c 'enumdrivers 3' \
   localhost | grep -B2 -A5 mydrivername
Printer Driver Info 3:
Version: [2]
Driver Name: [mydrivername]
Architecture: [Windows NT x86]
Driver Path: [\\kde-bitshop\print$\W32X86\2\cupsdrvr.dll]
Datafile: [\\kde-bitshop\print$\W32X86\2\mysmbtstprn.PPD]
Configfile: [\\kde-bitshop\print$\W32X86\2\cupsui.dll]
Helpfile: [\\kde-bitshop\print$\W32X86\2\cups.hlp]
```

   Remember, this command greps for the name you chose for the driver in step 6. This command must succeed before you can proceed.

9. Tell Samba which printer should use these driver files (**setdriver**).

```
root# rpcclient -Uroot%xxxx -c 'setdriver mysmbtstprn mydrivername' \
   localhost
Successfully set mysmbtstprn to driver mydrivername
```

   Since you can bind any printername (print queue) to any driver, this is a convenient way to setup many queues that use the same driver. You do not need to repeat all the previous steps for the setdriver command to succeed. The only preconditions are: **enumdrivers** must find the driver and **enumprinters** must find the printer.

10. **(Optional) Verify if Samba has recognized this association.**

```
root# rpcclient -Uroot%xxxx -c 'getprinter mysmbtstprn 2' localhost \
  | grep driver
drivername:[mydrivername]

root# rpcclient -Uroot%xxxx -c 'getprinter mysmbtstprn 2' localhost \
  | grep -C4 driv
servername:[\\kde-bitshop]
printername:[\\kde-bitshop\mysmbtstprn]
sharename:[mysmbtstprn]
portname:[Done]
drivername:[mydrivername]
comment:[mysmbtstprn]
location:[]
sepfile:[]
printprocessor:[winprint]

root# rpcclient -U root%xxxx -c 'getdriver mysmbtstprn' localhost
```

```
[Windows NT x86]
Printer Driver Info 3:
     Version: [2]
     Driver Name: [mydrivername]
     Architecture: [Windows NT x86]
     Driver Path: [\\kde-bitshop\print$\W32X86\2\cupsdrvr.dll]
     Datafile: [\\kde-bitshop\print$\W32X86\2\mysmbtstprn.PPD]
     Configfile: [\\kde-bitshop\print$\W32X86\2\cupsui.dll]
     Helpfile: [\\kde-bitshop\print$\W32X86\2\cups.hlp]
     Monitorname: []
     Defaultdatatype: [RAW]
     Monitorname: []
     Defaultdatatype: [RAW]

root# rpcclient -Uroot%xxxx -c 'enumprinters' localhost \
   | grep mysmbtstprn
     name:[\\kde-bitshop\mysmbtstprn]
     description:[\\kde-bitshop\mysmbtstprn,mydrivername,mysmbtstprn]
     comment:[mysmbtstprn]
```

Compare these results with the ones from steps 2 and 3. Every one of these commands show the driver is installed. Even the **enumprinters** command now lists the driver on the "*description*" line.

11. **(Optional) Tickle the driver into a correct device mode.** You certainly know how to install the driver on the client. In case you are not particularly familiar with Windows, here is a short recipe: Browse the Network Neighborhood, go to the Samba server, and look for the shares. You should see all shared Samba printers. Double-click on the one in question. The driver should get installed and the network connection set up. An alternate way is to open the **Printers (and Faxes)** folder, right-click on the printer in question and select **Connect** or **Install**. As a result, a new printer should have appeared in your client's local **Printers (and Faxes)** folder, named something like **printersharename on Sambahostname**. It is important that you execute this step as a Samba printer admin (as defined in `smb.conf`). Here is another method to do this on Windows XP. It uses a command line, which you may type into the "*DOS box*" (type root's smbpassword when prompted):

```
C:\> runas /netonly /user:root "rundll32 printui.dll,PrintUIEntry \
   /in /n \\sambaserver\mysmbtstprn"
```

Change any printer setting once (like changing **portrait** *to* **landscape**), click on **Apply**; change the setting back.

12. **Install the printer on a client (Point'n'Print).**

```
C:\> rundll32 printui.dll,PrintUIEntry /in /n \\sambaserver\mysmbtstprn
```

If it does not work it could be a permission problem with the *[print$]* share.

13. **(Optional) Print a test page.**

```
C:\> rundll32 printui.dll,PrintUIEntry /p /n "\\sambaserver\mysmbtstprn"
```

Then hit [TAB] five times, [ENTER] twice, [TAB] once and [ENTER] again and march to the printer.

14. **(Recommended) Study the test page.** Hmmm.... just kidding! By now you know everything about printer installations and you do not need to read a word. Just put it in a frame and bolt it to the wall with the heading "MY FIRST RPCCLIENT-INSTALLED PRINTER" — why not just throw it away!

15. **(Obligatory) Enjoy. Jump. Celebrate your success.**

```
root# echo "Cheeeeerioooooo! Success..." >> /var/log/samba/log.smbd
```

18.11.6 Troubleshooting Revisited

The setdriver command will fail, if in Samba's mind the queue is not already there. You had promising messages about the:

```
Printer Driver ABC successfully installed.
```

after the **adddriver** parts of the procedure? But you are also seeing a disappointing message like this one?

`result was NT_STATUS_UNSUCCESSFUL`

It is not good enough that you can see the queue in CUPS, using the **lpstat -p ir85wm** command. A bug in most recent versions of Samba prevents the proper update of the queuelist. The recognition of newly installed CUPS printers fails unless you restart Samba or send a HUP to all smbd processes. To verify if this is the reason why Samba does not execute the **setdriver** command successfully, check if Samba *"sees"* the printer:

```
root# rpcclient transmeta -N -U'root%xxxx' -c 'enumprinters 0'|grep ir85wm
        printername:[ir85wm]
```

An alternate command could be this:

```
root# rpcclient transmeta -N -U'root%secret' -c 'getprinter ir85wm'
        cmd = getprinter ir85wm
```

```
flags:[0x800000]
name:[\\transmeta\ir85wm]
description:[\\transmeta\ir85wm,ir85wm,DPD]
comment:[CUPS PostScript-Treiber for Windows NT/200x/XP]
```

By the way, you can use these commands, plus a few more, of course, to install drivers on remote Windows NT print servers too!

18.12 The Printing *.tdb Files

Some mystery is associated with the series of files with a tdb suffix appearing in every Samba installation. They are `connections.tdb`, `printing.tdb`, `share_info.tdb`, `ntdrivers.tdb`, `unexpected.tdb`, `brlock.tdb`, `locking.tdb`, `ntforms.tdb`, `messages.tdb` , `ntprinters.tdb`, `sessionid.tdb` and `secrets.tdb`. What is their purpose?

18.12.1 Trivial Database Files

A Windows NT (print) server keeps track of all information needed to serve its duty toward its clients by storing entries in the Windows registry. Client queries are answered by reading from the registry, Administrator or user configuration settings that are saved by writing into the registry. Samba and UNIX obviously do not have such a Registry. Samba instead keeps track of all client related information in a series of `*.tdb` files. (TDB = Trivial Data Base). These are often located in `/var/lib/samba/` or `/var/lock/samba/`. The printing related files are `ntprinters.tdb`, `printing.tdb`,`ntforms.tdb` and `ntdrivers.tdb`.

18.12.2 Binary Format

`*.tdb` files are not human readable. They are written in a binary format. "*Why not ASCII?*", you may ask. "*After all, ASCII configuration files are a good and proven tradition on UNIX.*" The reason for this design decision by the Samba team is mainly performance. Samba needs to be fast; it runs a separate **smbd** process for each client connection, in some environments many thousands of them. Some of these smbds might need to write-access the same `*.tdb` file *at the same time*. The file format of Samba's `*.tdb` files allows for this provision. Many smbd processes may write to the same `*.tdb` file at the same time. This wouldn't be possible with pure ASCII files.

18.12.3 Losing *.tdb Files

It is very important that all `*.tdb` files remain consistent over all write and read accesses. However, it may happen that these files *do* get corrupted. (A **kill -9 'pidof smbd'** while a write access is in progress could do the damage as well as a power interruption, etc.). In cases of trouble, a deletion of the old printing-related `*.tdb` files may be the only option. After that you need to re-create all print-related setup or you have made a backup of the `*.tdb` files in time.

18.12.4 Using tdbbackup

Samba ships with a little utility that helps the root user of your system to backup your `*.tdb` files. If you run it with no argument, it prints a usage message:

```
root# tdbbackup
 Usage: tdbbackup [options] <fname...>

 Version:3.0a
   -h            this help message
   -s suffix     set the backup suffix
   -v            verify mode (restore if corrupt)
```

Here is how I backed up my `printing.tdb` file:

```
root# ls
.              browse.dat      locking.tdb     ntdrivers.tdb printing.tdb
..             share_info.tdb  connections.tdb messages.tdb  ntforms.tdb
printing.tdbkp unexpected.tdb  brlock.tdb      gmon.out      namelist.debug
ntprinters.tdb sessionid.tdb

root# tdbbackup -s .bak printing.tdb
 printing.tdb : 135 records

root# ls -l printing.tdb*
 -rw-------    1 root     root         40960 May  2 03:44 printing.tdb
 -rw-------    1 root     root         40960 May  2 03:44 printing.tdb.bak
```

18.13 CUPS Print Drivers from Linuxprinting.org

CUPS ships with good support for HP LaserJet-type printers. You can install the generic driver as follows:

```
root# lpadmin -p laserjet4plus -v parallel:/dev/lp0 -E -m laserjet.ppd
```

The `-m` switch will retrieve the `laserjet.ppd` from the standard repository for not-yet-installed-PPDs, which CUPS typically stores in `/usr/share/cups/model`. Alternately, you may use `-P /path/to/your.ppd`.

The generic `laserjet.ppd,` however, does not support every special option for every LaserJet-compatible model. It constitutes a sort of "*least common denominator*" of all the models. If for some reason you must pay for the commercially available ESP Print Pro drivers, your first move should be to consult the database on `http://www.linuxprinting.org/printer_list.cgi`. Linuxprinting.org has excellent recommendations about which driver is best used for each printer. Its database is kept current by the tireless work of Till Kamppeter from MandrakeSoft, who is also the principal author of the **foomatic-rip** utility.

NOTE

The former **cupsomatic** concept is now being replaced by the new successor, a much more powerful **foomatic-rip**. **cupsomatic** is no longer maintained. Here is the new URL to the Foomatic-3.0 database: `http://www.linuxprinting.org/driver_list.cgi`. If you upgrade to **foomatic-rip**, remember to also upgrade to the new-style PPDs for your Foomatic-driven printers. foomatic-rip will not work with PPDs generated for the old **cupsomatic**. The new-style PPDs are 100% compliant to the Adobe PPD specification. They are also intended to be used by Samba and the cupsaddsmb utility, to provide the driver files for the Windows clients!

18.13.1 foomatic-rip and Foomatic Explained

Nowadays, most Linux distributions rely on the utilities of Linuxprinting.org to create their printing-related software (which, by the way, works on all UNIXes and on Mac OS X or Darwin, too). It is not known as well as it should be, that it also has a very end-user-friendly interface that allows for an easy update of drivers and PPDs for all supported models, all spoolers, all operating systems, and all package formats (because there is none). Its history goes back a few years.

Recently, Foomatic has achieved the astonishing milestone of 1000 listed[12] printer models. Linuxprinting.org keeps all the important facts about printer drivers, supported models and which options are available for the various driver/printer combinations in its Foomatic[13] database. Currently there are 245 drivers[14] in the database. Many drivers support various models, and many models may be driven by different drivers — its your choice!

18.13.1.1 690 "*Perfect*" Printers

At present, there are 690 devices dubbed as working perfectly, 181 mostly, 96 partially, and 46 are paperweights. Keeping in mind that most of these are non-PostScript models (PostScript printers are automatically supported by CUPS to perfection, by using their own manufacturer-provided Windows-PPD), and that a multifunctional device never qualifies as working perfectly if it does not also scan and copy and fax under GNU/Linux — then this is a truly astonishing achievement! Three years ago the number was not more than 500, and Linux or UNIX printing at the time wasn't anywhere near the quality it is today.

[12] http://www.linuxprinting.org/printer_list.cgi?make=Anyone

[13] http://www.linuxprinting.org/foomatic.html

[14] http://www.linuxprinting.org/driver_list.cgi

18.13.1.2 How the Printing HOWTO Started It All

A few years ago Grant Taylor[15] started it all. The roots of today's Linuxprinting.org are in the first Linux Printing HOWTO[16] that he authored. As a side-project to this document, which served many Linux users and admins to guide their first steps in this complicated and delicate setup (to a scientist, printing is "*applying a structured deposition of distinct patterns of ink or toner particles on paper substrates*", he started to build in a little Postgres database with information about the hardware and driver zoo that made up Linux printing of the time. This database became the core component of today's Foomatic collection of tools and data. In the meantime, it has moved to an XML representation of the data.

18.13.1.3 Foomatic's Strange Name

"*Why the funny name?*" you ask. When it really took off, around spring 2000, CUPS was far less popular than today, and most systems used LPD, LPRng or even PDQ to print. CUPS shipped with a few generic drivers (good for a few hundred different printer models). These didn't support many device-specific options. CUPS also shipped with its own built-in rasterization filter (`pstoraster`, derived from Ghostscript). On the other hand, CUPS provided brilliant support for *controlling* all printer options through standardized and well-defined PPD files (PostScript Printers Description files). Plus, CUPS was designed to be easily extensible.

Taylor already had in his database a respectable compilation of facts about many more printers and the Ghostscript "*drivers*" they run with. His idea, to generate PPDs from the database information and use them to make standard Ghostscript filters work within CUPS, proved to work very well. It also killed several birds with one stone:

- It made all current and future Ghostscript filter developments available for CUPS.
- It made available a lot of additional printer models to CUPS users (because often the traditional Ghostscript way of printing was the only one available).
- It gave all the advanced CUPS options (Web interface, GUI driver configurations) to users wanting (or needing) to use Ghostscript filters.

18.13.1.4 cupsomatic, pdqomatic, lpdomatic, directomatic

CUPS worked through a quickly-hacked up filter script named cupsomatic.[17] cupsomatic ran the printfile through Ghostscript, constructing automatically the rather complicated command line needed. It just needed to be copied into the CUPS system to make it work. To configure the way cupsomatic controls the Ghostscript rendering process, it needs a CUPS-PPD. This PPD is generated directly from the contents of the database. For CUPS and the respective printer/filter combo, another Perl script named CUPS-O-Matic did the PPD generation. After that was working, Taylor implemented within a few days a similar

[15] http://www2.picante.com:81/~gtaylor/
[16] http://www.linuxprinting.org/foomatic2.9/howto/
[17] http://www.linuxprinting.org/download.cgi?filename=cupsomatic&show=0

thing for two other spoolers. Names chosen for the config-generator scripts were PDQ-O-Matic[18] (for PDQ) and LPD-O-Matic[19] (for — you guessed it — LPD); the configuration here didn't use PPDs but other spooler-specific files.

From late summer of that year, Till Kamppeter[20] started to put work into the database. Kamppeter had been newly employed by MandrakeSoft[21] to convert its printing system over to CUPS, after they had seen his FLTK[22]-based XPP[23] (a GUI frontend to the CUPS lp-command). He added a huge amount of new information and new printers. He also developed the support for other spoolers, like PPR[24] (via ppromatic), GNUlpr[25] and LPRng[26] (both via an extended lpdomatic) and spoolerless printing (directomatic[27]).

So, to answer your question: "*Foomatic*" is the general name for all the overlapping code and data behind the "**omatic*" scripts. Foomatic, up to versions 2.0.x, required (ugly) Perl data structures attached to Linuxprinting.org PPDs for CUPS. It had a different "**omatic*" script for every spooler, as well as different printer configuration files.

18.13.1.5 The *Grand Unification* Achieved

This has all changed in Foomatic versions 2.9 (beta) and released as "*stable*" 3.0. It has now achieved the convergence of all *omatic scripts and is called the foomatic-rip.[28] This single script is the unification of the previously different spooler-specific *omatic scripts. foomatic-rip is used by all the different spoolers alike and because it can read PPDs (both the original PostScript printer PPDs and the Linuxprinting.org-generated ones), all of a sudden all supported spoolers can have the power of PPDs at their disposal. Users only need to plug foomatic-rip into their system. For users there is improved media type and source support — paper sizes and trays are easier to configure.

Also, the New Generation of Linuxprinting.org PPDs no longer contains Perl data structures. If you are a distro maintainer and have used the previous version of Foomatic, you may want to give the new one a spin, but remember to generate a new-version set of PPDs via the new foomatic-db-engine![29] Individual users just need to generate a single new PPD specific to their model by following the steps[30] outlined in the Foomatic tutorial or in this chapter. This new development is truly amazing.

foomatic-rip is a very clever wrapper around the need to run Ghostscript with a different syntax, options, device selections, and/or filters for each different printer or spooler. At the same time it can read the PPD associated with a print queue and modify the print job according to the user selections. Together with this comes the 100% compliance of the new

[18] http://www.linuxprinting.org/download.cgi?filename=lpdomatic&show=0
[19] http://www.linuxprinting.org/download.cgi?filename=lpdomatic&show=0
[20] http://www.linuxprinting.org/till/
[21] http://www.mandrakesoft.com/
[22] http://www.fltk.org/
[23] http://cups.sourceforge.net/xpp/
[24] http://ppr.sourceforge.net/
[25] http://sourceforge.net/projects/lpr/
[26] http://www.lprng.org/
[27] http://www.linuxprinting.org/download.cgi?filename=directomatic&show=0
[28] http://www.linuxprinting.org/foomatic2.9/download.cgi?filename=foomatic-rip&show=0
[29] http://www.linuxprinting.org/download/foomatic/foomatic-db-engine-3.0.0beta1.tar.gz
[30] http://www.linuxprinting.org/kpfeifle/LinuxKongress2002/Tutorial/II.Foomatic-User/II.tutorial-handout-foomatic-user.html

Foomatic PPDs with the Adobe spec. Some innovative features of the Foomatic concept may surprise users. It will support custom paper sizes for many printers and will support printing on media drawn from different paper trays within the same job (in both cases, even where there is no support for this from Windows-based vendor printer drivers).

18.13.1.6 Driver Development Outside

Most driver development itself does not happen within Linuxprinting.org. Drivers are written by independent maintainers. Linuxprinting.org just pools all the information and stores it in its database. In addition, it also provides the Foomatic glue to integrate the many drivers into any modern (or legacy) printing system known to the world.

Speaking of the different driver development groups, most of the work is currently done in three projects. These are:

- Omni[31] — a free software project by IBM that tries to convert their printer driver knowledge from good-ol' OS/2 times into a modern, modular, universal driver architecture for Linux/UNIX (still beta). This currently supports 437 models.
- HPIJS[32] — a free software project by HP to provide the support for their own range of models (very mature, printing in most cases is perfect and provides true photo quality). This currently supports 369 models.
- Gimp-Print[33] — a free software effort, started by Michael Sweet (also lead developer for CUPS), now directed by Robert Krawitz, which has achieved an amazing level of photo print quality (many Epson users swear that its quality is better than the vendor drivers provided by Epson for the Microsoft platforms). This currently supports 522 models.

18.13.1.7 Forums, Downloads, Tutorials, Howtos — also for Mac OS X and Commercial UNIX

Linuxprinting.org today is the one-stop shop to download printer drivers. Look for printer information and tutorials[34] or solve printing problems in its popular forums.[35] This forum it's not just for GNU/Linux users, but admins of commercial UNIX systems[36] are also going there, and the relatively new Mac OS X forum[37] has turned out to be one of the most frequented forums after only a few weeks.

Linuxprinting.org and the Foomatic driver wrappers around Ghostscript are now a standard toolchain for printing on all the important distros. Most of them also have CUPS underneath. While in recent years most printer data had been added by Kamppeter (who works at Mandrake), many additional contributions came from engineers with SuSE, RedHat,

[31] http://www-124.ibm.com/developerworks/oss/linux/projects/omni/
[32] http://hpinkjet.sf.net/
[33] http://gimp-print.sf.net/
[34] http://www.linuxprinting.org//kpfeifle/LinuxKongress2002/Tutorial/
[35] http://www.linuxprinting.org/newsportal/
[36] http://www.linuxprinting.org/macosx/
[37] http://www.linuxprinting.org/newsportal/thread.php3?name=linuxprinting.macosx.general

Connectiva, Debian, and others. Vendor-neutrality is an important goal of the Foomatic project.

NOTE

Till Kamppeter from MandrakeSoft is doing an excellent job in his spare time to maintain Linuxprinting.org and Foomatic. So if you use it often, please send him a note showing your appreciation.

18.13.1.8 Foomatic Database-Generated PPDs

The Foomatic database is an amazing piece of ingenuity in itself. Not only does it keep the printer and driver information, but it is organized in a way that it can generate PPD files on the fly from its internal XML-based datasets. While these PPDs are modelled to the Adobe specification of PostScript Printer Descriptions (PPDs), the Linuxprinting.org/Foomatic-PPDs do not normally drive PostScript printers. They are used to describe all the bells and whistles you could ring or blow on an Epson Stylus inkjet, or a HP Photosmart, or what-have-you. The main trick is one little additional line, not envisaged by the PPD specification, starting with the *`*cupsFilter`* keyword. It tells the CUPS daemon how to proceed with the PostScript print file (old-style Foomatic-PPDs named the cupsomatic filter script, while the new-style PPDs are now call foomatic-rip). This filter script calls Ghostscript on the host system (the recommended variant is ESP Ghostscript) to do the rendering work. foomatic-rip knows which filter or internal device setting it should ask from Ghostscript to convert the PostScript printjob into a raster format ready for the target device. This usage of PPDs to describe the options of non-PS printers was the invention of the CUPS developers. The rest is easy. GUI tools (like KDE's marvelous kprinter,[38] or the GNOME gtklp,[39] xpp and the CUPS Web interface) read the PPD as well and use this information to present the available settings to the user as an intuitive menu selection.

18.13.2 foomatic-rip and Foomatic-PPD Download and Installation

Here are the steps to install a foomatic-rip driven LaserJet 4 Plus-compatible printer in CUPS (note that recent distributions of SuSE, UnitedLinux and Mandrake may ship with a complete package of Foomatic-PPDs plus the **foomatic-rip** utility. Going directly to Linuxprinting.org ensures that you get the latest driver/PPD files):

- Open your browser at the Linuxprinting.org printer listpage.[40]
- Check the complete list of printers in the database.[41].
- Select your model and click on the link.

[38]http://printing.kde.org/overview/kprinter.phtml
[39]http://gtklp.sourceforge.net/
[40]http://www.linuxprinting.org/printer_list.cgi
[41]http://www.linuxprinting.org/printer_list.cgi?make=Anyone

- You'll arrive at a page listing all drivers working with this model (for all printers, there will always be *one* recommended driver. Try this one first).
- In our case (HP LaserJet 4 Plus), we'll arrive at the default driver for the HP-LaserJet 4 Plus.[42]
- The recommended driver is ljet4.
- Several links are provided here. You should visit them all if you are not familiar with the Linuxprinting.org database.
- There is a link to the database page for the ljet4.[43] On the driver's page, you'll find important and detailed information about how to use that driver within the various available spoolers.
- Another link may lead you to the homepage of the driver author or the driver.
- Important links are the ones that provide hints with setup instructions for CUPS, PDQ[44], LPD, LPRng and GNUlpr[45]) as well as PPR[46] or "*spooler-less*" printing.[47]
- You can view the PPD in your browser through this link: `http://www.linuxprinting.org/ppd-o-matic.cgi?driver=ljet4&printer=HP-LaserJet_4_Plus&show=1`
- Most importantly, you can also generate and download the PPD.[48]
- The PPD contains all the information needed to use our model and the driver; once installed, this works transparently for the user. Later you'll only need to choose resolution, paper size, and so on from the Web-based menu, or from the print dialog GUI, or from the command line.
- If you ended up on the drivers page[49] you can choose to use the "*PPD-O-Matic*" online PPD generator program.
- Select the exact model and check either **Download** or **Display PPD file** and click **Generate PPD file**.
- If you save the PPD file from the browser view, please do not use cut and paste (since it could possibly damage line endings and tabs, which makes the PPD likely to fail its duty), but use **Save as...** in your browser's menu. (It is best to use the **Download** option directly from the Web page).
- Another interesting part on each driver page is the **Show execution details** button. If you select your printer model and click on that button, a complete Ghostscript command line will be displayed, enumerating all options available for that combination of driver and printer model. This is a great way to "*learn Ghostscript by doing*". It is also an excellent cheat sheet for all experienced users who need to re-construct a good command line for that damn printing script, but can't remember the exact syntax.

[42] http://www.linuxprinting.org/show_printer.cgi?recnum=HP-LaserJet_4_Plus
[43] http://www.linuxprinting.org/show_driver.cgi?driver=ljet4
[44] http://www.linuxprinting.org/pdq-doc.html
[45] http://www.linuxprinting.org/lpd-doc.html
[46] http://www.linuxprinting.org/ppr-doc.html
[47] http://www.linuxprinting.org/direct-doc.html
[48] http://www.linuxprinting.org/ppd-o-matic.cgi?driver=ljet4&printer=HP-LaserJet_4_Plus&show=0
[49] http://www.linuxprinting.org/show_driver.cgi?driver=ljet4

- Some time during your visit to Linuxprinting.org, save the PPD to a suitable place on your harddisk, say `/path/to/my-printer.ppd` (if you prefer to install your printers with the help of the CUPS Web interface, save the PPD to the `/usr/share/cups/model/` path and restart cupsd).
- Then install the printer with a suitable command line, like this:

```
root# lpadmin -p laserjet4plus -v parallel:/dev/lp0 -E \
   -P path/to/my-printer.ppd
```

- For all the new-style "*Foomatic-PPDs*" from Linuxprinting.org, you also need a special CUPS filter named foomatic-rip.
- The foomatic-rip Perlscript itself also makes some interesting reading[50] because it is well documented by Kamppeter's inline comments (even non-Perl hackers will learn quite a bit about printing by reading it).
- Save foomatic-rip either directly in `/usr/lib/cups/filter/foomatic-rip` or somewhere in your $PATH (and remember to make it world-executable). Again, do not save by copy and paste but use the appropriate link or the **Save as...** menu item in your browser.
- If you save foomatic-rip in your $PATH, create a symlink:

```
root# cd /usr/lib/cups/filter/ ; ln -s 'which foomatic-rip'
```

  CUPS will discover this new available filter at startup after restarting cupsd.

Once you print to a print queue set up with the Foomatic-PPD, CUPS will insert the appropriate commands and comments into the resulting PostScript jobfile. foomatic-rip is able to read and act upon these and uses some specially encoded Foomatic comments embedded in the jobfile. These in turn are used to construct (transparently for you, the user) the complicated Ghostscript command line telling the printer driver exactly how the resulting raster data should look and which printer commands to embed into the data stream. You need:

- A "*foomatic+something*" PPD — but this is not enough to print with CUPS (it is only *one* important component).
- The *`foomatic-rip`* filter script (Perl) in `/usr/lib/cups/filters/`.
- Perl to make foomatic-rip run.
- Ghostscript (because it is doing the main work, controlled by the PPD/foomatic-rip combo) to produce the raster data fit for your printer model's consumption.
- Ghostscript *must* (depending on the driver/model) contain support for a certain device representing the selected driver for your model (as shown by **gs -h**).

[50] http://www.linuxprinting.org/foomatic2.9/download.cgi?filename=foomatic-rip&show=1

- foomatic-rip needs a new version of PPDs (PPD versions produced for cupsomatic do not work with foomatic-rip).

18.14 Page Accounting with CUPS

Often there are questions regarding print quotas where Samba users (that is, Windows clients) should not be able to print beyond a certain number of pages or data volume per day, week or month. This feature is dependent on the real print subsystem you're using. Samba's part is always to receive the job files from the clients (filtered *or* unfiltered) and hand it over to this printing subsystem.

Of course one could hack things with one's own scripts. But then there is CUPS. CUPS supports quotas that can be based on the size of jobs or on the number of pages or both, and span any time period you want.

18.14.1 Setting Up Quotas

This is an example command of how root would set a print quota in CUPS, assuming an existing printer named "*quotaprinter*":

```
root# lpadmin -p quotaprinter -o job-quota-period=604800 \
   -o job-k-limit=1024 -o job-page-limit=100
```

This would limit every single user to print 100 pages or 1024 KB of data (whichever comes first) within the last 604,800 seconds (= 1 week).

18.14.2 Correct and Incorrect Accounting

For CUPS to count correctly, the printfile needs to pass the CUPS pstops filter, otherwise it uses a dummy count of "*one*". Some print files do not pass it (e.g., image files) but then those are mostly one- page jobs anyway. This also means that proprietary drivers for the target printer running on the client computers and CUPS/Samba, which then spool these files as "*raw*" (i.e., leaving them untouched, not filtering them), will be counted as one-pagers too!

You need to send PostScript from the clients (i.e., run a PostScript driver there) to have the chance to get accounting done. If the printer is a non-PostScript model, you need to let CUPS do the job to convert the file to a print-ready format for the target printer. This is currently working for about a thousand different printer models. Linuxprinting has a driver list.[51]

18.14.3 Adobe and CUPS PostScript Drivers for Windows Clients

Before CUPS 1.1.16, your only option was to use the Adobe PostScript Driver on the Windows clients. The output of this driver was not always passed through the **pstops** filter

[51] http://www.linuxprinting.org/printer_list.cgi

on the CUPS/Samba side, and therefore was not counted correctly (the reason is that it often, depending on the PPD being used, wrote a PJL-header in front of the real PostScript which caused CUPS to skip **pstops** and go directly to the **pstoraster** stage).

From CUPS 1.1.16 onward, you can use the CUPS PostScript Driver for Windows NT/200x/XP clients (which is tagged in the download area of `http://www.cups.org/` as the `cups-samba-1.1.16.tar.gz` package). It does *not* work for Windows 9x/ME clients, but it guarantees:

- To not write a PJL-header.
- To still read and support all PJL-options named in the driver PPD with its own means.
- That the file will pass through the **pstops** filter on the CUPS/Samba server.
- To page-count correctly the print file.

You can read more about the setup of this combination in the man page for **cupsaddsmb** (which is only present with CUPS installed, and only current from CUPS 1.1.16).

18.14.4 The page_log File Syntax

These are the items CUPS logs in the `page_log` for every page of a job:

- Printer name
- User name
- Job ID
- Time of printing
- The page number
- The number of copies
- A billing information string (optional)
- The host that sent the job (included since version 1.1.19)

Here is an extract of my CUPS server's `page_log` file to illustrate the format and included items:

```
tec_IS2027 kurt 401 [22/Apr/2003:10:28:43 +0100] 1 3 #marketing 10.160.50.13
tec_IS2027 kurt 401 [22/Apr/2003:10:28:43 +0100] 2 3 #marketing 10.160.50.13
tec_IS2027 kurt 401 [22/Apr/2003:10:28:43 +0100] 3 3 #marketing 10.160.50.13
tec_IS2027 kurt 401 [22/Apr/2003:10:28:43 +0100] 4 3 #marketing 10.160.50.13
Dig9110 boss 402 [22/Apr/2003:10:33:22 +0100] 1 440 finance-dep 10.160.51.33
```

This was job ID *401*, printed on *tec_IS2027* by user *kurt*, a 64-page job printed in three copies and billed to *#marketing*, sent from IP address `10.160.50.13`. The next job had ID *402*, was sent by user *boss* from IP address `10.160.51.33`, printed from one page 440 copies and is set to be billed to *finance-dep*.

18.14.5 Possible Shortcomings

What flaws or shortcomings are there with this quota system?

- The ones named above (wrongly logged job in case of printer hardware failure, and so on).
- In reality, CUPS counts the job pages that are being processed in *software* (that is, going through the RIP) rather than the physical sheets successfully leaving the printing device. Thus if there is a jam while printing the fifth sheet out of a thousand and the job is aborted by the printer, the page count will still show the figure of a thousand for that job.
- All quotas are the same for all users (no flexibility to give the boss a higher quota than the clerk) and no support for groups.
- No means to read out the current balance or the "*used-up*" number of current quota.
- A user having used up 99 sheets of a 100 quota will still be able to send and print a thousand sheet job.
- A user being denied a job because of a filled-up quota does not get a meaningful error message from CUPS other than "*client-error-not-possible*".

18.14.6 Future Developments

This is the best system currently available, and there are huge improvements under development for CUPS 1.2:

- Page counting will go into the backends (these talk directly to the printer and will increase the count in sync with the actual printing process; thus, a jam at the fifth sheet will lead to a stop in the counting).
- Quotas will be handled more flexibly.
- Probably there will be support for users to inquire about their accounts in advance.
- Probably there will be support for some other tools around this topic.

18.15 Additional Material

A printer queue with *no* PPD associated to it is a "*raw*" printer and all files will go directly there as received by the spooler. The exceptions are file types `application/octet-stream` that need passthrough feature enabled. "*Raw*" queues do not do any filtering at all, they hand the file directly to the CUPS backend. This backend is responsible for sending the data to the device (as in the "*device URI*" notation: `lpd://`, `socket://`, `smb://`, `ipp://`, `http://`, `parallel:/`, `serial:/`, `usb:/`, and so on).

cupsomatic/Foomatic are *not* native CUPS drivers and they do not ship with CUPS. They are a third party add-on developed at Linuxprinting.org. As such, they are a brilliant hack to make all models (driven by Ghostscript drivers/filters in traditional spoolers) also work via CUPS, with the same (good or bad!) quality as in these other spoolers. `cupsomatic` is only a vehicle to execute a Ghostscript commandline at that stage in the CUPS filtering chain, where normally the native CUPS *pstoraster* filter would kick in. cupsomatic bypasses pstoraster, kidnaps the printfile from CUPS away and redirects it to go through Ghostscript. CUPS accepts this, because the associated cupsomatic/foomatic-PPD specifies:

```
*cupsFilter:  "application/vnd.cups-postscript 0 cupsomatic"
```

This line persuades CUPS to hand the file to cupsomatic, once it has successfully converted it to the MIME type *application/vnd.cups-postscript*. This conversion will not happen for Jobs arriving from Windows that are auto-typed *application/octet-stream*, with the according changes in `/etc/cups/mime.types` in place.

CUPS is widely configurable and flexible, even regarding its filtering mechanism. Another workaround in some situations would be to have in `/etc/cups/mime.types` entries as follows:

```
application/postscript            application/vnd.cups-raw  0  -
application/vnd.cups-postscript   application/vnd.cups-raw  0  -
```

This would prevent all PostScript files from being filtered (rather, they will through the virtual *nullfilter* denoted with "-"). This could only be useful for PS printers. If you want to print PS code on non-PS printers (provided they support ASCII text printing), an entry as follows could be useful:

```
*/*          application/vnd.cups-raw  0  -
```

and would effectively send *all* files to the backend without further processing.

You could have the following entry:

```
application/vnd.cups-postscript application/vnd.cups-raw 0 \
  my_PJL_stripping_filter
```

You will need to write a *my_PJL_stripping_filter* (which could be a shell script) that parses the PostScript and removes the unwanted PJL. This needs to conform to CUPS filter design (mainly, receive and pass the parameters printername, job-id, username, jobtitle, copies, print options and possibly the filename). It is installed as world executable into `/usr/lib/cups/filters/` and is called by CUPS if it encounters a MIME type *application/vnd.cups-postscript*.

CUPS can handle *-o job-hold-until=indefinite*. This keeps the job in the queue on hold. It will only be printed upon manual release by the printer operator. This is a requirement in many central reproduction departments, where a few operators manage the jobs of hundreds of users on some big machine, where no user is allowed to have direct access (such as when the operators often need to load the proper paper type before running the 10,000 page job requested by marketing for the mailing, and so on).

18.16 Auto-Deletion or Preservation of CUPS Spool Files

Samba print files pass through two spool directories. One is the incoming directory managed by Samba, (set in the *path* = /var/spool/samba directive in the *[printers]* section of `smb.conf`). The other is the spool directory of your UNIX print subsystem. For CUPS it is normally `/var/spool/cups/`, as set by the `cupsd.conf` directive `RequestRoot /var/spool/cups`.

18.16.1 CUPS Configuration Settings Explained

Some important parameter settings in the CUPS configuration file `cupsd.conf` are:

PreserveJobHistory Yes — This keeps some details of jobs in cupsd's mind (well it keeps the c12345, c12346, and so on, files in the CUPS spool directory, which do a similar job as the old-fashioned BSD-LPD control files). This is set to "*Yes*" as a default.

PreserveJobFiles Yes — This keeps the job files themselves in cupsd's mind (it keeps the d12345, d12346 etc. files in the CUPS spool directory). This is set to "*No*" as the CUPS default.

"MaxJobs 500" — This directive controls the maximum number of jobs that are kept in memory. Once the number of jobs reaches the limit, the oldest completed job is automatically purged from the system to make room for the new one. If all of the known jobs are still pending or active, then the new job will be rejected. Setting the maximum to 0 disables this functionality. The default setting is 0.

(There are also additional settings for *MaxJobsPerUser* and *MaxJobsPerPrinter*...)

18.16.2 Pre-Conditions

For everything to work as announced, you need to have three things:

- A Samba-smbd that is compiled against `libcups` (check on Linux by running `ldd 'which smbd'`).
- A Samba-`smb.conf` setting of *printing* = cups.
- Another Samba-`smb.conf` setting of *printcap* = cups.

NOTE

In this case, all other manually set printing-related commands (like *print command*, *lpq command*, *lprm command*, *lppause command* or *lpresume command*) are ignored and they should normally have no influence whatsoever on your printing.

18.16.3 Manual Configuration

If you want to do things manually, replace the *printing* = cups by *printing* = bsd. Then your manually set commands may work (I haven't tested this), and a *print command* = lp -d %P %s; rm %s" may do what you need.

18.17 Printing from CUPS to Windows Attached Printers

>From time to time the question arises, how can you print *to* a Windows attached printer *from* Samba? Normally the local connection from Windows host to printer would be done by USB or parallel cable, but this does not matter to Samba. From here only an SMB connection needs to be opened to the Windows host. Of course, this printer must be shared first. As you have learned by now, CUPS uses *backends* to talk to printers and other servers. To talk to Windows shared printers, you need to use the `smb` (surprise, surprise!) backend. Check if this is in the CUPS backend directory. This usually resides in `/usr/lib/cups/backend/`. You need to find an `smb` file there. It should be a symlink to `smbspool` and the file must exist and be executable:

```
root# ls -l /usr/lib/cups/backend/
total 253
drwxr-xr-x    3 root   root      720 Apr 30 19:04 .
drwxr-xr-x    6 root   root      125 Dec 19 17:13 ..
-rwxr-xr-x    1 root   root    10692 Feb 16 21:29 canon
-rwxr-xr-x    1 root   root    10692 Feb 16 21:29 epson
lrwxrwxrwx    1 root   root        3 Apr 17 22:50 http -> ipp
-rwxr-xr-x    1 root   root    17316 Apr 17 22:50 ipp
-rwxr-xr-x    1 root   root    15420 Apr 20 17:01 lpd
-rwxr-xr-x    1 root   root     8656 Apr 20 17:01 parallel
-rwxr-xr-x    1 root   root     2162 Mar 31 23:15 pdfdistiller
lrwxrwxrwx    1 root   root       25 Apr 30 19:04 ptal -> /usr/sbin/ptal-cups
-rwxr-xr-x    1 root   root     6284 Apr 20 17:01 scsi
lrwxrwxrwx    1 root   root       17 Apr  2 03:11 smb -> /usr/bin/smbspool
-rwxr-xr-x    1 root   root     7912 Apr 20 17:01 socket
-rwxr-xr-x    1 root   root     9012 Apr 20 17:01 usb

root# ls -l `which smbspool`
```

```
-rwxr-xr-x    1 root   root  563245 Dec 28 14:49 /usr/bin/smbspool
```

If this symlink does not exist, create it:

```
root# ln -s 'which smbspool' /usr/lib/cups/backend/smb
```

smbspool has been written by Mike Sweet from the CUPS folks. It is included and ships with Samba. It may also be used with print subsystems other than CUPS, to spool jobs to Windows printer shares. To set up printer *winprinter* on CUPS, you need to have a driver for it. Essentially this means to convert the print data on the CUPS/Samba host to a format that the printer can digest (the Windows host is unable to convert any files you may send). This also means you should be able to print to the printer if it were hooked directly at your Samba/CUPS host. For troubleshooting purposes, this is what you should do to determine if that part of the process chain is in order. Then proceed to fix the network connection/authentication to the Windows host, and so on.

To install a printer with the *smb* backend on CUPS, use this command:

```
root# lpadmin -p winprinter -v smb://WINDOWSNETBIOSNAME/printersharename \
  -P /path/to/PPD
```

The PPD must be able to direct CUPS to generate the print data for the target model. For PostScript printers, just use the PPD that would be used with the Windows NT PostScript driver. But what can you do if the printer is only accessible with a password? Or if the printer's host is part of another workgroup? This is provided for: You can include the required parameters as part of the `smb://` device-URI like this:

- `smb://WORKGROUP/WINDOWSNETBIOSNAME/printersharename`
- `smb://username:password@WORKGROUP/WINDOWSNETBIOSNAME/printersharename`
- `smb://username:password@WINDOWSNETBIOSNAME/printersharename`

Note that the device-URI will be visible in the process list of the Samba server (e.g., when someone uses the **ps -aux** command on Linux), even if the username and passwords are sanitized before they get written into the log files. So this is an inherently insecure option, however, it is the only one. Don't use it if you want to protect your passwords. Better share the printer in a way that does not require a password! Printing will only work if you have a working netbios name resolution up and running. Note that this is a feature of CUPS and you do not necessarily need to have smbd running.

18.18 More CUPS-Filtering Chains

The following diagrams reveal how CUPS handles print jobs.

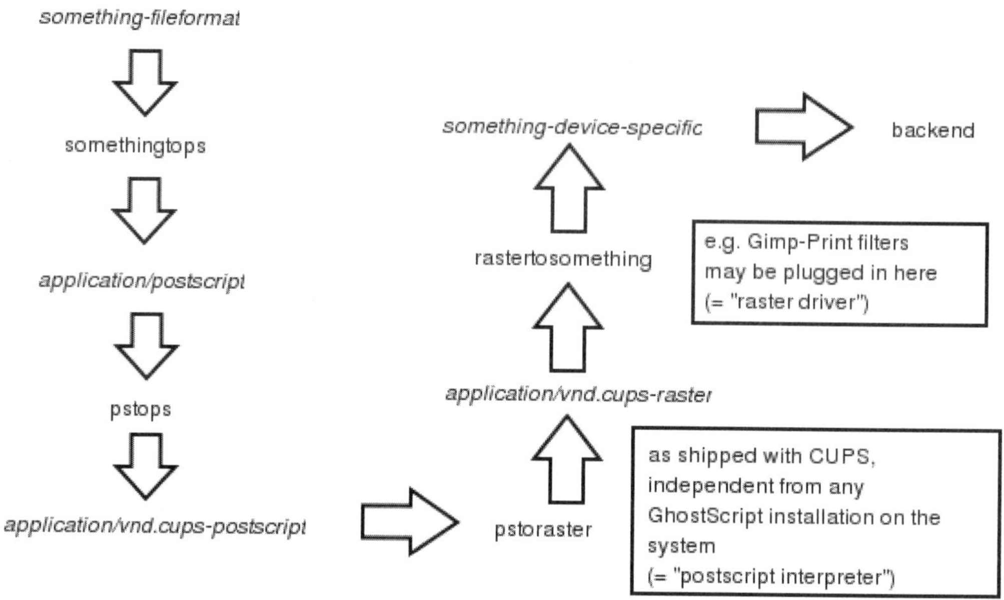

Figure 18.17. Filtering chain 1.

18.19 Common Errors

18.19.1 Windows 9x/ME Client Can't Install Driver

For Windows 9x/ME, clients require the printer names to be eight characters (or "*8 plus 3 chars suffix*") max; otherwise, the driver files will not get transferred when you want to download them from Samba.

18.19.2 "*cupsaddsmb*" Keeps Asking for Root Password in Never-ending Loop

Have you `security` = user? Have you used **smbpasswd** to give root a Samba account? You can do two things: open another terminal and execute **smbpasswd -a root** to create the account and continue entering the password into the first terminal. Or break out of the loop by pressing ENTER twice (without trying to type a password).

18.19.3 "*cupsaddsmb*" Errors

The use of "*cupsaddsmb*" gives "*No PPD file for printer...*" Message While PPD File Is Present. What might the problem be?

Have you enabled printer sharing on CUPS? This means: Do you have a *<Location /printers>....</Location>* section in CUPS server's `cupsd.conf` that does not deny access to the host you run "*cupsaddsmb*" from? It *could* be an issue if you use cupsaddsmb remotely, or if you use it with a `-h` parameter: `cupsaddsmb -H sambaserver -h cupsserver -v printername`.

Is your *TempDir* directive in `cupsd.conf` set to a valid value and is it writeable?

18.19.4 Client Can't Connect to Samba Printer

Use **smbstatus** to check which user you are from Samba's point of view. Do you have the privileges to write into the *[print$]* share?

18.19.5 New Account Reconnection from Windows 200x/XP Troubles

Once you are connected as the wrong user (for example, as `nobody`, which often occurs if you have *map to guest* = bad user), Windows Explorer will not accept an attempt to connect again as a different user. There will not be any byte transfered on the wire to Samba, but still you'll see a stupid error message that makes you think Samba has denied access. Use **smbstatus** to check for active connections. Kill the PIDs. You still can't reconnect and you get the dreaded `You can't connect with a second account from the same machine` message, as soon as you are trying. And you do not see any single byte arriving at Samba (see logs; use "*ethereal*") indicating a renewed connection attempt. Shut all Explorer Windows. This makes Windows forget what it has cached in its memory as established connections. Then reconnect as the right user. The best method is to use a DOS terminal window and *first* do `net use z: \\GANDALF\print$ /user:root`. Check with **smbstatus** that you are connected under a different account. Now open the **Printers** folder (on the Samba server in the **Network Neighborhood**), right-click on the printer in question and select **Connect...**

18.19.6 Avoid Being Connected to the Samba Server as the Wrong User

You see per **smbstatus** that you are connected as user nobody; while you want to be root or printeradmin. This is probably due to *map to guest* = bad user, which silently connects you under the guest account when you gave (maybe by accident) an incorrect username. Remove *map to guest*, if you want to prevent this.

18.19.7 Upgrading to CUPS Drivers from Adobe Drivers

This information came from a mailinglist posting regarding problems experienced when upgrading from Adobe drivers to CUPS drivers on Microsoft Windows NT/200x/XP Clients.

First delete all old Adobe-using printers. Then delete all old Adobe drivers. (On Windows 200x/XP, right-click in the background of **Printers** folder, select **Server Properties...**, select tab **Drivers** and delete here).

18.19.8 Can't Use "*cupsaddsmb*" on Samba Server Which Is a PDC

Do you use the "*naked*" root user name? Try to do it this way: `cupsaddsmb -U` *DOMAINNAME*`\\root -v` *printername*> (note the two backslashes: the first one is required to "*escape*" the second one).

18.19.9 Deleted Windows 200x Printer Driver Is Still Shown

Deleting a printer on the client will not delete the driver too (to verify, right-click on the white background of the **Printers** folder, select **Server Properties** and click on the **Drivers** tab). These same old drivers will be re-used when you try to install a printer with the same name. If you want to update to a new driver, delete the old ones first. Deletion is only possible if no other printer uses the same driver.

18.19.10 Windows 200x/XP "Local Security Policies"

Local Security Policies may not allow the installation of unsigned drivers. "*Local Security Policies*" may not allow the installation of printer drivers at all.

18.19.11 Administrator Cannot Install Printers for All Local Users

Windows XP handles SMB printers on a "*per-user*" basis. This means every user needs to install the printer himself. To have a printer available for everybody, you might want to use the built-in IPP client capabilities of WinXP. Add a printer with the print path of *http://cupsserver:631/printers/printername*. We're still looking into this one. Maybe a logon script could automatically install printers for all users.

18.19.12 Print Change Notify Functions on NT-clients

For print change, notify functions on NT++ clients. These need to run the **Server** service first (renamed to **File & Print Sharing for MS Networks** in XP).

18.19.13 WinXP-SP1

WinXP-SP1 introduced a Point and Print Restriction Policy (this restriction does not apply to "*Administrator*" or "*Power User*" groups of users). In Group Policy Object Editor, go to **User Configuration** -> **Administrative Templates** -> **Control Panel** -> **Printers**. The policy is automatically set to `Enabled` and the `Users can only Point and Print to machines in their Forest` . You probably need to change it to `Disabled` or `Users can only Point and Print to these servers` to make driver downloads from Samba possible.

18.19.14 Print Options for All Users Can't Be Set on Windows 200x/XP

How are you doing it? I bet the wrong way (it is not easy to find out, though). There are three different ways to bring you to a dialog that *seems* to set everything. All three dialogs *look* the same, yet only one of them does what you intend. You need to be Administrator or Print Administrator to do this for all users. Here is how I do in on XP:

A The first wrong way:

(a) Open the **Printers** folder.

(b) Right-click on the printer (**remoteprinter on cupshost**) and select in context menu **Printing Preferences...**

(c) Look at this dialog closely and remember what it looks like.

B The second wrong way:

(a) Open the **Printers** folder.

(b) Right-click on the printer (**remoteprinter on cupshost**) and select the context menu **Properties**.

(c) Click on the **General** tab.

(d) Click on the button **Printing Preferences...**

(e) A new dialog opens. Keep this dialog open and go back to the parent dialog.

C The third, and the correct way:

(a) Open the **Printers** folder.

(b) Click on the **Advanced** tab. (If everything is "*grayed out,*" then you are not logged in as a user with enough privileges).

(c) Click on the **Printing Defaults...** button.

(d) On any of the two new tabs, click on the **Advanced...** button.

(e) A new dialog opens. Compare this one to the other identical looking one from "*B.5*" or A.3".

Do you see any difference? I don't either. However, only the last one, which you arrived at with steps "*C.1.-6.*", will save any settings permanently and be the defaults for new users. If you want all clients to get the same defaults, you need to conduct these steps *as Administrator* (`printer admin` in `smb.conf`) *before* a client downloads the driver (the clients can later set their own *per-user defaults* by following the procedures *A* or *B* above).

18.19.15 Most Common Blunders in Driver Settings on Windows Clients

Don't use `Optimize for Speed`, but use `Optimize for Portability` instead (Adobe PS Driver). Don't use `Page Independence: No`: always settle with `Page Independence: Yes` (Microsoft PS Driver and CUPS PS Driver for Windows NT/200x/XP). If there are problems with fonts, use `Download as Softfont into printer` (Adobe PS Driver). For **TrueType Download Options** choose `Outline`. Use PostScript Level 2, if you are having trouble with a non-PS printer and if there is a choice.

18.19.16 cupsaddsmb Does Not Work with Newly Installed Printer

Symptom: The last command of **cupsaddsmb** does not complete successfully: **cmd = setdriver printername printername** result was NT_STATUS_UNSUCCESSFUL then possibly the printer was not yet recognized by Samba. Did it show up in Network Neighborhood? Did it show up i n **rpcclient hostname -c 'enumprinters'**? Restart smbd (or send a **kill -HUP** to all processes listed by **smbstatus** and try again.

18.19.17 Permissions on /var/spool/samba/ Get Reset After Each Reboot

Have you ever by accident set the CUPS spool directory to the same location? (*RequestRoot /var/spool/samba/* in `cupsd.conf` or the other way round: `/var/spool/cups/` is set as *path*> in the *[printers]* section). These *must* be different. Set

RequestRoot /var/spool/cups/ in `cupsd.conf` and *path* = /var/spool/samba in the *[printers]* section of `smb.conf`. Otherwise cupsd will sanitize permissions to its spool directory with each restart and printing will not work reliably.

18.19.18 Print Queue Called "*lp*" Mis-handles Print Jobs

In this case a print queue called "*lp*" intermittently swallows jobs and spits out completely different ones from what was sent.

It is a bad idea to name any printer "*lp*". This is the traditional UNIX name for the default printer. CUPS may be set up to do an automatic creation of Implicit Classes. This means, to group all printers with the same name to a pool of devices, and load-balancing the jobs across them in a round-robin fashion. Chances are high that someone else has a printer named "*lp*" too. You may receive his jobs and send your own to his device unwittingly. To have tight control over the printer names, set *BrowseShortNames No*. It will present any printer as *printername@cupshost* and then gives you better control over what may happen in a large networked environment.

18.19.19 Location of Adobe PostScript Driver Files for "*cupsaddsmb*"

Use **smbclient** to connect to any Windows box with a shared PostScript printer: **smbclient //windowsbox/print\$ -U guest**. You can navigate to the `W32X86/2` subdir to **mget ADOBE*** and other files or to `WIN40/0` to do the same. Another option is to download the `*.exe` packaged files from the Adobe Web site.

18.20 Overview of the CUPS Printing Processes

A complete overview of the CUPS printing processes can be found in Figure 18.19.

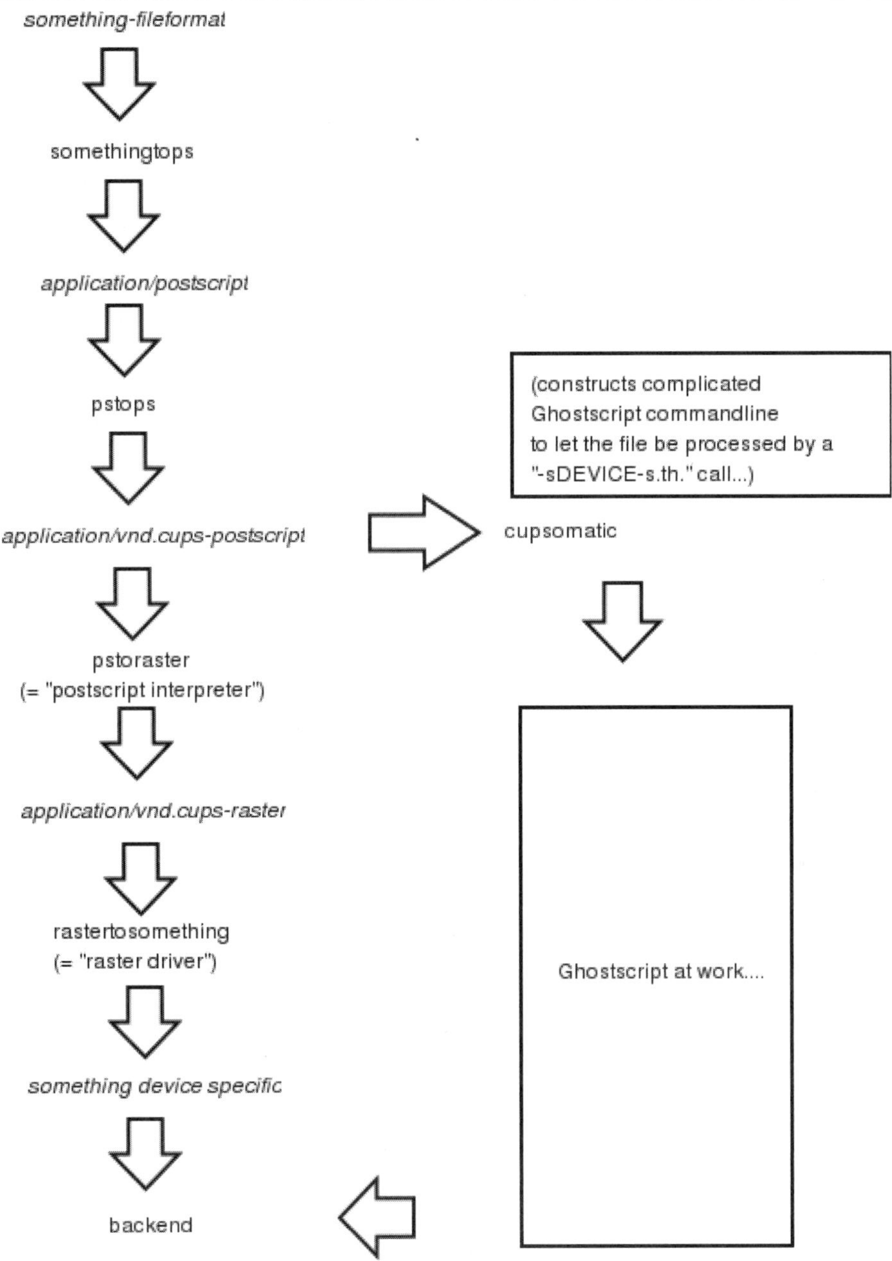
something-fileformat
somethingtops
application/postscript
pstops
application/vnd.cups-postscript
cupsomatic
(constructs complicated
Ghostscript commandline
to let the file be processed by a
"-sDEVICE-s.th." call...)
pstoraster
(= "postscript interpreter")
application/vnd.cups-raster
rastertosomething
(= "raster driver")
Ghostscript at work....
something device specific
backend
Note, that cupsomatic "kidnaps" the printfile after the
application/vnd.cups-postscript stage and deviates it gh
the CUPS-external, systemwide Ghostscript installation, bypassing the
"pstoraster" filter (therefore also bypassing the CUPS-raster-drivers
"rastertosomething", and hands the rasterized file directly to the CUPS
backend...

Figure 18.18. Filtering chain with cupsomatic

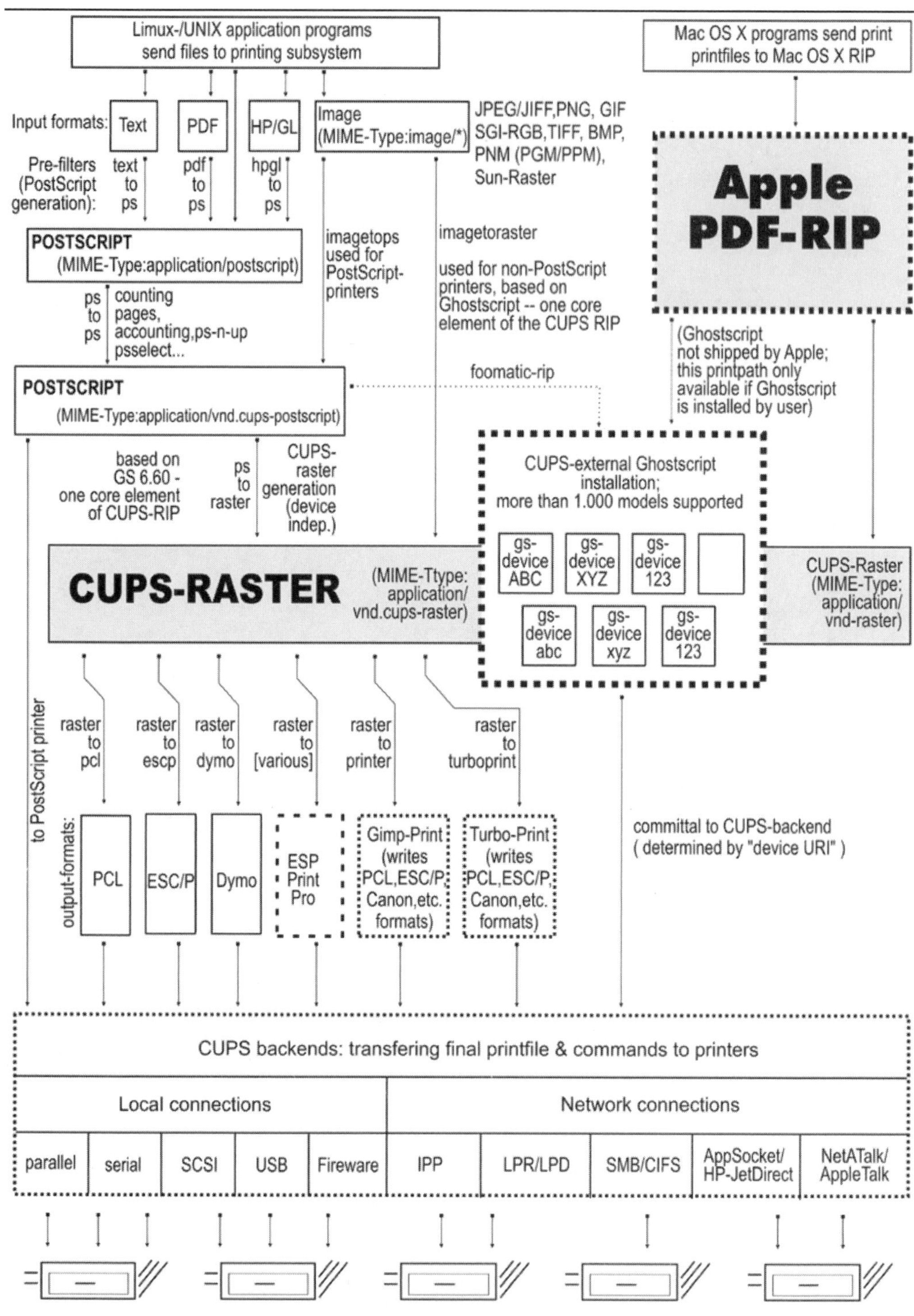

Figure 18.19. CUPS printing overview.

Chapter 19

STACKABLE VFS MODULES

19.1 Features and Benefits

Since Samba-3, there is support for stackable VFS (Virtual File System) modules. Samba passes each request to access the UNIX file system through the loaded VFS modules. This chapter covers all the modules that come with the Samba source and references to some external modules.

19.2 Discussion

If not supplied with your platform distribution binary Samba package you may have problems compiling these modules, as shared libraries are compiled and linked in different ways on different systems. They currently have been tested against GNU/Linux and IRIX.

To use the VFS modules, create a share similar to the one below. The important parameter is the *vfs objects* parameter where you can list one or more VFS modules by name. For example, to log all access to files and put deleted files in a recycle bin, see Example 19.1.

Example 19.1. smb.conf with VFS modules

```
[audit]
comment = Audited /data directory
path = /data
vfs objects = audit recycle
writeable = yes
browseable = yes
```

The modules are used in the order in which they are specified.

Samba will attempt to load modules from the `/lib` directory in the root directory of the Samba installation (usually `/usr/lib/samba/vfs` or `/usr/local/samba/lib/vfs`).

Some modules can be used twice for the same share. This can be done using a configuration similar to the one shown in Example 19.2.

Example 19.2. smb.conf with multiple VFS modules

```
[test]
comment = VFS TEST
path = /data
writeable = yes
browseable = yes
vfs objects = example:example1 example example:test
example1:  parameter = 1
example:  parameter = 5
test:  parameter = 7
```

19.3 Included Modules

19.3.1 audit

A simple module to audit file access to the syslog facility. The following operations are logged:

- share
- connect/disconnect
- directory opens/create/remove
- file open/close/rename/unlink/chmod

19.3.2 extd_audit

This module is identical with the **audit** module above except that it sends audit logs to both syslog as well as the **smbd** log files. The *log level* for this module is set in the `smb.conf` file.

Valid settings and the information that will be recorded are shown in Table 19.1.

Table 19.1. Extended Auditing Log Information

Log Level	Log Details - File and Directory Operations
0	Creation / Deletion
1	Create / Delete / Rename / Permission Changes
2	Create / Delete / Rename / Perm Change / Open / Close

19.3.3 fake_perms

This module was created to allow Roaming Profile files and directories to be set (on the Samba server under UNIX) as read only. This module will, if installed on the Profiles share, report to the client that the Profile files and directories are writable. This satisfies the client even though the files will never be overwritten as the client logs out or shuts down.

19.3.4 recycle

A Recycle Bin-like module. Where used, unlink calls will be intercepted and files moved to the recycle directory instead of being deleted. This gives the same effect as the **Recycle Bin** on Windows computers.

The **Recycle Bin** will not appear in Windows Explorer views of the network file system (share) nor on any mapped drive. Instead, a directory called `.recycle` will be automatically created when the first file is deleted. Users can recover files from the `.recycle` directory. If the *recycle:keeptree* has been specified, deleted files will be found in a path identical with that from which the file was deleted.

Supported options for the **recycle** module are as follow:

recycle:repository — Relative path of the directory where deleted files should be moved.

recycle:keeptree — Specifies whether the directory structure should be kept or if the files in the directory that is being deleted should be kept seperately in the recycle bin.

recycle:versions — If this option is set, two files with the same name that are deleted will both be kept in the recycle bin. Newer deleted versions of a file will be called "*Copy #x of filename*".

recycle:touch — Specifies whether a file's access date should be touched when the file is moved to the recycle bin.

recycle:maxsize — Files that are larger than the number of bytes specified by this parameter will not be put into the recycle bin.

recycle:exclude — List of files that should not be put into the recycle bin when deleted, but deleted in the regular way.

recycle:exclude_dir — Contains a list of directories. When files from these directories are deleted, they are not put into the recycle bin but are deleted in the regular way.

recycle:noversions — Opposite of *recycle:versions*. If both options are specified, this one takes precedence.

19.3.5 netatalk

A netatalk module will ease co-existence of Samba and netatalk file sharing services.

Advantages compared to the old netatalk module:

- Does not care about creating .AppleDouble forks, just keeps them in sync.

- If a share in `smb.conf` does not contain .AppleDouble item in hide or veto list, it will be added automatically.

19.4 VFS Modules Available Elsewhere

This section contains a listing of various other VFS modules that have been posted but do not currently reside in the Samba CVS tree for one reason or another (e.g., it is easy for the maintainer to have his or her own CVS tree).

No statements about the stability or functionality of any module should be implied due to its presence here.

19.4.1 DatabaseFS

URL: `http://www.css.tayloru.edu/~elorimer/databasefs/index.php`

By Eric Lorimer.[1]

I have created a VFS module that implements a fairly complete read-only filesystem. It presents information from a database as a filesystem in a modular and generic way to allow different databases to be used (originally designed for organizing MP3s under directories such as "*Artists*," "*Song Keywords*," and so on. I have since easily applied it to a student roster database.) The directory structure is stored in the database itself and the module makes no assumptions about the database structure beyond the table it requires to run.

Any feedback would be appreciated: comments, suggestions, patches, and so on. If nothing else, hopefully it might prove useful for someone else who wishes to create a virtual filesystem.

19.4.2 vscan

URL: `http://www.openantivirus.org/`

`samba-vscan` is a proof-of-concept module for Samba, which uses the VFS (virtual file system) features of Samba 2.2.x/3.0 alphaX. Of course, Samba has to be compiled with VFS support. `samba-vscan` supports various virus scanners and is maintained by Rainer Link.

[1] mailto:elorimer@css.tayloru.edu

Chapter 20

WINBIND: USE OF DOMAIN ACCOUNTS

20.1 Features and Benefits

Integration of UNIX and Microsoft Windows NT through a unified logon has been considered a "*holy grail*" in heterogeneous computing environments for a long time.

There is one other facility without which UNIX and Microsoft Windows network interoperability would suffer greatly. It is imperative that there be a mechanism for sharing files across UNIX systems and to be able to assign domain user and group ownerships with integrity.

winbind is a component of the Samba suite of programs that solves the unified logon problem. Winbind uses a UNIX implementation of Microsoft RPC calls, Pluggable Authentication Modules, and the Name Service Switch to allow Windows NT domain users to appear and operate as UNIX users on a UNIX machine. This chapter describes the Winbind system, explaining the functionality it provides, how it is configured, and how it works internally.

Winbind provides three separate functions:

- Authentication of user credentials (via PAM).
- Identity resolution (via NSS).
- Winbind maintains a database called winbind_idmap.tdb in which it stores mappings between UNIX UIDs / GIDs and NT SIDs. This mapping is used only for users and groups that do not have a local UID/GID. It stored the UID/GID allocated from the idmap uid/gid range that it has mapped to the NT SID. If *idmap backend* has been specified as ldapsam:url then instead of using a local mapping Winbind will obtain this information from the LDAP database.

NOTE

If **winbindd** is not running, smbd (which calls **winbindd**) will fall back to using purely local information from `/etc/passwd` and `/etc/group` and no dynamic mapping will be used.

20.2 Introduction

It is well known that UNIX and Microsoft Windows NT have different models for representing user and group information and use different technologies for implementing them. This fact has made it difficult to integrate the two systems in a satisfactory manner.

One common solution in use today has been to create identically named user accounts on both the UNIX and Windows systems and use the Samba suite of programs to provide file and print services between the two. This solution is far from perfect, however, as adding and deleting users on both sets of machines becomes a chore and two sets of passwords are required — both of which can lead to synchronization problems between the UNIX and Windows systems and confusion for users.

We divide the unified logon problem for UNIX machines into three smaller problems:

- Obtaining Windows NT user and group information.
- Authenticating Windows NT users.
- Password changing for Windows NT users.

Ideally, a prospective solution to the unified logon problem would satisfy all the above components without duplication of information on the UNIX machines and without creating additional tasks for the system administrator when maintaining users and groups on either system. The Winbind system provides a simple and elegant solution to all three components of the unified logon problem.

20.3 What Winbind Provides

Winbind unifies UNIX and Windows NT account management by allowing a UNIX box to become a full member of an NT domain. Once this is done the UNIX box will see NT users and groups as if they were "*native*" UNIX users and groups, allowing the NT domain to be used in much the same manner that NIS+ is used within UNIX-only environments.

The end result is that whenever any program on the UNIX machine asks the operating system to lookup a user or group name, the query will be resolved by asking the NT Domain Controller for the specified domain to do the lookup. Because Winbind hooks into the operating system at a low level (via the NSS name resolution modules in the C library), this redirection to the NT Domain Controller is completely transparent.

Users on the UNIX machine can then use NT user and group names as they would *"native"* UNIX names. They can chown files so they are owned by NT domain users or even login to the UNIX machine and run a UNIX X-Window session as a domain user.

The only obvious indication that Winbind is being used is that user and group names take the form `DOMAIN\user` and `DOMAIN\group`. This is necessary as it allows Winbind to determine that redirection to a Domain Controller is wanted for a particular lookup and which trusted domain is being referenced.

Additionally, Winbind provides an authentication service that hooks into the Pluggable Authentication Modules (PAM) system to provide authentication via an NT domain to any PAM-enabled applications. This capability solves the problem of synchronizing passwords between systems since all passwords are stored in a single location (on the Domain Controller).

20.3.1 Target Uses

Winbind is targeted at organizations that have an existing NT-based domain infrastructure into which they wish to put UNIX workstations or servers. Winbind will allow these organizations to deploy UNIX workstations without having to maintain a separate account infrastructure. This greatly simplifies the administrative overhead of deploying UNIX workstations into an NT-based organization.

Another interesting way in which we expect Winbind to be used is as a central part of UNIX-based appliances. Appliances that provide file and print services to Microsoft-based networks will be able to use Winbind to provide seamless integration of the appliance into the domain.

20.4 How Winbind Works

The Winbind system is designed around a client/server architecture. A long running **winbindd** daemon listens on a UNIX domain socket waiting for requests to arrive. These requests are generated by the NSS and PAM clients and is processed sequentially.

The technologies used to implement Winbind are described in detail below.

20.4.1 Microsoft Remote Procedure Calls

Over the last few years, efforts have been underway by various Samba Team members to decode various aspects of the Microsoft Remote Procedure Call (MSRPC) system. This system is used for most network-related operations between Windows NT machines including remote management, user authentication and print spooling. Although initially this work was done to aid the implementation of Primary Domain Controller (PDC) functionality in Samba, it has also yielded a body of code that can be used for other purposes.

Winbind uses various MSRPC calls to enumerate domain users and groups and to obtain detailed information about individual users or groups. Other MSRPC calls can be used to authenticate NT domain users and to change user passwords. By directly querying a

Windows PDC for user and group information, Winbind maps the NT account information onto UNIX user and group names.

20.4.2 Microsoft Active Directory Services

Since late 2001, Samba has gained the ability to interact with Microsoft Windows 2000 using its "*Native Mode*" protocols, rather than the NT4 RPC services. Using LDAP and Kerberos, a Domain Member running Winbind can enumerate users and groups in exactly the same way as a Windows 200x client would, and in so doing provide a much more efficient and effective Winbind implementation.

20.4.3 Name Service Switch

The Name Service Switch, or NSS, is a feature that is present in many UNIX operating systems. It allows system information such as hostnames, mail aliases and user information to be resolved from different sources. For example, a standalone UNIX workstation may resolve system information from a series of flat files stored on the local filesystem. A networked workstation may first attempt to resolve system information from local files, and then consult an NIS database for user information or a DNS server for hostname information.

The NSS application programming interface allows Winbind to present itself as a source of system information when resolving UNIX usernames and groups. Winbind uses this interface, and information obtained from a Windows NT server using MSRPC calls to provide a new source of account enumeration. Using standard UNIX library calls, one can enumerate the users and groups on a UNIX machine running Winbind and see all users and groups in a NT domain plus any trusted domain as though they were local users and groups.

The primary control file for NSS is `/etc/nsswitch.conf`. When a UNIX application makes a request to do a lookup, the C library looks in `/etc/nsswitch.conf` for a line that matches the service type being requested, for example the "*passwd*" service type is used when user or group names are looked up. This config line specifies which implementations of that service should be tried and in what order. If the passwd config line is:

```
passwd: files example
```

then the C library will first load a module called `/lib/libnss_files.so` followed by the module `/lib/libnss_example.so`. The C library will dynamically load each of these modules in turn and call resolver functions within the modules to try to resolve the request. Once the request is resolved, the C library returns the result to the application.

This NSS interface provides an easy way for Winbind to hook into the operating system. All that needs to be done is to put `libnss_winbind.so` in `/lib/` then add "*winbind*" into `/etc/nsswitch.conf` at the appropriate place. The C library will then call Winbind to resolve user and group names.

20.4.4 Pluggable Authentication Modules

Pluggable Authentication Modules, also known as PAM, is a system for abstracting authentication and authorization technologies. With a PAM module it is possible to specify different authentication methods for different system applications without having to recompile these applications. PAM is also useful for implementing a particular policy for authorization. For example, a system administrator may only allow console logins from users stored in the local password file but only allow users resolved from a NIS database to log in over the network.

Winbind uses the authentication management and password management PAM interface to integrate Windows NT users into a UNIX system. This allows Windows NT users to log in to a UNIX machine and be authenticated against a suitable Primary Domain Controller. These users can also change their passwords and have this change take effect directly on the Primary Domain Controller.

PAM is configured by providing control files in the directory `/etc/pam.d/` for each of the services that require authentication. When an authentication request is made by an application, the PAM code in the C library looks up this control file to determine what modules to load to do the authentication check and in what order. This interface makes adding a new authentication service for Winbind very easy. All that needs to be done is that the `pam_winbind.so` module is copied to `/lib/security/` and the PAM control files for relevant services are updated to allow authentication via Winbind. See the PAM documentation in Chapter 24, *PAM-Based Distributed Authentication* for more information.

20.4.5 User and Group ID Allocation

When a user or group is created under Windows NT/200x it is allocated a numerical relative identifier (RID). This is slightly different from UNIX which has a range of numbers that are used to identify users, and the same range in which to identify groups. It is Winbind's job to convert RIDs to UNIX ID numbers and vice versa. When Winbind is configured, it is given part of the UNIX user ID space and a part of the UNIX group ID space in which to store Windows NT users and groups. If a Windows NT user is resolved for the first time, it is allocated the next UNIX ID from the range. The same process applies for Windows NT groups. Over time, Winbind will have mapped all Windows NT users and groups to UNIX user IDs and group IDs.

The results of this mapping are stored persistently in an ID mapping database held in a tdb database). This ensures that RIDs are mapped to UNIX IDs in a consistent way.

20.4.6 Result Caching

An active system can generate a lot of user and group name lookups. To reduce the network cost of these lookups, Winbind uses a caching scheme based on the SAM sequence number supplied by NT Domain Controllers. User or group information returned by a PDC is cached by Winbind along with a sequence number also returned by the PDC. This sequence number is incremented by Windows NT whenever any user or group information is modified. If a cached entry has expired, the sequence number is requested from the PDC and compared against the sequence number of the cached entry. If the sequence numbers do not match, then the cached information is discarded and up-to-date information is requested directly from the PDC.

20.5 Installation and Configuration

20.5.1 Introduction

This section describes the procedures used to get Winbind up and running. Winbind is capable of providing access and authentication control for Windows Domain users through an NT or Windows 200x PDC for regular services, such as telnet and ftp, as well for Samba services.

- *Why should I do this?*

  This allows the Samba administrator to rely on the authentication mechanisms on the Windows NT/200x PDC for the authentication of Domain Members. Windows NT/200x users no longer need to have separate accounts on the Samba server.

- *Who should be reading this document?*

  This document is designed for system administrators. If you are implementing Samba on a file server and wish to (fairly easily) integrate existing Windows NT/200x users from your PDC onto the Samba server, this document is for you.

20.5.2 Requirements

If you have a Samba configuration file that you are currently using, *BACK IT UP!* If your system already uses PAM, *back up the `/etc/pam.d` directory contents!* If you haven't already made a boot disk, *MAKE ONE NOW!*

Messing with the PAM configuration files can make it nearly impossible to log in to your machine. That's why you want to be able to boot back into your machine in single user mode and restore your `/etc/pam.d` back to the original state they were in if you get frustrated with the way things are going.

The latest version of Samba-3 includes a functioning winbindd daemon. Please refer to the main Samba Web page[1] or, better yet, your closest Samba mirror site for instructions on downloading the source code.

To allow domain users the ability to access Samba shares and files, as well as potentially other services provided by your Samba machine, PAM must be set up properly on your machine. In order to compile the Winbind modules, you should have at least the PAM development libraries installed on your system. Please refer the PAM web site `http://www.kernel.org/pub/linux/libs/pam/`[2].

[1] http://samba.org/
[2] http://www.kernel.org/pub/linux/libs/pam/

20.5.3 Testing Things Out

Before starting, it is probably best to kill off all the Samba-related daemons running on your server. Kill off all smbd, nmbd, and winbindd processes that may be running. To use PAM, make sure that you have the standard PAM package that supplies the `/etc/pam.d` directory structure, including the PAM modules that are used by PAM-aware services, several pam libraries, and the `/usr/doc` and `/usr/man` entries for pam. Winbind built better in Samba if the pam-devel package is also installed. This package includes the header files needed to compile PAM-aware applications.

20.5.3.1 Configure nsswitch.conf and the Winbind Libraries on Linux and Solaris

PAM is a standard component of most current generation UNIX/Linux systems. Unfortunately, few systems install the `pam-devel` libraries that are needed to build PAM-enabled Samba. Additionally, Samba-3 may auto-install the Winbind files into their correct locations on your system, so before you get too far down the track be sure to check if the following configuration is really necessary. You may only need to configure `/etc/nsswitch.conf`.

The libraries needed to run the winbindd daemon through nsswitch need to be copied to their proper locations:

```
root# cp ../samba/source/nsswitch/libnss_winbind.so /lib
```

I also found it necessary to make the following symbolic link:

```
root# ln -s /lib/libnss_winbind.so /lib/libnss_winbind.so.2
```

And, in the case of Sun Solaris:

```
root# ln -s /usr/lib/libnss_winbind.so /usr/lib/libnss_winbind.so.1
root# ln -s /usr/lib/libnss_winbind.so /usr/lib/nss_winbind.so.1
root# ln -s /usr/lib/libnss_winbind.so /usr/lib/nss_winbind.so.2
```

Now, as root you need to edit `/etc/nsswitch.conf` to allow user and group entries to be visible from the winbindd daemon. My `/etc/nsswitch.conf` file look like this after editing:

```
passwd:     files winbind
shadow:     files
group:      files winbind
```

The libraries needed by the **winbindd** daemon will be automatically entered into the **ldconfig** cache the next time your system reboots, but it is faster (and you do not need to reboot) if you do it manually:

```
root#/sbin/ldconfig -v | grep winbind
```

This makes `libnss_winbind` available to winbindd and echos back a check to you.

20.5.3.2 NSS Winbind on AIX

(This section is only for those running AIX.)

The Winbind AIX identification module gets built as `libnss_winbind.so` in the nsswitch directory of the Samba source. This file can be copied to `/usr/lib/security`, and the AIX naming convention would indicate that it should be named WINBIND. A stanza like the following:

```
WINBIND:
        program = /usr/lib/security/WINBIND
        options = authonly
```

can then be added to `/usr/lib/security/methods.cfg`. This module only supports identification, but there have been success reports using the standard Winbind PAM module for authentication. Use caution configuring loadable authentication modules since you can make it impossible to logon to the system. More information about the AIX authentication module API can be found at "*Kernel Extensions and Device Support Programming Concepts for AIX*"in Chapter 18(John, there is no section like this in 18). Loadable Authentication Module Programming Interface[3] and more information on administering the modules can be found at "*System Management Guide: Operating System and Devices.*"[4]

20.5.3.3 Configure smb.conf

Several parameters are needed in the `smb.conf` file to control the behavior of winbindd. These are described in more detail in the winbindd(8) man page. My `smb.conf` file, as shown in Example 20.1, was modified to include the necessary entries in the [global] section.

20.5.3.4 Join the Samba Server to the PDC Domain

Enter the following command to make the Samba server join the PDC domain, where ***DOMAIN*** is the name of your Windows domain and ***Administrator*** is a domain user who has administrative privileges in the domain.

```
root#/usr/local/samba/bin/net rpc join -S PDC -U Administrator
```

The proper response to the command should be: "*Joined the domain DOMAIN*" where *DOMAIN* is your DOMAIN name.

20.5.3.5 Starting and Testing the winbindd Daemon

Eventually, you will want to modify your Samba startup script to automatically invoke the winbindd daemon when the other parts of Samba start, but it is possible to test out just

[3]http://publibn.boulder.ibm.com/doc_link/en_US/a_doc_lib/aixprggd/kernextc/sec_load_mod.htm
[4]http://publibn.boulder.ibm.com/doc_link/en_US/a_doc_lib/aixbman/baseadmn/iandaadmin.htm

Example 20.1. smb.conf for Winbind set-up

```
[global]
# separate domain and username with '+', like DOMAIN+username
winbind separator = +
# use uids from 10000 to 20000 for domain users
idmap uid = 10000-20000
# use gids from 10000 to 20000 for domain groups
winbind gid = 10000-20000
# allow enumeration of winbind users and groups
winbind enum users = yes
winbind enum groups = yes
# give winbind users a real shell (only needed if they have telnet access)
template homedir = /home/winnt/%D/%U
template shell = /bin/bash
```

the Winbind portion first. To start up Winbind services, enter the following command as root:

```
root#/usr/local/samba/bin/winbindd
```

NOTE

The above assumes that Samba has been installed in the `/usr/local/samba` directory tree. You may need to search for the location of Samba files if this is not the location of **winbindd** on your system.

Winbindd can now also run in "*dual daemon mode*". This will make it run as two processes. The first will answer all requests from the cache, thus making responses to clients faster. The other will update the cache for the query that the first has just responded. The advantage of this is that responses stay accurate and are faster. You can enable dual daemon mode by adding `-B` to the commandline:

```
root#/usr/local/samba/bin/winbindd -B
```

I'm always paranoid and like to make sure the daemon is really running.

```
root#ps -ae | grep winbindd
```

This command should produce output like this, if the daemon is running you would expect to see a report something like this:

```
3025 ?          00:00:00 winbindd
```

Now, for the real test, try to get some information about the users on your PDC:

```
root#/usr/local/samba/bin/wbinfo -u
```

This should echo back a list of users on your Windows users on your PDC. For example, I get the following response:

```
CEO+Administrator
CEO+burdell
CEO+Guest
CEO+jt-ad
CEO+krbtgt
CEO+TsInternetUser
```

Obviously, I have named my domain "*CEO*" and my ***winbind separator*** is "+".

You can do the same sort of thing to get group information from the PDC:

```
root# /usr/local/samba/bin/wbinfo -g
   CEO+Domain Admins
   CEO+Domain Users
   CEO+Domain Guests
   CEO+Domain Computers
   CEO+Domain Controllers
   CEO+Cert Publishers
   CEO+Schema Admins
   CEO+Enterprise Admins
   CEO+Group Policy Creator Owners
```

The function **getent** can now be used to get unified lists of both local and PDC users and groups. Try the following command:

```
root#getent passwd
```

You should get a list that looks like your `/etc/passwd` list followed by the domain users with their new UIDs, GIDs, home directories and default shells.

The same thing can be done for groups with the command:

```
root#getent group
```

20.5.3.6 Fix the init.d Startup Scripts

Linux The winbindd daemon needs to start up after the smbd and nmbd daemons are running. To accomplish this task, you need to modify the startup scripts of your system. They are located at `/etc/init.d/smb` in Red Hat Linux and they are located in `/etc/init.d/samba` in Debian Linux. Edit your script to add commands to invoke this daemon in the proper sequence. My startup script starts up smbd, nmbd, and winbindd from the `/usr/local/samba/bin` directory directly. The **start** function in the script looks like this:

```
start() {
        KIND="SMB"
        echo -n $"Starting $KIND services: "
        daemon /usr/local/samba/bin/smbd $SMBDOPTIONS
        RETVAL=$?
        echo
        KIND="NMB"
        echo -n $"Starting $KIND services: "
        daemon /usr/local/samba/bin/nmbd $NMBDOPTIONS
        RETVAL2=$?
        echo
        KIND="Winbind"
        echo -n $"Starting $KIND services: "
        daemon /usr/local/samba/bin/winbindd
        RETVAL3=$?
        echo
        [ $RETVAL -eq 0 -a $RETVAL2 -eq 0 -a $RETVAL3 -eq 0 ] && \
      touch /var/lock/subsys/smb || RETVAL=1
        return $RETVAL
}
```

If you would like to run winbindd in dual daemon mode, replace the line :

```
        daemon /usr/local/samba/bin/winbindd
```

in the example above with:

```
        daemon /usr/local/samba/bin/winbindd -B
```

.

The **stop** function has a corresponding entry to shut down the services and looks like this:

```
stop() {
        KIND="SMB"
        echo -n $"Shutting down $KIND services: "
        killproc smbd
        RETVAL=$?
        echo
        KIND="NMB"
        echo -n $"Shutting down $KIND services: "
        killproc nmbd
        RETVAL2=$?
        echo
```

```
        KIND="Winbind"
        echo -n $"Shutting down $KIND services: "
        killproc winbindd
        RETVAL3=$?
        [ $RETVAL -eq 0 -a $RETVAL2 -eq 0 -a $RETVAL3 -eq 0 ] && \
       rm -f /var/lock/subsys/smb
        echo ""
        return $RETVAL
}
```

Solaris Winbind does not work on Solaris 9, see Section 36.6.2 for details.

On Solaris, you need to modify the `/etc/init.d/samba.server` startup script. It usually only starts smbd and nmbd but should now start winbindd, too. If you have Samba installed in `/usr/local/samba/bin`, the file could contains something like this:

```
##
## samba.server
##

if [ ! -d /usr/bin ]
then                        # /usr not mounted
   exit
fi

killproc() {                # kill the named process(es)
   pid=`/usr/bin/ps -e |
        /usr/bin/grep -w $1 |
        /usr/bin/sed -e 's/^  *//' -e 's/ .*//'`
   [ "$pid" != "" ] && kill $pid
}

# Start/stop processes required for Samba server

case "$1" in

'start')
#
# Edit these lines to suit your installation (paths, workgroup, host)
#
echo Starting SMBD
   /usr/local/samba/bin/smbd -D -s \
   /usr/local/samba/smb.conf

echo Starting NMBD
   /usr/local/samba/bin/nmbd -D -l \
   /usr/local/samba/var/log -s /usr/local/samba/smb.conf
```

```
echo Starting Winbind Daemon
    /usr/local/samba/bin/winbindd
    ;;

'stop')
    killproc nmbd
    killproc smbd
    killproc winbindd
    ;;

*)
    echo "Usage: /etc/init.d/samba.server { start | stop }"
    ;;
esac
```

Again, if you would like to run Samba in dual daemon mode, replace:

```
/usr/local/samba/bin/winbindd
```

in the script above with:

```
/usr/local/samba/bin/winbindd -B
```

Restarting If you restart the smbd, nmbd, and winbindd daemons at this point, you should be able to connect to the Samba server as a Domain Member just as if you were a local user.

20.5.3.7 Configure Winbind and PAM

If you have made it this far, you know that **winbindd** and Samba are working together. If you want to use Winbind to provide authentication for other services, keep reading. The PAM configuration files need to be altered in this step. (Did you remember to make backups of your original `/etc/pam.d` files? If not, do it now.)

You will need a PAM module to use winbindd with these other services. This module will be compiled in the `../source/nsswitch` directory by invoking the command:

```
root#make nsswitch/pam_winbind.so
```

from the `../source` directory. The `pam_winbind.so` file should be copied to the location of your other PAM security modules. On my RedHat system, this was the `/lib/security` directory. On Solaris, the PAM security modules reside in `/usr/lib/security`.

```
root#cp ../samba/source/nsswitch/pam_winbind.so /lib/security
```

Linux/FreeBSD-specific PAM configuration The /etc/pam.d/samba file does not need to be changed. I just left this file as it was:

```
auth    required        /lib/security/pam_stack.so service=system-auth
account required        /lib/security/pam_stack.so service=system-auth
```

The other services that I modified to allow the use of Winbind as an authentication service were the normal login on the console (or a terminal session), telnet logins, and ftp service. In order to enable these services, you may first need to change the entries in /etc/xinetd.d (or /etc/inetd.conf). Red Hat Linux 7.1 and later uses the new xinetd.d structure, in this case you need to change the lines in /etc/xinetd.d/telnet and /etc/xinetd.d/wu-ftp from

```
enable = no
```

to:

```
enable = yes
```

For ftp services to work properly, you will also need to either have individual directories for the domain users already present on the server, or change the home directory template to a general directory for all domain users. These can be easily set using the smb.conf global entry *template homedir*.

The /etc/pam.d/ftp file can be changed to allow Winbind ftp access in a manner similar to the samba file. My /etc/pam.d/ftp file was changed to look like this:

```
auth       required     /lib/security/pam_listfile.so item=user sense=deny \
    file=/etc/ftpusers onerr=succeed
auth       sufficient   /lib/security/pam_winbind.so
auth       required     /lib/security/pam_stack.so service=system-auth
auth       required     /lib/security/pam_shells.so
account    sufficient   /lib/security/pam_winbind.so
account    required     /lib/security/pam_stack.so service=system-auth
session    required     /lib/security/pam_stack.so service=system-auth
```

The /etc/pam.d/login file can be changed nearly the same way. It now looks like this:

```
auth       required     /lib/security/pam_securetty.so
auth       sufficient   /lib/security/pam_winbind.so
auth       sufficient   /lib/security/pam_UNIX.so use_first_pass
auth       required     /lib/security/pam_stack.so service=system-auth
auth       required     /lib/security/pam_nologin.so
account    sufficient   /lib/security/pam_winbind.so
```

```
account    required    /lib/security/pam_stack.so service=system-auth
password   required    /lib/security/pam_stack.so service=system-auth
session    required    /lib/security/pam_stack.so service=system-auth
session    optional    /lib/security/pam_console.so
```

In this case, I added the

```
auth sufficient /lib/security/pam_winbind.so
```

lines as before, but also added the

```
required pam_securetty.so
```

above it, to disallow root logins over the network. I also added a

```
sufficient /lib/security/pam_unix.so use_first_pass
```

line after the **winbind.so** line to get rid of annoying double prompts for passwords.

Solaris-specific configuration The /etc/pam.conf needs to be changed. I changed this file so my Domain users can logon both locally as well as telnet. The following are the changes that I made. You can customize the pam.conf file as per your requirements, but be sure of those changes because in the worst case it will leave your system nearly impossible to boot.

```
#
#ident "@(#)pam.conf 1.14 99/09/16 SMI"
#
# Copyright (c) 1996-1999, Sun Microsystems, Inc.
# All Rights Reserved.
#
# PAM configuration
#
# Authentication management
#
login   auth required   /usr/lib/security/pam_winbind.so
login auth required  /usr/lib/security/$ISA/pam_UNIX.so.1 try_first_pass
login auth required  /usr/lib/security/$ISA/pam_dial_auth.so.1 try_first_pass
#
rlogin  auth sufficient /usr/lib/security/pam_winbind.so
rlogin  auth sufficient /usr/lib/security/$ISA/pam_rhosts_auth.so.1
rlogin auth required  /usr/lib/security/$ISA/pam_UNIX.so.1 try_first_pass
#
dtlogin auth sufficient /usr/lib/security/pam_winbind.so
dtlogin auth required  /usr/lib/security/$ISA/pam_UNIX.so.1 try_first_pass
#
rsh auth required /usr/lib/security/$ISA/pam_rhosts_auth.so.1
other   auth sufficient /usr/lib/security/pam_winbind.so
other auth required /usr/lib/security/$ISA/pam_UNIX.so.1 try_first_pass
#
```

```
# Account management
#
login   account sufficient      /usr/lib/security/pam_winbind.so
login account requisite /usr/lib/security/$ISA/pam_roles.so.1
login account required /usr/lib/security/$ISA/pam_UNIX.so.1
#
dtlogin account sufficient      /usr/lib/security/pam_winbind.so
dtlogin account requisite /usr/lib/security/$ISA/pam_roles.so.1
dtlogin account required /usr/lib/security/$ISA/pam_UNIX.so.1
#
other   account sufficient      /usr/lib/security/pam_winbind.so
other account requisite /usr/lib/security/$ISA/pam_roles.so.1
other account required /usr/lib/security/$ISA/pam_UNIX.so.1
#
# Session management
#
other session required /usr/lib/security/$ISA/pam_UNIX.so.1
#
# Password management
#
#other   password sufficient     /usr/lib/security/pam_winbind.so
other password required /usr/lib/security/$ISA/pam_UNIX.so.1
dtsession auth required /usr/lib/security/$ISA/pam_UNIX.so.1
#
# Support for Kerberos V5 authentication (uncomment to use Kerberos)
#
#rlogin auth optional /usr/lib/security/$ISA/pam_krb5.so.1 try_first_pass
#login auth optional /usr/lib/security/$ISA/pam_krb5.so.1 try_first_pass
#dtlogin auth optional /usr/lib/security/$ISA/pam_krb5.so.1 try_first_pass
#other auth optional /usr/lib/security/$ISA/pam_krb5.so.1 try_first_pass
#dtlogin account optional /usr/lib/security/$ISA/pam_krb5.so.1
#other account optional /usr/lib/security/$ISA/pam_krb5.so.1
#other session optional /usr/lib/security/$ISA/pam_krb5.so.1
#other password optional /usr/lib/security/$ISA/pam_krb5.so.1 try_first_pass
```

I also added a *try_first_pass* line after the `winbind.so` line to get rid of annoying double prompts for passwords.

Now restart your Samba and try connecting through your application that you configured in the pam.conf.

20.6 Conclusion

The Winbind system, through the use of the Name Service Switch, Pluggable Authentication Modules, and appropriate Microsoft RPC calls have allowed us to provide seamless integration of Microsoft Windows NT domain users on a UNIX system. The result is a great reduction in the administrative cost of running a mixed UNIX and NT network.

20.7 Common Errors

Winbind has a number of limitations in its current released version that we hope to overcome in future releases:

- Winbind is currently only available for the Linux, Solaris, AIX, and IRIX operating systems, although ports to other operating systems are certainly possible. For such ports to be feasible, we require the C library of the target operating system to support the Name Service Switch and Pluggable Authentication Modules systems. This is becoming more common as NSS and PAM gain support among UNIX vendors.
- The mappings of Windows NT RIDs to UNIX IDs is not made algorithmically and depends on the order in which unmapped users or groups are seen by Winbind. It may be difficult to recover the mappings of RID to UNIX ID mapping if the file containing this information is corrupted or destroyed.
- Currently the Winbind PAM module does not take into account possible workstation and logon time restrictions that may be set for Windows NT users, this is instead up to the PDC to enforce.

20.7.1 NSCD Problem Warning

WARNING

Do not under any circumstances run **nscd** on any system on which **winbindd** is running.

If **nscd** is running on the UNIX/Linux system, then even though NSSWITCH is correctly configured it will not be possible to resolve domain users and groups for file and directory controls.

20.7.2 Winbind Is Not Resolving Users and Groups

*"My `smb.conf` file is correctly configured. I have specified `idmap uid` = 12000, and `idmap gid` = 3000-3500 and **winbind** is running. When I do the following it all works fine."*

```
root# wbinfo -u
MIDEARTH+maryo
MIDEARTH+jackb
MIDEARTH+ameds
...
MIDEARTH+root

root# wbinfo -g
MIDEARTH+Domain Users
```

```
MIDEARTH+Domain Admins
MIDEARTH+Domain Guests
...
MIDEARTH+Accounts

root# getent passwd
root:x:0:0:root:/root:/bin/bash
bin:x:1:1:bin:/bin:/bin/bash
...
maryo:x:15000:15003:Mary Orville:/home/MIDEARTH/maryo:/bin/false
```

"But the following command just fails:

```
root# chown maryo a_file
chown: 'maryo': invalid user
```

This is driving me nuts! What can be wrong?"

Same problem as the one above. Your system is likely running **nscd**, the name service caching daemon. Shut it down, do not restart it! You will find your problem resolved.

Chapter 21

ADVANCED NETWORK MANAGEMENT

This section documents peripheral issues that are of great importance to network administrators who want to improve network resource access control, to automate the user environment and to make their lives a little easier.

21.1 Features and Benefits

Often the difference between a working network environment and a well appreciated one can best be measured by the *little things* that make everything work more harmoniously. A key part of every network environment solution is the ability to remotely manage MS Windows workstations, remotely access the Samba server, provide customized logon scripts, as well as other housekeeping activities that help to sustain more reliable network operations.

This chapter presents information on each of these areas. They are placed here, and not in other chapters, for ease of reference.

21.2 Remote Server Administration

"How do I get 'User Manager' and 'Server Manager'?"

Since I do not need to buy an NT4 Server, how do I get the 'User Manager for Domains' and the 'Server Manager'?

Microsoft distributes a version of these tools called `Nexus.exe` for installation on Windows 9x/Me systems. The tools set includes:

- Server Manager
- User Manager for Domains
- Event Viewer

Download the archived file at ftp://ftp.microsoft.com/Softlib/MSLFILES/NEXUS.EXE.

The Windows NT 4.0 version of the 'User Manager for Domains' and 'Server Manager' are available from Microsoft via ftp[1].

21.3 Remote Desktop Management

There are a number of possible remote desktop management solutions that range from free through costly. Do not let that put you off. Sometimes the most costly solution is the most cost effective. In any case, you will need to draw your own conclusions as to which is the best tool in your network environment.

21.3.1 Remote Management from NoMachine.Com

The following information was posted to the Samba mailing list at Apr 3 23:33:50 GMT 2003. It is presented in slightly edited form (with author details omitted for privacy reasons). The entire answer is reproduced below with some comments removed.

"*I have a wonderful Linux/Samba server running as pdc for a network. Now I would like to add remote desktop capabilities so users outside could login to the system and get their desktop up from home or another country.*"

"*Is there a way to accomplish this? Do I need a Windows Terminal Server? Do I need to configure it so it is a member of the domain or a BDC,PDC? Are there any hacks for MS Windows XP to enable remote login even if the computer is in a domain?*"

Answer provided: Check out the new offer from NoMachine, "*NX*" software: `http://www.nomachine.com/`.

It implements an easy-to-use interface to the Remote X protocol as well as incorporating VNC/RFB and rdesktop/RDP into it, but at a speed performance much better than anything you may have ever seen.

Remote X is not new at all, but what they did achieve successfully is a new way of compression and caching technologies that makes the thing fast enough to run even over slow modem/ISDN connections.

I could test drive their (public) Red Hat machine in Italy, over a loaded Internet connection, with enabled thumbnail previews in KDE konqueror which popped up immediately on "*mouse-over*". From inside that (remote X) session I started a rdesktop session on another, a Windows XP machine. To test the performance, I played Pinball. I am proud to announce that my score was 631750 points at first try.

NX performs better on my local LAN than any of the other "*pure*" connection methods I am using from time to time: TightVNC, rdesktop or Remote X. It is even faster than a direct crosslink connection between two nodes.

I even got sound playing from the Remote X app to my local boxes, and had a working "*copy'n'paste*" from an NX window (running a KDE session in Italy) to my Mozilla mailing agent. These guys are certainly doing something right!

[1] ftp://ftp.microsoft.com/Softlib/MSLFILES/SRVTOOLS.EXE

I recommend to test drive NX to anybody with a only a passing interest in remote computing `http://www.nomachine.com/testdrive.php`.

Just download the free of charge client software (available for Red Hat, SuSE, Debian and Windows) and be up and running within five minutes (they need to send you your account data, though, because you are assigned a real UNIX account on their testdrive.nomachine.com box.

They plan to get to the point were you can have NX application servers running as a cluster of nodes, and users simply start an NX session locally, and can select applications to run transparently (apps may even run on another NX node, but pretend to be on the same as used for initial login, because it displays in the same window. You also can run it fullscreen, and after a short time you forget that it is a remote session at all).

Now the best thing for last: All the core compression and caching technologies are released under the GPL and available as source code to anybody who wants to build on it! These technologies are working, albeit started from the command line only (and very inconvenient to use in order to get a fully running remote X session up and running.)

To answer your questions:

- You do not need to install a terminal server; XP has RDP support built in.
- NX is much cheaper than Citrix — and comparable in performance, probably faster.
- You do not need to hack XP — it just works.
- You log into the XP box from remote transparently (and I think there is no need to change anything to get a connection, even if authentication is against a domain).
- The NX core technologies are all Open Source and released under the GPL — you can now use a (very inconvenient) commandline at no cost, but you can buy a comfortable (proprietary) NX GUI frontend for money.
- NoMachine are encouraging and offering help to OSS/Free Software implementations for such a frontend too, even if it means competition to them (they have written to this effect even to the LTSP, KDE and GNOME developer mailing lists).

21.4 Network Logon Script Magic

There are several opportunities for creating a custom network startup configuration environment.

- No Logon Script.
- Simple universal Logon Script that applies to all users.
- Use of a conditional Logon Script that applies per user or per group attributes.
- Use of Samba's preexec and postexec functions on access to the NETLOGON share to create a custom logon script and then execute it.
- User of a tool such as KixStart.

The Samba source code tree includes two logon script generation/execution tools. See `examples` directory `genlogon` and `ntlogon` subdirectories.

The following listings are from the genlogon directory.

This is the `genlogon.pl` file:

```
#!/usr/bin/perl
#
# genlogon.pl
#
# Perl script to generate user logon scripts on the fly, when users
# connect from a Windows client. This script should be called from
# smb.conf with the %U, %G and %L parameters. I.e:
#
#       root preexec = genlogon.pl %U %G %L
#
# The script generated will perform
# the following:
#
# 1. Log the user connection to /var/log/samba/netlogon.log
# 2. Set the PC's time to the Linux server time (which is maintained
#    daily to the National Institute of Standard's Atomic clock on the
#    internet.
# 3. Connect the user's home drive to H: (H for Home).
# 4. Connect common drives that everyone uses.
# 5. Connect group-specific drives for certain user groups.
# 6. Connect user-specific drives for certain users.
# 7. Connect network printers.

# Log client connection
#($sec,$min,$hour,$mday,$mon,$year,$wday,$yday,$isdst) = localtime(time);
($sec,$min,$hour,$mday,$mon,$year,$wday,$yday,$isdst) = localtime(time);
open LOG, ">>/var/log/samba/netlogon.log";
print LOG "$mon/$mday/$year $hour:$min:$sec";
print LOG " - User $ARGV[0] logged into $ARGV[1]\n";
close LOG;

# Start generating logon script
open LOGON, ">/shared/netlogon/$ARGV[0].bat";
print LOGON "\@ECHO OFF\r\n";

# Connect shares just use by Software Development group
if ($ARGV[1] eq "SOFTDEV" || $ARGV[0] eq "softdev")
{
   print LOGON "NET USE M: \\\\$ARGV[2]\\SOURCE\r\n";
}
```

```
# Connect shares just use by Technical Support staff
if ($ARGV[1] eq "SUPPORT" || $ARGV[0] eq "support")
{
   print LOGON "NET USE S: \\\\$ARGV[2]\\SUPPORT\r\n";
}

# Connect shares just used by Administration staff
If ($ARGV[1] eq "ADMIN" || $ARGV[0] eq "admin")
{
   print LOGON "NET USE L: \\\\$ARGV[2]\\ADMIN\r\n";
   print LOGON "NET USE K: \\\\$ARGV[2]\\MKTING\r\n";
}

# Now connect Printers. We handle just two or three users a little
# differently, because they are the exceptions that have desktop
# printers on LPT1: - all other user's go to the LaserJet on the
# server.
if ($ARGV[0] eq 'jim'
     || $ARGV[0] eq 'yvonne')
{
   print LOGON "NET USE LPT2: \\\\$ARGV[2]\\LJET3\r\n";
   print LOGON "NET USE LPT3: \\\\$ARGV[2]\\FAXQ\r\n";
}
else
{
   print LOGON "NET USE LPT1: \\\\$ARGV[2]\\LJET3\r\n";
   print LOGON "NET USE LPT3: \\\\$ARGV[2]\\FAXQ\r\n";
}

# All done! Close the output file.
close LOGON;
```

Those wishing to use more elaborate or capable logon processing system should check out these sites:

- http://www.craigelachi.e.org/rhacer/ntlogon
- http://www.kixtart.org

21.4.1 Adding Printers without User Intervention

Printers may be added automatically during logon script processing through the use of:

```
C:\> rundll32 printui.dll,PrintUIEntry /?
```

See the documentation in the Microsoft knowledgebase article 189105.[2]

[2] http://support.microsoft.com/default.asp?scid=kb;en-us;189105

Chapter 22

SYSTEM AND ACCOUNT POLICIES

This chapter summarizes the current state of knowledge derived from personal practice and knowledge from Samba mailing list subscribers. Before reproduction of posted information, every effort has been made to validate the information given. Where additional information was uncovered through this validation it is provided also.

22.1 Features and Benefits

When MS Windows NT 3.5 was introduced, the hot new topic was the ability to implement Group Policies for users and groups. Then along came MS Windows NT4 and a few sites started to adopt this capability. How do we know that? By the number of "*booboos*" (or mistakes) administrators made and then requested help to resolve.

By the time that MS Windows 2000 and Active Directory was released, administrators got the message: Group Policies are a good thing! They can help reduce administrative costs and actually make happier users. But adoption of the true potential of MS Windows 200x Active Directory and Group Policy Objects (GPOs) for users and machines were picked up on rather slowly. This was obvious from the Samba mailing list as in 2000 and 2001 when there were few postings regarding GPOs and how to replicate them in a Samba environment.

Judging by the traffic volume since mid 2002, GPOs have become a standard part of the deployment in many sites. This chapter reviews techniques and methods that can be used to exploit opportunities for automation of control over user desktops and network client workstations.

A tool new to Samba — the **editreg** tool — may become an important part of the future Samba administrators' arsenal is described in this document.

22.2 Creating and Managing System Policies

Under MS Windows platforms, particularly those following the release of MS Windows NT4 and MS Windows 95, it is possible to create a type of file that would be placed in the NETLOGON share of a Domain Controller. As the client logs onto the network, this file is read and the contents initiate changes to the registry of the client machine. This file allows changes to be made to those parts of the registry that affect users, groups of users, or machines.

For MS Windows 9x/ME, this file must be called `Config.POL` and may be generated using a tool called `poledit.exe`, better known as the Policy Editor. The policy editor was provided on the Windows 98 installation CD, but disappeared again with the introduction of MS Windows Me (Millennium Edition). From comments of MS Windows network administrators, it would appear that this tool became a part of the MS Windows Me Resource Kit.

MS Windows NT4 Server products include the *System Policy Editor* under **Start -> Programs -> Administrative Tools**. For MS Windows NT4 and later clients, this file must be called `NTConfig.POL`.

New with the introduction of MS Windows 2000 was the Microsoft Management Console or MMC. This tool is the new wave in the ever-changing landscape of Microsoft methods for management of network access and security. Every new Microsoft product or technology seems to make the old rules obsolete and introduces newer and more complex tools and methods. To Microsoft's credit, the MMC does appear to be a step forward, but improved functionality comes at a great price.

Before embarking on the configuration of network and system policies, it is highly advisable to read the documentation available from Microsoft's Web site regarding Implementing Profiles and Policies in Windows NT 4.0[1] available from Microsoft. There are a large number of documents in addition to this old one that should also be read and understood. Try searching on the Microsoft Web site for "*Group Policies*".

What follows is a brief discussion with some helpful notes. The information provided here is incomplete — you are warned.

22.2.1 Windows 9x/ME Policies

You need the Windows 98 Group Policy Editor to set up Group Profiles under Windows 9x/ME. It can be found on the original full product Windows 98 installation CD under `tools/reskit/netadmin/poledit`. Install this using the Add/Remove Programs facility and then click on **Have Disk**.

Use the Group Policy Editor to create a policy file that specifies the location of user profiles and/or `My Documents`, and so on. Then save these settings in a file called `Config.POL` that needs to be placed in the root of the *[NETLOGON]* share. If Windows 98 is configured to log onto the Samba Domain, it will automatically read this file and update the Windows 9x/Me registry of the machine as it logs on.

Further details are covered in the Windows 98 Resource Kit documentation.

If you do not take the correct steps, then every so often Windows 9x/ME will check the integrity of the registry and restore its settings from the back-up copy of the registry it stores on each Windows 9x/ME machine. So, you will occasionally notice things changing back to the original settings.

[1]http://www.microsoft.com/ntserver/management/deployment/planguide/prof_policies.asp

Install the group policy handler for Windows 9x/Me to pick up Group Policies. Look on the Windows 98 CDROM in `\tools\reskit\netadmin\poledit`. Install group policies on a Windows 9x/Me client by double-clicking on `grouppol.inf`. Log off and on again a couple of times and see if Windows 98 picks up Group Policies. Unfortunately, this needs to be done on every Windows 9x/Me machine that uses Group Policies.

22.2.2 Windows NT4-Style Policy Files

To create or edit `ntconfig.pol` you must use the NT Server Policy Editor, **poledit.exe**, which is included with NT4 Server but not with NT Workstation. There is a Policy Editor on an NT4 Workstation but it is not suitable for creating domain policies. Furthermore, although the Windows 95 Policy Editor can be installed on an NT4 Workstation/Server, it will not work with NT clients. However, the files from the NT Server will run happily enough on an NT4 Workstation.

You need `poledit.exe`, `common.adm` and `winnt.adm`. It is convenient to put the two `*.adm` files in the `c:\winnt\inf` directory, which is where the binary will look for them unless told otherwise. This directory is normally "*hidden.*"

The Windows NT policy editor is also included with the Service Pack 3 (and later) for Windows NT 4.0. Extract the files using **servicepackname /x**, that's **Nt4sp6ai.exe /x** for service pack 6a. The Policy Editor, **poledit.exe**, and the associated template files (*.adm) should be extracted as well. It is also possible to downloaded the policy template files for Office97 and get a copy of the Policy Editor. Another possible location is with the Zero Administration Kit available for download from Microsoft.

22.2.2.1 Registry Spoiling

With NT4-style registry-based policy changes, a large number of settings are not automatically reversed as the user logs off. The settings that were in the `NTConfig.POL` file were applied to the client machine registry and apply to the hive key HKEY_LOCAL_MACHINE are permanent until explicitly reversed. This is known as tattooing. It can have serious consequences downstream and the administrator must be extremely careful not to lock out the ability to manage the machine at a later date.

22.2.3 MS Windows 200x/XP Professional Policies

Windows NT4 system policies allow the setting of registry parameters specific to users, groups and computers (client workstations) that are members of the NT4-style domain. Such policy files will work with MS Windows 200x/XP clients also.

New to MS Windows 2000, Microsoft recently introduced a style of group policy that confers a superset of capabilities compared with NT4-style policies. Obviously, the tool used to create them is different, and the mechanism for implementing them is much improved.

The older NT4-style registry-based policies are known as *Administrative Templates* in MS Windows 2000/XP Group Policy Objects (GPOs). The later includes the ability to set various security configurations, enforce Internet Explorer browser settings, change and redirect aspects of the users desktop (including the location of `My Documents` files (directory), as well as intrinsics of where menu items will appear in the Start menu). An additional new feature is the ability to make available particular software Windows applications to particular users and/or groups.

Remember, NT4 policy files are named `NTConfig.POL` and are stored in the root of the NETLOGON share on the Domain Controllers. A Windows NT4 user enters a username, password and selects the domain name to which the logon will attempt to take place. During the logon process, the client machine reads the `NTConfig.POL` file from the NETLOGON share on the authenticating server and modifies the local registry values according to the settings in this file.

Windows 200x GPOs are feature-rich. They are not stored in the NETLOGON share, but rather part of a Windows 200x policy file is stored in the Active Directory itself and the other part is stored in a shared (and replicated) volume called the SYSVOL folder. This folder is present on all Active Directory Domain Controllers. The part that is stored in the Active Directory itself is called the Group Policy Container (GPC), and the part that is stored in the replicated share called SYSVOL is known as the Group Policy Template (GPT).

With NT4 clients, the policy file is read and executed only as each user logs onto the network. MS Windows 200x policies are much more complex — GPOs are processed and applied at client machine startup (machine specific part) and when the user logs onto the network, the user-specific part is applied. In MS Windows 200x-style policy management, each machine and/or user may be subject to any number of concurrently applicable (and applied) policy sets (GPOs). Active Directory allows the administrator to also set filters over the policy settings. No such equivalent capability exists with NT4-style policy files.

22.2.3.1 Administration of Windows 200x/XP Policies

Instead of using the tool called The System Policy Editor, commonly called Poledit (from the executable name **poledit.exe**), GPOs are created and managed using a Microsoft Management Console (MMC) snap-in as follows:

1. Go to the Windows 200x/XP menu **Start->Programs->Administrative Tools** and select the MMC snap-in called **Active Directory Users and Computers**
2. Select the domain or organizational unit (OU) that you wish to manage, then right-click to open the context menu for that object, and select the **Properties**.
3. Left-click on the **Group Policy** tab, then left-click on the New tab. Type a name for the new policy you will create.
4. Left-click on the **Edit** tab to commence the steps needed to create the GPO.

All policy configuration options are controlled through the use of policy administrative templates. These files have an .adm extension, both in NT4 as well as in Windows 200x/XP. Beware, however, the .adm files are not interchangeable across NT4 and Windows 200x. The latter introduces many new features as well as extended definition capabilities. It is well beyond the scope of this documentation to explain how to program .adm files; for that the administrator is referred to the Microsoft Windows Resource Kit for your particular version of MS Windows.

NOTE

The MS Windows 2000 Resource Kit contains a tool called gpolmig.exe. This tool can be used to migrate an NT4 NTConfig.POL file into a Windows 200x style GPO. Be VERY careful how you use this powerful tool. Please refer to the resource kit manuals for specific usage information.

22.3 Managing Account/User Policies

Policies can define a specific user's settings or the settings for a group of users. The resulting policy file contains the registry settings for all users, groups, and computers that will be using the policy file. Separate policy files for each user, group, or computer are not necessary.

If you create a policy that will be automatically downloaded from validating Domain Controllers, you should name the file `NTConfig.POL`. As system administrator, you have the option of renaming the policy file and, by modifying the Windows NT-based workstation, directing the computer to update the policy from a manual path. You can do this by either manually changing the registry or by using the System Policy Editor. This can even be a local path such that each machine has its own policy file, but if a change is necessary to all machines, it must be made individually to each workstation.

When a Windows NT4/200x/XP machine logs onto the network, the client looks in the NETLOGON share on the authenticating domain controller for the presence of the NTConfig.POL file. If one exists it is downloaded, parsed and then applied to the user's part of the registry.

MS Windows 200x/XP clients that log onto an MS Windows Active Directory security domain may additionally acquire policy settings through Group Policy Objects (GPOs) that are defined and stored in Active Directory itself. The key benefit of using AS GPOs is that they impose no registry *spoiling* effect. This has considerable advantage compared with the use of `NTConfig.POL` (NT4) style policy updates.

In addition to user access controls that may be imposed or applied via system and/or group policies in a manner that works in conjunction with user profiles, the user management environment under MS Windows NT4/200x/XP allows per domain as well as per user account restrictions to be applied. Common restrictions that are frequently used include:

- Logon hours
- Password aging
- Permitted logon from certain machines only
- Account type (local or global)
- User rights

Samba-3.0.0 doe not yet implement all account controls that are common to MS Windows NT4/200x/XP. While it is possible to set many controls using the Domain User Manager for MS Windows NT4, only password expirey is functional today. Most of the remaining controls at this time have only stub routines that may eventually be completed to provide actual control. Do not be misled by the fact that a parameter can be set using the NT4 Domain User Manager or in the `NTConfig.POL`.

22.4 Management Tools

Anyone who wishes to create or manage Group Policies will need to be familiar with a number of tools. The following sections describe a few key tools that will help you to create a low maintenance user environment.

22.4.1 Samba Editreg Toolset

A new tool called **editreg** is under development. This tool can be used to edit registry files (called `NTUser.DAT`) that are stored in user and group profiles. `NTConfig.POL` files have the same structure as the `NTUser.DAT` file and can be edited using this tool. **editreg** is being built with the intent to enable `NTConfig.POL` files to be saved in text format and to permit the building of new `NTConfig.POL` files with extended capabilities. It is proving difficult to realize this capability, so do not be surprised if this feature does not materialize. Formal capabilities will be announced at the time that this tool is released for production use.

22.4.2 Windows NT4/200x

The tools that may be used to configure these types of controls from the MS Windows environment are: the NT4 User Manager for Domains, the NT4 System and Group Policy Editor, and the Registry Editor (regedt32.exe). Under MS Windows 200x/XP, this is done using the Microsoft Management Console (MMC) with appropriate "*snap-ins,*" the registry editor, and potentially also the NT4 System and Group Policy Editor.

22.4.3 Samba PDC

With a Samba Domain Controller, the new tools for managing user account and policy information include: **smbpasswd**, **pdbedit**, **net**, **rpcclient**. The administrator should read the man pages for these tools and become familiar with their use.

22.5 System Startup and Logon Processing Overview

The following attempts to document the order of processing the system and user policies following a system reboot and as part of the user logon:

1. Network starts, then Remote Procedure Call System Service (RPCSS) and Multiple Universal Naming Convention Provider (MUP) start.

2. Where Active Directory is involved, an ordered list of Group Policy Objects (GPOs) is downloaded and applied. The list may include GPOs that:
   - Apply to the location of machines in a Directory.
   - Apply only when settings have changed.
   - Depend on configuration of the scope of applicability: local, site, domain, organizational unit, and so on.

   No desktop user interface is presented until the above have been processed.
3. Execution of start-up scripts (hidden and synchronous by default).
4. A keyboard action to effect start of logon (Ctrl-Alt-Del).
5. User credentials are validated, user profile is loaded (depends on policy settings).
6. An ordered list of user GPOs is obtained. The list contents depends on what is configured in respect of:
   - Is the user a Domain Member, thus subject to particular policies?
   - Loopback enablement, and the state of the loopback policy (Merge or Replace).
   - Location of the Active Directory itself.
   - Has the list of GPOs changed? No processing is needed if not changed.
7. User Policies are applied from Active Directory. Note: There are several types.
8. Logon scripts are run. New to Windows 200x and Active Directory, logon scripts may be obtained based on Group Policy objects (hidden and executed synchronously). NT4-style logon scripts are then run in a normal window.
9. The User Interface as determined from the GPOs is presented. Note: In a Samba domain (like an NT4 Domain), machine (system) policies are applied at start-up; user policies are applied at logon.

22.6 Common Errors

Policy-related problems can be quite difficult to diagnose and even more difficult to rectify. The following collection demonstrates only basic issues.

22.6.1 Policy Does Not Work

"*We have created the* `Config.POL` *file and put it in the* NETLOGON *share. It has made no difference to our Win XP Pro machines, they just do not see it. It worked fine with Win 98 but does not work any longer since we upgraded to Win XP Pro. Any hints?*"

Policy files are not portable between Windows 9x/Me and MS Windows NT4/200x/XP-based platforms. You need to use the NT4 Group Policy Editor to create a file called `NTConfig.POL` so it is in the correct format for your MS Windows XP Pro clients.

Chapter 23

DESKTOP PROFILE MANAGEMENT

23.1 Features and Benefits

Roaming profiles are feared by some, hated by a few, loved by many, and a Godsend for some administrators.

Roaming profiles allow an administrator to make available a consistent user desktop as the user moves from one machine to another. This chapter provides much information regarding how to configure and manage roaming profiles.

While roaming profiles might sound like nirvana to some, they are a real and tangible problem to others. In particular, users of mobile computing tools, where often there may not be a sustained network connection, are often better served by purely local profiles. This chapter provides information to help the Samba administrator deal with those situations.

23.2 Roaming Profiles

WARNING

Roaming profiles support is different for Windows 9x/Me and Windows NT4/200x.

Before discussing how to configure roaming profiles, it is useful to see how Windows 9x/Me and Windows NT4/200x clients implement these features.

Windows 9x/Me clients send a NetUserGetInfo request to the server to get the user's profiles location. However, the response does not have room for a separate profiles location field, only the user's home share. This means that Windows 9x/Me profiles are restricted to being stored in the user's home directory.

Windows NT4/200x clients send a NetSAMLogon RPC request, which contains many fields including a separate field for the location of the user's profiles.

23.2.1 Samba Configuration for Profile Handling

This section documents how to configure Samba for MS Windows client profile support.

23.2.1.1 NT4/200x User Profiles

For example, to support Windows NT4/200x clients, set the followoing in the [global] section of the `smb.conf` file:

logon path = \\profileserver\profileshare\profilepath\%U\moreprofilepath

This is typically implemented like:

logon path = \\%L\Profiles\%u

where "*%L*" translates to the name of the Samba server and "*%u*" translates to the user name.

The default for this option is `\\%N\%U\profile`, namely `\\sambaserver\username\profile`. The `\\N%\%U` service is created automatically by the [homes] service. If you are using a Samba server for the profiles, you must make the share that is specified in the logon path browseable. Please refer to the man page for `smb.conf` in respect of the different semantics of "*%L*" and "*%N*", as well as "*%U*" and "*%u*".

NOTE

MS Windows NT/200x clients at times do not disconnect a connection to a server between logons. It is recommended to not use the *homes* meta-service name as part of the profile share path.

23.2.1.2 Windows 9x/Me User Profiles

To support Windows 9x/Me clients, you must use the *logon home* parameter. Samba has been fixed so `net use /home` now works as well and it, too, relies on the **logon home** parameter.

By using the logon home parameter, you are restricted to putting Windows 9x/Me profiles in the user's home directory. But wait! There is a trick you can use. If you set the following in the *[global]* section of your `smb.conf` file:

logon home = \\%L\%U\.profiles

then your Windows 9x/Me clients will dutifully put their clients in a subdirectory of your home directory called `.profiles` (making them hidden).

Not only that, but `net use /home` will also work because of a feature in Windows 9x/Me. It removes any directory stuff off the end of the home directory area and only uses the server and share portion. That is, it looks like you specified \\%L\%U for *logon home*.

23.2.1.3 Mixed Windows 9x/Me and Windows NT4/200x User Profiles

You can support profiles for Windows 9x and Windows NT clients by setting both the *logon home* and *logon path* parameters. For example:

```
logon home = \\%L\%u\.profiles
logon path = \\%L\profiles\%u
```

23.2.1.4 Disabling Roaming Profile Support

A question often asked is: *"How may I enforce use of local profiles?"* or *"How do I disable roaming profiles?"*

There are three ways of doing this:

In `smb.conf` — Affect the following settings and ALL clients will be forced to use a local profile: *logon home* and *logon path*

MS Windows Registry — By using the Microsoft Management Console gpedit.msc to instruct your MS Windows XP machine to use only a local profile. This, of course, modifies registry settings. The full path to the option is:

```
Local Computer Policy\
   Computer Configuration\
      Administrative Templates\
         System\
            User Profiles\

Disable: Only Allow Local User Profiles
Disable: Prevent Roaming Profile Change from Propagating to the Server
```

Change of Profile Type: — From the start menu right-click on **My Computer icon**, select **Properties**, click on the **User Profiles** tab, select the profile you wish to change from **Roaming** type to **Local**, and click on **Change Type**.

Consult the MS Windows registry guide for your particular MS Windows version for more information about which registry keys to change to enforce use of only local user profiles.

NOTE

The specifics of how to convert a local profile to a roaming profile, or a roaming profile to a local one vary according to the version of MS Windows you are running. Consult the Microsoft MS Windows Resource Kit for your version of Windows for specific information.

23.2.2 Windows Client Profile Configuration Information

23.2.2.1 Windows 9x/Me Profile Setup

When a user first logs in on Windows 9X, the file user.DAT is created, as are folders `Start Menu`, `Desktop`, `Programs`, and `Nethood`. These directories and their contents will be merged with the local versions stored in `c:\windows\profiles\username` on subsequent logins, taking the most recent from each. You will need to use the *[global]* options *preserve case* = yes, *short preserve case* = yes and *case sensitive* = no in order to maintain capital letters in shortcuts in any of the profile folders.

The `user.DAT` file contains all the user's preferences. If you wish to enforce a set of preferences, rename their `user.DAT` file to `user.MAN`, and deny them write access to this file.

1. On the Windows 9x/Me machine, go to **Control Panel** -> **Passwords** and select the **User Profiles** tab. Select the required level of roaming preferences. Press **OK**, but do not allow the computer to reboot.
2. On the Windows 9x/Me machine, go to **Control Panel** -> **Network** -> **Client for Microsoft Networks** -> **Preferences**. Select **Log on to NT Domain**. Then, ensure that the Primary Logon is **Client for Microsoft Networks**. Press **OK**, and this time allow the computer to reboot.

Under Windows 9x/ME, profiles are downloaded from the Primary Logon. If you have the Primary Logon as "*Client for Novell Networks*", then the profiles and logon script will be downloaded from your Novell Server. If you have the Primary Logon as "*Windows Logon*", then the profiles will be loaded from the local machine — a bit against the concept of roaming profiles, it would seem!

You will now find that the Microsoft Networks Login box contains `[user, password, domain]` instead of just `[user, password]`. Type in the Samba server's domain name (or any other domain known to exist, but bear in mind that the user will be authenticated against this domain and profiles downloaded from it, if that domain logon server supports it), user name and user's password.

Once the user has been successfully validated, the Windows 9x/Me machine will inform you that `The user has not logged on before` and asks you `Do you wish to save the user's preferences?` Select **Yes**.

Once the Windows 9x/Me client comes up with the desktop, you should be able to examine the contents of the directory specified in the *logon path* on the Samba server and verify that the `Desktop`, `Start Menu`, `Programs` and `Nethood` folders have been created.

These folders will be cached locally on the client, and updated when the user logs off (if you haven't made them read-only by then). You will find that if the user creates further folders or shortcut, that the client will merge the profile contents downloaded with the contents of the profile directory already on the local client, taking the newest folders and shortcut from each set.

If you have made the folders/files read-only on the Samba server, then you will get errors from the Windows 9x/Me machine on logon and logout as it attempts to merge the local and remote profile. Basically, if you have any errors reported by the Windows 9x/Me machine, check the UNIX file permissions and ownership rights on the profile directory contents, on the Samba server.

If you have problems creating user profiles, you can reset the user's local desktop cache, as shown below. When this user next logs in, the user will be told that he/she is logging in *"for the first time"*.

1. Instead of logging in under the [user, password, domain] dialog, press **escape**.
2. Run the **regedit.exe** program, and look in:

   `HKEY_LOCAL_MACHINE\Windows\CurrentVersion\ProfileList`

   You will find an entry for each user of ProfilePath. Note the contents of this key (likely to be `c:\windows\profiles\username`), then delete the key *ProfilePath* for the required user.
3. Exit the registry editor.
4. Search for the user's .PWL password-caching file in the `c:\windows` directory, and delete it.
5. Log off the Windows 9x/Me client.
6. Check the contents of the profile path (see *logon path* described above) and delete the `user.DAT` or `user.MAN` file for the user, making a backup if required.

WARNING

Before deleting the contents of the directory listed in the *ProfilePath* (this is likely to be `c:\windows\profiles\username`), ask the owner if they have any important files stored on their desktop or in their start menu. Delete the contents of the directory *ProfilePath* (making a backup if any of the files are needed).

This will have the effect of removing the local (read-only hidden system file) `user.DAT` in their profile directory, as well as the local *"desktop," "nethood," "start menu,"* and *"programs"* folders.

If all else fails, increase Samba's debug log levels to between 3 and 10, and/or run a packet sniffer program such as ethereal or **netmon.exe**, and look for error messages.

If you have access to an Windows NT4/200x server, then first set up roaming profiles and/or netlogons on the Windows NT4/200x server. Make a packet trace, or examine the example packet traces provided with Windows NT4/200x server, and see what the differences are with the equivalent Samba trace.

23.2.2.2 Windows NT4 Workstation

When a user first logs in to a Windows NT Workstation, the profile NTuser.DAT is created. The profile location can be now specified through the *logon path* parameter.

There is a parameter that is now available for use with NT Profiles: *logon drive*. This should be set to `H:` or any other drive, and should be used in conjunction with the new *logon home* parameter.

The entry for the NT4 profile is a directory not a file. The NT help on Profiles mentions that a directory is also created with a .PDS extension. The user, while logging in, must have write permission to create the full profile path (and the folder with the .PDS extension for those situations where it might be created.)

In the profile directory, Windows NT4 creates more folders than Windows 9x/Me. It creates `Application Data` and others, as well as `Desktop`, `Nethood`, `Start Menu`, and `Programs`. The profile itself is stored in a file `NTuser.DAT`. Nothing appears to be stored in the .PDS directory, and its purpose is currently unknown.

You can use the System Control Panel to copy a local profile onto a Samba server (see NT Help on Profiles; it is also capable of firing up the correct location in the System Control Panel for you). The NT Help file also mentions that renaming `NTuser.DAT` to `NTuser.MAN` turns a profile into a mandatory one.

The case of the profile is significant. The file must be called `NTuser.DAT` or, for a mandatory profile, `NTuser.MAN`.

23.2.2.3 Windows 2000/XP Professional

You must first convert the profile from a local profile to a domain profile on the MS Windows workstation as follows:

1. Log on as the *local* workstation administrator.
2. Right-click on the **My Computer** Icon, select **Properties**.
3. Click on the **User Profiles** tab.
4. Select the profile you wish to convert (click it once).
5. Click on the **Copy To** button.
6. In the **Permitted to use** box, click on the **Change** button.
7. Click on the **Look in** area that lists the machine name. When you click here, it will open up a selection box. Click on the domain to which the profile must be accessible.

> **NOTE**
>
> 
>
> You will need to log on if a logon box opens up. For example, connect as *DOMAIN*\root, password: *mypassword*.

8. To make the profile capable of being used by anyone, select "*Everyone*".
9. Click on **OK** and the Selection box will close.
10. Now click on **OK** to create the profile in the path you nominated.

Done. You now have a profile that can be edited using the Samba **profiles** tool.

> **NOTE**
>
> 
>
> Under Windows NT/200x, the use of mandatory profiles forces the use of MS Exchange storage of mail data and keeps it out of the desktop profile. That keeps desktop profiles from becoming unusable.

Windows XP Service Pack 1 There is a security check new to Windows XP (or maybe only Windows XP service pack 1). It can be disabled via a group policy in the Active Directory. The policy is called:

```
Computer Configuration\Administrative Templates\System\User Profiles\
Do not check for user ownership of Roaming Profile Foldersi
```

This should be set to `Enabled`.

Does the new version of Samba have an Active Directory analogue? If so, then you may be able to set the policy through this.

If you cannot set group policies in Samba, then you may be able to set the policy locally on each machine. If you want to try this, then do the following (N.B. I do not know for sure that this will work in the same way as a domain group policy):

1. On the XP workstation, log in with an Administrative account.
2. Click on **Start** -> **Run**.
3. Type **mmc**.
4. Click on **OK**.
5. A Microsoft Management Console should appear.
6. Click on **File** -> **Add/Remove Snap-in** -> **Add**.
7. Double-click on **Group Policy**.

8. Click on **Finish** -> **Close**.
9. Click on **OK**.
10. In the "*Console Root*" window expand **Local Computer Policy** -> **Computer Configuration** -> **Administrative Templates** -> **System** -> **User Profiles**.
11. Double-click on **Do not check for user ownership of Roaming Profile Folders**.
12. Select **Enabled**.
13. Click on **OK**.
14. Close the whole console. You do not need to save the settings (this refers to the console settings rather than the policies you have changed).
15. Reboot.

23.2.3 Sharing Profiles between W9x/Me and NT4/200x/XP Workstations

Sharing of desktop profiles between Windows versions is not recommended. Desktop profiles are an evolving phenomenon and profiles for later versions of MS Windows clients add features that may interfere with earlier versions of MS Windows clients. Probably the more salient reason to not mix profiles is that when logging off an earlier version of MS Windows, the older format of profile contents may overwrite information that belongs to the newer version resulting in loss of profile information content when that user logs on again with the newer version of MS Windows.

If you then want to share the same Start Menu/Desktop with W9x/Me, you will need to specify a common location for the profiles. The `smb.conf` parameters that need to be common are *logon path* and *logon home*.

If you have this set up correctly, you will find separate `user.DAT` and `NTuser.DAT` files in the same profile directory.

23.2.4 Profile Migration from Windows NT4/200x Server to Samba

There is nothing to stop you from specifying any path that you like for the location of users' profiles. Therefore, you could specify that the profile be stored on a Samba server, or any other SMB server, as long as that SMB server supports encrypted passwords.

23.2.4.1 Windows NT4 Profile Management Tools

Unfortunately, the Resource Kit information is specific to the version of MS Windows NT4/200x. The correct resource kit is required for each platform.

Here is a quick guide:

1. On your NT4 Domain Controller, right click on **My Computer**, then select the tab labeled **User Profiles**.
2. Select a user profile you want to migrate and click on it.

> **NOTE**
>
> 
>
> I am using the term "*migrate*" loosely. You can copy a profile to create a group profile. You can give the user *Everyone* rights to the profile you copy this to. That is what you need to do, since your Samba domain is not a member of a trust relationship with your NT4 PDC.

3. Click on the **Copy To** button.
4. In the box labeled **Copy Profile to** add your new path, e.g., `c:\temp\foobar`
5. Click on **Change** in the **Permitted to use** box.
6. Click on the group "*Everyone*", click on **OK**. This closes the "*choose user*" box.
7. Now click on **OK**.

Follow the above for every profile you need to migrate.

23.2.4.2 Side Bar Notes

You should obtain the SID of your NT4 domain. You can use smbpasswd to do this. Read the man page.

23.2.4.3 moveuser.exe

The Windows 200x professional resource kit has **moveuser.exe**. **moveuser.exe** changes the security of a profile from one user to another. This allows the account domain to change, and/or the user name to change.

This command is like the Samba **profiles** tool.

23.2.4.4 Get SID

You can identify the SID by using **GetSID.exe** from the Windows NT Server 4.0 Resource Kit.

Windows NT 4.0 stores the local profile information in the registry under the following key: `HKEY_LOCAL_MACHINE\SOFTWARE\Microsoft\Windows NT\CurrentVersion\ProfileList`

Under the ProfileList key, there will be subkeys named with the SIDs of the users who have logged on to this computer. (To find the profile information for the user whose locally cached profile you want to move, find the SID for the user with the **GetSID.exe** utility.) Inside the appropriate user's subkey, you will see a string value named *ProfileImagePath*.

23.3 Mandatory Profiles

A Mandatory Profile is a profile that the user does not have the ability to overwrite. During the user's session, it may be possible to change the desktop environment, however, as the user logs out all changes made will be lost. If it is desired to not allow the user any ability to change the desktop environment, then this must be done through policy settings. See the previous chapter.

NOTE

Under NO circumstances should the profile directory (or its contents) be made read-only as this may render the profile un-usable. Where it is essential to make a profile read-only within the UNIX file system, this can be done but then you absolutely must use the **fake-permissions** VFS module to instruct MS Windows NT/200x/XP clients that the Profile has write permission for the user. See Section 19.3.3.

For MS Windows NT4/200x/XP, the above method can also be used to create mandatory profiles. To convert a group profile into a mandatory profile, simply locate the `NTUser.DAT` file in the copied profile and rename it to `NTUser.MAN`.

For MS Windows 9x/ME, it is the `User.DAT` file that must be renamed to `User.MAN` to effect a mandatory profile.

23.4 Creating and Managing Group Profiles

Most organizations are arranged into departments. There is a nice benefit in this fact since usually most users in a department require the same desktop applications and the same desktop layout. MS Windows NT4/200x/XP will allow the use of Group Profiles. A Group Profile is a profile that is created first using a template (example) user. Then using the profile migration tool (see above), the profile is assigned access rights for the user group that needs to be given access to the group profile.

The next step is rather important. Instead of assigning a group profile to users (Using User Manager) on a "*per user*" basis, the group itself is assigned the now modified profile.

NOTE

Be careful with Group Profiles. If the user who is a member of a group also has a personal profile, then the result will be a fusion (merge) of the two.

23.5 Default Profile for Windows Users

MS Windows 9x/Me and NT4/200x/XP will use a default profile for any user for whom a profile does not already exist. Armed with a knowledge of where the default profile is located on the Windows workstation, and knowing which registry keys effect the path from which the default profile is created, it is possible to modify the default profile to one that has been optimized for the site. This has significant administrative advantages.

23.5.1 MS Windows 9x/Me

To enable default per use profiles in Windows 9x/ME, you can either use the Windows 98 System Policy Editor or change the registry directly.

To enable default per user profiles in Windows 9x/ME, launch the System Policy Editor, then select **File** -> **Open Registry**, next click on the **Local Computer** icon, click on **Windows 98 System**, select **User Profiles**, and click on the enable box. Remember to save the registry changes.

To modify the registry directly, launch the Registry Editor (**regedit.exe**) and select the hive `HKEY_LOCAL_MACHINE\Network\Logon`. Now add a DWORD type key with the name "*User Profiles,*" to enable user profiles to set the value to 1; to disable user profiles set it to 0.

23.5.1.1 User Profile Handling with Windows 9x/Me

When a user logs on to a Windows 9x/Me machine, the local profile path, `HKEY_LOCAL_MACHINE\Software\Microsoft\Windows\CurrentVersion\ProfileList`, is checked for an existing entry for that user.

If the user has an entry in this registry location, Windows 9x/Me checks for a locally cached version of the user profile. Windows 9x/Me also checks the user's home directory (or other specified directory if the location has been modified) on the server for the User Profile. If a profile exists in both locations, the newer of the two is used. If the User Profile exists on the server, but does not exist on the local machine, the profile on the server is downloaded and used. If the User Profile only exists on the local machine, that copy is used.

If a User Profile is not found in either location, the Default User Profile from the Windows 9x/Me machine is used and copied to a newly created folder for the logged on user. At log off, any changes that the user made are written to the user's local profile. If the user has a roaming profile, the changes are written to the user's profile on the server.

23.5.2 MS Windows NT4 Workstation

On MS Windows NT4, the default user profile is obtained from the location `%SystemRoot%\Profiles` which in a default installation will translate to `C:\Windows NT\Profiles`. Under this directory on a clean install there will be three (3) directories: `Administrator`, `All Users`, and `Default User`.

The `All Users` directory contains menu settings that are common across all system users. The `Default User` directory contains menu entries that are customizable per user depending on the profile settings chosen/created.

When a new user first logs onto an MS Windows NT4 machine, a new profile is created from:

- All Users settings.
- Default User settings (contains the default `NTUser.DAT` file).

When a user logs onto an MS Windows NT4 machine that is a member of a Microsoft security domain, the following steps are followed in respect of profile handling:

1. The users' account information that is obtained during the logon process contains the location of the users' desktop profile. The profile path may be local to the machine or it may be located on a network share. If there exists a profile at the location of the path from the user account, then this profile is copied to the location `%SystemRoot%\Profiles\%USERNAME%`. This profile then inherits the settings in the `All Users` profile in the `%SystemRoot%\Profiles` location.

2. If the user account has a profile path, but at its location a profile does not exist, then a new profile is created in the `%SystemRoot%\Profiles\%USERNAME%` directory from reading the `Default User` profile.

3. If the NETLOGON share on the authenticating server (logon server) contains a policy file (`NTConfig.POL`), then its contents are applied to the `NTUser.DAT` which is applied to the `HKEY_CURRENT_USER` part of the registry.

4. When the user logs out, if the profile is set to be a roaming profile it will be written out to the location of the profile. The `NTuser.DAT` file is then recreated from the contents of the `HKEY_CURRENT_USER` contents. Thus, should there not exist in the NETLOGON share an `NTConfig.POL` at the next logon, the effect of the previous `NTConfig.POL` will still be held in the profile. The effect of this is known as tattooing.

MS Windows NT4 profiles may be *local* or *roaming*. A local profile will stored in the `%SystemRoot%\Profiles\%USERNAME%` location. A roaming profile will also remain stored in the same way, unless the following registry key is created as shown:

```
HKEY_LOCAL_MACHINE\SYSTEM\Software\Microsoft\Windows NT\CurrentVersion\
winlogon\"DeleteRoamingCache"=dword:0000000
```

In this case, the local copy (in `%SystemRoot%\Profiles\%USERNAME%`) will be deleted on logout.

Under MS Windows NT4, default locations for common resources like `My Documents` may be redirected to a network share by modifying the following registry keys. These changes may be affected via use of the System Policy Editor. To do so may require that you create your own template extension for the policy editor to allow this to be done through the GUI. Another way to do this is by way of first creating a default user profile, then while logged in as that user, run **regedt32** to edit the key settings.

The Registry Hive key that affects the behavior of folders that are part of the default user profile are controlled by entries on Windows NT4 is:

```
HKEY_CURRENT_USER
  \Software
    \Microsoft
      \Windows
        \CurrentVersion
          \Explorer
            \User Shell Folders
```

The above hive key contains a list of automatically managed folders. The default entries are shown in Table 23.1.

Table 23.1. User Shell Folder Registry Keys Default Values

Name	**Default Value**
AppData	%USERPROFILE%\Application Data
Desktop	%USERPROFILE%\Desktop
Favorites	%USERPROFILE%\Favorites
NetHood	%USERPROFILE%\NetHood
PrintHood	%USERPROFILE%\PrintHood
Programs	%USERPROFILE%\Start Menu\Programs
Recent	%USERPROFILE%\Recent
SendTo	%USERPROFILE%\SendTo
Start Menu	%USERPROFILE%\Start Menu
Startup	%USERPROFILE%\Start Menu\Programs\Startup

The registry key that contains the location of the default profile settings is:

```
HKEY_LOCAL_MACHINE\SOFTWARE\Microsoft\Windows\CurrentVersion\Explorer\
User Shell Folders
```

The default entries are shown in Table 23.2.

Table 23.2. Defaults of Profile Settings Registry Keys

Common Desktop	%SystemRoot%\Profiles\All Users\Desktop
Common Programs	%SystemRoot%\Profiles\All Users\Programs
Common Start Menu	%SystemRoot%\Profiles\All Users\Start Menu
Common Startup	%SystemRoot%\Profiles\All Users\Start Menu\Programs\Startup

23.5.3 MS Windows 200x/XP

NOTE

MS Windows XP Home Edition does use default per user profiles, but cannot participate in domain security, cannot log onto an NT/ADS-style domain, and thus can obtain the profile only from itself. While there are benefits in doing this, the beauty of those MS Windows clients that can participate in domain logon processes allows the administrator to create a global default profile and enforce it through the use of Group Policy Objects (GPOs).

When a new user first logs onto an MS Windows 200x/XP machine, the default profile is obtained from `C:\Documents and Settings\Default User`. The administrator can modify or change the contents of this location and MS Windows 200x/XP will gladly use it. This is far from the optimum arrangement since it will involve copying a new default profile to every MS Windows 200x/XP client workstation.

When MS Windows 200x/XP participates in a domain security context, and if the default user profile is not found, then the client will search for a default profile in the NETLOGON share of the authenticating server. In MS Windows parlance, `%LOGONSERVER%\NETLOGON\Default User`, and if one exists there it will copy this to the workstation to the `C:\Documents and Settings\` under the Windows login name of the user.

NOTE

This path translates, in Samba parlance, to the `smb.conf` *[NETLOGON]* share. The directory should be created at the root of this share and must be called `Default Profile`.

If a default profile does not exist in this location, then MS Windows 200x/XP will use the local default profile.

On logging out, the users' desktop profile will be stored to the location specified in the registry settings that pertain to the user. If no specific policies have been created or passed to the client during the login process (as Samba does automatically), then the user's profile will be written to the local machine only under the path `C:\Documents and Settings\%USERNAME%`.

Those wishing to modify the default behavior can do so through these three methods:

- Modify the registry keys on the local machine manually and place the new default profile in the NETLOGON share root. This is not recommended as it is maintenance intensive.

- Create an NT4-style NTConfig.POL file that specified this behavior and locate this file in the root of the NETLOGON share along with the new default profile.
- Create a GPO that enforces this through Active Directory, and place the new default profile in the NETLOGON share.

The registry hive key that effects the behavior of folders that are part of the default user profile are controlled by entries on Windows 200x/XP is:

`HKEY_CURRENT_USER\Software\Microsoft\Windows\CurrentVersion\Explorer\User Shell Folders\`

The above hive key contains a list of automatically managed folders. The default entries are shown in Table 23.3

Table 23.3. Defaults of Default User Profile Paths Registry Keys

Name	Default Value
AppData	%USERPROFILE%\Application Data
Cache	%USERPROFILE%\Local Settings\Temporary Internet Files
Cookies	%USERPROFILE%\Cookies
Desktop	%USERPROFILE%\Desktop
Favorites	%USERPROFILE%\Favorites
History	%USERPROFILE%\Local Settings\History
Local AppData	%USERPROFILE%\Local Settings\Application Data
Local Settings	%USERPROFILE%\Local Settings
My Pictures	%USERPROFILE%\My Documents\My Pictures
NetHood	%USERPROFILE%\NetHood
Personal	%USERPROFILE%\My Documents
PrintHood	%USERPROFILE%\PrintHood
Programs	%USERPROFILE%\Start Menu\Programs
Recent	%USERPROFILE%\Recent
SendTo	%USERPROFILE%\SendTo
Start Menu	%USERPROFILE%\Start Menu
Startup	%USERPROFILE%\Start Menu\Programs\Startup
Templates	%USERPROFILE%\Templates

There is also an entry called "*Default*" that has no value set. The default entry is of type `REG_SZ`, all the others are of type `REG_EXPAND_SZ`.

It makes a huge difference to the speed of handling roaming user profiles if all the folders are stored on a dedicated location on a network server. This means that it will not be necessary to write the Outlook PST file over the network for every login and logout.

To set this to a network location, you could use the following examples:

`%LOGONSERVER%\%USERNAME%\Default Folders`

This would store the folders in the user's home directory under a directory called `Default Folders`. You could also use:

`\\`*`SambaServer`*`\`*`FolderShare`*`\%USERNAME%`

in which case the default folders will be stored in the server named *SambaServer* in the share called *FolderShare* under a directory that has the name of the MS Windows user as seen by the Linux/UNIX file system.

Please note that once you have created a default profile share, you MUST migrate a user's profile (default or custom) to it.

MS Windows 200x/XP profiles may be *Local* or *Roaming*. A roaming profile will be cached locally unless the following registry key is created:

```
HKEY_LOCAL_MACHINE\SYSTEM\Software\Microsoft\Windows NT\CurrentVersion\
  winlogon\"DeleteRoamingCache"=dword:00000001
```

In this case, the local cache copy will be deleted on logout.

23.6 Common Errors

The following are some typical errors, problems and questions that have been asked on the Samba mailing lists.

23.6.1 Configuring Roaming Profiles for a Few Users or Groups

With Samba-2.2.x, the choice you have is to enable or disable roaming profiles support. It is a global only setting. The default is to have roaming profiles and the default path will locate them in the user's home directory.

If disabled globally, then no one will have roaming profile ability. If enabled and you want it to apply only to certain machines, then on those machines on which roaming profile support is not wanted it is then necessary to disable roaming profile handling in the registry of each such machine.

With Samba-3, you can have a global profile setting in `smb.conf` and you can override this by per-user settings using the Domain User Manager (as with MS Windows NT4/ Win 200xx).

In any case, you can configure only one profile per user. That profile can be either:

- A profile unique to that user.
- A mandatory profile (one the user cannot change).
- A group profile (really should be mandatory, that is unchangable).

23.6.2 Cannot Use Roaming Profiles

A user requested the following: "*I do not want Roaming profiles to be implemented. I want to give users a local profile alone. Please help me, I am totally lost with this error. For the past two days I tried everything, I googled around but found no useful pointers. Please help me.*"

The choices are:

Local profiles — I know of no registry keys that will allow auto-deletion of LOCAL profiles on log out.

Roaming profiles — As a user logs onto the network, a centrally stored profile is copied to the workstation to form a local profile. This local profile will persist (remain on the workstation disk) unless a registry key is changed that will cause this profile to be automatically deleted on logout.

The roaming profile choices are:

Personal roaming profiles — These are typically stored in a profile share on a central (or conveniently located local) server.

Workstations cache (store) a local copy of the profile. This cached copy is used when the profile cannot be downloaded at next logon.

Group profiles — These are loaded from a central profile server.

Mandatory profiles — Mandatory profiles can be created for a user as well as for any group that a user is a member of. Mandatory profiles cannot be changed by ordinary users. Only the administrator can change or reconfigure a mandatory profile.

A Windows NT4/200x/XP profile can vary in size from 130KB to very large. Outlook PST files are most often part of the profile and can be many GB in size. On average (in a well controlled environment), roaming profile size of 2MB is a good rule of thumb to use for planning purposes. In an undisciplined environment, I have seen up to 2GB profiles. Users tend to complain when it takes an hour to log onto a workstation but they harvest the fruits of folly (and ignorance).

The point of all the above is to show that roaming profiles and good controls of how they can be changed as well as good discipline make up for a problem-free site.

Microsoft's answer to the PST problem is to store all email in an MS Exchange Server backend. This removes the need for a PST file.

Local profiles mean:

- If each machine is used by many users, then much local disk storage is needed for local profiles.
- Every workstation the user logs into has its own profile; these can be very different from machine to machine.

On the other hand, use of roaming profiles means:

- The network administrator can control the desktop environment of all users.
- Use of mandatory profiles drastically reduces network management overheads.
- In the long run, users will experience fewer problems.

23.6.3 Changing the Default Profile

"When the client logs onto the Domain Controller, it searches for a profile to download. Where do I put this default profile?"

First, the Samba server needs to be configured as a Domain Controller. This can be done by setting in `smb.conf`:

```
security = user
os level = 32 (or more)
domain logons = Yes
```

There must be a *[netlogon]* share that is world readable. It is a good idea to add a logon script to pre-set printer and drive connections. There is also a facility for automatically synchronizing the workstation time clock with that of the logon server (another good thing to do).

NOTE

To invoke auto-deletion of roaming profile from the local workstation cache (disk storage), use the Group Policy Editor to create a file called `NTConfig.POL` with the appropriate entries. This file needs to be located in the *netlogon* share root directory.

Windows clients need to be members of the domain. Workgroup machines do not use network logons so they do not interoperate with domain profiles.

For roaming profiles, add to `smb.conf`:

```
logon path = \\%N\profiles\%U
logon drive = H:
```

Chapter 24

PAM-BASED DISTRIBUTED AUTHENTICATION

This chapter should help you to deploy Winbind-based authentication on any PAM-enabled UNIX/Linux system. Winbind can be used to enable User-Level application access authentication from any MS Windows NT Domain, MS Windows 200x Active Directory-based domain, or any Samba-based domain environment. It will also help you to configure PAM-based local host access controls that are appropriate to your Samba configuration.

In addition to knowing how to configure Winbind into PAM, you will learn generic PAM management possibilities and in particular how to deploy tools like `pam_smbpass.so` to your advantage.

NOTE

The use of Winbind requires more than PAM configuration alone. Please refer to Chapter 20, *Winbind: Use of Domain Accounts*, for further information regarding Winbind.

24.1 Features and Benefits

A number of UNIX systems (e.g., Sun Solaris), as well as the xxxxBSD family and Linux, now utilize the Pluggable Authentication Modules (PAM) facility to provide all authentication, authorization and resource control services. Prior to the introduction of PAM, a decision to use an alternative to the system password database (`/etc/passwd`) would require the provision of alternatives for all programs that provide security services. Such a choice would involve provision of alternatives to programs such as: **login**, **passwd**, **chown**, and so on.

PAM provides a mechanism that disconnects these security programs from the underlying authentication/authorization infrastructure. PAM is configured by making appropriate modifications to one file `/etc/pam.conf` (Solaris), or by editing individual control files that are located in `/etc/pam.d`.

On PAM-enabled UNIX/Linux systems, it is an easy matter to configure the system to use any authentication backend so long as the appropriate dynamically loadable library modules are available for it. The backend may be local to the system, or may be centralized on a remote server.

PAM support modules are available for:

`/etc/passwd` — There are several PAM modules that interact with this standard UNIX user database. The most common are called: `pam_unix.so`, `pam_unix2.so`, `pam_pwdb.so` and `pam_userdb.so`.

Kerberos — The `pam_krb5.so` module allows the use of any Kerberos compliant server. This tool is used to access MIT Kerberos, Heimdal Kerberos, and potentially Microsoft Active Directory (if enabled).

LDAP — The `pam_ldap.so` module allows the use of any LDAP v2 or v3 compatible backend server. Commonly used LDAP backend servers include: OpenLDAP v2.0 and v2.1, Sun ONE iDentity server, Novell eDirectory server, Microsoft Active Directory.

NetWare Bindery — The `pam_ncp_auth.so` module allows authentication off any bindery-enabled NetWare Core Protocol-based server.

SMB Password — This module, called `pam_smbpass.so`, will allow user authentication off the passdb backend that is configured in the Samba `smb.conf` file.

SMB Server — The `pam_smb_auth.so` module is the original MS Windows networking authentication tool. This module has been somewhat outdated by the Winbind module.

Winbind — The `pam_winbind.so` module allows Samba to obtain authentication from any MS Windows Domain Controller. It can just as easily be used to authenticate users for access to any PAM-enabled application.

RADIUS — There is a PAM RADIUS (Remote Access Dial-In User Service) authentication module. In most cases, administrators will need to locate the source code for this tool and compile and install it themselves. RADIUS protocols are used by many routers and terminal servers.

Of the above, Samba provides the `pam_smbpasswd.so` and the `pam_winbind.so` modules alone.

Once configured, these permit a remarkable level of flexibility in the location and use of distributed Samba Domain Controllers that can provide wide area network bandwidth efficient authentication services for PAM-capable systems. In effect, this allows the deployment of centrally managed and maintained distributed authentication from a single-user account database.

24.2 Technical Discussion

PAM is designed to provide the system administrator with a great deal of flexibility in configuration of the privilege granting applications of their system. The local configuration of system security controlled by PAM is contained in one of two places: either the single system file, `/etc/pam.conf`, or the `/etc/pam.d/` directory.

24.2.1 PAM Configuration Syntax

In this section we discuss the correct syntax of and generic options respected by entries to these files. PAM-specific tokens in the configuration file are case insensitive. The module paths, however, are case sensitive since they indicate a file's name and reflect the case dependence of typical file systems. The case-sensitivity of the arguments to any given module is defined for each module in turn.

In addition to the lines described below, there are two special characters provided for the convenience of the system administrator: comments are preceded by a "#" and extend to the next end-of-line; also, module specification lines may be extended with a "\" escaped newline.

If the PAM authentication module (loadable link library file) is located in the default location, then it is not necessary to specify the path. In the case of Linux, the default location is `/lib/security`. If the module is located outside the default, then the path must be specified as:

```
auth  required  /other_path/pam_strange_module.so
```

24.2.1.1 Anatomy of /etc/pam.d Entries

The remaining information in this subsection was taken from the documentation of the Linux-PAM project. For more information on PAM, see The Official Linux-PAM home page.[1]

A general configuration line of the `/etc/pam.conf` file has the following form:

```
service-name   module-type   control-flag   module-path   args
```

Below, we explain the meaning of each of these tokens. The second (and more recently adopted) way of configuring Linux-PAM is via the contents of the `/etc/pam.d/` directory. Once we have explained the meaning of the above tokens, we will describe this method.

[1] http://ftp.kernel.org/pub/linux/libs/pam/

service-name — The name of the service associated with this entry. Frequently, the service name is the conventional name of the given application. For example, **ftpd**, **rlogind** and **su**, and so on.

There is a special service-name reserved for defining a default authentication mechanism. It has the name *OTHER* and may be specified in either lower- or upper-case characters. Note, when there is a module specified for a named service, the *OTHER* entries are ignored.

module-type — One of (currently) four types of module. The four types are as follows:

- *auth:* This module type provides two aspects of authenticating the user. It establishes that the user is who he claims to be by instructing the application to prompt the user for a password or other means of identification. Secondly, the module can grant group membership (independently of the `/etc/groups` file discussed above) or other privileges through its credential granting properties.
- *account:* This module performs non-authentication-based account management. It is typically used to restrict/permit access to a service based on the time of day, currently available system resources (maximum number of users) or perhaps the location of the applicant user "*root*" login only on the console.
- *session:* Primarily, this module is associated with doing things that need to be done for the user before and after they can be given service. Such things include the logging of information concerning the opening and closing of some data exchange with a user, mounting directories, and so on.
- *password:* This last module type is required for updating the authentication token associated with the user. Typically, there is one module for each "*challenge/response*" -based authentication *(auth)* module type.

control-flag — The control-flag is used to indicate how the PAM library will react to the success or failure of the module it is associated with. Since modules can be stacked (modules of the same type execute in series, one after another), the control-flags determine the relative importance of each module. The application is not made aware of the individual success or failure of modules listed in the `/etc/pam.conf` file. Instead, it receives a summary success or fail response from the Linux-PAM library. The order of execution of these modules is that of the entries in the `/etc/pam.conf` file; earlier entries are executed before later ones. As of Linux-PAM v0.60, this control-flag can be defined with one of two syntaxes.

The simpler (and historical) syntax for the control-flag is a single keyword defined to indicate the severity of concern associated with the success or failure of a specific module. There are four such keywords: *required, requisite, sufficient and optional.*

The Linux-PAM library interprets these keywords in the following manner:

- *required:* This indicates that the success of the module is required for the module-type facility to succeed. Failure of this module will not be apparent to the user until all of the remaining modules (of the same module-type) have been executed.

- *requisite:* Like required, however, in the case that such a module returns a failure, control is directly returned to the application. The return value is that associated with the first required or requisite module to fail. This flag can be used to protect against the possibility of a user getting the opportunity to enter a password over an unsafe medium. It is conceivable that such behavior might inform an attacker of valid accounts on a system. This possibility should be weighed against the not insignificant concerns of exposing a sensitive password in a hostile environment.

- *sufficient:* The success of this module is deemed *sufficient* to satisfy the Linux-PAM library that this module-type has succeeded in its purpose. In the event that no previous required module has failed, no more "*stacked*" modules of this type are invoked. (In this case, subsequent required modules are not invoked). A failure of this module is not deemed as fatal to satisfying the application that this module-type has succeeded.

- *optional:* As its name suggests, this control-flag marks the module as not being critical to the success or failure of the user's application for service. In general, Linux-PAM ignores such a module when determining if the module stack will succeed or fail. However, in the absence of any definite successes or failures of previous or subsequent stacked modules, this module will determine the nature of the response to the application. One example of this latter case, is when the other modules return something like PAM_IGNORE.

The more elaborate (newer) syntax is much more specific and gives the administrator a great deal of control over how the user is authenticated. This form of the control flag is delimited with square brackets and consists of a series of *value=action* tokens:

```
[value1=action1 value2=action2 ...]
```

Here, *value1* is one of the following return values:

```
success; open_err; symbol_err; service_err; system_err; buf_err;
perm_denied; auth_err; cred_insufficient; authinfo_unavail;
user_unknown; maxtries; new_authtok_reqd; acct_expired; session_err;
cred_unavail; cred_expired; cred_err; no_module_data; conv_err;
authtok_err; authtok_recover_err; authtok_lock_busy;
authtok_disable_aging; try_again; ignore; abort; authtok_expired;
module_unknown; bad_item; and default.
```

The last of these *(default)* can be used to set the action for those return values that are not explicitly defined.

The *action1* can be a positive integer or one of the following tokens: *ignore; ok;*

done; bad; die; and *reset*. A positive integer, J, when specified as the action, can be used to indicate that the next J modules of the current module-type will be skipped. In this way, the administrator can develop a moderately sophisticated stack of modules with a number of different paths of execution. Which path is taken can be determined by the reactions of individual modules.

- *ignore:* When used with a stack of modules, the module's return status will not contribute to the return code the application obtains.
- *bad:* This action indicates that the return code should be thought of as indicative of the module failing. If this module is the first in the stack to fail, its status value will be used for that of the whole stack.
- *die:* Equivalent to bad with the side effect of terminating the module stack and PAM immediately returning to the application.
- *ok:* This tells PAM that the administrator thinks this return code should contribute directly to the return code of the full stack of modules. In other words, if the former state of the stack would lead to a return of PAM_SUCCESS, the module's return code will override this value. Note, if the former state of the stack holds some value that is indicative of a modules failure, this *ok* value will not be used to override that value.
- *done:* Equivalent to *ok* with the side effect of terminating the module stack and PAM immediately returning to the application.
- *reset:* Clears all memory of the state of the module stack and starts again with the next stacked module.

Each of the four keywords: *required; requisite; sufficient;* and *optional*, have an equivalent expression in terms of the [...] syntax. They are as follows:

- *required* is equivalent to *[success=ok new_authtok_reqd=ok ignore=ignore default=bad]*.
- *requisite* is equivalent to *[success=ok new_authtok_reqd=ok ignore=ignore default=die]*.
- *sufficient* is equivalent to *[success=done new_authtok_reqd=done default=ignore]*.
- *optional* is equivalent to *[success=ok new_authtok_reqd=ok default=ignore]*.

Just to get a feel for the power of this new syntax, here is a taste of what you can do with it. With Linux-PAM-0.63, the notion of client plug-in agents was introduced. This is something that makes it possible for PAM to support machine-machine authentication using the transport protocol inherent to the client/server application. With the *[... value=action ...]* control syntax, it is possible for an application to be configured to support binary prompts with compliant clients, but to gracefully fall over into an alternative authentication mode for older, legacy applications.

module-path — The path-name of the dynamically loadable object file; the pluggable module itself. If the first character of the module path is "/", it is assumed to be a complete path. If this is not the case, the given module path is appended to the default module path: /lib/security (but see the notes above).

The arguments are a list of tokens that are passed to the module when it is invoked, much like arguments to a typical Linux shell command. Generally, valid arguments are optional and are specific to any given module. Invalid arguments are ignored by a module, however, when encountering an invalid argument, the module is required to write an error to syslog(3). For a list of generic options, see the next section.

If you wish to include spaces in an argument, you should surround that argument with square brackets. For example:

```
squid auth required pam_mysql.so user=passwd_query passwd=mada \
db=eminence [query=select user_name from internet_service where \
user_name=%u and password=PASSWORD(%p) and service=web_proxy]
```

When using this convention, you can include “[” characters inside the string, and if you wish to have a “]” character inside the string that will survive the argument parsing, you should use “\]”. In other words:

```
[..[..\]..]    -->    ..[..]..
```

Any line in one of the configuration files that is not formatted correctly will generally tend (erring on the side of caution) to make the authentication process fail. A corresponding error is written to the system log files with a call to syslog(3).

24.2.2 Example System Configurations

The following is an example /etc/pam.d/login configuration file. This example had all options uncommented and is probably not usable because it stacks many conditions before allowing successful completion of the login process. Essentially all conditions can be disabled by commenting them out, except the calls to pam_pwdb.so.

24.2.2.1 PAM: Original Login Config

```
#%PAM-1.0
# The PAM configuration file for the login service
#
auth        required    pam_securetty.so
auth        required    pam_nologin.so
# auth      required    pam_dialup.so
# auth      optional    pam_mail.so
auth        required    pam_pwdb.so shadow md5
# account   requisite   pam_time.so
account     required    pam_pwdb.so
```

```
session      required   pam_pwdb.so
# session    optional   pam_lastlog.so
# password   required   pam_cracklib.so retry=3
password     required   pam_pwdb.so shadow md5
```

24.2.2.2 PAM: Login Using pam_smbpass

PAM allows use of replaceable modules. Those available on a sample system include:

```
$/bin/ls /lib/security
```

```
pam_access.so    pam_ftp.so            pam_limits.so
pam_ncp_auth.so  pam_rhosts_auth.so    pam_stress.so
pam_cracklib.so  pam_group.so          pam_listfile.so
pam_nologin.so   pam_rootok.so         pam_tally.so
pam_deny.so      pam_issue.so          pam_mail.so
pam_permit.so    pam_securetty.so      pam_time.so
pam_dialup.so    pam_lastlog.so        pam_mkhomedir.so
pam_pwdb.so      pam_shells.so         pam_UNIX.so
pam_env.so       pam_ldap.so           pam_motd.so
pam_radius.so    pam_smbpass.so        pam_UNIX_acct.so
pam_wheel.so     pam_UNIX_auth.so      pam_UNIX_passwd.so
pam_userdb.so    pam_warn.so           pam_UNIX_session.so
```

The following example for the login program replaces the use of the `pam_pwdb.so` module that uses the system password database (`/etc/passwd`, `/etc/shadow`, `/etc/group`) with the module `pam_smbpass.so`, which uses the Samba database which contains the Microsoft MD4 encrypted password hashes. This database is stored in either `/usr/local/samba/private/smbpasswd`, `/etc/samba/smbpasswd`, or in `/etc/samba.d/smbpasswd`, depending on the Samba implementation for your UNIX/Linux system. The `pam_smbpass.so` module is provided by Samba version 2.2.1 or later. It can be compiled by specifying the `--with-pam_smbpass` options when running Samba's **configure** script. For more information on the `pam_smbpass` module, see the documentation in the `source/pam_smbpass` directory of the Samba source distribution.

```
#%PAM-1.0
# The PAM configuration file for the login service
#
auth       required   pam_smbpass.so nodelay
account    required   pam_smbpass.so nodelay
session    required   pam_smbpass.so nodelay
password   required   pam_smbpass.so nodelay
```

The following is the PAM configuration file for a particular Linux system. The default condition uses `pam_pwdb.so`.

```
#%PAM-1.0
# The PAM configuration file for the samba service
#
auth        required     pam_pwdb.so nullok nodelay shadow audit
account     required     pam_pwdb.so audit nodelay
session     required     pam_pwdb.so nodelay
password    required     pam_pwdb.so shadow md5
```

In the following example, the decision has been made to use the **smbpasswd** database even for basic Samba authentication. Such a decision could also be made for the **passwd** program and would thus allow the **smbpasswd** passwords to be changed using the **passwd** program:

```
#%PAM-1.0
# The PAM configuration file for the samba service
#
auth        required     pam_smbpass.so nodelay
account     required     pam_pwdb.so audit nodelay
session     required     pam_pwdb.so nodelay
password    required     pam_smbpass.so nodelay smbconf=/etc/samba.d/smb.conf
```

NOTE

PAM allows stacking of authentication mechanisms. It is also possible to pass information obtained within one PAM module through to the next module in the PAM stack. Please refer to the documentation for your particular system implementation for details regarding the specific capabilities of PAM in this environment. Some Linux implementations also provide the `pam_stack.so` module that allows all authentication to be configured in a single central file. The `pam_stack.so` method has some devoted followers on the basis that it allows for easier administration. As with all issues in life though, every decision makes trade-offs, so you may want to examine the PAM documentation for further helpful information.

24.2.3 smb.conf PAM Configuration

There is an option in `smb.conf` called *`obey pam restrictions`*. The following is from the online help for this option in SWAT;

When Samba is configured to enable PAM support (i.e., `--with-pam`), this parameter will control whether or not Samba should obey PAM's account and session management directives. The default behavior is to use PAM for cleartext authentication only and to ignore any account or session management. Samba always ignores PAM for authentication in the case of *encrypt passwords* = yes. The reason is that PAM modules cannot support the challenge/response authentication mechanism needed in the presence of SMB password encryption.

Default: *obey pam restrictions* = no

24.2.4 Remote CIFS Authentication Using winbindd.so

All operating systems depend on the provision of users credentials acceptable to the platform. UNIX requires the provision of a user identifier (UID) as well as a group identifier (GID). These are both simple integer type numbers that are obtained from a password backend such as `/etc/passwd`.

Users and groups on a Windows NT server are assigned a relative ID (RID) which is unique for the domain when the user or group is created. To convert the Windows NT user or group into a UNIX user or group, a mapping between RIDs and UNIX user and group IDs is required. This is one of the jobs that winbind performs.

As Winbind users and groups are resolved from a server, user and group IDs are allocated from a specified range. This is done on a first come, first served basis, although all existing users and groups will be mapped as soon as a client performs a user or group enumeration command. The allocated UNIX IDs are stored in a database file under the Samba lock directory and will be remembered.

The astute administrator will realize from this that the combination of `pam_smbpass.so`, **winbindd** and a distributed *passdb backend*, such as *ldap*, will allow the establishment of a centrally managed, distributed user/password database that can also be used by all PAM-aware (e.g., Linux) programs and applications. This arrangement can have particularly potent advantages compared with the use of Microsoft Active Directory Service (ADS) in so far as the reduction of wide area network authentication traffic.

WARNING

The RID to UNIX ID database is the only location where the user and group mappings are stored by **winbindd**. If this file is deleted or corrupted, there is no way for **winbindd** to determine which user and group IDs correspond to Windows NT user and group RIDs.

24.2.5 Password Synchronization Using pam_smbpass.so

`pam_smbpass` is a PAM module that can be used on conforming systems to keep the `smbpasswd` (Samba password) database in sync with the UNIX password file. PAM (Pluggable Authentication Modules) is an API supported under some UNIX operating systems, such as Solaris, HPUX and Linux, that provides a generic interface to authentication mechanisms.

This module authenticates a local `smbpasswd` user database. If you require support for authenticating against a remote SMB server, or if you are concerned about the presence of SUID root binaries on your system, it is recommended that you use `pam_winbind` instead.

Options recognized by this module are shown in Table 24.1.

Table 24.1. Options recognized by pam_smbpass

debug	log more debugging info.
audit	like debug, but also logs unknown usernames.
use_first_pass	do not prompt the user for passwords; take them from PAM_ items instead.
try_first_pass	try to get the password from a previous PAM module fall back to prompting the user.
use_authtok	like try_first_pass, but *fail* if the new PAM_AUTHTOK has not been previously set (intended for stacking password modules only).
not_set_pass	do not make passwords used by this module available to other modules.
nodelay	do not insert ~1 second delays on authentication failure.
nullok	null passwords are allowed.
nonull	null passwords are not allowed. Used to override the Samba configuration.
migrate	only meaningful in an "*auth*" context; used to update smbpasswd file with a password used for successful authentication.
smbconf=*file*	specify an alternate path to the `smb.conf` file.

The following are examples of the use of `pam_smbpass.so` in the format of Linux `/etc/pam.d/` files structure. Those wishing to implement this tool on other platforms will need to adapt this appropriately.

24.2.5.1 Password Synchronization Configuration

A sample PAM configuration that shows the use of pam_smbpass to make sure `private/smbpasswd` is kept in sync when `/etc/passwd (/etc/shadow)` is changed. Useful when an expired password might be changed by an application (such as **ssh**).

```
#%PAM-1.0
# password-sync
#
auth        requisite     pam_nologin.so
auth        required      pam_UNIX.so
account     required      pam_UNIX.so
password    requisite     pam_cracklib.so retry=3
password    requisite     pam_UNIX.so shadow md5 use_authtok try_first_pass
password    required      pam_smbpass.so nullok use_authtok try_first_pass
session     required      pam_UNIX.so
```

24.2.5.2 Password Migration Configuration

A sample PAM configuration that shows the use of pam_smbpass to migrate from plaintext to encrypted passwords for Samba. Unlike other methods, this can be used for users who have never connected to Samba shares: password migration takes place when users **ftp** in, login using **ssh**, pop their mail, and so on.

```
#%PAM-1.0
# password-migration
#
auth        requisite   pam_nologin.so
# pam_smbpass is called IF pam_UNIX succeeds.
auth        requisite   pam_UNIX.so
auth        optional    pam_smbpass.so migrate
account     required    pam_UNIX.so
password    requisite   pam_cracklib.so retry=3
password    requisite   pam_UNIX.so shadow md5 use_authtok try_first_pass
password    optional    pam_smbpass.so nullok use_authtok try_first_pass
session     required    pam_UNIX.so
```

24.2.5.3 Mature Password Configuration

A sample PAM configuration for a mature smbpasswd installation. private/smbpasswd is fully populated, and we consider it an error if the SMB password does not exist or does not match the UNIX password.

```
#%PAM-1.0
# password-mature
#
auth        requisite    pam_nologin.so
auth        required     pam_UNIX.so
account     required     pam_UNIX.so
password    requisite    pam_cracklib.so retry=3
password    requisite    pam_UNIX.so shadow md5 use_authtok try_first_pass
password    required     pam_smbpass.so use_authtok use_first_pass
session     required     pam_UNIX.so
```

24.2.5.4 Kerberos Password Integration Configuration

A sample PAM configuration that shows *pam_smbpass* used together with *pam_krb5*. This could be useful on a Samba PDC that is also a member of a Kerberos realm.

```
#%PAM-1.0
# kdc-pdc
```

```
#
auth        requisite   pam_nologin.so
auth        requisite   pam_krb5.so
auth        optional    pam_smbpass.so migrate
account     required    pam_krb5.so
password    requisite   pam_cracklib.so retry=3
password    optional    pam_smbpass.so nullok use_authtok try_first_pass
password    required    pam_krb5.so use_authtok try_first_pass
session     required    pam_krb5.so
```

24.3 Common Errors

PAM can be fickle and sensitive to configuration glitches. Here we look at a few cases from the Samba mailing list.

24.3.1 pam_winbind Problem

A user reported: I have the following PAM configuration:

```
auth required /lib/security/pam_securetty.so
auth sufficient /lib/security/pam_winbind.so
auth sufficient /lib/security/pam_UNIX.so use_first_pass nullok
auth required /lib/security/pam_stack.so service=system-auth
auth required /lib/security/pam_nologin.so
account required /lib/security/pam_stack.so service=system-auth
account required /lib/security/pam_winbind.so
password required /lib/security/pam_stack.so service=system-auth
```

When I open a new console with [ctrl][alt][F1], I can't log in with my user "*pitie*". I have tried with user "*scienceu+pitie*" also.

Answer: The problem may lie with your inclusion of *pam_stack.so service=system-auth*. That file often contains a lot of stuff that may duplicate what you are already doing. Try commenting out the *pam_stack* lines for *auth* and *account* and see if things work. If they do, look at /etc/pam.d/system-auth and copy only what you need from it into your /etc/pam.d/login file. Alternately, if you want all services to use Winbind, you can put the Winbind-specific stuff in /etc/pam.d/system-auth.

24.3.2 Winbind Is Not Resolving Users and Groups

"*My smb.conf file is correctly configured. I have specified idmap uid = 12000, and idmap gid = 3000-3500 and* ***winbind*** *is running. When I do the following it all works fine.*"

```
root# wbinfo -u
```

```
MIDEARTH+maryo
MIDEARTH+jackb
MIDEARTH+ameds
...
MIDEARTH+root

root# wbinfo -g
MIDEARTH+Domain Users
MIDEARTH+Domain Admins
MIDEARTH+Domain Guests
...
MIDEARTH+Accounts

root# getent passwd
root:x:0:0:root:/root:/bin/bash
bin:x:1:1:bin:/bin:/bin/bash
...
maryo:x:15000:15003:Mary Orville:/home/MIDEARTH/maryo:/bin/false
```

"But this command fails:"

```
root# chown maryo a_file
chown: 'maryo': invalid user
```

"This is driving me nuts! What can be wrong?"

Answer: Your system is likely running **nscd**, the name service caching daemon. Shut it down, do not restart it! You will find your problem resolved.

Chapter 25

INTEGRATING MS WINDOWS NETWORKS WITH SAMBA

This section deals with NetBIOS over TCP/IP name to IP address resolution. If your MS Windows clients are not configured to use NetBIOS over TCP/IP, then this section does not apply to your installation. If your installation involves the use of NetBIOS over TCP/IP then this section may help you to resolve networking problems.

NOTE

NetBIOS over TCP/IP has nothing to do with NetBEUI. NetBEUI is NetBIOS over Logical Link Control (LLC). On modern networks it is highly advised to not run NetBEUI at all. Note also there is no such thing as NetBEUI over TCP/IP — the existence of such a protocol is a complete and utter misapprehension.

25.1 Features and Benefits

Many MS Windows network administrators have never been exposed to basic TCP/IP networking as it is implemented in a UNIX/Linux operating system. Likewise, many UNIX and Linux administrators have not been exposed to the intricacies of MS Windows TCP/IP-based networking (and may have no desire to be either).

This chapter gives a short introduction to the basics of how a name can be resolved to its IP address for each operating system environment.

25.2 Background Information

Since the introduction of MS Windows 2000, it is possible to run MS Windows networking without the use of NetBIOS over TCP/IP. NetBIOS over TCP/IP uses UDP port 137 for NetBIOS name resolution and uses TCP port 139 for NetBIOS session services. When

NetBIOS over TCP/IP is disabled on MS Windows 2000 and later clients, then only the TCP port 445 will be used and the UDP port 137 and TCP port 139 will not.

> **NOTE**
>
> 
>
> When using Windows 2000 or later clients, if NetBIOS over TCP/IP is not disabled, then the client will use UDP port 137 (NetBIOS Name Service, also known as the Windows Internet Name Service or WINS), TCP port 139 and TCP port 445 (for actual file and print traffic).

When NetBIOS over TCP/IP is disabled, the use of DNS is essential. Most installations that disable NetBIOS over TCP/IP today use MS Active Directory Service (ADS). ADS requires Dynamic DNS with Service Resource Records (SRV RR) and with Incremental Zone Transfers (IXFR). Use of DHCP with ADS is recommended as a further means of maintaining central control over the client workstation network configuration.

25.3 Name Resolution in a Pure UNIX/Linux World

The key configuration files covered in this section are:

- `/etc/hosts`
- `/etc/resolv.conf`
- `/etc/host.conf`
- `/etc/nsswitch.conf`

25.3.1 /etc/hosts

This file contains a static list of IP addresses and names.

```
127.0.0.1   localhost localhost.localdomain
192.168.1.1 bigbox.quenya.org bigbox   alias4box
```

The purpose of `/etc/hosts` is to provide a name resolution mechanism so uses do not need to remember IP addresses.

Network packets that are sent over the physical network transport layer communicate not via IP addresses but rather using the Media Access Control address, or MAC address. IP addresses are currently 32 bits in length and are typically presented as four (4) decimal numbers that are separated by a dot (or period). For example, 168.192.1.1.

MAC Addresses use 48 bits (or 6 bytes) and are typically represented as two-digit hexadecimal numbers separated by colons: 40:8e:0a:12:34:56.

Every network interface must have a MAC address. Associated with a MAC address may be one or more IP addresses. There is no relationship between an IP address and a MAC address; all such assignments are arbitrary or discretionary in nature. At the most basic level, all network communications take place using MAC addressing. Since MAC addresses must be globally unique and generally remain fixed for any particular interface, the assignment of an IP address makes sense from a network management perspective. More than one IP address can be assigned per MAC address. One address must be the primary IP address — this is the address that will be returned in the ARP reply.

When a user or a process wants to communicate with another machine, the protocol implementation ensures that the "*machine name*" or "*host name*" is resolved to an IP address in a manner that is controlled by the TCP/IP configuration control files. The file `/etc/hosts` is one such file.

When the IP address of the destination interface has been determined, a protocol called ARP/RARP is used to identify the MAC address of the target interface. ARP stands for Address Resolution Protocol and is a broadcast-oriented method that uses User Datagram Protocol (UDP) to send a request to all interfaces on the local network segment using the all 1s MAC address. Network interfaces are programmed to respond to two MAC addresses only; their own unique address and the address ff:ff:ff:ff:ff:ff. The reply packet from an ARP request will contain the MAC address and the primary IP address for each interface.

The `/etc/hosts` file is foundational to all UNIX/Linux TCP/IP installations and as a minimum will contain the localhost and local network interface IP addresses and the primary names by which they are known within the local machine. This file helps to prime the pump so a basic level of name resolution can exist before any other method of name resolution becomes available.

25.3.2 /etc/resolv.conf

This file tells the name resolution libraries:

- The name of the domain to which the machine belongs.
- The name(s) of any domains that should be automatically searched when trying to resolve unqualified host names to their IP address.
- The name or IP address of available Domain Name Servers that may be asked to perform name-to-address translation lookups.

25.3.3 /etc/host.conf

`/etc/host.conf` is the primary means by which the setting in `/etc/resolv.conf` may be effected. It is a critical configuration file. This file controls the order by which name resolution may proceed. The typical structure is:

```
order hosts,bind
multi on
```

then both addresses should be returned. Please refer to the man page for `host.conf` for further details.

25.3.4 /etc/nsswitch.conf

This file controls the actual name resolution targets. The file typically has resolver object specifications as follows:

```
# /etc/nsswitch.conf
#
# Name Service Switch configuration file.
#

passwd:     compat
# Alternative entries for password authentication are:
# passwd:   compat files nis ldap winbind
shadow:     compat
group:      compat

hosts:      files nis dns
# Alternative entries for host name resolution are:
# hosts: files dns nis nis+ hesiod db compat ldap wins
networks:   nis files dns

ethers:     nis files
protocols:  nis files
rpc:    nis files
services:   nis files
```

Of course, each of these mechanisms requires that the appropriate facilities and/or services are correctly configured.

It should be noted that unless a network request/message must be sent, TCP/IP networks are silent. All TCP/IP communications assume a principal of speaking only when necessary.

Starting with version 2.2.0, Samba has Linux support for extensions to the name service switch infrastructure so Linux clients will be able to obtain resolution of MS Windows NetBIOS names to IP Addresses. To gain this functionality, Samba needs to be compiled with appropriate arguments to the make command (i.e., `make nsswitch/libnss_wins.so`). The resulting library should then be installed in the `/lib` directory and the *wins* parameter needs to be added to the "*hosts:*" line in the `/etc/nsswitch.conf` file. At this point, it will be possible to ping any MS Windows machine by its NetBIOS machine name, as long as that machine is within the workgroup to which both the Samba machine and the MS Windows machine belong.

25.4 Name Resolution as Used within MS Windows Networking

MS Windows networking is predicated about the name each machine is given. This name is known variously (and inconsistently) as the "*computer name,*" "*machine name,*" "*networking name,*" "*netbios name,*" or "*SMB name.*" All terms mean the same thing with the exception of "*netbios name*" that can also apply to the name of the workgroup or the domain name. The terms "*workgroup*" and "*domain*" are really just a simple name with which the machine is associated. All NetBIOS names are exactly 16 characters in length. The 16th character is reserved. It is used to store a one-byte value that indicates service level information for the NetBIOS name that is registered. A NetBIOS machine name is, therefore, registered for each service type that is provided by the client/server.

Table 25.1 and Table 25.2 list typical NetBIOS name/service type registrations.

Table 25.1. Unique NetBIOS Names

MACHINENAME<00>	Server Service is running on MACHINENAME
MACHINENAME<03>	Generic Machine Name (NetBIOS name)
MACHINENAME<20>	LanMan Server service is running on MACHINENAME
WORKGROUP<1b>	Domain Master Browser

Table 25.2. Group Names

WORKGROUP<03>	Generic Name registered by all members of WORKGROUP
WORKGROUP<1c>	Domain Controllers / Netlogon Servers
WORKGROUP<1d>	Local Master Browsers
WORKGROUP<1e>	Internet Name Resolvers

It should be noted that all NetBIOS machines register their own names as per the above. This is in vast contrast to TCP/IP installations where traditionally the system administrator will determine in the `/etc/hosts` or in the DNS database what names are associated with each IP address.

One further point of clarification should be noted. The `/etc/hosts` file and the DNS records do not provide the NetBIOS name type information that MS Windows clients depend on to locate the type of service that may be needed. An example of this is what happens when an MS Windows client wants to locate a domain logon server. It finds this service and the IP address of a server that provides it by performing a lookup (via a NetBIOS broadcast) for enumeration of all machines that have registered the name type *<1c>. A logon request is then sent to each IP address that is returned in the enumerated list of IP addresses. Whichever machine first replies, it then ends up providing the logon services.

The name "*workgroup*" or "*domain*" really can be confusing since these have the added significance of indicating what is the security architecture of the MS Windows network. The term "*workgroup*" indicates that the primary nature of the network environment is that of a peer-to-peer design. In a WORKGROUP, all machines are responsible for their own security, and generally such security is limited to the use of just a password (known as Share Level security). In most situations with peer-to-peer networking, the users who control their own machines will simply opt to have no security at all. It is possible to have

User Level Security in a WORKGROUP environment, thus requiring the use of a user name and a matching password.

MS Windows networking is thus predetermined to use machine names for all local and remote machine message passing. The protocol used is called Server Message Block (SMB) and this is implemented using the NetBIOS protocol (Network Basic Input Output System). NetBIOS can be encapsulated using LLC (Logical Link Control) protocol — in which case the resulting protocol is called NetBEUI (Network Basic Extended User Interface). NetBIOS can also be run over IPX (Internetworking Packet Exchange) protocol as used by Novell NetWare, and it can be run over TCP/IP protocols — in which case the resulting protocol is called NBT or NetBT, the NetBIOS over TCP/IP.

MS Windows machines use a complex array of name resolution mechanisms. Since we are primarily concerned with TCP/IP, this demonstration is limited to this area.

25.4.1 The NetBIOS Name Cache

All MS Windows machines employ an in-memory buffer in which is stored the NetBIOS names and IP addresses for all external machines that machine has communicated with over the past 10-15 minutes. It is more efficient to obtain an IP address for a machine from the local cache than it is to go through all the configured name resolution mechanisms.

If a machine whose name is in the local name cache has been shut down before the name had been expired and flushed from the cache, then an attempt to exchange a message with that machine will be subject to time-out delays. Its name is in the cache, so a name resolution lookup will succeed, but the machine cannot respond. This can be frustrating for users but is a characteristic of the protocol.

The MS Windows utility that allows examination of the NetBIOS name cache is called "*nbtstat*". The Samba equivalent of this is called **nmblookup**.

25.4.2 The LMHOSTS File

This file is usually located in MS Windows NT 4.0 or Windows 200x/XP in the directory `C:\WINNT\SYSTEM32\DRIVERS\ETC` and contains the IP Address and the machine name in matched pairs. The `LMHOSTS` file performs NetBIOS name to IP address mapping.

It typically looks like this:

```
# Copyright (c) 1998 Microsoft Corp.
#
# This is a sample LMHOSTS file used by the Microsoft Wins Client (NetBIOS
# over TCP/IP) stack for Windows98
#
# This file contains the mappings of IP addresses to NT computernames
# (NetBIOS) names. Each entry should be kept on an individual line.
# The IP address should be placed in the first column followed by the
# corresponding computername. The address and the computername
# should be separated by at least one space or tab. The "#" character
```

```
# is generally used to denote the start of a comment (see the exceptions
# below).
#
# This file is compatible with Microsoft LAN Manager 2.x TCP/IP lmhosts
# files and offers the following extensions:
#
#      #PRE
#      #DOM:<domain>
#      #INCLUDE <filename>
#      #BEGIN_ALTERNATE
#      #END_ALTERNATE
#      \0xnn (non-printing character support)
#
# Following any entry in the file with the characters "#PRE" will cause
# the entry to be preloaded into the name cache. By default, entries are
# not preloaded, but are parsed only after dynamic name resolution fails.
#
# Following an entry with the "#DOM:<domain>" tag will associate the
# entry with the domain specified by <domain>. This effects how the
# browser and logon services behave in TCP/IP environments. To preload
# the host name associated with #DOM entry, it is necessary to also add a
# #PRE to the line. The <domain> is always preloaded although it will not
# be shown when the name cache is viewed.
#
# Specifying "#INCLUDE <filename>" will force the RFC NetBIOS (NBT)
# software to seek the specified <filename> and parse it as if it were
# local. <filename> is generally a UNC-based name, allowing a
# centralized lmhosts file to be maintained on a server.
# It is ALWAYS necessary to provide a mapping for the IP address of the
# server prior to the #INCLUDE. This mapping must use the #PRE directive.
# In addition the share "public" in the example below must be in the
# LanManServer list of "NullSessionShares" in order for client machines to
# be able to read the lmhosts file successfully. This key is under
# \machine\system\currentcontrolset\services\lanmanserver\
# parameters\nullsessionshares
# in the registry. Simply add "public" to the list found there.
#
# The #BEGIN_ and #END_ALTERNATE keywords allow multiple #INCLUDE
# statements to be grouped together. Any single successful include
# will cause the group to succeed.
#
# Finally, non-printing characters can be embedded in mappings by
# first surrounding the NetBIOS name in quotations, then using the
# \0xnn notation to specify a hex value for a non-printing character.
#
# The following example illustrates all of these extensions:
#
# 102.54.94.97     rhino     #PRE #DOM:networking  #net group's DC
```

```
# 102.54.94.102    "appname  \0x14"       #special app server
# 102.54.94.123    popular   #PRE         #source server
# 102.54.94.117    localsrv  #PRE         #needed for the include
#
# #BEGIN_ALTERNATE
# #INCLUDE \\localsrv\public\lmhosts
# #INCLUDE \\rhino\public\lmhosts
# #END_ALTERNATE
#
# In the above example, the "appname" server contains a special
# character in its name, the "popular" and "localsrv" server names are
# preloaded, and the "rhino" server name is specified so it can be used
# to later #INCLUDE a centrally maintained lmhosts file if the "localsrv"
# system is unavailable.
#
# Note that the whole file is parsed including comments on each lookup,
# so keeping the number of comments to a minimum will improve performance.
# Therefore it is not advisable to simply add lmhosts file entries onto the
# end of this file.
```

25.4.3 HOSTS File

This file is usually located in MS Windows NT 4.0 or Windows 200x/XP in the directory `C:\WINNT\SYSTEM32\DRIVERS\ETC` and contains the IP Address and the IP hostname in matched pairs. It can be used by the name resolution infrastructure in MS Windows, depending on how the TCP/IP environment is configured. This file is in every way the equivalent of the UNIX/Linux `/etc/hosts` file.

25.4.4 DNS Lookup

This capability is configured in the TCP/IP setup area in the network configuration facility. If enabled, an elaborate name resolution sequence is followed, the precise nature of which is dependant on how the NetBIOS Node Type parameter is configured. A Node Type of 0 means that NetBIOS broadcast (over UDP broadcast) is used if the name that is the subject of a name lookup is not found in the NetBIOS name cache. If that fails then DNS, HOSTS and LMHOSTS are checked. If set to Node Type 8, then a NetBIOS Unicast (over UDP Unicast) is sent to the WINS Server to obtain a lookup before DNS, HOSTS, LMHOSTS, or broadcast lookup is used.

25.4.5 WINS Lookup

A WINS (Windows Internet Name Server) service is the equivalent of the rfc1001/1002 specified NBNS (NetBIOS Name Server). A WINS server stores the names and IP addresses that are registered by a Windows client if the TCP/IP setup has been given at least one WINS Server IP Address.

To configure Samba to be a WINS server, the following parameter needs to be added to the `smb.conf` file:

```
wins support = Yes
```

To configure Samba to use a WINS server, the following parameters are needed in the `smb.conf` file:

```
wins support = No
wins server = xxx.xxx.xxx.xxx
```

where *xxx.xxx.xxx.xxx* is the IP address of the WINS server.

For information about setting up Samba as a WINS server, read Chapter 9, *Network Browsing.*

25.5 Common Errors

TCP/IP network configuration problems find every network administrator sooner or later. The cause can be anything from keyboard mishaps, forgetfulness, simple mistakes, and carelessness. Of course, no one is ever deliberately careless!

25.5.1 Pinging Works Only in One Way

"I can ping my Samba server from Windows, but I cannot ping my Windows machine from the Samba server."

Answer: The Windows machine was at IP Address 192.168.1.2 with netmask 255.255.255.0, the Samba server (Linux) was at IP Address 192.168.1.130 with netmask 255.255.255.128. The machines were on a local network with no external connections.

Due to inconsistent netmasks, the Windows machine was on network 192.168.1.0/24, while the Samba server was on network 192.168.1.128/25 — logically a different network.

25.5.2 Very Slow Network Connections

A common cause of slow network response includes:

- Client is configured to use DNS and the DNS server is down.
- Client is configured to use remote DNS server, but the remote connection is down.
- Client is configured to use a WINS server, but there is no WINS server.
- Client is not configured to use a WINS server, but there is a WINS server.
- Firewall is filtering our DNS or WINS traffic.

25.5.3 Samba Server Name Change Problem

"The name of the Samba server was changed, Samba was restarted, Samba server cannot be pinged by new name from MS Windows NT4 Workstation, but it does still respond to ping using the old name. Why?"

From this description, three things are obvious:

- WINS is not in use, only broadcast-based name resolution is used.
- The Samba server was renamed and restarted within the last 10-15 minutes.
- The old Samba server name is still in the NetBIOS name cache on the MS Windows NT4 Workstation.

To find what names are present in the NetBIOS name cache on the MS Windows NT4 machine, open a **cmd** shell and then:

```
C:\> nbtstat -n

             NetBIOS Local Name Table

   Name                 Type          Status
------------------------------------------------
FRODO            <03>  UNIQUE      Registered
ADMINSTRATOR     <03>  UNIQUE      Registered
FRODO            <00>  UNIQUE      Registered
SARDON           <00>  GROUP       Registered
FRODO            <20>  UNIQUE      Registered
FRODO            <1F>  UNIQUE      Registered


C:\> nbtstat -c

            NetBIOS Remote Cache Name Table

   Name                 Type       Host Address     Life [sec]
--------------------------------------------------------------
GANDALF  <20>  UNIQUE      192.168.1.1          240

C:\>
```

In the above example, GANDALF is the Samba server and FRODO is the MS Windows NT4 Workstation. The first listing shows the contents of the Local Name Table (i.e., Identity information on the MS Windows workstation) and the second shows the NetBIOS name in the NetBIOS name cache. The name cache contains the remote machines known to this workstation.

Chapter 26

UNICODE/CHARSETS

26.1 Features and Benefits

Every industry eventually matures. One of the great areas of maturation is in the focus that has been given over the past decade to make it possible for anyone anywhere to use a computer. It has not always been that way, in fact, not so long ago it was common for software to be written for exclusive use in the country of origin.

Of all the effort that has been brought to bear on providing native language support for all computer users, the efforts of the Openi18n organization[1] is deserving of special mention.

Samba-2.x supported a single locale through a mechanism called *codepages*. Samba-3 is destined to become a truly trans-global file and printer-sharing platform.

26.2 What Are Charsets and Unicode?

Computers communicate in numbers. In texts, each number will be translated to a corresponding letter. The meaning that will be assigned to a certain number depends on the *character set (charset)* that is used.

A charset can be seen as a table that is used to translate numbers to letters. Not all computers use the same charset (there are charsets with German umlauts, Japanese characters, and so on). Usually a charset contains 256 characters, which means that storing a character with it takes exactly one byte.

There are also charsets that support even more characters, but those need twice as much storage space (or more). These charsets can contain **256 * 256 = 65536** characters, which is more than all possible characters one could think of. They are called multibyte charsets because they use more then one byte to store one character.

A standardized multibyte charset is unicode[2]. A big advantage of using a multibyte charset is that you only need one; there is no need to make sure two computers use the same charset when they are communicating.

---

[1] http://www.openi18n.org/

[2] http://www.unicode.org/

Old Windows clients use single-byte charsets, named *codepages*, by Microsoft. However, there is no support for negotiating the charset to be used in the SMB/CIFS protocol. Thus, you have to make sure you are using the same charset when talking to an older client. Newer clients (Windows NT, 200x, XP) talk unicode over the wire.

26.3 Samba and Charsets

As of Samba-3.0, Samba can (and will) talk unicode over the wire. Internally, Samba knows of three kinds of character sets:

unix charset — This is the charset used internally by your operating system. The default is `UTF-8`, which is fine for most systems, which covers all characters in all languages. The default in previous Samba releases was `ASCII`.

display charset — This is the charset Samba will use to print messages on your screen. It should generally be the same as the *unix charset*.

dos charset — This is the charset Samba uses when communicating with DOS and Windows 9x/Me clients. It will talk unicode to all newer clients. The default depends on the charsets you have installed on your system. Run **testparm -v | grep "dos charset"** to see what the default is on your system.

26.4 Conversion from Old Names

Because previous Samba versions did not do any charset conversion, characters in filenames are usually not correct in the UNIX charset but only for the local charset used by the DOS/Windows clients.

26.5 Japanese Charsets

Samba does not work correctly with Japanese charsets yet. Here are points of attention when setting it up:

- You should set *mangling method* = hash
- There are various iconv() implementations around and not all of them work equally well. glibc2's iconv() has a critical problem in CP932. libiconv-1.8 works with CP932 but still has some problems and does not work with EUC-JP.
- You should set *dos charset* = CP932, not Shift_JIS, SJIS.
- Currently only *UNIX charset* = CP932 will work (but still has some problems...) because of iconv() issues. *UNIX charset* = EUC-JP does not work well because of iconv() issues.
- Currently Samba-3.0 does not support *UNIX charset* = UTF8-MAC/CAP/HEX/JIS*.

More information (in Japanese) is available at: `http://www.atmarkit.co.jp/flinux/special/samba3/samba3a.html`.

26.6 Common Errors

26.6.1 CP850.so Can't Be Found

"Samba is complaining about a missing `CP850.so` file."

Answer: CP850 is the default ***dos charset***. The ***dos charset*** is used to convert data to the codepage used by your dos clients. If you do not have any dos clients, you can safely ignore this message.

CP850 should be supported by your local iconv implementation. Make sure you have all the required packages installed. If you compiled Samba from source, make sure to configure found iconv.

Chapter 27

BACKUP TECHNIQUES

27.1 Features and Benefits

The Samba project is over ten years old. During the early history of Samba, UNIX administrators were its key implementors. UNIX administrators will use UNIX system tools to backup UNIX system files. Over the past four years, an increasing number of Microsoft network administators have taken an interest in Samba. This is reflected in the questions about backup in general on the Samba mailing lists.

27.2 Discussion of Backup Solutions

During discussions at a Microsoft Windows training course, one of the pro-UNIX delegates stunned the class when he pointed out that Windows NT4 is so limiting compared with UNIX. He likened UNIX to a mechano set that has an unlimited number of tools that are simple, efficient, and, in combination, capable of achieving any desired outcome.

One of the Windows networking advocates retorted that if she wanted a mechano set, she would buy one. She made it clear that a complex single tool that does more than is needed but does it with a clear purpose and intent is preferred by some like her.

Please note that all information here is provided as is and without recommendation of fitness or suitability. The network administrator is strongly encouraged to perform due-diligence research before implementing any backup solution, whether free software or commercial.

A useful Web site I recently stumbled across that you might like to refer to is located at www.allmerchants.com.

The following three free software projects might also merit consideration.

27.2.1 BackupPC

BackupPC version 2.0.0 has been released on SourceForge.[1] New features include support for **rsync/rsyncd** and internationalization of the CGI interface (including English, French, Spanish, and German).

[1] http://backuppc.sourceforge.net

BackupPC is a high-performance Perl-based package for backing up Linux, UNIX or Windows PCs and laptops to a server's disk. BackupPC is highly configurable and easy to install and maintain. SMB (via smbclient), **tar** over **rsh/ssh** or **rsync/rsyncd** are used to extract client data.

Given the ever decreasing cost of disks and raid systems, it is now practical and cost effective to backup a large number of machines onto a server's local disk or network storage. This is what BackupPC does.

Key features are pooling of identical files (big savings in server disk space), compression, and a comprehensive CGI interface that allows users to browse backups and restore files.

BackupPC is free software distributed under a GNU GPL license. BackupPC runs on Linux/UNIX/freenix servers, and has been tested on Linux, UNIX, Windows 9x/ME, Windows 98, Windows 200x, Windows XP, and Mac OSX clients.

27.2.2 Rsync

rsync is a flexible program for efficiently copying files or directory trees.

rsync has many options to select which files will be copied and how they are to be transferred. It may be used as an alternative to **ftp, http, scp**, or **rcp**.

The rsync remote-update protocol allows rsync to transfer just the differences between two sets of files across the network link, using an efficient checksum-search algorithm described in the technical report that accompanies the rsync package.

Some of the additional features of rsync are:

- Support for copying links, devices, owners, groups, and permissions.
- Exclude and exclude-from options are similar to GNU tar.
- A CVS exclude mode for ignoring the same files that CVS would ignore.
- Can use any transparent remote shell, including rsh or ssh.
- Does not require root privileges.
- Pipelining of file transfers to minimize latency costs.
- Support for anonymous or authenticated rsync servers (ideal for mirroring).

27.2.3 Amanda

Amanda, the Advanced Maryland Automatic Network Disk Archiver, is a backup system that allows the administrator of a LAN to set up a single master backup server to back up multiple hosts to a single large capacity tape drive. Amanda uses native dump and/or GNU tar facilities and can back up a large number of workstations running multiple versions of UNIX. Recent versions can also use Samba to back up Microsoft Windows hosts.

For more information regarding Amanda, please check the www.amanda.org/ site.[2]

[2] http://www.amanda.org/

27.2.4 BOBS: Browseable Online Backup System

Browseable Online Backup System (BOBS) is a complete online backup system. Uses large disks for storing backups and lets users browse the files using a Web browser. Handles some special files like AppleDouble and icon files.

The home page for BOBS is located at bobs.sourceforge.net.[3]

[3] http://bobs.sourceforge.net/

Chapter 28

HIGH AVAILABILITY

28.1 Features and Benefits

Network administrators are often concerned about the availability of file and print services. Network users are inclined toward intolerance of the services they depend on to perform vital task responsibilities.

A sign in a computer room served to remind staff of their responsibilities. It read:

> All humans fail, in both great and small ways we fail continually. Machines fail too. Computers are machines that are managed by humans, the fallout from failure can be spectacular. Your responsibility is to deal with failure, to anticipate it and to eliminate it as far as is humanly and economically wise to achieve. Are your actions part of the problem or part of the solution?

If we are to deal with failure in a planned and productive manner, then first we must understand the problem. That is the purpose of this chapter.

Parenthetically, in the following discussion there are seeds of information on how to provision a network infrastructure against failure. Our purpose here is not to provide a lengthy dissertation on the subject of high availability. Additionally, we have made a conscious decision to not provide detailed working examples of high availability solutions; instead we present an overview of the issues in the hope that someone will rise to the challenge of providing a detailed document that is focused purely on presentation of the current state of knowledge and practice in high availability as it applies to the deployment of Samba and other CIFS/SMB technologies.

28.2 Technical Discussion

The following summary was part of a presentation by Jeremy Allison at the SambaXP 2003 conference that was held at Goettingen, Germany, in April 2003. Material has been added from other sources, but it was Jeremy who inspired the structure that follows.

28.2.1 The Ultimate Goal

All clustering technologies aim to achieve one or more of the following:

- Obtain the maximum affordable computational power.
- Obtain faster program execution.
- Deliver unstoppable services.
- Avert points of failure.
- Exact most effective utilization of resources.

A clustered file server ideally has the following properties:

- All clients can connect transparently to any server.
- A server can fail and clients are transparently reconnected to another server.
- All servers server out the same set of files.
- All file changes are immediately seen on all servers.
  - Requires a distributed file system.
- Infinite ability to scale by adding more servers or disks.

28.2.2 Why Is This So Hard?

In short, the problem is one of *state*.

- All TCP/IP connections are dependent on state information.

  The TCP connection involves a packet sequence number. This sequence number would need to be dynamically updated on all machines in the cluster to effect seemless TCP fail-over.

- CIFS/SMB (the Windows networking protocols) uses TCP connections.

  This means that from a basic design perspective, fail-over is not seriously considered.

  - All current SMB clusters are fail-over solutions — they rely on the clients to reconnect. They provide server fail-over, but clients can lose information due to a server failure.

- Servers keep state information about client connections.
  - CIFS/SMB involves a lot of state.
  - Every file open must be compared with other file opens to check share modes.

28.2.2.1 The Front-End Challenge

To make it possible for a cluster of file servers to appear as a single server that has one name and one IP address, the incoming TCP data streams from clients must be processed by the front end virtual server. This server must de-multiplex the incoming packets at the SMB protocol layer level and then feed the SMB packet to different servers in the cluster.

One could split all IPC$ connections and RPC calls to one server to handle printing and user lookup requirements. RPC Printing handles are shared between different IPC4 sessions — it is hard to split this across clustered servers!

Conceptually speaking, all other servers would then provide only file services. This is a simpler problem to concentrate on.

28.2.2.2 De-multiplexing SMB Requests

De-multiplexing of SMB requests requires knowledge of SMB state information, all of which must be held by the front-end *virtual* server. This is a perplexing and complicated problem to solve.

Windows XP and later have changed semantics so state information (vuid, tid, fid) must match for a successful operation. This makes things simpler than before and is a positive step forward.

SMB requests are sent by vuid to their associated server. No code exists today to affect this solution. This problem is conceptually similar to the problem of correctly handling, in Samba, requests from multiple requests from Windows 2000 Terminal Server. (John, this last sentence is garbled.)

One possibility is to start by exposing the server pool to clients directly. This could eliminate the de-mulitplexing step.

28.2.2.3 The Distributed File System Challenge

There exists many distributed file systems for UNIX and Linux.

Many could be adopted to backend our cluster, so long as awareness of SMB semantics is kept in mind (share modes, locking and oplock issues in particular). Common free distributed file systems include:

- NFS
- AFS
- OpenGFS
- Lustre

The server pool (cluster) can use any distributed file system backend if all SMB semantics are performed within this pool.

28.2.2.4 Restrictive Contraints on Distributed File Systems

Where a clustered server provides purely SMB services, oplock handling may be done within the server pool without imposing a need for this to be passed to the backend file system pool.

On the other hand, where the server pool also provides NFS or other file services, it will be essential that the implementation be oplock aware so it can interoperate with SMB services.

This is a significant challenge today. A failure to provide this will result in a significant loss of performance that will be sorely noted by users of Microsoft Windows clients.

Last, all state information must be shared across the server pool.

28.2.2.5 Server Pool Communications

Most backend file systems support POSIX file semantics. This makes it difficult to push SMB semantics back into the file system. POSIX locks have different properties and semantics from SMB locks.

All **smbd** processes in the server pool must of necessity communicate very quickly. For this, the current *tdb* file structure that Samba uses is not suitable for use across a network. Clustered **smbd**'s must use something else.

28.2.2.6 Server Pool Communications Demands

High speed inter-server communications in the server pool is a design prerequisite for a fully functional system. Possibilities for this include:

- Proprietary shared memory bus (example: Myrinet or SCI [Scalable Coherent Interface]). These are high cost items.
- Gigabit ethernet (now quite affordable).
- Raw ethernet framing (to bypass TCP and UDP overheads).

We have yet to identify metrics for performance demands to enable this to happen effectively.

28.2.2.7 Required Modifications to Samba

Samba needs to be significantly modified to work with a high-speed server inter-connect system to permit transparent fail-over clustering.

Particular functions inside Samba that will be affected include:

- The locking database, oplock notifications, and the share mode database.
- Failure semantics need to be defined. Samba behaves the same way as Windows. When oplock messages fail, a file open request is allowed, but this is potentially dangerous in a clustered environment. So how should inter-server pool failure semantics function and how should this be implemented?
- Should this be implemented using a point-to-point lock manager, or can this be done using multicast techniques?

28.2.3 A Simple Solution

Allowing fail-over servers to handle different functions within the exported file system removes the problem of requiring a distributed locking protocol.

If only one server is active in a pair, the need for high speed server interconnect is avoided. This allows the use of existing high availability solutions, instead of inventing a new one. This simpler solution comes at a price — the cost of which is the need to manage a more complex file name space. Since there is now not a single file system, administrators must remember where all services are located — a complexity not easily dealt with.

The *virtual server* is still needed to redirect requests to backend servers. Backend file space integrity is the responsibility of the administrator.

28.2.4 High Availability Server Products

Fail-over servers must communicate in order to handle resource fail-over. This is essential for high availaiblity services. The use of a dedicated heartbeat is a common technique to introduce some intelligence into the fail-over process. This is often done over a dedicated link (LAN or serial).

Many fail-over solutions (like Red Hat Cluster Manager, as well as Microsoft Wolfpack) can use a shared SCSI of Fiber Channel disk storage array for fail-over communication. Information regarding Red Hat high availability solutions for Samba may be obtained from: www.redhat.com.[1]

The Linux High Availability project is a resource worthy of consultation if your desire is to build a highly available Samba file server solution. Please consult the home page at www.linux-ha.org/.[2]

Front-end server complexity remains a challenge for high availability as it needs to deal gracefully with backend failures, while at the same time it needs to provide continuity of service to all network clients.

28.2.5 MS-DFS: The Poor Man's Cluster

MS-DFS links can be used to redirect clients to disparate backend servers. This pushes complexity back to the network client, something already included by Microsoft. MS-DFS creates the illusion of a simple, continuous file system name space, that even works at the file level.

Above all, at the cost of complexity of management, a distributed (pseudo-cluster) can be created using existing Samba functionality.

28.2.6 Conclusions

- Transparent SMB clustering is hard to do!
- Client fail-over is the best we can do today.
- Much more work is needed before a practical and managable high availability transparent cluster solution will be possible.

---

[1] http://www.redhat.com/docs/manuals/enterprise/RHEL-AS-2.1-Manual/cluster-manager/s1-service-samba.html

[2] http://www.linux-ha.org/

- MS-DFS can be used to create the illusion of a single transparent cluster.

Part IV

Migration and Updating

Chapter 29

UPGRADING FROM SAMBA-2.X TO SAMBA-3.0.0

This chapter deals exclusively with the differences between Samba-3.0.0 and Samba-2.2.8a. It points out where configuration parameters have changed, and provides a simple guide for the move from 2.2.x to 3.0.0.

29.1 Quick Migration Guide

Samba-3.0.0 default behavior should be approximately the same as Samba-2.2.x. The default behavior when the new parameter *passdb backend* is not defined in the `smb.conf` file provides the same default behviour as Samba-2.2.x with *encrypt passwords* = Yes, and will use the `smbpasswd` database.

So why say that *behavior should be approximately the same as Samba-2.2.x?* Because Samba-3.0.0 can negotiate new protocols, such as support for native Unicode, that may result in differing protocol code paths being taken. The new behavior under such circumstances is not exactly the same as the old one. The good news is that the domain and machine SIDs will be preserved across the upgrade.

If the Samba-2.2.x system was using an LDAP backend, and there is no time to update the LDAP database, then make sure that *passdb backend* = ldapsam_compat is specified in the `smb.conf` file. For the rest, behavior should remain more or less the same. At a later date, when there is time to implement a new Samba-3 compatible LDAP backend, it is possible to migrate the old LDAP database to the new one through use of the **pdbedit**. See Section 10.3.2.

29.2 New Features in Samba-3

The major new features are:

1. Active Directory support. This release is able to join an ADS realm as a member server and authenticate users using LDAP/kerberos.

2. Unicode support. Samba will now negotiate unicode on the wire and internally there is a much better infrastructure for multi-byte and unicode character sets.

3. New authentication system. The internal authentication system has been almost completely rewritten. Most of the changes are internal, but the new authoring system is also very configurable.

4. New filename mangling system. The filename mangling system has been completely rewritten. An internal database now stores mangling maps persistently.

5. New "*net*" command. A new "*net*" command has been added. It is somewhat similar to the "*net*" command in Windows. Eventually, we plan to replace a bunch of other utilities (such as smbpasswd) with subcommands in "*net*".

6. Samba now negotiates NT-style status32 codes on the wire. This considerably improves error handling.

7. Better Windows 200x/XP printing support including publishing printer attributes in Active Directory.

8. New loadable RPC modules for passdb backends and character sets.

9. New default dual-daemon winbindd support for better performance.

10. Support for migrating from a Windows NT 4.0 domain to a Samba domain and maintaining user, group and domain SIDs.

11. Support for establishing trust relationships with Windows NT 4.0 Domain Controllers.

12. Initial support for a distributed Winbind architecture using an LDAP directory for storing SID to UID/GID mappings.

13. Major updates to the Samba documentation tree.

14. Full support for client and server SMB signing to ensure compatibility with default Windows 2003 security settings.

Plus lots of other improvements!

29.3 Configuration Parameter Changes

This section contains a brief listing of changes to `smb.conf` options in the 3.0.0 release. Please refer to the smb.conf(5) man page for complete descriptions of new or modified parameters.

29.3.1 Removed Parameters

(Ordered Alphabetically):

- admin log
- alternate permissions
- character set

- client codepage
- code page directory
- coding system
- domain admin group
- domain guest group
- force unknown acl user
- nt smb support
- post script
- printer driver
- printer driver file
- printer driver location
- status
- stip dot
- total print jobs
- use rhosts
- valid chars
- vfs options

29.3.2 New Parameters

(New parameters have been grouped by function):

Remote Management

- abort shutdown script
- shutdown script

User and Group Account Management:

- add group script
- add machine script
- add user to group script
- algorithmic rid base
- delete group script
- delete user from group script
- passdb backend
- set primary group script

Authentication:

- auth methods
- realm

Protocol Options:

- client lanman auth
- client NTLMv2 auth
- client schannel
- client signing
- client use spnego
- disable netbios
- ntlm auth
- paranoid server security
- server schannel
- server signing
- smb ports
- use spnego

File Service:

- get quota command
- hide special files
- hide unwriteable files
- hostname lookups
- kernel change notify
- mangle prefix
- map acl inherit
- msdfs proxy
- set quota command
- use sendfile
- vfs objects

Printing:

- max reported print jobs

Unicode and Character Sets:

- display charset

- dos charset
- unicode
- UNIX charset

SID to UID/GID Mappings:

- idmap backend
- idmap gid
- idmap uid
- winbind enable local accounts
- winbind trusted domains only
- template primary group
- enable rid algorithm

LDAP:

- ldap delete dn
- ldap group suffix
- ldap idmap suffix
- ldap machine suffix
- ldap passwd sync
- ldap trust ids
- ldap user suffix

General Configuration:

- preload modules
- privatedir

29.3.3 Modified Parameters (Changes in Behavior):

- encrypt passwords (enabled by default)
- mangling method (set to hash2 by default)
- passwd chat
- passwd program
- password server
- restrict anonymous (integer value)
- security (new ads value)
- strict locking (enabled by default)

- winbind cache time (increased to 5 minutes)
- winbind uid (deprecated in favor of idmap uid)
- winbind gid (deprecated in favor of idmap gid)

29.4 New Functionality

29.4.1 Databases

This section contains brief descriptions of any new databases introduced in Samba-3. Please remember to backup your existing ${lock directory}/*tdb before upgrading to Samba-3. Samba will upgrade databases as they are opened (if necessary), but downgrading from 3.0 to 2.2 is an unsupported path.

The new tdb files are described in Table 29.1.

Table 29.1. TDB File Descriptions

Name	Description	Backup?
account_policy	User policy settings	yes
gencache	Generic caching db	no
group_mapping	Mapping table from Windows groups/SID to UNIX groups	yes
idmap	new ID map table from SIDS to UNIX UIDs/GIDs	yes
namecache	Name resolution cache entries	no
netlogon_unigrp	Cache of universal group membership obtained when operating as a member of a Windows domain	no
printing/*.tdb	Cached output from 'lpq command' created on a per print service basis	no
registry	Read-only Samba registry skeleton that provides support for exporting various db tables via the winreg RPCs	no

29.4.2 Changes in Behavior

The following issues are known changes in behavior between Samba-2.2 and Samba-3 that may affect certain installations of Samba.

1. When operating as a member of a Windows domain, Samba-2.2 would map any users authenticated by the remote DC to the "*guest account*" if a uid could not be obtained via the getpwnam() call. Samba-3 rejects the connection as NT_STATUS_LOGON_FAILURE. There is no current work around to re-establish the Samba-2.2 behavior.

2. When adding machines to a Samba-2.2 controlled domain, the "*add user script*" was used to create the UNIX identity of the Machine Trust Account. Samba-3 introduces a new "*add machine script*" that must be specified for this purpose. Samba-3 will not fall back to using the "*add user script*" in the absence of an "*add machine script*".

29.4.3 Charsets

You might experience problems with special characters when communicating with old DOS clients. Codepage support has changed in Samba-3. Read Chapter 26, *Unicode/Charsets*, for details.

29.4.4 Passdb Backends and Authentication

There have been a few new changes that Samba administrators should be aware of when moving to Samba-3.

1. Encrypted passwords have been enabled by default in order to interoperate better with out-of-the-box Windows client installations. This does mean that either (a) a Samba account must be created for each user, or (b) "*encrypt passwords = no*" must be explicitly defined in `smb.conf`.

2. Inclusion of new *`security`* = ads option for integration with an Active Directory domain using the native Windows Kerberos 5 and LDAP protocols.

Samba-3 also includes the possibility of setting up chains of authentication methods (*`auth methods`*) and account storage backends (*`passdb backend`*). Please refer to the `smb.conf` man page and Chapter 10, *Account Information Databases*, for details. While both parameters assume sane default values, it is likely that you will need to understand what the values actually mean in order to ensure Samba operates correctly.

Certain functions of the **smbpasswd** tool have been split between the new **smbpasswd** utility, the **net** tool and the new **pdbedit** utility. See the respective man pages for details.

29.4.5 LDAP

This section outlines the new features effecting Samba/LDAP integration.

29.4.5.1 New Schema

A new object class (sambaSamAccount) has been introduced to replace the old sambaAccount. This change aids us in the renaming of attributes to prevent clashes with attributes from other vendors. There is a conversion script (examples/LDAP/convertSambaAccount) to modify an LDIF file to the new schema.

Example:

```
$ ldapsearch .... -b "ou=people,dc=..." > old.ldif
$ convertSambaAccount <DOM SID> old.ldif new.ldif
```

The <DOM SID> can be obtained by running

```
$ net getlocalsid <DOMAINNAME>
```

on the Samba PDC as root.

The old sambaAccount schema may still be used by specifying the *ldapsam_compat* passdb backend. However, the sambaAccount and associated attributes have been moved to the historical section of the schema file and must be uncommented before use if needed. The Samba-2.2 object class declaration for a sambaAccount has not changed in the Samba-3 samba.schema file.

Other new object classes and their uses include:

- sambaDomain — domain information used to allocate RIDs for users and groups as necessary. The attributes are added in "*ldap suffix*" directory entry automatically if an idmap UID/GID range has been set and the "*ldapsam*" passdb backend has been selected.
- sambaGroupMapping — an object representing the relationship between a posixGroup and a Windows group/SID. These entries are stored in the "*ldap group suffix*" and managed by the "*net groupmap*" command.
- sambaUNIXIdPool — created in the "*ldap idmap suffix*" entry automatically and contains the next available "*idmap UID*" and "*idmap GID*".
- sambaIdmapEntry — object storing a mapping between a SID and a UNIX UID/GID. These objects are created by the idmap_ldap module as needed.

29.4.5.2 New Suffix for Searching

The following new smb.conf parameters have been added to aid in directing certain LDAP queries when *passdb backend = ldapsam://...* has been specified.

- ldap suffix — used to search for user and computer accounts.
- ldap user suffix — used to store user accounts.
- ldap machine suffix — used to store Machine Trust Accounts.
- ldap group suffix — location of posixGroup/sambaGroupMapping entries.
- ldap idmap suffix — location of sambaIdmapEntry objects.

If an *ldap suffix* is defined, it will be appended to all of the remaining sub-suffix parameters. In this case, the order of the suffix listings in smb.conf is important. Always place the *ldap suffix* first in the list.

Due to a limitation in Samba's `smb.conf` parsing, you should not surround the DNs with quotation marks.

29.4.5.3 IdMap LDAP Support

Samba-3 supports an ldap backend for the idmap subsystem. The following options inform Samba that the idmap table should be stored on the directory server onterose in the "ou=idmap,dc=quenya,dc=org" partition.

```
...
idmap backend = ldap:ldap://onterose/
ldap idmap suffix = ou=idmap,dc=quenya,dc=org
idmap uid = 40000-50000
idmap gid = 40000-50000
```

This configuration allows Winbind installations on multiple servers to share a UID/GID number space, thus avoiding the interoperability problems with NFS that were present in Samba-2.2.

Chapter 30

MIGRATION FROM NT4 PDC TO SAMBA-3 PDC

This is a rough guide to assist those wishing to migrate from NT4 Domain Control to Samba-3-based Domain Control.

30.1 Planning and Getting Started

In the IT world there is often a saying that all problems are encountered because of poor planning. The corollary to this saying is that not all problems can be anticipated and planned for. Then again, good planning will anticipate most show-stopper-type situations.

Those wishing to migrate from MS Windows NT4 Domain Control to a Samba-3 Domain Control environment would do well to develop a detailed migration plan. So here are a few pointers to help migration get under way.

30.1.1 Objectives

The key objective for most organizations will be to make the migration from MS Windows NT4 to Samba-3 Domain Control as painless as possible. One of the challenges you may experience in your migration process may well be one of convincing management that the new environment should remain in place. Many who have introduced open source technologies have experienced pressure to return to a Microsoft-based platform solution at the first sign of trouble.

Before attempting a migration to a Samba-3 controlled network, make every possible effort to gain all-round commitment to the change. Know precisely *why* the change is important for the organization. Possible motivations to make a change include:

- Improve network manageability.
- Obtain better user level functionality.
- Reduce network operating costs.
- Reduce exposure caused by Microsoft withdrawal of NT4 support.

- Avoid MS License 6 implications.
- Reduce organization's dependency on Microsoft.

Make sure everyone knows that Samba-3 is not MS Windows NT4. Samba-3 offers an alternative solution that is both different from MS Windows NT4 and offers advantages compared with it. Gain recognition that Samba-3 lacks many of the features that Microsoft has promoted as core values in migration from MS Windows NT4 to MS Windows 2000 and beyond (with or without Active Directory services).

What are the features that Samba-3 cannot provide?

- Active Directory Server.
- Group Policy Objects (in Active Directory).
- Machine Policy Objects.
- Logon Scripts in Active Directory.
- Software Application and Access Controls in Active Directory.

The features that Samba-3 does provide and that may be of compelling interest to your site include:

- Lower cost of ownership.
- Global availability of support with no strings attached.
- Dynamic SMB Servers (can run more than one SMB/CIFS server per UNIX/Linux system).
- Creation of on-the-fly logon scripts.
- Creation of on-the-fly Policy Files.
- Greater stability, reliability, performance and availability.
- Manageability via an ssh connection.
- Flexible choices of back-end authentication technologies (tdbsam, ldapsam, mysqlsam).
- Ability to implement a full single-sign-on architecture.
- Ability to distribute authentication systems for absolute minimum wide area network bandwidth demand.

Before migrating a network from MS Windows NT4 to Samba-3, consider all necessary factors. Users should be educated about changes they may experience so the change will be a welcome one and not become an obstacle to the work they need to do. The following are factors that will help ensure a successful migration:

30.1.1.1 Domain Layout

Samba-3 can be configured as a Domain Controller, a back-up Domain Controller (probably best called a secondary controller), a Domain Member, or as a stand-alone Server. The

Windows network security domain context should be sized and scoped before implementation. Particular attention needs to be paid to the location of the primary Domain Controller (PDC) as well as backup controllers (BDCs). One way in which Samba-3 differs from Microsoft technology is that if one chooses to use an LDAP authentication backend, then the same database can be used by several different domains. In a complex organization, there can be a single LDAP database, which itself can be distributed (have a master server and multiple slave servers) that can simultaneously serve multiple domains.

>From a design perspective, the number of users per server as well as the number of servers per domain should be scaled taking into consideration server capacity and network bandwidth.

A physical network segment may house several domains. Each may span multiple network segments. Where domains span routed network segments, consider and test the performance implications of the design and layout of a network. A centrally located Domain Controller that is designed to serve multiple routed network segments may result in severe performance problems. Check the response time (ping timing) between the remote segment and the PDC. If it's long (more than 100 ms), locate a backup controller (BDC) on the remote segment to serve as the local authentication and access control server.

30.1.1.2 Server Share and Directory Layout

There are cardinal rules to effective network design that cannot be broken with impunity. The most important rule: Simplicity is king in every well-controlled network. Every part of the infrastructure must be managed; the more complex it is, the greater will be the demand of keeping systems secure and functional.

Keep in mind the nature of how data must be shared. Physical disk space layout should be considered carefully. Some data must be backed up. The simpler the disk layout the easier it will be to keep track of backup needs. Identify what backup media will meet your needs; consider backup to tape, CD-ROM or (DVD-ROM), or other offline storage medium. Plan and implement for minimum maintenance. Leave nothing to chance in your design; above all, do not leave backups to chance: Backup, test, and validate every backup, create a disaster recovery plan and prove that it works.

Users should be grouped according to data access control needs. File and directory access is best controlled via group permissions and the use of the "*sticky bit*" on group controlled directories may substantially avoid file access complaints from Samba share users.

Inexperienced network administrators often attempt elaborate techniques to set access controls on files, directories, shares, as well as in share definitions. Keep your design and implementation simple and document your design extensively. Have others audit your documentation. Do not create a complex mess that your successor will not understand. Remember, job security through complex design and implementation may cause loss of operations and downtime to users as the new administrator learns to untangle your knots. Keep access controls simple and effective and make sure that users will never be interrupted by obtuse complexity.

30.1.1.3 Logon Scripts

Logon scripts can help to ensure that all users gain the share and printer connections they need.

Logon scripts can be created on-the-fly so all commands executed are specific to the rights and privileges granted to the user. The preferred controls should be affected through group membership so group information can be used to create a custom logon script using the *root preexec* parameters to the *NETLOGON* share.

Some sites prefer to use a tool such as **kixstart** to establish a controlled user environment. In any case, you may wish to do a Google search for logon script process controls. In particular, you may wish to explore the use of the Microsoft KnowledgeBase article KB189105 that deals with how to add printers without user intervention via the logon script process.

30.1.1.4 Profile Migration/Creation

User and Group Profiles may be migrated using the tools described in the section titled Desktop Profile Management.

Profiles may also be managed using the Samba-3 tool **profiles**. This tool allows the MS Windows NT-style security identifiers (SIDs) that are stored inside the profile `NTuser.DAT` file to be changed to the SID of the Samba-3 domain.

30.1.1.5 User and Group Accounts

It is possible to migrate all account settings from an MS Windows NT4 domain to Samba-3. Before attempting to migrate user and group accounts, it is STRONGLY advised to create in Samba-3 the groups that are present on the MS Windows NT4 domain *AND* to map them to suitable UNIX/Linux groups. By following this simple advice, all user and group attributes should migrate painlessly.

30.1.2 Steps in Migration Process

The approximate migration process is described below.

- You have an NT4 PDC that has the users, groups, policies and profiles to be migrated.
- Samba-3 set up as a DC with netlogon share, profile share, and so on. Configure the `smb.conf` file to fucntion as a BDC, i.e., *domain master = No*.

THE ACCOUNT MIGRATION PROCESS

1. Create a BDC account in the old NT4 domain for the Samba server using NT Server Manager.
   (a) Samba must not be running.
2. `net rpc join -S` *NT4PDC* `-w` *DOMNAME* `-U Administrator%`*passwd*
3. `net rpc vampire -S` *NT4PDC* `-U administrator%`*passwd*

4. pdbedit -L
   (a) Note — did the users migrate?
5. Now assign each of the UNIX groups to NT groups: (It may be useful to copy this text to a script called `initGroups.sh`)

```
#!/bin/bash
#### Keep this as a shell script for future re-use

# First assign well known domain global groups
net groupmap modify ntgroup="Domain Admins" unixgroup=root   rid=512
net groupmap modify ntgroup="Domain Users"  unixgroup=users  rid=513
net groupmap modify ntgroup="Domain Guests" unixgroup=nobody rid=514

# Now for our added domain global groups
net groupmap add ntgroup="Designers" unixgroup=designers type=d rid=3200
net groupmap add ntgroup="Engineers" unixgroup=engineers type=d rid=3210
net groupmap add ntgroup="QA Team"   unixgroup=qateam    type=d rid=3220
```

6. net groupmap list
   (a) Check that all groups are recognized.

Migrate all the profiles, then migrate all policy files.

30.2 Migration Options

Sites that wish to migrate from MS Windows NT4 Domain Control to a Samba-based solution generally fit into three basic categories. Table 30.1 shows the possibilities.

Table 30.1. The Three Major Site Types

Number of Users	Description
< 50	Want simple conversion with no pain.
50 - 250	Want new features, can manage some in-house complexity.
> 250	Solution/Implementation must scale well, complex needs. Cross-departmental decision process. Local expertise in most areas.

30.2.1 Planning for Success

There are three basic choices for sites that intend to migrate from MS Windows NT4 to Samba-3:

- Simple conversion (total replacement).
- Upgraded conversion (could be one of integration).
- Complete redesign (completely new solution).

Minimize down-stream problems by:

- Taking sufficient time.
- Avoiding Panic.
- Testing all assumptions.
- Testing the full roll-out program, including workstation deployment.

Table 30.2 lists the conversion choices given the type of migration being contemplated.

Table 30.2. Nature of the Conversion Choices

Simple	**Upgraded**	**Redesign**
Make use of minimal OS specific features.	Translate NT4 features to new host OS features.	Decide:
Move all accounts from NT4 into Samba-3	Copy and improve	Authentication regime (database location and access)
Make least number of operational changes	Make progressive improvements	Desktop management methods
Take least amount of time to migrate	Minimize user impact	Better control of Desktops/Users
Live versus isolated conversion	Maximize functionality	Identify Needs for: *Manageability, Scalability, Security, Availability*
Integrate Samba-3 then migrate while users are active, then change of control (swap out)	Take advantage of lower maintenance opportunity	

30.2.2 Samba-3 Implementation Choices

Authentication Database/Backend — Samba-3 can use an external authentication backend:

- Winbind (external Samba or NT4/200x server).
- External server could use Active Directory or NT4 Domain.
- Can use pam_mkhomedir.so to auto-create home dirs.
- Samba-3 can use a local authentication backend: *smbpasswd, tdbsam, ldapsam, mysqlsam*

Access Control Points — Samba permits Access Control Points to be set:

- On the share itself — using Share ACLs.
- On the file system — using UNIX permissions on files and directories.

Note: Can enable Posix ACLs in file system also.

- Through Samba share parameters — not recommended except as last resort.

Policies (migrate or create new ones) — Exercise great caution when affecting registry changes, use the right tool and be aware that changes made through NT4-style `NTConfig.POL` files can leave permanent changes.

- Using Group Policy Editor (NT4).
- Watch out for Tattoo effect.

User and Group Profiles — Platform-specific so use platform tool to change from a Local to a Roaming profile. Can use new profiles tool to change SIDs (`NTUser.DAT`).

Logon Scripts — Know how they work.

User and Group Mapping to UNIX/Linux — User and Group mapping code is new. Many problems have been experienced as network administrators who are familiar with Samba-2.2.x migrate to Samba-3. Carefully study the chapters that document the new password backend behavior and the new group mapping functionality.

- The *username map* facility may be needed.
- Use **net groupmap** to connect NT4 groups to UNIX groups.
- Use **pdbedit** to set/change user configuration.

  When migrating to LDAP backend, it may be easier to dump the initial LDAP database to LDIF, edit, then reload into LDAP.

OS Specific Scripts/Programs may be Needed — Every operating system has its peculiarities. These are the result of engineering decisions that were based on the experience of the designer, and may have side-effects that were not anticipated. Limitations that may bite the Windows network administrator include:

- Add/Delete Users: Note OS limits on size of name (Linux 8 chars) NT4 up to 254 chars.
- Add/Delete Machines: Applied only to Domain Members (Note: machine names may be limited to 16 characters).
- Use **net groupmap** to connect NT4 groups to UNIX groups.
- Add/Delete Groups: Note OS limits on size and nature. Linux limit is 16 char, no spaces and no upper case chars (**groupadd**).

Migration Tools — Domain Control (NT4 Style) Profiles, Policies, Access Controls, Security

- Samba: **net, rpcclient, smbpasswd, pdbedit, profiles.**
- Windows: **NT4 Domain User Manager, Server Manager (NEXUS)**

Chapter 31

SWAT — THE SAMBA WEB ADMINISTRATION TOOL

There are many and varied opinions regarding the usefulness of SWAT. No matter how hard one tries to produce the perfect configuration tool, it remains an object of personal taste. SWAT is a tool that will allow Web-based configuration of Samba. It has a wizard that may help to get Samba configured quickly, it has context-sensitive help on each `smb.conf` parameter, it provides for monitoring of current state of connection information, and it allows network-wide MS Windows network password management.

31.1 Features and Benefits

SWAT is a facility that is part of the Samba suite. The main executable is called **swat** and is invoked by the inter-networking super daemon. See Section 31.2.2 for details.

SWAT uses integral samba components to locate parameters supported by the particular version of Samba. Unlike tools and utilities that are external to Samba, SWAT is always up to date as known Samba parameters change. SWAT provides context-sensitive help for each configuration parameter, directly from **man** page entries.

There are network administrators who believe that it is a good idea to write systems documentation inside configuration files, and for them SWAT will aways be a nasty tool. SWAT does not store the configuration file in any intermediate form, rather, it stores only the parameter settings, so when SWAT writes the `smb.conf` file to disk, it will write only those parameters that are at other than the default settings. The result is that all comments, as well as parameters that are no longer supported, will be lost from the `smb.conf` file. Additionally, the parameters will be written back in internal ordering.

NOTE

Before using SWAT, please be warned — SWAT will completely replace your `smb.conf` with a fully-optimized file that has been stripped of all comments you might have placed there and only non-default settings will be written to the file.

31.2 Guidelines and Technical Tips

This section aims to unlock the dark secrets behind how SWAT may be made to work, may be made more secure, and how to solve Internationalization support problems.

31.2.1 Validate SWAT Installation

The very first step that should be taken before attempting to configure a host system for SWAT operation is to check that it is installed. This may seem a trivial point to some, however several Linux distributions do not install SWAT by default, even though they do ship an installable binary support package containing SWAT on the distribution media.

When you have configrmed that SWAT is installed it is necessary to validate that the installation includes the binary **swat** file as well as all the supporting text and Web files. A number of operating system distributions in the past have failed to include the necessary support files, evne though the **swat** binary executable file was installed.

Finally, when you are sure that SWAT has been fully installed, please check the SWAT has been enebled in the control file for the internetworking super-daemon (inetd or xinetd) that is used on your operating system platform.

31.2.1.1 Locating the swat File

To validate that SWAT is installed, first locate the **swat** binary file on the system. It may be found under the following directories:

`/usr/local/samba/bin` — the default Samba location.
`/usr/sbin` — the default location on most Linux systems.
`/opt/samba/bin`

The actual location is much dependant on the choice of the operating system vendor, or as determined by the administrator who compiled and installed Samba.

There are a number methods that may be used to locate the **swat** binary file. The following methods may be helpful:

If **swat** is in your current operating system search path it will be easy to find it. You can ask what are the command-line options for **swat** as shown here:

```
frodo:~ # swat -?
```

```
Usage: swat [OPTION...]
  -a, --disable-authentication          Disable authentication (demo mode)

Help options:
  -?, --help                            Show this help message
  --usage                               Display brief usage message

Common samba options:
  -d, --debuglevel=DEBUGLEVEL           Set debug level
  -s, --configfile=CONFIGFILE           Use alternative configuration file
  -l, --log-basename=LOGFILEBASE        Basename for log/debug files
  -V, --version                         Print version
```

31.2.1.2 Locating the SWAT Support Files

Now that you have found that **swat** is in the search path, it is easy to identify where the file is located. Here is another simple way this may be done:

```
frodo:~ # whereis swat
swat: /usr/sbin/swat /usr/share/man/man8/swat.8.gz
```

If the above measures fail to locate the **swat** binary, another approach is needed. The following may be used:

```
frodo:/ # find / -name swat -print
/etc/xinetd.d/swat
/usr/sbin/swat
/usr/share/samba/swat
frodo:/ #
```

This list shows that there is a control file for **xinetd**, the internetwork super-daemon that is installed on this server. The location of the SWAT binary file is `/usr/sbin/swat`, and the support files for it are located under the directory `/usr/share/samba/swat`.

We must now check where **swat** expects to find its support files. This can be done as follows:

```
frodo:/ # strings /usr/sbin/swat | grep "/swat"
/swat/
...
/usr/share/samba/swat
frodo:/ #
```

The `/usr/share/samba/swat/` entry shown in this listing is the location of the support files. You should verify that the support files exist under this directory. A sample list is as shown:

```
jht@frodo:/> find /usr/share/samba/swat -print
/usr/share/samba/swat
/usr/share/samba/swat/help
/usr/share/samba/swat/lang
/usr/share/samba/swat/lang/ja
/usr/share/samba/swat/lang/ja/help
/usr/share/samba/swat/lang/ja/help/welcome.html
/usr/share/samba/swat/lang/ja/images
/usr/share/samba/swat/lang/ja/images/home.gif
...
/usr/share/samba/swat/lang/ja/include
/usr/share/samba/swat/lang/ja/include/header.nocss.html
...
/usr/share/samba/swat/lang/tr
/usr/share/samba/swat/lang/tr/help
/usr/share/samba/swat/lang/tr/help/welcome.html
/usr/share/samba/swat/lang/tr/images
/usr/share/samba/swat/lang/tr/images/home.gif
...
/usr/share/samba/swat/lang/tr/include
/usr/share/samba/swat/lang/tr/include/header.html
/usr/share/samba/swat/using_samba
...
/usr/share/samba/swat/images
/usr/share/samba/swat/images/home.gif
...
/usr/share/samba/swat/include
/usr/share/samba/swat/include/footer.html
/usr/share/samba/swat/include/header.html
jht@frodo:/>
```

If the files needed are not available it will be necessary to obtain and install them before SWAT can be used.

31.2.2 Enabling SWAT for Use

SWAT should be installed to run via the network super-daemon. Depending on which system your UNIX/Linux system has, you will have either an **inetd**- or **xinetd**-based system.

The nature and location of the network super-daemon varies with the operating system implementation. The control file (or files) can be located in the file `/etc/inetd.conf` or in the directory `/etc/[x]inet[d].d` or similar.

The control entry for the older style file might be:

```
# swat is the Samba Web Administration Tool
swat stream tcp nowait.400 root /usr/sbin/swat swat
```

A control file for the newer style xinetd could be:

```
# default: off
# description: SWAT is the Samba Web Admin Tool. Use swat \
#              to configure your Samba server. To use SWAT, \
#              connect to port 901 with your favorite web browser.
service swat
{
   port    = 901
   socket_type     = stream
   wait    = no
   only_from = localhost
   user    = root
   server  = /usr/sbin/swat
   log_on_failure  += USERID
   disable = yes
}
```

Both of the above examples assume that the **swat** binary has been located in the `/usr/sbin` directory. In addition to the above, SWAT will use a directory access point from which it will load its Help files as well as other control information. The default location for this on most Linux systems is in the directory `/usr/share/samba/swat`. The default location using Samba defaults will be `/usr/local/samba/swat`.

Access to SWAT will prompt for a logon. If you log onto SWAT as any non-root user, the only permission allowed is to view certain aspects of configuration as well as access to the password change facility. The buttons that will be exposed to the non-root user are: **HOME**, **STATUS**, **VIEW**, **PASSWORD**. The only page that allows change capability in this case is **PASSWORD**.

As long as you log onto SWAT as the user *root*, you should obtain full change and commit ability. The buttons that will be exposed include: **HOME**, **GLOBALS**, **SHARES**, **PRINTERS**, **WIZARD**, **STATUS**, **VIEW**, **PASSWORD**.

31.2.3 Securing SWAT through SSL

Many people have asked about how to setup SWAT with SSL to allow for secure remote administration of Samba. Here is a method that works, courtesy of Markus Krieger.

Modifications to the SWAT setup are as follows:

1. Install OpenSSL.
2. Generate certificate and private key.

```
root# /usr/bin/openssl req -new -x509 -days 365 -nodes -config \
   /usr/share/doc/packages/stunnel/stunnel.cnf \
   -out /etc/stunnel/stunnel.pem -keyout /etc/stunnel/stunnel.pem
```

3. Remove swat-entry from [x]inetd.

4. Start **stunnel**.

```
root# stunnel -p /etc/stunnel/stunnel.pem -d 901 \
    -l /usr/local/samba/bin/swat swat
```

Afterward, simply connect to swat by using the URL `https://myhost:901`, accept the certificate and the SSL connection is up.

31.2.4 Enabling SWAT Internationalization Support

SWAT can be configured to display its messages to match the settings of the language configurations of your Web browser. It will be passed to SWAT in the Accept-Language header of the HTTP request.

To enable this feature:

- Install the proper **msg** files from the Samba `source/po` directory into $LIBDIR.
- Set the correct locale value for *display charset*.
- Set your browser's language setting.

The name of msg file is same as the language ID sent by the browser. For example en means "English", ja means "Japanese", fr means "French.

If you do not like some of messages, or there are no **msg** files for your locale, you can create them simply by copying the **en.msg** files to the dirertory for "*your language ID.msg*" and filling in proper strings to each "*msgstr*". For example, in `it.msg`, the **msg** file for the Italian locale, just set:

```
msgid "Set Default"
msgstr "Imposta Default"
```

and so on. If you find a mistake or create a new **msg** file, please email it to us so we will include this in the next release of Samba.

Note that if you enable this feature and the *display charset* is not matched to your browser's setting, the SWAT display may be corrupted. In a future version of Samba, SWAT will always display messages with UTF-8 encoding. You will then not need to set this `smb.conf` file parameter.

31.3 Overview and Quick Tour

SWAT is a tools that many be used to configure Samba, or just to obtain useful links to important reference materials such as the contents of this book, as well as other documents that have been found useful for solving Windows networking problems.

31.3.1 The SWAT Home Page

The SWAT title page provides access to the latest Samba documentation. The manual page for each Samba component is accessible from this page, as are the Samba HOWTO-Collection (this document) as well as the O'Reilly book "*Using Samba.*"

Administrators who wish to validate their Samba configuration may obtain useful information from the man pages for the diagnostic utilities. These are available from the SWAT home page also. One diagnostic tool that is not mentioned on this page, but that is particularly useful is **ethereal**.[1]

WARNING

SWAT can be configured to run in *demo* mode. This is not recommended as it runs SWAT without authentication and with full administrative ability. Allows changes to `smb.conf` as well as general operation with root privileges. The option that creates this ability is the -a flag to swat. *Do not use this in a production environment.*

31.3.2 Global Settings

The **GLOBALS** button will expose a page that allows configuration of the global parameters in `smb.conf`. There are two levels of exposure of the parameters:

- **Basic** — exposes common configuration options.
- **Advanced** — exposes configuration options needed in more complex environments.

To switch to other than **Basic** editing ability, click on **Advanced**. You may also do this by clicking on the radio button, then click on the **Commit Changes** button.

After making any changes to configuration parameters, make sure that you click on the **Commit Changes** button before moving to another area, otherwise your changes will be lost.

[1] http://www.ethereal.com/

NOTE

SWAT has context-sensitive help. To find out what each parameter is for, simply click on the **Help** link to the left of the configuration parameter.

31.3.3 Share Settings

To effect a currently configured share, simply click on the pull down button between the **Choose Share** and the **Delete Share** buttons, select the share you wish to operate on, then to edit the settings click on the **Choose Share** button. To delete the share, simply press the **Delete Share** button.

To create a new share, next to the button labeled **Create Share** enter into the text field the name of the share to be created, then click on the **Create Share** button.

31.3.4 Printers Settings

To affect a currently configured printer, simply click on the pull down button between the **Choose Printer** and the **Delete Printer** buttons, select the printer you wish to operate on, then to edit the settings click on the **Choose Printer** button. To delete the share, simply press the **Delete Printer** button.

To create a new printer, next to the button labeled **Create Printer** enter into the text field the name of the share to be created, then click on the **Create Printer** button.

31.3.5 The SWAT Wizard

The purpose if the SWAT Wizard is to help the Microsoft-knowledgeable network administrator to configure Samba with a minimum of effort.

The Wizard page provides a tool for rewriting the `smb.conf` file in fully optimized format. This will also happen if you press the **Commit** button. The two differ since the **Rewrite** button ignores any changes that may have been made, while the **Commit** button causes all changes to be affected.

The **Edit** button permits the editing (setting) of the minimal set of options that may be necessary to create a working Samba server.

Finally, there are a limited set of options that will determine what type of server Samba will be configured for, whether it will be a WINS server, participate as a WINS client, or operate with no WINS support. By clicking one button, you can elect to expose (or not) user home directories.

31.3.6 The Status Page

The status page serves a limited purpose. First, it allows control of the Samba daemons. The key daemons that create the Samba server environment are: smbd, nmbd, winbindd.

The daemons may be controlled individually or as a total group. Additionally, you may set an automatic screen refresh timing. As MS Windows clients interact with Samba, new smbd processes will be continually spawned. The auto-refresh facility will allow you to track the changing conditions with minimal effort.

Lastly, the Status page may be used to terminate specific smbd client connections in order to free files that may be locked.

31.3.7 The View Page

This page allows the administrator to view the optimized `smb.conf` file and, if you are particularly masochistic, will permit you also to see all possible global configuration parameters and their settings.

31.3.8 The Password Change Page

The Password Change page is a popular tool that allows the creation, deletion, deactivation, and reactivation of MS Windows networking users on the local machine. Alternately, you can use this tool to change a local password for a user account.

When logged in as a non-root account, the user will have to provide the old password as well as the new password (twice). When logged in as *root*, only the new password is required.

One popular use for this tool is to change user passwords across a range of remote MS Windows servers.

Part V

Troubleshooting

Chapter 32

THE SAMBA CHECKLIST

32.1 Introduction

This file contains a list of tests you can perform to validate your Samba server. It also tells you what the likely cause of the problem is if it fails any one of these steps. If it passes all these tests, then it is probably working fine.

You should do all the tests, in the order shown. We have tried to carefully choose them so later tests only use capabilities verified in the earlier tests. However, do not stop at the first error as there have been some instances when continuing with the tests has helped to solve a problem.

If you send one of the Samba mailing lists an email saying, "*it does not work*" and you have not followed this test procedure, you should not be surprised if your email is ignored.

32.2 Assumptions

In all of the tests, it is assumed you have a Samba server called BIGSERVER and a PC called ACLIENT both in workgroup TESTGROUP.

The procedure is similar for other types of clients.

It is also assumed you know the name of an available share in your `smb.conf`. I will assume this share is called *tmp*. You can add a *tmp* share like this by adding the lines shown in Example 32.1.

Example 32.1. smb.conf with [tmp] share

```
[tmp]
comment = temporary files
path = /tmp
read only = yes
```

NOTE

These tests assume version 3.0.0 or later of the Samba suite. Some commands shown did not exist in earlier versions.

Please pay attention to the error messages you receive. If any error message reports that your server is being unfriendly, you should first check that your IP name resolution is correctly set up. Make sure your `/etc/resolv.conf` file points to name servers that really do exist.

Also, if you do not have DNS server access for name resolution, please check that the settings for your `smb.conf` file results in **dns proxy = no**. The best way to check this is with **testparm smb.conf**.

It is helpful to monitor the log files during testing by using the **tail -F log_file_name** in a separate terminal console (use ctrl-alt-F1 through F6 or multiple terminals in X). Relevant log files can be found (for default installations) in `/usr/local/samba/var`. Also, connection logs from machines can be found here or possibly in `/var/log/samba`, depending on how or if you specified logging in your `smb.conf` file.

If you make changes to your `smb.conf` file while going through these test, remember to restart smbd and nmbd.

32.3 The Tests

DIAGNOSING YOUR SAMBA SERVER

1. In the directory in which you store your `smb.conf` file, run the command **testparm smb.conf**. If it reports any errors, then your `smb.conf` configuration file is faulty.

   NOTE

   Your `smb.conf` file may be located in: `/etc/samba` or in `/usr/local/samba/lib`.

2. Run the command **ping BIGSERVER** from the PC and **ping ACLIENT** from the UNIX box. If you do not get a valid response, then your TCP/IP software is not correctly installed. You will need to start a "*dos prompt*" window on the PC to run ping. If you get a message saying "*host not found*" or similar, then your DNS software or `/etc/hosts` file is not correctly setup. It is possible to run Samba without DNS entries for the server and client, but it is assumed you do have correct entries for the remainder of these tests. Another reason why ping might fail is if your host is running firewall software. You will need to relax the rules to let in the workstation in

question, perhaps by allowing access from another subnet (on Linux this is done via the appropriate firewall maintenance commands **ipchains** or **iptables**).

NOTE

Modern Linux distributions install ipchains/iptables by default. This is a common problem that is often overlooked.

If you wish to check what firewall rules may be present in a system under test, simply run **iptables -L -v** or if *ipchains*-based firewall rules are in use, **ipchains -L -v**. Here is a sample listing from a system that has an external ethernet interface (eth1) on which Samba is not active, and an internal (private network) interface (eth0) on which Samba is active:

```
frodo:~ # iptables -L -v
Chain INPUT (policy DROP 98496 packets, 12M bytes)
 pkts bytes target     prot opt in     out     source     destination
 187K  109M ACCEPT     all  --  lo     any     anywhere   anywhere
 892K  125M ACCEPT     all  --  eth0   any     anywhere   anywhere
1399K 1380M ACCEPT     all  --  eth1   any     anywhere   anywhere  \
               state RELATED,ESTABLISHED

Chain FORWARD (policy DROP 0 packets, 0 bytes)
 pkts bytes target     prot opt in     out     source     destination
 978K 1177M ACCEPT     all  --  eth1   eth0    anywhere   anywhere \
               state RELATED,ESTABLISHED
 658K   40M ACCEPT     all  --  eth0   eth1    anywhere   anywhere
    0     0 LOG        all  --  any    any     anywhere   anywhere \
               LOG level warning

Chain OUTPUT (policy ACCEPT 2875K packets, 1508M bytes)
 pkts bytes target     prot opt in     out     source     destination

Chain reject_func (0 references)
 pkts bytes target     prot opt in     out     source     destinat
```

3. Run the command: **smbclient -L BIGSERVER** on the UNIX box. You should get back a list of available shares. If you get an error message containing the string "*Bad password*", then you probably have either an incorrect ***hosts allow***, ***hosts deny*** or ***valid users*** line in your `smb.conf`, or your guest account is not valid. Check what your guest account is using testparm and temporarily remove any ***hosts allow***, ***hosts deny***, ***valid users*** or ***invalid users*** lines. If you get a message "*connection refused*" response, then the **smbd** server may not be running. If you installed it in `inetd.conf`, then you probably edited that file incorrectly. If you installed it as a

daemon, then check that it is running, and check that the netbios-ssn port is in a LISTEN state using **netstat -a**.

> **NOTE**
>
> 
>
> Some UNIX/Linux systems use **xinetd** in place of **inetd**. Check your system documentation for the location of the control files for your particular system implementation of the network super daemon.

If you get a message saying "*session request failed*", the server refused the connection. If it says "*Your server software is being unfriendly*", then it's probably because you have invalid command line parameters to smbd, or a similar fatal problem with the initial startup of smbd. Also check your config file (`smb.conf`) for syntax errors with testparm and that the various directories where Samba keeps its log and lock files exist. There are a number of reasons for which smbd may refuse or decline a session request. The most common of these involve one or more of the `smb.conf` file entries as shown in Example 32.2.

Example 32.2. Configuration for only allowing connections from a certain subnet

```
[globals]
...
hosts deny = ALL
hosts allow = xxx.xxx.xxx.xxx/yy
interfaces = eth0
bind interfaces only = Yes
...
```

In the above, no allowance has been made for any session requests that will automatically translate to the loopback adapter address 127.0.0.1. To solve this problem, change these lines as shown in Example 32.3.

Example 32.3. Configuration for allowing connections from a certain subnet and localhost

```
[globals]
...
hosts deny = ALL
hosts allow = xxx.xxx.xxx.xxx/yy 127.
interfaces = eth0 lo
...
```

Another common cause of these two errors is having something already running on port `139`, such as Samba (smbd is running from inetd already) or something like Digital's Pathworks. Check your `inetd.conf` file before trying to start smbd as a daemon — it can avoid a lot of frustration! And yet another possible cause for failure of this test is when the subnet mask and/or broadcast address settings are incorrect. Please check that the network interface IP Address/Broadcast Address/Subnet Mask settings are correct and that Samba has correctly noted these in the `log.nmbd` file.

4. Run the command: **nmblookup -B BIGSERVER __SAMBA__**. You should get back the IP address of your Samba server. If you do not, then nmbd is incorrectly installed. Check your `inetd.conf` if you run it from there, or that the daemon is running and listening to udp port 137. One common problem is that many inetd implementations can't take many parameters on the command line. If this is the case, then create a one-line script that contains the right parameters and run that from inetd.

5. Run the command: **nmblookup -B ACLIENT '*'** You should get the PC's IP address back. If you do not then the client software on the PC isn't installed correctly, or isn't started, or you got the name of the PC wrong. If ACLIENT does not resolve via DNS then use the IP address of the client in the above test.

6. Run the command: **nmblookup -d 2 '*'** This time we are trying the same as the previous test but are trying it via a broadcast to the default broadcast address. A number of NetBIOS/TCP/IP hosts on the network should respond, although Samba may not catch all of the responses in the short time it listens. You should see the "*got a positive name query response*" messages from several hosts. If this does not give a similar result to the previous test, then nmblookup isn't correctly getting your broadcast address through its automatic mechanism. In this case you should experiment with the ***`interfaces`*** option in `smb.conf` to manually configure your IP address, broadcast and netmask. If your PC and server aren't on the same subnet, then you will need to use the `-B` option to set the broadcast address to that of the PCs subnet. This test will probably fail if your subnet mask and broadcast address are not correct. (Refer to TEST 3 notes above).

7. Run the command: **smbclient //BIGSERVER/TMP**. You should then be prompted for a password. You should use the password of the account with which you are logged into the UNIX box. If you want to test with another account, then add the `-U accountname` option to the end of the command line. For example, **smbclient //bigserver/tmp -Ujohndoe**.

> NOTE
>
> 
>
> It is possible to specify the password along with the username as follows: **smbclient //bigserver/tmp -Ujohndoe%secret**.

Once you enter the password, you should get the `smb>` prompt. If you do not, then look at the error message. If it says "*invalid network name*", then the service *`tmp`* is not correctly setup in your `smb.conf`. If it says "*bad password*", then the likely causes are:

(a) You have shadow passwords (or some other password system) but didn't compile in support for them in smbd.

(b) Your ***valid users*** configuration is incorrect.

(c) You have a mixed case password and you haven't enabled the ***password level*** option at a high enough level.

(d) The ***path*** line in `smb.conf` is incorrect. Check it with testparm.

(e) You enabled password encryption but didn't map UNIX to Samba users. Run: **smbpasswd -a username**

Once connected, you should be able to use the commands **dir**, **get**, **put** and so on. Type **help command** for instructions. You should especially check that the amount of free disk space shown is correct when you type **dir**.

8. On the PC, type the command **net view \\BIGSERVER**. You will need to do this from within a dos prompt window. You should get back a list of shares available on the server. If you get a message "*network name not found*" or similar error, then netbios name resolution is not working. This is usually caused by a problem in **nmbd**. To overcome it, you could do one of the following (you only need to choose one of them):

   (a) Fixup the nmbd installation.

   (b) Add the IP address of BIGSERVER to the **wins server** box in the advanced TCP/IP setup on the PC.

   (c) Enable Windows name resolution via DNS in the advanced section of the TCP/IP setup.

   (d) Add BIGSERVER to your lmhosts file on the PC.

   If you get a message "*invalid network name*" or "*bad password error*", then apply the same fixes as for the **smbclient -L** test above. In particular, make sure your **hosts allow** line is correct (see the man pages). Also, do not overlook that fact that when the workstation requests the connection to the Samba server, it will attempt to connect using the name with which you logged onto your Windows machine. You need to make sure that an account exists on your Samba server with that exact same name and password. If you get a message "*specified computer is not receiving requests*" or similar, it probably means that the host is not contactable via TCP services. Check to see if the host is running TCP wrappers, and if so add an entry in the `hosts.allow` file for your client (or subnet, and so on.)

9. Run the command **net use x: \\BIGSERVER\TMP**. You should be prompted for a password, then you should get a `command completed successfully` message. If not, then your PC software is incorrectly installed or your `smb.conf` is incorrect. Make sure your ***hosts allow*** and other config lines in `smb.conf` are correct. It's also possible that the server can't work out what user name to connect you as. To see if this is the problem, add the line ***user*** = username to the *[tmp]* section of `smb.conf` where *username* is the username corresponding to the password you typed. If

you find this fixes things, you may need the username mapping option. It might also be the case that your client only sends encrypted passwords and you have *encrypt passwords* = no in `smb.conf`. Change this to "yes" to fix this.

10. Run the command **nmblookup -M** *testgroup* where *testgroup* is the name of the workgroup that your Samba server and Windows PCs belong to. You should get back the IP address of the master browser for that workgroup. If you do not, then the election process has failed. Wait a minute to see if it is just being slow, then try again. If it still fails after that, then look at the browsing options you have set in `smb.conf`. Make sure you have *preferred master* = yes to ensure that an election is held at startup.

11. >From file manager, try to browse the server. Your Samba server should appear in the browse list of your local workgroup (or the one you specified in `smb.conf`). You should be able to double click on the name of the server and get a list of shares. If you get the error message "*invalid password*", you are probably running Windows NT and it is refusing to browse a server that has no encrypted password capability and is in User Level Security mode. In this case, either set *security* = server and *password server* = Windows_NT_Machine in your `smb.conf` file, or make sure *encrypt passwords* is set to "*yes*".

Chapter 33

ANALYZING AND SOLVING SAMBA PROBLEMS

There are many sources of information available in the form of mailing lists, RFCs and documentation. The documentation that comes with the Samba distribution contains good explanations of general SMB topics such as browsing.

33.1 Diagnostics Tools

With SMB networking, it is often not immediately clear what the cause is of a certain problem. Samba itself provides rather useful information, but in some cases you might have to fall back to using a *sniffer*. A sniffer is a program that listens on your LAN, analyzes the data sent on it and displays it on the screen.

33.1.1 Debugging with Samba Itself

One of the best diagnostic tools for debugging problems is Samba itself. You can use the `-d option` for both smbd and nmbd to specify the ***debug level*** at which to run. See the man pages for **smbd, nmbd** and `smb.conf` for more information regarding debugging options. The debug level can range from 1 (the default) to 10 (100 for debugging passwords).

Another helpful method of debugging is to compile Samba using the **gcc -g** flag. This will include debug information in the binaries and allow you to attach gdb to the running **smbd/nmbd** process. To attach **gdb** to an **smbd** process for an NT workstation, first get the workstation to make the connection. Pressing ctrl-alt-delete and going down to the domain box is sufficient (at least, the first time you join the domain) to generate a *LsaEnumTrustedDomains*. Thereafter, the workstation maintains an open connection and there will be an smbd process running (assuming that you haven't set a really short smbd idle timeout). So, in between pressing **ctrl-alt-delete** and actually typing in your password, you can attach **gdb** and continue.

Some useful Samba commands worth investigating are:

```
$ testparm | more
```

```
$ smbclient -L //{netbios name of server}
```

33.1.2 Tcpdump

Tcpdump[1] was the first UNIX sniffer with SMB support. It is a command-line utility and now, its SMB support is somewhat lagging that of **ethereal** and **tethereal**.

33.1.3 Ethereal

Ethereal[2] is a graphical sniffer, available for both UNIX (Gtk) and Windows. Ethereal's SMB support is quite good.

For details on the use of **ethereal**, read the well-written Ethereal User Guide.

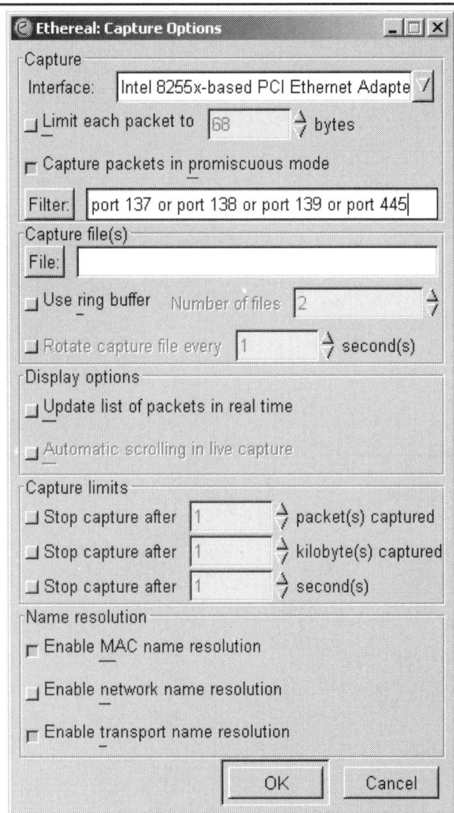

Figure 33.1. Starting a capture.

Listen for data on ports 137, 138, 139, and 445. For example, use the filter `port 137, port 138, port 139, or port 445` as seen in Figure 33.1.

[1] http://www.tcpdump.org/
[2] http://www.ethereal.com/

A console version of ethereal is available as well and is called **tethereal**.

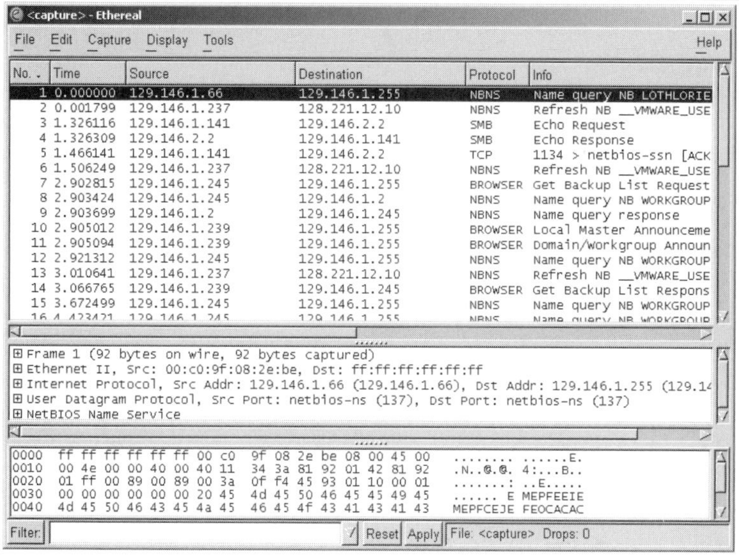

Figure 33.2. Main ethereal data window.

33.1.4 The Windows Network Monitor

For tracing things on Microsoft Windows NT, Network Monitor (aka Netmon) is available on Microsoft Developer Network CDs, the Windows NT Server install CD and the SMS CDs. The version of Netmon that ships with SMS allows for dumping packets between any two computers (i.e., placing the network interface in promiscuous mode). The version on the NT Server install CD will only allow monitoring of network traffic directed to the local NT box and broadcasts on the local subnet. Be aware that Ethereal can read and write Netmon formatted files.

33.1.4.1 Installing Network Monitor on an NT Workstation

Installing Netmon on an NT workstation requires a couple of steps. The following are instructions for installing Netmon V4.00.349, which comes with Microsoft Windows NT Server 4.0, on Microsoft Windows NT Workstation 4.0. The process should be similar for other versions of Windows NT version of Netmon. You will need both the Microsoft Windows NT Server 4.0 Install CD and the Workstation 4.0 Install CD.

Initially you will need to install Network Monitor Tools and Agent on the NT Server to do this:

- Go to **Start** -> **Settings** -> **Control Panel** -> **Network** -> **Services** -> **Add**.
- Select the **Network Monitor Tools and Agent** and click on **OK**.
- Click on **OK** on the Network Control Panel.

- Insert the Windows NT Server 4.0 install CD when prompted.

At this point, the Netmon files should exist in `%SYSTEMROOT%\System32\netmon\*.*`. Two subdirectories exist as well, `parsers\` which contains the necessary DLLs for parsing the Netmon packet dump, and `captures\`.

To install the Netmon tools on an NT Workstation, you will first need to install the Network Monitor Agent from the Workstation install CD.

- Go to **Start** -> **Settings** -> **Control Panel** -> **Network** -> **Services** -> **Add**.
- Select the **Network Monitor Agent**, click on **OK**.
- Click on **OK** in the Network Control Panel.
- Insert the Windows NT Workstation 4.0 install CD when prompted.

Now copy the files from the NT Server in `%SYSTEMROOT%\System32\netmon` to `%SYSTEMROOT%\System32\netmon` on the Workstation and set permissions as you deem appropriate for your site. You will need administrative rights on the NT box to run Netmon.

33.1.4.2 Installing Network Monitor on Windows 9x/Me

To install Netmon on Windows 9x/Me, install the Network Monitor Agent from the Windows 9x/Me CD (`\admin\nettools\netmon`). There is a readme file located with the Netmon driver files on the CD if you need information on how to do this. Copy the files from a working Netmon installation.

33.2 Useful URLs

- See how Scott Merrill simulates a BDC behavior at http://www.skippy.net/linux/smb-howto.html.
- FTP site for older SMB specs: ftp://ftp.microsoft.com/developr/drg/CIFS/

33.3 Getting Mailing List Help

There are a number of Samba-related mailing lists. Go to `http://samba.org`, click on your nearest mirror and then click on **Support** and next click on **Samba-related mailing lists**.

For questions relating to Samba TNG, go to http://www.samba-tng.org/. It has been requested that you do not post questions about Samba-TNG to the main-stream Samba lists.

If you do post a message to one of the lists, please observe the following guidelines :

- Always remember that the developers are volunteers, they are not paid and they never guarantee to produce a particular feature at a particular time. Any timelines are "*best guess*" and nothing more.

- Always mention what version of Samba you are using and what operating system it's running under. You should list the relevant sections of your `smb.conf` file, at least the options in *[global]* that affect PDC support.
- In addition to the version, if you obtained Samba via CVS, mention the date when you last checked it out.
- Try and make your questions clear and brief. Lots of long, convoluted questions get deleted before they are completely read! Do not post HTML encoded messages. Most people on mailing lists simply delete them.
- If you run one of those nifty "*I'm on holidays*" things when you are away, make sure its configured to not answer mailing list traffic. Auto-responses to mailing lists really irritate the thousands of people who end up having to deal with such bad netiquet bahavior.
- Don't cross post. Work out which is the best list to post to and see what happens. Do not post to both samba-ntdom and samba-technical. Many people active on the lists subscribe to more than one list and get annoyed to see the same message two or more times. Often someone will see a message and thinking it would be better dealt with on another list, will forward it on for you.
- You might include *partial* log files written at a debug level set to as much as 20. Please do not send the entire log but just enough to give the context of the error messages.
- If you have a complete Netmon trace (from the opening of the pipe to the error), you can send the *.CAP file as well.
- Please think carefully before attaching a document to an email. Consider pasting the relevant parts into the body of the message. The Samba mailing lists go to a huge number of people. Do they all need a copy of your `smb.conf` in their attach directory?

33.4 How to Get Off the Mailing Lists

To have your name removed from a Samba mailing list, go to the same place where you went to subscribe to it. Go to http://lists.samba.org, click on your nearest mirror, click on **Support** and then click on**Samba related mailing lists**.

Please do not post messages to the list asking to be removed. You will only be referred to the above address (unless that process failed in some way).

Chapter 34

REPORTING BUGS

34.1 Introduction

Please report bugs using Samba's Bugzilla[1] facilities and take the time to read this file before you submit a bug report. Also, check to see if it has changed between releases, as we may be changing the bug reporting mechanism at some point.

Please do as much as you can yourself to help track down the bug. Samba is maintained by a dedicated group of people who volunteer their time, skills and efforts. We receive far more mail than we can possibly answer, so you have a much higher chance of a response and a fix if you send us a "*developer friendly*" bug report that lets us fix it fast.

Do not assume that if you post the bug to the comp.protocols.smb newsgroup or the mailing list that we will read it. If you suspect that your problem is not a bug but a configuration problem, it is better to send it to the Samba mailing list, as there are thousands of other users on that list who may be able to help you.

You may also like to look though the recent mailing list archives, which are conveniently accessible on the Samba Web pages at `http://samba.org/samba/`.

34.2 General Information

Before submitting a bug report, check your config for silly errors. Look in your log files for obvious messages that tell you've misconfigured something. Run testparm to check your config file for correct syntax.

Have you looked through Chapter 32, *The Samba Checklist*? This is extremely important.

If you include part of a log file with your bug report, then be sure to annotate it with exactly what you were doing on the client at the time and exactly what the results were.

34.3 Debug Levels

If the bug has anything to do with Samba behaving incorrectly as a server (like refusing to open a file), then the log files will probably be quite useful. Depending on the problem, a

[1]https://bugzilla.samba.org/

log level of between 3 and 10 showing the problem may be appropriate. A higher level gives more detail, but may use too much disk space.

To set the debug level, use the *log level* in your `smb.conf`. You may also find it useful to set the log level higher for just one machine and keep separate logs for each machine. To do this, add the following lines to your main `smb.conf` file:

```
log level = 10
log file = /usr/local/samba/lib/log.%m
include = /usr/local/samba/lib/smb.conf.%m
```

and create a file `/usr/local/samba/lib/smb.conf.`*machine* where *machine* is the name of the client you wish to debug. In that file put any `smb.conf` commands you want, for example *log level* may be useful. This also allows you to experiment with different security systems, protocol levels and so on, on just one machine.

The `smb.conf` entry *log level* is synonymous with the parameter *debuglevel* that has been used in older versions of Samba and is being retained for backward compatibility of `smb.conf` files.

As the *log level* value is increased, you will record a significantly greater level of debugging information. For most debugging operations, you may not need a setting higher than `3`. Nearly all bugs can be tracked at a setting of `10`, but be prepared for a large volume of log data.

34.4 Internal Errors

If you get the message "*INTERNAL ERROR*" in your log files, it means that Samba got an unexpected signal while running. It is probably a segmentation fault and almost certainly means a bug in Samba (unless you have faulty hardware or system software).

If the message came from smbd, it will probably be accompanied by a message that details the last SMB message received by smbd. This information is often useful in tracking down the problem so please include it in your bug report.

You should also detail how to reproduce the problem, if possible. Please make this reasonably detailed.

You may also find that a core file appeared in a `corefiles` subdirectory of the directory where you keep your Samba log files. This file is the most useful tool for tracking down the bug. To use it, you do this:

```
$ gdb smbd core
```

adding appropriate paths to smbd and core so gdb can find them. If you do not have gdb, try `dbx`. Then within the debugger, use the command **where** to give a stack trace of where the problem occurred. Include this in your report.

If you know any assembly language, do a **disass** of the routine where the problem occurred (if its in a library routine, then disassemble the routine that called it) and try to work out exactly where the problem is by looking at the surrounding code. Even if you do not know assembly, including this information in the bug report can be useful.

34.5 Attaching to a Running Process

Unfortunately, some UNIXes (in particular some recent Linux kernels) refuse to dump a core file if the task has changed uid (which smbd does often). To debug with this sort of system, you could try to attach to the running process using `gdb smbd` *`PID`* where you get *`PID`* from smbstatus. Then use **`c`** to continue and try to cause the core dump using the client. The debugger should catch the fault and tell you where it occurred.

34.6 Patches

The best sort of bug report is one that includes a fix! If you send us patches, please use `diff -u` format if your version of diff supports it, otherwise use `diff -c4`. Make sure you do the diff against a clean version of the source and let me know exactly what version you used.

Part VI

Appendixes

Chapter 35

HOW TO COMPILE SAMBA

You can obtain the Samba source from the Samba Website.[1] To obtain a development version, you can download Samba from CVS or using **rsync**.

35.1 Access Samba Source Code via CVS

35.1.1 Introduction

Samba is developed in an open environment. Developers use Concurrent Versioning System (CVS) to "*checkin*" (also known as "*commit*") new source code. Samba's various CVS branches can be accessed via anonymous CVS using the instructions detailed in this chapter.

This chapter is a modified version of the instructions found at `http://samba.org/samba/cvs.html`

35.1.2 CVS Access to samba.org

The machine samba.org runs a publicly accessible CVS repository for access to the source code of several packages, including Samba, rsync, distcc, ccache, and jitterbug. There are two main ways of accessing the CVS server on this host:

35.1.2.1 Access via CVSweb

You can access the source code via your favorite WWW browser. This allows you to access the contents of individual files in the repository and also to look at the revision history and commit logs of individual files. You can also ask for a diff listing between any two versions on the repository.

Use the URL: `http://samba.org/cgi-bin/CVSweb`

[1] http://samba.org/

35.1.2.2 Access via CVS

You can also access the source code via a normal CVS client. This gives you much more control over what you can do with the repository and allows you to checkout whole source trees and keep them up-to-date via normal CVS commands. This is the preferred method of access if you are a developer and not just a casual browser.

To download the latest CVS source code, point your browser at the URL : `http://www.cyclic.com/`. and click on the "*How to get CVS*" link. CVS is free software under the GNU GPL (as is Samba). Note that there are several graphical CVS clients that provide a graphical interface to the sometimes mundane CVS commands. Links to theses clients are also available from the Cyclic Web site.

To gain access via anonymous CVS, use the following steps. For this example it is assumed that you want a copy of the Samba source code. For the other source code repositories on this system just substitute the correct package name.

RETRIEVING SAMBA USING CVS

1. Install a recent copy of CVS. All you really need is a copy of the CVS client binary.
2. Run the command: `cvs -d :pserver:cvs@samba.org:/cvsroot login`
3. When it asks you for a password, type `cvs`.
4. Run the command `cvs -d :pserver:CVS@samba.org:/cvsroot co samba`. This will create a directory called `samba` containing the latest Samba source code (i.e., the HEAD tagged CVS branch). This currently corresponds to the 3.0 development tree. CVS branches other then HEAD can be obtained by using the `-r` and defining a tag name. A list of branch tag names can be found on the "*Development*" page of the Samba Web site. A common request is to obtain the latest 3.0 release code. This could be done by using the following command: `cvs -d :pserver:cvs@samba.org:/cvsroot co -r SAMBA_3_0 samba`.
5. Whenever you want to merge in the latest code changes, use the following command from within the Samba directory: `cvs update -d -P`

35.2 Accessing the Samba Sources via rsync and ftp

pserver.samba.org also exports unpacked copies of most parts of the CVS tree at `ftp://pserver.samba.org/pub/unpacked` and also via anonymous rsync at `rsync://pserver.samba.org/ftp/unpacked/`. I recommend using rsync rather than ftp. See the rsync homepage for more info on rsync.

The disadvantage of the unpacked trees is that they do not support automatic merging of local changes like CVS does. **rsync** access is most convenient for an initial install.

35.3 Verifying Samba's PGP Signature

It is strongly recommended that you verify the PGP signature for any source file before installing it. Even if you're not downloading from a mirror site, verifying PGP signatures should be a standard reflex. Many people today use the GNU GPG toolset in place of PGP. GPG can substitute for PGP.

With that said, go ahead and download the following files:

```
$ wget http://us1.samba.org/samba/ftp/samba-2.2.8a.tar.asc
$ wget http://us1.samba.org/samba/ftp/samba-pubkey.asc
```

The first file is the PGP signature for the Samba source file; the other is the Samba public PGP key itself. Import the public PGP key with:

```
$ gpg --import samba-pubkey.asc
```

and verify the Samba source code integrity with:

```
$ gzip -d samba-2.2.8a.tar.gz
$ gpg --verify samba-2.2.8a.tar.asc
```

If you receive a message like, "*Good signature from Samba Distribution Verification Key...*" then all is well. The warnings about trust relationships can be ignored. An example of what you would not want to see would be:

```
gpg: BAD signature from Samba Distribution Verification Key
```

35.4 Building the Binaries

To build the binaries, first run the program `./configure` in the source directory. This should automatically configure Samba for your operating system. If you have unusual needs, then you may wish to run

```
root# ./configure --help
```

first to see what special options you can enable. Now execute `./configure` with any arguments it might need:

```
root# ./configure [... arguments ...]
```

Executing

```
root# make
```

will create the binaries. Once it is successfully compiled you can use

```
root# make install
```

to install the binaries and manual pages. You can separately install the binaries and/or man pages using

```
root# make installbin
```

and

```
root# make installman
```

Note that if you are upgrading from a previous version of Samba you might like to know that the old versions of the binaries will be renamed with an "*.old*" extension. You can go back to the previous version with

```
root# make revert
```

if you find this version a disaster!

35.4.1 Compiling Samba with Active Directory Support

In order to compile Samba with ADS support, you need to have installed on your system:

- The MIT or Heimdal kerberos development libraries (either install from the sources or use a package).
- The OpenLDAP development libraries.

If your kerberos libraries are in a non-standard location, then remember to add the configure option `--with-krb5=`*`DIR`*.

After you run configure, make sure that `include/config.h` it generates contain lines like this:

```
#define HAVE_KRB5 1
#define HAVE_LDAP 1
```

If it does not, configure did not find your KRB5 libraries or your LDAP libraries. Look in `config.log` to figure out why and fix it.

35.4.1.1 Installing the Required Packages for Debian

On Debian, you need to install the following packages:

- libkrb5-dev
- krb5-user

35.4.1.2 Installing the Required Packages for Red Hat Linux

On Red Hat Linux, this means you should have at least:

- krb5-workstation (for kinit)
- krb5-libs (for linking with)
- krb5-devel (because you are compiling from source)

in addition to the standard development environment.

If these files are not installed on your system, you should check the installation CDs to find which has them and install the files using your tool of choice. If in doubt about what tool to use, refer to the Red Hat Linux documentation.

35.4.1.3 SuSE Linux Package Requirements

SuSE Linux installs Heimdal packages that may be required to allow you to build binary packages. You should verify that the development libraries have been installed on your system.

SuSE Linux Samba RPMs support Kerberos. Please refer to the documentation for your SuSE Linux system for information regading SuSE Linux specific configuration. Additionally, SuSE are very active in the maintenance of Samba packages that provide the maximum capabilities that are available. You should consider using SuSE provided packages where they are available.

35.5 Starting the smbd and nmbd

You must choose to start smbd and nmbd either as daemons or from inetd. Don't try to do both! Either you can put them in `inetd.conf` and have them started on demand by inetd or xinetd, or you can start them as daemons either from the command line or in `/etc/rc.local`. See the man pages for details on the command line options. Take particular care to read the bit about what user you need to have to start Samba. In many cases, you must be root.

The main advantage of starting smbd and nmbd using the recommended daemon method is that they will respond slightly more quickly to an initial connection request.

35.5.1 Starting from inetd.conf

NOTE

The following will be different if you use NIS, NIS+ or LDAP to distribute services maps.

Look at your `/etc/services`. What is defined at port 139/tcp? If nothing is defined, then add a line like this:

```
netbios-ssn     139/tcp
```

Similarly for 137/udp, you should have an entry like:

```
netbios-ns  137/udp
```

Next, edit your `/etc/inetd.conf` and add two lines like this:

```
netbios-ssn stream tcp nowait root /usr/local/samba/bin/smbd smbd
netbios-ns dgram udp wait root /usr/local/samba/bin/nmbd nmbd
```

The exact syntax of `/etc/inetd.conf` varies between UNIXes. Look at the other entries in inetd.conf for a guide.

Some distributions use xinetd instead of inetd. Consult the xinetd manual for configuration information.

NOTE

Some UNIXes already have entries like netbios_ns (note the underscore) in `/etc/services`. You must edit `/etc/services` or `/etc/inetd.conf` to make them consistent.

NOTE

On many systems you may need to use the *interfaces* option in `smb.conf` to specify the IP address and netmask of your interfaces. Run ifconfig as root if you do not know what the broadcast is for your net. nmbd tries to determine it at run time, but fails on some UNIXes.

WARNING

Many UNIXes only accept around five parameters on the command line in `inetd.conf`. This means you shouldn't use spaces between the options and arguments, or you should use a script and start the script from **inetd**.

Restart inetd, perhaps just send it a HUP.

```
root# killall -HUP inetd
```

35.5.2 Alternative: Starting smbd as a Daemon

To start the server as a daemon, you should create a script something like this one, perhaps calling it `startsmb`.

```
#!/bin/sh
/usr/local/samba/bin/smbd -D
/usr/local/samba/bin/nmbd -D
```

Make it executable with **chmod +x startsmb**

You can then run **startsmb** by hand or execute it from `/etc/rc.local`.

To kill it, send a kill signal to the processes nmbd and smbd.

> NOTE
>
> 
>
> If you use the SVR4 style init system, you may like to look at the `examples/svr4-startup` script to make Samba fit into that system.

Chapter 36

PORTABILITY

Samba works on a wide range of platforms but the interface all the platforms provide is not always compatible. This chapter contains platform-specific information about compiling and using Samba.

36.1 HPUX

HP's implementation of supplementary groups is non-standard (for historical reasons). There are two group files, `/etc/group` and `/etc/logingroup`; the system maps UIDs to numbers using the former, but initgroups() reads the latter. Most system admins who know the ropes symlink `/etc/group` to `/etc/logingroup` (hard link does not work for reasons too obtuse to go into here). initgroups() will complain if one of the groups you're in in `/etc/logingroup` has what it considers to be an invalid ID, which means outside the range `[0..UID_MAX]`, where `UID_MAX` is (I think) 60000 currently on HP-UX. This precludes -2 and 65534, the usual `nobody` GIDs.

If you encounter this problem, make sure the programs that are failing to initgroups() are run as users, not in any groups with GIDs outside the allowed range.

This is documented in the HP manual pages under setgroups(2) and passwd(4).

On HP-UX you must use gcc or the HP ANSI compiler. The free compiler that comes with HP-UX is not ANSI compliant and cannot compile Samba.

36.2 SCO UNIX

If you run an old version of SCO UNIX, you may need to get important TCP/IP patches for Samba to work correctly. Without the patch, you may encounter corrupt data transfers using Samba.

The patch you need is UOD385 Connection Drivers SLS. It is available from SCO (ftp.sco.com, directory SLS, files uod385a.Z and uod385a.ltr.Z).

The information provided here refers to an old version of SCO UNIX. If you require binaries for more recent SCO UNIX products, please contact SCO to obtain packages that are ready to install. You should also verify with SCO that your platform is up-to-date for the binary packages you will install. This is important if you wish to avoid data corruption problems with your installation. To build Samba for SCO UNIX products may require significant patching of Samba source code. It is much easier to obtain binary packages directly from SCO.

36.3 DNIX

DNIX has a problem with seteuid() and setegid(). These routines are needed for Samba to work correctly, but they were left out of the DNIX C library for some reason.

For this reason Samba by default defines the macro NO_EID in the DNIX section of includes.h. This works around the problem in a limited way, but it is far from ideal, and some things still will not work right.

To fix the problem properly, you need to assemble the following two functions and then either add them to your C library or link them into Samba. Put the following in the file `setegid.s`:

```
        .globl  _setegid
_setegid:
        moveq   #47,d0
        movl    #100,a0
        moveq   #1,d1
        movl    4(sp),a1
        trap    #9
        bccs    1$
        jmp     cerror
1$:
        clrl    d0
        rts
```

Put this in the file `seteuid.s`:

```
        .globl  _seteuid
_seteuid:
        moveq   #47,d0
        movl    #100,a0
        moveq   #0,d1
        movl    4(sp),a1
        trap    #9
        bccs    1$
        jmp     cerror
1$:
        clrl    d0
        rts
```

After creating the above files, you then assemble them using

```
$ as seteuid.s
$ as setegid.s
```

that should produce the files `seteuid.o` and `setegid.o`

Then you need to add these to the LIBSM line in the DNIX section of the Samba Makefile. Your LIBSM line will then look something like this:

```
LIBSM = setegid.o seteuid.o -ln
```

You should then remove the line:

```
#define NO_EID
```

from the DNIX section of `includes.h`.

36.4 Red Hat Linux

By default during installation, some versions of Red Hat Linux add an entry to `/etc/hosts` as follows:

```
    127.0.0.1 loopback "hostname"."domainname"
```

This causes Samba to loop back onto the loopback interface. The result is that Samba fails to communicate correctly with the world and therefore may fail to correctly negotiate who is the master browse list holder and who is the master browser.

Corrective Action: Delete the entry after the word "loopback" in the line starting 127.0.0.1.

36.5 AIX

36.5.1 Sequential Read Ahead

Disabling Sequential Read Ahead using `vmtune -r 0` improves Samba performance significantly.

36.6 Solaris

36.6.1 Locking Improvements

Some people have been experiencing problems with F_SETLKW64/fcntl when running Samba on Solaris. The built-in file locking mechanism was not scalable. Performance would degrade to the point where processes would get into loops of trying to lock a file. It would try a lock, then fail, then try again. The lock attempt was failing before the grant was occurring. So the visible manifestation of this would be a handful of processes stealing all of the CPU, and when they were trussed they would be stuck if F_SETLKW64 loops.

Sun released patches for Solaris 2.6, 8, and 9. The patch for Solaris 7 has not been released yet.

The patch revision for 2.6 is 105181-34, for 8 is 108528-19 and for 9 is 112233-04.

After the install of these patches, it is recommended to reconfigure and rebuild Samba.

Thanks to Joe Meslovich for reporting this.

36.6.2 Winbind on Solaris 9

Nsswitch on Solaris 9 refuses to use the Winbind NSS module. This behavior is fixed by Sun in patch 113476-05, which as of March 2003, is not in any roll-up packages.

Chapter 37

SAMBA AND OTHER CIFS CLIENTS

This chapter contains client-specific information.

37.1 Macintosh Clients

Yes. Thursby[1] has a CIFS Client/Server called DAVE.[2] They test it against Windows 95, Windows NT /200x/XP and Samba for compatibility issues. At the time of this writing, DAVE was at version 4.1. Please refer to Thursby's Web site for more information regarding this product.

Alternatives — There are two free implementations of AppleTalk for several kinds of UNIX machines and several more commercial ones. These products allow you to run file services and print services natively to Macintosh users, with no additional support required on the Macintosh. The two free implementations are Netatalk,[3] and CAP.[4] What Samba offers MS Windows users, these packages offer to Macs. For more info on these packages, Samba, and Linux (and other UNIX-based systems), see http://www.eats.com/linux_mac_win.html.

Newer versions of the Macintosh (Mac OS X) include Samba.

37.2 OS2 Client

37.2.1 Configuring OS/2 Warp Connect or OS/2 Warp 4

Basically, you need three components:

- The File and Print Client (IBM Peer)
- TCP/IP (Internet support)
- The "*NetBIOS over TCP/IP*" driver (TCPBEUI)

[1] http://www.thursby.com/
[2] http://www.thursby.com/products/dave.html
[3] http://www.umich.edu/~rsug/netatalk/
[4] http://www.cs.mu.oz.au/appletalk/atalk.html

Installing the first two together with the base operating system on a blank system is explained in the Warp manual. If Warp has already been installed, but you now want to install the networking support, use the "*Selective Install for Networking*" object in the "*System Setup*" folder.

Adding the "*NetBIOS over TCP/IP*" driver is not described in the manual and just barely in the online documentation. Start **MPTS.EXE**, click on **OK**, click on **Configure LAPS** and click on **IBM OS/2 NETBIOS OVER TCP/IP** in **Protocols**. This line is then moved to **Current Configuration**. Select that line, click on **Change number** and increase it from 0 to 1. Save this configuration.

If the Samba server is not on your local subnet, you can optionally add IP names and addresses of these servers to the **Names List**, or specify a WINS server (NetBIOS Nameserver in IBM and RFC terminology). For Warp Connect, you may need to download an update for `IBM Peer` to bring it on the same level as Warp 4. See the Web page mentioned above.

37.2.2 Configuring Other Versions of OS/2

This sections deals with configuring OS/2 Warp 3 (not Connect), OS/2 1.2, 1.3 or 2.x.

You can use the free Microsoft LAN Manager 2.2c Client for OS/2 that is available from ftp://ftp.microsoft.com/BusSys/Clients/LANMAN.OS2/. In a nutshell, edit the file `\OS2VER` in the root directory of the OS/2 boot partition and add the lines:

```
20=setup.exe
20=netwksta.sys
20=netvdd.sys
```

before you install the client. Also, do not use the included NE2000 driver because it is buggy. Try the NE2000 or NS2000 driver from ftp://ftp.cdrom.com/pub/os2/network/ndis/ instead.

37.2.3 Printer Driver Download for OS/2 Clients

Create a share called *[PRINTDRV]* that is world-readable. Copy your OS/2 driver files there. The `.EA_` files must still be separate, so you will need to use the original install files and not copy an installed driver from an OS/2 system.

Install the NT driver first for that printer. Then, add to your `smb.conf` a parameter, *os2 driver map* = filename. Next, in the file specified by *filename*, map the name of the NT driver name to the OS/2 driver name as follows:

nt driver name = os2 driver name.device name, e.g.

HP LaserJet 5L = LASERJET.HP LaserJet 5L

You can have multiple drivers mapped in this file.

If you only specify the OS/2 driver name, and not the device name, the first attempt to download the driver will actually download the files, but the OS/2 client will tell you the driver is not available. On the second attempt, it will work. This is fixed simply by adding the device name to the mapping, after which it will work on the first attempt.

37.3 Windows for Workgroups

37.3.1 Latest TCP/IP Stack from Microsoft

Use the latest TCP/IP stack from Microsoft if you use Windows for Workgroups. The early TCP/IP stacks had lots of bugs.

Microsoft has released an incremental upgrade to their TCP/IP 32-bit VxD drivers. The latest release can be found on their ftp site at ftp.microsoft.com, located in `/peropsys/windows/public/tcpip/wfwt32.exe`. There is an update.txt file there that describes the problems that were fixed. New files include `WINSOCK.DLL`, `TELNET.EXE`, `WSOCK.386`, `VNBT.386`, `WSTCP.386`, `TRACERT.EXE`, `NETSTAT.EXE`, and `NBTSTAT.EXE`.

37.3.2 Delete .pwl Files After Password Change

Windows for Workgroups does a lousy job with passwords. When you change passwords on either the UNIX box or the PC, the safest thing to do is to delete the .pwl files in the Windows directory. The PC will complain about not finding the files, but will soon get over it, allowing you to enter the new password.

If you do not do this, you may find that Windows for Workgroups remembers and uses the old password, even if you told it a new one.

Often Windows for Workgroups will totally ignore a password you give it in a dialog box.

37.3.3 Configuring Windows for Workgroups Password Handling

There is a program call `admincfg.exe` on the last disk (disk 8) of the WFW 3.11 disk set. To install it, type `EXPAND A:\ADMINCFG.EX_ C:\WINDOWS\ADMINCFG.EXE`. Then add an icon for it via the Program Manager **New** Menu. This program allows you to control how WFW handles passwords, i.e., Disable Password Caching and so on. for use with *security* = user.

37.3.4 Password Case Sensitivity

Windows for Workgroups uppercases the password before sending it to the server. UNIX passwords can be case-sensitive though. Check the `smb.conf` information on *password level* to specify what characters Samba should try to uppercase when checking.

37.3.5 Use TCP/IP as Default Protocol

To support print queue reporting, you may find that you have to use TCP/IP as the default protocol under Windows for Workgroups. For some reason, if you leave NetBEUI as the default, it may break the print queue reporting on some systems. It is presumably a Windows for Workgroups bug.

37.3.6 Speed Improvement

Note that some people have found that setting *DefaultRcvWindow* in the *[MSTCP]* section of the `SYSTEM.INI` file under Windows for Workgroups to 3072 gives a big improvement.

My own experience with DefaultRcvWindow is that I get a much better performance with a large value (16384 or larger). Other people have reported that anything over 3072 slows things down enormously. One person even reported a speed drop of a factor of 30 when he went from 3072 to 8192.

37.4 Windows 95/98

When using Windows 95 OEM SR2, the following updates are recommended where Samba is being used. Please note that the above change will effect you once these updates have been installed.

There are more updates than the ones mentioned here. You are referred to the Microsoft Web site for all currently available updates to your specific version of Windows 95.

Kernel Update: KRNLUPD.EXE
Ping Fix: PINGUPD.EXE
RPC Update: RPCRTUPD.EXE
TCP/IP Update: VIPUPD.EXE
Redirector Update: VRDRUPD.EXE

Also, if using MS Outlook, it is desirable to install the **OLEUPD.EXE** fix. This fix may stop your machine from hanging for an extended period when exiting Outlook and you may notice a significant speedup when accessing network neighborhood services.

37.4.1 Speed Improvement

Configure the Windows 95 TCP/IP registry settings to give better performance. I use a program called **MTUSPEED.exe** that I got off the Internet. There are various other utilities of this type freely available.

37.5 Windows 2000 Service Pack 2

There are several annoyances with Windows 2000 SP2. One of which only appears when using a Samba server to host user profiles to Windows 2000 SP2 clients in a Windows domain. This assumes that Samba is a member of the domain, but the problem will most likely occur if it is not.

In order to serve profiles successfully to Windows 2000 SP2 clients (when not operating as a PDC), Samba must have *nt acl support* = no added to the file share which houses the roaming profiles. If this is not done, then the Windows 2000 SP2 client will complain about not being able to access the profile (Access Denied) and create multiple copies of it on disk (DOMAIN.user.001, DOMAIN.user.002, and so on). See the `smb.conf` man page for more details on this option. Also note that the *nt acl support* parameter was formally a global parameter in releases prior to Samba 2.2.2.

Example 37.1 provides a minimal profile share.

Example 37.1. Minimal profile share

```
[profile]
path = /export/profile
create mask = 0600
directory mask = 0700
nt acl support = no
read only = no
```

The reason for this bug is that the Windows 200x SP2 client copies the security descriptor for the profile that contains the Samba server's SID, and not the domain SID. The client compares the SID for SAMBA\user and realizes it is different from the one assigned to DOMAIN\user. Hence, the reason for the access denied message.

By disabling the *nt acl support* parameter, Samba will send the Windows 200x client a response to the QuerySecurityDescriptor trans2 call, which causes the client to set a default ACL for the profile. This default ACL includes:

DOMAIN\user "Full Control">

NOTE

This bug does not occur when using Winbind to create accounts on the Samba host for Domain users.

37.6 Windows NT 3.1

If you have problems communicating across routers with Windows NT 3.1 workstations, read this Microsoft Knowledge Base article.[5]

[5] http://support.microsoft.com/default.aspx?scid=kb;Q103765

Chapter 38

SAMBA PERFORMANCE TUNING

38.1 Comparisons

The Samba server uses TCP to talk to the client. Thus if you are trying to see if it performs well, you should really compare it to programs that use the same protocol. The most readily available programs for file transfer that use TCP are ftp or another TCP-based SMB server.

If you want to test against something like an NT or Windows for Workgroups server, then you will have to disable all but TCP on either the client or server. Otherwise, you may well be using a totally different protocol (such as NetBEUI) and comparisons may not be valid.

Generally, you should find that Samba performs similarly to ftp at raw transfer speed. It should perform quite a bit faster than NFS, although this depends on your system.

Several people have done comparisons between Samba and Novell, NFS or Windows NT. In some cases Samba performed the best, in others the worst. I suspect the biggest factor is not Samba versus some other system, but the hardware and drivers used on the various systems. Given similar hardware, Samba should certainly be competitive in speed with other systems.

38.2 Socket Options

There are a number of socket options that can greatly affect the performance of a TCP-based server like Samba.

The socket options that Samba uses are settable both on the command line with the `-O` option, or in the `smb.conf` file.

The *`socket options`* section of the `smb.conf` manual page describes how to set these and gives recommendations.

Getting the socket options correct can make a big difference to your performance, but getting them wrong can degrade it by just as much. The correct settings are very dependent on your local network.

The socket option TCP_NODELAY is the one that seems to make the biggest single difference for most networks. Many people report that adding *socket options* = TCP_NODELAY doubles the read performance of a Samba drive. The best explanation I have seen for this is that the Microsoft TCP/IP stack is slow in sending TCP ACKs.

38.3 Read Size

The option *read size* affects the overlap of disk reads/writes with network reads/writes. If the amount of data being transferred in several of the SMB commands (currently SMBwrite, SMBwriteX and SMBreadbraw) is larger than this value, then the server begins writing the data before it has received the whole packet from the network, or in the case of SMBreadbraw, it begins writing to the network before all the data has been read from disk.

This overlapping works best when the speeds of disk and network access are similar, having little effect when the speed of one is much greater than the other.

The default value is 16384, but little experimentation has been done as yet to determine the optimal value, and it is likely that the best value will vary greatly between systems anyway. A value over 65536 is pointless and will cause you to allocate memory unnecessarily.

38.4 Max Xmit

At startup the client and server negotiate a *maximum transmit* size, which limits the size of nearly all SMB commands. You can set the maximum size that Samba will negotiate using the *max xmit* option in `smb.conf`. Note that this is the maximum size of SMB requests that Samba will accept, but not the maximum size that the client will accept. The client maximum receive size is sent to Samba by the client and Samba honors this limit.

It defaults to 65536 bytes (the maximum), but it is possible that some clients may perform better with a smaller transmit unit. Trying values of less than 2048 is likely to cause severe problems. In most cases the default is the best option.

38.5 Log Level

If you set the log level (also known as *debug level*) higher than 2 then you may suffer a large drop in performance. This is because the server flushes the log file after each operation, which can be quite expensive.

38.6 Read Raw

The *read raw* operation is designed to be an optimized, low-latency file read operation. A server may choose to not support it, however, and Samba makes support for *read raw* optional, with it being enabled by default.

In some cases clients do not handle *read raw* very well and actually get lower performance using it than they get using the conventional read operations.

So you might like to try *read raw* = no and see what happens on your network. It might lower, raise or not effect your performance. Only testing can really tell.

38.7 Write Raw

The *write raw* operation is designed to be an optimized, low-latency file write operation. A server may choose to not support it, however, and Samba makes support for *write raw* optional, with it being enabled by default.

Some machines may find *write raw* slower than normal write, in which case you may wish to change this option.

38.8 Slow Logins

Slow logins are almost always due to the password checking time. Using the lowest practical *password level* will improve things.

38.9 Client Tuning

Often a speed problem can be traced to the client. The client (for example Windows for Workgroups) can often be tuned for better TCP performance. Check the sections on the various clients in Chapter 37, *Samba and Other CIFS Clients*.

38.10 Samba Performance Problem Due to Changing Linux Kernel

A user wrote the following to the mailing list:

I am running Gentoo on my server and Samba 2.2.8a. Recently I changed kernel version from `linux-2.4.19-gentoo-r10` to `linux-2.4.20-wolk4.0s`. And now I have a performance issue with Samba. Many of you will probably say, "*Move to vanilla sources!*" Well, I tried that and it didn't work. I have a 100mb LAN and two computers (Linux and Windows 2000). The Linux server shares directories with DivX files, the client (Windows 2000) plays them via LAN. Before when I was running the 2.4.19 kernel everything was fine, but now movies freeze and stop. I tried moving files between the server and Windows and it is terribly slow.

The answer he was given is:

Grab the mii-tool and check the duplex settings on the NIC. My guess is that it is a link layer issue, not an application layer problem. Also run ifconfig and verify that the framing error, collisions, and so on, look normal for ethernet.

38.11 Corrupt tdb Files

Our Samba PDC server has been hosting three TB of data to our 500+ users [Windows NT/XP] for the last three years using Samba without a problem. Today all shares went very slow. Also the main smbd kept spawning new processes so we had 1600+ running smbd's (normally we avg. 250). It crashed the SUN E3500 cluster twice. After a lot of searching, I decided to **rm /var/locks/*.tdb**. Happy again.

Question: Is there any method of keeping the *.tdb files in top condition or how can I detect early corruption?

Answer: Yes, run **tdbbackup** each time after stopping nmbd and before starting nmbd.

Question: What I also would like to mention is that the service latency seems a lot lower than before the locks cleanup. Any ideas on keeping it top notch?

Answer: Yes. Same answer as for previous question!

Chapter 39

DNS AND DHCP CONFIGURATION GUIDE

39.1 Features and Benefits

There are few subjects in the UNIX world that might raise as much contention as Domain Name System (DNS) and Dynamic Host Configuration Protocol (DHCP). Not all opinions held for or against particular implementations of DNS and DHCP are valid.

We live in a modern age where many information technology users demand mobility and freedom. Microsoft Windows users in particular expect to be able to plug their notebook computer into a network port and have things *"just work."*

UNIX administrators have a point. Many of the normative practices in the Microsoft Windows world at best border on bad practice from a security perspective. Microsoft Windows networking protocols allow workstations to arbitrarily register themselves on a network. Windows 2000 Active Directory registers entries in the DNS name space that are equally perplexing to UNIX administrators. Welcome to the new world!

The purpose of this chapter is to demonstrate the configuration of the Internet Software Consortiums (ISC) DNS and DHCP servers to provide dynamic services that are compatible with their equivalents in the Microsoft Windows 2000 Server products.

The purpose of this chapter is to provide no more than a working example of configuration files for both DNS and DHCP servers. The examples used match configuration examples used elsewhere in this document.

This chapter explicitly does not provide a tutorial, nor does it pretend to be a reference guide on DNS and DHCP, as this is well beyond the scope and intent of this document as a whole. Anyone who wants more detailed reference materials on DNS or DHCP should visit the ISC Web sites at http://www.isc.org. Those wanting a writen text might also be interested in the O'Reilly publications on these two subjects.

39.2 Example Configuration

The domain name system is to the Internet what water is to life. By it nearly all information resources (host names) are resolved to their Internet protocol (IP) address. Windows networking tried hard to avoid the complexities of DNS, but alas, DNS won. The alternative to DNS, the Windows Internet Name Service (WINS) an artifact of NetBIOS networking over the TCP/IP protocols, has demonstrated scalability problems as well as a flat non-hierachical name space that became unmanagable as the size and complexity of information technology networks grew.

WINS is a Microsoft implementation of the RFC1001/1002 NetBIOS Name Service (NBNS). It allows NetBIOS clients (like Microsoft Windows Machines) to register an arbitrary machine name that the administrator or user has chosen together with the IP address that the machine has been given. Through the use of WINS, network client machines could resolve machine names to their IP address.

The demand for an alternative to the limitations of NetBIOS networking finally drove Microsoft to use DNS and Active Directory. Microsoft's new implementation attempts to use DNS in a manner similar to the way that WINS is used for NetBIOS networking. Both WINS and Microsoft DNS rely on dynamic name registration.

Microsoft Windows clients can perform dynamic name registration to the DNS server on start-up. Alternately, where DHCP is used to assign workstation IP addresses, it is possible to register host names and their IP address by the DHCP server as soon as a client acknowledges an IP address lease. Lastly, Microsoft DNS can resolve hostnames via Microsoft WINS.

The following configurations demonstrate a simple insecure Dynamic DNS server and a simple DHCP server that matches the DNS configuration.

39.2.1 Dynamic DNS

The example DNS configuration is for a private network in the IP address space for network 192.168.1.0/24. The private class network address space is set forth in RFC1918.

It is assumed that this network will be situated behind a secure firewall. The files that follow work with ISC BIND version 9. BIND is the Berkely Internet Name Daemon. The following configuration files are offered:

The master configuration file for `/etc/named.conf` determines the location of all further configuration files used. The location and name of this file is specified in the start-up script that is part of the operating system.

```
# Quenya.Org configuration file

acl mynet {
   192.168.1.0/24;
   127.0.0.1;
};

options {

   directory "/var/named";
   listen-on-v6 { any; };
```

```
   notify no;
   forward first;
   forwarders {
      192.168.1.1;
      };
   auth-nxdomain yes;
   multiple-cnames yes;
   listen-on {
      mynet;
      };
};

# The following three zone definitions do not need any modification.
# The first one defines localhost while the second defines the
# reverse lookup for localhost. The last zone "." is the
# definition of the root name servers.

zone "localhost" in {
   type master;
   file "localhost.zone";
};

zone "0.0.127.in-addr.arpa" in {
   type master;
   file "127.0.0.zone";
};

zone "." in {
   type hint;
   file "root.hint";
};

# You can insert further zone records for your own domains below.

zone "quenya.org" {
   type master;
   file "/var/named/quenya.org.hosts";
   allow-query {
      mynet;
      };
   allow-transfer {
      mynet;
      };
   allow-update {
      mynet;
      };
   };
```

```
zone "1.168.192.in-addr.arpa" {
   type master;
   file "/var/named/192.168.1.0.rev";
   allow-query {
      mynet;
   };
   allow-transfer {
      mynet;
   };
   allow-update {
      mynet;
   };
};
```

The following files are all located in the directory /var/named. This is the /var/named/localhost.zone file:

```
$TTL 1W
@               IN SOA  @   root (
            42              ; serial (d. adams)
            2D              ; refresh
            4H              ; retry
            6W              ; expiry
            1W )            ; minimum

      IN NS            @
      IN A             127.0.0.1
```

The /var/named/127.0.0.zone file:

```
$TTL 1W
@               IN SOA          localhost.  root.localhost. (
            42              ; serial (d. adams)
            2D              ; refresh
            4H              ; retry
            6W              ; expiry
            1W )            ; minimum

            IN NS           localhost.
1               IN PTR          localhost.
```

The /var/named/quenya.org.host file:

```
$ORIGIN .
$TTL 38400      ; 10 hours 40 minutes
```

```
quenya.org      IN SOA  marvel.quenya.org. root.quenya.org. (
            2003021832 ; serial
            10800      ; refresh (3 hours)
            3600       ; retry (1 hour)
            604800     ; expire (1 week)
            38400      ; minimum (10 hours 40 minutes)
            )
         NS      marvel.quenya.org.
         MX      10 mail.quenya.org.
$ORIGIN quenya.org.
frodo                   A       192.168.1.1
marvel                  A       192.168.1.2
;
mail                    CNAME   marvel
www                     CNAME   marvel
```

The /var/named/192.168.1.0.rev file:

```
$ORIGIN .
$TTL 38400      ; 10 hours 40 minutes
1.168.192.in-addr.arpa  IN SOA  marvel.quenya.org. root.quenya.org. (
            2003021824 ; serial
            10800      ; refresh (3 hours)
            3600       ; retry (1 hour)
            604800     ; expire (1 week)
            38400      ; minimum (10 hours 40 minutes)
            )
         NS      marvel.quenya.org.
$ORIGIN 1.168.192.in-addr.arpa.
1                       PTR     frodo.quenya.org.
2                       PTR     marvel.quenya.org.
```

The above were copied from a fully working system. All dynamically registered entries have been removed. In addition to these files, BIND version 9 will create for each of the dynamic registration files a file that has a .jnl extension. Do not edit or tamper with the configuration files or with the .jnl files that are created.

39.2.2 DHCP Server

The following file is used with the ISC DHCP Server version 3. The file is located in /etc/dhcpd.conf:

```
ddns-updates on;
ddns-domainname "quenya.org";
option ntp-servers 192.168.1.2;
```

```
ddns-update-style ad-hoc;
allow unknown-clients;
default-lease-time 86400;
max-lease-time 172800;

option domain-name "quenya.org";
option domain-name-servers 192.168.1.2;
option netbios-name-servers 192.168.1.2;
option netbios-dd-server 192.168.1.2;
option netbios-node-type 8;

subnet 192.168.1.0 netmask 255.255.255.0 {
   range dynamic-bootp 192.168.1.60 192.168.1.254;
   option subnet-mask 255.255.255.0;
   option routers 192.168.1.2;
   allow unknown-clients;
}
```

In the above example, IP addresses between 192.168.1.1 and 192.168.1.59 are reserved for fixed address (commonly called `hard-wired`) IP addresses. The addresses between 192.168.1.60 and 192.168.1.254 are allocated for dynamic use.

Appendix A

MANUAL PAGES

This appendix contains most of the manual pages from the official Samba distribution. All manual pages have been written by members of the Samba Team[1].

A.1 smb.conf

SYNOPSIS

The `smb.conf` file is a configuration file for the Samba suite. `smb.conf` contains runtime configuration information for the Samba programs. The `smb.conf` file is designed to be configured and administered by the swat(8) program. The complete description of the file format and possible parameters held within are here for reference purposes.

FILE FORMAT

The file consists of sections and parameters. A section begins with the name of the section in square brackets and continues until the next section begins. Sections contain parameters of the form

name = *value*

The file is line-based - that is, each newline-terminated line represents either a comment, a section name or a parameter.

Section and parameter names are not case sensitive.

Only the first equals sign in a parameter is significant. Whitespace before or after the first equals sign is discarded. Leading, trailing and internal whitespace in section and parameter names is irrelevant. Leading and trailing whitespace in a parameter value is discarded. Internal whitespace within a parameter value is retained verbatim.

Any line beginning with a semicolon (";") or a hash ("#") character is ignored, as are lines containing only whitespace.

Any line ending in a "\" is continued on the next line in the customary UNIX fashion.

[1] http://samba.org/samba/team.html

The values following the equals sign in parameters are all either a string (no quotes needed) or a boolean, which may be given as yes/no, 0/1 or true/false. Case is not significant in boolean values, but is preserved in string values. Some items such as create modes are numeric.

SECTION DESCRIPTIONS

Each section in the configuration file (except for the [global] section) describes a shared resource (known as a "*share*"). The section name is the name of the shared resource and the parameters within the section define the shares attributes.

There are three special sections, [global], [homes] and [printers], which are described under *special sections*. The following notes apply to ordinary section descriptions.

A share consists of a directory to which access is being given plus a description of the access rights which are granted to the user of the service. Some housekeeping options are also specifiable.

Sections are either file share services (used by the client as an extension of their native file systems) or printable services (used by the client to access print services on the host running the server).

Sections may be designated *guest* services, in which case no password is required to access them. A specified UNIX *guest account* is used to define access privileges in this case.

Sections other than guest services will require a password to access them. The client provides the username. As older clients only provide passwords and not usernames, you may specify a list of usernames to check against the password using the "*user =*" option in the share definition. For modern clients such as Windows 95/98/ME/NT/2000, this should not be necessary.

The access rights granted by the server are masked by the access rights granted to the specified or guest UNIX user by the host system. The server does not grant more access than the host system grants.

The following sample section defines a file space share. The user has write access to the path `/home/bar`. The share is accessed via the share name "*foo*":

Example A.1.

```
[foo]
path = /home/bar
read only = read only = no
```

The following sample section defines a printable share. The share is read-only, but printable. That is, the only write access permitted is via calls to open, write to and close a spool file. The *guest ok* parameter means access will be permitted as the default guest user (specified elsewhere):

Example A.2.

```
[aprinter]
path = /usr/spool/public
read only = yes
printable = yes
guest ok = yes
```

SPECIAL SECTIONS

The [global] section

Parameters in this section apply to the server as a whole, or are defaults for sections that do not specifically define certain items. See the notes under PARAMETERS for more information.

The [homes] section

If a section called [homes] is included in the configuration file, services connecting clients to their home directories can be created on the fly by the server.

When the connection request is made, the existing sections are scanned. If a match is found, it is used. If no match is found, the requested section name is treated as a username and looked up in the local password file. If the name exists and the correct password has been given, a share is created by cloning the [homes] section.

Some modifications are then made to the newly created share:

- The share name is changed from homes to the located username.
- If no path was given, the path is set to the user's home directory.

If you decide to use a *path* = line in your [homes] section, you may find it useful to use the %S macro. For example :

```
path = /data/pchome/%S
```

is useful if you have different home directories for your PCs than for UNIX access.

This is a fast and simple way to give a large number of clients access to their home directories with a minimum of fuss.

A similar process occurs if the requested section name is "*homes*", except that the share name is not changed to that of the requesting user. This method of using the [homes] section works well if different users share a client PC.

The [homes] section can specify all the parameters a normal service section can specify, though some make more sense than others. The following is a typical and suitable [homes] section:

Example A.3.

```
[homes]
read only = no
```

An important point is that if guest access is specified in the [homes] section, all home directories will be visible to all clients *without a password*. In the very unlikely event that this is actually desirable, it is wise to also specify *read only access*.

The *browseable* flag for auto home directories will be inherited from the global browseable flag, not the [homes] browseable flag. This is useful as it means setting *browseable = no* in the [homes] section will hide the [homes] share but make any auto home directories visible.

The [printers] section

This section works like [homes], but for printers.

If a [printers] section occurs in the configuration file, users are able to connect to any printer specified in the local host's printcap file.

When a connection request is made, the existing sections are scanned. If a match is found, it is used. If no match is found, but a [homes] section exists, it is used as described above. Otherwise, the requested section name is treated as a printer name and the appropriate printcap file is scanned to see if the requested section name is a valid printer share name. If a match is found, a new printer share is created by cloning the [printers] section.

A few modifications are then made to the newly created share:

- The share name is set to the located printer name
- If no printer name was given, the printer name is set to the located printer name
- If the share does not permit guest access and no username was given, the username is set to the located printer name.

The [printers] service MUST be printable - if you specify otherwise, the server will refuse to load the configuration file.

Typically the path specified is that of a world-writeable spool directory with the sticky bit set on it. A typical [printers] entry looks like this:

Example A.4.

```
[printers]
path = /usr/spool/public
guest ok = yes
printable = yes
```

All aliases given for a printer in the printcap file are legitimate printer names as far as the server is concerned. If your printing subsystem doesn't work like that, you will have to set up a pseudo-printcap. This is a file consisting of one or more lines like this:

```
alias|alias|alias|alias...
```

Each alias should be an acceptable printer name for your printing subsystem. In the [global] section, specify the new file as your printcap. The server will only recognize names found in your pseudo-printcap, which of course can contain whatever aliases you like. The same technique could be used simply to limit access to a subset of your local printers.

An alias, by the way, is defined as any component of the first entry of a printcap record. Records are separated by newlines, components (if there are more than one) are separated by vertical bar symbols ("|").

NOTE

On SYSV systems which use lpstat to determine what printers are defined on the system you may be able to use "*printcap name = lpstat*" to automatically obtain a list of printers. See the "*printcap name*" option for more details.

PARAMETERS

Parameters define the specific attributes of sections.

Some parameters are specific to the [global] section (e.g., *security*). Some parameters are usable in all sections (e.g., *create mode*). All others are permissible only in normal sections. For the purposes of the following descriptions the [homes] and [printers] sections will be considered normal. The letter *G* in parentheses indicates that a parameter is specific to the [global] section. The letter *S* indicates that a parameter can be specified in a service specific section. All *S* parameters can also be specified in the [global] section - in which case they will define the default behavior for all services.

Parameters are arranged here in alphabetical order - this may not create best bedfellows, but at least you can find them! Where there are synonyms, the preferred synonym is described, others refer to the preferred synonym.

VARIABLE SUBSTITUTIONS

Many of the strings that are settable in the config file can take substitutions. For example the option "*path = /tmp/%u*" is interpreted as "*path = /tmp/john*" if the user connected with the username john.

These substitutions are mostly noted in the descriptions below, but there are some general substitutions which apply whenever they might be relevant. These are:

%U — session username (the username that the client wanted, not necessarily the same as the one they got).

%G — primary group name of %U.

%h — the Internet hostname that Samba is running on.

%m — the NetBIOS name of the client machine (very useful).

%L — the NetBIOS name of the server. This allows you to change your config based on what the client calls you. Your server can have a "*dual personality*".

This parameter is not available when Samba listens on port 445, as clients no longer send this information.

%M — the Internet name of the client machine.

%R — the selected protocol level after protocol negotiation. It can be one of CORE, COREPLUS, LANMAN1, LANMAN2 or NT1.

%d — The process id of the current server process.

%a — the architecture of the remote machine. Only some are recognized, and those may not be 100% reliable. It currently recognizes Samba, Windows for Workgroups, Windows 95, Windows NT and Windows 2000. Anything else will be known as "*UNKNOWN*". If it gets it wrong sending a level 3 log to samba@samba.org[2] should allow it to be fixed.

%I — The IP address of the client machine.

%T — the current date and time.

%D — Name of the domain or workgroup of the current user.

%$(*envvar*) — The value of the environment variable *envar*.

The following substitutes apply only to some configuration options (only those that are used when a connection has been established):

[2] mailto:samba@samba.org

%S — the name of the current service, if any.

%P — the root directory of the current service, if any.

%u — username of the current service, if any.

%g — primary group name of %u.

%H — the home directory of the user given by %u.

%N — the name of your NIS home directory server. This is obtained from your NIS auto.map entry. If you have not compiled Samba with the *–with-automount* option, this value will be the same as %L.

%p — the path of the service's home directory, obtained from your NIS auto.map entry. The NIS auto.map entry is split up as "*%N:%p*".

There are some quite creative things that can be done with these substitutions and other `smb.conf` options.

NAME MANGLING

Samba supports "*name mangling*" so that DOS and Windows clients can use files that don't conform to the 8.3 format. It can also be set to adjust the case of 8.3 format filenames.

There are several options that control the way mangling is performed, and they are grouped here rather than listed separately. For the defaults look at the output of the testparm program.

All of these options can be set separately for each service (or globally, of course).

The options are:

mangle case = yes/no — controls whether names that have characters that aren't of the "*default*" case are mangled. For example, if this is yes, a name like "*Mail*" will be mangled. Default *no.*

case sensitive = yes/no — controls whether filenames are case sensitive. If they aren't, Samba must do a filename search and match on passed names. Default *no.*

default case = upper/lower — controls what the default case is for new filenames. Default *lower.*

preserve case = yes/no — controls whether new files are created with the case that the client passes, or if they are forced to be the "*default*" case. Default *yes*.

short preserve case = yes/no — controls if new files which conform to 8.3 syntax, that is all in upper case and of suitable length, are created upper case, or if they are forced to be the "*default*" case. This option can be used with "*preserve case = yes*" to permit long filenames to retain their case, while short names are lowercased. Default *yes*.

By default, Samba 3.0 has the same semantics as a Windows NT server, in that it is case insensitive but case preserving.

NOTE ABOUT USERNAME/PASSWORD VALIDATION

There are a number of ways in which a user can connect to a service. The server uses the following steps in determining if it will allow a connection to a specified service. If all the steps fail, the connection request is rejected. However, if one of the steps succeeds, the following steps are not checked.

If the service is marked "*guest only = yes*" and the server is running with share-level security ("*security = share*", steps 1 to 5 are skipped.

1 If the client has passed a username/password pair and that username/password pair is validated by the UNIX system's password programs, the connection is made as that username. This includes the \\server\service%*username* method of passing a username.

2 If the client has previously registered a username with the system and now supplies a correct password for that username, the connection is allowed.

3 The client's NetBIOS name and any previously used usernames are checked against the supplied password. If they match, the connection is allowed as the corresponding user.

4 If the client has previously validated a username/password pair with the server and the client has passed the validation token, that username is used.

5 If a "*user =*" field is given in the `smb.conf` file for the service and the client has supplied a password, and that password matches (according to the UNIX system's password checking) with one of the usernames from the "*user =*" field, the connection is made as the username in the "*user =*" line. If one of the usernames in the "*user =*" list begins with a "@", that name expands to a list of names in the group of the same name.

6 If the service is a guest service, a connection is made as the username given in the "*guest account =*" for the service, irrespective of the supplied password.

EXPLANATION OF EACH PARAMETER

abort shutdown script (G) — *This parameter only exists in the HEAD cvs branch* This a full path name to a script called by smbd(8) that should stop a shutdown procedure issued by the *`shutdown script`*.

This command will be run as user.

Default: *None.*

Example: **abort shutdown script = /sbin/shutdown -c**

add group script (G) — This is the full pathname to a script that will be run *AS ROOT* by smbd(8) when a new group is requested. It will expand any *%g* to the group name passed. This script is only useful for installations using the Windows NT domain administration tools. The script is free to create a group with an arbitrary name to circumvent unix group name restrictions. In that case the script must print the numeric gid of the created group on stdout.

add machine script (G) — This is the full pathname to a script that will be run by smbd(8) when a machine is added to it's domain using the administrator username and password method.

This option is only required when using sam back-ends tied to the Unix uid method of RID calculation such as smbpasswd. This option is only available in Samba 3.0.

Default: **add machine script = <empty string>**

Example: **add machine script = /usr/sbin/adduser -n -g machines -c Machine -d /dev/null -s /bin/false %u**

addprinter command (G) — With the introduction of MS-RPC based printing support for Windows NT/2000 clients in Samba 2.2, The MS Add Printer Wizard (APW) icon is now also available in the "Printers..." folder displayed a share listing. The APW allows for printers to be add remotely to a Samba or Windows NT/2000 print server.

For a Samba host this means that the printer must be physically added to the underlying printing system. The *add printer command* defines a script to be run which will perform the necessary operations for adding the printer to the print system and to add the appropriate service definition to the `smb.conf` file in order that it can be shared by smbd(8).

The *addprinter command* is automatically invoked with the following parameter (in order):

- *printer name*
- *share name*
- *port name*
- *driver name*
- *location*
- *Windows 9x driver location*

All parameters are filled in from the PRINTER_INFO_2 structure sent by the Windows NT/2000 client with one exception. The "Windows 9x driver location" parameter is included for backwards compatibility only. The remaining fields in the structure are generated from answers to the APW questions.

Once the *addprinter command* has been executed, **smbd** will reparse the `smb.conf` to determine if the share defined by the APW exists. If the sharename is still invalid, then **smbd** will return an ACCESS_DENIED error to the client.

The "add printer command" program can output a single line of text, which Samba will set as the port the new printer is connected to. If this line isn't output, Samba won't reload its printer shares.

See also *deleteprinter command*, *printing*, *show add printer wizard*

Default: *none*

Example: **addprinter command = /usr/bin/addprinter**

add share command (G) — Samba 2.2.0 introduced the ability to dynamically add and delete shares via the Windows NT 4.0 Server Manager. The *add share command* is used to define an external program or script which will add a new service definition to `smb.conf`. In order to successfully execute the *add share command*, **smbd** requires that the administrator be connected using a root account (i.e. uid == 0).

When executed, **smbd** will automatically invoke the *add share command* with four parameters.

- *configFile* - the location of the global `smb.conf` file.
- *shareName* - the name of the new share.
- *pathName* - path to an **existing** directory on disk.
- *comment* - comment string to associate with the new share.

This parameter is only used for add file shares. To add printer shares, see the *addprinter command*.

See also *change share command*, *delete share command*.

Default: *none*

Example: **add share command = /usr/local/bin/addshare**

add user script (G) — This is the full pathname to a script that will be run *AS ROOT* by smbd(8) under special circumstances described below.

Normally, a Samba server requires that UNIX users are created for all users accessing files on this server. For sites that use Windows NT account databases as their primary user database creating these users and keeping the user list in sync with the Windows NT PDC is an onerous task. This option allows smbd to create the required UNIX users *ON DEMAND* when a user accesses the Samba server.

In order to use this option, smbd(8) must *NOT* be set to *security = share* and *add user script* must be set to a full pathname for a script that will create a UNIX user given one argument of *%u*, which expands into the UNIX user name to create.

When the Windows user attempts to access the Samba server, at login (session setup in the SMB protocol) time, smbd(8) contacts the `password server` and attempts to authenticate the given user with the given password. If the authentication succeeds then **smbd** attempts to find a UNIX user in the UNIX password database to map the Windows user into. If this lookup fails, and `add user script` is set then **smbd** will call the specified script *AS ROOT*, expanding any `%u` argument to be the user name to create.

If this script successfully creates the user then **smbd** will continue on as though the UNIX user already existed. In this way, UNIX users are dynamically created to match existing Windows NT accounts.

See also `security`, `password server`, `delete user script`.

Default: **add user script = <empty string>**

Example: **add user script = /usr/local/samba/bin/add_user %u**

add user to group script (G) — Full path to the script that will be called when a user is added to a group using the Windows NT domain administration tools. It will be run by smbd(8) *AS ROOT*. Any `%g` will be replaced with the group name and any `%u` will be replaced with the user name.

Default: **add user to group script =**

Example: **add user to group script = /usr/sbin/adduser %u %g**

admin users (S) — This is a list of users who will be granted administrative privileges on the share. This means that they will do all file operations as the super-user (root).

You should use this option very carefully, as any user in this list will be able to do anything they like on the share, irrespective of file permissions.

Default: *no admin users*

Example: **admin users = jason**

ads server (G) — If this option is specified, Samba does not try to figure out what ads server to use itself, but uses the specified ads server. Either one DNS name or IP address can be used.

Default: **ads server =**

Example: **ads server = 192.168.1.2**

algorithmic rid base (G) — This determines how Samba will use its algorithmic mapping from uids/gid to the RIDs needed to construct NT Security Identifiers.

Setting this option to a larger value could be useful to sites transitioning from WinNT and Win2k, as existing user and group rids would otherwise clash with sytem users etc.

All UIDs and GIDs must be able to be resolved into SIDs for the correct operation of ACLs on the server. As such the algorithmic mapping can't be 'turned off', but pushing it 'out of the way' should resolve the issues. Users and groups can then be assigned 'low' RIDs in arbitary-rid supporting backends.

Default: **algorithmic rid base = 1000**

Example: **algorithmic rid base = 100000**

allow hosts (S) — Synonym for *hosts allow*.

allow trusted domains (G) — This option only takes effect when the *security* option is set to `server`, `domain`, or `ads`. If it is set to no, then attempts to connect to a resource from a domain or workgroup other than the one which smbd is running in will fail, even if that domain is trusted by the remote server doing the authentication.

This is useful if you only want your Samba server to serve resources to users in the domain it is a member of. As an example, suppose that there are two domains DOMA and DOMB. DOMB is trusted by DOMA, which contains the Samba server. Under normal circumstances, a user with an account in DOMB can then access the resources of a UNIX account with the same account name on the Samba server even if they do not have an account in DOMA. This can make implementing a security boundary difficult.

Default: **allow trusted domains = yes**

announce as (G) — This specifies what type of server nmbd(8) will announce itself as, to a network neighborhood browse list. By default this is set to Windows NT. The valid options are : "NT Server" (which can also be written as "NT"), "NT Workstation", "Win95" or "WfW" meaning Windows NT Server, Windows NT Workstation, Windows 95 and Windows for Workgroups respectively. Do not change this parameter unless you have a specific need to stop Samba appearing as an NT server as this may prevent Samba servers from participating as browser servers correctly.

Default: **announce as = NT Server**

Example: **announce as = Win95**

announce version (G) — This specifies the major and minor version numbers that nmbd will use when announcing itself as a server. The default is 4.9. Do not change this parameter unless you have a specific need to set a Samba server to be a downlevel server.

Default: **announce version = 4.9**

Example: **announce version = 2.0**

auth methods (G) — This option allows the administrator to chose what authentication methods **smbd** will use when authenticating a user. This option defaults to sensible values based on *security*. This should be considered a developer option and used only in rare circumstances. In the majority (if not all) of production servers, the default setting should be adequate.

Each entry in the list attempts to authenticate the user in turn, until the user authenticates. In practice only one method will ever actually be able to complete the authentication.

Possible options include `guest` (anonymous access), `sam` (lookups in local list of accounts based on netbios name or domain name), `winbind` (relay authentication requests for remote users through winbindd), `ntdomain` (pre-winbindd method of authentication for remote domain users; deprecated in favour of winbind method), `trustdomain` (authenticate trusted users by contacting the remote DC directly from smbd; deprecated in favour of winbind method).

Default: **auth methods = <empty string>**

Example: **auth methods = guest sam winbind**

auto services (G) — This is a synonym for the *preload*.

available (S) — This parameter lets you "turn off" a service. If *available = no*, then *ALL* attempts to connect to the service will fail. Such failures are logged.

Default: **available = yes**

bind interfaces only (G) — This global parameter allows the Samba admin to limit what interfaces on a machine will serve SMB requests. It affects file service smbd(8) and name service nmbd(8) in a slightly different ways.

For name service it causes **nmbd** to bind to ports 137 and 138 on the interfaces listed in the interfaces parameter. **nmbd** also binds to the "all addresses" interface (0.0.0.0) on ports 137 and 138 for the purposes of reading broadcast messages. If this option is not set then **nmbd** will service name requests on all of these sockets. If *bind interfaces only* is set then **nmbd** will check the source address of any packets coming in on the broadcast sockets and discard any that don't match the broadcast addresses of the interfaces in the *interfaces* parameter list. As unicast packets are received on the other sockets it allows **nmbd** to refuse to serve names to machines that send packets that arrive through any interfaces not listed in the *interfaces* list. IP Source address spoofing does defeat this simple check, however, so it must not be used seriously as a security feature for **nmbd**.

For file service it causes smbd(8) to bind only to the interface list given in the interfaces parameter. This restricts the networks that **smbd** will serve to packets coming in those interfaces. Note that you should not use this parameter for machines that are serving PPP or other intermittent or non-broadcast network interfaces as it will not cope with non-permanent interfaces.

If *bind interfaces only* is set then unless the network address *127.0.0.1* is added to the *interfaces* parameter list smbpasswd(8) and swat(8) may not work as expected due to the reasons covered below.

To change a users SMB password, the **smbpasswd** by default connects to the *localhost - 127.0.0.1* address as an SMB client to issue the password change request. If *bind interfaces only* is set then unless the network address *127.0.0.1* is added to the *interfaces* parameter list then **smbpasswd** will fail to connect in it's default mode. **smbpasswd** can be forced to use the primary IP interface of the local host by using its smbpasswd(8) *-r remote machine* parameter, with *remote machine* set to the IP name of the primary interface of the local host.

The **swat** status page tries to connect with **smbd** and **nmbd** at the address *127.0.0.1* to determine if they are running. Not adding *127.0.0.1* will cause **smbd** and **nmbd** to always show "not running" even if they really are. This can prevent **swat** from starting/stopping/restarting **smbd** and **nmbd**.

Default: **bind interfaces only = no**

blocking locks (S) — This parameter controls the behavior of smbd(8) when given a request by a client to obtain a byte range lock on a region of an open file, and the request has a time limit associated with it.

If this parameter is set and the lock range requested cannot be immediately satisfied, samba will internally queue the lock request, and periodically attempt to obtain the lock until the timeout period expires.

If this parameter is set to no, then samba will behave as previous versions of Samba would and will fail the lock request immediately if the lock range cannot be obtained.

Default: **blocking locks = yes**

block size (S) — This parameter controls the behavior of smbd(8) when reporting disk free sizes. By default, this reports a disk block size of 1024 bytes.

Changing this parameter may have some effect on the efficiency of client writes, this is not yet confirmed. This parameter was added to allow advanced administrators to change it (usually to a higher value) and test the effect it has on client write performance without re-compiling the code. As this is an experimental option it may be removed in a future release.

Changing this option does not change the disk free reporting size, just the block size unit reported to the client.

browsable (S) — See the *browseable*.

browseable (S) — This controls whether this share is seen in the list of available shares in a net view and in the browse list.

Default: **browseable = yes**

browse list (G) — This controls whether smbd(8) will serve a browse list to a client doing a **NetServerEnum** call. Normally set to `yes`. You should never need to change this.

Default: **browse list = yes**

case sensitive (S) — See the discussion in the section NAME MANGLING.

Default: **case sensitive = no**

casesignames (S) — Synonym for case sensitive.

change notify timeout (G) — This SMB allows a client to tell a server to "watch" a particular directory for any changes and only reply to the SMB request when a change has occurred. Such constant scanning of a directory is expensive under UNIX, hence an smbd(8) daemon only performs such a scan on each requested directory once every *`change notify timeout`* seconds.

Default: **change notify timeout = 60**

Example: **change notify timeout = 300**

Would change the scan time to every 5 minutes.

change share command (G) — Samba 2.2.0 introduced the ability to dynamically add and delete shares via the Windows NT 4.0 Server Manager. The *`change share command`* is used to define an external program or script which will modify an existing service definition in `smb.conf`. In order to successfully execute the *`change share command`*, **smbd** requires that the administrator be connected using a root account (i.e. uid == 0).

When executed, **smbd** will automatically invoke the *`change share command`* with four parameters.

- *`configFile`* - the location of the global `smb.conf` file.
- *`shareName`* - the name of the new share.
- *`pathName`* - path to an **existing** directory on disk.
- *`comment`* - comment string to associate with the new share.

This parameter is only used modify existing file shares definitions. To modify printer shares, use the "Printers..." folder as seen when browsing the Samba host.

See also *`add share command`*, *`delete share command`*.

Default: *none*

Example: **change share command = /usr/local/bin/addshare**

client lanman auth (G) — This parameter determines whether or not smbclient(8) and other samba client tools will attempt to authenticate itself to servers using the weaker

LANMAN password hash. If disabled, only server which support NT password hashes (e.g. Windows NT/2000, Samba, etc... but not Windows 95/98) will be able to be connected from the Samba client.

The LANMAN encrypted response is easily broken, due to it's case-insensitive nature, and the choice of algorithm. Clients without Windows 95/98 servers are advised to disable this option.

Disabling this option will also disable the **client plaintext auth** option

Likewise, if the **client ntlmv2 auth** parameter is enabled, then only NTLMv2 logins will be attempted. Not all servers support NTLMv2, and most will require special configuration to us it.

Default : **client lanman auth = yes**

client ntlmv2 auth (G) — This parameter determines whether or not smbclient(8) will attempt to authenticate itself to servers using the NTLMv2 encrypted password response.

If enabled, only an NTLMv2 and LMv2 response (both much more secure than earlier versions) will be sent. Many servers (including NT4 < SP4, Win9x and Samba 2.2) are not compatible with NTLMv2.

Similarly, if enabled, NTLMv1, **client lanman auth** and **client plaintext auth** authentication will be disabled. This also disables share-level authentication.

If disabled, an NTLM response (and possibly a LANMAN response) will be sent by the client, depending on the value of **client lanman auth**.

Note that some sites (particularly those following 'best practice' security polices) only allow NTLMv2 responses, and not the weaker LM or NTLM.

Default : **client ntlmv2 auth = no**

client use spnego (G) — This variable controls controls whether samba clients will try to use Simple and Protected NEGOciation (as specified by rfc2478) with WindowsXP and Windows2000 servers to agree upon an authentication mechanism. SPNEGO client support for SMB Signing is currently broken, so you might want to turn this option off when operating with Windows 2003 domain controllers in particular.

Default: *client use spnego = yes*

comment (S) — This is a text field that is seen next to a share when a client does a queries the server, either via the network neighborhood or via **net view** to list what shares are available.

If you want to set the string that is displayed next to the machine name then see the *server string* parameter.

Default: *No comment string*

Example: **comment = Fred's Files**

config file (G) — This allows you to override the config file to use, instead of the default (usually `smb.conf`). There is a chicken and egg problem here as this option is set in the config file!

For this reason, if the name of the config file has changed when the parameters are loaded then it will reload them from the new config file.

This option takes the usual substitutions, which can be very useful.

If the config file doesn't exist then it won't be loaded (allowing you to special case the config files of just a few clients).

Example: **config file = /usr/local/samba/lib/smb.conf.%m**

copy (S) — This parameter allows you to "clone" service entries. The specified service is simply duplicated under the current service's name. Any parameters specified in the current section will override those in the section being copied.

This feature lets you set up a 'template' service and create similar services easily. Note that the service being copied must occur earlier in the configuration file than the service doing the copying.

Default: *no value*

Example: **copy = otherservice**

create mask (S) — A synonym for this parameter is *`create mode`*.

When a file is created, the necessary permissions are calculated according to the mapping from DOS modes to UNIX permissions, and the resulting UNIX mode is then bit-wise 'AND'ed with this parameter. This parameter may be thought of as a bit-wise MASK for the UNIX modes of a file. Any bit *not* set here will be removed from the modes set on a file when it is created.

The default value of this parameter removes the 'group' and 'other' write and execute bits from the UNIX modes.

Following this Samba will bit-wise 'OR' the UNIX mode created from this parameter with the value of the *`force create mode`* parameter which is set to 000 by default.

This parameter does not affect directory modes. See the parameter *`directory mode`* for details.

See also the *`force create mode`* parameter for forcing particular mode bits to be set on created files. See also the *`directory mode`* parameter for masking mode bits on created directories. See also the *`inherit permissions`* parameter.

Note that this parameter does not apply to permissions set by Windows NT/2000 ACL editors. If the administrator wishes to enforce a mask on access control lists also, they need to set the *`security mask`*.

Default: **create mask = 0744**

Example: **create mask = 0775**

create mode (S) — This is a synonym for *create mask*.

csc policy (S) — This stands for *client-side caching policy*, and specifies how clients capable of offline caching will cache the files in the share. The valid values are: manual, documents, programs, disable.

These values correspond to those used on Windows servers.

For example, shares containing roaming profiles can have offline caching disabled using **csc policy = disable**.

Default: **csc policy = manual**

Example: **csc policy = programs**

deadtime (G) — The value of the parameter (a decimal integer) represents the number of minutes of inactivity before a connection is considered dead, and it is disconnected. The deadtime only takes effect if the number of open files is zero.

This is useful to stop a server's resources being exhausted by a large number of inactive connections.

Most clients have an auto-reconnect feature when a connection is broken so in most cases this parameter should be transparent to users.

Using this parameter with a timeout of a few minutes is recommended for most systems.

A deadtime of zero indicates that no auto-disconnection should be performed.

Default: **deadtime = 0**

Example: **deadtime = 15**

debug hires timestamp (G) — Sometimes the timestamps in the log messages are needed with a resolution of higher that seconds, this boolean parameter adds microsecond resolution to the timestamp message header when turned on.

Note that the parameter *debug timestamp* must be on for this to have an effect.

Default: **debug hires timestamp = no**

debuglevel (G) — Synonym for *log level*.

debug pid (G) — When using only one log file for more then one forked smbd(8)-process there may be hard to follow which process outputs which message. This boolean parameter is adds the process-id to the timestamp message headers in the logfile when turned on.

Note that the parameter *debug timestamp* must be on for this to have an effect.

Default: **debug pid = no**

debug timestamp (G) — Samba debug log messages are timestamped by default. If you are running at a high *debug level* these timestamps can be distracting. This boolean parameter allows timestamping to be turned off.

Default: **debug timestamp = yes**

debug uid (G) — Samba is sometimes run as root and sometime run as the connected user, this boolean parameter inserts the current euid, egid, uid and gid to the timestamp message headers in the log file if turned on.

Note that the parameter *debug timestamp* must be on for this to have an effect.

Default: **debug uid = no**

default (G) — A synonym for *default service*.

default case (S) — See the section on NAME MANGLING. Also note the *short preserve case* parameter.

Default: **default case = lower**

default devmode (S) — This parameter is only applicable to printable services. When smbd is serving Printer Drivers to Windows NT/2k/XP clients, each printer on the Samba server has a Device Mode which defines things such as paper size and orientation and duplex settings. The device mode can only correctly be generated by the printer driver itself (which can only be executed on a Win32 platform). Because smbd is unable to execute the driver code to generate the device mode, the default behavior is to set this field to NULL.

Most problems with serving printer drivers to Windows NT/2k/XP clients can be traced to a problem with the generated device mode. Certain drivers will do things such as crashing the client's Explorer.exe with a NULL devmode. However, other printer drivers can cause the client's spooler service (spoolsv.exe) to die if the devmode was not created by the driver itself (i.e. smbd generates a default devmode).

This parameter should be used with care and tested with the printer driver in question. It is better to leave the device mode to NULL and let the Windows client set the correct values. Because drivers do not do this all the time, setting **default devmode = yes** will instruct smbd to generate a default one.

For more information on Windows NT/2k printing and Device Modes, see the MSDN documentation[3].

Default: **default devmode = no**

[3]http://msdn.microsoft.com/

default service (G) — This parameter specifies the name of a service which will be connected to if the service actually requested cannot be found. Note that the square brackets are *NOT* given in the parameter value (see example below).

There is no default value for this parameter. If this parameter is not given, attempting to connect to a nonexistent service results in an error.

Typically the default service would be a *guest ok*, *read-only* service.

Also note that the apparent service name will be changed to equal that of the requested service, this is very useful as it allows you to use macros like *%S* to make a wildcard service.

Note also that any "_" characters in the name of the service used in the default service will get mapped to a "/". This allows for interesting things.

Example:

```
[global]
   default service = pub

[pub]
   path = /%S
```

delete group script (G) — This is the full pathname to a script that will be run *AS ROOT* smbd(8) when a group is requested to be deleted. It will expand any *%g* to the group name passed. This script is only useful for installations using the Windows NT domain administration tools.

deleteprinter command (G) — With the introduction of MS-RPC based printer support for Windows NT/2000 clients in Samba 2.2, it is now possible to delete printer at run time by issuing the DeletePrinter() RPC call.

For a Samba host this means that the printer must be physically deleted from underlying printing system. The *deleteprinter command* defines a script to be run which will perform the necessary operations for removing the printer from the print system and from `smb.conf`.

The *deleteprinter command* is automatically called with only a single parameter: *"printer name"*.

Once the *deleteprinter command* has been executed, **smbd** will reparse the `smb.conf` to associated printer no longer exists. If the sharename is still valid, then **smbd** will return an ACCESS_DENIED error to the client.

See also *addprinter command*, *printing*, *show add printer wizard*

Default: *none*

Example: **deleteprinter command = /usr/bin/removeprinter**

delete readonly (S) — This parameter allows readonly files to be deleted. This is not normal DOS semantics, but is allowed by UNIX.

This option may be useful for running applications such as rcs, where UNIX file ownership prevents changing file permissions, and DOS semantics prevent deletion of a read only file.

Default: **delete readonly = no**

delete share command (G) — Samba 2.2.0 introduced the ability to dynamically add and delete shares via the Windows NT 4.0 Server Manager. The *delete share command* is used to define an external program or script which will remove an existing service definition from `smb.conf`. In order to successfully execute the *delete share command*, **smbd** requires that the administrator be connected using a root account (i.e. uid == 0).

When executed, **smbd** will automatically invoke the *delete share command* with two parameters.

- *configFile* - the location of the global `smb.conf` file.
- *shareName* - the name of the existing service.

This parameter is only used to remove file shares. To delete printer shares, see the *deleteprinter command*.

See also *add share command*, *change share command*.

Default: *none*

Example: **delete share command = /usr/local/bin/delshare**

delete user from group script (G) — Full path to the script that will be called when a user is removed from a group using the Windows NT domain administration tools. It will be run by smbd(8) *AS ROOT*. Any *%g* will be replaced with the group name and any *%u* will be replaced with the user name.

Default: **delete user from group script =**

Example: **delete user from group script = /usr/sbin/deluser %u %g**

delete user script (G) — This is the full pathname to a script that will be run by smbd(8) when managing users with remote RPC (NT) tools.

This script is called when a remote client removes a user from the server, normally using 'User Manager for Domains' or **rpcclient**.

This script should delete the given UNIX username.

Default: **delete user script = <empty string>**

Example: **delete user script = /usr/local/samba/bin/del_user %u**

delete veto files (S) — This option is used when Samba is attempting to delete a directory that contains one or more vetoed directories (see the *veto files* option). If this option is set to `no` (the default) then if a vetoed directory contains any non-vetoed files or directories then the directory delete will fail. This is usually what you want.

If this option is set to `yes`, then Samba will attempt to recursively delete any files and directories within the vetoed directory. This can be useful for integration with file serving systems such as NetAtalk which create meta-files within directories you might normally veto DOS/Windows users from seeing (e.g. `.AppleDouble`)

Setting **delete veto files = yes** allows these directories to be transparently deleted when the parent directory is deleted (so long as the user has permissions to do so).

See also the *veto files* parameter.

Default: **delete veto files = no**

deny hosts (S) — Synonym for *hosts deny*.

dfree command (G) — The *dfree command* setting should only be used on systems where a problem occurs with the internal disk space calculations. This has been known to happen with Ultrix, but may occur with other operating systems. The symptom that was seen was an error of "Abort Retry Ignore" at the end of each directory listing.

This setting allows the replacement of the internal routines to calculate the total disk space and amount available with an external routine. The example below gives a possible script that might fulfill this function.

The external program will be passed a single parameter indicating a directory in the filesystem being queried. This will typically consist of the string `./`. The script should return two integers in ASCII. The first should be the total disk space in blocks, and the second should be the number of available blocks. An optional third return value can give the block size in bytes. The default blocksize is 1024 bytes.

Note: Your script should *NOT* be setuid or setgid and should be owned by (and writeable only by) root!

Default: *By default internal routines for determining the disk capacity and remaining space will be used.*

Example: **dfree command = /usr/local/samba/bin/dfree**

Where the script dfree (which must be made executable) could be:

```
#!/bin/sh
df $1 | tail -1 | awk '{print $2" "$4}'
```

or perhaps (on Sys V based systems):

```
#!/bin/sh
/usr/bin/df -k $1 | tail -1 | awk '{print $3" "$5}'
```

Note that you may have to replace the command names with full path names on some systems.

directory (S) — Synonym for *path*.

directory mask (S) — This parameter is the octal modes which are used when converting DOS modes to UNIX modes when creating UNIX directories.

When a directory is created, the necessary permissions are calculated according to the mapping from DOS modes to UNIX permissions, and the resulting UNIX mode is then bit-wise 'AND'ed with this parameter. This parameter may be thought of as a bit-wise MASK for the UNIX modes of a directory. Any bit *not* set here will be removed from the modes set on a directory when it is created.

The default value of this parameter removes the 'group' and 'other' write bits from the UNIX mode, allowing only the user who owns the directory to modify it.

Following this Samba will bit-wise 'OR' the UNIX mode created from this parameter with the value of the *force directory mode* parameter. This parameter is set to 000 by default (i.e. no extra mode bits are added).

Note that this parameter does not apply to permissions set by Windows NT/2000 ACL editors. If the administrator wishes to enforce a mask on access control lists also, they need to set the *directory security mask*.

See the *force directory mode* parameter to cause particular mode bits to always be set on created directories.

See also the *create mode* parameter for masking mode bits on created files, and the *directory security mask* parameter.

Also refer to the *inherit permissions* parameter.

Default: **directory mask = 0755**

Example: **directory mask = 0775**

directory mode (S) — Synonym for *directory mask*

directory security mask (S) — This parameter controls what UNIX permission bits can be modified when a Windows NT client is manipulating the UNIX permission on a directory using the native NT security dialog box.

This parameter is applied as a mask (AND'ed with) to the changed permission bits, thus preventing any bits not in this mask from being modified. Essentially, zero bits in this mask may be treated as a set of bits the user is not allowed to change.

If not set explicitly this parameter is set to 0777 meaning a user is allowed to modify all the user/group/world permissions on a directory.

Note that users who can access the Samba server through other means can easily bypass this restriction, so it is primarily useful for standalone "appliance" systems. Administrators of most normal systems will probably want to leave it as the default of 0777.

See also the *force directory security mode*, *security mask*, *force security mode* parameters.

Default: **directory security mask = 0777**

Example: **directory security mask = 0700**

disable netbios (G) — Enabling this parameter will disable netbios support in Samba. Netbios is the only available form of browsing in all windows versions except for 2000 and XP.

NOTE

Note that clients that only support netbios won't be able to see your samba server when netbios support is disabled.

Default: **disable netbios = no**

Example: **disable netbios = yes**

disable spoolss (G) — Enabling this parameter will disable Samba's support for the SPOOLSS set of MS-RPC's and will yield identical behavior as Samba 2.0.x. Windows NT/2000 clients will downgrade to using Lanman style printing commands. Windows 9x/ME will be uneffected by the parameter. However, this will also disable the ability to upload printer drivers to a Samba server via the Windows NT Add Printer Wizard or by using the NT printer properties dialog window. It will also disable the capability of Windows NT/2000 clients to download print drivers from the Samba host upon demand. *Be very careful about enabling this parameter.*

See also use client driver

Default : **disable spoolss = no**

display charset (G) — Specifies the charset that samba will use to print messages to stdout and stderr and SWAT will use. Should generally be the same as the **unix charset**.

Default: **display charset = ASCII**

Example: **display charset = UTF8**

dns proxy (G) — Specifies that nmbd(8) when acting as a WINS server and finding that a NetBIOS name has not been registered, should treat the NetBIOS name word-for-word as a DNS name and do a lookup with the DNS server for that name on behalf of the name-querying client.

Note that the maximum length for a NetBIOS name is 15 characters, so the DNS name (or DNS alias) can likewise only be 15 characters, maximum.

nmbd spawns a second copy of itself to do the DNS name lookup requests, as doing a name lookup is a blocking action.

See also the parameter *wins support*.

Default: **dns proxy = yes**

domain logons (G) — If set to `yes`, the Samba server will serve Windows 95/98 Domain logons for the *workgroup* it is in. Samba 2.2 has limited capability to act as a domain controller for Windows NT 4 Domains. For more details on setting up this feature see the Samba-PDC-HOWTO included in the Samba documentation.

Default: **domain logons = no**

domain master (G) — Tell smbd(8) to enable WAN-wide browse list collation. Setting this option causes **nmbd** to claim a special domain specific NetBIOS name that identifies it as a domain master browser for its given *workgroup*. Local master browsers in the same *workgroup* on broadcast-isolated subnets will give this **nmbd** their local browse lists, and then ask smbd(8) for a complete copy of the browse list for the whole wide area network. Browser clients will then contact their local master browser, and will receive the domain-wide browse list, instead of just the list for their broadcast-isolated subnet.

Note that Windows NT Primary Domain Controllers expect to be able to claim this *workgroup* specific special NetBIOS name that identifies them as domain master browsers for that *workgroup* by default (i.e. there is no way to prevent a Windows NT PDC from attempting to do this). This means that if this parameter is set and **nmbd** claims the special name for a *workgroup* before a Windows NT PDC is able to do so then cross subnet browsing will behave strangely and may fail.

If **domain logons = yes**, then the default behavior is to enable the *domain master* parameter. If *domain logons* is not enabled (the default setting), then neither will *domain master* be enabled by default.

Default: **domain master = auto**

dont descend (S) — There are certain directories on some systems (e.g., the `/proc` tree under Linux) that are either not of interest to clients or are infinitely deep (recursive). This parameter allows you to specify a comma-delimited list of directories that the server should always show as empty.

Note that Samba can be very fussy about the exact format of the "dont descend" entries. For example you may need `./proc` instead of just `/proc`. Experimentation is the best policy :-)

Default: *none (i.e., all directories are OK to descend)*

Example: **dont descend = /proc,/dev**

dos charset (G) — DOS SMB clients assume the server has the same charset as they do. This option specifies which charset Samba should talk to DOS clients.

The default depends on which charsets you have installed. Samba tries to use charset 850 but falls back to ASCII in case it is not available. Run testparm(1) to check the default on your system.

dos filemode (S) — The default behavior in Samba is to provide UNIX-like behavior where only the owner of a file/directory is able to change the permissions on it. However, this behavior is often confusing to DOS/Windows users. Enabling this parameter allows a user who has write access to the file (by whatever means) to modify the permissions on it. Note that a user belonging to the group owning the file will not be allowed to change permissions if the group is only granted read access. Ownership of the file/directory is not changed, only the permissions are modified.

Default: **dos filemode = no**

dos filetime resolution (S) — Under the DOS and Windows FAT filesystem, the finest granularity on time resolution is two seconds. Setting this parameter for a share causes Samba to round the reported time down to the nearest two second boundary when a query call that requires one second resolution is made to smbd(8).

This option is mainly used as a compatibility option for Visual C++ when used against Samba shares. If oplocks are enabled on a share, Visual C++ uses two different time reading calls to check if a file has changed since it was last read. One of these calls uses a one-second granularity, the other uses a two second granularity. As the two second call rounds any odd second down, then if the file has a timestamp of an odd number of seconds then the two timestamps will not match and Visual C++ will keep reporting the file has changed. Setting this option causes the two timestamps to match, and Visual C++ is happy.

Default: **dos filetime resolution = no**

dos filetimes (S) — Under DOS and Windows, if a user can write to a file they can change the timestamp on it. Under POSIX semantics, only the owner of the file or root may change the timestamp. By default, Samba runs with POSIX semantics and refuses to change the timestamp on a file if the user **smbd** is acting on behalf of is not the file owner. Setting this option to `yes` allows DOS semantics and smbd(8) will change the file timestamp as DOS requires.

Default: **dos filetimes = no**

encrypt passwords (G) — This boolean controls whether encrypted passwords will be negotiated with the client. Note that Windows NT 4.0 SP3 and above and also Windows 98 will by default expect encrypted passwords unless a registry entry is changed. To use encrypted passwords in Samba see the chapter "User Database" in the Samba HOWTO Collection.

In order for encrypted passwords to work correctly smbd(8) must either have access to a local smbpasswd(5) file (see the smbpasswd(8) program for information on how to set up and maintain this file), or set the security = [server|domain|ads] parameter which causes **smbd** to authenticate against another server.

Default: **encrypt passwords = yes**

enhanced browsing (G) — This option enables a couple of enhancements to cross-subnet browse propagation that have been added in Samba but which are not standard in Microsoft implementations.

The first enhancement to browse propagation consists of a regular wildcard query to a Samba WINS server for all Domain Master Browsers, followed by a browse synchronization with each of the returned DMBs. The second enhancement consists of a regular randomised browse synchronization with all currently known DMBs.

You may wish to disable this option if you have a problem with empty workgroups not disappearing from browse lists. Due to the restrictions of the browse protocols these enhancements can cause a empty workgroup to stay around forever which can be annoying.

In general you should leave this option enabled as it makes cross-subnet browse propagation much more reliable.

Default: **enhanced browsing = yes**

enumports command (G) — The concept of a "port" is fairly foreign to UNIX hosts. Under Windows NT/2000 print servers, a port is associated with a port monitor and generally takes the form of a local port (i.e. LPT1:, COM1:, FILE:) or a remote port (i.e. LPD Port Monitor, etc...). By default, Samba has only one port defined–`"Samba Printer Port"`. Under Windows NT/2000, all printers must have a valid port name. If you wish to have a list of ports displayed (**smbd** does not use a port name for anything) other than the default `"Samba Printer Port"`, you can define *`enumports command`* to point to a program which should generate a list of ports, one per line, to standard output. This listing will then be used in response to the level 1 and 2 EnumPorts() RPC.

Default: *no enumports command*

Example: **enumports command = /usr/bin/listports**

exec (S) — This is a synonym for *`preexec`*.

fake directory create times (S) — NTFS and Windows VFAT file systems keep a create time for all files and directories. This is not the same as the ctime - status change time - that Unix keeps, so Samba by default reports the earliest of the various times Unix does keep. Setting this parameter for a share causes Samba to always report midnight 1-1-1980 as the create time for directories.

This option is mainly used as a compatibility option for Visual C++ when used against Samba shares. Visual C++ generated makefiles have the object directory as a dependency for each object file, and a make rule to create the directory. Also, when NMAKE compares timestamps it uses the creation time when examining a directory. Thus the object directory will be created if it does not exist, but once it does exist it will always have an earlier timestamp than the object files it contains.

However, Unix time semantics mean that the create time reported by Samba will be updated whenever a file is created or or deleted in the directory. NMAKE finds all object files in the object directory. The timestamp of the last one built is then compared to the timestamp of the object directory. If the directory's timestamp if newer, then all object files will be rebuilt. Enabling this option ensures directories always predate their contents and an NMAKE build will proceed as expected.

Default: **fake directory create times = no**

fake oplocks (S) — Oplocks are the way that SMB clients get permission from a server to locally cache file operations. If a server grants an oplock (opportunistic lock) then the client is free to assume that it is the only one accessing the file and it will aggressively cache file data. With some oplock types the client may even cache file open/close operations. This can give enormous performance benefits.

When you set **fake oplocks = yes**, smbd(8) will always grant oplock requests no matter how many clients are using the file.

It is generally much better to use the real *oplocks* support rather than this parameter.

If you enable this option on all read-only shares or shares that you know will only be accessed from one client at a time such as physically read-only media like CDROMs, you will see a big performance improvement on many operations. If you enable this option on shares where multiple clients may be accessing the files read-write at the same time you can get data corruption. Use this option carefully!

Default: **fake oplocks = no**

follow symlinks (S) — This parameter allows the Samba administrator to stop smbd(8) from following symbolic links in a particular share. Setting this parameter to `no` prevents any file or directory that is a symbolic link from being followed (the user will get an error). This option is very useful to stop users from adding a symbolic link to `/etc/passwd` in their home directory for instance. However it will slow filename lookups down slightly.

This option is enabled (i.e. **smbd** will follow symbolic links) by default.

Default: **follow symlinks = yes**

force create mode (S) — This parameter specifies a set of UNIX mode bit permissions that will *always* be set on a file created by Samba. This is done by bitwise 'OR'ing these bits onto the mode bits of a file that is being created or having its permissions changed. The default for this parameter is (in octal) 000. The modes in this parameter are bitwise 'OR'ed onto the file mode after the mask set in the *create mask* parameter is applied.

See also the parameter *create mask* for details on masking mode bits on files.

See also the *inherit permissions* parameter.

Default: **force create mode = 000**

Example: **force create mode = 0755**

would force all created files to have read and execute permissions set for 'group' and 'other' as well as the read/write/execute bits set for the 'user'.

force directory mode (S) — This parameter specifies a set of UNIX mode bit permissions that will *always* be set on a directory created by Samba. This is done by bitwise 'OR'ing these bits onto the mode bits of a directory that is being created. The default for this parameter is (in octal) 0000 which will not add any extra permission bits to a created directory. This operation is done after the mode mask in the parameter *directory mask* is applied.

See also the parameter *directory mask* for details on masking mode bits on created directories.

See also the *inherit permissions* parameter.

Default: **force directory mode = 000**

Example: **force directory mode = 0755**

would force all created directories to have read and execute permissions set for 'group' and 'other' as well as the read/write/execute bits set for the 'user'.

force directory security mode (S) — This parameter controls what UNIX permission bits can be modified when a Windows NT client is manipulating the UNIX permission on a directory using the native NT security dialog box.

This parameter is applied as a mask (OR'ed with) to the changed permission bits, thus forcing any bits in this mask that the user may have modified to be on. Essentially, one bits in this mask may be treated as a set of bits that, when modifying security on a directory, the user has always set to be 'on'.

If not set explicitly this parameter is 000, which allows a user to modify all the user/group/world permissions on a directory without restrictions.

Note that users who can access the Samba server through other means can easily bypass this restriction, so it is primarily useful for standalone "appliance" systems. Administrators of most normal systems will probably want to leave it set as 0000.

See also the *directory security mask*, *security mask*, *force security mode* parameters.

Default: **force directory security mode = 0**

Example: **force directory security mode = 700**

force group (S) — This specifies a UNIX group name that will be assigned as the default primary group for all users connecting to this service. This is useful for sharing files by ensuring that all access to files on service will use the named group for their permissions checking. Thus, by assigning permissions for this group to the files and directories within this service the Samba administrator can restrict or allow sharing of these files.

In Samba 2.0.5 and above this parameter has extended functionality in the following way. If the group name listed here has a '+' character prepended to it then the current user accessing the share only has the primary group default assigned to this group if they are already assigned as a member of that group. This allows an administrator to decide that only users who are already in a particular group will create files with group ownership set to that group. This gives a finer granularity of ownership assignment. For example, the setting `force group = +sys` means that only users who are already in group sys will have their default primary group assigned to sys when accessing this Samba share. All other users will retain their ordinary primary group.

If the *force user* parameter is also set the group specified in *force group* will override the primary group set in *force user*.

See also *force user*.

Default: *no forced group*

Example: **force group = agroup**

force security mode (S) — This parameter controls what UNIX permission bits can be modified when a Windows NT client is manipulating the UNIX permission on a file using the native NT security dialog box.

This parameter is applied as a mask (OR'ed with) to the changed permission bits, thus forcing any bits in this mask that the user may have modified to be on. Essentially, one bits in this mask may be treated as a set of bits that, when modifying security on a file, the user has always set to be 'on'.

If not set explicitly this parameter is set to 0, and allows a user to modify all the user/group/world permissions on a file, with no restrictions.

Note that users who can access the Samba server through other means can easily bypass this restriction, so it is primarily useful for standalone "appliance" systems. Administrators of most normal systems will probably want to leave this set to 0000.

See also the *force directory security mode*, *directory security mask*, *security mask* parameters.

Default: **force security mode = 0**

Example: **force security mode = 700**

force user (S) — This specifies a UNIX user name that will be assigned as the default user for all users connecting to this service. This is useful for sharing files. You should also use it carefully as using it incorrectly can cause security problems.

This user name only gets used once a connection is established. Thus clients still need to connect as a valid user and supply a valid password. Once connected, all file operations will be performed as the "forced user", no matter what username the client connected as. This can be very useful.

In Samba 2.0.5 and above this parameter also causes the primary group of the forced user to be used as the primary group for all file activity. Prior to 2.0.5 the primary group was left as the primary group of the connecting user (this was a bug).

See also *force group*

Default: *no forced user*

Example: **force user = auser**

fstype (S) — This parameter allows the administrator to configure the string that specifies the type of filesystem a share is using that is reported by smbd(8) when a client queries the filesystem type for a share. The default type is `NTFS` for compatibility with Windows NT but this can be changed to other strings such as `Samba` or `FAT` if required.

Default: **fstype = NTFS**

Example: **fstype = Samba**

getwd cache (G) — This is a tuning option. When this is enabled a caching algorithm will be used to reduce the time taken for getwd() calls. This can have a significant impact on performance, especially when the *wide links* parameter is set to `no`.

Default: **getwd cache = yes**

group (S) — Synonym for *force group*.

guest account (G,S) — This is a username which will be used for access to services which are specified as *guest ok* (see below). Whatever privileges this user has will be available to any client connecting to the guest service. Typically this user will exist in the password file, but will not have a valid login. The user account "ftp" is often a good choice for this parameter. If a username is specified in a given service, the specified username overrides this one.

One some systems the default guest account "nobody" may not be able to print. Use another account in this case. You should test this by trying to log in as your guest user (perhaps by using the **su -** command) and trying to print using the system print command such as **lpr(1)** or **lp(1)**.

This parameter does not accept % macros, because many parts of the system require this value to be constant for correct operation.

Default: *specified at compile time, usually "nobody"*

Example: **guest account = ftp**

guest ok (S) — If this parameter is **yes** for a service, then no password is required to connect to the service. Privileges will be those of the *guest account*.

This paramater nullifies the benifits of setting *restrict anonymous* = 2

See the section below on *security* for more information about this option.

Default: **guest ok = no**

guest only (S) — If this parameter is **yes** for a service, then only guest connections to the service are permitted. This parameter will have no effect if *guest ok* is not set for the service.

See the section below on *security* for more information about this option.

Default: **guest only = no**

hide dot files (S) — This is a boolean parameter that controls whether files starting with a dot appear as hidden files.

Default: **hide dot files = yes**

hide files (S) — This is a list of files or directories that are not visible but are accessible. The DOS 'hidden' attribute is applied to any files or directories that match.

Each entry in the list must be separated by a '/', which allows spaces to be included in the entry. '*' and '?' can be used to specify multiple files or directories as in DOS wildcards.

Each entry must be a Unix path, not a DOS path and must not include the Unix directory separator '/'.

Note that the case sensitivity option is applicable in hiding files.

Setting this parameter will affect the performance of Samba, as it will be forced to check all files and directories for a match as they are scanned.

See also *hide dot files*, *veto files* and *case sensitive*.

Default: *no file are hidden*

Example: **hide files = /.*/DesktopFolderDB/TrashFor%m/resource.frk/**

The above example is based on files that the Macintosh SMB client (DAVE) available from Thursby[4] creates for internal use, and also still hides all files beginning with a dot.

hide local users (G) — This parameter toggles the hiding of local UNIX users (root, wheel, floppy, etc) from remote clients.

Default: **hide local users = no**

hide special files (S) — This parameter prevents clients from seeing special files such as sockets, devices and fifo's in directory listings.

Default: **hide special files = no**

hide unreadable (S) — This parameter prevents clients from seeing the existance of files that cannot be read. Defaults to off.

Default: **hide unreadable = no**

hide unwriteable files (S) — This parameter prevents clients from seeing the existance of files that cannot be written to. Defaults to off. Note that unwriteable directories are shown as usual.

Default: **hide unwriteable = no**

homedir map (G) — If*nis homedir* is `yes`, and smbd(8) is also acting as a Win95/98 *logon server* then this parameter specifies the NIS (or YP) map from which the server for the user's home directory should be extracted. At present, only the Sun auto.home map format is understood. The form of the map is:

username server:/some/file/system

and the program will extract the servername from before the first ':'. There should probably be a better parsing system that copes with different map formats and also Amd (another automounter) maps.

> NOTE
>
> 
>
> A working NIS client is required on the system for this option to work.

See also *nis homedir*, *domain logons*.

Default: **homedir map = <empty string>**

[4] http://www.thursby.com

Example: **homedir map = amd.homedir**

host msdfs (G) — If set to yes, Samba will act as a Dfs server, and allow Dfs-aware clients to browse Dfs trees hosted on the server.

See also the *msdfs root* share level parameter. For more information on setting up a Dfs tree on Samba, refer to Chapter 16, *Hosting a Microsoft Distributed File System tree.*

Default: **host msdfs = no**

hostname lookups (G) — Specifies whether samba should use (expensive) hostname lookups or use the ip addresses instead. An example place where hostname lookups are currently used is when checking the **hosts deny** and **hosts allow**.

Default: **hostname lookups = yes**

Example: **hostname lookups = no**

hosts allow (S) — A synonym for this parameter is *allow hosts*.

This parameter is a comma, space, or tab delimited set of hosts which are permitted to access a service.

If specified in the [global] section then it will apply to all services, regardless of whether the individual service has a different setting.

You can specify the hosts by name or IP number. For example, you could restrict access to only the hosts on a Class C subnet with something like **allow hosts = 150.203.5.**. The full syntax of the list is described in the man page hosts_access(5) . Note that this man page may not be present on your system, so a brief description will be given here also.

Note that the localhost address 127.0.0.1 will always be allowed access unless specifically denied by a *hosts deny* option.

You can also specify hosts by network/netmask pairs and by netgroup names if your system supports netgroups. The *EXCEPT* keyword can also be used to limit a wildcard list. The following examples may provide some help:

Example 1: allow all IPs in 150.203.*.*; except one

hosts allow = 150.203. EXCEPT 150.203.6.66

Example 2: allow hosts that match the given network/netmask

hosts allow = 150.203.15.0/255.255.255.0

Example 3: allow a couple of hosts

hosts allow = lapland, arvidsjaur

Example 4: allow only hosts in NIS netgroup "foonet", but deny access from one particular host

hosts allow = @foonet

hosts deny = pirate

NOTE

Note that access still requires suitable user-level passwords.

See testparm(1) for a way of testing your host access to see if it does what you expect.

Default: *none (i.e., all hosts permitted access)*

Example: **allow hosts = 150.203.5. myhost.mynet.edu.au**

hosts deny (S) — The opposite of *hosts allow* - hosts listed here are *NOT* permitted access to services unless the specific services have their own lists to override this one. Where the lists conflict, the *allow* list takes precedence.

Default: *none (i.e., no hosts specifically excluded)*

Example: **hosts deny = 150.203.4. badhost.mynet.edu.au**

hosts equiv (G) — If this global parameter is a non-null string, it specifies the name of a file to read for the names of hosts and users who will be allowed access without specifying a password.

This is not be confused with *hosts allow* which is about hosts access to services and is more useful for guest services. *hosts equiv* may be useful for NT clients which will not supply passwords to Samba.

NOTE

The use of *hosts equiv* can be a major security hole. This is because you are trusting the PC to supply the correct username. It is very easy to get a PC to supply a false username. I recommend that the *hosts equiv* option be only used if you really know what you are doing, or perhaps on a home network where you trust your spouse and kids. And only if you *really* trust them :-).

Default: *no host equivalences*

Example: **hosts equiv = /etc/hosts.equiv**

idmap backend (G) — The purpose of the idmap backend parameter is to allow idmap to NOT use the local idmap tdb file to obtain SID to UID / GID mappings, but instead to obtain them from a common LDAP backend. This way all domain members and controllers will have the same UID and GID to SID mappings. This avoids the risk of UID / GID inconsistencies across UNIX / Linux systems that are sharing information over protocols other than SMB/CIFS (ie: NFS).

Default: **idmap backend = <empty string>**

Example: **idmap backend = ldap:ldap://ldapslave.example.com**

idmap gid (G) — The idmap gid parameter specifies the range of group ids that are allocated for the purpose of mapping UNX groups to NT group SIDs. This range of group ids should have no existing local or NIS groups within it as strange conflicts can occur otherwise.

The availability of an idmap gid range is essential for correct operation of all group mapping.

Default: **idmap gid = <empty string>**

Example: **idmap gid = 10000-20000**

idmap uid (G) — The idmap uid parameter specifies the range of user ids that are allocated for use in mapping UNIX users to NT user SIDs. This range of ids should have no existing local or NIS users within it as strange conflicts can occur otherwise.

Default: **idmap uid = <empty string>**

Example: **idmap uid = 10000-20000**

include (G) — This allows you to include one config file inside another. The file is included literally, as though typed in place.

It takes the standard substitutions, except *%u*, *%P* and *%S*.

Default: *no file included*

Example: **include = /usr/local/samba/lib/admin_smb.conf**

inherit acls (S) — This parameter can be used to ensure that if default acls exist on parent directories, they are always honored when creating a subdirectory. The default behavior is to use the mode specified when creating the directory. Enabling this option sets the mode to 0777, thus guaranteeing that default directory acls are propagated.

Default: **inherit acls = no**

inherit permissions (S) — The permissions on new files and directories are normally governed by *create mask*, *directory mask*, *force create mode* and *force directory mode* but the boolean inherit permissions parameter overrides this.

New directories inherit the mode of the parent directory, including bits such as setgid.

New files inherit their read/write bits from the parent directory. Their execute bits continue to be determined by *map archive*, *map hidden* and *map system* as usual.

Note that the setuid bit is *never* set via inheritance (the code explicitly prohibits this).

This can be particularly useful on large systems with many users, perhaps several thousand, to allow a single [homes] share to be used flexibly by each user.

See also *create mask*, *directory mask*, *force create mode* and *force directory mode*.

Default: **inherit permissions = no**

interfaces (G) — This option allows you to override the default network interfaces list that Samba will use for browsing, name registration and other NBT traffic. By default Samba will query the kernel for the list of all active interfaces and use any interfaces except 127.0.0.1 that are broadcast capable.

The option takes a list of interface strings. Each string can be in any of the following forms:

- a network interface name (such as eth0). This may include shell-like wildcards so eth* will match any interface starting with the substring "eth"
- an IP address. In this case the netmask is determined from the list of interfaces obtained from the kernel
- an IP/mask pair.
- a broadcast/mask pair.

The "mask" parameters can either be a bit length (such as 24 for a C class network) or a full netmask in dotted decimal form.

The "IP" parameters above can either be a full dotted decimal IP address or a hostname which will be looked up via the OS's normal hostname resolution mechanisms.

For example, the following line:

interfaces = eth0 192.168.2.10/24 192.168.3.10/255.255.255.0

would configure three network interfaces corresponding to the eth0 device and IP addresses 192.168.2.10 and 192.168.3.10. The netmasks of the latter two interfaces would be set to 255.255.255.0.

See also *bind interfaces only*.

Default: *all active interfaces except 127.0.0.1 that are broadcast capable*

invalid users (S) — This is a list of users that should not be allowed to login to this service. This is really a *paranoid* check to absolutely ensure an improper setting does not breach your security.

A name starting with a '@' is interpreted as an NIS netgroup first (if your system supports NIS), and then as a UNIX group if the name was not found in the NIS netgroup database.

A name starting with '+' is interpreted only by looking in the UNIX group database. A name starting with '&' is interpreted only by looking in the NIS netgroup database (this requires NIS to be working on your system). The characters '+' and '&' may be used at the start of the name in either order so the value *+&group* means check the UNIX group database, followed by the NIS netgroup database, and the value *&+group* means check the NIS netgroup database, followed by the UNIX group database (the same as the '@' prefix).

The current servicename is substituted for *%S*. This is useful in the [homes] section.

See also *valid users*.

Default: *no invalid users*

Example: **invalid users = root fred admin @wheel**

keepalive (G) — The value of the parameter (an integer) represents the number of seconds between *keepalive* packets. If this parameter is zero, no keepalive packets will be sent. Keepalive packets, if sent, allow the server to tell whether a client is still present and responding.

Keepalives should, in general, not be needed if the socket being used has the SO_KEEPALIVE attribute set on it (see *socket options*). Basically you should only use this option if you strike difficulties.

Default: **keepalive = 300**

Example: **keepalive = 600**

kernel oplocks (G) — For UNIXes that support kernel based *oplocks* (currently only IRIX and the Linux 2.4 kernel), this parameter allows the use of them to be turned on or off.

Kernel oplocks support allows Samba *oplocks* to be broken whenever a local UNIX process or NFS operation accesses a file that smbd(8) has oplocked. This allows complete data consistency between SMB/CIFS, NFS and local file access (and is a *very* cool feature :-).

This parameter defaults to on, but is translated to a no-op on systems that no not have the necessary kernel support. You should never need to touch this parameter.

See also the *oplocks* and *level2 oplocks* parameters.

Default: **kernel oplocks = yes**

lanman auth (G) — This parameter determines whether or not smbd(8) will attempt to authenticate users using the LANMAN password hash. If disabled, only clients which support NT password hashes (e.g. Windows NT/2000 clients, smbclient, etc... but not Windows 95/98 or the MS DOS network client) will be able to connect to the Samba host.

The LANMAN encrypted response is easily broken, due to it's case-insensitive nature, and the choice of algorithm. Servers without Windows 95/98 or MS DOS clients are advised to disable this option.

Unlike the **encypt passwords** option, this parameter cannot alter client behaviour, and the LANMAN response will still be sent over the network. See the **client lanman auth** to disable this for Samba's clients (such as smbclient)

If this option, and **ntlm auth** are both disabled, then only NTLMv2 logins will be permited. Not all clients support NTLMv2, and most will require special configuration to us it.

Default : **lanman auth = yes**

large readwrite (G) — This parameter determines whether or not smbd(8) supports the new 64k streaming read and write varient SMB requests introduced with Windows 2000. Note that due to Windows 2000 client redirector bugs this requires Samba to be running on a 64-bit capable operating system such as IRIX, Solaris or a Linux 2.4 kernel. Can improve performance by 10% with Windows 2000 clients. Defaults to on. Not as tested as some other Samba code paths.

Default: **large readwrite = yes**

ldap admin dn (G) — The *`ldap admin dn`* defines the Distinguished Name (DN) name used by Samba to contact the ldap server when retreiving user account information. The *`ldap admin dn`* is used in conjunction with the admin dn password stored in the `private/secrets.tdb` file. See the smbpasswd(8) man page for more information on how to accmplish this.

ldap delete dn (G) — This parameter specifies whether a delete operation in the ldapsam deletes the complete entry or only the attributes specific to Samba.

Default: *ldap delete dn = no*

ldap filter (G) — This parameter specifies the RFC 2254 compliant LDAP search filter. The default is to match the login name with the `uid` attribute for all entries matching the `sambaAccount` objectclass. Note that this filter should only return one entry.

Default: **ldap filter = (&(uid=%u)(objectclass=sambaAccount))**

ldap group suffix (G) — This parameters specifies the suffix that is used for groups when these are added to the LDAP directory. If this parameter is unset, the value of *`ldap suffix`* will be used instead.

Default: *none*

Example: *dc=samba,ou=Groups*

ldap idmap suffix (G) — This parameters specifies the suffix that is used when storing idmap mappings. If this parameter is unset, the value of *ldap suffix* will be used instead.

Default: *none*

Example: *ou=Idmap,dc=samba,dc=org*

ldap machine suffix (G) — It specifies where machines should be added to the ldap tree.

Default: *none*

ldap passwd sync (G) — This option is used to define whether or not Samba should sync the LDAP password with the NT and LM hashes for normal accounts (NOT for workstation, server or domain trusts) on a password change via SAMBA.

The *ldap passwd sync* can be set to one of three values:

- *Yes* = Try to update the LDAP, NT and LM passwords and update the pwdLastSet time.
- *No* = Update NT and LM passwords and update the pwdLastSet time.
- *Only* = Only update the LDAP password and let the LDAP server do the rest.

Default: **ldap passwd sync = no**

ldap server (G) — This parameter is only available if Samba has been configure to include the –**with-ldapsam** option at compile time.

This parameter should contain the FQDN of the ldap directory server which should be queried to locate user account information.

Default : **ldap server = localhost**

ldap ssl (G) — This option is used to define whether or not Samba should use SSL when connecting to the ldap server This is *NOT* related to Samba's previous SSL support which was enabled by specifying the –**with-ssl** option to the `configure` script.

The *ldap ssl* can be set to one of three values:

- *Off* = Never use SSL when querying the directory.
- *Start_tls* = Use the LDAPv3 StartTLS extended operation (RFC2830) for communicating with the directory server.
- *On* = Use SSL on the ldaps port when contacting the *ldap server*. Only available when the backwards-compatiblity –**with-ldapsam** option is specified to configure. See *passdb backend*

Default : **ldap ssl = start_tls**

ldap suffix (G) — Specifies where user and machine accounts are added to the tree. Can be overriden by **ldap user suffix** and **ldap machine suffix**. It also used as the base dn for all ldap searches.

Default: *none*

ldap user suffix (G) — This parameter specifies where users are added to the tree. If this parameter is not specified, the value from **ldap suffix**.

Default: *none*

level2 oplocks (S) — This parameter controls whether Samba supports level2 (read-only) oplocks on a share.

Level2, or read-only oplocks allow Windows NT clients that have an oplock on a file to downgrade from a read-write oplock to a read-only oplock once a second client opens the file (instead of releasing all oplocks on a second open, as in traditional, exclusive oplocks). This allows all openers of the file that support level2 oplocks to cache the file for read-ahead only (ie. they may not cache writes or lock requests) and increases performance for many accesses of files that are not commonly written (such as application .EXE files).

Once one of the clients which have a read-only oplock writes to the file all clients are notified (no reply is needed or waited for) and told to break their oplocks to "none" and delete any read-ahead caches.

It is recommended that this parameter be turned on to speed access to shared executables.

For more discussions on level2 oplocks see the CIFS spec.

Currently, if *kernel oplocks* are supported then level2 oplocks are not granted (even if this parameter is set to `yes`). Note also, the *oplocks* parameter must be set to `yes` on this share in order for this parameter to have any effect.

See also the *oplocks* and *kernel oplocks* parameters.

Default: **level2 oplocks = yes**

lm announce (G) — This parameter determines if nmbd(8) will produce Lanman announce broadcasts that are needed by OS/2 clients in order for them to see the Samba server in their browse list. This parameter can have three values, `yes`, `no`, or `auto`. The default is `auto`. If set to `no` Samba will never produce these broadcasts. If set to `yes` Samba will produce Lanman announce broadcasts at a frequency set by the parameter *lm interval*. If set to `auto` Samba will not send Lanman announce broadcasts by default but will listen for them. If it hears such a broadcast on the wire it will then start sending them at a frequency set by the parameter *lm interval*.

See also *lm interval*.

Default: **lm announce = auto**

Example: **lm announce = yes**

lm interval (G) — If Samba is set to produce Lanman announce broadcasts needed by OS/2 clients (see the *lm announce* parameter) then this parameter defines the frequency in seconds with which they will be made. If this is set to zero then no Lanman announcements will be made despite the setting of the *lm announce* parameter.

See also *lm announce*.

Default: **lm interval = 60**

Example: **lm interval = 120**

load printers (G) — A boolean variable that controls whether all printers in the printcap will be loaded for browsing by default. See the printers section for more details.

Default: **load printers = yes**

local master (G) — This option allows nmbd(8) to try and become a local master browser on a subnet. If set to `no` then **nmbd** will not attempt to become a local master browser on a subnet and will also lose in all browsing elections. By default this value is set to `yes`. Setting this value to `yes` doesn't mean that Samba will *become* the local master browser on a subnet, just that **nmbd** will *participate* in elections for local master browser.

Setting this value to `no` will cause **nmbd** *never* to become a local master browser.

Default: **local master = yes**

lock dir (G) — Synonym for *lock directory*.

lock directory (G) — This option specifies the directory where lock files will be placed. The lock files are used to implement the *max connections* option.

Default: **lock directory = ${prefix}/var/locks**

Example: **lock directory = /var/run/samba/locks**

locking (S) — This controls whether or not locking will be performed by the server in response to lock requests from the client.

If **locking = no**, all lock and unlock requests will appear to succeed and all lock queries will report that the file in question is available for locking.

If **locking = yes**, real locking will be performed by the server.

This option *may* be useful for read-only filesystems which *may* not need locking (such as CDROM drives), although setting this parameter of `no` is not really recommended even in this case.

Be careful about disabling locking either globally or in a specific service, as lack of locking may result in data corruption. You should never need to set this parameter.

Default: **locking = yes**

lock spin count (G) — This parameter controls the number of times that smbd should attempt to gain a byte range lock on the behalf of a client request. Experiments have shown that Windows 2k servers do not reply with a failure if the lock could not be immediately granted, but try a few more times in case the lock could later be aquired. This behavior is used to support PC database formats such as MS Access and FoxPro.

Default: **lock spin count = 3**

lock spin time (G) — The time in microseconds that smbd should pause before attempting to gain a failed lock. See *lock spin count* for more details.

Default: **lock spin time = 10**

log file (G) — This option allows you to override the name of the Samba log file (also known as the debug file).

This option takes the standard substitutions, allowing you to have separate log files for each user or machine.

Example: **log file = /usr/local/samba/var/log.%m**

log level (G) — The value of the parameter (a astring) allows the debug level (logging level) to be specified in the `smb.conf` file. This parameter has been extended since the 2.2.x series, now it allow to specify the debug level for multiple debug classes. This is to give greater flexibility in the configuration of the system.

The default will be the log level specified on the command line or level zero if none was specified.

Example: **log level = 3 passdb:5 auth:10 winbind:2**

logon drive (G) — This parameter specifies the local path to which the home directory will be connected (see *logon home*) and is only used by NT Workstations.

Note that this option is only useful if Samba is set up as a logon server.

Default: **logon drive = z:**

Example: **logon drive = h:**

logon home (G) — This parameter specifies the home directory location when a Win95/98 or NT Workstation logs into a Samba PDC. It allows you to do

```
C:\> NET USE H: /HOME
```

from a command prompt, for example.

This option takes the standard substitutions, allowing you to have separate logon scripts for each user or machine.

This parameter can be used with Win9X workstations to ensure that roaming profiles are stored in a subdirectory of the user's home directory. This is done in the following way:

logon home = \\%N\%U\profile

This tells Samba to return the above string, with substitutions made when a client requests the info, generally in a NetUserGetInfo request. Win9X clients truncate the info to \\server\share when a user does **net use /home** but use the whole string when dealing with profiles.

Note that in prior versions of Samba, the *logon path* was returned rather than *logon home*. This broke **net use /home** but allowed profiles outside the home directory. The current implementation is correct, and can be used for profiles if you use the above trick.

This option is only useful if Samba is set up as a logon server.

Default: **logon home = "\\%N\%U"**

Example: **logon home = "\\remote_smb_server\%U"**

logon path (G) — This parameter specifies the home directory where roaming profiles (NTuser.dat etc files for Windows NT) are stored. Contrary to previous versions of these manual pages, it has nothing to do with Win 9X roaming profiles. To find out how to handle roaming profiles for Win 9X system, see the *logon home* parameter.

This option takes the standard substitutions, allowing you to have separate logon scripts for each user or machine. It also specifies the directory from which the "Application Data", (`desktop`, `start menu`, `network neighborhood`, `programs` and other folders, and their contents, are loaded and displayed on your Windows NT client.

The share and the path must be readable by the user for the preferences and directories to be loaded onto the Windows NT client. The share must be writeable when the user logs in for the first time, in order that the Windows NT client can create the NTuser.dat and other directories.

Thereafter, the directories and any of the contents can, if required, be made read-only. It is not advisable that the NTuser.dat file be made read-only - rename it to NTuser.man to achieve the desired effect (a *MAN*datory profile).

Windows clients can sometimes maintain a connection to the [homes] share, even though there is no user logged in. Therefore, it is vital that the logon path does not include a reference to the homes share (i.e. setting this parameter to `\%N\%U\profile_path` will cause problems).

This option takes the standard substitutions, allowing you to have separate logon scripts for each user or machine.

Note that this option is only useful if Samba is set up as a logon server.

Default: **logon path = \\%N\%U\profile**

Example: **logon path = \\PROFILESERVER\PROFILE\%U**

logon script (G) — This parameter specifies the batch file (.bat) or NT command file (.cmd) to be downloaded and run on a machine when a user successfully logs in. The file must contain the DOS style CR/LF line endings. Using a DOS-style editor to create the file is recommended.

The script must be a relative path to the [netlogon] service. If the [netlogon] service specifies a *path* of `/usr/local/samba/netlogon`, and **logon script = STARTUP.BAT**, then the file that will be downloaded is:

`/usr/local/samba/netlogon/STARTUP.BAT`

The contents of the batch file are entirely your choice. A suggested command would be to add **NET TIME \\SERVER /SET /YES**, to force every machine to synchronize clocks with the same time server. Another use would be to add **NET USE U: \\SERVER\UTILS** for commonly used utilities, or

```
NET USE Q: \\SERVER\ISO9001_QA
```

for example.

Note that it is particularly important not to allow write access to the [netlogon] share, or to grant users write permission on the batch files in a secure environment, as this would allow the batch files to be arbitrarily modified and security to be breached.

This option takes the standard substitutions, allowing you to have separate logon scripts for each user or machine.

This option is only useful if Samba is set up as a logon server.

Default: *no logon script defined*

Example: **logon script = scripts\%U.bat**

lppause command (S) — This parameter specifies the command to be executed on the server host in order to stop printing or spooling a specific print job.

This command should be a program or script which takes a printer name and job number to pause the print job. One way of implementing this is by using job priorities, where jobs having a too low priority won't be sent to the printer.

If a *%p* is given then the printer name is put in its place. A *%j* is replaced with the job number (an integer). On HPUX (see *printing=hpux*), if the *-p%p* option is added to the lpq command, the job will show up with the correct status, i.e. if the job priority is lower than the set fence priority it will have the PAUSED status, whereas if the priority is equal or higher it will have the SPOOLED or PRINTING status.

Note that it is good practice to include the absolute path in the lppause command as the PATH may not be available to the server.

See also the *printing* parameter.

Default: Currently no default value is given to this string, unless the value of the *printing* parameter is `SYSV`, in which case the default is :

lp -i %p-%j -H hold

or if the value of the *printing* parameter is `SOFTQ`, then the default is:

qstat -s -j%j -h

Example for HPUX: **lppause command = /usr/bin/lpalt %p-%j -p0**

lpq cache time (G) — This controls how long lpq info will be cached for to prevent the **lpq** command being called too often. A separate cache is kept for each variation of the **lpq** command used by the system, so if you use different **lpq** commands for different users then they won't share cache information.

The cache files are stored in `/tmp/lpq.xxxx` where xxxx is a hash of the **lpq** command in use.

The default is 10 seconds, meaning that the cached results of a previous identical **lpq** command will be used if the cached data is less than 10 seconds old. A large value may be advisable if your **lpq** command is very slow.

A value of 0 will disable caching completely.

See also the *printing* parameter.

Default: **lpq cache time = 10**

Example: **lpq cache time = 30**

lpq command (S) — This parameter specifies the command to be executed on the server host in order to obtain **lpq**-style printer status information.

This command should be a program or script which takes a printer name as its only parameter and outputs printer status information.

Currently nine styles of printer status information are supported; BSD, AIX, LPRNG, PLP, SYSV, HPUX, QNX, CUPS, and SOFTQ. This covers most UNIX systems. You control which type is expected using the *printing* = option.

Some clients (notably Windows for Workgroups) may not correctly send the connection number for the printer they are requesting status information about. To get around this, the server reports on the first printer service connected to by the client. This only happens if the connection number sent is invalid.

If a *%p* is given then the printer name is put in its place. Otherwise it is placed at the end of the command.

Note that it is good practice to include the absolute path in the *lpq command* as the `$PATH` may not be available to the server. When compiled with the CUPS libraries, no *lpq command* is needed because smbd will make a library call to obtain the print queue listing.

See also the *printing* parameter.

Default: *depends on the setting of printing*

Example: **lpq command = /usr/bin/lpq -P%p**

lpresume command (S) — This parameter specifies the command to be executed on the server host in order to restart or continue printing or spooling a specific print job.

This command should be a program or script which takes a printer name and job number to resume the print job. See also the *lppause command* parameter.

If a *%p* is given then the printer name is put in its place. A *%j* is replaced with the job number (an integer).

Note that it is good practice to include the absolute path in the *lpresume command* as the PATH may not be available to the server.

See also the *printing* parameter.

Default: Currently no default value is given to this string, unless the value of the *printing* parameter is SYSV, in which case the default is :

lp -i %p-%j -H resume

or if the value of the *printing* parameter is SOFTQ, then the default is:

qstat -s -j%j -r

Example for HPUX: **lpresume command = /usr/bin/lpalt %p-%j -p2**

lprm command (S) — This parameter specifies the command to be executed on the server host in order to delete a print job.

This command should be a program or script which takes a printer name and job number, and deletes the print job.

If a *%p* is given then the printer name is put in its place. A *%j* is replaced with the job number (an integer).

Note that it is good practice to include the absolute path in the *lprm command* as the PATH may not be available to the server.

See also the *printing* parameter.

Default: *depends on the setting of printing*

Example 1: **lprm command = /usr/bin/lprm -P%p %j**

Example 2: **lprm command = /usr/bin/cancel %p-%j**

machine password timeout (G) — If a Samba server is a member of a Windows NT Domain (see the security = domain) parameter) then periodically a running smbd process will try and change the MACHINE ACCOUNT PASSWORD stored in the TDB called `private/secrets.tdb`. This parameter specifies how often this password will be changed, in seconds. The default is one week (expressed in seconds), the same as a Windows NT Domain member server.

See also smbpasswd(8), and the security = domain) parameter.

Default: **machine password timeout = 604800**

magic output (S) — This parameter specifies the name of a file which will contain output created by a magic script (see the *magic script* parameter below).

Warning: If two clients use the same *magic script* in the same directory the output file content is undefined.

Default: **magic output = <magic script name>.out**

Example: **magic output = myfile.txt**

magic script (S) — This parameter specifies the name of a file which, if opened, will be executed by the server when the file is closed. This allows a UNIX script to be sent to the Samba host and executed on behalf of the connected user.

Scripts executed in this way will be deleted upon completion assuming that the user has the appropriate level of privilege and the file permissions allow the deletion.

If the script generates output, output will be sent to the file specified by the *magic output* parameter (see above).

Note that some shells are unable to interpret scripts containing CR/LF instead of CR as the end-of-line marker. Magic scripts must be executable *as is* on the host, which for some hosts and some shells will require filtering at the DOS end.

Magic scripts are *EXPERIMENTAL* and should *NOT* be relied upon.

Default: *None. Magic scripts disabled.*

Example: **magic script = user.csh**

mangle case (S) — See the section on NAME MANGLING

Default: **mangle case = no**

mangled map (S) — This is for those who want to directly map UNIX file names which cannot be represented on Windows/DOS. The mangling of names is not always what is needed. In particular you may have documents with file extensions that differ between DOS and UNIX. For example, under UNIX it is common to use `.html` for HTML files, whereas under Windows/DOS `.htm` is more commonly used.

So to map `html` to `htm` you would use:

mangled map = (*.html *.htm)

One very useful case is to remove the annoying `;1` off the ends of filenames on some CDROMs (only visible under some UNIXes). To do this use a map of (*;1 *;).

Default: *no mangled map*

Example: **mangled map = (*;1 *;)**

mangled names (S) — This controls whether non-DOS names under UNIX should be mapped to DOS-compatible names ("mangled") and made visible, or whether non-DOS names should simply be ignored.

See the section on NAME MANGLING for details on how to control the mangling process.

If mangling is used then the mangling algorithm is as follows:

- The first (up to) five alphanumeric characters before the rightmost dot of the filename are preserved, forced to upper case, and appear as the first (up to) five characters of the mangled name.
- A tilde "~" is appended to the first part of the mangled name, followed by a two-character unique sequence, based on the original root name (i.e., the original filename minus its final extension). The final extension is included in the hash calculation only if it contains any upper case characters or is longer than three characters.

  Note that the character to use may be specified using the *mangling char* option, if you don't like '~'.
- The first three alphanumeric characters of the final extension are preserved, forced to upper case and appear as the extension of the mangled name. The final extension is defined as that part of the original filename after the rightmost dot. If there are no dots in the filename, the mangled name will have no extension (except in the case of "hidden files" - see below).
- Files whose UNIX name begins with a dot will be presented as DOS hidden files. The mangled name will be created as for other filenames, but with the leading dot removed and "___" as its extension regardless of actual original extension (that's three underscores).

The two-digit hash value consists of upper case alphanumeric characters.

This algorithm can cause name collisions only if files in a directory share the same first five alphanumeric characters. The probability of such a clash is 1/1300.

The name mangling (if enabled) allows a file to be copied between UNIX directories from Windows/DOS while retaining the long UNIX filename. UNIX files can be renamed to a new extension from Windows/DOS and will retain the same basename. Mangled names do not change between sessions.

Default: **mangled names = yes**

mangled stack (G) — This parameter controls the number of mangled names that should be cached in the Samba server smbd(8).

This stack is a list of recently mangled base names (extensions are only maintained if they are longer than 3 characters or contains upper case characters).

The larger this value, the more likely it is that mangled names can be successfully converted to correct long UNIX names. However, large stack sizes will slow most directory accesses. Smaller stacks save memory in the server (each stack element costs 256 bytes).

It is not possible to absolutely guarantee correct long filenames, so be prepared for some surprises!

Default: **mangled stack = 50**

Example: **mangled stack = 100**

mangle prefix (G) — controls the number of prefix characters from the original name used when generating the mangled names. A larger value will give a weaker hash and therefore more name collisions. The minimum value is 1 and the maximum value is 6.

mangle prefix is effective only when mangling method is hash2.

Default: **mangle prefix = 1**

Example: **mangle prefix = 4**

mangling char (S) — This controls what character is used as the *magic* character in name mangling. The default is a '~' but this may interfere with some software. Use this option to set it to whatever you prefer. This is effective only when mangling method is hash.

Default: **mangling char = ~**

Example: **mangling char = ^**

mangling method (G) — controls the algorithm used for the generating the mangled names. Can take two different values, "hash" and "hash2". "hash" is the default and is the algorithm that has been used in Samba for many years. "hash2" is a newer and considered a better algorithm (generates less collisions) in the names. However, many Win32 applications store the mangled names and so changing to the new algorithm must not be done lightly as these applications may break unless reinstalled.

Default: **mangling method = hash2**

Example: **mangling method = hash**

map acl inherit (S) — This boolean parameter controls whether smbd(8) will attempt to map the 'inherit' and 'protected' access control entry flags stored in Windows ACLs into an extended attribute called user.SAMBA_PAI. This parameter only takes effect if Samba is being run on a platform that supports extended attributes (Linux and IRIX so far) and allows the Windows 2000 ACL editor to correctly use inheritance with the Samba POSIX ACL mapping code.

Default: **map acl inherit = no**

map archive (S) — This controls whether the DOS archive attribute should be mapped to the UNIX owner execute bit. The DOS archive bit is set when a file has been modified since its last backup. One motivation for this option it to keep Samba/your PC from making any file it touches from becoming executable under UNIX. This can be quite annoying for shared source code, documents, etc...

Note that this requires the *create mask* parameter to be set such that owner execute bit is not masked out (i.e. it must include 100). See the parameter *create mask* for details.

Default: **map archive = yes**

map hidden (S) — This controls whether DOS style hidden files should be mapped to the UNIX world execute bit.

Note that this requires the *create mask* to be set such that the world execute bit is not masked out (i.e. it must include 001). See the parameter *create mask* for details.

Default: **map hidden = no**

map system (S) — This controls whether DOS style system files should be mapped to the UNIX group execute bit.

Note that this requires the *create mask* to be set such that the group execute bit is not masked out (i.e. it must include 010). See the parameter *create mask* for details.

Default: **map system = no**

map to guest (G) — This parameter is only useful in security modes other than *security = share* - i.e. `user`, `server`, and `domain`.

This parameter can take three different values, which tell smbd(8) what to do with user login requests that don't match a valid UNIX user in some way.

The three settings are :

- `Never` - Means user login requests with an invalid password are rejected. This is the default.
- `Bad User` - Means user logins with an invalid password are rejected, unless the username does not exist, in which case it is treated as a guest login and mapped into the *guest account*.
- `Bad Password` - Means user logins with an invalid password are treated as a guest login and mapped into the guest account. Note that this can cause problems as it means that any user incorrectly typing their password will be silently logged on as "guest" - and will not know the reason they cannot access files they think they should - there will have been no message given to them that they got their

password wrong. Helpdesk services will *hate* you if you set the *map to guest* parameter this way :-).

Note that this parameter is needed to set up "Guest" share services when using *security* modes other than share. This is because in these modes the name of the resource being requested is *not* sent to the server until after the server has successfully authenticated the client so the server cannot make authentication decisions at the correct time (connection to the share) for "Guest" shares.

For people familiar with the older Samba releases, this parameter maps to the old compile-time setting of the `GUEST_SESSSETUP` value in local.h.

Default: **map to guest = Never**

Example: **map to guest = Bad User**

max connections (S) — This option allows the number of simultaneous connections to a service to be limited. If *max connections* is greater than 0 then connections will be refused if this number of connections to the service are already open. A value of zero mean an unlimited number of connections may be made.

Record lock files are used to implement this feature. The lock files will be stored in the directory specified by the *lock directory* option.

Default: **max connections = 0**

Example: **max connections = 10**

max disk size (G) — This option allows you to put an upper limit on the apparent size of disks. If you set this option to 100 then all shares will appear to be not larger than 100 MB in size.

Note that this option does not limit the amount of data you can put on the disk. In the above case you could still store much more than 100 MB on the disk, but if a client ever asks for the amount of free disk space or the total disk size then the result will be bounded by the amount specified in *max disk size*.

This option is primarily useful to work around bugs in some pieces of software that can't handle very large disks, particularly disks over 1GB in size.

A *max disk size* of 0 means no limit.

Default: **max disk size = 0**

Example: **max disk size = 1000**

max log size (G) — This option (an integer in kilobytes) specifies the max size the log file should grow to. Samba periodically checks the size and if it is exceeded it will rename the file, adding a `.old` extension.

A size of 0 means no limit.

Default: **max log size = 5000**

Example: **max log size = 1000**

max mux (G) — This option controls the maximum number of outstanding simultaneous SMB operations that Samba tells the client it will allow. You should never need to set this parameter.

Default: **max mux = 50**

max open files (G) — This parameter limits the maximum number of open files that one smbd(8) file serving process may have open for a client at any one time. The default for this parameter is set very high (10,000) as Samba uses only one bit per unopened file.

The limit of the number of open files is usually set by the UNIX per-process file descriptor limit rather than this parameter so you should never need to touch this parameter.

Default: **max open files = 10000**

max print jobs (S) — This parameter limits the maximum number of jobs allowable in a Samba printer queue at any given moment. If this number is exceeded, smbd(8) will remote "Out of Space" to the client. See all *total print jobs*.

Default: **max print jobs = 1000**

Example: **max print jobs = 5000**

max protocol (G) — The value of the parameter (a string) is the highest protocol level that will be supported by the server.

Possible values are :

- `CORE`: Earliest version. No concept of user names.
- `COREPLUS`: Slight improvements on CORE for efficiency.
- `LANMAN1`: First *modern* version of the protocol. Long filename support.
- `LANMAN2`: Updates to Lanman1 protocol.
- `NT1`: Current up to date version of the protocol. Used by Windows NT. Known as CIFS.

Normally this option should not be set as the automatic negotiation phase in the SMB protocol takes care of choosing the appropriate protocol.

See also *min protocol*

Default: **max protocol = NT1**

Example: **max protocol = LANMAN1**

max reported print jobs (S) — This parameter limits the maximum number of jobs displayed in a port monitor for Samba printer queue at any given moment. If this number is exceeded, the excess jobs will not be shown. A value of zero means there is no limit on the number of print jobs reported. See all *total print jobs* and *max print jobs* parameters.

Default: **max reported print jobs = 0**

Example: **max reported print jobs = 1000**

max smbd processes (G) — This parameter limits the maximum number of smbd(8) processes concurrently running on a system and is intended as a stopgap to prevent degrading service to clients in the event that the server has insufficient resources to handle more than this number of connections. Remember that under normal operating conditions, each user will have an smbd(8) associated with him or her to handle connections to all shares from a given host.

Default: **max smbd processes = 0** ## no limit

Example: **max smbd processes = 1000**

max ttl (G) — This option tells nmbd(8) what the default 'time to live' of NetBIOS names should be (in seconds) when **nmbd** is requesting a name using either a broadcast packet or from a WINS server. You should never need to change this parameter. The default is 3 days.

Default: **max ttl = 259200**

max wins ttl (G) — This option tells smbd(8) when acting as a WINS server (*wins support = yes*) what the maximum 'time to live' of NetBIOS names that **nmbd** will grant will be (in seconds). You should never need to change this parameter. The default is 6 days (518400 seconds).

See also the *min wins ttl* parameter.

Default: **max wins ttl = 518400**

max xmit (G) — This option controls the maximum packet size that will be negotiated by Samba. The default is 65535, which is the maximum. In some cases you may find you get better performance with a smaller value. A value below 2048 is likely to cause problems.

Default: **max xmit = 65535**

Example: **max xmit = 8192**

message command (G) — This specifies what command to run when the server receives a WinPopup style message.

This would normally be a command that would deliver the message somehow. How this is to be done is up to your imagination.

An example is:

message command = csh -c 'xedit %s;rm %s' &

This delivers the message using **xedit**, then removes it afterwards. *NOTE THAT IT IS VERY IMPORTANT THAT THIS COMMAND RETURN IMMEDIATELY.* That's why I have the '&' on the end. If it doesn't return immediately then your PCs may freeze when sending messages (they should recover after 30 seconds, hopefully).

All messages are delivered as the global guest user. The command takes the standard substitutions, although *%u* won't work (*%U* may be better in this case).

Apart from the standard substitutions, some additional ones apply. In particular:

- *%s* = the filename containing the message.
- *%t* = the destination that the message was sent to (probably the server name).
- *%f* = who the message is from.

You could make this command send mail, or whatever else takes your fancy. Please let us know of any really interesting ideas you have.

Here's a way of sending the messages as mail to root:

message command = /bin/mail -s 'message from %f on %m' root < %s; rm %s

If you don't have a message command then the message won't be delivered and Samba will tell the sender there was an error. Unfortunately WfWg totally ignores the error code and carries on regardless, saying that the message was delivered.

If you want to silently delete it then try:

message command = rm %s

Default: *no message command*

Example: **message command = csh -c 'xedit %s; rm %s' &**

min passwd length (G) — Synonym for *min password length*.

min password length (G) — This option sets the minimum length in characters of a plaintext password that **smbd** will accept when performing UNIX password changing.

See also *unix password sync*, *passwd program* and *passwd chat debug*.

Default: **min password length = 5**

min print space (S) — This sets the minimum amount of free disk space that must be available before a user will be able to spool a print job. It is specified in kilobytes. The default is 0, which means a user can always spool a print job.

See also the *printing* parameter.

Default: **min print space = 0**

Example: **min print space = 2000**

min protocol (G) — The value of the parameter (a string) is the lowest SMB protocol dialect than Samba will support. Please refer to the *max protocol* parameter for a list of valid protocol names and a brief description of each. You may also wish to refer to the C source code in `source/smbd/negprot.c` for a listing of known protocol dialects supported by clients.

If you are viewing this parameter as a security measure, you should also refer to the *lanman auth* parameter. Otherwise, you should never need to change this parameter.

Default : **min protocol = CORE**

Example : **min protocol = NT1** # disable DOS clients

min wins ttl (G) — This option tells nmbd(8) when acting as a WINS server (*wins support = yes*) what the minimum 'time to live' of NetBIOS names that **nmbd** will grant will be (in seconds). You should never need to change this parameter. The default is 6 hours (21600 seconds).

Default: **min wins ttl = 21600**

msdfs proxy (S) — This parameter indicates that the share is a stand-in for another CIFS share whose location is specified by the value of the parameter. When clients attempt to connect to this share, they are redirected to the proxied share using the SMB-Dfs protocol.

Only Dfs roots can act as proxy shares. Take a look at the *msdfs root* and *host msdfs* options to find out how to set up a Dfs root share.

Example: **msdfs proxy = \\\\otherserver\\someshare**

msdfs root (S) — If set to `yes`, Samba treats the share as a Dfs root and allows clients to browse the distributed file system tree rooted at the share directory. Dfs links are specified in the share directory by symbolic links of the form `msdfs:serverA\\shareA,serverB\\shareB` and so on. For more information on setting up a Dfs tree on Samba, refer to Chapter 16, *Hosting a Microsoft Distributed File System tree.*

See also *host msdfs*

Default: **msdfs root = no**

name cache timeout (G) — Specifies the number of seconds it takes before entries in samba's hostname resolve cache time out. If the timeout is set to 0. the caching is disabled.

Default: **name cache timeout = 660**

Example: **name cache timeout = 0**

name resolve order (G) — This option is used by the programs in the Samba suite to determine what naming services to use and in what order to resolve host names to IP addresses. Its main purpose to is to control how netbios name resolution is performed. The option takes a space separated string of name resolution options.

The options are: "lmhosts", "host", "wins" and "bcast". They cause names to be resolved as follows:

- `lmhosts` : Lookup an IP address in the Samba lmhosts file. If the line in lmhosts has no name type attached to the NetBIOS name (see the lmhosts(5) for details) then any name type matches for lookup.
- `host` : Do a standard host name to IP address resolution, using the system `/etc/hosts`, NIS, or DNS lookups. This method of name resolution is operating system depended for instance on IRIX or Solaris this may be controlled by the `/etc/nsswitch.conf` file. Note that this method is used only if the NetBIOS name type being queried is the 0x20 (server) name type or 0x1c (domain controllers). The latter case is only useful for active directory domains and results in a DNS query for the SRV RR entry matching _ldap._tcp.domain.
- `wins` : Query a name with the IP address listed in the *wins server* parameter. If no WINS server has been specified this method will be ignored.
- `bcast` : Do a broadcast on each of the known local interfaces listed in the *interfaces* parameter. This is the least reliable of the name resolution methods as it depends on the target host being on a locally connected subnet.

Default: **name resolve order = lmhosts host wins bcast**

Example: **name resolve order = lmhosts bcast host**

This will cause the local lmhosts file to be examined first, followed by a broadcast attempt, followed by a normal system hostname lookup.

When Samba is functioning in ADS security mode (**security = ads**) it is advised to use following settings for *name resolve order*:

name resolve order = wins bcast

DC lookups will still be done via DNS, but fallbacks to netbios names will not inundate your DNS servers with needless querys for DOMAIN<0x1c> lookups.

netbios aliases (G) — This is a list of NetBIOS names that nmbd will advertise as additional names by which the Samba server is known. This allows one machine to appear in browse lists under multiple names. If a machine is acting as a browse server or logon server none of these names will be advertised as either browse server or logon servers, only the primary name of the machine will be advertised with these capabilities.

See also *netbios name*.

Default: *empty string (no additional names)*

Example: **netbios aliases = TEST TEST1 TEST2**

netbios name (G) — This sets the NetBIOS name by which a Samba server is known. By default it is the same as the first component of the host's DNS name. If a machine is a browse server or logon server this name (or the first component of the hosts DNS name) will be the name that these services are advertised under.

See also *netbios aliases*.

Default: *machine DNS name*

Example: **netbios name = MYNAME**

netbios scope (G) — This sets the NetBIOS scope that Samba will operate under. This should not be set unless every machine on your LAN also sets this value.

nis homedir (G) — Get the home share server from a NIS map. For UNIX systems that use an automounter, the user's home directory will often be mounted on a workstation on demand from a remote server.

When the Samba logon server is not the actual home directory server, but is mounting the home directories via NFS then two network hops would be required to access the users home directory if the logon server told the client to use itself as the SMB server for home directories (one over SMB and one over NFS). This can be very slow.

This option allows Samba to return the home share as being on a different server to the logon server and as long as a Samba daemon is running on the home directory server, it will be mounted on the Samba client directly from the directory server. When Samba is returning the home share to the client, it will consult the NIS map specified in *homedir map* and return the server listed there.

Note that for this option to work there must be a working NIS system and the Samba server with this option must also be a logon server.

Default: **nis homedir = no**

non unix account range (G) — The non unix account range parameter specifies the range of 'user ids' that are allocated by the various 'non unix account' passdb backends. These backends allow the storage of passwords for users who don't exist in /etc/passwd. This is most often used for machine account creation. This range of ids should have no existing local or NIS users within it as strange conflicts can occur otherwise.

> **NOTE**
>
> 
>
> These userids never appear on the system and Samba will never 'become' these users. They are used only to ensure that the algorithmic RID mapping does not conflict with normal users.

Default: **non unix account range = <empty string>**

Example: **non unix account range = 10000-20000**

nt acl support (S) — This boolean parameter controls whether smbd(8) will attempt to map UNIX permissions into Windows NT access control lists. This parameter was formally a global parameter in releases prior to 2.2.2.

Default: **nt acl support = yes**

ntlm auth (G) — This parameter determines whether or not smbd(8) will attempt to authenticate users using the NTLM encrypted password response. If disabled, either the lanman password hash or an NTLMv2 response will need to be sent by the client.

If this option, and **lanman auth** are both disabled, then only NTLMv2 logins will be permited. Not all clients support NTLMv2, and most will require special configuration to us it.

Default : **ntlm auth = yes**

nt pipe support (G) — This boolean parameter controls whether smbd(8) will allow Windows NT clients to connect to the NT SMB specific `IPC$` pipes. This is a developer debugging option and can be left alone.

Default: **nt pipe support = yes**

nt status support (G) — This boolean parameter controls whether smbd(8) will negotiate NT specific status support with Windows NT/2k/XP clients. This is a developer debugging option and should be left alone. If this option is set to `no` then Samba offers exactly the same DOS error codes that versions prior to Samba 2.2.3 reported.

You should not need to ever disable this parameter.

Default: **nt status support = yes**

null passwords (G) — Allow or disallow client access to accounts that have null passwords.

See also smbpasswd(5).

Default: **null passwords = no**

obey pam restrictions (G) — When Samba 3.0 is configured to enable PAM support (i.e. –with-pam), this parameter will control whether or not Samba should obey PAM's account and session management directives. The default behavior is to use PAM for clear text authentication only and to ignore any account or session management. Note that Samba always ignores PAM for authentication in the case of *encrypt passwords = yes*. The reason is that PAM modules cannot support the challenge/response authentication mechanism needed in the presence of SMB password encryption.

Default: **obey pam restrictions = no**

only guest (S) — A synonym for *guest only*.

only user (S) — This is a boolean option that controls whether connections with usernames not in the *user* list will be allowed. By default this option is disabled so that a client can supply a username to be used by the server. Enabling this parameter will force the server to only use the login names from the *user* list and is only really useful in share level security.

Note that this also means Samba won't try to deduce usernames from the service name. This can be annoying for the [homes] section. To get around this you could use **user = %S** which means your *user* list will be just the service name, which for home directories is the name of the user.

See also the *user* parameter.

Default: **only user = no**

oplock break wait time (G) — This is a tuning parameter added due to bugs in both Windows 9x and WinNT. If Samba responds to a client too quickly when that client issues an SMB that can cause an oplock break request, then the network client can fail and not respond to the break request. This tuning parameter (which is set in milliseconds) is the amount of time Samba will wait before sending an oplock break request to such (broken) clients.

DO NOT CHANGE THIS PARAMETER UNLESS YOU HAVE READ AND UNDERSTOOD THE SAMBA OPLOCK CODE.

Default: **oplock break wait time = 0**

oplock contention limit (S) — This is a *very* advanced smbd(8) tuning option to improve the efficiency of the granting of oplocks under multiple client contention for the same file.

In brief it specifies a number, which causes smbd(8)not to grant an oplock even when requested if the approximate number of clients contending for an oplock on the same file goes over this limit. This causes **smbd** to behave in a similar way to Windows NT.

DO NOT CHANGE THIS PARAMETER UNLESS YOU HAVE READ AND UNDERSTOOD THE SAMBA OPLOCK CODE.

Default: **oplock contention limit = 2**

oplocks (S) — This boolean option tells **smbd** whether to issue oplocks (opportunistic locks) to file open requests on this share. The oplock code can dramatically (approx. 30% or more) improve the speed of access to files on Samba servers. It allows the clients to aggressively cache files locally and you may want to disable this option for unreliable network environments (it is turned on by default in Windows NT Servers). For more information see the file `Speed.txt` in the Samba `docs/` directory.

Oplocks may be selectively turned off on certain files with a share. See the *veto oplock files* parameter. On some systems oplocks are recognized by the underlying operating system. This allows data synchronization between all access to oplocked files, whether it be via Samba or NFS or a local UNIX process. See the *kernel oplocks* parameter for details.

See also the *kernel oplocks* and *level2 oplocks* parameters.

Default: **oplocks = yes**

os2 driver map (G) — The parameter is used to define the absolute path to a file containing a mapping of Windows NT printer driver names to OS/2 printer driver names. The format is:

<nt driver name> = <os2 driver name>.<device name>

For example, a valid entry using the HP LaserJet 5 printer driver would appear as **HP LaserJet 5L = LASERJET.HP LaserJet 5L**.

The need for the file is due to the printer driver namespace problem described in Chapter 17, *Classical Printing Support*. For more details on OS/2 clients, please refer to Chapter 37, *Samba and Other CIFS Clients*.

Default: **os2 driver map = <empty string>**

os level (G) — This integer value controls what level Samba advertises itself as for browse elections. The value of this parameter determines whether nmbd(8) has a chance of becoming a local master browser for the *WORKGROUP* in the local broadcast area.

Note :By default, Samba will win a local master browsing election over all Microsoft operating systems except a Windows NT 4.0/2000 Domain Controller. This means that a misconfigured Samba host can effectively isolate a subnet for browsing purposes. See `BROWSING.txt` in the Samba `docs/` directory for details.

Default: **os level = 20**

Example: **os level = 65**

pam password change (G) — With the addition of better PAM support in Samba 2.2, this parameter, it is possible to use PAM's password change control flag for Samba. If enabled, then PAM will be used for password changes when requested by an SMB client instead of the program listed in *passwd program*. It should be possible to enable this without changing your *passwd chat* parameter for most setups.

Default: **pam password change = no**

panic action (G) — This is a Samba developer option that allows a system command to be called when either smbd(8) or smbd(8) crashes. This is usually used to draw attention to the fact that a problem occurred.

Default: **panic action = <empty string>**

Example: **panic action = "/bin/sleep 90000"**

paranoid server security (G) — Some version of NT 4.x allow non-guest users with a bad passowrd. When this option is enabled, samba will not use a broken NT 4.x server as password server, but instead complain to the logs and exit.

Disabling this option prevents Samba from making this check, which involves deliberatly attempting a bad logon to the remote server.

Default: **paranoid server security = yes**

passdb backend (G) — This option allows the administrator to chose which backends to retrieve and store passwords with. This allows (for example) both smbpasswd and tdbsam to be used without a recompile. Multiple backends can be specified, separated by spaces. The backends will be searched in the order they are specified. New users are always added to the first backend specified.

This parameter is in two parts, the backend's name, and a 'location' string that has meaning only to that particular backed. These are separated by a : character.

Available backends can include:

- **smbpasswd** - The default smbpasswd backend. Takes a path to the smbpasswd file as an optional argument.
- **tdbsam** - The TDB based password storage backend. Takes a path to the TDB as an optional argument (defaults to passdb.tdb in the *private dir* directory.
- **ldapsam** - The LDAP based passdb backend. Takes an LDAP URL as an optional argument (defaults to **ldap://localhost**)

  LDAP connections should be secured where possible. This may be done using either Start-TLS (see *ldap ssl*) or by specifying *ldaps://* in the URL argument.
- **nisplussam** - The NIS+ based passdb backend. Takes name NIS domain as an optional argument. Only works with sun NIS+ servers.

- **mysql** - The MySQL based passdb backend. Takes an identifier as argument. Read the Samba HOWTO Collection for configuration details.

Default: **passdb backend = smbpasswd**

Example: **passdb backend = tdbsam:/etc/samba/private/passdb.tdb smbpasswd:/etc/samba/smbpasswd**

Example: **passdb backend = ldapsam:ldaps://ldap.example.com**

Example: **passdb backend = mysql:my_plugin_args tdbsam**

passwd chat (G) — This string controls the *"chat"* conversation that takes places between smbd(8) and the local password changing program to change the user's password. The string describes a sequence of response-receive pairs that smbd(8) uses to determine what to send to the *passwd program* and what to expect back. If the expected output is not received then the password is not changed.

This chat sequence is often quite site specific, depending on what local methods are used for password control (such as NIS etc).

Note that this parameter only is only used if the *unix password sync* parameter is set to `yes`. This sequence is then called *AS ROOT* when the SMB password in the smbpasswd file is being changed, without access to the old password cleartext. This means that root must be able to reset the user's password without knowing the text of the previous password. In the presence of NIS/YP, this means that the passwd program must be executed on the NIS master.

The string can contain the macro *%n* which is substituted for the new password. The chat sequence can also contain the standard macros `\\n`, `\\r`, `\\t` and `\\s` to give line-feed, carriage-return, tab and space. The chat sequence string can also contain a '*' which matches any sequence of characters. Double quotes can be used to collect strings with spaces in them into a single string.

If the send string in any part of the chat sequence is a full stop ".", then no string is sent. Similarly, if the expect string is a full stop then no string is expected.

If the *pam password change* parameter is set to `yes`, the chat pairs may be matched in any order, and success is determined by the PAM result, not any particular output. The \n macro is ignored for PAM conversions.

See also *unix password sync*, *passwd program* ,*passwd chat debug* and *pam password change*.

Default: **passwd chat = *new*password* %n\\n *new*password* %n\\n *changed***

Example: **passwd chat = "*Enter OLD password*" %o\\n "*Enter NEW password*" %n\\n "*Reenter NEW password*" %n\\n "*Password changed*"**

passwd chat debug (G) — This boolean specifies if the passwd chat script parameter is run in *debug* mode. In this mode the strings passed to and received from the passwd chat are printed in the smbd(8) log with a *debug level* of 100. This is a dangerous option as it will allow plaintext passwords to be seen in the **smbd** log. It is available to help Samba admins debug their *passwd chat* scripts when calling the *passwd program* and should be turned off after this has been done. This option has no effect if the *pam password change* paramter is set. This parameter is off by default.

See also *passwd chat*, *pam password change*, *passwd program*.

Default: **passwd chat debug = no**

passwd program (G) — The name of a program that can be used to set UNIX user passwords. Any occurrences of *%u* will be replaced with the user name. The user name is checked for existence before calling the password changing program.

Also note that many passwd programs insist in *reasonable* passwords, such as a minimum length, or the inclusion of mixed case chars and digits. This can pose a problem as some clients (such as Windows for Workgroups) uppercase the password before sending it.

Note that if the *unix password sync* parameter is set to yes then this program is called *AS ROOT* before the SMB password in the smbpasswd file is changed. If this UNIX password change fails, then **smbd** will fail to change the SMB password also (this is by design).

If the *unix password sync* parameter is set this parameter *MUST USE ABSOLUTE PATHS* for *ALL* programs called, and must be examined for security implications. Note that by default *unix password sync* is set to no.

See also *unix password sync*.

Default: **passwd program = /bin/passwd**

Example: **passwd program = /sbin/npasswd %u**

password level (G) — Some client/server combinations have difficulty with mixed-case passwords. One offending client is Windows for Workgroups, which for some reason forces passwords to upper case when using the LANMAN1 protocol, but leaves them alone when using COREPLUS! Another problem child is the Windows 95/98 family of operating systems. These clients upper case clear text passwords even when NT LM 0.12 selected by the protocol negotiation request/response.

This parameter defines the maximum number of characters that may be upper case in passwords.

For example, say the password given was "FRED". If *password level* is set to 1, the following combinations would be tried if "FRED" failed:

"Fred", "fred", "fRed", "frEd","freD"

If *password level* was set to 2, the following combinations would also be tried:

"FRed", "FrEd", "FreD", "fREd", "fReD", "frED", ..

And so on.

The higher value this parameter is set to the more likely it is that a mixed case password will be matched against a single case password. However, you should be aware that use of this parameter reduces security and increases the time taken to process a new connection.

A value of zero will cause only two attempts to be made - the password as is and the password in all-lower case.

Default: **password level = 0**

Example: **password level = 4**

password server (G) — By specifying the name of another SMB server or Active Directory domain controller with this option, and using **security = [ads|domain|server]** it is possible to get Samba to to do all its username/password validation using a specific remote server.

This option sets the name or IP address of the password server to use. New syntax has been added to support defining the port to use when connecting to the server the case of an ADS realm. To define a port other than the default LDAP port of 389, add the port number using a colon after the name or IP address (e.g. 192.168.1.100:389). If you do not specify a port, Samba will use the standard LDAP port of tcp/389. Note that port numbers have no effect on password servers for Windows NT 4.0 domains or netbios connections.

If parameter is a name, it is looked up using the parameter *name resolve order* and so may resolved by any method and order described in that parameter.

The password server must be a machine capable of using the "LM1.2X002" or the "NT LM 0.12" protocol, and it must be in user level security mode.

NOTE

Using a password server means your UNIX box (running Samba) is only as secure as your password server. *DO NOT CHOOSE A PASSWORD SERVER THAT YOU DON'T COMPLETELY TRUST.*

Never point a Samba server at itself for password serving. This will cause a loop and could lock up your Samba server!

The name of the password server takes the standard substitutions, but probably the only useful one is *%m*, which means the Samba server will use the incoming client as the password server. If you use this then you better trust your clients, and you had better restrict them with hosts allow!

If the *security* parameter is set to `domain` or `ads`, then the list of machines in this option must be a list of Primary or Backup Domain controllers for the Domain or the character '*', as the Samba server is effectively in that domain, and will use cryptographically authenticated RPC calls to authenticate the user logging on. The advantage of using **security = domain** is that if you list several hosts in the *password server* option then **smbd** will try each in turn till it finds one that responds. This is useful in case your primary server goes down.

If the *password server* option is set to the character '*', then Samba will attempt to auto-locate the Primary or Backup Domain controllers to authenticate against by doing a query for the name `WORKGROUP<1C>` and then contacting each server returned in the list of IP addresses from the name resolution source.

If the list of servers contains both names/IP's and the '*' character, the list is treated as a list of preferred domain controllers, but an auto lookup of all remaining DC's will be added to the list as well. Samba will not attempt to optimize this list by locating the closest DC.

If the *security* parameter is set to `server`, then there are different restrictions that **security = domain** doesn't suffer from:

- You may list several password servers in the *password server* parameter, however if an **smbd** makes a connection to a password server, and then the password server fails, no more users will be able to be authenticated from this **smbd**. This is a restriction of the SMB/CIFS protocol when in **security = server** mode and cannot be fixed in Samba.
- If you are using a Windows NT server as your password server then you will have to ensure that your users are able to login from the Samba server, as when in **security = server** mode the network logon will appear to come from there rather than from the users workstation.

See also the *security* parameter.

Default: **password server = <empty string>**

Example: **password server = NT-PDC, NT-BDC1, NT-BDC2, ***

Example: **password server = windc.mydomain.com:389 192.168.1.101 ***

Example: **password server = ***

path (S) — This parameter specifies a directory to which the user of the service is to be given access. In the case of printable services, this is where print data will spool prior to being submitted to the host for printing.

For a printable service offering guest access, the service should be readonly and the path should be world-writeable and have the sticky bit set. This is not mandatory of course, but you probably won't get the results you expect if you do otherwise.

Any occurrences of *%u* in the path will be replaced with the UNIX username that the client is using on this connection. Any occurrences of *%m* will be replaced by the NetBIOS name of the machine they are connecting from. These replacements are very useful for setting up pseudo home directories for users.

Note that this path will be based on *root dir* if one was specified.

Default: *none*

Example: **path = /home/fred**

pid directory (G) — This option specifies the directory where pid files will be placed.

Default: **pid directory = ${prefix}/var/locks**

Example: **pid directory = /var/run/**

posix locking (S) — The smbd(8) daemon maintains an database of file locks obtained by SMB clients. The default behavior is to map this internal database to POSIX locks. This means that file locks obtained by SMB clients are consistent with those seen by POSIX compliant applications accessing the files via a non-SMB method (e.g. NFS or local file access). You should never need to disable this parameter.

Default: **posix locking = yes**

postexec (S) — This option specifies a command to be run whenever the service is disconnected. It takes the usual substitutions. The command may be run as the root on some systems.

An interesting example may be to unmount server resources:

postexec = /etc/umount /cdrom

See also *preexec*.

Default: *none (no command executed)*

Example: **postexec = echo \"%u disconnected from %S from %m (%I)\" >> /tmp/log**

preexec (S) — This option specifies a command to be run whenever the service is connected to. It takes the usual substitutions.

An interesting example is to send the users a welcome message every time they log in. Maybe a message of the day? Here is an example:

preexec = csh -c 'echo \"Welcome to %S!\" | /usr/local/samba/bin/smbclient -M %m -I %I' &

Of course, this could get annoying after a while :-)

See also *preexec close* and *postexec*.

Default: *none (no command executed)*

Example: **preexec = echo \"%u connected to %S from %m (%I)\" >> /tmp/log**

preexec close (S) — This boolean option controls whether a non-zero return code from *preexec* should close the service being connected to.

Default: **preexec close = no**

prefered master (G) — Synonym for *preferred master* for people who cannot spell :-).

preferred master (G) — This boolean parameter controls if nmbd(8) is a preferred master browser for its workgroup.

If this is set to `yes`, on startup, **nmbd** will force an election, and it will have a slight advantage in winning the election. It is recommended that this parameter is used in conjunction with *domain master* = **yes**, so that **nmbd** can guarantee becoming a domain master.

Use this option with caution, because if there are several hosts (whether Samba servers, Windows 95 or NT) that are preferred master browsers on the same subnet, they will each periodically and continuously attempt to become the local master browser. This will result in unnecessary broadcast traffic and reduced browsing capabilities.

See also *os level*.

Default: **preferred master = auto**

preload (G) — This is a list of services that you want to be automatically added to the browse lists. This is most useful for homes and printers services that would otherwise not be visible.

Note that if you just want all printers in your printcap file loaded then the *load printers* option is easier.

Default: *no preloaded services*

Example: **preload = fred lp colorlp**

preload modules (G) — This is a list of paths to modules that should be loaded into smbd before a client connects. This improves the speed of smbd when reacting to new connections somewhat.

Default: **preload modules =**

Example: **preload modules = /usr/lib/samba/passdb/mysql.so+++**

preserve case (S) — This controls if new filenames are created with the case that the client passes, or if they are forced to be the *default case*.

Default: **preserve case = yes**

See the section on NAME MANGLING for a fuller discussion.

printable (S) — If this parameter is `yes`, then clients may open, write to and submit spool files on the directory specified for the service.

Note that a printable service will ALWAYS allow writing to the service path (user privileges permitting) via the spooling of print data. The *read only* parameter controls only non-printing access to the resource.

Default: **printable = no**

printcap (G) — Synonym for *printcap name*.

printcap name (S) — This parameter may be used to override the compiled-in default printcap name used by the server (usually `/etc/printcap`). See the discussion of the [printers] section above for reasons why you might want to do this.

To use the CUPS printing interface set **printcap name = cups**. This should be supplemented by an addtional setting printing = cups in the [global] section. **printcap name = cups** will use the "dummy" printcap created by CUPS, as specified in your CUPS configuration file.

On System V systems that use **lpstat** to list available printers you can use **printcap name = lpstat** to automatically obtain lists of available printers. This is the default for systems that define SYSV at configure time in Samba (this includes most System V based systems). If *printcap name* is set to **lpstat** on these systems then Samba will launch **lpstat -v** and attempt to parse the output to obtain a printer list.

A minimal printcap file would look something like this:

```
print1|My Printer 1
print2|My Printer 2
print3|My Printer 3
print4|My Printer 4
print5|My Printer 5
```

where the '|' separates aliases of a printer. The fact that the second alias has a space in it gives a hint to Samba that it's a comment.

> NOTE
>
> 
>
> Under AIX the default printcap name is `/etc/qconfig`. Samba will assume the file is in AIX `qconfig` format if the string `qconfig` appears in the printcap filename.

Default: **printcap name = /etc/printcap**

Example: **printcap name = /etc/myprintcap**

print command (S) — After a print job has finished spooling to a service, this command will be used via a **system()** call to process the spool file. Typically the command specified will submit the spool file to the host's printing subsystem, but there is no requirement that this be the case. The server will not remove the spool file, so whatever command you specify should remove the spool file when it has been processed, otherwise you will need to manually remove old spool files.

The print command is simply a text string. It will be used verbatim after macro substitutions have been made:

%s, %f - the path to the spool file name

%p - the appropriate printer name

%J - the job name as transmitted by the client.

%c - The number of printed pages of the spooled job (if known).

%z - the size of the spooled print job (in bytes)

The print command *MUST* contain at least one occurrence of *%s* or *%f* - the *%p* is optional. At the time a job is submitted, if no printer name is supplied the *%p* will be silently removed from the printer command.

If specified in the [global] section, the print command given will be used for any printable service that does not have its own print command specified.

If there is neither a specified print command for a printable service nor a global print command, spool files will be created but not processed and (most importantly) not removed.

Note that printing may fail on some UNIXes from the `nobody` account. If this happens then create an alternative guest account that can print and set the *guest account* in the [global] section.

You can form quite complex print commands by realizing that they are just passed to a shell. For example the following will log a print job, print the file, then remove it. Note that ';' is the usual separator for command in shell scripts.

print command = echo Printing %s >> /tmp/print.log; lpr -P %p %s; rm %s

You may have to vary this command considerably depending on how you normally print files on your system. The default for the parameter varies depending on the setting of the *printing* parameter.

Default: For **printing = BSD, AIX, QNX, LPRNG or PLP :**

print command = lpr -r -P%p %s

For **printing = SYSV or HPUX :**

print command = lp -c -d%p %s; rm %s

For **printing = SOFTQ :**

print command = lp -d%p -s %s; rm %s

For printing = CUPS : If SAMBA is compiled against libcups, then printcap = cups uses the CUPS API to submit jobs, etc. Otherwise it maps to the System V commands with the -oraw option for printing, i.e. it uses **lp -c -d%p -oraw; rm %s**. With **printing = cups**, and if SAMBA is compiled against libcups, any manually set print command will be ignored.

Example: **print command = /usr/local/samba/bin/myprintscript %p %s**

printer (S) — Synonym for *printer name*.

printer admin (S) — This is a list of users that can do anything to printers via the remote administration interfaces offered by MS-RPC (usually using a NT workstation). Note that the root user always has admin rights.

Default: **printer admin = <empty string>**

Example: **printer admin = admin, @staff**

printer name (S) — This parameter specifies the name of the printer to which print jobs spooled through a printable service will be sent.

If specified in the [global] section, the printer name given will be used for any printable service that does not have its own printer name specified.

Default: *none (but may be lp on many systems)*

Example: **printer name = laserwriter**

printing (S) — This parameters controls how printer status information is interpreted on your system. It also affects the default values for the *print command*, *lpq command*, *lppause command*, *lpresume command*, and *lprm command* if specified in the [global] section.

Currently nine printing styles are supported. They are BSD, AIX, LPRNG, PLP, SYSV, HPUX, QNX, SOFTQ, and CUPS.

To see what the defaults are for the other print commands when using the various options use the testparm(1) program.

This option can be set on a per printer basis

See also the discussion in the [printers] section.

print ok (S) — Synonym for *printable*.

private dir (G) — This parameters defines the directory smbd will use for storing such files as smbpasswd and secrets.tdb.

Default :**private dir = ${prefix}/private**

profile acls (S) — This boolean parameter controls whether smbd(8) This boolean parameter was added to fix the problems that people have been having with storing user profiles on Samba shares from Windows 2000 or Windows XP clients. New versions of Windows 2000 or Windows XP service packs do security ACL checking on the owner and ability to write of the profile directory stored on a local workstation when copied from a Samba share.

When not in domain mode with winbindd then the security info copied onto the local workstation has no meaning to the logged in user (SID) on that workstation so the profile storing fails. Adding this parameter onto a share used for profile storage changes two things about the returned Windows ACL. Firstly it changes the owner and group owner of all reported files and directories to be BUILTIN\\Administrators, BUILTIN\\Users respectively (SIDs S-1-5-32-544, S-1-5-32-545). Secondly it adds an ACE entry of "Full Control" to the SID BUILTIN\\Users to every returned ACL. This will allow any Windows 2000 or XP workstation user to access the profile.

Note that if you have multiple users logging on to a workstation then in order to prevent them from being able to access each others profiles you must remove the "Bypass traverse checking" advanced user right. This will prevent access to other users profile directories as the top level profile directory (named after the user) is created by the workstation profile code and has an ACL restricting entry to the directory tree to the owning user.

Default: **profile acls = no**

protocol (G) — Synonym for *max protocol*.

public (S) — Synonym for *guest ok*.

queuepause command (S) — This parameter specifies the command to be executed on the server host in order to pause the printer queue.

This command should be a program or script which takes a printer name as its only parameter and stops the printer queue, such that no longer jobs are submitted to the printer.

This command is not supported by Windows for Workgroups, but can be issued from the Printers window under Windows 95 and NT.

If a *%p* is given then the printer name is put in its place. Otherwise it is placed at the end of the command.

Note that it is good practice to include the absolute path in the command as the PATH may not be available to the server.

Default: *depends on the setting of* *printing*

Example: **queuepause command = disable %p**

queueresume command (S) — This parameter specifies the command to be executed on the server host in order to resume the printer queue. It is the command to undo the behavior that is caused by the previous parameter (*queuepause command*).

This command should be a program or script which takes a printer name as its only parameter and resumes the printer queue, such that queued jobs are resubmitted to the printer.

This command is not supported by Windows for Workgroups, but can be issued from the Printers window under Windows 95 and NT.

If a *%p* is given then the printer name is put in its place. Otherwise it is placed at the end of the command.

Note that it is good practice to include the absolute path in the command as the PATH may not be available to the server.

Default: *depends on the setting of* *printing*

Example: **queuepause command = enable %p**

read bmpx (G) — This boolean parameter controls whether smbd(8) will support the "Read Block Multiplex" SMB. This is now rarely used and defaults to no. You should never need to set this parameter.

Default: **read bmpx = no**

read list (S) — This is a list of users that are given read-only access to a service. If the connecting user is in this list then they will not be given write access, no matter what the *read only* option is set to. The list can include group names using the syntax described in the *invalid users* parameter.

See also the *write list* parameter and the *invalid users* parameter.

Default: **read list = <empty string>**

Example: **read list = mary, @students**

read only (S) — An inverted synonym is *writeable*.

If this parameter is yes, then users of a service may not create or modify files in the service's directory.

Note that a printable service (**printable = yes**) will *ALWAYS* allow writing to the directory (user privileges permitting), but only via spooling operations.

Default: **read only = yes**

read raw (G) — This parameter controls whether or not the server will support the raw read SMB requests when transferring data to clients.

If enabled, raw reads allow reads of 65535 bytes in one packet. This typically provides a major performance benefit.

However, some clients either negotiate the allowable block size incorrectly or are incapable of supporting larger block sizes, and for these clients you may need to disable raw reads.

In general this parameter should be viewed as a system tuning tool and left severely alone. See also *write raw*.

Default: **read raw = yes**

read size (G) — The option *read size* affects the overlap of disk reads/writes with network reads/writes. If the amount of data being transferred in several of the SMB commands (currently SMBwrite, SMBwriteX and SMBreadbraw) is larger than this value then the server begins writing the data before it has received the whole packet from the network, or in the case of SMBreadbraw, it begins writing to the network before all the data has been read from disk.

This overlapping works best when the speeds of disk and network access are similar, having very little effect when the speed of one is much greater than the other.

The default value is 16384, but very little experimentation has been done yet to determine the optimal value, and it is likely that the best value will vary greatly between systems anyway. A value over 65536 is pointless and will cause you to allocate memory unnecessarily.

Default: **read size = 16384**

Example: **read size = 8192**

realm (G) — This option specifies the kerberos realm to use. The realm is used as the ADS equivalent of the NT4 **domain**. It is usually set to the DNS name of the kerberos server.

Default: **realm =**

Example: **realm = mysambabox.mycompany.com**

remote announce (G) — This option allows you to setup nmbd(8)to periodically announce itself to arbitrary IP addresses with an arbitrary workgroup name.

This is useful if you want your Samba server to appear in a remote workgroup for which the normal browse propagation rules don't work. The remote workgroup can be anywhere that you can send IP packets to.

For example:

remote announce = 192.168.2.255/SERVERS 192.168.4.255/STAFF

the above line would cause **nmbd** to announce itself to the two given IP addresses using the given workgroup names. If you leave out the workgroup name then the one given in the *workgroup* parameter is used instead.

The IP addresses you choose would normally be the broadcast addresses of the remote networks, but can also be the IP addresses of known browse masters if your network config is that stable.

See Chapter 9, *Network Browsing*.

Default: **remote announce = <empty string>**

remote browse sync (G) — This option allows you to setup nmbd(8) to periodically request synchronization of browse lists with the master browser of a Samba server that is on a remote segment. This option will allow you to gain browse lists for multiple workgroups across routed networks. This is done in a manner that does not work with any non-Samba servers.

This is useful if you want your Samba server and all local clients to appear in a remote workgroup for which the normal browse propagation rules don't work. The remote workgroup can be anywhere that you can send IP packets to.

For example:

remote browse sync = 192.168.2.255 192.168.4.255

the above line would cause **nmbd** to request the master browser on the specified subnets or addresses to synchronize their browse lists with the local server.

The IP addresses you choose would normally be the broadcast addresses of the remote networks, but can also be the IP addresses of known browse masters if your network config is that stable. If a machine IP address is given Samba makes NO attempt to validate that the remote machine is available, is listening, nor that it is in fact the browse master on its segment.

Default: **remote browse sync = <empty string>**

restrict anonymous (G) — The setting of this parameter determines whether user and group list information is returned for an anonymous connection. and mirrors the effects of the `HKEY_LOCAL_MACHINE\SYSTEM\CurrentControlSet \Control\LSA\RestrictAnonymous` registry key in Windows 2000 and Windows NT. When set to 0, user and group list information is returned to anyone who asks. When set to 1, only an authenticated user can retrive user and group list information. For the value 2, supported by Windows 2000/XP and Samba, no anonymous connections are allowed at all. This can break third party and Microsoft applications which expect to be allowed to perform operations anonymously.

The security advantage of using restrict anonymous = 1 is dubious, as user and group list information can be obtained using other means.

NOTE

The security advantage of using restrict anonymous = 2 is removed by setting *guest ok* = yes on any share.

Default: **restrict anonymous = 0**

root (G) — Synonym for *root directory"*.

root dir (G) — Synonym for *root directory"*.

root directory (G) — The server will **chroot()** (i.e. Change its root directory) to this directory on startup. This is not strictly necessary for secure operation. Even without it the server will deny access to files not in one of the service entries. It may also check for, and deny access to, soft links to other parts of the filesystem, or attempts to use ".." in file names to access other directories (depending on the setting of the *wide links* parameter).

Adding a *root directory* entry other than "/" adds an extra level of security, but at a price. It absolutely ensures that no access is given to files not in the sub-tree specified in the *root directory* option, *including* some files needed for complete operation of the server. To maintain full operability of the server you will need to mirror some system files into the *root directory* tree. In particular you will need to mirror `/etc/passwd` (or a subset of it), and any binaries or configuration files needed for printing (if required). The set of files that must be mirrored is operating system dependent.

Default: **root directory = /**

Example: **root directory = /homes/smb**

root postexec (S) — This is the same as the *postexec* parameter except that the command is run as root. This is useful for unmounting filesystems (such as CDROMs) after a connection is closed.

See also *postexec*.

Default: **root postexec = <empty string>**

root preexec (S) — This is the same as the *preexec* parameter except that the command is run as root. This is useful for mounting filesystems (such as CDROMs) when a connection is opened.

See also *preexec* and *preexec close*.

Default: **root preexec = <empty string>**

root preexec close (S) — This is the same as the *preexec close* parameter except that the command is run as root.

See also *preexec* and *preexec close*.

Default: **root preexec close = no**

security (G) — This option affects how clients respond to Samba and is one of the most important settings in the `smb.conf` file.

The option sets the "security mode bit" in replies to protocol negotiations with smbd(8) to turn share level security on or off. Clients decide based on this bit whether (and how) to transfer user and password information to the server.

The default is **security = user**, as this is the most common setting needed when talking to Windows 98 and Windows NT.

The alternatives are **security = share**, **security = server** or **security = domain**.

In versions of Samba prior to 2.0.0, the default was **security = share** mainly because that was the only option at one stage.

There is a bug in WfWg that has relevance to this setting. When in user or server level security a WfWg client will totally ignore the password you type in the "connect drive" dialog box. This makes it very difficult (if not impossible) to connect to a Samba service as anyone except the user that you are logged into WfWg as.

If your PCs use usernames that are the same as their usernames on the UNIX machine then you will want to use **security = user**. If you mostly use usernames that don't exist on the UNIX box then use **security = share**.

You should also use **security = share** if you want to mainly setup shares without a password (guest shares). This is commonly used for a shared printer server. It is more difficult to setup guest shares with **security = user**, see the *`map to guest`* parameter for details.

It is possible to use **smbd** in a *hybrid mode* where it is offers both user and share level security under different *`NetBIOS aliases`*.

The different settings will now be explained.

SECURITY = SHARE

When clients connect to a share level security server they need not log onto the server with a valid username and password before attempting to connect to a shared resource (although modern clients such as Windows 95/98 and Windows NT will send a logon request with a username but no password when talking to a **security = share** server). Instead, the clients send authentication information (passwords) on a per-share basis, at the time they attempt to connect to that share.

Note that **smbd** *ALWAYS* uses a valid UNIX user to act on behalf of the client, even in **security = share** level security.

As clients are not required to send a username to the server in share level security, **smbd** uses several techniques to determine the correct UNIX user to use on behalf of the client.

A list of possible UNIX usernames to match with the given client password is constructed using the following methods :

- If the *`guest only`* parameter is set, then all the other stages are missed and only the *`guest account`* username is checked.
- Is a username is sent with the share connection request, then this username (after mapping - see *`username map`*), is added as a potential username.

- If the client did a previous *logon* request (the SessionSetup SMB call) then the username sent in this SMB will be added as a potential username.
- The name of the service the client requested is added as a potential username.
- The NetBIOS name of the client is added to the list as a potential username.
- Any users on the *user* list are added as potential usernames.

If the *guest only* parameter is not set, then this list is then tried with the supplied password. The first user for whom the password matches will be used as the UNIX user.

If the *guest only* parameter is set, or no username can be determined then if the share is marked as available to the *guest account*, then this guest user will be used, otherwise access is denied.

Note that it can be *very* confusing in share-level security as to which UNIX username will eventually be used in granting access.

See also the section NOTE ABOUT USERNAME/PASSWORD VALIDATION.

SECURITY = USER

This is the default security setting in Samba 3.0. With user-level security a client must first "log-on" with a valid username and password (which can be mapped using the *username map* parameter). Encrypted passwords (see the *encrypted passwords* parameter) can also be used in this security mode. Parameters such as *user* and *guest only* if set are then applied and may change the UNIX user to use on this connection, but only after the user has been successfully authenticated.

Note that the name of the resource being requested is *not* sent to the server until after the server has successfully authenticated the client. This is why guest shares don't work in user level security without allowing the server to automatically map unknown users into the *guest account*. See the *map to guest* parameter for details on doing this.

See also the section NOTE ABOUT USERNAME/PASSWORD VALIDATION.

SECURITY = DOMAIN

This mode will only work correctly if net(8) has been used to add this machine into a Windows NT Domain. It expects the *encrypted passwords* parameter to be set to `yes`. In this mode Samba will try to validate the username/password by passing it to a Windows NT Primary or Backup Domain Controller, in exactly the same way that a Windows NT Server would do.

Note that a valid UNIX user must still exist as well as the account on the Domain Controller to allow Samba to have a valid UNIX account to map file access to.

Note that from the client's point of view **security = domain** is the same as **security = user**. It only affects how the server deals with the authentication, it does not in any way affect what the client sees.

Note that the name of the resource being requested is *not* sent to the server until after the server has successfully authenticated the client. This is why guest shares don't work in user level security without allowing the server to automatically map unknown users into the *guest account*. See the *map to guest* parameter for details on doing this.

See also the section NOTE ABOUT USERNAME/PASSWORD VALIDATION.

See also the *password server* parameter and the *encrypted passwords* parameter.

SECURITY = SERVER

In this mode Samba will try to validate the username/password by passing it to another SMB server, such as an NT box. If this fails it will revert to **security = user**. It expects the *encrypted passwords* parameter to be set to `yes`, unless the remote server does not support them. However note that if encrypted passwords have been negotiated then Samba cannot revert back to checking the UNIX password file, it must have a valid `smbpasswd` file to check users against. See the chapter about the User Database in the Samba HOWTO Collection for details on how to set this up.

NOTE

This mode of operation has significant pitfalls, due to the fact that is activly initiates a man-in-the-middle attack on the remote SMB server. In particular, this mode of operation can cause significant resource consuption on the PDC, as it must maintain an active connection for the duration of the user's session. Furthermore, if this connection is lost, there is no way to reestablish it, and futher authenticaions to the Samba server may fail. (From a single client, till it disconnects).

NOTE

From the client's point of view **security = server** is the same as **security = user**. It only affects how the server deals with the authentication, it does not in any way affect what the client sees.

Note that the name of the resource being requested is *not* sent to the server until after the server has successfully authenticated the client. This is why guest shares don't work in user level security without allowing the server to automatically map unknown users into the *guest account*. See the *map to guest* parameter for details on doing this.

See also the section NOTE ABOUT USERNAME/PASSWORD VALIDATION.

See also the *password server* parameter and the *encrypted passwords* parameter.

SECURITY = ADS

In this mode, Samba will act as a domain member in an ADS realm. To operate in this mode, the machine running Samba will need to have Kerberos installed and configured and Samba will need to be joined to the ADS realm using the net utility.

Note that this mode does NOT make Samba operate as a Active Directory Domain Controller.

Read the chapter about Domain Membership in the HOWTO for details.

See also the *ads server* parameter, the *realm* paramter and the *encrypted passwords* parameter.

Default: **security = USER**

Example: **security = DOMAIN**

security mask (S) — This parameter controls what UNIX permission bits can be modified when a Windows NT client is manipulating the UNIX permission on a file using the native NT security dialog box.

This parameter is applied as a mask (AND'ed with) to the changed permission bits, thus preventing any bits not in this mask from being modified. Essentially, zero bits in this mask may be treated as a set of bits the user is not allowed to change.

If not set explicitly this parameter is 0777, allowing a user to modify all the user, group or world permissions on a file.

Note that users who can access the Samba server through other means can easily bypass this restriction, so it is primarily useful for standalone "appliance" systems. Administrators of most normal systems will probably want to leave it set to 0777.

See also the *force directory security mode*, *directory security mask*, *force security mode* parameters.

Default: **security mask = 0777**

Example: **security mask = 0770**

server schannel (G) — This controls whether the server offers or even demands the use of the netlogon schannel. *server schannel = no* does not offer the schannel, *server schannel = auto* offers the schannel but does not enforce it, and *server schannel = yes* denies access if the client is not able to speak netlogon schannel. This is only the case for Windows NT4 before SP4.

Please note that with this set to *no* you will have to apply the WindowsXP requireSignOrSeal-Registry patch found in the docs/Registry subdirectory.

Default: **server schannel = auto**

Example: **server schannel = yes**

server string (G) — This controls what string will show up in the printer comment box in print manager and next to the IPC connection in **net view**. It can be any string that you wish to show to your users.

It also sets what will appear in browse lists next to the machine name.

A *%v* will be replaced with the Samba version number.

A *%h* will be replaced with the hostname.

Default: **server string = Samba %v**

Example: **server string = University of GNUs Samba Server**

set directory (S) — If **set directory = no**, then users of the service may not use the setdir command to change directory.

The **setdir** command is only implemented in the Digital Pathworks client. See the Pathworks documentation for details.

Default: **set directory = no**

set primary group script (G) — Thanks to the Posix subsystem in NT a Windows User has a primary group in addition to the auxiliary groups. This script sets the primary group in the unix userdatase when an administrator sets the primary group from the windows user manager or when fetching a SAM with **net rpc vampire**. *%u* will be replaced with the user whose primary group is to be set. *%g* will be replaced with the group to set.

Default: *No default value*

Example: **set primary group script = /usr/sbin/usermod -g '%g' '%u'**

share modes (S) — This enables or disables the honoring of the *share modes* during a file open. These modes are used by clients to gain exclusive read or write access to a file.

These open modes are not directly supported by UNIX, so they are simulated using shared memory, or lock files if your UNIX doesn't support shared memory (almost all do).

The share modes that are enabled by this option are `DENY_DOS`, `DENY_ALL`, `DENY_READ`, `DENY_WRITE`, `DENY_NONE` and `DENY_FCB`.

This option gives full share compatibility and enabled by default.

You should *NEVER* turn this parameter off as many Windows applications will break if you do so.

Default: **share modes = yes**

short preserve case (S) — This boolean parameter controls if new files which conform to 8.3 syntax, that is all in upper case and of suitable length, are created upper case, or if they are forced to be the *default case*. This option can be use with **preserve case = yes** to permit long filenames to retain their case, while short names are lowered.

See the section on NAME MANGLING.

Default: **short preserve case = yes**

show add printer wizard (G) — With the introduction of MS-RPC based printing support for Windows NT/2000 client in Samba 2.2, a "Printers..." folder will appear on Samba hosts in the share listing. Normally this folder will contain an icon for the MS Add Printer Wizard (APW). However, it is possible to disable this feature regardless of the level of privilege of the connected user.

Under normal circumstances, the Windows NT/2000 client will open a handle on the printer server with OpenPrinterEx() asking for Administrator privileges. If the user does not have administrative access on the print server (i.e is not root or a member of the *printer admin* group), the OpenPrinterEx() call fails and the client makes another open call with a request for a lower privilege level. This should succeed, however the APW icon will not be displayed.

Disabling the *show add printer wizard* parameter will always cause the OpenPrinterEx() on the server to fail. Thus the APW icon will never be displayed. *Note* :This does not prevent the same user from having administrative privilege on an individual printer.

See also *addprinter command*, *deleteprinter command*, *printer admin*

Default :**show add printer wizard = yes**

shutdown script (G) — *This parameter only exists in the HEAD cvs branch* This a full path name to a script called by smbd(8) that should start a shutdown procedure.

This command will be run as the user connected to the server.

%m %t %r %f parameters are expanded:

- *%m* will be substituted with the shutdown message sent to the server.
- *%t* will be substituted with the number of seconds to wait before effectively starting the shutdown procedure.
- *%r* will be substituted with the switch *-r*. It means reboot after shutdown for NT.
- *%f* will be substituted with the switch *-f*. It means force the shutdown even if applications do not respond for NT.

Default: *None.*

Example: **shutdown script = /usr/local/samba/sbin/shutdown %m %t %r %f**

Shutdown script example:

```
#!/bin/bash
```

```
$time=0
let "time/60"
let "time++"

/sbin/shutdown $3 $4 +$time $1 &
```

Shutdown does not return so we need to launch it in background.

See also *abort shutdown script*.

smb passwd file (G) — This option sets the path to the encrypted smbpasswd file. By default the path to the smbpasswd file is compiled into Samba.

Default: **smb passwd file = ${prefix}/private/smbpasswd**

Example: **smb passwd file = /etc/samba/smbpasswd**

smb ports (G) — Specifies which ports the server should listen on for SMB traffic.

Default: **smb ports = 445 139**

socket address (G) — This option allows you to control what address Samba will listen for connections on. This is used to support multiple virtual interfaces on the one server, each with a different configuration.

By default Samba will accept connections on any address.

Example: **socket address = 192.168.2.20**

socket options (G) — This option allows you to set socket options to be used when talking with the client.

Socket options are controls on the networking layer of the operating systems which allow the connection to be tuned.

This option will typically be used to tune your Samba server for optimal performance for your local network. There is no way that Samba can know what the optimal parameters are for your net, so you must experiment and choose them yourself. We strongly suggest you read the appropriate documentation for your operating system first (perhaps **man setsockopt** will help).

You may find that on some systems Samba will say "Unknown socket option" when you supply an option. This means you either incorrectly typed it or you need to add an include file to includes.h for your OS. If the latter is the case please send the patch to samba-technical@samba.org[5].

Any of the supported socket options may be combined in any way you like, as long as your OS allows it.

[5] mailto:samba-technical@samba.org

This is the list of socket options currently settable using this option:

- SO_KEEPALIVE
- SO_REUSEADDR
- SO_BROADCAST
- TCP_NODELAY
- IPTOS_LOWDELAY
- IPTOS_THROUGHPUT
- SO_SNDBUF *
- SO_RCVBUF *
- SO_SNDLOWAT *
- SO_RCVLOWAT *

Those marked with a '*' take an integer argument. The others can optionally take a 1 or 0 argument to enable or disable the option, by default they will be enabled if you don't specify 1 or 0.

To specify an argument use the syntax SOME_OPTION = VALUE for example **SO_SNDBUF = 8192**. Note that you must not have any spaces before or after the = sign.

If you are on a local network then a sensible option might be:

socket options = IPTOS_LOWDELAY

If you have a local network then you could try:

socket options = IPTOS_LOWDELAY TCP_NODELAY

If you are on a wide area network then perhaps try setting IPTOS_THROUGHPUT.

Note that several of the options may cause your Samba server to fail completely. Use these options with caution!

Default: **socket options = TCP_NODELAY**

Example: **socket options = IPTOS_LOWDELAY**

source environment (G) — This parameter causes Samba to set environment variables as per the content of the file named.

If the value of this parameter starts with a "|" character then Samba will treat that value as a pipe command to open and will set the environment variables from the output of the pipe.

The contents of the file or the output of the pipe should be formatted as the output of the standard Unix **env(1)** command. This is of the form:

Example environment entry:

SAMBA_NETBIOS_NAME = myhostname

Default: *No default value*

Examples: **source environment = |/etc/smb.conf.sh**

Example: **source environment = /usr/local/smb_env_vars**

stat cache (G) — This parameter determines if smbd(8) will use a cache in order to speed up case insensitive name mappings. You should never need to change this parameter.

Default: **stat cache = yes**

stat cache size (G) — This parameter determines the number of entries in the *stat cache*. You should never need to change this parameter.

Default: **stat cache size = 50**

strict allocate (S) — This is a boolean that controls the handling of disk space allocation in the server. When this is set to `yes` the server will change from UNIX behaviour of not committing real disk storage blocks when a file is extended to the Windows behaviour of actually forcing the disk system to allocate real storage blocks when a file is created or extended to be a given size. In UNIX terminology this means that Samba will stop creating sparse files. This can be slow on some systems.

When strict allocate is `no` the server does sparse disk block allocation when a file is extended.

Setting this to `yes` can help Samba return out of quota messages on systems that are restricting the disk quota of users.

Default: **strict allocate = no**

strict locking (S) — This is a boolean that controls the handling of file locking in the server. When this is set to `yes`, the server will check every read and write access for file locks, and deny access if locks exist. This can be slow on some systems.

When strict locking is disabled, the server performs file lock checks only when the client explicitly asks for them.

Well-behaved clients always ask for lock checks when it is important. So in the vast majority of cases, **strict locking = no** is preferable.

Default: **strict locking = no**

strict sync (S) — Many Windows applications (including the Windows 98 explorer shell) seem to confuse flushing buffer contents to disk with doing a sync to disk. Under UNIX, a sync call forces the process to be suspended until the kernel has ensured that all outstanding data in kernel disk buffers has been safely stored onto stable storage. This is very slow and should only be done rarely. Setting this parameter to `no` (the default) means that smbd(8) ignores the Windows applications requests for a sync call. There is only a possibility of losing data if the operating system itself that Samba is running on crashes, so there is little danger in this default setting. In addition, this fixes many performance problems that people have reported with the new Windows98 explorer shell file copies.

See also the *sync always* parameter.

Default: **strict sync = no**

strip dot (G) — This is a boolean that controls whether to strip trailing dots off UNIX filenames. This helps with some CDROMs that have filenames ending in a single dot.

Default: **strip dot = no**

sync always (S) — This is a boolean parameter that controls whether writes will always be written to stable storage before the write call returns. If this is `no` then the server will be guided by the client's request in each write call (clients can set a bit indicating that a particular write should be synchronous). If this is `yes` then every write will be followed by a **fsync()** call to ensure the data is written to disk. Note that the *strict sync* parameter must be set to `yes` in order for this parameter to have any affect.

See also the *strict sync* parameter.

Default: **sync always = no**

syslog (G) — This parameter maps how Samba debug messages are logged onto the system syslog logging levels. Samba debug level zero maps onto syslog `LOG_ERR`, debug level one maps onto `LOG_WARNING`, debug level two maps onto `LOG_NOTICE`, debug level three maps onto LOG_INFO. All higher levels are mapped to `LOG_DEBUG`.

This parameter sets the threshold for sending messages to syslog. Only messages with debug level less than this value will be sent to syslog.

Default: **syslog = 1**

syslog only (G) — If this parameter is set then Samba debug messages are logged into the system syslog only, and not to the debug log files.

Default: **syslog only = no**

template homedir (G) — When filling out the user information for a Windows NT user, the winbindd(8) daemon uses this parameter to fill in the home directory for that user. If the string *%D* is present it is substituted with the user's Windows NT domain name. If the string *%U* is present it is substituted with the user's Windows NT user name.

Default: **template homedir = /home/%D/%U**

template shell (G) — When filling out the user information for a Windows NT user, the winbindd(8) daemon uses this parameter to fill in the login shell for that user.

Default: **template shell = /bin/false**

time offset (G) — This parameter is a setting in minutes to add to the normal GMT to local time conversion. This is useful if you are serving a lot of PCs that have incorrect daylight saving time handling.

Default: **time offset = 0**

Example: **time offset = 60**

time server (G) — This parameter determines if nmbd(8) advertises itself as a time server to Windows clients.

Default: **time server = no**

timestamp logs (G) — Synonym for *debug timestamp*.

total print jobs (G) — This parameter accepts an integer value which defines a limit on the maximum number of print jobs that will be accepted system wide at any given time. If a print job is submitted by a client which will exceed this number, then smbd(8) will return an error indicating that no space is available on the server. The default value of 0 means that no such limit exists. This parameter can be used to prevent a server from exceeding its capacity and is designed as a printing throttle. See also *max print jobs*.

Default: **total print jobs = 0**

Example: **total print jobs = 5000**

unicode (G) — Specifies whether Samba should try to use unicode on the wire by default. Note: This does NOT mean that samba will assume that the unix machine uses unicode!

Default: **unicode = yes**

unix charset (G) — Specifies the charset the unix machine Samba runs on uses. Samba needs to know this in order to be able to convert text to the charsets other SMB clients use.

Default: **unix charset = UTF8**

Example: **unix charset = ASCII**

unix extensions (G) — This boolean parameter controls whether Samba implments the CIFS UNIX extensions, as defined by HP. These extensions enable Samba to better serve UNIX CIFS clients by supporting features such as symbolic links, hard links, etc... These extensions require a similarly enabled client, and are of no current use to Windows clients.

Default: **unix extensions = yes**

unix password sync (G) — This boolean parameter controls whether Samba attempts to synchronize the UNIX password with the SMB password when the encrypted SMB password in the smbpasswd file is changed. If this is set to `yes` the program specified in the *`passwd program`*parameter is called *AS ROOT* - to allow the new UNIX password to be set without access to the old UNIX password (as the SMB password change code has no access to the old password cleartext, only the new).

See also *`passwd program`*, *`passwd chat`*.

Default: **unix password sync = no**

update encrypted (G) — This boolean parameter allows a user logging on with a plaintext password to have their encrypted (hashed) password in the smbpasswd file to be updated automatically as they log on. This option allows a site to migrate from plaintext password authentication (users authenticate with plaintext password over the wire, and are checked against a UNIX account database) to encrypted password authentication (the SMB challenge/response authentication mechanism) without forcing all users to re-enter their passwords via smbpasswd at the time the change is made. This is a convenience option to allow the change over to encrypted passwords to be made over a longer period. Once all users have encrypted representations of their passwords in the smbpasswd file this parameter should be set to `no`.

In order for this parameter to work correctly the *`encrypt passwords`* parameter must be set to `no` when this parameter is set to `yes`.

Note that even when this parameter is set a user authenticating to **smbd** must still enter a valid password in order to connect correctly, and to update their hashed (smbpasswd) passwords.

Default: **update encrypted = no**

use client driver (S) — This parameter applies only to Windows NT/2000 clients. It has no effect on Windows 95/98/ME clients. When serving a printer to Windows NT/2000 clients without first installing a valid printer driver on the Samba host, the client will be required to install a local printer driver. From this point on, the client will treat the print as a local printer and not a network printer connection. This is much the same behavior that will occur when **disable spoolss = yes**.

The differentiating factor is that under normal circumstances, the NT/2000 client will attempt to open the network printer using MS-RPC. The problem is that because the client considers the printer to be local, it will attempt to issue the OpenPrinterEx() call requesting access rights associated with the logged on user. If the user possesses local administator rights but not root privilegde on the Samba host (often the case), the OpenPrinterEx() call will fail. The result is that the client will now display an "Access Denied; Unable to connect" message in the printer queue window (even though jobs may successfully be printed).

If this parameter is enabled for a printer, then any attempt to open the printer with the PRINTER_ACCESS_ADMINISTER right is mapped to PRINTER_ACCESS_USE instead. Thus allowing the OpenPrinterEx() call to succeed. *This parameter MUST not be able enabled on a print share which has valid print driver installed on the Samba server.*

See also *disable spoolss*

Default: **use client driver = no**

use mmap (G) — This global parameter determines if the tdb internals of Samba can depend on mmap working correctly on the running system. Samba requires a coherent mmap/read-write system memory cache. Currently only HPUX does not have such a coherent cache, and so this parameter is set to `no` by default on HPUX. On all other systems this parameter should be left alone. This parameter is provided to help the Samba developers track down problems with the tdb internal code.

Default: **use mmap = yes**

user (S) — Synonym for *username*.

username (S) — Multiple users may be specified in a comma-delimited list, in which case the supplied password will be tested against each username in turn (left to right).

The *username* line is needed only when the PC is unable to supply its own username. This is the case for the COREPLUS protocol or where your users have different WfWg usernames to UNIX usernames. In both these cases you may also be better using the \\server\share%user syntax instead.

The *username* line is not a great solution in many cases as it means Samba will try to validate the supplied password against each of the usernames in the *username* line in turn. This is slow and a bad idea for lots of users in case of duplicate passwords. You may get timeouts or security breaches using this parameter unwisely.

Samba relies on the underlying UNIX security. This parameter does not restrict who can login, it just offers hints to the Samba server as to what usernames might correspond to the supplied password. Users can login as whoever they please and they will be able to do no more damage than if they started a telnet session. The daemon runs as the user that they log in as, so they cannot do anything that user cannot do.

To restrict a service to a particular set of users you can use the *valid users* parameter.

If any of the usernames begin with a '@' then the name will be looked up first in the NIS netgroups list (if Samba is compiled with netgroup support), followed by a lookup in the UNIX groups database and will expand to a list of all users in the group of that name.

If any of the usernames begin with a '+' then the name will be looked up only in the UNIX groups database and will expand to a list of all users in the group of that name.

If any of the usernames begin with a '&' then the name will be looked up only in the NIS netgroups database (if Samba is compiled with netgroup support) and will expand to a list of all users in the netgroup group of that name.

Note that searching though a groups database can take quite some time, and some clients may time out during the search.

See the section NOTE ABOUT USERNAME/PASSWORD VALIDATION for more information on how this parameter determines access to the services.

Default: **The guest account if a guest service, else <empty string>.**

Examples:**username = fred, mary, jack, jane, @users, @pcgroup**

username level (G) — This option helps Samba to try and 'guess' at the real UNIX username, as many DOS clients send an all-uppercase username. By default Samba tries all lowercase, followed by the username with the first letter capitalized, and fails if the username is not found on the UNIX machine.

If this parameter is set to non-zero the behavior changes. This parameter is a number that specifies the number of uppercase combinations to try while trying to determine the UNIX user name. The higher the number the more combinations will be tried, but the slower the discovery of usernames will be. Use this parameter when you have strange usernames on your UNIX machine, such as `AstrangeUser`.

Default: **username level = 0**

Example: **username level = 5**

username map (G) — This option allows you to specify a file containing a mapping of usernames from the clients to the server. This can be used for several purposes. The most common is to map usernames that users use on DOS or Windows machines to those that the UNIX box uses. The other is to map multiple users to a single username so that they can more easily share files.

The map file is parsed line by line. Each line should contain a single UNIX username on the left then a '=' followed by a list of usernames on the right. The list of usernames on the right may contain names of the form @group in which case they will match any UNIX username in that group. The special client name '*' is a wildcard and matches any name. Each line of the map file may be up to 1023 characters long.

The file is processed on each line by taking the supplied username and comparing it with each username on the right hand side of the '=' signs. If the supplied name matches any of the names on the right hand side then it is replaced with the name on the left. Processing then continues with the next line.

If any line begins with a '#' or a ';' then it is ignored

If any line begins with an '!' then the processing will stop after that line if a mapping was done by the line. Otherwise mapping continues with every line being processed. Using '!' is most useful when you have a wildcard mapping line later in the file.

For example to map from the name `admin` or `administrator` to the UNIX name `root` you would use:

root = admin administrator

Or to map anyone in the UNIX group `system` to the UNIX name `sys` you would use:

sys = @system

You can have as many mappings as you like in a username map file.

If your system supports the NIS NETGROUP option then the netgroup database is checked before the `/etc/group` database for matching groups.

You can map Windows usernames that have spaces in them by using double quotes around the name. For example:

tridge = "Andrew Tridgell"

would map the windows username "Andrew Tridgell" to the unix username "tridge".

The following example would map mary and fred to the unix user sys, and map the rest to guest. Note the use of the '!' to tell Samba to stop processing if it gets a match on that line.

```
!sys = mary fred
guest = *
```

Note that the remapping is applied to all occurrences of usernames. Thus if you connect to \\server\fred and `fred` is remapped to `mary` then you will actually be connecting to \\server\mary and will need to supply a password suitable for `mary` not `fred`. The only exception to this is the username passed to the *password server* (if you have one). The password server will receive whatever username the client supplies without modification.

Also note that no reverse mapping is done. The main effect this has is with printing. Users who have been mapped may have trouble deleting print jobs as PrintManager under WfWg will think they don't own the print job.

Default: *no username map*

Example: **username map = /usr/local/samba/lib/users.map**

users (S) — Synonym for *username*.

use sendfile (S) — If this parameter is `yes`, and Samba was built with the –with-sendfile-support option, and the underlying operating system supports sendfile system call, then some SMB read calls (mainly ReadAndX and ReadRaw) will use the more efficient sendfile system call for files that are exclusively oplocked. This may make more efficient use of the system CPU's and cause Samba to be faster. This is off by default as it's effects are unknown as yet.

Default: **use sendfile = no**

use spnego (G) — This variable controls controls whether samba will try to use Simple and Protected NEGOciation (as specified by rfc2478) with WindowsXP and Windows2000 clients to agree upon an authentication mechanism. Unless further issues are discovered with our SPNEGO implementation, there is no reason this should ever be disabled.

Default: *use spnego = yes*

utmp (G) — This boolean parameter is only available if Samba has been configured and compiled with the option –**with-utmp**. If set to `yes` then Samba will attempt to add utmp or utmpx records (depending on the UNIX system) whenever a connection is made to a Samba server. Sites may use this to record the user connecting to a Samba share.

Due to the requirements of the utmp record, we are required to create a unique identifier for the incoming user. Enabling this option creates an n^2 algorithm to find this number. This may impede performance on large installations.

See also the *utmp directory* parameter.

Default: **utmp = no**

utmp directory (G) — This parameter is only available if Samba has been configured and compiled with the option –**with-utmp**. It specifies a directory pathname that is used to store the utmp or utmpx files (depending on the UNIX system) that record user connections to a Samba server. See also the *utmp* parameter. By default this is not set, meaning the system will use whatever utmp file the native system is set to use (usually `/var/run/utmp` on Linux).

Default: *no utmp directory*

Example: **utmp directory = /var/run/utmp**

-valid (S) — This parameter indicates whether a share is valid and thus can be used. When this parameter is set to false, the share will be in no way visible nor accessible.

This option should not be used by regular users but might be of help to developers. Samba uses this option internally to mark shares as deleted.

Default: *True*

valid users (S) — This is a list of users that should be allowed to login to this service. Names starting with '@', '+' and '&' are interpreted using the same rules as described in the *invalid users* parameter.

If this is empty (the default) then any user can login. If a username is in both this list and the *invalid users* list then access is denied for that user.

The current servicename is substituted for *%S*. This is useful in the [homes] section.

See also *invalid users*

Default: *No valid users list (anyone can login)*

Example: **valid users = greg, @pcusers**

veto files (S) — This is a list of files and directories that are neither visible nor accessible. Each entry in the list must be separated by a '/', which allows spaces to be included in the entry. '*' and '?' can be used to specify multiple files or directories as in DOS wildcards.

Each entry must be a unix path, not a DOS path and must *not* include the unix directory separator '/'.

Note that the *case sensitive* option is applicable in vetoing files.

One feature of the veto files parameter that it is important to be aware of is Samba's behaviour when trying to delete a directory. If a directory that is to be deleted contains nothing but veto files this deletion will *fail* unless you also set the *delete veto files* parameter to *yes*.

Setting this parameter will affect the performance of Samba, as it will be forced to check all files and directories for a match as they are scanned.

See also *hide files* and *case sensitive*.

Default: *No files or directories are vetoed.*

Examples:

```
; Veto any files containing the word Security,
; any ending in .tmp, and any directory containing the
; word root.
veto files = /*Security*/*.tmp/*root*/

; Veto the Apple specific files that a NetAtalk server
; creates.
veto files = /.AppleDouble/.bin/.AppleDesktop/Network Trash Folder/
```

veto oplock files (S) — This parameter is only valid when the *oplocks* parameter is turned on for a share. It allows the Samba administrator to selectively turn off the granting of oplocks on selected files that match a wildcarded list, similar to the wildcarded list used in the *veto files* parameter.

Default: *No files are vetoed for oplock grants*

You might want to do this on files that you know will be heavily contended for by clients. A good example of this is in the NetBench SMB benchmark program, which causes heavy client contention for files ending in `.SEM`. To cause Samba not to grant oplocks on these files you would use the line (either in the [global] section or in the section for the particular NetBench share :

Example: **veto oplock files = /*.SEM/**

vfs object (S) — Synonym for *vfs objects*.

vfs objects (S) — This parameter specifies the backend names which are used for Samba VFS I/O operations. By default, normal disk I/O operations are used but these can be overloaded with one or more VFS objects.

Default: *no value*

Example: **vfs objects = extd_audit recycle**

volume (S) — This allows you to override the volume label returned for a share. Useful for CDROMs with installation programs that insist on a particular volume label.

Default: *the name of the share*

wide links (S) — This parameter controls whether or not links in the UNIX file system may be followed by the server. Links that point to areas within the directory tree exported by the server are always allowed; this parameter controls access only to areas that are outside the directory tree being exported.

Note that setting this parameter can have a negative effect on your server performance due to the extra system calls that Samba has to do in order to perform the link checks.

Default: **wide links = yes**

winbind cache time (G) — This parameter specifies the number of seconds the winbindd(8) daemon will cache user and group information before querying a Windows NT server again.

Default: **winbind cache type = 300**

winbind enum groups (G) — On large installations using winbindd(8) it may be necessary to suppress the enumeration of groups through the **setgrent()**, **getgrent()** and **endgrent()** group of system calls. If the *winbind enum groups* parameter is `no`, calls to the **getgrent()** system call will not return any data.

Warning: Turning off group enumeration may cause some programs to behave oddly.

Default: **winbind enum groups = yes**

winbind enum users (G) — On large installations using winbindd(8) it may be necessary to suppress the enumeration of users through the **setpwent()**, **getpwent()** and **endpwent()** group of system calls. If the *winbind enum users* parameter is `no`, calls to the **getpwent** system call will not return any data.

Warning: Turning off user enumeration may cause some programs to behave oddly. For example, the finger program relies on having access to the full user list when searching for matching usernames.

Default: **winbind enum users = yes**

winbind gid (G) — This parameter is now an alias for **idmap gid**

The winbind gid parameter specifies the range of group ids that are allocated by the winbindd(8) daemon. This range of group ids should have no existing local or NIS groups within it as strange conflicts can occur otherwise.

Default: **winbind gid = <empty string>**

Example: **winbind gid = 10000-20000**

winbind separator (G) — This parameter allows an admin to define the character used when listing a username of the form of *`DOMAIN`**`user`*. This parameter is only applicable when using the `pam_winbind.so` and `nss_winbind.so` modules for UNIX services.

Please note that setting this parameter to + causes problems with group membership at least on glibc systems, as the character + is used as a special character for NIS in /etc/group.

Default: **winbind separator = '\\'**

Example: **winbind separator = +**

winbind uid (G) — This parameter is now an alias for **idmap uid**

The winbind gid parameter specifies the range of user ids that are allocated by the winbindd(8) daemon. This range of ids should have no existing local or NIS users within it as strange conflicts can occur otherwise.

Default: **winbind uid = <empty string>**

Example: **winbind uid = 10000-20000**

winbind use default domain (G) — This parameter specifies whether the winbindd(8) daemon should operate on users without domain component in their username. Users without a domain component are treated as is part of the winbindd server's own domain. While this does not benifit Windows users, it makes SSH, FTP and e-mail function in a way much closer to the way they would in a native unix system.

Default: **winbind use default domain = <no>**

Example: **winbind use default domain = yes**

wins hook (G) — When Samba is running as a WINS server this allows you to call an external program for all changes to the WINS database. The primary use for this option is to allow the dynamic update of external name resolution databases such as dynamic DNS.

The wins hook parameter specifies the name of a script or executable that will be called as follows:

wins_hook operation name nametype ttl IP_list

- The first argument is the operation and is one of "add", "delete", or "refresh". In most cases the operation can be ignored as the rest of the parameters provide sufficient information. Note that "refresh" may sometimes be called when the name has not previously been added, in that case it should be treated as an add.
- The second argument is the NetBIOS name. If the name is not a legal name then the wins hook is not called. Legal names contain only letters, digits, hyphens, underscores and periods.
- The third argument is the NetBIOS name type as a 2 digit hexadecimal number.
- The fourth argument is the TTL (time to live) for the name in seconds.
- The fifth and subsequent arguments are the IP addresses currently registered for that name. If this list is empty then the name should be deleted.

An example script that calls the BIND dynamic DNS update program **nsupdate** is provided in the examples directory of the Samba source code.

wins partners (G) — A space separated list of partners' IP addresses for WINS replication. WINS partners are always defined as push/pull partners as defining only one way WINS replication is unreliable. WINS replication is currently experimental and unreliable between samba servers.

Default: **wins partners =**

Example: **wins partners = 192.168.0.1 172.16.1.2**

wins proxy (G) — This is a boolean that controls if nmbd(8) will respond to broadcast name queries on behalf of other hosts. You may need to set this to `yes` for some older clients.

Default: **wins proxy = no**

wins server (G) — This specifies the IP address (or DNS name: IP address for preference) of the WINS server that nmbd(8) should register with. If you have a WINS server on your network then you should set this to the WINS server's IP.

You should point this at your WINS server if you have a multi-subnetted network.

If you want to work in multiple namespaces, you can give every wins server a 'tag'. For each tag, only one (working) server will be queried for a name. The tag should be seperated from the ip address by a colon.

> **NOTE**
>
> 
>
> You need to set up Samba to point to a WINS server if you have multiple subnets and wish cross-subnet browsing to work correctly.

See the Chapter 9, *Network Browsing.*

Default: *not enabled*

```
Example: wins server = mary:192.9.200.1 fred:192.168.3.199 \
                       mary:192.168.2.61
```

For this example when querying a certain name, 192.19.200.1 will be asked first and if that doesn't respond 192.168.2.61. If either of those doesn't know the wins name server 192.168.3.199 will be queried.

Example: **wins server = 192.9.200.1 192.168.2.61**

wins support (G) — This boolean controls if the nmbd(8) process in Samba will act as a WINS server. You should not set this to `yes` unless you have a multi-subnetted network and you wish a particular **nmbd** to be your WINS server. Note that you should *NEVER* set this to `yes` on more than one machine in your network.

Default: **wins support = no**

workgroup (G) — This controls what workgroup your server will appear to be in when queried by clients. Note that this parameter also controls the Domain name used with the **security = domain** setting.

Default: *set at compile time to WORKGROUP*

Example: **workgroup = MYGROUP**

writable (S) — Synonym for *`writeable`* for people who can't spell :-).

writeable (S) — Inverted synonym for *`read only`*.

write cache size (S) — If this integer parameter is set to non-zero value, Samba will create an in-memory cache for each oplocked file (it does *not* do this for non-oplocked files). All writes that the client does not request to be flushed directly to disk will be stored in this cache if possible. The cache is flushed onto disk when a write comes in whose offset would not fit into the cache or when the file is closed by the client. Reads for the file are also served from this cache if the data is stored within it.

This cache allows Samba to batch client writes into a more efficient write size for RAID disks (i.e. writes may be tuned to be the RAID stripe size) and can improve performance on systems where the disk subsystem is a bottleneck but there is free memory for userspace programs.

The integer parameter specifies the size of this cache (per oplocked file) in bytes.

Default: **write cache size = 0**

Example: **write cache size = 262144**

for a 256k cache size per file.

write list (S) — This is a list of users that are given read-write access to a service. If the connecting user is in this list then they will be given write access, no matter what the *read only* option is set to. The list can include group names using the @group syntax.

Note that if a user is in both the read list and the write list then they will be given write access.

See also the *read list* option.

Default: **write list = <empty string>**

Example: **write list = admin, root, @staff**

write ok (S) — Inverted synonym for *read only*.

write raw (G) — This parameter controls whether or not the server will support raw write SMB's when transferring data from clients. You should never need to change this parameter.

Default: **write raw = yes**

wtmp directory (G) — This parameter is only available if Samba has been configured and compiled with the option **–with-utmp**. It specifies a directory pathname that is used to store the wtmp or wtmpx files (depending on the UNIX system) that record user connections to a Samba server. The difference with the utmp directory is the fact that user info is kept after a user has logged out.

See also the *utmp* parameter. By default this is not set, meaning the system will use whatever utmp file the native system is set to use (usually `/var/run/wtmp` on Linux).

Default: *no wtmp directory*

Example: **wtmp directory = /var/log/wtmp**

WARNINGS

Although the configuration file permits service names to contain spaces, your client software may not. Spaces will be ignored in comparisons anyway, so it shouldn't be a problem - but be aware of the possibility.

On a similar note, many clients - especially DOS clients - limit service names to eight characters. smbd(8) has no such limitation, but attempts to connect from such clients will fail if they truncate the service names. For this reason you should probably keep your service names down to eight characters in length.

Use of the [homes] and [printers] special sections make life for an administrator easy, but the various combinations of default attributes can be tricky. Take extreme care when designing these sections. In particular, ensure that the permissions on spool directories are correct.

SEE ALSO

samba(7), smbpasswd(8), swat(8), smbd(8), nmbd(8), smbclient(1), nmblookup(1), testparm(1), testprns(1).

A.2 nmblookup

Synopsis

```
nmblookup [-M] [-R] [-S] [-r] [-A] [-h] [-B <broadcast address>] [-U
     <unicast address>] [-d <debug level>] [-s <smb config file>] [-i
     <NetBIOS scope>] [-T] [-f] name
```

DESCRIPTION

This tool is part of the Samba(7) suite.

nmblookup is used to query NetBIOS names and map them to IP addresses in a network using NetBIOS over TCP/IP queries. The options allow the name queries to be directed at a particular IP broadcast area or to a particular machine. All queries are done over UDP.

OPTIONS

-M — Searches for a master browser by looking up the NetBIOS name *name* with a type of `0x1d`. If *name* is "-" then it does a lookup on the special name `__MSBROWSE__`. Please note that in order to use the name "-", you need to make sure "-" isn't parsed as an argument, e.g. use : `nmblookup -M -- -`.

-R — Set the recursion desired bit in the packet to do a recursive lookup. This is used when sending a name query to a machine running a WINS server and the user wishes to query the names in the WINS server. If this bit is unset the normal (broadcast responding) NetBIOS processing code on a machine is used instead. See RFC1001, RFC1002 for details.

-S — Once the name query has returned an IP address then do a node status query as well. A node status query returns the NetBIOS names registered by a host.

-r — Try and bind to UDP port 137 to send and receive UDP datagrams. The reason for this option is a bug in Windows 95 where it ignores the source port of the requesting packet and only replies to UDP port 137. Unfortunately, on most UNIX systems root privilege is needed to bind to this port, and in addition, if the nmbd(8) daemon is running on this machine it also binds to this port.

-A — Interpret *name* as an IP Address and do a node status query on this address.

-n <primary NetBIOS name> — This option allows you to override the NetBIOS name that Samba uses for itself. This is identical to setting the *netbios name* parameter in the `smb.conf` file. However, a command line setting will take precedence over settings in `smb.conf`.

-i <scope> — This specifies a NetBIOS scope that **nmblookup** will use to communicate with when generating NetBIOS names. For details on the use of NetBIOS scopes, see rfc1001.txt and rfc1002.txt. NetBIOS scopes are *very* rarely used, only set this parameter if you are the system administrator in charge of all the NetBIOS systems you communicate with.

-W|–workgroup=domain — Set the SMB domain of the username. This overrides the default domain which is the domain defined in smb.conf. If the domain specified is the same as the servers NetBIOS name, it causes the client to log on using the servers local SAM (as opposed to the Domain SAM).

-O socket options — TCP socket options to set on the client socket. See the socket options parameter in the `smb.conf` manual page for the list of valid options.

-h|–help — Print a summary of command line options.

-B <broadcast address> — Send the query to the given broadcast address. Without this option the default behavior of nmblookup is to send the query to the broadcast address of the network interfaces as either auto-detected or defined in the *interfaces*[6] parameter of the smb.conf(5) file.

-U <unicast address> — Do a unicast query to the specified address or host *unicast address*. This option (along with the *-R* option) is needed to query a WINS server.

-V — Prints the program version number.

[6] smb.conf.5.html#INTERFACES

-s <configuration file> — The file specified contains the configuration details required by the server. The information in this file includes server-specific information such as what printcap file to use, as well as descriptions of all the services that the server is to provide. See `smb.conf` for more information. The default configuration file name is determined at compile time.

-d|–debug=debuglevel — *`debuglevel`* is an integer from 0 to 10. The default value if this parameter is not specified is zero.

The higher this value, the more detail will be logged to the log files about the activities of the server. At level 0, only critical errors and serious warnings will be logged. Level 1 is a reasonable level for day-to-day running - it generates a small amount of information about operations carried out.

Levels above 1 will generate considerable amounts of log data, and should only be used when investigating a problem. Levels above 3 are designed for use only by developers and generate HUGE amounts of log data, most of which is extremely cryptic.

Note that specifying this parameter here will override the *`log level`* parameter in the `smb.conf` file.

-l|–logfile=logbasename — File name for log/debug files. The extension `".client"` will be appended. The log file is never removed by the client.

-T — This causes any IP addresses found in the lookup to be looked up via a reverse DNS lookup into a DNS name, and printed out before each

IP address .... NetBIOS name

pair that is the normal output.

-f — Show which flags apply to the name that has been looked up. Possible answers are zero or more of: Response, Authoritative, Truncated, Recursion_Desired, Recursion_Available, Broadcast.

name — This is the NetBIOS name being queried. Depending upon the previous options this may be a NetBIOS name or IP address. If a NetBIOS name then the different name types may be specified by appending '#<type>' to the name. This name may also be '*', which will return all registered names within a broadcast area.

EXAMPLES

nmblookup can be used to query a WINS server (in the same way **nslookup** is used to query DNS servers). To query a WINS server, **nmblookup** must be called like this:

nmblookup -U server -R 'name'

For example, running :

nmblookup -U samba.org -R 'IRIX#1B'

would query the WINS server samba.org for the domain master browser (1B name type) for the IRIX workgroup.

SEE ALSO

nmbd(8), samba(7), and smb.conf(5).

A.3 rpcclient

Synopsis

```
rpcclient [-A authfile] [-c <command string>] [-d debuglevel] [-h] [-l
     logfile] [-N] [-s <smb config file>] [-U username[%password]] [-W
     workgroup] [-N] [-I destinationIP] server
```

DESCRIPTION

This tool is part of the Samba(7) suite.

rpcclient is a utility initially developed to test MS-RPC functionality in Samba itself. It has undergone several stages of development and stability. Many system administrators have now written scripts around it to manage Windows NT clients from their UNIX workstation.

OPTIONS

server — NetBIOS name of Server to which to connect. The server can be any SMB/CIFS server. The name is resolved using the *name resolve order* line from smb.conf(5).

-c|–command='command string' — execute semicolon separated commands (listed below))

-I IP-address — *IP address* is the address of the server to connect to. It should be specified in standard "a.b.c.d" notation.

Normally the client would attempt to locate a named SMB/CIFS server by looking it up via the NetBIOS name resolution mechanism described above in the *name resolve order* parameter above. Using this parameter will force the client to assume that the server is on the machine with the specified IP address and the NetBIOS name component of the resource being connected to will be ignored.

There is no default for this parameter. If not supplied, it will be determined automatically by the client as described above.

-V — Prints the program version number.

-s <configuration file> — The file specified contains the configuration details required by the server. The information in this file includes server-specific information such as what printcap file to use, as well as descriptions of all the services that the server is to provide. See `smb.conf` for more information. The default configuration file name is determined at compile time.

-d|–debug=debuglevel — *`debuglevel`* is an integer from 0 to 10. The default value if this parameter is not specified is zero.

The higher this value, the more detail will be logged to the log files about the activities of the server. At level 0, only critical errors and serious warnings will be logged. Level 1 is a reasonable level for day-to-day running - it generates a small amount of information about operations carried out.

Levels above 1 will generate considerable amounts of log data, and should only be used when investigating a problem. Levels above 3 are designed for use only by developers and generate HUGE amounts of log data, most of which is extremely cryptic.

Note that specifying this parameter here will override the *`log level`* parameter in the `smb.conf` file.

-l|–logfile=logbasename — File name for log/debug files. The extension `".client"` will be appended. The log file is never removed by the client.

-N — If specified, this parameter suppresses the normal password prompt from the client to the user. This is useful when accessing a service that does not require a password.

Unless a password is specified on the command line or this parameter is specified, the client will request a password.

-k — Try to authenticate with kerberos. Only useful in an Active Directory environment.

-A|–authfile=filename — This option allows you to specify a file from which to read the username and password used in the connection. The format of the file is

```
username = <value>
password = <value>
domain   = <value>
```

Make certain that the permissions on the file restrict access from unwanted users.

-U|–user=username[%password —] Sets the SMB username or username and password.

If %password is not specified, the user will be prompted. The client will first check the `USER` environment variable, then the `LOGNAME` variable and if either exists, the string is uppercased. If these environmental variables are not found, the username `GUEST` is used.

A third option is to use a credentials file which contains the plaintext of the username and password. This option is mainly provided for scripts where the admin does not wish to pass the credentials on the command line or via environment variables. If this method is used, make certain that the permissions on the file restrict access from unwanted users. See the *-A* for more details.

Be cautious about including passwords in scripts. Also, on many systems the command line of a running process may be seen via the **ps** command. To be safe always allow **rpcclient** to prompt for a password and type it in directly.

-n <primary NetBIOS name> — This option allows you to override the NetBIOS name that Samba uses for itself. This is identical to setting the *netbios name* parameter in the `smb.conf` file. However, a command line setting will take precedence over settings in `smb.conf`.

-i <scope> — This specifies a NetBIOS scope that **nmblookup** will use to communicate with when generating NetBIOS names. For details on the use of NetBIOS scopes, see rfc1001.txt and rfc1002.txt. NetBIOS scopes are *very* rarely used, only set this parameter if you are the system administrator in charge of all the NetBIOS systems you communicate with.

-W|–workgroup=domain — Set the SMB domain of the username. This overrides the default domain which is the domain defined in smb.conf. If the domain specified is the same as the servers NetBIOS name, it causes the client to log on using the servers local SAM (as opposed to the Domain SAM).

-O socket options — TCP socket options to set on the client socket. See the socket options parameter in the `smb.conf` manual page for the list of valid options.

-h|–help — Print a summary of command line options.

COMMANDS

LSARPC

lsaquery — Query info policy.

lookupsids — Resolve a list of SIDs to usernames.

lookupnames — Resolve a list of usernames to SIDs.

enumtrusts — Enumerate trusted domains.

enumprivs — Enumerate privileges.

getdispname — Get the privilege name.

lsaenumsid — Enumerate the LSA SIDS.

lsaenumprivsaccount — Enumerate the privileges of an SID.

lsaenumacctrights — Enumerate the rights of an SID.

lsaenumacctwithright — Enumerate accounts with a right.

lsaaddacctrights — Add rights to an account.

lsaremoveacctrights — Remove rights from an account.

lsalookupprivvalue — Get a privilege value given its name.

lsaquerysecobj — Query LSA security object.

LSARPC-DS

dsroledominfo — Get Primary Domain Information.

DFS

dfsexist — Query DFS support.

dfsadd — Add a DFS share.

dfsremove — Remove a DFS share.

dfsgetinfo — Query DFS share info.

dfsenum — Enumerate dfs shares.

REG

shutdown — Remote Shutdown.

abortshutdown — Abort Shutdown.

SRVSVC

srvinfo — Server query info.

netshareenum — Enumerate shares.

netfileenum — Enumerate open files.

netremotetod — Fetch remote time of day.

SAMR

queryuser — Query user info.

querygroup — Query group info.

queryusergroups — Query user groups.

querygroupmem — Query group membership.

queryaliasmem — Query alias membership.

querydispinfo — Query display info.

querydominfo — Query domain info.

enumdomusers — Enumerate domain users.

enumdomgroups — Enumerate domain groups.

enumalsgroups — Enumerate alias groups.

createdomuser — Create domain user.

samlookupnames — Look up names.

samlookuprids — Look up names.

deletedomuser — Delete domain user.

samquerysecobj — Query SAMR security object.

getdompwinfo — Retrieve domain password info.

lookupdomain — Look up domain.

SPOOLSS

adddriver <arch> <config> — Execute an AddPrinterDriver() RPC to install the printer driver information on the server. Note that the driver files should already exist in the directory returned by **getdriverdir**. Possible values for *arch* are the same as those for the **getdriverdir** command. The *config* parameter is defined as follows:

```
Long Printer Name:\
Driver File Name:\
Data File Name:\
Config File Name:\
Help File Name:\
Language Monitor Name:\
Default Data Type:\
Comma Separated list of Files
```

Any empty fields should be enter as the string "NULL".

Samba does not need to support the concept of Print Monitors since these only apply to local printers whose driver can make use of a bi-directional link for communication. This field should be "NULL". On a remote NT print server, the Print Monitor for a driver must already be installed prior to adding the driver or else the RPC will fail.

addprinter <printername> <sharename> <drivername> <port> — Add a printer on the remote server. This printer will be automatically shared. Be aware that the printer driver must already be installed on the server (see **adddriver**) and the *port* must be a valid port name (see **enumports**.

deldriver — Delete the specified printer driver for all architectures. This does not delete the actual driver files from the server, only the entry from the server's list of drivers.

enumdata — Enumerate all printer setting data stored on the server. On Windows NT clients, these values are stored in the registry, while Samba servers store them in the printers TDB. This command corresponds to the MS Platform SDK GetPrinterData() function (This command is currently unimplemented.)

enumdataex — Enumerate printer data for a key.

enumjobs <printer> — List the jobs and status of a given printer. This command corresponds to the MS Platform SDK EnumJobs() function.

enumkey — Enumerate printer keys.

enumports [level —] Executes an EnumPorts() call using the specified info level. Currently only info levels 1 and 2 are supported.

enumdrivers [level —] Execute an EnumPrinterDrivers() call. This lists the various installed printer drivers for all architectures. Refer to the MS Platform SDK documentation for more details of the various flags and calling options. Currently supported info levels are 1, 2, and 3.

enumprinters [level —] Execute an EnumPrinters() call. This lists the various installed and share printers. Refer to the MS Platform SDK documentation for more details of the various flags and calling options. Currently supported info levels are 1, 2 and 5.

getdata <printername> <valuename;> — Retrieve the data for a given printer setting. See the **enumdata** command for more information. This command corresponds to the GetPrinterData() MS Platform SDK function.

getdataex — Get printer driver data with keyname.

getdriver <printername> — Retrieve the printer driver information (such as driver file, config file, dependent files, etc...) for the given printer. This command corresponds to the GetPrinterDriver() MS Platform SDK function. Currently info level 1, 2, and 3 are supported.

getdriverdir <arch> — Execute a GetPrinterDriverDirectory() RPC to retrieve the SMB share name and subdirectory for storing printer driver files for a given architecture. Possible values for *arch* are "Windows 4.0" (for Windows 95/98), "Windows NT x86", "Windows NT PowerPC", "Windows Alpha_AXP", and "Windows NT R4000".

getprinter <printername> — Retrieve the current printer information. This command corresponds to the GetPrinter() MS Platform SDK function.

getprintprocdir — Get print processor directory.

openprinter <printername> — Execute an OpenPrinterEx() and ClosePrinter() RPC against a given printer.

setdriver <printername> <drivername> — Execute a SetPrinter() command to update the printer driver associated with an installed printer. The printer driver must already be correctly installed on the print server.

See also the **enumprinters** and **enumdrivers** commands for obtaining a list of of installed printers and drivers.

addform — Add form.

setform — Set form.

getform — Get form.

deleteform — Delete form.

enumforms — Enumerate form.

setprinter — Set printer comment.

setprinterdata — Set REG_SZ printer data.

rffpcnex — Rffpcnex test.

NETLOGON

logonctrl2 — Logon Control 2.

logonctrl — Logon Control.

samsync — Sam Synchronisation.

samdeltas — Query Sam Deltas.

samlogon — Sam Logon.

GENERAL COMMANDS

debuglevel — Set the current debug level used to log information.

help (?) — Print a listing of all known commands or extended help on a particular command.

quit (exit) — Exit **rpcclient**.

BUGS

rpcclient is designed as a developer testing tool and may not be robust in certain areas (such as command line parsing). It has been known to generate a core dump upon failures when invalid parameters where passed to the interpreter.

From Luke Leighton's original rpcclient man page:

WARNING

The MSRPC over SMB code has been developed from examining Network traces. No documentation is available from the original creators (Microsoft) on how MSRPC over SMB works, or how the individual MSRPC services work. Microsoft's implementation of these services has been demonstrated (and reported) to be a bit flaky in places.

The development of Samba's implementation is also a bit rough, and as more of the services are understood, it can even result in versions of smbd(8) and rpcclient(1) that are incompatible for some commands or services. Additionally, the developers are sending reports to Microsoft, and problems found or reported to Microsoft are fixed in Service Packs, which may result in incompatibilities.

A.4 smbcacls

Synopsis

```
smbcacls //server/share filename [-D acls] [-M acls] [-A acls] [-S acls]
     [-C name] [-G name] [-n] [-t] [-U username] [-h] [-d]
```

DESCRIPTION

This tool is part of the Samba(7) suite.

The **smbcacls** program manipulates NT Access Control Lists (ACLs) on SMB file shares.

OPTIONS

The following options are available to the **smbcacls** program. The format of ACLs is described in the section ACL FORMAT.

-A acls — Add the ACLs specified to the ACL list. Existing access control entries are unchanged.

-M acls — Modify the mask value (permissions) for the ACLs specified on the command line. An error will be printed for each ACL specified that was not already present in the ACL list.

-D acls — Delete any ACLs specified on the command line. An error will be printed for each ACL specified that was not already present in the ACL list.

-S acls — This command sets the ACLs on the file with only the ones specified on the command line. All other ACLs are erased. Note that the ACL specified must contain at least a revision, type, owner and group for the call to succeed.

-U username — Specifies a username used to connect to the specified service. The username may be of the form "username" in which case the user is prompted to enter in a password and the workgroup specified in the smb.conf(5) file is used, or "username%password" or "DOMAIN\username%password" and the password and workgroup names are used as provided.

-C name — The owner of a file or directory can be changed to the name given using the *-C* option. The name can be a sid in the form S-1-x-y-z or a name resolved against the server specified in the first argument.

This command is a shortcut for -M OWNER:name.

-G name — The group owner of a file or directory can be changed to the name given using the *-G* option. The name can be a sid in the form S-1-x-y-z or a name resolved against the server specified n the first argument.

This command is a shortcut for -M GROUP:name.

-n — This option displays all ACL information in numeric format. The default is to convert SIDs to names and ACE types and masks to a readable string format.

-t — Don't actually do anything, only validate the correctness of the arguments.

-h|–help — Print a summary of command line options.

-V — Prints the program version number.

-s <configuration file> — The file specified contains the configuration details required by the server. The information in this file includes server-specific information such as what printcap file to use, as well as descriptions of all the services that the server is to provide. See `smb.conf` for more information. The default configuration file name is determined at compile time.

-d|–debug=debuglevel — *debuglevel* is an integer from 0 to 10. The default value if this parameter is not specified is zero.

The higher this value, the more detail will be logged to the log files about the activities of the server. At level 0, only critical errors and serious warnings will be logged. Level 1 is a reasonable level for day-to-day running - it generates a small amount of information about operations carried out.

Levels above 1 will generate considerable amounts of log data, and should only be used when investigating a problem. Levels above 3 are designed for use only by developers and generate HUGE amounts of log data, most of which is extremely cryptic.

Note that specifying this parameter here will override the *log level* parameter in the `smb.conf` file.

-l|–logfile=logbasename — File name for log/debug files. The extension `".client"` will be appended. The log file is never removed by the client.

ACL FORMAT

The format of an ACL is one or more ACL entries separated by either commas or newlines. An ACL entry is one of the following:

```
REVISION:<revision number>
OWNER:<sid or name>
GROUP:<sid or name>
ACL:<sid or name>:<type>/<flags>/<mask>
```

The revision of the ACL specifies the internal Windows NT ACL revision for the security descriptor. If not specified it defaults to 1. Using values other than 1 may cause strange behaviour.

The owner and group specify the owner and group SIDs for the object. If a SID in the format CWS-1-x-y-z is specified this is used, otherwise the name specified is resolved using the server on which the file or directory resides.

ACLs specify permissions granted to the SID. This SID again can be specified in CWS-1-x-y-z format or as a name in which case it is resolved against the server on which the file or directory resides. The type, flags and mask values determine the type of access granted to the SID.

The type can be either 0 or 1 corresponding to ALLOWED or DENIED access to the SID. The flags values are generally zero for file ACLs and either 9 or 2 for directory ACLs. Some common flags are:

- `#define SEC_ACE_FLAG_OBJECT_INHERIT 0x1`
- `#define SEC_ACE_FLAG_CONTAINER_INHERIT 0x2`
- `#define SEC_ACE_FLAG_NO_PROPAGATE_INHERIT 0x4`
- `#define SEC_ACE_FLAG_INHERIT_ONLY 0x8`

At present flags can only be specified as decimal or hexadecimal values.

The mask is a value which expresses the access right granted to the SID. It can be given as a decimal or hexadecimal value, or by using one of the following text strings which map to the NT file permissions of the same name.

- *R* - Allow read access
- *W* - Allow write access
- *X* - Execute permission on the object
- *D* - Delete the object
- *P* - Change permissions
- *O* - Take ownership

The following combined permissions can be specified:

- *READ* - Equivalent to 'RX' permissions
- *CHANGE* - Equivalent to 'RXWD' permissions
- *FULL* - Equivalent to 'RWXDPO' permissions

EXIT STATUS

The **smbcacls** program sets the exit status depending on the success or otherwise of the operations performed. The exit status may be one of the following values.

If the operation succeeded, smbcacls returns and exit status of 0. If **smbcacls** couldn't connect to the specified server, or there was an error getting or setting the ACLs, an exit status of 1 is returned. If there was an error parsing any command line arguments, an exit status of 2 is returned.

A.5 smbclient

Synopsis

smbclient servicename [password] [-b <buffer size>] [-d debuglevel] [-D Directory] [-U username] [-W workgroup] [-M <netbios name>] [-m maxprotocol] [-A authfile] [-N] [-l logfile] [-L <netbios name>] [-I destinationIP] [-E] [-c <command string>] [-i scope] [-O <socket options>] [-p port] [-R <name resolve order>] [-s <smb config file>] [-T<c|x>IXFqgbNan] [-k]

DESCRIPTION

This tool is part of the Samba(7) suite.

smbclient is a client that can 'talk' to an SMB/CIFS server. It offers an interface similar to that of the ftp program (see ftp(1)). Operations include things like getting files from the server to the local machine, putting files from the local machine to the server, retrieving directory information from the server and so on.

OPTIONS

servicename — servicename is the name of the service you want to use on the server. A service name takes the form `//server/service` where *server* is the NetBIOS name of the SMB/CIFS server offering the desired service and *service* is the name of the service offered. Thus to connect to the service "printer" on the SMB/CIFS server "smbserver", you would use the servicename `//smbserver/printer`

The server name required is NOT necessarily the IP (DNS) host name of the server ! The name required is a NetBIOS server name, which may or may not be the same as the IP hostname of the machine running the server.

The server name is looked up according to either the *-R* parameter to **smbclient** or using the name resolve order parameter in the smb.conf(5) file, allowing an administrator to change the order and methods by which server names are looked up.

password — The password required to access the specified service on the specified server. If this parameter is supplied, the *-N* option (suppress password prompt) is assumed.

There is no default password. If no password is supplied on the command line (either by using this parameter or adding a password to the *-U* option (see below)) and the *-N* option is not specified, the client will prompt for a password, even if the desired service does not require one. (If no password is required, simply press ENTER to provide a null password.)

Some servers (including OS/2 and Windows for Workgroups) insist on an uppercase password. Lowercase or mixed case passwords may be rejected by these servers.

Be cautious about including passwords in scripts.

-R <name resolve order> — This option is used by the programs in the Samba suite to determine what naming services and in what order to resolve host names to IP addresses. The option takes a space-separated string of different name resolution options.

The options are :"lmhosts", "host", "wins" and "bcast". They cause names to be resolved as follows:

- `lmhosts`: Lookup an IP address in the Samba lmhosts file. If the line in lmhosts has no name type attached to the NetBIOS name (see the lmhosts(5) for details) then any name type matches for lookup.
- `host`: Do a standard host name to IP address resolution, using the system `/etc/hosts`, NIS, or DNS lookups. This method of name resolution is operating system dependent, for instance on IRIX or Solaris this may be controlled by the `/etc/nsswitch.conf` file). This method is only used if the NetBIOS name type being queried is the 0x20 (server) name type, otherwise it is ignored.
- `wins`: Query a name with the IP address listed in the *`wins server`* parameter. If no WINS server has been specified this method will be ignored.
- `bcast`: Do a broadcast on each of the known local interfaces listed in the *`interfaces`* parameter. This is the least reliable of the name resolution methods as it depends on the target host being on a locally connected subnet.

If this parameter is not set then the name resolve order defined in the smb.conf(5) file parameter (name resolve order) will be used.

The default order is lmhosts, host, wins, bcast and without this parameter or any entry in the *`name resolve order`* parameter of the smb.conf(5) file the name resolution methods will be attempted in this order.

-M NetBIOS name — This options allows you to send messages, using the "WinPopup" protocol, to another computer. Once a connection is established you then type your message, pressing ^D (control-D) to end.

If the receiving computer is running WinPopup the user will receive the message and probably a beep. If they are not running WinPopup the message will be lost, and no error message will occur.

The message is also automatically truncated if the message is over 1600 bytes, as this is the limit of the protocol.

One useful trick is to cat the message through **smbclient**. For example: **cat mymessage.txt | smbclient -M FRED** will send the message in the file `mymessage.txt` to the machine FRED.

You may also find the *`-U`* and *`-I`* options useful, as they allow you to control the FROM and TO parts of the message.

See the *`message command`* parameter in the smb.conf(5) for a description of how to handle incoming WinPopup messages in Samba.

Copy WinPopup into the startup group on your WfWg PCs if you want them to always be able to receive messages.

-p port — This number is the TCP port number that will be used when making connections to the server. The standard (well-known) TCP port number for an SMB/CIFS server is 139, which is the default.

-l logfilename — If specified, *logfilename* specifies a base filename into which operational data from the running client will be logged.

The default base name is specified at compile time.

The base name is used to generate actual log file names. For example, if the name specified was "log", the debug file would be `log.client`.

The log file generated is never removed by the client.

-h|–help — Print a summary of command line options.

-I IP-address — *IP address* is the address of the server to connect to. It should be specified in standard "a.b.c.d" notation.

Normally the client would attempt to locate a named SMB/CIFS server by looking it up via the NetBIOS name resolution mechanism described above in the *name resolve order* parameter above. Using this parameter will force the client to assume that the server is on the machine with the specified IP address and the NetBIOS name component of the resource being connected to will be ignored.

There is no default for this parameter. If not supplied, it will be determined automatically by the client as described above.

-E — This parameter causes the client to write messages to the standard error stream (stderr) rather than to the standard output stream.

By default, the client writes messages to standard output - typically the user's tty.

-L — This option allows you to look at what services are available on a server. You use it as **smbclient -L host** and a list should appear. The *-I* option may be useful if your NetBIOS names don't match your TCP/IP DNS host names or if you are trying to reach a host on another network.

-t terminal code — This option tells **smbclient** how to interpret filenames coming from the remote server. Usually Asian language multibyte UNIX implementations use different character sets than SMB/CIFS servers (*EUC* instead of *SJIS* for example). Setting this parameter will let **smbclient** convert between the UNIX filenames and the SMB filenames correctly. This option has not been seriously tested and may have some problems.

The terminal codes include CWsjis, CWeuc, CWjis7, CWjis8, CWjunet, CWhex, CWcap. This is not a complete list, check the Samba source code for the complete list.

-b buffersize — This option changes the transmit/send buffer size when getting or putting a file from/to the server. The default is 65520 bytes. Setting this value smaller (to 1200 bytes) has been observed to speed up file transfers to and from a Win9x server.

-V — Prints the program version number.

-s <configuration file> — The file specified contains the configuration details required by the server. The information in this file includes server-specific information such as what printcap file to use, as well as descriptions of all the services that the server is to provide. See `smb.conf` for more information. The default configuration file name is determined at compile time.

-d|–debug=debuglevel — *debuglevel* is an integer from 0 to 10. The default value if this parameter is not specified is zero.

The higher this value, the more detail will be logged to the log files about the activities of the server. At level 0, only critical errors and serious warnings will be logged. Level 1 is a reasonable level for day-to-day running - it generates a small amount of information about operations carried out.

Levels above 1 will generate considerable amounts of log data, and should only be used when investigating a problem. Levels above 3 are designed for use only by developers and generate HUGE amounts of log data, most of which is extremely cryptic.

Note that specifying this parameter here will override the *log level* parameter in the `smb.conf` file.

-l|–logfile=logbasename — File name for log/debug files. The extension `".client"` will be appended. The log file is never removed by the client.

-N — If specified, this parameter suppresses the normal password prompt from the client to the user. This is useful when accessing a service that does not require a password.

Unless a password is specified on the command line or this parameter is specified, the client will request a password.

-k — Try to authenticate with kerberos. Only useful in an Active Directory environment.

-A|–authfile=filename — This option allows you to specify a file from which to read the username and password used in the connection. The format of the file is

```
username = <value>
```

```
password = <value>
domain   = <value>
```

Make certain that the permissions on the file restrict access from unwanted users.

-U|–user=username[%password —] Sets the SMB username or username and password.

If %password is not specified, the user will be prompted. The client will first check the `USER` environment variable, then the `LOGNAME` variable and if either exists, the string is uppercased. If these environmental variables are not found, the username `GUEST` is used.

A third option is to use a credentials file which contains the plaintext of the username and password. This option is mainly provided for scripts where the admin does not wish to pass the credentials on the command line or via environment variables. If this method is used, make certain that the permissions on the file restrict access from unwanted users. See the *-A* for more details.

Be cautious about including passwords in scripts. Also, on many systems the command line of a running process may be seen via the **ps** command. To be safe always allow **rpcclient** to prompt for a password and type it in directly.

-n <primary NetBIOS name> — This option allows you to override the NetBIOS name that Samba uses for itself. This is identical to setting the ***netbios name*** parameter in the `smb.conf` file. However, a command line setting will take precedence over settings in `smb.conf`.

-i <scope> — This specifies a NetBIOS scope that **nmblookup** will use to communicate with when generating NetBIOS names. For details on the use of NetBIOS scopes, see rfc1001.txt and rfc1002.txt. NetBIOS scopes are *very* rarely used, only set this parameter if you are the system administrator in charge of all the NetBIOS systems you communicate with.

-W|–workgroup=domain — Set the SMB domain of the username. This overrides the default domain which is the domain defined in smb.conf. If the domain specified is the same as the servers NetBIOS name, it causes the client to log on using the servers local SAM (as opposed to the Domain SAM).

-O socket options — TCP socket options to set on the client socket. See the socket options parameter in the `smb.conf` manual page for the list of valid options.

-T tar options — smbclient may be used to create **tar(1)** compatible backups of all the files on an SMB/CIFS share. The secondary tar flags that can be given to this option are :

- `c` - Create a tar file on UNIX. Must be followed by the name of a tar file, tape device or "-" for standard output. If using standard output you must turn the log level to its lowest value -d0 to avoid corrupting your tar file. This flag is mutually exclusive with the *x* flag.

- *x* - Extract (restore) a local tar file back to a share. Unless the -D option is given, the tar files will be restored from the top level of the share. Must be followed by the name of the tar file, device or "-" for standard input. Mutually exclusive with the *c* flag. Restored files have their creation times (mtime) set to the date saved in the tar file. Directories currently do not get their creation dates restored properly.
- *I* - Include files and directories. Is the default behavior when filenames are specified above. Causes tar files to be included in an extract or create (and therefore everything else to be excluded). See example below. Filename globbing works in one of two ways. See r below.
- *X* - Exclude files and directories. Causes tar files to be excluded from an extract or create. See example below. Filename globbing works in one of two ways now. See *r* below.
- *b* - Blocksize. Must be followed by a valid (greater than zero) blocksize. Causes tar file to be written out in blocksize*TBLOCK (usually 512 byte) blocks.
- *g* - Incremental. Only back up files that have the archive bit set. Useful only with the *c* flag.
- *q* - Quiet. Keeps tar from printing diagnostics as it works. This is the same as tarmode quiet.
- *r* - Regular expression include or exclude. Uses regular expression matching for excluding or excluding files if compiled with HAVE_REGEX_H. However this mode can be very slow. If not compiled with HAVE_REGEX_H, does a limited wildcard match on '*' and '?'.
- *N* - Newer than. Must be followed by the name of a file whose date is compared against files found on the share during a create. Only files newer than the file specified are backed up to the tar file. Useful only with the *c* flag.
- *a* - Set archive bit. Causes the archive bit to be reset when a file is backed up. Useful with the *g* and *c* flags.

Tar Long File Names

smbclient's tar option now supports long file names both on backup and restore. However, the full path name of the file must be less than 1024 bytes. Also, when a tar archive is created, **smbclient**'s tar option places all files in the archive with relative names, not absolute names.

Tar Filenames

All file names can be given as DOS path names (with '\\' as the component separator) or as UNIX path names (with '/' as the component separator).

Examples

Restore from tar file `backup.tar` into myshare on mypc (no password on share).

smbclient //mypc/yshare "" -N -Tx backup.tar

Restore everything except `users/docs`

smbclient //mypc/myshare "" -N -TXx backup.tar users/docs

Create a tar file of the files beneath `users/docs`.

smbclient //mypc/myshare "" -N -Tc backup.tar users/docs

Create the same tar file as above, but now use a DOS path name.

smbclient //mypc/myshare "" -N -tc backup.tar users\edocs

Create a tar file of all the files and directories in the share.

smbclient //mypc/myshare "" -N -Tc backup.tar *

-D initial directory — Change to initial directory before starting. Probably only of any use with the tar -T option.

-c command string — command string is a semicolon-separated list of commands to be executed instead of prompting from stdin. *-N* is implied by `-c`.

This is particularly useful in scripts and for printing stdin to the server, e.g. **-c 'print -'**.

OPERATIONS

Once the client is running, the user is presented with a prompt :

`smb:\>`

The backslash ("\\") indicates the current working directory on the server, and will change if the current working directory is changed.

The prompt indicates that the client is ready and waiting to carry out a user command. Each command is a single word, optionally followed by parameters specific to that command. Command and parameters are space-delimited unless these notes specifically state otherwise. All commands are case-insensitive. Parameters to commands may or may not be case sensitive, depending on the command.

You can specify file names which have spaces in them by quoting the name with double quotes, for example "a long file name".

Parameters shown in square brackets (e.g., "[parameter]") are optional. If not given, the command will use suitable defaults. Parameters shown in angle brackets (e.g., "<parameter>") are required.

All commands operating on the server are actually performed by issuing a request to the server. Thus the behavior may vary from server to server, depending on how the server was implemented.

The commands available are given here in alphabetical order.

? [command —] If *command* is specified, the ? command will display a brief informative message about the specified command. If no command is specified, a list of available commands will be displayed.

! [shell command —] If *shell command* is specified, the ! command will execute a shell locally and run the specified shell command. If no command is specified, a local shell will be run.

altname file — The client will request that the server return the "alternate" name (the 8.3 name) for a file or directory.

cancel jobid0 [jobid1 ... [jobidN] —] The client will request that the server cancel the printjobs identified by the given numeric print job ids.

chmod file mode in octal — This command depends on the server supporting the CIFS UNIX extensions and will fail if the server does not. The client requests that the server change the UNIX permissions to the given octal mode, in standard UNIX format.

chown file uid gid — This command depends on the server supporting the CIFS UNIX extensions and will fail if the server does not. The client requests that the server change the UNIX user and group ownership to the given decimal values. There is currently no way to remotely look up the UNIX UID and GID values for a given name. This may be addressed in future versions of the CIFS UNIX extensions.

cd [directory name —] If "directory name" is specified, the current working directory on the server will be changed to the directory specified. This operation will fail if for any reason the specified directory is inaccessible.

If no directory name is specified, the current working directory on the server will be reported.

del <mask> — The client will request that the server attempt to delete all files matching *mask* from the current working directory on the server.

dir <mask> — A list of the files matching *mask* in the current working directory on the server will be retrieved from the server and displayed.

exit — Terminate the connection with the server and exit from the program.

get <remote file name> [local file name —] Copy the file called `remote file name` from the server to the machine running the client. If specified, name the local copy `local file name`. All transfers in **smbclient** are binary. See also the lowercase command.

help [command —] See the ? command above.

lcd [directory name —] If *directory name* is specified, the current working directory on the local machine will be changed to the directory specified. This operation will fail if for any reason the specified directory is inaccessible.

If no directory name is specified, the name of the current working directory on the local machine will be reported.

link source destination — This command depends on the server supporting the CIFS UNIX extensions and will fail if the server does not. The client requests that the server create a hard link between the source and destination files. The source file must not exist.

lowercase — Toggle lowercasing of filenames for the get and mget commands.

When lowercasing is toggled ON, local filenames are converted to lowercase when using the get and mget commands. This is often useful when copying (say) MSDOS files from a server, because lowercase filenames are the norm on UNIX systems.

ls <mask> — See the dir command above.

mask <mask> — This command allows the user to set up a mask which will be used during recursive operation of the mget and mput commands.

The masks specified to the mget and mput commands act as filters for directories rather than files when recursion is toggled ON.

The mask specified with the mask command is necessary to filter files within those directories. For example, if the mask specified in an mget command is "source*" and the mask specified with the mask command is "*.c" and recursion is toggled ON, the mget command will retrieve all files matching "*.c" in all directories below and including all directories matching "source*" in the current working directory.

The value for mask defaults to blank (equivalent to "*") and remains so until the mask command is used to change it. It retains the most recently specified value indefinitely. To avoid unexpected results it would be wise to change the value of mask back to "*" after using the mget or mput commands.

md <directory name> — See the mkdir command.

mget <mask> — Copy all files matching *mask* from the server to the machine running the client.

mask is interpreted differently during recursive operation and non-recursive operation - refer to the recurse and mask commands for more information. All transfers in **smbclient** are binary. See also the lowercase command.

mkdir <directory name> — Create a new directory on the server (user access privileges permitting) with the specified name.

mput <mask> — Copy all files matching ***mask*** in the current working directory on the local machine to the current working directory on the server.

mask is interpreted differently during recursive operation and non-recursive operation - refer to the recurse and mask commands for more information. All transfers in **smbclient** are binary.

print <file name> — Print the specified file from the local machine through a printable service on the server.

See also the printmode command.

printmode <graphics or text> — Set the print mode to suit either binary data (such as graphical information) or text. Subsequent print commands will use the currently set print mode.

prompt — Toggle prompting for filenames during operation of the mget and mput commands.

When toggled ON, the user will be prompted to confirm the transfer of each file during these commands. When toggled OFF, all specified files will be transferred without prompting.

put <local file name> [remote file name —] Copy the file called `local file name` from the machine running the client to the server. If specified, name the remote copy `remote file name`. All transfers in **smbclient** are binary. See also the lowercase command.

queue — Displays the print queue, showing the job id, name, size and current status.

quit — See the exit command.

rd <directory name> — See the rmdir command.

recurse — Toggle directory recursion for the commands mget and mput.

When toggled ON, these commands will process all directories in the source directory (i.e., the directory they are copying from) and will recurse into any that match the mask specified to the command. Only files that match the mask specified using the mask command will be retrieved. See also the mask command.

When recursion is toggled OFF, only files from the current working directory on the source machine that match the mask specified to the mget or mput commands will be copied, and any mask specified using the mask command will be ignored.

rm <mask> — Remove all files matching ***mask*** from the current working directory on the server.

rmdir <directory name> — Remove the specified directory (user access privileges permitting) from the server.

setmode <filename> <perm=[+|\- rsha> —] A version of the DOS attrib command to set file permissions. For example:

setmode myfile +r

would make myfile read only.

symlink source destination — This command depends on the server supporting the CIFS UNIX extensions and will fail if the server does not. The client requests that the server create a symbolic hard link between the source and destination files. The source file must not exist. The server will not create a link to any path that lies outside the currently connected share. This is enforced by the Samba server.

tar <c|x>[IXbgNa —] Performs a tar operation - see the *-T* command line option above. Behavior may be affected by the tarmode command (see below). Using g (incremental) and N (newer) will affect tarmode settings. Using the "-" option with tar x may not work - use the command line option instead.

blocksize <blocksize> — Blocksize. Must be followed by a valid (greater than zero) blocksize. Causes tar file to be written out in *blocksize**TBLOCK (usually 512 byte) blocks.

tarmode <full|inc|reset|noreset> — Changes tar's behavior with regard to archive bits. In full mode, tar will back up everything regardless of the archive bit setting (this is the default mode). In incremental mode, tar will only back up files with the archive bit set. In reset mode, tar will reset the archive bit on all files it backs up (implies read/write share).

NOTES

Some servers are fussy about the case of supplied usernames, passwords, share names (AKA service names) and machine names. If you fail to connect try giving all parameters in uppercase.

It is often necessary to use the -n option when connecting to some types of servers. For example OS/2 LanManager insists on a valid NetBIOS name being used, so you need to supply a valid name that would be known to the server.

smbclient supports long file names where the server supports the LANMAN2 protocol or above.

ENVIRONMENT VARIABLES

The variable USER may contain the username of the person using the client. This information is used only if the protocol level is high enough to support session-level passwords.

The variable PASSWD may contain the password of the person using the client. This information is used only if the protocol level is high enough to support session-level passwords.

The variable LIBSMB_PROG may contain the path, executed with system(), which the client should connect to instead of connecting to a server. This functionality is primarily intended as a development aid, and works best when using a LMHOSTS file

INSTALLATION

The location of the client program is a matter for individual system administrators. The following are thus suggestions only.

It is recommended that the smbclient software be installed in the /usr/local/samba/bin/ or /usr/samba/bin/ directory, this directory readable by all, writeable only by root. The client program itself should be executable by all. The client should *NOT* be setuid or setgid!

The client log files should be put in a directory readable and writeable only by the user.

To test the client, you will need to know the name of a running SMB/CIFS server. It is possible to run smbd(8) as an ordinary user - running that server as a daemon on a user-accessible port (typically any port number over 1024) would provide a suitable test server.

DIAGNOSTICS

Most diagnostics issued by the client are logged in a specified log file. The log file name is specified at compile time, but may be overridden on the command line.

The number and nature of diagnostics available depends on the debug level used by the client. If you have problems, set the debug level to 3 and peruse the log files.

A.6 net

Synopsis

```
net <ads|rap|rpc> [-h] [-w workgroup] [-W myworkgroup] [-U user] [-I
     ip-address] [-p port] [-n myname] [-s conffile] [-S server] [-l] [-P]
     [-D debuglevel]
```

DESCRIPTION

This tool is part of the Samba(7) suite.

The samba net utility is meant to work just like the net utility available for windows and DOS. The first argument should be used to specify the protocol to use when executing a certain command. ADS is used for ActiveDirectory, RAP is using for old (Win9x/NT3) clients and RPC can be used for NT4 and Windows 2000. If this argument is omitted, net will try to determine it automatically. Not all commands are available on all protocols.

OPTIONS

-h|–help — Print a summary of command line options.

-w target-workgroup — Sets target workgroup or domain. You have to specify either this option or the IP address or the name of a server.

-W workgroup — Sets client workgroup or domain

-U user — User name to use

-I ip-address — IP address of target server to use. You have to specify either this option or a target workgroup or a target server.

-p port — Port on the target server to connect to (usually 139 or 445). Defaults to trying 445 first, then 139.

-n <primary NetBIOS name> — This option allows you to override the NetBIOS name that Samba uses for itself. This is identical to setting the *netbios name* parameter in the `smb.conf` file. However, a command line setting will take precedence over settings in `smb.conf`.

-s <configuration file> — The file specified contains the configuration details required by the server. The information in this file includes server-specific information such as what printcap file to use, as well as descriptions of all the services that the server is to provide. See `smb.conf` for more information. The default configuration file name is determined at compile time.

-S server — Name of target server. You should specify either this option or a target workgroup or a target IP address.

-l — When listing data, give more information on each item.

-P — Make queries to the external server using the machine account of the local server.

-d|–debug=debuglevel — *debuglevel* is an integer from 0 to 10. The default value if this parameter is not specified is zero.

The higher this value, the more detail will be logged to the log files about the activities of the server. At level 0, only critical errors and serious warnings will be logged. Level 1 is a reasonable level for day-to-day running - it generates a small amount of information about operations carried out.

Levels above 1 will generate considerable amounts of log data, and should only be used when investigating a problem. Levels above 3 are designed for use only by developers and generate HUGE amounts of log data, most of which is extremely cryptic.

Note that specifying this parameter here will override the *log level* parameter in the `smb.conf` file.

COMMANDS

CHANGESECRETPW

This command allows the Samba machine account password to be set from an external application to a machine account password that has already been stored in Active Directory. DO NOT USE this command unless you know exactly what you are doing. The use of this command requires that the force flag (-f) be used also. There will be NO command prompt. Whatever information is piped into stdin, either by typing at the command line or otherwise, will be stored as the literal machine password. Do NOT use this without care and attention as it will overwrite a legitimate machine password without warning. YOU HAVE BEEN WARNED.

TIME

The **NET TIME** command allows you to view the time on a remote server or synchronise the time on the local server with the time on the remote server.

TIME Without any options, the **NET TIME** command displays the time on the remote server.

TIME SYSTEM Displays the time on the remote server in a format ready for **/bin/date**

TIME SET Tries to set the date and time of the local server to that on the remote server using **/bin/date**.

TIME ZONE Displays the timezone in hours from GMT on the remote computer.

[RPC|ADS] JOIN [TYPE] [-U username[%password]] [options]

Join a domain. If the account already exists on the server, and [TYPE] is MEMBER, the machine will attempt to join automatically. (Assuming that the machine has been created in server manager) Otherwise, a password will be prompted for, and a new account may be created.

[TYPE] may be PDC, BDC or MEMBER to specify the type of server joining the domain.

[RPC] OLDJOIN [options]

Join a domain. Use the OLDJOIN option to join the domain using the old style of domain joining - you need to create a trust account in server manager first.

[RPC|ADS] USER

[RPC|ADS] USER DELETE *target* Delete specified user

[RPC|ADS] USER LIST List all users

[RPC|ADS] USER INFO *target* List the domain groups of a the specified user.

[RPC|ADS] USER ADD *name* **[password] [-F user flags] [-C comment]** Add specified user.

[RPC|ADS] GROUP

[RPC|ADS] GROUP [misc options] [targets] List user groups.

[RPC|ADS] GROUP DELETE *name* **[misc. options]** Delete specified group.

[RPC|ADS] GROUP ADD *name* **[-C comment]** Create specified group.

[RAP|RPC] SHARE

[RAP|RPC] SHARE [misc. options] [targets] Enumerates all exported resources (network shares) on target server.

[RAP|RPC] SHARE ADD *name=serverpath* **[-C comment] [-M maxusers] [targets]** Adds a share from a server (makes the export active). Maxusers specifies the number of users that can be connected to the share simultaneously.

SHARE DELETE *sharenam* Delete specified share.

[RPC|RAP] FILE

[RPC|RAP] FILE List all open files on remote server.

[RPC|RAP] FILE CLOSE *fileid* Close file with specified *fileid* on remote server.

[RPC|RAP] FILE INFO *fileid* Print information on specified *fileid*. Currently listed are: file-id, username, locks, path, permissions.

[RAP|RPC] FILE USER

NOTE

Currently NOT implemented.

SESSION

RAP SESSION Without any other options, SESSION enumerates all active SMB/CIFS sessions on the target server.

RAP SESSION DELETE|CLOSE *CLIENT_NAME* Close the specified sessions.

RAP SESSION INFO *CLIENT_NAME* Give a list with all the open files in specified session.

RAP SERVER *DOMAIN*

List all servers in specified domain or workgroup. Defaults to local domain.

RAP DOMAIN

Lists all domains and workgroups visible on the current network.

RAP PRINTQ

RAP PRINTQ LIST *QUEUE_NAME* Lists the specified print queue and print jobs on the server. If the *QUEUE_NAME* is omitted, all queues are listed.

RAP PRINTQ DELETE *JOBID* Delete job with specified id.

RAP VALIDATE *user* [*password*]

Validate whether the specified user can log in to the remote server. If the password is not specified on the commandline, it will be prompted.

NOTE

Currently NOT implemented.

RAP GROUPMEMBER

RAP GROUPMEMBER LIST *GROUP* List all members of the specified group.

RAP GROUPMEMBER DELETE *GROUP USER* Delete member from group.

RAP GROUPMEMBER ADD *GROUP USER* Add member to group.

RAP ADMIN *command*

Execute the specified *command* on the remote server. Only works with OS/2 servers.

NOTE

Currently NOT implemented.

RAP SERVICE

RAP SERVICE START *NAME* **[arguments...]** Start the specified service on the remote server. Not implemented yet.

NOTE

Currently NOT implemented.

RAP SERVICE STOP Stop the specified service on the remote server.

NOTE

Currently NOT implemented.

RAP PASSWORD *USER OLDPASS NEWPASS*

Change password of *USER* from *OLDPASS* to *NEWPASS*.

LOOKUP

LOOKUP HOST *HOSTNAME* **[***TYPE***]** Lookup the IP address of the given host with the specified type (netbios suffix). The type defaults to 0x20 (workstation).

LOOKUP LDAP [*DOMAIN* Give IP address of LDAP server of specified *DOMAIN*. Defaults to local domain.

LOOKUP KDC [*REALM***]** Give IP address of KDC for the specified *REALM*. Defaults to local realm.

LOOKUP DC [*DOMAIN***]** Give IP's of Domain Controllers for specified *DOMAIN*. Defaults to local domain.

LOOKUP MASTER *DOMAIN* Give IP of master browser for specified *DOMAIN* or workgroup. Defaults to local domain.

CACHE

Samba uses a general caching interface called 'gencache'. It can be controlled using 'NET CACHE'.

All the timeout parameters support the suffixes:

s - Seconds
m - Minutes
h - Hours
d - Days
w - Weeks

CACHE ADD *key data time-out* Add specified key+data to the cache with the given timeout.

CACHE DEL *key* Delete key from the cache.

CACHE SET *key data time-out* Update data of existing cache entry.

CACHE SEARCH *PATTERN* Search for the specified pattern in the cache data.

CACHE LIST List all current items in the cache.

CACHE FLUSH Remove all the current items from the cache.

GETLOCALSID [DOMAIN]

Print the SID of the specified domain, or if the parameter is omitted, the SID of the domain the local server is in.

SETLOCALSID S-1-5-21-x-y-z

Sets domain sid for the local server to the specified SID.

GROUPMAP

Manage the mappings between Windows group SIDs and UNIX groups. Parameters take the for "parameter=value". Common options include:

- unixgroup - Name of the UNIX group
- ntgroup - Name of the Windows NT group (must be resolvable to a SID
- rid - Unsigned 32-bit integer
- sid - Full SID in the form of "S-1-..."
- type - Type of the group; either 'domain', 'local', or 'builtin'
- comment - Freeform text description of the group

GROUPMAP ADD Add a new group mapping entry

net groupmap add {rid=int|sid=string} unixgroup=string [type={domain|local|builtin}] [ntgroup=string] [comment=string]

GROUPMAP DELETE Delete a group mapping entry

net groupmap delete {ntgroup=string|sid=SID}

GROUPMAP MODIFY Update an existing group entry

net groupmap modify {ntgroup=string|sid=SID} [unixgroup=string] [comment=string]

GROUPMAP LIST List existing group mapping entries

net groupmap list [verbose] [ntgroup=string] [sid=SID]

MAXRID

Prints out the highest RID currently in use on the local server (by the active 'passdb backend').

RPC INFO

Print information about the domain of the remote server, such as domain name, domain sid and number of users and groups.

[RPC|ADS] TESTJOIN

Check whether participation in a domain is still valid.

[RPC|ADS] CHANGETRUSTPW

Force change of domain trust password.

RPC TRUSTDOM

RPC TRUSTDOM ADD *DOMAIN* Add a interdomain trust account for *DOMAIN* to the remote server.

RPC TRUSTDOM DEL *DOMAIM* Remove interdomain trust account for *DOMAIN* from the remote server.

NOTE

Currently NOT implemented.

RPC TRUSTDOM ESTABLISH *DOMAIN* Establish a trust relationship to a trusting domain. Interdomain account must already be created on the remote PDC.

RPC TRUSTDOM REVOKE *DOMAIN* Abandon relationship to trusted domain

RPC TRUSTDOM LIST List all current interdomain trust relationships.

RPC ABORTSHUTDOWN

Abort the shutdown of a remote server.

SHUTDOWN [-t timeout] [-r] [-f] [-C message]

Shut down the remote server.

-r — Reboot after shutdown.

-f — Force shutting down all applications.

-t timeout — Timeout before system will be shut down. An interactive user of the system can use this time to cancel the shutdown.

-C message — Display the specified message on the screen to announce the shutdown.

SAMDUMP

Print out sam database of remote server. You need to run this on either a BDC.

VAMPIRE

Export users, aliases and groups from remote server to local server. Can only be run an a BDC.

GETSID

Fetch domain SID and store it in the local `secrets.tdb`.

ADS LEAVE

Make the remote host leave the domain it is part of.

ADS STATUS

Print out status of machine account of the local machine in ADS. Prints out quite some debug info. Aimed at developers, regular users should use **NET ADS TESTJOIN**.

ADS PRINTER

ADS PRINTER INFO [*PRINTER*] [*SERVER*] Lookup info for *PRINTER* on *SERVER*. The printer name defaults to "*", the server name defaults to the local host.

ADS PRINTER PUBLISH *PRINTER* Publish specified printer using ADS.

ADS PRINTER REMOVE *PRINTER* Remove specified printer from ADS directory.

ADS SEARCH *EXPRESSION ATTRIBUTES...*

Perform a raw LDAP search on a ADS server and dump the results. The expression is a standard LDAP search expression, and the attributes are a list of LDAP fields to show in the results.

Example: `net ads search '(objectCategory=group)' sAMAccountName`

ADS DN *DN (attributes)*

Perform a raw LDAP search on a ADS server and dump the results. The DN standard LDAP DN, and the attributes are a list of LDAP fields to show in the result.

Example: `net ads dn 'CN=administrator,CN=Users,DC=my,DC=domain' SAMAccountName`

WORKGROUP

Print out workgroup name for specified kerberos realm.

HELP [COMMAND]

Gives usage information for the specified command.

A.7 nmbd

Synopsis

```
nmbd [-D] [-F] [-S] [-a] [-i] [-o] [-h] [-V] [-d <debug level>] [-H
     <lmhosts file>] [-l <log directory>] [-n <primary netbios name>] [-p
     <port number>] [-s <configuration file>]
```

DESCRIPTION

This program is part of the Samba(7) suite.

nmbd is a server that understands and can reply to NetBIOS over IP name service requests, like those produced by SMB/CIFS clients such as Windows 95/98/Me, Windows NT, Windows 2000, Windows XP and LanManager clients. It also participates in the browsing protocols which make up the Windows "Network Neighborhood" view.

SMB/CIFS clients, when they start up, may wish to locate an SMB/CIFS server. That is, they wish to know what IP number a specified host is using.

Among other services, **nmbd** will listen for such requests, and if its own NetBIOS name is specified it will respond with the IP number of the host it is running on. Its "own NetBIOS name" is by default the primary DNS name of the host it is running on, but this can be overridden with the *-n* option (see OPTIONS below). Thus **nmbd** will reply to broadcast queries for its own name(s). Additional names for **nmbd** to respond on can be set via parameters in the smb.conf(5) configuration file.

nmbd can also be used as a WINS (Windows Internet Name Server) server. What this basically means is that it will act as a WINS database server, creating a database from name registration requests that it receives and replying to queries from clients for these names.

In addition, **nmbd** can act as a WINS proxy, relaying broadcast queries from clients that do not understand how to talk the WINS protocol to a WINS server.

OPTIONS

-D — If specified, this parameter causes **nmbd** to operate as a daemon. That is, it detaches itself and runs in the background, fielding requests on the appropriate port. By default, **nmbd** will operate as a daemon if launched from a command shell. nmbd can also be operated from the **inetd** meta-daemon, although this is not recommended.

-F — If specified, this parameter causes the main **nmbd** process to not daemonize, i.e. double-fork and disassociate with the terminal. Child processes are still created as normal to service each connection request, but the main process does not exit. This operation mode is suitable for running **nmbd** under process supervisors such as **supervise** and **svscan** from Daniel J. Bernstein's **daemontools** package, or the AIX process monitor.

-S — If specified, this parameter causes **nmbd** to log to standard output rather than a file.

-i — If this parameter is specified it causes the server to run "interactively", not as a daemon, even if the server is executed on the command line of a shell. Setting this parameter negates the implicit daemon mode when run from the command line. **nmbd** also logs to standard output, as if the `-S` parameter had been given.

-h|–help — Print a summary of command line options.

-H <filename> — NetBIOS lmhosts file. The lmhosts file is a list of NetBIOS names to IP addresses that is loaded by the nmbd server and used via the name resolution mechanism *`name resolve order`* described in smb.conf(5) to resolve any NetBIOS name queries needed by the server. Note that the contents of this file are *NOT* used by **nmbd** to answer any name queries. Adding a line to this file affects name NetBIOS resolution from this host *ONLY*.

The default path to this file is compiled into Samba as part of the build process. Common defaults are `/usr/local/samba/lib/lmhosts`, `/usr/samba/lib/lmhosts` or `/etc/samba/lmhosts`. See the lmhosts(5) man page for details on the contents of this file.

-V — Prints the program version number.

-s <configuration file> — The file specified contains the configuration details required by the server. The information in this file includes server-specific information such as what printcap file to use, as well as descriptions of all the services that the server is to provide. See `smb.conf` for more information. The default configuration file name is determined at compile time.

-d|–debug=debuglevel — *debuglevel* is an integer from 0 to 10. The default value if this parameter is not specified is zero.

The higher this value, the more detail will be logged to the log files about the activities of the server. At level 0, only critical errors and serious warnings will be logged. Level 1 is a reasonable level for day-to-day running - it generates a small amount of information about operations carried out.

Levels above 1 will generate considerable amounts of log data, and should only be used when investigating a problem. Levels above 3 are designed for use only by developers and generate HUGE amounts of log data, most of which is extremely cryptic.

Note that specifying this parameter here will override the *log level* parameter in the `smb.conf` file.

-l|–logfile=logbasename — File name for log/debug files. The extension `".client"` will be appended. The log file is never removed by the client.

-p <UDP port number> — UDP port number is a positive integer value. This option changes the default UDP port number (normally 137) that **nmbd** responds to name queries on. Don't use this option unless you are an expert, in which case you won't need help!

FILES

`/etc/inetd.conf` — If the server is to be run by the **inetd** meta-daemon, this file must contain suitable startup information for the meta-daemon.

`/etc/rc` — or whatever initialization script your system uses.

If running the server as a daemon at startup, this file will need to contain an appropriate startup sequence for the server.

`/etc/services` — If running the server via the meta-daemon **inetd**, this file must contain a mapping of service name (e.g., netbios-ssn) to service port (e.g., 139) and protocol type (e.g., tcp).

`/usr/local/samba/lib/smb.conf` — This is the default location of the smb.conf(5) server configuration file. Other common places that systems install this file are `/usr/samba/lib/smb.conf` and `/etc/samba/smb.conf`.

When run as a WINS server (see the *`wins support`* parameter in the smb.conf(5) man page), **nmbd** will store the WINS database in the file `wins.dat` in the `var/locks` directory configured under wherever Samba was configured to install itself.

If **nmbd** is acting as a *browse master* (see the *`local master`* parameter in the smb.conf(5) man page, **nmbd** will store the browsing database in the file `browse.dat` in the `var/locks` directory configured under wherever Samba was configured to install itself.

SIGNALS

To shut down an **nmbd** process it is recommended that SIGKILL (-9) *NOT* be used, except as a last resort, as this may leave the name database in an inconsistent state. The correct way to terminate **nmbd** is to send it a SIGTERM (-15) signal and wait for it to die on its own.

nmbd will accept SIGHUP, which will cause it to dump out its namelists into the file `namelist.debug` in the `/usr/local/samba/var/locks` directory (or the `var/locks` directory configured under wherever Samba was configured to install itself). This will also cause **nmbd** to dump out its server database in the `log.nmb` file.

The debug log level of nmbd may be raised or lowered using smbcontrol(1) (SIGUSR[1|2] signals are no longer used since Samba 2.2). This is to allow transient problems to be diagnosed, whilst still running at a normally low log level.

SEE ALSO

inetd(8), smbd(8), smb.conf(5), smbclient(1), testparm(1), testprns(1), and the Internet RFC's `rfc1001.txt`, `rfc1002.txt`. In addition the CIFS (formerly SMB) specification is available as a link from the Web page http://samba.org/cifs/.

A.8 pdbedit

Synopsis

```
pdbedit [-L] [-v] [-w] [-u username] [-f fullname] [-h homedir] [-D drive]
    [-S script] [-p profile] [-a] [-m] [-r] [-x] [-i passdb-backend] [-e
    passdb-backend] [-b passdb-backend] [-g] [-d debuglevel] [-s
    configfile] [-P account-policy] [-C value] [-c account-control]
```

DESCRIPTION

This tool is part of the Samba(7) suite.

The pdbedit program is used to manage the users accounts stored in the sam database and can only be run by root.

The pdbedit tool uses the passdb modular interface and is independent from the kind of users database used (currently there are smbpasswd, ldap, nis+ and tdb based and more can be added without changing the tool).

There are five main ways to use pdbedit: adding a user account, removing a user account, modifing a user account, listing user accounts, importing users accounts.

OPTIONS

-L — This option lists all the user accounts present in the users database. This option prints a list of user/uid pairs separated by the ':' character.

Example: **pdbedit -L**

```
sorce:500:Simo Sorce
samba:45:Test User
```

-v — This option enables the verbose listing format. It causes pdbedit to list the users in the database, printing out the account fields in a descriptive format.

Example: **pdbedit -L -v**

```
---------------
username:       sorce
user ID/Group:  500/500
user RID/GRID:  2000/2001
Full Name:      Simo Sorce
Home Directory: \\BERSERKER\sorce
HomeDir Drive:  H:
Logon Script:   \\BERSERKER\netlogon\sorce.bat
Profile Path:   \\BERSERKER\profile
---------------
username:       samba
user ID/Group:  45/45
user RID/GRID:  1090/1091
Full Name:      Test User
Home Directory: \\BERSERKER\samba
HomeDir Drive:
Logon Script:
```

```
Profile Path:    \\BERSERKER\profile
```

-w — This option sets the "smbpasswd" listing format. It will make pdbedit list the users in the database, printing out the account fields in a format compatible with the `smbpasswd` file format. (see the smbpasswd(5) for details)

Example: **pdbedit -L -w**

```
sorce:500:508818B733CE64BEAAD3B435B51404EE \
   :D2A2418EFC466A8A0F6B1DBB5C3DB80C \
   :[UX          ]:LCT-00000000:
   samba:45:0F2B255F7B67A7A9AAD3B435B51404EE \
   :BC281CE3F53B6A5146629CD4751D3490 \
   :[UX          ]:LCT-3BFA1E8D:
```

-u username — This option specifies the username to be used for the operation requested (listing, adding, removing). It is *required* in add, remove and modify operations and *optional* in list operations.

-f fullname — This option can be used while adding or modifing a user account. It will specify the user's full name.

Example: **-f "Simo Sorce"**

-h homedir — This option can be used while adding or modifing a user account. It will specify the user's home directory network path.

Example: **-h "\\\\BERSERKER\\sorce"**

-D drive — This option can be used while adding or modifing a user account. It will specify the windows drive letter to be used to map the home directory.

Example: **-d "H:"**

-S script — This option can be used while adding or modifing a user account. It will specify the user's logon script path.

Example: **-s "\\\\BERSERKER\\netlogon\\sorce.bat"**

-p profile — This option can be used while adding or modifing a user account. It will specify the user's profile directory.

Example: **-p "\\\\BERSERKER\\netlogon"**

-G SID|rid — This option can be used while adding or modifying a user account. It will specify the users' new primary group SID (Security Identifier) or rid.

Example: **-G S-1-5-21-2447931902-1787058256-3961074038-1201**

-U SID|rid — This option can be used while adding or modifying a user account. It will specify the users' new SID (Security Identifier) or rid.

Example: **-U S-1-5-21-2447931902-1787058256-3961074038-5004**

-c account-control — This option can be used while adding or modifying a user account. It will specify the users' account control property. Possible flags that can be set are: N, D, H, L, X.

Example: **-c "[X]"**

-a — This option is used to add a user into the database. This command needs a user name specified with the -u switch. When adding a new user, pdbedit will also ask for the password to be used.

Example: **pdbedit -a -u sorce**

```
new password:
retype new password
```

-r — This option is used to modify an existing user in the database. This command needs a user name specified with the -u switch. Other options can be specified to modify the properties of the specified user. This flag is kept for backwards compatibility, but it is no longer necessary to specify it.

-m — This option may only be used in conjunction with the *-a* option. It will make pdbedit to add a machine trust account instead of a user account (-u username will provide the machine name).

Example: **pdbedit -a -m -u w2k-wks**

-x — This option causes pdbedit to delete an account from the database. It needs a username specified with the -u switch.

Example: **pdbedit -x -u bob**

-i passdb-backend — Use a different passdb backend to retrieve users than the one specified in smb.conf. Can be used to import data into your local user database.

This option will ease migration from one passdb backend to another.

Example: **pdbedit -i smbpasswd:/etc/smbpasswd.old**

-e passdb-backend — Exports all currently available users to the specified password database backend.

This option will ease migration from one passdb backend to another and will ease backing up.

Example: **pdbedit -e smbpasswd:/root/samba-users.backup**

-g — If you specify *-g*, then *-i in-backend -e out-backend* applies to the group mapping instead of the user database.

This option will ease migration from one passdb backend to another and will ease backing up.

-b passdb-backend — Use a different default passdb backend.

Example: **pdbedit -b xml:/root/pdb-backup.xml -l**

-P account-policy — Display an account policy

Valid policies are: minimum password age, reset count minutes, disconnect time, user must logon to change password, password history, lockout duration, min password length, maximum password age and bad lockout attempt.

Example: **pdbedit -P "bad lockout attempt"**

```
account policy value for bad lockout attempt is 0
```

-C account-policy-value — Sets an account policy to a specified value. This option may only be used in conjunction with the *-P* option.

Example: **pdbedit -P "bad lockout attempt" -C 3**

```
account policy value for bad lockout attempt was 0
account policy value for bad lockout attempt is now 3
```

-h|–help — Print a summary of command line options.

-V — Prints the program version number.

-s <configuration file> — The file specified contains the configuration details required by the server. The information in this file includes server-specific information such as what printcap file to use, as well as descriptions of all the services that the server is to provide. See `smb.conf` for more information. The default configuration file name is determined at compile time.

-d|–debug=debuglevel — *debuglevel* is an integer from 0 to 10. The default value if this parameter is not specified is zero.

The higher this value, the more detail will be logged to the log files about the activities of the server. At level 0, only critical errors and serious warnings will be logged. Level 1 is a reasonable level for day-to-day running - it generates a small amount of information about operations carried out.

Levels above 1 will generate considerable amounts of log data, and should only be used when investigating a problem. Levels above 3 are designed for use only by developers and generate HUGE amounts of log data, most of which is extremely cryptic.

Note that specifying this parameter here will override the *log level* parameter in the `smb.conf` file.

-l|–logfile=logbasename — File name for log/debug files. The extension `".client"` will be appended. The log file is never removed by the client.

NOTES

This command may be used only by root.

SEE ALSO

smbpasswd(5), samba(7)

A.9 smbcquotas

Synopsis

```
smbcquotas //server/share [-u user] [-L] [-F] [-S QUOTA_SET_COMMAND] [-n]
     [-t] [-v] [-d debuglevel] [-s configfile] [-l logfilebase] [-V] [-U
     username] [-N] [-k] [-A]
```

DESCRIPTION

This tool is part of the Samba(7) suite.

The **smbcquotas** program manipulates NT Quotas on SMB file shares.

OPTIONS

The following options are available to the **smbcquotas** program.

-u user — Specifies the user of whom the quotas are get or set. By default the current user's username will be used.

-L — Lists all quota records of the share.

-F — Show the share quota status and default limits.

-S QUOTA_SET_COMMAND — This command set/modify quotas for a user or on the share, depending on the QUOTA_SET_COMMAND parameter witch is described later

-n — This option displays all QUOTA information in numeric format. The default is to convert SIDs to names and QUOTA limits to a readable string format.

-t — Don't actually do anything, only validate the correctness of the arguments.

-v — Be verbose.

-h|–help — Print a summary of command line options.

-V — Prints the program version number.

-s <configuration file> — The file specified contains the configuration details required by the server. The information in this file includes server-specific information such as what printcap file to use, as well as descriptions of all the services that the server is to provide. See `smb.conf` for more information. The default configuration file name is determined at compile time.

-d|–debug=debuglevel — *debuglevel* is an integer from 0 to 10. The default value if this parameter is not specified is zero.

The higher this value, the more detail will be logged to the log files about the activities of the server. At level 0, only critical errors and serious warnings will be logged. Level 1 is a reasonable level for day-to-day running - it generates a small amount of information about operations carried out.

Levels above 1 will generate considerable amounts of log data, and should only be used when investigating a problem. Levels above 3 are designed for use only by developers and generate HUGE amounts of log data, most of which is extremely cryptic.

Note that specifying this parameter here will override the *log level* parameter in the `smb.conf` file.

-l|–logfile=logbasename — File name for log/debug files. The extension ".client" will be appended. The log file is never removed by the client.

-N — If specified, this parameter suppresses the normal password prompt from the client to the user. This is useful when accessing a service that does not require a password.

Unless a password is specified on the command line or this parameter is specified, the client will request a password.

-k — Try to authenticate with kerberos. Only useful in an Active Directory environment.

-A|–authfile=filename — This option allows you to specify a file from which to read the username and password used in the connection. The format of the file is

```
username = <value>
password = <value>
domain   = <value>
```

Make certain that the permissions on the file restrict access from unwanted users.

-U|–user=username[%password —] Sets the SMB username or username and password.

If %password is not specified, the user will be prompted. The client will first check the `USER` environment variable, then the `LOGNAME` variable and if either exists, the string is uppercased. If these environmental variables are not found, the username `GUEST` is used.

A third option is to use a credentials file which contains the plaintext of the username and password. This option is mainly provided for scripts where the admin does not wish to pass the credentials on the command line or via environment variables. If this method is used, make certain that the permissions on the file restrict access from unwanted users. See the *-A* for more details.

Be cautious about including passwords in scripts. Also, on many systems the command line of a running process may be seen via the **ps** command. To be safe always allow **rpcclient** to prompt for a password and type it in directly.

QUOTA_SET_COMAND

The format of an ACL is one or more ACL entries separated by either commas or newlines. An ACL entry is one of the following:

for user setting quotas for the specified by -u or the current username:

`UQLIM:<username><softlimit><hardlimit>`

for setting the share quota defaults limits:

```
FSQLIM:<softlimit><hardlimit>
```

for changing the share quota settings:

```
FSQFLAGS:QUOTA_ENABLED/DENY_DISK/LOG_SOFTLIMIT/LOG_HARD_LIMIT
```

EXIT STATUS

The **smbcquotas** program sets the exit status depending on the success or otherwise of the operations performed. The exit status may be one of the following values.

If the operation succeeded, smbcquotas returns an exit status of 0. If **smbcquotas** couldn't connect to the specified server, or when there was an error getting or setting the quota(s), an exit status of 1 is returned. If there was an error parsing any command line arguments, an exit status of 2 is returned.

A.10 smbd

Synopsis

```
smbd [-D] [-F] [-S] [-i] [-h] [-V] [-b] [-d <debug level>] [-l <log
     directory>] [-p <port number>] [-O <socket option>] [-s
     <configuration file>]
```

DESCRIPTION

This program is part of the Samba(7) suite.

smbd is the server daemon that provides filesharing and printing services to Windows clients. The server provides filespace and printer services to clients using the SMB (or CIFS) protocol. This is compatible with the LanManager protocol, and can service LanManager clients. These include MSCLIENT 3.0 for DOS, Windows for Workgroups, Windows 95/98/ME, Windows NT, Windows 2000, OS/2, DAVE for Macintosh, and smbfs for Linux.

An extensive description of the services that the server can provide is given in the man page for the configuration file controlling the attributes of those services (see smb.conf(5). This man page will not describe the services, but will concentrate on the administrative aspects of running the server.

Please note that there are significant security implications to running this server, and the smb.conf(5) manual page should be regarded as mandatory reading before proceeding with installation.

A session is created whenever a client requests one. Each client gets a copy of the server for each session. This copy then services all connections made by the client during that session. When all connections from its client are closed, the copy of the server for that client terminates.

The configuration file, and any files that it includes, are automatically reloaded every minute, if they change. You can force a reload by sending a SIGHUP to the server. Reloading the configuration file will not affect connections to any service that is already established. Either the user will have to disconnect from the service, or **smbd** killed and restarted.

OPTIONS

-D — If specified, this parameter causes the server to operate as a daemon. That is, it detaches itself and runs in the background, fielding requests on the appropriate port. Operating the server as a daemon is the recommended way of running **smbd** for servers that provide more than casual use file and print services. This switch is assumed if **smbd** is executed on the command line of a shell.

-F — If specified, this parameter causes the main **smbd** process to not daemonize, i.e. double-fork and disassociate with the terminal. Child processes are still created as normal to service each connection request, but the main process does not exit. This operation mode is suitable for running **smbd** under process supervisors such as **supervise** and **svscan** from Daniel J. Bernstein's **daemontools** package, or the AIX process monitor.

-S — If specified, this parameter causes **smbd** to log to standard output rather than a file.

-i — If this parameter is specified it causes the server to run "interactively", not as a daemon, even if the server is executed on the command line of a shell. Setting this parameter negates the implicit deamon mode when run from the command line. **smbd** also logs to standard output, as if the **-S** parameter had been given.

-V — Prints the program version number.

-s **<configuration file>** — The file specified contains the configuration details required by the server. The information in this file includes server-specific information such as what printcap file to use, as well as descriptions of all the services that the server is to provide. See `smb.conf` for more information. The default configuration file name is determined at compile time.

-d|–debug=debuglevel — *`debuglevel`* is an integer from 0 to 10. The default value if this parameter is not specified is zero.

The higher this value, the more detail will be logged to the log files about the activities of the server. At level 0, only critical errors and serious warnings will be logged. Level 1 is a reasonable level for day-to-day running - it generates a small amount of information about operations carried out.

Levels above 1 will generate considerable amounts of log data, and should only be used when investigating a problem. Levels above 3 are designed for use only by developers and generate HUGE amounts of log data, most of which is extremely cryptic.

Note that specifying this parameter here will override the *log level* parameter in the `smb.conf` file.

-l|–logfile=logbasename — File name for log/debug files. The extension `".client"` will be appended. The log file is never removed by the client.

-h|–help — Print a summary of command line options.

-b — Prints information about how Samba was built.

-l <log directory> — If specified, *log directory* specifies a log directory into which the "log.smbd" log file will be created for informational and debug messages from the running server. The log file generated is never removed by the server although its size may be controlled by the *max log size* option in the smb.conf(5) file. *Beware:* If the directory specified does not exist, **smbd** will log to the default debug log location defined at compile time.

The default log directory is specified at compile time.

-p <port number> — *port number* is a positive integer value. The default value if this parameter is not specified is 139.

This number is the port number that will be used when making connections to the server from client software. The standard (well-known) port number for the SMB over TCP is 139, hence the default. If you wish to run the server as an ordinary user rather than as root, most systems will require you to use a port number greater than 1024 - ask your system administrator for help if you are in this situation.

In order for the server to be useful by most clients, should you configure it on a port other than 139, you will require port redirection services on port 139, details of which are outlined in rfc1002.txt section 4.3.5.

This parameter is not normally specified except in the above situation.

FILES

`/etc/inetd.conf` — If the server is to be run by the **inetd** meta-daemon, this file must contain suitable startup information for the meta-daemon.

`/etc/rc` — or whatever initialization script your system uses).

If running the server as a daemon at startup, this file will need to contain an appropriate startup sequence for the server.

`/etc/services` — If running the server via the meta-daemon **inetd**, this file must contain a mapping of service name (e.g., netbios-ssn) to service port (e.g., 139) and protocol type (e.g., tcp).

`/usr/local/samba/lib/smb.conf` — This is the default location of the smb.conf(5) server configuration file. Other common places that systems install this file are `/usr/samba/lib/smb.conf` and `/etc/samba/smb.conf`.

This file describes all the services the server is to make available to clients. See smb.conf(5) for more information.

LIMITATIONS

On some systems **smbd** cannot change uid back to root after a setuid() call. Such systems are called trapdoor uid systems. If you have such a system, you will be unable to connect from a client (such as a PC) as two different users at once. Attempts to connect the second user will result in access denied or similar.

ENVIRONMENT VARIABLES

`PRINTER` — If no printer name is specified to printable services, most systems will use the value of this variable (or `lp` if this variable is not defined) as the name of the printer to use. This is not specific to the server, however.

PAM INTERACTION

Samba uses PAM for authentication (when presented with a plaintext password), for account checking (is this account disabled?) and for session management. The degree too which samba supports PAM is restricted by the limitations of the SMB protocol and the ***obey pam restrictions*** smb.conf(5) paramater. When this is set, the following restrictions apply:

- *Account Validation*: All accesses to a samba server are checked against PAM to see if the account is vaild, not disabled and is permitted to login at this time. This also applies to encrypted logins.
- *Session Management*: When not using share level secuirty, users must pass PAM's session checks before access is granted. Note however, that this is bypassed in share level secuirty. Note also that some older pam configuration files may need a line added for session support.

DIAGNOSTICS

Most diagnostics issued by the server are logged in a specified log file. The log file name is specified at compile time, but may be overridden on the command line.

The number and nature of diagnostics available depends on the debug level used by the server. If you have problems, set the debug level to 3 and peruse the log files.

Most messages are reasonably self-explanatory. Unfortunately, at the time this man page was created, there are too many diagnostics available in the source code to warrant describing each and every diagnostic. At this stage your best bet is still to grep the source code and inspect the conditions that gave rise to the diagnostics you are seeing.

SIGNALS

Sending the **smbd** a SIGHUP will cause it to reload its `smb.conf` configuration file within a short period of time.

To shut down a user's **smbd** process it is recommended that **SIGKILL (-9)** *NOT* be used, except as a last resort, as this may leave the shared memory area in an inconsistent state. The safe way to terminate an **smbd** is to send it a SIGTERM (-15) signal and wait for it to die on its own.

The debug log level of **smbd** may be raised or lowered using smbcontrol(1) program (SIGUSR[1|2] signals are no longer used since Samba 2.2). This is to allow transient problems to be diagnosed, whilst still running at a normally low log level.

Note that as the signal handlers send a debug write, they are not re-entrant in **smbd**. This you should wait until **smbd** is in a state of waiting for an incoming SMB before issuing them. It is possible to make the signal handlers safe by un-blocking the signals before the select call and re-blocking them after, however this would affect performance.

SEE ALSO

hosts_access(5), inetd(8), nmbd(8), smb.conf(5), smbclient(1), testparm(1), testprns(1), and the Internet RFC's `rfc1001.txt`, `rfc1002.txt`. In addition the CIFS (formerly SMB) specification is available as a link from the Web page http://samba.org/cifs/.

A.11 smbpasswd

Synopsis

`smbpasswd`

DESCRIPTION

This tool is part of the Samba(7) suite.

smbpasswd is the Samba encrypted password file. It contains the username, Unix user id and the SMB hashed passwords of the user, as well as account flag information and the time the password was last changed. This file format has been evolving with Samba and has had several different formats in the past.

FILE FORMAT

The format of the smbpasswd file used by Samba 2.2 is very similar to the familiar Unix `passwd(5)` file. It is an ASCII file containing one line for each user. Each field ithin each line is separated from the next by a colon. Any entry beginning with '#' is ignored. The smbpasswd file contains the following information for each user:

name — This is the user name. It must be a name that already exists in the standard UNIX passwd file.

uid — This is the UNIX uid. It must match the uid field for the same user entry in the standard UNIX passwd file. If this does not match then Samba will refuse to recognize this smbpasswd file entry as being valid for a user.

Lanman Password Hash — This is the LANMAN hash of the user's password, encoded as 32 hex digits. The LANMAN hash is created by DES encrypting a well known string with the user's password as the DES key. This is the same password used by Windows 95/98 machines. Note that this password hash is regarded as weak as it is vulnerable to dictionary attacks and if two users choose the same password this entry will be identical (i.e. the password is not "salted" as the UNIX password is). If the user has a null password this field will contain the characters "NO PASSWORD" as the start of the hex string. If the hex string is equal to 32 'X' characters then the user's account is marked as `disabled` and the user will not be able to log onto the Samba server.

WARNING! Due to the challenge-response nature of the SMB/CIFS authentication protocol, anyone with a knowledge of this password hash will be able to impersonate the user on the network. For this reason these hashes are known as *plain text equivalents* and must *NOT* be made available to anyone but the root user. To protect these passwords the smbpasswd file is placed in a directory with read and traverse access only to the root user and the smbpasswd file itself must be set to be read/write only by root, with no other access.

NT Password Hash — This is the Windows NT hash of the user's password, encoded as 32 hex digits. The Windows NT hash is created by taking the user's password as represented in 16-bit, little-endian UNICODE and then applying the MD4 (internet rfc1321) hashing algorithm to it.

This password hash is considered more secure than the LANMAN Password Hash as it preserves the case of the password and uses a much higher quality hashing algorithm. However, it is still the case that if two users choose the same password this entry will be identical (i.e. the password is not "salted" as the UNIX password is).

WARNING !!. Note that, due to the challenge-response nature of the SMB/CIFS authentication protocol, anyone with a knowledge of this password hash will be able to impersonate the user on the network. For this reason these hashes are known as *plain text equivalents* and must *NOT* be made available to anyone but the root user.

To protect these passwords the smbpasswd file is placed in a directory with read and traverse access only to the root user and the smbpasswd file itself must be set to be read/write only by root, with no other access.

Account Flags — This section contains flags that describe the attributes of the users account. In the Samba 2.2 release this field is bracketed by '[' and ']' characters and is always 13 characters in length (including the '[' and ']' characters). The contents of this field may be any of the following characters:

- *U* - This means this is a "User" account, i.e. an ordinary user. Only User and Workstation Trust accounts are currently supported in the smbpasswd file.
- *N* - This means the account has no password (the passwords in the fields LANMAN Password Hash and NT Password Hash are ignored). Note that this will only allow users to log on with no password if the ***null passwords*** parameter is set in the smb.conf(5) config file.
- *D* - This means the account is disabled and no SMB/CIFS logins will be allowed for this user.
- *W* - This means this account is a "Workstation Trust" account. This kind of account is used in the Samba PDC code stream to allow Windows NT Workstations and Servers to join a Domain hosted by a Samba PDC.

Other flags may be added as the code is extended in future. The rest of this field space is filled in with spaces.

Last Change Time — This field consists of the time the account was last modified. It consists of the characters 'LCT-' (standing for "Last Change Time") followed by a numeric encoding of the UNIX time in seconds since the epoch (1970) that the last change was made.

All other colon separated fields are ignored at this time.

SEE ALSO

smbpasswd(8), Samba(7), and the Internet RFC1321 for details on the MD4 algorithm.

A.12 smbpasswd

Synopsis

```
smbpasswd [-a] [-x] [-d] [-e] [-D debuglevel] [-n] [-r <remote machine>]
    [-R <name resolve order>] [-m] [-U username[%password]] [-h] [-s] [-w
    pass] [-i] [-L] [username]
```

DESCRIPTION

This tool is part of the Samba(7) suite.

The smbpasswd program has several different functions, depending on whether it is run by the *root* user or not. When run as a normal user it allows the user to change the password used for their SMB sessions on any machines that store SMB passwords.

By default (when run with no arguments) it will attempt to change the current user's SMB password on the local machine. This is similar to the way the **passwd(1)** program works. **smbpasswd** differs from how the passwd program works however in that it is not *setuid root* but works in a client-server mode and communicates with a locally running smbd(8). As a consequence in order for this to succeed the smbd daemon must be running on the local machine. On a UNIX machine the encrypted SMB passwords are usually stored in the smbpasswd(5) file.

When run by an ordinary user with no options, smbpasswd will prompt them for their old SMB password and then ask them for their new password twice, to ensure that the new password was typed correctly. No passwords will be echoed on the screen whilst being typed. If you have a blank SMB password (specified by the string "NO PASSWORD" in the smbpasswd file) then just press the <Enter> key when asked for your old password.

smbpasswd can also be used by a normal user to change their SMB password on remote machines, such as Windows NT Primary Domain Controllers. See the (`-r`) and `-U` options below.

When run by root, smbpasswd allows new users to be added and deleted in the smbpasswd file, as well as allows changes to the attributes of the user in this file to be made. When run by root, **smbpasswd** accesses the local smbpasswd file directly, thus enabling changes to be made even if smbd is not running.

OPTIONS

-a — This option specifies that the username following should be added to the local smbpasswd file, with the new password typed (type <Enter> for the old password). This option is ignored if the username following already exists in the smbpasswd file and it is treated like a regular change password command. Note that the default passdb backends require the user to already exist in the system password file (usually `/etc/passwd`), else the request to add the user will fail.

This option is only available when running smbpasswd as root.

-x — This option specifies that the username following should be deleted from the local smbpasswd file.

This option is only available when running smbpasswd as root.

-d — This option specifies that the username following should be `disabled` in the local smbpasswd file. This is done by writing a `'D'` flag into the account control space in the smbpasswd file. Once this is done all attempts to authenticate via SMB using this username will fail.

If the smbpasswd file is in the 'old' format (pre-Samba 2.0 format) there is no space in the user's password entry to write this information and the command will FAIL. See smbpasswd(5) for details on the 'old' and new password file formats.

This option is only available when running smbpasswd as root.

-e — This option specifies that the username following should be `enabled` in the local smbpasswd file, if the account was previously disabled. If the account was not disabled this option has no effect. Once the account is enabled then the user will be able to authenticate via SMB once again.

If the smbpasswd file is in the 'old' format, then **smbpasswd** will FAIL to enable the account. See smbpasswd(5) for details on the 'old' and new password file formats.

This option is only available when running smbpasswd as root.

-D debuglevel — *debuglevel* is an integer from 0 to 10. The default value if this parameter is not specified is zero.

The higher this value, the more detail will be logged to the log files about the activities of smbpasswd. At level 0, only critical errors and serious warnings will be logged.

Levels above 1 will generate considerable amounts of log data, and should only be used when investigating a problem. Levels above 3 are designed for use only by developers and generate HUGE amounts of log data, most of which is extremely cryptic.

-n — This option specifies that the username following should have their password set to null (i.e. a blank password) in the local smbpasswd file. This is done by writing the string "NO PASSWORD" as the first part of the first password stored in the smbpasswd file.

Note that to allow users to logon to a Samba server once the password has been set to "NO PASSWORD" in the smbpasswd file the administrator must set the following parameter in the [global] section of the `smb.conf` file :

null passwords = yes

This option is only available when running smbpasswd as root.

-r remote machine name — This option allows a user to specify what machine they wish to change their password on. Without this parameter smbpasswd defaults to the local host. The *remote machine name* is the NetBIOS name of the SMB/CIFS server to contact to attempt the password change. This name is resolved into an IP address using the standard name resolution mechanism in all programs of the Samba suite. See the *-R name resolve order* parameter for details on changing this resolving mechanism.

The username whose password is changed is that of the current UNIX logged on user. See the *-U username* parameter for details on changing the password for a different username.

Note that if changing a Windows NT Domain password the remote machine specified must be the Primary Domain Controller for the domain (Backup Domain Controllers only have a read-only copy of the user account database and will not allow the password change).

Note that Windows 95/98 do not have a real password database so it is not possible to change passwords specifying a Win95/98 machine as remote machine target.

-R name resolve order — This option allows the user of smbpasswd to determine what name resolution services to use when looking up the NetBIOS name of the host being connected to.

The options are :"lmhosts", "host", "wins" and "bcast". They cause names to be resolved as follows:

- `lmhosts`: Lookup an IP address in the Samba lmhosts file. If the line in lmhosts has no name type attached to the NetBIOS name (see the lmhosts(5) for details) then any name type matches for lookup.
- `host`: Do a standard host name to IP address resolution, using the system `/etc/hosts`, NIS, or DNS lookups. This method of name resolution is operating system depended for instance on IRIX or Solaris this may be controlled by the `/etc/nsswitch.conf` file). Note that this method is only used if the NetBIOS name type being queried is the 0x20 (server) name type, otherwise it is ignored.
- `wins`: Query a name with the IP address listed in the *wins server* parameter. If no WINS server has been specified this method will be ignored.
- `bcast`: Do a broadcast on each of the known local interfaces listed in the *interfaces* parameter. This is the least reliable of the name resolution methods as it depends on the target host being on a locally connected subnet.

The default order is **lmhosts, host, wins, bcast** and without this parameter or any entry in the smb.conf(5) file the name resolution methods will be attempted in this order.

-m — This option tells smbpasswd that the account being changed is a MACHINE account. Currently this is used when Samba is being used as an NT Primary Domain Controller.

This option is only available when running smbpasswd as root.

-U username — This option may only be used in conjunction with the *-r* option. When changing a password on a remote machine it allows the user to specify the user name on that machine whose password will be changed. It is present to allow users who have different user names on different systems to change these passwords.

-h — This option prints the help string for **smbpasswd**, selecting the correct one for running as root or as an ordinary user.

-s — This option causes smbpasswd to be silent (i.e. not issue prompts) and to read its old and new passwords from standard input, rather than from `/dev/tty` (like the **passwd(1)** program does). This option is to aid people writing scripts to drive smbpasswd

-w password — This parameter is only available if Samba has been configured to use the experimental –**with-ldapsam** option. The *-w* switch is used to specify the password to be used with the *ldap admin dn*. Note that the password is stored in the `secrets.tdb` and is keyed off of the admin's DN. This means that if the value of *ldap admin dn* ever changes, the password will need to be manually updated as well.

-i — This option tells smbpasswd that the account being changed is an interdomain trust account. Currently this is used when Samba is being used as an NT Primary Domain Controller. The account contains the info about another trusted domain.

This option is only available when running smbpasswd as root.

-L — Run in local mode.

username — This specifies the username for all of the *root only* options to operate on. Only root can specify this parameter as only root has the permission needed to modify attributes directly in the local smbpasswd file.

NOTES

Since **smbpasswd** works in client-server mode communicating with a local smbd for a non-root user then the smbd daemon must be running for this to work. A common problem is to add a restriction to the hosts that may access the **smbd** running on the local machine by specifying either *allow hosts* or *deny hosts* entry in the smb.conf(5) file and neglecting to allow "localhost" access to the smbd.

In addition, the smbpasswd command is only useful if Samba has been set up to use encrypted passwords.

SEE ALSO

smbpasswd(5), Samba(7).

A.13 smbstatus

Synopsis

```
smbstatus [-P] [-b] [-d <debug level>] [-v] [-L] [-B] [-p] [-S] [-s
    <configuration file>] [-u <username>]
```

DESCRIPTION

This tool is part of the Samba(7) suite.

smbstatus is a very simple program to list the current Samba connections.

OPTIONS

-P|–profile — If samba has been compiled with the profiling option, print only the contents of the profiling shared memory area.

-b|–brief — gives brief output.

-V — Prints the program version number.

-s <configuration file> — The file specified contains the configuration details required by the server. The information in this file includes server-specific information such as what printcap file to use, as well as descriptions of all the services that the server is to provide. See `smb.conf` for more information. The default configuration file name is determined at compile time.

-d|–debug=debuglevel — *debuglevel* is an integer from 0 to 10. The default value if this parameter is not specified is zero.

The higher this value, the more detail will be logged to the log files about the activities of the server. At level 0, only critical errors and serious warnings will be logged. Level 1 is a reasonable level for day-to-day running - it generates a small amount of information about operations carried out.

Levels above 1 will generate considerable amounts of log data, and should only be used when investigating a problem. Levels above 3 are designed for use only by developers and generate HUGE amounts of log data, most of which is extremely cryptic.

Note that specifying this parameter here will override the *log level* parameter in the `smb.conf` file.

-l|–logfile=logbasename — File name for log/debug files. The extension `".client"` will be appended. The log file is never removed by the client.

-v|–verbose — gives verbose output.

-L|–locks — causes smbstatus to only list locks.

-B|–byterange — causes smbstatus to include byte range locks.

-p|–processes — print a list of smbd(8) processes and exit. Useful for scripting.

-S|–shares — causes smbstatus to only list shares.

-h|–help — Print a summary of command line options.

-u|–user=<username> — selects information relevant to *username* only.

SEE ALSO

smbd(8) and smb.conf(5).

A.14 smbtree

Synopsis

`smbtree [-b] [-D] [-S]`

DESCRIPTION

This tool is part of the Samba(7) suite.

smbtree is a smb browser program in text mode. It is similar to the "Network Neighborhood" found on Windows computers. It prints a tree with all the known domains, the servers in those domains and the shares on the servers.

OPTIONS

-b — Query network nodes by sending requests as broadcasts instead of querying the (domain) master browser.

-D — Only print a list of all the domains known on broadcast or by the master browser

-S — Only print a list of all the domains and servers responding on broadcast or known by the master browser.

-V — Prints the program version number.

-s <configuration file> — The file specified contains the configuration details required by the server. The information in this file includes server-specific information such as what printcap file to use, as well as descriptions of all the services that the server is to provide. See `smb.conf` for more information. The default configuration file name is determined at compile time.

-d|–debug=debuglevel — *debuglevel* is an integer from 0 to 10. The default value if this parameter is not specified is zero.

The higher this value, the more detail will be logged to the log files about the activities of the server. At level 0, only critical errors and serious warnings will be logged. Level 1 is a reasonable level for day-to-day running - it generates a small amount of information about operations carried out.

Levels above 1 will generate considerable amounts of log data, and should only be used when investigating a problem. Levels above 3 are designed for use only by developers and generate HUGE amounts of log data, most of which is extremely cryptic.

Note that specifying this parameter here will override the *log level* parameter in the `smb.conf` file.

-l|–logfile=logbasename — File name for log/debug files. The extension `".client"` will be appended. The log file is never removed by the client.

-N — If specified, this parameter suppresses the normal password prompt from the client to the user. This is useful when accessing a service that does not require a password.

Unless a password is specified on the command line or this parameter is specified, the client will request a password.

-k — Try to authenticate with kerberos. Only useful in an Active Directory environment.

-A|–authfile=filename — This option allows you to specify a file from which to read the username and password used in the connection. The format of the file is

```
username = <value>
password = <value>
domain   = <value>
```

Make certain that the permissions on the file restrict access from unwanted users.

-U|–user=username[%password —] Sets the SMB username or username and password.

If %password is not specified, the user will be prompted. The client will first check the `USER` environment variable, then the `LOGNAME` variable and if either exists, the string is uppercased. If these environmental variables are not found, the username `GUEST` is used.

A third option is to use a credentials file which contains the plaintext of the username and password. This option is mainly provided for scripts where the admin does not wish to pass the credentials on the command line or via environment variables. If this method is used, make certain that the permissions on the file restrict access from unwanted users. See the *-A* for more details.

Be cautious about including passwords in scripts. Also, on many systems the command line of a running process may be seen via the **ps** command. To be safe always allow **rpcclient** to prompt for a password and type it in directly.

-h|–help — Print a summary of command line options.

A.15 testparm

Synopsis

```
testparm [-s] [-h] [-v] [-L <servername>] [-t <encoding>] config filename
    [hostname hostIP]
```

DESCRIPTION

This tool is part of the Samba(7) suite.

testparm is a very simple test program to check an smbd(8) configuration file for internal correctness. If this program reports no problems, you can use the configuration file with confidence that **smbd** will successfully load the configuration file.

Note that this is *NOT* a guarantee that the services specified in the configuration file will be available or will operate as expected.

If the optional host name and host IP address are specified on the command line, this test program will run through the service entries reporting whether the specified host has access to each service.

If **testparm** finds an error in the `smb.conf` file it returns an exit code of 1 to the calling program, else it returns an exit code of 0. This allows shell scripts to test the output from **testparm**.

OPTIONS

-s — Without this option, **testparm** will prompt for a carriage return after printing the service names and before dumping the service definitions.

-h|–help — Print a summary of command line options.

-V — Prints the program version number.

-L servername — Sets the value of the %L macro to *servername*. This is useful for testing include files specified with the %L macro.

-v — If this option is specified, testparm will also output all options that were not used in smb.conf(5) and are thus set to their defaults.

-t encoding — Output data in specified encoding.

configfilename — This is the name of the configuration file to check. If this parameter is not present then the default smb.conf(5) file will be checked.

hostname — If this parameter and the following are specified, then **testparm** will examine the *hosts allow* and *hosts deny* parameters in the smb.conf(5) file to determine if the hostname with this IP address would be allowed access to the **smbd** server. If this parameter is supplied, the hostIP parameter must also be supplied.

hostIP — This is the IP address of the host specified in the previous parameter. This address must be supplied if the hostname parameter is supplied.

FILES

smb.conf(5) — This is usually the name of the configuration file used by smbd(8).

DIAGNOSTICS

The program will issue a message saying whether the configuration file loaded OK or not. This message may be preceded by errors and warnings if the file did not load. If the file was loaded OK, the program then dumps all known service details to stdout.

SEE ALSO

smb.conf(5), smbd(8)

A.16 wbinfo

Synopsis

```
wbinfo [-u] [-g] [-N netbios-name] [-I ip] [-n name] [-s sid] [-U uid] [-G
     gid] [-S sid] [-Y sid] [-t] [-m] [--sequence] [-r user] [-a
     user%password] [--set-auth-user user%password] [--get-auth-user] [-p]
```

DESCRIPTION

This tool is part of the Samba(7) suite.

The **wbinfo** program queries and returns information created and used by the winbindd(8) daemon.

The winbindd(8) daemon must be configured and running for the **wbinfo** program to be able to return information.

OPTIONS

-u — This option will list all users available in the Windows NT domain for which the winbindd(8) daemon is operating in. Users in all trusted domains will also be listed. Note that this operation does not assign user ids to any users that have not already been seen by winbindd(8) .

-g — This option will list all groups available in the Windows NT domain for which the Samba(7) daemon is operating in. Groups in all trusted domains will also be listed. Note that this operation does not assign group ids to any groups that have not already been seen by winbindd(8).

-N name — The *-N* option queries winbindd(8) to query the WINS server for the IP address associated with the NetBIOS name specified by the *name* parameter.

-I ip — The *-I* option queries winbindd(8) to send a node status request to get the NetBIOS name associated with the IP address specified by the *ip* parameter.

-n name — The *-n* option queries winbindd(8) for the SID associated with the name specified. Domain names can be specified before the user name by using the winbind separator character. For example CWDOM1/Administrator refers to the Administrator user in the domain CWDOM1. If no domain is specified then the domain used is the one specified in the smb.conf(5) *workgroup* parameter.

-s sid — Use *-s* to resolve a SID to a name. This is the inverse of the *-n* option above. SIDs must be specified as ASCII strings in the traditional Microsoft format. For example, S-1-5-21-1455342024-3071081365-2475485837-500.

-U uid — Try to convert a UNIX user id to a Windows NT SID. If the uid specified does not refer to one within the idmap uid range then the operation will fail.

-G gid — Try to convert a UNIX group id to a Windows NT SID. If the gid specified does not refer to one within the idmap gid range then the operation will fail.

-S sid — Convert a SID to a UNIX user id. If the SID does not correspond to a UNIX user mapped by winbindd(8) then the operation will fail.

-Y sid — Convert a SID to a UNIX group id. If the SID does not correspond to a UNIX group mapped by winbindd(8) then the operation will fail.

-t — Verify that the workstation trust account created when the Samba server is added to the Windows NT domain is working.

-m — Produce a list of domains trusted by the Windows NT server winbindd(8) contacts when resolving names. This list does not include the Windows NT domain the server is a Primary Domain Controller for.

–sequence — Show sequence numbers of all known domains

-r username — Try to obtain the list of UNIX group ids to which the user belongs. This only works for users defined on a Domain Controller.

-a username%password — Attempt to authenticate a user via winbindd. This checks both authenticaion methods and reports its results.

–set-auth-user username%password — Store username and password used by winbindd during session setup to a domain controller. This enables winbindd to operate in a Windows 2000 domain with Restrict Anonymous turned on (a.k.a. Permissions compatiable with Windows 2000 servers only).

–get-auth-user — Print username and password used by winbindd during session setup to a domain controller. Username and password can be set using '-A'. Only available for root.

-p — Check whether winbindd is still alive. Prints out either 'succeeded' or 'failed'.

-V — Prints the program version number.

-h|–help — Print a summary of command line options.

EXIT STATUS

The wbinfo program returns 0 if the operation succeeded, or 1 if the operation failed. If the winbindd(8) daemon is not working **wbinfo** will always return failure.

SEE ALSO

winbindd(8)

A.17 winbindd

Synopsis

```
winbindd [-F] [-S] [-i] [-Y] [-d <debug level>] [-s <smb config file>]
     [-n]
```

DESCRIPTION

This program is part of the Samba(7) suite.

winbindd is a daemon that provides a service for the Name Service Switch capability that is present in most modern C libraries. The Name Service Switch allows user and system information to be obtained from different databases services such as NIS or DNS. The exact behaviour can be configured throught the `/etc/nsswitch.conf` file. Users and groups are allocated as they are resolved to a range of user and group ids specified by the administrator of the Samba system.

The service provided by **winbindd** is called 'winbind' and can be used to resolve user and group information from a Windows NT server. The service can also provide authentication services via an associated PAM module.

The `pam_winbind` module in the 2.2.2 release only supports the ***auth*** and ***account*** module-types. The latter simply performs a getpwnam() to verify that the system can obtain a uid for the user. If the `libnss_winbind` library has been correctly installed, this should always succeed.

The following nsswitch databases are implemented by the winbindd service:

hosts — User information traditionally stored in the `hosts(5)` file and used by **gethostbyname(3)** functions. Names are resolved through the WINS server or by broadcast.

passwd — User information traditionally stored in the `passwd(5)` file and used by **getpwent(3)** functions.

group — Group information traditionally stored in the `group(5)` file and used by **getgrent(3)** functions.

For example, the following simple configuration in the `/etc/nsswitch.conf` file can be used to initially resolve user and group information from `/etc/passwd` and `/etc/group` and then from the Windows NT server.

```
passwd:         files winbind
group:          files winbind
```

The following simple configuration in the `/etc/nsswitch.conf` file can be used to initially resolve hostnames from `/etc/hosts` and then from the WINS server.

OPTIONS

-F — If specified, this parameter causes the main **winbindd** process to not daemonize, i.e. double-fork and disassociate with the terminal. Child processes are still created as normal to service each connection request, but the main process does not exit. This operation mode is suitable for running **winbindd** under process supervisors such as **supervise** and **svscan** from Daniel J. Bernstein's **daemontools** package, or the AIX process monitor.

-S — If specified, this parameter causes **winbindd** to log to standard output rather than a file.

-V — Prints the program version number.

-s <configuration file> — The file specified contains the configuration details required by the server. The information in this file includes server-specific information such as what printcap file to use, as well as descriptions of all the services that the server is to provide. See `smb.conf` for more information. The default configuration file name is determined at compile time.

-d|–debug=debuglevel — *debuglevel* is an integer from 0 to 10. The default value if this parameter is not specified is zero.

The higher this value, the more detail will be logged to the log files about the activities of the server. At level 0, only critical errors and serious warnings will be logged. Level 1 is a reasonable level for day-to-day running - it generates a small amount of information about operations carried out.

Levels above 1 will generate considerable amounts of log data, and should only be used when investigating a problem. Levels above 3 are designed for use only by developers and generate HUGE amounts of log data, most of which is extremely cryptic.

Note that specifying this parameter here will override the *log level* parameter in the `smb.conf` file.

-l|–logfile=logbasename — File name for log/debug files. The extension `".client"` will be appended. The log file is never removed by the client.

-h|–help — Print a summary of command line options.

-i — Tells **winbindd** to not become a daemon and detach from the current terminal. This option is used by developers when interactive debugging of **winbindd** is required. **winbindd** also logs to standard output, as if the **-S** parameter had been given.

-n — Disable caching. This means winbindd will always have to wait for a response from the domain controller before it can respond to a client and this thus makes things slower. The results will however be more accurate, since results from the cache might not be up-to-date. This might also temporarily hang winbindd if the DC doesn't respond.

-Y — Single daemon mode. This means winbindd will run as a single process (the mode of operation in Samba-2.2). Winbindd's default behavior is to launch a child process that is responsible for updating expired cache entries.

NAME AND ID RESOLUTION

Users and groups on a Windows NT server are assigned a relative id (rid) which is unique for the domain when the user or group is created. To convert the Windows NT user or group into a UNIX user or group, a mapping between rids and UNIX user and group ids is required. This is one of the jobs that **winbindd** performs.

As winbindd users and groups are resolved from a server, user and group ids are allocated from a specified range. This is done on a first come, first served basis, although all existing users and groups will be mapped as soon as a client performs a user or group enumeration command. The allocated UNIX ids are stored in a database file under the Samba lock directory and will be remembered.

WARNING: The rid to UNIX id database is the only location where the user and group mappings are stored by winbindd. If this file is deleted or corrupted, there is no way for winbindd to determine which user and group ids correspond to Windows NT user and group rids.

CONFIGURATION

Configuration of the **winbindd** daemon is done through configuration parameters in the smb.conf(5) file. All parameters should be specified in the [global] section of smb.conf.

- *winbind separator*

- *idmap uid*
- *idmap gid*
- *winbind cache time*
- *winbind enum users*
- *winbind enum groups*
- *template homedir*
- *template shell*
- *winbind use default domain*

EXAMPLE SETUP

To setup winbindd for user and group lookups plus authentication from a domain controller use something like the following setup. This was tested on a RedHat 6.2 Linux box.

In `/etc/nsswitch.conf` put the following:

```
passwd:     files winbind
group:      files winbind
```

In `/etc/pam.d/*` replace the ***auth*** lines with something like this:

```
auth required  /lib/security/pam_securetty.so
auth required  /lib/security/pam_nologin.so
auth sufficient   /lib/security/pam_winbind.so
auth required   /lib/security/pam_pwdb.so use_first_pass shadow nullok
```

Note in particular the use of the ***sufficient*** keyword and the ***use_first_pass*** keyword.

Now replace the account lines with this:

account required /lib/security/pam_winbind.so

The next step is to join the domain. To do that use the **net** program like this:

net join -S PDC -U Administrator

The username after the *-U* can be any Domain user that has administrator privileges on the machine. Substitute the name or IP of your PDC for "PDC".

Next copy `libnss_winbind.so` to `/lib` and `pam_winbind.so` to `/lib/security`. A symbolic link needs to be made from `/lib/libnss_winbind.so` to `/lib/libnss_winbind.so.2`. If you are using an older version of glibc then the target of the link should be `/lib/libnss_winbind.so.1`.

Finally, setup a smb.conf(5) containing directives like the following:

```
[global]
   winbind separator = +
        winbind cache time = 10
        template shell = /bin/bash
        template homedir = /home/%D/%U
        idmap uid = 10000-20000
        idmap gid = 10000-20000
        workgroup = DOMAIN
        security = domain
        password server = *
```

Now start winbindd and you should find that your user and group database is expanded to include your NT users and groups, and that you can login to your UNIX box as a domain user, using the DOMAIN+user syntax for the username. You may wish to use the commands **getent passwd** and **getent group** to confirm the correct operation of winbindd.

NOTES

The following notes are useful when configuring and running **winbindd**:

nmbd(8) must be running on the local machine for **winbindd** to work. **winbindd** queries the list of trusted domains for the Windows NT server on startup and when a SIGHUP is received. Thus, for a running **winbindd** to become aware of new trust relationships between servers, it must be sent a SIGHUP signal.

PAM is really easy to misconfigure. Make sure you know what you are doing when modifying PAM configuration files. It is possible to set up PAM such that you can no longer log into your system.

If more than one UNIX machine is running **winbindd**, then in general the user and groups ids allocated by winbindd will not be the same. The user and group ids will only be valid for the local machine.

If the the Windows NT RID to UNIX user and group id mapping file is damaged or destroyed then the mappings will be lost.

SIGNALS

The following signals can be used to manipulate the **winbindd** daemon.

SIGHUP — Reload the smb.conf(5) file and apply any parameter changes to the running version of winbindd. This signal also clears any cached user and group information. The list of other domains trusted by winbindd is also reloaded.

SIGUSR1 — The SIGUSR1 signal will cause **winbindd** to write status information to the winbind log file including information about the number of user and group ids allocated by **winbindd**.

Log files are stored in the filename specified by the log file parameter.

FILES

`/etc/nsswitch.conf(5)` — Name service switch configuration file.

/tmp/.winbindd/pipe — The UNIX pipe over which clients communicate with the **winbindd** program. For security reasons, the winbind client will only attempt to connect to the winbindd daemon if both the `/tmp/.winbindd` directory and `/tmp/.winbindd/pipe` file are owned by root.

$LOCKDIR/winbindd_privilaged/pipe — The UNIX pipe over which "*privilaged*" clients communicate with the **winbindd** program. For security reasons, access to some winbindd functions - like those needed by the **ntlm_auth** utility - is restricted. By default, only users in the 'root' group will get this access, however the administrator may change the group permissions on $LOCKDIR/winbindd_privilaged to allow programs like 'squid' to use ntlm_auth. Note that the winbind client will only attempt to connect to the winbindd daemon if both the `$LOCKDIR/winbindd_privilaged` directory and `$LOCKDIR/winbindd_privilaged/pipe` file are owned by root.

/lib/libnss_winbind.so.X — Implementation of name service switch library.

$LOCKDIR/winbindd_idmap.tdb — Storage for the Windows NT rid to UNIX user and group id mapping. The lock directory is specified when Samba is initialy compiled using the *--with-lockdir* option. This directory is by default `/usr/local/samba/var/locks`.

$LOCKDIR/winbindd_cache.tdb — Storage for cached user and group information.

SEE ALSO

`nsswitch.conf(5)`, Samba(7), wbinfo(8), smb.conf(5)

Appendix B

THE GNU GENERAL PUBLIC LICENSE

Version 2, June 1991

Preamble

The licenses for most software are designed to take away your freedom to share and change it. By contrast, the GNU General Public License is intended to guarantee your freedom to share and change free software—to make sure the software is free for all its users. This General Public License applies to most of the Free Software Foundation's software and to any other program whose authors commit to using it. (Some other Free Software Foundation software is covered by the GNU Library General Public License instead.) You can apply it to your programs, too.

When we speak of free software, we are referring to freedom, not price. Our General Public Licenses are designed to make sure that you have the freedom to distribute copies of free software (and charge for this service if you wish), that you receive source code or can get it if you want it, that you can change the software or use pieces of it in new free programs; and that you know you can do these things.

To protect your rights, we need to make restrictions that forbid anyone to deny you these rights or to ask you to surrender the rights. These restrictions translate to certain responsibilities for you if you distribute copies of the software, or if you modify it.

For example, if you distribute copies of such a program, whether gratis or for a fee, you must give the recipients all the rights that you have. You must make sure that they, too, receive or can get the source code. And you must show them these terms so they know their rights.

We protect your rights with two steps: (1) copyright the software, and (2) offer you this license which gives you legal permission to copy, distribute and/or modify the software.

Also, for each author's protection and ours, we want to make certain that everyone understands that there is no warranty for this free software. If the software is modified by someone else and passed on, we want its recipients to know that what they have is not the original, so that any problems introduced by others will not reflect on the original authors' reputations.

Finally, any free program is threatened constantly by software patents. We wish to avoid the danger that redistributors of a free program will individually obtain patent licenses, in effect making the program proprietary. To prevent this, we have made it clear that any patent must be licensed for everyone's free use or not licensed at all.

The precise terms and conditions for copying, distribution and modification follow.

Terms and Conditions For Copying, Distribution and Modification

0. This License applies to any program or other work which contains a notice placed by the copyright holder saying it may be distributed under the terms of this General Public License. The "Program", below, refers to any such program or work, and a "work based on the Program" means either the Program or any derivative work under copyright law: that is to say, a work containing the Program or a portion of it, either verbatim or with modifications and/or translated into another language. (Hereinafter, translation is included without limitation in the term "modification".) Each licensee is addressed as "you".

   Activities other than copying, distribution and modification are not covered by this License; they are outside its scope. The act of running the Program is not restricted, and the output from the Program is covered only if its contents constitute a work based on the Program (independent of having been made by running the Program). Whether that is true depends on what the Program does.

1. You may copy and distribute verbatim copies of the Program's source code as you receive it, in any medium, provided that you conspicuously and appropriately publish on each copy an appropriate copyright notice and disclaimer of warranty; keep intact all the notices that refer to this License and to the absence of any warranty; and give any other recipients of the Program a copy of this License along with the Program.

   You may charge a fee for the physical act of transferring a copy, and you may at your option offer warranty protection in exchange for a fee.

2. You may modify your copy or copies of the Program or any portion of it, thus forming a work based on the Program, and copy and distribute such modifications or work under the terms of Section 1 above, provided that you also meet all of these conditions:

   (a) You must cause the modified files to carry prominent notices stating that you changed the files and the date of any change.

   (b) You must cause any work that you distribute or publish, that in whole or in part contains or is derived from the Program or any part thereof, to be licensed as a whole at no charge to all third parties under the terms of this License.

(c) If the modified program normally reads commands interactively when run, you must cause it, when started running for such interactive use in the most ordinary way, to print or display an announcement including an appropriate copyright notice and a notice that there is no warranty (or else, saying that you provide a warranty) and that users may redistribute the program under these conditions, and telling the user how to view a copy of this License. (Exception: if the Program itself is interactive but does not normally print such an announcement, your work based on the Program is not required to print an announcement.)

These requirements apply to the modified work as a whole. If identifiable sections of that work are not derived from the Program, and can be reasonably considered independent and separate works in themselves, then this License, and its terms, do not apply to those sections when you distribute them as separate works. But when you distribute the same sections as part of a whole which is a work based on the Program, the distribution of the whole must be on the terms of this License, whose permissions for other licensees extend to the entire whole, and thus to each and every part regardless of who wrote it.

Thus, it is not the intent of this section to claim rights or contest your rights to work written entirely by you; rather, the intent is to exercise the right to control the distribution of derivative or collective works based on the Program.

In addition, mere aggregation of another work not based on the Program with the Program (or with a work based on the Program) on a volume of a storage or distribution medium does not bring the other work under the scope of this License.

3. You may copy and distribute the Program (or a work based on it, under Section 2) in object code or executable form under the terms of Sections 1 and 2 above provided that you also do one of the following:

   (a) Accompany it with the complete corresponding machine-readable source code, which must be distributed under the terms of Sections 1 and 2 above on a medium customarily used for software interchange; or,

   (b) Accompany it with a written offer, valid for at least three years, to give any third party, for a charge no more than your cost of physically performing source distribution, a complete machine-readable copy of the corresponding source code, to be distributed under the terms of Sections 1 and 2 above on a medium customarily used for software interchange; or,

   (c) Accompany it with the information you received as to the offer to distribute corresponding source code. (This alternative is allowed only for noncommercial distribution and only if you received the program in object code or executable form with such an offer, in accord with Subsection b above.)

The source code for a work means the preferred form of the work for making modifications to it. For an executable work, complete source code means all the source code for all modules it contains, plus any associated interface definition files, plus the scripts used to control compilation and installation of the executable. However, as a special exception, the source code distributed need not include anything that is normally distributed (in either source or binary form) with the major components (compiler, kernel, and so on) of the operating system on which the executable runs, unless that component itself accompanies the executable.

If distribution of executable or object code is made by offering access to copy from a designated place, then offering equivalent access to copy the source code from the same place counts as distribution of the source code, even though third parties are not compelled to copy the source along with the object code.

4. You may not copy, modify, sublicense, or distribute the Program except as expressly provided under this License. Any attempt otherwise to copy, modify, sublicense or distribute the Program is void, and will automatically terminate your rights under this License. However, parties who have received copies, or rights, from you under this License will not have their licenses terminated so long as such parties remain in full compliance.

5. You are not required to accept this License, since you have not signed it. However, nothing else grants you permission to modify or distribute the Program or its derivative works. These actions are prohibited by law if you do not accept this License. Therefore, by modifying or distributing the Program (or any work based on the Program), you indicate your acceptance of this License to do so, and all its terms and conditions for copying, distributing or modifying the Program or works based on it.

6. Each time you redistribute the Program (or any work based on the Program), the recipient automatically receives a license from the original licensor to copy, distribute or modify the Program subject to these terms and conditions. You may not impose any further restrictions on the recipients' exercise of the rights granted herein. You are not responsible for enforcing compliance by third parties to this License.

7. If, as a consequence of a court judgment or allegation of patent infringement or for any other reason (not limited to patent issues), conditions are imposed on you (whether by court order, agreement or otherwise) that contradict the conditions of this License, they do not excuse you from the conditions of this License. If you cannot distribute so as to satisfy simultaneously your obligations under this License and any other pertinent obligations, then as a consequence you may not distribute the Program at all. For example, if a patent license would not permit royalty-free redistribution of the Program by all those who receive copies directly or indirectly through you, then the only way you could satisfy both it and this License would be to refrain entirely from distribution of the Program.

   If any portion of this section is held invalid or unenforceable under any particular circumstance, the balance of the section is intended to apply and the section as a whole is intended to apply in other circumstances.

   It is not the purpose of this section to induce you to infringe any patents or other property right claims or to contest validity of any such claims; this section has the sole purpose of protecting the integrity of the free software distribution system, which is implemented by public license practices. Many people have made generous contributions to the wide range of software distributed through that system in reliance on consistent application of that system; it is up to the author/donor to decide if he or she is willing to distribute software through any other system and a licensee cannot impose that choice.

This section is intended to make thoroughly clear what is believed to be a consequence of the rest of this License.

8. If the distribution and/or use of the Program is restricted in certain countries either by patents or by copyrighted interfaces, the original copyright holder who places the Program under this License may add an explicit geographical distribution limitation excluding those countries, so that distribution is permitted only in or among countries not thus excluded. In such case, this License incorporates the limitation as if written in the body of this License.

9. The Free Software Foundation may publish revised and/or new versions of the General Public License from time to time. Such new versions will be similar in spirit to the present version, but may differ in detail to address new problems or concerns.

   Each version is given a distinguishing version number. If the Program specifies a version number of this License which applies to it and "any later version", you have the option of following the terms and conditions either of that version or of any later version published by the Free Software Foundation. If the Program does not specify a version number of this License, you may choose any version ever published by the Free Software Foundation.

10. If you wish to incorporate parts of the Program into other free programs whose distribution conditions are different, write to the author to ask for permission. For software which is copyrighted by the Free Software Foundation, write to the Free Software Foundation; we sometimes make exceptions for this. Our decision will be guided by the two goals of preserving the free status of all derivatives of our free software and of promoting the sharing and reuse of software generally.

No Warranty

11. Because the program is licensed free of charge, there is no warranty for the program, to the extent permitted by applicable law. Except when otherwise stated in writing the copyright holders and/or other parties provide the program "as is" without warranty of any kind, either expressed or implied, including, but not limited to, the implied warranties of merchantability and fitness for a particular purpose. The entire risk as to the quality and performance of the program is with you. Should the program prove defective, you assume the cost of all necessary servicing, repair or correction.

12. In no event unless required by applicable law or agreed to in writing will any copyright holder, or any other party who may modify and/or redistribute the program as permitted above, be liable to you for damages, including any general, special, incidental or consequential damages arising out of the use or inability to use the program (including but not limited to loss of data or data being rendered inaccurate or losses sustained by you or third parties or a failure of the program to operate with any other programs), even if such holder or other party has been advised of the possibility of such damages.

End of Terms and Conditions

Appendix: How to Apply These Terms to Your New Programs

If you develop a new program, and you want it to be of the greatest possible use to the public, the best way to achieve this is to make it free software which everyone can redistribute and change under these terms.

To do so, attach the following notices to the program. It is safest to attach them to the start of each source file to most effectively convey the exclusion of warranty; and each file should have at least the "copyright" line and a pointer to where the full notice is found.

> one line to give the program's name and a brief idea of what it does.
> Copyright (C) yyyy name of author
>
> This program is free software; you can redistribute it and/or modify it under the terms of the GNU General Public License as published by the Free Software Foundation; either version 2 of the License, or (at your option) any later version.
>
> This program is distributed in the hope that it will be useful, but WITHOUT ANY WARRANTY; without even the implied warranty of MERCHANTABILITY or FITNESS FOR A PARTICULAR PURPOSE. See the GNU General Public License for more details.
>
> You should have received a copy of the GNU General Public License along with this program; if not, write to the Free Software Foundation, Inc., 59 Temple Place - Suite 330, Boston, MA 02111-1307, USA.

Also add information on how to contact you by electronic and paper mail.

If the program is interactive, make it output a short notice like this when it starts in an interactive mode:

> Gnomovision version 69, Copyright (C) yyyy name of author
> Gnomovision comes with ABSOLUTELY NO WARRANTY; for details type 'show w'.
> This is free software, and you are welcome to redistribute it under certain conditions; type 'show c' for details.

The hypothetical commands `show w` and `show c` should show the appropriate parts of the General Public License. Of course, the commands you use may be called something other than `show w` and `show c`; they could even be mouse-clicks or menu items—whatever suits your program.

You should also get your employer (if you work as a programmer) or your school, if any, to sign a "copyright disclaimer" for the program, if necessary. Here is a sample; alter the names:

> Yoyodyne, Inc., hereby disclaims all copyright interest in the program 'Gnomovision' (which makes passes at compilers) written by James Hacker.
>
> signature of Ty Coon, 1 April 1989
> Ty Coon, President of Vice

This General Public License does not permit incorporating your program into proprietary programs. If your program is a subroutine library, you may consider it more useful to permit linking proprietary applications with the library. If this is what you want to do, use the GNU Library General Public License instead of this License.

GLOSSARY

Access Control List (`ACL`)

A detailed list of permissions granted to users or groups with respect to file and network resource access. See Chapter 12, *File, Directory and Share Access Controls*, for details.

Active Directory Service (`ADS`)

A service unique to Microsoft Windows 200x servers that provides a centrally managed directory for management of user identities, and computer objects, as well as the permissions each user or computer may be granted to access distributed network resources. ADS uses Kerberos-based authentication and LDAP over Kerberos for directory access.

Common Internet File System (`CIFS`)

The new name for SMB. Microsoft renamed the SMB protocol to CIFS during the Internet hype in the nineties. At about the time that the SMB protocol was renamed to CIFS, an additional dialect of the SMB protocol was in development. The need for the deployment of the NetBIOS layer was also removed, thus paving the way for use of the SMB protocol natively over TCP/IP (known as NetBIOSless SMB or "*naked*" TCP transport).

Common UNIX Printing System (`CUPS`)

A recent implementation of a high capability printing system for UNIX developed by .[1] The design objective of CUPS was to provide a rich print processing system that has built-in intelligence that is capable of correctly rendering (processing) a file that is submitted for printing even if it was formatted for an entirely different printer.

Domain Master Browser (`DMB`)

The Domain Master Browser maintains a list of all the servers that have announced their services within a given workgroup or NT domain. See Section 9.4.1 for details.

Domain Name Service (`DNS`)

A protocol by which computer host names may be resolved to the matching IP address/es. DNS is implemented by the Berkeley Internet Name Daemon. There exists

[1] http://www.easysw.com/

a recent version of DNS that allows dynamic name registration by network clients or by a DHCP server. This recent protocol is known as Dynamic DNS (DDNS).

Dynamic Host Configuration Protocol (`DHCP`)

A protocol that was based on the BOOTP protocol that may be used to dynamically assign an IP address, from a reserved pool of addresses, to a network client or device. Additionally, DHCP may assign all network configuration settings and may be used to register a computer name and its address with a Dynamic DNS server.

Extended Metafile Format (`EMF`) An intermediate file format used by Microsoft Windows-based servers and clients. EMF files may be rendered into a page description language by a print processor.

Graphical Device Interface (`GDI`)

Device Independant format for printing used by Microsoft Windows. It is quite similar to what PostScript is for UNIX. Printing jobs are first generated in GDI and then converted to a device-specific format. See Section 18.4.1 for details.

Group IDentifier (`GID`)

The UNIX system Group Identifier; on olders systems a 32-bit unsigned integer and on newer systems an unsigned 64-bit integer. The GID is used in UNIX-like operating systems for all group level access control.

Internet Print Protocol (`IPP`)

An IETF standard for network printing. CUPS implements IPP.

Key Distribution Center (`KDC`)

The Kerberos authentication protocol makes use of security keys (also called a ticket) by which access to network resources is controlled. The issuing of Kerberos tickets is effected by a KDC.

NetBIOS Extended User Interface (`NetBEUI`)

Very simple network protocol invented by IBM and Microsoft. It is used to do NetBIOS over ethernet with low overhead. NetBEUI is a non-routable protocol.

Network Basic Input/Output System (`NetBIOS`)

NetBIOS is a simple application programming interface (API) invented in the eighties that allows programs to send data to certain network names. NetBIOS is always run over another network protocol such as IPX/SPX, TCP/IP, or Logical Link Control (LLC). NetBIOS run over LLC is best known as NetBEUI (The NetBIOS Extended User Interface — a complete misnomer!).

NetBT (`NBT`)

Protocol for transporting NetBIOS frames over TCP/IP. Uses ports 137, 138 and 139. NetBT is a fully routable protocol.

Local Master Browser (`LMB`)

The Local Master Browser maintains a list of all servers that have announced themselves within a given workgroup or NT domain on a particular broadcast isolated subnet. See Section 9.4.1 for details.

Printer Command Language (`PCL`)

A printer page description language that was developed by Hewlett Packard and is in common use today.

Portable Document Format (`PDF`) A highly compressed document format, based on postscript, used as a document distribution format that is supported by Web browsers as well as many applications. Adobe also distribute an application called "*acrobat*" which is a PDF reader.

Page Description Language (`PDL`)

A language for describing the layout and contents of a printed page. The best-known PDLs are Adobe PostScript and Hewlett-Packard PCL (Printer Control Language), both of which are used to control laser printers.

PostScript Printer Description (`PPD`)

PPD's specify and control options supported by postscript printers, such as duplexing, stapling, DPI, ... See also Section 18.4.4. PPD files can be read by printing applications to enable correct postscript page layout for a particular postscript printer.

Server Message Block (`SMB`)

SMB was the original name of the protocol 'spoken' by Samba. It was invented in the eighties by IBM and adopted and extended further by Microsoft. Microsoft renamed the protocol to CIFS during the Internet hype in the nineties.

User IDentifier (`UID`)

The UNIX system User Identifier; on olders systems a 32-bit unsigned integer and on newer systems an unsigned 64-bit integer. The UID is used in UNIX-like operating systems for all user level access control.

Universal Naming Convention (`UNC`)

A syntax for specifying the location of network resources (such as file shares). The UNC syntax was developed in the early days of MS DOS 3.x and is used internally by the SMB protocol.

SUBJECT INDEX

Prentice Hall Professional Technical Reference
http://www.phptr.com/
Prentice Hall PTR
InformIT
InformIT Online Books
Financial Times Prentice Hall
ft.com
PTG Interactive
Reuters
PRENTICE HALL PTR
TOMORROW'S SOLUTIONS FOR TODAY'S PROFESSIONALS
Prentice Hall Professional Technical Reference
Browse
Book Series
What's New
User Groups
Alliances
Special Sales
Contact Us
Search | Help | Home
Quick Search
PTR Favorites
Find a Bookstore
Book Series
Special Interests
Newsletters
Press Room
International
Best Sellers
Solutions Beyond the Book
Shopping Bag
Keep Up to Date with
PH PTR Online
We strive to stay on the cutting edge of what's happening in professional computer science and engineering. Here's a bit of what you'll find when you stop by www.phptr.com:
What's new at PHPTR? We don't just publish books for the professional community, we're a part of it. Check out our convention schedule, keep up with your favorite authors, and get the latest reviews and press releases on topics of interest to you.
Special interest areas offering our latest books, book series, features of the month, related links, and other useful information to help you get the job done.
User Groups Prentice Hall Professional Technical Reference's User Group Program helps volunteer, not-for-profit user groups provide their members with training and information about cutting-edge technology.
Companion Websites Our Companion Websites provide valuable solutions beyond the book. Here you can download the source code, get updates and corrections, chat with other users and the author about the book, or discover links to other websites on this topic.
Need to find a bookstore? Chances are, there's a bookseller near you that carries a broad selection of PTR titles. Locate a Magnet bookstore near you at www.phptr.com.
Subscribe today! Join PHPTR's monthly email newsletter! Want to be kept up-to-date on your area of interest? Choose a targeted category on our website, and we'll keep you informed of the latest PHPTR products, author events, reviews and conferences in your interest area.
Visit our mailroom to subscribe today! http://www.phptr.com/mail_lists

THE LITTLE GIANT BOOK OF

Joseph Rosenbloom

Illustrated by Sanford Hoffman

Sterling Publishing Co., Inc.

New York

Library of Congress Cataloging-in-Publication Data

Rosenbloom, Joseph
The little giant book of riddles / by Joseph Rosenbloom ; illustrated by Sanford Hoffman.
p cm.
Includes index.
Summary: A collection of hundreds of illustrated riddles, arranged in such categories as "Outlaws and Lawpersons," "Science for Dummies," and "Fun Food."
ISBN 0-8069-6100-7
1. Riddles, Juvenile. [1. Riddles. 2. Jokes.] I. Hoffman, Sanford, ill. II. Title.
PN6371.5.R6123 1996
818'.5402–dc20 96–9811
CIP
AC

10

Published by Sterling Publishing Company, Inc.
387 Park Avenue South, New York, N.Y. 10016
Material in this collection was adapted from *Mad Scientist: Riddles, Jokes , Fun*; *School's Out: Great Vacation Riddles & Jokes*; *Get Well Quick!* (published in paper as *Super Sick Jokes & Riddles*); *Wacky Insults & Terrible Jokes*; *Wild West Riddles & Jokes*, and *World's Best Sports Riddles & Jokes*, all © by Joseph Rosenbloom

Distributed in Canada by Sterling Publishing
℅ Canadian Manda Group, One Atlantic Avenue, Suite 105
Toronto, Ontario, Canada M6K 3E7
Distributed in Great Britain and Europe by Cassell PLC
Wellington House, 125 Strand, London WC2R 0BB, England
Distributed in Australia by Capricorn Link (Australia) Pty Ltd.
P.O. Box 6651, Baulkham Hills, Business Centre, NSW 2153, Australia
Manufactured in Canada

Sterling ISBN 0-8069-6100-7

CONTENTS

1. GO DIRECTLY TO JAIL

Why did the clock get arrested?
Because it struck twelve.

Why can't you keep a clock in jail?
Because time is always running out.

What hired killer never goes to jail?
An exterminator.

Why did Robin Hood steal from the rich?
Because the poor didn't have any money.

What is the difference between a jeweler and a jailer?

A jeweler sells watches; a jailer watches cells.

How many prisoners can you put into an empty cell?

One. After that the cell isn't empty anymore.

How are prisoners like astronauts?

Both are interested in outer space.

Why do prisoners like to eat a lot of sweets?

They are hoping to break out.

Why did the prisoner take a shower before he broke out of jail?

He wanted to make a clean getaway.

How is a person in jail like a sinking ship?

Both want to be bailed out.

Why are prisoners in jail the slowest talkers in the world?

They can spend 25 years on a single sentence.

How is an escaping prisoner like an airline pilot?

Both want safe flights.

Why was the picture sent to jail?

Because it was framed.

What did the police do in the shoe store?

Rounded up the sneakers and the loafers.

What kind of bars won't keep a prisoner in jail?

Chocolate bars.

What kind of party do prisoners like best?

A going-away party.

Why did the Sheriff arrest the chicken?
It used fowl language.

A prisoner was in jail. All he had in his cell was a piano. Yet, he managed to escape. How did he do it?

He played the piano until he found the right key.

Why were the clothespins arrested?
For holding up a pair of pants.

Why were the tennis players arrested?
Because they had racquets.

Why were the pair of old watches arrested?
Because they were two-timers.

Why were the walls arrested?
Because they were holding up the ceiling.

Why was the stale loaf of bread arrested?
It tried to get fresh.

Why was the deck of cards arrested?
The joker was wild.

Why is it dangerous to play cards in the jungle?
Because of all the cheetahs.

Why did the Sheriff arrest the cook?

For beating the eggs and whipping the cream.

When is a jail not on land and not on water?

When it is on fire.

What did the thief get for stealing the calendar?

Twelve months.

Why was the photographer arrested?

He shot his customers and then blew them up.

Why was the fisherman arrested?

For packing a rod.

What did the burglar get for robbing the rubber band factory?

A long stretch.

Why is it hard to keep a bank robbery secret?

Because so many people who work in the bank are tellers.

What kind of candy would a doomed prisoner like to have before he is hanged?

A Life Saver.

What did the comic say when the gangster stuffed a dirty piece of cloth in his mouth?

"That's an old gag."

What show do prisoners like to put on?
A cell-out (sell-out).

What do you call a sheep that hangs out with forty thieves?
Ali Baa Baa.

What do you call an elephant that hangs out with forty thieves?
Ali Babar.

What do you call someone who steals soap at camp?
A dirty crook.

How do hangmen keep up with current events?
They read the noose-paper.

What is the favorite sport of executioners?
Hang gliding.

What did the hangman give his wife for her birthday?

A choker.

What did the prisoner about to be hanged say when he was pardoned at the last minute?

"No noose is good noose."

2. OUTLAWS & LAWPERSONS

Why were outlaws the strongest men in the Old West?

They could hold up trains.

Why did outlaws sleep on the ground after they robbed a bank?

Because they wanted to lie low.

Who pulled the biggest holdup in history?
Atlas – he held up the whole world.

Which way did the varmint go when he stole the computer?

Data-way.

Who was the most famous cat in the Wild West?

Kit-ty Carson.

Who was the thirstiest outlaw in the West?

The one who drank Canada Dry.

How do you make a strawberry shake?

Introduce it to Jesse James.

Where do dead outlaws go on Saturday night?

To ghost towns.

What happened when the outlaw ran away with the circus?

The Sheriff made him bring it back.

What kind of cat chases outlaws?

A posse cat.

Who robbed stagecoaches and wore dirty clothes?

Messy James.

What has red bumps and is the fastest gun in the West?

Rootin' Tootin' Raspberry.

What is small, purple, and dangerous?

A grape with a six-shooter.

What is green and dangerous?

A thundering herd of pickles.

How did the Sheriff find the missing barber?

He combed the town.

Why did the Sheriff go to the barbecue?

He heard it was a place to have a steak out.

Why did the outlaw hold up the bakery?

He kneaded the dough.

What is the hardest thing to deal with in a poker game?

A greasy deck of cards.

Why did the outlaw steal the deck of cards?

He heard there were 13 diamonds in it.

Why didn't the sailors play cards?

Because the captain was standing on the deck.

What does a pickle say when it wants to enter a poker game?

"Dill me in."

Why did the outlaw hold up the river?

He heard it had two banks.

What is the difference between an outlaw and chocolate cake?

One hits the mark; the other hits the spot.

Why did the outlaw wear loud socks?

So his feet wouldn't fall asleep.

What lives in the ocean, has eight legs and is quick on the draw?

Billy the Squid.

Who is the toughest pickle in Dodge City?
Marshall Dill.

Where do most of the nuts in Dodge City hang out?
At the Hershey bar.

Who is the meanest goat in the West?
Billy the Kid.

What did the banana do when it saw Billy the Kid?
The banana split.

Why did Billy the Kid set Dodge City on fire?

So he could be the toast of the town.

What was Billy the Kid's favorite subject in school?

Triggernometry.

What kind of bandit steals cats?

A purr-snatcher.

Two outlaws robbed a bank. They decided to bury the money they stole. If it took two outlaws five days to dig a hole, how many days would it take them to dig half a hole?

None. You can't dig half a hole.

Why is a saltine like an outlaw?

Both are safecrackers.

What did the outlaw give his wife for her birthday?

A stole.

What is small and yellow and wears a mask?
The Lone Lemon.

What happened when the painter threw his pictures at the outlaw?
The outlaw had an art attack.

When is an outlaw neither left-handed nor right-handed?
When he is underhanded.

What do outlaws eat with their milk?
Crookies.

What kind of sweets do outlaws steal?
Hot chocolate.

What do you get if you cross a big bell and an outlaw?
A gongster.

Why did the gang of outlaws suddenly leave the restaurant?

Because they had finished eating.

What is the difference between an outlaw and a church bell?

One steals from the people; the other peals from the steeple.

Why couldn't the chuch steeple keep a secret?

Because the bell always tolled.

What would you get if you crossed Jesse James and a cow?

Better not try it. Jesse James doesn't like to be crossed.

What would you call a short, sunburned outlaw riding a horse?

Little Red Riding Hood.

Why wasn't the outlaw buried in the town cemetery?

Because he wasn't dead.

Why did the outlaw hold up air-conditioned banks?

To get cold cash.

When is a pistol like a young horse?
When it is a Colt.

How is a stolen pistol like a racing car?
They're both hot rods.

When is a gun unemployed?
When it is fired.

How come a duck won the shoot-out?
It was a quack shot.

What would happen if an ice cream cone picked a fight with Jesse James?
The ice cream cone would get licked.

What do you call Jesse James when he has the flu?
A sick shooter.

What is a shotgun?
A worn-out rifle.

What do you call a baby rifle?
A son-of-a-gun.

What would you get if you crossed a clock and a gun?

A ticks-shooter.

What is the safest way to talk with an outlaw?

By long distance.

Why did the outlaw carry a bottle of glue when he robbed the stagecoach?

So he could stick up the passengers.

Why did the outlaw brush his teeth with gunpowder?

So he could shoot his mouth off.

Why did the cowboy put a whistle in his ten-gallon hat?

So he could blow his top.

Why did the banana run from the outlaw?
Because it was yellow.

Why did the outlaw shoot the clock?
He was just killing time.

What happened to the outlaw who fell into the cement mixer?

He became a hardened criminal.

Did you hear about the stupid outlaw? When he saw a sign saying, "MAN WANTED FOR ROBBERY," he applied for the job.

What did the ten-gallon hat say to the outlaw?

"I've got you covered."

3. MEANWHILE... BACK AT THE RANCH

What cattle follow you wherever you go?
Your calves.

Where do calves eat?
In calf-eterias.

Where do cattle eat?
In re-steer-rants.

What key do cattle sing in?

Beef-flat (B-flat).

What did the bored cow say when she got up in the morning?

"Just an udder day."

What goes out black and comes in white?
A black cow in a snowstorm.

Can you spell COW in thirteen letters?
SEE O DOUBLE YOU.

Why don't most cows go to college?
Because not many graduate from high school.

Why did the cow go to the psychiatrist?
It had a fodder complex.

What do cows give after an earthquake?
Milk shakes.

What is the easiest way to keep milk from turning sour?
Leave it in the cow.

What is a calf after it is a year old?
Two years old.

Why don't cows have money?
Because the farmer milks them dry.

How is a political speech like a steer?

There is a point here and there and a lot of bull in between.

What do you call a cow that has lost its calf?

De-calf-inated.

Why is it better to own a cow than a bull?

Because a cow gives milk, but a bull always charges.

What is the best thing to do if a bull charges you?

Pay him.

What would you have if cattle fought each other?

Steer Wars.

How do cowboys drive steers?

With steer-ing wheels.

How do you make meatloaf?

Send a cow to the seashore.

When was beef at its highest?

When the cow jumped over the moon.

Where do cows go for entertainment?
To the moo-vies.

Where do bulls go to dance?
To the meatball.

Where do cows go to dance?
To a dis-cow-theque.

What music do cows like to dance to?
Cow-lypso (calypso) music.

There were three tomatoes on a shelf. Two were ripe and one was green. Which one is the cowboy?

The green one. The other two are redskins.

Why was the bowlegged cowboy fired?

Because he couldn't get his calves together.

What do you do with a green cowboy?

Wait until he ripens.

What kind of ponchos do Mexican cowboys wear on a rainy day?

Wet ones.

In what kind of home does the buffalo roam?

A very dirty one.

What has six legs and walks with only four?

A horse and rider.

What is the most important use for cowhide?

To keep the cow together.

When is it good manners to spit in a rancher's face?

When his mustache is on fire.

What has four legs and can see just as well from either end?

A horse with its eyes closed.

When is a horse not a horse?

When it turns into a stable.

How long should a horse's legs be?

Long enough to reach the ground.

What always follows a horse?

Its tail.

What kind of horse eats and drinks with its tail?

They all do. No horse takes off its tail to eat or drink.

What is the hardest thing about learning to ride a bucking horse?

The ground.

Why did the cowboy saddle up a porcupine?

So he wouldn't have to ride it bareback.

How much do you have to know to teach a horse tricks?

More than the horse.

An outlaw went on a trip on Friday, stayed three days, and came back on Friday. How was that possible?

His horse was named Friday.

What kind of horses frighten ranchers?

Nightmares.

Why are horses always poorly dressed?

Because they wear shoes but no socks.

What was the fastest way to ship small horses in the Old West?

By Pony Express.

What smells good and rides a horse?

The Cologne Ranger.

How do you make a horse float?

Take two scoops of ice cream, root beer – and add one horse.

What did the rancher see when he fell off his horse?

An all-star show.

Are horses good acrobats?

Yes, they can turn cartwheels.

What part of a cowboy's outfit is the saddest?

Blue jeans.

If dogs have fleas, what do sheep have?

Fleece.

What did the cowboy say when he wanted to get the sheep's attention?

"Hey, ewe!"

What does a sheep say when it has problems?

"Where there's a wool, there's a way."

What does a male sheep do when he gets angry?

He goes on a ram-page.

Why do sheep go into saloons?

To look for the baa-tender.

Why is a rodeo horse rich?
It has a million bucks.

Which part of a horse is the most important?
The mane (main) part.

Why is it hard to recognize horses from the back?

Because they are always switching their tails.

What did the horse say when it finished eating a bale of hay?

"That's the last straw!"

4. PLAY BALL!

What position do pigs play on a baseball team?

Short-slop.

What was a spider doing on the baseball team?

Catching flies.

When do monkeys play baseball?
In Ape-ril.

What kind of hit do you find in the zoo?
A lion drive.

What has 18 legs and catches flies?
A baseball team.

What has two gloves and four legs?
Two baseball players.

How do you hold a bat?
By the wings.

When is an umpire like a telephone operator?
When he makes a call.

What kind of umpires do you find at the North Pole?
Cold ones.

What's the difference between a rain barrel and a bad fielder?
One catches drops; the other drops catches.

Why did the umpire penalize the chicken?

For using fowl (foul) language.

What is the difference between an umpire and a pickpocket?

The umpire watches steals, the pickpocket steals watches.

How are tough teachers like umpires?

They penalize you for errors.

What does a skunk do when it disagrees with the umpire?

It raises a stink.

Why was night baseball invented?

Because bats like to sleep during the day.

Why did the ball player blink his eyes?

He needed batting practice.

What would you get if you crossed a lobster and a baseball player?

A pinch hitter.

Why was the mummy sent into the game as a pinch hitter?

With a mummy at bat, the game would be all tied up.

What would you get if Mickey Mantle married Betty Crocker?
Better batters.

What do you get if you cover a baseball field with sandpaper?
A diamond in the rough.

Why is a baseball field hot after the game?
Because all the fans have gone home.

Why can't turtles play baseball?
They can't run home.

How can you pitch a winning baseball game without ever throwing a ball?
Throw only strikes.

What happens when you hit a pop fly?
The same thing that happens when you hit a mom fly.

Where in a baseball stadium do the fans wear the whitest clothes?

In the bleachers.

What is the difference between a queen who likes to dance and a baseball player?

One throws balls, the other catches them.

What does an umpire do before he eats?

He brushes off the plate.

Where do coal diggers play baseball?

In the minor (miner) leagues.

Why couldn't Robin play baseball?

He forgot his bat, man.

Why don't baseball players join unions?

Because they don't like to be called out on strikes.

Where do great dragon baseball players go?

To the Hall of Flame.

What is the difference between someone who hits the ball but does not score – and someone who beats a chicken?

One fouls the hit, the other hits the fowl.

Why was the chef hired to coach the baseball team?

Because he knew how to handle a batter.

Which takes longer to run: from first base to second or from second base to third?

From second base to third, because there's a shortstop in the middle.

Why did the silly baseball fan take his car to the game?

He heard it was a long drive to center field.

Why couldn't the fans get soda pop at the double-header?

Because the home team lost the opener.

Why did the outlaw gang try to steal the baseball field?

Because it was the biggest diamond in the world.

Which baseball league has the most trees and shrubs?

The bush leagues.

Why is school like baseball?

The bell strikes one, two, three – and you're out!

Why did the baseball team sign up a two-headed monster?

To play double-headers.

What is the difference between a boy who is late for dinner and a baseball hit over the fence?

One runs home; the other is a home run.

What is the difference between a baseball player and a vampire?

One bats flies, the other flies bats.

What is the best way to get rid of flies?
Get good outfielders.

How can you make a fly ball?
Hit him with a bat.

Why was the piano tuner hired to play on the baseball team?

Because he had perfect pitch.

What song did the baseball player hum while he waited on third base?

"There's no place like home."

5. SCIENCE FOR DUMMIES

What did the robot say when it ran out of electricity?

"AC come, AC go."

What do you get when a robot's wires are reversed?

A lot of backtalk.

Why did the robot go to a psychiatrist?
It had a screw loose.

When is a robot like a surgeon?
When it operates on batteries.

What kind of doctor operates on Styrofoam robots?

A plastic surgeon.

What did the little electric robot say to its mother?

"I love you watts and watts."

What snacks should you serve robots at parties?

Assorted nuts.

Why did the scientist study electricity?

He wanted to keep up with current events.

What did the scientist get when she crossed an electric eel and a sponge?

Shock absorbers.

How did the scientist fix the robot gorilla?
With a monkey wrench.

What did the scientist get when he crossed an old car and a gorilla?
A greasy monkey.

What would you get if you crossed a breakfast drink and a monkey?

An orangu-tang.

What would you get if you crossed a stone and a shark?

Rockjaws.

What would you get if you crossed a Martian, a skunk, and an owl?

An animal that stinks to high heaven and doesn't give a hoot.

What is the most educated thing in the scientist's laboratory?

A thermometer, because it has so many degrees.

When is it best to buy a thermometer?

In the winter, when it is lower.

How do you use a thermometer to find the height of a building?

Lower the thermometer on a string from the top of the building to the ground. Then measure the length of the string.

Why did the scientist throw the thermometer out of the laboratory on a hot day?

He wanted to see the temperature drop.

What mysterious thing did the scientist see in the skillet?

An unidentified frying object.

What would you get if you crossed a parrot and an elephant?

Something that tells everything it remembers.

What do you get if you cross a parrot and an army man?

A parrot-trooper.

What would you get if you crossed a parrot and a bumblebee?

An animal that talks all the time about how busy it is.

What would you get if you crossed a parrot and a canary?

A bird that knows both the words and the music.

What did the scientist get when he crossed a cat and a parrot?

A purr-a-keet.

What does it mean when a barometer falls?

That whoever nailed it up didn't do such a good job.

What would you get if you crossed the Invisible Man and a cow?

Vanishing cream.

What dog likes to hang around scientists?

A laboratory retriever.

Why didn't the scientist need a pocket calculator?

Because he already knew how many pockets he had.

What did the scientist get when he made an exact duplicate of Texas?

A clone star state.

What did the scientist get when he crossed a chicken and a cow?

Roost beef.

What is the difference between electricity and lightning?

You have to pay for electricity.

If lightning strikes an orchestra, who is most likely to get hit?

The conductor.

What is yellow and goes "hmmmm"?
An electric lemon.

What is yellow and long and always points north?
A magnetic banana.

What would you get if you crossed a banana and a bell?
A banana you can peel more than once.

What would you get if you crossed a sheep and a banana?
A baa-nana.

What would you get if you crossed a banana and a banana?
A pair of slippers.

Why don't bananas ever get lonely?
Because they go around in bunches.

What kind of typewriter does Count Dracula use in his laboratory?

One that types blood.

What would you get if you crossed a porcupine and an alarm clock?

A stickler for punctuality.

What did the scientist get when he crossed a clock and a rooster?

An alarm cluck.

What's the quickest way to make oil boil?
Add the letter "B."

What did the scientist get when he crossed a chicken and a cement truck?
A hen that lays sidewalks.

What can you find in the center of gravity?
The letter "V."

Why can't you trust the law of gravity?
Because it always lets you down.

What weighs more – a pound of lead or a pound of feathers?
They weigh the same – one pound.

What can you measure that has no length, width, or thickness?
The temperature.

What did the scientist get when he crossed a frog and a soft drink?

Croak-a-cola.

What did the scientist get when he crossed an egg and a soft drink?

Yolk-a-cola.

The scientist invented a liquid that would dissolve anything it touched. He couldn't sell his invention, however. Why?

There was nothing in which he could put the liquid.

What happened when the scientist fell into the lens grinding machine?

He made a spectacle of himself.

What is H, I, J, K, L, M, N, O?

The formula for water – H to O.

What is the most important rule in chemistry?

Never lick the spoon.

How is an airplane like an atom bomb?

One drop and you're dead.

What did the scientist write on the robot's tombstone?

"RUST IN PEACE."

Who was the first nuclear scientist in history?

Eve – She knew all about atom (Adam).

Why did the scientist keep talking about the atom bomb?

He didn't want to drop the subject.

What do nuclear scientists argue about?

Whether splitting the atom was a wisecrack.

What is an atomic scientist's favorite snack?

Fission chips.

What is a hydrogen bomb?

Something that makes molehills out of mountains.

6. FUN FOOD

What snacks should you serve computer scientists at parties?
A byte of everything.

What kind of gum do chickens chew?
Chicklets.

What's the difference between a stupid person and a pizza?
One is easy to cheat and the other is cheesy to eat.

How many chickens does it take to serve ten people?

Chickens aren't good at serving. Better get waiters and waitresses.

How do you make an elephant sandwich?

First, get a very large loaf of bread...

Is chicken soup good for your health?

Not if you're a chicken.

How do you know that chickens love money?

They are always going, "Buck-buck! Buck-buck!"

What has bread on both sides and frightens easily.

A chicken sandwich.

What do dogs put on their pizza?

Mutts-arella.

What do ants put on their pizza?

Ant-chovies.

WHAT DO YOU LIKE WITH YOUR HAMBURGERS?

What do computer scientists like with their hamburgers?
Chips.

What do musicians like with their hamburgers?
Piccolos (pickle-o's).

What do spiders like with their hamburgers?
French flies.

What do cats put on their hamburgers?
Mouse-tard.

Why couldn't the hamburger talk?
The catsup got its tongue.

Is it a good idea to eat hamburgers on an empty stomach?
No, it's neater to eat them on a plate.

What music do young hamburgers like to listen to?
Rock 'n roll.

What's a burger's all-time favorite movie director?
Sizzle B. De Mille.

What does a lion eat when he goes to a restaurant?
The waiter.

What is the best thing to eat in a bathtub?
A sponge cake.

What do witches eat at cookouts?

Halloweenies (hollow wienies).

What smells best at a barbecue?
A pickle holding its breathe.

Where do you put a very smart hot dog?
On the honor roll.

How do you make a hot dog roll?
Tilt your plate.

What is the best way to talk to a hot dog?
Be frank.

What is a hot dog's favorite song?
"Franks for the memory..."

What did the hot dog say when it won the race?
"I'm a wiener!"

What is green and red all over?
A pickle holding its breath.

What is a geologist's favorite dessert?
Marble cake.

What sweets do geologists like?
Rock candy.

What does the funniest kid in camp have for breakfast?

Cream of Wit.

What do cowboys put on their pancakes?

Maple stirrup.

What did Mary have at the cookout?

Everyone knows that Mary had a little lamb.

Why did the frog like French fries?

Because it was a pota-toad.

What has four legs and flies?

A picnic table.

What kind of ant can break a picnic table with one blow?

A gi-ant.

What do ghouls drink at picnics?
Ice-ghoul lemonade.

What do frogs drink at picnics?
Crock-a-cola.

What do dogs drink at picnics?
Pupsi-cola.

7. HAVING A WONDERFUL TIME

Why is "H" the most popular letter of the alphabet?

It is the start of every holiday.

Why is it so hard to do nothing all summer?

Because you can't stop to rest.

Where does a Brontosaurus go for vacation?

To the dino-shore.

When are you most likely to dream about going away for the summer?

When you're asleep.

What is grey and wet and vacations in Florida?

A melted penguin.

What would you get if you crossed a bunch of bones and a week in Florida?

A skele-tan.

Why do skeletons always vacation alone?

Because they have no-body to go with.

Where do fish go for vacation?

Finland.

Where do songbirds go for vacation?

The Canary Islands.

What's black and white and hates to be touched?

A zebra just back from the beach.

Why is a cat on the beach like Christmas?

It has sandy claws (Santa Claus).

What is heavier in the summer than in the winter?

Traffic to the beach.

What did Cinderella wear when she went to the beach?

Glass flippers.

Where do race cars go swimming?

In the car pool.

Where do phantoms go swimming?

At the sea ghost.

Where do mummies go swimming?
In the Dead Sea.

What game do you play with fish?
Carps and robbers.

What do you get if a bunch of thieves dives into the swimming pool?

A crime wave.

Why aren't elephants allowed in the swimming pool?

Because they can't keep their trunks up.

Why did the tire need a vacation?
It couldn't take the pressure anymore.

Why did the three little pigs go on vacation?
Because their father was a boar.

Why isn't the moon a good place to go on vacation?
It lacks atmosphere.

Where do zombies go on vacation?
Club Dead.

Where does Lassie go on vacation?
Collie-fornia.

What is tan and has four legs and a trunk?
A mouse coming back from vacation.

What do octopuses take on camping trips?
Tent-acles.

How do you fix a torn tepee?
Apache here, Apache there.

Can you make a fire with one stick?
Yes, if it's a match.

What would you get if you crossed a rabbit and a stand-up comic?

A funny bunny that walks on its hind legs.

What is the difference between a rabbit that runs three miles a day and a so-so comedian?

One is a fit bunny, the other is a bit funny.

When are boxers like comedians?

When they have you in stitches.

What would you get if you crossed a banana and a clown?

Peels of laughter.

What ducks crack jokes?

Wise quackers.

How are comedians like surgeons?

They are both cut-ups.

What's a comedian's favorite motorcycle?

A Yama-haha.

A hiker went without sleep for seven days and wasn't tired. How come?

He slept at night.

How did Barbie help the chicken with its part in the camp play?

Barbie cued the chicken.

Where do you send a shoe in the summer?

To boot camp.

What do you call someone with a big red nose and purple hair who takes a plane from New York to Alaska?

A passenger.

What do you call a father who takes a plane to the North Pole?

A cold pop.

Why isn't an elephant allowed on a plane?

Because his trunk won't fit under the seat.

How do you get a mouse to fly?

Buy it an airline ticket.

How do dogs travel?

By mutt-a-cycle.

What is big and hairy and travels 1,200 miles an hour?

King Kongcorde.

Where do pilots keep their personal things?

In air pockets.

What people travel the most?

Romans.

What people travel the fastest?

Russians.

Why don't people visit Transylvania?

Because it is a terror-tory.

What egg travels to unknown places?

An eggs-plorer.

How do mice find their way when they travel?

With rod-ent maps.

Where do people leave their dogs when they go on vacation?

At the arf-anage (orphanage).

What's the difference between a dog with fleas and a person going on vacation?

One is going to itch and the other is itching to go.

HOW DO THEY TRAVEL?

How do rabbits travel?
By hareplane.

How does King Kong travel?
By hairyplane.

How do barbers travel?
By hairplane.

How do pizza pies travel?
By pie-cycle.

How do witches travel when they don't have a broom?
They witch-hike.

Where do wolves stay when they travel?

In a Howl-iday Inn.

What always follows a wolf when it travels?

Its tail.

What boat takes dentists on short trips?

The Tooth Ferry.

Why is travel by boat the cheapest way of getting around?

Because boats run on water.

When does a boat show affection?

When it hugs the shore.

What vegetable is dangerous to have in a boat?

A leek (leak).

What would you get if you crossed a lake in a boat that had a leak in it?

About halfway.

What happens when you hike across a stream and a river?

Your feet get wet.

How do robots cross a lake?
In a row-bot.

What do frogs wear on their feet in summer?
Open toad shoes.

What did the hiker say when he ran into a porcupine?

"Ouch!"

What kind of singers do you find in Yellowstone National Park?
Bear-itones.

How do bears walk?
In their bear feet.

8. ARE WE HAVING FUN YET?

What did the hiker say after being on safari for one week?

"Safari so good."

What did the hiker yell when he saw the avalanche?

"Here come the Rolling Stones!"

What musical key do you hear when a race car speeds through a coal mine?

A-flat miner (minor).

What musical instruments are best for catching fish?

Castanets.

What would you get if you crossed a sweet potato and a jazz musician?

Yam sessions.

What kind of music do you hear when you throw a stone into the lake?

Plunk rock.

What kind of car does a rich rock star drive?

A rock-n-Rolls Royce.

Why shouldn't you hit a famous composer?

He might hit you Bach.

Who leads a duck orchestra?

The con-duck-tor.

Why do so many orchestras have bad names?

Because they don't know how to conduct themselves.

What do you get when you cross a bag of cement, a stone, and a radio?

Hard rock music.

How is an actor in a hit show like a hockey player?

One sticks with a play, the other plays with a stick.

What sport is like a perfect score in the Olympics?

Tennis (ten is).

Why are waiters like tennis players?

They both have to know how to serve.

What is the difference between a tennis racket and a doughnut?

You can't dunk a tennis racket in a glass of milk.

Do vampires play tennis?

No, they prefer bat-minton.

What is the difference between a prince and a tennis ball?

One is heir to the throne, the other is thrown to the air.

Why weren't tennis players allowed in summer camp?

Because they made too much of a racket (racquet).

Why are mountain climbers curious?

Because they always want to take another peak (peek).

Why didn't the mountain climber hurt himself when he fell off the cliff?

Because he was wearing a light fall suit.

How can you climb Mount Everest without getting tired?

Be born on top.

If you go on a trek through the desert, what should you take along?

A thirst-aid kit.

If you were on a trek in the Sahara Desert, where would you get milk?

From the drome-dary (dairy).

Who is boss in the dairy?

The big cheese.

Why can't you play hide-and-seek with baby chickens?

Because they're always peeping.

Where do cows go to see art?

Moo-seums.

Why don't scarecrows have fun?

Because they are such stuffed shirts.

Why is Count Dracula like the Frankenstein monster?

Neither can play Ping-Pong.

Why didn't the cow want to play Ping-Pong?

She wasn't in the moo-d.

What did one toad say to the other toad?

"One more game of leapfrog – and I'll croak!"

What were the chickens doing in the health club?

Eggs-ercising.

What is green, has big eyes, and eats like a pig?

Kermit the Hog.

What is green, has big eyes, and is hard to see through?

Kermit the Fog.

What does a mechanical frog say?

"Robot, robot!"

What wallows in mud and carries colored eggs?

The Easter Piggie.

Why can't you play games with pigs?
Because they hog the ball.

When do pigs give their girlfriends presents?
On Valen-swine's Day.

What did the executioner do at Christmas?

He went sleighing (slaying).

What is the difference between little kids at Christmas and werewolves?

Werewolves have claws on their fingers; little kids at Christmas have claws (Claus) on their minds.

What would you get if you crossed a reindeer and a firefly?

Rudolph the Red-Nosed Firefly.

What did the hen do when it saw the large order of Kentucky Fried Chicken?

It kicked the bucket.

What would you get if you crossed some chocolate candy and a sheep?

A Hershey baa.

What lives on the bottom of the sea and is popular at Easter?

An oyster egg.

How can you drop an egg ten feet without breaking it?

Drop it eleven feet. It won't break for the first ten.

What does a parrot say on the Fourth of July?

"Polly wants a firecracker!"

What did one firecracker say to the other firecracker?

"My pop is bigger than your pop."

Where do geologists go to relax?

Rock concerts.

What is a geologist's favorite lullaby?
"Rock-a-bye, Baby."

How do you make notes out of stone?
Rearrange the letters.

Who is brown and hairy and fights forest fires?
A suntanned forest ranger who needs a shave.

How do hikers cross a patch of poison ivy?
They itch hike.

How do hikers dress on cold mornings?
Quickly.

Two campers were playing checkers. They played five games and each won the same number of games. How is this possible?
They played different people.

What would you get if you crossed a chicken and a television set?

A TV show that lays eggs.

Why are you like a shrub after a long hike?

Because you're bushed.

What is the worst thing you're likely to find in the camp kitchen?

The food.

How do they count muffins in the camp kitchen?

They have a roll call.

How do you greet a web-footed bird?

"What's up, duck?"

What is the difference between a football player and a duck?

You'll find one in a huddle and the other in a puddle.

How can you tell if there is a football team in your bathtub?

It's hard to close the shower curtain.

What did the football say to the player?
"I get a kick out of you."

What did the football say after the player threw it?
"You send me!"

What did the helmet say to the football player?
"You're putting me on!"

How do we know that football referees are happy?
Because they whistle while they work.

Who are the happiest people at the game?
The cheer leaders.

Which football player wears the biggest helmet?
The one with the biggest head.

Would you rather have a 300-pound football player attack you or a 300-pound wrestler?

I'd rather have them attack each other.

What did the football player say when he was hit by lightning?

"Got to glow now!"

9. BACK TO MOTHER NATURE

Which is lighter – the sun or the earth?
The sun, because it rises every morning.

Why is the moon like a dollar?
Because it has four quarters.

How many pieces of string does it take to reach the moon?
One, if it is long enough.

When is the moon heaviest?

When it is full.

Who was the first man in space?

The man in the moon.

Why did Humpty Dumpty have a great fall?

To make up for a boring summer.

Whose fault will it be if California falls into the ocean?

San Andreas fault.

What do you call a geologist who doesn't hear anything?

Stone deaf.

What happens to a small stone when it works up its courage?

It becomes a little boulder (bolder).

When are geologists unpopular?

When they are fault-finders.

When are geologists unhappy?

When people take them for granite.

What does a geologist have for breakfast?
Rock-n-roll.

What color is the wind?
Blew.

What gets harder to catch the faster you run?
Your breath.

What runs, but never gets out of breath?
Water.

What goes through water but doesn't get wet?
A ray of light.

How do you open the Great Lakes?
With the Florida Keys.

Why are there bridges over water?
So people won't step on the fishes.

Where is a lake deepest?
On the bottom.

What does one raindrop say to the other raindrop?
"My plop is bigger than your plop."

What is the difference between the North Pole and the South Pole?

All the difference in the world.

What did one pile of sand say to the other pile of sand?

"Dune anything tonight?"

Why did the river bend?

Because it saw the waterfall.

Where do rivers sleep?

In river beds.

When does a river flood?

When it gets too big for its bridges.

What body of water is a famous spy?

James Pond.

What doesn't get wetter – no matter how much it rains?

The ocean.

What's the most romantic part of the ocean?

The spot where buoy meets gull.

What did the ocean say to the beach?

"I'm not shore."

Who does the ocean date?

It goes out with the tide.

How do you cut the ocean in two?

With a sea-saw.

Why is the ocean so grouchy?

Because it has crabs all over its bottom.

What lives in the water and takes you anywhere you want to go?

A taxi crab.

When is the ocean friendliest?
When it waves.

Why did the ocean roar at the ships?
Because they crossed it so many times.

Why was the ocean arrested?
Because it beat upon the shore.

What do two oceans say when they meet after many years?
"Long time no sea."

What do you call a long series of hurricane names?
A gust (guest) list.

A camper fell out of a canoe into the middle of the lake. He neither swam nor sank. How could that be?
He floated.

What did one magnet say to the other magnet?

"You attract me."

What did one volcano say to the other volcano?

"I lava you."

What do you call it when Mother Nature crosses an earthquake and a forest fire?

Shake and Bake.

What's the longest distance you can see?

Down a road with telephone poles, because then you can see from pole to pole.

What is "mean temperature"?

Twenty degrees below zero when you don't have long underwear.

What is the best kind of letter to read on a hot day?

Fan mail.

What animal cools you off on a hot day?

A pup-sicle.

What makes bluebirds blue?

Their blue genes (jeans).

Who wears a black cape, flies through the night and wants to drink your flood?

A mosquito in a black cape.

What is green, has big eyes, and lives all alone in the pond?

Hermit the Frog.

What goes "Dit-dot-dot-croak, dit-dot-dot-croak"?

Morse toad.

What do you call a fat tree limb?

A porky twig.

What is the smartest tree in the forest?

Albert Pinestein.

What is a tree's favorite drink?

Root beer.

A Boy Scout climbed a tall pine tree to gather some acorns. He tried all morning, but couldn't get any. Why not?

Acorns don't grow on pine trees. They grow on oak trees.

What lives in the forest, is green, and pecks on trees?

Woody Wood Pickle.

10. BIRDS OF A FEATHER

What kind of geese come from Portugal?
Portu-geese.

Why did the hen sit on the axe?
She thought she could hatch-it.

Why did the muddy chicken cross the road twice?

Because she was a dirty double-crosser.

Why is a black chicken smarter than a white one?

A black chicken can lay white eggs, but a white chicken can't lay black eggs.

Which side of a chicken has the most feathers?

The outside.

Where can you find out more about chickens?

In a hen-cyclopedia.

Where can you find out more about ducks?

In a duck-tionary.

What book tells you about all the different kinds of owls?

Who's Whoo.

Why was the owl the hit of the talent show?

He was a h-owl!

What do you call it when it rains chickens and ducks?
Foul (fowl) weather.

What was the near-sighted chicken doing in the farmer's garden?
She was sitting on an eggplant.

What does a Spanish farmer say to his chickens?
"Oh, lay!" (Olé)

Why couldn't the chicken find her eggs?
She mislaid them.

What has four legs, is very long, and goes "quack, quack"?
A ducks-hund.

Why did the hens go on strike?
They refused to work for chicken feed.

Why did the hen sit down in the middle of the tennis court?

She wanted to lay it on the line.

What would you get if you crossed a chicken and an old timepiece?

A grandfather cluck.

How do chickens get out of their shell?

They eggs-it.

How can you make an egg run faster?

You egg it on.

11. HEALTHY, WEALTHY, & WEIRD

What doctor treats his patients like animals?

A vet.

When is a vet busiest?

When it rains cats and dogs.

When it rains cats and dogs, what does a vet step into?

Poodles.

What is worse than an elephant with an earache?

A giraffe with a sore throat.

Does a giraffe get a sore throat when its feet get wet?

Yes, but not until two weeks later.

What's the difference between a photocopier and a virus?

One makes facsimilies; the other makes sick families.

What's the difference between a bus driver and a cold?

One knows the stops; the other stops the nose.

What do they do with a cowboy whose voice is really hoarse?

They put a saddle on it.

Why did the doctor pour oil on his hands?

He wanted to be a smooth operator.

What does a polite doctor say when he is about to operate?

"May I cut in?"

Why can't you believe what doctors say?

Because they make MD (empty) promises.

How long should doctors practice medicine?

Until they get it right.

What is a medicine dropper?

A doctor with greasy fingers.

What did the doctor use to fix a broken heart?

Ticker tape.

What was the plumber doing in the operating room?

He was a drain surgeon.

Why do surgeons wear masks during operations?

So that, if they make a mistake, no one will know who did it.

Why did the dog see the doctor?

Because a stitch in time saves canine.

How can you tell when Count Dracula is catching cold?

From his coffin (coughin').

What happens when corn catches cold?

It gets an earache.

Where does a sneeze usually point?

At-choo (at you)!

What kind of paper is most like a sneeze?

A tissue.

What is the difference between a person with a cold and a strong wind?

One blows a sneeze; the other blows a breeze.

What is red, white, and blue, and convenient when you sneeze?

A hanky doodle dandy.

What should you say when the Statue of Liberty sneezes?

"God bless America."

You can never catch cold going up in an elevator. True or false?

True. You come down with a cold. You come up with a cure.

What would happen if you swallowed a dress?

You'd have a frock (frog) in your throat.

What sickness can't you talk about until it's cured?

Laryngitis.

What sickness did Bruce Lee get?

Kung flu.

What's the difference between a boxer and a person with a cold?

A boxer knows his blows; a person with a cold blows his nose.

What's the difference between ammonia and pneumonia?

Ammonia comes in bottles; pneumonia comes in chests.

What did one elevator say to the other elevator?

"I think I'm coming down with something."

How do you feel if you have a sore throat and fleas?

Hoarse and buggy.

What's the difference between a hill and a pill?

A hill is hard to get up; a pill is hard to get down.

What's the difference between a sick sailor and a blind man?

One can't go to sea; the other can't see to go.

What has 18 legs, red spots, and catches flies?

A baseball team with the measles.

Why is a catcher's glove like the measles?

Both are catching.

What's the best game to play when you've got the measles?

Hide-and-sick.

What game is dangerous to your mental health?

Marbles – if you lose them.

When was medicine first mentioned in the Bible?

It was when Moses received the two tablets.

What do you give an elk with indigestion?

Elk-a-Seltzer.

What did the farmer use to cure his sick hog?

Oinkment (ointment).

How is medicine packaged for astronauts?

In space capsules.

Why did the silly kid jump up and down?

The medicine label said, "Shake well."

Why did the silly kid swing on the chandelier?

Because her doctor told her to get some light exercise.

What do doctors give elephants to calm them down?

Trunk-quilizers.

How much did the elephant have to pay to the psychiatrist?

A hundred dollars for the hour and six hundred dollars for the couch.

Why did the wrestler go to the psychiatrist?
He couldn't get a grip on himself.

Why did the fencer go to the psychiatrist?
Because of her duel (dual) personality.

When someone comes to your door, what is the polite thing to do?
Vitamin (invite him in).

What contains the most vitamins?
A health food store.

What bee is necessary to your health?
Vitamin B.

On what day do you moan the most?
On Moan-day.

What kind of fish is in charge of an operating room?

The head sturgeon.

What would you do if you found yourself with water on the knee, water on the elbow, and water on the brain?

Turn off the shower.

What happened when the plastic surgeon stood too close to the fire?

He melted.

Who flies on a broom and carries a medicine bag?

A witch doctor.

What is the best way to get rid of demons?

Exorcise (exercise).

What kind of physician comes from Cairo?

A chiropractor.

What did the tree say to the tree surgeon?

"Leaf me alone!"

Why don't anteaters have infections?

Because they're filled with anty-bodies.

What gets 25 miles to a gallon of plasma?

A bloodmobile.

What should you do with a sick boat?

Take it to the dock (doc).

What does a doctor do with a sick zeppelin?

He tries to helium.

What's the difference between a sick person and seven days?

One is a weak one; the other is one week.

What's the difference between a dressmaker and a nurse?

One cuts the dresses; the other dresses the cuts.

Where is the best place to build offices for opticians and optometrists?

On a sight for sore eyes.

How is an eye doctor like a teacher?

They both test the pupils.

What do eye doctors sing when they test you?

"Oh, say can you see ..."

What did Old MacDonald see on the eye chart?

E-I-E-I-O.

What's the difference between a person with a terrible toothache and a rainy day?

One is roaring with pain; the other is pouring with rain.

Why do elephants in Alabama have to go to the dentist so often?

Because in Alabama Tuscaloosa (tusks are looser).

What is the best time to see a dentist?
Tooth-hurty (2:30).

Where does a rat go when it has a sore tooth?
To the rodent-ist.

What does a dentist say when you knock on his door?
"Gum on in!"

What did the tooth say to the dentist?
"Fill 'er up!"

What does the dentist do on his yacht?
Off-shore drilling.

When do most of your dollars go to the orthodontist?
When you have buck teeth.

12. GET WELL SOON!

Why did the chicken see the doctor?
It had people pox.

Why did the math book see the doctor?
It had problems.

What is grey, carries flowers, and cheers you when you're sick?

A get-well-ephant.

What relative will help you when you have an infection?

Anti-biotics.

WHAT DO THEY COME DOWN WITH?

What do dancers come down with?
Ballet-aches.

What do chimneys come down with?
The flu.

What does grass come down with?
Hay fever.

What do cabbages come down with?
Headaches.

What animal do you feel like when you have a fever?

A little otter (hotter).

Why don't rabbits multiply when they have colds?

Because they can't breed (breathe).

WHAT DO THEY COME DOWN WITH?

What do beekeepers come down with?
Hives.

What do motorcycle riders come down with?
Vroom-atism.

What does Mickey Mouse come down with?
Disney spells.

What do video cassettes come down with?
Tapeworms.

Where do they take care of sick parrots?

In a polyclinic.

Where do they send sick kangaroos?

To the hop-ital.

Where do they send sick ponies?
To the horse-pital.

Where do they send sick librarians?
To the hush-pital.

What do they give the sick insects?
To the wasp-ital.

What medical condition helps you run faster?

Athlete's foot.

What is a foot doctor's favorite song?

"There's No Business Like Toe Business."

What kind of X-rays do foot doctors make?

Foot-ographs.

What do you get if you add 13 hospital patients and 13 hospital patients?

Twenty sicks (26).

Why did the nurse tiptoe past the medicine cabinet?

She didn't want to wake the sleeping pills.

How does a sick pig get to the hospital?

In a ham-bulance.

Why aren't vampires welcome at the bloodmobile?

Because they only want to make withdrawals.

Why wasn't the chicken allowed to visit the hospital?

Because she fowled things up.

What would you get if your doctor became a vampire?

More blood tests than ever.

What kind of music do you hear when the nurse turns down your bed?

Sheet music.

What sign do you see in front of a dog hospital?

No Barking Zone."

How did the patient get to the hospital so fast?

Flu.

Why do you lie down on a hospital bed?

Because you can't lie up.

What kind of alligator do you find in a hospital?

An illigator.

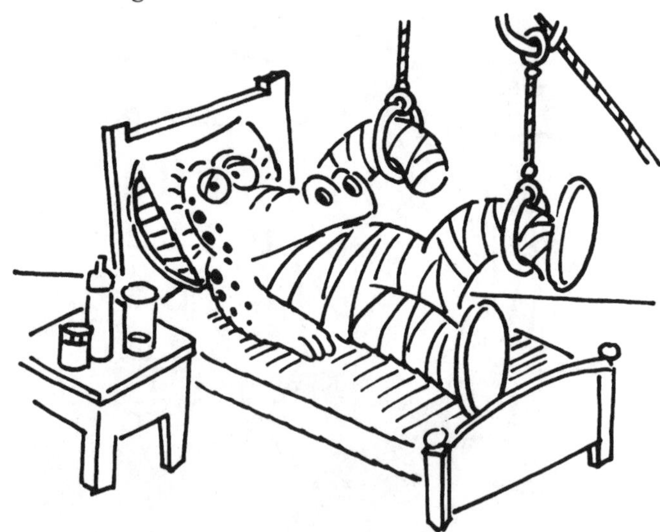

What has 15 letters, begins with an "A," ends with a "G," and means incredible pain?

"ARRRRRHHHHH-HHHG!"

Why do many mummies have high blood pressure?

Because they're so wound up.

What did the bed sheet say to the patient?

"Hold still – I've got you covered!"

Why were the bedcovers depressed?

Because the nurses turned them down.

Why did the doctor give up his practice?
He lost his patience.

When is it all right to belt a doctor?
When he gets in your car.

Why did the mother owl take her baby to the doctor?

Because it didn't give a hoot.

What do you tell a germ when it fools around?

"Don't bacilli!"

Why did the germ cross the microscope?

To get to the other side.

Which germ tastes best with pancakes?

Aunt Germ-ima.

What did one germ say to the other germ?

"You're making me sick."

Why do people with colds get plenty of exercise?

Their noses run.

Why was Whistler's mother in the hospital?

She went off her rocker.

Why did the turkey go to the hospital?
He had one foot in the grave-y.

What goes "Chit-chat, tick-tock, boom-gong"?

A sick clock.

GRADUATING FROM MEDICAL SCHOOL

What do you call a duck who graduates from medical school?

"Duck-tor."

What do you call a dog who graduates from medical school?

"Dog-tor."

What do you call a squid that graduates from medical school?

"Doc-topus."

What mental illness does Santa suffer from?

Claustrophobia.

What nationality is Santa Claus?

North Polish.

When do most doctors graduate from medical school?

In Doc-tober.

What do you have if your head feels hot, your feet are cold, and you see spots in front of your eyes?

You probably have a polka-dotted sock over your head.

WHAT'S THE PROBLEM?

What was Ronald McDonald's problem?

Fallen arches.

What was the Olympic athlete's problem?

Slipped discus.

What problems do you get from eating too much?

You get thick (sick) to your stomach.

What did the surgeon say to the patient as he sewed him up?

"That's enough out of you!"

13. FIGHTING WORDS

Who tells people where to get off and gets away with it?

A bus driver.

Who tells people where to go and gets away with it?

A travel agent.

How do cattle defend themselves?

They use cow-a-ti (karate).

How does a skunk defend itself?

Instinct.

Why did the karate expert wear a black belt?

To keep his pants up.

What tree is a karate champion?
Spruce Lee.

What is small, round and green, and knows karate?
Bruce Pea.

What do you get when a pea picks a fight with a boxer?

A black-eyed pea.

How do you shake hands with a judo expert?

Very carefully.

Why did the matador take judo lessons?

He wanted to learn how to throw the bull.

On the door of a school was a sign that read:

PLEASE DON'T KNOCK BEFORE ENTERING

What kind of school was it?

A karate school.

What would you get if a pig learned karate?

Pork chops.

What was the artist doing in the boxing ring?

They needed him in case the fight ended in a draw.

What is the difference between a winter day and a boxer who is down for the count?

One is cold out, and other is out cold.

The boxer was knocked out. What number did they use to revive him?

They brought him 2.

Why is a boxer's hand never larger than eleven inches long?

If it were twelve inches long, it would be a foot.

Why couldn't the boxer light the fire?

Because he lost all his matches.

What is the difference between a nail and a bad boxer?

One is knocked in; the other is knocked out.

Why did the boxer hit the grandfather clock?

The clock struck first.

Why did the clock strike first?

Because it was ticked off.

What would you get if a hen stepped into the ring with the heavyweight champion of the world?

Creamed chicken.

What kind of potatoes would you get if you stepped into the ring with the heavyweight champion of France?

French fright (fried).

What happened when the watch fought the heavyweight champion?

The watch took a licking, but kept on ticking.

Why is boxing the world champion like singing in a barbershop quartet?

Because if you don't look sharp, you'll be flat.

Why did the mother skunk take her baby to the doctor?

Because it was out of odor.

What four letters can you say to someone who has been in the ring with a professional wrestler?

R-U-O-K.

Where does a big, mean, 300-pound wrestler sit, when he gets on a bus?

Anywhere he wants to.

Who is the best fencer in the ocean?
The swordfish.

How do you tell a big, mean wrestler from a bunny rabbit?
You don't tell a big, mean wrestler anything.

What did the fencer say when he was defeated?

"Curses! Foiled again!"

What does a fencing master do at twelve o'clock?

He goes to lunge.

Why do mosquitoes bother people most late at night?

Because mosquitoes like a little bite before they go to sleep.

When are mosquitoes most annoying?

When they get under your skin.

Did you hear about the mosquito who went to Hollywood?

All she could get were bit parts.

What geometric figure is the most dangerous?
A firing line.

How do people feel after they've been shot by a six-shooter?
Holier.

How do you make a mothball?
Hit it.

What does a tick attack?
A tick attacks a toe (Tic-tac-toe).

Why was the insect kicked out of the national park?
Because it was a litterbug.

What did the lettuce say to the farmer?
"Lettuce alone!"

What two vegetables fight crime?
Beetman and Radish.

Why was the lamb punished?
Because it was baaaa-d!

14. THE NOT-SO-GREAT OUTDOORS

What is a mile high and spins?
A mountain top.

How do mountains hear?
With mountaineers (mountain ears).

Where do scientists raise magnets?
In magnetic fields.

What is yellow and always helpful?
A Boy Scout banana.

What grows fast and goes camping?
A Boy Sprout.

What do you call a deer with no eyes?
No-eye-deer (no idea).

What do you call a deer wiith no eyes and no legs?
Still no-eye-deer.

How do deer start a race?
They say, "Ready, set – DOE!"

Why did the horse sneeze?
It had a little colt.

What is the difference between a horse and the weather?
One is reined up, the other rains down.

What would happen if black widow spiders were as big as horses?
If one bit you, you could ride it to the hospital.

When you are trying to tell a story around the campfire, why don't you want goats to be there?

Because they're always butting in.

What do you call a very large moose?
Enor-moose.

What is the hardest thing about learning to ride wild horses?
The ground.

Why are wild horses rich?
They have a million bucks.

What do you get if you blow your hair dryer down a rabbit hole?

Hot cross bunnies.

A horse was tied to a 15-foot rope, but he walked 30 feet. How come?
The rope wasn't tied to anything.

Why was the little horse unhappy?
Because its mother always said, "Neigh!"

How do you say goodbye to a horse?
You say, "I've got to whoa now!"

How does a farmer cut his grass?
With a lawn moo-er.

How did the farmer find his missing cow?
He tractor down.

Why do cows live in barns?
They're too big for birdhouses.

What cow lives in a haunted farmhouse?

Cow-nt Dracula.

What is the difference between a berry farmer and a pirate?

The farmer treasures his berries; the pirate buries his treasures.

How do gardeners start a race?

They say "Ready, set – sow!"

How do farmers start a race?

They say "Ready, set – hoe!"

What goes in one ear and out the other?

A worm in a cornfield.

Why did the farmer take a hammer to bed with him?

So he could hit the hay.

Why do ranchers ride horses?

Because horses are too heavy to carry.

What do baby sweet potatoes sleep in?
Their yammies.

What do you get when a football team plays in your potato field?
Mashed potatoes.

What tables grow on a farm?
Vegetables.

What kind of pole do you have when five frogs sit on top of each other?

A toad-em pole.

What do you say when you meet a toad?

"Wart's new?"

Where do frogs shop?

Montgomery Wart.

Why are frogs so touchy?

Go near them and they croak.

What would you get if you crossed a rabbit and a frog?

A bunny ribbit.

What job did the frog take in the resort hotel?

Bell hop.

What animal uses a nutcracker?

A toothless squirrel.

How does a nut feel when a squirrel chews on it?

Nut so good.

What do you call a woodpecker with no beak?

A headbanger.

Why didn't the eagle get its hair wet when it went swimming?

It was a bald eagle.

What does an owl answer when you knock on its door?

"Whooooo is it?"

Why do owls call at night?

Because night rates are cheaper.

What detective stories do owls read?

Whoo-dunits.

How do we know that owls are smarter than chickens?

Have you ever heard of Kentucky Fried Owl?

What would you get if you crossed an owl and a goat?

A hootenanny.

What do you say to an intelligent firefly?

"For a little fellow, you're very bright."

Why did the farmer put an umbrella over the rabbit hutch?

He didn't want his hare to get wet.

What would you get if you crossed an insect and a hare?

Bugs Bunny.

What did the boy firefly say to the girl firefly?

"I glow for you."

Two flies were in the camp kitchen. Which one was the football player?

The one in the Sugar Bowl.

What time is it when the farmer looks at his bees?

Hive o'clock.

What do you call a young bee?

A babe-bee.

What is a bee's favorite song?

"Stinging in the Rain."

Why do bees hum?

Because they don't know the words.

What did the farmer get when he tried to reach the beehive?

A buzzy signal.

Why do bees have sticky hair?

Because there's honey on their combs.

Where do sheep go when they need a haircut?

To the baa-baa shop.

What is a sheep's favorite snack?
A bah-loney sandwich.

Why don't sheep have enough money to go on vacation?
The farmer is always fleecing them.

What is a fighter's favorite dog?
A boxer.

What is a bowler's favorite dog?
A setter.

What is a baseball player's favorite dog?
A good retriever that wears a muzzle, chases flies, and beats it for home when it sees the catcher.

What is a weightlifter's favorite dog?
A Siberian husky.

What is a chef's favorite dog?
A chow.

What is Hamlet's favorite dog?
A Great Dane.

Where do little dogs sleep when they go camping?

In pup tents.

What do you tell young dogs when they make noise outside your tent?

"Hush, puppies!"

Why didn't the witch like to sleep in a tent?

Because it didn't have a broom closet.

When does a camper go "Zzzz – meow – zzz – meow"?

When he's taking a catnap.

Why did the camper put his tent on the stove?

He wanted a home on the range.

What do people say about a mushroom who cracks great jokes?

"What a fungi!"

What is worse than having a snake in your sleeping bag?

Having two snakes in your sleeping bag.

What would you get if you crossed an animal from the Wild West and a duck?

Buffalo Bill.

What did the father buffalo say to his son when he went off to school?

"Bison!"

What do buffaloes celebrate every 100 years?

Their Bison-tennial.

What is the best thing to do if you find a skunk in your sleeping bag?

Sleep somewhere else.

What fur do you get from a skunk?

As fur as possible.

What do skunks have that no other animals have?

Baby skunks.

What did one skunk say to the other skunk when they met a group of hikers in the woods?

"Come – let us spray."

What would you get if you crossed a skunk and a boomerang?

A smell you couldn't get rid of.

Which skunk smells the worst?

The one with the cheapest perfume.

What would you get if you crossed a small bear and a skunk?

Winnie the Pooh.

What would you get if you crossed a small bear and a cow?

Winnie the Moo.

What is quicker than a fish?

The one who catches it.

How do you stop fish from smelling?

Hold their noses.

How do you communicate with a fish?

Drop it a line.

Why did the silly athlete bring a rod and reel to the football tryouts?

He heard they were looking for a tackle.

Why did Dr. Jekyll go to the beach?

To tan his hide (Hyde).

Why did Little Audrey tiptoe past the campers?

She didn't want to wake the sleeping bags.

What happened when the silly camper bought a sleeping bag?

He spent three weeks trying to wake it up.

How can you tell if there is an elephant in your sleeping bag?

By the smell of peanuts on his breath.

How do you make a bedroll?

Push it down a hill.

Why did the little boy put a frog in his sister's sleeping bag?

Because he couldn't find a snake.

How do moths swim?
They do the butterfly.

How do babies swim?
They do the crawl.

How can you tell when the ocean wants to meet you?

The tidal (tide'll) wave.

What did the banana say before it dived into the lake?

"I think I'll peel first."

How can you dive without getting wet?
Go sky diving.

How do little kids get to use the swimming pool?
They wade on line.

What do lawyers like to wear when they go swimming?
Bathing suits.

What do lawyers wear when they go running?
Briefs.

Why couldn't Batman go fishing?
Because Robin ate all the worms.

What is a worm's favorite opera?
Rigoletto (wriggle-etto).

What do you call a frightened scuba diver?

Chicken of the Sea.

Where did the seahorse live?
In the barn-acle.

What do you say to someone who falls off his surfboard because he's showing off?
"Surf's you right!"

How do boaters start a race?
They say, "Ready, set – row!"

Gladys went swimming. She saw a big shark, but she wasn't afraid. Why not?
Because it was a man-eating shark.

What is worse than seeing a shark's fins?
Seeing its tonsils.

What do sharks call swimmers?
Dinner!

Why is the ocean so angry?
It's been crossed.

How does a river say goodbye?
"Got to flow now!"

How does the ocean say goodbye?
"I'll be sea-ing you!"

How do you say goodbye to the ocean?
You wave.

15. FAR OUT!

What did the Martian say when he landed in a field of weeds?

"Take me to your weeder."

What is soft, white, and comes from Mars?

Martian-mallows.

Where do Martians leave their spaceships?
At parking meteors.

What does the sun do when it gets tired?
It sets a while.

How do you tune in to the sun?
With a sun dial.

What travels around the earth all year without using a drop of fuel?
The moon.

Who really likes to be down and out?
An airsick astronaut.

What is the secret of being a happy astronaut?
Never look down.

What is an astronaut's favorite meal?
Launch.

Why do astronauts wear bullet-proof vests?

To protect themselves against shooting stars.

What do astronauts do when they get angry?

They blast off.

What do astronauts do when they get dirty?

They take a meteor shower.

How do Martians drink their tea?

From flying saucers.

What do you call a person who is crazy about going into space?

An astro-nut.

What do Martians do in space when they get thirsty?

They drink from the Big Dipper.

How do you arrange for a trip to Mars?

You planet (plan it).

How do you get to the Planet of the Apes?

By banana boat.

What is round and purple and orbits the sun?

The Planet of the Grapes.

What is the noisiest planet?

Saturn, because it has so many rings.

What do you call an astronaut who is afraid of heights?

A failure.

Why was the astronaut wrong when he landed on the moon and reported there was no life there?

There was – with him on it.

What did the astronaut get when the rocket fell on his foot?

Mistletoe.

Why did the astronaut lie on the bed before he blasted off?

He wanted to count down.

How do you put a baby astronaut to sleep?

You rock-et.

What would you get if you crossed a 50-foot Martian and a 300-pound chicken?

The biggest cluck in the solar system.

What is covered with ribbons and bows and comes from outer space?
A gift-wrapped Martian.

Where do Martians go swimming?
In the galaxies (galax-seas).

What do little astronauts get when they do their homework?
Gold stars.

Why did the scientist pinch the waitress?
He wanted to see some flying saucers.

What would you get if you crossed a galaxy and a toad?
Star Warts.

16. DRIVING YOURSELF CRAZY

What looks like a snake, swims in water, and honks?

An automob-eel.

What kind of car do you drive in the fall?

An autumn-mobile.

What makes the road broad?

The letter "B."

What kind of car do rich cats drive?

Cat-illacs.

What kind of car do rich steers drive?

Cattle-acs.

CROSSING THE ROAD

Why did the chicken cross the road?
To get to the other side.

What was the farmer doing on the other side of the road?
Catching all the chickens who tried to cross the road.

What do you call a chicken that crosses the road without looking both ways?
Dead.

What has two arms, two wings, two tails, three heads, three bodies, and eight legs?

A cowboy on horseback holding a chicken.

CROSSING THE ROAD

Why did the turkey cross the road?
It was the chicken's day off.

Why did the elephant cross the road?
To prove he wasn't chicken.

Why did the one-handed man cross the road?
To get to the second-hand shop.

Why did the surfer cross the road?
To get to the other tide.

What has fur on the outside and feathers on the inside?

A chicken in a mink coat.

What kind of running means walking?
Running out of gas.

What is the difference between the back light of a car and a short story?
One is a tail light; the other is a light tale.

What equipment is standard on a crying car?
Windshield weepers.

Why do cars have such bad memories?
Because things go in one gear and out the other.

What songs do automobiles sing?
Car tunes (cartoons).

Where do old Volkswagens end up?
In the Old Volks (folks) Home.

What kind of car do toads drive?
Hop rods.

What is purple, very wrinkled, and goes SLAM-SLAM-SLAM-SLAM?
A four-door prune.

What would you get if you crossed an automobile, a dog, and a broom?

A car-pet sweeper.

How do you keep a dog from barking in the front seat of your car?

Make him sit in the back.

What is a good license plate for a sports car?

X L R 8.

What would you have if your car's motor was in flames?

A fire engine.

What would happen if a bunch of frogs sat in a no-parking zone?

They would get toad (towed) away.

What do you call an elephant hitchhiker?

A two-and-a-half-ton pickup.

What happened when the tire drove over the nail?

The nail knocked it flat.

What kind of car is good for carrying bunny rabbits?

A hutch-back.

What kind of car do hound dogs drive?

Houndas.

What do you call a car thief who steals Hondas?

A Honda-taker (undertaker).

What do Honda owners wear close to their skin?

Honda-wear.

What is the funniest car on the road?

A Jolkswagen.

Why did the motorist shoot his car?

He wanted to kill the motor.

CROSSING THE ROAD

Why did the duck cross the road?
Because the chicken retired and moved to Florida.

Why did the otter cross the road?
To get to the otter side.

Why did the turtle cross the road?
To get to the Shell station.

Why did the goose cross the road?
Because the light was green.

Why did the man put a rabbit in his gas tank?

Because he only used the car for short hops.

CROSSING THE ROAD

Why did the rabbit cross the road?
To get to the hopping mall.

Why did the cow cross the road?
To see its fodder.

Why did Dr. Jekyll cross the road?
To get to the other Hyde.

What was Count Dracula doing on the road?
Looking for the main artery.

What has four wheels and roars down the highway?

A lion on a skateboard.

A man drove 2,000 miles with his family without knowing he had a flat tire. How come?

It was his spare tire that was flat.

What is the laziest part of a car?

The wheels – they're always tired.

What did one car say to the other car?

"Well, strip my gears and call me shiftless!"

What did one car muffler say to the other car muffler?

"Boy, am I exhausted!"

What did the jack say to the car?

"Can I give you a lift?"

What has cities without houses, rivers without water, and forests without trees?

A road map.

CROSSING THE ROAD

Why did the dinosaur cross the road?

Because in those days they didn't have chickens.

Why was the elephant on the road?

Trying to trip the ants.

Why did the atoms cross the road?

I was time to split.

What would you get if you crossed the road with a bag of money?

Mugged!

What do police use to patrol the seashore?
A squid car.

What is the last thing a trapeze flyer wants to be?

The fall guy.

What goes "peckety-peck" and points north?

A magnetic chicken.

What do you say to a hitchhiking frog?

"Hop in!"

What do you say to a hitchhiking angel?

"Harp in!"

17. PLAYING AROUND

When was baseball first mentioned in the Bible?

In the opening words: "In the big inning (beginning)."

In which inning is the score always 0–0?

In the OP-inning (opening).

What inning is it when the Frankenstein monster steps up to bat?

The fright-inning (frightening).

What is the difference between a slow ball and a fast ball?

The difference between a lump on the head and a fractured skull.

Why did the pitcher let the baseball player walk?

He was too tired to run.

What would you get if you crossed a pitcher and the Invisible Man?

Pitching like no one has ever seen.

How are baseball players like song writers?

They're both interested in big hits.

Why didn't the Confederate soldier want to go to the ball game?

He heard the Yankees were playing.

What happens when baseball players get old?

They go batty.

Which three Rs must every cheerleader know?

RAH! RAH! RAH!

What color is a cheerleader?
Yell-ow.

What do cheerleaders like to drink?
Root beer.

What flavor ice cream do cheerleaders like best?
Rahs-berry.

What do cheerleaders have for breakfast?
Cheer-ios.

Why do elephants wear blue sneakers?
Because white ones get dirty too fast.

Why did the elephant go to the gym wearing Adidas?
His Reeboks were in the wash.

Where did the Loch Ness monster put on its sneakers?

In the loch-er room.

What kind of bell doesn't ring?

A dumbbell.

What do you call it when a weightlifter drops his dumbbell?

A power failure.

What happened when the weightlifter took a bath?

The police made him bring it back.

What shellfish lifts weights?

Mussels.

What grows on trees and can lift tremendous weights?

Hercu-leaves.

Why did the chicken lift weights?

She needed the eggs-ercise.

How do witches feel when they play croquet?

Wicket.

How can you tell witch twins apart?

It's not easy to tell which witch is which.

What goes GNIP-GNOP, GNIP-GNOP?
A Ping-Pong ball bouncing backwards.

How do you slice a Ping-Pong ball?
With a knife.

What do you do to a bad Ping-Pong ball?
Paddle it.

What can you serve, but never eat?
A tennis ball.

Why is tennis such a romantic sport?
Because every game starts with "Love."

Why are fish poor tennis players?
They don't like to get close to the net.

How does a tennis player sneeze?
"A-tennis shoe! A-tennis-shoe!"

Why are waiters good tennis players?

They know how to serve.

Who was the first tennis player in history?

Joseph, in the Bible, because he served in Pharaoh's court.

What is the quietest sport?
Bowling – you can usually hear a pin drop.

Why did all the bowling pins lie down?
They were on strike.

Why do great bowlers play slowly?
Because they have time to spare.

What do bowlers order when they go to a restaurant?
Spare ribs.

Why do they say bowling is good for teenagers?
Because it takes them off the streets and puts them in the alleys.

What did the executioner say to the bowling pins?
"I'll spare you this time."

What can you do with old bowling balls?

Give them to elephants to shoot marbles with.

What present does everyone kick about?
A soccer ball.

Why do soccer players do well in school?
They know how to use their heads.

What has 22 legs and goes CRUNCH CRUNCH CRUNCH?
A football team eating potato chips.

What do you call a monster that chases a whole football team?
Hungry!

How do you feel when a football team lands on you?
Very low.

Why don't skeletons play football?
Because they can't make body contact.

What kitten do you need when a football team tackles you?

A first-aid kit.

Why did the chicken cross the football field?

To score a touchdown.

Why did the silly kid bring a ladder to the ball game?

He wanted to shake hands with the Giants.

Why did giants do push-ups every morning?

To get their extra-size (exercise).

How many feet are there in a football field?

That depends on how many people are standing in it.

When is a football player like a judge?
When he sits on the bench.

What ghost haunts a football team?
The team spirit.

What do you do when a 400-pound football player breaks his big toe?

Call a big toe (tow) truck.

Why couldn't the football player make a phone call?

He couldn't find the receiver.

Why is a football receiver like measles?

Both are catching.

Why did the football coach date the watch?

He wanted to take time out.

Why didn't Cinderella get on the football team?

She had a pumpkin for a coach.

Why was the football coach unpopular?

He was rotten to the end.

What can't a coach ever say to a team of zombies?

"Look alive!"

How do you serve a football player his clam chowder?

In a soup-er bowl.

What happened when the egg got nasty with the football coach?

It was egg-spelled from the game.

What did the coach say when the whole team came down with the flu?

"Win a flu, lose a flu."

What happens when an egg sees a thrilling football game?

It gets egg-cited.

Why was the mayonnaise late for the game?

Because it was dressing.

What would you have to give up if you were the last person in the world?

Team sports.

If a basketball team were chasing a baseball team, what time would it be?

Five after nine.

How did the midget qualify for the basketball team?

He lied about his height.

Why are basketball players tall?

Because their heads are so far from their feet.

How do very tall basketball players greet each other?

They say, "Small world, isn't it?"

What do basketball players order when they go into a restaurant?

Chicken in the Basket.

What do basketball players read in their spare time?

Tall stories.

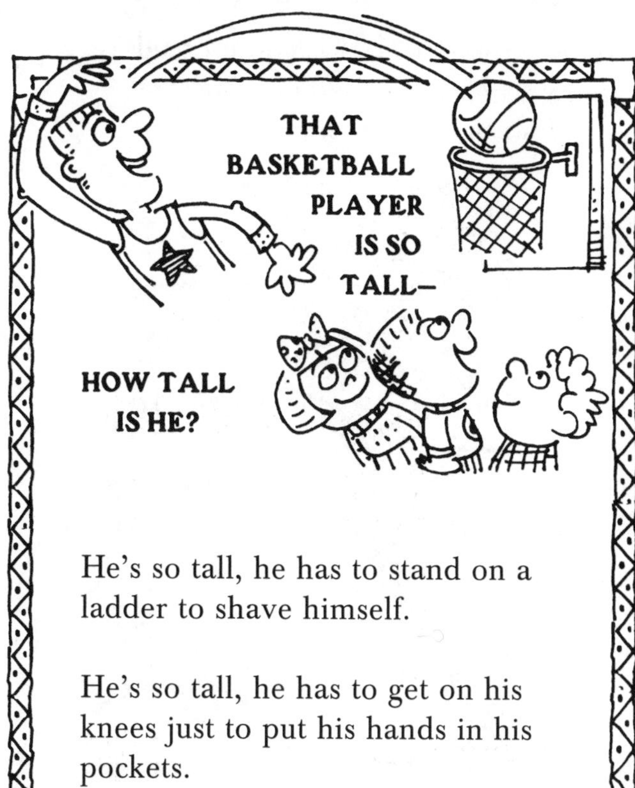

THAT
BASKETBALL
PLAYER
IS SO
TALL—
HOW TALL
IS HE?

He's so tall, he has to stand on a ladder to shave himself.

He's so tall, he has to get on his knees just to put his hands in his pockets.

What would you get if you crossed a basketball with a newborn snake?

A bouncing baby boa.

What is black and white and red all over?
A penguin that has done 100 push-ups.

What do you call a team of Czech basketball players whose games are called off?

Cancelled Czechs (checks).

Why is it hard for basketball players to be neat?

Because they dribble so much.

Why was the termite kicked off the basketball team?

It ate the backboard.

18. GO FOR THE GOLD

Where do judges go to relax?
To the tennis court.

Who keeps locomotives running?
The track coach.

Where do locomotives compete?
At the track meet.

What would you get if you crossed a computer programmer and an Olympic athlete?

A floppy disk-us thrower.

What happened when the discus thrower lost the tournament?

He became discus-ted.

What did the javelin say when it was thrown?

"Oh, spear me! Spear me!"

When does a broad jumper jump highest?

In a leap year.

When can you jump over three men without getting up?

In a checkers game.

Can any broad jumper jump higher than a house?

Yes, a house can't jump.

What kind of house weighs the least?

A lighthouse.

How do you make a lighthouse?
Use balsa wood.

What is brown and white and turns cartwheels?
A brown and white horse pulling a cart.

Why are Boy Scouts so great at gymnastics?
They're always doing good turns.

What season is it when you're on a trampoline?
Springtime.

How do you make fruit punch?
Give it boxing lessons.

Why did the monster give up boxing?
He didn't want to improve his looks.

What has two blades and breathes fire?

A dragon on ice skates.

What has two wings but doesn't fly?
A hockey team.

What position do monsters play on a hockey team?
Ghoulie.

When a hockey player goes to the barber, does he get a haircut?
No, he gets all of them cut.

What happens when a hockey player tastes a lemon?
He puck-ers up.

What do you need to play ice hockey?
Good ice sight.

What is a ski pro's favorite song?
"There's no business like snow business."

Why did the ski pro say he was an actor?

Because he broke his leg and was in a cast for six months.

What happens when skiers get old?
They go downhill.

Why did the cross-country skier wear only one boot?
He heard the snow was one foot deep.

What winter game do you learn in the fall?
Ice skating.

What is the hardest thing about learning to skate?
The ice.

What time is it when three skiers go ice skating?
Wintertime.

When are Olympic swimmers like babies?
When they do the crawl.

What is the only way a miser will swim?
Freestyle.

What do you say when you swim into kelp and it pulls you down?
"Kelp!"

A lemon and an orange were on a high diving board. The orange jumped, but the lemon didn't. Why?

The lemon was yellow.

What happened when the diver leaped 100 feet into a glass of root beer?

Nothing. It was a soft drink.

Why wouldn't the skeleton jump off the diving board?

It had no guts.

19. BROKEN DOWN ON THE INFORMATION SUPERHIGHWAY

What home computers grow on trees?
Apples.

What is a digital computer?
Someone who counts on his fingers.

What does a proud computer call his little kid?

A microchip off the old block.

What does a computer call its mother and father?

Mama and data.

Where do computers keep their money?

In memory banks.

Why did the silly kid put cheese in her computer?

She wanted to feed the mouse.

What kind of royal cat do you find in a computer?

A Sir Kit (circuit).

What would you get if you crossed a computer and a kangaroo?

I don't know what you would call it, but it would always jump to conclusions.

How do computer scientists sail?

On silicon chips (ships).

What did the scientist get when he crossed the mummy and a stopwatch?

An old-timer.

Why did the scientist like bargains?

Because he was 50% off himself.

What happened when the scientist threw an elastic band into the computer?

It gave snappy answers.

Why did the farmer put a computer in the hen house?

To make the chickens multiply faster.

What kind of feet do mathematicians have?

Square feet.

When will a mathematician die?

When his number is up.

What kind of beat do mathematicians like to dance to?

Logarithms.

Why do mosquitoes make great mathematicians?

Because they add to misery, subtract from pleasure, divide attention, and multiply rapidly.

Why are bacteria bad mathematicians?

Because they multiply by dividing.

What figures do the most walking?

Roman (roamin') numerals.

How is a telephone like arithmetic?
One mistake and you get the wrong number.

How do nuclear scientists relax?
They go fission.

What would you get if you crossed an elephant and a computer?

A ten-thousand-pound know-it-all.

What would you get if you crossed a midget and a computer?

A short circuit.

Why did the little computer go to the orthodontist?

To improve its byte.

What did the digital clock say to its mother?

"Look, Ma! No hands!"

The scientist came to his laboratory without a key and found all the windows and doors locked. How did he get in?

He ran round and round the building – until he was all in.

20. MOVING RIGHT ALONG

Why was Adam the best runner of all time?

Because he was first in the human race.

A cabbage, a faucet, and a tomato had a race. How did it go?

The cabbage was ahead (a head), the faucet was running, and the tomato tried to ketchup (catch up).

Why couldn't the orange finish the race?

It ran out of juice.

What race is like the Indianapolis 500 – but without the apples?

The Indian apple-less 500.

What happened when the race car driver slammed into a pile of IOUs?

He ran into debt.

What are two auto racers who drive the same car?

Vroom-mates.

How do fleas start a race?

The starter says,"One, two flea – go!"

How do fireflies start a race?

The starter says, "Ready, set – glow."

How do chickens start a race?

From scratch.

How would a vampire like to see a race finish?

Neck and neck.

Where do you go after you've jogged around a ship ten times?

To the poop deck.

What happens when long distance runners get old?

They go round the bend.

Which big punctuation mark is like a race?

A 40-yard dash.

What do joggers say when they leave you?

"So long – got to run!"

INDEX